The chemistry of
**organic germanium, tin
and lead compounds**

THE CHEMISTRY OF FUNCTIONAL GROUPS

A series of advanced treatises under the general editorship of
Professors Saul Patai and Zvi Rappoport

The chemistry of alkenes (2 volumes)
The chemistry of the carbonyl group (2 volumes)
The chemistry of the ether linkage
The chemistry of the amino group
The chemistry of the nitro and nitroso groups (2 parts)
The chemistry of carboxylic acids and esters
The chemistry of the carbon–nitrogen double bond
The chemistry of amides
The chemistry of the cyano group
The chemistry of the hydroxyl group (2 parts)
The chemistry of the azido group
The chemistry of acyl halides
The chemistry of the carbon–halogen bond (2 parts)
The chemistry of the quinonoid compounds (2 volumes, 4 parts)
The chemistry of the thiol group (2 parts)
The chemistry of the hydrazo, azo and azoxy groups (2 parts)
The chemistry of amidines and imidates (2 volumes)
The chemistry of cyanates and their thio derivatives (2 parts)
The chemistry of diazonium and diazo groups (2 parts)
The chemistry of the carbon–carbon triple bond (2 parts)
The chemistry of ketenes, allenes and related compounds (2 parts)
The chemistry of the sulphonium group (2 parts)
Supplement A: The chemistry of double-bonded functional groups (2 volumes, 4 parts)
Supplement B: The chemistry of acid derivatives (2 volumes, 4 parts)
Supplement C: The chemistry of triple-bonded functional groups (2 volumes, 3 parts)
Supplement D: The chemistry of halides, pseudo-halides and azides (2 volumes, 4 parts)
Supplement E: The chemistry of ethers, crown ethers, hydroxyl groups and their sulphur analogues (2 volumes, 3 parts)
Supplement F: The chemistry of amino, nitroso and nitro compounds and their derivatives (2 parts)
The chemistry of the metal–carbon bond (5 volumes)
The chemistry of peroxides
The chemistry of organic selenium and tellurium compounds (2 volumes)
The chemistry of the cyclopropyl group (2 parts)
The chemistry of sulphones and sulphoxides
The chemistry of organic silicon compounds (2 parts)
The chemistry of enones (2 parts)
The chemistry of sulphinic acids, esters and their derivatives
The chemistry of sulphenic acids and their derivatives
The chemistry of enols
The chemistry of organophosphorus compounds (3 volumes)
The chemistry of sulphonic acids, esters and their derivatives
The chemistry of alkanes and cycloalkanes
Supplement S: The chemistry of sulphur-containing functional groups
The chemistry of organic arsenic, antimony and bismuth compounds
The chemistry of enamines (2 parts)
The chemistry of organic germanium, tin and lead compounds

UPDATES

The chemistry of α-haloketones, α-haloaldehydes and α-haloimines
Nitrones, nitronates and nitroxides
Crown ethers and analogs
Cyclopropane derived reactive intermediates
Synthesis of carboxylic acids, esters and their derivatives
The silicon–heteroatom bond
Syntheses of lactones and lactams
The syntheses of sulphones, sulphoxides and cyclic sulphides

Patai's 1992 guide to the chemistry of functional groups — *Saul Patai*

C–Ge, C–Sn, C–Pb

The chemistry of
organic germanium, tin and lead compounds

Edited by

SAUL PATAI

The Hebrew University, Jerusalem

1995

JOHN WILEY & SONS

CHICHESTER–NEW YORK–BRISBANE–TORONTO–SINGAPORE

An Interscience® Publication

Other Wiley Editorial Offices

John Wiley & Sons, Inc., 605 Third Avenue,
New York, NY 10158-0012, USA

Jacaranda Wiley Ltd, 33 Park Road, Milton,
Queensland 4064, Australia

John Wiley & Sons (Canada) Ltd, 22 Worcester Road,
Rexdale, Ontario M9W 1L1, Canada

John Wiley & Sons (SEA) Pte Ltd, 37 Jalan Pemimpin #05-04,
Block B, Union Industrial Building, Singapore 2057

Library of Congress Cataloging-in-Publication Data

The chemistry of organic germanium, tin, and lead compounds / edited
by Saul Patai.
p.   cm. — (The chemistry of functional groups)
'An Interscience publication.'
Includes bibliographical references (p.   -   ) and index.
ISBN 0-471-94207-3 (alk. paper)
1. Organogermanium compounds.   2. Organotin compounds.
3. Organolead compounds.   I. Patai, Saul.   II. Series.
QD412.G5C47   1995
547.05′684 — dc20   95-19750
CIP

British Library Cataloguing in Publication Data

A catalogue record for this book is available from the British Library

ISBN 0 471 94207 3

Typeset in 9/10pt Times by Laser Words, Madras, India
Printed and bound in Great Britain by Biddles Ltd, Guildford, Surrey
This book is printed on acid-free paper responsibly manufactured from sustainable forestation, for
which at least two trees are planted for each one used for paper production.

Contributing authors

Harold Basch	Department of Chemistry, Bar-Ilan University, Ramat-Gan 52900, Israel
Carla Cauletti	Dipartimento di Chimica, Università di Roma 'La Sapienza', Piazzale Aldo Moro 5, 00185 Roma, Italy
Marvin Charton	Chemistry Department, School of Liberal Arts and Sciences, Pratt Institute, Brooklyn, New York 11205, USA
Peter J. Craig	Department of Chemistry, School of Applied Sciences, De Montfort University, The Gateway, Leicester, LE1 9BH, UK
J. T. van Elteren	Department of Chemistry, School of Applied Sciences, De Montfort University, The Gateway, Leicester, LE1 9BH, UK
Marcel Gielen	Faculty of Applied Sciences, Free University of Brussels, Room 8G512, Pleinlaan 2, B-1050 Brussels, Belgium
Charles M. Gordon	School of Chemical Sciences, Dublin City University, Dublin 9, Ireland
Sarina Grinberg	Institutes for Applied Research, Ben-Gurion University of the Negev, Beer-Sheva 84110, Israel
Tova Hoz	Department of Chemistry, Bar-Ilan University, Ramat-Gan 52900, Israel
L. M. Ignatovich	Latvian Institute of Organic Synthesis, Riga, LV 1006 Latvia
Jim Iley	Physical Organic Chemistry Research Group, Chemistry Department, The Open University, Milton Keynes, MK7 6AA, UK
Jill A. Jablonowski	Department of Chemistry and Biochemistry, University of South Carolina, Columbia, South Carolina 29208, USA
Helen Joly	Department of Chemistry, Laurentian University, Sudbury, Ontario P3E 2C6, Canada
Thomas M. Klapötke	Institut für Anorganische und Analytische Chemie, Technische Universität Berlin, Strasse des 17 Juni 135, D-10623 Berlin, Germany
Joel F. Liebman	Department of Chemistry and Biochemistry, University of Maryland, Baltimore County, 5401 Wilkens Avenue, Baltimore, Maryland 21228-5398, USA

Conor Long — School of Chemical Sciences, Dublin City University, Dublin 9, Ireland

E. Lukevics — Latvian Institute of Organic Synthesis, Riga, LV 1006 Latvia

Kenneth M. Mackay — School of Science and Technology, University of Waikato, P.B. 3105, Hamilton, New Zealand

Shigeru Maeda — Department of Applied Chemistry and Chemical Engineering, Faculty of Engineering, Kagoshima University, 1-21-40 Korimoto, Kagoshima 890, Japan

James A. Marshall — Department of Chemistry and Biochemistry, University of South Carolina, Columbia, South Carolina 29208, USA

Michael Michman — Department of Organic Chemistry, The Hebrew University of Jerusalem, Jerusalem 91904, Israel

Axel Schulz — Institut für Anorganische und Analytische Chemie, Technische Universität Berlin, Strasse des 17 Juni 135, D-10623 Berlin, Germany

Larry R. Sherman — Department of Chemistry, University of Scranton, Scranton, Pennsylvania 18519-4626, USA

José A. Martinho Simões — Departamento de Química, Faculdade de Ciências, Universidade de Lisboa, 1700 Lisboa, Portugal

Suzanne W. Slayden — Department of Chemistry, George Mason University, 4400 University Drive, Fairfax, Virginia 22030-4444, USA

Stefano Stranges — Dipartimento di Chimica, Università di Roma 'La Sapienza', Piazzale Aldo Moro 5, 00185 Roma, Italy

John M. Tsangaris — Department of Chemistry, University of Ioannina, GR-45100 Ioannina, Greece

Kenneth C. Westaway — Department of Chemistry, Laurentian University, Sudbury, Ontario P3E 2C6, Canada

Rudolph Willem — Faculty of Applied Sciences, Free University of Brussels, Room 8G512, Pleinlaan 2, B-1050 Brussels, Belgium

Jacob Zabicky — Institutes for Applied Research, Ben-Gurion University of the Negev, Beer-Sheva 84110, Israel

Foreword

As was the case with the volume *The chemistry of organic arsenic, antimony and bismuth compounds*, published in 1994, it was clear that the set of five volumes describing organometallic compounds (edited by Professor Frank R. Hartley) did not deal in sufficient depth with organic compounds of germanium, tin and lead. Hence we decided to publish the present volume, which we hope will be a useful and worthwhile addition to the series *The Chemistry of Functional Groups*. In this volume the authors' literature search extended in most cases up to the end of 1994.

The following chapters unfortunately did not materialize: Mass spectra; NMR and Mössbauer spectroscopy; Organic Ge, Sn and Pb compounds as synthones; Ge, Sn and Pb analogs of radicals and of carbenes; and Rearrangements. Moreover, the volume does not contain a 'classical' chapter on biochemistry, although much of the relevant material is included in the chapter on environmental methylation of Ge, Sn and Pb and in the chapter on the toxicity of organogermanium compounds, in the chapter on organotin toxicology and also in the chapter on safety and environmental effects.

I hope that the above shortcomings will be amended in one of the forthcoming supplementary volumes of the series.

I will be indebted to readers who will bring to my attention mistakes or omissions in this or in any other volume of the series.

Jerusalem
May 1995

SAUL PATAI

The Chemistry of Functional Groups
Preface to the series

The series 'The Chemistry of Functional Groups' was originally planned to cover in each volume all aspects of the chemistry of one of the important functional groups in organic chemistry. The emphasis is laid on the preparation, properties and reactions of the functional group treated and on the effects which it exerts both in the immediate vicinity of the group in question and in the whole molecule.

A voluntary restriction on the treatment of the various functional groups in these volumes is that material included in easily and generally available secondary or tertiary sources, such as Chemical Reviews. Quarterly Reviews, Organic Reactions, various 'Advances' and 'Progress' series and in textbooks (i.e. in books which are usually found in the chemical libraries of most universities and research institutes), should not, as a rule, be repeated in detail, unless it is necessary for the balanced treatment of the topic. Therefore each of the authors is asked not to give an encyclopaedic coverage of his subject, but to concentrate on the most important recent developments and mainly on material that has not been adequately covered by reviews or other secondary sources by the time of writing of the chapter, and to address himself to a reader who is assumed to be at a fairly advanced postgraduate level.

It is realized that no plan can be devised for a volume that would give a complete coverage of the field with no overlap between chapters, while at the same time preserving the readability of the text. The Editors set themselves the goal of attaining reasonable coverage with moderate overlap, with a minimum of cross-references between the chapters. In this manner, sufficient freedom is given to the authors to produce readable quasi-monographic chapters.

The general plan of each volume includes the following main sections:

(a) An introductory chapter deals with the general and theoretical aspects of the group.

(b) Chapters discuss the characterization and characteristics of the functional groups, i.e. qualitative and quantitative methods of determination including chemical and physical methods, MS, UV, IR, NMR, ESR and PES — as well as activating and directive effects exerted by the group, and its basicity, acidity and complex-forming ability.

(c) One or more chapters deal with the formation of the functional group in question, either from other groups already present in the molecule or by introducing the new group directly or indirectly. This is usually followed by a description of the synthetic uses of the group, including its reactions, transformations and rearrangements.

(d) Additional chapters deal with special topics such as electrochemistry, photochemistry, radiation chemistry, thermochemistry, syntheses and uses of isotopically labelled compounds, as well as with biochemistry, pharmacology and toxicology. Whenever applicable, unique chapters relevant only to single functional groups are also included (e.g. 'Polyethers', 'Tetraaminoethylenes' or 'Siloxanes').

This plan entails that the breadth, depth and thought-provoking nature of each chapter will differ with the views and inclinations of the authors and the presentation will necessarily be somewhat uneven. Moreover, a serious problem is caused by authors who deliver their manuscript late or not at all. In order to overcome this problem at least to some extent, some volumes may be published without giving consideration to the originally planned logical order of the chapters.

Since the beginning of the Series in 1964, two main developments have occurred. The first of these is the publication of supplementary volumes which contain material relating to several kindred functional groups (Supplements A, B, C, D, E, F and S). The second ramification is the publication of a series of 'Updates', which contain in each volume selected and related chapters, reprinted in the original form in which they were published, together with an extensive updating of the subjects, if possible, by the authors of the original chapters. A complete list of all above mentioned volumes published to date will be found on the page opposite the inner title page of this book. Unfortunately, the publication of the 'Updates' has been discontinued for economic reasons.

Advice or criticism regarding the plan and execution of this series will be welcomed by the Editors.

The publication of this series would never have been started, let alone continued, without the support of many persons in Israel and overseas, including colleagues, friends and family. The efficient and patient co-operation of staff-members of the publisher also rendered us invaluable aid. Our sincere thanks are due to all of them.

The Hebrew University   SAUL PATAI
Jerusalem, Israel   ZVI RAPPOPORT

Contents

List of abbreviations used

Ac	acetyl (MeCO)
acac	acetylacetone
Ad	adamantyl
AIBN	azoisobutyronitrile
Alk	alkyl
All	allyl
An	anisyl
Ar	aryl
Bz	benzoyl (C_6H_5CO)
Bu	butyl (also t-Bu or Bu^t)
CD	circular dichroism
CI	chemical ionization
CIDNP	chemically induced dynamic nuclear polarization
CNDO	complete neglect of differential overlap
Cp	η^5-cyclopentadienyl
Cp*	η^5-pentamethylcyclopentadienyl
DABCO	1,4-diazabicyclo[2.2.2]octane
DBN	1,5-diazabicyclo[4.3.0]non-5-ene
DBU	1,8-diazabicyclo[5.4.0]undec-7-ene
DIBAH	diisobutylaluminium hydride
DME	1,2-dimethoxyethane
DMF	N,N-dimethylformamide
DMSO	dimethyl sulphoxide
ee	enantiomeric excess
EI	electron impact
ESCA	electron spectroscopy for chemical analysis
ESR	electron spin resonance
Et	ethyl
eV	electron volt

Fc	ferrocenyl
FD	field desorption
FI	field ionization
FT	Fourier transform
Fu	furyl(OC_4H_3)

| GLC | gas liquid chromatography |

Hex	hexyl(C_6H_{13})
c-Hex	cyclohexyl(C_6H_{11})
HMPA	hexamethylphosphortriamide
HOMO	highest occupied molecular orbital
HPLC	high performance liquid chromatography

i-	iso
Ip	ionization potential
IR	infrared
ICR	ion cyclotron resonance

LAH	lithium aluminium hydride
LCAO	linear combination of atomic orbitals
LDA	lithium diisopropylamide
LUMO	lowest unoccupied molecular orbital

M	metal
M	parent molecule
MCPBA	*m*-chloroperbenzoic acid
Me	methyl
MNDO	modified neglect of diatomic overlap
MS	mass spectrum

n	normal
Naph	naphthyl
NBS	*N*-bromosuccinimide
NCS	*N*-chlorosuccinimide
NMR	nuclear magnetic resonance

Pc	phthalocyanine
Pen	pentyl(C_5H_{11})
Pip	piperidyl($C_5H_{10}N$)
Ph	phenyl
ppm	parts per million
Pr	propyl (also *i*-Pr or Pr^i)
PTC	phase transfer catalysis or phase transfer conditions
Pyr	pyridyl (C_5H_4N)

R	any radical
RT	room temperature
s-	secondary
SET	single electron transfer
SOMO	singly occupied molecular orbital
t-	tertiary
TCNE	tetracyanoethylene
TFA	trifluoroacetic acid
THF	tetrahydrofuran
Thi	thienyl(SC_4H_3)
TLC	thin layer chromatography
TMEDA	tetramethylethylene diamine
TMS	trimethylsilyl or tetramethylsilane
Tol	tolyl(MeC_6H_4)
Tos or Ts	tosyl(p-toluenesulphonyl)
Trityl	triphenylmethyl(Ph_3C)
Xyl	xylyl($Me_2C_6H_3$)

In addition, entries in the 'List of Radical Names' in *IUPAC Nomenclature of Organic Chemistry*, 1979 Edition. Pergamon Press, Oxford, 1979, p. 305–322, will also be used in their unabbreviated forms, both in the text and in formulae instead of explicitly drawn structures.

CHAPTER **1**

The nature of the C−M bond (M = Ge, Sn, Pb)

HAROLD BASCH and TOVA HOZ

Department of Chemistry, Bar-Ilan University, Ramat-Gan 52900, Israel
Fax: +(972)-3-535-1250;   e-mail: HBASCH@MANGO.CC.BIU.AC.IL

The chemistry of organic germanium, tin and lead compounds
Edited by S. Patai   © 1995 John Wiley & Sons Ltd

I. INTRODUCTION

The nature of the carbon–M bond as a function of the metal (M) atoms Ge, Sn and Pb has been traditionally described using differences in the atomic properties of these atoms to explain trends in molecular bonding characteristics such as bond distances, angles and energy properties. Emphasis has been on a comparison of properties contrasting behavior relative to carbon and silicon bonding to C, and among the metals themselves. The importance of relativistic effects in determining the properties of the heavier metal–ligand bonds has also been extensively addressed.

The ability of the lightest of the Group 14 atoms, the carbon atom, to bind in so many ways with carbon and with other atoms in the Periodic Table attracts extensive comparison with the analogous compounds of Si, Ge, Sn and Pb, both real and hypothetical. The wider the comparison, the greater the opportunity to gain insight into the secrets of chemical binding involving the Group 14 atoms, and to detect the nuances that differentiate their properties. Some of the causes of the differences are large, obvious and consistent. Other causes are more subtle and difficult to identify. A combination of contrary trends can effectively mask their individual characters when the individual effects are small.

The most obvious property to examine for trends and their causes is geometric structure. Historically, bond lengths and bond angles in molecules were used to elucidate electronic structure trends and construct descriptions of chemical bonding[1]. The major obstacle hindering this approach is the general lack of a sufficiently large number and variety of experimentally known molecular structures. Happily, recent developments in *ab initio* electronic structure theory have provided chemists with the tools for accurately calculating geometric structures for ever-increasing sizes of molecules[2]. At the same time, developments in relativistic effective core potentials (RCEP)[3] have allowed the incorporation of both direct and indirect radial scaling effects due to relativistic properties of the core electrons in the heavier atoms into the electronic structure description of their valence electrons. As has been known for some time already, certain differences in chemical properties in going down a column in the Periodic Table can be attributed to relativistic effects in the heavier atoms[4].

Therefore, the common approach to building a bonding description of these type compounds is to combine the few experimentally known geometric structures with a larger number of theoretically calculated geometries to infer bonding properties and trends in simple Group 14 compounds. It is, however, first necessary to identify those atomic properties which distinguish the various Group 14 atoms and which can contribute to differences in the properties of their corresponding compounds. Of course, the same properties which can be used to explain trends in geometric structure could also be used for energy properties, such as bond dissociation energies. However, energy properties typically involve both an initial state and a final state, where the energetics of the process depend on the difference in properties between the two states, both involving the same Group 14 atom. Trends in energy properties as a function of atom then involve another differencing step. This can be more subtle and difficult than treating just geometry, which involves only one state. In addition, theoretical methods for calculating geometry are more developed and reliable than for energy difference properties.

In this review we will first discuss the atomic properties that are expected to be relevant to trends in molecular structure and bonding for compounds of the Group 14 atoms[5]. Reference will be made mainly to atomic radii[6] and atomic orbital energies[6−9]. The resultant conclusions will contribute to interpreting trends in the geometric structures of small molecules having the generic formula $XH_3−Y$, where X = C, Si, Ge, Sn and Pb, and Y is one of the 53 substituents ranging from Y = H to Y = $C(O)OCH_3$. The calculated $XH_3−Y$ bond energies will also be presented and analyzed. The generated data will allow other derivative thermodynamic quantities for simple generic-type chemical reactions involving the Group 14 atom compounds to be calculated. The $XH_3−Y$ molecules are restricted to those having a formal single bond between the Group 14 atom X and the direct bonding atom of the Y group.

II. ATOMIC PROPERTIES

The atomic properties of most relevance to determining the structure and energies of molecular compounds have been identified and discussed[4,5]. The values of these properties are collected in Table 1[6−16]. The ground-state electronic configuration of the Group 14 atoms is $[core]ns^2np^2$, with n = 2,3,4,5 and 6 for C, Si Ge, Sn and Pb, respectively. In L−S coupling the electronic ground state has the term symbol ^3P. Relativistic effects are very large for the heaviest atom, lead, with a spin−orbit coupling in the thousands of cm^{-1}[17]. The splitting of the valence $np_{1/2}$ and $np_{3/2}$ spinors can affect molecular binding through their different spatial and energetic interactions with other atoms, even in closed-shell electronic states[18]. For simplicity, spin−orbit averaged values for calculated properties are shown in Table 1 for discussing trends and making comparisons.

The trend in orbital energy values for the valence ns and np atomic orbitals going down the Group 14 column is shown in Table 1. The orbital energies are taken from the numerical atomic Dirac−Fock compilation of Desclaux[6] and these open-shell systems do not rigorously obey Koopmans' Theorem[19]. As such, besides the other approximations inherent to Koopmans' Theorem, these orbital energies can only give a rough measure of values and trends in the atomic orbital ionization energies. In any event, these numbers show an interesting behavior which must reflect fundamental underlying effects. The (absolute value) np orbital energy is seen to decrease steadily, if not uniformly, with increasing atom size. There is a relatively very large energy gap between the carbon and silicon atoms, small gaps among the Si−Ge and Sn−Pb pairs, and a somewhat larger energy gap between the Ge and Sn orbital energies. The valence ns atomic orbital energy, on the other hand, is seen to have a sawtooth, alternating behavior in going down the Group 14 column[20]. As with the np atomic orbital, there is a very large energy decrease between carbon and silicon, but from Si the orbital energies alternately increase and decrease. Again, the differences between Si and Ge and between Sn and Pb are small, while the gap between Ge and Sn is larger.

The experimental ionization energies[7−9] in Table 1 show similar trends; ns ionization alternates, while np ionization decreases until Pb, where it increases slightly. Again, there is a large energy gap between carbon and silicon. The calculated atomic radii ($\langle r \rangle$) for the Desclaux orbitals[6] in Table 1 mirror the general behavior of the orbital and ionization energies: sawtooth for ns and uniformly increasing for np, where the np values for Sn and Pb are almost equal.

Although it is not completely clear which definition of each property is most appropriate for discussing molecular bonding (i.e. with or without spin−orbit averaging, radial maxima or expectation values, choice of final state for ionization energy, etc.) the general trends seem to be roughly independent of definition. The size and energies of the two valence atomic orbitals, which properties should be very important for the atom's

TABLE 1.   Properties of the Group 14 atoms

Atom	C	Si	Ge	Sn	Pb
n	2	3	4	5	6
Orbital energy[a]					
ns	−19.39	−14.84	−15.52	−13.88	−15.41
np[b]	−11.07	−7.57	−7.29	−6.71	−6.48
Ionization energy[c]					
ns[b,d]	16.60	13.64	14.43	13.49	16.04
np[b,e]	11.26	8.15	7.90	7.39	7.53
Electron affinity[f]	1.26	1.39	1.23	1.11	0.36
Polarizability[g]	1.76	5.38	6.07	7.7	6.8
Electronegativity[h]					
Mulliken[i]	1.92	1.46	1.40	1.30	1.21
Pauling[j]	2.55	1.90	2.01	1.96	2.33
Allen[k]	2.28	1.76	1.81	1.68	1.91
Atomic radius[l]					
ns	1.58	2.20	2.19	2.48	2.39
np[b]	1.74	2.79	2.88	3.22	3.22

[a] In eV; from Reference 6.
[b] Spin-orbit averaged.
[c] In eV; from References 7–9.
[d] For the process, $ns^2np^2(^3P) \rightarrow ns^1np^2(^4P)$.
[e] For the process, $ns^2np^2(^3P) \rightarrow ns^2np^1(^2P)$.
[f] In eV; from References 10 and 11.
[g] In 10^{-24} cm^3; from References 12 and 13; dipole polarizability.
[h] Relative to hydrogen = 2.20.
[i] Average of np atomic ionization energy and electron affinity. Data from appropriate lines in this Table. See Reference 14.
[j] Pauling scale (Reference 1) as calculated in Reference 15.
[k] Weighted average of ns and np ionization energies from the appropriate lines in this Table. See Reference 16.
[l] $\langle r \rangle$ in au; from Reference 6.

chemical behavior, generally show different trends for ns and np. The ns energy alternates with increasing atom size while the np energy generally decreases steadily, at least until Pb. The result is a nonuniform trend in energy gap between the ns and np atomic orbitals which can affect the degree of ns–np hybridization in chemical bonds involving the Group 14 atoms, and, thereby, the chemical behavior of their molecular compounds. On the other hand, the ns–np atomic radius ($\langle r \rangle$) difference increases steadily with atomic number, with a particularly large change between carbon and silicon. This difference can also affect the degree of ns–np hybridization through the (radial) overlap which controls the bonding effectiveness of resultant hybrid valence orbitals. We can therefore anticipate a somewhat complex, somewhat alternating chemical behavior going down the Group 14 column of the Periodic Table[1,4,16,20].

Another property which anticipates these trends is the electronegativity, also shown for several definitions in Table 1. Pauling's empirical electronegativity scale based on bond energies, as updated by Allerd[15], shows a sawtooth behavior, with predictable chemical consequences. Electronegativity is used to correlate a vast number of chemical and physical properties. Allen's revised definition of electronegativity[16] as the average configuration energy of the valence ns and np electrons also shows the alternating behavior with atomic number in the Group 14 column, as expected from the above discussion of orbital and ionization energies. The Mulliken definition[14], based on just the np atomic orbital ionization energy and the corresponding electron affinity, does not show the sawtooth behavior, and must be considered deficient for neglecting the effect of the ns atomic orbital on chemical behavior. The Mulliken scale also defines a higher electronegativity for hydrogen relative to carbon[15].

The source of the differential behavior between the ns and np atomic orbitals in going down the Group 14 column of the Periodic Table can be attributed to a combination of screening and relativistic effects, both of which preferentially stabilize the ns atomic orbital[4,16]. Filling the first transition series affects germanium this way through incomplete screening of its 4s atomic orbital which gives it a higher effective nuclear charge. Filling the first lanthanide series analogously stabilizes the Pb 6s atomic orbital through incomplete screening, which is further enhanced by relativistic effects[4,20]. Although incomplete screening and relativistic terms also affect the np atomic orbital, the stabilization is stronger for the ns atomic orbital because of its nonzero charge density at the nucleus.

The dipole polarizability term for the atoms (in Table 1) shows the usual gap between carbon and silicon, increasing values for Si $\rightarrow$ Sn and a decrease at lead. This is another reason to expect somewhat unusual behavior for lead compounds compared to the lighter metals.

The role of d-type orbitals is not addressed in Table 1. This subject has been addressed for second-row atoms in previous reviews[21,22] which contain many references to this subject. It is very difficult to define the energy and radius of the outer-sphere d-type orbitals (nd) in isolated atoms since they are not occupied in the ground state. Rather than make use of some excited state definition, we prefer to postpone a discussion of this subject until after an inspection of the calculated results on the molecular compounds.

III. CALCULATIONAL METHODS

Using atomic properties alone for predictive capabilities with regard to the geometric and electronic structure of molecules is often insufficient. Except for weakly bound systems, the chemical bond is more than just a perturbation of the electronic structure of atoms. Molecular properties determined experimentally have been used to infer the electronic structure description of simple systems, from which predictions are made for more complicated molecules using group property and additivity concepts. This empirical approach has been used very extensively in identifying and defining the determining factors in the geometric and electronic structure of molecules. These latter are then used in a predictive mode for unknown systems. The opposite approach is to use *ab initio* quantum chemical calculations to determine everything. The disadvantages in the latter methodology is that no intuitive understanding is derived from the purely mechanical calculational process which can be used for chemical systems that are too large for the *ab initio* machinery.

In this review we will try to combine the best of both approaches. On the one hand, there are very little experimental data for the Group 14 compounds for the atoms below Si. For simple molecular compounds of Ge, Sn and Pb, *ab initio* methods can be used to generate an 'experimental' database from which the electronic structure properties of such compounds can be inferred. Hopefully, the principles learned from this reference set of molecules can then be applied to larger systems. Although not the subject of this chapter, the corresponding carbon and silicon[21] systems are also examined to help elucidate trends in properties going down the complete Group 14 column of the Periodic Table and for general comparison purposes. More experimental information is available for the corresponding carbon and silicon compounds so that these can also be used to evaluate the accuracy of the calculated properties of the germanium, tin and lead database set.

The *ab initio* methods and approach used here are similar to that reported in previous studies[22–24]. The geometries of a generic set of XH_3Y molecules were determined calculationally. X is any one of the Group 14 atoms (carbon, silicon, germanium, tin and lead) and Y is any of the substituent groups, F, AlH_2, BH_2, SH, Br, H, C≡CH, PH_2,

NH_2, SCH_3, Cl, NO, ON, C(O)H, SeCN, NCSe, C(O)F, C(O)NH_2, ONO_2, NCS, SCN, CH_2CH_3, C(O)OH, NO_2, ONO, PC, CP, NCO, OCN, CN, NC, OCH_3, CH=CH_2, NNN, OH, CH_3, SiH_3, GeH_3, SnH_3, PbH_3, CF_3, C(O)OCH_3, OC(O)CH_3, PO, OP, C(O)Cl, OF, $OSiH_3$, C(O)CH_3, PO_2, OPO, OPO_2 and OS(O)OH. The Y substituents are written where attachment to X is through the leftmost atom. Attachment to X alternately by different atoms of the Y group gives rise to the possibility of linkage isomerism for the Y group. The plethora of bonding possibilities with respect to type of atom, attachment site, substitution and conformation should combine to give a balanced and comprehensive picture of the chemical bonding situation in these systems.

The geometries of the XH_3Y molecules were optimized at the MP2 (Moeller–Plesset to second order) level[2] using compact effective potentials (CEP) for the atoms in the first two rows of the Periodic Table (B−F and Al−Cl)[25] and their relativistic analogs (RCEP) for the main group atoms below the second row[26]. The RCEP are generated from Dirac–Fock all-electron relativistic atomic orbitals[6,27] and therefore implicitly include the indirect relativistic effects of the core electrons on the radial distribution of the valence electrons[18]. This could be particularly important for the lead atom. The effective potentials or pseudopotentials replace the chemically inactive core electrons.

The valence electron Gaussian basis sets were taken from the respective CEP[25] and RCEP[26] tabulations. The published basis sets show a valence atomic orbital splitting that can be denoted as (R)CEP-N1G, where $N = 3$ for first- and second-row atoms, and $N = 4$ for the heavier main group elements. This type basis set is generically called double-zeta (DZ) for historical reasons connected to Slater orbital (exponential-type) basis sets. In these calculations the valence DZ distributions were converted to triple-zeta (TZ) by splitting off the smallest exponent Gaussian member of the contracted (N) set, to give the (R)CEP-K11 valence atomic orbital distribution ($K = 2$ for first- and second-row atoms and $K = 3$ for beyond). The valence TZ Gaussian basis set for each atom was augmented by a double (D) set of d-type polarization (DP) functions (all 6 components) taken from the GAMESS tabulation[28,29] as follows. The reported[29] single Gaussian d-type polarization function was converted to DP form by scaling the single Gaussian exponent (α) by 1.4α and 0.4α to form two distinct d-type polarization functions. Both the single and double set of single Gaussians exponents are displayed in Table 2. The valence TZ hydrogen atom basis set was taken from the GAUSSIAN92[30] code as the 311G group, and augmented by a single Gaussian p-type polarization function with exponent 0.9. Overall, this basis set is denoted TZDP. All geometry optimizations were carried out in this basis set at the MP2 level (denoted MP2/TZDP or MP2/CEP-TZDP) using the GAUSSIAN92[30] .set of computer programs.

The extended basis sets are necessary to describe the adaptation of the atom to the molecular environment. Experience has shown[31] that the major effect on the radial extent of each atom in a molecule is in the bonding region. A frozen atomic orbital basis set is unable to provide the differential flexibility required in the short, intermediate and long-range radial distances from the nucleus to accurately describe the electron density changes in the molecule. The valence TZ basis set has that flexibility. Analogously, Magnusson[32] has recently discussed the effect of angular polarization functions on the inner and outer parts of the valence atomic orbitals of the main group elements The different polarization needs in the different regions of space about each atom in the molecule leads to the use of a double set of d-type basis function.

The MP2 level calculation is the first step beyond the Hartree–Fock (HF) level[33,34], and is thereby defined as a post-Hartree–Fock method. Theory predicts[35] and actual calculations have shown[2] that HF level calculated geometries generally give bond distances that are too short compared to experiment for normal covalent bonds. For these cases, MP2 level optimized geometries give better agreement with experiment[36]. The same is

TABLE 2. Polarization and diffuse Gaussian exponents[a]

Atom	Polarization single α	Polarization double[b] α_1	α_2	Diffuse single Gaussian
H	0.9000			0.03237
B	0.7000	0.9800	0.2800	0.02559
C	0.7500	1.050	0.3000	0.03691
N	0.8000	1.120	0.3200	0.05171
O	0.8500	1.190	0.3400	0.06181
F	0.9000	1.260	0.3600	0.07461
Al	0.3250	0.4550	0.1300	0.01691
Si	0.3950	0.5530	0.1580	0.02324
P	0.4650	0.6510	0.1860	0.02919
S	0.5420	0.7588	0.2168	0.03461
Cl	0.6000	0.8400	0.2400	0.04395
Ge	0.2460	0.3444	0.0984	0.02132
Se	0.3200	0.4480	0.1280	0.02934
Br	0.3600	0.5040	0.1440	0.03574
Sn	0.1830	0.2562	0.0732	0.01858
Pb	0.1640	0.2296	0.0656	0.01574

[a] Except for H, the polarization functions are d-type and the diffuse functions are sp-type. For the hydrogen atom polarization is p-type and diffuse is s-type.
[b] $\alpha_1 = 1.4\alpha$; $\alpha_2 = 0.4\alpha$; values of α are from Ref. 29.

true for vibrational frequencies[37]. The reason for the improved description of the normal covalent bond at the post-HF level is the improved description of the incipient homolytic bond dissociation process (i.e. reduced ionicity) at the MP2 level compared to HF in the neighborhood of the equilibrium bond distances. The resultant geometry optimized bond lengths are listed in Tables 3–8.

TABLE 3. Bond distances (in Å) not involving Group 14 atoms[a]

Compound	B−H[b] Al−H[b]	N−H[b] P−H[b]	O−H S−H	N=N	N≡N	N−O P−O	N=O P=O
CH_3NO_2							1.241[b]
SiH_3NO_2							1.252[b]
GeH_3NO_2							1.250[b]
SnH_3NO_2							1.251[b]
PbH_3NO_2							1.248[b]
CH_3ONO						1.411	1.203
SiH_3ONO						1.481	1.180
GeH_3ONO						1.374	1.216
SnH_3ONO						1.338	1.231
PbH_3ONO						1.306	1.245
CH_3OH			0.967				
SiH_3OH			0.965				
GeH_3OH			0.968				
SnH_3OH			0.968				
PbH_3OH			0.971				
CH_3NNN				1.246	1.159		
SiH_3NNN				1.233	1.168		

(continued overleaf)

TABLE 3.   (*continued*)

Compound	B−H[b] Al−H[b]	N−H[b] P−H[b]	O−H S−H	N=N	N≡N	N−O P−O	N=O P=O
GeH$_3$NNN				1.237	1.169		
SnH$_3$NNN				1.234	1.173		
PbH$_3$NNN				1.242	1.175		
CH$_3$ONO$_2$						1.422	1.217
							1.223[b]
SiH$_3$ONO$_2$						1.405	1.214
							1.232[b]
GeH$_3$ONO$_2$						1.388	1.218
							1.236[b]
SnH$_3$ONO$_2$						1.366	1.220
							1.247[b]
PbH$_3$ONO$_2$						1.345	1.223
							1.259[b]
CH$_3$NO							1.233
SiH$_3$NO							1.252
GeH$_3$NO							1.240
SnH$_3$NO							1.238
PbH$_3$NO							1.224
CH$_3$ON							1.251
SiH$_3$ON[c]							—
GeH$_3$ON							1.277
SnH$_3$ON							1.285
PbH$_3$ON							1.257
CH$_3$SH			1.320				
SiH$_3$SH			1.322				
GeH$_3$SH			1.322				
SnH$_3$SH			1.322				
PbH$_3$SH			1.321				
CH$_3$C(O)OH			0.976				
SiH$_3$C(O)OH			0.980				
GeH$_3$C(O)OH			0.980				
SnH$_3$C(O)OH			0.981				
PbH$_3$C(O)OH			0.981				
CH$_3$NH$_2$		1.019					
SiH$_3$NH$_2$		1.014					
GeH$_3$NH$_2$		1.018					
SnH$_3$NH$_2$		1.019					
PbH$_3$NH$_2$		1.023					
CH$_3$PH$_2$		1.412					
SiH$_3$PH$_2$		1.413					
GeH$_3$PH$_2$		1.413					
SnH$_3$PH$_2$		1.414					
PbH$_3$PH$_2$		1.413					
CH$_3$BH$_2$	1.193						
SiH$_3$BH$_2$	1.190						
GeH$_3$BH$_2$	1.189						
SnH$_3$BH$_2$	1.189						
PbH$_3$BH$_2$	1.186						
CH$_3$AlH$_2$	1.584						
SiH$_3$AlH$_2$	1.582						
GeH$_3$AlH$_2$	1.581						
SnH$_3$AlH$_2$	1.582						
PbH$_3$AlH$_2$	1.579						
CH$_3$C(O)NH$_2$		1.011					

TABLE 3. (*continued*)

Compound	B—H[b] Al—H[b]	N—H[b] P—H[b]	O—H S—H	N=N	N≡N	N—O P—O	N=O P=O
				Bond type			
$SiH_3C(O)NH_2$		1.012					
$GeH_3C(O)NH_2$		1.012					
$SnH_3C(O)NH_2$		1.013					
$PbH_3C(O)NH_2$		1.013					
CH_3PO						1.513	
SiH_3PO						1.527	
GeH_3PO						1.524	
SnH_3PO						1.524	
PbH_3PO						1.520	
CH_3OP					1.621		
SiH_3OP					1.640		
GeH_3OP					1.631		
SnH_3OP					1.631		
PbH_3OP					1.624		
CH_3PO_2						1.483[b]	
SiH_3PO_2						1.488[b]	
GeH_3PO_2						1.489[b]	
SnH_3PO_2						1.491[b]	
PbH_3PO_2						1.491[b]	
CH_3OPO					1.627	1.499	
SiH_3OPO					1.627	1.499	
GeH_3OPO					1.615	1.503	
SnH_3OPO					1.602	1.509	
PbH_3OPO					1.582	1.523	
CH_3OF					1.455[d]		
SiH_3OF					1.465[d]		
GeH_3OF					1.464[d]		
SnH_3OF					1.470[d]		
PbH_3OF					1.473[d]		
$CH_3OS(O)OH$			0.979	1.630[e]		1.659[f]	1.464[g]
$SiH_3OS(O)OH$			0.980	1.633[e]		1.641[f]	1.465[g]
$GeH_3OS(O)OH$			0.980	1.611[e]		1.652[f]	1.470[g]
$SnH_3OS(O)OH$			0.980	1.590[e]		1.649[f]	1.482[g]
$PbH_3OS(O)OH$			0.979	1.564[e]		1.651[f]	1.500[g]
CH_3OPO_2						1.592	1.476 1.482[h]
SiH_3OPO_2						1.586	1.475 1.484[h]
GeH_3OPO_2						1.580	1.477 1.487[h]
SnH_3OPO_2						1.572	1.478 1.492[h]
PbH_3OPO_2						1.563	1.478 1.501[h]

[a] MP2 optimized geometries in the TZDP basis set.
[b] Averaged.
[c] Not calculated.
[d] O—F.
[e] (X) O—S.
[f] S—O(H).
[g] S=O.
[h] Facing X.

TABLE 4. Bond distances (in Å) involving carbon[a]

Compound	C–H[b]	C–C	C=C[c]	C–N C–P	C=N[c] C=P[c]	C–O C–S C–Se	C=O[c] C=S[c] C=Se[c]	C–F C–Cl C–Br	C–B C–Al
CH_3OCH_3	1.095					1.423			
SiH_3OCH_3	1.093					1.434			
GeH_3OCH_3	1.094					1.433			
SnH_3OCH_3	1.095					1.430			
PbH_3OCH_3	1.097					1.432			
CH_3NO_2	1.087			1.496					
CH_3ONO	1.090					1.446			
CH_3OH	1.093					1.433			
CH_3Cl	1.086							1.796	
CH_3CP	1.093	1.471			1.571				
SiH_3CP					1.577				
GeH_3CP					1.577				
SnH_3CP					1.578				
PbH_3CP					1.578				
CH_3PC	1.090			1.859	1.639				
SiH_3PC					1.636				
GeH_3PC					1.635				
SnH_3PC					1.633				
PbH_3PC					1.633				
CH_3CH_3	1.092	1.533							
SiH_3CH_3	1.092								
GeH_3CH_3	1.091								
SnH_3CH_3	1.091								
PbH_3CH_3	1.088								
CH_4	1.089								
CH_3CHCH_2	1.093 1.088[d] 1.085[e]	1.503	1.339						
SiH_3CHCH_2	1.089[d] 1.086[e]		1.346						
GeH_3CHCH_2	1.088[d] 1.086[e]		1.343						
SnH_3CHCH_2	1.089[d] 1.087[e]		1.345						
PbH_3CHCH_2	1.086[d] 1.087[e]		1.341						
CH_3NNN	1.090			1.487					
CH_3CN	1.090	1.466			1.174				
SiH_3CN					1.178				
GeH_3CN					1.178				
SnH_3CN					1.179				
PbH_3CN					1.180				
CH_3NC	1.089			1.432	1.185				
SiH_3NC					1.189				
GeH_3NC					1.189				
SnH_3NC					1.190				
PbH_3NC					1.190				
CH_3SCN	1.089				1.181	1.812	1.684		
SiH_3SCN					1.181		1.685		

TABLE 4. (*continued*)

Compound	C–H[b]	C–C	C=C[c]	C–N / C–P	C=N[c] / C=P[c]	C–O / C–S / C–Se	C=O[c] / C=S[c] / C=Se[c]	C–F / C–Cl / C–Br	C–B / C–Al
GeH₃SCN					1.182		1.683		
SnH₃SCN					1.183		1.682		
PbH₃SCN					1.184		1.680		
CH₃NCS	1.091			1.444	1.204		1.571		
SiH₃NCS					1.204		1.566		
GeH₃NCS					1.208		1.569		
SnH₃NCS					1.202		1.574		
PbH₃NCS					1.218		1.574		
CH₃OCN	1.089				1.180	1.465	1.303		
SiH₃OCN					1.180		1.299		
GeH₃OCN					1.182		1.296		
SnH₃OCN					1.183		1.292		
PbH₃OCN					1.185		1.291		
CH₃NCO	1.090			1.457	1.216		1.186		
SiH₃NCO					1.213		1.183		
GeH₃NCO					1.217		1.185		
SnH₃NCO					1.213		1.188		
PbH₃NCO					1.226		1.189		
CH₃SCH₃	1.091					1.795			
SiH₃SCH₃	1.090					1.817			
GeH₃SCH₃	1.090					1.815			
SnH₃SCH₃	1.089					1.819			
PbH₃SCH₃	1.090					1.817			
CH₃CH₂CH₃	1.093	1.531							
	1.095[e]	1.531							
SiH₃CH₂CH₃	1.095[e]	1.540							
	1.093								
GeH₃CH₂CH₃	1.095[e]	1.536							
	1.094								
SnH₃CH₂CH₃	1.094	1.537							
PbH₃CH₂CH₃	1.091[e]	1.531							
	1.094								
CH₃CF₃	1.089	1.503						1.355	
SiH₃CF₃								1.361	
GeH₃CF₃								1.361	
SnH₃CF₃								1.364	
PbH₃CF₃								1.359	
CH₃CCH	1.092	1.467							
	1.063[d]		1.217						
SiH₃CCH	1.066		1.225						
GeH₃CCH	1.065		1.224						
SnH₃CCH	1.066		1.226						
PbH₃CCH	1.066		1.225						
CH₃ONO₂	1.089					1.446			
CH₃F	1.090							1.400	
CH₃NO	1.092				1.484				
CH₃ON	1.088					1.500			
CH₃C(O)H	1.092	1.507					1.221		

(*continued overleaf*)

TABLE 4.   (*continued*)

Compound	C–H[b]	C–C	C=C[c]	C–N C–P	C=N[c] C=P[c]	C–O C–S C–Se	C=O[c] C=S[c] C=Se[c]	C–F C–Cl C–Br	C–B C–Al
	1.109[d]								
SiH$_3$C(O)H	1.114						1.231		
GeH$_3$C(O)H	1.114						1.226		
SnH$_3$C(O)H	1.115						1.227		
PbH$_3$C(O)H	1.112						1.221		
CH$_3$Br	1.086							1.944	
CH$_3$SeCN	1.088				1.181	1.961	1.849		
SiH$_3$SeCN					1.182		1.851		
GeH$_3$SeCN					1.182		1.849		
SnH$_3$SeCN					1.182		1.850		
PbH$_3$SeCN					1.183		1.847		
CH$_3$NCSe	1.091		1.440		1.201		1.736		
SiH$_3$NCSe					1.202		1.718		
GeH$_3$NCSe					1.199		1.725		
SnH$_3$NCSe					1.199		1.729		
PbH$_3$NCSe					1.197		1.735		
CH$_3$SH	1.088					1.806			
CH$_3$C(O)OH	1.090	1.506				1.368	1.216		
SiH$_3$C(O)OH						1.369	1.220		
GeH$_3$C(O)OH						1.370	1.217		
SnH$_3$C(O)OH						1.374	1.217		
PbH$_3$C(O)OH						1.369	1.213		
CH$_3$C(O)F	1.090	1.498					1.194	1.378	
SiH$_3$C(O)F							1.195	1.397	
GeH$_3$C(O)F							1.194	1.395	
SnH$_3$C(O)F							1.194	1.405	
PbH$_3$C(O)F							1.190	1.401	
CH$_3$NH$_2$	1.093			1.477					
CH$_3$PH$_2$	·1.091			1.862					
CH$_3$BH$_2$	1.095								1.562
CH$_3$AlH$_2$	1.094								1.963
CH$_3$C(O)NH$_2$	1.091	1.519		1.381			1.227		
SiH$_3$C(O)NH$_2$				1.374			1.236		
GeH$_3$C(O)NH$_2$				1.374			1.231		
SnH$_3$C(O)NH$_2$				1.375			1.231		
PbH$_3$C(O)NH$_2$				1.371			1.225		
CH$_3$C(O)OCH$_3$	1.090	1.509				1.361	1.218		
	1.089					1.448[f]			
SiH$_3$C(O)OCH$_3$	1.088					1.363	1.223		
						1.454[f]			
GeH$_3$C(O)OCH$_3$	1.088					1.365	1.219		
						1.454[f]			
SnH$_3$C(O)OCH$_3$	1.088					1.369	1.219		
						1.455[f]			
PbH$_3$C(O)OCH$_3$	1.088					1.363	1.214		
						1.458[f]			
SiH$_3$OC(O)CH$_3$	1.090	1.505				1.363	1.220		
GeH$_3$OC(O)CH$_3$	1.090	1.508				1.351	1.225		
SnH$_3$OC(O)CH$_3$	1.090	1.507				1.333	1.236		

TABLE 4. (*continued*)

	Bond type								
Compound	C—H[b]	C—C	C=C[c]	C—N C—P	C=N[c] C=P[c]	C—O C—S C—Se	C=O[c] C=S[c] C=Se[c]	C—F C—Cl C—Br	C—B C—Al
PbH₃OC(O)CH₃	1.090	1.508				1.314	1.249		
CH₃PO	1.094			1.848					
CH₃OP	1.090					1.459			
CH₃C(O)Cl	1.090	1.502					1.195	1.823	
SiH₃C(O)Cl							1.192	1.875	
GeH₃C(O)Cl							1.188	1.903	
SnH₃C(O)Cl							1.188	1.903	
PbH₃C(O)Cl							1.187	1.885	
CH₃C(O)CH₃	1.092	1.518[b]					1.226		
SiH₃C(O)CH₃	1.094	1.515					1.234		
GeH₃C(O)CH₃	1.093	1.517					1.228		
SnH₃C(O)CH₃	1.094	1.517					1.227		
PbH₃C(O)CH₃	1.093	1.510					1.220		
CH₃PO₂	1.089			1.811					
CH₃OPO	1.089					1.457			
CH₃OF	1.092					1.423			
CH₃OS(O)OH	1.089					1.453			
CH₃OS(O)OH	1.087					1.465			

[a] MP2 optimized geometries in the TZDP basis set.
[b] Average bond lengths.
[c] Double or triple bond.
[d] CH group.
[e] CH₂ group.
[f] O—CH₃.

TABLE 5. Bond distances (in Å) involving silicon[a]

	Bond type					
Compound	Si—H[b]	Si—C Si—Si Si—Ge Si—Sn Si—Pb	Si—N Si—P	Si—O Si—S Si—Se	Si—F Si—Cl Si—Br	Si—B Si—Al
SiH₃OCH₃	1.475			1.660		
SiH₃NO₂	1.465		1.847			
SiH₃ONO	1.471			1.705		
SiH₃OH	1.473			1.665		
SiH₃Cl	1.468				2.070	
SiH₃CP	1.473	1.853				
SiH₃PC	1.468		2.279			
SiH₃CH₃	1.477	1.879				
SiH₃SiH₃	1.475	2.347				
SiH₃GeH₃	1.474	2.389				
SiH₃SnH₃	1.475	2.579				
SiH₃PbH₃	1.473	2.584				
SiH₄	1.472					

(*continued overleaf*)

TABLE 5. (*continued*)

Compound	Si–H[b]	Si–C / Si–Si / Si–Ge / Si–Sn / Si–Pb	Si–N / Si–P	Si–O / Si–S / Si–Se	Si–F / Si–Cl / Si–Br	Si–B / Si–Al
SiH_3CHCH_2	1.476	1.868				
SiH_3NNN	1.471		1.763			
SiH_3CN	1.468	1.860				
SiH_3NC	1.467		1.768			
SiH_3SCN	1.468			2.165		
SiH_3NCS	1.470		1.738			
SiH_3OCN	1.466			1.715		
SiH_3NCO	1.470		1.736			
SiH_3SCH_3	1.474			2.116		
$SiH_3CH_2CH_3$	1.477	1.882				
SiH_3CF_3	1.469	1.929				
SiH_3CCH	1.472	1.838				
SiH_3ONO_2	1.467			1.729		
SiH_3F	1.469				1.619	
SiH_3NO	1.471		1.853			
SiH_3ON^c	—			—		
$SiH_3C(O)H$	1.473	1.923				
SiH_3Br	1.467				2.234	
SiH_3SeCN	1.468			2.320		
SiH_3NCSe	1.470		1.751			
SiH_3SH	1.471			2.129		
$SiH_3C(O)OH$	1.470	1.922				
$SiH_3C(O)F$	1.468	1.921				
SiH_3PH_2	1.474		2.266			
SiH_3NH_2	1.477		1.739			
SiH_3BH_2	1.478					2.022
SiH_3AlH_2	1.479					2.486
$SiH_3C(O)NH_2$	1.473	1.922				
$SiH_3C(O)OCH_3$	1.471	1.920				
$SiH_3OC(O)CH_3$	1.469			1.710		
SiH_3PO	1.476		2.356			
SiH_3OP	1.469			1.715		
$SiH_3C(O)Cl$	1.468	1.920				
$SiH_3C(O)CH_3$	1.474	1.933				
SiH_3PO_2	1.467		2.285			
SiH_3OPO	1.468			1.707		
SiH_3OF	1.469			1.709		
SiH_3OSiH_3	1.474			1.656		
GeH_3OSiH_3	1.476			1.654		
SnH_3OSiH_3	1.478			1.648		
PbH_3OSiH_3	1.480			1.650		
$SiH_3OS(O)OH$	1.469			1.707		
SiH_3OPO_2	1.466			1.716		

[a] MP2 optimized geometries in the TZDP basis set.
[b] Average bond lengths.
[c] Not calculated.

TABLE 6. Bond distances (in Å) involving germanium[a]

Compound	Ge—H[b]	Ge—C / Ge—Si / Ge—Ge / Ge—Sn / Ge—Pb	Ge—N / Ge—P	Ge—O / Ge—S / Ge—Se	Ge—F / Ge—Cl / Ge—Br	Ge—B / Ge—Al
GeH_3OCH_3	1.521			1.779		
GeH_3NO_2	1.510		1.944			
GeH_3ONO	1.514			1.860		
GeH_3OH	1.519			1.783		
GeH_3Cl	1.513				2.171	
GeH_3CP	1.518	1.930				
GeH_3PC	1.513		2.363			
GeH_3CH_3	1.524	1.955				
GeH_3SiH_3	1.522	2.389				
GeH_3GeH_3	1.522	2.427				
GeH_3SnH_3	1.522	2.610				
GeH_3PbH_3	1.520	2.621				
GeH_4	1.518					
GeH_3CHCH_2	1.522	1.943				
GeH_3NNN	1.515		1.876			
GeH_3CN	1.511	1.937				
GeH_3NC	1.510		1.861			
GeH_3SCN	1.513			2.259		
GeH_3NCS	1.513		1.847			
GeH_3OCN	1.510			1.841		
GeH_3NCO	1.515		1.844			
GeH_3SCH_3	1.520			2.206		
$GeH_3CH_2CH_3$	1.525	1.961				
GeH_3CF_3	1.514	2.000				
GeH_3CCH	1.517	1.913				
GeH_3ONO_2	1.511			1.851		
GeH_3F	1.514				1.738	
GeH_3NO	1.519		1.972			
GeH_3ON	1.512			1.897		
$GeH_3C(O)H$	1.520	2.000				
GeH_3Br	1.513				2.321	
GeH_3SeCN	1.514			2.398		
GeH_3NCSe	1.513		1.846			
GeH_3SH	1.517			2.221		
$GeH_3C(O)OH$	1.516	1.994				
$GeH_3C(O)F$	1.513	1.992				
GeH_3NH_2	1.522		1.848			
GeH_3PH_2	1.521		2.333			
GeH_3BH_2	1.525					2.072
GeH_3AlH_2	1.528					2.504
$GeH_3C(O)NH_2$	1.520	1.997				
$GeH_3C(O)OCH_3$	1.517	1.993				
$GeH_3OC(O)CH_3$	1.513			1.834		
GeH_3PO	1.524		2.418			
GeH_3OP	1.514			1.840		
$GeH_3C(O)Cl$	1.513	1.998				
$GeH_3C(O)CH_3$	1.521	2.011				
GeH_3PO_2	1.513		2.348			

(*continued overleaf*)

TABLE 6. (continued)

Compound	Ge−H[b]	Ge−C Ge−Si Ge−Ge Ge−Sn Ge−Pb	Ge−N Ge−P	Ge−O Ge−S Ge−Se	Ge−F Ge−Cl Ge−Br	Ge−B Ge−Al
			Bond type			
GeH_3OPO	1.512			1.838		
GeH_3OF	1.515			1.822		
GeH_3OSiH_3	1.518			1.779		
$GeH_3OS(O)OH$	1.512			1.838		
GeH_3OPO_2	1.510			1.848		

[a] MP2 optimized geometries in the CEP-TZDP basis set.
[b] Average bond lengths.

TABLE 7. Bond distances (in Å) involving tin[a]

Compound	Sn−H[b]	Sn−C Sn−Si Sn−Ge Sn−Sn Sn−Pb	Sn−N Sn−P	Sn−O Sn−S Sn−Se	Sn−F Sn−Cl Sn−Br	Sn−B Sn−Al
			Bond type			
SnH_3OCH_3	1.691			1.959		
SnH_3NO_2	1.680		2.131			
SnH_3ONO	1.683			2.076		
SnH_3OH	1.689			1.963		
SnH_3Cl	1.682				2.350	
SnH_3CP	1.688	2.110				
SnH_3PC	1.682		2.545			
SnH_3CH_3	1.695	2.140				
SnH_3SiH_3	1.693	2.579				
SnH_3GeH_3	1.693	2.610				
SnH_3SnH_3	1.693	2.785				
SnH_3PbH_3	1.692	2.797				
SnH_4	1.689					
SnH_3CHCH_2	1.692	2.126				
SnH_3NNN	1.684		2.056			
SnH_3CN	1.681	2.117				
SnH_3NC	1.680		2.039			
SnH_3SCN	1.681			2.448		
SnH_3NCS	1.682		2.017			
SnH_3OCN	1.679			2.023		
SnH_3NCO	1.684		2.015			
SnH_3SCH_3	1.690			2.389		
$SnH_3CH_2CH_3$	1.696	2.148				
SnH_3CF_3	1.684	2.192				
SnH_3CCH	1.687	2.088				
SnH_3ONO_2	1.680			2.051		
SnH_3F	1.684				1.910	
SnH_3NO	1.690		2.189			
SnH_3ON	1.682			2.077		
$SnH_3C(O)H$	1.690	2.197				
SnH_3Br	1.684				2.496	

TABLE 7.   (*continued*)

Compound	Sn—H[b]	Sn—C Sn—Si Sn—Ge Sn—Sn Sn—Pb	Sn—N Sn—P	Sn—O Sn—S Sn—Se	Sn—F Sn—Cl Sn—Br	Sn—B Sn—Al
				Bond type		
SnH_3SeCN	1.683			2.584		
SnH_3NCSe	1.682		2.024			
SnH_3SH	1.687			2.402		
$SnH_3C(O)OH$	1.686	2.182				
$SnH_3C(O)F$	1.683	2.178				
SnH_3NH_2	1.692		2.031			
SnH_3PH_2	1.691		2.521			
SnH_3BH_2	1.696					2.264
SnH_3AlH_2	1.698					2.696
$SnH_3C(O)NH_2$	1.691	2.186				
$SnH_3C(O)OCH_3$	1.687	2.180				
$SnH_3OC(O)CH_3$	1.684			2.039		
SnH_3PO	1.694		2.629			
SnH_3OP	1.684			2.028		
$SnH_3C(O)Cl$	1.683	2.189				
$SnH_3C(O)CH_3$	1.692	2.210				
SnH_3PO_2	1.682		2.540			
SnH_3OPO	1.681			2.040		
SnH_3OF	1.684			2.010		
SnH_3OSiH_3	1.687			1.959		
$SnH_3OS(O)OH$	1.682			2.042		
SnH_3OPO_2	1.679			2.042		

[a] MP2 optimized geometries in the TZDP basis set.
[b] Average bond lengths.

TABLE 8.   Bond distances (in Å) involving lead[a]

Compound	Pb—H[b]	Pb—C Pb—Si Pb—Ge Pb—Sn Pb—Pb	Pb—N Pb—P	Pb—O Pb—S Pb—Se	Pb—F Pb—Cl Pb—Br	Pb—B Pb—Al
				Bond type		
PbH_3OCH_3	1.735			2.064		
PbH_3NO_2	1.724		2.180			
PbH_3ONO	1.726			2.206		
PbH_3OH	1.733			2.070		
PbH_3Cl	1.726				2.414	
PbH_3CP	1.732	2.169				
PbH_3PC	1.726		2.598			
PbH_3CH_3	1.742	2.181				
PbH_3SiH_3	1.743	2.584				
PbH_3GeH_3	1.743	2.621				
PbH_3SnH_3	1.744	2.797				
PbH_3PbH_3	1.742	2.812				
PbH_4	1.735					

(*continued overleaf*)

TABLE 8. (*continued*)

		Bond type				
	Pb–H[b]	Pb–C Pb–Si Pb–Ge Pb–Sn Pb–Pb	Pb–N Pb–P	Pb–O Pb–S Pb–Se	Pb–F Pb–Cl Pb–Br	Pb–B Pb–Al
Compound						
PbH$_3$CHCH$_2$	1.742	2.167				
PbH$_3$NNN	1.728		2.157			
PbH$_3$CN	1.724	2.180				
PbH$_3$NC	1.721		2.130			
PbH$_3$SCN	1.726			2.497		
PbH$_3$NCS	1.725		2.139			
PbH$_3$OCN	1.721			2.134		
PbH$_3$NCO	1.726		2.126			
PbH$_3$SCH$_3$	1.737			2.433		
PbH$_3$CH$_2$CH$_3$	1.746	2.182				
PbH$_3$CF$_3$	1.730	2.206				
PbH$_3$CCH	1.730	2.144				
PbH$_3$ONO$_2$	1.726			2.168		
PbH$_3$F	1.726				2.026	
PbH$_3$NO	1.744		2.264			
PbH$_3$ON	1.732			2.194		
PbH$_3$C(O)H	1.739	2.231				
PbH$_3$Br	1.728				2.549	
PbH$_3$SeCN	1.728			2.624		
PbH$_3$NCSe	1.724		2.117			
PbH$_3$SH	1.733			2.451		
PbH$_3$C(O)OH	1.733	2.210				
PbH$_3$C(O)F	1.728	2.212				
PbH$_3$NH$_2$	1.738		2.123			
PbH$_3$PH$_2$	1.741		2.546			
PbH$_3$BH$_2$	1.745					2.279
PbH$_3$AlH$_2$	1.750					2.696
PbH$_3$C(O)NH$_2$	1.740	2.216				
PbH$_3$C(O)OCH$_3$	1.735	2.208				
PbH$_3$OC(O)CH$_3$	1.732			2.174		
PbH$_3$PO	1.747		2.639			
PbH$_3$OP	1.728			2.131		
PbH$_3$C(O)Cl	1.728	2.230				
PbH$_3$C(O)CH$_3$	1.743	2.238				
PbH$_3$PO$_2$	1.727		2.563			
PbH$_3$OPO	1.728			2.194		
PbH$_3$OF	1.728			2.100		
PbH$_3$OSiH$_3$	1.730			2.075		
PbH$_3$OS(O)OH	1.729			2.184		
PbH$_3$OPO$_2$	1.723			2.179		

[a] MP2 optimized geometries in the TZDP basis set.
[b] Average bond lengths.

The MP2/TZDP optimized structures were then used to calculate the stationary state geometry force constants and harmonic vibrational frequencies, also at the MP2 level. These results serve several purposes. Firstly, they test that the calculated geometry is really an energy minimum by showing all real frequencies in the normal coordinate analysis. Secondly, they provide values of the zero-point energy (ZPE) that can be used to convert the total electronic energy differences to thermodynamic enthalpies that take into account zero-point vibrational energy differences in chemical reactions. Thirdly, they provide values of vibrational frequencies in the XH_3Y series that can be compared for atom and substituent effects. This information contributes another dimension to the analysis of the electronic structure description of the bonding in these systems.

Another property obtained at the MP2 level using the relaxed MP2 densities calculated as energy derivatives[38,39] are the Mulliken populations and atomic charges. A comprehensive discussion of the whole topic of population analyses has recently been given[40]. The specific deficiencies of the Mulliken partitioning of the basis function space in the wave function charge distribution is well known[41]. Clearly, great care must be taken in interpreting trends in structure properties based on derived atomic charges alone. The individual MP2 atomic charges were summed to calculate group charges for all the XH_3 and Y substituents. These are shown in Table 9. The individual s, p and (five component) d contributions to the atomic populations at the MP2 level for the X atoms in all the XH_3Y molecules are found in Table 10. The individual atomic charges for all the atoms in XH_3Y are tabulated in Tables 11–15. The MP2/TZDP level dipole moments are tabulated in Table 16.

TABLE 9. Mulliken charges on the H and XH_3 groups in RY^a

Y\R	H	CH_3	SiH_3	GeH_3	SnH_3	PbH_3
F	0.412	0.327	0.649	0.622	0.736	0.669
AlH_2	−0.080	−0.221	0.065	0.174	0.235	0.193
BH_2	−0.027	−0.171	0.141	0.131	0.132	0.136
SH	0.165	0.036	0.319	0.350	0.398	0.455
Br	0.139	−0.050	0.161	0.267	0.368	0.452
H	0.	−0.147	0.075	0.060	0.089	0.101
CCH	0.191	0.071	0.447	0.332	0.457	0.444
PH_2	0.051	−0.126	0.137	0.171	0.207	0.243
NH_2	0.257	0.094	0.358	0.342	0.466	0.420
SCH_3	0.138	0.003	0.288	0.312	0.374	0.417
Cl	0.211	0.066	0.357	0.388	0.464	0.524
NO	0.226	0.133	0.394	0.333	0.391	0.318
ON	0.319	0.174	b	0.493	0.587	0.493
C(O)H	0.098	0.049	0.277	0.229	0.279	0.274
SeCN	0.104	−0.057	0.141	0.240	0.391	0.454
NCSe	0.304	0.235	0.549	0.474	0.587	0.563
C(O)F	0.130	0.091	0.336	0.281	0.344	0.355
$C(O)NH_2$	0.079	0.013	0.244	0.181	0.235	0.217
ONO_2	0.306	0.139	0.496	0.494	0.593	0.584
NCS	0.302	0.218	0.511	0.482	0.597	0.594
SCN	0.187	0.086	0.391	0.434	0.500	0.564
CH_2CH_3	0.138	0.140	0.256	0.234	0.310	0.294
C(O)OH	0.115	0.059	0.311	0.254	0.307	0.311
NO_2	0.267	0.188	0.478	0.445	0.535	0.526
ONO	0.267	0.111	0.425	0.430	0.506	0.484
PC	0.145	0.037	0.365	0.379	0.383	0.445
CP	0.183	0.087	0.387	0.317	0.381	0.409
NCO	0.299	0.194	0.494	0.465	0.600	0.568
OCN	0.364	0.234	0.604	0.615	0.715	0.696
CN	0.266	0.241	0.512	0.465	0.506	0.528
NC	0.334	0.309	0.553	0.502	0.605	0.579
OCH_3	0.289	0.080	0.408	0.421	0.566	0.499

(*continued overleaf*)

TABLE 9. (*continued*)

Y\R	H	CH_3	SiH_3	GeH_3	SnH_3	PbH_3
$CHCH_2$	0.125	0.007	0.294	0.248	0.318	0.333
NNN	0.288	0.171	0.445	0.435	0.553	0.547
OH	0.303	0.070	0.388	0.379	0.526	0.468
CH_3	0.147	0.	0.273	0.260	0.330	0.329
SiH_3	−0.075	−0.273	0.	0.068	0.139	0.129
GeH_3	−0.060	−0.260	−0.068	0.	0.128	0.094
SnH_3	−0.089	−0.330	−0.139	−0.128	0.	−0.089
PbH_3	−0.101	−0.329	−0.129	−0.094	0.089	0.
CF_3	0.103	0.085	0.285	0.222	0.283	0.288
$C(O)OCH_3$	0.117	0.051	0.312	0.243	0.360	0.376
$OC(O)CH_3$	0.307	0.138	0.479	0.575	0.686	0.640
PO	−0.034	−0.117	0.083	0.088	0.124	0.106
OP	0.355	0.171	0.496	0.497	0.622	0.587
C(O)Cl	0.153	0.103	0.363	0.311	0.384	0.407
$C(O)CH_3$	0.079	0.016	0.231	0.166	0.215	0.200
PO_2	0.065	−0.043	0.248	0.265	0.305	0.324
OPO	0.325	0.149	0.479	0.479	0.583	0.555
OF	0.319	0.126	0.480	0.476	0.595	0.570
$OSiH_3$	0.388	0.110	0.430	0.455	0.626	0.582
OS(O)OH	0.347	0.147	0.480	0.484	0.588	0.575
OPO_2	0.359	0.199	0.544	0.564	0.683	0.687

[a] All results are at the geometry optimized MP2/TZDP level. Connectivity is to the leftmost atom in Y.
[b] Not calculated.

TABLE 10. MP2/TZDP Mulliken atomic orbital populations for X in XH_3Y^a

	C			Si			Ge			Sn			Pb		
X = Y	s	p	d	s	p	d	s	p	d	s	p	d	s	p	d
F	1.23	2.71	0.12	1.07	1.79	0.32	1.20	1.85	0.22	1.16	1.75	0.19	1.32	1.70	0.12
H	1.20	3.23	0.06	1.16	2.26	0.23	1.28	2.34	0.16	1.23	2.25	0.16	1.35	2.16	0.10
Cl	1.23	3.01	0.11	1.14	2.00	0.31	1.26	2.03	0.21	1.22	1.94	0.19	1.35	1.84	0.11
Br	1.25	3.14	0.11	1.21	2.12	0.32	1.31	2.12	0.21	1.25	2.02	0.19	1.38	1.92	0.10
OH	1.25	2.84	0.12	1.08	1.89	0.33	1.20	1.95	0.23	1.15	1.83	0.20	1.31	1.79	0.12
SH	1.21	3.09	0.10	1.17	2.10	0.30	1.29	2.16	0.21	1.23	2.06	0.20	1.36	1.97	0.11
NH_2	1.24	2.95	0.10	1.11	2.01	0.31	1.21	2.06	0.21	1.16	1.93	0.19	1.31	1.88	0.10
PH_2	1.22	3.24	0.08	1.17	2.21	0.25	1.27	2.24	0.18	1.22	2.14	0.17	1.35	2.04	0.08
BH_2	1.19	3.25	0.07	1.13	2.28	0.21	1.26	2.33	0.14	1.21	2.25	0.14	1.34	2.17	0.06
AlH_2	1.17	3.41	0.06	1.14	2.40	0.19	1.20	2.39	0.13	1.16	2.29	0.13	1.31	2.12	0.06
CH_3	1.18	3.06	0.06	1.14	2.11	0.26	1.25	2.16	0.17	1.19	2.06	0.16	1.32	1.98	0.08
SiH_3	1.22	3.36	0.06	1.18	2.38	0.22	1.27	2.38	0.15	1.21	2.27	0.15	1.36	2.15	0.08
GeH_3	1.22	3.37	0.07	1.22	2.42	0.22	1.31	2.42	0.15	1.23	2.32	0.16	1.40	2.17	0.08
SnH_3	1.25	3.43	0.06	1.22	2.50	0.22	1.30	2.52	0.14	1.23	2.41	0.14	1.42	2.26	0.08
PbH_3	1.27	3.42	0.06	1.24	2.51	0.22	1.30	2.55	0.15	1.23	2.44	0.15	1.41	2.30	0.08
CN	1.17	3.08	0.08	1.14	2.10	0.25	1.25	2.14	0.17	1.19	2.02	0.15	1.33	1.94	0.08
NC	1.32	2.92	0.10	1.15	1.97	0.28	1.23	1.99	0.18	1.18	1.87	0.15	1.33	1.83	0.09
NO	1.26	2.98	0.08	1.17	2.02	0.25	1.29	2.07	0.17	1.25	1.99	0.16	1.40	1.93	0.10
ON	1.27	2.85	0.07	—	—	—	1.29	1.94	0.20	1.23	1.84	0.20	1.41	1.81	0.12
PO	1.20	3.29	0.07	1.19	2.30	0.22	1.31	2.34	0.15	1.26	2.23	0.15	1.41	2.12	0.09
OP	1.26	2.84	0.10	1.10	1.88	0.31	1.24	1.92	0.22	1.20	1.82	0.21	1.37	1.76	0.12
CP	1.13	3.06	0.10	1.18	2.11	0.27	1.28	2.16	0.18	1.21	2.04	0.15	1.33	1.96	0.07
PC	1.21	3.20	0.08	1.18	2.16	0.26	1.28	2.17	0.17	1.23	2.06	0.17	1.38	1.97	0.09
OF	1.25	2.82	0.11	1.11	1.87	0.30	1.24	1.92	0.21	1.19	1.82	0.19	1.34	1.75	0.12

TABLE 10. (*continued*)

X =	C			Si			Ge			Sn			Pb		
Y	s	p	d	s	p	d	s	p	d	s	p	d	s	p	d
CCH	1.24	3.06	0.09	1.21	2.10	0.27	1.32	2.15	0.18	1.26	2.03	0.16	1.35	1.93	0.08
C(O)H	1.15	3.11	0.07	1.20	2.17	0.24	1.31	2.23	0.17	1.24	2.12	0.16	1.37	2.04	0.09
CHCH$_2$	1.19	3.09	0.08	1.18	2.13	0.26	1.28	2.20	0.18	1.21	2.10	0.17	1.33	2.01	0.09
OCH$_3$	1.25	2.85	0.12	1.08	1.89	0.33	1.21	1.93	0.22	1.16	1.82	0.19	1.33	1.78	0.10
SCH$_3$	1.21	3.12	0.10	1.17	2.11	0.30	1.29	2.17	0.21	1.23	2.08	0.21	1.37	1.98	0.11
OSiH$_3$	1.26	2.86	0.12	1.09	1.90	0.34	1.20	1.94	0.23	1.15	1.82	0.19	1.33	1.79	0.11
CH$_2$CH$_3$	1.19	3.09	0.09	1.14	2.12	0.26	1.25	2.18	0.17	1.19	2.09	0.16	1.33	2.00	0.08
OCN	1.27	2.84	0.11	1.12	1.88	0.29	1.25	1.91	0.19	1.21	1.82	0.16	1.37	1.77	0.09
NCO	1.28	2.92	0.10	1.13	1.96	0.30	1.22	1.99	0.20	1.17	1.85	0.17	1.33	1.82	0.09
SCN	1.22	3.10	0.09	1.19	2.09	0.29	1.30	2.15	0.20	1.25	2.05	0.20	1.39	1.95	0.11
NCS	1.31	2.91	0.10	1.15	1.96	0.30	1.23	1.98	0.19	1.17	1.84	0.15	1.34	1.83	0.09
SeCN	1.24	3.22	0.10	1.23	2.20	0.29	1.32	2.20	0.20	1.25	2.10	0.18	1.39	1.98	0.09
NCSE	1.31	2.91	0.10	1.13	1.97	0.29	1.21	1.97	0.17	1.16	1.86	0.13	1.31	1.82	0.07
NNN	1.25	2.94	0.10	1.14	1.99	0.31	1.25	2.03	0.21	1.20	1.92	0.19	1.36	1.86	0.11
ONO	1.25	2.89	0.11	1.12	1.92	0.30	1.25	1.99	0.20	1.21	1.91	0.18	1.35	1.84	0.11
NO$_2$	1.29	2.97	0.08	1.17	2.02	0.27	1.27	2.06	0.20	1.22	1.95	0.18	1.36	1.88	0.12
OPO	1.26	2.86	0.11	1.10	1.91	0.30	1.23	1.96	0.21	1.19	1.87	0.20	1.36	1.78	0.13
PO$_2$	1.22	3.26	0.07	1.21	2.26	0.24	1.33	2.30	0.17	1.27	2.20	0.18	1.39	2.11	0.11
C(O)F	1.18	3.12	0.07	1.19	2.16	0.24	1.29	2.23	0.18	1.23	2.12	0.17	1.35	2.04	0.10
C(O)Cl	1.18	3.10	0.07	1.20	2.14	0.25	1.30	2.19	0.18	1.24	2.08	0.18	1.36	2.00	0.10
C(O)OH	1.19	2.12	0.07	1.19	2.16	0.24	1.29	2.21	0.17	1.23	2.11	0.17	1.35	2.03	0.10
C(O)NH$_2$	1.20	3.14	0.06	1.20	2.18	0.24	1.30	2.24	0.18	1.24	2.15	0.18	1.36	2.07	0.11
C(O)CH$_3$	1.17	3.14	0.07	1.20	2.19	0.24	1.32	2.25	0.17	1.26	2.17	0.17	1.38	2.09	0.09
CF$_3$	1.20	3.12	0.07	1.20	2.18	0.25	1.30	2.26	0.18	1.24	2.16	0.18	1.36	2.10	0.11
ONO$_2$	1.26	2.86	0.11	1.11	1.89	0.30	1.24	1.93	0.21	1.20	1.84	0.19	1.35	1.76	0.12
OPO$_2$	1.26	2.84	0.11	1.11	1.88	0.30	1.23	1.92	0.20	1.20	1.84	0.19	1.36	1.74	0.12
C(O)OCH$_3$	1.20	3.11	0.07	1.20	2.15	0.24	1.30	2.21	0.17	1.24	2.11	0.17	1.35	1.96	0.09
OC(O)CH$_3$	1.26	2.87	0.11	1.10	1.90	0.30	1.23	1.94	0.21	1.19	1.86	0.20	1.34	1.78	0.12
OS(O)OH	1.26	2.89	0.11	1.09	1.91	0.31	1.23	1.95	0.22	1.19	1.86	0.20	1.34	1.77	0.13

[a] Geometry optimized at the MP2/TZDP level, 5 d-type distribution. Connectivity is to the leftmost atom in Y.

TABLE 11. Mulliken atomic charges calculated at the MP2 level for carbon compounds[a]

	Atom type						
Compound	H(C)[b]	B Al	C	N P	O S Se	F Cl Br	H (A)
CH$_3$OCH$_3$	0.128		−0.305		−0.161		
CH$_3$NO$_2$	0.180		−0.353	−0.479	0.143 0.148		
CH$_3$ONO	0.156		−0.358	−0.407	0.045 0.250[c]		
CH$_3$OH	0.128		−0.314		−0.359		0.289 (O)
CH$_3$Cl	0.171		−0.448			−0.066	
CH$_3$CP	0.181		−0.456	0.168			
CH$_3$PC	0.198		−0.556[f] −0.559	0.523			
CH$_3$CH$_3$	0.138		−0.415				
CH$_4$	0.147		−0.590				

(*continued overleaf*)

TABLE 11. (*continued*)

Compound	H(C)[b]	B Al	C	N P	O S Se	F Cl Br	H (A)
CH$_3$CHCH$_2$	0.149[f]		−0.440[f]				
	0.109		−0.076				
	0.119[g]		−0.278[g]				
CH$_3$NNN	0.160		−0.309	−0.273			
				0.325[h]			
				−0.223			
CH$_3$CN	0.194		−0.341[f]	−0.535			
			0.294				
CH$_3$NC	0.181		−0.235[f]	−0.372			
			0.063				
CH$_3$SCN	0.187		−0.475[f]	−0.516	0.214		
			0.216				
CH$_3$NCS	0.170		−0.294[f]	−0.688	−0.062		
			0.532				
CH$_3$OCN	0.169		−0.274[f]	−0.536	0.014		
			0.288				
CH$_3$NCO	0.164		−0.297[f]	−0.542	−0.058		
			0.407				
CH$_3$SCH$_3$	0.160		−0.486		−0.005		
CH$_2$CH$_2$CH$_3$	0.140[f]		−0.423[f]				
	0.130[g]		−0.255[g]				
CH$_3$CF$_3$	0.171		−0.429[f]			−0.234	
			0.619				
CH$_3$CCH	0.176[f]		−0.457[f]				
	0.170[j]		0.113				
			−0.354[j]				
CH$_3$NO$_2$	0.162		−0.347	−0.630	0.004		
					0.263[c]		
					0.224[c,k]		
CH$_3$F	0.126		−0.050			−0.327	
CH$_3$NO	0.150		−0.286	−0.287	0.122		
CH$_3$ON	0.152		−0.282	0.320	−0.494		
CH$_3$C(O)H	0.160[f]		−0.432[f]				
	0.079[c]		−0.015[c]		−0.113		
CH$_3$Br	0.180		−0.590			0.050	
CH$_3$SeCN	0.188		−0.623[f]	−0.455	0.515		
			−0.003				
CH$_3$NCSe	0.174		−0.288[f]	−0.571	0.135		
			0.201				
CH$_3$SH	0.165		−0.458		−0.174		0.138 (S)
CH$_3$C(O)OH	0.174		−0.462[f]		−0.140[c]		0.307 (O)
			0.005[c]		−0.230[e]		
CH$_3$C(O)F	0.180		−0.451[f]			−0.256	
			0.244[c]		−0.079		
CH$_3$NH$_2$	0.134		−0.309	−0.574			0.240 (N)
CH$_3$PH$_2$	0.158		−0.602	0.066			0.030 (P)
CH$_3$BH$_2$	0.155	0.251	−0.636				0.040 (B)
CH$_3$AlH$_2$	0.155	0.403	−0.686				0.091 (Al)
CH$_3$C(O)NH$_2$	0.166		−0.484[f]	−0.436			0.244 (N)
			0.072[c]		−0.138		
CH$_3$C(O)OCH$_3$	0.170		−0.460[f]		−0.142		

TABLE 11.   (*continued*)

Compound	H(C)[b]	B Al	C	N P	O S Se	F Cl Br	H (A)
	0.151[i]		0.048[c] −0.316[f]		−0.094[c]		
CH$_3$PO	0.169		−0.625	0.514	−0.397		
CH$_3$OP	0.150		−0.278	0.160	−0.332		
CH$_3$C(O)Cl	0.183		−0.446[f] −0.010[c]		−0.017	−0.077	
CH$_3$C(O)CH$_3$	0.161		−0.467[f] −0.057[c]		−0.090		
CH$_3$PO$_2$	0.198		−0.638	0.900	−0.429[b]		
CH$_3$OPO	0.160		−0.331	0.612	−0.342 −0.387[c]		
CH$_3$OF	0.151		−0.325		0.069	−0.195	
CH$_3$OS(O)OH	0.161		−0.336		−0.173 0.383[d] −0.293[c] −0.386[e]		0.323 (O)
CH$_3$OPO$_2$	0.166		−0.298	0.943	−0.330 −0.406[b,c]		

[a] From MP2 optimized geometries in the TZDP basis set.
[b] Average.
[c] Doubly-bonded atom.
[d] Sulfur.
[e] Hydroxyl.
[f] Methyl.
[g] Methylene.
[h] Middle nitrogen atom.
[i] Methoxy.
[j] Triple bond.
[k] Facing X.

TABLE 12.   Mulliken atomic charges calculated at the MP2 level for silicon compounds[a]

Compound	Si	H(Si)[b]	B Al	C Si Ge Sn Pb	N P	O S Se	F Cl Br	H(A)
SiH$_3$OCH$_3$	0.685	−0.092		−0.313		−0.518		0.141 (C)
SiH$_3$NO$_2$	0.584	−0.035			−0.817	0.167 −0.172		
SiH$_3$ONO	0.639	−0.071			−0.430	−0.445 −0.450[c]		
SiH$_3$OH	0.676	−0.096				−0.700		0.312 (O)
SiH$_3$Cl	0.543	−0.062					−0.357	
SiH$_3$CP	0.575	−0.063		−0.613	−0.226			
SiH$_3$PC	0.419	−0.018		−0.589	−0.224			

(*continued overleaf*)

TABLE 12.   (*continued*)

Compound	Si	H(Si)[b]	B / Al	C / Si / Ge / Sn / Pb	N / P	O / S / Se	F / Cl / Br	H(A)
SiH_3CH_3	0.526	−0.084		−0.757				0.162 (C)
SiH_3SiH_3	0.193	−0.064						
SiH_3GeH_3	0.114	−0.061						
SiH_3SnH_3	0.039	−0.060						
SiH_3PbH_3	0.021	−0.050						
SiH_4	0.300	−0.075						
SiH_3CHCH_2	0.537	−0.081		−0.392 −0.275[f]				0.116 (C) 0.129 (C)
SiH_3NNN	0.634	−0.063			−0.558 −0.330[g] −0.217			
SiH_3CN	0.640	−0.043		−0.095	−0.417			
SiH_3NC	0.703	−0.050		−0.058	−0.496			
SiH_3SCN	0.492	−0.034		0.236	−0.513	−0.114		
SiH_3NCS	0.686	−0.058		0.613	−1.064	−0.061		
SiH_3OCN	0.742	−0.046		0.293	−0.548	−0.350		
SiH_3NCO	0.697	−0.067		0.438	−0.872	−0.060		
SiH_3SCH_3	0.486	−0.066		−0.491		−0.318		0.173 (C)
$SiH_3CH_2CH_3$	0.504	−0.083		−0.558[f] −0.454				0.151 (C) 0.151 (C)
SiH_3CF_3	0.416	−0.044		−0.408			−0.231	
SiH_3CCH	0.647	−0.067		−0.415[h] −0.217				0.184 (C)
SiH_3ONO_2	0.662	−0.055			−0.597	−0.375 0.282[c] 0.195[c,g]		
SiH_3F	0.905	−0.085					−0.648	
$SiH_3C(O)H$	0.449	−0.057		−0.290		−0.077		0.091 (C)
SiH_3NO	0.559	−0.055			−0.615	0.221		
SiH_3ON^d	—	—			—	—		
SiH_3Br	0.318	−0.052					−0.161	
SiH_3SeCN	0.234	−0.031		−0.008	−0.457	0.309		
SiH_3NCSe	0.875	−0.080		−0.065	−0.933	0.234		
SiH_3SH	0.511	−0.064				−0.480		0.160 (S)
$SiH_3C(O)OH$	0.470	−0.053		−0.284		−0.107[c] −0.215		0.294 (O)
$SiH_3C(O)F$	0.466	−0.043		−0.053		−0.025	−0.258	
SiH_3NH_2	0.651	−0.095			−0.864			0.253 (N)
SiH_3PH_2	0.330	−0.064			−0.234			0.149 (P)
SiH_3BH_2	0.341	−0.067	0.116					−0.012 (B)
SiH_3AlH_2	0.280	−0.071	0.075					−0.070 (Al)
$SiH_3C(O)NH_2$	0.438	−0.065		−0.216	−0.391	−0.120		0.241 (N)
$SiH_3C(O)OCH_3$	0.483	−0.057		−0.272[c] −0.324		−0.102[c] −0.077		0.152 (C)
$SiH_3OC(O)CH_3$	0.689	−0.070		−0.096[c] −0.464		−0.481 −0.151[c]		0.173 (C)
SiH_3PO	0.239	−0.058			0.303	−0.386		
SiH_3OP	0.693	−0.066			0.226	−0.722		
$SiH_3C(O)Cl$	0.472	−0.036		−0.273		0.045	−0.135	

TABLE 12. (*continued*)

				Atom type				
	Si	H(Si)[b]	B Al	C Si Ge Sn Pb	N P	O S Se	F Cl Br	H(A)
Compound								
$SiH_3C(O)CH_3$	0.415	−0.061		−0.204[c] −0.459		−0.054		0.162 (C)
SiH_3PO_2	0.310	−0.021			0.645	−0.446[b]		
SiH_3OPO	0.671	−0.064			0.643	−0.713 −0.408[c]		
SiH_3OF	0.668	−0.063				−0.292	−0.189	
SiH_3OSiH_3	0.689	−0.086				−0.859		
$SiH_3OS(O)OH$	0.686	−0.069				−0.533 −0.376[e] −0.293[c] −0.355[i]		
SiH_3OPO_2	0.695	−0.050			−0.938	−0.398 −0.400[c,j]		0.324 (O)

[a] From MP2 optimized geometries in the TZDP basis set.
[b] Average.
[c] Doubly-bonded atom.
[d] Not calculated.
[e] Sulfur.
[f] Methylene.
[g] Middle nitrogen atom.
[h] Attached to Si.
[i] Hydroxyl.
[j] Facing Si.

TABLE 13. Mulliken atomic charges calculated at the MP2 level for germanium compounds[a]

				Atom type				
	Ge	H(Ge)[b]	B Al	C Si Ge Sn Pb	N P	O S Se	F Cl Br	H (A)
Compound								
GeH_3OCH_3	0.617	−0.065		−0.329		−0.506		0.138 (C)
GeH_3NO_2	0.491	−0.016			−0.766	0.160 0.161		
GeH_3ONO	0.531	−0.034			−0.384	−0.249 0.203[c]		
GeH_3OH	0.589	−0.070				−0.682		0.303 (O)
GeH_3Cl	0.498	−0.037					−0.388	
GeH_3CP	0.441	−0.042		−0.548	0.231			
GeH_3PC	0.380	−0.000		−0.594	0.214			
GeH_3CH_3	0.472	−0.071		−0.758				0.166 (C)
GeH_3SiH_3	0.213	−0.048						
GeH_3GeH_3	0.136	−0.045						
GeH_3SnH_3	0.001	−0.043						
GeH_3PbH_3	0.008	−0.034						

(*continued overleaf*)

TABLE 13.   (*continued*)

Compound	Ge	H(Ge)[b]	B Al	C Si Ge Sn Pb	N P	O S Se	F Cl Br	H (A)
GeH$_4$	0.241	−0.060						
GeH$_3$CHCH$_2$	0.438	−0.063		−0.341				0.122 (C)
				−0.287^e				0.129 (C)
GeH$_3$NNN	0.543	−0.036			−0.513			
					−0.319^f			
					−0.242			
GeH$_3$CN	0.535	−0.023		−0.111	−0.353			
GeH$_3$NC	0.580	−0.026		−0.054	−0.449			
GeH$_3$SCN	0.470	−0.012		0.215	−0.512	−0.138		
GeH$_3$NCS	0.580	−0.032		0.517	−0.922	−0.077		
GeH$_3$OCN	0.667	−0.017		0.250	−0.321	−0.545		
GeH$_3$NCO	0.589	−0.041		0.379	−0.785	−0.059		
GeH$_3$SCH$_3$	0.451	−0.046		−0.510		−0.326		0.177 (C)
GeH$_2$CH$_2$CH$_3$	0.488	−0.084		−0.504^e				0.140 (C)
				−0.421				0.137 (C)
GeH$_3$CF$_3$	0.306	−0.028		0.462			−0.228	
GeH$_3$CCH	0.467	−0.045		−0.423^g				0.191 (C)
				−0.099				
GeH$_3$ONO$_2$	0.576	−0.027			−0.583	−0.319		
						−0.248^c		
						−0.160^{c,g}		
GeH$_3$F	0.796	−0.058					−0.622	
GeH$_3$C(O)H	0.356	−0.042		−0.281		−0.055		0.107 (C)
GeH$_3$NO	0.454	−0.040			−0.589	0.256		
GeH$_3$ON	0.569	−0.025			−0.449	−0.043		
GeH$_3$Br	0.354	−0.029					−0.267	
GeH$_3$SeCN	0.274	−0.011		−0.002	−0.460	0.222		
GeH$_3$NCSe	0.587	−0.038		0.018	−0.723	0.230		
GeH$_3$SH	0.549	−0.055				−0.528		0.144 (S)
GeH$_3$C(O)OH	0.368	−0.038		−0.264		−0.079^c		0.293 (O)
						−0.204		
GeH$_3$C(O)F	0.440	−0.031		−0.131		−0.139	−0.338	
GeH$_3$NH$_2$	0.567	−0.074			−0.853			0.255 (N)
GeH$_3$PH$_2$	0.319	−0.049			−0.276			0.052 (P)
GeH$_3$BH$_2$	0.296	−0.055	−0.005					−0.122 (B)
GeH$_3$AlH$_2$	0.339	−0.054	−0.069					−0.035 (Al)
GeH$_3$C(O)NH$_2$	0.331	−0.050		−0.183	−0.390	−0.094		0.243 (N)
GeH$_3$C(O)OCH$_3$	0.370	−0.042		−0.236^c		−0.072^c		
				−0.327		−0.071		0.154 (C)
GeH$_3$OC(O)CH$_3$	0.597	−0.042		−0.091^c		−0.437		
				−0.468		−0.169^c		0.171 (C)
GeH$_3$PO	0.208	−0.040			−0.293	−0.381		
GeH$_3$OP	0.611	−0.038			−0.187	−0.683		
GeH$_3$C(O)Cl	0.372	−0.020		−0.254		−0.063	−0.120	
GeH$_3$C(O)CH$_3$	0.301	−0.045		−0.158^c		−0.039		0.168 (C)
				−0.474				
Ge$_3$PO$_2$	0.280	−0.005			−0.609	−0.437		
Ge$_3$OPO	0.582	−0.034			−0.582	−0.678		
						−0.427^c		

TABLE 13. (*continued*)

Compound	Ge	H(Ge)[b]	B / Al	C / Si / Ge / Sn / Pb	N / P	O / S / Se	F / Cl / Br	H (A)
GeH₃OF	0.588	−0.038				−0.281	−0.195	
GeH₃OSiH₃	0.625	−0.057		0.671		−0.845		0.094 (Si)
GeH₃O(O)OH	0.597	−0.045				−0.487		
						−0.355[d]		
						−0.310[e]		
						−0.364[h]		0.321 (O)
GeH₃OPO₂	0.627	−0.021			−0.909	−0.652[g]		
						−0.405[c]		
						−0.415[c,g]		

[a] From MP2 optimized geometries in the TZDP basis set.
[b] Average.
[c] Doubly-bonded atom.
[d] Sulfur.
[e] Methylene.
[f] Middle nitrogen atom.
[g] Facing Ge.
[h] Hydroxyl.

TABLE 14. Mulliken atomic charges calculated at the MP2 level for tin compounds[a]

Compound	Sn	H(Sn)[b]	B / Al	C / Si / Ge / Sn / Pb	N / P	O / S / Se	F / Cl / Br	H (A)
SnH₃OCH₃	0.803	−0.079		−0.333		−0.628		0.132 (C)
SnH₃NO₂	0.627	−0.030			−0.856	−0.154		
						−0.166		
SnH₃ONO	0.645	−0.047			−0.396	−0.244		
						−0.134[c]		
SnH₃OH	0.782	−0.085				−0.821		0.295 (O)
SnH₃Cl	0.623	−0.053					−0.464	
SnH₃CP	0.586	−0.068		−0.620	−0.238			
SnH₃PC	0.447	−0.021		−0.573	−0.189			
SnH₃CH₃	0.607	−0.092		−0.806				0.159 (C)
SnH₃SiH₃	0.357	−0.072						
SnH₃GeH₃	0.332	−0.068						
SnH₃SnH₃	0.194	−0.065						
SnH₃PbH₃	0.261	−0.058						
SnH₄	0.427	−0.107						
SnH₃CHCH₂	0.566	−0.089		−0.410				0.117 (C)
				−0.286[e]				0.130 (C)
SnH₃NNN	0.697	−0.048			−0.655			
					−0.363[f]			

(*continued overleaf*)

TABLE 14.   (*continued*)

Compound	Sn	H(Sn)^b	B / Al	C / Si / Ge / Sn / Pb	N / P	O / S / Se	F / Cl / Br	H (A)
					−0.261			
SnH$_3$CN	0.645	−0.046		−0.216	−0.290			
SnH$_3$NC	0.746	−0.047		−0.068	−0.537			
SnH$_3$SCN	0.580	−0.026		0.200	−0.511	−0.189		
SnH$_3$NCS	0.750	−0.051		0.624	−1.116	−0.104		
SnH$_3$OCN	0.804	−0.030		0.233	−0.551	−0.397		
SnH$_3$NCO	0.777	−0.059		0.346	−0.875	−0.071		
SnH$_3$SCH$_3$	0.561	−0.062		−0.510		−0.382		0.172 (C)
SnH$_3$CH$_2$CH$_3$	0.570	−0.086		−0.607^e				0.149 (C)
				−0.460				0.153 (C)
SnH$_3$CF$_3$	0.425	−0.047		−0.376			−0.220	
SnH$_3$CCH	0.676	−0.073		−0.522^g				0.183 (C)
				−0.118				
SnH$_3$ONO$_2$	0.710	−0.039			−0.558	−0.359		
						0.237^c		
						0.086^c		
SnH$_3$F	0.961	−0.075					−0.736	
SnH$_3$C(O)H	0.484	−0.068		−0.336		−0.049		0.106 (C)
SnH$_3$NO	0.582	−0.064			−0.655	0.264		
SnH$_3$ON	0.715	−0.042			−0.463	−0.124		
SnH$_3$Br	0.503	−0.045					−0.368	
SnH$_3$SeCN	0.470	−0.027		−0.017	−0.459	0.085		
SnH$_3$NCSe	0.762	−0.058		0.019	−0.828	0.223		
SnH$_3$SH	0.586	−0.063				−0.553		0.155 (S)
SnH$_3$C(O)OH	0.486	−0.060		−0.327		−0.067^c		
						−0.201		0.288 (O)
SnH$_3$C(O)F	0.489	−0.048		−0.103		0.024	−0.264	
SnH$_3$NH$_2$	0.742	−0.092			−0.960			0.247 (N)
SnH$_3$PH$_2$	0.423	−0.072			−0.306			0.049 (P)
SnH$_3$BH$_2$	0.384	−0.084	−0.002					−0.135 (B)
SnH$_3$AlH$_2$	0.589	−0.107	−0.104					−0.083 (Al)
SnH$_3$C(O)NH$_2$	0.445	−0.067		−0.257	−0.366	−0.242		0.242 (N)
SnH$_3$C(O)OCH$_3$	0.494	−0.064		−0.313^c		−0.057^c		
				−0.329		−0.064		0.154 (C)
SnH$_3$OC(O)CH$_3$	0.711	−0.053		−0.137^c		−0.504		
				−0.473		−0.226^c		0.171 (C)
SnH$_3$PO	0.323	−0.066			0.254	−0.378		
SnH$_3$OP	0.771	−0.050			0.141	−0.763		
SnH$_3$C(O)Cl	0.493	−0.036		−0.323		0.088	−0.149	
SnH$_3$C(O)CH$_3$	0.413	−0.066		−0.225^c		−0.036		0.165 (C)
				−0.453				
SnH$_3$PO$_2$	0.385	−0.026			0.560	−0.432^c		
SnH$_3$OPO	0.714	−0.044			−0.713	−0.749		
						−0.455^c		
SnH$_3$OF	0.762	−0.056				−0.399	−0.196	
SnH$_3$OSiH$_3$	0.835	−0.070		0.641		−0.968		−0.100 (Si)
SnH$_3$OS(O)OH	0.727	−0.046				−0.548		
						−0.347^d		
						−0.346^c		

TABLE 14.   (*continued*)

				Atom type			
Compound	Sn	H(Sn)[b]	B Al	C Si Ge Sn Pb	N P	O S Se	F Cl Br	H (A)
SnH$_3$OPO$_2$	0.771	−0.029			0.886	−0.360[h] −0.728[g] −0.404[c] −0.437[c,g]		0.319 (O)

[a] From MP2 optimized geometries in the TZDP basis set.
[b] Average.
[c] Doubly-bonded atom.
[d] Sulfur.
[e] Methylene.
[f] Middle nitrogen atom.
[g] Facing Sn.
[h] Hydroxyl.

TABLE 15.   Mulliken atomic charges calculated at the MP2 level for lead compounds[a]

				Atom type			
Compound	Pb	H(Pb)[b]	B Al	C Si Ge Sn Pb	N P	O S Se	F Cl Br	H (A)
PbH$_3$OCH$_3$	0.765	−0.089		−0.357		−0.547		0.135 (C)
PbH$_3$NO$_2$	0.652	−0.042			−0.861	0.165 0.171		
PbH$_3$ONO	0.658	−0.058			−0.412	−0.134 0.062[c]		
PbH$_3$OH	0.750	−0.094				−0.751		0.284 (O)
PbH$_3$Cl	0.708	−0.061					−0.524	
PbH$_3$CP	0.645	−0.079		−0.634	0.225			
PbH$_3$PC	0.537	−0.031		−0.574	0.129			
PbH$_3$CH$_3$	0.660	−0.110		−0.838				0.169 (C)
PbH$_3$SiH$_3$	0.404	−0.092						
PbH$_3$GeH$_3$	0.355	−0.087						
PbH$_3$SnH$_3$	0.162	−0.084						
PbH$_3$PbH$_3$	0.224	−0.075						
PbH$_4$	0.404	−0.101						
PbH$_3$CHCH$_2$	0.635	−0.100		−0.458 −0.267[e]				0.129 (C) 0.131 (C)
PbH$_3$NNN	0.707	−0.053			−0.583 0.313[f] −0.277			
PbH$_3$CN	0.688	−0.053		−0.203	−0.326			
PbH$_3$NC	0.726	−0.049		−0.055	−0.524			
PbH$_3$SCN	0.665	−0.034		0.188	−0.507	−0.246		

(*continued overleaf*)

TABLE 15.   (*continued*)

Compound	Pb	H(Pb)[b]	B Al	C Si Ge Sn Pb	N P	O S Se	F Cl Br	H (A)
PbH$_3$NCS	0.744	−0.050		−0.350	−0.840	−0.103		
PbH$_3$OCN	0.788	−0.030		−0.180	−0.545	−0.332		
PbH$_3$NCO	0.751	−0.061		−0.280	−0.811	−0.037		
PbH$_3$SCH$_3$	0.652	−0.075		−0.517		−0.426		0.175 (C)
PbH$_3$CH$_2$CH$_3$	0.616	−0.107		−0.641^e −0.459				0.164 (C) 0.162 (C)
PbH$_3$CF$_3$	0.472	−0.062		−0.332			−0.206	
PbH$_3$CCH	0.693	−0.083		−0.425^g −0.201				0.181 (C)
PbH$_3$ONO$_2$	0.728	−0.048			0.561	−0.267 −0.227^c −0.016^{c,g}		
PbH$_3$F	0.913	−0.081					−0.669	
PbH$_3$C(O)H	0.530	−0.085		−0.364		−0.034		0.124 (C)
PbH$_3$NO	0.577	−0.086			−0.612	−0.294		
PbH$_3$ON	0.663	−0.057			−0.498	0.005		
PbH$_3$Br	0.614	−0.054					−0.452	
PbH$_3$SeCN	0.556	−0.035		−0.035	−0.455	0.037		
PbH$_3$NCSe	0.738	−0.058		−0.024	−0.735	−0.195		
PbH$_3$SH	0.681	−0.076				−0.607		0.152 (O)
PbH$_3$C(O)OH	0.538	−0.075		−0.386		−0.039^c −0.176		0.289 (O)
PbH$_3$C(O)F	0.643	−0.071		−0.021		−0.091	−0.318	
PbH$_3$NH$_2$	0.734	−0.104			−0.925			0.252 (N)
PbH$_3$PH$_2$	0.513	−0.090			−0.355			0.056 (P)
PbH$_3$BH$_2$	0.447	−0.103	−0.178					0.021 (B)
PbH$_3$AlH$_2$	0.498	−0.101	−0.076					−0.059 (Al)
PbH$_3$C(O)NH$_2$	0.481	−0.097		−0.301	−0.354	−0.054		0.256 (N)
PbH$_3$C(O)OCH$_3$	0.667	−0.097		−0.238^c −0.247		−0.174^c −0.152		0.145 (C)
PbH$_3$OC(O)CH$_3$	0.722	−0.067		0.136^c −0.472		−0.420 −0.279^c		0.171 (C)
PbH$_3$PO	0.362	−0.085			0.255	−0.361		
PbH$_3$OP	0.758	−0.057			0.106	−0.693		
PbH$_3$C(O)Cl	0.556	−0.050		−0.383		−0.099	−0.123	
PbH$_3$C(O)CH$_3$	0.458	−0.086		−0.262^c −0.448		−0.008		0.172 (C)
PbH$_3$PO$_2$	0.435	−0.037			0.512	−0.418		
PbH$_3$OPO	0.718	−0.054			0.625	−0.665 −0.516^c		
PbH$_3$OF	0.770	−0.067				−0.363	−0.207	
PbH$_3$OSiH$_3$	0.798	−0.072		0.622		−0.892		−0.104 (Si)
PbH$_3$OS(O)OH	0.748	−0.058				−0.474 −0.335^d −0.396^c −0.357^h		0.317 (O)

TABLE 15. (*continued*)

Compound	Pb	H(Pb)[b]	B Al	C Si Ge Sn Pb	N P	O S Se	F Cl Br	H (A)
				Atom type				
PbH$_3$OPO$_2$	0.791	−0.035			0.875	−0.666[g] −0.417[c] −0.479[c,g]		

[a] From MP2 optimized geometries in the TZDP basis set.
[b] Average.
[c] Doubly-bonded atom.
[d] Sulfur.
[e] Methylene.
[f] Middle nitrogen atoms.
[g] Facing Pb.
[h] Hydroxyl.

TABLE 16. MP2/TZDP calculated dipole moments (in *D*) for RY[a]

Y\R =		H	CH$_3$	SiH$_3$	GeH$_3$	SnH$_3$	PbH$_3$
F	0.	1.843	1.908	1.374	2.157	2.590	3.433
AlH$_2$	0.449	0.	0.552	0.390	0.587	0.725	1.321
BH$_2$	0.470	0.	0.552	0.391	0.536	0.719	1.123
SH	0.907	1.173	1.605	1.188	1.515	1.738	2.029
Br	0.	0.854	1.845	1.466	2.088	2.456	3.030
H	0.	0.	0.	0.	0.	0.	0.
CCH	0.742	0.	0.730	0.328	0.119	0.287	0.649
PH$_2$	0.631	0.656	1.142	0.716	0.733	0.742	0.644
NH$_2$	1.939	1.673	1.435	1.283	1.176	1.127	1.350
SCH$_3$	1.661	1.604	1.626	1.462	1.536	1.630	1.773
Cl	0.	1.234	1.990	1.501	2.241	2.693	3.359
NO	0.177	1.695	2.279	2.188	2.087	2.094	1.386
ON	0.177	2.700	3.326	b	3.680	3.961	4.017
C(O)H	1.551	2.318	2.670	2.170	2.116	2.095	1.856
SeCN	3.755	3.743	4.264	4.146	4.412	4.640	4.627
NCSe	3.755	1.980	3.730	1.817	3.102	3.745	5.052
C(O)F	0.675	2.047	2.854	2.752	2.860	3.092	2.993
C(O)NH$_2$	3.552	3.851	3.651	3.430	3.377	3.257	3.152
ONO$_2$	0.	2.125	2.933	2.401	3.344	3.685	4.216
NCS	2.915	1.991	3.435	2.606	3.607	4.721	4.604
SCN	2.915	3.478	4.129	4.031	4.456	4.835	4.944
CH$_2$CH$_3$	0.286	0.	0.086	0.775	0.748	0.748	1.007
C(O)OH	1.909	1.369	1.568	1.314	1.299	1.344	1.290
NO$_2$	0.166	2.537	3.397	3.566	3.973	4.448	4.601
ONO	0.166	1.481	2.132	1.274	2.458	2.725	3.006
PC	1.020	3.537	5.104	5.491	5.974	6.494	6.912
CP	1.020	0.534	1.608	0.408	0.835	1.016	1.327
NCO	0.488	2.036	2.855	2.100	2.956	3.645	4.052
OCN	0.488	3.853	4.499	4.570	5.456	6.206	6.735
CN	2.186	3.013	3.892	3.512	3.964	4.279	4.736
NC	2.186	3.283	4.127	3.540	4.367	4.841	5.723

(*continued overleaf*)

TABLE 16. (*continued*)

Y\R =		H	CH_3	SiH_3	GeH_3	SnH_3	PbH_3
OCH_3	1.918	1.705	1.327	1.216	1.226	1.386	1.903
$CHCH_2$	0.695	0.	0.332	0.662	0.525	0.470	0.603
NNN	0.	1.827	2.356	2.052	2.870	3.552	4.010
OH	1.699	1.948	1.705	1.353	1.507	1.670	2.177
CH_3	0.	0.	0.	0.684	0.614	0.596	0.719
SiH_3	0.040	0.	0.684	0.	0.144	0.187	0.622
GeH_3	0.078	0.	0.614	0.144	0.	0.067	0.491
SnH_3	0.286	0.	0.596	0.187	0.067	0.	0.480
PbH_3	0.543	0.	0.719	0.622	0.491	0.481	0.
CF_3	0.361	1.615	2.285	2.316	2.373	2.612	2.305
$C(O)OCH_3$	2.618	1.764	1.708	1.605	1.556	1.524	1.642
$OC(O)CH_3$	b	1.568	1.708	1.621	1.386	0.931	0.403
PO	1.903	2.239	2.796	2.165	2.041	1.970	1.754
OP	1.903	1.682	2.222	1.502	2.449	2.999	3.730
C(O)Cl	0.259	1.815	2.729	2.844	3.011	3.490	3.363
$C(O)CH_3$	2.354	2.670	2.830	2.410	2.310	2.240	2.172
PO_2	1.347	2.525	3.906	4.096	4.293	4.659	4.703
OPO	1.347	1.483	1.782	1.782	2.232	2.185	1.685
OF	0.175	1.902	2.296	1.940	2.565	2.986	3.430
$OSiH_3$	1.576	1.353	1.216	0.434	1.235	1.857	2.698
OS(O)OH	2.610	1.550	2.009	1.474	2.235	2.375	2.641
OPO_2	0.	3.051	3.987	3.646	4.649	5.117	5.211

a At the MP2/TZDP geometry optimized structures. Connectivity is to the leftmost atom in Y.
b Not calculated.

Another property calculated in the XH_3-Y series is the $X-Y$ bond dissociation energy. For this purpose the MP2 optimized geometries for the XH_3 and Y (doublet spin) radicals were obtained using the unrestricted HF (UHF) method. For comparison to experiment, the electronic energy differences for the reaction

$$XH_3-Y \longrightarrow XH_3\cdot + Y\cdot \qquad (1)$$

were converted to enthalpy changes by correcting the electronic energy differences for vibrational, rotational and translational motions at 298 K and adding $\Delta PV = RT\,\Delta n = +0.6$ kcal mol^{-1}[42,43]. Unfortunately, the UHF method does not give an exact eigenfunction of the spin-squared operator (S^2) and the resultant wave function, upon which the MP2 method is based, is contaminated by higher spin states[44]. If the calculated $\langle S^2 \rangle$ value is close to the formally exact 0.75 value for a spin doublet electronic state, then the properties of the UMP2 wave functions (including energy) are not expected to be significantly affected by higher-order corrections. However, where spin contamination is significant $(\langle S^2 \rangle > 0.75)$ in the starting UHF wave function, the UMPn $(n = 2, 3, \ldots)$ series converges slowly[45] and UMPn theory may give poor results for lower values of n.

A question that arises in using calculated values of ZPE is that of scaling. Experience[2,46] has shown that HF calculated harmonic vibrational frequencies are usually too large, and that scaling by an approximate factor of 0.89 generally brings them in good agreement with experiment. For MP2 calculated frequencies the best fitted scale factor for ZPE is about 0.96[47]. For the thermodynamic quantities calculated here no scaling of the ZPE was imposed. Firstly, a 4% change in a quantity that ranges from 1–6 kcal mol^{-1} as the difference between products and reactants in an enthalpy of reaction is negligible at the level of accuracy in these calculations. Secondly, uniform scaling does not guarantee uniform accuracy and the possible errors introduced into the ZPE by scaling could be larger than just using the unscaled values[48] for this purpose.

Another consideration in calculating bond dissociation energies is basis set superposition error (BSSE). BSSE is defined as the energy resulting from basis functions on each fragment contributing to the wave function of the other fragment at the equilibrium geometry of the molecule in a situation of zero interaction. BSSE is due to incompleteness of the Gaussian basis set for each fragment and will tend to enhance the calculated binding energy. A comprehensive discussion of BSSE has recently been given[49]. One of the advantages of using the CEP and RCEP methods to eliminate the core electrons is that it simultaneously reduces the BSSE. It has been shown[50] in weakly bonding situations that the core electron energy is a large part of the BSSE in all-electron calculations. It is not entirely clear how to define or treat BSSE in a strong bonding situation as in the XH_3-Y series. Therefore, BSSE has not been corrected for in this work.

In order to obtain more accurate and reliable values of the calculated bond dissociation energies[51,52] a set of diffuse sp-type functions was added to every atom except hydrogen, where only an additional s-type function was added. The exponents for these diffuse single Gaussian basis functions were obtained by using the ratio of the smallest two valence orbital exponents of each atom to calculate the diffuse member. The resultant extended basis set is denoted as TZDP++, and the exponents themselves are shown in Table 2. UMPn ($n = 2-4$) energies in the TZDP++ basis set were calculated at the UMP2 geometries optimized in the TZDP basis set (MPn/TZDP++//MP2/TZDP) and these were used for all the thermodynamic quantities calculated here. The accuracy of the MP4 level calculations has recently been discussed[53]. The HF and UMPn energies in the extended basis set are tabulated in Tables 17-23. The HY systems were included in these tabulations for completeness. They can be used to generate exchange reaction energies, for example. Calculated zero point energies, without (Table 24) and with (Table 25) temperature corrections for vibrational, rotational and translational motion, are also listed. Finally, the calculated XH_3-Y bond dissociation energies are shown in Tables 26-30.

TABLE 17. Molecular energies (in au) using the extended basis set (TZDP++) for radical species Y^a

Y	UHF	UMP2	PUMP2	UMP3	UMP4(SDTQ)	$\langle S^2 \rangle^b$
F	−23.799767	−23.963379	−23.964854	−23.973230	23.977973	0.7541
AlH$_2$	−3.024891	−3.081404	−3.081837	−3.094274	−3.098228	0.7541
BH$_2$	−3.765041	−3.839410	−3.839922	−3.854314	−3.858453	0.7534
SH	−10.671916	−10.803917	−10.805888	−10.824419	−10.829840	0.7626
Br	−13.130021	−13.239777	−13.241001	−13.253485	−13.256107	0.7569
H	−0.499284	−0.499819	−0.499819	−0.499819	−0.499819	0.7500
CCH	−11.404284	−11.628401	−11.645072	−11.644639	−11.661129	1.0367
PH$_2$	−7.496066	−7.614740	−7.616986	−7.636813	−7.643479	0.7692
NH$_2$	−10.805899	−10.970744	−10.972941	−10.987004	−10.993379	0.7601
SCH$_3$	−17.332429	−17.622118	−17.624170	−17.653382	−17.666810	0.7633
Cl	−14.655493	−14.789473	−14.807147	−14.810726	−14.810726	0.7585
NO	−25.327962	−25.682713	−25.687133	−25.681919	−25.710639	0.7796
C(O)H	−21.718911	−22.050010	−22.052610	−22.051547	−22.078809	0.7665
SeCN	−24.303562	−24.705910	−24.713728	−24.719332	−24.750130	0.8465
C(O)F	−45.013030	−45.546199	−45.549424	−45.539685	−45.579650	0.7758
C(O)NH$_2$	−32.018545	−32.534551	−32.536739	−32.539828	−32.575860	0.7630
ONO$_2$	−56.520359	−57.391503	−57.399005	−57.340797	−57.437639	0.8009
NCS	−25.259021	−25.675797	−25.687509	−25.695076	−25.726215	0.8923

(*continued overleaf*)

TABLE 17. *(continued)*

Y	UHF	UMP2	PUMP2	UMP3	UMP4(SDTQ)	$\langle S^2 \rangle^b$
CH_2CH_3	-13.854792	-14.152235	-14.154466	-14.183008	-14.195935	0.7637
C(O)OH	-37.442483	-37.976993	-37.979141	-37.974615	-38.013598	0.7642
NO_2	-40.946624	-41.554124	$-41\,558246$	-41.528762	-41.592847	0.7729
CP	-11.728528	-11.918498	-11.950155	-11.941271	-11.958843	1.5150
NCO	-30.834362	-31.305166	-31.313103	-31.311167	-31.347083	0.8245
CN	-15.069386	-15.326086	-15.343146	-15.334044	-15.357188	1.0128
OCH_3	-22.893201	-23.218905	-23.221254	-23.243824	-23.258934	0.7600
$CHCH_2$	-12.646539	-12.906259	-12.915173	-12.930969	-12.945404	0.9056
NNN	-28.958473	-29.449702	-29.465278	-29.453518	-29.496005	0.9056
OH	-16.232659	-16.406306	-16.408227	-16.418653	-16.424830	0.7573
CH_3	-7.191172	-7.333512	-7.335122	-7.352005	-7.357752	0.7621
SiH_3	-5.453401	-5.552897	-5.553452	-5.573041	-5.578951	0.7545
GeH_3	-5.412748	-5.509628	-5.510117	-5.528667	-5.534239	0.7537
SnH_3	-4.982343	-5.068931	-5.069256	-5.086815	-5.092287	0.7527
PbH_3	-5.026597	-5.114296	-5.114812	-5.131943	-5.137628	0.7541
CF_3	-77.056795	-77.805102	-77.806004	-77.802113	-77.843933	0.7543
$C(O)OCH_3$	-44.093096	-44.781817	-44.783954	-44.790605	-44.839330	0.7638
$OC(O)CH_3{}^c$	-44.132522	-44.815447	-44.817479	-44.819523	-44.871642	0.7621
PO	-22.038067	-22.361240	-22.363863	-22.359775	-22.393254	0.7709
C(O)Cl	-35.839677	-36.325336	-36.329559	-36.333762	-36.370096	0.7842
$C(O)CH_3$	-28.395781	-28.880609	-28.882872	-28.894038	-28.928588	0.7642
PO_2	-37.751092	-38.298398	-38.301842	-38.281682	-38.335961	0.7739
OF	-39.399257	-39.783029	-39.786408	-39.789509	-39.813737	0.7669
$OSiH_3$	-21.189909	-21.463675	-21.465636	-21.491555	-21.504921	0.7582
OS(O)OH	-57.705906	-58.497428	-58.502017	-58.486883	-58.548781	0.7820
OPO_2	-53.422076	-54.186484	-54.199172	-54.164330	-54.236633	0.8577

[a] Geometry optimized at the UMP2/TZDP level. Connectivity is to the leftmost atom in Y.
[b] UHF.
[c] Dissociative to $CO_2 + CH_3 \cdot$.

TABLE 18. Molecular energies (in au) for HY using the extended basis set (TZDP++)[a]

Y	RHF	MP2	MP3	MP4(SDTQ)
F	-24.452342	-24.683656	-24.683633	-24.693973
AlH_2	-3.631450	-3.709127	-3.725719	-3.730307
BH_2	-4.404355	-4.506177	-4.524671	-4.529558
SH	-11.281142	-11.448032	-11.469406	-11.475702
Br	-13.731209	-13.876241	-13.890675	-13.894456
H	-1.132632	-1.160186	-1.165898	-1.167349
CCH	-12.074921	-12.360204	-12.371830	-12.391149
PH_2	-8.096276	-8.243400	-8.267926	-8.275383
NH_2	-11.436526	-11.647833	-11.660319	-11.668954
SCH_3	-17.937449	-18.260876	-18.292940	-18.307217
Cl	-15.276903	-15.450150	-15.467284	-15.471667
NO	-25.864064	-26.260593	-26.261086	-26.291134
ON	-25.814715	-26.183972	-26.193043	-26.220124
C(O)H	-22.337551	-22.692569	-22.699191	-22.723399
SeCN	-24.884498	-25.333407	-25.341429	-25.376956
NCSe	-24.889023	-25.342992	-25.347509	-25.386392
C(O)F	-45.649649	-46.210398	-46.207307	-46.245434
$C(O)NH_2$	-32.644192	-33.186720	-33.194996	-33.229560
ONO_2	-57.206448	-58.045643	-58.023827	-58.093632
NCS	-25.862768	-26.336725	-26.345206	-26.382565

TABLE 18.   (*continued*)

Y	RHF	MP2	MP3	MP4(SDTQ)
SCN	−25.851791	−26.322363	−26.333280	−26.368990
CH$_2$CH$_3$	−14.486156	−14.819477	−14.850783	−14.865143
C(O)OH	−38.077587	−38.639665	−38.640536	−38.677945
NO$_2$	−41.538865	−42.168923	−42.150443	−42.205942
ONO	−41.560793	−42.175124	−42.166270	−42.216989
PC	−12.223216	−12.521366	−12.524252	−12.567352
CP	−12.371599	−12.648760	−12.659575	−12.684360
NCO	−31.470753	−31.994017	−31.988318	−32.030544
OCN	−31.429772	−31.954280	−31.951994	−31.989896
CN	−15.738360	−16.058196	−16.059274	−16.085141
NC	−15.723541	−16.029383	−16.037105	−16.060007
OCH$_3$	−23.515024	−23.896613	−23.913906	−23.932469
CHCH$_2$	−13.290608	−13.594953	−13.619318	−13.634937
NNN	−29.550298	−30.115785	−30.099699	−30.153583
OH	−16.869632	−17.100831	−17.105463	−17.115725
CH$_3$	−7.826270	−8.004758	−8.024477	−8.031724
SiH$_3$	−6.073864	−6.196006	−6.219496	−6.226089
GeH$_3$	−6.021489	−6.140564	−6.162901	−6.169222
SnH$_3$	−5.579618	−5.686694	−5.707949	−5.714181
PbH$_3$	−5.611904	−5.720058	−5.740939	−5.747365
CF$_3$	−77.698718	−78.478827	−78.478712	−78.520528
C(O)OCH$_3$	−44.726582	−45.443233	−45.455270	−45.502296
OC(O)CH$_3$	−44.752211	−45.469966	−45.482311	−45.527101
PO	−22.615863	−22.959869	−22.963788	−22.994903
OP	−22.574606	−22.902057	−22.915701	−22.938832
C(O)Cl	−36.460491	−36.973344	−36.984706	−37.018155
C(O)CH$_3$	−29.103578	−29.524241	−29.541740	−29.573767
PO$_2$	−38.362980	−38.923978	−38.913273	−38.961713
OPO	−38.390486	−38.945786	−38.942763	−38.986229
OF	−40.009426	−40.450039	−40.451656	−40.477937
OSiH$_3$	−21.831277	−22.161945	−22.181055	−22.198363
OS(O)OH	−58.333600	−59.142740	−59.140702	−59.196886
OPO$_2$	−54.109809	−54.874232	−54.861154	−54.918264

[a] Geometry optimized at the MP2/TZDP level. Connectivity is to the leftmost atom in Y.

TABLE 19.   Molecular energies (in au) for CH$_3$Y using the extended basis set (TZDP++)[a]

Y	RHF	MP2	MP3	MP4(SDTQ)
F	−31.102136	−31.479763	−31.493198	−31.512339
AlH$_2$	−10.307622	−10.547333	−10.576089	−10.587297
BH$_2$	−11.083433	−11.343741	−11.373706	−11.385160
SH	−17.937457	−18.260877	−18.292942	−18.307218
Br	−20.389256	−20.690675	−20.715658	−20.728147
H	−7.826270	−8.004758	−8.024477	−8.031724
CCH	−18.746487	−19.187546	−19.209267	−19.236538
PH$_2$	−14.760571	−15.066140	−15.101496	−15.116204
NH$_2$	−18.085132	−18.449212	−18.473690	−18.490041
SCH$_3$	−24.595932	−25.078782	−25.120679	−25.143412
Cl	−21.931749	−22.258545	−22.286633	−22.299372
NO	−32.534153	−33.085066	−33.095955	−33.135141
ON	−32.470626	−32.998533	−33.017031	−33.055332

(*continued overleaf*)

TABLE 19.   (*continued*)

Y	RHF	MP2	MP3	MP4(SDTQ)
C(O)H	−29.013578	−29.524241	−29.541740	−29.573767
SeCN	−31.549486	−32.158505	−32.176142	−32.220448
NCSe	−31.550323	−32.158950	−32.172993	−32.220456
C(O)F	−52.326393	−53.042222	−53.050673	−53.096317
C(O)NH$_2$	−39.315755	−40.014819	−40.034510	−40.076645
ONO$_2$	−63.856981	−64.852183	−64.840424	−64.921090
NCS	−32.520862	−33.150706	−33.168248	−33.214921
SCN	−32.514375	−33.143262	−33.163681	−33.208164
CH$_2$CH$_3$	−21.147475	−21.638431	−21.680634	−21.702550
C(O)OH	−44.752211	−45.469966	−45.482311	−45.527101
NO$_2$	−48.209065	−48.991726	−48.984840	−49.048667
ONO	−48.210692	−48.982487	−48.983073	−49.045209
PC	−18.899590	−19.354690	−19.366572	−19.418229
CP	−19.044620	−19.479720	−19.498530	−19.532936
NCO	−38.122587	−38.801943	−38.805779	−38.857808
OCN	−38.081630	−38.759085	−38.767520	−38.815308
CN	−22.416723	−22.890389	−22.902143	−22.935771
NC	−22.386037	−22.847818	−22.864830	−22.897071
OCH$_3$	−30.165088	−30.700900	−30.729734	−30.757240
CHCH$_2$	−19.956081	−20.418046	−20.452891	−20.476111
NNN	−36.202626	−36.926761	−36.918342	−36.983907
OH	−23.515024	−23.896613	−23.913906	−23.932469
CH$_3$	−14.486156	−14.819477	−14.850783	−14.865143
CF$_3$	−84.372937	−85.309760	*b*	*b*
C(O)OCH$_3$	−51.400251	−52.273122	*b*	*b*
PO	−29.292329	−29.785224	−29.801113	−29.838252
OP	−29.223723	−29.707424	−29.731257	−29.764942
C(O)Cl	−43.136427	−43.806820	−43.828254	−43.870107
C(O)CH$_3$	−35.686116	−36.354343	−36.382632	−36.422531
PO$_2$	−45.045912	−45.762117	−45.763597	−45.818355
OPO	−45.038256	−45.748534	−45.756221	−45.809836
OF	−46.668017	−47.261734	−47.274542	−47.310633
OSiH$_3$	−28.476222	−28.960014	−28.991128	−29.017370
OS(O)OH	−64.982251	−65.948210	*b*	*b*
OPO$_2$	−60.758359	−61.676477	−61.674685	−61.741609

a Geometry optimized at the MP2/TZDP level. Connectivity is to the leftmost atom in Y.
b Not calculated.

TABLE 20.   Molecular energies (in au) for SiH$_3$Y using the extended basis set (TZDP++)*a*

Y	RHF	MP2	MP3	MP4(SDTQ)
F	−29.432804	−29.760775	−29.775655	−29.793391
AlH$_2$	−8.557795	−8.739108	−8.774057	−8.784599
BH$_2$	−9.320913	−9.526048	−9.562345	−9.573191
SH	−16.227464	−16.496612	−16.532239	−16.544237
Br	−18.692744	−18.937565	−18.966077	−18.975614
H	−6.073864	−6.196006	−6.219496	−6.226089
CCH	−17.015104	−17.405845	−17.431193	−17.458063
PH$_2$	−13.025926	−13.277783	−13.317770	−13.330959
NH$_2$	−16.379691	−16.692385	−16.719528	−16.734433
SCH$_3$	−22.882167	−23.310796	−23.356339	−23.377029
Cl	−20.240451	−20.512604	−20.543812	−20.553745
NO	−30.787257	−31.288095	−31.301806	−31.340743

TABLE 20.   (*continued*)

Y	RHF	MP2	MP3	MP4(SDTQ)
ON[b]	—	—	—	—
C(O)H	−27.253486	−27.714118	−27.735983	−27.767984
SeCN	−29.833023	−30.386955	−30.408119	−30.450546
NCSe	−29.842248	−30.389342	−30.410438	−30.453330
C(O)F	−50.569624	−51.238133	−51.249253	−51.296837
C(O)NH$_2$	−37.559678	−38.211053	−38.233531	−38.277125
ONO$_2$	−62.163538	−63.106634	−63.098272	−63.176470
NCS	−30.829928	−31.404441	−31.426226	−31.470086
SCN	−30.799662	−31.375590	−31.399119	−31.441843
CH$_2$CH$_3$	−19.405420	−19.845506	−19.891746	−19.912650
C(O)OH	−42.994839	−43.664948	−43.680090	−43.726568
NO$_2$	−46.475089	−47.213208	−47.206376	−47.273278
ONO	−46.512591	−47.234343	−47.237327	−47.298197
PC	−17.185007	−17.578921	−17.597133	−17.644922
CP	−17.309578	−17.693506	−17.716439	−17.749747
NCO	−36.434480	−37.058257	−37.066295	−37.115175
OCN	−36.394243	−37.020794	−37.031619	−37.077714
CN	−20.680986	−21.104978	−21.120090	−21.153615
NC	−20.687006	−21.093035	−21.114683	−21.144932
OCH$_3$	−28.476222	−28.960014	−28.991128	−29.017370
CHCH$_2$	−18.214761	−18.625192	−18.664217	−18.686568
NNN	−34.497158	−35.175840	−35.165379	−35.232920
OH	−21.831277	−22.161945	−22.181055	−22.198363
CH$_3$	−12.748670	−13.030880	−13.066205	−13.079386
SiH$_3$	−10.996230	−11.222440	−11.263395	−11.275869
CF$_3$	−82.608461	−83.496483	*b*	*b*
C(O)OCH$_3$	−49.643273	−50.469066	*b*	*b*
OC(O)CH$_3$	−49.710384	−50.531856	*b*	*b*
PO	−27.535564	−27.985573	−28.004413	−28.044063
OP	−27.532023	−27.962471	−27.989389	−28.020756
C(O)Cl	−41.383059	−42.004571	−42.029246	−42.072182
C(O)CH$_3$	−33.927071	−34.546407	−34.578269	−34.618996
PO$_2$	−43.295283	−43.963323	−43.966721	−44.024709
OPO	−43.353735	−44.011521	−44.022179	−44.073746
OF	−44.966072	−45.509394	−45.524418	−45.559754
OSiH$_3$	−26.800078	−27.230809	−27.264568	−27.288834
OS(O)OH	−63.295310	−64.208432	*b*	*b*
OPO$_2$	−59.073836	−59.940106	−59.941078	−60.005981

[a] Geometry optimized at the MP2/TZDP level. Connectivity is to the leftmost atom in Y.
[b] Not calculated.

TABLE 21.   Molecular energies (in au) for GeH$_3$Y using the extended basis set (TZDP++)[a]

Y	RHF	MP2	MP3	MP4(SDTQ)
F	−29.356140	−29.684927	−29.697422	−29.716587
AlH$_2$	−8.514675	−8.692881	−8.726467	−8.736584
BH$_2$	−9.271984	−9.474020	−9.508929	−9.519472
SH	−16.172332	−16.438877	−16.473597	−16.485361
Br	−18.639589	−18.881428	−18.909168	−18.918329
H	−6.021489	−6.140564	−6.162901	−6.169222
CCH	−16.956116	−17.344637	−17.368614	−17.395614
PH$_2$	−12.976427	−13.225345	−13.264276	−13.277175

(*continued overleaf*)

TABLE 21. (*continued*)

Y	RHF	MP2	MP3	MP4(SDTQ)
NH_2	−16.310955	−16.623291	−16.648699	−16.664280
SCH_3	−22.826554	−23.252782	−23.297381	−23.317894
Cl	−20.183179	−20.452735	−20.483087	−20.492782
NO	−30.730505	−31.230540	−31.242720	−31.282523
ON	−30.697271	−31.174105	−31.192457	−31.230844
C(O)H	−27.199835	−27.658364	−27.678757	−27.710879
SeCN	−29.782990	−30.334271	−30.354384	−30.396551
NCSe	−29.776623	−30.322580	−30.341820	−30.385219
C(O)F	−50.515196	−51.181665	−51.191394	−51.238927
$C(O)NH_2$	−37.505485	−38.154829	−38.175882	−38.219438
ONO_2	−62.098583	−63.040332	−63.030085	−63.108855
NCS	−30.762845	−31.337221	−31.356741	−31.401722
SCN	−30.746660	−31.320278	−31.342667	−31.385241
CH_2CH_3	−19.349288	−19.787040	−19.831870	−19.852839
C(O)OH	−42.940779	−43.608752	−43.622564	−43.668833
NO_2	−46.419174	−47.154524	−47.146470	−47.213272
ONO	−46.448386	−47.168224	−47.169076	−47.230738
PC	−17.135924	−17.527616	−17.544334	−17.592212
CP	−17.251197	−17.633532	−17.654294	−17.688590
NCO	−36.366633	−36.990375	−36.996285	−37.046218
OCN	−36.327052	−36.953888	−36.962215	−37.009554
CN	−20.624170	−21.045604	−21.059451	−21.093026
NC	−20.621989	−21.028374	−21.047696	−21.078737
OCH_3	−28.400217	−28.885723	−28.914258	−28.941960
$CHCH_2$	−18.157337	−18.565821	−18.603366	−18.625823
NNN	−34.431592	−35.110314	−35.097379	−35.166464
OH	−21.755991	−22.087839	−22.104520	−22.123121
CH_3	−12.692173	−12.971848	−13.005848	−13.019034
SiH_3	−10.949936	−11.173113	−11.212732	−11.224821
GeH_3	−10.904062	−11.124099	−11.162458	−11.174173
CF_3	−82.553290	−83.439525	*b*	*b*
$C(O)OCH_3$	−49.589230	−50.413075	*b*	*b*
$OC(O)CH_3$	−49.641112	−50.463876	*b*	*b*
OP	−27.489299	−27.936728	−27.954406	−27.993740
PO	−27.461778	−27.894651	−27.918023	−27.951789
C(O)Cl	−41.328916	−41.948311	−41.971528	−42.014476
$C(O)CH_3$	−33.873938	−34.491195	−34.521729	−34.562303
PO_2	−43.246632	−43.912669	−43.914698	−43.972560
OPO	−43.284384	−43.943897	−43.951453	−44.004826
OF	−44.897923	−45.440591	−45.453521	−45.490016
$OSiH_3$	−26.723224	−27.155763	−27.186823	−27.212760
OS(O)OH	−63.226566	−64.140728	*b*	*b*
OPO_2	−59.006204	−59.873498	−59.871896	−59.938114

[a] Geometry optimized at the MP2/TZDP level. Connectivity is to the leftmost atom in Y.
[b] Not calculated.

TABLE 22. Molecular energies (in au) for SnH_3Y using the extended basis set $(TZDP++)^a$

Y	RHF	MP2	MP3	MP4(SDTQ)
F	−28.915420	−29.235862	−29.246727	−29.266458
AlH_2	−8.079296	−8.287355	−8.277375	−8.287355
BH_2	−8.830512	−9.020757	−9.054971	−9.065432
SH	−15.737039	−15.992544	−16.026451	−16.037881

TABLE 22. (*continued*)

Y	RHF	MP2	MP3	MP4(SDTQ)
Br	−18.208313	−18.438547	−18.465586	−18.474291
H	−5.579618	−5.686694	−5.707949	−5.714181
CCH	−16.516372	−16.894853	−16.918006	−16.945038
PH$_2$	−12.540065	−12.777637	−12.815756	−12.828408
NH$_2$	−15.866167	−16.169784	−16.193719	−16.209661
SCH$_3$	−22.389596	−22.804701	−22.848855	−22.868784
Cl	−19.750111	−20.008631	−20.038209	−20.047508
NO	−30.286501	−30.777414	−30.788482	−30.828918
ON	−30.255720	−30.726074	−30.741954	−30.781693
C(O)H	−26.755777	−27.204207	−27.223628	−27.255935
SeCN	−29.349766	−29.890006	−29.909258	−29.951194
NCSe	−29.339093	−29.875108	−29.893312	−29.936757
C(O)F	−50.073056	−50.730186	−50.738674	−50.786690
C(O)NH$_2$	−37.062170	−37.701866	−37.721783	−37.765667
ONO$_2$	−61.663641	−62.594532	−62.583748	−62.661932
NCS	−30.327054	−30.890518	−30.909566	−30.953808
SCN	−30.312082	−30.875157	−30.896640	−30.939013
CH$_2$CH$_3$	−18.904445	−19.332091	−19.375915	−19.396974
C(O)OH	−42.498257	−43.156684	−43.169336	−43.215951
NO$_2$	−45.980564	−46.705720	−46.696670	−46.763679
ONO	−46.011782	−46.722862	−46.721580	−46.784380
PC	−16.706605	−17.085393	−17.102024	−17.148787
CP	−16.811147	−17.183318	−17.202974	−17.237560
NCO	−35.929449	−36.542944	−36.548026	−36.597668
OCN	−35.889668	−36.507173	−36.514153	−36.561790
CN	−20.185779	−20.596873	−20.609945	−20.643467
NC	−20.184892	−20.581832	−20.600081	−20.631164
OCH$_3$	−27.955095	−28.432078	−28.458946	−28.487292
CHCH$_2$	−17.713381	−18.111929	−18.148439	−18.171006
NNN	−33.990819	−34.661954	−34.646348	−34.716872
OH	−21.312147	−21.635888	−21.650718	−21.670010
CH$_3$	−12.248533	−12.518040	−12.551031	−12.564247
SiH$_3$	−10.512254	−10.723470	−10.762314	−10.774290
GeH$_3$	−10.467883	−10.675836	−10.713397	−10.724955
SnH$_3$	−10.033955	−10.229195	−10.265896	−10.277263
CF$_3$	−82.108501	−82.985369	*b*	*b*
C(O)OCH$_3$	−49.146485	−49.960790	*b*	*b*
OC(O)CH$_3$	−49.203044	−50.018014	*b*	*b*
PO	−27.050478	−27.486726	−27.503527	−27.543011
OP	−27.020497	−27.445767	−27.466835	−27.501915
C(O)Cl	−40.887215	−41.496118	−41.518216	−41.561450
C(O)CH$_3$	−33.430322	−34.037762	−34.067173	−34.108098
PO$_2$	−42.810244	−43.466357	−43.467091	−43.525434
OPO	−42.846820	−43.498354	−43.503998	−43.558227
OF	−44.456829	−44.989524	−45.001159	−45.038356
OSiH$_3$	−26.281097	−26.705125	−26.734506	−26.761026
OS(O)OH	−62.790477	−63.695560	*b*	*b*
OPO$_2$	−58.569587	−59.428197	−59.425197	−59.491728

[a] Geometry optimized at the MP2/TZDP level. Connectivity is to the leftmost atom in Y.
[b] Not calculated.

TABLE 23. Molecular energies (in au) for PbH_3Y using the extended basis set (TZDP++)[a]

Y	RHF	MP2	MP3	MP4(SDTQ)
F	−28.946194	−29.271594	−29.280468	−29.302517
AlH$_2$	−8.120860	−8.287194	−8.319228	−8.329319
BH$_2$	−8.871541	−9.062798	−9.096179	−9.106839
SH	−15.778830	−16.036015	−16.069366	−16.081067
Br	−18.251266	−18.482858	−18.509454	−18.518381
H	−5.611904	−5.720058	−5.740939	−5.747365
CCH	−16.550182	−16.930863	−16.952990	−16.980737
PH$_2$	−12.583323	−12.822338	−12.859813	−12.872684
NH$_2$	−15.902605	−16.209877	−16.232593	−16.249642
SCH$_3$	−22.432404	−22.850735	−22.893841	−22.914543
Cl	−19.792129	−20.052228	−20.081310	−20.090889
NO	−30.332622	−30.827351	−30.837145	−30.879106
ON	−30.299724	−30.775465	−30.789501	−30.831002
C(O)H	−26.798393	−27.248988	−27.267366	−27.300283
SeCN	−29.393775	−29.936798	−29.955010	−29.997406
NCSe	−29.371515	−29.910167	−29.926960	−29.971475
C(O)F	−50.114529	−50.773947	−50.781384	−50.829899
C(O)NH$_2$	−37.105979	−37.748110	−37.766816	−37.811270
ONO$_2$	−61.712422	−62.646053	−62.633967	−62.713353
NCS	−30.361080	−30.929144	−30.946131	−30.992518
SCN	−30.356374	−30.922148	−30.942556	−30.985427
CH$_2$CH$_3$	−18.949860	−19.378828	−19.421484	−19.443000
C(O)OH	−42.540771	−43.201669	−43.213157	−43.260338
NO$_2$	−46.026770	−46.753308	−46.743393	−46.810900
ONO	−46.060805	−46.776545	−46.772398	−46.838238
PC	−16.747734	−17.128456	−17.144179	−17.191732
CP	−16.846265	−17.221353	−17.239188	−17.275320
NCO	−35.963542	−36.581524	−36.584789	−36.636289
OCN	−35.929686	−36.551827	−36.556744	−36.606234
CN	−20.220929	−20.633977	−20.646198	−20.680377
NC	−20.218708	−20.619394	−20.635947	−20.668257
OCH$_3$	−27.989262	−28.472786	−28.497269	−28.527863
CHCH$_2$	−17.754297	−18.154924	−18.190315	−18.213444
NNN	−34.029168	−34.704924	−34.687224	−34.759679
OH	−21.344308	−21.673462	−21.686261	−21.707504
CH$_3$	−12.291486	−12.562206	−12.594321	−12.607939
SiH$_3$	−10.553930	−10.766274	−10.804353	−10.816448
GeH$_3$	−10.509746	−10.718762	−10.755611	−10.767293
SnH$_3$	−10.075357	−10.271694	−10.307739	−10.319226
PbH$_3$	−10.117001	−10.314322	−10.349714	−10.361316
CF$_3$	−82.153154	−83.032061	*b*	*b*
C(O)OCH$_3$	−49.189408	−50.006217	*b*	*b*
OC(O)CH$_3$	−49.248926	−50.069376	*b*	*b*
PO	−27.095101	−27.533461	−27.549447	−27.589318
OP	−27.059241	−27.490253	−27.508513	−27.546471
C(O)Cl	−40.929707	−41.541389	−41.562422	−41.606165
C(O)CH$_3$	−33.474947	−34.084845	−34.112902	−34.154540
PO$_2$	−42.851645	−43.510268	−43.509948	−43.568893
OPO	−42.891059	−43.549001	−43.551985	−43.609000
OF	−44.498977	−45.034707	−45.044930	−45.083967
OSiH$_3$	−26.313248	−26.743205	−26.770285	−26.799071
OS(O)OH	−62.837510	−63.747596	*b*	*b*
OPO$_2$	−58.613526	−59.477789	−59.472739	−59.541243

[a] Geometry optimized at the MP2/TZDP level. Connectivity is to the leftmost atom in Y.
[b] Not calculated.

TABLE 24.   MP2 level zero point energies (ZPE, in kcal mol^{-1}) for RY^a

Y		H	CH$_3$	SiH$_3$	GeH$_3$	SnH$_3$	PbH$_3$
F	0.	5.9	25.1	17.7	16.4	14.3	13.6
AlH$_2$	6.5	11.7	30.5	23.1	22.8	21.1	20.4
BH$_2$	9.3	16.9	35.6	28.1	27.1	25.2	24.6
SH	4.0	9.9	29.6	22.1	20.9	18.9	18.2
Br	0.	3.9	23.8	16.7	15.5	13.6	13.0
H	0.	6.5	28.6	20.2	19.1	16.8	16.1
CCH	10.9	16.4	34.9	27.3	26.2	24.2	23.5
PH$_2$	8.8	15.6	34.9	27.3	26.3	24.3	23.6
NH$_2$	12.2	21.8	40.7	32.3	31.2	28.9	28.3
SCH$_3$	23.5	29.6	48.4	41.1	39.9	37.9	37.3
Cl	0.	4.4	24.2	17.0	15.8	13.9	13.3
NO	5.2	8.6	27.3	19.4	18.1	16.0	15.2
ON	5.2	8.7	27.4	b	18.4	16.3	15.2
C(O)H	8.2	16.8	35.1	27.0	25.8	23.7	23.1
SeCN	6.1	9.4	28.7	21.4	20.3	18.4	17.8
NCSe	6.1	11.1	29.9	22.7	21.4	19.4	18.7
C(O)F	5.1	13.0	30.9	22.8	21.6	19.5	18.9
C(O)NH$_2$	20.4	28.7	46.5	38.4	37.2	35.1	34.5
ONO$_2$	11.3	16.5	34.5	26.8	25.5	23.5	22.9
NCS	5.7	11.8	30.5	23.0	21.8	19.7	19.0
SCN	5.7	10.5	29.6	22.2	21.0	19.1	18.5
CH$_2$CH$_3$	37.9	47.6	65.8	57.5	56.3	54.2	53.6
C(O)OH	12.9	21.2	39.0	30.9	29.6	27.5	26.9
NO$_2$	5.6	13.8	31.7	23.4	22.2	20.0	19.3
ONO	5.6	12.4	30.5	22.1	21.6	19.7	19.3
PC	1.8	5.5	25.1	18.1	17.0	15.2	14.6
CP	1.8	8.3	26.7	19.3	18.1	16.1	15.5
NCO	7.0	13.2	32.0	24.6	23.3	21.2	20.5
OCN	7.0	13.1	31.6	23.9	22.6	20.5	19.9
CN	4.1	9.9	28.5	20.9	17.7	17.0	19.7
NC	4.1	9.5	28.5	20.8	19.6	17.5	16.7
OCH$_3$	23.8	32.6	50.8	43.0	41.6	39.4	38.7
CHCH$_2$	24.0	32.2	50.3	42.5	41.3	39.2	38.5
NNN	5.9	13.2	31.8	24.1	22.8	20.7	20.0
OH	5.4	13.5	32.6	24.7	23.4	21.1	20.4
CH$_3$	18.9	28.6	47.6	39.2	38.1	35.9	35.3
SiH$_3$	13.8	20.2	39.2	31.7	30.6	28.8	28.1
GeH$_3$	13.0	19.1	38.1	30.6	29.6	27.7	27.1
SnH$_3$	11.5	16.8	35.9	28.8	27.7	25.9	25.2
PbH$_3$	10.8	16.1	35.3	28.1	27.1	25.2	24.6
CF$_3$	7.6	16.0	33.2	24.8	23.6	21.5	20.9
C(O)OCH$_3$	31.2	39.3	56.8	48.8	47.6	45.5	44.9
OC(O)CH$_3$	26.1^c	39.0	56.8	49.1	47.8	45.7	45.1
PO	1.5	6.2	24.8	17.8	16.7	14.9	14.2
OP	1.5	8.2	26.8	19.1	18.0	15.7	15.0
C(O)Cl	4.1	11.9	29.9	21.9	20.7	18.6	17.9
C(O)CH$_3$	27.4	35.1	52.9	44.9	43.7	41.6	41.0
PO$_2$	4.2	10.2	28.6	21.2	20.1	18.2	17.6
OPO	4.2	11.0	29.3	21.6	20.3	18.3	17.8
OF	2.1	8.7	27.1	19.4	18.1	16.0	15.4
OSiH$_3$	16.9	24.7	43.0	35.4	34.1	31.9	31.2
OS(O)OH	14.1	20.3	38.7	31.0	29.7	27.7	27.0
OPO$_2$	9.2	14.4	32.4	24.8	23.5	21.4	20.9

^{a} Geometry optimized at the MP2/TZDP level. Connectivity is to the leftmost atom in Y.
^{b} Not calculated.
^{c} Dissociated to CO$_2$ and CH$_3\cdot$.

TABLE 25. MP2 level zero point energies and temperature corrections to 298 K (in kcal mol^{-1}) for RY[a]

				R			
Y		H	CH$_3$	SiH$_3$	GeH$_3$	SnH$_3$	PbH$_3$
F	0.	7.4	26.9	19.7	18.5	16.6	16.0
AlH$_2$	8.3	13.7	33.5	27.1	26.2	24.7	24.1
BH$_2$	11.1	18.7	38.0	31.0	30.1	28.4	27.8
SH	5.5	11.7	31.8	24.7	23.7	21.9	21.3
Br	0.	5.4	25.7	18.9	17.9	16.1	15.6
H	0.	8.0	30.4	22.1	21.0	18.9	18.2
CCH	12.4	18.2	37.4	30.3	29.3	27.4	26.9
PH$_2$	10.6	17.5	37.3	30.1	29.2	27.4	26.9
NH$_2$	14.0	23.6	42.9	34.9	33.8	31.8	31.2
SCH$_3$	25.4	31.8	51.4	44.6	43.5	41.8	41.2
Cl	0.	5.9	26.1	19.1	18.1	16.3	15.8
NO	6.7	10.4	29.7	22.3	21.1	19.3	18.6
ON	6.7	10.5	29.8	b	21.3	19.5	18.6
C(O)H	10.0	18.6	37.5	29.9	28.9	27.1	25.0
SeCN	8.1	11.9	32.0	25.2	24.2	22.5	22.0
NCSe	8.1	13.4	33.2	26.2	25.1	23.3	22.7
C(O)F	7.0	14.9	33.6	26.1	25.1	23.2	22.7
C(O)NH$_2$	22.4	30.7	49.6	42.1	41.0	39.2	38.7
ONO$_2$	13.3	18.8	37.6	30.2	29.1	27.3	26.8
NCS	7.6	13.9	33.7	26.5	25.4	23.6	22.9
SCN	7.6	12.8	32.7	25.7	24.7	23.0	22.5
CH$_2$CH$_3$	40.4	49.7	68.6	60.9	59.8	57.9	57.4
C(O)OH	14.9	23.2	41.8	34.3	33.2	31.3	30.8
NO$_2$	7.4	15.6	34.4	26.7	25.6	23.7	23.1
ONO	7.4	14.4	33.3	25.6	24.8	23.2	22.8
PC	3.2	7.4	27.9	21.4	20.4	18.8	18.3
CP	3.2	10.0	29.2	22.2	21.2	19.4	18.8
NCO	8.7	15.3	35.0	27.8	26.6	24.8	24.1
OCN	8.7	15.3	34.5	27.2	26.0	24.1	23.6
CN	5.5	11.5	30.8	23.6	22.6	20.7	20.2
NC	5.5	11.4	30.9	23.6	22.5	20.7	20.1
OCH$_3$	25.7	34.7	53.5	46.3	44.9	43.0	42.3
CHCH$_2$	26.0	34.1	52.9	45.6	44.5	42.6	42.0
NNN	7.6	15.3	34.6	27.3	26.1	24.3	23.6
OH	6.9	15.2	34.7	27.1	26.0	23.9	23.3
CH$_3$	20.9	30.4	49.7	41.8	40.8	38.8	38.4
SiH$_3$	15.7	22.1	41.8	34.8	33.9	32.2	31.7
GeH$_3$	14.9	21.0	40.8	33.9	33.0	31.3	30.7
SnH$_3$	13.4	18.9	38.8	32.2	31.3	29.7	29.1
PbH$_3$	12.8	18.2	38.4	31.7	30.7	29.1	28.6
CF$_3$	9.8	18.2	36.2	28.5	27.5	25.6	25.2
C(O)OCH$_3$	34.1	42.1	60.7	53.2	52.1	50.3	49.8
OC(O)CH$_3$	29.4[c]	41.8	60.7	53.3	52.1	50.3	49.7
PO	3.0	8.1	27.6	21.0	20.1	18.5	17.9
OP	3.0	10.0	29.3	22.0	20.9	18.9	18.3
C(O)Cl	6.2	14.0	32.8	25.4	24.3	22.5	22.0
C(O)CH$_3$	29.9	37.5	56.3	48.8	47.8	46.0	45.4
PO$_2$	6.2	12.3	31.8	24.9	24.0	22.3	21.8
OPO	6.2	13.2	32.5	25.1	24.0	22.1	21.6

TABLE 25. (*continued*)

Y		H	CH$_3$	SiH$_3$	GeH$_3$	SnH$_3$	PbH$_3$
					R		
OF	3.6	10.5	29.4	22.1	21.0	19.1	18.6
OSiH$_3$	19.1	27.1	46.3	39.2	37.9	36.0	35.4
OS(O)OH	16.7	23.4	42.6	35.3	34.1	32.3	31.7
OPO$_2$	11.5	17.0	36.1	28.8	27.6	25.7	25.2

[a] Geometry optimized at the MP2/TZDP level. Connectivity is to the leftmost atom in Y. The complete term is $\Delta E_{vib}{}^{298} + \Delta E_{rot}{}^{298} + \Delta E_{tr}{}^{298}$.
[b] Not calculated.
[c] Dissociated to CO$_2$ + CH$_3$·.

TABLE 26. Bond dissociation energies (in kcal mol^{-1}) for CH$_3$—Y → CH$_3$· + Y·[a]

Y	HF	MP2	PMP2	MP3	MP4	Experimental
F	64.4	109.4	107.7	100.0	105.4	112.8[b], 109.9[e]
AlH$_2$	53.8	79.4	78.1	77.8	78.7	
BH$_2$	74.4	101.8	100.5	99.6	100.6	
SH	41.9	72.7	70.4	68.3	70.3	74.6[b]
Br	38.5	69.5	67.7	64.9	67.5	70.6[b]
H	76.0	98.7	97.7	99.4	100.4	104.7[b]
CCH	91.3	138.1	126.6	129.9	133.1	125.3[b]
PH$_2$	40.8	68.8	66.4	71.3	66.9	72.1[b]
NH$_2$	47.9	83.6	81.2	77.1	79.8	84.7[b]
SCH$_3$	40.9	72.8	70.2	67.8	70.1	73.7[b], 77.6[c]
Cl	48.8	80.5	78.4	75.4	77.5	83.5[b]
NO	7.9	41.7	37.9	37.4	40.4	39.6[b]
ON	−32.0	−12.7	−16.5	−12.2	−9.8	
C(O)H	58.9	82.3	79.7	80.7	80.1	83.4[b]
SeCN	32.0	72.3	66.4	63.4	68.2	
NCSe	31.3	71.4	65.5	64.4	67.0	
C(O)F	71.6	96.9	93.8	94.7	94.6	98.9[b]
C(O)NH$_2$	60.8	86.4	84.0	83.8	84.1	
ONO$_2$	88.7	77.0	71.3	89.8	76.1	79.3[b]
NCS	39.7	84.1	75.8	71.4	77.6	
SCN	36.7	80.5	76.3	69.6	74.3	
CH$_2$CH$_3$	57.0	89.1	86.7	84.7	86.7	87.8[b]
C(O)OH	69.0	94.7	92.3	92.3	92.3	84.9[b]
NO$_2$	39.0	59.6	56.0	59.6	55.8	60.6[b]
ONO	41.3	55.1	51.5	59.8	55.0	58.6[b], 60.8[c]
PC	−12.0	61.2	40.4	42.8	60.6	
CP	73.9	138.4	117.5	124.3	131.3	
NCO	56.1	97.6	91.7	84.7	91.2	103.9[b]
OCN	30.9	71.3	65.3	61.2	65.0	
CN	94.2	141.0	129.3	131.8	134.8	121.8[b]
NC	74.8	114.2	102.5	108.3	110.4	97.8[b]
OCH$_3$	44.3	86.9	84.4	77.7	81.9	83.1[b]
CHCH$_2$	68.9	106.5	99.9	101.2	103.1	101.8[b], 98.1[f]

(*continued overleaf*)

TABLE 26. (*continued*)

Y	HF	MP2	PMP2	MP3	MP4	Experimental
NNN	27.7	84.6	73.8	65.3	76.2	80.0[b]
OH	50.9	92.1	89.9	83.6	87.8	92.5[b]
CH$_3$	57.8	88.4	86.3	84.8	86.6	89.8[b], 85.8[h]
CF$_3$	73.5	102.5	100.9	d	d	102.2[b]
C(O)OCH$_3$	67.7	94.1	91.7	d	d	92.5[b]
OC(O)CH$_3$	38.2	68.1	65.8	d	d	83.3[b]
PO	36.5	53.7	52.0	53.0	51.6	
OP	−8.3	3.2	0.5	7.4	3.9	
C(O)Cl	61.2	87.8	84.1	84.3	84.2	
C(O)CH$_3$	57.3	83.1	80.1	80.8	80.6	84.4[b], 81.8[e]
PO$_2$	60.9	77.6	74.4	77.4	74.1	
OPO	55.4	68.4	65.2	72.1	68.1	
OF	44.4	86.8	83.7	79.2	83.0	81.9[b,g]
OSiH$_3$	54.0	95.9	94.2	86.9	90.8	
OSO$_2$H	49.0	69.2	65.3	d	d	
OPO$_2$	88.0	95.1	86.1	96.3	89.3	

[a] All results are at the MP2/TZDP++//MP2/TZDP level, corrected for vibration, rotation and translation at 298 K from Table 25. An additional correction of $\Delta PV = RT\,\Delta n = 0.6$ kcal mol^{-1} is made for the reaction itself. Connectivity is to the leftmost atom in Y.
[b] Calculated from enthalpies of formation tabulated in References 13 and 224.
[c] From Reference 225.
[d] Not calculated.
[e] From Reference 232.
[f] Enthalpy of formation for C$_2$H$_3$· from References 226 and 227.
[g] Calculated ΔH_f (CH$_3$OF) from Reference 230.
[h] From Reference 231.

TABLE 27. Bond dissociation energies (in kcal mol^{-1}) for $SiH_3-Y \rightarrow SiH_3\cdot + Y\cdot$ [a]

Y	HF	MP2	PMP2	MP3	MP4	Experimental
F	109.3	150.0	148.7	140.5	145.0	155.9[b], 150.1[c,f]
AlH$_2$	47.4	63.3	62.6	64.5	64.9	
BH$_2$	61.7	81.3	80.7	85.3	85.8	57.9[b,f]
SH	61.2	84.8	83.2	81.7	82.1	70(?)[d]
Br	66.0	88.6	87.2	85.0	85.6	92.1[b], 88.4[c]
H	69.9	84.1	83.8	86.2	86.6	90.5[b,c]
CCH	97.2	139.3	128.5	132.4	135.2	128.7[b,j]
PH$_2$	44.8	65.9	64.2	64.5	64.9	
NH$_2$	70.9	101.3	99.6	95.5	97.1	100[d], 103.2[f]
SCH$_3$	57.6	82.3	80.7	78.6	79.5	
Cl	79.8	104.0	102.7	99.9	100.2	107.8[c]
NO	4.4	33.6	30.5	30.1	32.8	
ON[e]	—	—	—	—	—	
C(O)H	47.3	66.2	64.2	66.3	65.6	
SeCN	46.9	79.6	74.4	71.8	75.4	
NCSe	51.7	80.1	74.9	72.0	76.2	
C(O)F	62.0	84.4	82.1	82.9	83.9	
C(O)NH$_2$	51.7	74.2	72.4	72.3	73.4	
ONO$_2$	118.5	101.2	96.1	115.1	99.7	
NCS	71.1	107.7	100.0	96.6	100.9	
SCN	52.9	90.4	82.7	80.4	84.0	
CH$_2$CH$_3$	56.8	83.9	82.1	85.7	82.2	47.4[b], 82.6[f]
C(O)OH	59.0	81.7	79.6	80.0	81.0	

TABLE 27. (*continued*)

Y	HF	MP2	PMP2	MP3	MP4	Experimental
NO_2	44.1	63.6	60.7	62.6	60.7	
ONO	68.7	78.0	75.0	83.2	77.4	
PC	0.	65.6	45.4	50.1	65.3	
CP	77.4	136.7	116.5	124.1	130.3	
NCO	89.3	122.8	122.5	111.5	115.9	
OCN	64.6	99.9	94.6	64.1	93.0	
CN	97.5	140.0	129.0	131.9	134.7	
NC	101.2	132.5	121.5	128.5	129.2	
OCH_3	77.0	113.8	112.0	105.1	108.3	
$CHCH_2$	68.8	100.9	94.9	97.2	98.5	117.1^b, 113.4^g
						96.0^k
NNN	50.1	105.3	95.2	83.7	95.7	
OH	87.2	123.3	121.8	114.9	118.2	128^d, 122.4^f
						124.8^h, 121.9^i
CH_3	60.7	86.1	84.7	84.0	84.9	$88.2^{b,c}$
SiH_3	53.3	70.4	69.7	74.2	71.2	$73.8^{b,c}$
CF_3	62.3	87.5	83.6	*e*	*e*	
$C(O)OCH_3$	57.9	81.5	79.8	*e*	*e*	
$OC(O)CH_3$	70.5	95.0	93.3	*e*	*e*	
PO	26.0	43.1	41.1	43.2	43.4	
OP	22.7	27.6	25.6	32.8	27.8	
C(O)Cl	53.6	76.4	76.3	73.9	74.4	
$C(O)CH_3$	46.3	68.2	66.5	67.2	67.3	69.8^b
PO_2	54.6	67.9	65.4	67.9	66.5	
OPO	93.5	97.9	98.0	102.5	97.1	
OF	69.0	106.7	104.2	99.4	102.6	
$OSiH_3$	94.6	130.6	129.1	121.7	124.8	
OS(O)OH	83.0	96.9	93.7	*e*	*e*	
OPO_2	123.5	125.0	117.3	126.8	118.5	

^{a} All results are at the MP2/TZDP++//MP2/TZDP level, corrected for vibration, rotation and translation at 298 K from Table 25. An additional correction of $\Delta PV = RT\,\Delta n = 0.6$ kcal mol^{-1} is made for the reaction itself. Connectivity is to the leftmost atom in Y.
^{b} Calculated from the enthalpies of formation tabulated in References 13 and 224.
^{c} From Reference 234.
^{d} From Reference 21.
^{e} Not calculated.
^{f} From enthalpy of formation calculated in Reference 229.
^{g} Enthalpy of formation of C_2H_3 from References 226 and 227.
^{h} Calculated value from Reference 228.
^{i} Calculated value from Reference 92.
^{j} Calculated enthalpy of formation from Reference 219.
^{k} Calculated value from Reference 233.

TABLE 28. Bond dissociation energies (in kcal mol^{-1}) for $GeH_3-Y \rightarrow GeH_3\cdot + Y.^a$

Y	HF	MP2	PMP2	MP3	MP4
F	87.1	130.0	128.7	119.7	125.2
AlH_2	45.9	61.5	60.9	65.0	65.3
BH_2	55.6	74.9	78.4	75.5	76.1
SH	52.3	75.9	74.4	72.9	73.4
Br	58.4	80.4	79.4	77.3	77.9

(*continued overleaf*)

TABLE 28. (*continued*)

Y	HF	MP2	PMP2	MP3	MP4
H	62.8	76.8	76.5	78.8	79.3
CCH	85.9	128.2	117.5	121.2	124.3
PH$_2$	41.1	62.1	60.3	60.7	61.1
NH$_2$	53.6	85.4	83.7	79.2	81.5
SCH$_3$	48.5	73.4	71.8	69.8	70.7
Cl	69.5	93.8	91.8	89.8	90.2
NO	−5.3	25.1	22.0	21.3	24.7
ON	−26.2	−10.3	−13.4	−10.3	−7.7
C(O)H	39.4	58.6	56.6	58.4	58.0
SeCN	41.2	73.9	69.9	66.2	71.0
NCSe	36.3	65.7	60.5	57.4	61.8
C(O)F	53.5	76.4	74.0	74.6	75.9
C(O)NH$_2$	43.5	66.3	64.7	64.3	65.5
ONO$_2$	103.5	87.0	82.0	100.5	83.8
NCS	54.9	93.0	85.3	81.2	86.3
SCN	45.4	83.0	75.4	73.0	76.7
CH$_2$CH$_3$	47.4	74.6	72.9	71.5	73.1
C(O)OH	50.9	73.8	71.8	72.0	73.1
NO$_2$	34.8	54.3	54.7	56.5	51.4
ONO	54.0	63.7	60.8	68.2	63.1
PC	−5.1	60.7	40.6	45.0	62.8
CP	66.5	126.4	106.2	113.2	120.2
NCO	72.6	107.8	102.5	95.8	101.1
OCN	48.4	85.6	80.2	75.0	78.7
CN	87.5	130.1	119.1	121.9	124.9
NC	60.7	119.4	108.4	116.7	116.0
OCH$_3$	55.5	94.9	93.2	85.3	89.7
CHCH$_2$	58.5	91.1	85.2	87.2	88.7
NNN	34.9	91.7	81.7	72.9	82.5
OH	65.8	104.3	102.8	95.0	99.3
CH$_3$	51.0	76.4	75.0	74.2	75.3
SiH$_3$	49.9	66.7	66.0	67.0	67.3
GeH$_3$	46.7	63.2	62.6	63.4	63.7
CF$_3$	50.4	76.1	75.2	*b*	*b*
C(O)OCH$_3$	49.8	73.8	72.2	*b*	*b*
OC(O)CH$_3$	52.9	80.0	78.3	*b*	*b*
PO	22.5	39.7	37.8	39.8	40.0
OP	4.5	12.5	10.6	16.2	12.8
C(O)Cl	45.4	68.5	65.6	65.9	66.5
C(O)CH$_3$	38.6	61.0	52.2	59.7	60.0
PO$_2$	49.7	63.4	60.9	65.5	61.9
OPO	73.3	85.3	80.5	86.2	82.2
OF	52.0	90.9	88.5	83.0	87.2
OSiH$_3$	72.4	111.2	110.0	101.2	108.9
OS(O)OH	65.8	82.0	78.8	*b*	*b*
OPO$_2$	106.9	110.7	102.4	111.7	104.3

^{a} All results are at the MP2/TZDP++//MP2/TZDP level, corrected for vibration, rotation and translation at 298 K from Table 28. An additional correction of $\Delta PV = RT\,\Delta n = 0.6$ kcal mol^{-1} is made for the reaction itself. Connectivity is to the leftmost atom in Y.
^{b} Not calculated.

TABLE 29. Bond separation energies (in kcal mol^{-1}) for $SnH_3-Y \rightarrow SnH_3\cdot + Y\cdot$ [a]

Y	HF	MP2	PMP2	MP3	MP4
F	81.1	125.1	124.1	114.5	120.5
AlH_2	42.8	56.7	56.3	58.0	58.4
BH_2	48.9	67.2	66.7	68.1	68.7
SH	49.5	72.7	71.3	69.9	70.2
Br	58.1	79.4	78.4	76.5	76.9
H	56.3	69.1	68.9	71.2	71.7
CCH	80.4	122.9	112.3	116.1	119.2
PH_2	35.9	56.2	54.6	55.0	55.3
NH_2	45.1	77.8	76.3	75.8	74.0
SCH_3	44.6	68.9	67.4	65.8	66.4
Cl	68.2	92.0	90.8	88.2	88.4
NO	−13.5	17.6	14.6	13.8	17.7
ON	−33.1	−14.8	−17.8	−15.6	−12.1
C(O)H	31.1	50.4	48.6	50.4	50.1
SeCN	39.7	71.9	66.8	64.3	67.9
NCSe	32.2	61.7	56.6	53.5	58.0
C(O)F	46.5	70.0	67.8	68.2	69.8
$C(O)NH_2$	35.7	58.9	57.4	56.9	58.4
ONO_2	101.0	84.1	79.2	98.0	82.8
NCS	51.8	89.5	81.9	78.1	82.9
SCN	43.0	80.4	72.9	70.6	73.2
CH_2CH_3	38.7	66.1	64.5	63.1	64.7
C(O)OH	43.7	67.1	65.1	65.3	66.7
NO_2	30.1	49.6	46.8	48.6	47.0
ONO	50.2	60.8	58.0	64.7	60.5
PC	−4.1	59.9	39.8	44.8	59.7
CP	60.7	120.7	100.7	107.5	114.8
NCO	68.6	103.9	98.7	92.1	97.2
OCN	44.4	82.1	76.9	71.5	75.4
CN	82.9	125.6	114.6	117.5	120.5
NC	82.4	116.0	105.1	111.3	112.8
OCH_3	46.6	87.2	85.5	77.2	82.1
$CHCH_2$	50.4	83.2	77.4	79.4	81.1
NNN	28.7	87.2	77.3	63.8	78.0
OH	58.0	97.8	96.4	88.1	92.9
CH_3	43.2	68.6	67.4	66.5	67.8
SiH_3	45.5	61.3	60.7	61.8	62.2
GeH_3	43.3	58.6	58.1	59.0	59.4
SnH_3	41.2	55.0	54.6	55.6	55.9
CF_3	41.7	68.1	67.3	b	b
$C(O)OCH_3$	42.9	67.4	65.8	b	b
$OC(O)CH_3$	48.4	77.0	75.5	b	b
PO	17.4	34.0	32.1	35.7	34.6
OP	−1.9	7.8	6.0	10.7	8.3
C(O)Cl	38.6	61.6	58.8	59.0	60.0
$C(O)CH_3$	30.7	53.3	51.6	52.1	52.6
PO_2	46.1	62.1	60.0	60.4	61.0
OPO	69.2	80.3	78.0	83.1	80.0
OF	45.7	84.8	82.5	76.8	81.5
$OSiH_3$	68.3	105.4	103.9	95.1	99.9
OS(O)OH	62.5	79.3	76.4	b	b
OPO_2	103.4	108.2	100.1	109.0	102.0

[a] All results are at the MP2/TZDP++//MP2/TZDP level, corrected for vibration, rotation and translation at 298 K from Table 28. An additional correction of $\Delta PV = RT\Delta n = 0.6$ kcal mol^{-1} is made for the reaction itself. Connectivity is to the leftmost atom in Y.
[b] Not calculated.

TABLE 30.   Bond dissociation energies (in kcal mol^{-1}) for $PbH_3-Y \rightarrow PbH_3\cdot + Y\cdot$[a]

Y	HF	MP2	PMP2	MP3	MP4
F	72.6	119.1	118.2	107.4	114.7
AlH$_2$	41.1	55.0	54.4	56.0	56.2
BH$_2$	46.8	65.2	64.8	65.7	66.2
SH	48.0	71.5	70.3	68.5	68.9
Br	57.2	78.6	77.5	75.6	76.0
H	48.8	61.7	61.4	63.7	64.2
CCH	73.8	117.0	104.5	109.6	113.1
PH$_2$	37.0	55.6	53.9	54.2	54.6
NH$_2$	40.2	74.5	72.8	67.5	70.6
SCH$_3$	43.6	69.3	67.7	65.7	66.7
Cl	66.6	90.8	92.4	86.8	87.0
NO	−14.1	18.7	15.6	14.3	19.1
ON	−32.9	−12.0	−15.1	−13.8	−9.3
C(O)H	31.6	51.5	51.8	51.0	51.0
SeCN	39.4	72.7	68.5	64.6	68.3
NCSe	24.8	55.3	50.0	46.3	51.3
C(O)F	44.7	68.9	66.5	66.6	68.4
C(O)NH$_2$	35.3	59.4	57.7	56.7	58.5
ONO$_2$	103.7	87.9	82.9	101.1	86.5
NCS	73.2	85.4	77.7	72.8	78.8
SCN	70.7	81.4	73.7	71.0	74.8
CH$_2$CH$_3$	39.4	66.9	65.1	63.2	65.0
C(O)OH	42.5	66.8	67.8	64.4	66.0
NO$_2$	31.3	51.0	48.1	49.4	48.0
ONO	53.0	65.8	62.9	68.1	65.6
PC	−6.3	58.3	38.1	42.8	58.1
CP	54.8	116.1	95.9	101.9	110.0
NCO	62.4	99.7	94.4	86.9	93.1
OCN	41.6	81.6	76.3	69.8	74.8
CN	77.1	120.2	109.2	111.8	115.1
NC	75.8	111.1	100.1	105.4	107.6
OCH$_3$	68.2	84.4	82.6	73.0	79.2
CHCH$_2$	48.3	81.7	75.8	77.3	79.2
NNN	25.1	85.8	75.7	61.3	76.5
OH	50.4	92.9	91.4	82.1	88.0
CH$_3$	42.2	67.7	66.4	65.2	66.5
SiH$_3$	43.8	59.6	58.9	59.8	60.1
GeH$_3$	41.8	57.1	56.5	57.2	57.5
SnH$_3$	39.4	53.2	81.3	53.5	53.7
PbH$_3$	37.6	51.4	50.7	51.5	51.6
CF$_3$	41.8	68.7	67.8	b	b
C(O)OCH$_3$	41.4	66.8	65.1	b	b
OC(O)CH$_3$	49.5	80.7	79.1	b	b
PO	17.0	33.8	31.9	34.1	36.7
OP	−5.9	6.7	4.8	8.0	7.3
C(O)Cl	37.4	61.5	58.5	58.3	59.4
C(O)CH$_3$	30.9	54.3	52.6	52.4	53.3
PO$_2$	44.2	59.0	56.5	58.2	57.6
OPO	69.1	85.5	81.0	84.8	83.0
OF	44.3	84.6	82.2	75.9	81.6
OSiH$_3$	57.8	100.8	99.2	89.2	95.3
OS(O)OH	64.3	83.7	80.5	b	b
OPO$_2$	103.1	110.8	102.5	110.4	104.5

[a] All results are at the MP2/TZDP++//MP2/TZDP level, corrected for vibration, rotation and translation at 298 K from Table 28. An additional correction of $\Delta PV = RT \Delta n = 0.6$ kcal mol^{-1} is made for the reaction itself. Connectivity is to the leftmost atom in Y.
[b] Not calculated.

IV. STRUCTURES

A. XH$_4$

The XH$_4$ series, with X = C,Si,Ge,Sn and Pb, has been studied extensively both theoretically[54-56] and experimentally[57,58]. These simplest hydrocarbon analogues have served as benchmarks for the newest, relativistic *ab initio* methods[54-56]. The calculated (experimental) values are 1.089 (1.084–1.087), 1.472 (1.471–1.473), 1.518 (1.514–1.516), 1.689 (1.691–1.694) and 1.735 (1.754[59]), for methane, silane, germane, stannane and plumbane, respectively, in Tables 4–8. The 'experimental' PbH$_4$ value is extrapolated from a combination of theoretical and observed bond lengths[59]. Both the calculated and experimental sets of bond lengths are for equilibrium structures so that a direct comparison of calculated and experimental is valid.

The range of gas phase experimental values comes from different fitting procedures of the raw spectroscopic numbers. The combination of different distance parameter definitions (equilibrium, ground vibrational state, thermal averaging, effective positions, etc.), different experimental methods of structure determination (microwave, infrared and Raman spectroscopies, electron diffraction) and different model fitting procedures for each method or combinations of them, can result in differences of several hundredths of an Angstrom among the experimental values themselves for bond lengths. For comparison between theory and experiment we would like to have the experimental r_e values which correspond to the potential energy minimum and are the quantities actually calculated. Unfortunately, r_e values are generally not available experimentally. Other definitions relate to the zero-point vibrational levels and thermal averaging which, because of vibrational anharmonicity, are usually larger than r_e[60]. Because of the shorter amplitudes of vibration in a particular X—A bond as the atoms get heavier, the difference between r_e and the other definitions should get smaller in going down the Periodic Table. On the other hand, the X—A force constant may decrease in magnitude as the atoms get heavier, which will increase the amplitude of vibration and cancel the mass effect. A detailed analysis of every comparison between calculated and experimental bond distance is not possible here. Comparisons between theory and measured will therefore be carried out indiscriminately. However, it should always be kept in mind that, often, the same, exactly defined quantities are not being compared[60].

It is interesting to see if any definite conclusions can be derived about the accuracy of the calculated bond lengths from the comparison for the XH$_4$ series between theoretical and experimental. Generally, differences or errors can arise from inadequacies in the CEP, basis set, theoretical level (MP2) or treatment of relativistic effects. In general, bond lengths calculated at the MP2 level are expected to be longer than experimental distances[2,61]. This effect should decrease with increasing atomic number, as the importance of correlation diminishes as one goes down the Periodic Table. The one-component, quasi-relativistic method used here is known to give shorter X—H equilibrium bond lengths than the full four-component Fock–Dirac methods[55,56,62]. This effect should increase with atom size. The frozen core approximation inherent to the CEP method[3,25,26] is expected to give calculated r_e values that are too long compared to all-electron results at the same basis set and theoretical level. Since core polarization[64,65] is expected to increase with core size, this relaxation effect may cause the geometry errors to increase with atomic number. The effect of basis set size is more difficult to judge, but even smaller basis sets have been shown to be sufficiently large to give reasonably accurate equilibrium geometric structures[2,61,63] at the MP2 level for singly bonded atoms. There is also a synergistic interaction between the basis set size and correlation treatment, and a larger basis set is sometimes needed for the correlation calculation. The general conclusion obtained is that the calculated equilibrium bond lengths should be larger than the experimental r_e values for singly-bonded atoms not involving hydrogen. Different degrees of error cancellation in specific cases, however, may give other results. For example, in electron-deficient or strongly polar (ionic) bonds, where there are low-lying bonding or empty orbitals, the MP2 method could exaggerate the degree of bonding or back charge-transfer. In this case the equilibrium bond lengths could be calculated short, or at least cancel part of the expected overestimation. For X–hydrogen atom bonds the CEP and correlation effects are expected to be smaller, but the relativistic trend is the same as for X–atom (not hydrogen) single bonds, so that the X–H bond length may, by error cancellation, be calculated with reasonably good accuracy for r_e.

Table 9 shows the MP2/TZDP calculated Mulliken charges on the XH_3 groups in the series of XH_3–Y compounds. For Y = H the four ligands around X are, of course, equivalent. For X = C the charge on the methyl group is negative, showing the usual Mulliken population result that the methyl group is more electronegative than the hydrogen atom. Bader's *Atoms in Molecules* atomic charges[40,66] show reversed polarity. In any event, the net charge on XH_3 increases with Z except for a dip at the X = Ge atom[67]. This behavior pattern does not follow any simple trend in atomic properties shown in Table 2, although the equivalence of the four hydrogen atom ligands may level such trends. The ns ionization energies and orbital radii in Table 2 show dips at X = Ge ($n = 4$) and X = Pb ($n = 6$) which has been termed the 'inert pair effect'[68,69], and attributed to relativistic contraction and stabilization[69]. However, this effect is predicted to be stronger for Pb, whereas the XH_3 group charges in Table 9 show no decrease for X = Pb.

The distribution of electrons in the X atoms among the s-, p- and (five components) d-type basis functions is shown in Table 10. For Y = H the switch in polarity between XH_3 and Y at X = Si is reflected in the depletion of electrons in the p-type atomic orbitals of X with increasing size. At the same time, the s-type population does not vary very much so that the p/s ratio decreases steadily from carbon (p/s = 2.69) to lead (p/s = 1.60). This steady decrease in s–p hybridization ratio going down a column of the Periodic Table of the main group elements has been noted and explained previously[5,70,71] based on decreased overlap between the ns and np atomic orbitals with increasing atomic number. The numbers in Table 10 seem to indicate that the decreasing hybridization ratio with increasing atom size is mainly due to a steady depletion of electron density from the np atomic orbitals in going down the Group 14 column of the Periodic Table.

Other interesting aspects of Table 10 are the trends in s-type, p-type and (five component) d-type orbital populations of the central X atom. The d-orbital occupancies are expected to represent the degree of angular polarization of the X–H and X–Y bonds[32]. These will certainly be affected by the relative d-orbital energies in the atom X, which are hard to define from the isolated atom data because of the arbitrary charge, spin and occupancy choices that need to be made. The results in Table 10 show that, after the usual jump from carbon to silicon, the d-orbital populations generally decrease steadily from silicon to lead. The reduced importance of the d-type functions with increased atomic number may represent a decreased need for angular polarization about the central atom X as the X–H and X–R bond distances increase. The Mulliken atomic charges in Tables 12–15 (using six d-type orbitals) on X alternate from silicon to lead, with the above-noted dips for Ge and Pb. The s-orbital populations also follow the alternating pattern shown by the ns atomic orbital ionization energies in Table 1. This combination of trends results in the largest s-orbital population being calculated for the X = Pb atom.

B. XH₃A

The XH_3–A (A = F, Cl and Br) systems have been studied and analyzed extensively experimentally[72–84]. In this series of molecules, gas-phase geometric structures are known for all the members except SnH_3F and all the plumbyl halides (PbH_3-A). The MP2/TZDP calculated bond distances are shown in Tables 4-8. Comparison with the experimental gas-phase geometries (r_e where available) shows the difference between theory and measurement for the X−A(halide) bond lengths generally increasing with both the sizes of X and A. This result is not completely straightforward because the comparisons are not always uniform with corresponding levels of structure information. The largest error is for SnH_3-Br (0.027 Å)[83]. The error in the X−H bond length is typically smaller and there is a discernible trend of the X−H bond length being underestimated as X gets heavier. The X−F bond distances for X = Si[77] and Ge[79] are particularly well described, with the CH_3F[72] C−F bond length having a larger error (0.017 Å). CH_3F has the shortest X−A bond distance in this group and the fluorine atom lone-pair electrons crowding the C−H bonding electron pairs possibly contribute to the larger error for the C−F bond.

After a large positive jump in value from carbon to silicon, the Mulliken charges on the XH_3 groups show uniform increases for the bromide series (Table 9), a sharp zigzag pattern for the fluorides and intermediate behavior for the chlorides, in going from X = C to X = Pb. The alternating pattern would agree with such corresponding trends in atomic properties (Table 1), which are, apparently, emphasized by more electronegative ligands and more ionic bonds. The calculated dipole moments (Table 16) decrease from carbon to silicon for all the halides and then increase almost uniformly in going to lead. The dipole moments, therefore, show no correlation with the Mulliken group charges on XH_3, as might have been expected for such simple systems. The trend in the calculated HXA bond angles (not shown) is generally for them to decrease uniformly from about 108-109° for X = C to 104-105° for X = Pb. The experimental values generally agree with these trends. Since after carbon the charge on the hydrogen atoms in the XH_3 group is negative (Tables 12-15), reducing the HXA angle would tend to increase the dipole moment. A reduced HXA angle can also be interpreted as increasing ionicity of the XH_3-A bond as the XH_3 group tends to the planar cation structure[70]. This, again, is not completely consistent with all the calculated trends in XH_3 group charges (Table 9), although, except for the peak at germanium, it does approximately correlate with the X atom np populations (Table 10). An alternative explanation is that the Mulliken populations become increasingly unreliable as the bond becomes more ionic[85].

The ns, np and nd orbital occupancies in Table 10 for the halides behave as expected. The ns populations vary as their atomic ionization energies (Table 1), the np populations decrease uniformly (C → Pb) except for the inflection at Ge, and the nd populations decrease from Si → Pb and stay relatively constant and insensitive to the net atomic charge on X (Tables 10-15). The p/s ratios computed from the numbers in Table 10 decrease smoothly for X increasing in size, and the halogen decreasing in size. These latter results are consistent with Kutzelnigg's and von Schleyer's analyses[5,70,71] regarding the relative importance of the ns and np atomic orbitals in the Group 14 column compounds as a function of substituent.

C. XH$_3$AH

The two representatives of this grouping studied here are AH = OH and SH. Of this set, gas-phase molecular structures are available only for methanol[86,87] and methyl mercaptan[88], although both silanol (SiH_3OH) and silyl mercaptan (SiH_3SH) have been studied theoretically[89-92]. All the XH$_3$AH molecules here have staggered equilibrium geometries with an X−H bond *trans* to A−H across X−A. While the calculated C−O distance (Table 4) is 0.012 Å longer than the measured r_s value[86], which should be close to r_e[60], the MP2/TZDP r_e C−S bond length is 0.013 Å less than the observed r_0[88]. The experimental r_e value must then be within several thousandths of an Å of the calculated C–S r_e value in CH_3SH. The outstanding difference between XH_3OH and XH_3SH in geometric characteristic is in the ∠XAH. For X(O, S)H the calculated angles are C(107.2°, 96.9°), Si(116.6°, 95.5°), Ge(112.4°, 94.9°), Sn(113.6°, 94.8°) and Pb(107.8°, 92.6°). Recent studies[70,93] have shown that bond angles in such systems are determined primarily by a subtle balance between ionicity, hybridization effects and stabilizing bonding–antibonding/nonbonding–antibonding interactions between electron pairs, or bonds located *trans* across an intervening bond. The *trans* HXA angles are C(106.4°, 106.0°), Si(105.5°, 104.8°), Ge(104.3°, 104.5°), Sn(103.5°, 104.4°) and Pb(102.7°, 104.6°). The hyperconjugative effects are less important when the intervening bond involves one or both atoms beyond the first row of the Periodic Table. Thus, ∠XAH for A = sulfur and X = silicon or heavier is mainly determined by other effects. The large increase in ∠XOH from carbon to silicon, whereas the corresponding change for ∠XSH is actually a 1.4° decrease, signals a change in mechanism involving ionicity and dependence on the X–A bond length. From Si → Pb the ∠XAH generally decreases. The *trans* and *cis* HXA angles differ by 4–6° in each molecule. This leads to the ca 3° tilt angle observed for CH_3OH[87]. Both *cis* and *trans* HXA angles decrease with increasing atomic number of X as the XH_3–AH bond apparently becomes more ionic and XH_3 more planar.

The group charges for XH_3OH (X = C → Pb) show the same zigzag pattern (Table 9) as in XH_3F, while XH_3SH behaves as the other halides discussed. Analogously, the calculated dipole moments (Table 16) for both XH_3OH and XH_3SH decrease uniformly with increasing atomic number of X, after the initial decrease in going from X = carbon to silicon. In this latter case, the increased equilibrium bond length apparently dominates the increase in ionicity and more subtle bond angle change effects. Table 11 shows the same type of s, p and d orbital distributions and trends commented upon already for the previous Y substituents in the XH_3Y compounds.

An interesting question is the influence of the remote atom X on the ionicity of the A–H bond[91]. This effect can be reflected in the atomic charge on the hydrogen atom, the A–H bond length and the A–H stretch frequency. Table 3 shows a slight increase in the O–H bond length with increased size of X starting from X = Si. The H(O) atomic charges, however, from Tables 11–15 show no systematic changes. The MP2/TZDP calculated O–H harmonic stretch frequencies cluster about 3854 (±ca 40) cm^{-1} with PbH_3OH lowest at 3820 cm^{-1}. Thus, at least the bond lengths and vibrational frequencies show PbH_3O–H to be more ionic than the other XH_3O–H members of the series. The XH_3SH series shows no such trends for any of these properties.

D. XH₃AH₂

The four members of this series have A = N,P,B and Al. The silyl compounds have been discussed by Apeloig[21]. Only for CH_3NH_2[93,94], CH_3PH_2[95,96] and SiH_3PH_2[97,98] are there experimental gas-phase geometric structures. These molecules have staggered conformations, as expected, in an ethane-type geometry where a lone pair of electrons on N and P replace the C–H bonding pair of electrons in the hydrocarbons. For these systems the calculated r_e (Tables 3–5) and experimental average zero-point bond distances (r_0) typically agree to within a few hundredths of an ångstrom, and the bond angles are very close. For example, the measured Si–P bond length is 2.249 Å[97] compared to the calculated (Table 5) value of 2.266 Å. The P–H distance is underestimated by 0.007[98]–0.025 Å[97]. The observed dipole moment (Table 16) of 1.142D for CH_3PH_2 agrees nicely with the experimental 1.100D[96]. All the members of the XH_3NH_2 and

XH$_3$PH$_2$ (X = C, Si, Ge, Sn and Pb) set were calculated in the staggered geometric configuration and all the optimized structures were found to have real harmonic vibrational frequencies, indicative of equilibrium geometries. The major difference between the A = N and A = P structures is in the HAX angle which, characteristically, has a smaller (by 12–15°) value for A = P than for A = N. These results, of course, parallel the geometric structures of NH$_3$ and PH$_3$[2].

The XH$_3$BH$_2$ series, however, was (MP2/TZDP) found to have a quasi-staggered conformation, where the XBH$_2$ moiety is very nearly planar and perpendicular to one HXB plane. This staggered-planar geometric structure was not considered previously for the X = C and Si members[89]. CH$_3$AlH$_2$ also shows the staggered-planar structure. However, the other XH$_3$AlH$_2$ molecules with X = Si → Pb all have the eclipsed-planar (where a X—H eclipses an Al—H) equilibrium geometry, which was found previously[89] to be a transition state in the rotational barrier profile. The results here actually show the staggered-planar structure to be a transition state (one imaginary frequency) in the rotation profile for the heavier XH$_3$AlH$_2$ species. However, the (MP2/TZDP optimized) calculated energy differences are all found to be less than 0.1 kcal mol^{-1}, so that bending-rotation motion in these systems is essentially unhindered by a barrier, as was also concluded previously[89]. The calculated harmonic rotation frequency is below 37 cm^{-1} for the whole XH$_3$AlH$_2$ series. Both BH$_3$ and AlH$_3$ have planar equilibrium geometries[2].

Table 3 shows some interesting trends in the A—H bond lengths as a function of X. For the amines, the N—H bond length first decreases in going from carbon to silicon, but then increases steadily from Si → Pb. This behavior exactly parallels the calculated N—H harmonic stretch frequencies which first increase from the pair of values (3507 cm^{-1} and 3595 cm^{-1}) for X = C in going to Si (3572 cm^{-1} and 3667 cm^{-1}) and then decrease steadily from Si to Pb (3474 cm^{-1} and 3572 cm^{-1}). It thus seems that the N—H bond becomes weaker in going from Si → Pb. Another correlation is with the ∠HNH, which increases from 105.2° (C) to 108.7° (Si) and then decreases steadily to 104.6° (Pb). The trends from Si → Pb can, perhaps, be explained by the increased charge localization on the NH$_2$ group as the X—N bond distance increases with higher atomic number. Table 9 shows the general charge increase on NH$_2$ as the X atom gets longer, but with a superimposed zigzag pattern. The ∠HNH bond angle in NH$_2^-$ is calculated to be 99.4°. The jump in ∠HNH bond angle from carbon to silicon could be related to the increased charge density in the valence p shell of the nitrogen atom. This electron transfer increases its 2p/2s hybridization ratio which favors a more tetrahedral angle. In going Si → Pb, the p/s ratio on nitrogen decreases as the 2s atomic orbital is increasingly populated and the bonding involves more atomic 2p character. The result is a weaker N—H bond and a closing ∠HNH.

The P—H bond lengths in the phosphines, on the other hand, in Table 3 show no clear trend with increasing atomic number of the X atom. The P—H harmonic stretch frequencies are found to hardly change from their values in X = C (2475 cm^{-1} and 2482 cm^{-1}). The corresponding P—H vibrational energies for X = Pb are 2472 cm^{-1} and 2482 cm^{-1}. The ∠HPH angle does increase from 93.6° to 93.9° in going from carbon to silicon and then decreases steadily to 92.9° for lead, but these are much smaller changes than for the amine series.

For the boranes the two B—H harmonic stretch frequencies increase monotonically from methyl (2629 cm^{-1}, 2707 cm^{-1}) to plumbyl (2664 cm^{-1} and 2773 cm^{-1}). The B—H bond length (Table 4) decreases steadily from methylborane to plumbylborane but the HBH angle shows no consistent change with atom size (X). The calculated ∠BH$_2$ angle ranges from 117.4° (Si) to 120.1° (Pb) while the corresponding angle in BH$_2^-$ is 100.3°. Clearly, as can also be seen from the group charges in Table 9, BH$_2$ in the XH$_3$BH$_2$ compounds is not very negative. From the trends in bond length and B—H vibrational frequency it seems that the B−H bond strengthens with increasing atomic number of the X atom.

In the alanes the Al−H bond length (Table 4) decreases steadily from 1.584 Å (X = C) to 1.579 Å (X = Pb), the ∠HAlH angle remains at about 120°, but like in the XH_3BH_2 series, the largest value is for X = Pb (121.0°). The (MP2/TZDP) calculated equilibrium ∠HAlH in AlH_2^- is 93.1°, which is much smaller than in the XH_3 alanes. The harmonic stretch frequencies for Al−H stay within a narrow range, varying from 1927 cm^{-1} and 1939 cm^{-1} for carbon, to 1929 cm^{-1} and 1951 cm^{-1} for lead. In general, the group charges on BH_2 and AlH_2 in Table 9 and the dipole moments in Table 16 indicate that in the XH_3AH_2 series, A = boron and aluminum look much the same from an electronic-structure point of view. Thus, for example, the charge density about the B and Al atoms is very anisotropic due to the electron-deficient nature of these atoms, where the 'hole' is found perpendicular to the XAH_2 plane.

E. XH_3AH_3

The members of this group include A = X′ = C, Si, Ge, Sn and Pb so that all combinations of the group 14 binary hydrides are treated. Experimentally determined gas-phase geometries are available for CH_3CH_3[99−101], SiH_3CH_3[102,103], SiH_3SiH_3[60,104], GeH_3CH_3[105], GeH_3SiH_3[106], GeH_3GeH_3[107] and SnH_3CH_3[108]. A number of the $XH_3X′H_3$ systems have also been studied theoretically[2,21,89,109−112]. All the $XH_3X′H_3$ molecules were calculated in the staggered conformation. The microwave r_e values of Harmony[101] for C_2H_6 are slightly smaller (by 0.008 Å for C−C and 0.003 Å for C−H) than the calculated r_e bond distances in Table 4. The Si−C and Si−Si equilibrium bond lengths in SiH_3CH_3 and Si_2H_6, respectively, are both calculated, at most, ca 0.022 Å larger than experiment[60,102−104], while the error in the Si−H distance (Table 5) is much smaller and in the other direction. For the $GeH_3X′H_3$ group, with X′ = C, Si and Ge, the calculated Ge−C bond length is 0.01 Å too large[105], Ge−Si is ca 0.03 Å too large[106] and the Ge−Ge distance is 0.024 Å greater than experiment[107]. The comparisons with experimental r_e values, which are not available, could somewhat increase these differences for the X−X′ bond lengths. The calculated Si−H, Ge−H and Sn−H bond distances are consistently smaller than the experimental lengths, which are also not r_e values. Thus the calculated and experimental r_e bond distances for X−H are probably closer. The largest difference with r_s[60] is for Ge_2H_6 at 0.017 Å. The calculated (Table 16) dipole moment of 0.614 Å for methylgermane agrees very well with the 0.635 Å experimental value[105].

The calculated C−H equilibrium bond distances in XH_3CH_3 as a function of X (Table 3) show decreasing values as X becomes heavier. This would seem to indicate an increasingly stronger C−H bond. The same picture emerges from the calculated harmonic vibrational frequencies for the C−H mode, which increase steadily from X = Si to X = Pb. The C−H bond length, as well as all the X−H bond lengths, should be affected by several factors simultaneously. Schleyer and coworkers[111] have emphasized the importance of both vicinal and geminal interactions between bonding X−H and antibonding X−H molecular orbitals in determining the rotational barriers in these ethane-type structures. Such interactions are expected to both stabilize the bond but also increase its equilibrium bond length. Both effects will decrease with increased C−X distance as X gets heavier. This mechanism, therefore, seems to explain satisfactorily the calculated decrease in the C−H bond length with increased size of X′ in $CH_3X′H_3$.

However, in the $PbH_3X′H_3$ series, for example, where the hyperconjugative effect must be smallest because of the large Pb−X′ distances, a different trend is found. Here, the Pb–H equilibrium bond length (Table 8) increases from $X' = C$ to $X' = Sn$, and then decreases for $X' = Pb$. The Pb–H harmonic stretch frequency also decreases from $X' = C$ to $X' = Sn$, and then increases for $X = Pb$. The increased Pb–H bond length and decreased vibrational frequency, indicative of a weakened Pb–H bond, may be due to the decrease in charge on PbH_3 (Table 9) in going from $X' = C$ to $X' = Pb$. The greater the charge, the more contracted the valence shell about the lead atom, if because of the lower p/s ratio (Table 10) or because of a purely electrostatic affect[70] on the atomic radius. The decreasing charge on PbH_3 with increasing size of X' in $PbH_3X'H_3$ will increase the Pb np character of the Pb–H bond, which increases both its length and strength due to the longer range and enhanced overlap with the hydrogen atom 1s orbital. The consistent shortening of the $X'–H$ bond in $PbH_3X'H_3$ compared to the other $XH_3X'H_3$ members of these series for each X atom may be related to the nature of the Pb–X' bond in a way that is not obvious here.

F. XH₃AB

The molecules in this category include AB = CN, NC, CP, PC, NO, ON, PO, OP and OF, where connectivity to the XH_3 group is to the leftmost atom in AB. This grouping spans a range of internal ligand bonding types: A≡B (CN and CP), A=B (NC, NO, PC and PO), A–B (OF) and some indeterminate types (ON and OP). The possibility of linkage isomerism in binding to XH_3 also probes the fate of the unpaired spin which, in the electronic ground state of the dissociated $XH_3 \cdot$ +AB· radicals, may actually be located on the B atom. In such a case, the migration of the spin to the A atom to form a covalent bond with the XH_3 radical requires a reorganization of the charge and spin densities in AB during the bond-formation process which may require energy and give rise to a barrier. By examining both linkage isomers of AB we are certain to find the spin migration phenomenon in one of them.

In the XH_3AB series, experimental gas-phase geometries are available for CH_3CP[113,114], CH_3CN[115,116], CH_3NC[117], CH_3NO[118], SiH_3CN[119] and GeH_3CN[120]. The AB = CN, NC, CP and PC members of this group are (XAB) linear while the AB = NO, ON, PO, OP and OF members are (XAB) bent. If we compare the experimental[119,120] and calculated (Tables 5 and 6) bond lengths for SiH_3CN and GeH_3CN, then we find the usual results: the MP2/TZDP optimized Si–C and C≡N bond lengths are too long (by 0.010 Å and 0.017 Å, respectively) while the Si–H bond length is calculated slightly short (by 0.020 Å and 0.004 Å, respectively). Similar results are found for CH_3CN (References 115 and 116 and Table 4). Triple bonds are not described completely accurately at the MP2 level[51,121,122], even with the most extended basis sets. The bond length errors for the triply-bonded species are therefore expected to be larger than for singly-bonded atoms, and that is what is generally found in these comparisons. The C≡N bond length is overestimated by ca 0.02 Å in all three XH_3CN compounds (X = C, Si and Ge). The calculated dipole moment in Table 16 for GeH_3CN (3.964D) and for acetonitrile (3.892D), however, agrees well with the respective 3.99D[120] and 3.94D[123] experimental values. Analogously, good agreement is obtained for CH_3NO where the (Table 16) calculated value of 2.279D is very close to the measured 2.320D[118], and for CH_3CP (calculated[Table 16] = 1.608D, observed[113] = 1.499D).

The C≡N bond length in $XH_3C≡N$ (Table 4) increases as X gets heavier. The C≡N harmonic stretch frequency decreases correspondingly. A reasonable explanation of these calculated results is increased charge transfer from XH_3 to CN as X increases in size. The XH_3 group charges in Table 9, however, show an alternating, zigzag pattern of group charges with X, as is usually found with the more electronegative ligands (like F) in all the $XH_3–Y$ molecules.

A corresponding comparison can be made for the X−C bond length, where the nitriles are compared to the $H_3X−CH_3$ species. The interesting result is that the X−C bond length is significantly smaller in $XH_3−CN$ than in $XH_3−CH_3$ (Tables 4-8), with the differences starting at 0.067 Å for X = C and decreasing with increasing size of X until for X = Pb (Table 8) the difference in calculated bond lengths is only 0.001 Å. The shorter X−C bond length in $XH_3−CN$ can be interpreted in terms of back-bonding from $CN^{(−)}$ to the $XH_3^{(+)}$ unit which gives the X−C bond in $H_3X−CN$ a degree of double-bond character. Since double-bond character is sensitive to bond length in these systems[124,125] the degree of X−C double-bond character will decrease with increasing size of X and concomitant X−C distance.

Comparing the X−H bond length between the XH_3CH_3 and XH_3CN systems (Tables 4-8) shows the latter consistently shorter, with the difference increasing from 0.002 Å for X = C to 0.018 Å for X = Pb. The correlation here seems to be with increased charge on XH_3 in going down the Group 14 column which 'tightens' the X−H bond. The X−H harmonic stretch frequencies are also consistently larger for the XH_3CN group compared to the XH_3CH_3 series for the given X, with the difference roughly increasing with the size of X. Thus, the back-bonding that shortens the X−C bond does not seem to have a weakening effect on the X−H bond.

The C−N bond length in the isomeric XH_3NC series is slightly longer than the C−N bond length in the corresponding members of XH_3CN (Table 4), but only by about 0.01 Å. Since the C−N double bond length is generally found[60] to be some 0.05-0.06 Å longer than the triple bond length, the C−N bond in the $XH_3N−C$ series must have close to triple-bond character. The calculated harmonic C−N stretch frequencies in $XH_3N−C$ are calculated to be only some 1.5-3.0% smaller than for $XH_3C−N$ (frequency ca 2100 cm^{-1}), with the difference roughly decreasing with increasing size of X. The corresponding H−X equilibrium bond length in XH_3NC is uniformly shorter than in XH_3NH_2, as was found for CH_3CN, with the same dependence on increasing size of X.

The X−N bond lengths in $XH_3−NC$ (Tables 4-8) can be compared with the corresponding $XH_3−NH_2$ species for each atom X. When this is done, it is found that only for the X = C member is the X−C bond length calculated smaller in the isocyanide than in the amine. For X = Si → Pb the X−N equilibrium distance is predicted to be larger in the isocyanide compared to the amine, with the difference decreasing with increasing size of X. The $XH_3−NC$ harmonic stretch frequencies show the same behavior as the $XH_3−CN$ vibrational modes, decreasing in energy with increasing size of X, as expected. However, the $XH_3−NC$ frequencies are uniformly larger than the corresponding $XH_3−CN$ frequencies. However, as will be seen later in this chapter, the $XH_3−NC$ bond dissociation energies are uniformly lower than the corresponding values for $XH_3−CN$, but larger than for $XH_3−NH_2$ (Tables 26-30). The XH_3 group charges (Table 9) in the isocyanide compounds are uniformly larger than for the corresponding cyanide moieties, as are also the calculated dipole moments (Table 16). The conclusion from all these numbers seems to be that the CN → XH_3 back-bonding mechanism active in CH_3CN is weaker in CH_3NC with a strong X−N distance dependence. The $XH_3−NC$ bond distance for X = Si → Pb is longer than would be expected based on electrostatics (large negative charges on N; see Tables 12-15) and hybridization arguments.

In XH_3CP the X−C bond is, again, consistently shorter than for $XH_3−CH_3$, with the difference decreasing with increasing size of X. Comparing the $XH_3−PC$ bond length with $XH_3−PH_2$ shows a marginally shorter C−P bond in 2-phosphapropyne relative to methyl phosphine. However, for X = Si → Pb, the X−P bond in XH_3PC is longer than the corresponding bond in $XH_3−PH_2$. The behavior of the $XH_3−NC$ and $XH_3−PC$ bond lengths as a function of atom X should be investigated further.

The XH_3C-P and XH_3P-C bond lengths in Table 4 differ by about 0.07 Å consistently, showing that in XH_3PC the P–C bond is probably very close to double-bond character. The calculated C–P harmonic vibrational frequencies are substantially larger for CH_3CP relative to CH_3PC. The difference ranges from 461 cm^{-1} for X = C (1532 cm^{-1} in 1-phosphapropyne) to 246 cm^{-1} for X = Pb (1344 cm^{-1} in PbH_3CP). The dipole moments (Table 16) for XH_3PC are substantially larger than for XH_3CP, contrary to the same comparison for the cyanides where XH_3CN and XH_3NC have similar dipole moments for a given atom X. All these results reflect the different C–P bonding in the isomeric 1-phospha- and 2-phosphapropynes and their heavier Group 14 analogs. The H–X bond lengths are consistently shorter (within 0.002 Å) in the phospha compounds (–CP and –PC) compared to the phosphines, just as in the cyanides (–CN and –NC). Again, the difference increases with the size of the X atom.

In the nitroso series (XH_3NO) the corresponding X–N bonds are longer than for either XH_3NH_2 or XH_3NC, and increase with the size of X until, for Pb, the difference is 0.14 Å (Tables 4–8). The nitroso radical has a doublet-pi ground state[126] with the radical electron localized mainly on the nitrogen atom. The XNO angle in XH_3NO is calculated to be between 112° and 113° in the whole series, except for X = Pb which is about 4° smaller; reflecting the tendency in the whole series for homolytic dissociation to an NO π radical. This also formally leaves a lone pair of electrons in-plane on the nitrogen atom which may exert a repulsive interaction on the XH_3 group. Table 9 shows that $XH_3 \rightarrow$ NO charge transfer is not large compared even to XH_3NH_2. Free NO, with its radical electron in an antibonding pi molecular orbital, should tend to donate electron density. The individual nitrogen atom atomic charges in Tables 11–15 seem to be unusually large and may be unreliable. They are typically more negative than the oxygen atoms, which is counterintuitive, and their large values may reflect a highly asymmetric charge distribution about the nitrogen atom[40]. The XH_3 and Y fragment group charges should be more reliable, especially as the X–A distance increases. Thus, NO $\rightarrow XH_3$ backbonding should be expected to both relieve charge separation and strengthen the NO bond.

The N=O equilibrium bond length in XH_3NO (Table 3) is short for the methyl compound and then jumps to its largest value for nitrosilane. For X = Si $\rightarrow$ Pb the NO bond length decreases steadily. The N=O harmonic stretch frequency is relatively constant (±15 cm^{-1}) at about 1412 cm^{-1} for the silane to plumbane members of the series. The C–N distance in CH_3NO is only 0.007 Å larger than in CH_3NH_2. Together, these numbers indicate a weakening of the X–N bond and a slight strengthening of the N=O bond as the X atom gets larger. The average X–H bond length in XH_3NO is calculated to fall between the corresponding values in XH_3NC and XH_3NH_2, except for X = Pb where it has the largest value of all three plumbane molecules. XH_3 adopts a C_s skeletal structure with a X–H bond eclipsed with the N=O group. This results in inequivalent HXN angles. Interestingly enough, the out-of-plane HXN angles are consistently smaller than the in-plane HXN angle except for X = Pb, where the ordering is reversed. This correlates with the small PbNO bond angle (103.0°). Clearly, in PbH_3NO there is a special interaction between the in-plane hydrogen and oxygen atoms. The atomic charges (Table 15) and drop in dipole moment (Table 16) in going to the nitrosoplumbane compound are consistent with this conclusion.

The X–P distance (Tables 4–8) in XH_3PO is shortest for X = C and longest for X = Si $\rightarrow$ Pb compared to XH_3PC and XH_3PH_2. As for the nitroso series, this shows that different (covalent vs ionic) bonding mechanisms are operative in these cases. The P=O distance is also shortest for X = C, jumps to largest for X = Si and decreases steadily to X = Pb in the XH_3PO series. The X–H bond length is also consistently larger for XH_3PO compared with XH_3PC and XH_3PH_2. The XPO angle is consistently smaller in XH_3PO compared to XH_3NO, as expected, and is 107.5° for X = C and 101.0° ±0.9° for X = Si $\rightarrow$ Pb. The in-plane HXP angle is significantly smaller for X = Pb, indicating a H...O interaction, as in the PbH_3NO case. The amount of valence p character in X for XH_3PO is larger than in XH_3NO (Table 10), especially for X = C. This should selectively enhance the bonding interactions involving the X atom in the XH_3PO series, and especially for CH_3PO where covalent bonding is the dominant mechanism. The XH_3ON, XH_3OP and XH_3OF series will be compared with XH_3OH. The X—O bond length (Tables 4–9) is shortest in CH_3OF and longest in CH_3OP. For X = Si → Pb the shortest X—O distance is found consistently in XH_3OH. An equilibrium geometric structure could not be MP2/TZDP calculated for SnH_3ON. The longest X—O bond lengths among these molecular groups are in the OP series for X = Si and Ge, and in ON for X = Sn and Pb. In the free OF and OH radicals, the unpaired spin resides mainly on the oxygen atom and their covalent interaction and bond formation with CH_3 is unhindered electronically. For NO and PO the nitrogen atom is the major home of the radical electron and the oxygen atom localized radical is a spin-migrated excited state, which could even lead to metastable XH_3—OP and XH_3—ON species relative to the ground-state asymptotic dissociation radicals. As the bonding mechanism becomes more ionic, the original location of the spin in AB is less relevant. The P=O and N=O harmonic stretch frequencies are consistently lower in the —OP and —ON sets compared to the corresponding numbers in the —NO and —PO groups, where these can be identified in the harmonic vibrational analysis. The equilibrium XOP angle is consistently larger than XON by $9° ±4°$. A larger bond angle is usually interpreted as indicating a smaller np/ns ratio in the oxygen atom binding.

The bond angle observation can be related to a comparison of the NO and PO bond lengths in the respective linkage isomers. In XH_3ON the NO distance is only some 0.02–0.05 Å longer than in the corresponding member of XH_3NO. In XH_3OP, however, the increase spans 0.08–0.11 Å, making the OP bond of essentially single bond order. Looking ahead at XH_3—O—N=O and XH_3—O—P=O we can see the difference between single- and double-bond character in the N—O and P—O bonds (Table 3). Thus, the N—O equilibrium distance in XH_3O—N is 0.03–0.06 Å larger than in XH_3ON=O and XH_3N=O, while XH_3O—NO is 0.25–0.30 Å longer than XH_3ON=O. Thus, both XH_3N—O and XH_3O—N qualify as being stretched double bonds. However, XH_3P=O is 0.02–0.04 Å longer than XH_3OP=O while XH_3O—P spans the same bond length range as XH_3O—PO and is 0.11–0.15 Å longer than XH_3OP=O. The P—O bond in XH_3O—P is therefore of single-bond character.

G. XH₃ABH

There are two groups of this generic type: XH_3C=CH and $XH_3C(O)H$. Experimental gas-phase geometries have been reported for CH_3CCH[127,128], SiH_3CCH[129], GeH_3CCH[130], $CH_3C(O)H$[131,132] and a theoretical structure for $SiH_3C(O)H$[133]. The gas-phase structure of the permethyl-substituted $(CH_3)_3SnCCH$[134] and photoelectron spectrum of $(CH_3)_3PbCCH$[135] have also been reported and analyzed. The most relevant comparison between measured and calculated values here is for the germylacetylene compound. For this molecule, the Ge—C and C≡C bond lengths are both calculated ca 0.02 Å larger than experiment[130] and the equilibrium Ge—H bond distance is calculated 0.004 Å short. These results are consistent with previous such comparisons above. The calculated dipole moment of GeH_3CCH (Table 16) is $0.119D$, compared to the measured[130] $0.136D$. Analogously, $CH_3C(O)H$ has a reported dipole moment of $2.75D$[119] compared to the Table 16 calculated value of $2.670D$. The corresponding theoretical (experimental) dipole moments of the silyl and methyl acetylenes are $0.328D$ ($0.316D$) and $0.730D$ ($0.780D$), respectively. The corresponding comparison for methyl, silyl and germyl chlorides is $1.990D$ ($1.871D$), $1.501D$ ($1.303D$) and $2.241D$ ($2.124D$), respectively. For CH_3GeH_3, CH_3SiH_3 and GeH_3SiH_3 the numbers are $0.614D$ ($0.635D$),

0.684D (0.735D) and 0.144D (*ca* 0.1D). The measured values are all collected in Reference 130. These comparisons show that the MP2/TZDP optimized wave functions follow the electronic structure changes in the Group 14 series with different substituents rather accurately.

The acetylinic C−H bond length in the XH$_3$CCH series (Table 4) is found to be essentially independent of the Group 14 atom X. The C≡C equilibrium bond distance is also calculated to be independent of X for X = Si → Pb. The group charges in Table 9 also show the charge on XH$_3$ to be essentially constant for the four heavier atoms and somewhat different from methylacetylene. All the members of the XH$_3$CCH group are linear (C_{3v} symmetry) and therefore the cylindrical symmetry of the C≡C triple bond is preserved. The calculated C≡C harmonic bond stretch frequency is predicted to drop from 2130 cm^{-1} for X = C to *ca* 2023 cm^{-1} ($\pm$10 cm^{-1}) for the Si, Ge, Sn and Pb compounds. The acetylinic C−H vibrational frequency changes even less from 3482 cm^{-1} for the methyl compound to an average *ca* 3460 cm^{-1} ($\pm$5 cm^{-1}) for the heavier atoms. The C−H stretch frequencies of the methyl group are much lower at 3159 cm^{-1} (*e* symmetry) and 3078 cm^{-1} (a_1 symmetry).

For the XH$_3$C(O)H series, Table 9 shows almost no difference in the amount of charge transferred from XH$_3$ to C(O)H in going from formylsilane to formylplumbane. Consistent with these results, the calculated dipole moments in Table 16 show gradually decreasing values from X = Si → Pb, as the X−C equilibrium bond length increases. However, contrary to the XH$_3$CCH group, in XH$_3$C(O)H the C−H and C=O equilibrium distances do depend on the atom X even in the X = Si → Pb group (Table 4). Complicating factors in rationalizing these differences are the additional degrees of freedom afforded by the HCX and OCX angles and the rotational conformation about the X−C single bond in the formyl set. The equilibrium conformation found here for all the XH$_3$C(O)H structures have the C=O bond eclipsed with one of the X−H bonds. The HXC(O)H skeleton is planar and the molecules have C_s symmetry. These conformations have been properly characterized as having only real harmonic vibrational frequencies for all X (C → Pb). For X = C this is also the experimental result[60,131,132], although a different theoretical equilibrium conformation has been reported for SiH$_3$C(O)H[133].

In any event, using the consistent C_s structure, the formyl C−H bond in XH$_3$C(O)H (Table 4) is shortest for acetaldehyde (1.109 Å), is 0.005–0.006 Å longer for X = Si → Sn and then decreases to 1.112 Å for the lead compound. Analogously, the equilibrium C=O bond length is calculated at 1.221 Å for X = C, increases by 0.006–0.010 Å for X = Si → Sn and decreases back to 1.221 Å in formylplumbane. The formyl C−H and carbonyl harmonic stretch frequencies show the same trends as their respective bond lengths. For C−H (C=O) the vibrational energies are: X = C, 2942 cm^{-1} (1745 cm^{-1}); X = Si, 2868 cm^{-1} (1643 cm^{-1}); X = Ge, 2870 cm^{-1} (1657 cm^{-1}); X = Sn, 2853 cm^{-1} (1640 cm^{-1}); X = Pb, 2882 cm^{-1} (1660 cm^{-1}). The conclusion seems to be that the C−H and C=O bonds are stronger in the plumbane compound than in the formyl-silane, -germane or -stannane compounds. A hyperconjugative effect strengthening the in-plane X−H bond in the planar *trans* H−X−C−H dihedral conformation[70] can be found both in its shorter equilibrium bond length and higher harmonic vibrational frequency, relative to the corresponding property values of the other two out-of-plane X−H bonds. Presumably, there is a mutual effect on the formyl C−H bond along the series which is strongest for X = C because of the short C−C bond distance. The C=O bond is also expected to be strongest in the acetaldehyde member of the series due to electroneutrality between the CH$_3$ and C(O)H fragments.

The increased strength of the C−H and C=O bonds in PbH$_3$C(O)H relative to the X = Si → Sn compounds could be related to the nature of the X−C bond in these compounds. Examining Tables 4–8 for both the XH$_3$CCH and XH$_3$C(O)H shows that, relative to

XH_3CH_3, XH_3CN and XH_3CP, the X—C bond length is shortest in the acetylene series (within 0.001 Å for X = C) and longest in the formyl series (except for acetaldehyde). The negative charge on the proximate carbon atom in the more polarizable CCH group is larger than in CN (Tables 11–15). Therefore, in the more ionic X = Si → Pb members of their respective XH_3—Y series, the attractive X—C ionic interaction should be stronger, although this argument does not work well for Y = CP. For X = C, the dominant covalent bonding dictates that the C—C bond length in CH_3—Y is mainly determined by the hybridization ratio on the Y group carbon atom, where the smaller the ratio the shorter the C—C distance. Hence, Y = CH_3 > C(O)H > CCH is the order for the equilibrium C—C bond length in CH_3—Y. In these respects, there seems to be nothing unusual about $PbH_3C(O)H$ relative to the X = Si → Sn members of this series, except the smaller np/ns ratio for X = Pb (Table 10), which is true for all the XH_3Y series. Roughly speaking, the average value equilibrium H(formyl)CX angle increases (from 115.4° for X = C), and ∠OCX decreases (from 124.5° in X = C) with increasing size of X. Again, there seems to be no strong difference in these trends for X = Pb.

H. XH_3ABH_3

The only member of the XH_3AHBH_2 generic grouping reviewed here is $XH_3CH=CH_2$. Experimental gas-phase geometries have been reported for propene[136,137], vinylsilane[138] and vinylgermane[139]. All members of this series are found to be planar (C_s symmetry) where an X—H of the XH_3 group eclipses a C—H of the terminal CH_2 group and is staggered to the C—H bond of the central carbon atom. These are also the experimental conformations, where reported[136–139]. The measured (calculated—Table 16) dipole moments are $0.366D$[136] ($0.332D$), $0.657D$[138] ($0.662D$) and $0.50D$[139] ($0.525D$), respectively, for X = C, Si and Ge in $XH_3CH=CH_2$. Agreement here is seen to be very good. Comparison of the geometric structural parameters for the germyl compound shows the Ge—C equilibrium bond length is calculated (Table 6) to be too large (ca 0.02Å), as in previous cases. The calculated Si—C distance in vinylsilane is also calculated (Table 5) longer than experiment by about the same amount. The C=C distances in all three vinyl compounds, as well as the C—C distance in propene, are very close to their experimentally measured values. The small 'tilt' of the methyl group, defined as the angle between the approximate C_3 axis of the methyl group and the line of the C—C bond, is also well represented in these calculations. Although Tables 4–8 report only average X—H bond lengths, these actually have slightly different values for the in-plane and out-of-plane bonds which are accompanied by different HXC angles and X—H bond lengths for the two types of X—H bonds. This results in the approximate 'tilt' angle described above.

The MP2/TZDP optimized C=C equilibrium bond length in ethylene is 1.337 Å. Going to propene, the methyl group increases this distance to 1.339 Å (Table 4) and the Si → Sn atoms increase it even more to almost 0.01 Å longer than in ethylene. In $PbH_3CH=CH_2$ the double bond length is only 0.004 Å longer than in ethylene. The terminal methylene C—H bond length is insensitive to the nature of X in $XH_3CH=CH_2$. However, the middle carbon C—H bond distance shows the same decrease from the Si, Ge and Sn group to Pb as was calculated in the $XH_3C(O)H$ series. The equilibrium C—H distance in ethylene is calculated to be 1.084 Å, which is close to the terminal methylene C—H distances in propene, where the middle carbon C—H bond length is a larger 1.088 Å. In going to X = Pb from propene in the $XH_3CH=CH_2$ set, the two types of C—H distances converge to essentially a common value. This shows the decreased influence of the XH_3 group as the X—C distance increases. This effect is found also in the C—H harmonic vibrational frequencies of the middle and end carbon methylene groups. For X = C, the two terminal carbon C—H frequencies are 3165 cm^{-1} and 3266 cm^{-1}, and the middle C—H is 3177 cm^{-1}. For X = Pb, the respective stretch energies are 3140 cm^{-1}, 3231 cm^{-1} and 3190 cm^{-1}. Clearly, in going from X = carbon to X = lead the middle C—H vibrational frequency migrates to become the average of the two terminal values. This trend is not linear, however, and for the intermediate Si, Ge and Sn compounds the central carbon C—H stretch energy decreases to 3165 ± 4 cm^{-1}, keeping within 20 cm^{-1} of the lower energy terminal C—H frequency.

The C=C harmonic vibrational frequency is calculated at 1671 cm^{-1} in free ethylene and is infrared (IR) forbidden. Its IR intensity is therefore expected to remain low in the vinyl series of compounds. The C=C stretch energy is calculated to be 1687 cm^{-1} in propene and then decline to 1629 ± 4 cm^{-1} for X = Si → Pb. As in the equilibrium bond distance, there is also a very small counter-trend change in the vibrational frequency going from X = Sn to X = Pb that indicates a slight strengthening of the C=C bond.

The X—C bond lengths in the XH$_3$CH=CH$_2$ series are consistently somewhat shorter than in the corresponding XH$_3$CH$_3$ set (Tables 4–8). This difference is expected just on the basis of the lower s-type character in the sp^2 hybrid of the vinyl carbon group compared to the sp^3 hybrid in the methyl substituent. In this regard, the X—C bond length in XH$_3$—Y for Y = CHCH$_2$ lies somewhere between Y = CCH (sp hybridization) and Y = C(O)H, where the gap between CH=CH$_2$ and C≡CH is largest for X = C, diminishes with increasing size of X and is smallest for X = Pb. An examination of the internal in-plane angles of the XH$_3$CH=CH$_2$ series shows the major change being a swing of the XH$_3$ group towards the terminal carbon atom, as X gets larger, with the *cis*-HCC angle increasing slightly in an accommodating fashion to the motion of the XH$_3$ group. The picture here is not completely uniform and, in places, the PbH$_3$CH=CH$_2$ molecule shows counter-trend values of the geometric parameters. These bond angle changes may, in part, be driven by an attempt to minimize energetically unfavorable dipole moments which, in simple terms, represent the separation of charge. Thus, although the group charges (Table 9) do not change much or uniformly with the size of X (from Si), the dipole moments (Table 16) do not increase uniformly as would be expected from the steady increase in XH$_3$—C$_2$H$_3$ bond length with increasing size of X. This could indicate the effect of the bond angle changes.

The other groups of XH$_3$ABH$_3$ molecules reviewed here include XH$_3$OCH$_3$, XH$_3$OSiH$_3$ (X = Si → Pb) and XH$_3$SCH$_3$. In these sets, experimental gas-phase geometries have been reported for dimethyl ether[140,141], methyl silyl ether[142], dimethyl sulfide[143], methyl silyl sulfide[144] and disiloxane (SiH$_3$OSiH$_3$)[145]. The Si—O bond and the Si—O—X bond angle have aroused considerable interest in relation to the catalytic properties of zeolites and their structural analogs. The result has been a large number of theoretical studies on these type systems[146–152]. The most common geometric conformation for all the XH$_3$AX'H$_3$ molecules (where X' = C or Si) is C$_{2v}$ (for X = X') or quasi-C$_{2v}$ (for X ≠ X') symmetry, where each XH$_3$ or X'H$_3'$ group is staggered with respect to the opposite A—X' or A—X bond. This conformation is called staggered-staggered (*ss*) and has a planar H—X—A—X'—H' skeleton in a zigzag conformation. Rotation by 180° about either the X—A or X'—A bonds gives the *se* (*e* = eclipsed) conformation. The same rotation about both the X—A and X'—A bonds of 180° each from *ss* gives the *ee* conformation. All the XH$_3$OCH$_3$ molecules reviewed here were optimized in the *ss* structure and found to have all real harmonic vibrational frequencies and, therefore, to be equilibrium geometries. The *se* structure for CH$_3$OCH$_3$ was found to be a transition state at 2.5 kcal mol^{-1} higher energy. The calculated dipole moment of 1.327D (Table 16) of dimethyl ether in the *ss* conformation agrees very well with the 1.302D experimental value[136]. In the XH$_3$OSiH$_3$ set, for disiloxane the *ss* conformer was found to be a transition state and the equilibrium geometry was calculated to have the *ee* structure. As also found by others[150], the energies of these two conformers are essentially the same. The calculated dipole moment for CH_3OSiH_3 (1.216D) also agrees well with the experimental 1.38D value[144]. GeH_3OSiH_3 and SnH_3OSiH_3 have equilibrium *ss* configurations, although, again, the *ss* and *ee* structures are essentially energetically degenerate. The PbH_3OSiH_3 molecule optimized to a slightly twisted *ss* structure, where a set of terminal Pb−H and Si−H bonds are not aligned with the PbOSi plane or with each other. All the XH_3SCH_3 molecules have the *ss* conformation.

For SiH_3OSiH_3, comparing the calculated structural parameters in Table 5 with experiment[145] shows the usual result that the MP2/TZDP optimized Si−O bond is too long (by *ca* 0.02 Å) and the Si−H bond length is too short (by *ca* 0.01 Å). The SiOSi angle is underestimated by 5.5° (calculated = 138.6° and measured = 144.1°). The combination of a larger Si−O distance and a smaller SiOSi angle gives a Si...Si distance of 3.097 Å, which is almost the same as the electron diffraction value of 3.107 Å[145]. A similar comparison between theory and experiment for SiH_3OCH_3[142] again shows the Si−C and Si−O bond lengths too large (by 0.005 Å and 0.02 Å, respectively), but the SiOC angle in good agreement (observed = 120.6°, calculated = 120.1°). A recent calculation of the GeH_3OSiH_3 geometric structure[151] using density functional theory shows similar structural parameters to those obtained here, although the methods and basis sets are very different. The Ge−O, Si−O distances and GeOSi angle calculated here (best published[151]) are 1.779 Å (1.786 Å), 1.660 Å (1.655 Å) and 123.9° (131.6°). The major discrepancy is in the latter angle, which is also the most difficult to calculate accurately. For SiH_3SCH_3 the experimental (1.819 Å[144]) and calculated (1.817 Å, Table 4) C−S bond lengths agree. Similarly, in dimethyl sulfide the C−S bond distances agree to 0.007 Å, with the experimental[143] value being larger. The Si−C distance in methyl silyl sulfide is measured at 2.134 Å[144] and calculated at 2.116 Å. It thus appears that MP2/TZDP slightly underestimates the C−S and Si−S bond length values. The CSC and SiSC angles are observed at 98.8° and 98.3°. The corresponding computed values are 98.3° and 98.7°, respectively. To summarize, the XAX′ angle situation, the X = C, Si, A = O, S and X′ = C cases show excellent agreement between theory and experiment. As X and X′ become heavier, we can expect the bending potential to become flatter and the exact determination of the bending angle more difficult.

The two major geometric parameters of interest in these systems are the X−A bond length and XAX′ bond angle. Examining Tables 4−8 shows that the X−O bond in the XH_3OCH_3 and XH_3OSiH_3 series have the shortest X−O bond lengths in the Tables for all X. This removes the claim of an unusually large Si−O bond distance in disiloxane[21]. The calculated equilibrium geometries show the XH_3SCH_3 series members also have the shortest X−S bond lengths compared to XH_3SH and (looking ahead) XH_3SCN, which are the only X−S bearing entries in Tables 4−8. The S−C bond distance in the XH_3SCH_3 (X = Si → Pb) set are, however, longer than in CH_3SCN, probably because of back-bonding in the latter.

It has long been noted[140] that the XH_3 group in almost all the XH_3−Y systems, where Y is a nonlinear substituent, has its local C_3 rotation axis tilted a few degrees away from the X−A bond axis, where A is the attached atom in Y. The XH_3 group hydrogen atoms divide into two categories, H_s (in-plane) and H_a (out-of-plane). In the XH_3OCH_3, XH_3OSiH_3 and XH_3SCH_3 sets the H_aXA (A = O, S) angle is always larger than the H_sXA angle by 3−4° for X = C, decreasing to 2−3° for X = Pb. At the same time, both HXA angles are decreasing steadily as X gets larger. The larger H_aXA angles are explained by steric repulsion effects between the XH_3 and $X'H_3'$ groups in the *ss* conformation. The flattening of the XH_3 group as X gets heavier is probably due to a combination of the increasingly cationic nature of the XH_3 group and its tendency to be flatter due to the decreasing np/ns hybridization ratio on X (Table 10). The X−H_a and

X—H_s distances are also calculated (and observed)[140,143,144] to be different, with X—H_a slightly longer than X—H_s, consistent with the H_aXA angle being larger than $\angle H_s$XA. This differential in the H—X bond lengths and corresponding HXA angles involving the XH$_3$ group almost always decreases and almost disappears by PbH$_3$Y.

The most striking feature of the XH$_3$AX′H$_3'$ group of molecules is the variation in the XAX′ angle as a function of the three atoms X, A and X′[146]. The calculated trends here are as follows. In the XH$_3$OCH$_3$ series $\angle$XOC has the values (X) 110.4° (C), 120.1° (Si), 116.5° (Ge), 118.9° (Sn) and 110.0° (Pb). For XH$_3$OSiH$_3$ the corresponding angles are 120.1° (C), 138.6° (Si), 131.6° (Ge), 134.6° (Sn) and 123.2° (Pb). The calculated XSC angles in XH$_3$SCH$_3$ are 98.2° (C), 98.7° (Si), 98.3° (Ge), 99.3° (Sn) and 93.4° (Pb). A commonly used argument to explain the variation of the XOX′ bond angle in these systems has been the involvement of back-bonding from the oxygen atom lone-pair electrons to the XH$_3$ groups to give the X—O bond a certain degree of double-bond character. A concomitant result of this back-donation is a widening of the XOX′ angle to enhance the interaction through better alignment of the lone-pair electrons with the XH$_3$ group. A tendency to linearize the XOX′ angle will also cause the X—O sigma bond to take on a more valence s-type character, which reinforces the shortening of the X—O bond length to enhance the partial π bond. Clearly, because of the more ionic (XH$_3 \rightarrow$ O) nature of the X—O bond for X = Si $\rightarrow$ Pb, this mechanism will be stronger for the heavier X atoms. Thus, the XOC angle in XH$_3$OCH$_3$ jumps from 110.4° in dimethyl ether to 120.1° in methyl silyl ether. For the heavier X atoms the degree of partial double-bond character in the X—O bond should decrease due to the increasing X—O bond distance. The result is a drift of the XOC angle to lower values as X increases in size from Si. In the XH$_3$OSiH$_3$ series, the calculated XOSi angle jumps from 120.1° to 138.6°, with an accompanying decrease in the Si—O bond length from 1.660 Å to 1.656 Å (Table 5). As the X atom gets larger (from X = Si) the XOSi angle again drifts lower, but does not approach the X = C values (as in the XOC case) for the X = Pb compound because of electrostatic repulsion between the XH$_3$ (X>Si) and SiH$_3$ groups. In PbH$_3$OSiH$_3$ the Pb—O bond length is 0.11 Å longer than in PbH$_3$OCH$_3$. In the XH$_3$SCH$_3$ series, hybridization at the central sulfur atom is a dominant effect and the XSC angle is in the characteristic 90°–100° range for all X. Partial double-bond character, steric and electrostatic effects seem to be less important here (compared to the ethers) due to the larger X—C and C—S bond distances. The X—S bond for Y = SCH$_3$ is still somewhat shorter (Tables 4–8) and the XSA bond a bit wider than for Y = SH. It would be interesting to extend the orbital interaction analysis arguments recently described[146] to these heavier atom systems.

I. XH$_3$ABH$_5$

The only member of this group is the XH$_3$CH$_2$CH$_3$ set. Experimental gas-phase geometries have been reported for propane[60,153], ethylsilane[60], ethylgermane[154] and ethylstannane[155]. The calculated dipole moments (Table 16) for both GeH$_3$C$_2$H$_5$ and SnH$_3$C$_2$H$_5$ of 0.748D agree closely with the experimental values of 0.76D[154] and 0.99D[155], respectively. The equilibrium Ge—C bond length (Table 6) is larger than observed by 0.012 Å, the C—C bond is underestimated by 0.009 Å and Ge—H is calculated too large by 0.003 Å. The measured structural parameters refer to microwave r_s values[154]. For ethylstannane[155], the Sn—C bond length is overestimated (Table 7) by 0.005 Å and the C—C distance is calculated too short by 0.016 Å, although the experimental uncertainty for this structural parameter is ±0.025 Å. The geometric structure of the XH$_3$C$_2$H$_5$ set has the conformation described for the XH$_3$OCH$_3$ series, but with the central methylene hydrogen atoms also simultaneously staggered with the in-plane X—H and C—H bonds. Thus, the XH$_3$CH$_2$C and XCH$_2$CH$_3$ fragment units both have the staggered ethane-like conformation. For the silane member, the Si−C bond length (Table 5) is too large compared with experiment[60], C−C agrees exactly and Si−H is underestimated by 0.003–0.006 Å. In propane, the calculated and experimental C−C bond lengths are very close[60,153].

The reported vibrational spectra and normal coordinate analysis[154] of $GeH_3C_2H_5$ also allows a comparison of the calculated harmonic frequencies with experiment. The three methyl C−H stretch frequencies are calculated to have an average value of 3114 cm^{-1} with a spread of 86 cm^{-1}. The corresponding measured energies average to 2936 cm^{-1} with a (highest–lowest) splitting of 82 cm^{-1}. Thus, the MP2/TZDP vibrational energies for this mode have to be scaled by 94.3% to coincide with experiment. This scaling is similar to other reported experimental/calculated ratios for MP2 frequencies in an extended basis set[47]. For the central methylene group the calculated C−H stretch average is 3096 cm^{-1} with a spread of 47 cm^{-1}. The experimental values are 2950 cm^{-1} (95.3%) with a 39 cm^{-1} splitting. Finally, for Ge−H the calculated stretch vibration energies average to 2172 cm^{-1} with an 8 cm^{-1} distribution, while the measured quantities are 2081 cm^{-1} (95.8%) spanning 6 cm^{-1}. The C−C stretch is calculated at 1059 cm^{-1} and observed at 1030 cm^{-1} (97.3%). For ethylstannane[155] the analogous comparison between experiment and theory shows the CH_3 stretch modes spanning 2960–2884 cm^{-1} vs 3146–3059 cm^{-1} (94.2%), CH_2 is found at 2967 and 2934 cm^{-1} vs 3123 and 3076 cm^{-1} calculated (95.2%) and the C−C stretch is observed at 1018 cm^{-1} and calculated at 1045 cm^{-1} (97.4%). The Sn−H stretch mode is measured at ca 1870 cm^{-1} compared to the average theoretical energy of 1958 cm^{-1} (95.5%). The respective scaling factors for the two compounds are virtually identical.

We may note several trends in the geometric parameters of the $XH_3C_2H_5$ set and in comparison with other CH_3Y groups. As expected, the X−C bond length (Table 4) increases steadily in the series Y = CCH, $CHCH_2$ and CH_2CH_3 as the hybridization on the attached carbon atom increases its 2p/2s hybridization ratio, for all X. The C−C bond in Y = C_2H_5 jumps by 0.09 Å in going from X = C to X = Si because of the $XH_3 \rightarrow$ Y charge transfer and then decreases to the propane value for X = Pb. The trend is not uniform with increasing size of X since part of these changes are due to steric effects which decrease as the X−C bond lengthens. Thus, increasing charge transfer in going from X = C → Pb lengthens the C−C bond in Y = C_2H_5 while decreasing steric repulsion due to the increasing X−C bond distance allows the C−C bond to shorten. The calculated C−C stretch frequency shows little variation with X. Starting at 1075 cm^{-1} for X = C this vibrational energy remains at 1052 ±7 cm^{-1} for the four heavier Group 14 atoms. The X−C bond lengths in $XH_3CH_2CH_3$ (Tables 4–8) are about the same as in XH_3CH_3, showing the loss of the back-bonding effect noted in the XH_3OCH_3 series due to the substitution of the methylene group for the oxygen atom.

The XCC angle in the $XH_3CH_2CH_3$ set shows none of the widening effects found in the corresponding ethers. The angle values are 112.0°, 113.2°, 112.8°, 114.3° and 110.8°, respectively, for X = C, Si, Ge, Sn and Pb. Their magnitudes are somewhat larger than tetrahedral, as expected. It is not clear whether the small zigzag variations in these calculated equilibrium angles, which result in ethylstannane having the largest XCC angle, are significant because of the soft bending potential. For very soft bending modes, as will be found in the coming set (XH_3ABC) of compounds, the equilibrium bend angle can be strongly coupled to other geometric parameters so that small variations in the relevant bond distance/angles can influence the value of the bend angle significantly. This point was not investigated further here. Ethylplumbane is calculated to have the smallest XCC angle. This interesting result appears not infrequently for the XAB angle in these type systems.

J. XH₃ABC

There are thirteen sets of this XH_3Y grouping reviewed here, where A, B and C are nonhydrogen atoms from the first three rows of the Periodic Table. The XH_3ABC category divides into three subgroups: (1) $Y = N_3$, OCN, NCO, SCN, NCS, SeCN and NCSe (the pseudohalides) having either a linear ($\angle ABC = 180°$) or quasilinear ($\angle ABC > 170°$) chain structure, (2) ONO and OPO having a strongly bent ABC geometry and (3) NO_2, PO_2, C(O)F and C(O)Cl with a forked configuration and the attachment to X being through the middle atom of the triatomic Y substituent. The XH_3ABC general group offers a rich variety of attaching atoms, linkage isomerism, rearrangement isomerism and a wide range of bending angles for comparing structural properties. The Y groups also have different internal bonding characteristics in terms of combinations of single-, double- and triply-bonded atoms which can affect their interactions with the XH_3 group.

In the quasi-linear grouping a serious structural feature revolves about the question of how statically linear is the ABC chain in the equilibrium geometry. Certainly, even where the $\angle ABC$ lies between 170° and 180°, any barrier height at the linear configuration must be only in the tens of cm^{-1} range[23]. Experimental gas-phase geometries for the XH_3–azide set have been reported for CH_3N_3[156], SiH_3N_3[157] and GeH_3N_3[158,159] and discussed comparatively with other like systems both experimentally[160] and theoretically[161]. The prototypical hydrazoic acid (HN_3) and similar systems have been discussed previously[23]. All the members of the XH_3NNN set are calculated to have the same C_s structure, independent of X. The in-plane $H_sXN_\alpha N_\beta N_\gamma$ atoms are calculated to form a zigzag, trans structure with $\angle N_\alpha N_\beta N_\gamma = 172.4°$ (X = C), 173.6° (X = Si), 173.6° (X = Ge), 174.1° (X = Sn) and 174.4° (X = Pb). The XH_3-N_α and $N_\beta N_\gamma$ bonds are trans to each other across the $N_\beta N_\gamma$ bond. Experimentally, the $N_\alpha N_\beta N_\gamma$ angle has been estimated at 173.1° (X = C)[156b] and 171.5° (X = Ge)[159]. The measured dipole moment of GeH_3N_3 is $2.579D$[159], which is to be compared with the calculated $2.870D$ in Table 16. Looking at the observed r_a parameters[60], the MP2 calculated $Ge-N_\alpha$ distance (Table 6) is too large by 0.021 Å and the average Ge–H bond length is too small by 0.017 Å. The C_3 (pseudo-)rotation axis of the XH_3 group is predicted to be tilted relative to the $Ge-N_\alpha$ bond axis by a few degrees for all X. Thus, the difference between the H_sXN_α and H_aXN_α (H_a = out-of-plane hydrogen atoms) angles runs from 4.9° in the methyl compound to 3.0° in plumbyl azide, and reaching 5.4° and 5.6°, respectively, for X = Ge and Sn. Experimentally, these two angles have been found to differ by 4.2° in CH_3N_3[156b]. The Si–N_α distance is calculated too long (Table 5) by 0.044 Å compared to an electron diffraction study[147]. This difference is larger than expected based on previous comparisons. Considering that a comparison with a measured r_e value is expected to widen the gap suggests that the observed structure needs to be refined.

The $XN_\alpha N_\beta$ angle in the azides has been commented upon[157] as being smaller than is usually found in comparable chain XH_3ABC molecules, where A is a first-row atom. The measured (calculated) angle values for X = C, Si and Ge, respectively, are $113.8°$[155] ($113.6°$), $123.8°$[157] ($113.6°$) and $116°$–$119°$[159] ($119.0°$). Finally, the central nitrogen atom (N_β) in the azides is formally hypervalent, $-N_\alpha=N_\beta\equiv N_\gamma$. This predicts different N–N bond lengths for the two nitrogen–nitrogen bonds. The measured structural parameters give (double-bond minus triple-bond) differences of 0.086 Å[156a]–0.094 Å[156b], 0.179 Å[157] and 0.11 Å[158]–0.12 Å[159] for the methyl, silyl and germyl azides, respectively. The corresponding calculated values (Tables 4–6) are 0.086 Å, 0.065 Å and 0.068 Å. The discrepancy here is large for the SiH_3 and GeH_3 compounds.

The OCN ligand can attach to the XH_3 group to form either the cyanate ($-$OCN) or isocyanate ($-$NCO) species. Experimental gas-phase geometric structures have been published for CH_3OCN^{162}, $CH_3NCO^{156a,163}$, $SiH_3NCO^{164-166}$ and $GeH_3NCO^{167,168}$. A number of papers have also analyzed the chain XH_3ABC systems theoretically[161,169-172]. Both XH_3OCN and XH_3NCO have the *trans* C_s structure, except for PbH_3OCN which adopts the *cis* configuration (by 0.2°). The calculated OCN (NCO) angles are 178.2° (171.9°), 177.9° (175.8°), 178.4° (174.8°), 178.8° (176.0°) and 179.8° (174.8°) for the respective methyl, silyl, germyl, stannyl and plumbyl compounds. The known or assumed experimental values are 177° $(CH_3OCN)^{162}$, 170.3° $(CH_3NCO)^{163}$ and 173.8° $(GeH_3NCO)^{173}$, the latter from an X-ray analysis of the solid phase. The MP2 optimized geometry of GeH_3NCO overestimates the Ge–N bond length by 0.018 $Å^{168}$ or 0.013 $Å^{167}$, and underestimates the (average) Ge–H equilibrium distance by 0.005 $Å^{167}$ or 0.017 $Å^{164}$. The average HGeN and GeNC angles are observed at 108.3° and 142.2°, respectively, compared to the corresponding 107.2° and 138.2° calculated values. Recent all-electron (AE) calculations of germyl isocyanate at the MP2 level using valence double-[170] or triple-[161] zeta + single polarization basis sets obtained equilibrium NCO bend angles of $176.7°^{170}$ and $180°^{161}$, GeNC angles of $153.4°^{170}$ and $180°^{161}$ and average HGeN angles of $108.6°^{170}$ and $108.2°^{161}$. The largest discrepancy between the calculated angles is for the GeNC angle where the experimental value is significantly closer to the RCEP results reported here. Analogously, the experimental, AE and CEP values for the equilibrium SiNC angle in the isocyanate are $159.6°^{166}$, $180°^{171}$ and $156.3°^{161}$, and 146.6°, respectively. The MP2/CEP-TZDP calculated Si–N bond length (Table 5) is greater than experiment[164-166] by 0.033 Å and the calculated average Si–H bond length is too small by 0.02 Å. For CH_3NCO the experimental, AE and MP2/CEP-TZDP values for the CNC angle are $135.6°^{163}$, $138°^{161,174}$ and 134.0°, respectively. In CH_3OCN the analogous comparison for the COC angle gives $113.3°^{162}$, $113.5°^{174}$ and 113.0°, respectively. The larger basis set used in the CEP calculations for both X atoms and the relativistic effects included in the effective core potential (RCEP) for germanium seem to contribute to a better description of the equilibrium XAB angle for X = Si and Ge. However, although the geometric parameters in Table 4 agree well with the experimental structures, the calculated dipole moment for CH_3OCN (Table 16) of 4.499D is somewhat larger than the measured $4.26D^{162}$.

The calculated XH_3SCN and XH_3NCS equilibrium bond distances are shown in Tables 4–8. Experimental gas-phase geometric structures are available for CH_3SCN^{175}, $CH_3NCS^{156a,176,177}$, SiH_3NCS^{165} and GeH_3NCS^{178}. The analogous AE study was also carried out on the methyl, silyl and germyl thiocyanates and isothiocyanates[161,169-171]. In contrast to the above comparison of the dipole moment for CH_3OCN, the corresponding comparison between the calculated (Table 16) and observed[175] dipole moment of CH_3SCN shows their agreeing to 0.001D. The calculated SCN (NCS) angles for the XH_3SCN and XH_3NCS compounds are 178.5° (174.8°), 178.7° (178.7°), 179.1° (176.7°), 179.4° (179.8°) and 178.9° (176.5°), for X = C, Si, Ge, Sn and Pb, respectively. CH_3NCS is predicted to have the largest bend angle away from 180° and this has been estimated experimentally at 6.2°. In general, the $-N=N\equiv N$, $-Z-C\equiv N$ and $-N=C=Z$ (Z = O, S and Se) substituents are expected to be linear because of the combinations of internal double and triple bonds which give optimum interatom bonding for the linear structure. The bent XNC, XZC and XNN angles are consistent with the single-bond attachment to X. Thus, the calculated XSC (XNC) bond angles in the thio- and isothio-cyanates are 98.3° (144.4°), 95.6° (166.8°), 95.3° (149.4°), 94.3° (178.9°) and 89.0° (123.7°) for X = C, Si, Ge, Sn and Pb, respectively. The XSC angle is characteristically in the 90°–100° range. The experimental value for CH_3SCN is $99.0°^{175}$, which is 0.7° within the calculated value. However, the XH_3NCS set shows alternating XNC angle values with X = Si near-linear (within 13°), stannyl isothiocyanate actually linear (within 1.1°), while PbH_3NCS shows the smallest XNC angle. The observed XNC bond angle values are $141.6°^{156a}$ or $147.7°^{176}$ for X = C, and $163.8°^{164}$ or $180°^{178}$ for X = Si. The microwave spectrum of GeH_3NCS is considered to be consistent with a C_{3v} point group symmetry[179], having a linear GeNCS chain structure. Both AE studies at the MP2 level[161,170] predict a linear geometry, but a previous infrared spectroscopy study suggested a GeNC angle of $156°^{180}$, which is more consistent with the results reported here. This system warrants further study.

The calculated Ge—N bond length (Table 6) in GeNCS is 0.03 longer than experiment $(r_s)^{178}$ and Ge—H is calculated 0.007 Å shorter than assumed in the microwave+spectroscopy analysis of the structure[178]. The N=C bond length is underestimated by 0.062 Å (Table 4) relative to the measured value. This very large discrepancy is consistent with the calculated–measured difference in GeNC bond angles noted above. The C=S bond distance is calculated too small by only 0.027 Å. In SiH_3NCS the calculated (Table 5) Si—N distance is large by 0.024 Å^{164}–0.066 Å^{178}, N=C is short by 0.007 Å^{164}–0.016 Å^{178} and C=S is long by 0.006 Å^{164} or short by 0.008 Å^{178} relative to the observed value. The MP2 level AE calculation[171] also shows a 0.06 Å discrepancy with experiment for the Si—N bond distance. On the other hand, the calculated dipole moment (Table 16) of $2.606D$ is not far from the $2.38D$ value[178], despite the 16.2° difference in SiNC angles and 0.066 Å discrepancy in Si—N bond distance. These two discrepancies may be related and a re-evaluation of the SiH_3NCS microwave + infrared spectra to take into account quasi-linearity with a small barrier to inversion should be undertaken. For methyl isothiocyanate, all the calculated (Table 4) and experimental[155,176,177] bond-length values are within 0.025 Å of each other for the microwave studies.

For the XH_3SeCN and XH_3NCSe sets, experimental gas-phase geometries are available only for CH_3SeCN^{181} and $CH_3NCSe^{182,183}$. The calculated XSeC (XNC) angles are 95.0° (148.9°), 91.7° (179.7°), 91.8° (179.8°), 90.4° (179.9°) and 84.9° (179.2°) for X = C, Si, Ge, Sn and Pb, respectively. The selenocyanates are predicted to have SeCN angles $\geqslant$ 178.8°, where the equal sign is for the methyl compound. The NCSe angle is calculated to be 175.5° for X = C and 180° for all the heavier Group 14 atoms. Experimentally[181], the CSeC angle in methyl selenocyanate is found to be 96.0°, within 1° of calculated, and the SeCN angle is 179.4°. For CH_3NCSe the CNC angle is measured at $157.0°^{182}$–$161.7°^{183}$, for an assumed linear NCSe chain. In the XH_3SeCN set the AE calculational survey gives a XSeC bend angle of 97.9° for methyl[169], 95.7° for silyl[171] and 99.6° for germyl[170], while for XH_3NCSe the corresponding XNC angles were found to be 160°, 180° and 180°, respectively.

We can summarize the situation with respect to the structural parameters in the pseudohalides as follows. The XH_3—ZCN molecules, where Z is one of the chalconides (O, S, Se), are very bent at the Z atom, with the bend angle XZC being progressively smaller with increasing size of Z. The barrier to inversion through the linear structure is very high (thousands of cm^{-1}) and increases with the size of the chalconide[169]. PbH_3SCN and PbH_3SeCN are calculated to have PbZN angles that are smaller than 90°, possibly due to an attractive interaction between the lead and end nitrogen atoms. In this regard it should be noted that, contrary to all the others, both PbH_3SCN and PbH_3SeCN have the *cis* conformation relative to the Z—C bond, which is conducive to such an attractive interaction.

The C—N bond in the XH_3ZCN molecules (Table 4) are all about 1.181–1.185 Å, which is only several thousands Å larger than in CH_3CN. Thus, the triple-bond character of the CN group here is well preserved and other bond structures, such as $XH_3-\overset{-}{Z}=C=\overset{+}{N}$, do not contribute to any significant degree. The calculated $C\equiv N$ vibrational frequency is calculated at 2198 cm^{-1} in CH_3OCN, increases by 8 cm^{-1} in the Si compound and then decreases steadily through Ge and Sn to 2167 cm^{-1} in PbH_3OCN. In the XH_3SCN and XH_3SeCN sets the $C\equiv N$ energy decreases from 2102 cm^{-1} for X = C to 2068 cm^{-1} for X = Pb, and from 2079 cm^{-1} for X = C to 2054 cm^{-1} for X = Pb, respectively. The $C\equiv N$ vibrational frequencies in CH_3OCN are very similar to those in CH_3CN.

The CZ−CN bond lengths, however, are calculated to be significantly shorter than the C−ZCN distances, by 0.11–0.16 Å (Table 4), with the difference being in the order O>S>Se[169]. Inspection of Tables 4–8 shows that the Z−CN bond length is intermediate between single- and double-bond length. The XZ−CN distance also drifts to shorter values as X gets larger. Presumably, this trend approximately correlates with the extent of $XH_3 \rightarrow ZCN$ charge transfer. The shortening of the Z−CN bond therefore arises from $Z \rightarrow CN$ back-bonding[181], which is expected to have the Z dependence shown above. The X−O and X−S bond lengths behave alike. The X−Z bond length in XH_3ZCN is 0.01–0.07 Å longer than in the XH_3ZH and XH_3ZCH_3 reference compounds, for both Z = oxygen and sulfur, where the difference is smallest for X = carbon and roughly increases with the size of the X atom. This trend possibly reflects the increased relative p-character contribution of the X atom in going down the Periodic Table. As expected, the HXZ angle decreases with increasing size of X and decreasing size of Z, as these trends approximately correlate with increasing ionicity of the X−Z bond.

The XH_3NCZ (Z = O, S, Se) series of molecules have received much more attention than their XH_3ZCN linkage isomeric counterparts. The XH_3NNN set will be included here in the comparisons. The XNN/XNC angles are calculated to be generally bent, where the bending angle has a zigzag dependence on the size of the atom X. The smallest angles are found for carbon and lead. The XNN/XNC angles increase in the order Se > S > O > N. Thus, for Se, all the XNC angles are above 179°, except for CH_3NCSe (148.9°), and in the XH_3NCS set only the Sn compound has a near-180° SnCS angle. The greater bending angle in the azides has been explained[155,157] based on the all-bent valence bond structures,

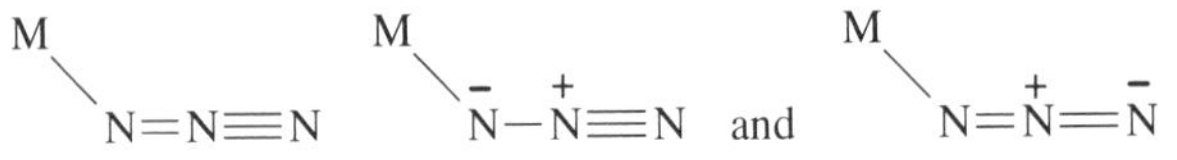

$$
\begin{array}{ccc}
\text{M} & \text{M} & \text{M} \\
\diagdown & \diagdown & \diagdown \\
\text{N=N}\equiv\text{N} & \overset{-}{\text{N}}-\overset{+}{\text{N}}\equiv\text{N} \quad\text{and} & \text{N}=\overset{+}{\text{N}}=\overset{-}{\text{N}} \quad \text{with M} = \text{XH}_3 \\
\textbf{(a)} & \textbf{(b)} & \textbf{(c)}
\end{array}
$$

In contrast, for structures XH_3NCZ, two of the possible structures are linear

$$
\text{H}_3\overset{-}{\text{X}}=\overset{+}{\text{N}}=\text{C}=\text{Z}, \quad \text{H}_3\text{X}-\overset{+}{\text{N}}\equiv\text{C}-\overset{-}{\text{Z}} \quad \text{and only one is bent} \quad (\text{H}_3\text{X}-\text{N}=\text{C}=\text{Z})^{169,178}
$$

$$
\textbf{(d)} \qquad\qquad \textbf{(e)} \qquad\qquad\qquad\qquad\qquad \textbf{(f)}
$$

All the NNN and NCZ bond angles are near-linear, with the former being the most bent in the 172° to 174° range. The smaller XNN angles probably induce a larger distortion in the NNN chain through interaction with the XH_3 group hydrogen atoms. It should be noted that in the XH_3NCZ series the lowest-energy vibrational frequencies involve low-energy (<67 cm^{-1} for all cases) coupled CH_3 rotation and XNCZ bending modes.

The N−C bond length in the XH_3NCZ sets is seen in Table 4 to be in the 1.20–1.23 Å range, where the dependence on Z is O>S>Se. Firstly, the bond distance size is somewhat larger than triple-bond length which indicates a possible contribution from structure **d** above. The calculated dependence of the equilibrium C−N bond distance on Z is consistent with the importance of **e**, where of all the chalconides the oxygen atom is best able to accommodate the negative charge. The C−N harmonic stretch frequencies in $XH_3N−CZ$

generally decrease in going from X = C to X = Pb, except for Ge → Sn when Z = O or S. The vibrational energy decrease is particularly large at X = Pb, a change of −63 cm^{-1} relative to Sn for Z = O or S, for example. An outstanding characteristic of the C−N harmonic motion in the XH$_3$NCZ series is the predicted large infrared intensity for all X and Z.

The C−O distance in the XH$_3$NCO set is calculated to be shorter than other formally doubly-bonded C=O atom pairs (Table 4), which is not consistent with the single bond C−O structure in **e**. The C=O distances in XH$_3$C(O)H, for example, are in the 1.22–1.23 Å range, while in XH$_3$NC=O they are in the 1.18–1.19 Å range. Analogously, the S−C distances in XH$_3$S−CN are all about 1.68 Å, much shorter already than the single-bond 1.81 Å S−C bond length in CH$_3$−SH and CH$_3$−SCN, for example. The S−C distance in the XH$_3$NC−S set is an even shorter 1.57 Å for all X. A C≡O triple-bonded structure is expected from a

$$H_3\overset{-}{X}=N-C\equiv\overset{+}{O}$$
(g)

type structure, which does not affect the XNC angle (relative to **f** above) and does give both a slightly elongated C−N bond length and reduced O−C distance. Clearly, no single valence bond structure is adequate to describe the XH$_3$NCZ series of molecules[167]. The Z−CN harmonic stretch frequencies decrease strongly in going from X = C → Pb. These seem to be moderately mixed with the corresponding X−NCZ mode, which explains the strong X dependence.

The X−N bond distances (Tables 4–8) in XH$_3$NNN and XH$_3$NCZ show the azide being consistently the larger. This is probably because of the smaller XNN bend angle which can introduce steric repulsion effects, and shows that structures with X=N double bonds like the linear H$_3\overset{-}{X}=\overset{+}{N}=N=N$ are not important. An analogous conclusion has been drawn about the XH$_3$NCZ series[167]. In the NCZ series the X−N bond lengths are generally comparable to those in XH$_3$−NH$_2$, so that structures like **d** above probably are not very important. A contrary conclusion has been reached for SiH$_3$NCS[178] based on the measured dipole moment.

The XH$_3$C(O)F and XH$_3$C(O)Cl series have C_s geometries, where the in-plane X−H$_s$ bond eclipses the carbonyl bond across X−C. Experimental gas-phase geometries are available only for the carbon member of each series[184,185]. In the comparison between calculated and experimental values both the C−F and C−Cl bonds are overestimated by *ca* 0.02 Å, C=O is calculated too large by *ca* 0.01 Å and *ca* 0.005 Å, respectively, and both C−C bonds are within 0.007 Å of experiment. The calculated angles all agree within 1°.

The FCO and ClCO angles are all within less than 1° of 120°, independent of X, with no obvious trends with regard to the nature of the X atom. The HXC angles are relatively isotropic and therefore the tilt angles are very small and, naturally, decrease in magnitude as X gets larger and the X−C distance widens. The WCX (W = halogen) angle is largest for X = C (109.9° for F and 111.2° for Cl) and decreases irregularly to 106.2° and 107.7° for X = Pb, respectively. The complementary OCX angle increases from 129.9° (F) and 128.3° (Cl) for X = C to 133.3° (F) and 131.4° (Cl) in the PbH$_3$C(O)W molecules. The larger OCX angle may be due to an optimum orientation of the unpaired electron molecular orbital in the C(O)W radical. This point requires further investigation.

The C=O bond length in both XH$_3$C(O)F and XH$_3$C(O)Cl (Table 4) decreases steadily in going from X = carbon to X = lead. At the same time, the C−F and C−Cl distances increase irregularly from X = C to X = Sn and then decrease in going to X = Pb. The calculated harmonic stretch frequencies for the C=O mode also show a mild decrease

from carbon to germanium, but then the vibrational energies increase somewhat in going
to the Pb compounds. Although the carbonyl bond lengths in the chlorides are generally
shorter than in the fluorides, the opposite trend is found in the vibrational frequencies,
which are higher in the fluorides. These mixed trends indicate that a combination of
steric and electronic factors govern these bond-length changes. By comparison with other
carbonyl systems (Tables 4–8) the X—C bond distances in the $XH_3C(O)W$ series here
show no unusual characteristics.

The last group in this series contains XH_3NO_2 and XH_3PO_2 (forked), and their respective linkage isomers, XH_3ONO and XH_3OPO (chain-bent). CH_3NO_2 and CH_3ONO have
been characterized experimentally in the gas phase[186,187]. Both CH_3NO_2/CH_3ONO and
SiH_3NO_2/SiH_3ONO have been discussed theoretically[188–190]. The chain-bent structure
of both XH_3ONO and XH_3OPO can, in principle, have four isomers for a geometry of
C_s symmetry, according to the conformation of the in-plane H_s—X—O—N—O chain. In
practice, two main isomer types (using NO_2, for example) are identified:

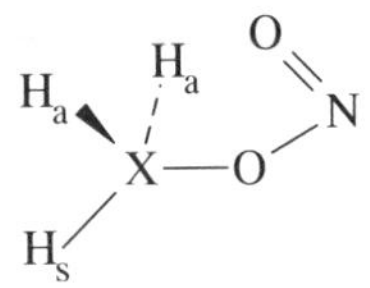

which differ in the ONO orientation. Both structural isomers have the staggered conformation of the XH_3 group with the terminal NO groups. For CH_3ONO[188] and SiH_3ONO[189] it
has been found that the *cis* conformation is marginally more stable. The results obtained
here at the optimized MP2/TZDP level are as follows. XH_3ONO adopts the *cis* conformation for X = C, Ge, Sn and Pb. However, for X = Si the *trans* conformation is
lower in energy (by 0.25 kcal mol^{-1}) at this level. The properties corresponding to the
lowest-energy structure in each case are listed in the Tables. Thus, for example, there is
a clear discontinuity in the values of the dipole moments in Table 16 for the XH_3ONO
set for X = Si. An interesting feature of the *cis* structure for the XH_3ONO set (except
SiH_3ONO) is the gradual swing of the ONO group as X gets larger to bring the terminal oxygen atom towards the X atom. In *cis*-PbH_3ONO the bound Pb—O distance is
2.21 Å while the 'nonbonded' Pb...O distance is 2.49 Å, to approximately form a four-
membered ring. This motion is accomplished by a gradual reduction of the OXH_s angle
from 104.3° for X = C to 90.6° for X = Pb, and the XON angle from 113.9° (X = C)
to 102.8° (X = Pb). Simultaneously, the XO—N distance decreases from 1.411 Å to
1.306 Å, N=O increases from 1.203 Å to 1.245 Å (Table 3) and the N=O harmonic
stretch frequency decreases from 1547 cm^{-1} to 1428 cm^{-1}. The preferred stability of
the *cis* XH_3ONO conformation is clearly driven by the X...O interaction as X increases
in size. For PbH_3ONO the optimized MP2/TZDP energy difference between the *cis* and
trans conformations is 9.4 kcal mol^{-1}. Actually, for CH_3ONO the results here predict
the *trans*-eclipsed equilibrium conformation to be next highest in energy after the *trans*-
staggered structure, where staggered and eclipsed refer to the relative orientations of the
C—H_s and O—N bonds.

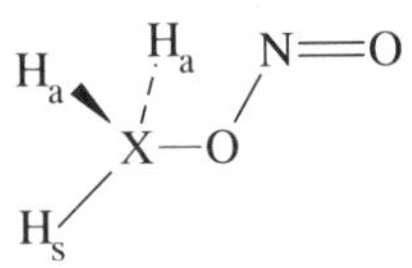

For XH_3OPO a slightly different result is obtained. CH_3OPO is MP2/TZDP optimized
to have the *cis*-eclipsed conformation as most stable, while the other XH_3OPO members
adopt the ordinary *cis*-staggered equilibrium geometry. Here, again, from Si → Pb the
H_sXO angle decreases (from 104.6° to 91.4°), ∠XOP decreases from 125.8° to 104.1° and
the Pb...O distance is 2.62 Å compared to the Pb—O bond length of 2.19 Å (Table 8). The
P=O harmonic vibrational energy decreases from 1199 cm^{-1} in SiH_3OPO to 1132 cm^{-1}
in PbH_3OPO.

The XH$_3$NO$_2$ and XH$_3$PO$_2$ series all have the planar C_s symmetry structure where one N=O/P=O bond eclipses a X$-$H$_s$ and the other N=O/P=O bond is staggered with the pair of X$-$H$_a$ bonds. The resultant asymmetry induced in the ONX and OPX angles is just a few degrees, on the average, for all the X atoms. The two N=O bond distances differ at most by 0.003 Å and the two P=O bond lengths by even less for a given atom. The NO$_2$ and PO$_2$ symmetric and antisymmetric stretch frequencies decrease uniformly from X = C to X = Sn. For example, the two vibrational energies are 1397 cm^{-1} and 1738 cm^{-1} for CH$_3$NO$_2$, and 1120 cm^{-1} and 1407 cm^{-1} in CH$_3$PO$_2$. These pairs of stretch modes are calculated to be 1326 cm^{-1} and 1638 cm^{-1} in PbH$_3$NO$_2$, and 1077 cm^{-1} and 1369 cm^{-1} in PbH$_3$PO$_2$. The values for X = Pb are very similar to those for X = Sn, sometimes a bit larger and sometimes smaller. The calculated ONO angle hovers about 125° for all X, while the OPO angle is centered about 133°. The P=O bonds are expected to be more ionic (or semipolar) than the N=O bonds (Tables 11$-$15) and this could account for the larger OPO angle.

K. XH$_3$ABC H

The only member of this group reviewed here is the XH$_3$C(O)OH set. An experimental gas-phase geometric structure is available only for CH$_3$C(O)OH[191]. The most stable conformer has the in-plane C$-$H$_s$ and O$-$H bonds eclipsing the C=O bond in C_s symmetry. The C(O)OH fragment always has the *syn* conformation[192] and the methyl group can adopt two rotameric configurations of the C$-$H$_s$ relative to C=O, eclipsed and staggered. In XH$_3$C(O)OH, X = C, Ge, Sn and Pb adopts the eclipsed (*e*) conformation and SiH$_3$C(O)OH favors the staggered (*s*) conformation. The calculated energy difference between the *e* and *s* forms for CH$_3$C(O)OH is 0.4 kcal mol^{-1} (140 cm^{-1}) in favor of *s*. For SiH$_3$C(O)OH the *s* form is more stable by only *ca* 0.1 kcal mol^{-1} (32 cm^{-1}) and the *s* conformer is predicted to be a transition state in the methyl rotation mode rotating the *e* conformation back into itself. The energy differences refer to MP2/TZDP optimized geometries for each conformer. In GeH$_3$C(O)OH, SnH$_3$C(O)OH and PbH$_3$C(O)OH the *s* conformer is more stable by 18 cm^{-1}, 25 cm^{-1} and 63 cm^{-1}, respectively. The relative stability of the *e* conformation in these last three molecules cannot be due to (C=)O ... H(X) interaction since both participating atoms have negative atomic charges (Tables 12$-$15).

It is interesting to compare experimental and theoretical geometric structural parameters for the CH$_3$C(O)OH case to demonstrate the methyl group asymmetry in these type systems. The calculated (measured[191]) bond lengths are: C$-$C, 1.506 Å (1.503 Å); C$-$O, 1.368 Å (1.352 Å), C=O, 1.216 Å (1.205 Å); O$-$H, 0.976 Å (0.971 Å); C$-$H$_s$, 1.087 Å (1.088 Å); C$-$H$_a$, 1.091 Å (1.094 Å). The angles are: C$-$C$-$O, 110.9° (111.7°); C$-$C=O, 126.5° (125.4°); COH, 105.1° (105.4°), CCH$_s$, 109.5° (109.7°); CCH$_a$, (out-of-plane), 109.6° (109.5°); H$_a$CH$_a$, 107.7° (107.6°); H$_s$CH$_a$, 110.3° (110.3°). The theoretical description of the asymmetry of the methyl group due to interactions with the Y substituent is seen here to agree closely with experiment.

The COH angle in the XH$_3$C(O)OH starts at 105.1° in CH$_3$C(O)OH, increases to 105.6° for X = Si, Ge and Sn (to within 0.1°) and then increases to 106.3° for PbH$_3$C(O)OH. The general trends for changes in the X$-$C=O and X$-$C$-$O angles in going down the Group 14 column is mixed. The former tends to increase (Ge → Pb) and the latter decreases (except for Si) with increasing size of X. The different equilibrium conformation for SiH$_3$C(O)OH makes it difficult to interpret the initial large decrease in $\angle$X$-$C=O (from 126.5° to 123.0°) and increase in X$-$C$-$O (from 110.9° to 114.7°) in going from X = C to X = Si. The narrowest range of angles is for O$-$C=O which is in the 122.0°$-$122.6° range for C to Sn and jumps to 123.7° for Pb. The almost constant value of the calculated dipole moment in Table 16 seems to indicate that the gradual swing of the C(O)OH group away from $C-H_s$ resultant from the above angle changes acts to moderate the magnitude of the usual dipole moment increase as X gets heavier. Rehybridization at the carbon atom should also be a factor. The degree of charge transfer (Table 9) also does not seem to depend strongly on the nature of X in these systems beyond methyl.

The C=O bond length in the $XH_3C(O)OH$ (Table 4) set is consistently larger (by *ca* 0.02–0.03 Å) than in $XH_3C(O)F$ and $XH_3C(O)Cl$ and slightly smaller (by *ca* 0.005–0.01 Å) than in $XH_3C(O)H$ for each X. The calculated C=O harmonic stretch frequencies follow the same trends. The differences are significant. For example, in the lead compounds, the calculated C=O vibrational energies in $PbH_3C(O)Z$ are: Z = H, 1660 cm^{-1}; OH, 1746 cm^{-1}; F, 1828 cm^{-1} and Cl, 1812 cm^{-1}. These, of course, have to be multiplied by *ca* 0.95 for comparison with experiment[47]. The O–H bond length (Table 3) in $XH_3C(O)OH$ is consistently somewhat longer than in XH_3OH (by *ca* 0.006–0.015 Å). The difference is smallest for X = C, largest for X = Si and decreases steadily to X = Pb. The calculated O–H stretch frequencies are correspondingly lower in $XH_3C(O)OH$ compared to XH_3OH by an almost uniform 200 cm^{-1}. The C–O(H) bond length in $XH_3C(O)OH$ is consistently and significantly shorter than in normal singly-bonded C–O (Table 4). The reason for these trends in the O–H and C–O(H) bonds is certainly connected to possible additional valence bond structures for the C(O)OH group[192,193] that involve its enhanced acidity.

The X–C bond lengths in the $XH_3C(O)OH$ set are slightly shorter than in the $XH_3C(O)H$ series (Tables 4–8), with the decrease ranging from 0.001 Å (X = C) to 0.021 Å (X = Pb) in a relatively uniform manner. In this sense, the corresponding X–C distances in $XH_3C(O)OH$ are very similar to those in $XH_3C(O)F$, and probably for the same reasons. The X–H equilibrium bond distances (Tables 4–8) span a narrow range of values for a given X. These, however, consistently show the ordering of X–H bond lengths to be $XH_3C(O)H > XH_3C(O)OH > XH_3C(O)Cl, XH_3C(O)F$. All three calculated harmonic stretch frequencies for the X–H mode show the same ordering, where the Z = OH, Cl and F compounds are relatively close at higher energies, and separate from the formyl set. The difference between the two sets revolves about the 20 cm^{-1} to 30 cm^{-1} range.

L. XH_3ABCH_2

The only set in this grouping reviewed here is $XH_3C(O)NH_2$. An experimental gas-phase geometry, determined by electron diffraction, is available for $CH_3C(O)NH_2$[60,194]. The agreement between calculation (Table 4) and experiment for the bond lengths is within 0.007 Å for the C–C, C–N and C–O bond lengths. The bond angles agree to within 0.6°. The equilibrium conformation of acetamide has the eclipsed (*e*) structure, where the C=O bond is *cis* to the in-plane $C-H_s$ bond. The $H_sC(O)N$ skeleton has approximate C_s symmetry and the NH_2 group is slightly pyramidal. The $H_\alpha NCO$ dihedral angle is calculated to be 15.8° and the $H_\beta NCO$ angle is 160.7°. The alternate staggered (*s*) conformation, with the methyl $C-H_s$ bond located *trans* to the C=O bond, is a transition state, at only 29 cm^{-1} higher energy. The *s* configuration has a possible steric repulsion interaction between the H_s atom and an amine hydrogen (H_β). For X = Si → Pb in $XH_3C(O)NH_2$, however, the *s* conformation is calculated to be the equilibrium geometry at 388 cm^{-1}, 253 cm^{-1}, 227 cm^{-1} and 186 cm^{-1} lower energies than the *e* conformation for X = Si, Ge, Sn and Pb, respectively. In all these latter cases, the *e* conformation is a transition state. The lowest-energy vibrational mode for the $XH_3C(O)NH_2$ series has frequencies of 52 cm^{-1}, 98 cm^{-1}, 79 cm^{-1}, 69 cm^{-1} and 63 cm^{-1}, respectively, for X = C → Pb, each in its own MP2/TZDP optimized equilibrium structure. This motion involves principally methyl group rotation. The particular conformational preference of a given X atom member of this set must be a combination of steric and electronic factors. The former favors the e conformation and the latter stabilizes the s geometry, while both decrease in importance as the X−C bond lengthens.

The trends in the XCO and XCN angles in $XH_3C(O)NH_2$ are similar to those found in $XH_3C(O)OH$. Comparing the acids and the amides, the XC=O angle increases slowly with X and XCN decreases with X (Si → Pb). The XC=O angle in the amide is 3–7° smaller than in the acid, XC−N is 5–6° larger in the amide than XC−O is in the acid and the O−C=O and N−C=O angles are very similar, including the increase in going Sn → Pb. Except for the C and Si compounds, the amide and the acid adopt different conformations. Also, apparently, because of the different structure and the different mode of binding, $CH_3C(O)NH_2$ does not conform to the general trends and has the largest XC=O (123.2°) and smallest XCN (114.7°) angles of the $XH_3C(O)NH_2$ set. The calculated dipole moments in the $XH_3C(O)NH_2$ set (Table 16) drift to lower values as X increases in size.

The C=O bond length in $XH_3C(O)NH_2$ is uniformly slightly larger than in the $XH_3C(O)OH$ and $XH_3C(O)H$ sets (Table 4). The harmonic stretch frequencies in the amines, however, are still calculated to be somewhat larger than in the aldehydes. For example, the C=O vibrational frequency in $PbH_3C(O)NH_2$ is 1717 cm^{-1}, which is larger than in $PbH_3C(O)H$. It should be noted that all the C=O modes are predicted to have high infrared intensities. Wiberg and coworkers[195] have discussed the nature of substituent effects in aliphatic carbonyls and this will be used to summarize trends in the properties of the C=O group in the coming section. The N−H bond length in the $XH_3C(O)NH_2$ set is calculated to be uniformly shorter than in the corresponding XH_3NH_2 set. The N−H harmonic stretch frequencies are larger in $XH_3C(O)NH_2$ than in XH_3NH_2 by variable amounts. The differences are smallest for Si (9 cm^{-1} and 54 cm^{-1}) and largest for C (78 cm^{-1} and 126 cm^{-1}) and Pb (64 cm^{-1} and 132 cm^{-1}). These trends are opposite to those found for the O−H bond in comparing $XH_3C(O)OH$ and XH_3OH. The HNH angle is 11–13° wider in the $XH_3C(O)NH_2$ set compared to XH_3NH_2. A clue to the behavior of the NH bond must lie in the short C−N distance found for all the members of the $XH_3C(O)NH_2$ set, which is reduced by ca 0.1 Å relative to a normal C−N single bond distance, as in CH_3NH_2, for example (Table 4). The corresponding reduction in the C−O(H) length in the acid relative to ethanol is only ca 0.06 Å.

The X−C distances in $XH_3C(O)NH_2$ are larger than in the previous carbonyl sets (formyl, fluoroformyl, chloroformyl and carboxyl) discussed above (Tables 4–8). The X−H distances in the carboxamides are also generally larger than in the previous carbonyl sets, although the X−H harmonic stretch frequencies are very similar between $XH_3C(O)NH_2$ and $XH_3C(O)H$. The Mulliken group charges for XH_3 (Table 9) show lower values for both the carboxamides and the aldehydes relative to the acids, fluoroformyl and chloroformyl compounds, for a given X. The X atom charges in Tables 11–15 show the same general trends. The higher the charge on the X atom, the shorter the X−H and X−C bond lengths, probably due to a radial contraction effect of the central atom.

M. XH_3ABCH_3

The members of this set all have the $XH_3C(O)CH_3$ formula. Only for dimethyl ketone has a gas-phase geometry been reported[196]. There are two important conformations relevant to this acetyl series within the planar $H_sXC(O)C'H'_s$ skeleton structure (C_s). $CH_3C(O)CH_3$ is reported[196] and calculated here to have both in-plane methyl hydrogen atoms (C−H_s and C′−H'_s) eclipsed (e) with the carbonyl oxygen atom. The molecules with X = Ge, Sn and Pb also adopt this $e−e'$ conformation as their lowest-energy equilibrium structure. Acetylsilane ($SiH_3C(O)CH_3$) prefers the $s-e'$ (s = staggered) conformation where $Si-H_s$ is *trans* to the C=O bond across the Si–C bond and $C-H_s$ is eclipsed with the C=O bond. The alternate $e-e'$ configuration in the silane is calculated to be a transition state (one imaginary frequency) only 42 cm^{-1} above $s-e'$, where all conformations are MP2 geometry optimized in the TZDP basis set. The lowest-energy harmonic vibrational frequency in the normal-mode analysis of the $s-e'$ silane structure is 45 cm^{-1} involving essentially rotation of the SH_3 group towards the $e-e'$ conformation. There is also an $e-s'$ conformation possible, where $Si-H_s$ is eclipsed and $C-H_s$ is staggered with the C=O bond. Its geometry-optimized energy is 367 cm^{-1} (1 kcal mol^{-1}) above $s-e'$. Given the small energy difference calculated between the $e-e'$ and $s-e'$ conformations and previous experience[195], the true ground-state geometry for the silane remains to be definitively determined. In this review the lowest energy ($s-e'$) structure is adopted. In X = C, the $e-s'$ (= $s-e'$) structure is 247 cm^{-1} (0.7 kcal mol^{-1}) above $e-e'$, compared to the experimental estimate[197] of 0.8 kcal mol^{-1}. The lowest-energy harmonic frequency in the equilibrium $e-e'$ geometry is 41 cm^{-1}. This latter motion is comprised of the synchronous rotation of the two methyl groups in opposite-sense directions. Acetylgermane also has a higher energy $e-e'$ configuration at 356 cm^{-1} (1 kcal mol^{-1}) above $e-s'$. The equilibrium $e-e'$ structure has a lowest-energy harmonic frequency of 31 cm^{-1} involving principally a rotation of the GeH_3 group to give the $s'-e$ geometry. The corresponding methyl rotation towards $e-s'$ has a frequency of 135 cm^{-1}. For $SnH_3C(O)CH_3$ the $e-s'$ conformation is 420 cm^{-1} (1.2 kcal mol^{-1}) above $e-e'$. In the latter geometry the SnH_3 rotation frequency is calculated to be 22 cm^{-1} going towards the $s-e'$ configuration, which is found to be a transition state in the $e-e'$ ↔ $e-e'$ interconversion. The methyl rotation frequency in the stannane towards $e-s'$ is at 142 cm^{-1}. Finally, for X = Pb, the lowest-energy vibrational frequency in the equilibrium $e-e'$ structure is calculated to be 39 cm^{-1}, for essentially PbH_3 rotation. The $s-e'$ conformation is 483 cm^{-1} (1.4 kcal mol^{-1}) above $e-e'$ and the methyl rotation frequency in the $e-e'$ ground equilibrium structure towards $e-s'$ is at 153 cm^{-1}.

If we focus just on the electronic energy differences between the $e-s'$ and $e-e'$ stationary states in the $XH_3C(O)CH_3$ set, then there is a gradual increase in the energy gap as X get heavier. The difference between these two conformations is in the rotation of a methyl group by 60°, and the energy difference is the rotation barrier. It has been shown[111] in the ethane-like XH_3CH_3 series that the dominant interactions that determine the preferred stability of the staggered conformations are hyperconjugative. In the RC(O)R' carbonyl compounds the stability of the structure is determined by the electrostatic interaction of the R and R' substituents with the carbonyl group. The preferred relative stability of the $e-e'$ geometry could be a combination of electrostatics, hyperconjugation, hybridization and induced dipole stabilization[198–200]. Most interaction mechanisms favor $e-e'$ over $e-s'$. As X gets larger, and the X–C(O) bond distance increases, the negative charge on oxygen gets smaller while the positive charge on the methyl hydrogen atoms does not change much (Tables 12–15). This would actually indicate a decreased preferred electrostatic stability for $e-e'$ with increasing size of X. Thus, the other mechanisms[198–200] must determine the ground state $e-e'$ equilibrium geometry conformation. An indication that the electronic structure is well described comes from the calculated dipole moment of $CH_3C(O)CH_3$ of 2.830D (Table 16), which agrees very well with the experimental value of 2.90D[201].

The C=O bond lengths in $XH_3C(O)CH_3$ follow the general pattern found for the carbonyl compounds (Table 4): an increase in distance in going from X = C to X = Si and then a gradual decrease with increasing size X until X = Pb which has the shortest

C=O bond length in the set. The C=O distances themselves for a given X conform to the general rule[195] that the more electronegative the substituent Z in $XH_3C(O)Z$, the shorter the C=O bond length. Both the C=O distances and harmonic stretch vibrational frequencies are similar for Z = H and Z = CH_3. The C=O vibrational energies for Z = CH_3, like those for Z = H, are lower than for Z = F, Cl, OH and NH_2, decreasing in that order.

The CC=O bond angle oscillates between 121.8° and 121.5° for X = C → Sn and jumps to 124.0° for X = Pb. ∠XC=O decreases from 121.8° for X = C to 118.4° for X = Pb, except for the X = Si which is 116.3°. The smaller XC=O angle for the silicon compound could be related to the methyl group being staggered with the C=O bond. The XCC bond angle is in the 116° to 119° range, except for Si where it complements ∠XC=O and is calculated to be 122.2°.

The C−C single bond in $XH_3C(O)CH_3$ (Table 4) generally decreases in length as X increases in size. This is generally interpreted as indicating a degree of increased bond character between the atoms, which in simple valence bond language would simultaneously involve a structure with reduced bond order in C=O. Since the C=O bond length also shortens as X increases, changes in bond orders cannot explain both trends simultaneously. Two other possible explanations involve increased positive charge on the carbonyl carbon atom or its rehybridization to include more s character in the sigma framework. The former is not supported by the atomic charges shown in Tables 11–15. Therefore, the hybridization argument[195,199] should be further examined for its ability to rationalize the bond length trends involving the carbonyl carbon atom as a function of X. In general, the C−C bond length in $XH_3C(O)CH_3$ is longer than in the previous members of the $XH_3C(O)Z$ sets (Table 4), probably because of steric repulsion effects. The equilibrium X−C bond length in the acetyl series is also the longest of the other $XH_3C(O)Z$ members for a given X atom, presumably for the same reason.

N. XH_3ABCD

There are two series of molecules in this category: the XH_3CF_3 set and the XH_3ONO_2, XH_3OPO_2 sets. For the trifluoromethyl group there are experimental gas-phase geometries for 1,1,1-trifluoroethane[202,203], trifluoromethylsilane[204] and trifluoromethylgermane[205]. In the XH_3CF_3 set all the members were calculated in the staggered conformation and these were found to be equilibrium geometries. Comparing calculation (Tables 4–6) with experiment for CH_3CF_3[202–205] we find the C−C distance overestimated by ca 0.01 Å. The CCF angle is calculated (observed) at 112.1° (112.3°) and ∠CCH is 109.0° (109.2°). In these comparisons the latest analysis of the microwave, infrared and electron diffraction results have been quoted[205]. For SiH_3CF_3 the Si−F distance is overestimated by 0.006 Å and the C−F bond length is in error by 0.013 Å relative to average zero-point level geometric parameters[204]. In the germane compound the MP2/TZDP calculated Ge−C bond distance is 0.003 Å larger than experiment[205], C−F is 0.009 Å too long and the average Ge−H distance is overestimated by 0.015 Å. The bond angles agree closely.

In spite of fluorine being a very electronegative atom, the XH_3 group is increasingly electropositive as X gets larger, and the XH_3CF_3 geometry stays relatively constant in the staggered ethane-like conformation. Thus the group charges in Table 9 and the calculated dipole moments in Table 16 show no striking increases in going down the Group 14 column of the Periodic Table, especially from X = Si. The HXC and FCX angles also do not show much variation as a function of X. Their calculated values are (∠HXC) 107.0° ±0.2° for Si → Pb and 109.0° for X = C, and (∠FCX) 112.1° ±0.6° for C → Pb, with no clear trends. However, the individual atomic charges in Tables 11–15 show that their distribution does not lead to reinforcing dipoles and that the local $X-H$, $C-F$ and $X-C$ moments can combine to partially cancel each other out.

The $C-F$ bond distance in the XH_3CF_3 series is shorter than in both CH_3F and the $XH_3C(O)F$ set (Table 4). The $C-F$ harmonic stretch frequencies are also uniformly larger in the trifluoro than in the fluoromethyl series for corresponding X atoms. The FCF angles in the trifluoro series are a relatively small $106.7° \pm 0.4°$ (compared to tetrahedral) for all X. The $C-F$ bond shortening could be attributed to a radial contraction of the carbon atom and electrostatic attraction between $C^{\delta+}$ and $F^{\delta-}$. In the bond angles the opposite effect is found for the HXH angles which are larger than tetrahedral, especially for $X = Si \rightarrow Pb$. Here, the larger angle can be explained by the tendency of the XH_3^+ group to be planar. One possible explanation for the small FCF angles would be a hyperconjugative interaction between the fluorine lone pairs of electrons and the $C-F$ σ^* molecular orbitals. This latter effect should tend to lengthen the $C-F$ bond which seems to be the opposite to what is found here. Another possibility is the contribution of ionic bond structures (CF^-) which involve simultaneously a higher bond order of the carbon with another fluorine atom ($C=F^+$). The resultant instantaneous charges could create an attractive interaction between the fluorine atoms that close the FCF angles somewhat[202]. Trifluoromethane also has small ($103.8°$) FCF angles[60]. In molecular orbital terms, this interaction is between the fluorine lone-pair (π) electrons and the $C-F$ σ^* orbitals[206]. It must then be that the shortening effect due to contraction about the carbon atom would be even larger were it not for the lone-pair σ^* interaction, which tends to lengthen the $C-F$ bond, and partially cancels the shortening effect.

The $X-C$ bond lengths in XH_3CF_3 are typically on the long side for a single bond connection between the respective atoms (Tables 4–8) for $X = Si \rightarrow Pb$, and on the short side for $X = C$. Here the comparison is with tetracoordinated carbon atoms. For the silane to plumbane compounds the longer $X-C$ bonds are probably due to the atoms involved having the same-sign charges (Tables 12–15). For the carbon compound the two carbon atoms are oppositely charged (Table 11).

All the members of the XH_3ONO_2 and XH_3OPO_2 series of molecules are calculated to have C_s symmetry equilibrium geometries with only the two methyl $C-H_a$ bonds symmetrically located out-of-plane. The $H_sCON=O_\alpha$ chain is arranged in a zigzag pattern and the $N=O_\beta$ bond is staggered with respect to the two $C-H_a$ bonds facing $XH_2\cdot$. This is also the experimental result for CH_3ONO_2[207]. The comparison between measured and calculated values (Tables 3 and 4) shows that the $C-O$, $O-N$, $N=O_\alpha$ and $N=O_\beta$ bond lengths are MP2/TZDP calculated too large by 0.009 Å, 0.020 Å, 0.009 Å and 0.018 Å, respectively. N=O double bonds are particularly troublesome in NO_3[208] but the results here for the bond distances in methyl nitrate are in the same error range as the other single and double bonds involving first-row atoms (Tables 3 and 4). The CON, ONO_α, ONO_β and $O_\alpha NO_\beta$ angles agree within $0.8°$ between theory and experiment[207]. The (Table 16) calculated dipole moment of $2.933D$ for CH_3ONO_2 agrees very well with the measured $3.081D$ value[207].

A particularly interesting feature of these geometries are the large $O_\alpha NO_\beta$ angles. Their calculated values are $130.2°$ (C), $129.5°$ (Si), $128.6°$ (Ge), $127.4°$ (Sn) and $126.1°$ (Pb). The corresponding $O_\alpha PO_\beta$ angles in XH_3OPO_2 are $133.7°$ (C), $133.4°$ (Si), $132.3°$ (Ge), $131.4°$ (Sn) and $130.5°$ (Pb). These large $O_\alpha NO_\beta$ angles are probably due to lone-pair electron steric and atom-centered electrostatic repulsion between the O_α and O_β atoms. The decrease in NO_2 angle with increasing size of X could reflect the decreasing charges on these oxygen atoms (Tables 11–15). However, this decrease is observed also for the XH_3OPO_2 series where the atomic charges on O_α and O_β are not getting smaller with increasing size of X (Tables 11–15). An alternative explanation to the decreasing $O_\alpha NO_\beta$ and $O_\alpha PO_\beta$ angle values with X lies in the gradual development of a bonding interaction between X and O_β to form, finally, for PbH_3ONO_2 and PbH_3OPO_2 four-membered $PbONO_\beta$ and $PbOPO_\beta$ rings. The $Pb-O_\beta$ distances for the nitrates are 2.53 Å (C), 2.66 Å (Si), 2.70 Å (Ge), 2.69 Å (Sn) and 2.55 Å (Pb). The corresponding values in the XH_3OPO_2 set are 2.90 Å (C), 3.12 Å (Si), 3.10 Å (Ge), 3.10 Å (Sn) and 2.75 Å (Pb). These can be compared to the normal X—O bond lengths in Tables 4–8. Other geometric and electronic structural parameters also show the increasing X—O_β interaction in Si → Pb. A similar result was obtained for the XH_3ONO and XH_3OPO sets above. One result of this interaction is a decrease in the $O_\alpha NO_\beta$ and $O_\alpha PO_\beta$ angles as, for example, the near-symmetric four-membered ring is formed with similar $O_\alpha NO_\beta$ and $O_\alpha NO$ pairs of angles. The generally larger $O_\alpha PO_\beta$ angles relative to $O_\alpha NO_\beta$ for each X is due to the larger charges on the oxygen atoms in the phosphite compounds, due to the semipolar nature of the P=O bond compared to the more covalent nature of the N=O bond[5,22].

Another consequence of the X...O_β attraction is the decreasing value of the H_sXO angle (102.8°, 99.5°, 98.0°, 95.8° and 91.7° for X = C, Si, Ge, Sn and Pb, respectively, in XH_3ONO_2 and 104.7°, 103.4°, 100.9°, 99.1° and 93.8° in the phosphite compounds) which follows the swing of the ONO_2 and OPO_2 groups relative to the X—O bonds. The XON (XOP) angles decrease correspondingly from 113.3° (123.8°) for X = Si to 102.5° (105.4°) for X = Pb. The XON and XOP angle values for the covalent X = C case are, however, 111.9° and 118.4°, which are smaller than for X = Si. The increased asymmetry in the N=O_α and N=O_β bond lengths and oxygen atom charges in the Tables, and similarly for P=O_α and P=O_β, are more indications of the ring-forming tendency of these molecules in going down the column of the Periodic Table. The X—H bond lengths also show increasingly different values for X—H_s and X—H_a as X gets larger.

Both the group charges in Table 9 and the calculated dipole moments in Table 16 agree that the bonding becomes increasingly more ionic in going from carbon to lead. The phosphites show consistently more charge transfer from XH_3 than the nitrates. These results must be connected to the nature of the central O—N and O—P bonds, which decrease steadily in length from silicon to lead (Table 3). These bond distances are consistently shorter than in the corresponding XH_3ONO and XH_3OPO molecules. The higher valencies of the nitrogen and phosphorus atoms in the $-ONO_2$ and $-OPO_2$ substituents compared to $-ONO$ and $-OPO$ would tend to shorten all bonds to the central N and P atoms. The increased ionicity of the XH_3-Y (Y = ONO_2 and OPO_2) bond is expected to enhance

valence bond structures such as $H_3X^+ O=N^{+}\!\!\begin{smallmatrix} O^- \\ \diagup \\ \diagdown \\ O^- \end{smallmatrix}$ and the analogous bonding arrangement

in XH_3OPO_2. Thus, the double-bond character of the (X)O—N and (X)O—P bonds is expected to increase in weight as the ionicity of the XH_3-Y bond increases. Since the P=O bond has intrinsically more of a semipolar nature than N=O, the above bonding structure should affect the (X)O—P bond to a lesser extent that the (X)O—N bond. The N=O_β and P=O_β bonds facing the X atom lengthen with increasing size of X, as expected from the X...O_β interaction. However, the remote N=O_α and P=O_α bonds do not change much, or increase slightly, as a function of X. The N=O and P=O harmonic stretch frequencies are calculated (before scaling[47]) to fall in the 1799 cm^{-1} (C) to 1679 cm^{-1} (Pb) range for the antisymmetric mode, and from 1275 cm^{-1} (C) to 1306 cm^{-1} (Pb) for the symmetric motion in XH_3ONO_2. The corresponding energies for the phosphite are 1417 cm^{-1} (C) to 1375 cm^{-1} (Pb) for the antisymmetric stretch, and 1157 cm^{-1} (Si) to 1142 cm^{-1} (Pb) for the symmetric mode. These parallel the trends in the N=O and P=O bond lengths, although there is no strong bond localization of the vibrational motions.

The X−O distances in both the nitrate and phosphite sets (Tables 4–8) are on the long side compared to the other X−O bonds reviewed here. The differences are smallest for X = C in the nitrate. The X...O$_\beta$ interaction should tend to lengthen X−O. The C−O(N) bond should be better described than C−O(P) because of the latter's higher O−P ionic character. For the same reason, the other X−O(P) bonds should be shorter in the phosphite compounds compared to the nitrates. These expectations are realized in the calculated relative X−O bond lengths (Tables 4–9) except for the lead compounds.

O. XH₃ABCDH

The XH₃OS(O)OH set has the following geometric structure: The H$_s$XO−S chain forms a *trans* conformation plane with the two X−H$_a$ bonds located symmetrically above and below the plane according to C_s symmetry. The *syn* O$_\alpha$=S−O$_\beta$H group is also approximately planar and roughly perpendicular to the O−S bond. The orientation of the O$_\alpha$=SO$_\beta$H plane to the H$_s$XO−S plane changes in going from X = C to X = Pb to make the X...O$_\alpha$ (=S) distance progressively smaller. Thus, the O=S−OX dihedral angle decreases (in absolute value) from −42.4° (X = C), through −16.3° (Si), −19.4° (Ge) and −10.4° (Sn) to −9.6° (Pb). The X...O (=S) distance decreases, in parallel, from 2.91 Å, through 2.95 Å, 2.95 Å, 2.83 Å and 2.55 Å, respectively. In PbH₃OS(O$_\alpha$)O$_\beta$H the PbO−S=O$_\alpha$ group forms an approximate four-membered ring, where the equilibrium Pb−O(−S) distance is 2.18 Å (Table 8), and the PbO−S=O$_\alpha$ skeleton is now approximately planar in a *syn* conformation. The (H)O$_\beta$S−OX dihedral angles are 66.3°, 93.5°, 89.7°, 98.3° and 98.2°, respectively, for X = C → Pb. The HO$_\beta$S angle stays relatively constant at 105.7 ± 0.5° for the Si → Pb compounds, after an initial 104.4° for X = C. The (H)O$_\beta$S=O$_\alpha$ angle also remains relatively constant around 105 ± 1° in going down the Group 14 column, and ∠O−S−O$_\beta$, initially 5° smaller in the methyl compound, is only 0.8° smaller for X = Pb. The large changes, as usual in these oxy compounds, are in the H$_s$XO angle, which decreases progressively from 105.0° (C) to 89.8° (Pb). The H$_a$XO angles, however, stay steady at about 110° ± 1°. The trend towards planarity in the XH₃ group is expressed in the motion of H$_s$ because of the approach of the O$_\alpha$(=S) atom to the H$_a$ side of X. The XO−S angle also decreases from 116.2° (C) to 104.9° (Pb) as the quasi four-membered ring tends to closure; the same behavior is followed by the O−S=O$_\alpha$ angle (from 108.9° for X = C to 102.6° for X = Pb). The result of the relative orientation of the sulfur−oxygen bonds in different directions is that, although the XH₃ → OS(O)OH charge transfer is large (Table 9), the dipole moments are not correspondingly large.

The O−H bond length in the XH₃OS(O)OH set (Table 3) is relatively unchanged at about 3689 cm^{-1} as a function of the very remote X atom. This invariance is also reflected in the O−H stretch frequency, which spans only ±9 cm^{-1} for all five compounds. The S−O$_\beta$ (H) length varies between 1.64–1.66 Å with no apparent trend. The S=O bond length increases steadily from X = C to Pb, in accord with its increased coordination to the X atom, while (X)O−S decreases steadily from X = C to Pb. Thus the two sulfur−oxygen bonds that are remote from increased oxygen coordination to the X atom approach each other in bond length.

The X−O single bond lengths in Tables 4–8 for XH₃OS(O)OH show a variable relationship to that found in the other Y groups in XH₃Y compounds having such a bond. For the heavier X atoms the additional X...O interaction for Y = ONO, OPO, ONO₂, OPO₂, OS(O)OH and, as will be seen in the next set, OC(O)CH₃ makes their directly bonded X−O distance longer than usual (Tables 7 and 8). The exception here is for Y = ON, which almost always has the longest X−O distance for all X. For the lighter X atoms the range of X−O single-bond distances is narrower. For example, for X = C, the equilibrium C−O bond length is calculated to have values from 1.423 Å (CH_3-OCH_3) to 1.500 Å (CH_3-ON). The C−O bond distance in $CH_3-OS(O)OH$ is closer to the lower end of this range at 1.453 Å (Table 4). For X = Si, the first of the more ionic structures, the S−O range is from 1.660 Å (CH_3-OSiH_3) to 1.729 Å (CH_3-ONO_2), while the value for $SiH_3-OS(O)OH$ is 1.707 Å. In the plumbyl compound the Pb−O range is 2.064 Å (CH_3-OPbH_3) to 2.194 Å (CH_3-ON, CH_3-OPO), and the bond length in the sulfite is 2.184 Å. Thus, in going from X = C to Pb the X−O single-bond length migrates from the shorter part to the longer part of the respective ranges.

P. XH_3ABCDH_3

Two sets of molecules belonging to this generic type were examined: $XH_3C(O)OCH_3$ and $CH_3C(O)OXH_3$. The X = C members of each set are the same molecule. The methyl[209], silyl[210] and germyl[211] acetates have had their gas-phase geometric structures determined. All the members of the acetate set are calculated to have C_s symmetry, with the planar H_sCCOXH_s' skeleton in a zigzag conformation. The O=CCH$_s$ group lies in the same plane with the *syn* or eclipsed (*e*) conformation, while O=COXH$_s'$ has staggered (*s*) C=O and X−H$_s'$ bonds. This is called the *e−s* conformation, where the second description refers to the XH_3 group. The X = C, Ge, Sn and Pb members of the $XH_3C(O)OCH_3$ set are also found to have *e−s* equilibrium geometric structures, but $SiH_3C(O)OCH_3$ is calculated to have the silyl rotated, *s−s* configuration. Here, the first descriptor refers to the XH_3 group.

In the acetate set the *s−s* conformation, which has the $H_sCC=O$ group in the methyl rotated *anti* or staggered position, is MP2/TZDP calculated to be 53 cm^{-1} (C), 210 cm^{-1} (Si), 160 cm^{-1} (Ge), 132 cm^{-1} (Sn) and 72 cm^{-1} (Pb) above ground state *e−s*, each in their gradient optimized geometries. In $CH_3C(O)OCH_3$, the only acetate tested, the *s−s* conformation is a transition state. In the ground *e−s* geometry the lowest-energy harmonic frequency is 36 cm^{-1}, corresponding to a *e−s* → *s−s* transformation. In the higher-energy *s−s* configuration the imaginary frequency is, coincidentally, the same as the 53 cm^{-1} total energy difference with the ground state and corresponds to the *s−s* → *e−s* transformation. Also in $CH_3C(O)OCH_3$, the geometry-optimized *e−e* conformation is found to be 409 cm^{-1} above *e−s* and a transition state. Its imaginary frequency of 124 cm^{-1} corresponds to a *e−e* → *e−s* rotation of the methoxy methyl group. Consistent with this result the second lowest-energy harmonic vibrational frequency in the ground state *e−s* methyl acetate is 160 cm^{-1} and rotates XH_3 to take conformer *e−s* to *e−e*. Thus in the $CH_3C(O)OXH_3$ set, the *syn* → *anti* transformation by methyl rotation is energetically preferred to XH_3 rotation. The *s−s* conformation is low in energy for X = C (53 cm^{-1}) and larger for Si → Pb, but decreasing in that order.

In contrast to the uniform configuration of the acetate series, $SiH_3C(O)OCH_3$ is calculated to have a *s−s* equilibrium ground-state geometric structure. The optimized *e−s* conformation is only 30 cm^{-1} higher. The lowest-energy harmonic vibrational frequency in the *s−s* configuration is 40 cm^{-1}, corresponding to the *s−s* → *e−s* rotation. The geometry-optimized *e−e* configuration is 421 cm^{-1} above *s−s*. For X = Ge, Sn and Pb in $XH_3C(O)OCH_3$ the optimized *s−s* structures are 17 cm^{-1}, 29 cm^{-1} and 67 cm^{-1} above *e−s* in total energy, respectively. The harmonic vibrational frequencies for the *e−s* → *s−s* motion, corresponding to a rotation of the XH_3 group, is 23 cm^{-1}, 30 cm^{-1} and 42 cm^{-1} for these same Group 14 atoms. Considering that this energy difference is negative for Si (i.e. *s−s* is more stable than *e−s*) shows that the XH_3 rotation barrier increases steadily from Si as X gets heavier, even though the X−C distance is increasing. The considerations discussed with regard to the acetyl series ($XH_3C(O)CH_3$) in this context are therefore probably also applicable here.

The acetate set has an increasing interaction between the carbonyl oxygen atom and X as the latter gets larger. By X = Pb the already familiar four-membered ring is formed, with one long Pb...O(=C) distance. For X = C → Pb this X...O distance is 2.63 Å, 2.79 Å, 2.79 Å, 2.71 Å and 2.52 Å, respectively. The progressive formation of the ring is reflected in the steady decrease of the C–OX angle from 115.6° (Si) to 98.8° (Pb), and the $H_sXO(-C)$ angle from 105.1° (C) to 92.0° (Pb). The calculated C=O harmonic stretch frequency also decreases steadily from 1772 cm^{-1} (C) to 1638 cm^{-1} (Pb). Experimentally[209], the C=O stretch in the methyl compound has been identified at 1769 cm^{-1}. Analogously, the C–C stretch is calculated at 642 cm^{-1} and measured at 636 cm^{-1}. In the $XH_3C(O)OCH_3$ series the C=O frequency is calculated to decrease from X = C to X = Si, but thereafter oscillate mildly about 1714 cm^{-1}. In this set there is no evidence of additional X...atom interactions.

The experimental gas-phase geometry of methyl acetate[209] also shows a planar heavy atom skeleton with the *syn* or *e–s* conformation. The calculated (experimental) r_e values are r_e(C=O)=1.218 Å (1.205 Å), r_e[(O=)C–O]=1.361 Å (1.359 Å), r_e[(H$_3$)C–O]=1.448 Å (1.458 Å) and r_e(C–C)=1.509 Å (1.506 Å) (Table 4). The important angles are calculated (observed) to be ∠COC = 113.7° (116.4), ∠OCO = 123.5° (123.0°), ∠CC–O = 110.5° (111.4°) and ∠H$_s$C–O=105.1° (103.1°). Silyl acetate is also observed[210] to adopt the *e–s* configuration. It was also determined that the best value of the O=C–O–Si dihedral angle is *ca* 10°, although a planar skeleton with large-amplitude torsional motion could not be ruled out. The MP2/TZDP optimized Si–O distance at 1.710 Å (Table 5) is larger than the experimental value of 1.685 Å. The COSi angle is calculated at 115.6° and measured at 116.5°. The crucial (C=)O...Si distance reported[210] at 2.795 Å is calculated here at 2.791 Å. Germyl acetate is also found[211] to have an *e–s* geometric structure, where a dihedral angle of 20° about the C–C bond gives the best fit to the electron diffraction pattern. The Ge–O distance observed at 1.830 Å is calculated (Table 6) here to be 1.834 Å. The experimental COGe angle is 113.0°, compared to the calculated 112.7°. The unusual shortness of the nonbonded Ge...O(=C) distance at 2.84 Å was commented upon in the electron diffraction study. Thus, the incipient X...O(=C) interaction leading to a four-membered ring with Pb is already detected in the germanium member of the acetate series. The small dipole moments in Table 16 show further effects of the incipient ring formation.

The $XH_3C(O)OCH_3$ set shows no unusual structural features and the internal angles are relatively insensitive to the nature of the X atom. The C=O bond length, in general, shortens from X = Si → Pb and is correspondingly shorter than in $XH_3C(O)CH_3$, probably because of the more electronegative OCH$_3$ substituent. The C–OCH$_3$ bond length is consistently shorter (by 0.09–0.10 Å) than O–CH$_3$, as expected, due to the different hybridization on the carbon atoms. The C–OCH$_3$ lengths are also somewhat consistently smaller than the C–OH distance in the acids (Table 4), possible due to steric effects which are expected to decrease as the X–C bond lengthens. The (XH$_3$)O–C bond distance in the acetates decreases going down the Group 14 column due to the tendency towards ring formation which will push the C–O and C=O bond lengths towards equal values. The C–C bond length in the acetates is similar to that found in the general carbonyl series, $CH_3C(O)Z$. There seems to be a rough correlation here between the shortness of the C–C distance and the electronegativity of Z, although steric and resonance factors may also be active. The X–C equilibrium bond length in the $XH_3C(O)OCH_3$ set is calculated to be on the short end of the scale compared to the other carbonyl series. Again, the electronegativity of the OCH$_3$ group seems to be the dominant factor.

V. BOND DISSOCIATION ENERGIES

As mentioned in Section III, Tables 17–23 contain the HF/ and MPn/TZDP++//MP2/TZDP total energies for the Y radical species, the HY molecules and the XH_3Y compounds with X = C, Si, Ge, Sn and Pb. For the open-shell radical species the UHF and UMPn methods were used[2,30]. The unrestricted Hartree–Fock wave function sometimes gives a poor expectation value of the spin-squared operator ($\langle S^2 \rangle$) and this can influence molecular properties[212]. For the doublet spin radical species in Table 17 the exact $\langle S^2 \rangle$ value is 0.75 [$s(s+1) = 0.75$, $s = 1/2$] and a larger calculated value indicates contamination of the ground-state electronic wave function with higher spin states. The degree of contamination is proportional to the deviation of $\langle S^2 \rangle$ from 0.75[213]. For example, a calculated $\langle S^2 \rangle$ value of 0.80 implies a mixture of 98.3% spin-doublet and 1.7% spin-quartet, if the latter were the only contaminant. Similarly $\langle S^2 \rangle = 0.90$ implies 95% $S = 1/2$ and 5% $S = 3/2$. For $\langle S^2 \rangle = 1.00$ the mixture is 91.7% and 8.3%. Projection techniques can be used to annihilate the contaminating higher-spin components, both at the UHF[213] and UMP2[214] levels. Table 17 shows the calculated UHF $\langle S^2 \rangle$ values and the projected UMP2 (PUMP2) energies for each radical. The largest deviations from the exact 0.75 value are for systems where the radical electron is found on atoms involved in multiple bonding. This is consistent with the observation that spin contamination in UHF implies the need for a multiconfigurational representation of the wave function[215,216]. The larger the initial (UHF) spin contamination the slower the UMPn expansion converges for property values[217,218]. This will be seen here for the calculated bond dissociation energy (BDE) values in Tables 26–30, for the process described in equation 1. A spin contamination of even 5% can lead to large differences in energetic processes.

The radicals with large deviations from $\langle S^2 \rangle = 0.75$ will have proportionately larger differences between their UMP2 and PUMP2 energies. The XH_3Y molecules are all closed-shell and have exact $\langle S^2 \rangle = 0$ values. Therefore, they have no energy difference between the MP2 and PMP2 levels of theory. Since the XH_3 radicals all have very-near 0.7500 values of $\langle S^2 \rangle$, the difference between the MP2 values of the XH_3–Y BDEs is in the $\langle S^2 \rangle$ nature of the Y radical. These relationships can be used to define corrections to the XH_3–Y BDEs at various UMPn levels that depend only on the $\langle S^2 \rangle$ nature of Y[219,220].

There are a number of problematics in calculating the XH_3–Y BDE. For a number of radical species in Table 17, the unpaired electron does not reside on the same atom as the attachment site to XH_3. This is always true for at least one of two possible linkage isomers, such as XH_3NO_2 and XH_3ONO. The NO_2 radical has ca 50% of its unpaired spin localized on the nitrogen atom. Other interesting cases are the C_2H_3[221] and CCH radicals[222] with large $\langle S^2 \rangle$ values whose unpaired spins are in-plane[222]. The ONO_2 or NO_3 radical is planar-symmetric (D_{3h})[223] with the unpaired spin evenly distributed on the three oxygen atoms, in-plane. The charge distribution in the XH_3ONO_2 molecules will certainly be different. If the ground-state radical species has inadvertently not been calculated for equation 1, then the BDEs in Tables 26–30 will be too large by the corresponding differences in energy between the true ground and the calculated electronic states.

The molecular energies for the MP2/TZDP HY species are given in Table 18 and for the XH_3Y compounds in Tables 19–23 for HF/ and MPn/TZDP++//MP2/TZDP level calculations. Table 18 is not used here to calculate BDEs but the tabulation is presented for the sake of completeness. It can, for example, be used in reactions of the type

$$XH_3Y + H_2 \longrightarrow XH_4 + HY \qquad (2)$$

or

$$XH_3Y + CH_4 \longrightarrow XH_3CH_3 + HY \qquad (3)$$

as measures of the relative XH_3-Y bond strength, without having to worry about the purity of the spin state in the radical species. Other measures of the relative XH_3-Y bond strengths without using the HY or radical species energies can be obtained, for example, from the following reaction, where the enthalpy of reaction 4 is $\Delta H\,(4)$:

$$XH_3Y + CH_4 \longrightarrow XH_4 + CH_3Y \qquad (4)$$

There is no intention of dealing here extensively with the relative energies of isomers in these energy Tables, and to compare them to experiment for the lighter Group 14 members where there is a greater possibility of finding experimental results or other calculations. Rather, the focus here will be, briefly, on the energy trends from HY and CH_3Y to PbH_3Y as the Group 14 atom gets heavier. Thus, for example, at the MP4(SDTQ) level, HNO is calculated to be more stable than HON by 45 kcal mol^{-1}. This difference becomes 50 kcal mol^{-1} for CH_3NO relative to CH_3ON and could not be determined for SiH_3NO/ON. For the heavier X atoms, XH_3NO is more stable than XH_3ON by 32 kcal mol^{-1} (Ge) and 30 kcal mol^{-1} (Sn, Pb). An analogous comparison between the $-NO_2$ and $-ONO$ forms shows $-ONO$ more stable by 7, -2, 16, 11, 13 and 17 kcal mol^{-1} at the MP4 level, for H, CH_3, SiH_3, GeH_3 and PbH_3, respectively, where the negative difference means that the $-NO_2$ form is lower in energy. It should be recalled that PbH_3ONO has a quasi-four-membered ring with one longer Pb...O distance which stabilizes the $-ONO$ form. For the $-PO$ versus $-OP$ coordination, the former linkage isomer is 35, 46, 15, 26, 26 and 27 kcal mol^{-1} more stable, respectfully, than H, CH_3, SiH_3, GeH_3, SnH_3 and PbH_3 attachment to $-OP$. For the $-OPO/-PO_2$ comparison, $-OPO$ is favored by 15, -5, 8, 20, 21 and 25 kcal mol^{-1}. Again, as in the case for $-ONO$, PbH_3OPO forms a quasi-four-membered ring which is expected to stabilize this isomeric form. Thus, in the last two cases, except for the methyl compound, preferred coordination is in the $-ONO$ and $-OPO$ modes, and not through the nitrogen or phosphorus atoms which carry most of the unpaired spin in the free radicals. For NO and PO the preferred attachment site is also the major location of the unpaired spin in the free radical. The difference in linkage stabilities may be due to the higher positive charge on the central N and P atoms in the dioxides and additional interaction with the remote oxygen atom in the $-ONO$ and $-OPO$ attachment modes. In both cases, the silicon compound behaves differently.

For the $-QCN$ versus $-NCQ$ compounds, with Q = O, S and Se, the preferred orientation is $-NCO$ by 26, 27, 24, 23, 23 and 19 kcal mol^{-1}, $-NCS$ by 9, 4, 18, 10, 9 and 4 kcal mol^{-1} and $-NCSe$ by 6, 0, 2, -7, -9 and -16 kcal mol^{-1} for attachment to H, CH_3, SiH_3, GeH_3, SnH_3 and PbH_3, respectively. The last negative values, of course, represent a preferred $-SeCN$ attachment. The QCN radicals have the unpaired electron residing primarily in a (symmetry broken) pi molecular orbital. As indicated by the deviation of their $\langle S^2 \rangle$ values from 0.75, these radicals can show significant spin polarization at the UHF level. At the UMP2 level only SCN and SeCN show significant spin polarization. While for OCN the UMP2 atomic spin populations are all positive, the nitrogen spin populations are -0.35 and -0.74, respectively, in SCN and SeCN. This makes it very difficult to correlate the preferred attachment site with the location of the unpaired spin population. The UMP2 calculated atomic charges on Q for the three QCN radicals all have the same value of about -0.5, so this substituent property cannot be used to correlate trends in linkage going Q = O $\rightarrow$ S $\rightarrow$ Se. The increased preference for Q end coordination compared to the N site as Q gets heavier must, then, be due to other properties such as polarizability and hybridization.

The total electronic energies tabulated in Tables 17–23 were used to calculate bond dissociation energies according to the process described in equation 1. The BDEs were corrected to 298 K for vibrational (ZPE), rotational and translational motion differences between reactant and products according to standard formulas[2,92]. An additional factor of $\Delta PV = RT \Delta n = 0.6$ kcal mol^{-1} was added for conversion from energy terms to enthalpy[43]. These correction terms together usually add up to just a few kcal mol^{-1} and become smaller as the X atom in XH_3Y becomes heavier. Since ZPE, especially, is larger for XH_3Y than for the sum of the asymptotic fragments, the total correction term reduces the purely electronic BDE. The dominant ZPE energies alone for all the species are listed in Table 24 and the total temperature corrected motion + enthalpy conversion terms are enumerated in Table 25. One particular species which requires special mention is the $OC(O)CH_3$ radical which, at the UMP2/TZDP level, dissociates spontaneously to CO_2 + $CH_3\cdot$. Therefore, for BDEs involving the $OC(O)CH_3$ radical asymptote, the motion correction terms are particularly large (ca 7–10 kcal mol^{-1}) mainly due to the ZPE differences between the $XH_3OC(O)CH_3$ compounds and the $XH_3 + CH_3 + CO_2$ asymptotes.

Experimental values for the BDEs of $CH_3−Y$ and $SiH_3−Y$ for comparison with the calculated values were taken either from tabulated heats of formation, ΔH_f° (298 K)[13,195,219,224−230] or bond dissociation energies[21,219,231−237]. There are not enough known BDEs for the corresponding Ge, Sn and Pb compounds to tabulate. Some of the experimental numbers carry large uncertainties. Thermodynamic quantities determined by different experimental methods sometimes differ significantly, such as for the vinyl (C_2H_3) radical[226,227]. The experimental BDE values listed in Tables 26 and 27 are based primarily on the heats of formation for reactants and products (equation 1) listed in References 13 and 224. Additional values are included where an experimental determination or a high-quality calculation gives a different BDE.

We will briefly review the comparison between theoretical and experimental BDEs for the methyl and silyl compounds in order to have some idea of the expected accuracy for the germyl, stannyl and plumbyl compounds. The MP4/TZDP++//MP2/TZDP level electronic energies in Tables 17–23, and the resultant BDEs in Tables 26–30, are similar to the G2(MP2)[237] level of the G2[42] method of Pople and coworkers. Generally, we expect the BDEs to be underestimated by such methods, since any lack of completeness in the basis set and theoretical level should generate a larger error in the compound relative to the two radical fragments. One exception to this expectation is to be found in those cases where the radical fragment has a higher bond order between two atoms than in its associated compounds with $XH_3\cdot$. Another, more relevant exception to the underestimation of equation 1 BDEs is when the radical fragment is poorly described theoretically.

Examination of Table 26, and ignoring differences of 1 kcal mol^{-1} and less, shows that for Y = CCH·, C(O)OH·, CN·, NC· and CHCH$_2$· the calculated $CH_3−Y$ BDEs at the MP4 level are substantially above the quoted experimental values. For the CCH·, CN· (NC·) and CHCH$_2$· cases the large UHF $\langle S^2 \rangle$ values in Table 17 indicate that these radical species are not being described sufficiently accurately. For C(O)OH· the UHF $\langle S^2 \rangle$ value is reasonable and the calculated $CH_3−C(O)OH$ BDE seems to have converged, going PMP2 → MP3 → MP4, but the experimental value is significantly smaller than theory. Hence the tabulated 'experimental' BDE, which is based on reported ΔH_f° values[13,224], must be in error, probably due to an inaccurate ΔH_f° for COOH[13]. Other radical species that have large deviations from 0.75 for the UHF $\langle S^2 \rangle$ quantity in Table 17, like SeCN·, SCN· and CP·, have no reported experimental BDEs for the methyl compound with which to compare. Their calculated BDEs are also expected to be overestimated relative to experiment. For $CH_3−NNN$ the calculated BDE shows a strong alternation with the n-value level in the MPn progression. The MP4 value, however, is still below experiment, contrary to expectations from the relatively large UHF $\langle S^2 \rangle$ value of 0.9056 in Table 17 for the NNN radical. The tabulated ΔH_f° value for NNN.[13] carries an uncertainty of ± 5 kcal mol^{-1}. The lower (-5) limit puts the 'experimental' BDE at 75 kcal mol^{-1}, which is more in line with expectations and experience with the calculated values. The ΔH_f° value for CH$_3$N$_3$ might also need re-examination. Another calculated BDE that is inconsistent with expectations is that for CH$_3$—NCO. Although the UHF $\langle S^2 \rangle$ value for NCO is 0.8245, the experimental BDE is substantially larger than the calculated values. The ΔH_f° values for NCO· and CH$_3$NCO need to be verified. The reported CH$_3$—Y BDE in Table 26 of 81.9 kcal mol^{-1} for Y = OF is based on a calculated ΔH_f° of -21 kcal mol^{-1} for CH$_3$—OF[230]. Although the UHF $\langle S^2 \rangle$ value for OF is a respectable 0.7669 (Table 17), the calculated MPn BDEs are still larger than 'experimental'. This would indicate that ΔH_f° for OF is larger than the tabulated 26.0 kcal mol^{-1}[13] and/or ΔH_f° for CH$_3$—OF is more negative than the calculated -21 kcal mol^{-1}[230].

There are also experimental BDE values listed in older tabulations based on spectroscopic or the known values of ΔH_f° at that time. Where these differ significantly from the more current tabulations[13,224,232] they have also been included for comparison purposes. However, the BDEs based on the most recent ΔH_f° values are presumably to be favored.

For the silyl compounds (Table 27) high-quality calculated BDE values have also been included with the experimentally derived quantities. Here, the uncertainties and distribution of experimental values are larger than for the methyl compounds. The calculated BDEs in this work and corresponding comparison experience with the methyl compounds can, perhaps, be used to narrow some of the uncertainties in the silyl compounds. For almost all the germaniun, tin and lead compounds in Tables 28–30, the calculated BDEs are the only information available to date.

If we examine the MPn series values of the BDEs in Table 26, we find that the PMP2 stabilities are smaller than the MP2 energies, as expected from the lower asymptote radical fragment energies due to projection of the high-spin component. The difference in MP2 and PMP2 energies are then actually a measure of spin contamination in the two asymptotic radicals in equation 1. The MP2–PMP2 difference in BDEs can then be used as a correction term to adjust the presumably more accurate MP4 values. These ideas have been used by a number of researchers[219,220] to obtain improved thermodynamic quantities for open-shell molecular systems. This method is just one of several possibilities that fall under the category of scaled energies[238,239]. An alternate approach to avoiding the vagaries of the $\langle S^2 \rangle$ problem is to use equations 2–4 to calculate relative binding energies. An additional refinement is to use the experimental BDE values of the known CH$_3$Y species in the following variant of equation 4:

$$\text{BDE(XH}_3\text{—Y)} = \text{BDE(CH}_3\text{—Y)} + \text{BDE(XH}_3\text{—H)} - \text{BDE(CH}_3\text{—H)} - \Delta H\,(4) \qquad (5)$$

where the experimental or best theoretical values of the quantities on the right-hand side of equation 5 are used, as available.

However, the values in Tables 26–30 are directly calculated BDEs, according to equation 1. Taking only those molecules for which there are reliable experimental information (Table 26), and excluding the problematic $\langle S^2 \rangle$ cases (CCH·, COOH·, NCO·, CN·, NC·, CHCH$_2$· and N$_3$·) and the decomposed radical (CH$_3$CO$_2$·), the average MP2 error for CH$_3$—Y is 2.0 kcal mol^{-1} (21 cases) and for MP4 is 3.8 kcal mol^{-1} (19 cases). It thus seems that the particular choice of basis set and level of calculation used here gives the best cancellation of errors for the MP2 method.

The accuracy of the calculated silyl BDEs (Table 27) are more difficult to judge because of the large uncertainties in the ΔH_f° values. For SiH$_3$—BH$_2$, the experimental BDE, based on a calculated ΔH_f° for SiH$_3$BH$_2$[229] and a ΔH_f° value of 48 ± 15 kcal mol^{-1}[224] for BH$_2$·

is some $23-28$ kcal mol^{-1} smaller than calculated, and therefore suspect. The reported BDE for SiH$_3$–SH is also low. For CCH· the calculated BDEs are, as expected, too large. SiH$_3$–CH$_2$CH$_3$ has two reported $\Delta H_f°$ values, 27 kcal mol^{-1}[224] and -8.2 kcal mol^{-1}[229], which differ by ca 35 kcal mol^{-1}. The calculated BDEs show the latter value to be more correct. For CHCH$_2$· the calculated silyl BDEs are, unexpectedly, smaller than experiment, although agreeing with another high-level calculation[233]. The $\Delta H_f°$ value for C$_2$H$_3$· is in doubt[226,227] and the experimental BDE could be as low as 108.8 kcal mol^{-1}, using ΔH_f(C$_2$H$_3$·)=63.4 kcal mol^{-1}[224]. Gathering the remaining nonproblematic BDEs in Table 27 (Y = F, Br, H, NH$_2$, Cl, CH$_3$CH$_3$, OH, CH$_3$, SiH$_3$ and C(O)CH$_3$) gives an average (absolute) MP2 error of 2.1 kcal mol^{-1} and an average MP4 error of 3.6 kcal mol^{-1}. In this comparison, where there was a choice of experimental BDE, the one closest to the calculated values was chosen for the averages. This selectivity does not affect the relative results between MP2 and MP4, and the outcome here, like for the CH$_3$–Y comparison above, is that the MP2 energy is generally closer to experiment.

We can now examine trends in BDE values in going down the Group 14 column of the Periodic Table. These trends divide into three groupings of Y substituents. In the first category (I-A) the BDE for XH$_3$–Y decreases steadily from X = C to X = Pb. In the second category (II-A) there is a significant jump in binding energy from X = C to X = Si, after which the BDE decreases steadily to X = Pb. The third category has either of the two characteristics of I and II, with the additional increase of the PbH$_3$–Y BDE relative to X = Sn. We will call these I-B and II-B, respectively. The numbers and references for the ensuing discussion are all found in Tables 26–30.

A. XH$_4$

The XH$_3$–H BDE decreases steadily as X gets heavier (class I-A). For CH$_4$ the MP2 error is larger than for MP4. The same is found for SiH$_4$. In both cases, MP2 is ca 6 kcal mol^{-1} too low and MP4 is about 4 kcal mol^{-1} underestimated, compared to experiment. Correcting the calculated MP2 and MP4 energies in Tables 27–30 by +6 and +4 kcal mol^{-1}, respectively, gives predicted BDEs for GeH$_3$–H, SnH$_3$–H and PbH$_3$–H of ca 83 kcal mol^{-1}, 75–76 kcal mol^{-1} and ca 68 kcal mol^{-1}, respectively. The extrapolated stability for GeH$_3$–H agrees very well with the reported experimental value of 82.7 kcal mol^{-1}[240].

B. XH$_3$A

Experimental XH$_3$–A BDEs are available for three halogens (F, Cl and Br), with X = C and Si. The stability for XH$_3$–F jumps substantially from X = C to X = Si (Type II-A) and this change is confirmed experimentally. The trends in comparing calculated with experimental are the same: MP2 is underestimated by 3.4 kcal mol^{-1} for CH$_3$–F and by 5.9 kcal mol^{-1} for SiH$_3$–F. For MP4 the respective errors are larger: 7.4 and 10.9 kcal mol^{-1}. If the calculated MP2 BDEs are corrected by ca 5 kcal mol^{-1} and the MP4 BDEs by ca 9 kcal mol^{-1}, then the predicted energy for GeH$_3$–F is 134–135 kcal mol^{-1}, for SnH$_3$–F is ca 130 kcal mol^{-1} and for PbH$_3$–F is ca 124 kcal mol^{-1}. For CH$_3$–Cl (SiH$_3$–Cl) the MP2 error is 3.0 (3.8) kcal mol^{-1} and the MP4 underestimation is 6.0 (7.6) kcal mol^{-1}. Using an MP2 (MP4) correction factor of 3.4 (6.8) kcal mol^{-1} gives adjusted BDE values of ca 97 kcal mol^{-1} for GeH$_3$–Cl, ca 95 kcal mol^{-1} for SnH$_3$–Cl and ca 94 kcal mol^{-1} for PbH$_3$–Cl. For CH$_3$Br a problem arises because of the relatively large uncertainty in the $\Delta H_f°$(CH$_3$Br)[224] value of

-19 ± 4 kcal mol^{-1}. The lower limit value gives a stability that agrees with Walsh[234]. Using the unshifted enthalpy of formation gives MP2 errors of 1.1 and 3.5 kcal mol^{-1} for CH$_3$–Br and SiH$_3$–Br, respectively, compared to experiment. For MP4 the corresponding numbers are 3.1 and 6.5 kcal mol^{-1}. Using 2.3 and 4.8 kcal mol^{-1} as MP2 and MP4 correction factors, the predicted BDE for GeH$_3$–Br is *ca* 83 kcal mol^{-1}, for SnH$_3$–Br is *ca* 72 kcal mol^{-1} and for PbH$_3$–Br is *ca* 81 kcal mol^{-1}.

C. XH$_3$AH

The two series with A = O and A = S both conform to the group II-A behavior. Stabilities for CH$_3$–OH, CH$_3$–SH and SiH$_3$–OH can be used for comparison purposes. The silyl compound values are more uncertain than those of the methyl group. For SiH$_3$–OH the calculated 124.8 kcal mol^{-1} BDE[228] is adopted as the experimental value, while for SiH$_3$–SH the quoted *ca* 70 kcal mol^{-1} energy[21] is too far from the calculated values to be useful for comparative purposes. On this basis, the MP2 errors for CH$_3$–OH and CH$_3$–SH are 0.4 and 1.5 kcal mol^{-1}, respectively. The corresponding MP4 errors are 4.7 and 6.6 kcal mol^{-1}. Using 1.5 (MP2) and 5.7 (MP4) kcal mol^{-1} adjustment factors gives predicted BDEs of 105-106 kcal mol^{-1} for GeH$_3$–OH, *ca* 99 kcal mole for SnH$_3$–OH and *ca* 94 kcal mol^{-1} for PbH$_3$–OH. For XH$_3$–SH the MP2 (MP4) shift is 1.9 (4.3) kcal mol^{-1}. Using these numbers to adjust the higher XH$_3$–OH stabilities gives (in kcal mol^{-1}) *ca* 86-87 for SiH$_3$–SH, *ca* 78 for GeH$_3$–SH, *ca* 75 for SnH$_3$–SH and *ca* 73 for PbH$_3$–SH.

D. XH$_3$AH$_2$

In this category are found the A = N, P, B and Al substituents. The amine set (XH$_3$–NH$_2$) behaves as II-A together with the XH$_3$–A and XH$_3$–AH sets, while the phosphorus, borine and alane series conform to class I-A, like XH$_3$–H. For the reference BDE of SiH$_3$–NH$_2$ the value based on the calculated heat of formation of SiH$_3$NH$_2$[229] is adopted. For that choice, the MP2 (MP4) error in CH$_3$–NH$_2$ stability is 1.1 (4.9) kcal mol^{-1} and for SiH$_3$–NH$_2$ is 1.9 (6.1) kcal mole. Using adjustment energies of 1.5 kcal mol^{-1} for MP2 and 5.5 kcal mol^{-1} for MP4 gives predicted BDEs for GeH$_3$–NH$_2$ of *ca* 87 kcal mol^{-1}, for SnH$_3$–NH$_2$ of 79–80 kcal mol^{-1} and for PbH$_3$–NH$_2$ of *ca* 76 kcal mol^{-1}. For XH$_3$–PH$_2$ only the methyl stability is used with an MP2 error of 3.3 kcal mol^{-1} and an MP4 underestimation of 5.4 kcal mol^{-1}. Using these numbers as energy shift values gives projected BDEs of 69–70 kcal mol^{-1} for SiH$_3$–PH$_2$, 65–66 kcal mol^{-1} for GeH$_3$–PH$_2$, 60–61 kcal mol^{-1} for SnH$_3$–PH$_2$ and 59–60 kcal mol^{-1} for PbH$_3$–PH$_2$. For the XH$_3$–BH$_2$ and XH$_3$–AlH$_2$ sets there is no reliable experimental stability information. However, the comparisons until here seem to show that a *ca* 4 kcal mol^{-1} upward adjustment of the average of MP2 and MP4 BDEs generally gives values closer to experiment than the raw, untreated MP2 and MP4 energies shown in Tables 26–30. On this basis, the XH$_3$–BH$_2$ and XH$_3$–AlH$_2$ stabilities (in kcal mol^{-1}) are calculated to be *ca* 105 and 83 (X = C), 87–88 and *ca* 68 (X = Si), 79–80 and *ca* 67 (X = Ge), *ca* 72 and 61–62 (X = Sn) and *ca* 70 and *ca* 60 (X = Pb), respectively.

E. XH$_3$–AH$_3$

Experimental BDEs are available for CH$_3$–CH$_3$, SiH$_3$–CH$_3$ and SiH$_3$–SiH$_3$ in Tables 26 and 27. MP2/MP4 underestimates the CH$_3$–CH$_3$ stability by 1.4/3.2 kcal mol^{-1}, the

SiH_3-CH_3 energy by $2.1/3.3$ kcal mol^{-1} and the SiH_3-SiH_3 BDE by $3.4/2.6$ kcal mol^{-1}. The stabilities of this series of ethane analogs has recently been discussed theoretically[112,241] using the LDF method. Their calculated BDEs for CH_3-CH_3 (86.8 kcal mol^{-1}) and SiH_3-CH_3 70.0 kcal mol^{-1})[241] are very close to the theoretical values here. However, for SiH_3-CH_3 the difference between the LDF method[241] and Table 27, and with experiment, is larger. In general, the homogeneous XH_3-XH_3 BDEs agree very well (within ca 2 kcal mol^{-1}) between the two calculational methods, except for PbH_3-PbH_3 where the difference is about 7 kcal mol^{-1}. In contrast, for the XH_3-CH_3 (X $\neq$ C) members' stabilities the difference between previous[241] and current (Tables 27–30) calculated binding energies ranges up to ca 16 kcal mol^{-1} for PbH_3-CH_3, increasing with the size of X. The calculated stabilities in Tables 26–30 are relatively well converged with respect to the value of n in the MPn series of energies. It should therefore be possible to extrapolate corrections to the calculated BDEs and obtain good estimates of the true bond energies. Adding 4 kcal mol^{-1} to the average MP2/MP4 calculated BDEs for GeH_3-GeH_3 gives a predicted ca 65 kcal mol^{-1}, which agrees well with the experimental 66.0 kcal mol^{-1} value[240]. In this manner, the other projected stabilities (in parenthesis, kcal mol^{-1}) are: GeH_3-CH_3 (ca 80), SnH_3-CH_3 (ca 72), PbH_3-CH_3 (ca 71), GeH_3-SiH_3 (ca 71), SnH_3-SiH_3 (ca 66), PbH_3-SiH_3 (ca 64), SnH_3-GeH_3 (ca 63), SnH_3-SnH_3 (ca 59.5), PbH_3-GeH_3 (ca 61.3), PbH_3-SnH_3 (ca 57.5) and PbH_3-PbH_3 (ca 55.5).

F. XH$_3$AB

The AB members of this series are AB = CN, NC, NO, ON, CP, PC, PO, OP and OF. The substituents CN, CP and PO have BDEs that behave as the more covalent class I-A. NC, PC, OP and OF conform to the more ionic class II-A, and NO behaves as class I-B. The latter has PbH_3-NO more stable than SnH_3-NO. The structural features of PbH_3-NO indicated an attractive interaction between PbH_3 and the oxygen atom. The class I-A CN and CP compounds have very similar BDEs for the X = C and X = Si members, indicating some intermediate behavior between classes I and II. The XH_3-ON set is not thermodynamically bound for any X and XH_3-OP is very weakly bound. The CN and CP radicals have large UHF $\langle S^2 \rangle$ values and therefore have exaggerated theoretical XH_3-CP and XH_3-CN stabilities. The CH_3-CN BDE can be compared with experiment which shows that MP2 is 19.2 kcal mol^{-1} and MP4 is 13.0 kcal mol^{-1} larger than experiment (Table 26). This ca 6 kcal mol^{-1} difference between MP2 and MP4 in CH_3-CN decreases to ca 5 kcal mol^{-1} for SiH_3-CN to PbH_3-CN (Tables 27–30). Applying a ca 13 kcal mol^{-1} correction uniformly to all the MP4 energies of the nitriles gives predicted XH_3-CN stabilities of ca 122 (Si), ca 112 (Ge), ca 107.5 (Sn) and ca 102 (Pb) kcal mol^{-1}. The XH_3-NC set is also strongly bound. The experimental BDE for CH_3-NC is 24 kcal mol^{-1} less than for CH_3-CN and this difference is exactly reproduced by the ca 24 kcal mol^{-1} difference in calculated MP4 binding energies (Table 26). Applying the same ca 13 kcal mol^{-1} correction to the MP4 binding energies of the other members of the XH_3-NC set gives adjusted BDEs of ca 116 (Si), ca 103 (Ge), ca 100 (Sn) and ca 95 (Pb) kcal mol^{-1}. It should be noted that because of the different behaviors of the –CN and –NC sets with X atom substitution in XH_3, the SiH_3-NC binding energy is calculated to be larger than for SiH_3-CN. CH_3-NO has a small (ca 40 kcal mol^{-1}) measured binding energy which is very close to both the MP2 and MP4 calculated stabilities. As noted above, the CH_3-OF experimental BDE, based on measured and calculated heats of formation, is anomalously smaller than the corresponding calculated values. The theoretical XH_3-OF binding energies remain large for Si $\rightarrow$ Pb. The XH_3-CP

BDEs are calculated to be very high due to the large, incorrect UHF $\langle S^2 \rangle$ value of the CP radical (Table 17). One possible way of estimating an energy adjustment term to correct for the error in $\langle S^2 \rangle$ is to use the difference in the PMP2 and MP2 BDE values to correct the MP4 stabilities. For CH_3-CN this gives a ca 12 kcal mol^{-1} shift in energy and ca 11 kcal mol^{-1} for SiH_3-CN to PbH_3-CN. These numbers are close to the 13 kcal mol^{-1} correction deduced by comparison to experiment for CH_3-CN and already applied above, generally for XH_3-CN. For XH_3-CP, however, the PMP2–MP2 correction term is ca 21 kcal mol^{-1} and that decreases to $\sim$ 20 kcal mol^{-1} for Si $\rightarrow$ Pb. Applying a uniform 22 kcal mol^{-1} reduction of the MP4 stabilities of XH_3-CP gives BDEs of ca 109 (C), ca 108 (Si), ca 98 (Ge), ca 93 (Sn) and ca 88 (Pb) kcal mol^{-1}. With this same size correction, the linkage isomer XH_3-PC binding energies are all below 50 kcal mol^{-1}.

G. XH₃ABH

As noted above, the calculated BDEs for the XH_3-CCH set suffer in accuracy from the high UHF $\langle S^2 \rangle$ value for the CCH radical (Table 17). The MP2 stability for CH_3-CCH is ca 13 kcal mol^{-1} too high at the MP2 level and ca 8 kcal mol^{-1} too large at the MP4 level. The PMP2 energy (Table 26) is only 1.3 kcal mol^{-1} over experiment. The differences for the silicon compounds (Table 27) are smaller. Applying the shift factors to the respective MP2, PMP2 and MP4 bond dissociation energies of the other XH_3-CCH compounds gives the following projected stabilities (in kcal mol^{-1}): ca 116 (Ge), ca 111 (Sn) and ca 105 (Pb). The C(O)H radical, on the other hand, has a UHF $\langle S^2 \rangle$ value that is close to 0.75. The MP2 calculated BDE (Table 26) is 1.1 kcal mol^{-1} too low and MP4 is 3.3 kcal mol^{-1} smaller than the experimentally derived stability of $CH_3-C(O)H$. If the 1.1 and 3.3 kcal mol^{-1} factors are added to the respective MP2 and MP4 binding energies of $XH_3-C(O)H$ in Tables 27–30, the following predicted BDEs (in kcal mol^{-1}) are obtained: 67–69 (Si), 60–61 (Ge), 52–53 (Sn) and 53–54 (Pb). The C(O)H substituent behaves as a class I-B group.

H. XH₃ABH₃

The comparison between theory and experiment for XH_3-CHCH_2 is doubly difficult, both because of the large UHF $\langle S^2 \rangle$ value for the vinyl radical (Table 17) and because of the uncertainty in its experimental enthalpy of formation[226,227]. The CH_3-CHCH_2 binding energy is based on this enthalpy value. The preferred experimental value is $\Delta H_f°$ $(C_2H_3\cdot)$ = 68 kcal mol^{-1}[226], which leads to a derived experimental BDE of 98.1 kcal mol^{-1} for CH_3-CHCH_2. On this basis, the MP2 energy is 8.4 kcal mol^{-1} and MP4 is 5.0 kcal mol^{-1} too high relative to experiment. Applying these correction terms to the other members of the XH_3-CHCH_2 set projects BDEs (in kcal mol^{-1}) of 92–94 (Si), 83–84 (Ge), 75–76 (Sn) and 73–74 (Pb). The experimental BDEs of CH_3-OCH_3 and CH_3-SCH_3 are known (Table 26). The MP2 energy is too large by 3.8 kcal mol^{-1} and MP4 underestimates by 1.2 kcal mol^{-1} relative to experiment. Applying these adjustment factors to the other XH_3-OCH_3 members gives the following predicted BDEs (in kcal mol^{-1}); ca 110 (Si), ca 91 (Ge), ca 83 (Sn) and ca 80.5 (Pb). Analogously, using the 73.7 kcal mol^{-1} experimental BDE for CH_3-SCH_3, the MP2 stability is 0.9 kcal mol^{-1} too low and MP4 is 3.6 kcal mol^{-1} too small compared to experiment. Using these numbers as respective MP2 and MP4 correction factors for the other XH_3-SCH_3 molecules leads to extrapolated BDEs (in kcal mol^{-1}) of ca 83 (Si), ca 74 (Ge), ca 70 (Sn) and ca 70 (Pb). The XH_3-OSiH_3 set (X = Si $\rightarrow$ Pb) has larger

BDEs than the corresponding members of the XH_3-OCH_3 and XH_3-SCH_3 sets, probably because of the very basic oxygen atom in the $OSiH_3$ fragment. SiH_3-OSiH_3 (disiloxane) has one of the highest BDEs of any compound in Tables 26–30.

I. XH₃ABH₅

The MP2 calculated BDE for $CH_3-CH_2CH_3$ is 1.3 kcal mol^{-1} above the measured value and MP4 is 1.1 kcal mol^{-1} smaller than the experimental stability. Averaging MP2 and MP4 calculated binding energies for the other members of the $XH_3-CH_3CH_3$ set gives BDEs of ca 83.0 (Si), ca 73.8 (Ge), ca 65.4 (Sn) and ca 65.9 (Pb) kcal mol^{-1}. The best available experimental BDE for $SiH_3-CH_2CH_3$ combines experimental and theoretical heats of formation to give a stability of 82.6 kcal mol^{-1} (Table 27). This agrees very well with the above 83 kcal mol^{-1} predicted value.

J. XH₃ABC

The azide radical (NNN·) has a calculated UHF $\langle S^2 \rangle$ of 0.9056 (Table 17). Consequently, the UMPn binding energies would be expected to be greater than observed. The derived experimental BDE of 80 kcal mol^{-1} (Table 26) has an uncertainty of at least ± 5 kcal mol^{-1} attached to it, based on the measured enthalpies of formation[13,224]. Another consideration is the theoretical difficulty in describing multiply bonded systems such as $CH_3-N=N\equiv N$. It is therefore difficult to formulate a scheme for estimating the heavier atom XH_3-NNN binding energies from the calculated MPn values. Of the six, $-OCN$, $-NCO$, $-SCN$, $-NCS$, $-SeCN$, and $-NCSe$ substituents, only the CH_3-NCO experimental BDE is listed in Table 26. Although the UHF $\langle S^2 \rangle$ value for NCO is above 0.8, the calculated stabilities are less than the experimental value. If the correction energies of 6.3 (MP2) and 12.7 (MP4) kcal mol^{-1} from the CH_3-NCO case are applied to the other XH_3-NCO molecules, then their predicted BDEs are (in kcal mol^{-1}): ca 129 (Si), ca 114 (Ge), ca 110 (Sn) and ca 106 (Pb). The substituents $-NNN$, $-OCN$, $-NCO$, $-NCS$ and $-NCSe$ behave as class II-A groups. For $-SCN$ and $-SeCN$, the BDEs increase from Sn to Pb (class II-B), for the reasons noted in the discussion on the structural features.

The ligands $-C(O)F$ and $-C(O)Cl$ behave as class I-A groups, as do all the Y substituents attached to XH_3 through a low-electronegative atom (like carbon). In $CH_3-C(O)F$, MP2 is 2.0 and MP4 is 4.3 kcal mol^{-1} below the experimentally derived BDE. Using these numbers as correction factors for the other X atom compounds with $-C(O)F$ gives BDEs (in kcal mol^{-1}) of ca 87 (Si), ca 79 (Ge), ca 73 (Sn) and ca 72 (Pb). The BDEs for CH_3-NO_2 and CH_3-ONO are listed in Table 26. For XH_3NO_2 the MP2 (1.0 kcal mol^{-1}) and MP4 (4.8 kcal mol^{-1}) correction terms derived from CH_3-NO_2 can be applied to the respective MP2 and MP4 BDEs for the other XH_3-NO_2 compounds to give predicted binding energies (in kcal mol^{-1}) of ca 65 (Si), 55–56 (Ge), 51–52 (Sn) and 52–53 (Pb). The analogous CH_3-ONO difference energies are 3.5 (MP2) and 3.6 (MP4) kcal mol^{-1} compared to experiment. Applying these shifts to XH_3-ONO gives (in kcal mol^{-1}) ca 81 (Si), ca 67 (Ge), ca 64 (Sn) and 68–69 (Pb). The XH_3-ONO and XH_3-OPO BDEs behave like class II-B, as expected from Y substituents with highly electronegative atoms attached to X. The increase in stability from Sn to Pb is the result of the above discussed Pb . . . O interaction in these systems.

K. XH₃ABCH

The $CH_3-C(O)OH$ BDE can be derived from tabulated ΔH_f° values for the constituent components of reaction 1 to give 84.9 kcal mol^{-1}. This energy is 9–10 kcal mol^{-1}

smaller than the calculated MPn BDEs (Table 26) which seem to be nicely converged with n. There is no reasonable explanation for such a large discrepancy between theory and experiment at this level. Support for the higher theoretical values comes from the $CH_3-C(O)OCH_3$ BDE, which is experimentally estimated at 92.5 kcal mol^{-1}. Substituting a methyl group for a hydrogen atom at such a remote position from the cleaved C–C bond should not affect the C–C BDE to a significant extent. It must be that the quoted $\Delta H_f°[C(O)OH]$ of -53.3 kcal mol^{-1}[13] is in error by 7–8 kcal mol^{-1}.

L. XH$_3$ABC H$_2$

The BDEs for $XH_3-C(O)NH_2$ are calculated to be uniformly lower than in $XH_3C(O)OH$. This trend is consistent both with the lower calculated C–C stretch frequency in $CH_3C(O)NH_2$ (851 cm^{-1}) relative to $CH_3C(O)OH$ (862 cm^{-1}), and with the lower electronegativity of $-NH_2$ relative to $-OH$.

M. XH$_3$ABC H$_3$

The calculated BDEs of the members of the $XH_3-C(O)CH_3$ set are uniformly lower than the respective binding energies of the $-C(O)NH_2$ and $-C(O)OH$ sets, as expected from electronegativity arguments. The MP2 calculated stability is 1.3 kcal mol^{-1} below and MP4 is 3.8 kcal mol^{-1} smaller than the experimentally derived BDE of $CH_3-C(O)CH_3$. Using these numbers as correction terms to the other members of the set gives predicted $XH_3-C(O)CH_3$ binding energies (in kcal mol^{-1}) of ca 70 (Si),˙ca 62 (Ge), ca 55 (Sn) and ca 56 (Pb). The derived experimental BDE for $SiH_3-C(O)CH_3$ (Table 27) is 69.8 kcal mol^{-1}, in excellent agreement with the above 70 kcal mol^{-1} projected value from using the $CH_3-C(O)CH_3$ shift energies. In $SiH_3-C(O)CH_3$ the MP4 value is only 2.5 kcal mol^{-1} below experiment and this has already been taken into account in the above extrapolated BDE values for X = Ge, Sn and Pb.

N. XH$_3$ABC D

Experimental BDE values in this catagory are listed in Table 26 for $-ONO_2$ and $-CF_3$. The calculated UHF $\langle S^2 \rangle$ value for the NO_3 radical (Table 17) is 0.8009 and the MP2–PMP2 energy difference for CH_3-ONO_2, for example, is ca 6 kcal mol^{-1}. The MPn binding energies as a function of n clearly have not converged for any of the XH_3 groups (Tables 26–30), but the patterns are the same: similar MP2 and MP4 energies, and 14–15 kcal mol^{-1} higher MP3 stabilities. The consistency of these trends can, perhaps, be used to extrapolate the heavier XH_3 group BDEs with NO_3. For CH_3-ONO_2 the MP2 binding energy is 2.3 and MP4 is 3.2 kcal mol^{-1} below experiment. Using these numbers to adjust the corresponding energies for the other XH_3-ONO_2 BDEs gives (in kcal mol^{-1}): ca 103 (Si), ca 88 (Ge), ca 86 (Sn) and ca 90 (Pb). Both XH_3-ONO_2 and XH_3-OPO_2 are class II-B systems because of the additional Pb . . . O interaction. For the XH_3-CF_3 set there is an experimentally derived BDE only for CH_3-CF_3. The MP2 energy is only 0.3 kcal mol^{-1} above experiment, so that the MP2 calculated BDEs for the other XH_3-CF_3 compounds should be close to their respective experimental stabilities.

O. XH$_3$ABC DH$_3$

The derived experimental $CH_3-C(O)OCH_3$ binding energy is 92.5 kcal mol^{-1}. The calculated MP2 BDE is 1.6 kcal mol^{-1} above that value. Reducing the MP2 binding energies for the other $XH_3-C(O)OCH_3$ species by 1.6 kcal mol^{-1} gives projected BDEs of ca 80 (Si), ca 72 (Ge), ca 66 (Sn) and ca 65 (Pb) kcal mol^{-1}. However, the MP2 calculated BDE for $CH_3-OC(O)CH_3$ is 17.5 kcal mol^{-1} smaller than the derived experimental value. The $CH_3CO_2\cdot$ radical was found to be unstable to dissociation to give $CO_2 + CH_3\cdot$. This, in fact, is also the experimental thermodynamic result obtained by using the $\Delta H_f°$ values of the 3 species involved[224]. The $C(O)OCH_3$ radical must then be metastable. The exothermicity of the $C(O)OCH_3\cdot \rightarrow CO_2+CH_3\cdot$ reaction is 18.7 kcal mol^{-1} experimentally[13,224] and 21.1 (MP2) or 20.3 (MP4) kcal mol^{-1} theoretically from the energies in Table 17. There must also be a metastable CH_3CO_2 radical with an (experimental) exothermicity of 9.5 kcal mol^{-1}[13,224], but it was not found by the present MP2/TZDP geometry optimization. The $XH_3-C(O)OCH_3$ BDEs show class I-A behavior, while $XH_3-OC(O)CH_3$ is class II-B. In the latter cases there is a Pb . . . O stabilizing interaction.

VI. REFERENCES

1. L. Pauling, *The Nature of the Chemical Bond*, 3rd ed., Cornell University Press, Ithaca, 1960.
2. W. J. Hehre, L. Radom, P. v. R. Schleyer and J. A. Pople, *Ab Initio Molecular Orbital Theory*, Wiley-Interscience, New York, 1986.
3. L. Kahn, P. Baybutt and D. G. Truhlar, *J. Chem. Phys.*, **65**, 3826 (1976).
4. P. Pyykko, *Chem. Rev.*, **88**, 563 (1988).
5. W. Kutzelnigg, *Angew. Chem., Int. Ed. Engl.*, **23**, 272 (1984).
6. J. P. Desclaux, *At. Data Nucl. Data Tables*, **12**, 311 (1973).
7. C. E. Moore, Atomic Energy Levels, NSRDS-NBS 35/V.I, U.S. Government Printing Office, Washington, D.C. 20402, 1971.
8. C. E. Moore, Atomic Energy Levels, NSRDS-NBS 35/V.II, U.S. Government Printing Office, Washington, D.C. 20402, 1971.
9. C. E. Moore, Atomic Energy Levels, NSRDS-NBS 35/V.III, U.S. Government Printing Office, Washington, D.C. 20402, 1971.
10. H. Hotop and W. C. Lineberger, *J. Phys. Chem. Ref. Data*, **14**, 731 (1985).
11. T. M. Miller, A. E. S. Miller and W. C. Lineberger, *Phys. Rev. A*, **33**, 3558 (1986).
12. T. M. Miller and B. Bederson, *Adv. At. Mol. Phys.*, **13**, 1 (1977).
13. D. R. Lide (Ed.), *CRC Handbook of Chemistry and Physics*, 74th ed., CRC Press, Boca Raton, 1993.
14. R. S. Mulliken, *J. Chem. Phys.*, **2**, 782 (1934).
15. A. L. Allerd, *J. Inorg. Nucl. Chem.*, **17**, 215 (1961).
16. L. C. Allen, *J. Am. Chem. Soc.*, **111**, 9003 (1989); *J. Am. Chem. Soc.*, **114**, 1510 (1992).
17. H. Basch, W. J. Stevens and M. Krauss, *J. Chem. Phys.*, **74**, 2416 (1981).
18. K. G. Dyall, *J. Chem. Phys.*, **98**, 2191 (1993) and references cited therein.
19. T. Koopmans, *Physica*, **1**, 104 (1933).
20. P. Schwerdtfeger, G. A. Heath, M. Dolg and M. A. Bennett, *J. Am. Chem. Soc.*, **114**, 7518 (1992).
21. Y. Apeloig, in *The Chemistry of Organic Silicon Compounds* (Eds. S. Patai and Z. Rappoport), Wiley, Chichester, 1989.
22. T. Hoz and H. Basch, in *Supplement S: The Chemistry of Sulphur Containing Functional Groups* (Eds. S. Patai and Z. Rappoport), Wiley, Chichester, 1993.
23. H. Basch and T. Hoz, in *Supplement B: The Chemistry of Acid Derivatives, Vol. 2* (Ed. S. Patai), Wiley, Chichester, 1992.
24. H. Basch and T. Hoz, in *Supplement C: The Chemistry of the Triple-bonded Functional Groups, Vol. 2* (Ed. S. Patai), Wiley, Chichester, 1994.
25. W. J. Stevens, H. Basch and M. Krauss, *J. Chem. Phys.*, **81**, 6026 (1984).
26. W. J. Stevens, M. Krauss, H. Basch and P. G. Jasien, *Can. J. Chem.*, **70**, 612 (1992).
27. J. P. Desclaux, *Comput. Phys. Commun.*, **9**, 31 (1975).
28. M. W. Schmidt, K. K. Baldridge, J. A. Boatz, S. T. Elbert, M. S. Gordon, J. H. Jensen, S. Koseki, N. Matsunaga, K. A. Nguyen, S. Su, T. L. Windus, M. Dupuis and J. A. Montgomery, Jr., *J. Comput. Chem.*, **14**, 1347 (1993).
29. GAMESS User's Guide, Section 4, p. 10.

30. GAUSSIAN92, Revision A. M. J. Frisch, G. W. Trucks, M. Head-Gordon, P. M. W. Gill, W. Wong, J. B. Foresman, H. B. Schlegel, M. A. Robb, E. S. Replogle, R. Gomperts, J. L. Andres, K. Raghavachari, J. S. Binkley, C. Gonzales, R. L. Martin, D. J. Fox, D. J. Defrees, J. Baker, J. J. P. Stewart and J. A. Pople, Gaussian, Inc., Pittsburgh PA 15213, 1992.

31. M. Krauss, *J. Res. Nat. Bur. Stand., Sec. A*, **68**, 635 (1964).

32. E. Magnusson, *J. Am. Chem. Soc.*, **115**, 1051 (1993).

33. J. A. Pople, J. S. Binkley and R. Seeger, *Int. J. Quantum Chem., Symp.*, **10**, 1 (1976).

34. J. S. Binkley and J. A. Pople, *Int. J. Quantum Chem.*, **9**, 229 (1975).

35. J. C. Slater, *Quantum Theory of Molecules and Solids*, Volume 1, McGraw-Hill, New York, 1963.

36. D. J. DeFrees, B. A. Levi, S. K. Pollack, W. J. Hehre, J. S. Binkley and J. A. Pople, *J. Am. Chem. Soc.*, **101**, 4085 (1975).

37. J. F. Stanton, J. Gauss and R. J. Bartlett, *Chem. Phys. Lett.*, **195**, 194 (1992).

38. J. A. Pople, M. Head-Gordon and K. Raghavachari, *J. Chem. Phys.*, **87**, 5968 (1987).

39. K. B. Wiberg, C. M. Hadad, T. J. LePage, C. M. Breneman and M. J. Frisch, *J. Phys. Chem.*, **96**, 671 (1992).

40. S. M. Bachrach, in *Reviews in Computational Chemistry*, Volume V (Eds. K. B. Lipkowitz and D. B. Boyd), Chap. 3, VCH Publ., New York, 1994.

41. R. S. Mulliken and P. Politzer, *J. Chem. Phys.*, **55**, 5135 (1971).

42. W. J. Hehre, R. Ditchfield, L. Radom and J. A. Pople, *J. Am. Chem. Soc.*, **92**, 4796 (1970).

43. L. C. Snyder and H. Basch, *J. Am. Chem. Soc.*, **91**, 2189 (1969).

44. See, for example, H. B. Schlegel, *J. Chem. Phys.*, **84**, 4530 (1986).

45. N. C. Handy, P. J. Knowles and K. Somasundram, *Theor. Chim. Acta*, **68**, 87 (1985).

46. L. A. Curtiss, K. Raghavachari, G. W. Trucks and J. A. Pople, *J. Chem. Phys.*, **94**, 7221 (1991).

47. J. A. Pople, A. P. Scott, M. W. Wong and L. Radom, *Isr. J. Chem.*, **33**, 345 (1993).

48. R. S. Grev, C. L. Janssen and H. F. Schaefer III, *J. Chem. Phys.*, **95**, 5128 (1991).

49. E. R. Davidson and S. J. Chakravorty, *Chem. Phys. Lett.*, **217**, 48 (1994).

50. P. G. Jasien and W. J. Stevens, *J. Chem. Phys.*, **84**, 3271 (1986).

51. M. J. Frisch, J. A. Pople and J. S. Binkley, *J. Chem. Phys.*, **80**, 3265 (1984).

52. J. A. Pople, M. Head-Gordon, D. J. Fox, K. Raghavachari and L. A. Curtiss, *J. Chem. Phys.*, **90**, 5622 (1989).

53. L. A. Curtiss, K. Raghavachari and J. A. Pople, *Chem. Phys. Lett.*, **214**, 183 (1993).

54. J. Almlof and K. Faegri, Jr., *Theor. Chim. Acta*, **69**, 438 (1986).

55. K. G. Dyall, P. R. Taylor, K. Faegri, Jr. and H. Parteidge, *J. Chem. Phys.*, **95**, 2583 (1991).

56. O. Visser, L. Visscher, P. J. C. Aerts and W. C. Nieuwpoort, *Theor. Chim. Acta*, **81**, 405 (1992); see also: M. Dolg, W. Kuchle, H. Stoll, H. Preuss and P. Schwerdtfeger, *Mol. Phys.*, **74**, 1265 (1991) and P. Schwerdtfeger, H. Silberbach and B. Miehlich, *J. Chem. Phys.*, **90**, 762 (1989).

57. E. Hirota, *J. Mol. Spectrosc.*, **77**, 213 (1979).

58. K. Ohno, H. Matsuura, Y. Endo and E. Hirota, *J. Mol. Spectrosc.*, **118**, 1 (1986).

59. J. P. Desclaux and P. Pykko, *Chem. Phys. Lett.*, **29**, 534 (1974).

60. Landolt-Bornstein, *Numerical Data and Functional Relationships in Science and Technology*, New Series (Ed. K. -H. Hellwege), Group II, Vol. 7, *Structure Data of Free Polyatomic Molecules* (Eds. J. H. Callomon, E. Hirota, K. Kuchitsu, W. F. Lafferty, A. G. Maki and C. S. Pote; K. - H. Hellwege and A. M. Hellwege); Springer-Verlag, Berlin, 1976; M. Hargittai and I. Hargittai, *Int. J. Quantum Chem.*, **44**, 1057 (1992).

61. R. J. Bartlett and J. F. Stanton, in *Reviews in Computational Chemistry*, Vol. 5 (Eds. K. B. Lipkowitz and D. B. Boyd), Chap. 2, VCH Publ., New York, 1994.

62. H. Basch and W. J. Stevens, unpublished calculations.

63. D. J. Defrees, B. A. Levi, S. K. Pollack, W. J. Hehre, J. S. Binkley and J. A. Pople, *J. Am. Chem. Soc.*, **101**, 4085 (1979).

64. W. Muller and W. Meyer, *J. Chem. Phys.*, **80**, 3311 (1984).

65. M. Krauss and W. J. Stevens, *J. Chem. Phys.*, **93**, 4236 (1990).

66. R. F. W. Bader, *Atoms in Molecules: A Quantum Theory*, Clarendon Press, Oxford, 1990 and references cited therein.

67. M. T. Carroll, M. S. Gordon and T. L. Windus, *Inorg. Chem.*, **31**, 825 (1992).

68. P. Pykko, *Chem. Phys. Lett.*, **156**, 337 (1989); **162**, 349 (1989).

69. P. Pykko and Y. Zhao, *J. Phys. Chem.*, **94**, 7753 (1990).

70. M. Kaupp and P. v. R. Schleyer, *J. Am. Chem. Soc.*, **115**, 1061 (1993).

71. W. Kutzelnigg, *J. Mol. Struct.*, **169**, 403 (1988).
72. T. Egawa, S. Yamamoto, M. Nakata and K. Kuchitsu, *J. Mol. Struct.*, **156**, 213 (1987).
73. D. F. Eggers, Jr., *J. Mol. Struct.*, **31**, 367 (1976).
74. M. Imachi, T. Tanaka and E. Hirota, *J. Mol. Spectrosc.*, **63**, 265 (1976).
75. P. Jensen, S. Brodersen and G. Guelachvili, *J. Mol. Spectrosc.*, **88**, 378 (1981).
76. G. Graner, *J. Mol. Spectrosc.*, **90**, 394 (1981).
77. A. G. Robiette, C. Georghiou and J. G. Baker, *J. Mol. Spectrosc.*, **63**, 391 (1976).
78. R. Kewley, P. M. McKinney and A. G. Robiette, *J. Mol. Spectrosc.*, **34**, 390 (1970).
79. M. Le Guennec, W. Chen, G. Woldarczak and J. Demaison, *J. Mol. Spectrosc.*, **150**, 493 (1991).
80. J. Demaison, G. Woldarczak and J. Burie, *J. Mol. Spectrosc.*, **140**, 322 (1990).
81. S. Cradock, D. C. McKean and M. W. MacKenzie, *J. Mol. Struct.*, **74**, 265 (1981).
82. L. C. Krisher, R. A. Gsell and J. M. Bellama, *J. Chem. Phys.*, **54**, 2287 (1971).
83. S. N. Wolf, L. C. Krisher and R. A. Gsell, *J. Chem. Phys.*, **54**, 4605 (1971).
84. W. Schneider and W. Thiel, *J. Chem. Phys.*, **86**, 923 (1987).
85. A. E. Reed, R. B. Weinstock and F. Weinhold, *J. Chem. Phys.*, **83**, 735 (1985).
86. M. C. L. Gerry, R. M. Lees and G. Winnewisser, *J. Mol. Spectrosc.*, **61**, 231 (1976).
87. T. Iijima, *J. Mol. Struct.*, **212**, 137 (1989).
88. T. Kojima, *J. Phys. Soc. Jpn.*, **15**, 1284 (1960).
89. B. T. Luke, J. A. Pople, M. B. Krogh-Jesperson, Y. Apeloig, J. Chandrasekhar and P. v. R. Schleyer, *J. Am. Chem. Soc.*, **108**, 260 (1986).
90. J. Sauer and R. Ahlrichs, *J. Chem. Phys.*, **93**, 2575 (1990).
91. M. S. Stave and J. B. Nicholas, *J. Phys. Chem.*, **97**, 9630 (1993).
92. D. J. Lucas, L. A. Curtiss and J. A. Pople, *J. Chem. Phys.*, **99**, 6697 (1993).
93. T. Iijima, *Bull. Chem. Soc. Jpn.*, **59**, 853 (1986); T. Iijima, H. Jimbo and M. Taguchi, *J. Mol. Struct.*, **144**, 381 (1986).
94. M. Kreglewski, *J. Mol. Spectrosc.*, **133**, 10 (1989).
95. L. S. Bartell, *J. Chem. Phys.*, **32**, 832 (1960).
96. T. Kojima, E. L. Breig and C. C. Lin, *J. Chem. Phys.*, **35**, 2139 (1961).
97. C. Glidwell, P. M. Pinder, A. G. Roberts and G. M. Sheldrick, *J. Chem. Soc., Dalton Trans.*, 1402 (1973).
98. J. R. Durig, Y. S. Li, M. M. Chen and J. D. Odom, *J. Mol. Spectrosc.*, **59**, 74 (1976).
99. L. S. Bartell and H. K. Higginbotham, *J. Chem. Phys.*, **42**, 851 (1965).
100. T. Iijima, *Bull. Chem. Soc. Jpn.*, **46**, 2311 (1973).
101. M. D. Harmony, *J. Chem. Phys.*, **93**, 7522 (1990).
102. A. C. Bond and L. O. Brockway, *J. Am. Chem. Soc.*, **76**, 3312 (1954).
103. M. Wong and I. Ozier, *J. Mol. Spectrosc.*, **102**, 89 (1983).
104. B. Beagley, A. R. Conrad, J. M. Freeman, J. J. Monaghan and B. G. Norton, *J. Mol. Struct.*, **11**, 371 (1972).
105. V. W. Laurie, *J. Chem. Phys.*, **30**, 1210 (1959).
106. A. P. Cox and R. Varma, *J. Chem. Phys.*, **46**, 2007 (1967).
107. B. Beagley and J. J. Monaghan, *Trans. Faraday Soc.*, **66**, 2745 (1970).
108. J. R. Durig, C. M. Whang and G. M. Attia, *J. Mol. Spectrosc.*, **108**, 240 (1984).
109. J. F. Sanz and A. Marquez, *J. Phys. Chem.*, **93**, 7328 (1989).
110. B. J. DeLeeuw and H. F. Schaefer III, *J. Chem. Educ.*, **69**, 441 (1992).
111. P. v. R. Schleyer, M. Kaupp, F. Hampel, M. Bremer and K. Mislow, *J. Am. Chem. Soc.*, **114**, 6791 (1992).
112. T. A. Hein, W. Thiel and T. J. Lee, *J. Phys. Chem.*, **97**, 4381 (1993).
113. H. W. Kroto, J. F. Nixon and N. Pc. Simmons, *J. Mol. Spectrosc.*, **77**, 270 (1979).
114. J. C. T. R. Burckett-St. Laurent, T. A. Cooper, H. W. Kroto, J. F. Nixon, O. Ohashi and K. Ohno, *J. Mol. Struct.*, **79**, 215 (1982).
115. K. Karakida, T. Fukuyama and K. Kuchitsu, *Bull. Chem. Soc. Jpn.*, **47**, 299 (1974).
116. J. Demaison, A. Dubrulle, D. Boucher and J. Burie, *J. Mol. Spectrosc.*, **76**, 1 (1979).
117. L. Halonen and I. M. Mills, J. Mol. Spectrosc., **73**, 494 (1978).
118. P. H. Turner and A. P. Cox, *J. Chem Soc.*, *Faraday* Trans. 2, **74**, 533 (1978).
119. P. D. Blair, A. J. Blake, R. W. Cockman, S. Cradock, E. A. V. Ebsworth and D. W. H. Rankin, *J. Mol. Struct.*, **193**, 279 (1989).
120. R. Varma and K. S. Buckton, *J. Chem. Phys.*, **46**, 1565 (1967).
121. H. Basch, S. Hoz, M. Goldberg and L. Games, *Israel J. Chem.*, **31**, 335 (1991).

122. W. D. Laidig, P. Saxe and R. J. Bartlett, *J. Chem. Phys.*, **86**, 887 (1986).
123. S. N. Ghosh, R. Trambarulo and W. Gordy, *J. Chem. Phys.*, **21**, 308 (1953).
124. G. Trinquier and J. -C. Barthelat, *J. Am. Chem. Soc.*, **112**, 9121 (1990).
125. R. S. Grev, *Adv. Organomet. Chem.*, **33**, 125 (1991).
126. K. Huber and G. Herzberg, *Constants of Diatomic Molecules*, Van Nostrand, New York, 1979.
127. A Bauer, D. Boucher, J. Burie, J. Demaison and A. Dubrulle, *J. Phys. Chem. Ref. Data*, **8**, 537 (1979).
128. H. Tam, I. An and J. A. Roberts, *J. Mol. Spectrosc.*, **135**, 349 (1989).
129. C. A. Brookman, S. Cradock, D. W. H. Rankin, N. Robertson and P. Vefghi, *J. Mol. Struct.*, **216**, 191 (1990).
130. E. C. Thomas and V. W. Laurie, *J. Chem. Phys.*, **44**, 2602 (1966).
131. T. Iijima and S. Tsuchiya, *J. Mol. Spectrosc.*, **44**, 88 (1972).
132. P. Nosberger, A. Bauder and Hs. H. Gunthard, *Chem. Phys.*, **1**, 418 (1973).
133. H. B. Schlegel and P. N. Skancke, *J. Am. Chem. Soc.*, **115**, 10916 (1993).
134. L. S. Khaikin, V. P. Novikov, V. S. Zavgorodnii and A. A. Petrov, *J. Mol. Struct.*, **39**, 91 (1977).
135. M. V. Adreocci, M. Bossa, C. Cauletti, S. Stranges, B. Wrackmeyer and K. Horschler, *Inorg. Chim. Acta*, **162**, 83 (1989).
136. D. R. Lide, Jr., and D. Christensen, *J. Chem. Phys.*, **35**, 1374 (1961).
137. I. Tokue, T. Fukuyama and K. Kuchitsu, *J. Mol. Struct.*, **17**, 207 (1973).
138. Y. Shiki, A. Hasegawa and M. Hayashi, *J. Mol. Struct.*, **78**, 185 (1982).
139. J. R. Durig, K. L. Kizer and Y. S. Li, *J. Am. Chem. Soc.*, **96**, 7400 (1974).
140. U. Blukis, P. H. Kasai and R. J. Myers, *J. Chem. Phys.*, **38**, 2753 (1963).
141. T. Kamagawa, M. Takemura, S. Konaka and M. Kimura, *J. Mol. Struct.*, **125**, 131 (1984).
142. C. Glidewell, D. W. H. Rankin, A. G. Robiette, G. M. Sheldrick, B. Beagley and J. M. Freeman, *J. Mol. Struct.*, **5**, 417 (1970).
143. M. Hayashi, N. Nakata and S. Miyazaki, *J. Mol. Spectrosc.*, **135**, 270 (1989).
144. J. Nakagawa, Y. Shiki and M. Hayashi, *J. Mol. Spectrosc.*, **122**, 1 (1987).
145. A. Almenningen, O. Bastiansen, V. Ewing, K. Hedberg and M. Traetteberg, *Acta Chem. Scand.*, **17**, 2455 (1963).
146. S. Shambayati, J. F. Blake, S. G. Wierschke, W. L. Jorgensen and S. L. Schreiber, *J. Am. Chem. Soc.*, **112**, 697 (1990).
147. Y. Apeloig, D. Arad and Z. Rappoport, *J. Am. Chem. Soc.*, **112**, 9131 (1990).
148. L. A. Curtiss, H. Brand, J. B. Nicholas and L. E. Iton, *Chem. Phys. Lett.*, **184**, 215 (1991).
149. J. F. Blake and W. L. Jorgensen, *J. Org. Chem.*, **56**, 6052 (1991).
150. B. T. Luke, *J. Phys. Chem.*, **97**, 7505 (1993).
151. M. S. Stave and J. B. Nicholas, *J. Phys. Chem.*, **97**, 9630 (1993).
152. I. S. Ignatyev, *J. Mol. Struct.*, **172**, 139 (1988).
153. T. Iijima, *Bull. Chem. Soc. Jpn.*, **45**, 1291 (1972).
154. J. R. During, A. D. Lopata and P. Groner, *J. Chem. Phys.*, **66**, 1888 (1977).
155. J. R. Durig, Y. S. Li, J. F. Sullivan, J. S. Church and C. B. Bradley, *J. Chem. Phys.*, **78**, 1046 (1983).
156. (a) D. W. W. Anderson, D. W. H. Rankin and A. Robertson, *J. Mol. Struct.*, **14**, 385 (1972).
    (b) N. Heineking and M. C. L. Gerry, *Z. Naturforsch.*, **44a**, 669 (1989).
157. C. Glidewell and A. G. Robiette, *Chem. Phys. Lett.*, **28**, 290 (1974).
158. J. D. Murdoch and D. W. H. Rankin, *J. Chem. Soc., Chem. Commun.*, 748 (1972).
159. P. Groner, G. M. Attia, A. B. Mohamad, J. F. Sullivan, Y. S. Li and J. R. Durig, *J. Chem. Phys.*, **91**, 1434 (1989).
160. A. Hammel, H. V. Volden, A. Haaland, J. Weidlein and R. Reischmann, *J. Organomet. Chem.*, **408**, 35 (1991).
161. M. H. Palmer and M. F. Guest, *Chem. Phys. Lett.*, **196**, 183 (1992).
162. T. Sakaizumi, H. Mure, O. Ohashi and I. Yamaguchi, *J. Mol. Spectrosc.*, **140**, 62 (1990).
163. J. Koput, *J. Mol. Spectrosc.*, **115**, 131 (1986).
164. C. Glidewell, A. G. Robiette and G. M. Sheldrick, *Chem. Phys. Lett.*, **16**, 526 (1972).
165. J. A. Duckett, A. G. Robiette and M. C. L. Gerry, *J. Mol. Spectrosc.*, **90**, 374 (1981).
166. M. Krglewski and P. Jensen, *J. Mol. Spectrosc.*, **103**, 312 (1984).
167. J. D. Murdoch, D. W. H. Rankin and B. Beagley, *J. Mol. Struct.*, **31**, 291 (1976).
168. S. Cradock, J. R. Durig, A. B. Mohamad, J. F. Sullivan and J. Koput, *J. Mol. Spectrosc.*, **128**, 68 (1988).

169. T. Pasinszki, T. Veszpremi and M. Feher, *Chem. Phys. Lett.*, **189**, 245 (1992).
170. M. Feher, T. Pasinski and T. Veszpremi, *Chem. Phys. Lett.*, **205**, 123 (1993).
171. M. Feher, T. Pasiszki and T. Veszpremi, *J. Phys. Chem.*, **97**, 1538 (1993).
172. M. Feher, T. Pasinszki and T. Veszpremi, *J. Am. Chem. Soc.*, **115**, 1500 (1993).
173. M. J. Barrow, E. A. V. Ebsworth and M. M. Harding, *J. Chem. Soc., Dalton Trans.*, 1838 (1980).
174. H. -G. Mack and H. Oberhammer, *Chem. Phys. Lett.*, **157**, 436 (1989).
175. H. Dreizler, H. D. Rudolph and H. Schleser, *Z. Naturforsch.*, **25a**, 1643 (1971).
176. J. Koput, *J. Mol. Spectrosc.*, **118**, 189 (1986).
177. M. Kreglewski, *Chem. Phys. Lett.*, **112**, 275 (1984).
178. K. F. Dossel and D. H. Sutter, *Z. Naturforsch.*, **32a**, 478 (1977); **34a**, 482 (1979).
179. J. R. Durig, Y. S. Li and J. F. Sullivan, *J. Chem. Phys.*, **71**, 1041 (1979).
180. G. Davidson, L. A. Woodward, K. M. Mackay and P. Robinson, *Spectrochim. Acta*, **23A**, 2383 (1967).
181. T. Sakaizumi, M. Obata, K. Takahashi, E. Sakaki, Y. Takeuchi, O. Ohashi and I. Yamaguchi, *Bull. Chem. Soc. Jpn.*, **59**, 3791 (1986).
182. T. Sakaizumi, A. Yasukawa, H. Miyamoto, O. Ohashi and I. Yamaguchi, *Bull. Chem. Soc. Jpn.*, **59**, 1614 (1986).
183. J. Koput, F. Stroh and M. Winnewisser, *J. Mol. Spectrosc.*, **140**, 31 (1990).
184. S. Tsuchiya, *J. Mol. Struct.*, **22**, 77 (1974).
185. S. Tsuchiya and T. Iljima, *J. Mol. Struct.*, **13**, 327 (1972).
186. A. P. Cox and S. Waring, *J. Chem. Soc., Faraday Trans. 2*, **68**, 1060 (1972).
187. P. H. Turner, M. J. Corkill and A. P. Cox, *J. Phys. Chem.*, **83**, 1473 (1979).
188. M. L. McKee, *J. Am. Chem. Soc.*, **108**, 5784 (1986) and many subsequent papers.
189. T. J. Packwood and M. Page, *Chem. Phys. Lett.*, **216**, 180 (1993).
190. H. Basch and S. Hoz, to be published.
191. B. P. van Eijck and E. van Zoren, *J. Mol. Spectrosc.*, **111**, 138 (1985).
192. Y. Li and K. N. Houk, *J. Am. Chem. Soc.*, **111**, 4505 (1989).
193. H. Basch and W. J. Stevens, *J. Am. Chem. Soc.*, **113**, 95 (1991).
194. M. Kitano and K. Kuchitsu, *Bull. Chem. Soc. Jpn.*, **46**, 3048 (1973).
195. K. B. Wiberg, C. M. Hadad, P. R. Rablen and J. Cioslowski, *J. Am. Chem. Soc.*, **114**, 8644 (1992).
196. T. Iijima, *Bull. Chem. Soc. Jpn.*, **45**, 3526 (1972).
197. L. Pierce and R. J. Nelson, *J. Mol. Spectrosc.*, **18**, 344 (1965).
198. K. B. Wiberg and E. J. Martin, *J. Am. Chem. Soc.*, **107**, 5035 (1985).
199. K. B. Wiberg, *J. Am. Chem. Soc.*, **107**, 5035 (1985).
200. K. B. Wiberg and K. E. Laidig, *J. Am. Chem. Soc.*, **109**, 5935 (1987).
201. J. D. Swalen and C. C. Costain, *J. Chem. Phys.*, **31**, 1562 (1959).
202. W. F. Edgell, G. B. Miller and J. W. Amy, *J. Am. Chem. Soc.*, **79**, 2391 (1957).
203. B. Beagley, M. O. Jones and M. A. Zanjanchi, *J. Mol. Struct.*, **56**, 215 (1979).
204. H. Beckers, H. Burger, R. Eugen, B. Rempfer and H. Oberhammer, *J. Mol. Struct.*, **140**, 281 (1986).
205. J. F. Sullivan, C. M. Whang, J. R. Durig, H. Burger, R. Eugen and S. Cradock, *J. Mol. Struct.*, **223**, 457 (1990).
206. W. Cherry, N. Epiotis and W. T. Borden, *Acc. Chem. Res.*, **10**, 167 (1977).
207. A. P. Cox and S. Waring, *Trans. Faraday Soc.*, **67**, 3441 (1971).
208. A. Stirling, I. Papai, J. Mink and D. R. Salahub, *J. Chem. Phys.*, **100**, 2910 (1994) and references cited therein.
209. W. Pyckhout, C. V. Alsenoy and H. J. Geise, *J. Mol. Struct.*, **144**, 265 (1986).
210. M. J. Barrow, S. Cradock, E. A. V. Ebsworth and D. W. H. Rankin, *J. Chem. Soc., Dalton Trans.*, 1988 (1981).
211. E. A. V. Ebsworth, C. M. Huntley and D. W. H. Rankin, *J. Organomet. Chem.*, **281**, 63 (1985).
212. P. O. Lowdin, *Phys. Rev.*, **97**, 1509 (1955).
213. C. Sosa and H. B. Schlegel, *Int. J. Quantum Chem., Quantum Chem. Symp.*, **21**, 267 (1987).
214. C. Gonzales, C. Sosa and H. B. Schlegel, *J. Phys. Chem.*, **93**, 2435 (1989).
215. N. C. Handy, M. -D. Su, J. Coffin and R. D. Amos, *J. Chem. Phys.*, **93**, 4123 (1990).
216. J. Baker, *J. Chem. Phys.*, **91**, 1789 (1989).
217. N. C. Handy, P. J. Knowles and K. Somasundrum, *Theor. Chim. Acta*, **68**, 87 (1985).
218. P. M. W. Gill, J. A. Pople, L. Radom and R. H. Nobes, *J. Chem. Phys.*, **89**, 7307 (1988).

219. M. D. Allendorf and C. F. Melius, *J. Phys. Chem.*, **96**, 428 (1992).
220. M. D. Allendorf and C. F. Melius, *J. Phys. Chem.*, **97**, 720 (1993).
221. L. A. Curtiss and J. A. Pople, *J. Chem. Phys.*, **88**, 7405 (1988).
222. L. A. Curtiss and J. A. Pople, *J. Chem. Phys.*, **91**, 2420 (1989).
223. A. Stirling, I. Papai, J. Mink and D. R. Salahub, *J. Chem. Phys.*, **100**, 2910 (1994).
224. S. G. Lias, J. E. Bartmess, J. F. Liebman, J. L. Holmes, R. D. Levin and W. G. Mallard, *J. Phys. Chem. Ref. Data*, **17**, Supplement No. 1 (1988).
225. K. B. Wiberg, L. S. Crocker and K. P. Morgan, *J. Am. Chem. Soc.*, **113**, 3447 (1991).
226. J. H. Kiefer, S. S. Sidhu, R. D. Kern, K. Xie, H. Chen and L. B. Harding, *Combust. Sci. Technol.*, **82**, 101 (1992).
227. J. L. Holmes, *Int. J. Mass Spectrom. Ion Processes*, **118/119**, *381* (1992).
228. C. L. Darling and H. B. Schlegel, *J. Phys. Chem.*, **97**, 8207 (1993).
229. G. Leroy, M. Sana, C. Wilante and D. R. Temsamani, *J. Mol. Struct.*, **259**, 369 (1992).
230. G. Leroy, M. Sana, C. Wilante and M. -J. van Zieleghem, *J. Mol. Struct.*, **247**, 199 (1991).
231. D. F. McMillen and D. M. Golden, *Annu. Rev. Phys. Chem.*, **33**, 493 (1982).
232. D. Griller, J. M. Kanabus-Kaminska and A. Maccoll, *J. Mol. Struct.*, **163**, 125 (1988).
233. K. B. Wiberg and P. R. Rablen, *J. Am. Chem. Soc.*, **115**, 9234 (1993).
234. R. Walsh, in *The Chemistry of Organic Silicon Compounds* (Eds. S. Patai and Z. Rappoport), Wiley, New York, 1989.
235. J. J. W. McDouall, H. B. Schlegel and J. S. Francisco, *J. Am. Chem. Soc.*, **111**, 4622 (1989).
236. M. -D. Su and H. B. Schlegel, *J. Phys. Chem.*, **97**, 9981 (1993).
237. L. A. Curtiss, K. Raghavachari and J. A. Pople, *J. Chem. Phys.*, **98**, 1293 (1993).
238. M. S. Gordon and D. G. Truhlar, *J. Am. Chem. Soc.*, **108**, 5412 (1986).
239. L. Pardo, J. R. Banfelder and R. Osman, *J. Am. Chem. Soc.*, **114**, 2382 (1992).
240. M. J. Almond, A. M. Doncaster, P. N. Noble and R. Walsh, *J. Am. Chem. Soc.*, **104**, 4717 (1982).
241. H. Jacobsen and T. Ziegler, *J. Am. Chem. Soc.*, **116**, 3667 (1994).

CHAPTER 2

Structural aspects of compounds containing C−E (E = Ge, Sn, Pb) bonds

KENNETH M. MACKAY

School of Science and Technology, University of Waikato, P.B. 3105, Hamilton, New Zealand
Fax: +(64)-7-838-4218

The chemistry of organic germanium, tin and lead compounds
Edited by S. Patai © 1995 John Wiley & Sons Ltd

LIST OF ABBREVIATIONS

Some of the more striking chemistry reported here depends on the presence of bulky ligands which will be denoted by the following abbreviations:

t-Bu or Bu^t	$-CMe_3$	Dmp	$2,6,-Me_2C_6H_3$
Nep	$-CH_2CMe_3$	Dep	$2,6-Et_2C_6H_3$
TMS	$-SiMe_3$	Dip	$2,6-(iso\text{-}Pr)_2C_6H_3$
Tsi	$-C(SiMe_3)_3$	Mes	$2,4,6-Me_3C_6H_2$
Bsi	$-CH(SiMe_3)_2$	Tip	$2,4,6-(iso\text{-}Pr)_3C_6H_2$
Sim	$-CH_2SiMe_3$	Ar(f) or Ar^f	$2,4,6-(CF_3)_3C_6H_2$
Tb	$2,4,6-[(Me_3Si)_2CH]_3C_6H_2$	Ttb	$2,4,6-(t\text{-}Bu)_3C_6H_2$

I. INTRODUCTION

The classical consensus rationalized differences between C and the heavier Group 14 elements as due mainly to carbon's ability to form

(i) strong element–element bonds giving homonuclear chains and rings, and

(ii) π overlap of p orbitals (in more current terminology) and hence alkenes, alkynes, aromatic compounds and the like.

Distinctive heavy-element properties were also recognized. In particular,

(iii) the existence of a distinct (II) state for Ge and its increasing stability through Sn to Pb,

(iv) the ability of the heavier elements to expand their coordination number above 4, allowing types of compound inaccessible to C, producing different structures for formally analogous compounds (e.g. halogen-bridged polymers for R_2EX_2) and providing lower-energy intermediates and hence major kinetic effects.

This review of the structural properties of organo-germanes, -stannanes, and -plumbanes covers only compounds whose structures have been determined by X-ray crystallography or other absolute methods. A much wider range of structures has been well established by spectroscopic methods. Advances in crystallography in the last ten years has made the method much more accessible, so that the determined structures are, by and large, a representative cross section of current chemistry. Much of the inspiration for study of Ge, Sn and Pb chemistry in recent years has been to test the classical generalizations. Thus our focus is to

(a) cover the carbon comparisons and contrasts outlined in (i) to (iv) above, and

(b) go beyond classical ideas especially in multiple bonding, divalent organometallics and more complex structures.

It is possible only to touch on heavier Group 14 organics in other actively developing fields such as metalla(car)boranes or in transition metal chemistry.

The picture is painted against a background of the classical chemistry of the tetravalent (IV) state. There are many valuable baseline reviews[1,2] covering to the early 1980s, and only a few classical references are quoted from the earlier period. Recent reviews[3] are relevant. For comparison with Si, recent summary references[4] usefully cover to 1988/9. The degree of interest has varied widely between the different Group 14 elements. Because of the accessibility, economic importance and greater amenability to older methods, there are about three times as many organotin structures in the Cambridge data base as there are organogermanium ones, which in turn has some five times as many entries as lead. In the last five years, interest in germanium has increased relatively though still less than tin, while organolead chemistry remains a minor field.

While the generalizations of an earlier period were invaluable in the initial structuring of the chemistry of Main Group elements, we now see that they provided only a skeleton. The present picture of Group 14, as of the rest, is one of a rich chemistry of each individual element, a subtly varying relationship between them, and sufficient unexpected and at present unique behaviour to indicate that further development will be exciting and complex.

II. SIMPLE TETRAVALENT SPECIES

A. Basic Bond Lengths

Mononuclear ER_4 and simple four-coordinate compounds of E(IV) states are the baseline for viewing the other coordination numbers, the effect of bulky ligands, bonds to other E or metals, E(II) compounds, multiple bonds and other phenomena discussed in later sections. Basic parameters for some simple compounds are presented in Table 1, taken from the gas-phase data summarized by Molloy and Zuckerman[5] and Haaland[6]. These data show the unperturbed molecules in the gas phase and provide the base for

Kenneth M. Mackay

TABLE 1. Bond distances (pm) and angles (deg) in simple tetravalent species

Compound	E—C	E—H or E—X	Angles or notes
Me$_4$Ge	194.5		CGeC = 109.6
Me$_3$GeH^a	194.7(6)	153.2	CGeC = 109.6
			CGeH = 109.3
Me$_2$GeH$_2$^{a}	195.0(10)		CGeC = 110.0
MeGeH$_3$^{a}	194.5	152.9	HGeH = 109.3
GeH$_4$^{b}		152.5	HGeH = 109.5
CH$_3$—CH$_2$GeH$_3$	194.9	152.2	CGeH = 109.7
CH$_2$=CHGeH$_3$^{a}	192.6(12)	152.0	CGeH = 109.7
CH≡CGeH$_3$^{a}	189.6	152.1	HGeH = 109.9
(σ-C$_5$H$_5$)GeH$_3$	196.5	153	planar C$_5$ ring
			GeC—ring angle = 64°
Me$_2$GeF$_2$	192.8	173.9	
MeGeF$_3$	190.4(9)	171.4	
Me$_3$GeCl^a	194.0	217.0	
Me$_2$GeCl$_2$	192.6	215.5	Ge—Cl = 214.3 in another study
MeGeCl$_3$	189.3(10)	213.2	
GeCl$_4$		211.3	
Me$_3$GeBr^a	193.6	232.3	
Me$_2$GeBr$_2$	191.1(12)	230.3	
MeGeBr$_3$	189	227.6	
GeBr$_4$		227.2	
Ge(CF$_3$)$_4$	198.9		
Me$_4$Sn	214.4		
Me$_3$SnH	214.9	171(7)	
Me$_2$SnH$_2$	215.3	168.0(15)	
MeSnH$_3$^{a}	214.3	170(2)	
SnH$_4$^{b}		171.1	HSnH = 109.5
Me$_3$SnC≡CH	214.1 (Me) and 212.6(8)		
Me$_3$SnC≡CSnMe$_3$	212.7 (Me) and 209.5		
(CH$_2$=CH)$_4$Sn	211.6		CSnC = 109.7
(CH≡C)$_4$Sn	206.7		
(CH≡C)$_3$SnI	206.0(6)	264.5	
(F$_3$CC≡C)$_4$Sn	207.0		
Ph$_4$Sn	216.8		
Me$_3$SnCl	210.6(6)	235.1	CSnC = 115
			CSnCl = 103
Me$_2$SnCl$_2$	210.9	232.7	
MeSnCl$_3$	210.4(16)	230.4	
SnCl$_4$		228.1	
Sn(CF$_3$)$_4$	216.8		
Me$_4$Pb	223.8		

All by electron diffraction except ^{a} microwave, ^{b} vibrational-rotational spectroscopy. Standard deviations are 5 or less in the units of the last digit, except where given.

comparison for more complex structures. Data vary because (i) different structural methods determine different measures related to geometrical parameters, (ii) techniques have improved with time and (iii) experimental problems, such as the stability of compounds, affect data quality. The large majority of bond-length determinations are available to 0.1 pm and a difference of 1 pm is commonly 2–3 σ when there are no experimental difficulties. Thus slight changes — e.g. in E—C with the number of E—H bonds — are not significant at this order of accuracy.

With H or alkyl as ligands, the bond lengths, and thus the E radius, remain constant within 1 pm for Ge and Sn. The reduction in M—C for alkene and for a single

TABLE 2.   Standard bond lengths (pm) in four-coordinate compounds of Ge, Sn or Pb[a]

E =	Ge	Sn	Pb
E−C(alkyl)	194.5	214	225
E−H	152.5	170	
E−F	171	196	201
E−O(ether)	176	194	
E−N(amine)	185	204	
E−Cl	215	235	243
E−S(sulphide)	223	243	249
E−P	236	254	262
E−Br	232	249	265
E−Se	238	255	
E−I	250	270	
E−Te		275	

[a] Adding further electronegative ligands reduces E−C and increases E−X by some 2 pm for each. A similar reduction occurs in E−C in going to sp^2 and then to sp C.

alkyne essentially matches the established differences between sp^3, sp^2 and sp C. The tetra-ethynes and the distannyl ethyne show further reductions. Increasing halogen substitution reduces both the E−C and the E−X bonds, as expected from the inductive effect of the electronegative ligands.

The gas-phase values are often a little larger than found for solid state structures. However, taking Table 1 data together with solid state values from simple compounds, we may generalize that 'normal' bond lengths would be within 2 pm of the values in Table 2.

B. Alkyls and Related Compounds

Based on these generalizations, several studies of alkyl species are of interest.

(i) The 158K X-ray structure[7] for Me_4Sn shows the molecule lying on a three-fold axis with three Sn−C lengths of 213.8(6) pm and one of 210.2(8) pm compared with the gas-phase value of 214.4(3) pm. Angles are essentially tetrahedral. While the differences are only marginally significant, the distortion does match those indicated by inelastic neutron scattering and spectroscopic studies on EMe_4 solids.

(ii) Crystal structures of the alkynyl-tin compounds, $Sn(C\equiv CR)_4$ for R = CMe_3 and $SiMe_3$[8], are part of a study of alkynyl exchange which included spectroscopic characterization of a wide range of species. Deviations from ideal tetrahedral configuration at Sn are slight with CSnC angles of 107.6° to 112.9° in the two compounds. Sn−C bonds averaged 207.4 pm (R = CMe_3) and 207.9 pm (R = $SiMe_3$), shorter than in Table 1. There was no short intermolecular distance.

(iii) There are few structures reported for lead compounds and the only crystal structure for a tetraalkyl is that[9] for tetrakis(2-chlorobenzyl)lead. The four benzyl CH_2 groups form a slightly flattened tetrahedron with Pb−C lengths of 226 pm and CPbC angles ranging from 105 to 115°.

(iv) The calculated[10] SCF-optimized geometries for Me_nPbF_{4-n} species show shortening with increasing electronegative ligands with a Pb−C length of 222.7 pm for n = 3, diminishing to 220.2 and 219.8 pm for n = 2 and 1, respectively. The CPbC angles are enlarged to 116.4° in the trimethyl and 134.8° in the dimethyl. As n decreases, the increasing positive charge on lead leads to an increase in the s orbital contribution giving the structural changes. Energy calculations show that Me_4Pb is stable with respect to Me_2Pb

and Pb and Me_3PbF is stable with respect to loss of CH_3F. In contrast, lead tetrahalides are unstable with respect to PbX_2 and MeF elimination from $MePbF_3$ is exothermic. Thus the at-first-sight unexpected stability of Pb(IV) in tetra-and tri-alkyls also reflects the change in s and p contributions to the bonding.

(v) The structures of two $Ge-CF_3$ oxides are of interest. $(CF_3)_2GeO^{11}$ is a linear polymer with $Ge-C = 198.5$ pm [similar to $Ge(CF_3)_4$], $Ge-O$ distances in the chain essentially equal at 173.1 and 175.1 pm and chain angles of 133° at O and 104° at Ge. An electron diffraction structure[12] of $[(CF_3)_3Ge]_2O$ shows a similar $Ge-CF_3$ distance (200.4 pm), which is lengthened compared with the methyl analogue, while $Ge-O$ (172.4 pm) is shortened. The GeOGe angle of 151.5° continues the widening trend seen[6] for the GeH_3 (126.5°) and $GeMe_3$ (141.0°) analogues. These angles are wider than for C counterparts, but not as open as for Si. Such changes are now ascribed[13] to bond polarity differences, together with the variation with E of the s–p energy gap, rather than by the older idea of π acceptance of the lone pair by the nd orbital.

(vi) Germacyclohexanes and polycyclic analogues have been extensively studied[14], with a strong input from NMR methods. The crystal structure of 1,3,5-trigermacyclohexane and the electron diffraction structures[15] of germacyclohexane, GeC_5H_{12}, and the dimethyl analogue $Me_2GeC_5H_{10}$ show the molecules are in the chair form with $Ge-H = 153.0$ pm and $Ge-C = 195.6$ pm ($Ge-CH_3$ assumed $= Ge-CH_2$).

(vii) The compounds $P_7(EMe_3)_3$ (**1**) form one of the few series of more complex compounds[16] which contain all the Group 14 elements. The sensitivity to oxygen gives rather high standard errors. The parent heptaphosphane, P_7H_3, structure can be envisaged as a P_4 tetrahedron with PH inserted into three $P-P$ edges. In all members, the non-substituted P_3 face has longer $P-P$ bonds than those involving $P(EMe_3)$. The $E-C$ bonds are all normal while $P-E$ lengths (pm) are 228.8 (Si), 235.5 (Ge), 253.9 (Sn) and 261.7 (Pb), which fit with Table 2 values.

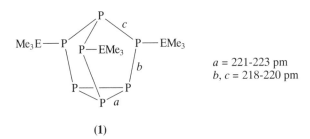

$a = 221\text{-}223$ pm
$b, c = 218\text{-}220$ pm

(**1**)

(viii) The distibanes, $(Me_3E)_2SbSb(EMe_3)_2$ (**2**), are reported for E = Si, Ge and Sn. The Ge–Me and Sn–Me lengths are normal[17] and the configuration is *trans* with angles at Sb of 92–93°. There are short intermolecular Sb...Sb contacts of 386 pm at $-110\,°C$ (Ge) and 381.1 pm at $-120\,°C$ (Sn) which increase 4–6 pm at room temperature.

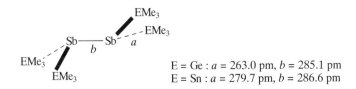

E = Ge : $a = 263.0$ pm, $b = 285.1$ pm
E = Sn : $a = 279.7$ pm, $b = 286.6$ pm

(**2**)

(ix) In $Me_3GeN(SO_2Me)_2$ (**3**), the two independent molecules[18] give tetrahedral Ge with essentially identical Ge−C values [192.3(2) to 193.6(3) pm], an unusually long Ge−N and planar nitrogen. The Pb analogue gives a chain structure by additional Pb−O bonding.

$$Me_3Ge \overset{a}{\underset{\displaystyle b}{\overset{\displaystyle b}{\text{—}}}} N \overset{SO_2Me}{\underset{SO_2Me}{}}$$

a = 198.4 pm
ab = 121.1°
bb = 117.3°

(**3**)

(x) The Me_3Ge group, in contrast to Me_3Si, allows a nearly-coplanar PhN_4Ph skeleton in the tetrazene, $Ph(Me_3Ge)NN=NN(GeMe_3)Ph$[19] maximizing π interaction. The configuration at Ge is tetrahedral (angles 106.1–113.1°, GeC = 194.0 pm, GeN = 191.2 pm) and the molecule has C_{2h} symmetry. Further potential chemistry is indicated by the existence of a SiN_4 ring.

C. Aryls

With larger ligands, we look for signs of steric distortion.

1. Basic tetraaryls

It is noticeable in Table 1 that the gas-phase structure for Ph_4Sn shows a Sn−C length of 216.8 pm, rather than a value reflecting sp^2 carbon, and this may imply a steric effect. Interestingly, the old crystal structure — determined by photographic methods — of Ph_4Pb has been redetermined by two groups recently[20,21]. The molecule is S_4 and contracted along this axis, giving four smaller and two larger CPbC angles, differing by 2.2° (average of the two determinations). Each phenyl ring is twisted by an angle α of 59° with respect to its CPbC plane, so that the conformation of each EPh_3 group is propeller-like. This S_4 configuration is found[21] for most Ar_4E structures (Ar = Ph, o-, m-, p-Tol, C_6F_5: E = Si to Pb, also CPh_4), although there are three different space groups. Twist angles α are 50–60°. The difference between the two larger and four smaller angles increases from Ge to Pb, is small for R = Ph or m-Tol, intermediate for o-Tol, and as large[22] as 7.4° (Sn) and 8.8° (Pb) in the $(p$-Tol$)_4E$ species. Even $Ph_3Si(p$-Tol) deviates[23] only slightly from idealized S_4 with all Si−C equal within 2σ and with PhSiTol giving one of the two larger 111° angles and two of the four smaller ones, again statistically indistinguishable from the PhSiPh angles. The only exceptions to S_4 were $(p$-Tol$)_4E$, E = Si or Ge, with a tetrahedron lengthened along one axis. This brings only three of the aryl groups closer than in the tetrahedron, as opposed to all four in the S_4 structures. While the E−C bond lengths are statistically indistinguishable, the three angles involving the extended axis average 110.3(1)° while the other three average 108.6(1)°. The mean bond angles at E are tetrahedral in all these molecules and the mean E−C bond lengths are listed in Table 3.

Within the set of values in Table 3, there is little indication of steric differences between the aryls, with the slightly longer o-tolyl bonds barely if at all significant. The *ortho*-methyl does adjust by moving away from the E group in-plane (Figure 1). The E−C(aryl) lengths are longer than the sp^3 C−E value (let alone the sp^2 one) for Ge but similar, or slightly shorter, for Sn and Pb, indicating steric adjustment in the germanium compounds. Many of these structural features carry over into the perphenyl catenated compounds discussed

TABLE 3. Mean E—C bond lengths (pm) in Ar_4E compounds determined crystallographically[a,b]

E/Ar	Ph	o-Tol	m-Tol	p-Tol	C_6F_5
Si	187.2	—	—	187.3	186.6
Ge	195.4	196.6	—	194.8	195.7
Sn	214.3	215.2	215.0	214.7	212.6
Pb	219.8	222.9	219.1	221.7	—

[a] Data from References 20–23 and references cited therein.
[b] Note some Sn—C bond lengths in highly hindered compounds of 218–225 pm quoted in H. Grützmacher, W. Deck, H. Pritzkow and M. Sander, *Angew. Chem., Int. Ed. Engl.*, **33**, 456 (1994).

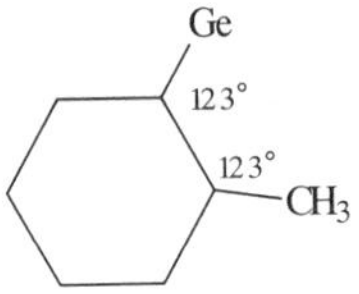

FIGURE 1. In-plane distortion in $(o\text{-Tol})_4Ge$

later. Many of these studies[20–23] also detail NMR and other spectroscopic parameters for a wider range of tetra-aryls including xylyls and mixed-ligand species and develop systematic relationships between structures and chemical shifts or coupling constants.

While not referring directly to Group 14, a review[24] on quantification of steric effects tabulates steric effect data such as cone angles for PR_3 and AsR_3 species, which are useful comparisons for SiR_3 and GeR_3 substituents.

2. Further aryl compounds

(i) The structure[25] of the fully substituted benzene, $(Me_3Ge)_6C_6$, shows somewhat lengthened Ge—C(aryl) bonds (197.6–199.8 pm) and Ge—C(methyl) bonds (192.3–195.8 pm) which are normal. As with organic analogues, C_6R_6, the structure relieves steric strain by (a) a slight puckering of the C_6 ring into a flattened chair, (b) a marginally significant lengthening of C—C to 141.8(13) pm from 139 pm in benzene and (c) substantial bending by 22° of the Me_3Ge groups out of the idealized plane of the molecule — alternatively above and below. This reduces the GeC(aryl)C(aryl) angles by about 3°. The angles at Ge are substantially distorted from tetrahedral (103–123°) while the benzene ring angles stay at 120°. Force field calculations indicate similar structural properties for the three $(Me_3E)_6C_6$ molecules with distortions increasing from E = C to Si to Ge.

(ii) In $Sn(p\text{-}C_6H_4OEt)_4$[26] the strain is relieved by one OEt group twisting out of plane, thus destroying the S_4 symmetry at Sn, though the tin coordination remains close to tetrahedral (Sn—C = 212.6–213.5 pm, CSnC averaging 109.4° with a spread 105–112°).

(iii) All the molecules $[Ph_{4-x}Sn\{CH_2P(S)Ph_2\}_x]$, x = 1 to 4, remain tetrahedral[27] with no indication of S coordination (Sn—Ph = 213.4 to 214.6 pm, Sn—C = 215.3 to 218.1 pm over the series, angles 104.5–112.4°). By contrast, the P=O analogue forms a dimer involving Sn...O coordination.

(iv) An indication of the intramolecular steric pressures is given by triphenylvinyltin[28] where the Sn—C bonds are all equal and normal, averaging 213 pm. The reduced strain is shown by the average value of the twist angle α of 51.6°, a reduction of 7° compared with Ph_4Sn, and with a larger change for the group remote from vinyl. The vinyl group is twisted 5°, moving its H atoms away from the nearest phenyl. In (E)-but-2-enyltriphenyltin[29] similar effects occur. The structure is very regular with the Sn–C bond (216.4 pm) a little longer than Sn–Ph (213.6 pm) and PhSnPh angles (108.8°) slightly smaller than PhSnC (110.2°).

(v) Reaction[30] of the tin chlorides with o-xylenediyls $[C_6H_4(CHR)_2]^{2-}$ (R = H or $SiMe_3$) gave $C_6H_4(CHR)_2SnPh_2$ or $[C_6H_4(CHR)_2]_2Sn]$ containing the $[C_6H_4(CHR)_2Sn]$ bicyclic unit with a CCC(R)C(R)Sn five-membered ring and the second with $spiro$-Sn. In all these compounds, Sn–Ph = 213 pm, Sn–C in the 5-ring averaged 214.5 pm and all bonds in the compounds with the disilyl ligand were 1–2 pm longer. A more detailed discussion of fold angles and steric effects is given. A stannylene produced $[C_6H_4(CHR)_2Sn]_4$ with an Sn_4 ring discussed later.

(vi) In an examination of the Lewis base behaviour of $(Ph_3Sn)_2O$ towards R_3M (M = Al, Ga) a dimeric product $[Me_2Al(\mu\text{-}OSnPh_3)]_2$ was identified[31], linked through a four-membered Al–O–Al–O ring. The configuration at Sn is normal (Sn–C = 211.8 pm, Sn–O = 198.4 pm, angles somewhat scattered, CSnC at 110–116° and OSnC at 104–108°). The $SnPh_3$ groups are $trans$ across the four-membered ring which shows an unusually short Al...Al distance.

(vii) A perchlorinated aryl[32] $(C_6Cl_5)_3GeCl$ has Ge–C = 197.7 pm, slightly lengthened compared with Table 3, indicating some strain. The Ge–Cl bond is normal at 215.8 pm.

D. Other Four-coordinated Species

(i) *Li compounds*. A direct E–Li bond is found in $Ph_3ELi(PMDETA)$ (E = Sn^{33}, Pb^{34}) with a trigonal structure (CSnC averaging 96.1°, CPbC = 95°, CSnLi 120.7°); Sn–C = 219.5 pm, Pb–C = 229.5 pm, Sn–Li = 287.2 pm and Pb–Li = 285.8 pm). In the trimer[35] $[(Me_3Ge)_2NLi]_3$ a Me_3Ge unit of normal geometry is present bonded to the interesting $(LiN)_3$ six-membered ring.

(ii) *Halides etc with large substituents*. While organogermyl halides are 4-coordinate, the majority of tin and lead halides associate through halogen bridges, usually to give five- or six-coordinate E in polymeric species. For tin, the triphenyl halides (X = Cl, Br, I) and also Ph_2SnCl_2 are mononuclear, or only slightly distorted. In particular, the combination[36] of Ph or Mes and I is sufficiently demanding that Ph_3SnI and Mes_3SnI are close to idealized trigonal symmetry at Sn. The phenyl compound fills the lattice closely while the Mes groups are sufficiently bulky to leave gaps for chloroform or toluene solvates. Further adaptation is shown by the bending away of the o-Me groups in the plane of the phenyl rings to give (Sn)CCMe angles of 123° (compare Figure 1). Within a relatively large uncertainty, the Sn–Ph distances are normal at 212 pm while the Sn–Mes values which range from 216–218 pm in the two solvates are a little longer than Sn–o-Tol (Table 3). Sn–I is 269.9 pm in the triphenyl and 275.1 pm in the trimesityls, comparable with other simple compounds (Table 2). Similarly, $ISn(2\text{-}MeOC_6H_4)_3$ is tetrahedral[37] with Sn–I = 271.3 pm and regular trigonal angles with CSnC averaging 113° and with CSnI, 107°. Even for fluorides, bulky ligands give mononuclear species[38,39] where, in R_3SnF, values for Sn–F are 196.1 pm (R = Mes), 196.5 pm [R = $C(SiMe_2Ph)_3$]. Similarly, Mes_3SnOH^{38} and $Mes_3GeNH_2^{40}$ are monomers.

(iii) *Tied bicyclopentadienyls*. Several structures were reported where R_2Si groups tie together two cp rings in metal cyclopentadienides. Later, $Me_2Ge(C_5Me_4)_2LnX$ compounds were characterized[41] for X a $(\mu\text{-}Cl)_2Li$ unit (Ln = Sm, Lu) or a bulky $CHSiMe_3$ group (Ln = Ho). The Ge–Me lengths (195 pm) were consistently about 2 pm shorter than Ge–cp while the MeGeMe angles of 103° averaged 5° wider than the cpGecp ones.

Compared with the unbridged analogues, the centroid cpLncp angles were contracted 3–9°, some 3° less than the effect of a Me_2Si tie. See also a similar tied zirconocene[42] as a selective polymerization catalyst.

(iv) *E-boranes and -carboranes*. Group 14 atoms have been bonded to a range of polyboranes. The unusual apical substitution in square-pyramidal pentaborane was found[43] for $(1\text{-}Ph_2ClSn)B_5H_8$ with normal Sn–Ph (214.3 pm) and Sn–Cl (238.3 pm) and Sn–B = 218.9 pm. The Ph_3Sn analogue also gave spectroscopic evidence for apical substitution, but readily rearranged into the form with Ph_3Sn bridging one base edge with normal Sn–Ph and Sn–B values of 213.8 and 246.7 pm. Long Sn–C bonds (283 pm), from a terpyridyl coordinated Sn, complete a *closo*-stannadicarbaheptborane[44]. Similarly, a ferrocenylmethylamine–Pb group completes a *closo*-plumbadicarbaheptborane[45] with Pb–C = 277.4 pm, Pb–B = 253 pm and the Pb–N(ligand) bond is bent away with Pb–N = 267 pm. Analogous Ge and Sn species are similar[46]. The $B_{11}E$ icosahedron E = Ge, Sn or Pb is readily synthesized in $[B_{11}H_{11}E]^{2-}$. The structure[47] of the derivative $[B_{11}H_{11}SnMe]^-$ shows the Sn–C bond bent slightly away from the idealized five-fold axis with Sn–Me = 210.5 pm and Sn–B = 229.5 pm. In the related[48] $[B_9H_9C_2Me_2GeCl]^-$ the Ge–Cl (237 pm) is bent away and the Ge–C distances are long at 254 pm, properly seen as π-bonded dicarbollyl ligands and Ge(II) (compare Section V.B.3). Similarly, the bent-sandwich[49] in an $Sn(B_4C_2)_2$ structure shows Sn(II) (section V.C.1). Here Sn completes two *closo*-heptacarboranes and the angle is 142° at Sn (average values, Sn–C = 270 pm and Sn–B = 238 pm).

(v) *Ph$_n$E compounds*. The Ph_3E group often supports otherwise fragile structures. One case is polysulphane chains[50] as in $(Ph_3Ge)-S-S-S-(GePh_3)$ and the cyclohexyl analogue. These compounds are relatively manageable, although they do deposit S on recrystallization. The configurations at Ge are tetrahedral and propeller-like and staggered with respect to the second GeR_3 group. The Ge–Ph length may be a little short at 190(4) pm and there is a normal Ge–C of 197.1(15) pm in the cyclohexyl. This difference matches the sp^2/sp^3 difference and suggests modest lengthening in the cyclohexyl compared with methyls. The PhGePh angles open to an average 111°, indicating relief of steric strain. The Ge–S and S–S distances average 227 pm and 202 pm, respectively, which are normal within the error limits. The angles at S are relatively wide, averaging 105° for GeSS and 109° for SSS. More recently, the Si analogue has been reported[51] together with the tetrasulphane, $(Ph_3Si)-S-S-S-S-(SiPh_3)$ where the inner S–S is similar at 202.5 pm with the longer outer S–S bonds averaging 206.2 pm.

The Ge–N–N–W chain found in the complexes *trans*-$[WI(NNGeR_3)(PR_3)_4]^{52}$ (**4**) results from the novel coupling reaction of a coordinated dinitrogen molecule with R_3GeCl

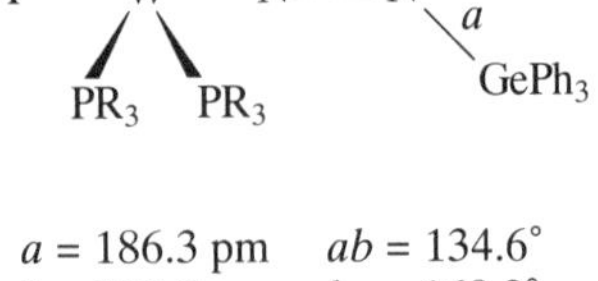

$$R_3P \diagdown \qquad \diagup PR_3$$

$$I\text{———}\overset{\diagup\;\diagdown}{W}\overset{c}{-}N\overset{b}{=\!=}N\diagdown_a$$

$$\diagup \quad \diagdown \qquad\qquad GePh_3$$

$$PR_3 \quad PR_3$$

a = 186.3 pm ab = 134.6°
b = 124.8 pm bc = 168.2°
c = 180.9 pm

(4)

in the presence of NaI. For R = Me, the product is extremely water sensitive and less stable than the Si analogues. However, the Ph analogue was much more tractable. The structure shows a normal GePh$_3$ group with bond lengths 195.0–196.7 pm and CGeC angles of 106.7–111.7°. The Ge−N length is also in the normal range for GeN single bonds. Thus attention focusses on the W−N−N unit which is formally double bonded. The NNGe angle of 135° is unusually wide, probably reflecting repulsion between the phenyl groups and the substituents on the *cis* phosphanes.

Tri- and di-phenylgermyl groups support Ge−S−CS−SR units in trithiocarbonate (trixanthate) derivatives, although these readily eliminate CS$_2$. The crystal structure[53] of Ph$_2$Ge[S$_2$C(S−i-Pr)]$_2$ shows expected Ge-Ph parameters (GeC = 193.3 pm, CGeC = 118.1°), an acute SGeS angle of 87.3° and slightly long Ge−S at 228.1 pm. These values are all a little more extreme than found[54] in the corresponding dithiocarbonates (xanthate) Ph$_2$Ge[S$_2$C(O−i-Pr)]$_2$ where SGeS = 103.2°, CGeC = 115.6° and GeS = 225.1 pm, while the smaller OMe analogue has a more acute SGeS at 93.4°, with CGeC = 117.1° and GeS = 226.2 pm. In these and similar structures of mono-, di-, tri- and tetra-thiocarbonate and other sulphur–germanium derivatives[55], there is no indication of expansion of the germanium coordination above 4, in contrast to the tin analogues where the remote O or S bonds back to give six-coordination. Similarly, Ph$_3$SnSC$_6$H$_4$(2-NH$_2$) is tetrahedral[56] despite the potential N donor (Sn−C = 211.7 pm, Sn−S = 243.6 pm).

Formation of 1,3,2-diazagermine was identified[57] by a crystal study of the novel eight-membered 1,5,2-diazagermocine ring which resulted from acetylene insertion.

(vi) *Bulky groups in β positions.* While bulky groups directly bonded to E have substantial and often exciting effects, E−CH$_2$−R or similar links suffice to reduce steric effects at E to a minimum, while the overall structure will optimize packing of the R groups. A simple example[58] is EtGe(SePh)$_3$ where Ge-C = 195.2 pm and two GeSe bonds are slightly longer (235.5 pm) and enclose a smaller angle (98.85°) than the third (234.8 pm and 107–112°).

In (ArCMe$_2$CH$_2$)$_3$PbBr[59] where Ar is the bulky 3,5-di-t-butylphenyl, the three −CH$_2$CMe$_2$Ar units are in a 'Manx legs' or pinwheel arrangement which minimizes interactions. The crystal was poor, but within larger error limits the configuration at Pb is tetrahedral with Pb−C = 216(5) pm and Pb−Br = 265.6 pm.

A further example results from a diene cyclization in the presence of Ph$_3$SnH, where Sn−H addition followed by cyclization yielded[60] Ph$_3$SnR where R is a *cis* fused 2-oxabicyclo[3.3.0]octan-7-ol unit linked to Ph$_3$Sn by a CH$_2$ group (5). The Sn−Ph distances are normal, averaging 213.3 pm, while Sn−CH$_2$ is slightly long at 216.6 pm. The angles at Sn vary from 104° to 116° with the PhSnPh angles averaging 3° less than the PhSn−CH$_2$ ones. The triphenyltin unit lies in the concave face of the bicycle which, though larger than Ph, is more distant by one bond, so the steric effect at tin is minor.

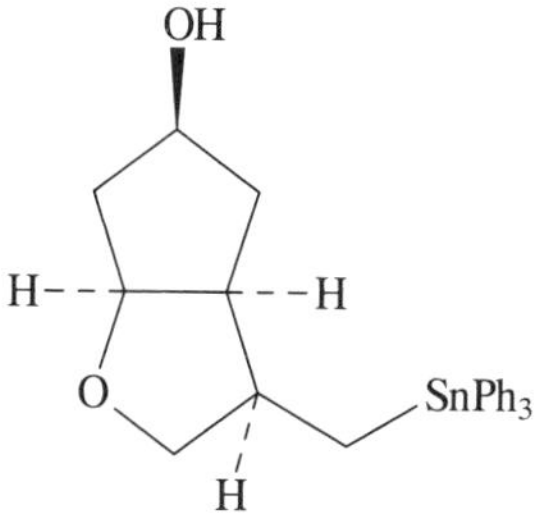

(5)

Two benzyl and a C_3 chain ligand allow one bulky tms group to be accommodated in a fairly regular tetrahedron in $(PhCH_2)_2Sn(tms)R$ [R $=$ $CH_2CH-C(SiMe_3)_2$][61] where Sn–C values are all in the same range at 224 pm (tms), 219 pm (R), 224 and 221 pm (benzyl). Similarly, angles range from $106.4-114.5°$. Replacing L by $OSiMe_3$ allows Sn–C to shorten a little to 217 pm and Sn–O is 193 pm.

An attempt[62] to prepare a π-bonded product using pentaphenylcyclopentadienyl yielded instead the sigma-bonded, bulky ligand product $Ph_5C_5GeCl_3$ with normal values for GeC $=$ 197.5 pm and GeCl $=$ 213.1 pm.

(vii) *Oxy-E chain.* Structures are well established for Group 14 element–oxygen compounds which are cyclic, multi-ring or cage, but simple chain examples are rare. In $Ph_3Sn-O-SiPh_2-O-SiPh_2-O-SnPh_3$, the Sn–Ph and Si–Ph lengths[63] are normal at 213 and 189 pm while the PhSnPh angles range from 111.3 to $114.4°$. Sn–O is 194.5 pm and Si–O averages 160.5 pm. The SiOSi angles ($165.4°$) are at the high end of the range of observed values while the SnOSi values of $140.0°$ and $142.7°$ are similar to the $144.2°$ of $Ph_3SnOSiPh_3$.

III. UNSTRAINED RINGS CONTAINING FOUR-COORDINATE ISOLATED E

Three-membered rings which are highly strained usually derive from studies on divalent E or E=E species and are discussed separately. $(RR'EX)_n$ rings X $=$ O,S,Se,Te make up a common class of four-coordinate E species. For oxides, while rings as large as 16 members ($n = 8$) are known for Si, only smaller molecules have so far been characterized for Ge and Sn with 6 and 8 ($n = 3,4$) as the commonest sizes for larger rings. There are also many examples of 4-rings ($n = 2$). Bulky substituents allow isolation of such molecules whereas the common products with small ligands are isoluble polymers.

A. Rings Formed by Hydrogen Bonding

Many 4-valent E compounds are found as cyclic compounds. The least perturbation of the geometry at E is probably found in the case[64] of Ph_3GeOH which forms a cyclic tetramer by OH...O hydrogen bonding. Here, there are four Ph_3GeOH units in the cell and the Ge atoms are tetrahedral with Ge–C $=$ 191.4–195.5 pm and Ge–O averaging 179.1 pm, respectively somewhat shorter and longer than in simpler analogues. The four O atoms form a flattened tetrahedron with H-bonded contacts ranging from 260.4 pm to 265.7 pm. For the other elements E, Ph_3SiOH is isomorphous with the Ge compound while Ph_3SnOH and Ph_3PbOH have long-established zigzag chain structures with planar EPh_3 groups linked by OH with a trigonal bipyramidal coordination at E.

Bulky substituents are necessary to allow isolation of a dihydroxide. In t-$Bu_2Ge(OH)_2$ the Ge[65] is tetrahedral, distorted by a wide CGeC angle of $122.5°$ reflecting the bulky groups and with $OGeO = 103.5°$. The Ge–O distance of 178.0 pm is normal and Ge–C at 196.9 pm is quite short for a t-butyl. The molecules are linked into a chain by hydrogen bonds alternatively 270 pm and 283 pm and the chains are further linked into a chain of rings by 273 pm hydrogen bonds. Dehydration gives a trimeric cyclic oxide.

B. Six-ring Compounds, $(RR'EY)_3$, Y $=$ O, S, Se, Te and Others

The following set of oxides provides a useful series of six-rings for comparison of substituent effects. Where R $=$ R' and is bulky, six-membered rings are planar (**6a**) as in $(t$-$Bu_2GeO)_3$ formed above[65] and the Sn analogues[66] with R $=$ t-Bu or t-amyl. The stannylene $Ar(f)_2Sn$, discussed later, readily forms a similar cyclic oxide[67] and use of a

similar bulky aryl [Ar = Dep] earlier[68] allowed the isolation of a trimeric oxide en route to the tristannane, also discussed later.
Parameters may be summarized as follows:

ER_2	$GeBu_2{}^t$	$SnBu_2{}^t$	$SnAm_2{}^t$	$SnAr(f)_2$	$SnDep_2$
E–C (pm)	200.5	219.4	227	220.3	213.8/216.9
E–O (pm)	178.1	196.5	196.1	193.1	194.5
CEC (deg)	114.4	119.5	115/121	110.1	
OEO (deg)	107.0	106.9	106.1	104.3	
EOE (deg)	133.0	133.1	134.1	135.5	136.2

Values are normal allowing for the bulky groups and the effect of the CF_3 substituents, and the angles at O are interestingly similar. In one example with two different substituents on Sn, $[Me(Tsi)SnO]_3$, the ring is non-planar[69] with one Sn and the *trans* O displaced from the plane of the other four in a flattened boat. The Sn–Me values average 215 pm while Sn–Tsi was very variable. Sn–O and OSnO values matched the planar rings, but the angles reversed in size with OSnO at 105° and two CSnC at 118° with the third at 111°.

The trimeric rings, $(R_2E)_3Y_3$, involving Y = S, Se or Te, exist for most combinations and for a range of R, even R = $MeO_2CCH_2CH_2$[70], although bulky R favours cyclic dimers (e.g. R = *t*-Bu — see below) or chains (R = *i*-Pr). The common form of these E_3Y_3 rings is a twisted boat configuration with a C_2 axis passing through one E and the opposite Y (**6b**). For $[Me_2SnS]_3$ in the tetragonal form[71], the bond to the unique Sn, which has no close intermolecular contacts, is slightly shorter [240.5(3) pm compared with 242.2(3) pm] than those to the two other Sn atoms which have external contacts in the range 388–410 pm.

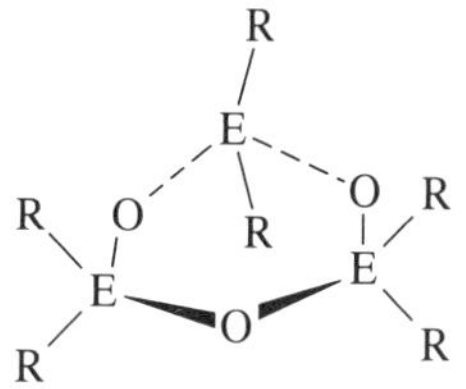

(6a) 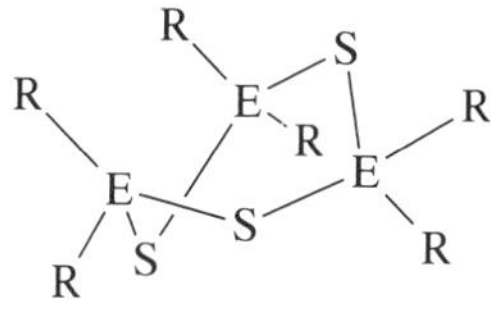   (6b)

With the larger Y = Te, such contacts disappear and the two crystal modifications[72] of $[Me_2SnTe]_3$ and for the analogue[73] $[(Me_3SiCH_2)_2SnTe]_3$ show all Sn–Te matching at 272.8–274.1 pm. These rings have angles at Te around 97° while the sulphur rings show wider angles averaging 103°.

With larger ligands, as for $[Ph_2ES]_3$ with E = Sn or Pb[74], there is no indication of additional contacts and the E–Ph and E–S parameters are within the normal range. NMR studies show exchange between Sn and Pb groups in solution, giving the possible mixed-E rings.

Other six-rings. A different six-ring is near-planar $-SnR_2SSnR_2NSN-$ (R = *t*-Bu) with Sn–R = 216 pm, RSnR = 120° and ring distances of Sn–S = 241 pm, Sn–N = 204 pm and SN = 151 pm. A second product[75] had two five-rings $-SnSNSN-$, longer bonds (SnC = 221 pm, SnN = 215 pm, SnS = 262 pm) linked by a SnSNSN four-ring (SnN = 233 pm).

A mixed-member six-ring, $-GeMe_2SeCH_2CMe_2CH_2Se-$, has a symmetrical twisted boat configuration[76] with $GeSe = 235$ pm, $GeMe = 196$ pm, $CGeC = 111.5°$ and $SeGeSe = 103.2°$.

A reaction of $[Me_2GeS]_3$ with a Zr-Ph[77], probably via an insertion of $Me_2Ge=S$, gives $Cp_2ZrSGeMe_2(o\text{-}C_6H_4)$ with a $-ZrSGeCC-$ ring $[Ge-C = 195.8$ pm (ring), 194.4 pm (Me), $Ge-S = 221.7$ pm].

C. Eight-ring Oxygen Compounds

A wide range of eight-membered rings involving Si is known but those with heavier E are rarer. A 1994 paper[78] presents a classification of eight-ring structural types. The structures of 42 rings had been reported by that date with Si or Ge and O, N or C, and these fell into 9 types ranging from essentially planar through species with 1,2,3 and 4 atoms out of the plane of the rest to a fully irregular case. However, only the two structures of Figure 2 were common, the 'extended chair' and the tub. This paper also reports the structure of the mixed-E eight-ring $[R_2GeOR'_2SiO]_2$ ($R = Et$, $R' = Ph$) which is 'extended chair' and isotypical with the analogue where $R = Me$. This was characterized earlier[79] along with the molecule with $R = R' = Ph$ where the Si and Ge are disordered over the ring positions. The parameters of the Me and Et analogues are very similar with Ge-C relatively short at 191 pm average and other values normal. The angles at Ge differ substantially with $OGeO = 104.6°$ and $CGeC = 118.5°$ while those at Si are closer at $CSiC = 109.9°$ and $OSiO = 112.9°$. The angles at O range from $136°$ to $143°$. A similar picture emerges from the values in the perphenyl analogue. The latter parameters also correlate with the earlier study[80] of $[Ph_2GeO]_4$ — an eight-ring with $E = E' = Ge$ which adopts the tub configuration.

The eight-ring[81] with $R = Cl$, $R' = t\text{-}Bu$ resulting from the reaction of $GeCl_4$ with $t\text{-}Bu_2Si(OH)_2$ may also be noted. This ring is planar and the angles at Ge are closer to tetrahedral with $ClGeCl = 108°$ and $OGeO = 112°$; GeO (169.8 pm) and GeCl (209.8 pm) are short. The most striking feature is that angles at one pair of opposite oxygens are nearly linear (174.8°) while the other GeOSi values are 142.9°. When $SnCl_4$ was reacted with the silane, the analogous eight-ring did not form but instead a bicycloheptane with 3O and 3Cl atoms forming a distorted octahedron at Sn.

A further interesting structure[82] combines an eight-ring with a germacyclohexane in the cyclic tetramer $[(C_5GeH_{11})O]_4$. The ring is in the tub configuration with normal Ge-C (192.9-194.7 pm) and Ge-O (176.5-178.4 pm) while the GeOGe angles average 125.8°. There are two slightly different chair configurations for the GeC_5 rings with larger CCC angles (113.2°-116.0°) and reduced $CGeC$ (105.1°-107.0°).

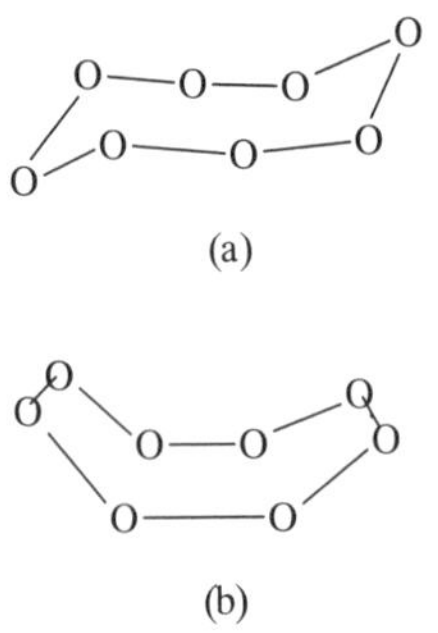

(a)

(b)

FIGURE 2. Common shapes for E_4O_4 rings

D. Seven-membered Rings

By reacting a digermane, it was possible to prepare[83] the novel octaphenyl-1,2-digerma-4,6-disila-3,5,7-trioxa seven-membered ring analogous to the eight-rings. The ring has a flattened twisted chair conformation with normal values for GeO (177.7 pm), SiO (161.5 pm) and OSiO (111.3°). The two Ge–C bonds at each Ge differ a little [193.9(2) and 195.5(2)] while CGeC and OGeGe are close to tetrahedral. The GeOSi angles are a little wider than in the eight-ring at 144.4° while SiOSi at 157.9° is within the range found in Si_4O_4 rings.

A seven-ring containing only one E atom, C_6Sn, in dichlorodihydrodibenzostannepine[84], shows normal Sn–C (210.2 pm) and Sn–Cl (234 pm) and the large ring allows near-tetrahedral coordination with angles at Sn in the range 103°–119°.

E. Four-membered Rings, [RR′EY]₂, Y = O, S, Se, Te

The symmetrical −EYEY− ring is common, while a few cases of the −EEYY− iso-meric form are also known. In one example[85] reaction of the digermenes $Ar_2Ge=GeAr_2$ (Ar = Dep or Dip) with oxygen adds O–O across the double bond giving the unsymmet-ric ring, and this product photolyses to the GeOGeO isomer which can also be formed from the digeroxirane. The structures of the Dep species show normal Ge–Ar distances (195.1–197.0 pm) and the aryl groups are arranged roughly helically about each Ge. In the unsymmetric compound, a two-fold axis bisects Ge–Ge (length 244.1 pm) and O–O (147 pm) and the trapezoidal ring is twisted from planarity with a Ge–O–O–Ge torsion angle of 19.5° (GeGeO = 74.1°, GeOO = 103.9°). The symmetric ring is a slightly puck-ered square (dihedral angles 8.5°) with GeO = 181.7 pm, OGeO = 87.6° and GeOGe = 92.1°. A similar pair of isomers is known for Si. Related five- and six-membered rings[86] with similar parameters were produced by similar chemistry with folded structures and formulae, −$Ar_2GeOOGeAr_2CH_2$− and −$Ar_2GeOOGeAr_2CH=CPh$− (Ar = Dep).

The four-membered [R_2EY]₂ ring is common for heavier Y and with bulky R. Members of the [Bu_2^tSnY]₂ series (Y = S,Se,Te) have been known to be cyclic dimers since 1975, and Ge analogues a little later[87]. The rings are planar[88] (7) with Sn–C and the CSnC angle normal for t-Bu substituents.

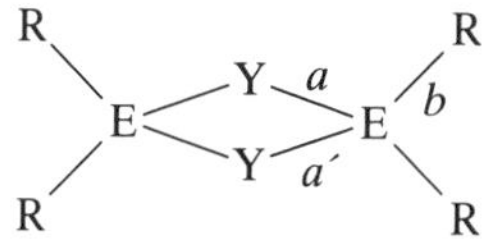

E = Sn		
Y = S	Se	Te
a = 243	255	276
b = 221	218	220
aa = 86°	83°	80°
bb = 117°	115°	117°
$aa′$ = 94°	98°	100°

(7)

A further interaction with a donor within the ligand may also favour the four-membered ring, as in [MeN(CH₂CH₂CH₂)₂SnS]₂ (8) where[89] there is a central, rectangular planar [SnS]₂ unit with a long Sn-N bond and the coordination at tin is nearer 5 than 4 in a distorted trigonal bipyramid.

A similar case[90], this time with a rectangular distortion of the [SnS]₂ geometry, is caused by weak coordination of 2N at each Sn in [(Me₂NCH₂CH₂CH₂)₂Sn(Ph)S]₂ to give irregular six-coordination at Sn. The two Sn–S distances are 244 and 248 pm and the two Sn...N values are 281 pm (*trans* to the shorter Sn–S) and 316 pm. The CSnC angle is 137° and SSnS is 93°.

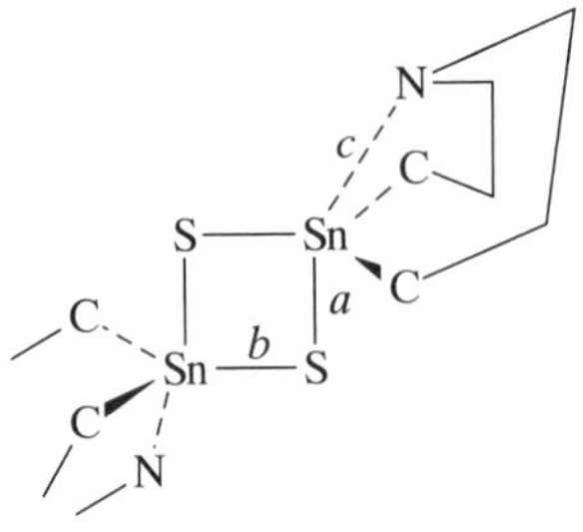

$a = 252$ pm
$b = 239$ pm
$c = 255$ pm
$ab = 87°$
$CSnC = 120°$

(8)

A fuller NMR study[91] of exchange in $[Ar_2ES]_2$, for $E = Sn$, Pb, demonstrated redistribution over combinations of Ar and Sn or Pb. When the four-membered $[Bu^t_2SnS]_2$ was included in the exchange, mixed element products were identified with both four- and six-rings, such as $[(Ph_2PbS)(Bu^t_2SnS)_x]$ for $x = 1$ or 2.

Since the 1980s, the focus on bulky ligands, ER_2 compounds and double-bonded species has led to syntheses of four-membered rings as dimers of divalent entities. The stannene $(Tip)_2Sn=CR_2$ (CR_2 = fluorenylidene) slowly dimerises to give a planar 1,3-distannacyclobutane ring[92] with long Sn–C bonds (229.0 pm), angles at Sn of 88° and at C of 92°. The Sn–Tip bonds (221 pm) are normal for a large ligand. These ring bonds compare with values[93] around 219 pm in less-hindered stanncyclobutanes.

The reaction[94] of $E(NR_2)_2$ with $HSCR_3$ [$R = C(SiMe_3)_3$] gave a number of rearranged species including both isomers of $[(R_2N)(PhCH_2)GeS]_2$ and $[Pb(NR_2)\mu\text{-}SCR_3]_2$ from $E = Ge$, Pb, respectively. The structure of the *cis* isomer showed a buckled $[GeS]_2$ ring with average parameters: $GeS = 224.4$ pm, $GeC = 196$ pm, $GeN = 182$ pm and ring angles of 84.2° at S and 95.5° at Ge, with planar N. The Pb_2S_2 ring was quite distorted with PbS distances of 288 and 274 pm, PbN values of 207 and 235 pm, and ring angles of 76.3 and 81.2° at the lead atoms and 90.0 at S. Although there is no direct Ge–C bond, it is worth noting the $[(R_2N)_2GeTe]_2$ ring which was produced in a continuation of this study[95] with Ge–Te averaging 259.6 pm and ring angles of 85.6° at Te and 94.5° at Ge.

In a closer study of potential doubly-bonded Sn=Y compounds, structural characterization[96] was reported of a number of these captured as dimers. The study used the bulky, but disk-like, Mes or Tip group as one ligand and the more spherical Tb as the second ligand on Sn. Isomers with *cis* and *trans* configurations of these ligands across the $[SnY]_2$ ring were characterized. In the *trans* configuration, the balance of steric effects gave a completely

TABLE 4.   Parameters (pm and deg) for *cis* or *trans* $[ArTbSnY]_2$ compounds[a]

Parameters	*trans*	*trans*	*trans*	*cis*	*cis*
Ar/Y	Mes/S	Mes/Se	Tip/Se	Mes/S	Mes/Se
Sn–C(Tb)	219	219.2	219	219.2	218.8
Sn–C(Ar)	218	218	221	218.5	217.3
Sn–Y	243.3	256	256.7	243.8/246.3	255.7/258.6
ArSnTb	114.1	113.3	125.8/111.1	114.2	114.3
YSnY	90.84	91.70	90.3	88.33	89.09
SnYSn	89.1	88.3	89.35	84.34	82.88
Fold angles					
about Sn...Sn	0	0	8.95	40.9	41.3
about Y...Y	0	0	9.02	39.8	43.6

[a] Data from Reference 96.

planar, nearly square ring while the *cis* isomers were distinctly buckled (Table 4). In the same paper, the Sn=Y was also captured by adding across an alkyne to give $-SnC=CY-$ four-membered rings whose structures were determined for the Tb(Tip)Sn species. The Sn/aryl parameters were similar to those in Table 4, with ArSnAr 122.9° and 121.3° in the S and Se rings, respectively. The rings were slightly folded by 7.3° (S) and 10.7° (Se) and other parameters were Sn–S = 265.1, SC = 173, CC = 133 and SnC = 217 pm with angles at the ring atoms of 66.5° (SSnC), 73.1° (SnSC), 121° (SCC) and 98.6° (SnCC) in the S compound and Sn–Se = 274.6, SeC = 200, CC = 133 and SnC = 230 pm, 74.3° (SSnC), 64.3° (SnSeC), 134° (SeCC) and 87° (SnCC) in the Se ring.

F. Four-membered Rings with Y = Group 15 Atom

Addition of the germene, $Mes_2 Ge=CR_2$ (CR_2 = fluorenylidene), to a diazo compound gave[97] the unusual four-membered digermazane ring (9) in place of the expected three-membered one. The angles across the non-planar ring diagonals are 34° across N⋯N and 38° across Ge⋯Ge. The Ge–N and Ge–C distances are normal for large substituent products.

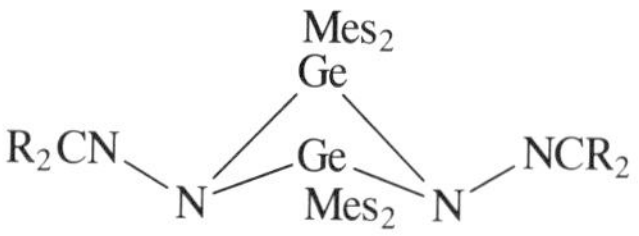

GeN = 189.5 pm
GeMes = 197 pm
GeNGe = 94.5°
NGeN = 79.6°

(9)

A similar SnNSnN ring but with further exocyclic $N-SnR_x$ units (10) was found in stannylamine studies[98]. The ring fold angle is 28° and there is a very weak interaction between exocyclic Cl and ring Sn. The exocyclic Sn are in the distorted trigonal bipyramidal configuration commonly found for five-coordination (see below).

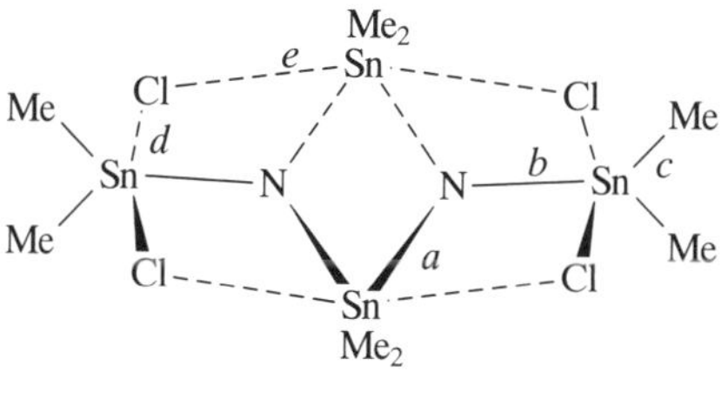

a = 215.6–217.9 pm
b = 208.8–212.2 pm
c = 212 pm
d = 255.7–261.5 pm
dd = 162.4°–165.0°
e = 313 pm

(10)

The *t*-Bu groups lie nicely in square planar $[Bu^t_2SnPH]_2$[99] with regular Sn–P (254.4 pm), normally extended Sn–C (220.1 pm), CSnC = 114° and ring angles of 82° at P and 98° at Sn. $[Bu^t(Mes)GePH]_2$ is formed with related three-membered rings[100] (Section VII.E) and has similar geometry (Ge–P = 234.6 pm, Ge–C = 200.8 pm, Ge–Mes = 198.0 pm, angle at P = 86.8°, at Ge = 62.1°).

Starting from a diarsanylsilane, the four-ring, $-R_2SiAsHGeR'_2AsH-$, was synthesized and converted with loss of H_2 and formation of As–As to two EAs_2 rings sharing the

As—As edge, similarly to the phosphorus system. The cyclo-[(Mes)(t-Bu)GeAsH(Tip)(t-Bu)SiAsH] ring[101] is buckled at 23.5° with the two t-Bu groups trans: GeAs = 245.5 pm, SiAs = 239.3 pm, ring angles 84.0° (As), 92.2° (Ge) and 95.3° (Si).

G. More Complex Ring Species

Other interesting four-membered rings include the —GeCCO— rings with various substituents on C which result from the addition of C=O groups across Ge=C in the germene $Mes_3Ge=CR_2$ where CR_2 = fluorenylidene[102]. The rings are folded with angles at Ge of 75° and Ge–O = 183 pm, Ge–C = 203–207 pm while Ge–Mes = 196–204 pm.

The GeCNS ring resulted similarly[103] from a dichlorogermylene reaction with an adamantylimidethione. In the ring, angles are close to 90°, Ge–C = 200.8 pm and Ge–N = 181.3 pm.

The unusual GeNBN ring in $Me_2Ge(NBu^t)_2BCl$[104] has a ring angle at Ge of 75° and normal Ge–C (192.8 pm) and GeN (186.3 pm).

The five-membered SnOSnC=C ring (11b) is formed[105] by the oxidation of the Sn—Sn bond of the SnSnCC ring (11a) discussed later.

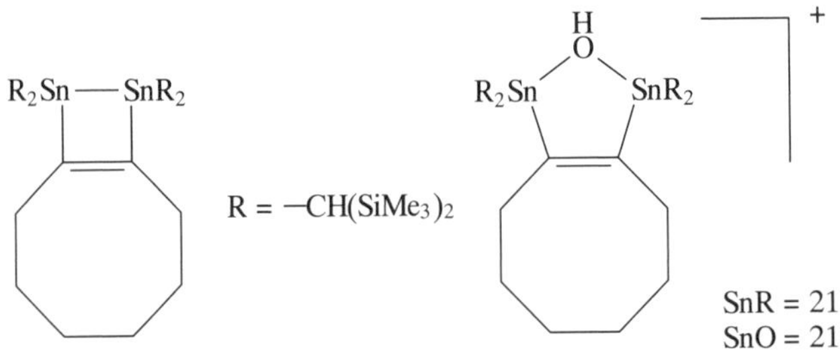

R_2Sn —— SnR_2

R = —CH(SiMe₃)₂

$$R = —CH(SiMe_3)_2$$

SnR = 215.5-216.8 pm
SnO = 210.4
SnC(ring) = 216.6 pm
C=C = 134.7 pm

(11a)   (11b)

Five-membered EY_4 rings have been established[106] for E = Si, Ge, Sn and Y = S or Se. They are supported by very large ligands on E including Mes and Tb (= 2, 4, 6 − {(Me₃Si)₂CH}₃C₆H₂) and formation is more easy from Si to Ge to Sn. Similar structures are reported for $Tb(Mes)EY_4$ (for E = Sn and Y = S,Se; E = Ge, Y = Se and E = Si, Y = S) with a distorted half-chair configuration for the EY_4 ring. There are small variations in Y−Y in the rings which are also slightly asymmetrically bonded to E (e.g. SnSe = 254.9 and 260.3 pm; SnS = 243.8 and 248.1 pm; GeSe = 240.9 and 242.1 pm; SiS = 215.5 and 222.4 pm). Ge−Tb is 6 pm longer than a normal Ge−Mes of 195 pm but the Sn−C values show little variation.

Reaction of dimethylgermylene with a 1,3-diene gives a germacyclopentane with 3,4-alkene substituents[107]. Lengths and angles are normal with Ge–C(ring) = 195.0 pm, the ring angle at Ge=92°, Ge–Me = 192.9 pm, and MeGeMe = 113°. A diphenylgermanium in the 3-position of a bicyclo-octane has less constricted geometry[108] [Ge–Ph = 195 pm, Ge–C(ring) = 194 pm, ring angle at Ge = 103°, PhGePh = 108.4°].

A spiro-Sn completing two stannacyclopentadiene rings[109] has twisted ring structures due to the bulky substituents; Sn–C 215–216.5 pm and the ring angles at Sn are 85° and the non-ring ones, 117°.

The earlier structures[1,2,3,5] of sesquisulphides and selenides, $(RE)_4X_6$, were of the adamantane or P_4O_6 type. The alternative arrangement was found for RE = t-BuGe,

X = S, when mild reaction conditions were used. $(Bu^tGe)_4S_6$ has a structure[110] with two parallel and nearly planar Ge_2S_2 four-rings linked by Ge–S–Ge bridges into a complex eight-membered ring. The molecule is of high, D_{2h}, symmetry and converts into the adamantane isomer on heating. The ring GeS distances are marginally longer at 224.3(8) pm than the bridge ones [221.6(3) pm]; Ge–C = 200 pm and the ring angles are 83.2° at S and 96.7° at Ge. The GeSGe angle for the bridges between the rings is 108.5°.

Rings involving partly halogenated Ge give open, linked-ring, structures[87]. In $(RGe)_6O_8Cl_2$ (R = t-Bu) an eight-ring $(RGeO)_4$ has two pairs of GeOGe further bridged by OGe(RCl)O, forming two six-rings sharing a common GeOGe with the eight-ring. Bond lengths are in normal ranges (Ge–C = 194–196 pm; Ge–O = 173–177 pm; Ge–Cl = 215.4 pm). The GeOGe angles in the six-ring are 128.9°–130.5° and all the remaining angles are close to tetrahedral. With Y = S or Se, $(RGe)_4Y_5Cl_2$ is formed which is two six rings sharing a common RGeYRGe bridge with the Cl on the non-bridging Ge. Geometries at Ge are similar with slightly lengthened GeC (196–200 pm) and GeCl (217.8 pm) and bonds to Y varying with substitution: –RGe-S = 222.5 pm and RClGe–S = 221.3 pm and the corresponding Ge–Se values, 235.5 pm and 233.7 pm.

Hydrolysis of the trichloride yields $(RGe)_6O_9$ (R = t-Bu, i-Pr, Mes, C_6H_{11}) which have a cage structure[87,111] where two $(GeO)_3$ six-rings are linked by GeOGe bonds creating three additional eight-rings. All the Ge–O bonds fall in the range 171–179 pm, average 175.6 pm, and Ge–C ranges from 190.2 pm to 194.8 pm.

IV. ELEMENT(IV) SPECIES WITH COORDINATION NUMBERS ABOVE FOUR

Even with bulky ligands, there is always the possibility of an additional action when a potential donor atom is present in a suitable position in a substituent. We find a continuum from non-coordination, as for the remote S or O atoms in the xanthates and thioxanthates above, through weak interaction as seen in some of the ring compounds, to unequal interactions giving irregular 5- or 6- coordination, and finally equal interactions giving regular structures. Higher coordination numbers may also be achieved by use of lone pairs on directly bonded ligands, as in the well-established polymeric structures of organotin halides. Lewis acidity or acceptor power clearly increases with the size of E and with the number of electronegative substituents. Thus expanded coordination is least likely with Ge and with four E–C bonds.

While Si and Ge have a significant 5- and 6-coordinate chemistry in their organo compounds, the steep increase in size from Ge to Sn allows a more extensive chemistry for Sn, including both the II and the IV state and extending to coordination numbers above 6. For lead, high coordination chemistry is also developed with ligands such as bi- or polydenate O, N or S species but very limited by the paucity of organo-lead compounds. Non-organic complex chemistry often provides a valuable indication of potential structures[112].

Interest in coordination numbers above four stems in part from the idea that an $(n + 1)$ coordination, involving one or more donor atoms, acts as a model for the reaction intermediate for attack on an n-coordinate centre.

A. Tin in Coordination Numbers Above Four

Since the dominant higher-coordination chemistry is that of tin, we survey this first. The range of structural studies is so wide that it is possible only to generalize and look at recent and representative examples. Intramolecular coordination in organotin compounds has been extensively reviewed up to 1992[113].

As the number of organic groups bonded to tin increases, the acceptor power decreases. Thus we find the frequency of different coordination numbers varies broadly as follows:

Number of Sn—C	Coordination frequency
4	$4 > 5 \gg 6$
3	$5 > 4 > 6$
2	$6 \approx 5 > 4 > 7, 8$
1	$6 > 5 > 4 \approx 7, 8$

For five-coordination, the usual geometry is trigonal bipyramidal. This varies from fully symmetrical, as in Ph_3SnF, to extremely distorted. At the limit the shape can equally well be described as capped tetrahedral or distorted square pyramid.

The bonds to the axial positions in these five-coordinate species are distinctly longer and weaker than the equatorial bonds, and often differ significantly in length, even to formally equivalent atoms. At the limit, the distinction between a long fifth axial bond and a slightly perturbed tetrahedron becomes arbitrary. Where the three equatorial substituents are the same and reasonably symmetrical, any bonding interaction shows up more clearly in the equatorial angles which increase towards 120°. The axial positions are normally occupied by the most electronegative substituents. An assessment of normal axial bond lengths for such substituents is given in Table 5, to act as a basis for judging interactions.

TABLE 5.   Representative tin—element bond lengths in different coordinations (pm)

Bond	4 or 5 (equatorial)	5 (axial) or 6	Sum of van der Waals radii
Sn—O	194	212	360
Sn—N	204	215	365
Sn—Cl	235	245	385
Sn—S	243	250	390
Sn—Br	249	265	395
Sn—I	268	283	410

In interpreting such values, both the experimental uncertainties and the empirical basis of van der Waals radii need to be taken into account.

Apart from mononuclear species (e.g. **14, 32b, 33**), five-coordination is found in chain polymers (as Figures 3 and 4) where divalent atoms, usually halogen or oxygen, or else bidentate ligands, bridge neighbouring Sn. Cyclic dimers, trimers, and larger species up to hexamers have also been seen (e.g. **19, 23, 25, 26, 29**). Bidentate ligands also form a variety of ring sizes (commonly 4-membered **22b** or 5-membered **15, 16** and sometimes 6-membered), asymmetric bonds are common and the ligand bite angles also contribute to the distortions. Some examples are established of tridentate and more complex ligands (Figure 6; see later). Where bonding is to C and an electronegative atom, the Sn—C bonds are equatorial as far as possible (**14**). In a few cases, alternative donor atoms are available: the usual preference is for $O > N > S$.

Most of these comments also apply to octahedral complexes, which tend to show longer bonds also. Again, there are examples ranging from near-regular to shapes better described as distorted tetrahedral with two long interactions.

1. Tetraorgano compounds

Although the tin atom is a poorer acceptor than in the presence of more electronegative substituents, a number of higher-coordinate tetraorganotin compounds are now established where a donor group is present in a suitable position.

We note first the interesting structure[114] of $Me_2Sn(CPr^i=CEtBEt_2C\equiv CPr^i)$ **(12)** which contains Sn formally 5-coordinate to carbon — alternatively 4-coordinate with the $C\equiv C$ triple-bonded unit bonded sideways on to Sn in a π-ethyne mode. The coordination at Sn can be envisaged as a pyramidal Me_2CSn^+ ion (angles $114°-119°$) with the Sn apex pointing at the triple bond with the Sn–alkyne distances 15% longer than Sn–C sigma bonds, reflecting the π interaction. The geometry remains close to the free alkyne, and the angles at the C atoms $170°-175°$. Note also the closely related $Sn[C(R)=C(R)B(Et_2)C\equiv CR]_2$[115] where Sn is normally bonded to two C (Sn–C = 212 pm) and coordinated asymmetrically to two alkyne groups (Sn–C 235.3 and 251.2 pm), with the two Sn–C and the midpoints of the triple bonds lying tetrahedrally and the whole giving tin 6-coordinated to C!

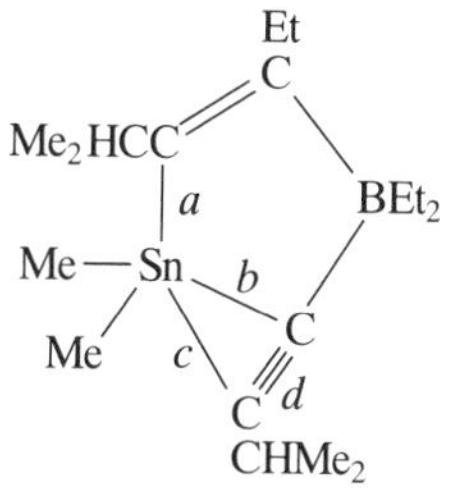

(12)

A further study[116] incorporating a donor atom in the group bonded to $C^{(1)}$ produced structures for three C_5SnD species [QD = $CNMe_2$ **(13a,b)** or C=COMe **(13c)** with varying substituents R,R$'$]. In these molecules, the sigma donor and the pi effects compete, and varying flexibility is present. The data (Table 6) show retention of the basic pyramidal Me_2SnC geometry but significant changes in the Sn–alkyne interactions. For all three cases, the tin–alkyne interaction weakens with a further 5% increase in the distances: for the two N donors, the SnC distances become more nearly equal but with the less-constrained O donor the difference in b and c is very large. These papers also provide spectroscopic data illustrating the sigma/pi competition.

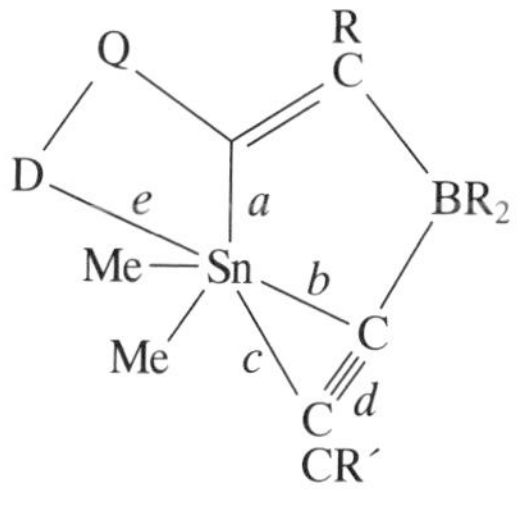

(13)

(a)	**(b)**	**(c)**
QD = CH_2NMe_2	QD = CH_2NMe_2	QD = CH=CHOMe
R = Et	R = i-Pr	R = i-Pr
R$'$ = NMe_2	R$'$ = NMe_2	R$'$ = CH=CHOMe

TABLE 6.   Data (pm and deg) for π-bonded tin alkynyl species

	12	13a	13b	13c
Sn–Me	213.2	213.8	213.3	211.2
MeSnMe	113.9	111.8	112.5	113.6
a	211.6	210.4	211.4	212.7
b	233.9	262.6	255.4	237.3
c	252.3	260.4	258.9	266.1
d	121.3	122.0	121.2	122.1
e		252.1	248.1	259.2
bc	28.6	27.0	27.3	27.3
B–C≡C	170.1	177.8	174.9	163.5
C≡C–C	174.0	171.0	169.6	179.2
ae		61.8	62.1	74.0

The other reported R_4SnD structures **14a, b, c** all have O or N donor atoms as part of a rigid unit also bonded through a Sn–C bond and with the donor atom occupying one of the apex positions in a trigonal bipyramid while the Sn–C is equatorial. Three alkyls or aryls occupy the remaining positions. We include in this group an example, with an optically active R′ where the R″ group is replaced by H. A final variant has a diptych structure with an axial N bonded via two equatorial C–Sn.

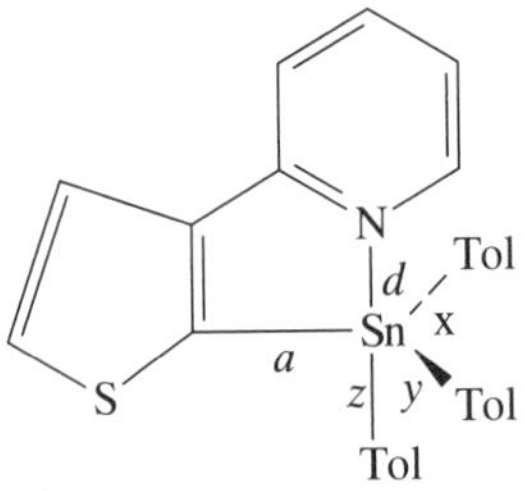

(a)

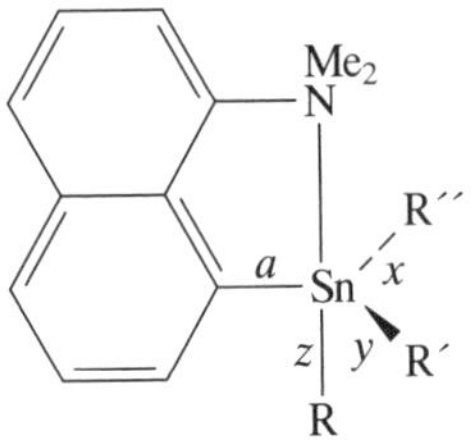

(b) R = Ph = R′ = R″
(f) R = Me, R′ = (−)-Menthyl
     R″ = H

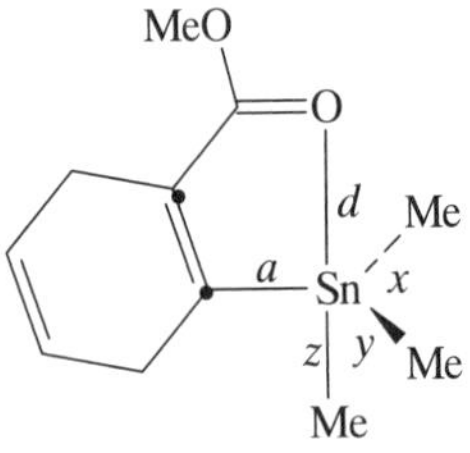

(c) 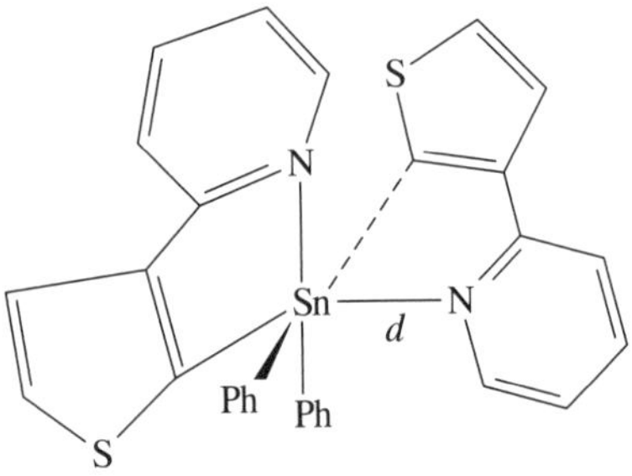   (g)

(14)

Data for these R_4SnD species are collected in Table 7, which also includes the one example of six-coordinate R_4SnD_2 (**14g**). Note that **d** in Table 7 is Ph_3Sn bonded to OH in a similar environment to **c** and **e** is the MeSn diptych (neither is illustrated).

TABLE 7.   Structural parameters (pm and deg) for R_4SnD compounds (14)

Compound	a	$x = y$	z	d^a	xy	$xz = yz$	ad	dz
a[118]	215.3	214.5	218.3	284.1	113.7	103.8	65.8	166.6
b[119] b	213.7	214.8	218.3	288.4	114.4	102.1	68.8	168.5
c[120]	217.7	213.6	215.0	271.8(0)	113.1	105.4	68.8	172.4
d[121]	213	214	219	277(0)	105	104	67	168
e[122]	215.1 –217.4	d	221.4	262.4	113.6	105.2	d	179.6
f[123] c	215.1 –216.3	218.4	212.2	288.5 –293.1	108	106.5	67.3	162.6
R_4SnD_2 example								
g[124]				265.0				

^{a} Values are Sn–N except where shown.
^{b} For the analogous bromide, SnN shortens to 249.6 pm, SnBr = 266.7 pm, Sn/C parameters remain similar, CSnBr angles are reduced and CSnN angles increased[119].
^{c} Two independent molecules, significant variations shown. Sn–H = 152 to 163 pm.
^{d} This structure is approximately symmetric around the NSnMe axis, so $a = x = y$ etc.

A tetraorganotin member, Me_3Sn-1,4-cyclohexadiene-COOMe, is included with the triorganotin series summarized in Table 8 (see later).

As H is a very similar ligand to R, the chiral t-butyl(8-dimethylaminonaphthyl)(−)menthyltin hydride[120] is included with the tetraorganics. A weak Sn...N interaction (SnN = 290.3 pm) perturbs the shape towards trigonal bipyramidal (Sn–H = 157 pm, Sn–Bu = 221.2 pm, Sn–Nt = 215.7 pm, Sn–Menthyl = 218.3 pm).

An example[117] of marginal 5-coordination in $R_3RSn(...D)$ involves a flexible, but potentially chelating, substituent in $(tol)_3Sn\{C_4H_2S\text{-}C_3(Me_2)H_2NO\}$, where the O of the oxazolinyl substituent on the thienyl ring points at the Sn at a distance of 297.7 pm. The Sn–C(tolyl) distances average 213.4 pm and Sn–C(thienyl) is 215.1 pm while the CSnC angles range from 105° to 113°.

2. Triorgano compounds

Replacing one Sn–R bond by an electronegative substituent greatly enhances the acceptor power and a wide range of complexes are known containing R_3Sn units.

a. Halogen compounds. One series is closely related to the R_4SnD compounds of Table 7 where the Sn–C axial apex is replaced by Sn–X in 14. The detailed summary[125] reveals an inverse correlation between the Sn–D length and both the Sn–X distance and the average XSnC angles. The validity of the simply-based correlations is striking as the ranges of values are wide: Sn–N = 237.2–267.4 pm; Sn–O = 228.7–272.0 pm; Sn–Cl = 243.2–261.3 pm; Sn–Br = 258.8–274.8 pm and single cases with Sn–F = 197.4 pm and Sn–I = 283.0 pm.

Whereas for simple R, R_3SnX compounds with the heavier halides are only weakly coordinated, similar R_3SnF show a strong interaction. Ph_3SnF has an exact trigonal bipyramidal structure[126] with axial Sn–F = 214.6 pm, and Sn–Ph = 211.5 pm. The overall structure is rod-like, as also found for R = CH_2Ph and CH_2SiMe_3 and, in contrast to the zigzag chains, bent at SnFSn in Me_3SnF and other small alkyls (and for the heavier halogens). Very large ligands such as Mes and trimethylsilylmethyl give mononuclear monofluorides[38].

The anion [Me₃SnF₂]⁻ has been isolated with an unusual diphosphacyclobutane counterion[127]. The structure is a regular trigonal bipyramid within 1.8° with Sn−C = 214.4 pm. The Sn−F bonds are axial and reported as 259.6 and 260.7 pm, which seem exceptionally long.

In the crystal, Me₃SnCl has longer bonds to C and Cl than the gas-phase values of Table 1 (210.6 and 243.4 pm) and a further Cl, *trans* to the bonded one, at 325.9 pm showing incipient 5-coordination. Structures of the [Me₃SnCl₂]⁻ anion with different counterions have been determined and reviewed[128] and all show axial Cl−Sn−Cl in trigonal bipyramids of varying regularity. A half-way stage to five-coordination occurs in [tmpH₂]⁺[Me₃SnCl₂]⁻ with Sn−Cl distances of 245.4 pm and 303.4 pm but involving hydrogen bonding. With a large counterion, [K(18-crown-6)]⁺, near-ideal geometry results (Sn−C = 210.6−211.3 pm; CSnC = 118.8°−120.6°; Sn−Cl = 261.8 pm; ClSnCl = 179.2°). The structure is a chain of cation−anion interactions. The [Me₃SnC(NC)]⁻ ion is similar (Sn−N = 265.4 pm).

[R₃SnBr₂]⁻ (R = Ph or *p*-C₆H₄SMe) are near-regular trigonal bipyramids[129] with axial Br (Sn−Br = 273.1−279.1 pm; BrSnBr = 175.3°) but with packing differences resulting from the change in R. (The paper gives a valuable comparison of Sn−Cl and Sn−Br values.)

When one R group carries a donor atom, mononuclear species result, usually with X and the donor in apical positions in the trigonal bipyramid. The core structure with axial ClSnO and equatorial SnC₃ is well represented[130]. Values fall in the ranges Sn−Cl = 247−253 pm; Sn−O = 230−239 pm and Sn−C around 211 pm. ClSnO angles lie in the range 175°−179°, despite varying steric demands by the R groups. Similar geometry is shown by heavier analogues, as in a OSnBr species[131] with Sn−Br = 257.4 pm while a SSnCl[132] species has a much weaker Sn−S interaction (Sn−S = 319.5 pm, Sn−Cl = 244.2 pm and SSnCl = 171°).

Orthorhombic Ph₃SnNCS[133] shows a zigzag chain polymer as found previously for R₃SnNCX, with similar geometry at Sn but varying in SnNC and CSSn angles: NSnS = 174°, Sn−N = 226 pm, Sn···S = 290.4 pm and SnC = 213 pm with angles close to 120°.

An unusual species is Ph₃SnCl·H₂O[134] which occurs as 1:1 addition complexes with metal salcylaldimine complexes and is stabilized by hydrogen bonding (R₃SnX do not react with Schiff base complexes). In two structures, the geometry at Sn is near-regular with ClSnOH₂ = 178.7−179.2°, CSnC within 4° of 120°, SnC normal, Sn−Cl = 248−250 pm and Sn−OH₂ = 233.3 and 241.8 pm. In [(PhCH₂)₃(Cl)SnOOCR′][135] where R is the betaine, Ph₃P⁺CH₂CH₂CO₂⁻, this has a fairly long coordination to Sn (Sn−O = 225.2 pm) *trans* to Cl (Sn−Cl also long at 259.2 pm) to complete a trigonal bipyramid with normal, near-regular equatorial SnC₃ coordination.

The C₃SnXY geometry is also found, but involving a chelate ring as in the set[136] of molecules (15), Me₂SnX (1,4-cyclohexadiene-COOMe) for X = Cl, Br, I and also Me. When X = F, a further very weak interaction occurs (Sn···F = 364.1 pm) to form a six-coordinate dimer with an unsymmetric Sn−F···Sn−F··· four-membered ring. Parameters are listed in Table 8.

The presence of large groups and a chelate ring gives[137] Ph₂SnI(CH₂CH₂CH₂OH) distorted trigonal bipyramid configuration with a long Sn−I (285.7 pm) at 168° to Sn−O (248.7 pm) and normal Sn−C and Sn−Ph. Heating drives off PhH and gives the dimeric [PhSnI(CH₂CH₂CH₂O)]₂ with similar geometry at tin, an SnOSnO ring with μ³-O and shorter Sn−I (277.6 pm) *trans* across the ring. With a very bulky butadiene unit forming the potential chelate, a very weak fifth interaction occurs[138] in Ph₂SnBr(CPh=CPhCPh=CPhX) where X is axial to the Br (Sn−Br = 268.2 pm, BrSnI = 168.5°) but at a very long distance of 388.4 pm (X = I) or 383.8 pm (X = Br) and outside bonding for X = Cl(428 pm).

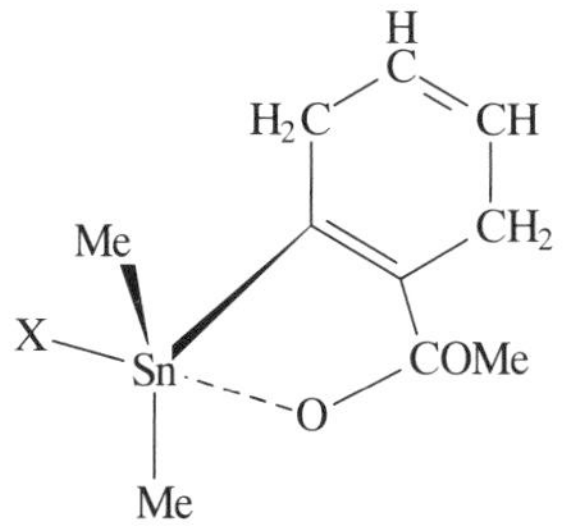

(15)

TABLE 8.   Parameters (pm and deg) for the cyclohexadiene compounds, $XSnMe_2C_6H_6COOMe$ (15)

	X = F	X = Cl	X = Br	X = I	X = Me
Sn–X	197.4	243.2	258.8	283.0	215.0
Sn–C(aryl)	215	215.1	215.8	215.5	217.7
Sn–Me	213	212	213	212	213
Sn–O	252	247	247	239	278
XSnMe	103; 94	99	99	99	106
XSnC(aryl)	93	99	100	99	104
C(aryl)SnMe	113, 124	118	118	117.5	112
MeSnMe	119	118	117	119	116

A similar chelate ring to **15** is seen in a chlorodimethyltin(benzenesulphonyl pyrro-lidide)[139] where the geometry is similar to the chloride above except the axial OSn is longer at 252.9 pm. In a further similar series[140] **16** (next page) the changes are rung on the halogen and the donor atom as part of a study of the effect of systematic variations on geometry and packing. The three compounds are isomorphous with a half-chair conformations and the CSnPY atoms nearly coplanar. The geometry at Sn is the standard trigonal bipyramidal one with the parameters as given in the accompanying table.

	(16a)	(16b)	(16c)
	X = F, Y = O	X = Cl, Y = Se	X = Br, Y = S
Sn–Me	212.4	212.5	212.6/215.5
Sn–C	215.8	214.5	215.8
Sn–X	203.5	250.0	265.0
Sn–Y	245.4	302.2	287.2
CSnC (av)	119.2	119.6	119.6
XSnY	172.3	173.2	175.1

This single-chelate structure is also seen[141] with the optically active substituent, [8-(dimethylamino)naphthyl]-(–)-menthylmethyltin bromide] where the amine-N coordinates to an axial position opposite the Br with the three Sn–C bonds equatorial. The ligand bite is only 74°, NSnBr is bent to 168.1° and the Sn is displaced 19 pm from the C_3 plane towards Br. [Note, the ligand bite is the relatively fixed angle demanded by a small or otherwise rigid bidentate ligand] The SnC distances are within the normal range and SnBr is 264 pm. The Sn–N bond is long at 255 pm, showing a weak interaction, but the structure is definitely closer to tbp than to distorted tetrahedral. The opposite view[142] is

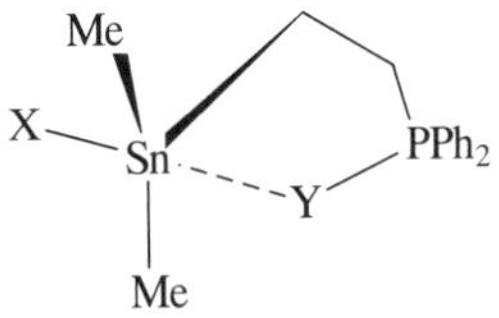

Me
\
X—\—Sn· - - - - ⟍PPh₂
      | ⟍ - - - ⟍Y
      Me

(a) X = F, Y = O
(b) X = Cl, Y = Se
(c) X = Br, Y = S

(16)

taken of $(PhMe_2Si)_3CSnMe_2X$ where the giant ligand causes distortion, but the interaction with the second O when X = NO_3 is very slight (Sn–O = 292.8 pm; compare the bonded Sn–O = 209.5 pm). (A second structure showed no close approach to the S for X = NCS.)

b. Oxygen donors. The coordination unit, R_3SnO_2 with a trigonal SnC_3 plane and apical oxygens, is a common one where a range of behaviours is found.

When X in R_3EX is OH^-, the common configuration of most $R_3Sn(OH)$ is polymeric with axial oxygens bridging. Thus for R = Et^{143}, OSnO = 177.9°, SnO = 215.6 and 224.4 pm, Sn–C = 210 pm and CSnC values range from 117° to 125°. Similar parameters were reported for R = Ph, but with a bigger difference between the SnO values.

Single coordination units are found where the O ligands are large, as in $Ph_3SnLL'^{144}$ for L = diphenylcyclopropenone and L' an isothiazolone or the similar saccharine. OSnO is near-linear and CSnC adjust to steric factors in the range 114–134°, SnO(ketone) = 237–241 pm, Sn–O(hetero-ring) = 216 pm, Sn–C = 211–216 pm.

Such a configuration is also found[145] for $[Me_3SnOC_6H_3Me_2]$ and for $[Me_3Sn\{\mu\text{-}(MeSO_2)_2N\}$ $Me_3Sn(\mu\text{-}OH]_n$. The latter consists of chains with the building unit of Figure 3, linked in pairs by O—H···N hydrogen bonding. Both the tin sites are nearly regular trigonal bipyramids with normal Sn—C bonds around 211 pm and CSnC angles in the 117°–120° range, normal axial Sn–OH distances averaging 212.1 pm, and the Sn(OH)Sn angle is 138°. Axial–equatorial angles are around 97° and the OSnO angle is bent a little at 174°. For the interaction with the anion, the Sn–O distance is 253 pm at 178 K, increasing 5 pm at 296 K. These values, taken with CPMAS[119]Sn data and calculations, serve to identify the energy barriers to $2\pi/3$ propeller-like jumps of the Me_3Sn unit.

In $Ph_3SnOBu^i\cdot Bu^iOH$, a similar core is found[146] and the molecules are linked in a chain by hydrogen bonding between the OBu of one molecule and the coordinated BuOH of the next (OH...O = 177°, O...O = 268.7 pm). The two Sn–O interactions are quite different (Sn–OBu = 206.5 pm: Sn–OH(Bu) = 255.0 pm) but OSnO (173.5°). The asymmetry is also reflected in the CSnC angles of 112°, 114° and 131°. This structure

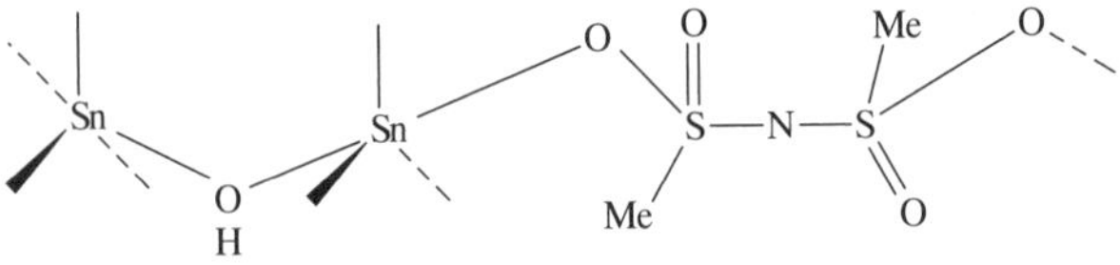

FIGURE 3. Repeat unit in $[Me_3Sn(OH)SnMe_3]^+[(MeSO_2)_2N]^-$

contrasts with the classical structure of the methoxide, where μ-OMe groups formed a chain of Me_3Sn units linked by apical O while the butanol molecule leads to an analogous but differently constructed O_2SnC_3. The phenyl groups are propeller-wise around a rather distorted trigonal plane (Sn–C = 212.2–214.9 pm).

Many carboxylates[147] form a chain polymer via Sn–O–CR–O–Sn bridging. Such a structure was found[148] for trimethyltin(2-furancarboxylate). The carboxylate at the shorter axial Sn–O distance (219.1 pm) has its second oxygen at 317.5 pm which is an extremely weak interaction. The more distant axial Sn–O is at 243.0 pm, and the OSnO angle is 172.4°. The ring oxygen is non-interacting at 358 pm. Similarly, tin is five-coordinate in triphenyltin glyoxalate O-methyloxime[149] despite the presence of the alternative O and N potential donor atoms. The Sn–Ph bonds averaged 214.2 pm while the PhSnPh angles were irregular at 110.0°, 115.6° and 134.0°: Sn–O (218.5 and 236.7 pm), OSnO (173.2°). Similar values hold for bis(triphenyltin)citraconate[150] [intermolecular Sn–O(carboxylate) = 239.7 pm] and Bu_3Sn(uracilacetate)[151] (axial acetate O at 212.4 pm; weaker interaction with uracil O from a different ligand at 266.9 pm).

A unique macrocyclic tetramer with framework $[SnOCO]_4$ was formed[152] for the difluorobenzoate $\{n - Bu_3SnO_2C(2,6-F_2C_6H_3)\}_4$. The coordination at Sn is standard (Sn–O = 218.6 pm and 251.4 pm, OSnO = 175.2°) with normal Sn–C lengths and variable equatorial angles from 115° to 126°. The tetramer is square (S_4 symmetry) with an O–Sn–O along each edge and a carboxylate OCO unit turning each corner. The core structure is essentially that of **26c** without the central μ^3-O and with the sides straightened out.

The presence of a large substituent R supports chelation by a carboxylate group as in Ph_3SnO_2CR, where R is a disubstituted phenylazobenzene, phenoxybenzoato or 2-thiophene[153]. For the former, the structure at Sn is a distorted trigonal bipyramid with one phenyl near-axial (PhSnO angle of 145°); the SnOCO ring of the chelate is planar with OSnO = 53.6° and Sn–O distances of 208.9 pm and 266.0 pm. The SnPh bonds average 214 pm and angles 112°. The thiophene product has an even weaker interaction with the second O (Sn–O = 207.6 and 276.8 pm) so the structure is closer to tetrahedral, while in $Ph_3SnO_2CC_6H_4(2-OPh)$[154] the Sn–O distances are even more disparate at 205.3 and 283.2 pm. In a further, unusual, example[155] R = $CpM(L)(PPh_3)$ [M(L) = Fe(CO) or Re(NO)] the bite is larger (57.4°) and SnO values are closer (212.3 and 234.2 pm).

The polyfunctionality of oxalate gives a range of structures. $[(Ph_3Sn)_3ox_2]^-$ (Figure 4) has[156] a *cis* trigonal bipyramidal $[Ph_3Snox]^-$ anion with a chelate $-SnOCCO-$ ring linked through a third oxalate O axially to $[(Ph_3Sn)_2ox]$. In this unit the Sn atoms are linked by a twisted oxalate group using a *trans* pair of O atoms. The fourth O of the *cis* oxalate links to a neighbouring molecule (Sn–O = 245.1 pm). In the chelate, Sn–O

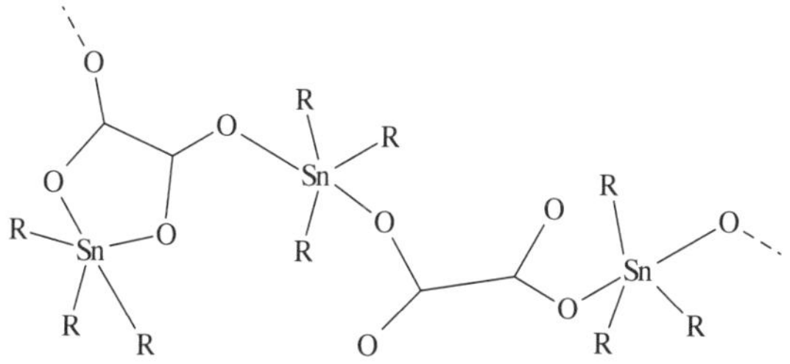

FIGURE 4. The structure of polymeric $[Ph_3Snox]^-[Ph_3Sn)_2ox]$

values are 211.3 and 235.2 pm, and the apical bridge is 233.2 pm while the bidentate bridging oxalate distances are 213.8 and 218.7 pm. The coordination at the Sn involving the chelate O has the unusual *trans*-CSnO arrangement (angle $= 156°$), while the other two Sn atoms have the normal *trans*-OSnO (angles $174°$ and $179°$). [$(C_6H_{11})_3Sn]_2(\mu$-ox) has a similar structure[157] and geometry at Sn, in this case with the Sn atoms alternately *cis* chelated and *trans* bridged so all the oxalates are tetradentate.

In a Ph$_3$Sn succinate/dimethylformamide system[158], Sn environments are *trans*. A dinuclear tin–succinate–tin unit (Sn–O $= 214.5$ pm) is terminated by one donor formamide per tin (Sn–O $= 240.4$ pm) and links through the second O on each carboxylate (Sn–O $= 244.4$ pm) to a formamide-free tin–succinate–tin unit (Sn–O $= 211.3$ pm) and this tetramer is the repeat unit in the 3-D network. Illustrative of many of these structural themes is a study[159] using arylphenoxide ligands, with a potential chelating O, together with amine donors, to form 5- and 6-coordinate mono- and di-organic tin species including SnOSnO bridged dinuclear compounds and an example of all-*trans* octahedral $C_2SnN_2O_2$ coordination.

c. Nitrogen donors. Similar structural themes are found for N donors.

The cation [Me$_3$Sn(NH$_3$)$_2$]$^+$, as the [N(SO$_2$Me)$_2$]$^-$ salt, is a near-regular trigonal bipyramid[160] with Sn–N $= 232.8$ and 238.3 pm, Sn–C 212.1 pm, NSnN $= 179.2°$, CSnC within $2°$ of $120°$ and NSnC in the range 86.8–$93.2°$.

A chain cation is formed by coordination of dipyridyl in [Me$_3$Sn(μ-4,4′-bpy)-]$_n^{n+}$; in regular trigonal bipyramidal geometry[161] Sn–N $= 241.6$ pm, SnC $= 212.3$ pm, NSnN $= 176.8°$, CSnN $= 87°$–$94°$ and CSnC $= 116°$–$124°$.

A SnN$_2$ chelate is found[162] in R$_3$SiN=N–N(SnMe$_3$)SiR$_3$ where the third N of the triazene link also coordinates to Sn (Sn–N $= 256$ pm, compare directly bonded Sn–N $= 221$ pm and NSnN $= 54.3°$). The coordination is a distorted trigonal bipyramid with axial N and C (Sn–C $= 230$ pm, CSnC $= 165°$). The other two methyls are also irregular (Sn–C $= 203$ and 225 pm, CSnC $= 90°$).

N and O occupy the *trans* axial positions in triphenyl[2-(4-pyridylthio)acetato]tin (Sn–O $= 215$ pm, Sn–N $= 250$ pm, OSnN $= 172°$) and in the glycolic acid adduct of triphenyltinsaccharine (Sn–O $= 241$ pm, Sn–N $= 224$ pm, OSnN $= 176°$), but with reversed strengths of interaction. The SnC geometry is standard[163]. In Ph$_3$SnOC(O)C$_5$H$_4$N[164], effectively only one carboxylate O coordinates (Sn–O $= 213.7$ pm; Sn$\cdots$O $= 327.1$ pm) and the molecules are linked by pyridine N–Sn (256.8 pm) giving axial OSnN $= 173.1°$, Sn displaced from the C$_3$ plane towards O (CSnC $= 118°$).

In triphenyl compounds, quite symmetric structures result with larger donors. In the DMF adduct of triphenyltinsaccharin[165], the three phenyls lie almost symmetrically in the equatorial plane (CSnC from $116°$–$124°$; Sn–C 212.5 pm) and the axial O from the formamide and N from the benzisothiazolone are near-linear axial (NSnO $= 176°$; Sn–N $= 224.2$ pm, Sn–O $= 240.2$ pm).

d. Sulphur donors. In an older study, chelation by a disulphur ligand was found[166] in Ph$_3$SnS$_2$CN(CH$_2$)$_4$ with asymmetric SnS (248.1 and 291.9 pm). In contrast, Ph$_3$Sn[Ph(S)C=C(SCH$_3$)Ph] has[167] almost perfectly tetrahedral Sn (Sn–S $= 242.5$ pm) and the SCH$_3$ group is non-interacting, although coordination in the Ph$_2$Sn analogue does involve this S weakly (see below and **22b**).

Ph$_3$SnSD structures [SD $=$ S(CH$_2$)$_2$NH$_2$ or SC$_5$H$_4$NO] involve chelate donors with S and N or S and O. The structures[168] are *cis* with S equatorial (Sn–S respectively 242.6 pm and 249.4 pm) and one phenyl axial with O or N (Sn–C $= 217.0$ pm, Sn–N $= 264.7$ pm, CSnN $= 169°$; Sn–C $= 217.1$ pm, Sn–O $= 236.4$ pm, CSnO $= 163°$).

A similar five-membered ring involving Sn–O (226.1 pm) and Sn–S (257.7 pm) but with the unusual square pyramidal geometry occurs[169] in $(PhCH_2)_3SnSD$ where SD = 2-thiolatopyridine-N-oxide. One benzyl is axial (Sn–C = 216.7 pm) and the Sn is above the basal plane [axial–Sn–basal angles of 102° (to S), 100° (to O) and 110° (to each C)]. The *trans* base–Sn–base angles are 140°–147°, and the OSnS bite is 73°. While distorted, the square pyramid description is the closest.

3. Diorgano compounds

a. Halogen compounds. The simplest structures for diorganotin species are those of the polyhalide ions. Among the trihalides, $[Ph_2SnCl_3]^-$, like dialkyl trichlorides reported earlier, is a near-regular trigonal bipyramid[170] in its Et_4N^+ salt. The phenyls are at 115.6° and 213.3 pm while the axial Sn–Cl = 252.7 pm and 174°. The equatorial Sn–Cl is shorter at 237.8 pm. Among the tetrahalides is $[Me_2SnF_4]^{2-}$, determined[171] as the NH_4^+ salt. The cation is a *trans* octahedron with all angles within 2° of 90°, and with Sn–C = 210.9 pm and Sn–F = 212.7 pm, respectively shorter and longer than the tetrahedral values. Each cation is linked to four different anions by N–H...F bonding. A similar *trans* octahedral structure is found[172] for $[Ph_2SnCl_4]^{2-}$ in the presence of a large counterion (Sn–Ph = 215.7 pm; Sn–Cl = 256.3 and 260.3 pm). This is comparable to earlier $[R_2SnX_4]^{2-}$ determinations.

A second fluoro-anion, $[Me_4Sn_2F_5]^-$, shows a similar *trans* octahedral arrangement[173], but now with two different Sn environments and bridged into a complex structure (17). The core structure is a chain of *trans* and zigzag –F–Sn–F–Sn–links (FSnF = 180°; SnFSn = 150°). The remaining F link successive pairs of chain Sn centres to a third Sn forming a succession of $[SnF]_3$ rings on alternate sides of the chain. The remaining two F on this third Sn are terminal and normal length as are the different Sn–C bonds. The CSnC angle on the bridging Sn is reduced and angles involving the terminal F open to 94°. This structure is thus part way to the long-established one of Me_2SnF_2 where all the F atoms are equatorial and bridging, giving a sheet structure.

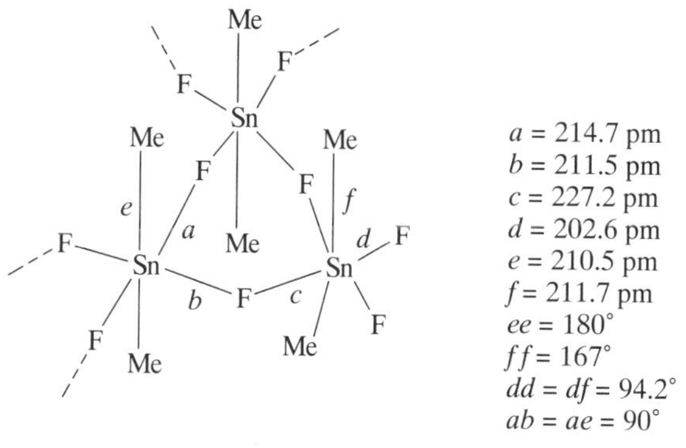

$a = 214.7$ pm
$b = 211.5$ pm
$c = 227.2$ pm
$d = 202.6$ pm
$e = 210.5$ pm
$f = 211.7$ pm
$ee = 180°$
$ff = 167°$
$dd = df = 94.2°$
$ab = ae = 90°$

(17)

The $[Me_4Sn_2Cl_6]^{2-}$ is a simple dimer[174] with Cl-bridged, edge-sharing octahedra. The methyls are *trans* (Sn–C = 208 pm, CSnC = 167°), terminal Sn–Cl = 247–52 pm and bridging Sn–Cl is 290 pm.

For the parent molecules, R_2SnCl_2 species show long Sn...Cl interactions producing (4 plus 1; R = C_6H_{11})[175] or (4 plus 2; R = Me, Et, Bu) coordination[176]. Typically,

Sn$-$C is normal, RSnR $= 125°-135°$, Sn$-$Cl distances are 236$-$240 pm while Sn...Cl is around 350 pm. Both coordinations are linked into zigzag chains which are linear at Sn (Cl$-$Sn...Cl 170$-$175°) and bent at Cl (Sn$-$Cl...Sn around 105°). The cyclohexylbromide shows similar features (Sn$-$Br $= 250$ pm; Sn$\cdots$Br $= 377$ pm). While the long Sn$\cdots$Cl or Br distances are only 10% less than the van der Waals radii, the interpretation in terms of very weak additional coordination seems justified. By comparison, in truly 4-coordinate (biphenylyl)$_2$SnCl$_2$, Sn$-$Cl $= 238.6$ pm and all the intermolecular Sn/Cl distances exceed 600 pm, and in Et$_2$SnI$_2$ the extramolecular Sn/I is 428 pm, distinctly outside the van der Waals limit. In the end the matter is one of interpretation[175]. When R$_2$ = Ph$_2$ or MePh, tetrameric units are found containing both 5- and 6-coordinate Sn.

While the dominant species in solution is probably the 1:1 adduct, diorganodihalotin, donor adducts are normally isolated as the less soluble 1:2 products, R$_2$SnX$_2$·2D. A recent study[177] of the system with R $= C_6H_4F$, D $=$ thiirane-1-oxide, C$_2$H$_4$SO, shows O donation to give an octahedron with all angles close to 90°, *trans* R (213.7 pm) and *cis* Cl (247 pm) and D (Sn$-$O in the longer range at 232 pm). For R $=$ Me (210 pm), D $=$ imidazoline (Sn$-$O $= 242$ pm) and X $=$ Cl (249 pm)[178] the same geometry adjusts to a somewhat longer bond to O. Similarly, when the donor atoms are part of the ligands as in the Schiff-base compound[179] [Cl$_2$Sn(C$_6$H$_3$MeNtol)$_2$], two long *cis* Sn$-$N bonds (275 and 286 pm) complete a distorted octahedron with normal Sn$-$C (211 pm) and Sn$-$Cl (238 pm).

With larger substituents, (C$_6$H$_{11}$)$_2$SnBr$_2$D$_2$ (D $=$ pyrazole or imidazole) form[180] all-*trans* octahedra with Sn$-$Br $= 270-80$ pm, Sn$-$N 236$-$239 pm and Sn$-$C averaging 216 pm. Using a substituted imidazole, D $=$ imidazolinethione bonding through S, the product Bu$_2$SnCl$_2$D is a 1:1 complex with the standard trigonal bipyramidal configuration. The smaller methyl analogue, Me$_2$SnCl$_2$D[181], forms a dimer by very weak chloride bridges. The S (249.5 pm) and the Me groups (212 pm) occupy approximately equatorial trigonal bipyramidal sites (CSnC $= 144°$, SSnC $= 107°$) and the shorter Cl$-$Sn$-$Cl are nearly linear (178°, 254 and 267 pm) while the dimer involves a sixth Sn...Cl interaction of 359 pm.

The *t*-Bu$_2$SnF group bonded to phosphorus ylides[182] is weakly coordinated axially to the BF$_4^-$ counterion (SnF $= 278-285$ pm) giving trigonal bipyramid geometry with the other axial Sn$-$F $= 197-203$ pm, and all the equatorial Sn$-$C around 218 pm. A symmetric four-membered SnCSnC ring (Sn$-$C $= 224$ pm; angles 84.8° at Sn and 95.3° at C) was also characterized with longer Sn$-$Bu of 222 pm.

The polyfunctional ligand di-2-pyridylketone 2-aminobenzoylhydrazone (Hdpa) reacts with Ph$_2$SnCl$_2$ to give two products[183], [Ph$_2$SnCl$_2$Hdpa] (**18**) and, with loss of PhH, the deprotonated ligand product [PhSnCl$_2$dpa]. In **18** Sn$-$C and Sn$-$Cl bonds are normal, and three weaker Sn$-$O or N bonds give a distorted pentagonal bipyramid. In the second, these bonds become much stronger (SnO $= 220$ pm, SnN $= 216$ and 214 pm) and the Sn$-$Cl bonds are *trans* (170°; length 246 pm).

Structures are established for a variety of diorganotin monohalides. With a highly demanding ligand Ar^f, a singly-bridged oxide (Ar$_2^f$SnCl)$_2$O resulted[184] from the oxidation of Ar$_2^f$Sn^{II} in the presence of Cl$^-$. Bond lengths are appropriate for 4-coordination with a large ligand (Sn$-$C $= 219.3$ pm; Sn$-$Cl $= 231.5$ pm; Sn$-$O $= 193.7$ pm) and SnOSn is 144.8°.

One set of structures which covers the standard bidentate ligand monomer and dimer structures (Figure 5) is the benzoxathiostannole series, [(DD′)SnMe$_2$X]$^-$, where DD′ is (C$_6$H$_4$OS) and X $=$ F, Cl or I[185]. Here, Sn$-$C (211$-$216 pm) and Sn$-$S (242$-$245 pm) lie in the normal range and occupy the equatorial positions. The equatorial angles become increasingly distorted from 118°$-$122° in the fluoride to 108° (SSnS) and 140° (CSnC) in the iodide. Sn$-$O is normal for the axial position (212$-$219 pm) while the Sn$-$X values

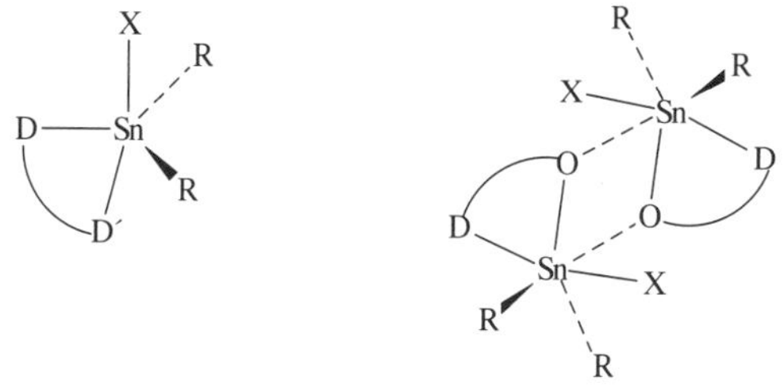

(a)   (b)

FIGURE 5. Common configurations of diorganotin compounds with bidenate ligand (a) mononuclear and (b) SnOSnO four-membered ring dimer

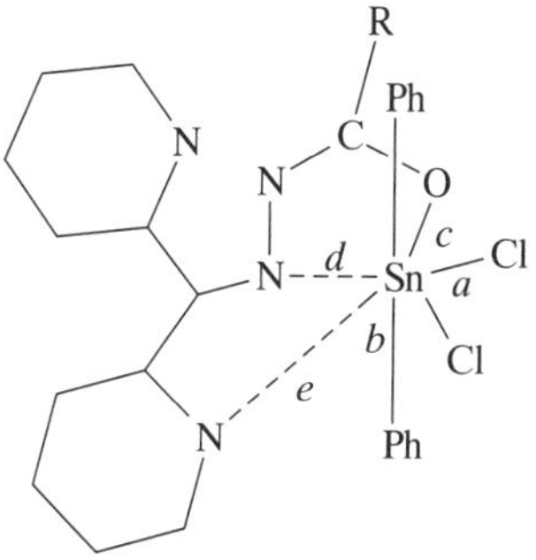

$a = 249$ pm $aa = 88.6°$
$b = 212$ pm $bb = 163.8°$
$c = 243$ pm $ab = 88°$
$d = 275$ pm $ac = 75°$
$e = 270$ pm

(18)

204 pm (F), 256 pm (Cl) and 323 pm (I) become increasingly extended from F to I. The iodide adopts the dimeric form with the longer Sn–O distance of 264.6 pm completing an irregular octahedron with *trans* axial methyls.

An exciting result was found[186] for the amine $(Me_2SnCl)_3N$ (19) which has a planar Sn_3N skeleton (Sn–N = 198 pm, SnNSn = 120°; Sn–C = 207 pm) stabilized by slightly unsymmetric and quite long Sn–Cl–Sn bridges (the Sn–Cl distances average 272 pm and 280 pm). Calculations suggest a p orbital on N and planarity resulting from stabilization produced by the bridges, rather than any Sn–N π effect. $(Me_3Sn)_2N$ (Dpp) also shows[98] planar N (Sn–N = 204.4 pm, SnNSn = 125°) with no π interaction.

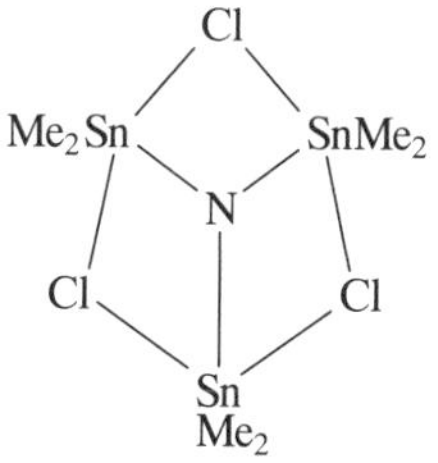

(19)

b. *Oxygen donors*. The tetrahydrated cation, $[Me_2Sn(H_2O)_4]^{2+}$, has been isolated[187], stabilized by the large benzene-1,2-disulphonic imide anion and by hydrogen bonding from the water ligands to the anion O atoms. The structure is almost regular *trans* octahedral (normal Sn–C 208.8 pm; CSnC = 179.5°) and a small range of Sn–O values (221–226 pm; angles involving O within 2° of 90°).

In the $[Ag(PPh_3)_4]^+$ salt[188], the anion $[Ph_2Sn(NO_3)_2X]^-$ contains X = Cl or NO_3 disordered over an equatorial site of a bipyramid with axial phenyls (Sn–C = 211 pm, CSnC = 164°). The ordered nitrates are equatorial and asymmetrically bidentate (Sn–O = 234 and 255 pm). The remaining equatorial site is 50/50 occupied by Cl (Sn–Cl = 257 pm), giving a pentagonal bipyramid, or by NO_3 (Sn–O = 224 and 243 pm) giving an example of rare eight-coordination in a hexagonal bipyramid, encouraged by the tight nitrate bite angle of 50–54°.

(i) *Carboxylates.* In diorgantin carboxylates in general, $[R_2Sn(O_2CR')_2]$ commonly[189] show a distorted octahedron where two short Sn–O and two longer Sn–O lie in a plane with R lying over the weaker Sn–O (20). Some typical parameters are listed in Table 9. Oxalato-complexes form similar, usually *cis* substituted, octahedra (21).

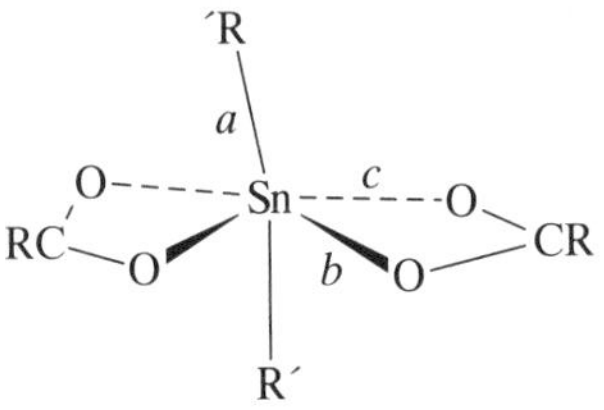

(20)

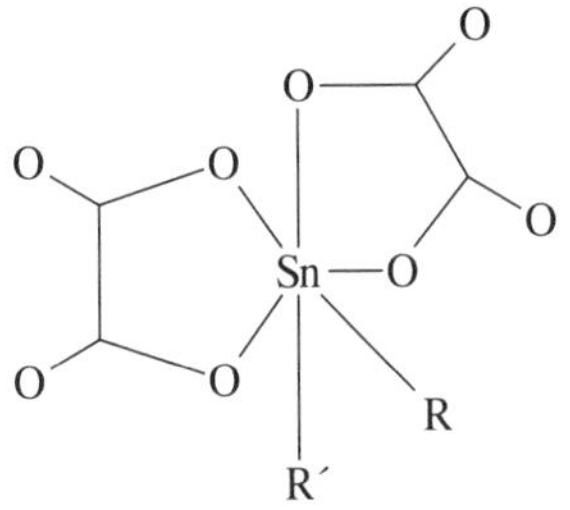

(21)

The values in Table 9 show short Sn–O values corresponding to full bonds for axial substitution, while the longer bonds are still significant interactions. Interestingly, the S in the case of R' = CH_2SPh showed no sign of coordinating and the standard structure was

TABLE 9. Parameters (pm and deg) of some representative $R'_2Sn(OOCR)_2$ species (20)

R' and R	SnC (a)	SnO (b)	Sn...O (c)	CSnC (aa)	OSn...O (bc)
Me, Me[190]	209.8	210.6	259.3	136	55
Bu, benzoate[189]	211	211	255	145	
Bu, CEt$_2$COO[191(a)]	212	215	249	159, 145	56
Me, p-NH$_2$C$_6$H$_4^{192}$	210	209	255	135	55

[a] Average over different phases.

observed[193]. In the related acetate anion[190] [Me$_2$Sn(OAc)$_3$]$^-$ the coordination is pentagonal bipyramidal with two unsymmetric bidentate acetates (SnO of 228 and 252 pm), the third acetate coordinating through one O and the Me groups axial. In contrast, dimethyldiformatotin, Me$_2$Sn(OOCH)$_2$ is a sheet polymer[194] with linear MeSnMe moieties (angle 179.7°; quite short Sn–C = 209.7 and 211.6 pm) and each formate bridges a pair of Sn atoms, but with all Sn–O bonds equal at 224.7 pm. The coordination at tin is almost regular square bipyramidal with angles ranging from 84°–94°. This structure is very similar to the long-known Me$_2$SnF$_2$ where Sn–C is even shorter at 208 pm.

In [Bu$_2$Sn(OOCCEt$_2$COO)]$_x$, the second carboxylate links[191] to the neighbouring Sn to give an extended chain. When the compound reacts with the diamine, H$_2$N(CH$_2$)$_3$NH(CH$_2$)$_3$Si(OMe)$_3$, a similar linear polymer results but the N displaces the weak Sn–O (Sn–N = 228.4 and 238.4 pm) leaving monodentate carboxylates (Sn–O = 214.5 pm and Sn...O = 350 pm which is non-bonding]. The Sn–C lengths average 215.5 pm and the groups are *cis* in a less distorted octahedron (CSnC = 110°).

Two picolinates, R$_2$Sn(OOCC$_5$H$_4$N)$_2$ with R = Me or Ph, have been reported. The diphenyl compound[195] adopts the *cis* octahedral structure, compare **22a**, but with the N and only one O of the carboxylate coordinating. The two SnN bonds (228.4 pm) are *cis* (angle 80.3°) and the OSnO alignment is distorted *trans* at 153° and Sn–O is quite strong (Sn–O = 209.5 pm). The two phenyls are at 105° (Sn–C = 212.8 pm). In contrast, in the longer-known Me analogue[196] a second O of the carboxylate bridges to a second Sn, giving overall pentagonal bipyramidal coordination with the Me groups axial. This same paper gives one of the few *trans* octahedral examples, Me$_2$Sn(kojate)$_2$, though very distorted with MeSnMe = 148°.

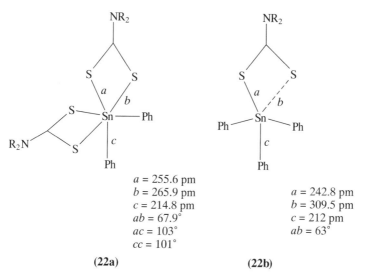

a = 255.6 pm
b = 265.9 pm
c = 214.8 pm
ab = 67.9°
ac = 103°
cc = 101°

a = 242.8 pm
b = 309.5 pm
c = 212 pm
ab = 63°

(22a)   (22b)

(ii) Oxides and hydroxides. Diorganotin oxides and hydroxides are found in a variety of structures. While mononuclear species are rare and depend on shielding of OH groups, dinuclear and polynuclear species of increasing complexity have been identified.

Dinuclear species. Diorganostannoxane carboxylate hydroxides often form dimers joined by a rectangular SnOSnO unit, as in [Ph$_2$Sn(OOCCCl$_3$)(OH)]$_2$ **(23)**. The Sn–OH bonds are asymmetric (b = 202.4 pm, c = 216.7 pm). Axial Sn–OC (a = 215.6 pm), Sn–Ph (d, e = 211 pm) and PhSnPh (133.9°) follow the normal pattern. There is hydrogen bonding between OH and carboxylate (OH...O = 264 pm)[197].

$$
\begin{array}{c}
\text{Cl}_3\text{COO} \quad \overset{H}{\underset{}{O}} \quad \overset{Ph}{/} \\
a \diagdown \underset{\text{Sn}}{} \underset{c}{\diagup} \quad \text{Sn} \text{----- Ph} \\
\text{Ph} \diagup d \diagup e \diagdown \quad \text{O} \diagdown \text{O} \\
\text{Ph} \quad \overset{}{\underset{H}{O}} \diagdown \text{O--C} \diagup \\
\text{O--C} \\
\text{Cl}_3
\end{array}
$$

(23)

A similar Sn(OH)Sn(OH) ring is found[198] in the R_2Sn(OH)$_2$ dimer where R is a substituted ferrocene bonded to Sn through a ring C. The dimer forms via Sn(OH)...Sn interactions where the Sn–O distances are 202 and 223 pm. The ferrocene also contains an adjacent $-CH_2NMe_2$ which, surprisingly, does not coordinate to Sn, probably because they are involved in strong hydrogen bridges. Two Sn–C bonds (214 pm average) and an O are equatorial.

When the acid group is a thiophosphate[199], only the O coordinates. The hydroxy-bridged dimer [^{t}Bu$_2$Sn(μ-OH){O(S)P(OEt)$_2$)}]$_2$ has a rectangular bridge (Sn–O = 201.8 and 219.7 pm) and has, as axial groups, the phosphate (Sn–OP = 212.1 pm) and OH (OSnO = 156°). The CSnC angle is 124° and the Sn–C lengths are normal for a t-Bu at 216 pm.

In L = 2,6-pyridinedicarboxylate compounds[200], L is a tridentate ligand through the N and one O from each carboxylate. In the [R_2SnL(H$_2$O)]$_2$ dimer the SnOSnO forms a rectangle and these two O, with the ligand donor atoms, lie in a nearly planar pentagon. The SnR$_2$ units (R = Me, Et, Bu, Ph) lie in axial positions (CSnC 166°) completing a pentagonal bipyramid with normal bond lengths (Sn–O = 218–247 pm and 260–278 pm; Sn–N = 228 pm, Sn–C = 204–211 pm).

In [Ph$_2$SnCl(μ-CH$_2$)P(O)Ph$_2$]$_2$[27] the binuclear ring is an 8-membered twisted chair [SnCPO]$_2$ with a near-symmetrical trigonal bipyramidal configuration (axial ClSnO = 170.7°; Sn–Cl = 251.1 pm, Sn–O = 229.1 pm; equatorial CSnC angles 119.1°–120.2°, SnC 213–217 pm).

The *ortho*-substituted dibenzoates, Bu$_2$Sn(OOCC$_6$H$_3$X$_2$)$_2$ (X = OH or Cl), show[201] two types of weak additional coordination. Each carboxylate has one strong (Sn–O = 211 pm) and one weaker (Sn–O = 260 pm) bond to Sn, and the two Sn–C bonds are normal (212 pm) forming a bicapped tetrahedron overall. These six-coordinate units are then weakly linked into dimers through SnO...SnO... interactions (Sn...O = 338 pm; compare van der Waals sum of 370 pm)]. The interaction is qualitatively that of (**30**) but minor.

The β-diketones of substituted pyrazolonates [O–O] form[202] two 6-membered –SnOCCCO– rings in all-*cis* R$_2$Sn[O–O]$_2$ octahedral configurations. One ring is half-chair and the other is boat. There is one short (210–215 pm) and one long (234–246 pm) Sn–O bond from each and Sn–C is normal at around 210 pm for n-alkyl and 220 pm for t-Bu.

Trinuclear species. When salicylaldoxime reacted[203] with dibutyltin oxide, the complex product (**24**) contained three tin atoms, two with SnC$_2$O$_3$ coordination in trigonal bipyramidal (Sn^5) and one with SnC$_2$O$_3$N$_2$ links in pentagonal bipyramidal (Sn^7) configurations. The Sn–C distances are normal at 210–218 pm, the CSn^{7}C angle is 160° and, at the 5-coordinate centres, 125–129°. The triply-bridging O is 202 pm from Sn^5 and 214 pm from Sn^7. The Sn^7–OAr distances are longer, 226 pm to μ-O and 268 pm to μ^3-O, and the Sn^7–N distances show the reciprocal asymmetry (267 pm and 229 pm).

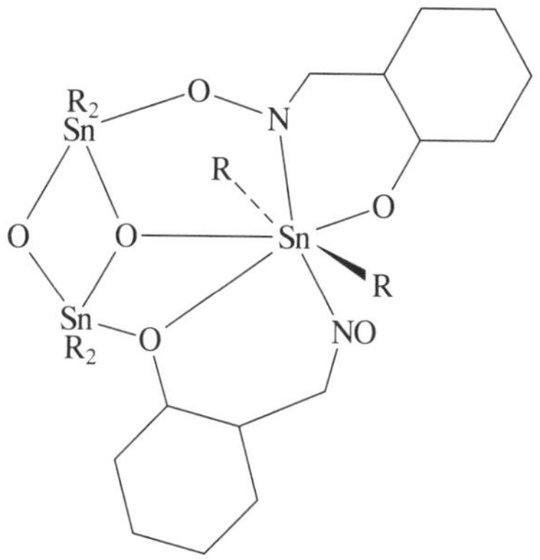

(24)

Tetranuclear and larger species. There are four tin and four oxygen centres in the three-ring oxide–hydroxide structure which was found[204] among the products of hydrolysis of chlorides. The central SnOSnO shares edges with outer SnOSnOH rings **(25)**. The Sn_4O_4 core is only slightly stepped, and the R substituents form an envelope above and below. Bond lengths (Table 10) are relatively short for apical Sn–O and Sn–OH showing strong interactions and are similar to those of earlier studies.

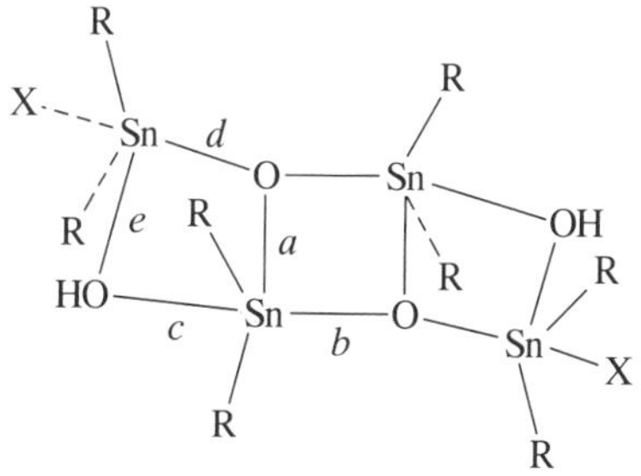

(25)

Oxide species containing four tin centres are found in a number of variants on the compact structure based on a central SnOSnO ring sharing edges with four other rings shown in **(26)**. Typical parameters for the varieties are given in Table 11.

The thiophosphate[199] oxide, related to the dimeric hydroxide above, $[Me_2Sn(\mu\text{-}OH)\{O(S)P(OEt_2)\}]_2O$, has the compact multi-ring structure of **(26a)** [R = Me, Z =

TABLE 10. Bond lengths (pm) in the triple-ring oxide–hydroxides of structure **(25)**

R;X	a	b	c	d	e	Other features
Ph; Cl[a]	204.8	212.1	213.8	202.1	214.9	Sn–Cl = 249.5 pm; OH is hydrogen-bonded to C_3H_7NO
Ph; Cl	204.8	211.7	217.1	202.2	218.0	Sn–Cl = 248.0 pm

[a]DMF adduct.

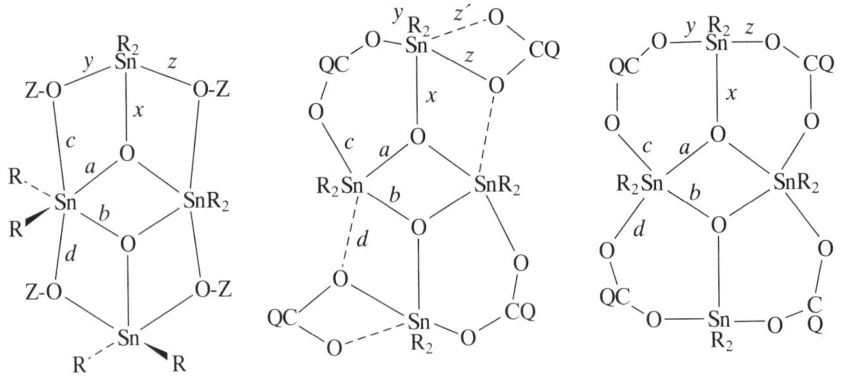

(a)   (b)   (c)

(26)

TABLE 11. Bond lengths (pm) in the complex oxides related to (26)

	$26a^{199}$	$26b^{192}$	$26b^{206}$	$26b^{207}$	$26c^{206}$	$26b^{192}$ variant
a	201.0	202.9	203.6	205.8	211.0	203.6
b	203.4	216.2	217.6	211.4	208.8	216.6
c	260	225.4	222.6	227.5	235.3	220.6
d	277	287.7	286.3	279.3	233.0	331.5
x	198.9	200.7	202.4	201.7	200.5	200.9
y	225	226.0	228.0	225.2	222.8	268.8
z	221	215.4	216.2	220.4	224.3	210.4
z'		290.9	274.6	298.4		257.3
						293.5
Sn–C (av)	209	211	214	213	211	213

(EtO)$_2$P(S)] with short oxide bonds and phosphate oxygen weakly bridging to the second tin. This relates to the carboxylate structure of (26b) where the additional bridges are weaker.

Related to the dimethyltinaminobenzoate of Table 9 is a pair of oxides with interesting structures[192]. The oxide formed by *ortho*-aminobenzoate, [(Me$_2$SnOOCC$_6$H$_4$-*o*-NH$_2$)$_2$O]$_2$ is a dimer which contains a central SnOSnO bridge and further bridges to two outer Sn atoms (26b) (R = Me, Q = C$_6$H$_4$-*o*-NH$_2$). This has a short equatorial or a longer axial bond between the different Sn and the oxide O. The carboxylates are alternately mono- and bi-dentate with bonds to the carboxylate O of the short and longer axial types. The monodentate carboxylate also has two very long additional interactions. Thus both types of Sn are trigonal bipyramidal with a weak sixth interaction. In the *para*-amino analogue, all the carboxylates become monodentate (26b variant in Table 11). That is, bonds y in 26b are broken. Additional weak interactions (294–332 pm) occur between Sn and both the formally bonded and the formally non-bonded O. A further example of 26b [Q = C$_5$H$_4$(SMe)N; R = Et] is found in the thiopyridinecarboxylate oxide dimer [Et$_2$Sn(2-MeS-3-C$_5$H$_4$NCO$_2$)$_2$O]$_2$[207]. Yet a further variant, 26c, has all four corners bridged by the O–CQ–O unit. This structure is found[206] for [(Me$_2$Sn(*t*-BuCO$_2$)$_2$O]$_2$ (Q = Bu^t; R = Me), while in the Et$_2$Sn analogue, the structure changes to 26b.

A related structure was isolated[91] during the preparation of cyclic sulphides from the halides. This contains two SnOSnSH rings which are fused to a central SnOSnO ring (27), a molecule of formula $[(H_{11}C_6)_2Sn]_4O_2(SH)_2Cl_2$. The rings form a planar ladder and there is a weak interaction between the central Sn and the Cl on the outer Sn giving a structure of five linked rings comparable to (26). (Note that SH and Cl are not unambiguously distinguished by X-rays.) The Sn–SH bonds are much longer than ring Sn–S ones, and the Sn–O and Sn–Cl values are also lengthened while Sn–C, with a range 213–218 pm covering both tin sites, are normal.

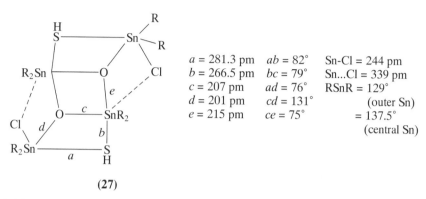

a = 281.3 pm ab = 82°   Sn-Cl = 244 pm
b = 266.5 pm bc = 79°   Sn...Cl = 339 pm
c = 207 pm ad = 76°   RSnR = 129°
d = 201 pm cd = 131°   (outer Sn)
e = 215 pm ce = 75°   = 137.5°
                                (central Sn)

(27)

While carboxylate oxides and salicylic acids form normal SnOSnO bridged dimers as in $[\{R_2Sn(OOCCH_2SPh)_2\}O]_2$[193], when thiosalicylic acid was treated with dibutyltin oxide, a cyclic hexamer resulted[208], with a 24-membered ring $[Sn-O-C=O]_6$. The three different Sn atoms are in very similar trigonal bipyramidal configurations with apical O from two carboxylates (OSnO = 176°) and equatorial n-Bu and S (angles 131°, 119° and 109°). The main bond lengths show little variation (SnC = 215 pm, SnO = 222 pm, SnS = 241 pm). In addition, there is a more remote oxygen at 307, 311 or 316 pm from the three Sn sites.

c. Sulphur donors. Diorganotin dithiolates give both *cis* and *trans* octahedra, depending on the bulk of R and R′, and the SnS₂ coordination may be symmetric or asymmetric. In $[(t\text{-}Bu)_2Sn(S_2CNMe_2)_2]$ coordination drops to 5, with one dithiolate becoming monodentate (Sn–S = 257 pm) and the bidentate unit very asymmetric (Sn–S = 248.9 pm and 279.5 pm). Five-coordination is also found for triorganotin compounds like $R_3Sn(S_2CNR_2')$ and the asymmetry is substantial, so that the fifth coordination is weak (average Sn–S = 246 pm and 325 pm depending on R and R′). Bulky R′ give four-coordination, as in $Ph_3Sn(S_2COPr^i)$ or $Ph_3Sn(S_2POEt_2)$, which have monodentate thiolates. A similar theme appears[209] in dithiocarbamate structures, investigated for bioactivity. The diorganotin, $Ph_2Sn(S_2CNEt_2)_2$ (22a) is an octahedron with modest asymmetry in Sn–S in contrast to the triorganotin analogue, $Ph_3Sn(S_2CNEt_2)$ (22b), which is distorted tetrahedral with a weak fifth interaction (CSnC angles 104°, 106° and 121°). A very asymmetric version of (22a) is found[167] in $Ph_2Sn[(Ph(S)C=C(SMe)Ph]_2$ where the Sn–S distances are 245.2 pm and 330.8 pm in the chelates. The morpholinocarbdithioato compound[210], $Bu_2Sn[S_2C(NC_4H_8)_2]_2$, has a structure similar to (22a) with much more asymmetric Sn–S lengths of 252.5 and 300.1 pm.

A further comparison of closely related compounds is provided by the related chlorides, $Ph_2(Cl)Sn(S_2CNR_2'')$, where R″ is Et[211] or C_6H_{11}[212]. These have structures similar to 22b but with reduced asymmetry in Sn–S (244.5 and 271.6 pm for R″ = Et; 244.0 and

265.7 pm for the cyclohexyl). The Sn−Cl bonds are the same at 243.9 pm and the Cl and the longer S occupy axial sites in a very distorted trigonal bipyramid (ClSnS angles of 158° and 154°, respectively). The t-Bu analogues of (22a) and of the monochloride show similar features. There is thus a subtle interplay between substituent size and geometrical features, especially in the asymmetry of the SnS_2 coordination.

The dithiosquarate[213] ligand bonds to Me_2Sn through the S atoms, presumably because of the more comfortable bite angle (89°). There are four strong bonds which are close to tetrahedral, allowing for the smaller chelate angle (Sn−C = 210 pm, Sn−S = 252 pm; CSnC = 135°, CSnS = 105°−108°). However, one O from each squarate bonds weakly to a neighbouring Sn in cis positions (268−288 pm), giving a very distorted (4 strong plus 2 weak) octahedron. In the Me_2SO adduct, the O (234.5 pm) completes a trigonal bipyramid with similar SnS and SnC values and there is an even weaker squarate O−Sn interaction of 294 pm.

The alternative choice between O and S as strong and weaker ligand is found[214] in $Me_2Sn(OSPPh_2)_2$ where the P=O bonds strongly (Sn−O = 206 pm) and the P=S completes the trigonal bipyramid with a much weaker bond (Sn...S = 294 pm; OSn...S = 163°) linking the molecules into a zigzag chain.

d. Nitrogen donors. Diorganotin compounds with relatively weak Sn−N bonds are of interest as anticancer agents. $[R_2Sn(bipym)Cl_2]$ (bipym = bipyrimidine; R = Ph, Bu) are mononuclear[215] with bidentate bipym and *trans* R [CSnC = 175° (Bu) and 169° (Ph); all Sn−C = 214 pm]. The Sn−N (240 pm) and Sn−Cl (246 pm) bonds are about 5 pm shorter in the phenyl compound than in the butyl one, making the latter the better anti-tumour prospect.

In the dimethylbipyridyl[216], $(C_5H_9)_2SnCl_2 \cdot bipyMe_2$ (28), the ClSnCl angle of 107.5° complements the bite angle of the chelate. The Sn−C distance is normal at 215.5 pm and the CSnC angle is 175°. The cyclopentyls exert some steric effect which shows up in Sn−N bonds of 243.5 pm, compared with 236 pm in $Ph_2SnCl_2 \cdot bipy$. Similar geometry was found in earlier determinations of $R_2Z_2Sn(L-L)$ compounds containing a range of bidentate L-L groups.

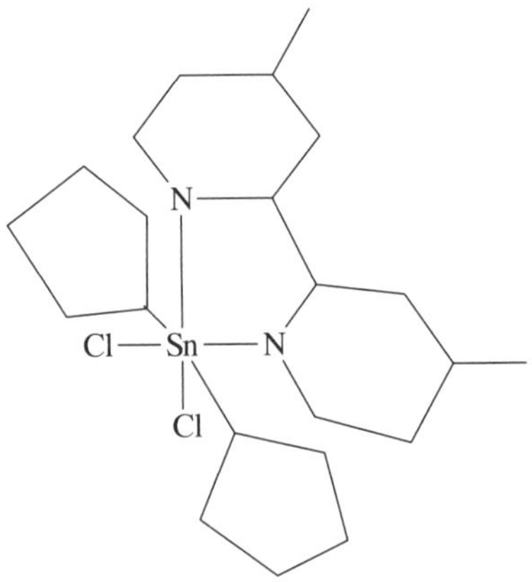

(28)

A related geometry results from the use[217] of 2,6-pyridine dicarboxylate in $Ph_2SnC_7H_3NO_4 \cdot H_2O$. The two phenyl groups are at normal distances and *trans* (CSnC = 172°). The tridentate ligand occupies three sites in the equatorial plane and the coordinated

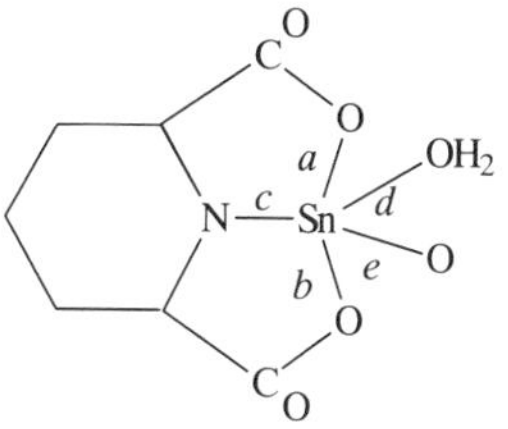

$$a = 221.4 \text{ pm} \quad ac = 69.7°$$
$$b = 244.6 \text{ pm} \quad bc = 65.9°$$
$$c = 233.8 \text{ pm} \quad ad = 73.6°$$
$$d = 227.1 \text{ pm} \quad be = 72.7°$$
$$e = 238.0 \text{ pm} \quad de = 78.3°$$

FIGURE 6. Equatorial plane of pentagonal bipyramidal $Ph_2SnC_7H_3NO_4\cdot H_2O$. Phenyl groups lie above and below this plane

water fills a fourth. Interestingly, a carboxylate O from a neighbouring molecule (Sn–O = 238 pm) is also coordinated to give 7-coordination, pentagonal bipyramidal geometry at Sn and an overall zigzag chain polymer. The equatorial pentagon (Figure 6) is irregular but essentially planar. The compound is one of a family which shows significant anti-tumour activity[218]. Compare also analogous[219] N-(2-hydroxyethyl), six-coordinate potential anti-tumour species.

The larger tridentate[220] benzoylsalicylahydrazone ligand forms three strong bonds to give the usual distorted trigonal bipyramid geometry. The ligand is near-planar and fills the two axial (SnO = 213 pm) and one equatorial (SnN = 217 pm) site in the dibutyltin derivative (SnC = 212 pm).

The N-(2-mercaptophenyl)-4-oxo-2-pentylideneamine molecule acts as a tridentate ligand bonding through three different donors — S, N and O. The $LSnPh_2$ product[221] has the usual distorted trigonal bipyramid with equatorial N and C.

Dipeptides. Interest in diorganotindipeptides stem from their antileukaemic activity. More recent structures[222] with various Sn–R include GlyMet, GlyGly, GlyAla, GlyVal. All show very similar trigonal bipyramids around Sn with normal bond lengths for the configuration. The tridentate dipeptide is usually nearly planar with the O and the amino N (225–230 pm) in axial positions and the peptide N (around 210 pm) with the two organic groups equatorial. In [$Et_2Sn(GlyHis)$]·$MeOH$[223], there are two tin environments. One is the standard trigonal bipyramid, as above, but in the second, the coordination at Sn is increased to 6 by a long bond to a ring N (Sn–N = 279 pm) of the histidine on the trigonal bipyramidal molecule. This lies in the equatorial plane of the 2C and the peptide N (lengthened 3 pm) with a reduction in angles.

e. General. Reaction of [t-Bu_2SnO]$_3$ with $RB(OH)_2$ (R = Ph or Mes) gives two product types[224] with structures shown in (**29**), both interesting variants on the trigonal bipyramid theme.

Several other recent papers further illustrate the above themes, such as tetrahedron plus two weak interactions[225]; trigonal bipyramids[226], with chelate rings[227]; a tridentate ring[228], including $Ph_2Sn(CH_2CH_2COOMe)I$ with SnI = 281 pm[229], and bridging into a linear polymer[230]; all *trans* octahedron[231], *cis* octahedron[232] and finally octahedral with chelate rings[233].

4. Mono-organo compounds

For mono-organotin, five- and six-coordination are often found[234]. Five-coordinate geometry verging on the square pyramid is sometimes observed as in various $RSnL_2$ compounds[235] for L a bidentate S ligand. Seven-coordination is also represented, usually

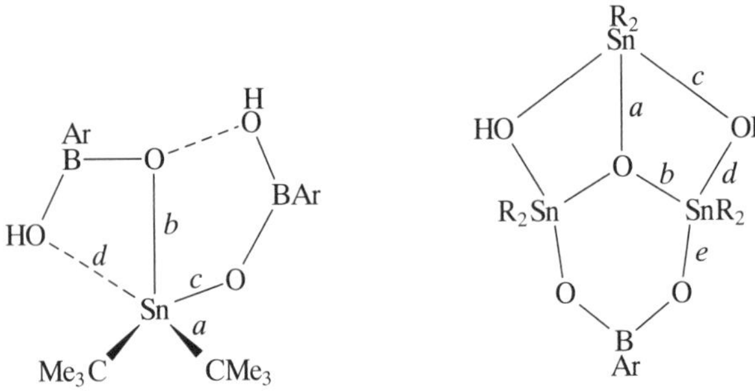

a = 217 pm
b = 197 pm
c = 199 pm
d = 275 pm
aa = 133°
cd = 146°

a = 210 pm
b = 206 pm
c = 217 pm
d = 233 pm
e = 204 pm
SnC = 218 pm

(29)

in pentagonal bipyramidal geometry. In 2-mercapto-pyridine (SPy) or -pyrimidine (SPym) complexes[236] p-TolSn(SPy)$_3$ and BuSn(SPym)$_3$ have R and the S of one ligand in axial positions (Sn–Bu = 213.1 pm and Sn–SPym = 244.5 pm; Sn–Tol = 211.9 pm, Sn–SPy = 248.6 pm). In the equatorial plane two ligands are coordinated with Sn–S bonds cis and longer than the axial (Sn–S = 250–256 pm) and the three N atoms complete the pentagon (Sn–N = 244–262 pm). The axial CSnS angle is 152°–156°.

a. Halogen compounds. Structures of [RSnCl$_5$]$^-$ ions have been reported[237] for R = Et or Ph with various counter-ions. The Sn is displaced towards R (angles average 94°) and the $trans$ Sn–Cl (241 pm for Et, 247 pm for Ph) is shorter than the cis values and the difference is greater in the Et anion (252 pm) than for R = Ph (249 pm). As hydrogen bonding to counter-ions occurs, it is unclear whether there is any specific $trans$ effect.

With a non-demanding substituent a dimeric anion is found[238] for [BuSn(OH)Cl$_2$· H$_2$O]$_2$$^{2-}$ with very distorted octahedral Sn and an $-$Sn(OH)Sn(OH)$-$ ring. One Cl (SnCl = 242.7 pm) and the Bu (213.2 pm) are $trans$ to the bridging OH (Sn–OH = 203.8 pm and 212.2 pm; CSnO = 160°, ClSnO = 172°) and the remaining two sites are occupied by Cl $trans$ to the water molecule (Sn–Cl = 248.5 pm, Sn–OH$_2$ = 221.6 pm). The water is hydrogen-bonded to the counter-ion.

RSnX$_3$ compounds may retain four-coordination with large R. In an interesting series where R is large and has available donor atoms, higher coordinations are found. For the case[239] of the polyfunctional ligand R = [2,6-(MeO)$_2$C$_6$H$_3$]$_3$C we find seven-coordination—to the C and 3X atoms (X = F, Cl, Br) and to one methoxy O on each of the three 2,6-dimethoxyphenyl groups. The coordinated oxygens are $anti$ to the halogens, giving approximately a trigonal antiprism with the Sn–C bond capping. This capped octahedron contrasts with the more common pentagonal bipyramidal geometry found for seven-coordination. The rings are twisted regularly in a propeller configuration. In contrast to the five-coordinate diorganotin compounds of Table 8, bond lengths increase with increasing halogen size (Table 12) and angles become more variable—for example, the CSnBr angles are 114.6°, 117.4° and 128.3°. All the bonds are lengthened, reflecting the high coordination, and the Sn–O distances denote a relatively weak but still substantial interaction. NMR studies explored exchange of OMe and ring rotation.

TABLE 12.   Parameters (pm and deg) for $[2,6\text{-}(MeO)_2C_6H_3]_3CSnX_3$ compounds

	X = F	X = Cl^a	X = Br
Sn—C	222	226	231
Sn—O	240–246	256–264	259–276
Sn—X	196	239	253
XSnX	94	92–99	94–100
OSnO	108–113	100–111	96–110
CSnX	119–124	114–126	114–128

^{a} Over the two independent molecules in the unit cell.

A less obvious route to mono-organotin species is to start with $SnCl_2$ and add L and Cl by reaction with LHgCl. Such a study[240] produced octahedrally coordinated $(LSnCl_2O)_2$ and $[LSnCl_2(OH)]_2$ dimers [L is a (phenylazo)phenyl species bonding through C—Sn and N—Sn] with SnOSnO or Sn(OH)Sn(OH) (30) four-membered rings. Unfortunately, the crystal for the oxide was disordered as a comparison of ring shapes would be interesting. In the $[Sn(OH)]_2$ ring, the Sn—O distances are quite asymmetric and relatively short.

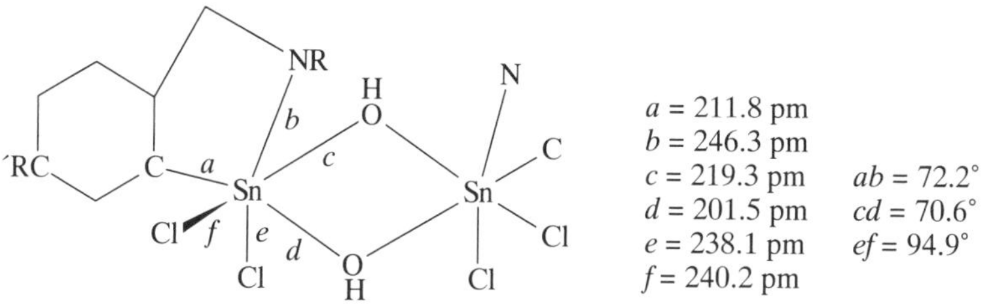

$a = 211.8$ pm
$b = 246.3$ pm
$c = 219.3$ pm $ab = 72.2°$
$d = 201.5$ pm $cd = 70.6°$
$e = 238.1$ pm $ef = 94.9°$
$f = 240.2$ pm

(30)

A similar $-SnO(R)SnO(R)-$ ring is found for $Pr^iSn(OPr^i)_3$, which is a dimer with five-coordinate Sn^{241}. The ring is asymmetric (Sn—O = 202.6 and 227.3 pm) and the latter is axial on Sn with the longer external O (OSnO = 158°); Sn—C = 217.7 pm.

Hydrolysis of trihalides gives complex cagelike structures[242], with an interesting example[243] involving the encapsulation of a Na^+ ion in $[(iso\text{-}PrSn)_{12}O_4(OH)_{24}]^{4+}$ $[Ag_7I_{11}]^{4-}\cdot NaCl\cdot H_2O\cdot 10DMSO$. The structure consists of 12 SnO_6 octahedra sharing edges and corners with 4 μ^3-O (Sn—O = 204.3–218.5 pm; Sn—OH = 199.1–221.9 pm), and with outward pointing Sn—R. Trimers of octahedra are located tetrahedrally around the central cavity holding the Na^+.

The pair of molecules Ar_2SnXY (31) illustrate nicely[244] the balance between five- and six-coordination. The ligand Ar is the 8-dimethylnaphthyl group where the side-chain N is in a potential chelating position. When X = Y = Br, both Sn—N distances are equal and correspond to a relatively long bond and distorted octahedral coordination. In contrast, when X = Me and Y = I, one Sn—N is shorter (at 253 pm) but the other is so long that the interaction is very weak.

b. Carboxylates. $[PhCH_2Sn(Ox)_2(OOCPh)]^{2-}$, R = $C_6H_5CH_2$; R' = $OC(O)C_6H_5$) has cis octahedral[245] bidentate oxalates whose remote oxygens are hydrogen bonded to the cations. The benzyl (Sn—C = 214.5 pm) and one O of the benzoate ion (Sn—O = 207.2 pm) are cis and complete the octahedron. The coordination to Sn of each oxalate

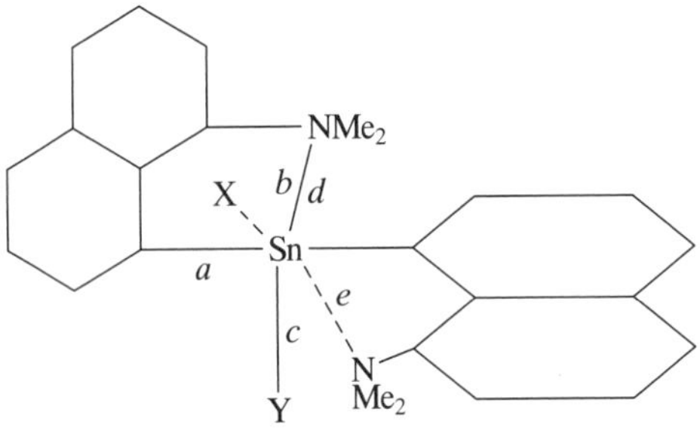

X = Me	X = Br
Y = I	Y = Br
a = 215 pm	a = 212 pm
b = 221 pm	b = 261 pm
c = 295 pm	c = 263 pm
d = 253 pm	d = 262 pm
e = 310 pm	e = 264 pm

(31)

is slightly asymmetric, and the group *trans* to benzyl is a little closer (Sn–O values 207.3 and 210.4 pm and 212.2 and 215.8 pm, respectively). The similar diorgano oxalate, $[Ph_2Sn(Ox)_2]^{2-}$ (see **21**, R = C_6H_5 = R'), has similar *cis* geometry with slightly longer and more asymmetric Sn–O (212.1 and 219.7 pm), normal Sn–Ph lengths and with CSnC = 106.2°.

c. Sulfur donors. Coordination geometry similar to that of (**22a**) is found in various $(S_2)_2SnRX$ species. Typical is the oxypropyl (R = CH_2CH_2COOMe; X = Cl, SS = Me_2NCS_2)[246] with the chelate Sn–S fairly symmetric (253.5 and 263.5 pm), SnCl = 245.8 pm and other geometry standard. Using the pyrrole derivative[247], SS = $C_4H_4NCS_2$, the PhSnCl derivative also gives a structure similar to (**22a**) with longer SnS (258 and 265 pm). The same study produced a $Me_2Sn(SS)_2$ derivative with much more asymmetry (Sn–S = 252 and 301 pm) and $Sn(SS)_4$ with one symmetric and one near-monodentate ligand (SnS = 241 and 319 pm).

Seven-coordination is found among bidentate 1.1-dithiolate complexes[248] of mono-organotin, $[RSn(S_2CY)_3]$, where R is alkyl or phenyl and Y is commonly NR'_2 or OR'. The distorted pentagonal bipyramids have R and one S in axial sites (average values Sn–S = 249 pm, CSnS = 165°) with the remaining S equatorial and more distant (Sn–S = 260–282 pm).

d. Nitrogen donors. The $CSnN_3Cl_2$ geometry of $PhCl_2SnL$ for L = pyrazolylborate is distorted octahedral[249] with a long Sn–C (223 pm), variable Sn–N (221.8, 224.1 and 228.9 pm) and SnCl (242.0 and 243.8 pm) governed largely by the geometry of the tridentate ligand.

The pair[250] of sulphinylimidazole complexes (**32a** and **32b**) nicely illustrate the delicate balance of acceptor power for $MeSnCl_3$ versus Me_2SnCl_2. The trichloride has a moderately long Sn–N bond and a clear Sn–O link giving distorted octahedral coordination. The dichloride has longer Sn–N and Sn–Cl bonds in axial positions in a trigonal bipyramid and the Sn...O distance is now 275 pm, which indicates only a very weak interaction, though shorter than the van der Waals sum. A pyridine analogue of **32a** showed similar geometry but with Sn–N = 231 pm.

B. Coordination Numbers Above Four for Ge and Pb

1. Germanium

a. Conventional complexes. The ion $[(CF_3)_3GeF_2]^{-}$[251] is a regular trigonal bipyramid (Ge–F = 183.5 pm, Ge–CF_3 = 200.0 pm), while the mono- or bis-trifluoromethyls give octahedral fluoro-anions.

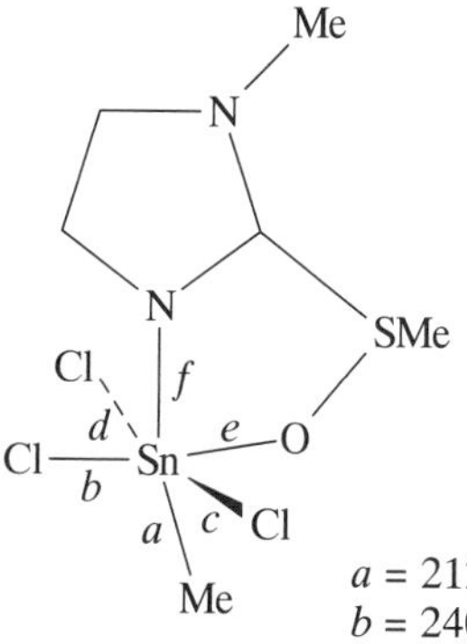

(32a)

a = 212.2 pm
b = 240.3 pm
c = 245.4 pm
d = 247.7 pm
e = 226.2 pm
f = 219.4 pm
ab = 96.5°
ef = 75.3°

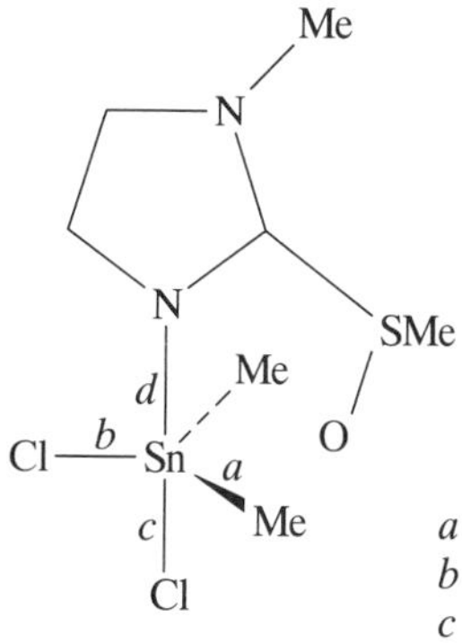

(32b)

a = 210.2 pm
b = 241.7 pm
c = 255.0 pm
d = 241.1 pm
cd = 175.7°
aa = 154.5°
ab = 103°

Hexafluorocumyl alcohol (= LH) gives a spirogermanacycle, L_2Ge, with two –GeCCCO– rings and distorted tetrahedral coordination (Ge–C = 189.8 pm, Ge–O = 178.6 pm, CGeC = 138°, OGeO = 110°). This adds nucleophiles to give five-coordinate anions, such as $[L_2GeBu]^-$ in a relatively regular trigonal bipyramid with axial oxygens (GeO = 198.9 pm, OGeO = 173.8°) and three equatorial carbons [Ge–C = 194.7 pm (Bu), 193.0 and 197.2 pm; CGeC = 119°–121°][252].

$Me_2GeX(S_2CNMe_2)$ (X = Cl, Br, I)[253] show trigonal bipyramidal geometry with axial X and S (XGeS = 195°–161°). The second Ge–S is equatorial, and much shorter (Table 13).

The tridentate dipeptides, as L = glycyl-L-methionate, give standard trigonal bipyramid coordination in Me_2GeL[254] with axial OGeN (amino) angle of 161.8° and Ge–O = 200.8 pm, Ge–N = 210.3 pm and equatorial peptide N (Ge–N = 188.9 pm) and methyls (Ge–C = 191 pm). The geometry is very similar to that found for the glygly analogue and also for corresponding tin dipeptides.

In mono-organogermanes, a systematic study[255] exemplifies work on factors influencing the variation between trigonal bipyramid and square pyramid geometries. This used *ortho-* disubstituted benzenes to provide (D...D) units, in $[RGe(D_2C_6X_4)_2]^-[Et_4N]^+$ for: (A) R = Me, D = O, X = H; (B) R = Me, D = S, X = H; (C) R = Ph, D = O, X = H; (D) R = Me, D = O, X = Cl and (E) R = Ph, D = O, X = Cl. The molecules are trigonal bipyramids, with increasing distortion from (A) to (B) to (C), while (D) and (E) are close to square pyramids (Table 14). Interestingly, the Et_3NH^+ salt of (C) is

TABLE 13.   Bond lengths (pm) in $Me_2GeX(S_2CNMe_2)^a$

	X = Cl	X = Br	X = I
Ge–Me	192.7	189/196	191.8
Ge–S(equat)	225.4	222.2/225.2	225.5
Ge–S(ax)	289.6	281.7/284	268.5
Ge–X	225.1	242	271.2

^{a}Two independent molecules for bromide.

TABLE 14.   Parameters (pm and deg) for $RGe(D_2)_2^-$ ions[a]

(1) Trigonal bipyramid geometry (mean values)

	(A) Me, O	(B) Me, S	(C) Ph, O
Ge—C	194.6	197	193.6
Ge—D(ax)	192.0	239.8	189.0
Ge—D(eq)	183.4	227.3	184.6
DGeD(ax,ax)	166.3	165.2	160.1
CGeD(eq)	123.1	130.4	134.0
CGeD(ax)	96.8	97.1	100
DGeD(eq,ax)	86.8	86.8	86.1

(2) Square pyramid values

	(D) Me, O	(E) Ph, O
Ge—C	190.7	193.0
Ge—D	187.2–189.0	185.9–188.4
DGeD (trans)	142.5, 149.4	143.0, 150.3
CGeD	105.2–109.4	104.6–109.3
DGeD (cis)	84.4–85.7	85.0–85.8

[a] For (A), (B), (C), (D), (E) — see text.

closer to square pyramid as hydrogen bonding to the cation lengthens two of the Ge—O bonds.

Six-coordination is found in L = lactamo-N-methyl compounds $L_2GeCl(OSO_2CF_3)$ where L bonds through C and N and structures for NC_x ring sizes with $x = 4,5,6$ were determined[256]. Only one Cl could be replaced by triflate. The structure is distorted all-*trans* with OGeO = $167°–173.8°$ (increasing with x) and Ge–O = 198–205 pm; CGeC = $108°–113°$ and Ge–C = 192–197 pm. The third axis contains Ge—Cl (213–217 pm) and the weakly bonded triflate (Ge–O = 302–336 pm) with OGeCl = $165°–169°$. The coordination is clearly 5 plus 1, but the angles match the distorted octahedron description reasonably.

A porphyrin $Me_2Ge(dpb)$ shows considerable cyctotoxic effects[257]. The six-coordinate structure shows GeC = 199 pm and GeN = 202–203 pm with a MeGe plane angle of $86.9°$.

An example of a weak interaction is seen[258] in $Mes_3GeNHCOBu^t$ which has normal values for GeN (189.9 pm) and Ge–mes (198.6 pm) but irregular angles (two NGeC at $101.7°$ and one = $111.6°$; two CGeC at $117.4°$, and one = $106.3°$). The Ge...O distance is 316 pm, a little shorter than the van der Waals radii sum, and the direction involves the larger angles, suggesting a weak interaction, and in accord with a study[259] which identifies weak to modest five-coordination over a range of Ge–O distances of 323 pm to 248 pm.

The search for reaction models led to the synthesis[260] of the Ar_3GeH species where donation by side-chain N offers the possibility of seven-coordination (Figure 7). The structure of the germanium triphenyl derivative shows three Ge—Ar bonds and the Ge—H in approximately tetrahedral array with Ge–C = 195.9 pm, Ge–H = 158 pm, mean ArGeAr angles of $106.7°$ and ArGeH averaging $112°$. In addition, each N is *trans* to an aryl (average NGeC = $174°$) at an average N···Ge distance of 305 pm, some 7 pm shorter than the van der Waals radii sum. Thus the Ge has four strong and three weak interactions. Similar properties are found for related complexes, including a silicon analogue.

b. Germatranes and related species. The classical reaction of triethanolamine (**33**) with $RSiCl_3$ to give *silatrane* with E = Si, D = O) was intriguing because the possibility of

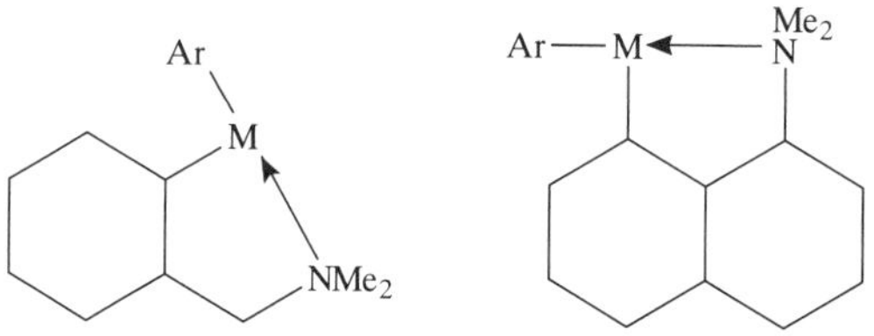

FIGURE 7. Schematic coordination of Ar—N ligands giving high coordination Ge and Si species

transannular donation from N to Si raised options for modification of Si–R reactions and involved the, at the time, relatively rare trigonal bipyramid coordination of Si. Additional interest was provided by the biological activity of silatranes—again, an unusual feature of Si chemistry then current.

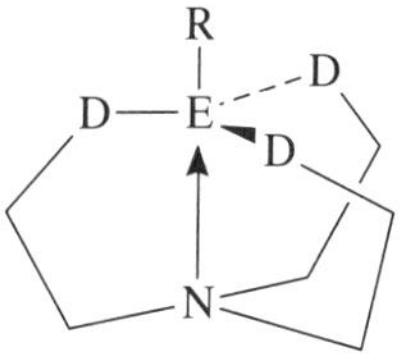

(33)

The name was taken up and the *atrane* root is now widely and conveniently used to designate the total class. For compounds where the O atoms are replaced by NR, the term *azatrane* is used, and the terms *proatrane* and *quasiatrane* have been suggested for molecules where the E...N interaction across the ring is non-existent (E...N distance equal to van der Waals sum) or weak, respectively. The whole class of atranes is now very extensive, involving E from transition metal groups as well as main groups. The full field is discussed by Verkade[261] with emphasis on the effect of the E...N interaction on the chemistry. Within the organometallic subset[262] where E = Ge, Sn and R is organic, germatranes were extensively studied from the 60s, partly because of the possibility of biological activity for germanium, though it is now found that such activity is relatively weak[263]. The Ge–N transannular bond is typically 213–224 pm long—a relatively weak interaction (compare Tables 2 and 5). Recent work has developed azatrane chemistry of Ge[264] and Sn[265]. The structure of the phenyltin compound **33**, ER = SnPh, D = NH, shows two independent molecules with average values Sn–C = 216.4 pm, Sn–NH = 206 pm and transannular Sn–N = 242 pm.

2. Lead

Shortening from four-coordination is seen in the Pb–Ph lengths (215.3–218.9 pm) in Ph₃PbBr.OPPh₃[266]. The shape is near-regular (Pb–Br = 275.4 pm, Pb–O = 255.6 pm, OPbBr = 176.9°, CPbC = 118.5°–120°). Similarly, shorter Pb–CH₃ (218.8 pm) is seen in Me₃PbN(SO₂Me)[18] in which, in contrast to the Ge analogue, there is an intermolecular interaction (Pb–O = 265 pm, NPbO = 169°) to give a chain of trigonal bipyramidal Pb units. Similar features were found for (μ-NNN)PbMe₃[267]. A shorter Pb–CH₃ value of 217 pm is found for (2-furanato)trimethyllead[268] where a planar PbMe₃ unit linked into a chain by slightly unsymmetrically bridging carboxylate units (Pb–O distances of 235.3 pm and 253.4 pm). The distance to the formally non-bonded carboxylate O is 317 pm and there is a further Pb...O interaction of 355 pm to the furan ring O. Similar structures and values are reported[269] in other trimethyl- and triphenyl-lead carboxylates. The Pb–CH$_3$ distance reduces again[270] to 214.2 pm in dimethylleadbis(4-pyrimidinecarboxylate), presumably six-coordinated.

Simple six-coordination is seen[271] in [PhPbCl$_5$]$^{2-}$. Ph$_2$Pb(NCS)$_2$.2HMPA shows regular *trans*-octahedral geometry[272] (CPbC, NPbN and OPbO all 180°; PbC = 215 pm, Pb–O = 234 pm, PbN = 242 pm). *Trans*-octahedral geometry is also found in Ph$_2$Pb{S$_2$P(OCH$_2$Ph)$_2$}$_2$[273] with axial Ph (angle 165°, length 219.5 pm) and two four-ring asymmetric chelates (PbS = 270 and 295 pm). In contrast, the triorganics Ph$_3$PbS$_2$P(OCH$_2$Et)$_2$ and Ph$_3$PbSPh are tetrahedral with normal Pb–C (220 pm) and shorter Pb–S of 252–255 pm.

Seven-coordination is found in Ph$_2$PbL.H$_2$O[274] where L = 2,6-pyridinedicarboxylate. The phenyls are axial (Pb–C = 214.3 pm, CPbC = 172.8°) and the central pentagon contains one tridentate L (Pb–O = 237.4 and 257.1 pm, Pb–N = 245.2 pm), the water molecule (Pb–O = 247.2 pm) and the second O from the carboxylate of a neighbouring molecule (Pb–O = 251.4 pm). The bridging gives chains which are probably connected by hydrogen bonding involving the water and carboxylate. Axial-equatorial angles are in the range 85.5°–94.8°, so the pentagonal bipyramid is fairly regular. A mono-organic also gives seven-coordinate Pb in 2-XC$_6$H$_4$Pb(OAc)$_3$ for X = Me or Cl[275]. The acetates are bidentate with asymmetric Pb–O (averaging 221 and 247 pm). The aryl and one O are axial (Pb–C = 218 pm, Pb–O = 220 pm, CPbO = 150°) and the remaining five Pb–O are equatorial. A more complex structure is found for [Ph$_2$Pb(OAc)$_3$]$_2$.H$_2$O[276] where the two phenyls are axial (CPbC = 169°) and five O atoms are coplanar with Pb to give a pentagonal bipyramid. The PbO$_5$ coordination differs for the two Pb nuclei. Each has one bidentate asymmetric acetate (Pb–O = 233 and 251 pm) coordinated only to it. A second acetate is involved in holding the two centres together in different ways. One (Pb–O = 232 and 264 pm) also bridges to the second Pb (272 pm). The other (Pb–O = 233 and 265 pm) links back by hydrogen bonding to the water which is bonded to the first Pb (Pb–OH$_2$ = 258 pm). The Pb–Ph distances are 198 pm on the first Pb and longer at 217 pm on the second. Thus the molecule has two pentagonal bipyramids linked by one apex and a hydrogen bond.

An older study[277] established 8-coordination as a hexagonal bipyramid in [Ph$_2$Pb(OOCMe)$_3$]$^-$.

V. ORGANO-E(IV) COMPOUNDS OF TRANSITION METALS

It is possible only to sketch lightly this major field. By 1981[1,2], it was well-established that R$_3$E groups could bond to any transition metal and such bonds were most readily regarded as two-electron two-centre bonds. One or two of the R groups could be replaced by H, X or other functional groups. A further significant class had R$_2$E groups bridging a M—M bond or the edge of a cluster, or else forming an —EMEM— four-membered ring. In all such compounds the geometry at E is basically tetrahedral and the E—C bond parameters fall within the normal range. The number of reported structures is distinctly less for E = Pb than for Sn or Ge. Transition metal–heavy Main Group chemistry is reviewed to 1990 and–tin chemistry to 1988[278].

An indicative group of more recent studies includes structures involving Ph$_3$Ge—Yb[279], Ph$_3$Sn—Ti (with 7-coordinate Ti)[280], Ph$_3$Sn—Zr, Hf (involving 8-coordinate M)[281], Ph$_3$Pb—V[282], Ph$_2$SnCl—Cr[283], Ph$_3$Sn—Mo[284] an interesting chain in (Ph$_3$Ge—W—)$_2$[285], germylene-W including Cl(cp)Ge—W[286], BuSnCl$_2$—W[287], a C$_4$Ge ring bonded to Mn,

Fe or Co[288], $Ph_2XGe-Re$[289], various Ge–Fe compounds[290], Me_3Sn and Me_3Pb-Fe or Co[291], $(Ph_2RSn)_2-Fe$[292], $(Et_2Pb-Fe)_2$ ring[293], $Me_2(SH)Sn-Ru$[294], tol_3Ge-Os[295], Me_3Sn-Co[296], $(F_5C_6)_2Ge-Ni$[297], an $R_2SnCCNi$ ring[298], $(Me_2ClGe)_2-Pt$ from a digermane[299] and $MeCl_2Sn-Pt(IV)$[300].

In some cases E is involved in further interactions. Thus additional M–X–E bridging gives five-coordinate Sn in species involving $RCl_2Sn(...Cl)$-Mo[301] and $RCl_2Sn(...S)$-Mo[302], possible Sn–H–Fe 3-centre bonds in a $Ph_3Sn-FeH_3$ species[303]. A chlorotetraphenylgermole is characterized as a Ge–Fe species[304]. A phenylgermole shows different coordination to two Co atoms[305] in η^1 and η^3 modes.

Insertion of the germylene or stannylene, $E(\mu\text{-NBu}^t)_2SiMe_2$, into the Fe–Me bond of $cpFe(CO)_2Me$ gives[306] $cpFe(CO)_2E(Me)(\mu\text{-NBu-}t)_2SiMe_2$ with four-coordinate E and normal E–Me (Sn–C = 214.8 pm, Ge–C = 197.7 pm; MeEFe angles 112°). Similar insertions into the dimer give inorganic analogues with two EFe bonds. An unusual $-(R_3P)_2Pt-SnAr_2-O-SnAr_2-O$ – five-membered ring is formed[67b] in the reaction with a cyclostannoxane [$Ar = Ar(f)$] and the platinum dioxygen complex: Sn–Pt = 262.8 pm, Sn–O averages 195 pm and the Sn–C bonds are 224 pm.

More complex M_xE_y structures are also found, including raft species as in R_2Sn bridging Ru–Ru in a triangle of triangles[307], and R_2Sn bridging Ir in a raft with the sail up[308]; a Main Group cluster $(R_2Ge)_3Bi_2$ further bonded to Pt[309]; the product from $Tb(Tip)SnS_4$ with $Os_3(CO)_{12}$ to give a cluster[310] including a Sn–Os bond (SnOs = 266.2 pm, SnTb = 219.3 pm, SnTip = 216.5 pm) linked also by a μ^3-S to a second Os (SnS = 248.9 pm). In $\{[cp^*Rb(\eta\text{-}C_6H_5)]_4Ge\}^{+}$[311], each ring of tetraphenylgermane is π-bonded to a Ru. When mono-organo compounds are used[312], the RE unit can act as apex of open or closed clusters as in the $RECo_3$ trigonal pyramid (closed for E = Ge, open for E = Sn), the $(RGe)_2M_4$ square bipyramid (M = Co, Fe) and nets of linked EM_2 (E = Ge, Sn; M = Fe, Co) triangles terminated by RE or R_2E groups[313].

VI. COMPOUNDS CONTAINING E–E OR E–E′ BONDS: E = Ge, Sn OR Pb; E′ = Si, Ge, Sn OR Pb

A. Introduction

A range of syntheses is available, but the two commonest are:

(a) combination of radicals formed by pyrolysis, photolysis, plasma or silent electric discharge;

(b) coupling, most commonly from E–X bonds (X = electronegative group such as halide) using alkali metals. Specific bonds, especially E–E′ ones, may be formed from the reaction of E–X with $E'-M^+$ species such as R_3EK.

The classical perception of restricted chains in E_nH_{2n+2}, has been overturned in the second half of this century and there is now no upper limit to the straight or branched chain compounds for E = Si, Ge or Sn, alone or in combination. Rings with from 4 to 7 of these E atoms were also stable, together with indications of much larger species. Three-membered rings are characterized but much less stable. Limits to E–E formation are apparent only for lead[314], where structurally characterized compounds are restricted to R_6Pb_2 though preparations are reported for R_8Pb_3 and $Pb(PbR_3)_4$.

A few early vapour-phase structures (both ED and MW) and the first wave of X-ray studies allowed Dräger and colleagues[315] to list 15 structures with Ge–Ge bonds in 1983. Molloy and Zuckerman[5] could nevertheless quote the limited number of polygermanes and the dominance of ditins in their list of structures as showing a decreasing tendency to catenation down the group. By the mid-80s, there were over 100 Si–Si structures, with about half that number of Sn—Sn, one third Ge—Ge, and few Pb—Pb structures with a similar low proportion of mixed E—E' structures, and these numbers have increased markedly over the last ten years. An even larger range and variety of structures and isomers is indicated spectroscopically including Sn Mossbauer and Sn[119/7], Pb[209] and even Ge[73] NMR to reinforce H[1] and vibrational methods.

Compounds containing mixed E—E' bonds have been increasingly studied and a number of crystal structures have appeared in the last five years—interestingly, only three years ago Sheldrick[4a] could report only one determination of the Ge—Si bond length and that from silylgermane.

In the last few years, a further exciting development has been the characterization of prismanes, cubanes and similar clusters paralleling the similar interst in organic chemistry—though we do not yet have a buckminsterfullerene.

B. Dinuclear Compounds, E_2R_6 and Similar Species

Table 15 lists bond parameters for the simple digermanes, distannanes, diplumbanes and mixed dinuclear species. Some disilanes and the gas-phase value for Ge_2H_6 are included for comparison. The set of Si—Ge examples has increased since Sheldrick's review, and all the other E—E' combinations are now represented.

The E—ligand bond lengths, including those for bulky ligands, match those found in simple mononuclear compounds (Tables 2 and 3). The E—E bond lengths may be compared with values for tetrahedral E in the elements (pm):

C—C	Si—Si	Ge—Ge	Sn—Sn	Pb—Pb
154.45	235.2	245.0	281.0	(350 in 12-coordination)

Determinations in Table 15 cover some twenty-five years, and a number of studies suffered from poor crystals, so the initial level of comparison is at the level of a few picometers. In a number of cases, disorder between E and E' vitiated detailed discussion, for example of the potentially interesting and well-represented series of Ar_6EE' molecules. For simple ligands, the E—E bond lengths fall into quite tight ranges, slightly shorter than in the elements, and the E—E' values interpolate.

Steric and electron donation properties of the ligands strikingly effect the t-Bu species, especially $Si_2Bu_6^t$ where Si—Si is 35 pm larger than the mean, and lengthening persists for all $E_2Bu_6^t$ species with even $Pb_2Bu_6^t$ showing an increase of some 5 pm.

Steric effects are seen for other larger ligands. For Ph, an impact is seen for Si, where Si_2Ph_6 has not been crystallized on its own but in $Si_2Ph_6 \cdot Pb_2Ph_6$, and Si—Si is some 20 pm longer than the Si—Si norm. The value for $Me_3SiSiPh_3$ is normal, indicating a steric rather than an electronic effect. For dinuclear species of the heavier elements, including Si—Ge, the effect of simple aryls is much less marked and bonds in the hexaphenyls and p-tolyls fall close to Table 3 values, but the hexa*ortho*-tolyl derivatives do show an increase.

At a more precise level, bond lengths reflect expected electronic effects; thus the Ge—Ge distance of 241.3 pm in $(PhGeCl_2)_2$ is the shortest value reported from the crystal structure of an unconstrained digermane as expected for electron-withdrawing ligands, but the effect is small, noting that Ge—Ge in $Ph_3GeGeMe_3$ is only 241.8 pm, and that the gas-phase electron diffraction value for Ge_2H_6 is 240.3 pm. Detailed comparisons need to be undertaken with caution unless a closely related series is studied under similar conditions.

One such study is that of Pannell and colleagues[348] who have related the changing crystal structure of the $Me_3EE'Ph_3$ series to change of E and E'. Five of the reported structures (E = E' = Si; E = Si, E' = Ge; E = Ge, E' = Si) crystallized in a triclinic space group and were isomorphous while the other two (E = Ge, E' = Sn and

TABLE 15. Parameters (pm and deg) for dinuclear species E_2R_6 and $R_3EE'R_3'$

Compound	d (E–E)	d (E–R/X)	Ref.	Other parameters	Notes
Disilanes					
Si_2H_6	233.1	Si–H = 149.2	316	HSiSi = 101.3	ED
Si_2Me_6	234.0	Si–Me = 187.7		MeSiMe = 108.4	XRD & ED
$Ph_3SiSiMe_3$	235.5	Si–Me = 186.2		SiSiMe = 109.9	
		Si–Ph = 188.6		SiSiPh = 110.3	
Si_2Ph_6	251.9	Si–Ph = 189.2	317	SiSiPh = 107.3	Cocryst with Ph_6Pb_2
$Si_2(Bu-t)_6$	269.7	Bu^t = 199	318		Longest reported Si–Si
Digermanes					
Ge_2H_6	240.3		319		ED
Ge_2Ph_6	243.7	Ge–Ph = 194.8–196.3	320	triclinic mod	
$Ge_2Ph_6{\cdot}2C_6H_6$	244.6	Ge–Ph = 195.8–197.1	321	rhombohedral	Molecules are achiral bipropellers (S_6 symmetry)
$Ge_2(p\text{-}Tol)_6$	242.3	Me = 194.3	322		
$Ph_3GeGeMe_3$	241.8	Ph = 195.7	323		
$Ge_2(Bu-t)_6$	270.5 and 271.4	C = 204.7–208.5 and 205.6–212.3	324	2 mols in unit cell	Longest Ge–Ge and Ge–C known so far;
$(PhCl_2Ge)_2$	241.3		325		Ge–Ge in $Ph_3Ge_2Cl_3$ is most readily cleaved of all mixed Cl/Ph digermanes
Silylgermanes					
H_3GeSiH_3	235.7		326		MW
$Me_3GeSiPh_3$	239.4	Ge–Me 195.8	327		
		Si–Ph 188.5			

(continued overleaf)

TABLE 15. (continued)

Compound	d (E–E)	d (E–R/X)	Ref.	Other parameters	Notes
Ph$_3$GeSiMe$_3$	238.4	Ge–Ph 195.8, Si–Me 186.3	328, 329	CSiC = 110.1°, CGeC = 107.9°	
Ph$_3$GeSiMe$_2$L, L = Fe(CO)$_2$cp	240.5	Si–Me 187.7–188.5, Ge–Ph 195.9–196.6	328	Fe–Si = 232.8pm, GeSiFe = 118.1°	Compare disilane[330]
Germylstannanes					
Me$_3$GeSnPh$_3$	260.1	Ge–Me = 194.7–197.3, Sn–Ph = 214.3–216.2	323, 331		Note Ge–Sn correction from Ref.[323]
Ph$_3$GeSnMe$_3$	261.1	Ge–Ph = 192.4–194.0, Sn–Me = 213.7–219.1	323, 331		Note Ge–Sn correction from Ref.[323]
Ph$_3$GeSnPh$_3$	260.6		332		
Germylplumbanes					
Ph$_3$GePbPh$_3$	262.3	Ge–Ph = 197.3–210.5, Pb–Ph = 212.9–218.8	317		Ge and Pb 50/50 disordered over sites
Ph$_3$GePb(p-Tol)$_3$	264.2	Ge–Ph = 194, Pb–Tol = 222	333	CGeC = 109.4, CPbC = 105.5	
(p-Tol)$_3$GePb(p-Tol)$_3$	259.9	Ge–Ph = 208, Pb–Tol = 216	333	CGeC = 105.0, CPbC = 104.7	Ge and Pb 63/37 disordered over sites
Distannanes					
Sn$_2$Ph$_6$	278.0 (a), 275.9 (b)	Sn–Ph = 216.8 (a), 214.1, 216.8, 226.8 (b)	334		(a) (b) Two independent molecules in unit cell
Sn$_2$Ph$_6$·2C$_6$H$_6$	276.66	Sn–Ph = 213.7	335		Benzene molecules perpendicular to C$_3$ axis of distannane

Compound	Sn–Sn	Sn–X	Ref.	Angles	Comments
Sn$_2$p-Tol$_6$	A 277.7, B 277.8	Sn–Tol A 214.4–215.5, B 213.9	336		A from CHCl$_3$ and B from C$_6$H$_6$ are homeotypes with differences of parameters
Sn$_2$o-Tol$_6$	288.3	Sn–Tol 216.1–217.2	337	CSnC = 105°–106°: SnSnC = 112°–113°	
Sn$_2$Bu$^t_6$	289.4	Sn–C = 222–226	338	CSnC = 106.4°–109.2° SnSnC = 110.6°–112.0°	
Sn$_2$Bz$_6$	282.3	Sn–C = 219.1	338	CSnC = 103.2°: SnSnC = 115.2°	
[ArBu$^t_2$Sn]$_2$ Ar = Tip	303.4	Sn–Ar = 224.0, Sn–Bu^t = 224.6 and 227.2	339	ArSnBu^t = 99°–120° Bu^{t}SnBu^t = 103.8°	Very long Sn–Sn, irregular angles, near-eclipsed with syn aryl groups
[Ar$_2$MeSn]$_2$ Ar = Tip	282.9	Sn–Ar = 217–219, Sn–Me = 222	338	ArSnAr = 116° ArSnMe = 103°	
[Ar$_2$BrSn]$_2$ Ar = Tip	284.1	Sn–Ar = 213–221, Sn–Br = 254.6	338	ArSnAr = 103°–108° BrSnSn = 99.5°	
[SnMe$_2$Cl]$_2$ (with intermolecular Cl...Sn links)	277.0	Sn–Me 214, Sn...Cl 244.5, Sn...Cl 324.0, 329.2	340	CSnC = 113.9° SnSnCl = 120.5° SnSnCl = 100.8° SnSn...Cl = 77.8°	Intermolecular Sn...Cl links give double chain with distorted trigonal bipyramidal Sn Cl trans across Sn–Sn
[(Me$_3$SiCH$_2$)$_2$ClSn]$_2$	284.4	Sn–C = 219, Sn–Cl = 236.5	341	CSnC = 109.2°	

(continued overleaf)

TABLE 15. (continued)

Compound	d (E–E)	d (E–R/X)	Ref.	Other parameters	Notes
Plumbylstannanes					
$Ph_3SnPbPh_3$	280.9 and 284.8	(Sn,Pb)–Ph = 221	317	2 mols in unit cell	Sn and Pb 50% disordered over E sites
$p\text{-}Tol_3SnPbp\text{-}Tol_3$	281.3	M-Tol = 216–218	336	MMC = 113° CMC = 107°	Sn and Pb 1:1 disordered
$o\text{-}Tol_3SnPbo\text{-}Tol_3$	284.5	M-Tol = 220.9–222.4	337		Sn and Pb disordered
Diplumbanes					
Pb_2Me_6	288		342		ED
Pb_2Ph_6	284.8 and 283.9	Pb–Ph = 220.5–225.3	343	CPbC = 108.7° (av) CPbPb = 110.2° (av)	Two independent molecules in unit cell
	284.6	Pb–Ph = 222.2	314		Cocryst with Si_2Ph_6
$Pb_2(p\text{-}Tol)_6$	285.1	Pb–Tol = 217.8–221.8	336	CPbC = 106.4° PbPbC = 112.4°	
$Pb_2(p\text{-}Tol)_3Ph_3$	283 and 286	Pb–C = 223	344	$R_3PbLi + R'_3PbCl$ gives range of products from migrations of R, R'	2 disordered independent molecules in unit cell; substituent distribution $Ph(p\text{-}Tol)_2Pb\text{-}Pb(p\text{-}Tol)Ph_2$
$Pb_2o\text{-}Tol_6$	289.5	Pb–Tol = 224.2–224.9	337, 345	CPbC = 105.7° PbPbC = 113°	
$Pb_2Bu^t_6$	293.7		337		
$Pb_2(C_6H_{11})_6$	287.6		346		
$Ph_3PbPb(Tsi)Ph_2$	290.8	Ph = 225.0 av Pb–C(Tsi) = 228.0	347	PbPbC angles 118°–121° in the PhPbPbTsi plane, 100°–105° out of plane	

E = Sn, E′ = Ge) crystallized in a closely related orthorhombic group. The structures consist of chains of molecules fitting head-to-tail and with the triphenyl groups of one chain fitting the clefts of the neighbouring ones, with adjacent chains oriented in opposite directions for the triclinic structure and in the same direction for the orthorhombic one. Pannell found the missing link compound, $Ph_3GeGeMe_3$, was isostructural with the triclinic group, defining more closely where the morphotropic change occurs. In addition, the two well-defined pairs of values for the isomers — Si–Ge = 239.4(1) and 238.4(1) and Ge–Sn = 261.1(1) and 260.1(1) — does give a firm indication of the electronic effects. Relationships between structure, force constants and NMR properties are discussed for

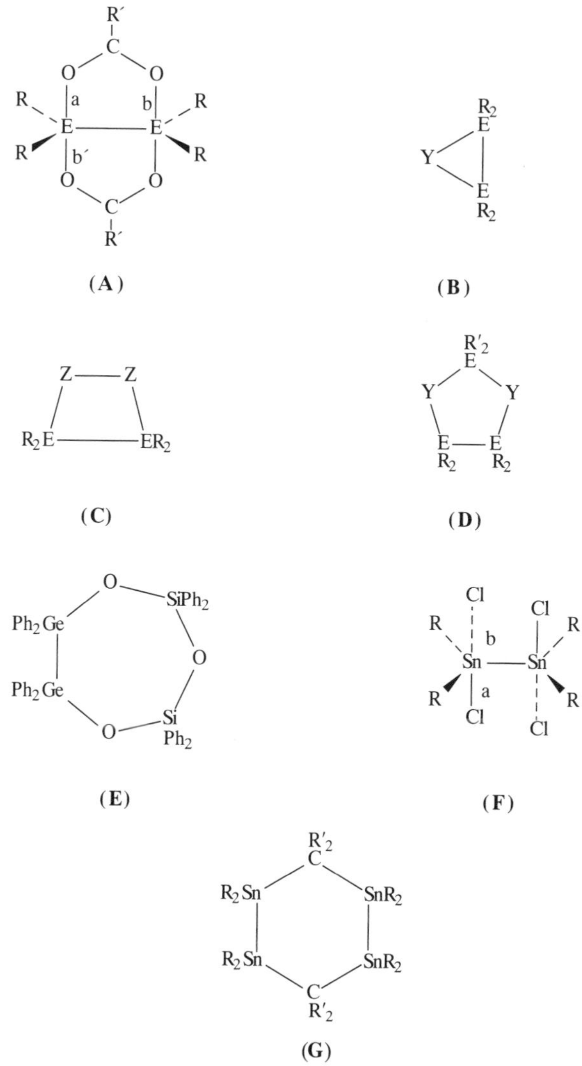

FIGURE 8. Representations of more complex structures containing E–E bonds

the now-extensive range of Ar_6EE' compounds[349]. These studies illustrate the potential for refined analysis in favourable cases.

When E−E is part of a larger structure, it is of interest to see how far the bond length accommodates to constraints. Table 16 collects E−E data for bonds in more complex situations, as part of larger rings or with increased coordination, and Figure 8 (structures A to G) summarizes some geometries.

The E−E bonds can accommodate substantially to imposed constraints, as in the classic case of the acetates, $[R'_2SnO_2CR]_2$, where the two O−CR−O groups bridge the Sn−Sn bond giving 5-coordinate Sn A in the usual trigonal bipyramid configuration with axial oxygens with non-linear OEO. The Sn–Sn distance is reduced some 8 pm, and varies a little with R, in both cases of $R' = Me$ or Ph, presumably reflecting the bite of the bridging acid groups. The germanium analogue, $[Ph'_2GeO_2CCl_3]_2$, shows a similar though smaller contraction.

Similarly, Sn−Sn in $(ClSnMe_2)_2$, is short at 277.0 pm and there is relatively weak intermolecular Sn. . . Cl interaction, giving a double chain and trigonal bipyramidal coordination. Similar geometry is found with $MeN(CH_2CH_2CH_2)_2$ which bonds through two C and N in the Sn−Sn linked dimer.

An interesting case is provided by the 1,2,4,5-tetrastannacyclohexanes (Figure 8, G). Of the four compounds which have been crystallographically characterized, two are in the boat form and two in the chair. Where tin bears 2 Me groups, the octamethyl[367] and decamethyl[366] compounds are boat while the dodecamethyl[365] is chair, explicable by the increasing 1,4 axial interaction between, respectively, the two CH_2, CHMe and CMe_2 groups. For these three, the SnCSn angle decreases steadily with increasing methyl number from 116° to 111°. However, when $SnMe_2$ is replaced by $SnPh_2$, the chair form is found in the octaphenyl species[369], showing that interactions between groups on the Sn are also important in determining the stereochemistry.

Contractions are seen when single Ge−Ge bonds are found in GeGeX three-membered rings B, where X is a small group like CH_2, NPh or S, even with bulky ligands on Ge. Where X = Te, or in larger rings C, D or E, the Ge−Ge values revert to the normal range. In complex cyclobutenes[376] D, Ge−Ge remains normal at 245.9 pm and Sn−Sn at 280.3 pm.

The Sn−Sn bond in the azadistannirane B is unusually short while most of the other ring values are normal. The enlarging effect of t-butyl and polysubstituted phenyls persists in the rings.

An unusual Ge−Sn situation arises in the product from the reaction where two molecules of the germylene, $cp^*Ge(Bsi)$, add to $(Me_3Sn)_2C=N_2$ to give $[cp^*Ge(Bsi)(SnMe_3)]NNC[cp^*Ge(Bsi)(SnMe_3)]$[377], where the two Ge−Sn distances are equal (264.8 and 265.2 pm) as are Ge−cp^* (205.6 pm). The Ge−C(Bsi) vary slightly with the longer (198.4 pm) matching the longer chain bond (GeC = 193.5 pm) and the other (196.9 pm) on the N-bonded Ge (GeN = 188.3 pm). The linking chain is linear (CNN = 174.2°) and has GeCN = 150° and NNGe = 130 pm.

C. Polynuclear Compounds

1. General

Polysilanes, whose heavier analogues have not yet been structurally characterized, include:

(a) Highly sterically crowded molecules[378] like the octasilane $[(Me_3Si)_3Si]_2$ and other species with multiple $(Me_3Si)_3Si$ ('supersilyl') or similar ligands, including the novel Tl−Tl bonded species[379] $[\{Me_3Si)_3Si\}_2Tl]_2$.

TABLE 16. Parameters (pm and deg) for more complex structures involving E–E or E–E' bonds

Compound	d(E–E)	d(E–R/X)	Ref.	Structure	Values
Carboxylates bridging E–E					
[(Cl₃CCO₂)Ph₂Ge]₂	239.3	Ge–Ph = 193.5 and 194.9	350	A	$a = 207.3$ $b = 231.4$ $ab' = 175.4$
[SnMe₂O₂CR]₂ (a) R = CH₂Cl (b) R = CCl₃ (c) R = CF₃	269.2 (a) 271.1 (b) 270.7 (c)		351	A	
[SnPh₂O₂CR]₂ (a) R = CH₃ (b) R = CCl₃ (c) R = CF₃	269.4 (a) 271.1 (b) 271.9 (c)	Sn–Ph = 211.0–213.8	352	A	$a = 226.0$ (a), 229.5 (b) and (c), $b = 227.6$(a), 232.2 (b), 231.7 (c) $ab' = 168.2–168.9$
Ge–Ge units in larger rings					
Ar₂Ge–GeAr₂ Ar = Mes	237.6	Ge–S = 226.3, 227.7	353	B ER = GeAr Y = S	One long Ge–Mes bond GeSGe = 63.1°
Ar₂Ge–GeAr₂ Ar = Dep	238.7	Ge–Ar = 3 of 194.7–195.3, 201.5	354	B ER = GeAr Y = S	
Ar₂Ge–GeAr₂ Ar = Dep	239.8		354	B ER = GeAr Y = Se	
Ar₂Ge–GeAr₂ Ar = Dep	243.5	Ge–Te = 259.7 Ge–Ar = 199.3–200	355	B ER = GeAr Y = Te	GeTeGe = 55.9° Aryl groups slightly disordered
Ar₂Ge–GeAr₂ Ar = Dep	237.9	Ge–CH₂ = 197.0 Ge–Ar = 197.5, 198.4	356	B ER = GeAr Y = CH₂	
Ar₂Ge–GeAr₂ Ar = Dep	237.9	Ge–N = 187.6 Ge–Ar = 197.1, 198.6	356	B ER = GeAr Y = NPh	GeGeNC(Ph) skeleton is planar

(continued overleaf)

TABLE 16. (continued)

Compound	d(E–E)	d(E–R/X)	Ref.	Structure	Values
Ar₂Ge–GeAr₂ Ar = Dep	244.1	Ge–O = 185.7 Ge–Ar = 195.8, 197.0 O–O = 147 OGeGe = 74.1° OOGe = 103.9°	357	**C** ER₂ = GeAr₂ Z = O	GeGeOO ring C_2 and non-planar; GeOO/OOGe torsion = 19.5°
Ph₂Ge–GePh₂	242.1	Ge–Ph = 193.9–195.7 Ge–P = 233.2–234.7	358	**C** Z = PBu^t	
Ph₂Ge–GePh₂	243.0	Ge–Ph = 193–197 Ge–CMe₃ = 200 Ge–O = 178.9	359	**D** E = Ge R = Ph R′ = Bu^t Y = O	
Ph₂Ge–GePh₂	245.86	Ge–Ph = 193.9–195.9 Ge–O = 177.7	360	**E**	Angles at Ge 108° – 111° Ge–O–Si = 143.2° – 145.3°
Ph₂Ge–GePh₂ in pergerma-dioxane	244.8	Ge–Ph = 193.4–194.9 Ge–O = 178–179	361		Chair conformation, C_{2h}
Ph₂Ge–GePh₂	241.5	Ge–Ph = 194.2–196.7 Ge–S = 236–237	361	**D** E = Ge R = R′ = Ph Y = S	Half-chair conformation, C_2
Me₂Ge–GeMe₂	243.8	Ge-phane 194.1–194.6	362	**F** [22] paracyclophane bridged by 2 Ge–Ge	
Sn–Sn [SnMe₂Cl]₂ (with intermolecular Cl...Sn links giving double chains	277.0	Sn–Me = 214	363	**F**	a = 244.5 b = 324.0, 329.2 ab = 178.1 CSnC = 113.9°: SnSnC = 120.5° SnSnCl = 100.8°

Compound	Sn–Sn	Sn–C	Ref	Description	Configuration / Notes
[(Me₃SiCH₂)₂Sn]₂	Sn–Sn = 276.4	Sn–C = 220.7	364		SnSn...Cl = 77.8° Formally Sn=Sn
Sn units in larger rings					
Me₂Sn—SnMe₂	277.53	Sn–C(ring) = 217.7, 216.2; Sn–Me = 215.0–216.8	365	in G, R = Me, CR′₂ = CMe₂	Chair configuration SnCSn = 111°
Me₂Sn—SnMe₂	277.5, 276.6		366	in G, R = Me, CR′₂ = CHMe	Boat configuration SnCSn = 112.2°
Me₂Sn—SnMe₂	278.0, 279.1	Sn–C(ring) = 213–216; Sn–Me = 211–218	367	in G, R = Me, CR′₂ = CH₂	Boat configuration SnCSn = 115.5°
R₂Sn—SnR₂ R = (Me₃Si)₂CH	281.7	Sn–C(ring) = 220; Sn–C(R) = 220.3–222.6	368	In the cyclobutene **11a**	Distannacyclobutene ring is planar with CSnSn = 70.7° and SnCC = 107.6° and 110.9°
Ph₂Sn—SnPh₂	278.3	Sn–C(ring) = 215.0	369	In G, R = Ph	Flattened chair configuration SnCSn = 120.6°
Ar₂Sn—SnAr₂ Ar = Tip	284.0	Sn–Ph = 213.9–216.4; Sn–C(ring) = 217.4, 220.2; Sn–Me = 215.0–217.7	370	CR′₂ = CH₂ **C**; ER₂ = SnAr₂; Z–Z = CPh=CH **C**	
Ar₂Sn—SnAr₂ Ar = Tip	288.9	Sn–S = 248.0; Sn–Ar = 218.5	371	ER₂ = SnAr₂; Z = S **B**	
Ar₂Sn—SnAr₂ Ar = Tip	282.7	Sn–Te = 280.0, 277.9	371	ER₂ = SnAr₂; Y = Te **B**	
Ar₂Sn—SnAr₂ Ar = Ar(f)	270.9	Sn–Ar = 215.9–219.1; Sn–N = 207.6; Sn–Ar = 230.1–231.6	372	ER₂ = SnAr₂; Y = N(Mes) **B**	

(continued overleaf)

154

TABLE 16. (continued)

Compound	d(E–E)	d(E–R/X)	Ref.	Structure	Values
Bu^t_2Sn–$SnBu^t_2$	288.2	Sn–S = 240–243 Sn–C = 218–223 SnSnS = 108.1° SnSnS = 107.9° SSnS = 115.3°	373	**D** $ER_2 = ER'_2 =$ $SnBu^t_2$ Y = S	Planar ring
Bu^t_2Sn–$SnBu^t_2$	287.5	Sn–Se = 251.9–253.1 Sn–C = 217–227 SnSnSe = 105.8° SnSeSn = 106.4° SeSnSe = 115.1°	373	**D** $ER_2 = ER'_2 =$ $SnBu^t_2$ Y = Se	Planar ring
Bu^t_2Sn–$SnBu^t_2$	284.0	Sn–Te = 273.8 Sn–C = 219 SnSnTe = 105.0–106.2° SnTeSn = 101.9° TeSnTe = 114.1°	373	**D** $ER_2 = ER'_2 =$ $SnBu^t_2$ Y = Te	Puckered ring, C_{2v} Two independent molecules with different pucker angles
R_2SnSnR_2 R = Bsi	287.8, 289.4	Sn–C = 220.3 Sn–P = 254.4	374	**C** $ER_2 = SnR_2$ Z–Z = P=CBu^t	Planar ring SnSnC = 76.1° SnSnP = 74.8° SnPC = 94.5° SnCP = 114.5°
$[ClSnL]_2$ L = MeN[(CH$_2$)$_3$]$_2$	283.1	Sn–C = 214.5 and 216.7 Sn–N = 244.7	375	Two NC$_3$Sn rings per Sn	ClSnN = 168.9° CSnC = 118.7° CSnSn = 113.4°

(b) Other highly substituted/bulky ligand species, such as a cyclopentadiene ring with four disilane substituents, in the Lappert[380] issue of *Journal of Organometallic Chemistry*.

(c) [Si(H)CH$_2$Ph]$_6$, a cyclohexane analogue with a nearly flat ring[381].

(d) A more extended series of E-paracyclophanes including [nn] and [nnn] Si$_2$ bridges and a heptasilane bridge[382] (Me$_2$Si)$_7$(C$_6$H$_4$).

(e) Sheldrick[4a] includes cases such as [SiMe$_2$]$_n$ with n greater than 6; $n = 7$, 13 and 16 are listed and a wider range of very long bonds as in Bu^t substituted trisilanes.

As for dinuclear compounds, perphenyl compounds represent the largest single group of structurally characterized polynuclear species. Straight-chain and monocyclic alkane analogues are established for up to 6 Ge or Sn atoms (Tables 17 and 18). Although there are no crystal data for analogues of the larger silicon rings, like (Me$_2$Si)$_{13 \text{ or } 17}$, spectroscopic evidence for larger units is strong. A significant number of interesting structures involve sterically demanding ligands like mesityl or t-butyl, usually stemming from work on R$_2$E species. The α,ω-dihalides in the list reflect ring-closure/ring-cleavage preparations and reactions. Mixed-element chains, usually three-membered, are well represented.

Lengths of E−R and E−X bonds in polynuclear compounds match those seen in simpler compounds. The average E−E and E−E′ values, excluding very bulky ligands, tend to be a little higher than the dinuclear values. There is a high proportion of polyphenyl species, whose structures are dominated by the need to minimize phenyl−phenyl interactions with propeller orientations of EPh$_3$ units and adjacent units conforming. For Ge−Ge particularly, there are several cases where there are two independent molecules in the unit cell, or where there is no crystal symmetry constraint on chains and rings, and these indicate a 0.5–3 pm variability in 'independent' Ge−Ge values with identical ligands. Differences between central and outer Ge−Ge bonds in the longer chains are also of this order. The difference in Ge−Ge between Ge$_6$Me$_{12}$ and Ge$_6$Ph$_{12}$ is striking.

2. Highly branched species

A further, and undoubtedly increasing, class are those compounds like [{(Me$_3$Si)$_3$Si}$_2$GeCl]$_2$ which can be viewed either as a highly branched mixed element chain with 18 E elements, or as a dinuclear species with bulky group 14 ligands.

An anion, [Ge(SiMe$_3$)$_3$]$^-$, has been isolated[407] as a lithium-crown salt and shows a pyramidal structure (Ge−Si = 236.7 pm, SiGeSi = 102.6°). As expected, the angle is more acute than for R$_3$Si$^-$ ions but less so than for Ph$_3$ELi (E = Sn, Pb) (Section II.D).

Series with changing E allow useful insight, as in the growing set (Me$_3$Si)$_3$E−E(SiMe$_3$)$_3$. The known members are prepared by elimination/coupling reactions of the Li salts. Various attempts at forming the E = C member have failed, probably because of excessive steric interaction between the two ends of the molecule. The Si, Ge and Sn compounds are established. (Me$_3$Si)$_3$GeGe(SiMe$_3$)$_3$ is disordered but crystal structures of the Si and Sn species are reported. The structure[408] of (Me$_3$Si)$_3$SiSi(SiMe$_3$)$_3$ shows long Si−Si bonds (240 pm for the central bond, 239 pm for the outer ones) and other features consistent with appreciable steric strain. The Sn analogue (Table 17) shows a normal Sn−Sn bond but with dihedral angle distortions away from D$_{3d}$ suggesting there is still some, but much less, need for steric accommodation.

The structural parameters of E−E′(SiMe$_3$)$_3$ species have been calculated[409] for E = C, Si, Ge or Sn and E′ = (a) Si and (b) C. Cone angles decrease in the (a) series from 199° for E = C through 182° (Si), 175° (Ge) to 168° for E = Sn. The (b) set cone angles are each some 20° larger. The structure cited in Table 17 for {(Me$_3$Si)$_3$Si}$_2$SnCl$_2$ shows a good match with calculation with an experimental cone angle of 166°; the calculated Si−Sn distance of 161 pm matches the experimental value also. The effect of the larger cone angle is well illustrated by the attempt to prepare the Ge analogue, [(Me$_3$Si)$_3$Si]$_2$GeCl$_2$ (angle

TABLE 17. Parameters (pm and deg) for polynuclear chain species

Compound	d(E–E)	d(E–R/X)	Ref.	Notes
Polygermanes				
Ge$_3$Ph$_8$	243.8, 244.1	Ge–Ph = 193.5–198.6 [196.0 av]	383	Si–Si = 239.4 in Si$_3$Ph$_8$ [384]
(Ph$_3$Ge)$_2$GeMe$_2$	242.9	Ge–Ph = 195.8, Ge–Me = 194.4	385	GeGeGe = 120.3°, C$_2$ molecular symmetry
(ClPh$_2$Ge)$_2$GePh$_2$	(a) 241.9, 243.7, (b) 241.3, 242.3	Ge–Ph = 195, Ge–Cl = 219.1	386	Two conformations (a) *anti–gauche* (b) *gauche–gauche*
Ge$_4$Ph$_{10}$·2C$_6$H$_6$	246.3 (outer), 246.1 (inner)	Ge–Ph = 196.3–197.7 [196.8 av]	383	
1,4-Cl$_2$Ge$_4$Ph$_8$	245.0 (outer), 244.2 (inner)	Ge–Ph = 197 av, Ge–Cl = 213.2	386	All-*anti*, centrosymmetric
1,4-I$_2$Ge$_4$Ph$_8$	245.1 (outer), 245.9 (inner)	Ge–Ph = 193.5–196.1, Ge–I = 255.9	387	All-*trans*, fully staggered C$_i$ molecular symmetry
Ge$_5$Ph$_{12}$	244.3 (outer), 247.6 (inner)	Ge–Ph = 195 av	388	antiperiplanar–anticlinal
Silylgermanes				
Ge(SiH$_3$)$_4$	237.0	Ge–H = 170	389	ED Si–H = 149.7
Mes$_3$GeGe(H)MesSiEt$_3$	Ge–Ge = 252.7, Ge–Si = 244.0	Ge–C 200.5–204.6	390	
Ph$_3$GeSiPh$_2$SiPh$_3$	Ge–Si = 234.9, Si–Si = 230.2	Ge–Ph etc. no data	384	Si/Ge 55/45 disorder over end positions
Ph$_3$GeSiPh$_2$GePh$_3$	Ge–Si = 240.3	Ge–Ph etc. no data	384	
Ph$_3$SiGePh$_2$SiPh$_3$	Ge–Si = 241.8, 242.5		384	Si/Ge 60/40 disorder over end positions
Ph$_3$GeGePh$_2$SiPh$_3$	Ge–Ge = 243.1, Ge–Si = 235.0		384	
Ph$_3$Ge(SiPh$_2$)$_2$GePh$_3$	Ge–Si = 242.4	Si–Si = 237.9	391	GeSiSi = 116.5° (i) Crowding at Ge accommodated by twisting of rings.
(Me$_3$Si)$_3$SiGePh$_3$	Ge–Si = 241.6	Si–Si = 236.6	392	(ii) Calculations give Ge–Si = 242.8, Si–Si = 238.2

Compound			Ref.	Notes
Ph$_3$Ge(SiPh$_2$)$_4$GePh$_3$	Ge–Si = 242.9	Si–Si = 241.0, 240.1	391	GeSiSi = 118.2°, SiSiSi = 116.1°, All-*trans* chain
[[(Me$_3$Si)$_3$Si]$_2$GeCl$_2$]$_2$	Ge–Ge = 242.1, Ge–Si = 251.4	Ge–Cl = 215.8, 228.8	393	Si–Si = 202.8, 245.0, 261.9, GeGeSi = 122°, Crystal highly disordered, large uncertainties (2.8 to 5.2) in Ge–Cl, Ge–Si, Si–Si bond lengths
Germylstannanes (Ph$_3$Sn)$_2$GePh$_2$	Ge–Sn isomorphous with silylstannane analogue		394	Crystals not suitable for full analysis
Cp*[CH(SiMe$_3$)$_2$](Me$_3$Sn)GeN=N=CGe(SnMe$_3$)Cp*[CH(SiMe$_3$)$_2$]	Ge–Sn = 265.0	Ge–CHSi$_2$ = 198.4, 196.9; Ge–Cp* = 205.6; Ge–CNN = 193.5; Ge–NNC = 188.3	395	Ge–C=N = 150°, C=N=N = 174°, N=N–Ge = 130°
Silylgermylstannane [(Me$_3$Si)$_3$Ge]$_2$SnCl$_2$	Ge–Sn = 262.6, 263.6, Ge–Si = 238.8–239.7	Sn–Cl = 238.5	396	ClSnCl = 98.8°, GeSnGe = 142.1°, Angles at Ge and Si 106°–112°
Polystannanes SnBu$^t_2$(SnPh$_3$)$_2$	279.8	Sn–C = 222–230	397	SnSnSn = 107°
Sn$_3$Bu$^t_8$	296.6		338	SnSnSn = 122.1°, CSnSn = 108°–115°, CSnC = 106°

(continued overleaf)

TABLE 17. (continued)

Compound	d(E–E)	d(E–R/X)	Ref.	Notes
$Sn(L)_2(SnPh_3)_2$ $L = Me_2NCH_2CH_2CH_2$	277.5	Sn–C = 212–216	398	SnSnSn = 111.3° No Sn–N interaction; all-*trans* chain
$(SnBu_2^t SnPh_3)_2$	282.5 (outer) 286.8 (inner)		397	SnSnSn = 118°
$1,4\text{-}Sn_4Bu_8^t Br_2$	288.7 (A) 288.7 (B) (outer) 292.8 (A) 290.4 (B) (inner)	Sn–C = 220–224 (A); 223–226 (B) Sn–Br = 255.1	338	SnSnSn = 117.8° BrSnSn = 99.8° (A) 97.2° (B) CSnC = 106° A,B different molecules in unit cell
$1,4\text{-}Sn_4Bu_8^t I_2$	289.5 (outer) 292.4 (inner)	Sn–C = 220–226 Sn–I = 275.3	395	SnSnSn = 117.8° ISnSn = 100.8° CSnC = 109° All-*trans* chain
$1,4\text{-}Sn_4Bu_8^t(SPh)_2$	290.9 (outer) 292.1 (inner)	Sn–C = 220–225 Sn–S = 246.6	338	SnSnSn = 118.3° SSnSn = 92.5° CSnC = 107.4°
$[(Me_3Sn)_3Sn]_2Ln$ $(THF)_4$ Ln = (a) Sm, (b) Yb, (c) Yb (second form)	(a) 275.4–280.0 (b) 277.5–281.8 (c) 279.8–280.3	(a) Sn–Sm = 339.7 (b) Sn–Yb = 329.4 (c) Sn–Yb = 330.0 Sn–C = 213 (a), 216 (b), 219 (c)	399	(a) and (b) Isostructural and orthorhombic, crystals decayed in X-ray, R = 0.068–0.083 (c) Tetragonal, more stable, R = 0.043
$[(Me_3Sn)_3SnMo$ $(NMe_2)_2]_2$	276.8–277.9	Sn–C = 214.3–222.6 Sn–Mo = 277.4–278.3	400	MoMo (triple bond) = 220.1
$SnBu_2^t(SnBu_2^t SnPh_3)_2$	283.7 (outer) 291.2 (inner)		397	SnSnSn (central) = 119° SnSnSn (outer) = 114°
$(SnBu_2^t SnBu_2^t SnPh_3)_2$	284.5 (outer)		397	SnSnSn (inner) = 121° SnSnSn (outer) = 114°

	293 (inner) 296.6 (central)			Central bond is longest known Sn–Sn in polystannanes
Silylstannanes				
Ph₃SnSiPh₂SnPh₃	Sn–Si = 257.5	Sn–Ph = 213–219 Si–Ph = 191	394	SnSiSn = 118.5° C₃SnSiC₂SnC₃ skeleton C₂
Ph₃Sn(SiPh₂)₂SnPh₃	Sn–Si = 259.1	Si–Si = 236.5	391	SnSiSi = 116.2° All-*trans* chain
Ph₃Sn(SiPh₂)₄SnPh₃	Sn–Si = 259.3	Si–Si = 238.7	391	SnSiSi = 116.1° SiSiSi = 116.4° All-*trans* chain
[(Me₃Si)₃Si]₂SnCl₂	Sn–Si = 259.7, 260.4	Si–Si = 235.0–236.5 Sn–Cl = 243.3	401	ClSnCl = 99.1° SiSnSi = 142° SiSnCl = 99°–106° Compare Si₃Ge analogue above
[(Me₃Si)₃Si]₂Sn(II). ClLi(THF)₃	Sn–Si = 266.6, 268.1		402	Sn...Cl = 275.4, SiSn...Cl = 92.1°–95.5°, SiSnSi = 114.2° Pyramidal at Sn
[(Me₃Si)₃Sn]₂	Sn–Sn = 278.9, Sn–Si = 260.9–261.1	Si–C = 190,	403	First Sn(II)-Si bond found SiSnSn = 110°–112° SiSnSi = 106°–109° Me₃Si groups disordered
(Me₃Sn)₃SiW(CO)₅⁻	Sn–Si = 256.1–258.4,	W–Si = 265.2	404	SnSiSn = 100.4°–103.9° Quotes unpublished work Sn–Si = 256.9 pm in Cp₂Zr(Cl) analogue
Polyplumbanes				
Pb₃Ph₈, Pb₅Ph₁₂				Indicated, not characterized

(continued overleaf)

TABLE 17. (continued)

Compound	d(E–E)	d(E–R/X)	Ref.	Notes
$Pb_3(C_6H_{11})_8$			315	PbPbPb = 113.0°
Silylplumbanes				
$(Me_3Si)_3SiPbPh_3$	Pb–Si = 260 (calc)		405	Calc values PbSiSi = 109.9° SiSiSi = 109.0° Compound characterized spectroscopically
$[(Me_3Si)_3SiPbPh_2]_2$	Pb–Pb = 291.1, Pb–Si = 264.8,	Si–Si = 233.7–235.7 Si–C = 184.4–191.0 Pb–C = 225.5	405	PbPbSi = 122.4° SiSiSi = 111.0° av PbSiSi = 107.9° av
$Pb(SiMe_3)_4$	No crystal data		406	

TABLE 18. Parameters (pm and deg) for cyclic polynuclear compounds

Compound	d (E–E)	d (E–R/X)	Ref.	Notes
Cyclic polygermanes				
cyclo-Ge$_3$Bu$^t_6$	256.3	Ge–C = 205.6	324	D_3 symmetry
cyclo-Ge$_3$Mes$_6\cdot$2CH$_2$Cl$_2$	253.7	Ge–C = 200.3–202.5	412	GeGeGe = 60.0° ± 0.1°
cyclo-Ge$_3$(2,6-Me$_2$C$_6$H$_3$)$_6$	254.1	Ge–C = 197.7–201.5	413	GeGeGe = 60.0° ± 0.1°
cyclo-Ge$_3$(2,6-Et$_2$C$_6$H$_3$)$_6$ {cis, cis}	259.0	Ge–C = 200	414	
cyclo-[Ge(Mes)(Dmp)]$_3$	253.1–256.4	Ge–C = 198.2–200.6	414	
cyclo-Ge$_4$R$_8$ R = Me, Pr, Sim, Tol	No structure		415	Spectroscopically characterized
cyclo-Ge$_4$Ph$_8$	245.8–247.2 246.5 av		416	Square near-planar ring, 3.9° pucker
cyclo-[(GeClBu^t)$_4$	245.5–247.1	Ge–C = 195.1–197.6	417, 418	Ring puckered, dihedral angle 21°. Cl all-*trans* and pseudoaxial, Bu^t pseudoequatorial
1,3-(Bu-t)$_2$bis(1,2(μ-R$_2$)) cyclo-Ge$_4$ R = (Me$_3$Si)C(PMe$_2$)$_2$	248.9 (unbridged) 252.9 (bridged)	Ge–C = 201.8 Ge(1)–P = 241.3 Ge(2)–P = 236.3	419	
cyclo-Ge$_5$Ph$_{10}$	244.0–247.3	Ge–C = 196.0–198.6	420	Conformation of ring between C$_s$ and C$_2$;
cyclo-Ge$_5$Ph$_{10}\cdot$C$_6$H$_6$	243.8–247.3	Ge–C = 192.8–199.3	421	crystal benzene disordered
cyclo-Ge$_6$Me$_{12}$	236.5–237.7	Ge–C = 195 av (ax = eq)	422	triclinic
cyclo-Ge$_6$Ph$_{12}\cdot$7C$_6$H$_6$	245.7	Ge–Ph = 197.3 av (ax = eq)	423	Flattened chair; also known solvent-free and as monoclinic 2 benzene solvate
cyclo-Ge$_6$Ph$_{12}\cdot$2C$_6$H$_5$CH$_3$	246.3	Ge–Ph = 196.9 av (ax = eq)	424	Monoclinic; C$_i$ similar to above, toluenes above and below ring, steepening axial phenyls and lengthening Ge–Ge

(continued overleaf)

162

TABLE 18. (continued)

Compound	d (E—E)	d (E-R/X)	Ref.	Notes
Hetero-rings containing Ge—Ge				
(Ph₂Ge)₄S	244.3–245.4	Ge–C = 193.7–198.7 Ge–S = 223.0, 225.0	425	Near-planar twist conformation, C₂
(Ph₂Ge)₄Se	244.8 243.7 Ge–Ge(Se)	Ge–C = 195.1–196.9 Ge–Se = 237.3	411	GeGeGe = 105.8° GeSeGe = 106.3° GeGeSe = 110.4°
(Buᵗ₂Ge)₃Se (Buᵗ₂Ge)₃Te	256.8 258.4	Ge–Se = 239.4 Ge–Te = 259.0	426 429	Planar ring with long Ge–Ge planar ring with long Ge–Ge Ring angles 88°–92°
1,2,3-(Me₂Ge)₃CH=CR R = 2-ethylnylphenyl	241.1, 240.6	Ge–Me = 192–195 Ge–C (ring) = 197.9, 195.4	427	Ge₃C₂ ring nearly planar; C=C = 130
cyclo-(Me₂Ge)₃BNLCR NL = piperidine CR = fluorenylidene (containing a GeGeGeBC ring)	240.2, 243.5	Ge–Me = 194–199 Ge–C(ring) = 207.6 Ge–B = 216.5	428	GeGeGe = 89.4° GeGeB = 103.5° GeBC = 112.9° BCGe = 108.0° CGeGe = 102.7°
Cyclo-Silylgermanes				
cyclo-Ge₂SiMes₆·2CH₂Cl₂ (isostructural with Ge₃ analogue listed above)	Ge–Ge(Si) = 250.8	Ge(Si)–C = 201.1	412	3-fold disorder of Si/Ge; taking Ge–Ge from Ge₃Mes₆ gives Ge–Si = 249.3
cyclo-Si₂Ge(TMS)₆	Ge–Si = 239.1 (ring) 235.6 (ext)	Si–Si = 237.7 (ring) 236.6 (ext)	407	Mean exocyclic SiSi(Ge)Si angles 105.5° 3-fold disorder of Si/Ge
cyclo-[(Prⁱ₂Si)₃Ge(Sim)₂]	Ge–Si = 245.2, 246.2	Si–Si = 238.0, 239.1 Ge–C = 198.9	430	H GeSiSi = 88.5°, 89.0° SiSiSi = 90.4° SiGeSi = 87.1° Puckered ring, angle 24°
cyclo-[((BuᵗCH₂)₂Si)₃GeR₂] R = Sim	Ge–Si = 242.7, 246.1	Si–Si = 239.3 Ge–C = 198.9	430	H GeSiSi = 86.4°, 87.2°

Compound	Distances (pm)	No.	Comments
cyclo-[(Me₃Si)₃SiGeCl]₄·C₆H₆	Ge–Ge = 250.9, 255.8, Ge–Si = 244.6, 247.4, Ge–Cl = 221.6	396	SiSiSi = 88.7°, SiGeSi = 86.4°, Puckered ring, angle 36° I
	Si–Si = 236.4–238.1		GeGeGe = 88.1°–89.3°, Non-planar with internal torsions 17.2°. Alternate Cl and Si(SiMe₃)₃ groups on opposites sides of ring, and latter twist to minimize strain
Cyclic polystannanes			
cyclo-(Ar₂Sn)₃ Ar = Dep	2 × 285.5, 287.0	431	Sn–C = 216.7–220.3, Isosceles triangle, angles 60.3° and 2 × 59.8°
cyclo-(Bu^t₂Sn)₄	288.7	432	Sn–C = 222–227, SnSnSn = 90.0°, CSnC = 105°
cyclo-(Am^t₂Sn)₄	291.4–292.3	432	Sn–C = 226–229, C₂ₕ, square planar Sn₄, SnSnSn = 89.0°, CSnC = 105.3°–106.4°, Sn₄ ring puckered by about 20°
cyclo-[(Me₃SiCH₂)₂Sn]₄	283.9, 283.4, 285.2	433, 434	Sn–C = 219.1–220.6, Sn–C = 209, 215, approx. C₂ₕ, square Sn₄, SnSnSn = 88.2°, Tetramer of the stannaidene LLSn^II
cyclo-[(L-L)₄Sn₄] L-L = meso-[CH(SiMe₃)]₂C₆H₃			
cyclo-(SnPh₂)₆	278.1 (A), 278.6–279.1 (B)	435	Sn–Ph = 213.3–217.3, CSnC = 105.2°–107.1° (A), 99.8°–107.1° (B), SnSnSn = 108.3°–121° (A), 107.4°–116.0° (B), Distorted chair with different distortions in the independent molecules A and B

(continued overleaf)

TABLE 18. (*continued*)

Compound	d (E–E)	d (E-R/X)	Ref.	Notes
cyclo-(SnPh₂)₆·2C₆H₅Me	278.0	Sn–Ph = 215.0	436	CSnC = 106.7° SnSnSn = 112.5° Chair conformation
cyclo-(SnBz₂)₆·DMF	279.2–281.1	Sn–C = 218.9–222.0	435	CSnC = 104.2°–107.1° SnSnSn = 105.5°–111.7° (A) 108.4°–112.9° (B) chair conformation; values for independent molecules A, B largely very similar
Hetero-rings containing Sn–Sn				
(Bu₂ᵗSn)₄S	288.4–289.8 (A) 286.4–288.6 (B)	Sn–C = 208–224 (A) 214–238 (B) Sn–S = 240.7–242.8	437	SnSnSn = 100.5° SnSnS = 106.7°–108.4° SnSSn = 116.6° (A), (B) differ in twist angles
(Bu₂ᵗSn)₄Se	288	Sn–Se = 252–255	437	SnSnSn = 102° SnSnSe = 108.5° SnSeSn = 112.8°
(Bu₂ᵗSn)₄Te	286.9–288.9	Sn–C = 219–227 Sn–Te = 274.3	437	SnSnSn = 103.9° SnSnTe = 109.4° SnTeSn = 107.4°

175°), which gave instead, by cleaving one Ge−Si, the digermane $[(Me_3Si)_3SiGeCl_2]_2$ and the cyclotetragermane $[(Me_3Si)_3SiGeCl]_4$, listed in Table 17. Note the very similar structure of $[Bu^tGeCl]_4$. Further indications of the extent of steric effects come from the lead[410] compounds $(Me_3Si)_3SiPbPh_3$ and $[(Me_3Si)_3SiPbPh_2]_2$. Molecular mechanics calculations on the former indicated very little distortion, in contrast to lighter congeners, while the diplumbane shows distinct lengthening of the Pb−Pb and Pb−C bonds compared with Pb_2Ph_6. Thus the steric effects of $(Me_3Si)_3Si$ fade, but do not disappear even for lead species.

The ligand series has been extended to E′ = Sn, including the transition metal complexes cited in the Table, and just recently to E′ = Ge. This work[396] produced $\{(Me_3Si)_3Ge\}_2SnCl_2$ where the cone angle at 168° was indistinguishable from that of the E′ = Si analogue. The full E−E′(E″Me_3)_3 set, with E, E′, E″ all permutations of the group 14 elements, thus offers an accessible series (apart from those of high Pb content) of chemically similar species with a range of cone angles covering some 70° from 216° down. These will undoubtedly lead to further detailed insight into steric effects on products and structures.

3. Rings

Three-membered rings are intimately linked with divalent and double-bonded species and are discussed separately below.

For four-membered rings, both planar and puckered rings occur. Whether folding occurs is the result of the balance of the E−E distance and the steric requirements of the ligands. Examples in Table 18 exhibit these features. A borderline case is Ge_4Ph_8 where the folding is very slight. The $GeSi_3$ rings listed (Figure 9, **H**) fold less markedly than the Si_4 analogues. Marked folding is also seen in rings with different substituents of different bulk on E like $[ClGeBu^t]_4$, where the tert-butyl groups are on alternate sides of the ring (Figure 9, **I**) and the ring puckers and the groups twist to reduce interactions.

4. Clusters

Forming further E−E bonds in E_n chains and rings creates double and multiple rings and further steps in this process culminate in the formation of nets, open clusters and finally closed polyhedral solids like prismanes and cubanes. The challenge and aesthetic attraction of such platonic and ideal solid structures was felt earlier by carbon chemists with the first report, that of cubane[438(a)], appearing in 1964. Only in the last five years has this challenge been met for the heavier Group 14 elements but development has been striking.[438(b)]

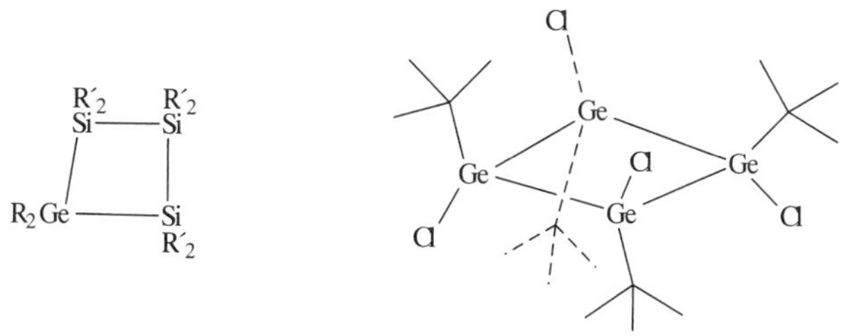

(H)   (I)

FIGURE 9. Schematic structures for polynuclear rings

First, however, we should note briefly a second class of clusters, represented mainly by the E_n^{x-} anions. While the first class of clusters is electron-precise, the second class is usually electron-deficient, best treated by multi-centred cluster orbitals. While there are not yet organometallic examples, this class adds a further dimension to the self-bonding properties of E elements. These were first indicated by highly coloured solutions formed when the elements were added to sodium solutions in liquid ammonia, e.g. a 9/4 ratio of Pb to Na atoms suggested a $[Pb_9{}^{4-}]_n$ cluster. Such large anions could not be isolated and structurally characterized until the advent of polydentate O and N ligands to give large coordinated cations to stabilize the solids. A range of heavier Main Group cluster ions has been characterized, largely by the work of Corbett[439] and his colleagues There is a recent report[440] of a substituted cluster of this type — $[(CO)_5CrSn]_6{}^{2-}$ containing a regular octahedron of tin atoms and isolated as the K[2.2.2]cryptand salt. Since the $Sn-Cr(CO)_5$ unit is isolobal with $Sn-CH_3$, there is no formal reason why organometallic clusters of this second class should not be preparable.

Theoretical studies in 1985 and later[441] indicated that a range of $(EH)_n$ structures should be stable, and that the preferred shape was the n-prismane. Tetrahedrane, E_4H_4, with only triangular faces has similar strain energies for C, Si and Ge but, while the strain energy increases for C compounds from tetrahedrane to prismane to cubane, the reverse pattern is found for Si, Ge, Sn and Pb — matching the effects in the corresponding four-membered rings, $[EH_2]_4$. The first syntheses of Si, Ge and Sn prismanes were reported almost simultaneously in 1988–9.

Present syntheses are relatively undirected, giving low yields of the characterized products and undoubtedly other interesting species in the mixture. No doubt similar compounds were among the long-known 'ER$_2$' and 'ER' species, with identifications like hexaphenylhexagermabenzene, formed in reductions of less bulky REX$_3$ species.

Figure 10 (J to N) shows some of the products structurally characterized to date. In reductive dehalogenation of monogermanes, products depend on R with RGeCl$_3$ giving $[RGe]_6$ J for R $=$ CH(SiMe$_3$)$_2$[442] or $[RGe]_8$ K for R $=$ CEt$_2$Me or 2,6-Et$_2$C$_6$H$_3$[443]. For Si, RSiX$_3$ gives the cubane K for R $=$ t-BuMe$_2$Si[444], CMe$_2$CHMe$_2$[445] and 2,6-Et$_2$C$_6$H$_3$[443].

An alternative product, L, the tetracyclo[3.3.0$^{2.7}$0$^{3.6}$]octasilane or germane, which may be seen as the $trans$ isomer of the penultimate intermediate en $route$ to the cubane, was isolated from a similar dehalogenation of Bu^{t}EX$_2$EX$_2$Bu^t (E $=$ Ge, X $=$ Br; E $=$ Si, X $=$ Cl) with Li naphthallide. The silicon reaction was characteristically complex[446] with several products mostly in small yields — L (5%) and including a cyclotetrasilane (20%) and the analogous tricyclo[2.2.0.0$^{2.5}$]hexasilane (6%) which was also structurally characterized. The Ge system[447,448] was much cleaner with 50% of L and no cyclogermane. The enantiomers of L were paired in the unit cell, and it was unusually stable to heat and to air and moisture. The separation of the last two halogens presumably precludes the final step to the cubane.

An alternative synthesis was used in tin studies. Thermolysis of the cyclotristannane, $[(2,6-Et_2C_6H_3)_2Sn]_3$, at 200 °C, gave[449] not only the cubane K, but also pentagonal prismane M and the propellane N, which may be seen as three triangles sharing a common edge. These products derive from the initial stannylene, R$_2$Sn, by aryl transfer, part of which appears as the distannane Sn_2R_6.

Typical parameters found for representative $[n]$ prismane clusters are given below.

	ER	Ref.	E–E, n-face	E–E, square	Notes
J	GeCH(SiMe$_3$)$_2$	442	258.0	252.1	angles 60° and 90°
K	GeC$_6$H$_4$Et$_2$	443	247.8–250.3	same	angles 88.9–91.1°
M	SnC$_6$H$_4$Et$_2$	449	285.6	285.6	angles 108° 90°

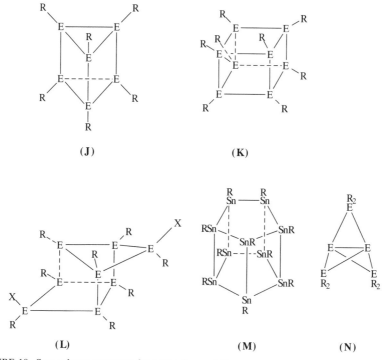

(J)   (K)

(L)   (M)   (N)

FIGURE 10. Some cluster structures for germanium and tin

The structures are essentially ideal ones. The Sn–Sn distance in tin **K** is 285.4 pm. The values indicate lengthening of Ge–Ge in the trigonal prism, but little effect in the cubanes or **K**. This is in accord with calculations of strain energies for the [n] prismanes[450] which show strain energy decreasing from $n = 3$ to 4 to 5, then rising so that $n = 8$ is less stable than $n = 3$. Calculated bond lengths in the $n = 5$, 6, 8 and 12 prismanes differ little with n, averaging 252 for Ge–Ge and 288 for Sn–Sn with the square faces less than 1 pm longer apart from $n = 12$, where the difference rises to 2 pm for Sn. The same calculation shows the pagodane [edges 249–253 (Ge) and 285–290 (Sn)] is less strained and the dodechedron (250.3 and 286.2, respectively) even less so, which presents a nice challenge to synthesis. The same papers also examine other structures including possible aromatics (D_{3d} more stable than planar) and tetrahedranes (where the lead member is less unstable relative to the lighter elements than for other structures).

Germanium provides two examples of the tetracyclooctagermane **L**, with R = t-Bu and X = Cl or Br; Ge–Ge bonds range from 244–254 pm, with those involving GeX the shortest.

Tin forms the propellane **N** (R = dep), which has normal outer bonds (284–287 pm) and a long central bond (Sn–Sn = 336.7 pm), 10% longer than any other Sn–Sn. Calculations[451] show that the central bonding interactions decrease on descending the group. A tetracyclic heptastannane derivative R_8Sn_7[452] may be seen as the propellane with two of the wingtips linked by an Sn–Sn bridge. The central Sn–Sn remains long (334.9 pm), the bridge Sn–Sn = 290.2 pm and the remaining values range from 282 to 286 pm. An earlier study[453] produced [P(SnMe_3)_3]_2 which has two trigonal PSn_3 units joined by Sn–Sn edges—but with no central bond (P...P = 518 pm)—not a propellane but a nicely symmetric cage. A simpler relation[454] had a $(PSn_2Me_4)_2$ six-membered ring with the two P linked by a fifth $SnMe_2$ group. In these molecules, Sn–Sn = 279 pm and Sn–P = 251–253 pm. The first bicyclic hexastannane[455], $Dep_{10}Sn_6$, consists of two four-membered, puckered rings sharing a common edge at a dihedral angle of 132°. The shared edge is the shortest (281.8 pm) by a little over the rest (283.5–293.1 pm).

Recent reports of further structures for silicon indicate further likely developments for the heavier elements. Most striking is the synthesis of the tetrahedrane, Si_4R_4, despite the unfavourable strain energy, by use for R of the 'supersilyl' highly bulky $SiBu_3^t$ ligand. There are two slightly distorted tetradranes in the unit cell with Si–Si lengths ranging from 231.5–234.1 pm (calc for Si_4H_4, 231.4 pm). Exocyclic Si–Si was 235.5–236.5 while Si–C was 191.7–196.2 pm; compare with Si–Si = 268.5 pm in $Si_2Bu_6^t$ which co-crystallizes[456]. Also to be noted is the mixed-element cubane[457], $(t\text{-BuSiP})_4$, the analogue of K with alternating Si and P in a distorted cage which can be described as a large P_4 tetrahedron interpenetrating a smaller Si_4 one.

Finally, we note two larger clusters containing Ge. There is a distorted rhombohedron structure in the dimer[458], $\{t\text{-BuGe}[As(SiR_3)_3Li]_3\}_2$, where the Ge is bonded to 3 As (244.6 pm) and the Et (199.8 pm) in a regular tetrahedral array. The two $GeAs_3$ groups are linked by six Li triply bridging between As. The high symmetry $Ni_9(GeEt)_6(CO)_8$[459] has a cube of NiCO face-bridged by GeEt and with a central Ni, the first such structure with Group 14 face bridges.

VII. DIVALENT SPECIES AND LOWER VALENT CLUSTERS

A. General

For all the elements E = Si to Pb, the simple divalent ER_2 molecules have the lone pair in a single orbital (singlet S) and this orbital has increasing s character for heavier E, so that the bond angle in simple ER_2 species approaches 90°. Such relatively closed angles persist through to higher coordination numbers, i.e. bonds tend to be bent away from the lone-pair direction. Thus we find complexes of divalent species, especially those of tin(II), which retain the lone-pair site and become trigonal or square pyramidal for the addition of 1 or 3 additional bonds with the lone pair completing the tetrahedral or octahedral configuration. For two added bonds, the commonest configuration is the C_{2v} one, which results from the lone pair occupying an equatorial position in a trigonal bipyramid. Similar remarks apply to REX species. A second manifestation is in formally π-bonded systems, as in Ecp_2.

R_2E and REX are found in combined forms:

(a) as formally double-bonded dimers, $R_2E{=}ER_2$,

(b) formally doubly bonded to other elements, such as $R_2E{=}CR_2'$, $R_2E{=}PR'$ or $R_2E{=}ML_n$,

(c) as reaction products with unsaturated systems, for example after reaction with A=B as three-membered EAB rings; some larger ring products formed similarly, such as EEOO and EECC, have been included in Section VI.B,

(d) by polymerization into rings or clusters, discussed in Section VI.C,

(e) as complexes with donors, $R_2E{\cdot}D_x$.

Such products are often formed reversibly and so also act as sources of R_2E or REX.

In the following sections, we examine R_2E and these further products. The role of Group 14 ER_2 in transition metal chemistry has been reviewed by Lappert and Rowe[460] and recent advances in bivalent organic chemistry by Lappert[461].

B. R$_2$Ge Species

In contrast to silicon, germanium has a well-established though limited chemistry of inorganic compounds in the +II state which are of reasonable thermal stability though usually air-sensitive. Divalent organogermanium(II) species known at present fall into three groups:

(a) Simple R$_2$Ge species, which are now well-established by chemical means, have been isolated in low-temperature matrices, and have calculated spectroscopic properties which give respectable matches to observations on the matrix species. These species are of transient existence at ambient temperatures.

(b) Related R$_2$Ge species with bulky ligands R whose higher stability has allowed structural characterization in the solid.

(c) π-Bonded systems, stable at room temperature, exemplified by germanocene.

In addition, there are neutral and ionic complexes of Ge(II) with coordination numbers from three upwards of variable stability.

1. Small-ligand germylenes

Dimethylgermylene, Me$_2$Ge, and similar species have long been postulated as a reaction intermediate and the evidence for such compounds is now very firm and ably reviewed by Neumann[462,463].

In brief summary, germylenes are usually produced by thermolysis or photolysis of a precursor and identified by the products, either oligomers or trapped species. Precursors are either strained organic derivatives like 7-germabenzonornadiene, or strained or weakly-bonded inorganic compounds like cyclotrigermanes, R$_2$Ge(SeR')$_2$ or germanium mercurials. Characterization includes products from insertion into sigma bonds like C—halogen or E—H, addition to alkenes, alkynes, dienes and the like, and isolation of oligomers. Some products containing Ge—Ge bonds, including three-membered rings, arise from the intermediate formation of the digermene, R$_2$Ge=Ge$_2$. The chemical evidence, supplemented by matrix and theoretical studies, gives very satisfactory evidence for the singlet germylene and not the triplet, nor alternative reaction paths such as those with radical intermediates. As Neumann warns, however, 'Nevertheless, some UFO's (unidentified factual objects) are still crawling through the landscape of photolytic germylene generation.'!

2. Germylene structures

Theoretical calculations of R$_2$Ge give bond parameters and vibrational frequencies. For example, Me$_2$Ge has calculated GeC = 202 pm and CGeC = 98° with the Ge–C stretches at 560 cm^{-1} (A$_1$) and 497 cm^{-1} (B$_1$). As the latter agree reasonably with experimental values from matrix-isolated species (527 and 541 cm^{-1}), the structural values are probably good indications.

Some of the doubly-bonded R$_2$Ge=GeR$_2$ compounds dissociate in solution (e.g. R = Bsi by Lappert, see Section VIII). The gas-phase electron diffraction of (Bsi)$_2$Ge[464] shows Ge–C = 203.8 pm and CGeC = 107°. Other determined germylene structures are of nonorganic species and show angles ranging from 86° to 111°[461]. Isolation of R$_2$Ge was first reported for the even more hindered R = 2,4,6-Bu$^t_3$C$_6$H$_2$. To produce a germylene stable enough to allow an X-ray structure required the addition of a further SiMe$_3$ group to Lappert's compound in (Tsi)Ge(Bsi)[465]. The Ge–C(SiMe$_3$)$_3$ length = 206.7 pm, Ge–CH(SiMe$_3$)$_2$ = 201.2 pm and the CGeC angle = 111.3°. The additional silyl prevents dimerization (shortest Ge...Ge = 570 pm). Going the other way, the compound[464] with one silyl less than Lappert's — (Sim)Ge(Bsi) — decomposes at −20°C.

A kinetically stabilized diarylgermylene, Tb(Tip)Ge, is also stable in hexane solution with no tendency to dimerize. The structure has not yet been measured but it was characterized as the base-free mononuclear transition metal complex formed with the reactive $M(CO)_5$. THF adduct ($M = $ Mo,W). The crystal structure of the W complex[466] shows Ge=W $= 259.3$ pm, Ge–Tb $= 198.8$ pm, Ge–Tip $= 199.9$ pm, CGeC $= 108.4°$, TbGeW $= 138.9°$ and TipGeW $= 112.2°$. Thus the Ge is pyramidal and the structure is obviously the result of a balance of steric repulsions between the carbonyls, Tb, and Tip. Older work on similar Ge=W compounds, $RR'Ge=W(CO)_5$ with $RR' = Cp^*(Cl)$ or $(Bsi)_2$[467], show Ge=W lengths of 251.1 and 263.2 pm, respectively, indicating the combined steric and electronic effects.

3. Germanocenes and related compounds

Dicyclopentadienylgermanium(II), $(C_5H_5)_2Ge$, was the first germanium π complex to be synthesized, by Curtis and Scibelli in 1973 followed five years later by the monomethylcyclopentadiene analogue. Through the 1980s, a rapid and substantial development of this branch of germanium(II) chemistry has ensued with a major contribution from Jutzi and his group[468].

Common syntheses are by coupling the dihalide with the anion and varying ratios may yield the mono- or di-substituted product:

$$GeX_2 + C_5R_5^- = (C_5R_5)_2Ge \text{ or } (C_5R_5)GeX$$

Disproportionation between $(C_5R_5)_2Ge$ and GeX_2, coupling with different anion or removing one cp ring gives further product types such as $(C_5R_5)GeR''$, $(C_5R_5)Ge(C_5R_5')$ or $[Me_5C_5Ge]^+ GeCl_3^-$.

These germanium(II) organometallics are thermally stable though sometimes air-sensitive. The larger the substituted cp, generally the more stable. The structures (Figure 11)[469] indicate Ge(II) configurations with a lone pair. In $(C_5R_5)_2Ge$ (Figure 11a) the rings are not parallel (D_{5h} or D_{5d}) but are instead bent towards each other (C_{2v}) by a combination of bending the angle at Ge made by the centroids of the rings with small tilts of the ring planes to produce an angle α between the planes of the two rings which lies in the range $34°–50°$. In $(C_5R_5)GeR$ (or X) the angle at Ge is around $110°$ and the Ge is displaced from above the ring centre towards an edge, and is thus effectively dihapto (Figure 11c). The Ge–ring distances vary from $210–225$ pm to the centroid in Figure 11a or b (pentahapto structures) to differences in GeC distances up to 60 pm for asymmetric, dihapto, placings.

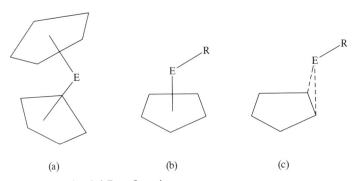

(a)   (b)   (c)

FIGURE 11. Common π-bonded E configurations

The bent structure for germanocene may be understood on qualitative MO grounds by noting that the highest filled orbital in the co-planar D_{5d} structure is the antibonding a_{1g} orbital formed by the out-of-phase combination of Ge 4s with the rings. Of the remaining six orbitals, the higher are two pairs of degenerate orbitals which are π on the rings and non-bonding or only weakly bonding, while the lowest are a pair of sigma bonding orbitals involving the Ge s and p_z orbitals. Bending to a C_{2v} configuration makes little net change in the energies of these six orbitals, but does allow mixing of Ge p character into the highest orbital, creating a lone pair pointing away from the rings. Thus this interaction with the rings goes from antibonding to weakly bonding, substantially stabilizing the bent configuration.

In the cation, with one ring removed there is an outward pointing hybrid holding the lone pair with a weakly bonding interaction with the ring, together with two p orbitals to the π bond. Finally, the cpGeR molecules can be envisaged as a Ge–R unit with the Ge using an sp^2 hybrid to bond to R, another to sigma bond to the ring and the third for the lone pair. This leaves a p orbital for π bonding to the ring. *Ab initio* MO calculations give good matches to the structural parameters and ionization energies of cpGeR and cpGe$^+$.

The mixed ligand cp*GeTsi shows a more dihapto interaction (GeC of cp* values 224.6 and 229.4 and two of 260 pm) with Ge–C(Si)$_3$ = 213.5 pm. The angle at Ge of the mid-point of the C–C bond and C is 118°. Similar geometry if found for cp*GeTtb with GeC(cp*) 322.5 and 230.4 pm and Ge–C(Ttb) = 208.7 pm with the angle reduced to 101°[470].

When GeCl$_2$ reacts with (Me$_4$C$_5$H)SiMe$_2$(Me$_4$C$_5$)$^-$, it gives a Ge cation with a penta-hapto link to the dienyl (GeC = 224–231 pm) and a dihapto one to the diene (GeC = 318 and 334 pm, the rest at 372–416 pm). The coordination is completed by a long interaction to 2Cl (324 and 339 pm) of the GeCl$_3^-$ counterion. A similar Sn product was formed and isolated as a dimer with two SnFBFSn links through the BF$_4^-$ counterion (SnC values averaging 250 pm in the penthapto, 322 and 390 in the dihapto and Sn–F = 274 and 282 pm)[471].

The indenyl complex cp*Ge(1,3-(SiMe$_3$)$_2$C$_9$H$_5$)[472] has approximately pentahapto bonding from both rings (Ge–C for cp* values of 233.2, 233.1, 238.6, 244.5 and 241.5 pm; for indenyl, 273.2, 276.7, 272.5, 270.4 269.0 pm; centroid–Ge–centroid angle = 163°).

The lone pair can be used for further bonding, especially to transition metals. The early examples were Cr or W carbonyls, Me$_5$C$_5$GeL $\longrightarrow$ W(CO)$_5$ where L = Cl, Me, CH(SiMe$_3$)$_2$ or N(SiMe$_3$)$_2$. Crystal structures show these species retain the dihapto coordination of ring to Ge with a reduction of the ring–Ge–R angle from 110° to 101° (L = Cl) or 102° (L = CH(SiMe$_3$)$_2$) and very similar Ge–L distances of 224 pm for the chloride and 199 pm for the disilylmethyl. The W(CO)$_5$ group lies over the ring with Ge–W = 263 pm and the ring–Ge–W angle around 135°.

A further related class of π complexes is that where the 5-membered ring is provided by the CB$_4$, C$_2$B$_3$ or CP (or As) B$_3$ face of a *nido*undecacarborane or heterocarborane. Examples are found with Ge–R or with an outward-pointing lone pair. While there has been no crystal structure report, the structure with the Ge completing the *closo*-dodecacarborane structure is compatible with B^{11} and other NMR data. 2-carba, 2,3-dicarba, 2,4-dicarba, 2,3-phospha(or arsa)carba and 2,4-phosphacarba skeletons are all represented.

C. R₂Sn(II) Species

Calculated parameters for Me$_2$Sn are Sn–C = 220.3 pm and CSnC = 95.3°.

Zuckerman and colleagues[473] postulated the first monomeric stannylene on the basis of Mossbauer results on [2,6-(CF$_3$)$_2$C$_6$H$_3$]$_2$Sn in 1981. Ten years later the first crystal structure was reported, of the related trisubstituted ring analogue[474], bis[2,4,6-tris(trifluoromethyl)phenyl]stannylene, $(Ar^f)_2Sn$. The Sn–C distance is 228.1 pm and the CSnC angle is 98.3°. The two nearest ortho-F...Sn distances average 267 pm while the closest intermolecular Sn...F is 351 pm; compare the van der Waals sum of 364 pm. Thus there is significant intermolecular interaction but little intermolecular effect. The barrier to oligomerization is probably provided in part sterically and in part by weak intramolecular donation from F to Sn.

This solid state structure is comparable with the first stannylene gas-phase structure reported[475] by Lappert. In $[(Me_3Si)_2CH]_2Sn$, the Sn–C distances were 222 pm and the CSnC angle was 97(2)°, essentially equivalent to the above.

$Ar(f)_2Sn$ reacts with a cumulene with transfer of one Ar(f) group to form $Ar(f)SnC(PPh_3)C(O)Ar(f)$[476] where the Sn is two-coordinate to C and the CSnC(P)C skeleton is planar with CSnC = 108.4°, SnCC = 113.3° and SnCP = 135.2°. The Sn–Ar(f) distance (225 pm) is normal for Ar–Sn but the Sn–C(P) length of 212 pm is shorter than a single bond (Table 1 and 2) though longer than the value calculated for $H_2Sn=CH_2$ (198.2 pm) and matches a picture of a multiple bond delocalized over the Sn–C–C–O frame. A similar study[477] with C=N–R yielded $Ar(f)_2SnCNR$ with Sn–CN = 239.7 pm, Sn–Ar(f) = 231 pm, SnCN = 154° and angles at Sn of 83°, 103° and 105°.

The stannylene $(Bsi)_2Sn$ gives two Zr–Sn bonds in $cp_2Zr(SnBsi_2)_2$[478]; Sn–Zr = 287.2 pm, Sn–C = 224 pm and there is no tin-tin bond (Sn...Sn = 424 pm). The geometry at Sn is distorted planar with CSnC = 102.4° and the two ZrSnC angles are widely different at 136° and 121°.

Alkyltin(II) compounds formed[479] using the anion $R^- = C_5H_4N[C(SiMe_3)_2]^-$ which bonds to the Sn through C and N in a four-membered chelate. For SnR_2, there are three independent molecules giving Sn–N = 238.4–244.9 pm, Sn–C = 233.4–237.7 pm and NSnC at 60° with the angles between the two ligands NSnN = 141° and CSnC = 125°. In RSnCl (Sn–Cl = 244.3 pm) and $RSnN(SiMe_3)_2$ [Sn–N(amine) = 214.4 pm] the ring SnC and SnN and angle values are similar.

1. Stannocenes and related compounds

Tin(II) has a similar[468], though richer, π complex chemistry to germanium. $(C_5H_5)_2Sn$ was synthesized in 1956 followed by the methylcyclopentadiene analogue in 1959. For two decades these were the only stable organometallic tin(II) species known. Through the 1980s, many other examples were added including those with Me_3Si- and Me_3Sn-substituted rings. A spectacular example was the synthesis of $(Ph_5C_5)_2Sn$ by Zuckerman and colleagues in 1984, later followed by the Ge and Pb analogues[480]. Unlike the others, this has parallel rings — the first group 14 ferrocene-type structure. With less bulky substituents, the structures are the bent ones found also for the germanocenes. The Sn atom tends to be off-centre and the structure thus tends towards a dihapto form and the individual Sn–C distances vary in a 10% range upwards from 253 pm. It is interesting that the differences for the two independent molecules in the unit cell of $(C_5H_5)_2Sn$ are quite significant — in particular, the displacement from the centroid of the Sn-ring perpendicular is 21 pm in one and 36 pm in the other, corresponding to a 5–10% difference in the distance to the ring and in the angle between the ring planes, α. This illustrates the flexibility of these molecules which also shows up in the spectroscopic observations.

Reaction[481] of cp_2Sn with further cp^- produces $[cp_3Sn]^-$ which has three η^3 bonded cp rings arranged trigonally around Sn (centroid–Sn = 257.1 pm, centroid–Sn–centroid = 118.8°). The Sn–C distances going stepwise round the ring are 262.1, 280.5, 309.2, 312.6 and 286.2 pm. A similar reaction replaced cp to form an $[SnR_3]^-$ anion when R is the much larger fluorenyl, to give a product which is clearly pyramidal (Sn–C = 234.6 pm, CSnC = 91.3°–102.9°). A similar species with the $[Na\{(Me_2NCH_2CH_2)_2NMe\}]^+$ counterion sees one of the cp rings bridging between the Sn and the cation with Sn–cp = 273.3 pm and cp–Na = 255 pm (compare the other two non-bridged cp–Sn at 254 pm)[482]. Reaction of cp_2Sn with $LiN=C(NMe_2)_2$ gave a four-membered ring[483] $[cpSnN=C(NMe_2)_2]_2$ where the SnNSnN is centrosymmetric with almost equal sides (Sn–N = 219.6 and 218.5 pm; angle at Sn 75.4°, at N 104.6°). The cp rings are trihapto and *trans* across the central ring (Sn–centroid = 243.2 pm).

An η^6 coordination of Ar = Me_6C_6 is found[484] in $[ArSnCl(AlCl_4)]_4$ where the seven-coordinate Sn also has 6 Cl bridges to 2 Al and a second Sn in a centrosymmetric tetramer. Bis-arene coordination from triptycene (R = $C_{20}H_{14}$) completes a distorted octahedron[485] around Sn in the dimer $[RClSnCl_2AlCl_2]_2$ with Sn bridged by two Cl to the second Sn (257.8 and 268.0 pm) and by two Cl to the Al (280.7 and 307.0 pm). Sn–centroid distances are 290 and 338 pm. A tris-arene interaction with Sn was achieved[486] by using the paracyclophane (p-$C_6H_4CH_2CH_2)_3$=R, in $[RSn]^{2+}$ with $AlCl_4^-$ counterions. There is a weak interaction (Sn–Cl = 307.3 pm) with one ion. The Sn–centroid distances are slightly varied because of this (253, 262 and 266 pm) but the Sn(centroid)$_3$ configuration is close to trigonal planar. The same study produced a germanium(II) product, $[RGeCl]^+$, with Ge–centroid distances averaging 274 pm and GeCl = 222.4 pm.

2. Sn(II) complexes

The same type of bidentate ligand bonding through C and N that gives higher coordination tetraorganotin complexes (Section IV.A.1) also yields Sn(II) compounds[113]. Thus $Sn(NC)_2$ coordination is found for alkylamine-substituted phenyl or naphthyl ligands, or for trimethylsilyl substituted pyridine. Coordinations include shapes derived from both the trigonal bipyramid with equatorial lone pair and square pyramid with axial lone pair. Tridentate ligands bonding through two N and C give similar products LSnCl or LSnR with equatorial Cl or R in shapes including equatorial lone pairs.

D. Pb(II) Species

The first structure[487] of a stable diaryllead depended, like the SnAr$_2$ above, on use of the tris(trifluoromethyl)phenyl ligand. In $[2,4,6-(CF_3)_3C_6H_2]_2Pb$, the Pb–C length was 236.6 pm and the CPbC angle was 94.5°. As for the Sn analogue, there were close intramolecular lead–fluorine interactions with the shortest pair of Pb...F distances averaging 279 pm.

While plumbocene has been known since 1966, little recent work is reported. The diazaplumbocene[488] $[2,5-(Bu^t)_2C_4NH_2]_2Pb$ has ring planes tilted at 143° and Pb–C distances in the range 260.4–283.2 pm and Pb–N = 287.8 and 285.4 pm.

E. Three-membered Rings

Three-membered rings are linked to lower valent compounds by the relation (E, E′, E″ = Si, Ge, Sn)

$$R_2E + R_2'E'=E''R''_2 = cyclo - R_2ER_2'E'R''_2E''$$

This reaction is a common basis for ring synthesis and may be reversed by photolysis, or often by heating, making the rings convenient sources of the -ylenes and the -enes. Such reactions are reviewed by Weidenbruch[489].

(a) Cyclopropane analogues, $(ER_2)_3$ for $E = Si, Ge, Sn$. These were seen to be inaccessible up to about 1980 because of the expected high reactivity of the strained ring. Various synthetic routes failed. For example, reduction of dihalides or similar species, R_2EX_2, yielded $(R_2E)_n$ only for $n = 4$ or more and the larger rings extruded R_2E to form the $(n - 1)$ ring for $n = 5,6$ etc but not for $n = 4$. Such reactions involved R = Me, or other small substituents, and the key to forming three-membered rings was the use of bulky R groups. Thus, Masamune and his colleagues prepared[490] the first cyclotrisilane, Si_3R_6, by using R = Dmp. Analogues with R = Mes, t-Bu and similar groups followed[491] quickly, including the striking persilylated member[492], the nonasilane $[(Et_3Si)_2Si]_3$. A similar strategy soon produced the cyclotri-germanes and -stannanes listed in Table 18. Preferred reducing agents include alkyllithiums, Li or Na naphthalenide, or Mg metal. While mechanisms are not proven, a reasonable and common postulate is formation of R_2E, then $R_2E=ER_2$, followed by addition of the two, or a similar process. Photolysis of the rings yield $R_2E=ER_2$.

The first mixed-E rings were not reported until the 1990s. Coupling[493] $[ClMes_2Ge]_2$ with Cl_2Mes_2Si produced the $SiGe_2$ ring, while reaction[494] of $GeCl_2$ with tris(trimethylsilyl)silyllithium gave a $GeSi_2$ ring in $[(Me_3Si)_2Ge]_3$, the analogue of the above nonasilane.

Structural details of these E_3 and E_2E' rings are given in Table 18. By comparison with the chains and larger rings, these all show $E-E$ bonds some 10 pm longer than the norm. These extensions reflect the bulky ligand rather than an adjustment to ring size. Thus (comparing the sterically similar ligands 2,6-dimethylphenyl and 2,4,6-trimethylphenyl) the $Si-Si$ bond length[495] of 240.7 pm in Si_3Dmp_6 and the $Ge-Ge$ value in Ge_3Mes_6 give a good prediction of $Ge-Si$ in $SiGe_2Mes_6$. Another comparison is with the linear $Mes_3GeGeMes(H)SiEt_3$ where both the $Ge-Ge$ and $Ge-Si$ bonds are lengthened despite the presence of the non-demanding $Ge-H$ bond. Similar lengthening of bonds between E atoms with large substituents is seen in larger rings.

When the trigermane values are compared with the corresponding trisilanes (compare the invaluable review by Tsumuraya and colleagues[496]), it is seen that the $Ge-Ge$ bond length in $(t$-Bu$)_6Ge_3$ falls into the same range as for other substituents while the $Si-Si$ value of 255.1 pm for $(t$-Bu$)_6Si_3$ substantially exceeds the $Si-Si$ range of 238-244 pm found for ligands like Dmp, Mes and neopentyl.

So far there is only one reported cyclotristannane structure, Sn_3dep_6, and no mixed ring containing tin. The closely related 2,4,6-triisopropylphenyl tristannane has been shown[497] to be in thermal equilibrium with the distannene over a wide temperature range. Not unexpectedly, there are no lead compounds of this class. It is an interesting sidelight[498] on ring stability that the dominant mode of fragmentation in electron impact mass spectroscopy is the loss of R as alkene, leaving the ring apparently intact for E = Si, Ge and Sn.

(b) Three-membered species with EC_2 and E_2C rings. A further significant group are three-membered rings containing C which were also once thought to be too reactive to isolate. Syntheses usually involve divalent or unsaturated species, as in the formation of EC_2 rings by addition of R_2E to alkynes or alkenes.

Recent interest in the structures has led to the crystallographic characterization of a significant proportion of the possible types. Known cyclopropane analogues, '-iranes', include E_2C, $EE'C$ and EC_2 rings while cyclopropene analogues, '-irenes', are found for the EC_2 composition. Most structures are of rings with bulky substituents on E reflecting syntheses from stabilized ER_2, but this is not a precondition for stability, as shown by the report of a dimethylgermirene.

The first structure reports were of germirene and stannirene in the 1980s following closely the first silirene. In a pair of germirene[499] and stannirene[500] structures, the

C=C ring edge is shared with a seven-membered sulphur-containing ring in a 4-thia-8-germa(stanna)bicyclo[5.1.0]octene. The first saturated germirane rings were reported ten years later[501], along with a germirene with a second type of bicyclic structure — this time a diazagermabicycloheptene. In most of the reported structures there are bulky R = CH(SiMe$_3$)$_2$ groups on E, which stabilized the germ(stann)ylene reagent.

Compounds were air-sensitive and some decomposed in the X-ray beam, so the data are of variable quality. Within the error limits, the Ge—C bond lengths in the two germirane rings match at 196 pm, while the two germirenes are distinctly shorter at 191 pm. The external Ge—C bonds are normal and the external CGeC angle is wide, ranging from 114° to 126° in different compounds. Table 19 shows the critical ring CC bond lengths and CGeC angles.

The C=C length is similar for E = Ge and Sn while the angle at E is smallest for Sn. This angle is markedly smaller than in E$_3$ compounds, an indication that the bonding at E is adjusting to CC and EC bond length limitations (compare, for example, the pioneering[503] CNDO calculations on C$_2$X). There may be a trend from mols 2 to 4 to 3 reflecting delocalization from the external double bonds in 4, but the large error makes

FIGURE 12. Core structure for Table 19

TABLE 19. Some germirene rings

Compound[a]	CC length (pm)	CEC angle (deg)	ECC angle (deg)	
1	133.1	40.5	70.3,69.2	
2	139	43.2	68.4	C$_2$ axis
3	151,162	45,47.7	68,71	2 mols in unit cell
4	146	44	66.7,68.3	
5	134.0	36.6	71.7	ext CSnC = 110°
Some calculated values[502]	CC	GeC	GeSi	GeGe
GeC$_2$H$_6$	154.3	194.4		
Ge$_2$CH$_6$		199.6		237.3
GeSiCH$_6$		201.3 (SiC = 188.7)	231.2	
Ge$_3$H$_6$				246.2
Ge$_2$SiH$_6$			240.2	246.0
GeSi$_2$H$_6$			239.9 (SiSi = 233.6)	

[a] Compounds (compare Figure 12):
1. R = Me, A = CMe$_2$, Z = CH$_2$CH$_2$.
2. R = Bsi, A = CMe$_2$, Z is NN shared with a CONPhCO ring.
3. R = Bsi, A = CO, Z = NPh.
4. A is =CMe$_2$, no Z.
5. Sn analogue of 1.

this uncertain. For the tin compound[504], NMR studies suggest the R_2Sn addition to the alkyne is reversible and the system is in further equilibrium with R_4Sn_2.

Using *ab initio* calculations[505], the stabilities to decomposition to $:EH_2 + H_2E'=E''H_2$ were evaluated for the ten rings containing combinations of E, E', E'' = C, Si and Ge. The germirane ring, GeC_2, is substantially the least stable with a value only two-thirds of the remaining Ge-containing rings, which are all quite similar, and about half of those for Si rings. The tin analogues should be even less stable while plumbirane is almost certainly thermodynamically unstable. Such results match the experimental experience, where it took some 10 years after the first cyclotrigermane and germirene syntheses before germiranes appeared, and then the rings were supported by the C substituents. Similarly, while cyclotristannanes and a stannirene are known, there is not yet a stannirane.

Addition of a CH_2 group to the digermene formed by photolysis of the cyclotrigermane, $Ge_3(Dep)_6$, using diazomethane, gave the first structurally characterized digermirane[506], $(Dep)_4Ge_2CH_2$. The external Ge—C bonds were normal at 198.4 and 197.5 pm and the angle between them was 112.6°. The phenyl rings are all twisted some 65° in the same sense. Within the three-membered ring, values are close to the calculated values, with GeGe = 237.9 pm and GeC = 197.0 pm; the angle at C is 74.3° and at Ge, 52.9°.

(c) Hetero-rings containing elements from other groups. There are several examples of E_2X rings where X comes from Group 15 or 16. In a similar reaction to the digermirane formation, reaction of a digermene with phenylazide gave[506] the azadigermiridine, Ge_2N ring. Such reactions earlier produced Ge_2X rings for X = S, Se, Te as in Table 16. Similar syntheses have given Sn_2N and Sn_2Te rings. Ge—Ge is normal length in the telluradigermirane and shortens progressively through Se and S to the N and C rings; the GeXGe angle increases in parallel.

A substantial number of GeXY rings have been identified spectroscopically, but crystallographic determinations are rarer and include the following:

(i) The GeSC ring in 3-(2-adamantyl)-2,2-dimesityl-1,2-thiagermirane is formed[507,508] by the addition of germylene to the C=S bond of adamantanethione. The two crystallographically independent rings are very similar with Ge—S = 222.2 pm, Ge—C = 197.1 pm and SC = 187.9 pm. The ring angles are 52.9° at Ge, 70.4° at C and 56.8° at S. The Ge—Mes lengths are normal at 195.4–196.9 pm. These values interpolate reasonably with those in digermirane and sulphadigermirane. This ring is air-stable but easily oxidized with the insertion of an oxygen and formation of S=O to give the four-membered GeOS(O)C ring, probably via the thiagermirane S-oxide. Two of the GeC distances are little altered at 195.9 pm (ring) and 197.1 pm (Mes) but the second Ge—Mes lengthens to 203.3 pm. The ring is folded 24.3° across the GeS diagonal.

The analogous SnCS thiastannirane ring was formed[509] using the stannylene $(Bsi)_2Sn$ to react with $Bu_2^t=C=S$. The ring has SnS = 243.8 pm, SnC = 212 pm and CS = 179 pm and angles of 46° at Sn, 58° at S and 77° at C. The Sn tolerates a more acute angle than the Ge or Si analogues. The external Sn—C averages 217 pm and the CSnC angle is 116°.

(ii) A GePS ring is found in $Mes_2GeSPAr$ (Ar = 2,4,6-tris-t-butylphenyl) formed[510] by S insertion into Ge=P. The ring parameters are GeP = 231.6 pm, GeS = 222.7 pm, SP = 217.6 pm; angles are 57.2° at Ge, 59.3° at P and 63.5° at S; Ge—Mes = 196.1 pm.

(iii) While single GePP rings have been identified spectroscopically, the crystal data[511] are for the linked pair of GePP triangles which share a P—P edge in the 1,3-diphospha-2,4-digermabicyclo[1.1.0]butane, $[t-Bu(Mes)Ge]_2P_2$. The GeP length is 231.1 pm, Ge—Bu is 199.0 pm, Ge—Mes is 197.2 pm, all normal values, while PP at 238.3 pm is 7% longer than in simple diphosphines. The ring angles are 58.9° at P and 62.1° at Ge, and the molecule is bent across the PP edge shown by the GePGe angle of 86.8°. The two Mes groups are in pseudo-axial positions, also describable as *cis* to the double ring. We might include here the data from the same study for the corresponding four-membered ring cyclo-[t-Bu(Mes)GePH]$_2$ where there is no P–P bond. Here the Ge–C distances are the same but GeP is longer at 234.7 pm, and the ring is planar with angles of 95.3° at Ge and 84.8° at P. The pairs of organic groups are in *trans* positions across the ring. Replacement of P–H via Li by HgR, then photolysis to remove 2HgR, forms the transannular P–P bond of the bicycle. A related product in the system contains two Ge$_2$P$_2$ rings linked through a linear P–Hg–P bond and with a terminal P–Hg–t-Bu group on the second P. Here the Ge–P (Hg bridge) length is 234.2 pm while Ge–P (Hg terminal) is a little longer at 236.6 pm. The rings are slightly bent across the Ge...Ge diagonal and the Mes groups are now *cis* across the four-membered ring. The –Ge(Bsi)$_2$P=CBu^t– ring has also been established[512].

VIII. DOUBLY BONDED GERMANIUM, TIN AND LEAD

Early ideas of doubly bonded Si or Sn were dispersed when Kipping and his contemporaries demonstrated in the 1920s that the E$_2$R$_4$ species were not alkene analogues but small rings. Some fifty years later, the discovery by Lappert and his colleagues of genuine R$_4$E$_2$ species reopened the whole question and led to an intensive period of experimental and theoretical examination of the concept of an E=E double bond.

It is first worth noting some calculated parameters[513] for the radical cations Me$_6$E$_2$$^+$ (E–E = 265, 257, 342 and 305 pm and E–C = 183, 192, 212 and 217 pm, respectively, for E = Si, Ge, Sn and Pb) which warn of the interplay of orbital energies down the Group. Double bonds involving transition metals and unsubstituted Ge, Sn and Pb are noteworthy[514].

From the 1970s, the presence of intermediates with double bonds to Ge was well established by chemical studies which identified the species by recovery of its polymer, or by use of a trapping agent. Satgé and coworkers[515–517] have written a valuable and intriguing series of reviews which shows the development of the field over the years. Thus, Et$_2$Ge=CH$_2$ was established by identification of the dimer, 1,3-digermacyclobutane, and the germacyclohexene by trapping by dimethyldibutadiene. Interestingly Me$_2$Ge=C(Me)GeMe$_3$ formed the digermane head-to-head dimer Me$_3$GeCHMeGeMe$_2$GeMe$_2$C(=CH$_2$)GeMe$_3$. Many similar studies established transient R$_2$Ge=O, R$_2$Ge=S, R$_2$Ge=NR, R$_2$Ge=PR′ and R$_2$Ge=BiR′ as well as a range of examples of R$_2$Ge=CR$_2'$ and several digermenes R$_2$Ge=GeR$_2$, but much more limited Sn and Pb chemistry. The excellent 1991 review[496] and a further survey in 1993[518] of the heavier Group 14 and 15 element pπ interactions show that the major emphasis of work is still on Si. In their 1994 review[515] Satgé and colleagues list some 15 E=E and E=X (X=N, P and S mainly) systems which are established with stable species, although there are also noticeable gaps such as Ge=Sn or any As or Se species. Structural studies so far cover a much narrower sub-set.

In 1976, Lappert and his colleagues[519] reported the preparation of E$_2$R$_4$ [R = Bsi] species for E = Ge, Sn and Pb. The crystal structure defined R$_2$SnSnR$_2$, while the Ge and Pb compounds were identified as analogues by comparing properties, with the assignment of a strong Raman band at 300 cm^{-1} to the Ge=Ge stretching frequency. The compounds dissociated to ER$_2$ in solution. This discovery initiated an interesting decade of effort through the 1980s when the set of structurally characterized ethene analogues was extended to include examples of silene (R$_2$Si=CR$_2'$), disilene (R$_2$Si=SiR$_2$), germene (R$_2$Ge=CR$_2'$), germasilene (R$_2$Si=GeR$_2'$) as well as further digermenes (R$_2$Ge=GeR$_2$). The silicon species and theoretical approaches have been reviewed by Sheldrick[4a] to 1989.

For the digermenes, Lappert later reported[520] the full crystal structure of R$_2$Ge=GeR$_2$, R = Bsi and a refinement of the tin structure while Masamune has characterized the analog

TABLE 20.   Structural parameters of doubly bonded species

(A) Compounds of formula $R_2E=ER_2$ (see Figure 13B)

E	R(a)	E–E (pm)	E–C (pm)	Fold (x)	Ref.
C	Mes	136.4	149.4	0	531
Si	Mes	216.0	187.1	18	532
	Dep	214.0	188.2	0	533
	Bu^t, Mes	214.3	188.4	0	531
Ge	Dep	221.3	196.2	12	521
	Mes, Dip	230.1	198 (av)	36	521
	Bsi	234.7	201.0	32	520
Sn	Bsi	276.8	221.6	41	520

(B) Compounds of formula $R_2E=E'R'_2$

E	R	E'R'	E–E'	E–C	Ref.
Ge	Ar(1)	fluorenylidene	180.3	193.6	522
Ge	Mes	PMes	213.8		533

where $R = Dep^{521}$ and also the unsymmetrically substituted RR'Ge=GeRR'(**36**)^{519} with $R = Dip$ and $R' = Mes$. Data are included in Table 20.

Not until 1987 were the first characterized germenes reported^{522}. In $Mes_2Ge=CR_2$ (CR_2 = fluorenylidine) the Ge=C bond is 180.1 and 180.6 pm in two independent molecules which is clearly shortened over normal single bonds (compare with Tables 1 and 2). This shortening is even more marked compared with the bond in $Mes_2GeH-CHR_2$, where Ge–C at 201 pm is unusually long, showing the intense steric pressures. Similarly there is a distinct shortening in Ge–Mes to 193.1 and 194.4 pm in $Mes_2Ge=CR_2$ compared with 196.5 pm in $Mes_2GeH-CHR_2$. These values are all in accord with expectation for sp^2 Ge.

The 1976 structure of $R_2Sn=SnR_2$, $R = Bsi$ showed that, while formally an ethene analogue, the conformation at Sn was pyramidal, not planar, and the Sn_2C_4 core adopted a *trans*-folded C_{2h} structure instead of the D_{2h} ethene form. This was reflected in *ab initio* calculations by Trinquier and colleagues following pioneer work by Gowenlock and Hunter. These researchers found that, while $H_2Ge=CH_2$ was planar C_{2v}^{523} like ethene and silene, the planar formulation of digermene^{524} was about 15 kJ mol^{-1} less stable than the *trans* folded form. In this, $H_2Ge=GeH_2$ was the exception as all other $H_2Ge=X$ systems examined (X = O, S, NH, PH as well as CH_2) were planar like the Si analogues. Lappert and others reached similar conclusions from *ab initio* calculations for digermene^{522} and for distannene^{525}. Relativistic extended Huckel calculations have been performed^{526} for digermene, distannene and diplumbene, as well as MNDO calculations^{527} for distannene. These, and other contributions to the burst of theoretical work through the 1980s, were ably summarized by Grev^{528} in 1991.

The results of calculations on E_2H_4 are worth summarizing (Figure 13), and were discussed in detail by Sheldrick^{4a}. Rearrangement barriers from $H_2Ge=EH_2$ to H_3EGeH have also been evaluated^{529}. Three structures are found among the most stable forms: the planar D_{2h} structure (A) of ethene, the *trans* folded C_{2h} structure (B), and the H_3EEH isomer (C) involving one divalent E atom and a singlet lone pair. There are also several excited states including the triplet state of (C) and a twisted diradical (D) (where the EH_2

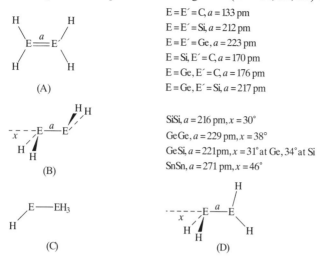

$E = E' = C, a = 133\,\text{pm}$
$E = E' = Si, a = 212\,\text{pm}$
$E = E' = Ge, a = 223\,\text{pm}$
$E = Si, E' = C, a = 170\,\text{pm}$
$E = Ge, E' = C, a = 176\,\text{pm}$
$E = Ge, E' = Si, a = 217\,\text{pm}$

$SiSi, a = 216\,\text{pm}, x = 30°$
$GeGe, a = 229\,\text{pm}, x = 38°$
$GeSi, a = 221\text{pm}, x = 31° \text{ at Ge}, 34° \text{ at Si}$
$SnSn, a = 271\,\text{pm}, x = 46°$

FIGURE 13. Some calculated structures for E_2H_4 species

group is planar for E = C but pyramidal for E = Si, Ge, Sn, Pb). The singlet and triplet forms of (D) are of similar energy.

For ethene, all other isomers are $250-300\,\text{kJ mol}^{-1}$ less stable than planar (A). However, in disilene the folded structure (B) is just below (A) and becomes increasingly stable relative to (A) as E gets heavier. Although isomer (C), methylmethylene, is $300\,\text{kJ mol}^{-1}$ less stable than (A), this gap drops to about 30 kJ for E = Si while for E = Ge, isomer (C) actually lies between (A) and (B). The *trans* folded structure (B) is also the most stable for Sn_2H_4. $EE'H_4$ species vary in a similar way. Although the gap decreases enormously compared with C_2H_4, the planar, C_{2v} form of silene (A) is some $17\,\text{kJ mol}^{-1}$ more stable than methylsilene (C) and this order is actually reversed for germene, reflecting the increased relative stability of Ge(II). The twisted diradical (D) forms are some 130 kJ higher for each element, with other structures including the methylene isomers of (C) higher still. The species corresponding to (B) with pyramidal $Si(Ge)H_2$ was apparently not considered in these studies. Germylsilene resembles germene in having the silyl-germylene isomer (C) as the most stable form at 25 kJ below (B). However, it is closer to disilene than digermene in the small size of the gap between (B) and (A).

The established R_4E_2 structures (Table 20) may be compared with the ER_2 monomers (previous section). They agree with the calculations on the hydride models in showing the predicted stable structure: the *trans* folded structure (B), with the fold angle a good match to the calculated one. However, the substituents on E are extremely bulky and have varying electronic properties, so distortions are found. In several cases the fold angle is substantially diminished, but the groups twist and often the two E–R distances also differ. Thus, aspects of both (B) and (D) isomers are present. All the E=E bonds show shortening over the single-bond values (compare Section VI.B and Table 15) by 2–12%, tending to decrease for larger E. Despite the variation in observed single bonds, this contraction is significant.

The first crystal structure of a germaphosphene[530] was of $Mes_2Ge=P(2,4,6-\text{tri}tert\text{butylphenyl})$ formed by dehydrofluorination. The Ge=P length was some 20 pm shorter than Ge–P values at 213.8 pm. The molecule shows an almost planar double-bond configuration with an angle at P of 107.5°. The Ge–Mes distances were normal and the (Mes)Ge(Mes) angle is 112.9°.

Lappert's compounds E_2R_4 readily dissociated to ER_2 monomers in solution, emphasizing the relationship and leading to the very natural simple description of these $CH(SiMe_3)_2$ compounds as a double donor–acceptor complexes. For Dip, Dep and mes species which are more stable in solution, the structures are still close to those of the Lappert compounds.

IX. REFERENCES

1.   G. Wilkinson, F. G. A. Stone and E. W. Abel (Eds.), *Comprehensive Organometallic Chemistry*, Pergamon Press, London, 1982. In particular Chapters 10, *Germanium*, P. Rivière, M. Rivière-Baudet and J. Satgè, p.399; 11, *Tin*, A. G. Davies and P. J. Smith, p.519; 12, *Lead*, P. G. Harrison, p.629; 43, *Compounds with Bonds between a Transition Metal and Either Silicon, Germanium, Tin or Lead*, K. M. Mackay and B. K. Nicholson, p.1043; and *Index of Structures Determined by Diffraction Methods*, M. I. Bruce, p.1209. See also the corresponding sections in: E. W. Abel, F. G. A. Stone and G. Wilkinson (Eds.), *Comprehensive Organometallic Chemistry II*, Pergamon Press, London, 1995.
     G. Wilkinson, R. D. Gillard and J. A. McCleverty (Eds.), *Comprehensive Coordination Chemistry*, Pergamon Press, London, 1987; particularly Section 12.2 by P. G. Harrison and T. Kikabbau, and Chapter 26 by P. G. Harrison.
     Recent volumes of the *Gmelin Handbook*: *Organogermanium—Part 5, Ge—O Compounds* and *Organotin*, Vol.20, Springer-Verlag, Berlin, 1993.
2.   S. Patai (Ed.), *The Chemistry of Functional Groups, Supplement A: The Chemistry of Double-bonded Functional Groups*, Vol.2. (esp. pp.1–52) Wiley, Chichester, 1989; F. Hartley and S. Patai (Eds.), *The Chemistry of the Metal–carbon Bond*, Wiley, Chichester, 1982; P. A. Cusack, P. J. Smith, J. D. Donaldson and S. M. Grimes, *A Bibliography of X-Ray Crystal Structures of Tin Compounds*, International Tin Research Institute Publication 588, 1981; Organotin Cluster Chemistry, R. R. Holmes, Acct. Chem. Research, **22**, 190 (1989).
3.   Ge, Sn and Pb Conference Reports: VI (Brussels), *Main Group Metal Chemistry*, **XII**, 277; VII (Riga), *Frontiers of Organo-germanium, -tin and -lead Chemistry*, (Eds. E. Lukevics and L. Ignatovich), Latvian Institute of Organic Synthesis, 1993; VIII (Tokyo), *Main Group Metal Chemistry*, **XVII**, 1; see also: J. T. B. H. Jastrzebski and G. van Koten, 'Intramolecular Coordination in Organotin Chemistry', *Adv. Organomet. Chem.*, **35**, 241 (1993); V. G. K. Das, N. S. Weng and M. Gielen (Eds.), *Chemistry and Technology of Silicon and Tin*, Oxford University Press, 1992; K. C. Molloy, 'Organotin Heterocycles', *Adv. Organomet. Chem.*, **33**, 171 (1991); 'Group 14 Metalloles', 1. J. Dubac, A. Laporterie and G. Manuel; 2. E. Colomer, R. J. Corriu and M. Lheureux, *Chem. Rev.*, **90**, 215 and 265 (1990); P. G. Harrison *Chemistry of Tin*, Blackie, London, 1989; L. Omae, 'Organotin Chemistry', *J. Organomet. Chem. Library*, **21**, 238 (1989); V. G. Kumar Das, L. K. Mun and N. S. Weng, 'C-Heterocyclic Tin(IV) Compounds', Main Group Metal Chemistry, **XI**, 251 (1988).
4.   (a) W. S. Sheldrick, Chap. 3, 'Structural Chemistry of Organic Silicon Compounds', in The Chemistry of Organic Silicon Compounds (Eds. S. Patai and Z. Rappoport), Wiley, Chichester, 1989, p.227.
     (b) E. Lukevics, O. Pudova and R. Sturkovich, *Molecular Structure of Organosilicon Compounds*, English edn., Ellis Horwood, Chichester, 1989.
5.   K. C. Molloy and J. J. Zuckerman, *Adv. Inorg. Chem. Radiochem.*, **27**, 113, 1983.
6.   A. Haaland, Chap. 8 in *Stereochemical Applications of Gas Phase Electron Diffraction, Part B* (Eds. I and M. Hargittai), VCH, Stuttgart, 1988, p.328.
7.   B. Krebs, G. Henkel and M. Dartmann, *Acta Cryst.*, **C45**, 1010 (1989).
8.   B. Wrackmeyer, G. Kehr, D. Wettinger and W. Milius, *Main Group Met. Chem.*, **XVI**, 445 (1993).
9.   U. Fahrenkampf, M. Schürmann and F. Huber, *Acta Cryst.*, **C49**, 1066 (1993).
10. M. Kaupp and P. von R. Schleyer, *Angew. Chem., Int. Ed. Engl.*, **31**, 1224 (1992).
11. A. Haas, H. -J. Kutsch and C. Kruger, *Chem. Ber.*, **122**, 271 (1989).
12. R. Eujen, F. E. Laufs and H. Oberhammer, *Z. Anorg. Allg. Chem.*, **561**, 82 (1988).
13. For a fuller discussion, see Reference 4a, pp.242–245.
14. Y. Takeuchi, K. Ogawa, G. Manuel, R. Boukherroub and I. Zicmane, *Main Group Metal Chem.*, **XVII**, 121 (1994).
15. Q. Shen, S. Rhodes, Y. Takeuchi and K. Tanaka, *Organometallics*, **11**, 1752 (1992); H. Schmidbaur, J. Rott, G. Reber and G. Muller, *Z. Naturforsch.*, **B43**, 727 (1988).

2. Structural aspects of compounds containing C–E (E = Ge, Sn, Pb) bonds 181

16. G. Fritz, K. D. Hoppe, W. Hönle, D. Weber, C. Mujica, V. Manriquez and H. G. von Schnering, *J. Organomet. Chem.*, **249**, 63 (1983).
17. S. Roller, M. Dräger, H. J. Breunig, M. Ates and S. Gülec, *J. Organomet. Chem.*, **378**, 327 (1989); **329**, 319 (1987); G. Becker, M. Meiser, O. Mundt and J. Weidlein, *Z. Anorg. Allg. Chem.*, **569**, 62 (1989).
18. A. Blaschette, T. Hamann, A. Michalides and P. G. Jones, *J. Organomet. Chem.*, **456**, 49 (1993).
19. G. A. Miller, S. W. Lee and W. C. Trogler, *Organometallics*, **8**, 738 (1989).
20. H. Preut and F. Huber, *Acta Cryst.*, **C49**, 1372 (1993).
21. C. Schneider-Koglin, B. Mathiasch and M. Dräger, *J. Organomet. Chem.*, **469**, 25 (1994).
22. M. Charissé, S. Roller and M. Dräger, *J. Organomet. Chem.*, **427**, 23 (1992).
23. M. Charissé, V. Gauthey and M. Dräger, *J. Organomet. Chem.*, **448**, 47 (1993).
24. D. White and N. J. Colville, *Adv. Organomet. Chem.*, **36**, 95 (1994).
25. W. Weissensteiner, I. I. Schuster, J. F. Blount and K. Mislow, *J. Am. Chem. Soc.*, **108**, 6664 (1986).
26. I. Wharf and M. G. Simard, *Acta Cryst.*, **C47**, 1314 (1991).
27. J. P. Fackler Jr., G. Garzon, R. A. Kresinski, H. H. Murray III and R. G. Raptis, *Polyhedron*, **13**, 1705 (1994).
28. F. Theobald and B. Trimaille, *J. Organomet. Chem.*, **267**, 143 (1984).
29. M. T. Ahmet, A. Houlton, C. S. Frampton, J. R. Miller, R. M. G. Roberts, J. Silver and B. Yavari, *J. Chem. Soc., Dalton Trans.*, 3085 (1993).
30. M. F. Lappert, W-P. Leung, C. L. Ralston, B. W. Skelton and A. H. White, *J. Chem. Soc., Dalton Trans.*, 775 (1992).
31. S. U. Ghazi, R. Kumar, M. J. Heeg and J. P. Oliver, *Inorg. Chem.*, **33**, 411 (1994).
32. L. Fajari, L. Julia, J. Riera, E. Molins and C. Miravitlles, *J. Organomet. Chem.*, **363**, 31 (1989).
33. D. Reed, D. Stalke and D. S. Wright, *Angew. Chem., Int. Ed. Engl.*, **30**, 1459 (1991).
34. D. R. Armstrong, M. G. Davidson, D. Moncreiff, D. Stalke and D. S. Wright, *J. Chem. Soc., Chem. Commun.*, 1413 (1982).
35. M. Rannenberg, H-D. Hausen and J. Weidlein, *J. Organomet. Chem.*, **376**, C27 (1989).
36. M. G. Simard and I. Wharf, *Acta Cryst.*, **C50**, 397 (1994).
37. R. A. Howie, J-N. Ross, J. L. Wardell and J. N. Low, *Acta Cryst.*, **C50**, 229 (1994).
38. H. Reuter and H. Puff, *J. Organomet. Chem.*, **379**, 223 (1989).
39. S. S. Al-Juaid, S,M, Dhaher, C. Eaborn, P. B. Hitchcock and J. D. Smith, *J. Organomet. Chem.*, **325**, 117 (1987).
40. M. Riviere-Baudet, A. Morere, J. F. Britten and M. Onyszchuk, *J. Organomet. Chem.*, **423**, C5 (1992).
41. H. Schumann, L. Esser, J. Loebel and A. Dietrich, *Organometallics*, **10**, 2585 (1991).
42. Y-X, Chen, M. D. Rausch and J. C. W. Chien, *Organometallics*, **13**, 748 (1994).
43. D. K. Srivastava, N. P. Rath and L. Barton, *Organometallics*, **11**, 2263 (1992).
44. U. Siriwardane and N. S. Hosmane, *Acta Cryst.*, **C44**, 1572 (1988).
45. H. Zhang, L. Jia and N. S. Hosmane, *Acta Cryst.*, **C49**, 791 (1993).
46. N. S. Hosmane, K-J Lu H. Zhang, J. A. Maguire, L. Jia and R. D. Barreto, *Organometallics*, **11**, 2458 (1992).
47. R. W. Chapman, J. G. Kester, K. Folting, W. E. Streib and L. J. Todd, *Inorg. Chem.*, **31**, 979 (1992).
48. P. Jutzi, D. Wegener, H-G. Stammler, A. Karaulov and M. B. Hursthouse, *Inorg. Chim. Acta*, **198–200**, 369 (1992).
49. L. Jia, H. Zhang and N. S. Hosmane, *Organometallics*, **11**, 2957 (1992).
50. F. Brisse, M. Vanier, M. J. Oliver, Y. Gareau and K. Stelliou, *Organometallics*, **2**, 878 (1983).
51. R. Minkwitz, A. Kornath and H. Preut, *Z. Anorg. Allg. Chem.*, **620**, 981 (1994).
52. H. Oshita, Y. Mizobe and M. Hidai, *J. Organomet. Chem.*, **456**, 213 (1993).
53. J. E. Drake and J. Yang, *Inorg. Chem.*, **33**, 854 (1994).
54. J. E. Drake, A. G. Mislankar and M. L. Y. Wong, *Inorg. Chem.*, **30**, 2174 (1991); J. E. Drake, A. B. Sarker and M. L. Y. Wong, *Inorg. Chem.*, **29**, 785 (1990).
55. J. E. Drake, A. G. Mislankar and J. Yang, *Inorg. Chem.*, **31**, 5543 (1992) and references cited therein.
56. S. W. Ng, V. G. Kumar Das, F. L. Lee, E. J. Gabe and F. E. Smith, *Acta Cryst.*, **C45**, 1294 (1989).

57. J. Barluenga, M. Tomas, A. Ballesteros, J. S. Kong, S. G. Granda and P. Pertierra, *Organometallics*, **11**, 2348 (1992).

58. H. J. Gysling and H. R. Luss, *Organometallics*, **8**, 363 (1989).

59. R. Okazaki, K. Shibata, and N. Tokitoh, *Tetrahedron Lett.*, **32**, 6601 (1991).

60. S. Hanessian, Y. L. Bennani and R. Léger, *Acta Cryst.*, **C46**, 701 (1990).

61. P. Brown, M. F. Mahon and K. C. Molloy, *J. Chem. Soc., Dalton Trans.*, 2643 (1990).

62. C. Janiak, M. Schwichtenberg and F. E. Hahn, *J. Organomet. Chem.*, **365**, 37 (1989).

63. B. J. Brisdon, M. F. Mahon, K. C. Molloy and P. J. Schofield, *J. Organomet. Chem.*, **465**, 145 (1994).

64. G. Ferguson, J. F. Gallagher, D. Murphy, T. R. Spalding, C. Glidewell and H. D. Holden, *Acta Cryst.*, **C48**, 1228 (1992).

65. H. Puff, S. Franken, W. Schuh and W. Schwab, *J. Organomet. Chem.*, **254**, 33 (1983).

66. H. Puff, W. Schuh, R. Sievers, W. Wald and R. Zimmer, *J. Organomet. Chem.*, **260**, 271 (1984).

67. (a) J. F. Van der Maelen Uria, M. Belay, F. T. Edelmann and G. M. Sheldrick, *Acta Cryst.*, **C50**, 403 (1994).
    (b) H. Grützmacher and H. Pritzkow, *Chem. Ber.*, **126**, 2409 (1993).

68. S. Masamune, L. Sita and D. L. Williams, *J. Am. Chem. Soc.*, **105**, 630 (1983).

69. V. K. Belsky, N. N. Zemlyansky, I. V. Borisova, N. D. Kolosova and I. P. Beletskaya, *J. Organomet. Chem.*, **254**, 189 (1983).

70. O. -S. Jung, J. H. Jeong and Y. S. Sohn, *Polyhedron*, **8**, 2557 (1989).

71. E. R. T. Tiekink, *Main Group Metal Chem.*, **XVI**, 65 (1993).

72. R. J. Batchelor, F. B. Einstein and C. H. W. Jones, *Acta Cryst.*, **C45**, 1813 (1989).

73. F. B. Einstein, I. D. Gay, C. H. W. Jones, A. Riesen and R. D. Sharma, *Acta Cryst.*, **C49**, 470 (1993).

74. B. M. Schmidt and M. Dräger, *J. Organomet. Chem.*, **399**, 63 (1990); A. J. Edwards and B. F. Hoskins, *Acta Cryst.*, **C46**, 1397 (1990).

75. D. Hänssgen, M. Jansen, W. Assenmacher and H. Salz, *J. Organomet. Chem.*, **445**, 61 (1993).

76. S. Tomada, M. Shimoda, M. Sanami, Y. Takeuchi and Y. Iitaka, *J. Chem. Soc., Chem. Commun.*, 1304 (1989).

77. J. Bodiguel, P. Meunier, M. M. Kubicki, P. Richard and B. Gautheron, *Organometallics*, **11**, 1423 (1992).

78. M. Akkurt, T. R. Kök, P. Faleschini, L. Randaccio, H. Puff and W. Schuh, *J. Organomet. Chem.*, **470**, 59 (1994).

79. H. Puff, M. P. Böckmann, T. R. Kök and W. Schuh, *J. Organomet. Chem.*, **268**, 197 (1984).

80. L. Ross and M. Dräger, *Z. Naturforsch.*, **B39**, 868 (1984).

81. A. Mazzah, A. Haoudi-Mazzah, M. Noltemeyer and H. W. Roesky, *Z. Anorg. Allg. Chem.*, **604**, 93 (1991).

82. K. Ogawa, K. Tanaka, S. Yoshimura, Y. Takeuchi, J. Chien, Y. Kai and N. Kasai, *Acta Cryst.*, **C47**, 2558 (1991).

83. H. Puff, T. R. Kök, P. Nauroth and W. Schuh, *J. Organomet. Chem.*, **281**, 141 (1985).

84. H. Preut, J. Koch and F. Huber, *Acta Cryst.*, **C46**, 2088 (1990).

85. S. Masamune, S. A. Batcheller, J. Park, W. M. Davis, O. Yamashita, Y. Ohta and Y. Kabe, *J. Am. Chem. Soc.*, **111**, 1888 (1989).

86. M. Kako, T. Akasaka and W. Ando, *J. Chem. Soc., Dalton Trans.*, 457, 458 (1992).

87. e.g. H. Puff, K. Braun, S. Franken, T. R. Kök and W. Schuh, *J. Organomet. Chem.*, **335**, 167 (1987); M. Wojnowska, M. Noltemeyer, H-J. Füllgrabe and A Meller, *J. Organomet. Chem.*, **228**, 229 (1982).

88. H. Puff, G. Bertram, B. Ebeling, M. Franken, R. Gattermayer, R. Hundt, W. Schuh and R. Zimmer, *J. Organomet. Chem.*, **379**, 235 (1989).

89. B. Schmidt, M. Dräger and K. Jurkschat, *J. Organomet. Chem.*, **410**, 43 (1991).

90. K. Jurkschat, S. van Dreumel, G. Dyson, D. Dakternieks, T. J. Barstow, M. E. Smith and M. Dräger, *Polyhedron*, **11**, 2747 (1992).

91. O. R. Flöck and M. Dräger, *Organometallics*, **12**, 4623 (1993).

92. G. Anselme, J-P. Declercq, A. Dubourg, H. Ranaivonjatovo, J. Escudié and C. Couret, *J. Organomet. Chem.*, **458**, 49, (1993).

93. M. Weidenbruch, K. Schäfers, J. Schlaefke, K. Peters and H. G. von Schnering, *J. Organomet. Chem.*, **415**, 343 (1991).

2. Structural aspects of compounds containing C–E (E = Ge, Sn, Pb) bonds   183

94. P. B. Hitchcock, H. A. Jasim, R. E. Kelly and M. F. Lappert, *J. Chem. Soc., Chem. Commun.*, 1776 (1985).
95. P. B. Hitchcock, H. A. Jasim, M. F. Lappert, W-P. Leung, A. K. Rai and R. E. Taylor, *Polyhedron.*, **10**, 1203 (1991).
96. Y. Matsuhashi, N. Tokitoh, R. Okazaki and M. Goto, *Organometallics*, **12**, 2573 (1993).
97. M. Lazraq, C. Couret, J. P. Declercq, A. Dubourg, J. Escudie and M. Riviere-Baudet, *Organometallics*, **9**, 845 (1990).
98. S. Diemer, H. Nöth, K. Polborn and W. Storch, *Chem. Ber.*, **125**, 389 (1992).
99. D. Hänssgen, H. Aldenhoven and M. Nieger, *Chem. Ber.*, **123**, 1837 (1990).
100. M. Dreiss, H. Pritzkow and U. Winkler, *Chem. Ber.*, **125**, 1541 (1992).
101. M. Dreiss and H. Pritzkow, *Chem. Ber.*, **127**, 477 (1994).
102. M. Lazraq, G. Couret, J. Escudie, J. Satge and M. Dräger, *Organometallics*, **10**, 1771 (1991).
103. A. May, H. W. Roesky, R. Herbst-Irmer, S. Freitag and G. M. Sheldrick, *Organometallics*, **11**, 15 (1992).
104. K. -H. van Bonn, P. Schreyer, P. Paetzold and R. Boese, *Chem. Ber.*, **121**, 1045 (1988).
105. L. R. Sita, I. Kinoshita and S. P. Lee, *Organometallics*, **9**, 1644 (1990).
106. Y. Matsuhashi, N. Tokitoh, R. Okazaki, M. Goto and S. Nagase, *Organometallics*, **12**, 1351 (1993); N. Tokitoh, T. Matsumoto and R. Okazaki, *Tetrahedron Lett.*, **33**, 2531 (1992); N. Tokitoh, H. Suzuki, T. Matsumoto, Y. Matsuhashi, R. Okazaki and M. Goto, *J. Am. Chem. Soc.*, **113**, 7047 (1991).
107. H. Preut, S. Wienken and W. P. Neumann, *Acta Cryst.*, **C49**, 184 (1993).
108. A. G. Sommese, S. E. Cremer, J. A. Campbell and M. R. Thompson, *Organometallics*, **9**, 1784 (1990).
109. B. Wrackmeyer, G. Kehr and R. Boese, *Chem. Ber.*, **125**, 643 (1992).
110. W. Ando, T. Kadowaki, Y. Kabe and M. Ishii, *Angew. Chem., Int. Ed. Engl.*, **31**, 59 (1992).
111. H. Puff, K. Braun, S. Franken, T. R. Kök and W. Schuh, *J. Organomet. Chem.*, **349**, 293 (1988); H. Puff, S. Franken and W. Schuh, *J. Organomet. Chem.*, **256**, 23 (1983).
112. See, for example, M. Vieth, M. Nötzel, I. Stahl and V. Huch, *Z. Anorg. Allg. Chem.*, **620**, 1264 (1994) and references cited therein.
113. J. T. B. H. Jastrzebski and G. van Koten, *Adv. Organomet. Chem.*, **35**, 241 (1993) in an article dedicated to G. J. M. van der Kerk.
114. B. Wrackmeyer, S. Kundler and R. Boese, *Chem. Ber.*, **126**, 1361 (1993).
115. B. Wrackmeyer, G. Kehr and R. Boese, *Angew. Chem., Int. Ed. Engl.*, **30**, 1370 (1991).
116. B. Wrackmeyer, S. Kundler, W. Milius and R. Boese, *Chem. Ber.*, **127**, 333 (1994).
117. S. Selvaratnam, K. M. Lo and V. G. K. Das, *J. Organomet. Chem.*, **464**, 143 (1994).
118. V. G. Kumar Das, L. K. Mun, C. Wei, S. J. Blunden and T. C. W. Mak, *J. Organomet. Chem.*, **322**, 163 (1987).
119. J. T. B. H. Jastrzebiski, J. Boersma, P. M. Esch and G. van Koten, *Organometallics*, **10**, 930 (1991).
120. B. Jousseaume, P. Villeneuve, M. Dräger, S. Roller and J. M. Chezeau, *J. Organomet. Chem.*, **349**, C1 (1988).
121. H. Pan, R. Willem, J. Meunier-Piret and M. Gielen, *Organometallics*, **9**, 2199 (1990).
122. K. Jurkschat, A. Tzschach and J. Meunier-Piret, *J. Organomet. Chem.*, **315**, 45 (1986).
123. H. Schumann, B. C. Wasserman and F. E. Hahn, *Organometallics*, **11**, 2803 (1992).
124. V. G. Kumar Das, L. K. Mun, C. Wei and T. C. W. Mak, *Organometallics*, **6**, 10 (1987).
125. See Table 1 in Reference 113.
126. D. Tudela, E. Gutiérrez-Puebla and A. Monge, *J. Chem. Soc., Dalton Trans.*, 1069 (1992).
127. L. Heuer, L. Ernst, R. Schmutzler, and D. Schomburg, *Angew. Chem., Int. Ed. Engl.*, **28**, 1507 (1989).
128. S. E. Johnson and C. B. Knobler, *Organometallics*, **11**, 3684 (1992).
129. I. Wharf and M. G. Simard, *Acta Cryst.*, **C47**, 1605 (1991).
130. S-G. Teoh, S-B. Teo, G-Y. Yeap, H-K Fun and P. J. Smith *J. Organomet. Chem.*, **454**, 73 (1993); S. W. Ng and V. G. Kumar Das, *Acta Cryst.*, **C48**, 1839 (1992); H-K. Fun, S-B. Teo, S-G. Teoh, and G-Y. Yeap, *Acta Cryst.*, **C47**, 1824 (1991).
131. C. Wei, N. W. Kong, V. G. K. Das, G. B. Jameson and R. J. Butcher, *Acta Cryst.*, **C46**, 2034 (1990).
132. J. Cox, S. M. S. V. Doidge-Harrison, I. W. Nowell, R. A. Howie, J. L. Wardell and J. M. Wigzell, *Acta Cryst.*, **C46**, 1015 (1990).

133. L. E. Khoo, X-M. Chen and T. C. W. Mak, *Acta Cryst.*, **C47**, 2647 (1991).
134. N. Clarke, D. Cunningham, T. Higgins, P. McArdle, J. McGinley and M. O'Gara, *J. Organomet. Chem.*, **469**, 33 (1994).
135. S. W. Ng and V. G. Kumar Das, *Main Group Metal Chem.*, **XVI**, 81 (1993).
136. U. Kolb, M. Dräger and B. Jousseaume, *Organometallics*, **10**, 2737 (1991).
137. A. R. Forrester, S. J. Garden, R. A. Howie and J. L. Wardell, *J. Chem. Soc., Dalton Trans.*, 2615 (1992).
138. C. R. A. Muchmore and M. J. Heeg, *Acta Cryst.*, **C46**, 1743 (1990).
139. H. Preut, C. Wicenec and W. P. Neumann, *Acta Cryst.*, **C48**, 366 (1992).
140. H. Preut, B. Godry and T. Mitchell, *Acta Cryst.*, **C48**, 1491; 1834, 1894 (1992).
141. H. Schumann, B. C. Wassermann and J. Pickardt, *Organometallics*, **12**, 3051 (1993).
142. S. S. Al-Juaid, M. Al-Rawi, C. Eaborn, P. B. Hitchcock and J. D. Smith, *J. Organomet. Chem.*, **446**, 161 (1993).
143. G. B. Deacon, E. Lawrenz, K. T. Nelson and E. R. T. Tiekink, *Main Group Metal Chem.*, **XVI**, 265 (1993).
144. S. W. Ng, V. G. K. Das, W-H. Yip and T. C. W. Mak, *J. Organomet. Chem.*, **456**, 181 (1993).
145. M. Suzuki, I-H. Son, R. Noyori and H. Masuda, *Organometallics*, **9**, 3043 (1990); J. Kümmerlen, I. Lange, W. Milius, A. Sebald and A. Blaschette, *Organometallics*, **12**, 3541 (1993); J. Kümmerlen, A. Sebald and E. Sendermann, *Organometallics*, **13**, 802 (1994).
146. H. Reuter and D. Schröder, *Acta Cryst.*, **C49**, 954 (1993).
147. E. R. T. Tiekink, *Appl. Organomet. Chem.*, **5**, 1 (1991).
148. E. R. T. Tiekink, G. K. Sandhu and S. P. Verma, *Acta Cryst.*, **C45**, 1810 (1989).
149. K. M. Lo, S. W. Ng, C. Wei and V. G. Kumar Das, *Acta Cryst.*, **C48**, 1657 (1992).
150. A. Samuel-Lewis, P. J. Smith, J. H. Aupers, D. Hampson and D. C. Povey, *J. Organomet. Chem.*, **437**, 131 (1992).
151. E. Kellö, V. Vrabel, V. Rattay, J. Sivy and J. Kozísek, *Acta Cryst.*, **C48**, 51 (1992).
152. M. Gielen, A. El Khloufi, M. Biesemans, F. Kayser, R. Willem, B. Mahieu, D. Maes, J. N. Lisgarten, L. Wyns, A. Moreira, T. K. Chattopadhay and R. A. Palmer, *Organometallics*, **13**, 2849 (1994).
153. B. C. Das, G. Biswas, B. B. Maji, K. L. Ghatak, S. N. Ganguly, Y. Iitaka and A. Banerjee, *Acta Cryst.*, **C49**, 216 (1993); S. W. Ng, V. G. Kumar Das, F. van Meurs, J. D. Schagen and L. H. Straver, *Acta Cryst.*, **C45**, 568 (1989).
154. S. W. Ng, V. G. Kumar Das, W. H. Yip and T. C. W. Mak, *Acta Cryst.*, **C47**, 1593 (1991).
155. D. H. Gibson, J. F. Richardson and T-S. Ong, *Acta Cryst.*, **C47**, 259 (1991).
156. S. W. Ng and V. G. Kumar Das, *J. Organomet. Chem.*, **456**, 175 (1993).
157. S. W. Ng, V. G. Kumar Das, S-L. Li and T. C. W. Mak, *J. Organomet. Chem.*, **467**, 47 (1994).
158. S. W. Ng and V. G. Kumar Das, *Acta Cryst.*, **C49**, 754 (1993).
159. G. D. Smith, V. M. Visciglio, P. E. Fanwick and I. P. Rothwell, *Organometallics*, **11**, 1064 (1992).
160. A. Blaschette, I. Hippel, J. Krahl, E. Wieland, P. G. Jones and A. Sebald, *J. Organomet. Chem.*, **437**, 279 (1992).
161. I. Lange, E. Wieland, P. G. Jones and A. Blaschette, *J. Organomet. Chem.*, **458**, 57 (1993).
162. N. Wiberg, P. Karampatses, E. Kühnel, M. Veith and V. Huch, *Z. Anorg. Allg. Chem.*, **562**, 91 (1988).
163. S. W. Ng and V. G. Kumar Das, *Acta Cryst.*, **C48**, 2025 (1992); S. W. Ng, C. Wei, V. G. Kumar Das and T. C. W. Mak, *J. Organomet. Chem.*, **379**, 247 (1989).
164. S. W. Ng, V. G. Kumar Das, F. van Meurs, J. D. Schagen and L. H. Straver, *Acta Cryst.*, **C45**, 570 (1989).
165. S. W. Ng, A. J. Kuthubutheen, Z. Arifin, C. Wei and V. G. K. Das, *J. Organomet. Chem.*, **403**, 101 (1991).
166. E. M. Holt, F. A. K. Nasser, A. Wilson Jr. and J. J. Zuckerman, *Organometallics*, **4**, 2073 (1985).
167. G. N. Schrauzer, R. K. Chadha, C. Zhang and H. K. Reddy, *Chem. Ber.*, **126**, 2367 (1993).
168. B. D. James, R. J. Magee, W. C. Patalinghug, B. W. Skelton and A. H. White, *J. Organomet. Chem.*, **467**, 51 (1994).
169. S. W. Ng, C. Wei, V. G. Kumar Das and T. C. W. Mak, *J. Organomet. Chem.*, **326**, C61 (1987).
170. E. G. Martinéz, A. S. González, A. Castiñeiras, J. S. Casas and J. Sordo, *J. Organomet. Chem.*, **469**, 41 (1994).

2. Structural aspects of compounds containing C–E (E = Ge, Sn, Pb) bonds   185

171. D. Tudela, *J. Organomet. Chem.*, **471**, 63 (1994).
172. S-G. Teoh, S-B. Teo and G-Y. Yeap, *Polyhedron*, **11**, 2351 (1992).
173. T. H. Lambertsen, P. G. Jones and R. Schmutzler, *Polyhedron*, **11**, 331 (1992).
174. S-G. Teoh, S-B. Teo and G-Y. Yeap, *J. Organomet. Chem.*, **439**, 139 (1992).
175. P. Ganis, G. Valle, D. Furlani and G. Tagliavini, *J. Organomet. Chem.*, **302**, 165 (1986).
176. J. F. Sawyer, *Acta Cryst.*, **C44**, 633 (1988); compare also NMR studies in D. Dakternieks and H. Zhu, *Organometallics*, **11**, 3820 (1992).
177. W. A. Schenk, A. Khadra and C. Burschka, *J. Organomet. Chem.*, **468**, 75 (1994).
178. P. Tavridou, U. Russo, G. Valle and D. Kovala-Demertzi, *J. Organomet. Chem.*, **460**, C16 (1993).
179. K. Ding, Y. Wu, Y. Wang, Y. Zhu and L. Yang, *J. Organomet. Chem.*, **463**, 77 (1993).
180. G. Bandoli, A. Dolmella, V. Peruzzo and G. Plazzogna, *J. Organomet. Chem.*, **452**, 47 (1993).
181. E. G. Martinéz, A. S. González, J. S. Casas, J. Sordo, G. Valle and U. Russo *J. Organomet. Chem.*, **453**, 47 (1993).
182. H. Grützmacher and H. Pritzkow, *Organometallics*, **10**, 938 (1991).
183. S. Ianelli, M. Orcesi, C. Pelizzi, G. Pelizzi and G. Predieri, *J. Organomet. Chem.*, **451**, 59 (1993).
184. S. Brooker, F. T. Edelmann and D. Stalke, *Acta Cryst.*, **C47**, 2527 (1991).
185. J. F. Vollano, R. O. Day and R. R. Holmes, *Organometallics*, **3**, 750 (1984).
186. C. Kober, J. Kroner and W. Storch, *Angew. Chem., Int. Ed. Engl.*, **32**, 1608 (1993).
187. I. Hippel, P. G. Jones and A. Blaschette, *J. Organomet. Chem.*, **448**, 63 (1993).
188. C. Pelizzi, G. Pelizzi, and P. Tarasconi, *J. Organomet. Chem.*, **277**, 29 (1984).
189. E. R. T. Tiekink, *Appl. Organomet. Chem.*, **5**, 1 (1991).
190. T. P. Lockhart, J. C. Calabrese and F. Davidson, *Organometallics*, **6**, 2479 (1987).
191. J. H. Wengrovius and M. F. Garbauskas, *Organometallics*, **11**, 1334 (1992); *Acta Cryst.*, **C47**, 1969 (1991).
192. V. Chandrasekhar, R. O. Day, J. M. Holmes and R. R. Holmes, *Inorg. Chem.*, **27**, 958 (1988).
193. G. K. Sandhu, N. Sharma and E. R. T. Tiekink, *J. Organomet. Chem.*, **403**, 119 (1991).
194. F. Mistry, S. J. Rettig, J. Trotter and F. Aubke, *Acta Cryst.*, **C46**, 2091 (1990).
195. M. Gielen, M. Boûâlam and E. R. T. Tiekink, *Main Group Metal Chem.*, **XVI**, 251 (1993).
196. T. P. Lockhart and F. Davidson, *Organometallics*, **6**, 2471 (1987).
197. N. W. Alcock and S. M. Roe, *Acta Cryst.*, **C50**, 227 (1994).
198. K. Jurkschat, C. Krüger and J. Meunier-Piret, *Main Group Metal Chem.*, **XV**, 61 (1992).
199. V. B. Mokal, V. K. Jain and E. R. T. Tiekink, *J. Organomet., Chem.*, **471**, 53 (1994).
200. M. Gielen, M. Acheddad and E. R. T. Tiekink, *Main Group Metal Chem.*, **XVI**, 367 (1993); F. Huber, H. Preut, E. Hoffmann and M. Gielen, *Acta Cryst.*, **C45**, 51 (1989).
201. S. P. Narula, S. K. Bharadwaj, Y. Sharda, R. O. Day, L. Howe and R. R. Holmes, *Organometallics*, **11**, 2206 (1992).
202. B. Bovio, A. Cingolani, F. Marchetti and C. Pettinari, *J. Organomet. Chem.*, **458**, 39 (1993); C. Pettinari, G. Rafaiani, G. G. Lobbia, A. Lorenzotti, F. Bonati and B. Bovio, *J. Organomet. Chem.*, **405**, 75 (1991).
203. F. Kayser, M. Biesemans, M. Boûâlam, E. R. T. Tiekink, A. El Khloufi, J. Meunier-Piret, A. Bouhdid, K. Jurschat, M. Gielen and R. Willem, *Organometallics*, **13**, 1098 (1994).
204. R. A. Kresinski, R. J. Staples and J. P. Fackler, *Acta Cryst.*, **C50**, 40 (1994); E. R. T. Tiekink, *Acta Cryst.*, **C47**, 661 (1991).
205. V. B. Mokal, V. K. Jain and E. R. T. Tiekink, *J. Organomet. Chem.*, **471**, 53 (1994).
206. C. Vatsa, V. K. Jain, T. Kesavadas and E. R. T. Tiekink, *J. Organomet. Chem.*, **408**, 157 (1991).
207. M. Gielen, A. El Khloufi, M. Biesemans, R. Willem and J. Meunier-Piret, *Polyhedron*, **11**, 1861 (1992).
208. J. Meunier-Piret, M. Boûâlam, R. Willem and M. Gielen, *Main Group Metal Chem.*, **XVI**, 329 (1993).
209. J. M. Hook, B. M. Linahan, R. L. Taylor, E. R. T. Tiekink, L. van Gorkom and L. K. Webster, *Main Group Metal Chem.*, **XVII**, 293 (1994).
210. V. Vrabel and E. Kellö, *Acta Cryst.*, **C49**, 873, (1993).
211. D. Dakternieks, H. Zhu, D. Masi and C. Mealli, *Inorg. Chem.*, **31**, 3601 (1992).
212. T. S. B. Baul and E. R. T. Tiekink, *Main Group Metal Chem.*, **XVI**, 201 (1993).
213. G. Engel and G. Mattern, *Z. Anorg. Allg. Chem.*, **620**, 723 (1994).
214. C. Silvestru, I. Haiduc, F. Caruso, M. Rossi, B. Mahieu and M. Gielen, *J. Organomet. Chem.*, **448**, 75 (1993).

215. F. Caruso, M. Giomini, A. M. Giulani and E. Rivarola, *J. Organomet. Chem.*, **466**, 69 (1994); J. Costamagna, J. Canales, J. Vargas, M. Camalli, F. Caruso and E. Rivarola, *Pure Appl. Chem.*, **65**, 1521 (1993).

216. S. W. Ng, L. K. Mun and V. G. Kumar Das, *Main Group Metal Chem.*, **XVI**, 101 (1993).

217. M. Gielen, E. Joosen, T. Mancilla, K. Jurschat, R. Willem, C. Roobol, J. Bernheim, G. Atassi, F. Huber, E. Hoffmann, H. Preut and B. Mahieu, *Main Group Metal Chem.*, **X**, 147 (1987).

218. M. Gielen, *Main Group Metal Chem.*, **XVII**, 1 (1994).

219. A. Meriem, R. Willem, J. Meunier-Piret and M. Gielen, *Main Group Metal Chem.*, **XII**, 187 (1989).

220. Z.-K. Yu, S.-H. Wang, Z.-Y. Yang, X.-M. Liu and N.-H. Yu, *J. Organomet. Chem.*, **447**, 189 (1993).

221. R. C. Okechukwu, H-K Fun, S-G. Teoh, S-B. Teo and K. Chinnakali, *Acta Cryst.*, **C49**, 368 (1993).

222. H. Preut, B. Mundus, F. Huber and R. Barbieri, *Acta Cryst.*, **C45**, 728 (1989); **C42**, 536 (1986).

223. H. Preut, M. Vornefeld and F. Huber, *Acta Cryst.*, **C47**, 264 (1991).

224. P. Brown, M. F. Mahon and K. C. Molloy, *J. Chem. Soc., Dalton Trans.*, 3503 (1992).

225. M. Bouâlam, J. Meunier-Piret, M. Biesemans, R. Willem and M. Gielen, *Inorg. Chim. Acta*, **198–200**, 249 (1992); P. Harston, R. A. Howie, J. L. Wardell, S. M. S. V. Doidge-Harrison and P. J. Cox, *Acta Cryst.*, **C48**, 279 (1992).

226. H. K. Fun, S. -B. Teo, S. -G. Teoh and G. -Y. Yeap, *Acta Cryst.*, **47**, 1845 (1991); M. Wada, T. Fujii, S. Iijima, S. Hayase and T. Erabi, *J. Organomet. Chem.*, **445**, 65 (1993).

227. E. Kellö, V. Vrábel, A. Lycka and J. Sivy, *Acta Cryst.*, **C49**, 1493 (1993); G. Ossig, A. Meller, S. Freitag, H. Herbst-Irmer and G. M. Sheldrick, *Chem. Ber.*, **126**, 2247 (1993); H. Preut, C. Wicenec and W. P. Neumann, *Acta Cryst.*, **C47**, 2214 (1991); O. S. Jung, J. H. Jeong and Y. S. Sohn, *Organometallics*, **10**, 2217 (1991); A. R. Forrester, R. A. Howie, J-N. Ross, J. N. Low and J. L. Wardell, *Main Group Metal Chem.*, **XIV**, 293 (1991).

228. M. M. Amini, M. J. Heeg, R. W. Taylor, J. J. Zuckerman and S. W. Ng, *Main Group Metal Chem.*, **XVI**, 415 (1993).

229. P. Harston, R. A. Howie, G. P. McQuillan, J. L. Wardell, E. Zanetti, S. M. S. V. Doidge-Harrison, N. S. Stewart and P. J. Cox, *Polyhedron*, **10**, 1085 (1991).

230. C. Silvestru, I. Haiduc, F. Caruso, M. Rossi, B. Mahieu and M. Gielen, *J. Organomet. Chem.*, **448**, 75 (1993); P. A. Bates, M. B. Hursthouse, A. G. Davies and S. D. Slater, *J. Organomet. Chem.*, **325**, 129 (1987).

231. H. K. Fun, S. -B. Teo, S. -G. Teoh, G. -Y. Yeap and T. -S. Yeoh, *Acta Cryst.*, **C47**, 1602 (1991).

232. C. Wei, N. W. Kong, V. G. K. Das, G. B. Jameson and R. J. Butcher, *Acta Cryst.*, **C45**, 861 (1989).

233. S. W. Ng, *Acta Cryst.*, **C49**, 753 (1993); B. Bovio, A. Cingolani, F. Marchetti and C. Pettinari, *J. Organomet. Chem.*, **458**, 39 (1993); V. Vrábel, J. Lokaj, E. Kellö, V. Rattay, A. C. Batsanov and Yu. T. Struchkov, *Acta Cryst.*, **C48**, 627, 633 (1992); A. A. S. El-Khaldy, Y. P. Singh, R. Bohra, R. K. Mehrotra and G. Srivastava, *Main Group Metal Chem.*, **XIV**, 319 (1991).

234. D. Dakternieks, G. Dyson, K. Jurschat, R. Tozer and E. R. T. Tiekink, *J. Organomet. Chem.*, **458**, 29 (1993); V. Peruzzo, G. Plazzogna and G. Bandoli, *J. Organomet. Chem.*, **415**, 335 (1991).

235. S. M. S. V. Doidge-Harrison, R. A. Howie, J. T. S. Irvine, G. M. Spencer and J. L. Wardell, *J. Organomet. Chem.*, **436**, 23 (1992).

236. M. Schürmann and F. Huber, *Acta Cryst.*, **C50**, 206 (1994); R. Schmiedgen, F. Huber and M. Schürmann, *Acta Cryst.*, **C50**, 391 (1994).

237. P. Storck and A. Weiss, *Acta Cryst.*, **C46**, 767 (1990); *Ber. Bunsenges. Phys. Chem.*, **93**, 454 (1989).

238. W. Chen, *J. Organomet. Chem.*, **471**, 69 (1994); compare S-B. Teo, S-G. Teoh, R. C. Okechukwu and H. K. Fun, *J. Organomet. Chem.*, **454**, 67 (1993).

239. S. Dostal, S. J. Stoudt, P. Fanwick, W. F. Sereatan, B. Kahr and J. E. Jackson, *Organometallics*, **12**, 2284 (1993).

240. J. Vicente, M-T. Chicote, M-del-C. Ramirez-de-Arallano and P. G. Jones, *J. Chem. Soc., Dalton Trans.*, 1839 (1992).

241. H. Reuter and D. Schröder, *J. Organomet. Chem.*, **455**, 83 (1993).

242. R. R. Holmes, R. O. Day, K. C. K. Swamy, C. G. Schmid, S. D. Burton and J. M. Holmes, *Main Group Metal Chem.*, **XII**, 291 (1989).

243. H. Reuter, *Angew. Chem., Int. Ed. Engl.*, **30**, 1482 (1991).

244. J. T. B. H. Jastrzebski, P. A. van der Schaaf, J. Boersma, G. van Koten, D. J. A. de Ridder and D. Heijdenrijk, *Organometallics*, **11**, 1521 (1992); J. T. B. H. Jastrzebski, P. A. van der Schaaf, J. Boersma, G. van Koten, M. de Wit, Y. Wang, D. Heijdenrijk and C. H. Stam, *J. Organomet. Chem.*, **407**, 301 (1991).
245. S. W. Ng and V. G. Kumar Das, *Main Group Metal Chem.*, **XVI**, 87, 95 (1993).
246. O. S. Jung, J. H. Jeong and Y. S. Sohn, *Acta Cryst.*, **C46**, 31 (1990).
247. N. Seth, V. D. Gupta, H. Nöth and M. Thomann, *Chem. Ber.*, **125**, 1523 (1992).
248. E. R. T. Tiekink, *Main Group Metal Chem.*, **XV**, 161 (1992); **XVI**, 129 (1993).
249. G. G. Lobbia, S. Calogero, B. Bovio and P. Cecchi, *J. Organomet. Chem.*, **440**, 27 (1992).
250. G. F. de Sousa, C. A. L. Filgueiras, M. Y. Darensbourg and J. H. Reibenspies, *Inorg. Chem.*, **31**, 3044 (1992).
251. D. J. Brauer, J. Wilke and R. Eujen, *J. Organomet. Chem.*, **316**, 261 (1986).
252. S. E. Denmark, R. T. Roberts, G. Dai-Ho and S. Wilson, *Organometallics*, **9**, 3015 (1990).
253. R. K. Chadha, J. E. Drake, A. B. Sarkar and M. L. Y. Wong, *Acta Cryst.*, **C45**, 37 (1989).
254. H. Preut, M. Vornefeld and F. Huber, *Acta Cryst.*, **C45**, 1504 (1989); M. Vornefeld, F. Huber, H. Preut and H. Brunner, *Appl. Organomet. Chem.*, **3**, 177 (1989).
255. R. R. Holmes, R. O. Day, A. C. Sau and J. M. Holmes, *Inorg. Chem.*, **25**, 600 (1986); R. R. Holmes, R. O. Day, A. C. Sau, C. A. Poutasse and J. M. Holmes, *Inorg. Chem.*, **25**, 607 (1986).
256. Yu. I. Baukov, A. G. Shipov, L. S. Smirnova, E. P. Kramarova, S. Yu. Bylikin, Yu. E. Ovchinnikov and Yu. T. Struchkov, *J. Organomet. Chem.*, **461**, 39 (1993).
257. T. K. Miyamoto, *Main Group Metal Chem.*, **XVII**, 145 (1994).
258. A. Morère, M. Rivière-Baudet, J. F. Britten and M. Onyszchuk, *Main Group Metal Chem.*, **XV**, 141 (1992).
259. A. O. Mozzchukhin, A. A. Macharashvili, V. E. Shklover, Yu. T. Struchkov, A. V. Shipov, V. N. Sergeev, S. A. Artamkin, S. V. Pestunovich and Yu. I. Baukov, *J. Organomet. Chem.*, **408**, 305 (1991).
260. C. Brelière, F. Carré, R. J. P. Corriu, G. Royo, M. W. C. Man and J. Lapasset, *Organometallics*, **13**, 307 (1994); C. Brelière, F. Carré, R. J. P. Corriu and G. Royo, *Organometallics*, **7**, 1006 (1988).
261. J. G. Verkade, *Acc. Chem. Res.*, **26**, 483 (1993).
262. S. N. Tandura, M. G. Voronkov and N. V. Alekseev, *Top. Curr. Chem.*, **131**, 99 (1986); M. G. Voronkov, V. Dyakov and S. V. Kirpichenko, *J. Organomet. Chem.*, **233**, 1 (1982).
263. E. Lukevics and L. Ignatovich, *Main Group Metal Chem.*, **XVII**, 133 (1994).
264. Y. Wan and J. G. Verkade, *Inorg. Chem.*, **32**, 79 (1993).
265. W. Plass and J. G. Verkade, *Inorg. Chem.*, **32**, 5145, 5153 (1993).
266. H. J. Eppley, J. L. Ealy, C. H. Yoder, J. N. Spencer and A. L. Rheingold, *J. Organomet. Chem.*, **431**, 133 (1992).
267. J. Muller, U. Muller, A. Loss, J. Lorbeth, H. Donath and W. Massa, *Z. Naturforsch.*, **B40**, 1320 (1985).
268. H. Preut, P. Röhm and F. Huber, *Acta Cryst.*, **C42**, 657 (1986).
269. G. Fehlberg-Sternemann, dissertation, Univ. of Dortmund (1992), cited reference 9; A. Glowacki, F. Huber and H. Preut, *J. Organomet. Chem.*, **306**, 9 (1986).
270. E. Hoffmann, dissertation, Univ. of Dortmund (1988), cited reference 9.
271. A. V. Yaysenko, S. V. Medvedev and L. A. Aslanov, *Zh. Strukt. Khim.*, **30**, 1192, (1989).
272. M. Onyszchuk, I. Wharf, M. Simard and A. L. Beaychamp, *J. Organomet. Chem.*, **326**, 25 (1987).
273. M. C. Begley, C. Gaffney, P. G. Harrison and A. Steel, *J. Organomet. Chem.*, **289**, 281 (1985).
274. H. Preut, F. Huber and E. Hoffmann, *Acta Cryst.*, **C44**, 755 (1988).
275. F. Huber, H. Preut, D. Scholz and M. Schürmann, *J. Organomet. Chem.*, **441**, 227 (1992).
276. C. Gaffney, P. G. Harrison and T. J. King, *J. Chem. Soc., Dalton Trans.*, 1061 (1982).
277. N. W. Alcock, *J. Chem. Soc., Dalton Trans.*, 1189 (1972).
278. N. A. Compton, R. J. Errington and N. C. Norman, *Adv. Organomet. Chem.*, **31**, 91 (1990); M. S. Holt, W. L. Wilson and J. H. Nelson. *Chem. Rev.*, **89**, 11 (1989).
279. M. N. Bochkarev, I. M. Penyagini, L. N. Zakharov, Yu. N. Rad'kov, E. A. Fedorova, S. Ya. Khorshev and Yu. T. Struchkov, *J. Organomet. Chem.*, **378**, 363 (1989).
280. J. E. Ellis and P. Yuen, *Inorg. Chem.*, **32**, 4998 (1993).

281. J. E. Ellis, K-M. Chi, A-J. DiMaio, S. R. Frerichs, J. R. Stenzel, A. L. Rheingold and B. S. Haggerty, *Angew. Chem., Int. Ed. Engl.*, **30**, 194 (1991).
282. F. Calderazzo, G. Pampaloni, G. Pellizi and F. Vitali, *Polyhedron*, **7**, 2039 (1988).
283. S. Mock and U. Schubert, *Chem. Ber.*, **126**, 2591 (1993).
284. D. Miguel, J. A. Pérez-Martinez, V. Riera and S. Garcia-Granda, *Organometallics*, **12**, 2888 (1993); S. Seebald, B. Mayer and U. Schubert, *J. Organomet. Chem.*, **462**, 225 (1993).
285. M. H. Chisholm, G. J. Gama and I. P. Parkin, *Polyhedron*, **12**, 961 (1993); **10**, 1215 (1991).
286. P. Jutzi, B. Hampel, M. B. Waterhouse and A. J. Howes, *J. Organomet. Chem.*, **299**, 19 (1986); P. Jutzi, B. Hampel, K. Stroppel, C. Kruger, K. Angermund and P. Hofman, *Chem. Ber.*, **118**, 2789 (1985).
287. D. Miguel, J. A. Pérez-Martinez, V. Riera and S. Garcia-Granda, *J. Organomet. Chem.*, **455**, 121 (1993).
288. D. Lei, M. J. Hampden-Smith, E. N. Duesler and J. C. Huffman, *Inorg. Chem.*, **29**, 795 (1990); D. Lei, M. J. Hampden-Smith, J. W. G. Garvey and J. C. Huffman, *J. Chem. Soc., Dalton Trans.*, 2449 (1991).
289. K. E. Lee, A. M. Arif and J. A. Gladysz, *Organometallics*, **10**, 751 (1991).
290. H. K. Sharma, F. Cervantes-Lee and K. H. Pannell, *J. Organomet. Chem.*, **409**, 321 (1991); G. Barsuaskas, D. Lei, M. J. Hampden-Smith and E. N. Duesler, *Polyhedron*, **9**, 773 (1990).
291. U. Behrens, A. K. Brimah, T. M. Soliman, R. D. Fischer, D. C. Apperley, N. A. Davies and R. K. Harris, *Organometallics*, **11**, 1718 (1992).
292. P. Braunstein, M. Knorr, M. Stampfer, A. DeCian and J. Fischer, *J. Chem.Soc., Dalton Trans.*, 117 (1994); M. Akita, T. Oku, M. Tanaka and Y. Moro-Oka, *Organometallics*, **10**, 3080 (1991).
293. C. Campbell and L. J. Farrugia, *Acta Cryst.*, **C45**, 1817 (1989).
294. G. R. Clarke, K. R. Flower, W. R. Roper, C. M. Salter and L. J. Wright, *Organometallics*, **12**, 3810 (1993).
295. G. R. Clarke, K. R. Flower, C. E. F. Rickard, W. R. Roper, D. M. Salter and L. J. Wright, *J. Organomet. Chem.*, **462**, 331 (1993).
296. C. Loubser, J. L. M. Dillen and S. Lotz, *Polyhedron*, **10**, 2535 (1991).
297. L. V. Pankratov, V. l. Nevodchikov, L. N. Zakharov, M. N. Bochkarev, l. V. Zdanovich, V. N. Latyaeva, A. N. Lineva, A. S. Batsanov and Yu. T. Struchkov, *J. Organomet. Chem.*, **429**, 13 (1992).
298. C. Pluta, K. R. Pörschke, l. Ortmann and C. Krüger, *Chem. Ber.*, **125**, 103 (1992).
299. H. Yamashita, T. Kobayashi, M. Tanaka, J. A. Samuels and W. E. Streib, *Organometallics*, **11**, 2330 (1992).
300. W. J. J. Smeets, A. L. Spek, J. A. M. van Beek and G. van Koten, *Acta Cryst.*, **C48**, 745 (1992).
301. M. Cano, M. Panizo, J. A. Campo, J. Tornero and N. Menéndez, *J. Organomet. Chem.*, **463**, 121 (1993).
302. G. Barrado, D. Miguel, J. A. Pérez-Martinez, V. Riera and S. Garcia-Granda, *J. Organomet. Chem.*, **466**, 147 (1994); **463**, 127 (1993).
303. U. Schubert, S. Gilbert and S. Mock, *Chem. Ber.*, **125**, 835 (1993).
304. F. Meier-Brocks and E. Weiss, *J. Organomet. Chem.*, **453**, 33 (1993).
305. P. Dufour, M-J. Menu, M. Dartiguenave, Y. Dartiguenave and J. Dubac, *Organometallics*, **10**, 1645 (1991).
306. M. Veith, L. Stahl and V. Huch, *Organometallics*, **12**, 1914 (1993).
307. C. J. Cardin, D. J. Cardin, G. A. Lawless, J. M. Power, M. B. Power and M. B. Hursthouse, *J. Organomet. Chem.*, **325**, 203 (1987).
308. C. J. Cardin, M. B. Power and D. J. Cardin, *J. Organomet. Chem.*, **462**, C27 (1993).
309. M. N. Bochkarev, G. A. Razuvaev, L. N. Zakharov and Yu. T. Struchkov, *J. Organomet. Chem.*, **199**, 205 (1980).
310. N. Tokitoh, Y. Matsuhashi and R. Okazaki, *Organometallics*, **12**, 2894 (1993).
311. P. J. Fagan, M. D. Ward and J. C. Calbrese, *J. Am. Chem. Soc.*, **112**, 3540 (1990).
312. K. H. Whitmire, *J. Coord. Chem.*, **17**, 95 (1988); *J. Cluster Sci.*, **2**, 231 (1991); W. A. Herrmann, *Angew. Chem., Int. Ed. Engl.*, **25**, 56 (1986).
313. S. K. Lee, K. M. Mackay and B. K. Nicholson, *J. Chem. Soc., Dalton Trans.*, 715 (1993); S. K. Lee, K. M. Mackay, B. K. Nicholson and M. Service, *J. Chem. Soc., Dalton Trans.*, 1709 (1992); S. G. Anema, S. K. Lee, K. M. Mackay and B. K. Nicholson *Organomet. Chem.*, **444**, 211 (1993) and references cited therein.
314. K. M. Mackay and R. Watt, *Organomet. Chem. Rev., A*, **4**, 137 (1969).

315. M. Dräger, L. Ross and D. Simon, *Rev. Silicon, Germanium, Tin Lead Compd.*, **7**, 299 (1983).
316. Silicon data from Table 17 in Reference 4b.
317. N. Kleiner and M. Dräger, *J. Organomet. Chem.*, **270**, 151 (1984).
318. N. Wiberg, H. Schuster, A. Simon and K. Peters, *Angew. Chem., Int. Ed. Engl.*, **25**, 79, (1986).
319. L. Pauling, A. W. Laubergayer and J. L. Hoard, *J. Am. Chem. Soc.*, **60**, 605 (1938).
320. M. Dräger and L. Ross, *Z. Anorg. Allg. Chem.*, **460**, 207 (1980).
321. M. Dräger and L. Ross, *Z. Anorg. Allg. Chem.*, **469**, 115 (1980).
322. S. Roller, Dissertation, Mainz (1988), cited in Reference 333.
323. L. Párkányi, A. Kalman, S. Sharma, D. M. Nolen and K. H. Pannell, *Inorg. Chem.*, **33**, 180 (1994).
324. M. Weidenbruch, F-T. Grimm, M. Herrndorf, A. Schäfer, K. Peters and H. G. von Schnering, *J. Organomet. Chem.*, **341**, 335 (1988).
325. K. Häberle and M. Dräger, *Z. Naturforsch. B*, **42**, 323 (1987).
326. A. P. Cox and R. Varma, *J. Chem. Phys.*, **46**, 2007 (1967).
327. K. H. Pannell, R. N. Kapoor, R. Raptis, L. Párkányi and V. Fülöp, *J. Organomet. Chem.*, **384**, 41 (1990).
328. L. Párkányi, C. Hernandez and K. H. Pannell, *J. Organomet. Chem.*, **301**, 145 (1986).
329. J. Drahnak, R. West and J. C. Calabrase, *J. Organomet. Chem.*, **198**, 55 (1980).
330. L. Párkányi, K. H. Pannell and C. Hernandez, *J. Organomet. Chem.*, **252**, 127 (1983).
331. K. H. Pannell, L. Párkányi, H. Sharma and F. Cervantes-Lee, *Inorg. Chem.*, **31**, 522 (1992).
332. S. Adams, Dissertation, Mainz (1987), cited in Reference 333.
333. H. -J. Koglin, K. Behrends and M. Dräger, *Organometallics*, **13**, 2733 (1994).
334. H. Preut, H-J. Haupt and F. Huber, *Z. Anorg. Allg. Chem.*, **396**, 81 (1973).
335. H. Piana, U. Kirchgassner and U. Schubert, *Chem. Ber.*, **124**, 743 (1991).
336. C. Schneider and M. Dräger, *J. Organomet. Chem.*, **415**, 349 (1991).
337. C. Schneider-Koglin, K. Behrends and M. Dräger, *J. Organomet. Chem.*, **448**, 29 (1993).
338. H. Puff, B. Breuer, G. Gehrke-Brinkmann, P. Kind, H. Reuter, W. Schuh, W. Wald and G. Weidenbrück, *J. Organomet. Chem.*, **363**, 265 (1989).
339. M. Weidenbruch, J. Schlaefke, K. Peters and H. G. von Schnering, *J. Organomet. Chem.*, **414**, 319 (1991).
340. S. Adams, M. Dräger and B. Mathiasch, *Z. Anorg. Allg. Chem.*, **532**, 81 (1986).
341. D. E. Goldberg, D. H. Harris, M. F. Lappert and K. M. Thomas, *J. Chem. Soc., Chem. Commun.*, 261 (1976).
342. H. A. Skinner and L. E. Sutton, *Trans. Faraday Soc.*, **36**, 1209 (1940).
343. H. Preut and F. Huber, *Z. Anorg. Allg. Chem.*, **419**, 92 (1976).
344. N. Kleiner and M. Dräger, *J. Organomet. Chem.*, **293**, 323 (1985).
345. See A. Sebald and R. K. Harris, *Organometallics*, **9**, 2096 (1990) for Pb CPMAS study.
346. N. Kleiner and M. Dräger, *Z. Naturforsch., Teil B*, **40**, 477 (1985).
347. S. P. Mallela, J. Myrczek, l. Bernal and R. A. Geanangel, *J. Chem. Soc., Dalton Trans.*, 2891 (1993).
348. L. Parkanyi, A. Kalman, S. Sharma, D. M. Nolen and K. H. Pannell, *Inorg. Chem.*, **33**, 180 (1994).
349. H. -J. Koglin, K. Behrends and M. Dräger, *Organometallics*, **13**, 2733 (1994).
350. D. Simon, K. Häberle and M. Dräger, *J. Organomet. Chem.*, **267**, 133 (1984).
351. Older structures are summarized in Reference 338.
352. S. Adams, M. Dräger and B. Mathiasch, *J. Organomet. Chem.*, **326**, 173 (1987).
353. T. Tsumuraya, S. Sato and W. Ando, *Organometallics*, **7**, 2015, (1988).
354. S. A. Batcheller, thesis, cited in T. Tsumuraya, S. A. Batcheller and S. Masamune, *Angew. Chem., Int. Ed. Engl.*, **30**, 902 (1991).
355. T. Tsumuraya, S. Sato and W. Ando, *J. Chem. Soc., Chem. Commun.*, 1159, (1990).
356. T. Tsumuraya, S. Sato and W. Ando, *Organometallics*, **9**, 2061, (1990).
357. S. Masamune, S. A. Batcheller, J. Park, W. M. Davis, O. Yamashita, Y. Ohta and Y. Kabe, *J. Am. Chem. Soc.*, **111**, 1888, (1989).
358. K-F. Tebbe and R. Fröhlich, *Z. Anorg. Allg. Chem.*, **506**, 27 (1983).
359. H. Puff, H. Heisig, W. Schuh and W. Schwab, *J. Organomet. Chem.*, **303**, 343 (1986).
360. H. Puff, T. R. Kok, P. Nauroth and W. Schuh, *J. Organomet. Chem.*, **281**, 141 (1985).
361. M. Dräger and K. Häberle, *J. Organomet. Chem.*, **280**, 183 (1985).
362. A. Sekiguchi, T. Yatabe, C. Kabuto and H. Sakurai, *J. Organomet. Chem.*, **390**, C27 (1990).

363. S. Adams, M. Dräger and B. Mathiasch, *Z. Anorg. Allg. Chem.*, **532**, 81 (1986).
364. D. E. Goldberg, D. H. Harris, M. F. Lappert and K. M. Thomas, *J. Chem. Soc., Chem. Commun.*, 261 (1976).
365. H. Preut and T. N. Mitchell, *Acta Cryst.*, **C47**, 951 (1991).
366. H. Preut, P. Bleckmann, T. N. Mitchell and B. Fabische, *Acta Cryst.*, **C40**, 370 (1984).
367. H. Preut and T. N. Mitchell, *Acta Cryst.*, **C45**, 35 (1989).
368. L. R. Sita, l. Kinoshita and S. P. Lee, *Organometallics*, **9**, 1644 (1990).
369. J. Meunier-Piret, M. Van Meerssche, M. Gielen and K. Jurkschat, *J. Organomet. Chem.*, **252**, 289 (1983).
370. M. Weidenbruch, A. Schafer, H. Kilian, S. Pohl, W. Saak and H. Marsmann, *Chem. Ber.*, **125**, 563 (1992).
371. A. Schafer, M. Weidenbruch, W. Saak, S. Pohl and H. Marsmann, *Angew. Chem., Int. Ed. Engl.*, **30**, 834, 962 (1991).
372. H. Grutzmacher and H. Pritzkow, *Angew. Chem., Int. Ed. Engl.*, **30**, 1017 (1991).
373. H. Puff, B. Breuer, W. Schuh, R. Sievers and R. Zimmer, *J. Organomet. Chem.*, **332**, 279 (1987).
374. A. H. Cowley, S. W. Hall, C. M. Nunn and J. M. Power, *Angew. Chem., Int. Ed. Engl.*, **27**, 838 (1988).
375. K. Jurkschat, A. Tzschach, C. Mügge, J. Piret-Meunier, M. van Meerssche, G. van Binst, C. Wynants, M. Gielen and R. Willem, *Organometallics*, **7**, 593 (1988).
376. A. Krebs, A. Jacobsen-Bauer, E. Haupt, M. Vieth and V. Huch, *Angew. Chem., Int. Ed. Engl.*, **28**, 603 (1989).
377. C. Leue, R. Réau, B. Neumann, H-G. Stammler, P. Jutzi and G. Bertrand, *Organometallics*, **13**, 436 (1994).
378. H. Bock J. Meuret and H. Schödel., *Chem. Ber.*, **B126**, 2227 (1993); H. Bock, J. Meuret and K. Ruppert *J. Organomet. Chem.*, **445**, 19 (1993); H. Bock, J. Meuret, C. Näther and K. Ruppert *Angew. Chem., Int. Ed. Engl.*, **32**, 414 (1993).
379. S. Henkel, K. W. Klinkhammer and W. Schwarz, *Angew. Chem., Int. Ed. Engl.*, **33**, 681 (1994).
380. *J. Organomet. Chem.*, **462** (1993).
381. H. Li, l. A. Butler and J. F. Harrod, *Organometallics*, **12**, 4533, (1993).
382. W. Ando, T. Tsumuraya and Y. Kabe, *Angew. Chem., Int. Ed. Engl.*, **29**, 778 (1990).
383. S. Roller, D. Simon and M. Dräger, *J. Organomet. Chem.*, **301**, 27 (1986); see also K. Haberle and M. Dräger, *Z. Anorg. Allg. Chem.*, **551**, 116 (1987) for extensive NMR data on phenylpolygermanes.
384. M. Charissé, M, Mathes, D. Simon and M. Dräger, *J. Organomet. Chem.*, **445**, 39 (1993).
385. M. Dräger and D. Simon, *J. Organomet. Chem.*, **306**, 183 (1986).
386. K. Häberle and M. Dräger, *J. Organomet. Chem.*, **312**, 155 (1986).
387. M. Dräger and D. Simon, *Z. Anorg. Allg. Chem.*, **472**, 120 (1981).
388. S. Roller and M. Drager, *J. Organomet. Chem.*, **316**, 57 (1986).
389. T. Lobreyer, H. Oberhammer and W. Sundermeyer, *Angew. Chem., Int. Ed. Engl.*, **32**, 586 (1993).
390. K. M. Baines, J. A. Cooke and J. J. Vittal, *J. Chem. Soc., Chem. Commun.*, 1484 (1992).
391. M. Dräger and M. Mathes, unpublished (1989).
392. S. P. Mallela, M. A. Ghuman and R. A. Geanangel, *Inorg. Chim. Acta*, **202**, 211 (1992).
393. S. P. Mallela and R. A. Geanangel, *Inorg. Chem.*, **30**, 1480 (1991).
394. S. Adams and M. Dräger, *J. Organomet. Chem.*, **323**, 11 (1987).
395. S. Adams and M. Dräger, *J. Organomet. Chem.*, **288**, 295 (1985).
396. S. P. Mallela and R. A. Geanangel, *Inorg. Chem.*, **33**, 1115 (1994).
397. S. Adams and M. Dräger, *Angew. Chem., Int. Ed. Engl.*, **26**, 1255 (1987).
398. D. Schollmeyer, H. Hartung, C. Mreftani-Klaus and K. Jurkschat, *Acta Cryst.*, **C47**, 2365 (1991).
399. L. N. Bochkarev, O. V. Grachev, N. E. Molosnova, S. F. Zhiltsov, L. N. Zakharov, G. K. Fukin, A. l. Yanovsky and Y. T. Struchkov, *J. Organomet. Chem.*, **443**, C26 (1993).
400. M. H. Chisholm, H-T. Chiu, K. Folting and J. C. Huffman, *Inorg. Chem.*, **23**, 4097 (1984).
401. S. P. Mallela and R. A. Geanangel, *Inorg. Chem.*, **29**, 3525 (1990).
402. A. M. Arif, A. H. Cowley and T. M. Elkins, *J. Organomet. Chem.*, **325**, C11 (1987).
403. S. P. Mallela and R. A. Geanangel, *Inorg. Chem.*, **32**, 5623 (1993).
404. R. H. Heyn and T. D. Tilley, *Inorg. Chem.*, **29**, 4051 (1990).
405. S. P. Mallela and R. A. Geanangel, *Inorg. Chem.*, **32**, 602 (1993).

406. L. Rosch and U Stark, *Angew. Chem., Int. Ed. Engl.*, **22**, 557 (1983).
407. A. Heine and D. Stalke, *Angew. Chem., Int. Ed. Engl.*, **33**, 113 (1994).
408. H. Bock, J. Meuret and K. Ruppert, *J. Organomet. Chem.*, **445**, 19 (1993).
409. M. Aggarwal and R. A. Geanangel, *Main Group Metal Chem.*, **XIV**, 263 (1991).
410. S. P. Mallela and R. A. Geanangel, *Inorg. Chem.*, **32**, 602 (1993).
411. L. Ross and M. Dräger, *Z. Anorg. Allg. Chem.*, **472**, 109 (1981).
412. K. M. Baines, J. A. Cooke, N. C. Paine and J. J. Vittal, *Organometallics*, **11**, 1408 (1992).
413. S. Masamune, Y. Hanzawa and D. J. Williams, *J. Am. Chem. Soc.*, **104**, 6136 (1982).
414. T. Tsumuraya, S. A. Batcheller and S. Masamune, *Angew.Chem., Int. Ed. Engl.*, **30**, 902 (1991).
415. H. Sakurai, K. Sakamoto and M. Kira, *Chem. Lett.*, 1379 (1984).
416. W. Ando and T. Tsumuraya, *J. Chem. Soc., Chem. Commun.*, 1514 (1987).
417. L. Ross and M. Dräger, *J. Organomet. Chem.*, **199**, 195 (1980).
418. H. Sakurai and A. Sekiguchi, in *Frontiers of Organogermanium, -Tin, and -Lead Chemistry* (Eds. E. Lukevics and L. Ignatovich), Latvian Inst. of Organic Synthesis, Riga, 1993, p.102; A. Sekiguchi, T. Yatabe, H. Naito, C. Kabuto and H. Sakurai, *Chem. Lett.*, 1697 (1992).
419. H. H. Karsch, G. Baumgartner and S. Gamper, *J. Organomet. Chem.*, **462**, C3 (1993).
420. L. Ross, Thesis, Mainz (1980); see also K. Haberle and M. Dräger, *Z. Anorg. Allg. Chem.*, **551**, 116 (1987) for extensive NMR data on phenylpolygermanes.
421. L. Ross and M. Dräger, *Z. Anorg. Allg. Chem.*, **519**, 225 (1984).
422. W. Jensen, R. Jacobson and J. Benson, *Cryst. Struct. Commun.*, **4**, 299 (1975).
423. M. Dräger, L. Ross and D. Simon, *Z. Anorg. Allg. Chem.*, **466**, 145 (1980).
424. M. Dräger and L. Ross, *Z. Anorg. Allg. Chem.*, **476**, 95 (1981).
425. L. Ross and M. Dräger, *J. Organomet. Chem.*, **194**, 23 (1980).
426. M. Weidenbruch, A. Ritschl, K. Peters and H. G. von Schnering, *J. Organomet. Chem.*, **438**, 39 (1992).
427. H. Preut, M. P. Weisbeck and W. P. Neumann, *Acta Cryst.*, **C49**, 182 (1993).
428. E. P. Mayer, H. Nöth, W. Rattay and U. Wietelman, *Chem. Ber.*, **125**, 401 (1992).
429. M. Weidenbruch, A. Ritschl, K. Peters and H. G. von Schnering, *J. Organometal. Chem.*, **437**, C_{25} (1992).
430. H. Suzuki, K. Okabe, R. Kato, N. Sato, Y. Fukuda, H. Watanabe and M. Goto, *Organometallics*, **12**, 4833 (1993); H. Suzuki, Y. Fukuda, N. Sato, H. Ohmori, M. Goto and H. Watanabe, *Chem., Lett.*, 853 (1991).
431. S. Masamune, L. Sita and D. L. Williams, *J. Am. Chem. Soc.*, **105**, 630 (1983).
432. H. Puff, C. Bach, W. Schuh and R. Zimmer, *J. Organomet. Chem.*, **312**, 313 (1986).
433. V. K. Belsky, N. N. Zemlyansky, N. D. Kolosova and l. V. Borisova, *J. Organomet. Chem.*, **215**, 41 (1981).
434. M. F. Lappert, W-P.-Leung, C. L. Raston, B. W. Skelton and A. H. White, *J. Chem. Soc., Dalton Trans.*, 775 (1992); M. F. Lappert, W-P.-Leung, C. L. Raston, A. J. Thorne, B. W. Skelton and A. H. White, *J. Organomet. Chem.*, **233**, C28 (1982).
435. H. Puff, C. Bach, H. Reuter and W. Schuh, *J. Organomet. Chem.*, **277**, 17 (1984); D. H. Olson and R. E. Rundle, *Inorg. Chem.*, **2**, 1310 (1963).
436. M. Dräger, B. Mathiasch, L. Ross and M. Ross, *Z. Anorg.Allg.Chem.*, **506**, 99 (1983).
437. H. Puff, A. Bongartz, W. Schuh and R. Zimmer, *J. Organomet. Chem.*, **248**, 61 (1983).
438. (a) P. E. Eaton and T. W. Cole Jr., *J. Am. Chem. Soc.*, **86**, 3157 (1964); (b), L. R. Sita, *Acct. Chem. Research*, **27**, 191 (1994).
439. J. D. Corbett, *Chem. Rev.*, **85**, 383 (1985).
440. B. Schiemenz and G. Huttner, *Angew. Chem., Int. Ed. Engl.*, **32**, 297 (1993).
441. S. Nagase, *Angew. Chem., Int. Ed. Engl.*, **28**, 329 (1989); *Polyhedron*, **10**, 1299 (1991); A. F. Sax and R. Janoschek, *Angew. Chem., Int. Ed. Engl.*, **25**, 651 (1986).
442. A. Sekiguchi, C. Kabuto and H. Sakurai, *Angew. Chem., Int. Ed. Engl.*, **28**, 55 (1989).
443. A. Sekiguchi, T. Yatabe, H. Kamatani, C. Kabuto and H. Sakurai, *J. Am. Chem. Soc.*, **114**, 6260 (1992).
444. H. Matsumoto, K. Higuchi, Y. Hoshino, H. Koike and Y. Nagai, *J. Chem. Soc., Chem. Commun.*, 1083 (1988).
445. H. Matsumoto, K. Higuchi, S. Kyushin and M. Goto, *Angew. Chem., Int. Ed. Engl.*, **31**, 1354 (1991).
446. Y. Kabe, M. Kuroda, Y. Honda, O. Yamashita, T. Kawase and S. Masamune, *Angew. Chem., Int. Ed. Engl.*, **27**, 1725 (1988).

447. M. Weidenbruch, F. -T. Grimm, S. Pohl and W. Saak, *Angew. Chem., Int. Ed. Engl.*, **28**, 198 (1989).
448. A. Sekiguchi, H. Naito, H. Nameki, K. Ebata, C. Kabuto and H. Sakurai, *J. Organomet. Chem.*, **368**, C1 (1989).
449. L. R. Sita and I. Konoshita, *Organometallics*, **9**, 2865 (1990); *J. Am. Chem. Soc.*, **113**, 1856 (1991); L. R. Sita and R. D. Bickerstaff, *J. Am. Chem. Soc.*, **111**, 6454 (1989).
450. S. Nagase, K. Kobayashi and T. Kudo, *Main Group Metal Chem.*, **XVII**, 171 (1994).
451. M. S. Gordon, K. A. Nguyen and M. T. Carroll, *Polyhedron*, **10**, 1247 (1991).
452. L. R. Sita and I. Kinoshita, *J. Am. Chem. Soc.*, **114**, 7024 (1992).
453. M. Dräger and B. Mathiasch, *Angew. Chem., Int. Ed. Engl.*, **20**, 1029 (1980).
454. B. Mathiasch and M. Dräger, *Angew. Chem., Int. Ed. Engl.*, **17**, 767 (1978).
455. L. R. Sita and R. D. Bickerstaff, *J. Am. Chem. Soc.*, **111**, 3769 (1989).
456. N. Wiberg, C. M. M. Finger and K. Polborn, *Angew. Chem., Int. Ed. Engl.*, **32**, 1054 (1993).
457. M. Baudler, G. Scholz, K-F. Tebbe and M. Feher, *Angew. Chem., Int. Ed. Engl.*, **28**, 340 (1989).
458. L. Zsolnai, G. Huttner and M. Driess, *Angew. Chem., Int. Ed. Engl.*, **32**, 1439 (1993).
459. J. P. Zebrowski, R. K. Hayashi, A. Bjarnason and L. F. Dahl, *J. Am. Chem. Soc.*, **114**, 3121 (1992).
460. M. F. Lappert and R. S. Rowe, *Coord. Chem. Rev.*, **100**, 267 (1990).
461. M. F. Lappert, *Main Group Metal Chem.*, **XVII**, 183 (1994).
462. W. P. Neumann, *Chem. Rev.*, **91**, 311 (1991).
463. W. P. Neumann, M. P. Weisbeck and S. Wienken, *Main Group Metal Chem.*, **XVII**, 151 (1994).
464. T. Fjeldberg, A. Haaland, B. E. R. Schilling, P. B. Hitchcock, M. F. Lappert and A. J. Thorne, *J. Chem. Soc., Dalton Trans.*, 1551 (1986).
465. P. Jutzi, A. Becker, H. G. Stammler and B. Neumann, *Organometallics*, **10**, 1647 (1991).
466. N. Tokitoh, K. Manmaru and R. Okazaki, *Organometallics*, **13**, 167 (1994).
467. P. Jutzi, B. Hampel, B. Hursthouse and A. J. Howes, *J. Organomet. Chem.*, **299**, 19 (1986); P. Jutzi, B. Hampel, K. Stroppel, C. Krüger, K. Angermund and P. Hofmann, *Chem. Ber.*, **118**, 2789 (1985).
468. P. Jutzi, *Adv. Organomet. Chem.*, **26**, 217 (1986).
469. L. Fernholt, A. Haaland, P. Jutzi, F. X. Kohl and R. Seip, *Acta Chem. Scand.*, **A38**, 211 (1984); J. Almlöf, L. Fernholt, K. Faegri, A. Haaland, B. E. R. Schilling, R. Seip and K. Taugbøl, *Acta Chem. Scand.*, **A37**, 131 (1984); M. Grenz, E. Hahn, W. -W. du Mont and J. Pickardt, *Angew. Chem., Int. Ed. Engl.*, **23**, 61 (1984); H. Schumann, C. Janiak, E. Hahn, J. Loebel and J. J. Zuckerman, *Angew. Chem., Int. Ed. Engl.*, **24**, 773 (1985).
470. P. Jutzi, A. Becker, C. Leue, H. G. Stammler, B. Neumann, M. B. Hursthouse and A. Karaulov, *Organometallics*, **10**, 3838 (1991).
471. F. X. Kohl, R. Dickbreder, P. Jutzi, G. Muller and B. Huber, *Chem. Ber.*, **122**, 871 (1989).
472. A. H. Cowley, M. A. Mardones, S. Avendaño, E. Román, J. M. Manriquez and C. J. Carrano, *Polyhedron*, **12**, 125 (1993).
473. M. P. Bigwood, P. J. Corvan and J. J. Zuckerman, *J. Am. Chem. Soc.*, **103**, 7643 (1981).
474. H. Grützmacher, H. Pritzkow and F. T. Edelmann, *Organometallics*, **10**, 23 (1991).
475. T. Fjeldberg, A. Haaland, B. E. R. Schilling, P. B. Hitchcock, M. F. Lappert and A. J. Thorne, *J. Chem. Soc., Dalton Trans.*, 1551 (1986).
476. H. Grützmacher, W. Deck, H. Pritzkow and M. Sander, *Angew. Chem., Int. Ed. Engl.*, **33**, 456 (1994).
477. H. Grützmacher, S. Freitag, R. Herbst-Irmer and G. S. Sheldrick, *Angew. Chem., Int. Ed. Engl.*, **31**, 437 (1992).
478. W. E. Piers, R. M. Whittal, G. Ferguson, J. F. Gallagher, R. D. J. Froese, H. J. Stronks and P. H. Krygsman, *Organometallics*, **11**, 4015, (1992).
479. B. S. Jolly, M. F. Lappert, L. M. Engelhardt, A. H. White and C. L. Raston, *J. Chem. Soc., Dalton Trans.*, 2653 (1993).
480. C. Janiak, H. Schumann, C. Stader, B. Wrackmeyer and J. J. Zuckerman, *Chem. Ber.*, **121**, 1745 (1988).
481. A. J. Edwards, M. A. Paver, P. R. Raithby, C. A. Russell, D. Stalke, A. Steiner and D. S. Wright, *J. Chem. Soc., Dalton Trans.*, 1465 (1993).
482. M. G. Davidson, D. Stalke and D. S. Wright, *Angew. Chem., Int. Ed. Engl.*, **31**, 1226 (1992).
483. D. Stalke, M. A. Paver and D. S. Wright, *Angew. Chem., Int. Ed. Engl.*, **32**, 428 (1993).

2. Structural aspects of compounds containing C–E (E = Ge, Sn, Pb) bonds   193

484. H. Schmidbaur, T. Probst, B. Huber, G. Müller and C. Krüger, *J. Organomet. Chem.*, **365**, 53 (1989).
485. H. Schmidbaur, T. Probst and O. Steigelmann, *Organometallics*, **10**, 3176 (1991).
486. T. Probst, O. Steigelmann, J. Riede and H. Schmidbaur, *Angew. Chem., Int. Ed. Engl.*, **29**, 1397 (1990).
487. S. Brooker, J-K. Buijink and F. T. Edelmann, *Organometallics*, **10**, 25 (1991).
488. N. Kuhn, G. Henkel and S. Stubenrauch, *Angew. Chem., Int. Ed. Engl.*, **31**, 778 (1992).
489. M. Weidenbruch, *Main Group Metal Chem.*, **XVII**, 9 (1994).
490. S. Masamune, Y. Hanzawa, S. Murakami, T. Bally and J. F. Blount, *J. Am. Chem. Soc.*, **104**, 1150 (1982).
491. T. Tsumuraya, S. A. Batcheller and S. Masamune, *Angew.Chem., Int. Ed. Engl.*, **30**, 904 (1991).
492. H. Matsumoto, A. Sakamoto and Y. Nagai, *J. Chem. Soc., Chem. Commun.*, 1786 (1986).
493. K. M. Baines, J. A. Cooke, N. C. Paine and J. J. Vittal, *Organometallics*, **11**, 1408 (1992).
494. A. Heine and D. Stalke, *Angew. Chem., Int. Ed. Engl.*, **33**, 113 (1994).
495. S. Masamune, Y. Hanzawa, S. Murikami, T. Bally and M. F. Blount, *J. Am. Chem. Soc.*, **104**, 1150 (1982).
496. T. Tsumuraya, S. A. Batcheller and S. Masamune, *Angew.Chem., Int. Ed. Engl.*, **30**, 902 (1991).
497. S. Masamune and L. R. Sita, *J. Am. Chem. Soc.*, **107**, 6390 (1985).
498. T. Tsumuraya, S. A. Batcheller and S. Masamune, *Angew.Chem., Int. Ed. Engl.*, **30**, 907 (1991).
499. M. P. Egorov, S. P. Kolesnikov, Yu. T. Struchkov, M. Yu. Antipin, S. V. Sereda and O. M. Nefedov, *J. Organomet. Chem.*, **290**, C27 (1985); *Izv. Akad. Nauk SSSR, Ser. Khim.*, (**4**), 959 (1985).
500. L. R. Sita and R. D. Bickerstaff, *J. Am. Chem. Soc.*, **110**, 5208 (1988).
501. W. Ando, H. Ohgaki and Y. Kabe, *Angew. Chem., Int. Ed. Engl.*, **33**, 659 (1994).
502. D. Horner, R. S. Grev and H. F. Schaeffer III, *J. Am. Chem. Soc.*, **114**, 2093 (1992).
503. G. L. Delker, Y. Wang, G. D. Stucky, R. L. Lambert, C. K. Haas and D. Seyferth, *J. Am. Chem. Soc.*, **98**, 1779 (1976).
504. L. R. Sita and R. D. Bickerstaff, *J. Am. Chem. Soc.*, **110**, 5208 (1988).
505. D. Horner, R. S. Grev and H. F. Schaeffer III, *J. Am. Chem. Soc.*, **114**, 2093 (1992).
506. T. Tsumaraya, S. Sato and W. Ando, *Organometallics*, **9**, 2061 (1990).
507. T. Tsumaraya, and W. Ando, *Organometallics*, **8**, 1467 (1989).
508. T. Tsumaraya, S. Sato and W. Ando, *Organometallics*, **8**, 161 (1989).
509. T. Ohtaki, Y. Kabe and W. Ando, *Organometallics*, **12**, 4 (1993).
510. M. Andrianarison, C. Couret, J-P. Declerq, A. Dubourg, J. Escudie, H. Ranaivonjatavo and J. Satgé, *Organometallics*, **7**, 1545 (1988).
511. M. Driess, H. Pritzkow and U. Winkler, *Chem. Ber.*, **125**, 1541 (1992).
512. A. H. Cowley, S. W. Hall, C. M. Nunn and J. M. Power, *J. Chem. Soc., Chem. Commun.*, 753 (1988).
513. C. Glidewell, *J. Organomet. Chem.*, **461**, 15 (1993).
514. W. A. Herrmann, *Angew. Chem., Int. Ed. Engl.*, **25**, 56 (1986).
515. J. Escudié, C. Couret, H. Ranaivonjatovo, G. Anselme, G. Delpon-Lacaze, M,-A Chaudon, A. Kandri Roti and J. Satgé, *Main Group Metal Chem.*, **XVII**, 33 (1994).
516. J. Barrau, J. Escudié and J. Satgé, *Chem. Rev.*, **90**, 283 (1990).
517. J. Satgé, *Adv. Organomet. Chem.*, **21**, 241 (1982).
518. N. C. Norman, *Polyhedron*, **12**, 2341 (1993).
519. D. E. Goldberg, D. H. Harris, M. F. Lappert and K. M. Thomas, *J. Chem. Soc., Chem. Commun.*, 21 (1976); P. J. Davidson, D. H. Harris and M. F. Lappert, *J. Chem. Soc., Dalton Trans.*, 2268 (1976); J. D. Cotton, P. J. Davidson, M. F. Lappert, J. D. Donaldson and J. Silver, *J. Chem. Soc., Dalton Trans.*, 2286 (1976).
520. P. B. Hitchcock, M. F. Lappert, S. J. Miles and A. J. Thorne, *J. Chem. Soc., Chem. Commun.*, 480 (1984); D. E. Goldberg, P. B. Hitchcock, M. F. Lappert, K. M. Thomas, A. J. Thorne, T. Fjeldberg, A. Haaland and B. E. R. Schilling, *J. Chem. Soc., Dalton Trans.*, 2387 (1986).
521. J. T. Snow, S. Murakami, S. Masamune and D. J. Williams, *Tetrahedron Lett.*, **25**, 4191 (1984); S. A. Batcheller, T. Tsumuraya, O. Tempkin, W. M. Davis and S. Masamune, *J. Am. Chem. Soc.*, **112**, 9394 (1990).
522. M. Lazraq, J. Escudie, C. Couret, J. Satgé, M. Dräger and R. Dammel, *Angew. Chem., Int. Ed. Engl.*, **27**, 828 (1988); H. Meyer, G. Baum, W. Massa and A. Berndt, *Angew. Chem., Int. Ed. Engl.*, **26**, 798 (1987).

523. G. Trinquier, J. C. Barthelat and J. Satge, *J. Am. Chem. Soc.*, **104**, 5931 (1982); B. G. Gowenlock and J. A. Hunter, *J. Organomet. Chem.*, **140**, 265 (1977); **111**, 171 (1976).

524. G. Trinquier, J. -P. Malrieu and P. Riviere *J. Am. Chem. Soc.*, **104**, 4592 (1982); T. Fjeldberg, A. Haaland, M. F. Lappert, B. E. R. Schilling, R. Seip and A. J. Thorne, *J. Chem. Soc., Chem. Commun.*, 1407 (1982); S. Nagase and T. Kudo, *J. Mol. Struct. Theochem.*, **103**, 35 (1983).

525. T. Fjeldberg, A. Haaland, B. E. R. Schilling, H. V. Volden, M. F. Lappert and A. J. Thorne, *J. Organomet. Chem.*, **280**, C43 (1985).

526. J. T. Gleghorn and N. D. A. Hammond, *Chem. Phys. Lett.*, **105**, 621 (1984).

527. M. J. S. Dewar, G. L. Grady, D. R. Kuhn and K. M. Merz, *J. Am. Chem. Soc.*, **106**, 6773 (1984).

528. R. S. Grev, *Adv. Organomet. Chem.*, **33**, 125, (1991).

529. R. S. Grev and H. F. Schaeffer III, *Organometallics*, **11**, 3489 (1992).

530. M. Dräger, J. Escudie, C. Couret, H. Ranaivonjatovo and J. Satgé, *Organometallics*, **7**, 1010 (1988).

531. J. F. Blount, K. Mislow and J. Jacobus, *Acta Cryst.*, **A28**, S12 (1972).

532. M. J. Fink, M. J. Michalczyk, K. J. Haller, R. West and J. Michl, *Organometallics*, **3**, 793 (1984).

533. S. Masamune, S. Murakami, J. T. Snow, H. Tobita and D. J. Williams, *Organometallics*, **3**, 333 (1984).

CHAPTER 3

Stereochemistry and conformation of organogermanium, organotin and organolead compounds

JAMES A. MARSHALL and JILL A. JABLONOWSKI

Department of Chemistry and Biochemistry, University of South Carolina, Columbia, SC 29208, USA
Fax: 803-777-9385;   e-mail: marshall@chem.chem.scarolina.edu

The chemistry of organic germanium, tin and lead compounds
Edited by S. Patai   © 1995 John Wiley & Sons Ltd

I. INTRODUCTION

It is the purpose of this chapter to summarize what is currently known about the stereo-chemistry and conformation of organogermanium, tin and lead compounds. Coverage is selective rather than exhaustive. The first section deals with compounds in which substitution by four different groups causes the metal atom to be stereogenic. We have limited our discussion to those cases in which at least three of the four substituents are alkyl or aryl. In this section we also briefly discuss pentacoordinated triorgano halostannanes.

The second section examines organogermanium, tin and lead compounds in which chirality resides in the organic group attached to the metal center. We have organized this section according to the nature of the chiral organic substituent which in turn reflects the method of synthesis. We have only briefly noted applications of the foregoing compounds as reagents for organic synthesis.

II. CHIRAL ORGANOMETALLICS OF TYPE $R^1R^2R^3R^4M$

A. Germanium

1. Synthesis

The first chiral nonracemic germanes were prepared from the tetraphenyl derivative through a series of successive electrophilic and nucleophilic substitutions as illustrated in Scheme 1. Brook and Peddle were able to separate the diastereomeric $(-)$-menthyloxy derivatives **1** and **2** by fractional crystallization[1]. Treatment of each diastereomer with LiAlH$_4$ afforded the $(+)$ and $(-)$ enantiomeric hydrides R-**3** and S-**3**, respectively,

The assignment of absolute configuration to these hydrides was based on their mixture melting point behavior with the analogous silanes of known absolute configuration. Thus, R-**3** admixed with its R-$(+)$ sila analogue behaved as a solid solution with a melting range of $1\,°C$ or less over a range of concentrations. Admixture of R-**3** with the S-$(-)$ sila analogue resulted in mp ranges of $10-25\,°C$. Additional support for the assigned configuration was secured through application of Brewster's rules of atomic asymmetry[2].

An independent and virtually simultaneous report by Eaborn and coworkers described the preparation of germyl hydride (**5**), the ethyl analogue of **3**, by an analogous route (equation 1)[3]. However, only one of the diastereomeric menthyloxy derivatives (**4**) could be separated by crystallization; the other was not readily purified. Assignment of absolute configuration was not made for **5** but, based on the sign of rotation, it is most likely R.

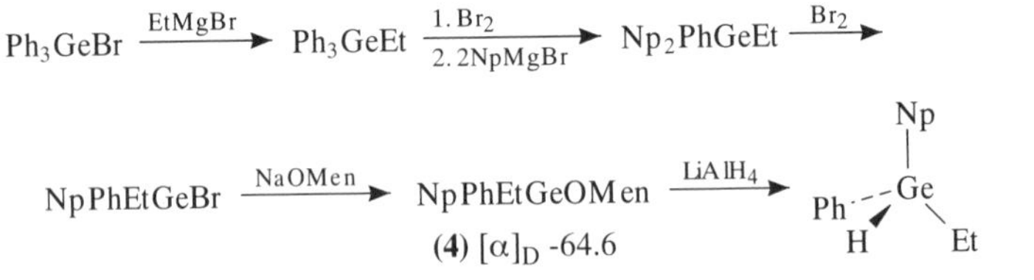

$$Ph_3GeBr \xrightarrow{EtMgBr} Ph_3GeEt \xrightarrow[2.\,2NpMgBr]{1.\,Br_2} Np_2PhGeEt \xrightarrow{Br_2}$$

$$NpPhEtGeBr \xrightarrow{NaOMen} NpPhEtGeOMen \xrightarrow{LiAlH_4} \underset{H\quad\quad Et}{\overset{Np}{Ph\diagdown\!\!-\!\!Ge\diagup}} \tag{1}$$

(4) $[\alpha]_D$ -64.6

(R)-**(5)** $[\alpha]_D$ 23.6

The isopropyl homologue R-**6** of the chiral germyl hydrides **3** and **5** was prepared by Carré and Corriu as outlined in equation 2[4]. The route closely parallels that of the previous investigators.

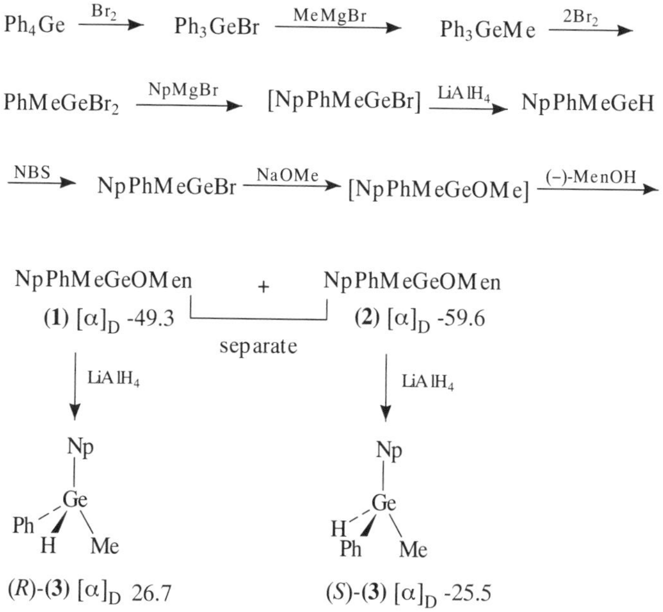

$Ph_4Ge \xrightarrow{Br_2} Ph_3GeBr \xrightarrow{MeMgBr} Ph_3GeMe \xrightarrow{2Br_2}$

$PhMeGeBr_2 \xrightarrow{NpMgBr} [NpPhMeGeBr] \xrightarrow{LiAlH_4} NpPhMeGeH$

$\xrightarrow{NBS} NpPhMeGeBr \xrightarrow{NaOMe} [NpPhMeGeOMe] \xrightarrow{(-)-MenOH}$

$$\underset{\substack{(\mathbf{1})\ [\alpha]_D\ -49.3}}{NpPhMeGeOMen} \quad + \quad \underset{\substack{(\mathbf{2})\ [\alpha]_D\ -59.6}}{NpPhMeGeOMen}$$

separate

| LiAlH₄ | LiAlH₄ |

(R)-$(\mathbf{3})\ [\alpha]_D\ 26.7$ (S)-$(\mathbf{3})\ [\alpha]_D\ -25.5$

SCHEME 1. Np = 1-naphthyl, Men = $(-)$-menthyl, $(-)$-MenOH =

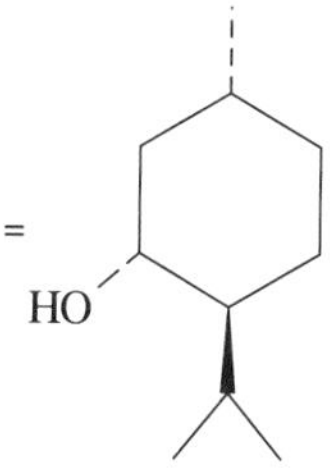

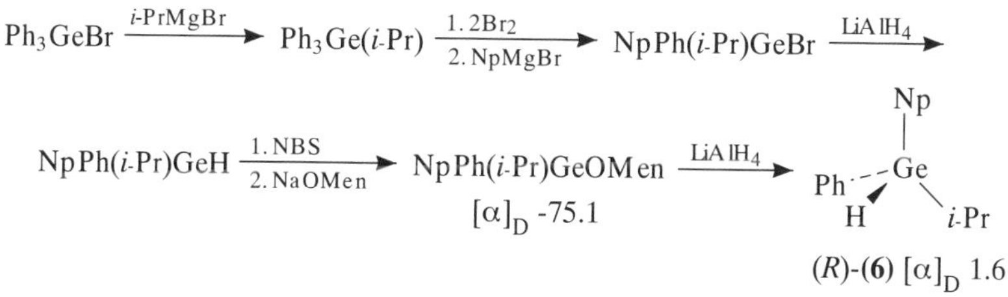

$Ph_3GeBr \xrightarrow{i\text{-}PrMgBr} Ph_3Ge(i\text{-}Pr) \xrightarrow[2.\ NpMgBr]{1.\ 2Br_2} NpPh(i\text{-}Pr)GeBr \xrightarrow{LiAlH_4}$

$NpPh(i\text{-}Pr)GeH \xrightarrow[2.\ NaOMen]{1.\ NBS} \underset{[\alpha]_D\ -75.1}{NpPh(i\text{-}Pr)GeOMen} \xrightarrow{LiAlH_4}$ (2)

(R)-$(\mathbf{6})\ [\alpha]_D\ 1.6$

The absolute configuration of germane R-**6** was determined by conversion to the methylated derivative S-**8** via the chlorogermane S-**7** as depicted in Scheme 2[4]. The enantiomeric germane R-**8** was prepared analogously from the known methylgermyl hydride R-**3**. It is assumed that chlorination of the hydrides R-**6** and R-**3** proceeds with retention and the reaction of chlorogermanes S-**7** and S-**9** with MeLi and i-PrLi proceeds with inversion of configuration.

Recently, Terunuma and coworkers reported a novel synthesis of nonracemic tetraorgano germanes containing an amine substituent. Reaction of diazomethane with $GeCl_4$ led to the (chloromethyl)germyl chloride, which could be sequentially substituted by various Grignard reagents (equation 3)[5]. Ammonolysis of the triorgano chloromethylgermanes

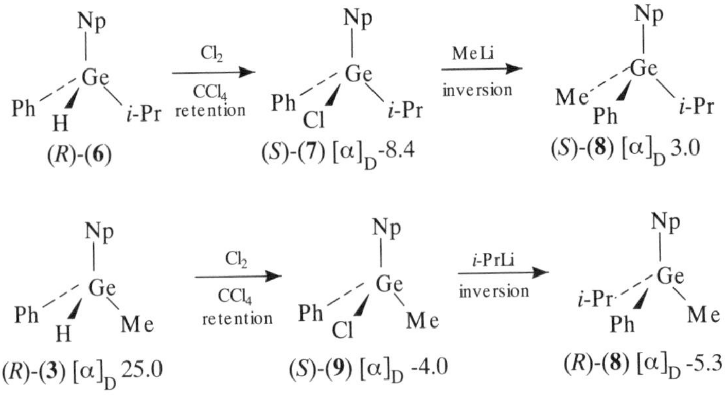

$$\text{(R)-(6)} \xrightarrow[\text{retention}]{\underset{CCl_4}{Cl_2}} \text{(S)-(7)} [\alpha]_D -8.4 \xrightarrow[\text{inversion}]{MeLi} \text{(S)-(8)} [\alpha]_D 3.0$$

$$\text{(R)-(3)} [\alpha]_D 25.0 \xrightarrow[\text{retention}]{\underset{CCl_4}{Cl_2}} \text{(S)-(9)} [\alpha]_D -4.0 \xrightarrow[\text{inversion}]{i\text{-PrLi}} \text{(R)-(8)} [\alpha]_D -5.3$$

SCHEME 2. Np = 1-naphthyl

led to the amines, which could be resolved as salts of α-phenylpropionic or tartaric acid. The ee of the resulting amines was shown to be > 90% by ^{1}H NMR analysis of the Mosher amides.

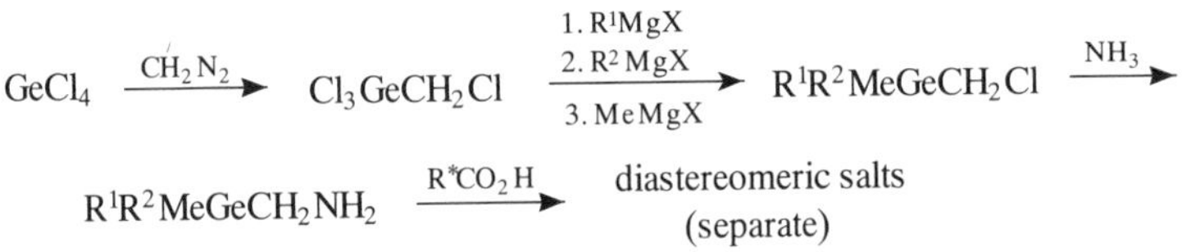

$$GeCl_4 \xrightarrow{CH_2N_2} Cl_3GeCH_2Cl \xrightarrow[\substack{1.\,R^1MgX \\ 2.\,R^2MgX \\ 3.\,MeMgX}]{} R^1R^2MeGeCH_2Cl \xrightarrow{NH_3}$$

$$R^1R^2MeGeCH_2NH_2 \xrightarrow{R^*CO_2H} \begin{array}{c} \text{diastereomeric salts} \\ \text{(separate)} \end{array} \qquad (3)$$

R^1	R^2
Ph	o-tolyl
Ph	Bn
o-tolyl	Bn
p-MeOC$_6$H$_4$	Bn

In a report describing the first enzymatic synthesis of a chiral nonracemic tetraorgano germane, Tacke and coworkers subjected the prochiral *cis*-hydroxymethyl derivative (10) to acetylation catalyzed by pig liver esterase (Scheme 3)[6]. The resulting monoacetate (11) was shown to be of 55% ee through ^{1}H NMR analysis of the Mosher ester derivative.

2. Interconversions — Walden cycles

Lithiation of the germyl hydride S-3 with BuLi in ether and subsequent protonolysis led to recovered hydride of $[\alpha]_D - 18.5$ (86% retention)[2]. It is assumed that both steps proceed with predominant retention of configuration. Carboxylation of the lithio derivative afforded the carboxylic acid R-12, also with retention of configuration (Scheme 4). The corresponding methyl ester underwent decarbonylation upon heating to afford the methoxy derivative R-15, which was subsequently reduced with LiAlH$_4$ to the (−)-hydride S-3. Thus the decarbonylation and reduction steps must both proceed with retention or inversion of configuration. From a consideration of Brewster's rules and by mixture

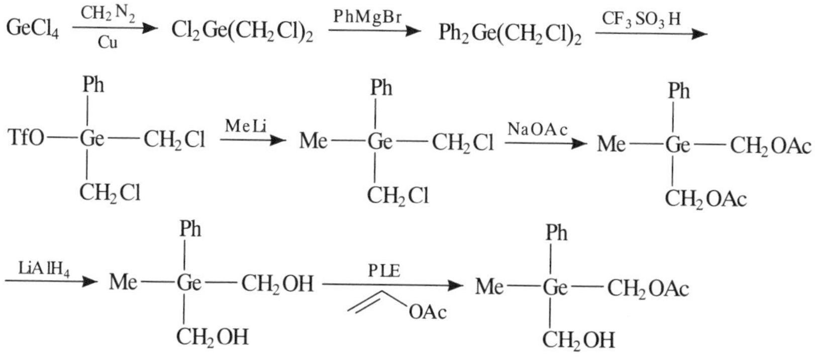

$$GeCl_4 \xrightarrow[\text{Cu}]{CH_2N_2} Cl_2Ge(CH_2Cl)_2 \xrightarrow{PhMgBr} Ph_2Ge(CH_2Cl)_2 \xrightarrow{CF_3SO_3H}$$

TfO—Ge(Ph)(CH₂Cl)—CH₂Cl $\xrightarrow{MeLi}$ Me—Ge(Ph)(CH₂Cl)—CH₂Cl $\xrightarrow{NaOAc}$ Me—Ge(Ph)(CH₂OAc)—CH₂OAc

$\xrightarrow{LiAlH_4}$ Me—Ge(Ph)(CH₂OH)—CH₂OH $\xrightarrow[\diagup OAc]{PLE}$ Me—Ge(Ph)(CH₂OAc)—CH₂OH

(10)   **(11)** $[\alpha]_D$ -11.4

SCHEME 3. PLE = pig liver esterase, Tf = CF_3SO_2

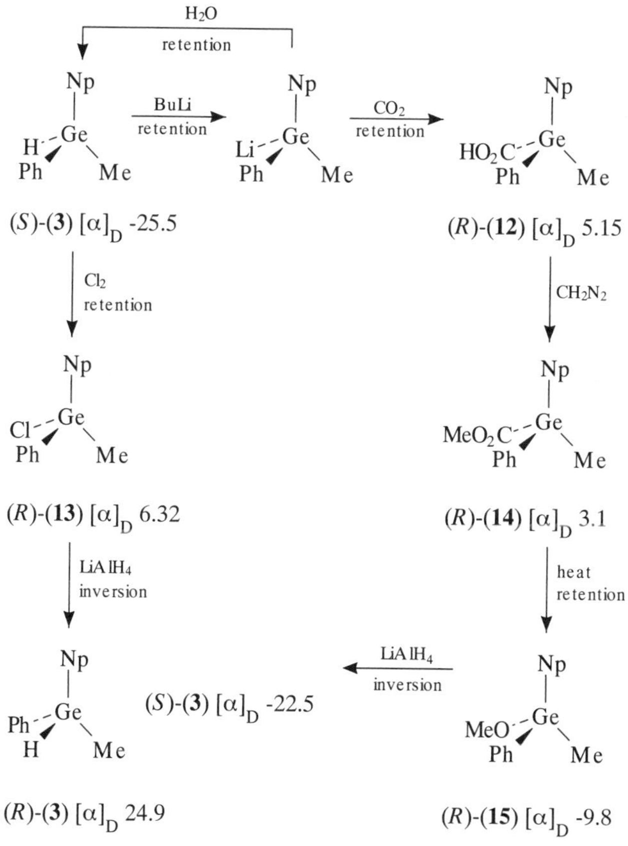

H₂O / retention

Np—Ge(H)(Ph)—Me $\xrightarrow[\text{retention}]{BuLi}$ Np—Ge(Li)(Ph)—Me $\xrightarrow[\text{retention}]{CO_2}$ Np—Ge(HO₂C)(Ph)—Me

(S)-**(3)** $[\alpha]_D$ -25.5   (R)-**(12)** $[\alpha]_D$ 5.15

$\downarrow$ Cl₂ retention $\downarrow$ CH₂N₂

Np—Ge(Cl)(Ph)—Me   Np—Ge(MeO₂C)(Ph)—Me

(R)-**(13)** $[\alpha]_D$ 6.32   (R)-**(14)** $[\alpha]_D$ 3.1

$\downarrow$ LiAlH₄ inversion $\downarrow$ heat retention

Np—Ge(Ph)(H)—Me   Np—Ge(MeO)(Ph)—Me

(S)-**(3)** $[\alpha]_D$ -22.5 $\xleftarrow[\text{inversion}]{LiAlH_4}$

(R)-**(3)** $[\alpha]_D$ 24.9   (R)-**(15)** $[\alpha]_D$ -9.8

SCHEME 4. Np = 1-naphthyl

melting point determination with the sila analogue of the methoxy derivative R-**15**, it was concluded that both of these reactions proceed with retention.

In a second Walden cycle, germane S-**3** was converted to the chloro derivative R-**13**, which was reduced by $LiAlH_4$ to the enantiomeric germane R-**3**[2]. The configuration of the chloride R-**13** was assigned by mixture melting point with the known R sila analogue and from consideration of Brewster's rules. Thus, chlorination must proceed with retention and hydride reduction with inversion of configuration.

The free radical chlorination of germane R-**3** was studied in some detail by Sakurai and coworkers[7]. They showed that the intermediate germyl radical abstracts Cl atoms from CCl_4 more rapidly than it inverts when the latter is employed as solvent. Progressive dilution with cyclohexane gave rise to increasing amounts of the racemic chloride. An activation energy of 5.7 kcal mol^{-1} was estimated for the inversion process (equation 4). Attempted preparation of a nonracemic bromogermane through bromination of the hydride R-**5** in CCl_4 led to racemic product[3].

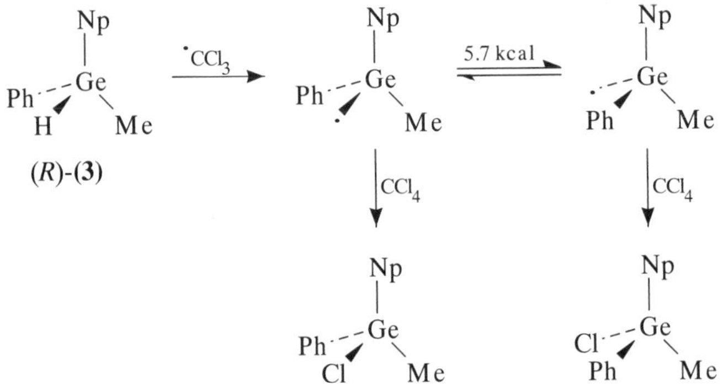

$$\begin{array}{ccc}
\underset{(R)\text{-}(3)}{\overset{\displaystyle Np}{\underset{\displaystyle H \quad Me}{Ph\text{-}Ge}}} & \xrightarrow{\cdot CCl_3} & \underset{}{\overset{\displaystyle Np}{\underset{\displaystyle \cdot \quad Me}{Ph\text{-}Ge}}} & \underset{}{\overset{5.7\,kcal}{\rightleftharpoons}} & \underset{}{\overset{\displaystyle Np}{\underset{\displaystyle Ph \quad Me}{\cdot\text{-}Ge}}}
\end{array}$$

$$\downarrow CCl_4 \qquad\qquad \downarrow CCl_4 \qquad\qquad (4)$$

$$\underset{(S)\text{-}(13)}{\overset{\displaystyle Np}{\underset{\displaystyle Cl \quad Me}{Ph\text{-}Ge}}} \qquad\qquad \underset{(R)\text{-}(13)}{\overset{\displaystyle Np}{\underset{\displaystyle Ph \quad Me}{Cl\text{-}Ge}}}$$

Corriu and coworkers noted that chlorogermanes such as **7** racemize when allowed to stand in certain donor solvents (Table 1)[8]. Nondonor solvents such as benzene were ineffective in this regard, and racemization was appreciably retarded in mixtures of benzene

TABLE 1.   Racemization of chlorogermanes in various solvents

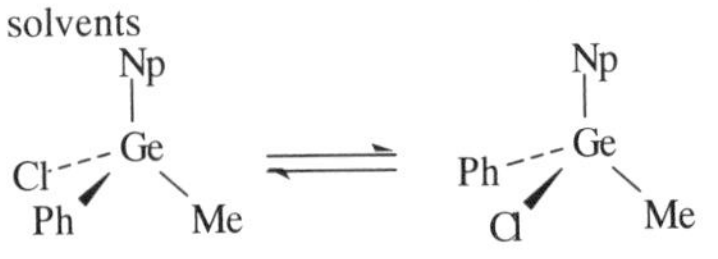

$$\underset{(R)\text{-}(7)}{\overset{\displaystyle Np}{\underset{\displaystyle Ph \quad Me}{Cl\text{-}Ge}}} \rightleftharpoons \underset{(S)\text{-}(7)}{\overset{\displaystyle Np}{\underset{\displaystyle Cl \quad Me}{Ph\text{-}Ge}}}$$

	Half-life in minutes	
Solvent	R = Me	R = i-Pr
THF	1.5	20
HOAc	1.5	55
Ac_2O	0	0.5
$MeOCH_2CH_2OMe$	0	0.5
EtOAc	3	90
$PhCO_2Et$	0	36

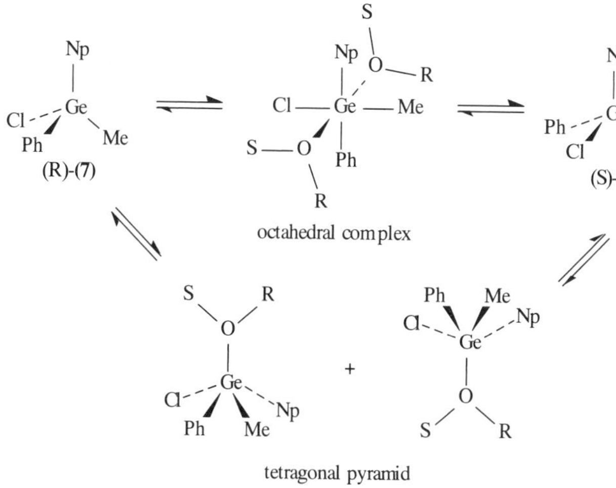

FIGURE 1. Solvent participation in the racemization of chiral chlorogermanes

and donor solvents. A mechanism involving a hexacoordinated octahedral germane was proposed for these racemizations (Figure 1). It should be noted that racemization could also proceed through a tetragonal pyramid intermediate.

Alkylations of the germyllithium derived from germane *R*-5 were found to proceed with retention or inversion depending upon the nature of the alkyl halide employed (Scheme 5)[9]. Iodides gave rise to inverted products, whereas the corresponding chlorides or bromides yielded products with retention of configuration at Ge.

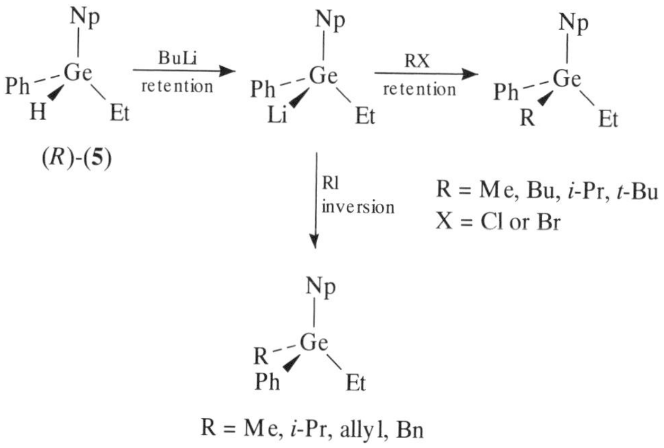

R = Me, *i*-Pr, allyl, Bn

SCHEME 5. Np = 1-naphthyl

The assignment of configuration to these products was made by analogy with the R = CH$_3$ system[9]. In that case, the two methylated enantiomers were compared with an authentic sample prepared from the known germyl hydride *R*-5 via the chloro derivative as outlined in equation 5.

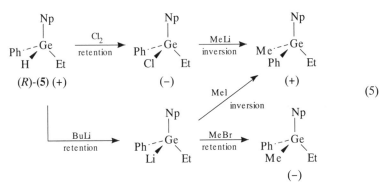

FIGURE 2. Proposed pathway for alkylation of germyllithium reagents with alkyl iodides

It was suggested that iodides undergo metal–halogen exchange with the germyllithium reagent, followed by S_N2 attack of the resulting organolithium on the iodogermane (Figure 2)[9]. In keeping with this proposal, lithiation of germane R-5 with BuLi in the presence of MeI leads to appreciable amounts of the butylated germane[9]. Lithiation in the presence of MeBr, in contrast, gives the methylated product and only a trace of butyl adduct.

An alternative route to chiral tetraorganogermanes entails additions of organometallic reagents to chiral alkoxygermanes. With chlorogermanes, such reactions are known to proceed with inversion[9,10]. However, the steric course of the alkoxy displacement differs for alkyl- and allyllithium reagents[10]. The former displace the OR group with retention, whereas inversion is seen with the latter (Scheme 6). Thus, reaction of allyllithium with the menthyloxy derivative R-19 leads to the product R-17 of negative rotation. Hydrogenation of this product affords the levorotatory propyl derivative R-18. Direct displacement of the menthyloxy group in R-19 by propyllithium leads to the dextrorotatory propyl derivative S-18. S-18 is also produced by treatment of the chlorogermane S-7 with propyllithium, a reaction presumed to proceed with inversion of configuration. It therefore follows that the reaction of allyllithium with the menthyloxygermane R-19 occurs with inversion. The methoxygermane R-16 behaves analogously but affords the substitution product R-17 of lower ee. This may reflect a lower ee of the methoxy derivative R-16 or a decreased stereospecificity for such displacements on R-16 vs R-19.

It should be noted that the hydrogenolysis of methoxygermane R-16 by LiAlH$_4$ proceeds with retention of configuration[9].

3. Digermanes

Displacements on halogermanes and alkoxygermanes by germyllithium species have been found to occur with inversion at the electrophilic and retention at the nucleophilic germanium center (Schemes 7 and 8)[11]. Thus, the known menthyloxygermane S-19, upon hydrogenolysis with LiAlH$_4$ (retention) and chlorination (retention), affords the chloroger-mane S-7. Reaction with triphenylgermyllithium yields the levorotatory digermane R-20.

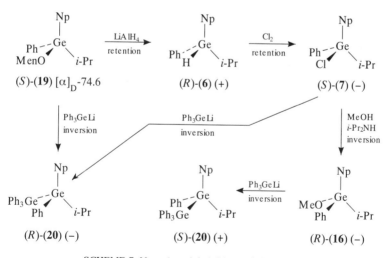

SCHEME 6. Np = 1-naphthyl, Men = (−)-menthyl

SCHEME 7. Np = 1-naphthyl, Men = (−)-menthyl

This same germane was secured through displacement of the menthyloxy grouping of S-19 with Ph₃GeLi. Methanolysis of chlorogermane S-7 proceeds with inversion to afford the levorotatory methoxygermane R-16. Reaction with Ph₃GeLi leads to dextrorotatory digermane S-20. Accordingly, displacements on chloro and alkoxygermanes S-7, R-16, and S-19 by Ph₃GeLi must proceed with inversion.

The stereochemical fate of the nucleophilic germanium component of these reactions was probed as outlined in Scheme 8[11]. Hydrogenolysis of the known menthyloxygermane S-19 afforded the hydride R-6 with retention of configuration. Lithiation and subsequent reaction with Ph$_3$GeBr led to the digermane S-20 of positive rotation. On the other hand, displacement of the menthyloxygermane S-19 with Ph$_3$GeLi, a reaction known to proceed with inversion, afforded digermane R-20 of negative rotation. Therefore, the germyllithium R-21 derived from hydride R-6 must retain its configuration in the reaction leading to digermane S-20.

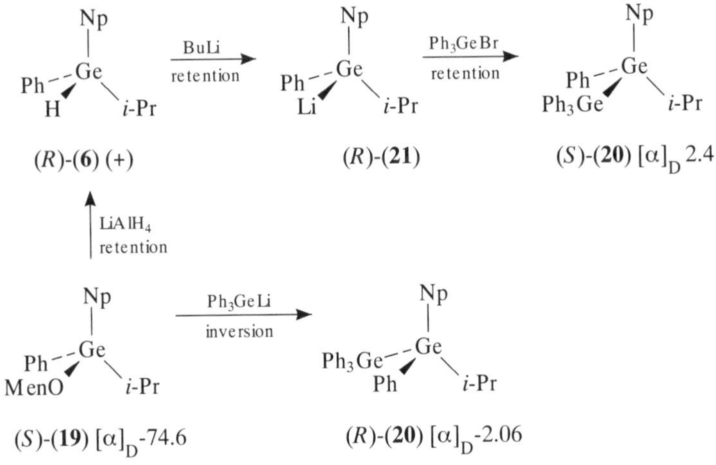

Np
|
Ph--Ge
H i-Pr

$\xrightarrow[\text{retention}]{\text{BuLi}}$

Np
|
Ph--Ge
Li i-Pr

$\xrightarrow[\text{retention}]{\text{Ph}_3\text{GeBr}}$

Np
|
Ph--Ge
Ph$_3$Ge i-Pr

(R)-(6) (+)   (R)-(21)   (S)-(20) $[\alpha]_D$ 2.4

$\uparrow$ LiAlH$_4$
retention

Np
|
Ph--Ge
MenO i-Pr

$\xrightarrow[\text{inversion}]{\text{Ph}_3\text{GeLi}}$

Np
|
Ph$_3$Ge--Ge
Ph i-Pr

(S)-(19) $[\alpha]_D$-74.6   (R)-(20) $[\alpha]_D$-2.06

SCHEME 8. Np = 1-naphthyl, Men = $(-)$-menthyl

Additional support for these conclusions was obtained as outlined in Scheme 9[11]. Accordingly, the lithiogermane R-21, prepared from the known menthyloxygermane S-19 by hydrogenolysis (retention) and lithiation (retention), was treated with the known menthyloxygermane R-19 to afford the dextrorotatory digermane S,S-22. Hydrogenolysis and lithiation of menthyloxygermane R-19 gave rise to the lithiogermane S-21, whose reaction with menthyloxygermane S-19 afforded the levorotatory digermane R,R-22. As expected, combination of lithiogermane S-21 with menthyloxygermane R-19 yielded the inactive digermane, *meso*-22.

4. Other reactions

The hydrogermylation of phenylacetylene has been shown to proceed with retention at germanium[12]. Addition of the germyl hydride R-3 in the presence of a Pt or Rh catalyst led to the adduct S-23 as the major product. The enantiomer R-23 was prepared by addition of *trans-β*-styryllithium to chlorogermane S-7 prepared as shown in Scheme 10. The isopropyl analogues of S-23 and R-23 were similarly prepared from hydride R-6[12].

In a study designed to test the feasibility of a germanium aza Brook rearrangement, the N-benzyl derivative 25 of amine 24 was treated with BuLi[13]. After subsequent addition of water, the hydride 28 was obtained as the sole Ge-containing product (Scheme 11).

Treatment of hydride 28 with dichlorocarbene, a process known to proceed with retention of configuration[14], afforded the insertion product 27. Hydrogenolysis with Bu$_3$SnH led to the monochloro derivative 26, which afforded the amine 24 of nearly identical

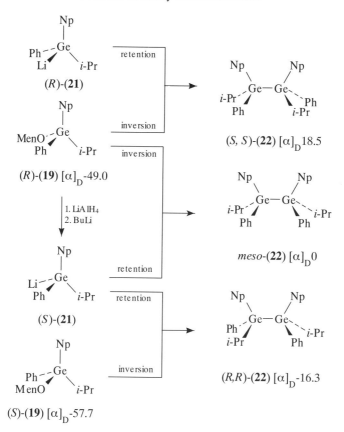

SCHEME 9. Np = 1-naphthyl, Men = (−)-menthyl

SCHEME 10. Np = 1-naphthyl

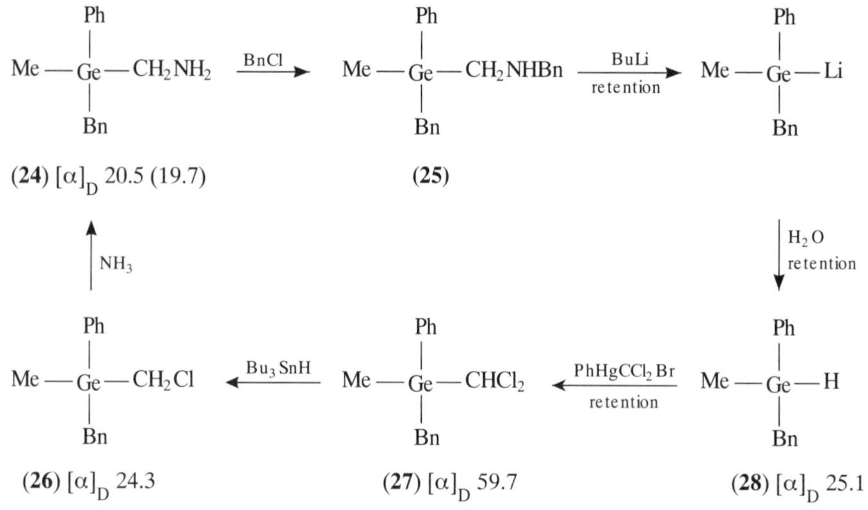

Ph
|
Me —Ge— CH$_2$NH$_2$ $\xrightarrow{\text{BnCl}}$ Me —Ge— CH$_2$NHBn $\xrightarrow[\text{retention}]{\text{BuLi}}$ Me —Ge— Li
| | |
Bn   Bn   Bn

(24) [α]$_D$ 20.5 (19.7)   (25)

↑ NH$_3$ | H$_2$O
                                                                                 ↓ retention

Ph   Ph   Ph
| | |
Me —Ge— CH$_2$Cl $\xleftarrow{\text{Bu}_3\text{SnH}}$ Me —Ge— CHCl$_2$ $\xleftarrow[\text{retention}]{\text{PhHgCCl}_2\text{Br}}$ Me —Ge— H
| | |
Bn   Bn   Bn

(26) [α]$_D$ 24.3   (27) [α]$_D$ 59.7   (28) [α]$_D$ 25.1

SCHEME 11

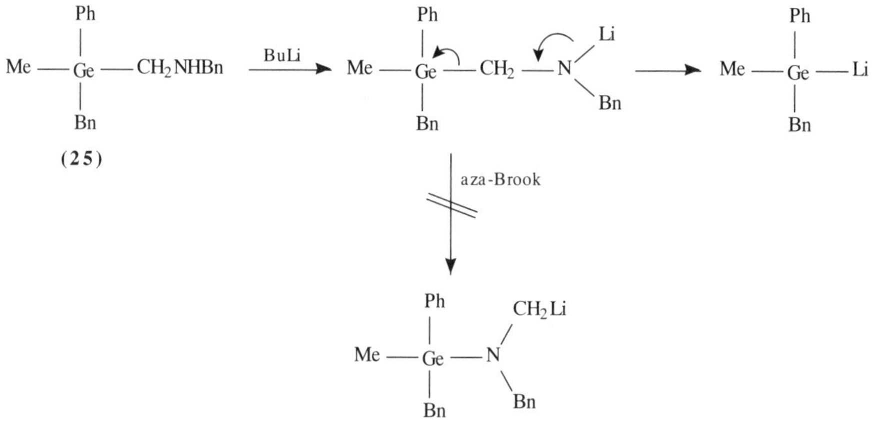

Ph   Ph   Ph
| | |
Me —Ge— CH$_2$NHBn $\xrightarrow{\text{BuLi}}$ Me —Ge— CH$_2$—N—Li → Me —Ge— Li
| | \Bn |
Bn   Bn   Bn

(25)

                                           | aza-Brook
                                           ‖
                                           ↓

                              Ph   CH$_2$Li
                              | /
                         Me —Ge— N
                              | \
                              Bn   Bn

FIGURE 3. Possible reaction pathway for Ge–C bond cleavage in amine 25

rotation to that of the starting sample. Thus, cleavage of the Ge–C bond of germane **25** must proceed with retention of configuration. A possible reaction pathway is shown in Figure 3.

B. Tin

1. Configurational stability

The successful preparation of nonracemic tetraorganogermanes stimulated investigations on the synthesis of related stannanes. By the late 1960s, it was apparent from ^{1}H NMR studies that chiral stannanes should be capable of existence and that their stability

Me^a Me^b X   Me   Me^a Me^b Me   X

Ph   Sn   Ph   Sn

Ph   Ph   Ph   Ph

(S)-(29) X = Cl   (R)-(29) X = Cl
(R)-(30) X = H   (S)-(30) X = H

FIGURE 4. Configurational stability of chiral organostannanes according to ^{1}H NMR analysis of diastereotopic methyl signals

would depend upon the substituents on the tin atom[15]. From chemical shift differences of diastereotopic protons in appropriately substituted triorgano halo- and hydridostannanes such as **29** and **30**, it proved possible to study stereochemical inversion as a function of temperature and solvent (Figure 4)[16].

It was found that the geminal methyl substituents of chlorostannane **29** are nonequivalent in nonpolar solvents such as benzene, toluene or CCl$_4$ at concentrations less than 0.2 M. At higher concentration peak coalescence was observed, indicating rapid interconversion of the enantiomers on the NMR time scale. The addition of DMSO, acetone or HCl also caused coalescence, even in nonpolar media. It was concluded that inversion at tin in **29** was occurring by self-association in nonpolar solvents or through ligand addition. In each case, a transient five-coordinated Sn species is a likely intermediate.

The stannyl hydride **30**, in contrast, showed nonequivalence of the geminal methyls for the neat sample up to 222 °C. Furthermore, no noticeable line broadening was observed in DMSO–dioxane at 160 °C. Thus, it may be concluded that the hydride **30** is configurationally stable while the chloride **29** is not.

2. Synthesis

In 1973, two laboratories announced the preparation of tetraorganotin compounds with measurable optical rotations[17,18]. In one of these, the diastereomeric menthyl esters, secured through addition of racemic NpArMeSnH to (−)-menthyl acrylate, were separated by column chromatography and purified by recrystallization to afford a pure diastereomer of [α]$_D$ − 24 (Scheme 12)[17]. Addition of MeMgI to this ester led to the chiral stannane **32** of high optical purity but unknown configuration.

The second synthesis involved preparation of the (−)-menthyloxystannane **33** and subsequent displacement of the menthyloxy grouping with benzylmagnesium chloride to afford stannane **34**, [α]$_D$ 4.6, of unknown configuration and enantiomeric purity (equation 6)[18].

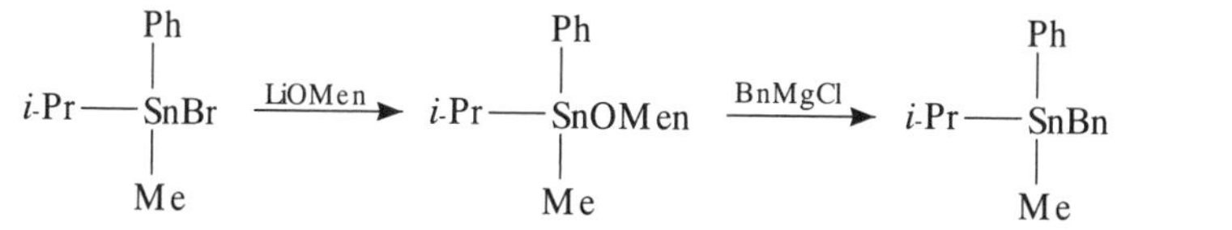

$$i\text{-}Pr \overset{Ph}{\underset{Me}{—Sn Br}} \xrightarrow{LiOMen} i\text{-}Pr \overset{Ph}{\underset{Me}{—SnOMen}} \xrightarrow{BnMgCl} i\text{-}Pr \overset{Ph}{\underset{Me}{—SnBn}} \quad (6)$$

(33)   (34) [α]$_D$ 4.6

The use of a chiral alcohol to prepare diastereomeric alkoxystannanes from racemic triorganostannyl halides, then displacement with a Grignard reagent, constitutes a general route to nonracemic tetraorganostannanes. Chinconine has proven particularly effective as the chiral alcohol (equation 7)[19].

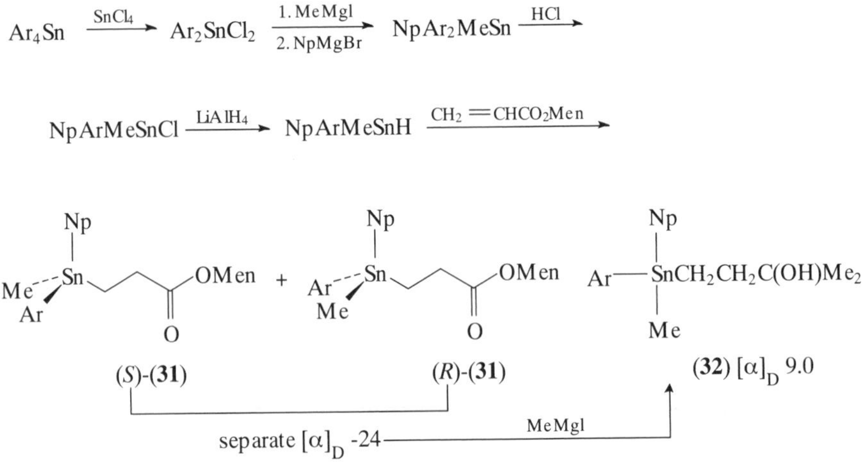

$$Ar_4Sn \xrightarrow{SnCl_4} Ar_2SnCl_2 \xrightarrow[2.\,NpMgBr]{1.\,MeMgl} NpAr_2MeSn \xrightarrow{HCl}$$

$$NpArMeSnCl \xrightarrow{LiAlH_4} NpArMeSnH \xrightarrow{CH_2=CHCO_2Men}$$

(S)-(31)   (R)-(31)   (32) $[\alpha]_D$ 9.0

separate $[\alpha]_D$ -24 $\xleftarrow{\qquad\qquad MeMgl \qquad\qquad}$

SCHEME 12. Np = 1-naphthyl, Men = (−)-menthyl, Ar = p-MeOC$_6$H$_4$

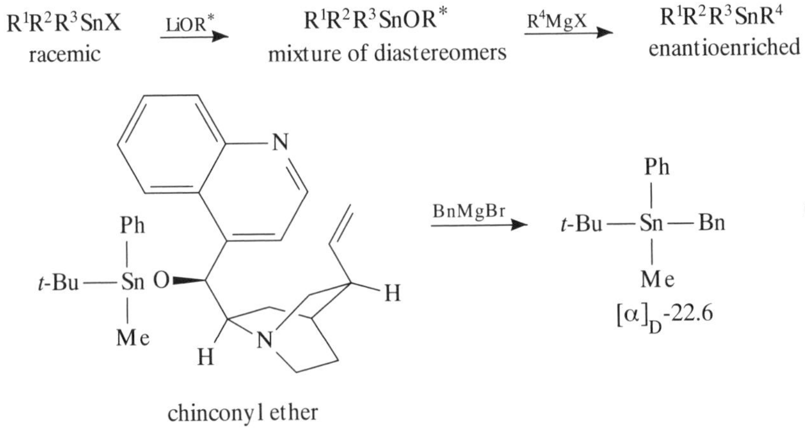

$$R^1R^2R^3SnX \xrightarrow{LiOR^*} R^1R^2R^3SnOR^* \xrightarrow{R^4MgX} R^1R^2R^3SnR^4$$

racemic   mixture of diastereomers   enantioenriched

$$\xrightarrow{BnMgBr} \quad t\text{-Bu}-\overset{\overset{\displaystyle Ph}{|}}{\underset{\underset{\displaystyle Me}{|}}{Sn}}-Bn \qquad (7)$$

$$[\alpha]_D\text{-}22.6$$

chinconyl ether

Enantionmerically enriched pairs of p-stannylated benzoic acids (+)/(−)-**35** and (+)/(−)-**36** were obtained by resolution with brucine or strychnine (Scheme 13)[20]. Again, neither the configuration nor the enantiomeric enrichment could be determined.

A series of nonracemic triorganotin hydrides (**37**–**39**) was prepared by reduction of racemic halostannane precursors with a chiral alkoxy hydride reagent (Scheme 14)[21].

Stannane **37** lost 50% of its optical activity when allowed to stand as a 0.2 M solution in benzene for 17 days. Addition of AIBN to the solution at 80 °C caused complete racemization after 30 min. With added hydroquinone, the benzene solution at 80 °C showed no decrease in rotation after 2 h. It was thus concluded that racemization proceeds by homolysis.

Racemization of stannane **37** was also observed in donor solvents as shown in Table 2. Although stable for short periods in phenethylamine and acetonitrile, stannane **37** is appreciably racemized in DMSO, HMPA and especially MeOH. Pentacoordinated tin species are likely intermediates. Interestingly, exposure to HMPA leads to significant formation

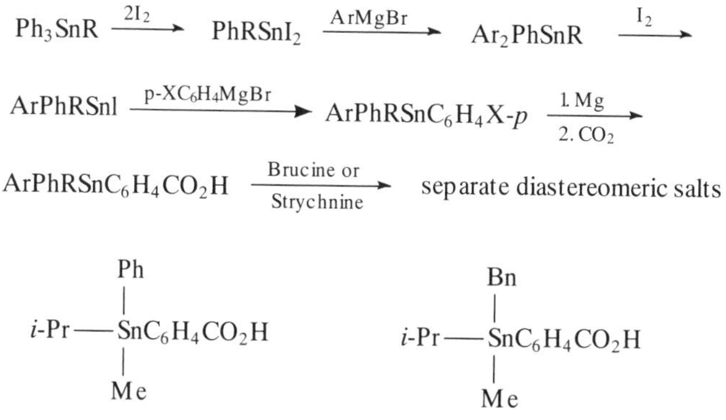

$$Ph_3SnR \xrightarrow{2I_2} PhRSnI_2 \xrightarrow{ArMgBr} Ar_2PhSnR \xrightarrow{I_2}$$

$$ArPhRSnI \xrightarrow{p\text{-}XC_6H_4MgBr} ArPhRSnC_6H_4X\text{-}p \xrightarrow[2.\,CO_2]{1.\,Mg}$$

$$ArPhRSnC_6H_4CO_2H \xrightarrow[\text{Strychnine}]{\text{Brucine or}} \text{separate diastereomeric salts}$$

$$i\text{-Pr}\overset{\displaystyle Ph}{\underset{\displaystyle Me}{\overline{}Sn}}C_6H_4CO_2H \qquad\qquad i\text{-Pr}\overset{\displaystyle Bn}{\underset{\displaystyle Me}{\overline{}Sn}}C_6H_4CO_2H$$

(**35**) $[\alpha]_D$ 0.37 and -0.50   (**36**) $[\alpha]_D$ 0.73 and -0.60

SCHEME 13. Ar = p-MeOC$_6$H$_4$

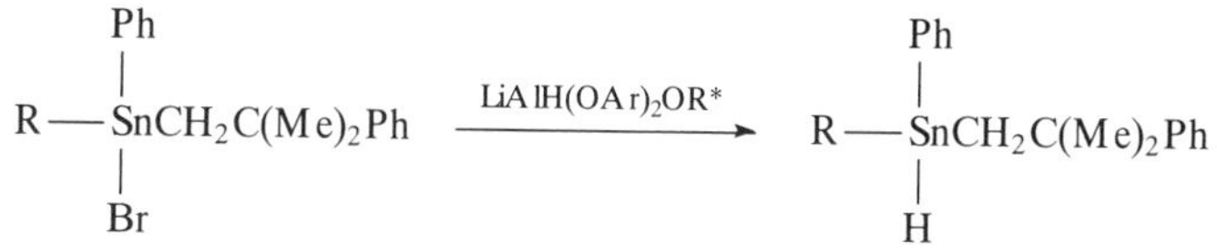

$$R\overset{\displaystyle Ph}{\underset{\displaystyle Br}{\overline{}Sn}}CH_2C(Me)_2Ph \xrightarrow{LiAlH(OAr)_2OR^*} R\overset{\displaystyle Ph}{\underset{\displaystyle H}{\overline{}Sn}}CH_2C(Me)_2Ph$$

(**37**) R = Me $[\alpha]_{365}$ 6.4

(**38**) R = t-Bu $[\alpha]_{365}$ -3.5

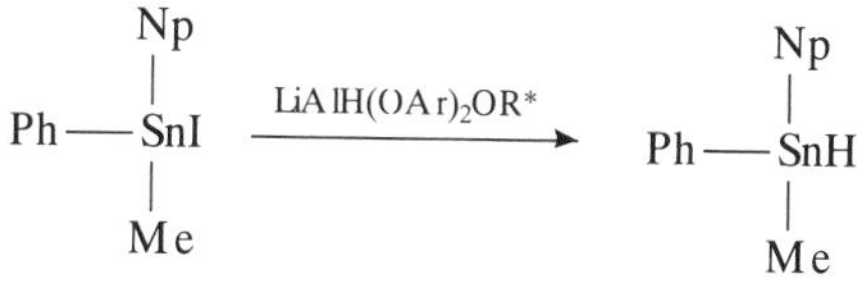

$$Ph\overset{\displaystyle Np}{\underset{\displaystyle Me}{\overline{}Sn}}I \xrightarrow{LiAlH(OAr)_2OR^*} Ph\overset{\displaystyle Np}{\underset{\displaystyle Me}{\overline{}Sn}}H$$

(**39**) $[\alpha]_{365}$ 2.0

SCHEME 14. Ar = 3,5-Me$_2$C$_6$H$_3$, R* = (−)-Me$_2$NCH(Me)CH(Ph)

of the racemic hexaorganodistannane product. Hydrides **38** and **39** were found to be less readily racemized in benzene than **37**.

Hydride **38** is converted to the methyl derivative **40** upon treatment with diazomethane in the presence of Cu (equation 8)[21]. Presumably this insertion reaction proceeds with retention of configuration.

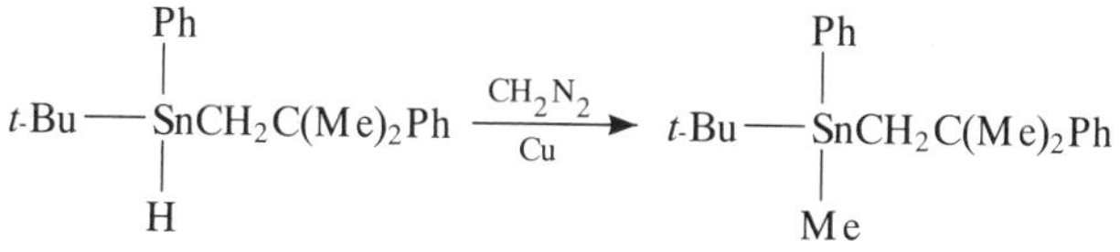

$$t\text{-Bu}\overset{\displaystyle Ph}{\underset{\displaystyle H}{\overline{}Sn}}CH_2C(Me)_2Ph \xrightarrow[Cu]{CH_2N_2} t\text{-Bu}\overset{\displaystyle Ph}{\underset{\displaystyle Me}{\overline{}Sn}}CH_2C(Me)_2Ph \qquad (8)$$

(**38**) $[\alpha]_{365}$ -0.9   (**40**) $[\alpha]_{365}$ 1.5

TABLE 2. Racemization of stannane **37**

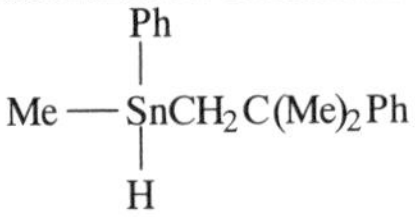

$$Me - \underset{\underset{H}{|}}{\overset{\overset{Ph}{|}}{Sn}} CH_2 C(Me)_2 Ph$$

(37) $[\alpha]_{365}$ 6.4

Solvent	t (h)	Racemization (%)
PhCHMeNH$_2$	2	0
MeCN	1	0
DMSO	1	51
HMPA	1	90
MeOH	<1	100

Finally, it should be noted that several chiral tetraorganostannanes have been partially resolved by column chromatography on microcrystalline cellulose triacetate[22].

3. Interconversions

The reaction of hydride **37** with CCl$_4$ affords the chlorostannane **41**. When this reaction was followed by ORD, an initial increase in optical rotation was observed until 65–70% conversion — at which point the rotation gradually decreased to zero following first-order kinetics (equation 9)[21].

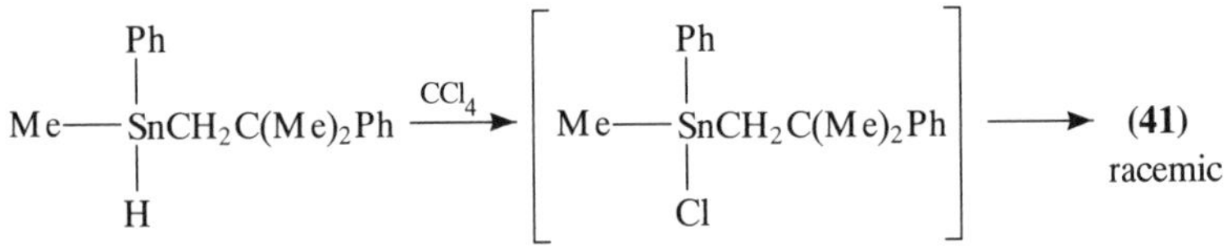

$$Me - \underset{\underset{H}{|}}{\overset{\overset{Ph}{|}}{Sn}} CH_2 C(Me)_2 Ph \xrightarrow{CCl_4} \left[Me - \underset{\underset{Cl}{|}}{\overset{\overset{Ph}{|}}{Sn}} CH_2 C(Me)_2 Ph \right] \longrightarrow \underset{racemic}{\textbf{(41)}} \qquad (9)$$

(37) $[\alpha]_{365}$ 14   (+)-**(41)** $[\alpha]_{365}$ 28

Chlorination does not occur in the presence of hydroquinone, hence a radical process is implicated. It is inferred that the stannyl radical retains configuration en route to chloride (+)-**41**, which then racemizes by a dissociative process. It is estimated that racemization of chloride (+)-**41** in CCl$_4$ at 0.18 M concentration requires about 10 min at room temperature.

The deuteride (+)-**37D** undergoes exchange at 40 °C with Ph$_3$SnH, affording the hydride (+)-**37** with an estimated 93% retention of configuration (equation 10)[21,23]. A chiral radical intermediate is probably involved here as well. ESR studies of hindered triarylstannyl radicals suggest a nonplanar arrangement with a 14–15° out-of-plane angle for the aryl substituents[24].

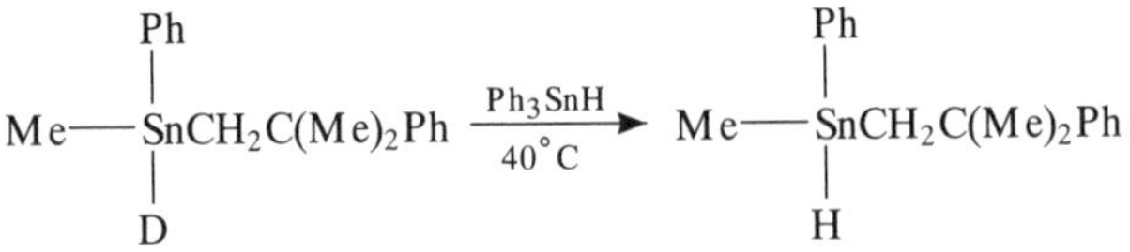

$$Me - \underset{\underset{D}{|}}{\overset{\overset{Ph}{|}}{Sn}} CH_2 C(Me)_2 Ph \xrightarrow[40°C]{Ph_3SnH} Me - \underset{\underset{H}{|}}{\overset{\overset{Ph}{|}}{Sn}} CH_2 C(Me)_2 Ph \qquad (10)$$

(37D) $[\alpha]_{365}$ 5.8   **(37)** $[\alpha]_{365}$ 5.0

TABLE 3. Stereochemistry of nucleophilic substitution at tin

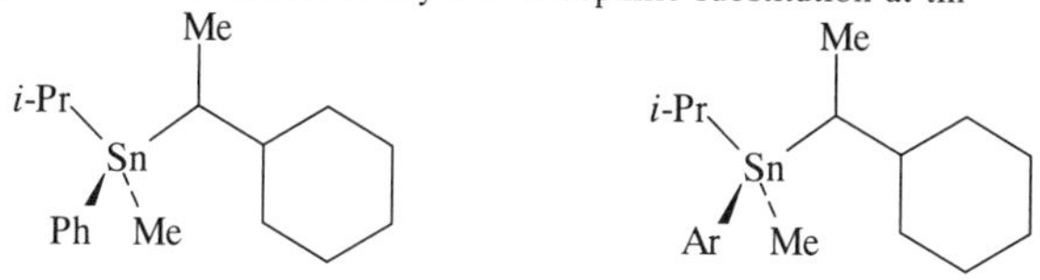

(R^*)-(**44**) $p\text{-}CH_3C_6H_4Li$ → (R^*)-(**45**)

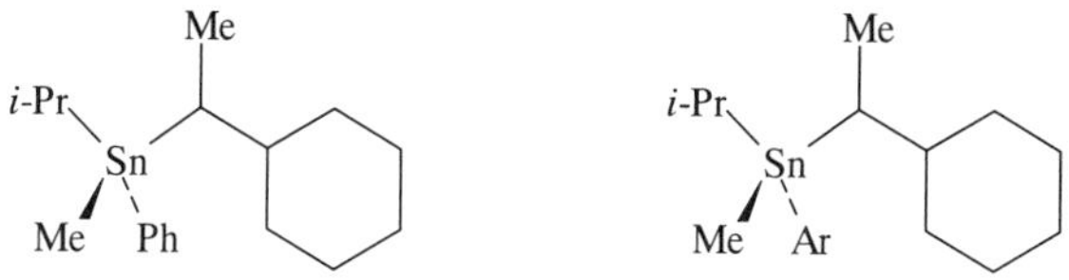

(S^*)-(**44**) (S^*)-(**45**)

$(R^*):(S^*)$-(**44**)	Solvent	$T(^\circ C)$	$(R^*):(S^*)$-(**45**)
6:1	Et$_2$O	27	6:1
4:1	THF	−75	1.1:1

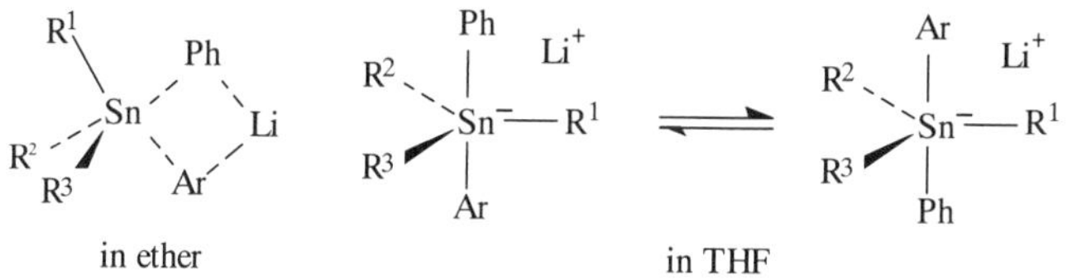

in ether   in THF

FIGURE 5. Ligand exchange mechanisms in tetraorganostannanes

Reich and coworkers have studied the exchange of organolithium species with tetraorganostannanes and found that reaction of the bicyclic stannane **42** with CD$_3$Li proceeds with retention in ether (equation 11). Known mixtures of diastereomeric acyclic stannanes R^*-**44** and S^*-**44**, of unknown configuration at tin, underwent Ph/p-tolyl exchange with clean retention in ether upon treatment with p-tolyllithium. In THF, however, racemization at tin was observed (Table 3). In the former case, a configurationally stable four-center transition state is proposed. Reaction in THF is thought to proceed through a long-lived configurationally labile pentacoordinated ate complex (Figure 5).

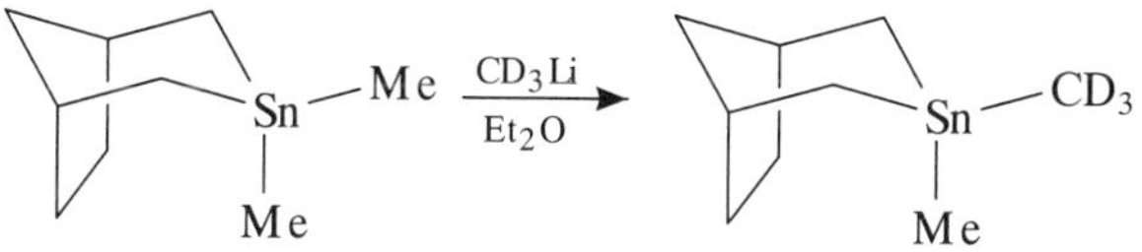

(11)

(**42**)   (**43**)

4. Distannanes

Upon exposure to 10% Pd/C in pentane, the nonracemic stannyl hydride **37** is converted to a mixture of meso and enantioenriched distannanes **46**[25]. Dimerization of **37** can also be achieved by treatment with $LiAlH_4$ or Me_2Hg, but these reactions lead to mixtures of meso and racemic distannanes (equation 12).

$$
\underset{\substack{| \\ H \\ \textbf{(37)} \; [\alpha]_{365} \; 13.2}}{\overset{\substack{Ph \\ |}}{Me\!\!-\!\!SnCH_2C(Me)_2Ph}}
\xrightarrow{\;Pd/C\;}
\underset{\substack{| \\ Ph \\ \textbf{(46)} \; [\alpha]_{365} \, \text{-}28.9}}{\overset{\substack{Ph \\ | \\ Me\!\!-\!\!SnCH_2C(Me)_2Ph \\ | \\ Me\!\!-\!\!SnCH_2C(Me)_2Ph}}{}}
\tag{12}
$$

5. Pentacoordinated triorgano halostannanes

The configurational stability of triorganotin halides is considerably enhanced by the presence of an amine ligand that can coordinate intramolecularly with the tin atom. This was demonstrated by analysis of the ^{1}H NMR spectrum of the stannyl bromide **47** depicted in Figure 6[26]. Below 30 °C, both the N-methyl and the benzylic protons are diastereotopic; above 30 °C, the N-methyl signals coalesce, indicating rapid dissociation to S-**48** or R-**48**.

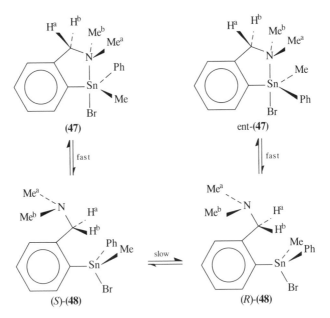

FIGURE 6. Inversion of pentacoordinated triorganotin halides **47** and ent-**47** according to ^{1}H NMR analysis

TABLE 4. Equilibration of diastereomeric pentacoordinated tri-organo tin halides

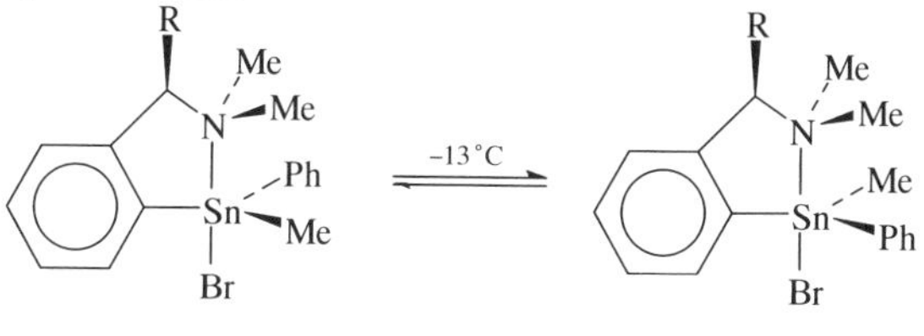

(49)   **(50)**

R	49:50
Me	60:40
Et	60:40
i-Pr	80:20
t-Bu	>98:2

However, the benzylic protons remain nonequivalent to at least 125 °C, suggesting that interconversion of S-**48** and R-**48** is slow on the NMR time scale.

The diastereomeric pentacoordinated bromostannanes **49** and **50** were found to equilibrate at −13 °C in toluene to a mixture of diastereomers[27]. The composition of the equilibrium mixture is dependent upon the R substituent (Table 4). As the size of this group is increased, the equilibrium is shifted in favor of the sterically less congested diastereomer **49**.

Similarly, the (−)-menthyl substituted stannyl bromides **51** and **53**, whose diastereomeric purity was confirmed by single crystal X-ray structure analysis, were found to slowly equilibrate to nearly 1:1 mixtures of diastereoisomers on standing in solution (equation 13)[28].

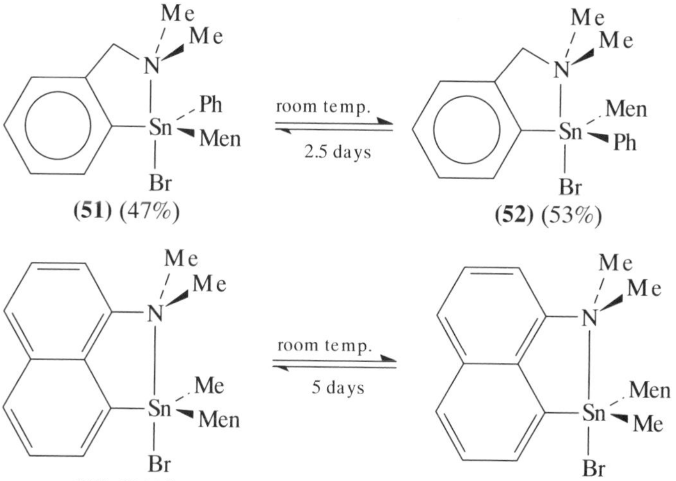

(51) (47%)   **(52)** (53%)

(53) (46%)   **(54)** (54%)   (13)

C. Lead

To date, few chiral organolead compounds have been prepared[29]. The route outlined in Scheme 15 for the preparation of plumbane **55** is similar to the approach employed for the pentacoordinated organostannanes **47**–**54** described in the previous section. The X-ray crystal structure of **55** resembles that of the diphenyltin analogue but the structures are not isomorphous[30]. The configurational stability of **55** could not be determined because both the *N*-methyl and benzylic H signals appear as singlets in the ^{1}H NMR spectrum. This is attributed, at least in part, to the small disymmetry of **55**. Presumably, replacing one of the aryl groupings with an alkyl or a naphthyl substituent would enable studies comparable to those described for the analogous stannanes to be conducted.

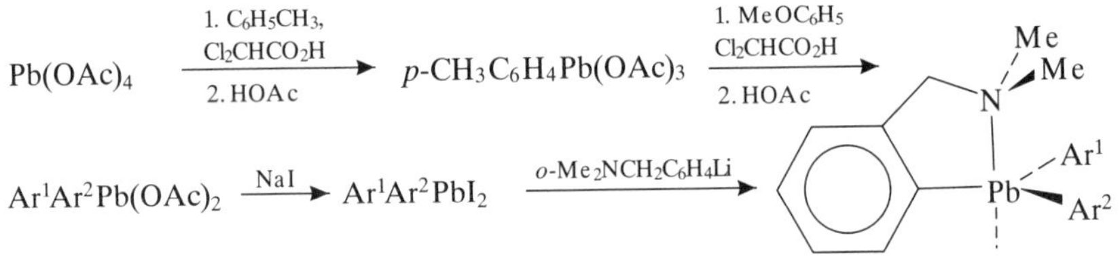
1. C6H5CH3,
Cl2CHCO2H
2. HOAc
Pb(OAc)4
p-CH3C6H4Pb(OAc)3
1. MeOC6H5
Cl2CHCO2H
2. HOAc
Ar1Ar2Pb(OAc)2
NaI
Ar1Ar2PbI2
o-Me2NCH2C6H4Li
Me
Me
N
Pb
Ar1
Ar2

(55)

SCHEME 15. $Ar^1 = p\text{-}MeC_6H_4$, $Ar^2 = p\text{-}MeOC_6H_4$

III. GERMACYCLOPENTANES AND CYCLOHEXANES

A. Cyclopentanes

A *cis*/*trans* mixture of 1,2-dimethylgermacyclopentanes **56** and **57** was prepared as outlined in Scheme 16[32]. The isomeric germanes are stable and can be separated by spinning band distillation. Stereochemical assignments were made by comparison of the chemical shifts of the methyl carbons in the ^{13}C NMR spectra with those of the analogous dimethylcyclopentanes and silacyclopentanes.

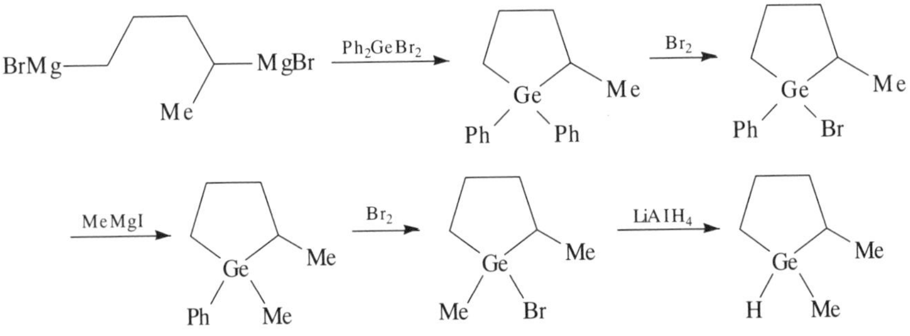
BrMg
MgBr
Me
Ph2GeBr2
Ge
Ph
Ph
Me
Br2
Ge
Ph
Br
Me
MeMgI
Ge
Ph
Me
Me
Br2
Ge
Me
Br
Me
LiAlH4
Ge
H
Me
Me

(56) *trans*
(57) *cis*

SCHEME 16

Hydrogermylations by these hydrides were found to proceed with retention of stereochemistry at Ge (equation 14)[32].

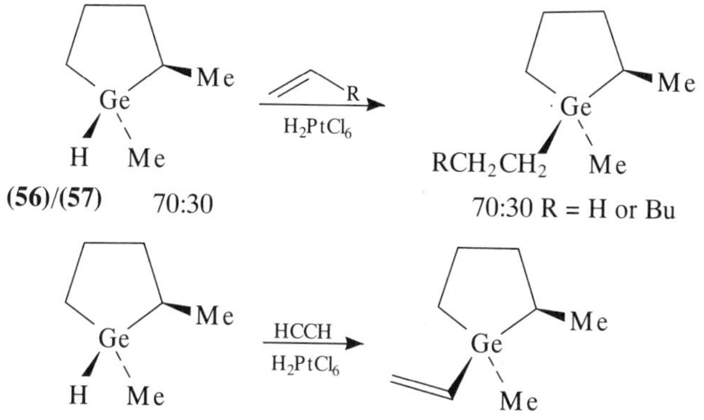

(56)/(57)   70:30   70:30 R = H or Bu

(14)

(56)/(57)   70:30   70:30

Chlorination with Cl_2, CCl_4, SO_2Cl_2 or $ClCH_2OMe$ also took place with predominant retention of configuration[33]. Reduction of the chlorogermanes with lithium aluminum hydride proceeded mainly with inversion to complete the Walden cycle (equation 15). Bromination, on the other hand, led to a nearly 1:1 mixture of *cis* and *trans* bromogermanes.

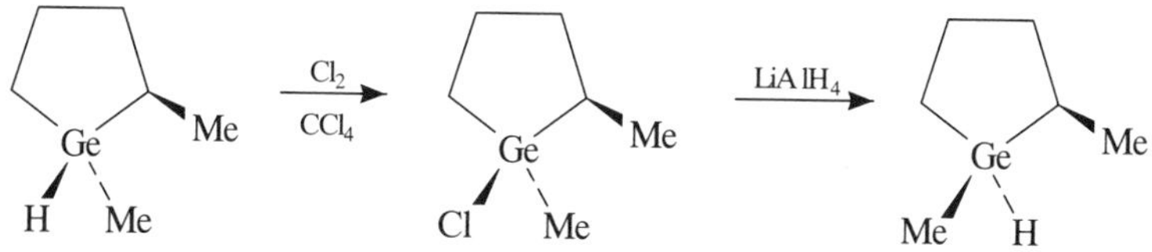

(56)/(57)   80:20   70:30   (57)/(56)   65:35

A series of studies was carried out on various 1,2-dimethylgermacyclopentanes bearing electronegative substituents at germanium in order to determine the steric course of substitution reactions by various hydrides and Grignard or organolithium reagents (Figure 7)[34]. It was found that inversion was most favored for the chlorogermanes, whereas the amino and phosphino derivatives underwent substitution with predominant retention. Of the hydrides, $LiBH_4$ most favored inversion while DIBAH tended to give mainly retention. Allylmagnesium bromide showed the highest preference for inversion; butyllithium tended toward retention. The overall trends are comparable to those observed with chiral silanes.

Mechanisms involving pentacoordinated germanes were proposed for these substitutions (Figure 8).

B. Cyclohexanes

Takeuchi and coworkers prepared a series of methyl and phenyl substituted germacyclohexanes to evaluate conformational preferences of these substituents (Scheme 17)[35,36]. Based on analysis of ^{13}C NMR spectra and molecular mechanics calculations, they concluded that a C-methyl prefers the equatorial orientation by *ca* 1.4 kcal mol^{-1} but the Ge-methyl substituent actually shows a slight intrinsic preference for the axial orientation (Figure 9). A similar conclusion was reached for the Ge-phenyl substituent.

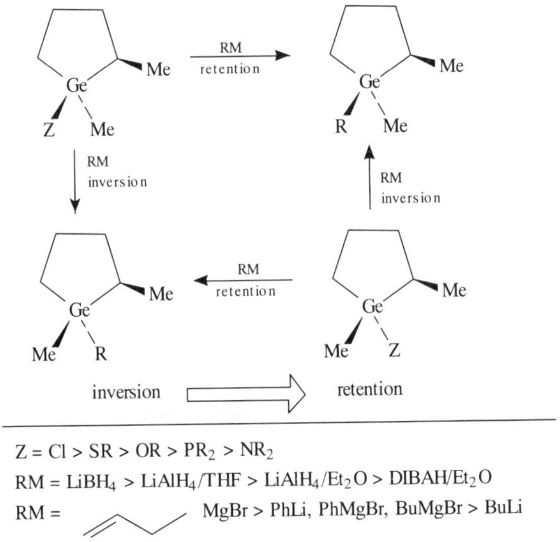

$Z = Cl > SR > OR > PR_2 > NR_2$

$RM = LiBH_4 > LiAlH_4/THF > LiAlH_4/Et_2O > DIBAH/Et_2O$

$RM = $ ⟋⟍⟋ $MgBr > PhLi, PhMgBr, BuMgBr > BuLi$

FIGURE 7. Trends in nucleophilic displacements on germacyclopentanes

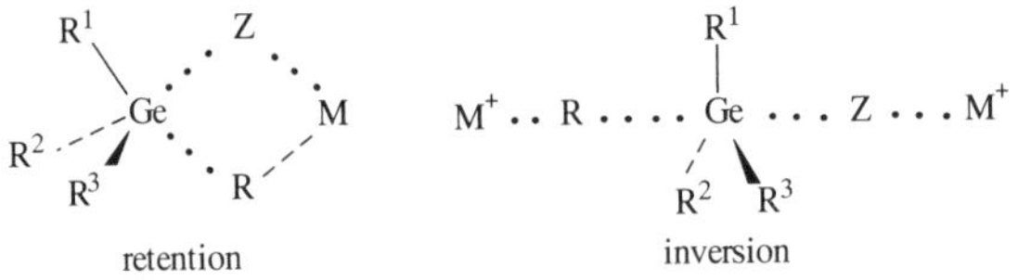

FIGURE 8. S_N mechanisms in triorganogermyl Z systems

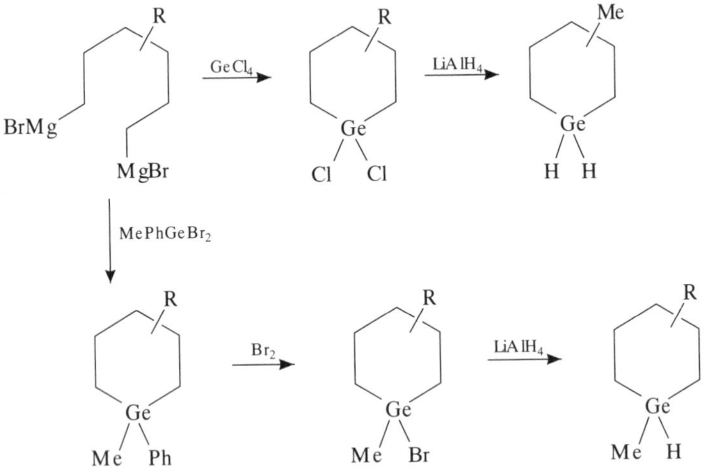

SCHEME 17. R = H or Me

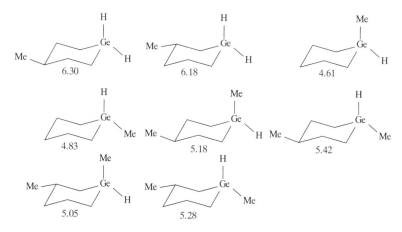

FIGURE 9. Calculated energies (kcal mol^{-1}) of methyl and dimethylgermacyclohexanes

IV. ORGANOMETALLICS OF TYPE R*MR₃

A. Chiral Acyclic Systems

In the first systematic study on nucleophilic substitutions of chiral halides by Group IV metal anions, Jensen and Davis showed that (S)-2-bromobutane is converted to the (R)-2-triphenylmetal product with predominant inversion at the carbon center (Table 5)[37]. Replacement of the phenyl substituents by alkyl groups was possible through sequential brominolysis and reaction of the derived stannyl bromides with a Grignard reagent (equation 16). Subsequently, Pereyre and coworkers employed the foregoing Grignard sequence to prepare several trialkyl(s-butyl)stannanes (equation 17)[38]. They also developed an alternative synthesis of more hindered trialkyl derivatives (equation 18).

$$\text{SnPh}_3 \xrightarrow{2\,\text{Br}_2} \text{Ph}\diagdown\text{SnBr}_2 \xrightarrow{2\,\text{RMgBr}} \text{Ph}\diagdown\text{SnR}_2 \xrightarrow{\text{Br}_2}$$

$$(16)$$

$$\text{Br}\diagdown\text{SnR}_2 \xrightarrow{\text{RMgBr}} \text{SnR}_3$$

$$\text{SnPh}_3 \xrightarrow{3\,\text{Br}_2} \text{SnBr}_3 \xrightarrow{\text{RMgX}} \text{SnR}_3$$

$$(17)$$

R = i-Pr, Bu, i-Bu, s-Bu

$$\text{SnPh}_3 \xrightarrow[\text{MeOH}]{\text{HCl}} \text{Cl}\diagdown\text{SnPh}_2 \xrightarrow{\text{RMgX}} \text{R}\diagdown\text{SnPh}_2 \xrightarrow{\text{repeat 2X}} \text{SnR}_3$$

R = Me₃CCH₂, Et₂CH

$$(18)$$

TABLE 5. S_N2 displacements on (S)-2-bromobutane by Ph$_3$MNa

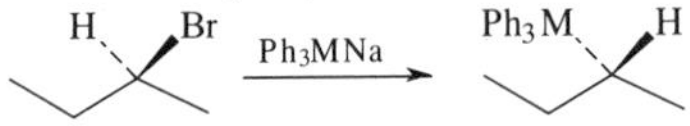

M	$[\alpha]_D$	Inversion (%)
Ge	−2.33	67
Sn	−3.54	88
Pb	−4.16	67

Displacement of 2-chlorobutane by trimethyl- and tributylstannylsodium was found to proceed with nearly complete inversion of configuration (equation 19). Displacements on 2-octyl tosylate, chloride and bromide by Me$_3$SnLi and Me$_3$SnNa afforded inverted product with varying degrees of enantiomeric purity, depending on reaction conditions (equation 20)[39]. Addition of the bromide to solutions of the stannane gave product of 34–83% ee. Selectivity was better for Li than for Na. Higher selectivity was also observed at lower temperatures. Addition of the Me$_3$SnM to the bromide, on the other hand, resulted in product of 93–100% ee. The tosylate and chloride displacements were relatively insensitive to reaction conditions. Product of 90% ee or greater was secured irrespective of conditions. It is suggested that a free radical process can occur with the bromide but not with the chloride or tosylate.

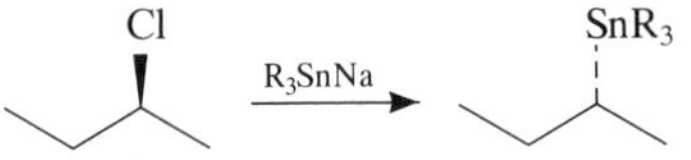

$$R = Me, Bu$$ (19)

(20)

X = OTs, Cl, Br

The corresponding reaction with Ph$_3$SnM proceeds with complete inversion regardless of addition mode (equation 21). In this case, radical reactions appear unimportant. Displacement of the tosylate by a trimethyltin cuprate was also found to take place by inversion.

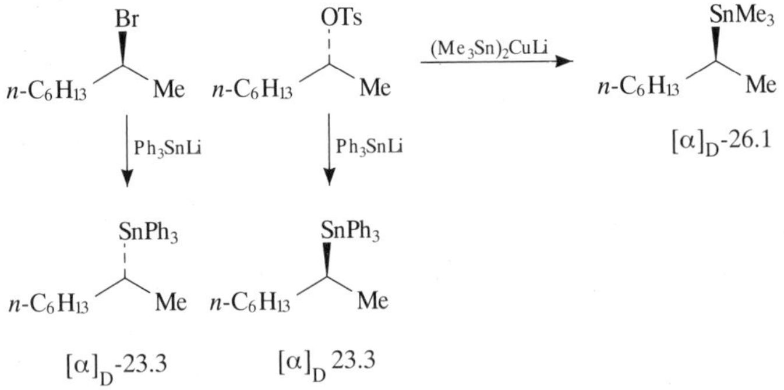

(21)

A comparison of displacements on (R)-2-octyl tosylate, chloride and bromide by Ph$_3$CLi, Ph$_3$SiLi, Ph$_3$GeLi and Ph$_3$SnLi is summarized in Table 6[40]. In each case, the S_N2 product is formed with inversion. The basicity of the triphenylmethyl anion caused

TABLE 6. S_N2 displacements on 2-octyl derivatives by Ph_3MLi

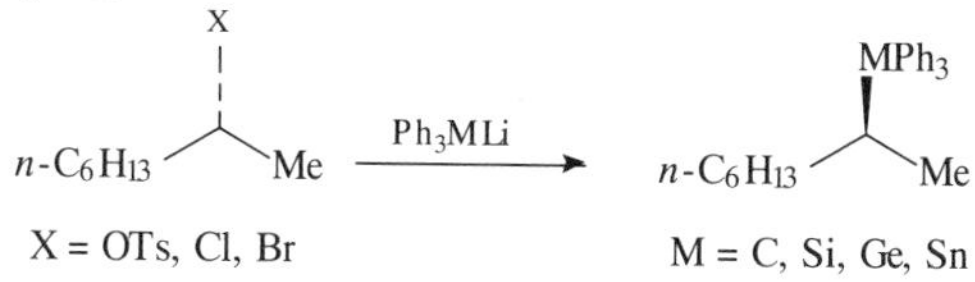

$$n\text{-}C_6H_{13} \overset{X}{\underset{Me}{\diagup}} \quad \xrightarrow{Ph_3MLi} \quad n\text{-}C_6H_{13} \overset{MPh_3}{\underset{Me}{\diagup}}$$

X = OTs, Cl, Br   M = C, Si, Ge, Sn

Rate for X = Cl	Yield of S_N2 product
M = C ~ Si > Ge > Sn	M = C; Cl > OTs > Br[a]
	M = Si; Cl > OTs > Br[b]
	M = Ge; Br > OTs > Cl
	M = Sn; Br > OTs > Cl

[a] Mainly elimation product.
[b] Mainly n-octane.

extensive elimination. The triphenylsilyl anion effected hydrogenolysis to n-octane by an unknown pathway. Reactions of the triphenylgermyl and triphenylstannyl anions gave mainly the S_N2 products. In each case, the bromide proved more reactive than the tosylate or chloride.

B. Cyclohexyl Systems

Traylor and coworkers examined S_N displacements of some cyclohexyl derivatives with Me_3SnLi to determine the stereochemistry of such reactions (Scheme 18)[41]. cis-4-t-Butylcyclohexyl bromide underwent substitution with retention of configuration, as did 1-bromoadamantane and 1-chlorocamphane[42]. However, trans-4-t-butylcyclohexyl tosylate was found to react with inversion to give the cis product 56. The trans isomer 57 was prepared by addition of Me_3SnCl to 4-t-butylcyclohexylmagnesium bromide. The two isomers showed distinctive differences for the Me_3Sn protons in the ^1H NMR spectra. A mechanism involving halogen–metal interchange was proposed for the retention pathway. Somewhat conflicting results were reported by Kitching and coworkers[43]. They found that cis-4-methyl- and cis-4-t-butylcyclohexyl bromide gave roughly 2:1 mixtures of trans and cis substitution products with Me_3SnLi, suggesting predominant inversion of configuration (Scheme 19).

Interestingly, the analogous reaction with Me_3GeLi proceeded mainly with retention to give a ca 70:30 mixture of cis and trans substitution products. The inverted products were assumed to arise through backside displacement of bromide. A two-step metal–halogen exchange mechanism was proposed for the cis products, but the possibility of electron transfer leading to radical intermediates could not be ruled out.

Both Me_3GeLi and Me_3SnLi react with cyclohexene oxide to give the trans product (equation 22)[44]. These reactions most likely proceed by S_N2 mechanisms.

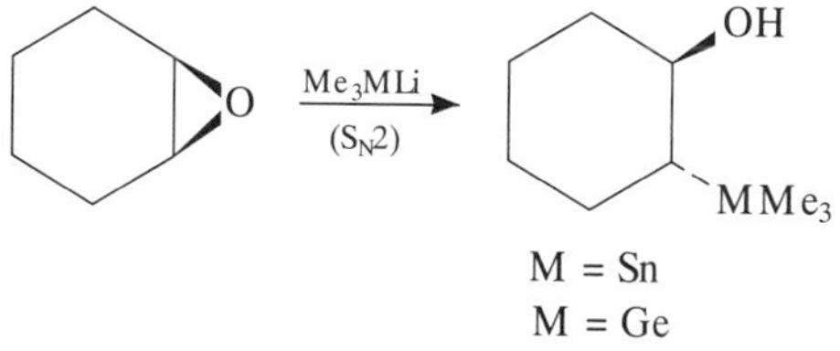

$$\text{(cyclohexene oxide)} \xrightarrow[\text{(S_N2)}]{Me_3MLi} \text{(trans product with OH and MMe}_3\text{)}$$

M = Sn
M = Ge

(22)

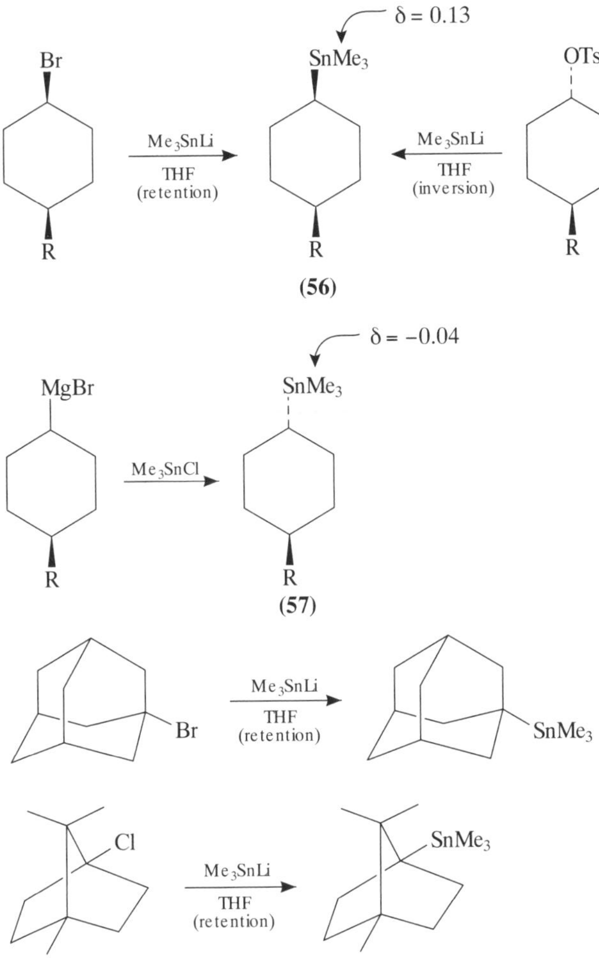

SCHEME 18. R = t-Bu

Evidence for a free radical pathway in the foregoing cycloalkyl bromide reactions was secured by San Filippo and coworkers, who found that *cis* and *trans*-4-t-butylcyclohexyl bromide afford nearly identical product mixtures with Me_3SnLi, Me_3SnNa and Me_3SnK under a given set of conditions (Table 7)[45]. Of the various combinations examined, that of *cis*-bromide and Me_3SnLi appeared to be most favorable to S_N2 displacement.

Reactions of the corresponding tosylates with Me_3SnLi proceeded in low yield, but products of inverted configuration were produced. The *cis* and *trans* chlorides led to extensive loss of configuration at carbon with a slight preference for inversion.

In support of a radical pathway for such reactions, both cyclopropylcarbinyl bromide and iodide gave rise to appreciable homoallylic substitution product with the foregoing metallostannanes. In contrast, the corresponding chloride and tosylate gave only the unrearranged product.

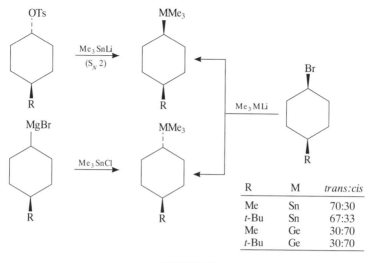

SCHEME 19

R	M	trans:cis
Me	Sn	70:30
t-Bu	Sn	67:33
Me	Ge	30:70
t-Bu	Ge	30:70

TABLE 7. Reaction of *cis*- and *trans*-4-*t*-butylcyclohexyl bromides with Me₃SnM reagents

cis-bromide

M	Yield (%)	0 °C trans:cis	−70 °C trans:cis
Li	76	70:30	82:18
Na	79	61:39	64:36
K	71	65:35	68:32

trans-bromide

M	Yield (%)	0 °C trans:cis	−70 °C trans:cis
Li	86	74:26	70:30
Na	59	64:36	64:36
K	88	64:36	65:35

An analogous set of experiments with *cis*- and *trans*-4-methylcyclohexyl bromide and the germyl nucleophiles Me₃GeLi and Ph₃GeLi revealed a similar scenario (Table 8)[6]. In this case, the *cis* products were favored. Reaction of these germyllithium reagents with 6-bromo-1-heptene led to mixtures of S_N2 product and (2-methylcyclopentyl)methylgermanes. The latter are products of free radical cyclization.

TABLE 8.   Conversion of *cis*- and *trans*-cyclohexyl bromides to germane derivatives

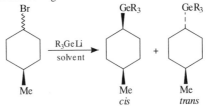

cis-bromide		
R	Solvent	*cis*:*trans*
Me	HMPA	70:30
Ph	THF	55:45

trans-bromide		
R	Solvent	*cis*:*trans*
Me	HMPA	70:30
Ph	THF	66:34

These studies were subsequently extended to the analogous stannanes and the 3-methyl and 2-methylcyclohexyl bromides (Table 9)[47]. The Me₃SnLi reactions, as expected, afforded similar ratios of isomeric product from a given set of stereoisomeric bromides. However, Ph₃SnLi led to clean inversion with 4-methyl and *trans*-3-methylcyclohexyl bromides. Evidently, the Ph₃Sn radical is not easily formed under these conditions.

The presence of Me₃Sn and Me₃Ge radicals was ascertained by reactions of the lithio reagents with cyclopropylcarbinyl bromide and 6-bromo-1-heptene, whereupon products of free radical reactions were produced along with those from direct displacement.

TABLE 9.   Stereochemistry of substitution reactions on methylcyclohexyl bromides by Me₃SnLi and Me₃GeLi

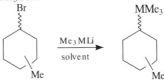

Bromide	M	Solvent	*trans*:*cis*
trans-1,4	Sn	THF	68:32
cis-1,4	Sn	THF	70:30
trans-1,3	Sn	THF	24:76
cis-1,3	Sn	THF	33:67
trans-1,2	Sn	THF	85:15
cis-1,2	Sn	THF	90:10
trans-1,4	Ge	HMPA	27:73
cis-1,4	Ge	HMPA	30:70

C. Allylic Systems

Displacements on allylic chlorides by stannyl and germyllithium reagents have been found to proceed with high regio- and stereoselectivity. Thus, *cis-* and *trans-*5-methyl-3-chlorocyclohexene, as unequal mixtures of isomers, afforded like mixtures of inverted substitution products upon treatment with Ph_3SnLi or Me_3SnLi (Scheme 20)[48,49]. Evaluation of the $^{13}C/^{119}Sn$ coupling constant revealed a strong preference for the axial SnR_3 conformer in the *trans* product[49]. A consideration of substituent *A*-values ($CH_3 = 1.7$, $Me_3Sn = 1.0$, $Ph_3Sn = 1.5$) is in accord with this finding—particularly in the case of Me_3Sn. In addition, it is proposed that σ-π interactions should favor an axial carbon–tin bond. Nonetheless, the *cis* stannanes were found to prefer the depicted diequatorial conformation. Accordingly, such orbital interactions must make only modest contributions to the conformational preference of the stannyl substituent.

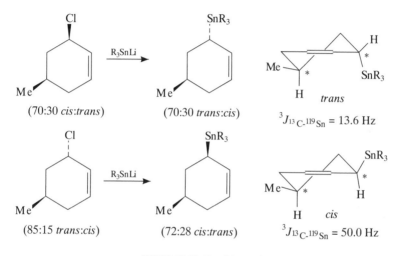

$^3J_{^{13}C-^{119}Sn} = 13.6$ Hz

$^3J_{^{13}C-^{119}Sn} = 50.0$ Hz

SCHEME 20. R = Me or Ph

The question of regiochemistry was addressed by displacements on deuterated 5-methyl-3-chlorocyclohexenes as exemplified in Scheme 21[49]. These experiments were rather complex, as reactions were performed on *cis/trans* mixtures of chlorides containing deuterium at both C-1 and C-5. However, careful analysis of the resulting product mixtures by ^1H, ^2H, ^{13}C and ^{119}Sn NMR showed that virtually all products were the result of S_N2 displacement[49].

Reactions of the isomeric 5-methyl-3-chlorocyclohexenes with Me_3GeLi (Scheme 22) were complicated by the formation of by-products resulting from double bond migration **58** and disproportionation **59** and **60**[49]. These side reactions could be minimized by careful control of conditions. The predominant process was found to be S_N2 displacement for both isomers, with the *trans* chloride showing somewhat higher specificity than the *cis* isomer.

D. β-Organometallo Ketones

Recently, it has been shown that Me_3GeLi undergoes conjugate addition to cyclohexenones to give β-Me_3Ge cyclohexanones (Scheme 23)[51]. The intermediate enolate can be trapped with MeI to afford the *trans* methylated product.

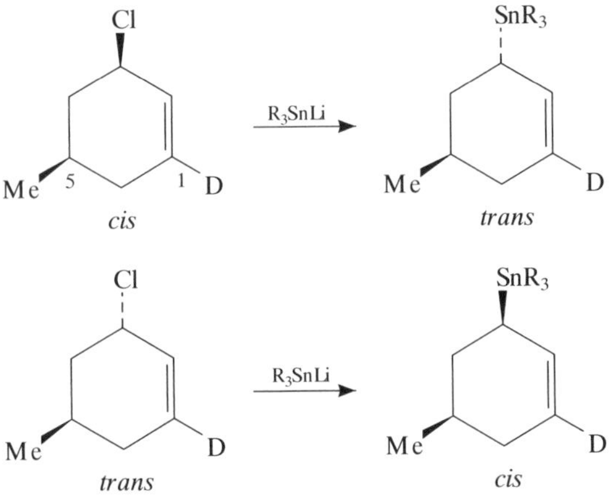

SCHEME 21. R = Me or Ph

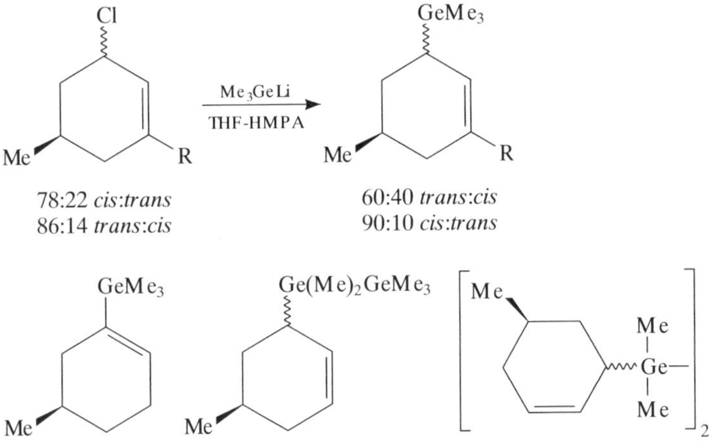

78:22 *cis:trans*   60:40 *trans:cis*
86:14 *trans:cis*   90:10 *cis:trans*

(58)   (59)   (60)

SCHEME 22. R = H or Me

It has been known for some time that trialkylstannyllithium reagents readily afford 1,4-adducts with conjugated ketones[52]. Kitching and coworkers found that Me₃SnLi adds exclusively 1,4 to 5-methyl-2-cyclohexen-1-one[53]. The resulting product is an 82:18 mixture of *trans* and *cis* isomers (Scheme 24). By analysis of vicinal $^{13}C-^{119}Sn$ coupling constants, they estimated the *trans* product exists as a 60:40 mixture favoring the axial CH_3 conformer. This finding is unexpected in view of the *A*-values for CH_3 vs Me_3Sn. It is suggested that the equatorial C—Sn bond experiences a stabilizing orbital interaction with the carbonyl grouping. The higher-order cyanocuprate, $Bu_3Sn(Bu)Cu(CN)Li_2$, also gives 1,4-adducts with conjugated ketones[54]. The reagent is highly selective for transfer of the Bu_3Sn moiety. Addition proceeds from the less hindered face of the double bond (equation 23).

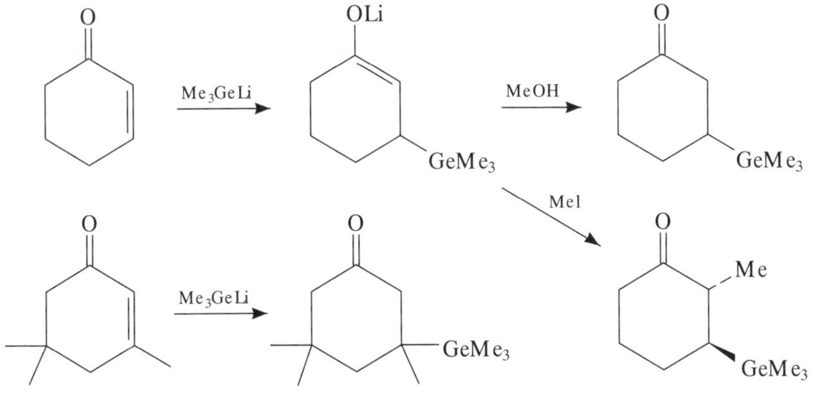

SCHEME 23

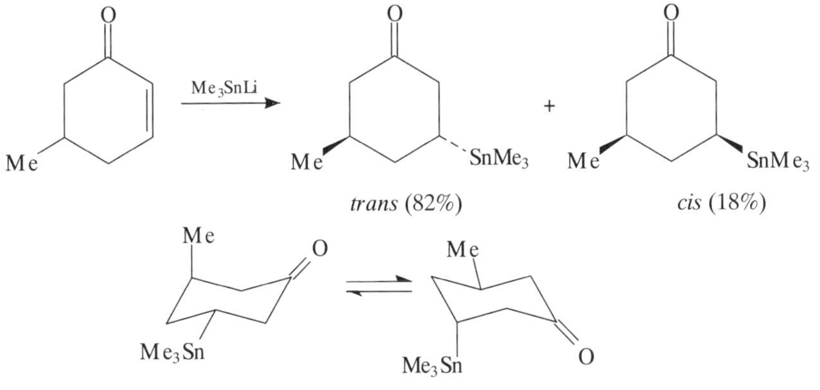

trans (82%)   *cis* (18%)

60:40

SCHEME 24

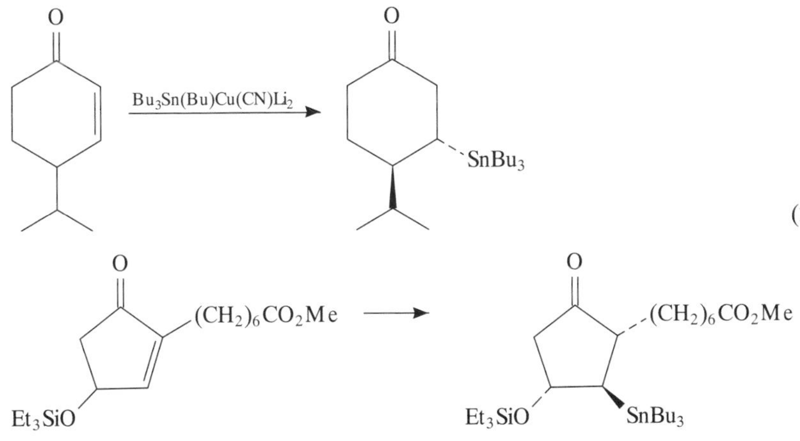

(23)

As might be expected, R_3PbLi compounds also react with enones by 1,4- addition (equation 24)[55].

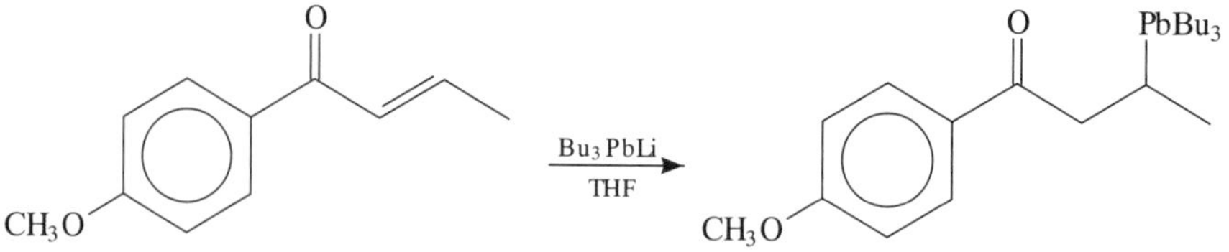

(24)

Tetraalkylstannanes show little affinity for Lewis base donor atoms. But when one of the alkyl substituents is replaced by an electron withdrawing group such as halogen, interactions with even weakly basic donor atoms become favorable. Thus, the cis-3-trialkylstannyl cyclohexyl benzoate 61 exists in the diequatorial conformation exclusively (Figure 10)[56]. However, the halostannane 62 adopts the diaxial arrangement. In this conformation, the ester oxygen forms a pentacoordinated donor–acceptor complex with the tin atom. Evidence for this interaction was obtained from the $^{13}C-^{119}Sn$ coupling constants and ^{119}Sn chemical shift data, with additional confirmation from single crystal X-ray structure analysis of the chloro derivative (X = Cl). The ^{13}C NMR spectrum of this compound showed nonequivalence of the two CH_3 carbons at $-28\,^{\circ}C$. These signals coalesced at $4\,^{\circ}C$. A dissociation–inversion process is postulated, with an activation energy of 13.8 kcal mol^{-1}.

As expected, both trans analogues 63 and 64 of stannanes 61 and 62 adopt the chair e/a conformation. The halostannane 64 shows no tendency toward intra- or intermolecular association.

The foregoing stabilizing 1,3-diaxial interaction was shown to have potentially useful applications for stereochemical control of addition reactions[56]. The β-trimethylstannyl cyclohexenone ketal 65 affords a nearly 1:1 mixture of isomeric cis-diols 66 and 67 when hydroxylated with OsO_4 (equation 25). However, the chlorostannane 68 upon hydroxylation with OsO_4, then Sn methylation, yields a 94:6 mixture favoring the α,α,-diol 66. Evidently, the conformational change induced by the 1,3-diaxial donor–acceptor

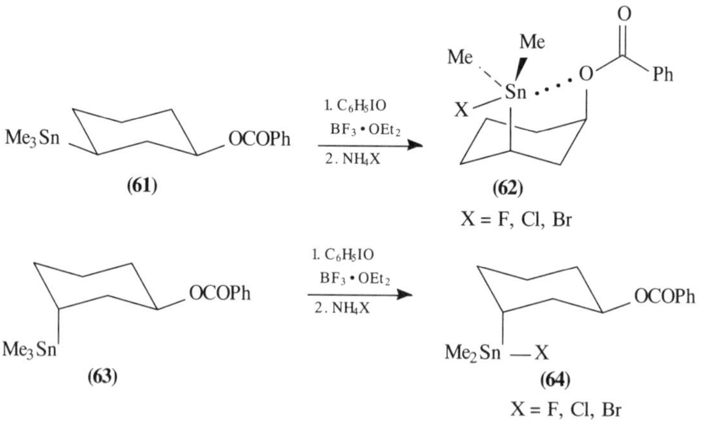

FIGURE 10. Preferred conformations of cis and trans 3-trialkyl and 3-dialkylhalostannyl cyclohexyl benzoates

interaction in chlorostannane **68** effectively blocks the β-face of the double bond, as shown.

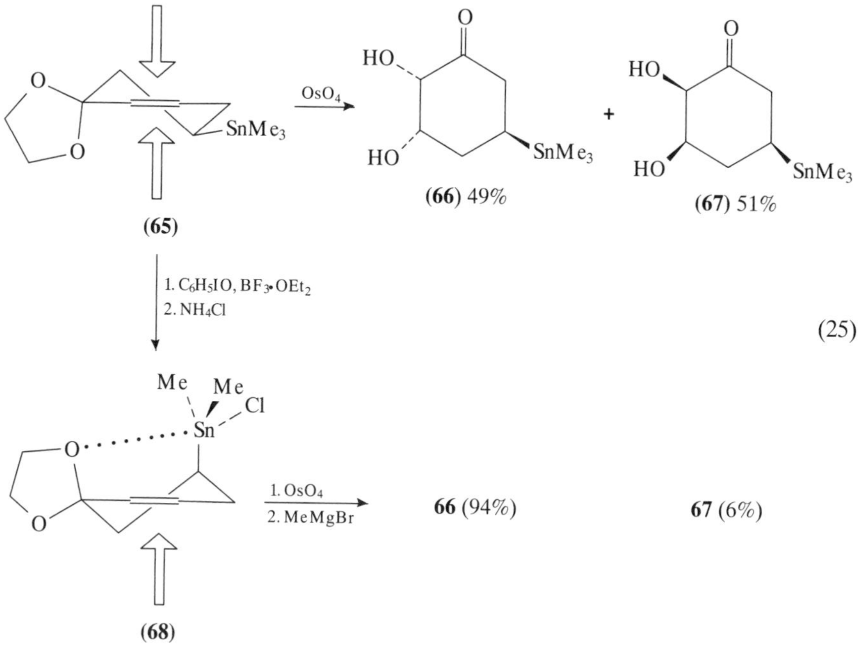
SnMe3
OsO4
HO
HO
SnMe3
+
HO
HO
SnMe3
(65)
(66) 49%
(67) 51%
1. C6H5IO, BF3•OEt2
2. NH4Cl
Me
Me
Cl
Sn
1. OsO4
2. MeMgBr
66 (94%)
67 (6%)
(68)

(25)

E. α-, β- and γ-Organometallo Esters

Crotonic esters and certain homologues, when converted to their enolates with LDA and treated with stannyl and germyl chlorides, afford the γ-metallo derivatives (Table 10)[57]. In contrast, silylation of these enolates leads to the O-silyl derivatives. Interestingly, the halostannane derivatives show a strong preference for the (Z) geometry suggestive of a donor–acceptor interaction between the carbonyl oxygen and the electropositive tin atom,

TABLE 10. Preparation of allylmetallics from conjugated esters

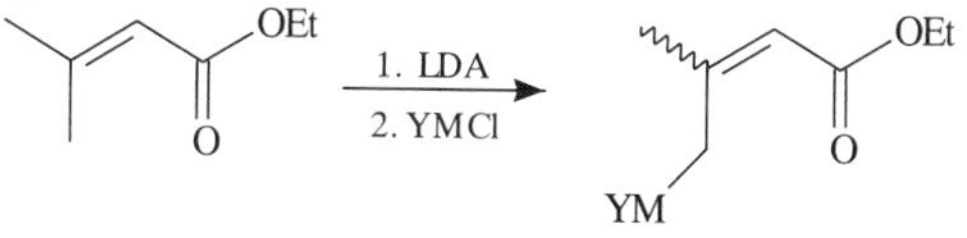
OEt
O
1. LDA
2. YMCl
OEt
O
YM

Y	M	Yield (%)	Z:E
Bu_3	Sn	83	87:13
Bu_2Cl	Sn	75	>95:5
$BuCl_2$	Sn	44	>95:5
Cl_3	Sn	68	>95:5
Ph_3	Ge	50	83:17

as noted above. The regiochemistry of germylation depends upon the structure of the ester and the nature of the germyl chloride. Thus Me_3GeCl adds α to ethyl β-methylcrotonate (Scheme 25) but Ph_3GeCl adds to the γ-position (Table 10). The enolate of ethyl tiglate, on the other hand, gives only γ-addition with Me_3GeBr (Scheme 25).

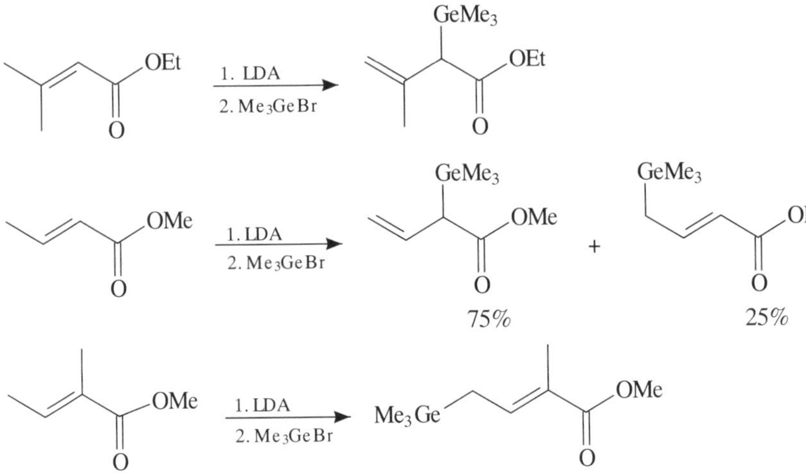

SCHEME 25

Hydrostannation of conjugated esters and nitriles leads to β-stannylated derivatives by a free radical mechanism (equation 26)[58]. The additions are highly stereoselective, but the relative stereochemistry of the adducts was not determined. Since both (E)- and (Z)-alkenoates gave rise to the same stereoisomeric adduct, it can be concluded that a stepwise process is involved.

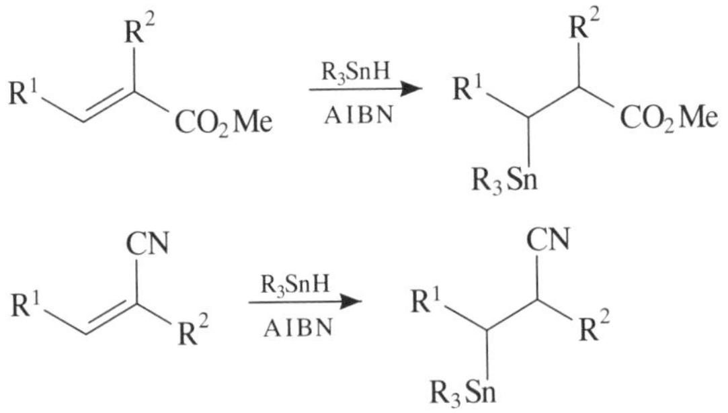

(26)

F. α- and β-Oxygenated Organometallics

The first chiral α-oxygenated stannanes were prepared by Still[59]. Addition of Bu_3SnLi to α-methyl-β-phenylpropionaldehyde followed by MOMCl led to a separable 1:1 mixture of syn and anti alkoxy stannanes (Scheme 26). Lithiation with n-BuLi and addition of acetone gave the respective adducts with overall retention of stereochemistry. Thus, it is implied that the intermediate α-alkoxy lithio derivatives retain their configuration.

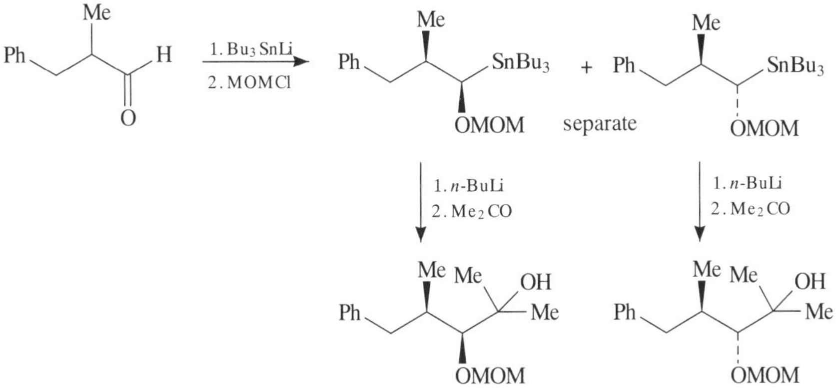

SCHEME 26. MOM = MeOCH$_2$

α-Methyl-β-(benzyloxymethoxy)propionaldehyde (68) afforded an 89:11 mixture of *anti* and *syn* adducts by the analogous sequence (equation 27)[59]. Lithiation followed by methylation yielded an 8:1 mixture of the methylated products.

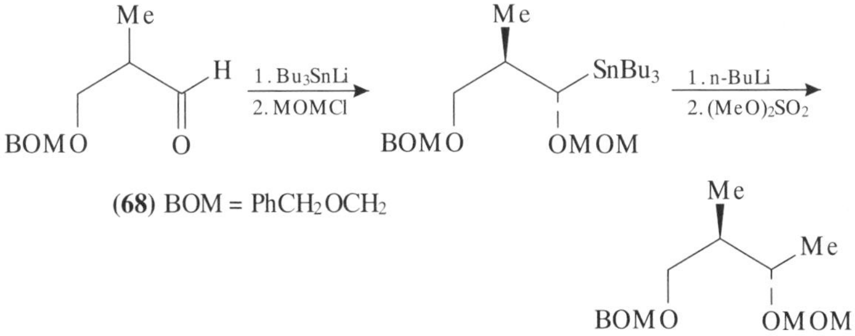

(27)

(68) BOM = PhCH$_2$OCH$_2$

It was further shown that the nonracemic stannane 70 can be prepared from the Mosher ester derivative 69 through reductive cleavage and BOM protection (equation 28). Lithiation and methylation gave the nonracemic ether with retention of configuration[59].

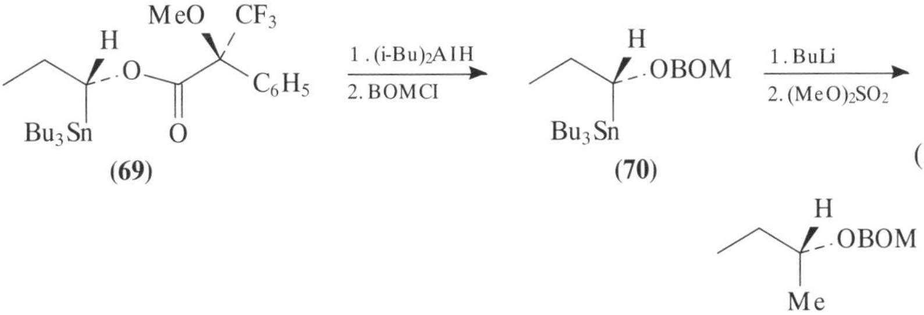

(28)

The nonracemic α-menthyloxymethoxy allylic stannanes 71 and 72 were obtained from crotonaldehyde by addition of Bu$_3$SnLi and etherification with ($-$)-menthyloxymethyl chloride (Scheme 27)[60]. The diastereomeric ethers **71** and **72** are separable by column chromatography. It was found that on heating with mainly aryl aldehydes, these allylic stannanes give the *anti* adducts, stereospecifically. A cyclic S_E2' transition state was proposed.

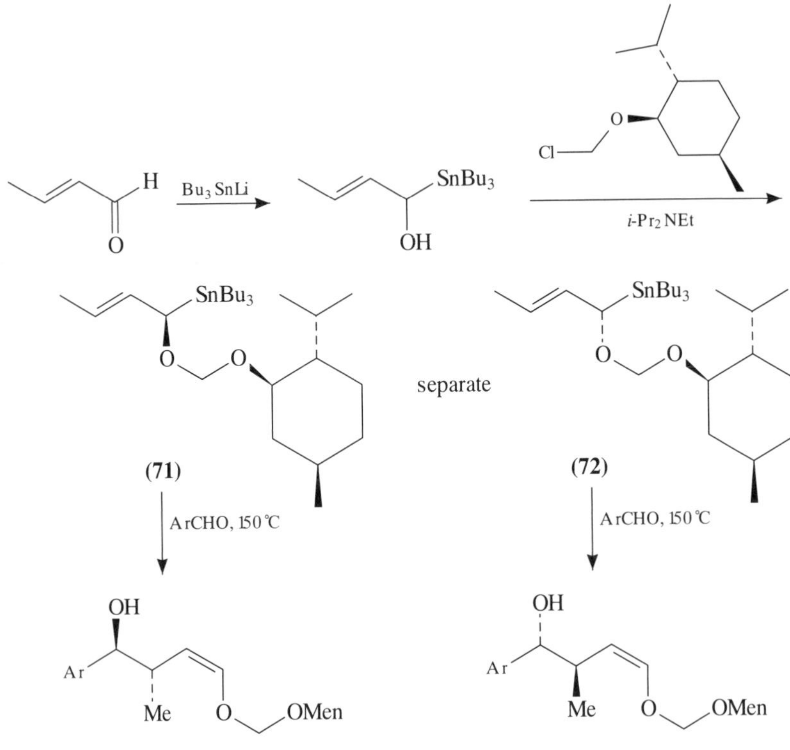

SCHEME 27. Men = (−)-menthyl

An alternative route to nonracemic α-alkoxy stannanes entails the reduction of acyl stannanes with chiral hydrides[61,62]. Accordingly, conjugated stannyl enones yield (S)-α-alkoxy allylic stannanes by reduction with (R)-(+)-BINAL-H. As expected, (S)-(−)-BINAL-H gives rise to the enantiomeric (R)-α-alkoxy allylic stannanes (equation 29)[61]. Upon treatment with Lewis acids, these stannanes undergo a stereospecific *anti* 1,3-isomerization to the (Z)-γ-alkoxy allylic stannanes[61].

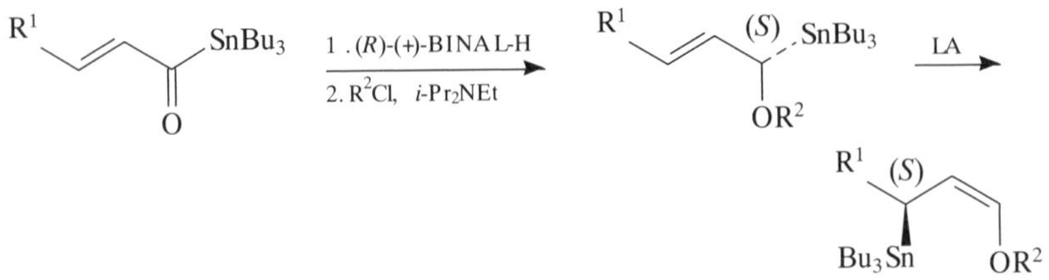

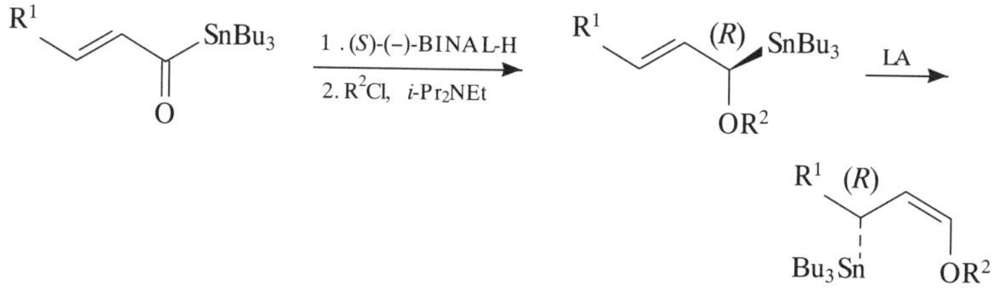

$$R^1 \xrightarrow[O]{} SnBu_3 \quad \xrightarrow[\text{2. } R^2Cl,\ i\text{-Pr}_2NEt]{\text{1.}(S)\text{-}(-)\text{-BINAL-H}} \quad R^1 \xrightarrow[OR^2]{(R)} SnBu_3 \xrightarrow{LA}$$

$$R^1 \xrightarrow[Bu_3Sn \quad OR^2]{(R)} \tag{29}$$

$$R^1 = Me, Bu, c\text{-}C_6H_{11}; \quad R^2 = MeOCH_2, BnOCH_2; \quad LA = BF_3 \cdot OEt_2, Bu_3SnOTf, LiClO_4$$

γ-Alkoxy and silyloxy allylic stannanes have been used as chiral reagents to prepare nonracemic 1,2-diol derivatives (equation 30)[64].

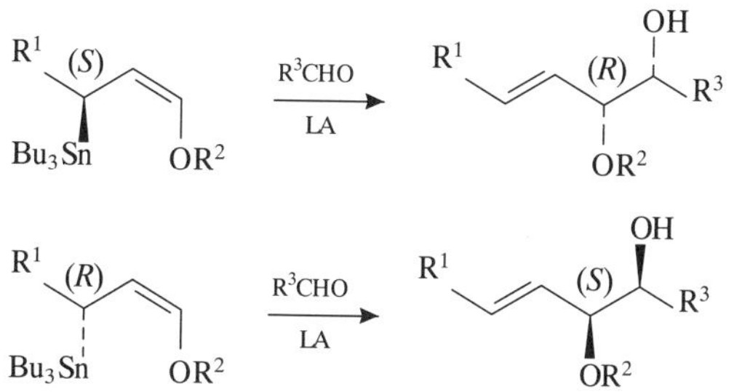

$$R^1 \xrightarrow[Bu_3Sn \quad OR^2]{(S)} \quad \xrightarrow[LA]{R^3CHO} \quad R^1 \xrightarrow[OR^2]{(R)} \overset{OH}{\underset{}{}} R^3$$

$$R^1 \xrightarrow[Bu_3Sn \quad OR^2]{(R)} \quad \xrightarrow[LA]{R^3CHO} \quad R^1 \xrightarrow[OR^2]{(S)} \overset{OH}{\underset{}{}} R^3 \tag{30}$$

A second route to nonracemic γ-oxygenated allylic stannanes utilizes an enantioselective deprotonation of allylic carbamates by BuLi in the presence of $(-)$-sparteine. The configurationally stable α-lithio carbamate intermediate undergoes enantioselective S_E2' reaction with Bu_3SnCl and Me_3SnCl (Scheme 28)[65]. Once formed, the γ-carbamoyloxy stannanes can be inverted by successive lithiation with s-BuLi and stannation with R_3SnCl (Scheme 29)[65]. The former reaction proceeds with S_E2' retention and the latter by S_E2' inversion. Nonracemic allylic carbamates can also be used to prepare chiral stannanes. Deprotonation with s-BuLi·TMEDA proceeds stereospecifically with retention (Scheme 29)[65].

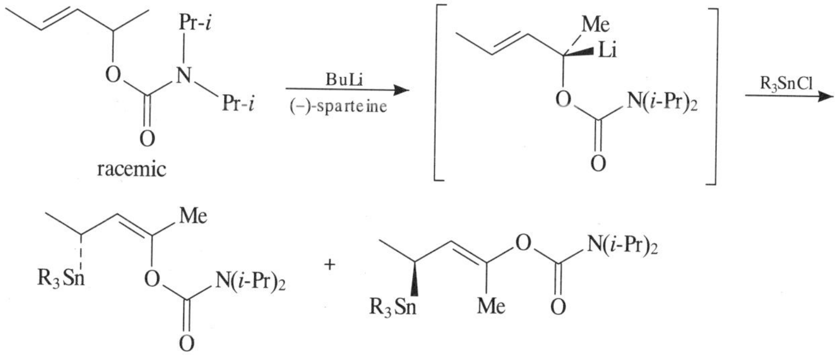

R = Bu; 87:13 (80% ee)
R = Me; 92:8 (60% ee)

SCHEME 28

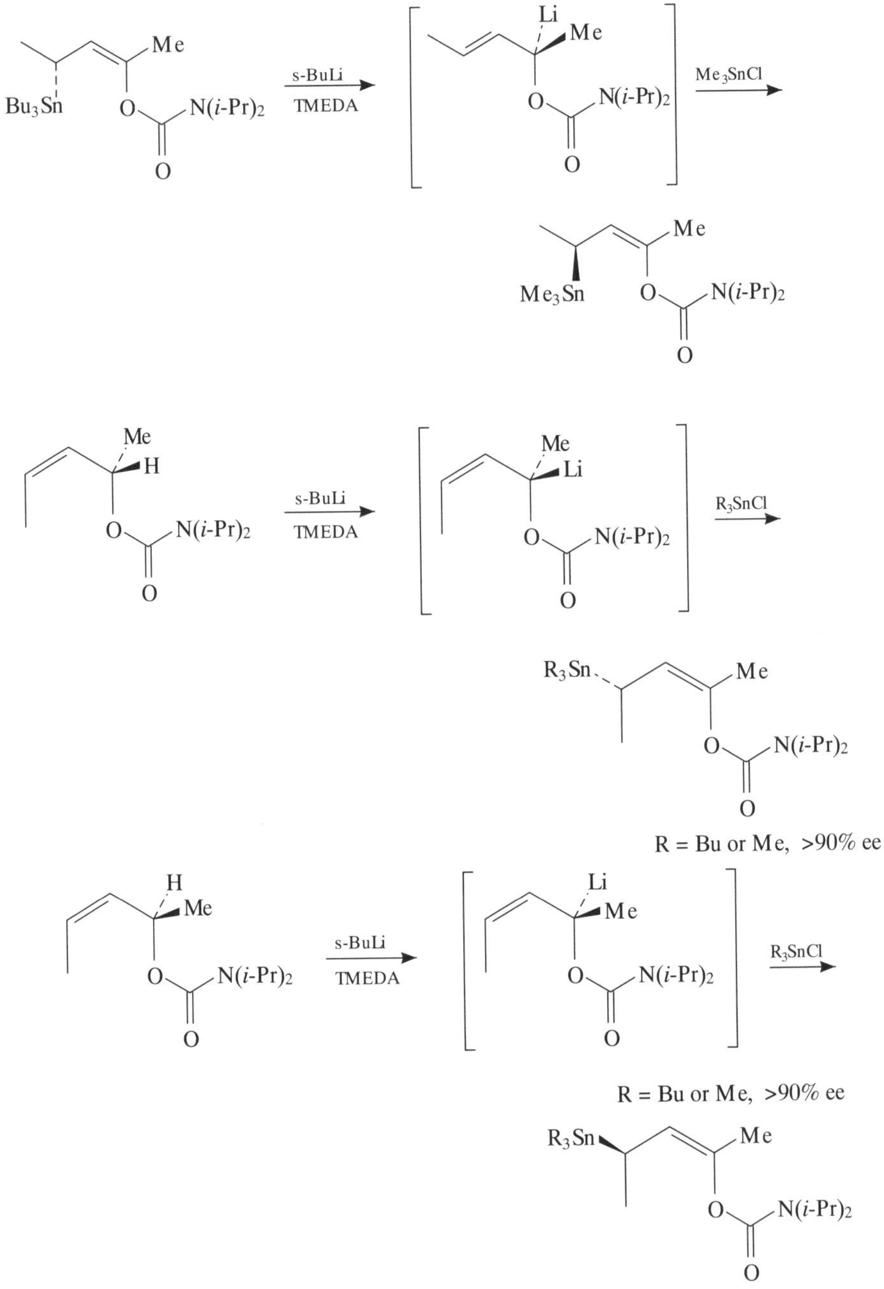

SCHEME 29

A third route to nonracemic α-alkoxy and α-hydroxy stannanes employs the chiral acetal **73** prepared from (R,R)-2,4-pentanediol (Scheme 30)[66]. Addition of various Grignard reagents to this acetal in the presence of TiCl$_4$ results in selective displacement yielding (S)-α-alkoxy stannanes. The corresponding α-hydroxy derivatives can be obtained after oxidation and mild base treatment. Organocuprates can also be employed to cleave this acetal but with somewhat lower selectivity[67].

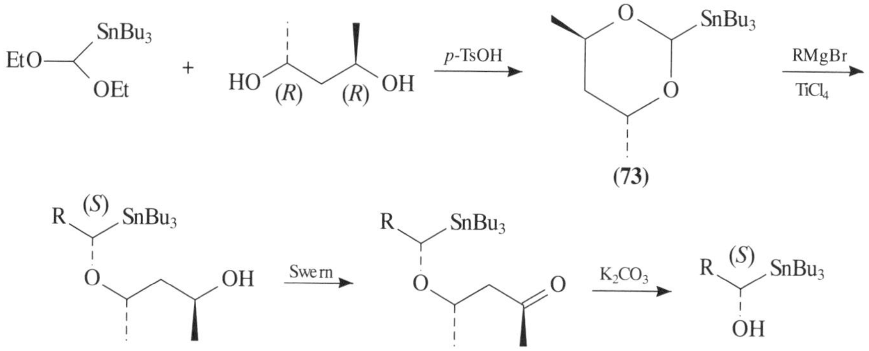

R	d.r
Et	>95:5
Bu	92:8
CH$_2$=CHCH$_2$	>95:5
C$_6$H$_5$	60:40
Me	65:35

SCHEME 30

Addition of Bu$_3$SnLi or Me$_3$SnLi to 4-t-butylcyclohexanone affords mixtures of *trans* and *cis* adducts in ratios that depend on reaction conditions (Table 11)[68]. In THF, a 93:7 mixture is obtained with both reagents. This ratio is thought to represent the thermodynamic distribution — the axial stannane being favored. In ether, the *cis* isomer predominates, suggesting a kinetic preference for equatorial addition. Each of the two isomers can be lithiated with BuLi. Subsequent treatment with alkyl halides or carbonyl compounds affords the substituted alkoxy cyclohexanes with retention of stereochemistry.

α-Hydroxy alkylstannanes can be transformed into the α-halo derivatives under neutral or mildly basic conditions (Table 12). Upon treatment with BuLi, these halostannanes are converted to dimeric olefins[69].

The mesylate derivatives of nonracemic α-hydroxystannanes undergo S$_N$2 displacement by alkoxides to afford the inverted α-alkoxy derivatives (equation 31)[70].

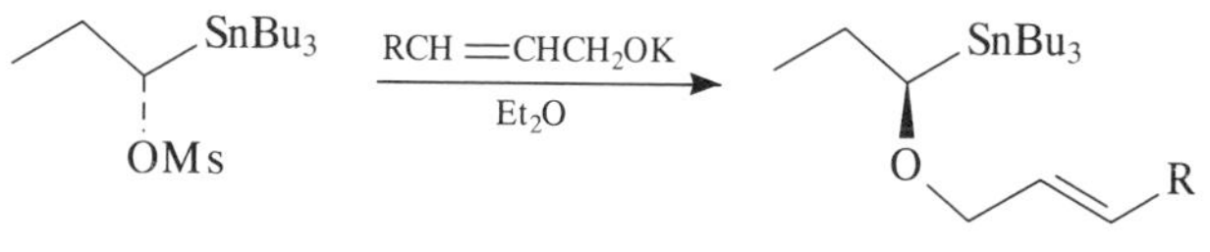

$$R = H, Me \tag{31}$$

A series of β-hydroxy organometal derivatives was prepared by reaction of the triphenylmetallo anion with various epoxides (equation 32)[71]. Upon treatment with acidic

TABLE 11.   Additions of trialklystannyllithium reagents to 4-t-butylcyclohexanone

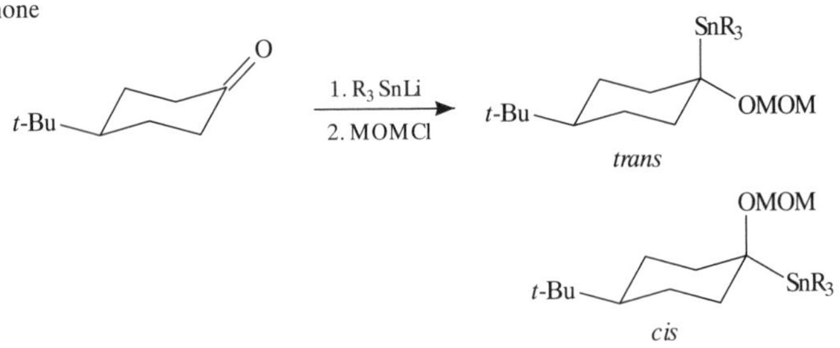

R	Solvent	T($^\circ$C)	trans:cis
Bu	THF	− 78	93:7
Bu	THF	− 40	93:7
Me	THF	− 78	93:7
Bu	ether	− 78	45:55
Me	ether	− 78	45:55
Bu	ether	−100	25:75
Bu	DME	− 40	50:50

TABLE 12.   Conversion of α-hydroxy stannanes to α-halo stannanes

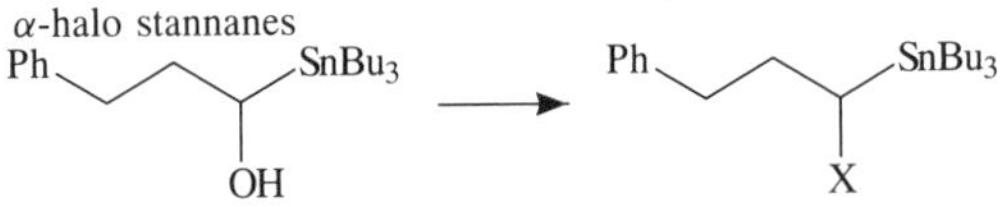

Reagent	X	Yield (%)
p-TsCl, pyridine	Cl	65
Ph$_3$P, DEAD, CH$_2$Cl$_2$	Cl	65
Ph$_3$P, CBr$_4$, CH$_2$Cl$_2$	Br	100
Ph$_3$P, DEAD, CH$_3$I	I	60

methanol, a stereospecific *anti* elimination to the olefin was observed, with relative rates of M = Pb > Sn ≫ Ge > Si for a given substitution pattern.

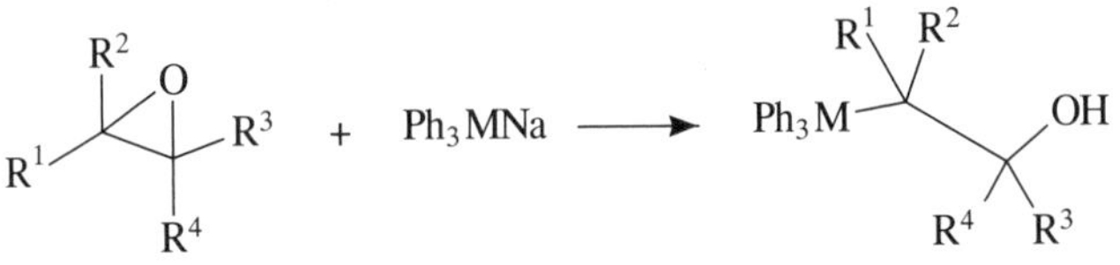

$$R^1\text{–}R^4 = \text{H or Me} \qquad \text{M = Si, Ge, Sn, Pb}$$

β-Hydroxy stannanes can be prepared by cleavage of epoxides with Bu$_3$SnLi or cleavage of epoxy stannanes with organocuprates (equation 33)[72]. The two methods are stereochemically complementary. The higher order cyanocuprate, Bu$_3$Sn(Bu)Cu(CN)Li$_2$, also affords β-hydroxy stannanes by reaction with epoxides[54].

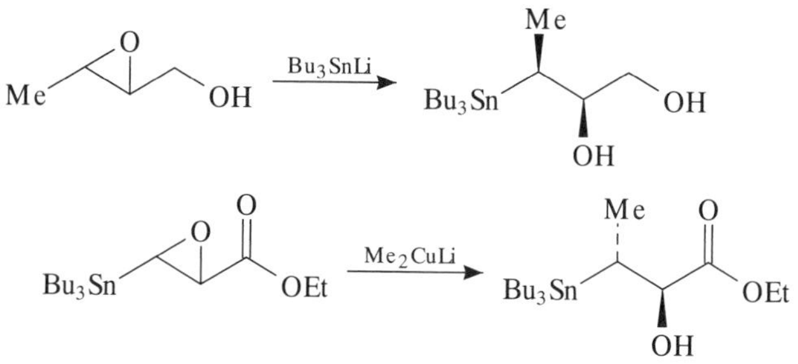

(33)

Thermolysis of the *syn* and *anti* β-acetoxy stannanes, obtained by addition of Bu₃SnLi to *trans*- or *cis*- 1,2-dimethyloxirane and subsequent acetylation, led to (Z)- or (E)-2-butene, respectively, by a stereospecific *anti* process (equation 34).[73]. It is postulated that a hyperconjugative interaction of the C—Sn bond facilitates departure of the acetate group in these acyclic systems. The trimethylstannyl and triphenylstannyl analogues likewise undergo *anti* elimination.

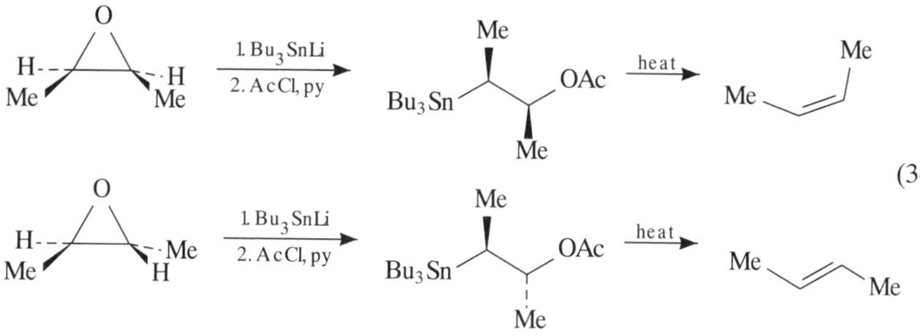

(34)

Certain δ- and ε-oxygenated allylic stannanes have been found to transmetallate with SnCl₄ to give chiral pentacoordinated chloro stannane intermediates which add stereoselectively to aldehydes (Scheme 31)[74]. These reactions proceed with net 1,5-and 1,6-asymmetric induction.

Enantioenriched α-alkoxyorganolead compounds have been prepared through lithiation of stannane precursors and trapping of the lithiated species with a triorganolead halide (equation 35)[75]. In the presence of TiCl₄, these plumbanes add to aldehydes to afford monoprotected *syn*-1,2-diols. The use of BF₃·OEt₂ as a Lewis acid promoter leads mainly to the *anti* adducts (Table 13)[70].

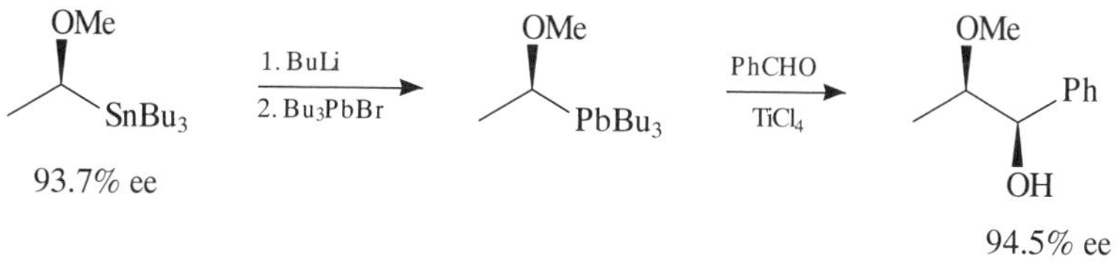

(35)

G. α-Aminostannanes

Upon treatment with dibenzyl iminodicarbonate under Mitsunobu conditions, nonracemic α-hydroxy stannanes are converted to α-imino stannanes with inversion of

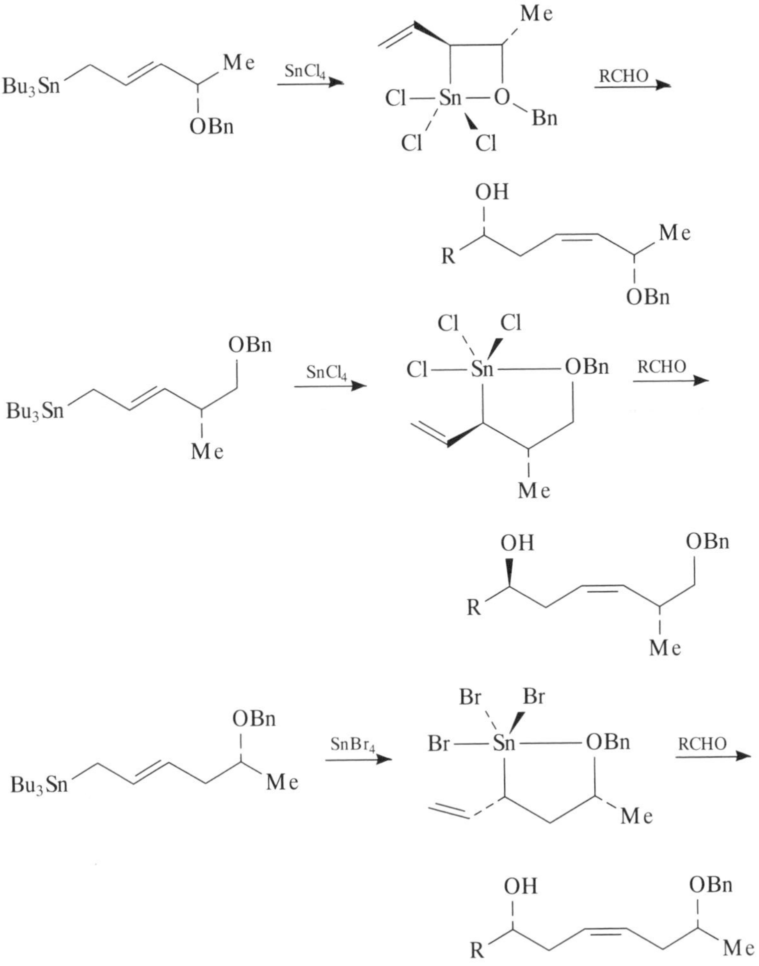

SCHEME 31

configuration (equation 36)[76]. The reaction can also be effected with phthalimide to yield the phthalimido analogues.

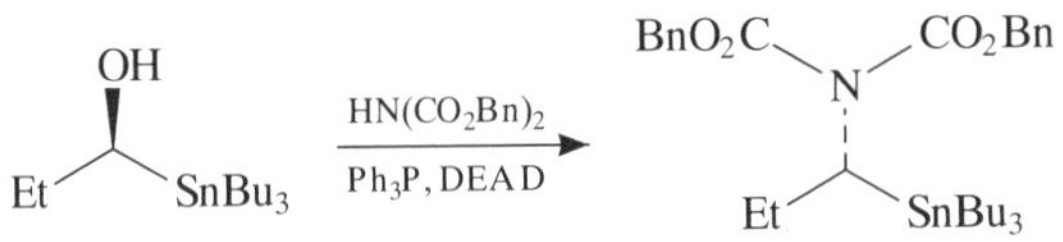

$$(36)$$

An alternative synthesis of nonracemic α-amino stannanes is outlined in equation 37[77]. The diastereomeric stannanes, obtained by sequential lithiation and stannylation of the starting nonracemic piperidinooxazoline, can be separated by chromatography. Subsequent removal of the chiral auxiliary and N-methylation completes the synthesis.

TABLE 13. Additions of α-methoxy organoplumbanes to aldehydes

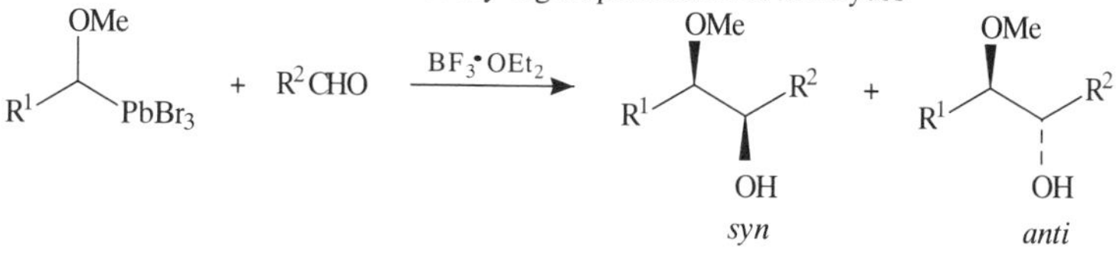

R^1	R^2	syn:anti
C_7H_{15}	C_6H_5	$99:1^a$
C_7H_{15}	C_7H_{15}	39:61
c-C_6H_{11}	C_7H_{15}	30:70
c-C_6H_{11}	c-C_6H_{11}	11:89

^{a} TiCl$_4$ was employed as the Lewis acid.

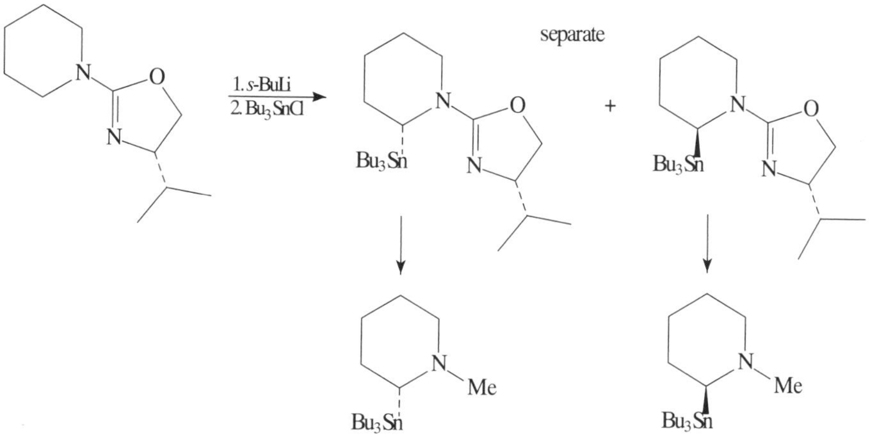

(37)

V. PROPARGYL/ALLENYL SYSTEMS

Chiral allenylstannanes can be prepared by S_N2' displacement of propargylic halides sulfinates or sulfonates with tin cuprates (Table 14)[78]. The nonracemic propargylic mesylate (74) afforded a nonracemic allene, $[\alpha]_D$ -570, whose configuration was assigned by application of Brewster's rules (equation 38)[78]. Displacements on the steroidal mesylates 75 and 76 afforded the allenic products with complete inversion of configuration (Scheme 32)[78].

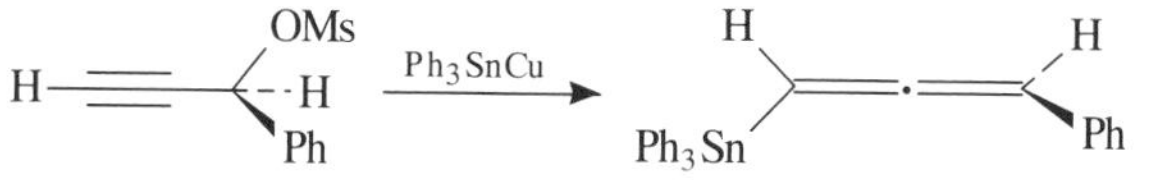

(74)

(38)

An alternative route to allenyl stannanes involves organocuprate displacements on propargylic chlorides bearing an alkynyl Ph$_3$Sn substituent (equation 39)[79]. Interestingly, transmetallation by attack of the cuprate on the tin substituent is not observed in these systems. A parallel strategy can be employed for allenylgermanes (equation 39)[79]. The

TABLE 14.　Synthesis of allenylstannanes by S_N2' displacements

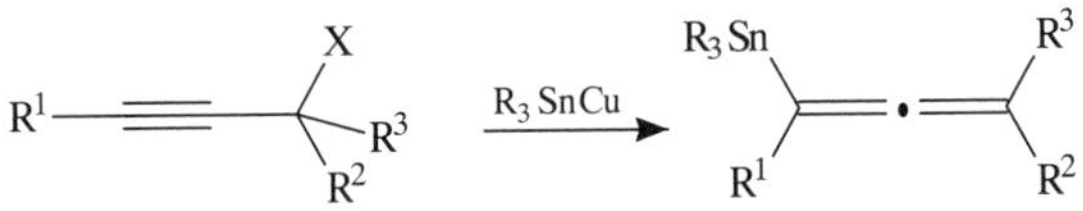

R^1	R^2	R^3	R	X	Yield (%)
H	H	H	Ph	Br	90
H	H	H	Ph	OMs	90
H	H	Me	Ph	OAc	80
H	H	Me	Ph	OMs	80
H	H	Ph	Ph	OMs	90
H	Me	Me	Ph	OMs	95
H	H	Pr	Me	OMs	90
H	H	t-Bu	Ph	OMs	80
Me	Me	Me	Ph	OMs	90
Me	H	Pr	Ph	OMs	56

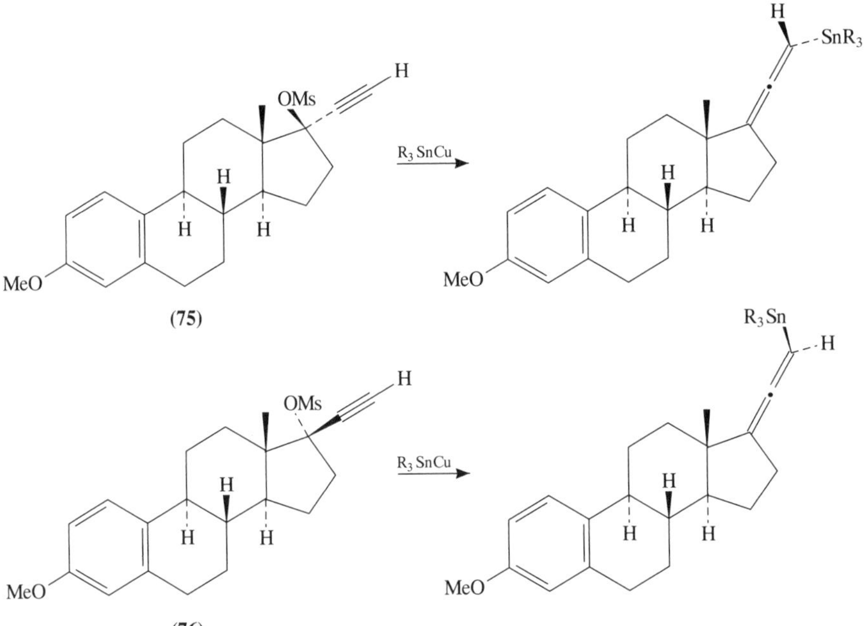

SCHEME 32. R = Ph or Me

nonracemic steroidal mesylate **77** reacts with inversion to yield the allenylgermane derivative (equation 40)[79].

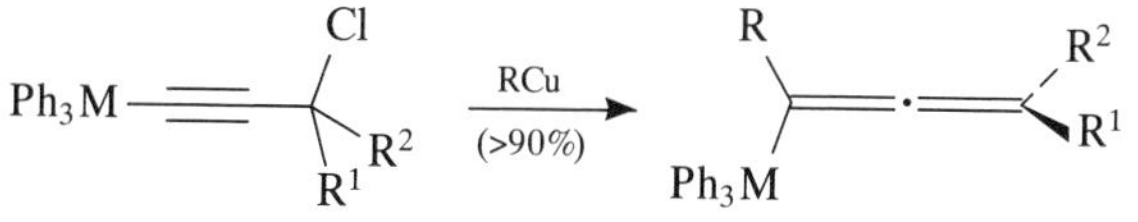

$$M = Ge \text{ or } Sn$$
$$R^1, R^2 = Me \text{ or } H$$
$$R = Et, i\text{-}Pr, t\text{-}Bu,$$

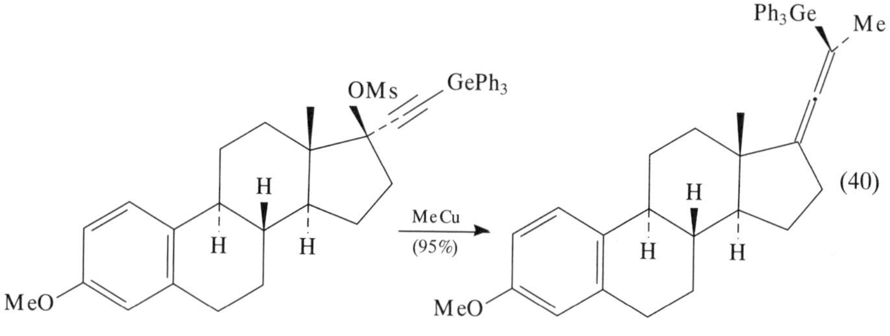

(39)

(40)

(**77**)

$$R^1 = Et, R^2 = C_6H_{13}$$

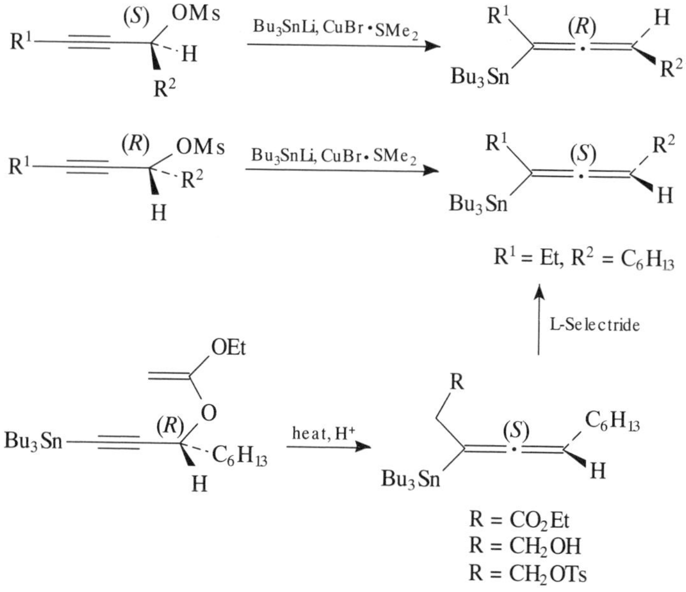

SCHEME 33

Various nonracemic allenylstannanes have been prepared from nonracemic propargylic mesylates and $(Bu_3Sn)_2CuLi$. The stereochemistry of the displacement was shown to be *anti* by correlation with an allenic stannane prepared through Claisen orthoester rearrangement of a propargylic alcohol of known configuration (Scheme 33)[80].

The foregoing allenylstannanes participate in stereospecific S_E2' additions to α-branched aldehydes to afford homopropargylic alcohols with a high degree of diastereoselectivity (Scheme 34)[80]. In the presence of $SnCl_4$, the allenylstannanes undergo S_E2' exchange with inversion[74]. The intermediate propargylic trichlorostannanes afford allenic alcohols upon addition to aldehydes (Scheme 35)[81]. If allowed to equilibrate, the

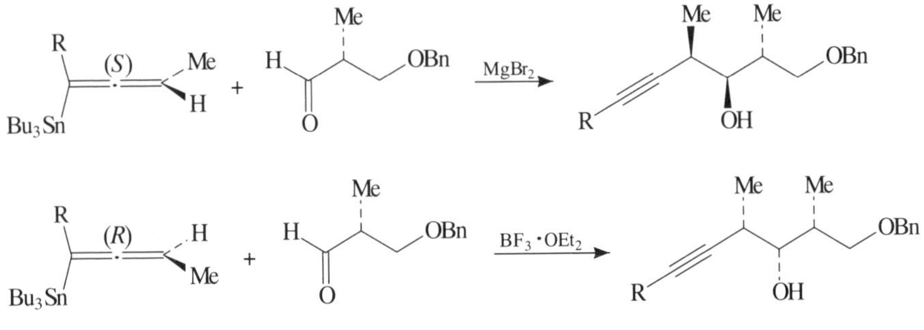

SCHEME 34. R = CH_2OAc or Et

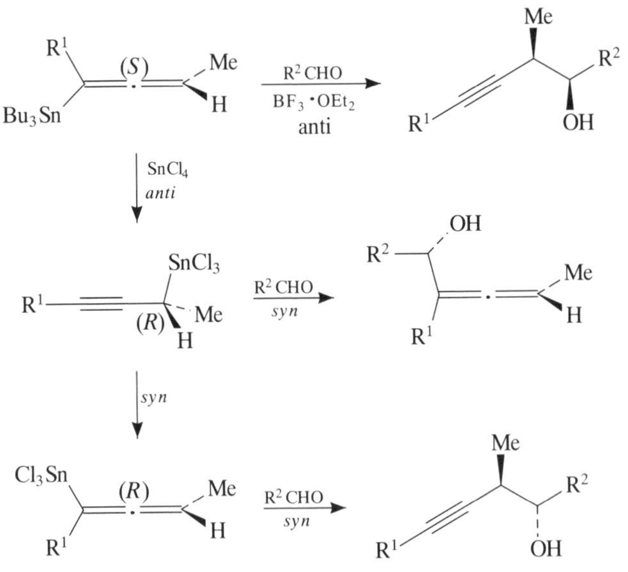

SCHEME 35. $R^1 = CH_2OAc$, C_7H_{15}; $R^2 = i\text{-Pr}$, $BnOCH_2CH(CH_3)_2$

propargylic trichlorostannanes isomerize to allenic isomers with retention. These stannanes react with aldehydes giving *anti* homopropargylic alcohols stereoselectively and in high yield[81].

VI. REFERENCES

1. A. G. Brook and G. J. D. Peddle, *J. Am. Chem. Soc.*, **85**, 1869 (1963).
2. A. G. Brook, G. J. D. Peddle, *J. Am. Chem. Soc.*, **85**, 2338 (1963); C. Eaborn, R. E. E. Hill and P. Simpson, *J. Organomet. Chem.*, **37**, 251 (1972); **37**, 267 (1972).
3. R. W. Bott, C. Eaborn and I. D. Varma, *Chem. Ind.*, 614 (1963); C. Eaborn, P. Simpson and I. D. Varma, *J. Chem. Soc. (A)*, 1133 (1966).
4. F. Carré and R. Corriu, *J. Organomet. Chem.*, **25**, 395 (1970).
5. D. Terunuma, H. Kizaki, T. Sato, K. Masuo and H. Nohira, *Bull. Chem. Soc. Jpn.*, **66**, 664 (1993).
6. R. Tacke, S. A. Wagner and J. Sperlich, *Chem. Ber.*, **127**, 639 (1994).
7. H. Sakurai and K. Mochida, *J. Chem. Soc., Chem. Commun.*, 1581 (1971); K. Mochida, T. Yamauchi and H. Sakurai, *Bull. Chem. Soc. Jpn.*, **62**, 1982 (1989).
8. F. H. Carré, R. J. P. Corriu and R. B. Thomassin, *J. Chem. Soc., Chem. Commun.*, 560 (1968); F. Carré, R. Corriu and M. Leard, *J. Organomet. Chem.*, **24**, 101 (1970).
9. C. Eaborn, R. E. E. Hill and P. Simpson, *J. Organomet. Chem.*, **37**, 275 (1972); C. Eaborn, R. E. E. Hill, P. Simpson, A. G. Brook and D. MacRae, *J. Organomet. Chem.*, **15**, 241 (1968). These reductions show a solvent and concentration dependence: A. Jean and M. Lequan, *Tetrahedron Lett.*, 1517 (1970).
10. F. Carré and R. Corriu, *J. Organomet. Chem.*, **65**, 343 (1974).
11. F. Carré and R. Corriu, *J. Organomet. Chem.*, **65**, 349 (1974).
12. R. J. P. Corriu and J. J. E. Moreau, *J. Chem. Soc., Chem. Commun.*, 812 (1971). Olefins show analogous behavior: R. J. P. Corriu and J. J. E. Moreau, *J. Organomet. Chem.*, **40**, 55 (1972).
13. D. Terunuma, K. Masuo, H. Kizaki and H. Nohira, *Bull. Chem. Soc. Jpn.*, **67**, 160 (1994).
14. A. G. Brook, J. M. Duff and D. G. Anderson, *J. Am. Chem. Soc.*, **92**, 7567 (1970).
15. For a comprehensive review, see M. Gielen, *Top. Curr. Chem.*, **104**, 57 (1982).
16. G. J. D. Peddle and G. Redl, *J. Am. Chem. Soc.*, **92**, 365 (1970).
17. M. Gielen, *Acc. Chem. Res.*, **6**, 198 (1973); M. Gielen and H. Moakhtar-Jamai, *Bull. Soc. Chim. Belg.*, **84**, 197 (1975).
18. U. Folli, D. Iarossi and F. Taddei, *J. Chem. Soc., Perkin Trans. 2*, 638 (1973); R. M. Lequan and M. Lequan, *J. Organomet. Chem.*, **202**, C99 (1980).
19. R. M. Lequan and M. Lequan, *Tetrahedron Lett.*, **22**, 1323 (1981).
20. M. Lequan, F. Meganem and Y. Besace, *J. Organomet. Chem.*, **131**, 231 (1977).
21. M. Gielen, S. Simon, Y. Tondeur, M. Van de Steen, C. Hoogzand and I. Van den Eynde, *Israel J. Chem.*, **15**, 74 (1976); M. Gielen and Y. Tondeur, *J. Organomet. Chem.*, **169**, 265 (1979).
22. I. Vanden Eynde, M. Gielen, G. Stühler and A. Mannschreck, *Polyhedron*, **1**, 1 (1982).
23. M. Gielen, *Pure Appl. Chem.*, **52**, 657 (1980).
24. M. Lehnig, H. -U. Buschhaus, W. P. Neumann and T. Apoussidis, *Bull. Soc. Chim. Belg.*, **89**, 907 (1980).
25. H. J. Reich, J. P. Borst, M. B. Coplien and N. H. Phillips, *J. Am. Chem. Soc.*, **114**, 6577 (1992).
26. M. Gielen and Y. Tondeur, *J. Chem. Soc., Chem. Commun.*, 81 (1978). For a review, see J. T. B. H. Jastrzebski and G. Van Koten, *Adv. Organomet. Chem.*, **35**, 241 (1993).
27. G. van Koten and G. J. Noltes, *J. Am. Chem. Soc.*, **98**, 5393 (1976).
28. J. T. B. H. Jastrzebski, J. Boersma and G. Van Koten, *J. Organomet. Chem.*, **413**, 43 (1991).
29. H. Schumann, B. C. Wassermann and F. E. Hahn, *Organometallics*, **11**, 2803 (1992); H. Schumann, B. C. Wassermann and J. Pickardt, *Organometallics*, **12**, 3051 (1993).
30. H. O. van der Kooi, J. Wolters and A. van der Gen, *Recl. Trav. Chim. Pays-Bas*, **98**, 353 (1979).
31. H. O. van der Kooi, W. H. den Brinker and A. J. de Kok, *Acta Crystallogr.*, **C41**, 869 (1985).
32. J. Dubac, P. Mazerolles, M. Joly and F. Piau, *J. Organomet. Chem.*, **127**, C69 (1977).
33. J. Dubac, P. Mazerolles, M. Joly and J. Cavezzan, *J. Organomet. Chem.*, **165**, 163 (1979).

34. J. Dubac, J. Cavezzan, A. Laporterie and P. Mazerolles, *J. Organomet. Chem.*, **209**, 25 (1981); J. Dubac, J. Cavezzan, A. Laporterie and P. Mazerolles, *J. Organomet. Chem.*, **197**, C15 (1980).

35. Y. Takeuchi, M. Shimoda, K. Tanaka, S. Tomodo, K. Ogawa and H. Suzuki, *J. Chem. Soc., Perkin Trans. 2*, 7 (1988).

36. Y. Takeuchi, K. Tanaka, T. Harazono, K. Ogawa and S. Yoshimura, *Tetrahedron*, **44**, 7531 (1988).

37. F. R. Jensen and D. D. Davis, *J. Am. Chem. Soc.*, **93**, 4047 (1971).

38. A. Rahm, M. Pereyre, M. Petraud and B. Barbe, *J. Organomet. Chem.*, **139**, 49 (1977).

39. J. San Filippo, Jr. and J. Silbermann, *J. Am. Chem. Soc.*, **103**, 5588 (1981).

40. K. -W. Lee and J. San Filippo, Jr., *Organometallics*, **1**, 1496 (1982).

41. G. S. Koermer, M. L. Hall and T. G. Traylor, *J. Am. Chem. Soc.*, **94**, 7205 (1972).

42. For analogous findings on stannylations: (a) of cyclopropyl halides, see K. Sisido, S. Kozima and K. Takizawa, *Tetrahedron Lett.*, 33 (1967); (b) of 7-halonorbornenes, see H. G. Kuivila, J. L. Considine and J. D. Kennedy, *J. Am. Chem. Soc.*, **94**, 7206 (1972).

43. W. Kitching, H. Olszowy, J. Waugh and D. Doddrell, *J. Org. Chem.*, **43**, 898 (1978).

44. Ph$_3$SnLi gives an analogous result: R. H. Fish and B. M. Broline, *J. Organomet. Chem.*, **136**, C41 (1977).

45. J. San Filippo, Jr., J. Silbermann and P. J. Fagan, *J. Am. Chem. Soc.*, **100**, 4834 (1978).

46. W. Kitching, H. Olszowy and K. Harvey, *J. Org. Chem.*, **46**, 2423 (1981).

47. W. Kitching, H. A. Olszowy and K. Harvey, *J. Org. Chem.*, **47**, 1893 (1982).

48. J. P. Quintard, M. Degueil-Castaing, G. Dumartin, A. Rahm and M. Pereyre, *J. Chem. Soc., Chem. Commun.*, 1004 (1980).

49. G. Wickham, D. Young and W. Kitching, *J. Org. Chem.*, **47**, 4884 (1982).

50. The crystal structure of Ph$_3$CH$_2$CH=CH$_2$ shows a dihedral angle for Sn−C−C=C of 108° and 97° for two independent molecules in the unit cell. P. Ganis, D. Furlani, D. Marton, G. Tagliavini and G. Valle, *J. Organomet. Chem.*, **293**, 207 (1985).

51. K. Mochida and M. Nanba, *Polyhedron*, **13**, 915 (1994).

52. W. C. Still, *J. Am. Chem. Soc.*, **99**, 4836 (1977); J. Hudec, *J. Chem. Soc., Perkin Trans. 1*, 1020 (1975).

53. G. Wickham, H. A. Olszowy and W. Kitching, *J. Org. Chem.*, **47**, 3788 (1982).

54. B. H. Lipshutz, E. L. Ellsworth, S. H. Dimock and D. C. Reuter, *Tetrahedron Lett.*, **30**, 2065 (1989).

55. I. Suzuki, T. Furuta and Y. Yamamoto, *J. Organomet. Chem.*, **443**, C6 (1993).

56. M. Ochiai, S. Iwaki, T. Ukita, Y. Matsuura, M. Shiro and Y. Nagao, *J. Am. Chem. Soc.*, **110**, 4606 (1988).

57. Y. Yamamoto, S. Hatsuya and J. -I. Yamada, *J. Org. Chem.*, **55**, 3118 (1990).

58. J. C. Podesta, A. B. Chopa and A. D. Ayala, *J. Organomet. Chem.*, **212**, 163 (1981).

59. W. C. Still and C. Sreekumar, *J. Am. Chem. Soc.*, **102**, 1201 (1980).

60. V. J. Jephcote, A. J. Pratt and E. J. Thomas, *J. Chem. Commun.*, 800 (1984); *J. Chem. Soc., Perkin Trans. 1*, 1529 (1989).

61. J. A. Marshall and W. Y. Gung, *Tetrahedron Lett.*, **29**, 1657 (1988); J. A. Marshall, G. S. Welmaker and B. W. Gung, *J. Am. Chem. Soc.*, **113**, 647 (1991).

62. P. C. -M. Chan and J. M. Chong, *J. Org. Chem.*, **53**, 5584 (1988).

63. J. A. Marshall and W. Y. Gung, *Tetrahedron Lett.*, **30**, 2183 (1989); **30**, 7349 (1989); J. A. Marshall and G. S. Welmaker, *Tetrahedron Lett.*, **32**, 2101 (1991); J. P. Quintard, G. Dumartin, B. Elissondo, A. Rahm and M. Pereyre, *Tetrahedron*, **45**, 1017 (1989).

64. J. A. Marshall and S. Beaudoin, *J. Org. Chem.*, **59**, 6614 (1994); J. A. Marshall, B. M. Seletsky and P. S. Coan, *J. Org. Chem.*, **59**, 5139 (1994); J. A. Marshall and G. S. Welmaker, *J. Org. Chem.*, **59**, 4122 (1994); J. A. Marshall, B. M. Seletsky and G. P. Luke, *J. Org. Chem.*, **59**, 3413 (1994); J. A. Marshall, *Chemtracts—Org. Chem.*, 75 (1992).

65. O. Zschage, J. R. Schwark, T. Kramer and D. Hoppe, *Tetrahedron*, **48**, 8377 (1992).

66. K. Tomooka, T. Igarashi and T. Nakai, *Tetrahedron Lett.*, **35**, 1913 (1994).

67. J. -L. Parrain, J. -C. Cintrat and J. P. Quintard, *J. Organomet. Chem.*, **437**, C19 (1992).

68. J. S. Sawyer, A. Kucerovy, T. L. Macdonald and G. J. McGarvey, *J. Am. Chem. Soc.*, **110**, 842 (1988).

69. Y. Torisawa, M. Shibasaki and S. Ikegami, *Tetrahedron Lett.*, **22**, 2397 (1981).

70. K. Tomooka, T. Igarashi, M. Watanabe and T. Nakai, *Tetrahedron Lett.*, **33**, 5795 (1992).

71. D. D. Davis and C. E. Gray, *J. Organomet. Chem.*, **18**, P1 (1969).

72. J. M. Chong and E. K. Mar, *J. Org. Chem.*, **57**, 46 (1992).

73. B. Jousseaume, N. Noiret, M. Pereyre, J. -M. Frances and M. Petraud, *Organometallics*, **11**, 3910 (1992).
74. For an overview, see E. J. Thomas, *Chemtracts—Org. Chem.*, 207 (1994).
75. Y. Yamamoto, *Chemtracts—Org. Chem.*, 255 (1991).
76. J. M. Chong and S. B. Park, *J. Org. Chem.*, **57**, 2220 (1992).
77. R. E. Gawley and Q. Zhang, *J. Am. Chem. Soc.*, **115**, 7515 (1993).
78. K. Ruitenberg, H. Westmijze, J. Meijer, C. J. Elsevier and P. Vermeer, *J. Organomet. Chem.*, **241**, 417 (1983).
79. K. Ruitenberg, H. Westmijze, H. Kleijn and P. Vermeer, *J. Organomet. Chem.*, **277**, 227 (1984).
80. J. A. Marshall and X. -J. Wang, *J. Org. Chem.*, **55**, 6246 (1990); **56**, 3211 (1991); **57**, 1242 (1992).
81. J. A. Marshall and J. Perkins, *J. Org. Chem.*, **59**, 3509 (1994).

CHAPTER 4

Thermochemistry of organometallic compounds of germanium, tin and lead

JOSÉ A. MARTINHO SIMÕES

Departamento de Química, Faculdade de Ciências, Universidade de Lisboa, 1700 Lisboa, Portugal
Fax: (+1)351-1-7599404; e-mail: fjams@cc.fc.ul.pt

JOEL F. LIEBMAN

Department of Chemistry and Biochemistry, University of Maryland, Baltimore County, 5401 Wilkens Avenue, Baltimore, Maryland 21228-5398, USA
Fax: (+1)410-455-2608; e-mail: jliebman@umbc2.umbc.edu

and

SUZANNE W. SLAYDEN

Department of Chemistry, George Mason University, 4400 University Drive, Fairfax, Virginia 22030-4444, USA
Fax: (+1)703-993-1055; e-mail: sslayden@gmu.edu

The chemistry of organic germanium, tin and lead compounds
Edited by S. Patai © 1995 John Wiley & Sons Ltd

245

I. GERMANIUM COMPOUNDS

A. Enthalpies of Formation

A casual user of thermochemical data for germanium organometallic compounds will be surprised when comparing the information provided by the two main comprehensive reviews on the subject, one by Pilcher and Skinner[1] and the other by Tel'noi and Rabinovich[2] (Table 1). Although the latter review is only about two years older, the number of enthalpies of formation listed for organogermanium compounds is about twice that given in Pilcher and Skinner's survey. In addition, examination of recommended data in both papers reveals that some values of the enthalpies of formation differ considerably, despite the fact that they rely on the same original references. This unfortunate state of affairs is mainly due to the interpretation of the experimental results obtained from static-bomb combustion calorimetry, a technique which was used to probe the thermochemistry of most of the organogermanium compounds. As noted by Pilcher[3], the structure assigned to the solid combustion product GeO_2 (amorphous, hexagonal or tetragonal) has a significant influence on the final enthalpy of formation values: the enthalpies of transition $GeO_2(am) \longrightarrow GeO_2(c, hex)$ and $GeO_2(am) \longrightarrow GeO_2(c, tet)$ are -15.7 and -41.1 kJ mol^{-1}, respectively. Therefore, the discrepancies in Table 1 result from the fact that the Manchester group recalculated all the static-bomb results on the basis of amorphous GeO_2^4, whereas the Gorki thermochemists in their 1980 review have probably considered a hexagonal state for germanium dioxide. In addition to the data quoted from the two aforementioned surveys, Table 1 contains values taken from Sussex–NPL Tables[5] (where the amorphous state for GeO_2 was assumed) and from an NBS compilation[6]. More recent results are also included[7-13].

TABLE 1.   Standard enthalpies of formation of organogermanium compounds (kJ mol^{-1})^a

Compound	ΔH_f° (l/c)	Method/Ref.^{b}	ΔH_v° or $\Delta H_s^{\circ C}$	ΔH_f° (g)
GeH$_4$, g		RC/6		90.8 ± 2.1
Ge$_2$H$_6$, l	137.3	RC/6	25.0	162.3 ± 1.3
Ge$_3$H$_8$, l	193.7	RC/6	33.1	226.8
Me$_2$Ge=CH$_2$, g		IMR/7		46^d
GeMe$_4$, l	−98.3 ± 9.4	SB/1	27.3 ± 0.4^e	−71.0 ± 9.4
	−117.2 ± 8.4	SB/2		−89.9 ± 8.4
	−131.1 ± 8.3	SB/8		−103.8 ± 8.3
Me$_3$Ge(t-Bu)		IMR/9		−232.8
Et$_3$GeH, l	−167.4 ± 8.4	SB/2	41.9	−125.5 ± 12.6
GeEt$_4$, l	−206.4 ± 7.5	RB/1	45.7 ± 0.4^e	−160.7 ± 7.5
	−213.4 ± 8.4	SB,RB/2		−167.7 ± 8.4
GePr$_4$, l	−288.3 ± 4.9	SB/1	61.5 ± 4.2	−226.8 ± 6.5
	−305.4 ± 4.2	SB/2		−243.9 ± 5.9
GeBu$_4$, l	−384.9 ± 4.2	SB/2	50.2	−334.7 ± 4.2
Ge(i-Bu)$_4$, l	−464.4 ± 4.2	SB/2	46.0	−418.4 ± 4.2
Ph$_2$GeH$_2$, l	224 ± 3.3	SB/10		
Ph$_2$Ge(C≡CH)$_2$, c	656.9 ± 8.4	SB/2	133.9	790.8 ± 8.4
	658 ± 8.4	SB/11		791.9 ± 8.4
Ph$_2$Ge(CH$_2$)$_4$, c	34.3 ± 10.0	RB/12	104.6 ± 2.8	138.9 ± 10.4
Ph$_3$GeCH=CH$_2$, c	263.9 ± 7.0	RB/13	98.7 ± 1.6	362.6 ± 7.2
GePh$_4$,c	288.6 ± 23.6	RB/1	156.9 ± 4.2	445.5 ± 23.9
	280.3 ± 12.6	SB,RB/2		437.2 ± 13.3
Ph$_3$GeC≡CPh, c	471.4 ± 7.9	RB/13	107.5 ± 1.5	578.9 ± 8.0
GeBz$_4$, c	223.3 ± 11.9	RB/1	168.6 ± 8.4	391.9 ± 14.6
	217.6 ± 8.4	RB/2	184.1	401.7 ± 12.6

TABLE 1. (*continued*)

Compound	ΔH_f° (l/c)	Method/Ref.[b]	ΔH_v° or $\Delta H_s^{\circ C}$	ΔH_f° (g)
Ge(PhCH=CH)$_4$, c	832.6 ± 12.6	SB/2		

, c	866.1 ± 8.4	SB/2		
	868 ± 8.0	SB/11		
Me$_3$GeOEt, l	−397.5 ± 8.4	RC/2	33.5	−364.0 ± 12.6
Et$_3$GeOO(*t*-Bu), l	−486.2 ± 7.0	SB/1	43.5 ± 4.2	−442.7 ± 8.1
	−502.1 ± 4.2	SB/2		−458.6 ± 5.9
Ge(OMe)$_4$, l	−842.2 ± 4.6	SB/5	40.2 ± 0.4	−802.0 ± 4.6
	−861.9 ± 4.2	SB/2		−821.7 ± 4.2
Ge(OEt)$_4$, l	−1006.9 ± 4.8	SB/5	43.1 ± 0.4	−963.8 ± 4.8
	−1025.1 ± 4.2	SB/2		−982.0 ± 4.2
Ge(OPh)$_4$, l	−458.1 ± 6.7	SB/5	37.4 ± 0.4	−420.7 ± 6.7
	−472.8 ± 4.2	SB/2		−435.4 ± 4.2
Me$_3$GeNMe$_2$, l	−154.8 ± 8.4	RC/2	33.5	−121.3 ± 12.6
Et$_3$GeNEt$_2$, l	−342.5 ± 6.1	SB/1	46.0 ± 4.8	−296.5 ± 7.8
	−359.8 ± 4.2	SB/2		−313.8 ± 6.4
Me$_3$GeN$_3$, l	117.2 ± 4.2	RC/2	37.6	154.8 ± 8.4
Me$_3$GeSBu, l	−272.0 ± 8.4	RC/2	41.9	−230.1 ± 12.6
Me$_3$GeCl, l	−301.2 ± 8.4	RC/2	33.4	−267.8 ± 12.6
Me$_3$GeBr, l	−255.2 ± 8.4	RC/2	37.6	−217.6 ± 12.6
(GeMe$_3$)$_2$		IMR/9		−261.3
(GeEt$_3$)$_2$, l	−372.9 ± 11.9	SB/1	62.8 ± 2.1	−310.2 ± 12.1
	−410.0 ± 8.4	SB/2		−347.2 ± 8.7
(GePh$_3$)$_2$, c	453.7 ± 14.2	RB/1	209.2 ± 4.2	662.9 ± 14.8
	447.7 ± 8.4	RB/2		656.9 ± 9.4
(GeEt$_3$)$_2$O, l	−611.3 ± 11.9	SB/1	58.6 ± 4.2	−552.7 ± 12.6
	−631.8 ± 8.4	SB/2		−573.2 ± 9.4
(GePh$_3$)$_2$O, c	161.1 ± 10.6	RB/13	98.0 ± 1.5	259.1 ± 10.7
(Me$_2$GeO)$_4$, c	−1514.5 ± 25.9	SB/1	68.2 ± 4.2	−1446.3 ± 26.2
	−1585.7 ± 8.4	SB/2		−1517.5 ± 9.4
(Et$_2$GeO)$_4$, c	−1519.3 ± 27.6	SB/1		
	−1589.9 ± 20.9	SB/2		
Me$_3$GeSiMe$_3$		IMR/9		−310.1
Me$_3$GeSnMe$_3$		IMR/9		−165.9
(GeEt$_3$)$_2$Hg, l	−101.0 ± 9.5	SB/1	62.8 ± 4.2	−38.2 ± 10.4
	−133.9 ± 4.2	SB/2		−71.1 ± 5.9
[Ge(*i*-Pr)$_3$]$_2$Hg, l	−273.4 ± 9.7	SB/1	54.4 ± 4.2	−219.0 ± 10.6
	−309.6 ± 4.2	SB/2		−255.2 ± 5.9

[a] See text for a detailed explanation of the tabulated enthalpies of formation.

[b] IMR = ion–molecule reactions; RB = rotating-bomb combustion calorimetry; RC = reaction calorimetry; SB = static-bomb combustion calorimetry.

[c] The enthalpies of vaporization or sublimation recommended in References 1 and 2 are in agreement, except in the case of GeBz$_4$.

[d] The value relies on the proton affinity value of NH$_3$, which is taken here as 854 kJ mol^{-1}.

[e] Values from M. H. Abraham and R. J. Irving, *J. Chem. Thermodyn.*, **12**, 539 (1980).

Because of the controversy surrounding the use of static-bomb calorimetry for determining enthalpies of formation of organogermanium compounds[1,2], the reliability of most data in Table 1 cannot be fully assessed. It is, however, possible to discuss generally some of the results.

The best experimental determination for ΔH_f°(GeMe$_4$, l) is probably the one obtained by Long and Pulford who used a highly purified sample and made a careful analysis of the final states (GeO$_2$, c, hex) of the static-bomb combustion experiments[8]. The values for ΔH_f°(GeEt$_4$, l) and ΔH_f°(GePh$_4$, l) were derived from rotating-bomb measurements (GeO$_2$, c, hex) and these are preferred to the ones recommended by Tel'noi and Rabinovich[2], which are averages of static and rotating-bomb results. Although the two values listed for ΔH_f°(GeBz$_4$, c) are both derived from the same original reference (GeO$_2$, c, hex), it is not clear from the review articles how the authors corrected the results. The two values listed for the enthalpy of formation of Ge(n-Pr)$_4$ are both corrected from the same original reference (GeO$_2$, c, hex) but the value given from Reference 1 has been further corrected to refer to amorphous GeO$_2$.

Assessing the remaining data in Table 1, particularly those that were determined from static-bomb calorimetry experiments, is difficult. For example, the errors due to the uncertainty in the germanium dioxide state will be proportional to the number of germanium atoms in the molecule. It is therefore not surprising that the enthalpies of formation of the compounds (GeR$_3$)$_2$, (GeR$_3$)$_2$O, (R$_2$GeO)$_4$ and (R$_3$Ge)$_2$Hg (Table 1) given by Pilcher and Skinner[1] differ considerably from those in Tel'noi and Rabinovich's review[2].

A classic method[14] for examining the thermochemical regularity of an organic homologous series is plotting the standard molar enthalpies of formation versus the number of carbon atoms in the compounds. The linear relationship may be expressed as equation 1 where all the enthalpies of formation are in either the gaseous or a condensed phase, α is the slope, β is the y-intercept and n_c is the number of carbon atoms in the compound.

$$\Delta H_f^\circ = \alpha \bullet [n_c] + \beta \qquad (1)$$

In general, more reliable constants are obtained[14] when $n_c \geqslant 4$ (per alkyl substituent), but in the present analysis there are not enough data to fulfill this condition. Accordingly, when the preferred gaseous enthalpy-of-formation values for tetraethyl, tetra-n-propyl and tetra-n-butyl germanium compounds (GeO$_2$, c, hex) are subjected to a least-squares analysis[15], an excellent straight line results ($r^2 = 0.99985$, $\alpha = -21.92 \pm 0.53$, $\beta = 16.68 \pm 7.4$). The methylene increment, α, is not too different from the purported 'universal' methylene increment of -20.6 kJ mol^{-1} derived from the homologous n-alkanes and is within the range of variation observed for functionally substituted alkanes. The resulting straight line for the liquid-phase enthalpies of formation of these same germanium derivatives is also fairly good ($r^2 = 0.9988$, $\alpha = -21.66 \pm 1.62$, $\beta = -40.71 \pm 21.79$). A notable aspect of the α values for the gaseous- and liquid-phase germanium series is their near identity. However, a lack of regularity in the enthalpies of vaporization for these species in Table 1 shows they cannot all be correct. For other homologous series[14] the liquid-phase α values are generally 4–5 kJ mol^{-1} more negative, consonant with an earlier analysis[16].

For all homologous series studied, the enthalpy of formation of the methyl substituted species deviates from the otherwise linear relationship established by the higher members of the series[14]. The direction[17] of deviation depends on the electronegativity of the attached atom relative to carbon: for atoms of greater electronegativity, the methyl substituted species has a more positive enthalpy of formation than calculated from equation 1 (a positive deviation) and for atoms of lower electronegativity, the methyl substituted species has a more negative enthalpy of formation than calculated (a negative deviation). The absolute magnitude of the deviation generally increases with increasing electronegativity[18] or increasing electropositivity (relative to carbon) of the attached atom.

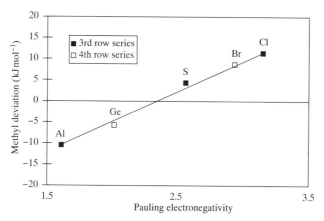

FIGURE 1. The relationships between the methyl deviation, $\delta(CH_3-Z)$, and Pauling electronegativity of the bonded Z atom in various homologous functional group series, $CH_3-(CH_2)_m-Z$. See References 17b and 18

We have shown[19] that there are separate but nearly parallel linear relationships between the methyl deviations, $\delta(CH_3-Z)$, of methyl substituted species CH_3-Z and the electronegativity of the atom in Z bonded to carbon for functionality in the second row of the periodic table ($Z = -OH, -NH_2, -CH_3, -B<$) and in the third row of the periodic table ($Z = -Cl, -SH, -Al<$). If the methyl deviations for CH_3Br and $(CH_3)_4Ge$, both containing fourth-row atoms, are now calculated and the results plotted versus the Pauling electronegativity, the two new points fall on the line established by the third-row series (Figure 1).

Another simple method that has been used for assessing the data for many families of compounds, say ML_n, consists in plotting $\Delta H_f^\circ(ML_n)$ versus $\Delta H_f^\circ(LH)$, with ML_n and LH in either their standard reference states (their stable physical states at 298.15 K and 1 bar) or in the gas phase[20]. It has been observed that many[21] of the above plots which involve reliable thermochemical data define excellent straight lines. This empirical linear relationship may be expressed as equation 2.

$$\Delta H_f^\circ(ML_n) = a[\Delta H_f^\circ(LH)] + b \tag{2}$$

The meaning of this observation is seen by considering Scheme 1. $\Delta H^\circ(3)$ and $\Delta H^\circ(4)$ are the enthalpies of hypothetical reactions of a 'family' of compounds ML_n, where reactants and products are in the standard reference states and in the gas phase, respectively, and ΔH_V° are vaporization or sublimation enthalpies.

$$ML_n \text{ (rs)} + n\text{HY (rs)} \xrightarrow{\Delta H^\circ(3)} MY_n \text{ (rs)} + n\text{LH (rs)} \tag{3}$$

$$\Big\downarrow \Delta H_{V1}^\circ \qquad \Big\downarrow n\Delta H_{V2}^\circ \qquad \Big\downarrow \Delta H_{V3}^\circ \qquad \Big\downarrow n\Delta H_{V4}^\circ$$

$$ML_n \text{ (g)} + n\text{HY (g)} \xrightarrow{\Delta H^\circ(4)} MY_n \text{ (g)} + n\text{LH (g)} \tag{4}$$

SCHEME 1

Observation of the empirically linear equation 2 for the gas-phase enthalpy-of-formation data implies that $\Delta H°(3)$ is constant for the series of compounds $ML_n{}^{20a,b}$. $\Delta H°(3)$ can also be expressed in terms of the bond dissociation enthalpies (equation 5) by again using Scheme 1.

$$\Delta H°(3) = n[\bar{D}(M-L) - D(L-H)] + [\Delta H°_{V1} - n\Delta H°_{V4}] + n\Delta H°_{V2}$$

$$+ [nD(H-Y) - n\bar{D}(M-Y) - \Delta H°_{V3}] \qquad (5)$$

The third bracketed term in equation 5 is constant for a given Y, implying that $\{n[\bar{D}(M-L) - D(L-H)] + [\Delta H°_{V1} - n\Delta H°_{V4}]\}$ is also constant for the series of compounds that obey the linear correlation. It is therefore very likely that $\bar{D}(M-L)$ and $D(L-H)$ follow nearly parallel trends. Obviously, if the linear correlation holds for reactants and products in the gas phase, then $\Delta H°(4)$ is constant and so will be $[\bar{D}(M-L) - D(L-H)]$.

The application of the above method to the gaseous germanium tetra-alkyls is shown in Figure 2. The linear fit involving the preferred values for tetraethyl, tetra-n-propyl, tetra-n-butyl and also tetrabenzyl germanium species (GeO2, c, hex) and the enthalpies of formation of the corresponding alkanes RH[22] leads to ($r^2 = 0.99996$):

$$\Delta H°_f(GeR_4, g) = (4.118 \pm 0.017)\Delta H°_f(RH, g) + (184.6 \pm 1.7) \qquad (6)$$

While the slope is slightly different from the ideal value of 4 (equal to the number of ligands bonded to the germanium atom), the statistical behavior of the correlation is good. The methyl species' enthalpies of formation are not expected to fit this correlation because tetramethyl germanium and methane deviate by different magnitudes from the linear relationships of their respective homologous series in equation 1. If enthalpy-of-formation values are calculated from equation 1 for Me4Ge (-71.0 kJ mol^{-1}) and CH4 (-63.8 kJ mol^{-1})[23] and then used in the correlation of equation 2, the fit is very good ($r^2 = 0.9998$, $a = 4.118 \pm 0.03$, $b = 186.1 \pm 2.71$).

We now turn to the calculation of bond enthalpy terms. In a first approximation, the enthalpy required to break all the chemical bonds in, say, gaseous GeR4, the so-called

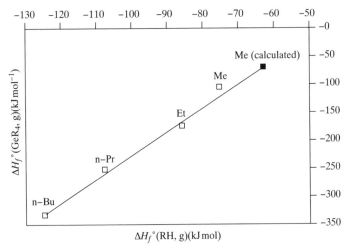

FIGURE 2. Enthalpies of formation of gaseous germanium tetra-alkyls versus the enthalpies of formation of the corresponding hydrocarbons. The point Me (cal'c.) is for the calculated values of $\Delta H°_f(CH_4)$ and $\Delta H°_f(GeMe_4)$. See text

enthalpy of atomization[14], can be identified with a sum of terms (E), each one representing the enthalpy of a given bond. For example,

$$\Delta H^{\circ}_{atom}(\text{GeMe}_4) = 4E\,(\text{Ge}-\text{C}) + 12E\,(\text{C}-\text{H})_p$$

$$\Delta H^{\circ}_{atom}(\text{GePr}_4) = 4E\,(\text{Ge}-\text{C}) + 8E\,(\text{C}-\text{H})_s^{\text{Ge}} + 8E\,(\text{C}-\text{H})_s + 8E\,(\text{C}-\text{C}) + 12E\,(\text{C}-\text{H})_p$$

$$\Delta H^{\circ}_{atom}(\text{GePh}_4) = 4E\,(\text{Ge}-\text{C}_b) + 24E\,(\text{C}_b-\text{C}_b) + 20E\,(\text{C}_b-\text{H})$$

$$\Delta H^{\circ}_{atom}(\text{GeBz}_4) = 4E\,(\text{Ge}-\text{C}) + 8E\,(\text{C}-\text{H})_s^{\text{Ge}} + 4E\,(\text{C}-\text{C}_b)$$
$$+ 24E\,(\text{C}_b-\text{C}_b) + 20E\,(\text{C}_b-\text{H})$$

where the subscripts p and s indicate primary and secondary carbon atoms, respectively, b refers to a carbon from a phenyl group and the superscript Ge means that the carbon is bonded to a germanium atom (a detailed description of this 'scheme', developed by Laidler, may be found elsewhere)[1,14]. As the appropriate terms for hydrocarbons are available, the selected enthalpies of formation in Table 1 can be taken together with the enthalpies of formation of the gaseous elements to derive $E\,(\text{Ge}-\text{C})$, $E\,(\text{Ge}-\text{H})_s^{\text{Ge}}$ and $E\,(\text{Ge}-\text{C}_b)$[24]. This exercise leads to $E\,(\text{Ge}-\text{C}) = 257.09$ kJ mol^{-1} (from GeMe$_4$), $E\,(\text{Ge}-\text{C}_b) = 262.91$ kJ mol^{-1} and to slightly different values for $E\,(\text{Ge}-\text{H})_s^{\text{Ge}}$, according to the germanium compound used as source (kJ mol^{-1}): 404.22 kJ mol^{-1} (GeEt$_4$), 402.19 kJ mol^{-1} (GePr$_4$) and 402.69 kJ mol^{-1} (GeBz)$_4$. Let us take the values for $E\,(\text{Ge}-\text{C})$ and $E\,(\text{Ge}-\text{C}_b)$, together with $E\,(\text{Ge}-\text{H})_s^{\text{Ge}} = 402.5$ kJ mol^{-1}, and predict the enthalpy of atomization of the five-membered metallacycle Ph$_2$Ge(CH$_2$)$_4$. The value obtained, 15575 kJ mol^{-1}, is 54 kJ mol^{-1} *lower* that the experimental result, 15629 kJ mol^{-1}, calculated from the enthalpy of formation in Table 1, i.e. the compound is predicted to be *less* stable than it actually is. Strain in the metallacycle, for which there is no evidence from the X-ray structure[9], would imply an opposite trend. Therefore, the selected bond terms are apparently unsuitable to address this issue. Part of the problem is probably caused by the value of $E\,(\text{Ge}-\text{C}_b)$ which was transferred from GePh$_4$ to Ph$_2$Ge(CH$_2$)$_4$. Steric hindrance in tetraphenyl germanium will make $E\,(\text{Ge}-\text{C}_b)$ smaller than in the metallacycle compound, as indicated by data for tin compounds (see below): it is observed that $E\,(\text{Sn}-\text{C}_b)$ in SnPh$_4$ is about 15 kJ mol^{-1} smaller than in Me$_3$SnPh[25]. If $E\,(\text{Ge}-\text{C}_b)$ used above (269.21 kJ mol^{-1}) is increased by that amount, the calculated atomization enthalpy will be closer to the experimental value (but still 24 kJ mol^{-1} lower).

The calculation of other bond enthalpy terms, such as $E\,(\text{Ge}-\text{Ge})$, $E\,(\text{Ge}-\text{O})$, $E\,(\text{Ge}-\text{N})$ and $E\,(\text{Ge}-\text{S})$, can be made from data in Table 1. However, due to the above-mentioned controversy involving most of the data obtained with static-bomb combustion calorimeters, we refrain from tabulating those terms.

B. Bond Dissociation Enthalpies

Photoacoustic calorimetry (PAC) studies by Clark and Griller have provided significant insights into the bonding energetics of organogermanium compounds[26]. The main conclusion from those studies was that alkyl and aryl substituents have a rather small influence on Ge$-$H bond dissociation enthalpies in germanium hydrides (Table 2). The almost negligible effect of alkylation on $D\,(\text{Ge}-\text{H})$ had been observed for the analogous silanes, where for example $D\,(\text{Si}-\text{H})$ in SiH$_4$ (378 $\pm$ kJ mol^{-1}) is indistinguishable from $D\,(\text{Si}-\text{H})$ in Me$_3$SiH[27,28]. However, Griller and coworkers have also found that the silicon$-$hydrogen bonds are considerably weakened by silylation, e.g. $D\,[(\text{Me}_3\text{Si})_3\text{Si}-\text{H}]$

TABLE 2. Bond dissociation enthalpies in organogermanium compounds ($kJ\,mol^{-1}$)

Compound	Method/Ref.[a]	$D(M-L)^b$
Ge−H	PIMS/32b	$> 225^{c,d}$
HGe−H	PIMS/32b	$< 288^{c,e}$
H_2Ge-H	PIMS/32b	$> 236^{c,f}$
H_3Ge-H	KG/32c	346 ± 10
	PIMS/32b	$< 358^{c,g}$
Me_3Ge-H	KG/32a	340 ± 10
	PAC/26	345.6 ± 2.1^h
Me_3Ge-Me	/32d	347 ± 17
	VLPP/32e	339 ± 13
Et_3Ge-H	PAC/26	348.5 ± 2.5^h
Bu_3Ge-H	PAC/26	349.8 ± 2.5^h
$PhGeH_2-H$	PAC/26	335.6 ± 2.9^h
Ph_2GeH-H	PAC/26	336.8 ± 2.9^h
Ph_3Ge-H	PAC/26	339.7 ± 3.3^h
$IGeH_2-H$	KG/32f	332 ± 10
$Mes_3Ge-GeMes_3^i$	ES/38	87 ± 8
$Tep_3Ge-GeTep_3^j$	ES/38	44 ± 8

[a] ES = equilibrium in solution; KG = kinetics in the gas phase; PAC = photoacoustic calorimetry; PIMS = photoionization mass spectrometry; VLPP = very low pressure pyrolysis.
[b] Values at ~ 298 K, unless indicated otherwise.
[c] Value at 0 K.
[d] Authors recommend 264 $kJ\,mol^{-1}$.
[e] Authors recommend 277 $kJ\,mol^{-1}$.
[f] Authors recommend 247 $kJ\,mol^{-1}$.
[g] Authors recommend 343 ± 8 $kJ\,mol^{-1}$.
[h] The value reported in the original paper is 4.2 $kJ\,mol^{-1}$ smaller because the authors have used different auxiliary data to derive the equation relating the experimentally measured quantity with the bond dissociation enthalpy.
[i] Mes = mesityl.
[j] Tep = 2,4,6-triethylphenyl.

is about 48 $kJ\,mol^{-1}$ smaller than $D(H_3Si-H)^{29}$, so that it is possible that germylation will reduce $D(Ge-H)$. Regarding the aryl substitution effect, it seems that it is smaller for germanium than for silicon: $D(Ph_3Si-H)$ is about 26 $kJ\,mol^{-1}$ lower than $D(H_3Si-H)^{30}$.

The PAC results reported in Table 2 were obtained in benzene or isooctane solution and may therefore be affected by solvation. The available evidence, however, indicates that these solvation phenomena do not disturb significantly the energetics of the bond cleavages[31], so that the PAC results can be compared with values derived from gas-phase experiments[32]. Indeed, it is noted in Table 2 that there is satisfactory agreement between $D(Me_3Ge-H)$ obtained by PAC and from a gas-phase kinetic study by Doncaster and Walsh[32a].

One of the interesting features in Table 2 regards the values of $D(Me_3Ge-H)$ and $D(Me_3Ge-Me)$. Despite the uncertainties that affect the latter, these bond dissociation enthalpies are remarkably similar, again resembling what is observed for the analogous silicon compounds[27]. Accepting $D(Me_3Ge-Me) = 339 \pm 13$ $kJ\,mol^{-1}$ and taking the selected enthalpy of formation of $GeMe_4$ in Table 1 together with the enthalpy of formation of methyl radical[33], one obtains $\Delta H_f^\circ(GeMe_3, g) = 88 \pm 15$ $kJ\,mol^{-1}$.

The enthalpy of formation of trimethylgermanyl radical could also be derived from the (probably more accurate) PAC value for $D(\mathrm{Me_3Ge-H})$, if the enthalpy of formation of $\mathrm{Me_3GeH}$ were available. Although no experimental value of this quantity has been reported, we trust that it can be estimated as $-55.2\ \mathrm{kJ\,mol^{-1}}$ by using the Laidler scheme mentioned above, namely the terms $E(\mathrm{Ge-C}) = 257.09\ \mathrm{kJ\,mol^{-1}}$, $E(\mathrm{Ge-H}) = 289.55\ \mathrm{kJ\,mol^{-1}}$ and $E(\mathrm{C-H})_p = 411.26\ \mathrm{kJ\,mol^{-1}}$ [24,34,35]. The estimated value for $\Delta H_f^\circ(\mathrm{Me_3GeH, g})$ implies $\Delta H_f^\circ(\mathrm{GeMe_3, g}) = 72.4\ \mathrm{kJ\,mol^{-1}}$ [or $D(\mathrm{Me_3Ge-Me}) = 323\ \mathrm{kJ\,mol^{-1}}$], which is almost within the error bar assigned to the above $88 \pm 15\ \mathrm{kJ\,mol^{-1}}$, but it is nevertheless significantly lower. Unfortunately, the discrepancy cannot be settled with the present data. However, since the selected enthalpy of formation of $\mathrm{GeMe_4}$ affects both methods of deriving $\Delta H_f^\circ(\mathrm{GeMe_3, g})$, albeit with different weights, the disagreement strongly suggests that either $D(\mathrm{Me_3Ge-Me}) = 339\ \mathrm{kJ\,mol^{-1}}$ is an upper limit or, less likely, $D(\mathrm{Me_3Ge-H})$ is too low. In other words, it is possible that, unlike the case of silanes, the germanium–hydrogen bond is 15–20 units stronger than the germanium–methyl bond. Incidentally, this would be in line with a general trend observed for transition metal-hydrides and -alkyls, where the differences $D(\mathrm{M-H}) - D(\mathrm{M-Me})$ are smaller for more electropositive M[36].

The PAC result for $D(\mathrm{Et_3Ge-H})$, $348.5\ \mathrm{kJ\,mol^{-1}}$, together with $\Delta H_f^\circ(\mathrm{Et_3GeH, g})$ from Table 1, affords $\Delta H_f^\circ(\mathrm{GeEt_3, g}) = 5 \pm 13\ \mathrm{kJ\,mol^{-1}}$, which in turn leads to $D(\mathrm{Et_3Ge-Et}) = 285\ \mathrm{kJ\,mol^{-1}}$[33], by using $\Delta H_f^\circ(\mathrm{GeEt_4, g}) = -160.7\ \mathrm{kJ\,mol^{-1}}$. This germanium–carbon bond dissociation enthalpy is, however, about $25\ \mathrm{kJ\,mol^{-1}}$ lower than expected on the basis of two assumptions: (1) As suggested above, the difference $D(\mathrm{Ge-H}) - D(\mathrm{Ge-Me})$ is ca $20\ \mathrm{kJ\,mol^{-1}}$; (2) the approximate relationship $D(\mathrm{R_3Ge-Me}) - D(\mathrm{R_3Ge-Et}) \approx D(\mathrm{Me-H}) - D(\mathrm{Et-H}) = 18\ \mathrm{kJ\,mol^{-1}}$ holds for the compounds under discussion[36]. One may invoke a substantial error in the experimental value of $\Delta H_f^\circ(\mathrm{Et_3GeH, g})$ to explain the apparent discrepancy. The use of Laidler terms, including $E(\mathrm{Ge-C}) = 257.09\ \mathrm{kJ\,mol^{-1}}$, $E(\mathrm{Ge-H})_s^{\mathrm{Ge}} = 402.5\ \mathrm{kJ\,mol^{-1}}$ and $E(\mathrm{Ge-H}) = 289.55\ \mathrm{kJ\,mol^{-1}}$ [24,34] yield $\Delta H_f^\circ(\mathrm{Et_3GeH, g}) = -87.5\ \mathrm{kJ\,mol^{-1}}$, a result which is well above the experimental one in Table 1 and leads to $\Delta H_f^\circ(\mathrm{GeEt_3, g}) = 43\ \mathrm{kJ\,mol^{-1}}$ and to the more sensible (but probably too high) value $D(\mathrm{Et_3Ge-Et}) = 323\ \mathrm{kJ\,mol^{-1}}$.

The enthalpy of formation of $\mathrm{GeH_4}$ shown in Table 1 and $D(\mathrm{H_3Ge-H})$ in Table 2 yield $\Delta H_f^\circ(\mathrm{GeH_3, g}) = 219 \pm 10\ \mathrm{kJ\,mol^{-1}}$. This value, together with $\Delta H_f^\circ(\mathrm{Ge_2H_6})$ from Table 1, gives $D(\mathrm{H_3Ge-GeH_3}) = 275 \pm 15\ \mathrm{kJ\,mol^{-1}}$. A similar exercise involving $\mathrm{GeEt_4}$ and $\mathrm{Ge_2Et_6}$ leads to $D(\mathrm{Et_3Ge-GeEt_3}) = 396\ \mathrm{kJ\,mol^{-1}}$ or $320\ \mathrm{kJ\,mol^{-1}}$ according to the value adopted for $\Delta H_f^\circ(\mathrm{GeEt_3, g})$, $43\ \mathrm{kJ\,mol^{-1}}$ or $5\ \mathrm{kJ\,mol^{-1}}$, respectively. Although the latter result, based on $\Delta H_f^\circ(\mathrm{GeEt_3, g}) = 5\ \mathrm{kJ\,mol^{-1}}$, is closer to $D(\mathrm{H_3Ge-GeH_3})$, and therefore looks more reasonable (see below), it is likely that the above calculation is affected by a large error in $\Delta H_f^\circ(\mathrm{Ge_2Et_6, g})$. In fact, the use of the appropriate Laidler terms[24], including $E(\mathrm{Ge-Ge}) = 162.40\ \mathrm{kJ\,mol^{-1}}$ [obtained from $\Delta H_f^\circ(\mathrm{Ge_2H_6, g})$ and $E(\mathrm{Ge-H})$[34]], $E(\mathrm{Ge-H})_s^{\mathrm{Ge}} = 402.5\ \mathrm{kJ\,mol^{-1}}$, and $E(\mathrm{Ge-C}) = 257.09\ \mathrm{kJ\,mol^{-1}}$, yield $\Delta H_f^\circ(\mathrm{Ge_2Et_6, g}) = -194.3\ \mathrm{kJ\,mol^{-1}}$, much higher than the experimental result in Table 1. This estimated enthalpy of formation of hexamethyldigermane implies $D(\mathrm{Et_3Ge-GeEt_3}) = 280\ \mathrm{kJ\,mol^{-1}}$, quite close to $D(\mathrm{H_3Ge-GeH_3})$.

The comparison of $D(\mathrm{H_3Si-SiH_3})$ with $D(\mathrm{Me_3Si-SiMe_3})$ has been recently addressed in detail by Pilcher and coworkers[37]. Although the question has not been settled, the authors argue that those bond dissociation enthalpies 'may be closer together than previously thought', and that 'this lack of methyl group substituent effect on the Si–Si dissociation enthalpies would be consistent with the constancy of Si—H dissociation enthalpies in the methylsilanes'[28]. These conclusions may well apply for Ge_2Et_6 and Ge_2H_6, supporting the similarity between $D(H_3Ge-GeH_3) = 275 \pm 15$ kJ mol^{-1} and the value found for $D(Et_3Ge-GeEt_3)$, 280 kJ mol^{-1}.

In contrast to the relatively high Ge—Ge bond dissociation enthalpies that have just been discussed, the two last values of $D(Ge-Ge)$ presented in Table 2, obtained from van't Hoff plots involving the equilibrium $(GeR_3)_2 = 2GeR_3$ in solution[38], are rather low. This has been attributed to the steric effects caused by the bulky substituents in the phenyl rings[38]. Unfortunately, the obvious comparison with $D(Ph_3Ge-GePh_3)$ is not obtainable due to the lack of an experimental value of $\Delta H_f^{\circ}(Ph_3GeH, g)$, which, together with $D(Ph_3Ge-H)$ in Table 2, would afford $\Delta H_f^{\circ}(Ph_3Ge, g)$. It is possible, however, to make an estimate of the missing enthalpy of formation, having in mind the discussion in the previous section about the Laidler term for Ge—C_b. Recall that it was shown that $E(Ge-C_b)$ in $GePh_4$ should be a lower limit and that in a strain-free molecule the value should be increased by ca 15 kJ mol^{-1}, i.e. $E(Ge-C_b) \approx 284$ kJ mol^{-1}. Accepting this value (as an upper limit) and using the necessary Laidler terms[24,34], one obtains $\Delta H_f^{\circ}(Ph_3GeH, g) = 312$ kJ mol^{-1} and $\Delta H_f^{\circ}(Ph_3Ge, g) = 434$ kJ mol^{-1}. Finally, the enthalpy of formation of hexaphenyldigermane from Table 1 (662.9 kJ mol^{-1}) implies $D(Ph_3Ge-GePh_3) = 205$ kJ mol^{-1}. Assuming a strain-free Ge—Ge bond in this molecule[38], one half of the difference $D(Ph_3Ge-GePh_3) - D(R_3Ge-GeR_3) \approx$ 75 kJ mol^{-1} (R = alkyl) should reflect the resonance stabilization of the $GePh_3$ radical. We recall, however, the small effect of arylation on Ge—H bond dissociation enthalpies in the hydrides (Table 2), which indicates a negligible stabilization of that radical. Therefore, if the above estimated values for $D(Ph_3Ge-GePh_3)$ and $D(R_3Ge-GeR_3)$ are correct, one is forced to conclude that $(GePh_3)_2$ is also considerably destabilized by steric effects, although these are rather more pronounced in the case of the two molecules with bulky substituents in the phenyl rings.

II. TIN COMPOUNDS

A. Enthalpies of Formation

With very few exceptions[39-42], the available enthalpies of formation of organotin compounds have been compiled in the reviews by Pilcher and Skinner[1] and by Tel'noi and Rabinovich[2] (Table 3). Nevertheless, in contrast to what has been noted for the germanium molecules, the disagreements between the values recommended by the two groups are now usually smaller than 4 kJ mol^{-1}, i.e. within the respective experimental uncertainty intervals. This is not surprising because static-bomb combustion calorimetry, the method used to probe the thermochemistry of most of the compounds in Table 3, usually yields well characterized products; there is general recognition that the results obtained with this technique are reliable[3].

Let us use some of the methods applied to the germanium compounds to assess a few values from Table 3. A plot of the three gaseous enthalpies of formation for tetraethyl, tetra-n-propyl and tetra-n-butyl tin species versus the number of carbon atoms in the compound (equation 1) shows that probably at least one of them is inaccurate. In the liquid phase there is an additional enthalpy of formation, that for tetra-n-hexyl tin. A plot of the liquid enthalpies of formation versus total carbon number shows that the enthalpy of formation for tetraethyl tin is an outlier and the remaining three points define a fairly good straight line[15] ($r^2 = 0.99953$, $\alpha = -21.77 \pm 0.80$, $\beta = 47.78 \pm 12.39$). If

TABLE 3. Standard enthalpies of formation of organotin compounds ($kJ\,mol^{-1}$)

Compound	ΔH_f° (l/c)	Method/Ref.[a]	ΔH_v° or $\Delta H_s^{\circ b}$	ΔH_f° (g)
SnH₄, g		RC/6		162.8 ± 2.1
Me₂SnH₂, l	58.6 ± 4.2	SB/2	29.3	87.9 ± 4.2
Me₂Sn=CH₂, g		IMR/7		130^c
Me₂SnEt₂, l	-121.3 ± 20.9	SB/2	41.8	-79.5 ± 20.9
Me₃SnH, l	-8.4 ± 4.2	SB/2	33.5	25.1 ± 4.2
SnMe₄, l	-52.3 ± 1.7	SB/1	32.0 ± 0.8^d	-20.3 ± 1.9
Me₃SnCH=CH₂, l	54.4 ± 13.3	RC/1	37.2 ± 2.1	91.7 ± 13.4
Me₃SnEt, l	-67.2 ± 2.4	SB/1	37.7 ± 1.7	-29.5 ± 2.9
Me₃(i-PrSn), l	-87.4 ± 4.3	SB/1	40.6 ± 2.1	-46.8 ± 4.8
Me₃(t-BuSn), c	-121.1 ± 4.5	SB/1	54.0 ± 4.2	-67.1 ± 6.2
Me₃(t-BuSn), g		IMR/9		-104.1
Me₃SnPh, l	60.8 ± 3.1	RC/1	52.3 ± 4.2	113.1 ± 5.2
Me₃SnBz, l	26.3 ± 3.9	RC/1	56.5 ± 4.2	82.8 ± 5.7
Et₂SnH₂, l	12.6 ± 4.2	SB/2	29.2	41.8 ± 4.2
Et₃SnH, l	-46.0 ± 8.4	SB/2	46.0	0 ± 8.4
Et₃SnMe, l	-225.9 ± 4.2	SB/2		
Et₃SnCH=CH₂, l	36.2 ± 3.2	SB/1,2	51.7	87.9 ± 8.4
SnEt₄, l	-95.9 ± 2.5	SB/1	50.6 ± 0.2^d	-45.3 ± 2.5
Sn(CH=CH₂)₄, l	300.8 ± 7.6	SB/1,2	50.7	351.5 ± 8.4
Sn(CH₂CH=CH₂)₄, l	-170.2 ± 7.3	SB/1		
Pr₃SnH, l	-133.9 ± 4.2	RC/2		
SnPr₄, l	-211.3 ± 5.3	SB/1	65.4 ± 2.5^d	-145.9 ± 5.9
Sn(i-Pr)₄, l	-188.0 ± 5.6	SB/1	64.9 ± 4.2	-123.1 ± 7.0
Et₂SnBu₂, l	-225.9 ± 8.4	SB/2		
Bu₃SnH, l	-205.0 ± 4.2	RC/2		
SnBu₄, l	-302.1 ± 3.7	SB/1	82.8 ± 2.1	-219.2 ± 4.2
Sn(i-Bu)₄, l	-330.9 ± 6.3	SB/1		
Ph₂Sn(CH₂)₄, c	195.0 ± 5.2	SB/39	106.8 ± 5.5	301.8 ± 7.6
Ph₂Sn(CH₂)₅, l	214.0 ± 6.8	SB/39	75.0 ± 1.5	289.0 ± 7.0
Ph₃SnMe		IMR/40		406 ± 29
Ph₃SnEt		IMR/40		381 ± 29
Ph₃SnCH=CH₂, c	411.4 ± 6.6	SB/41	114.1 ± 2.5	525.5 ± 7.1
Ph₃SnC≡CPh, c	596.2 ± 7.8	SB/41	137.6 ± 2.0	733.8 ± 8.1
SnPh₄, c	411.6 ± 3.7	??RB/1	161.1 ± 4.2	572.7 ± 5.6
SnHex₄, l	-468.6 ± 12.6	SB/2		
Sn(cy-Hex)₄, c	-364.7 ± 28.6	SB/1		
Me₃SnOH, c	-380.7 ± 4.2	RC/2	62.7	-318.0 ± 8.4
Me₃SnOEt, l	-305.4 ± 4.2	RC/2	41.8	-263.6 ± 8.4
Ph₃SnOH, c	50.2 ± 8.4	SB/2	129.7	179.9 ± 8.4
Me₃SnOCOPh, c	-491.7 ± 8.5	SB/1		
Et₃SnOO(t-Bu), l	-421.3 ± 16.0	SB/1	48.8 ± 2.1	-372.5 ± 16.1
Et₃SnOCOPh, c	-575.8 ± 4.5	SB/1		
Et₃SnOOC(Me)₂Ph, l	-285.6 ± 8.6	SB/1	56.5 ± 2.1	-229.1 ± 8.9
Me₃SnNMe₂, l	-58.6 ± 4.2	RC/2	37.7	-20.9 ± 8.4
Et₃SnNEt₂, l	-210.3 ± 4.4	SB/1	50.2 ± 4.2	-160.1 ± 6.1
Me₃SnN₃, c	188.3 ± 4.2	RC/2	62.7	251.0 ± 8.4
Bu₃SnNCO, l	-368.2 ± 4.2	SB/2		
Me₃SnSBu, l	-196.6 ± 4.2	RC/2	41.8	-154.8 ± 8.4
Me₃SnCl, l	-213.0 ± 7.1	RC/1		
Me₃SnBr, l	-185.5 ± 3.5	RC/1	47.3 ± 4.2	-138.2 ± 5.5
Me₃SnI, l	-130.6 ± 3.9	RC/1	48.1 ± 4.2	-82.5 ± 5.7

(*continued overleaf*)

TABLE 3. (continued)

Compound	ΔH_f° (l/c)	Method/Ref.[a]	ΔH_v° or $\Delta H_s^{\circ b}$	ΔH_f° (g)
Bu$_3$SnBr, l	-356.2 ± 1.3	RC/1	83.7 ± 12.6	-272.5 ± 12.6
Ph$_3$SnBr, c	188.0 ± 7.7	RC/1		
Ph$_3$SnI, g		IMR/40		381 ± 29
Ph$_3$SnSPh, g		IMR/40		431
Me$_2$SnCl$_2$, l	-330.3 ± 7.7	RC/1		
Ph$_2$SnBr$_2$, c	-19.5 ± 6.8	RC/1		
MeSnCl$_3$, l	-443.1 ± 7.6	RC/1		
(SnMe$_3$)$_2$, l	-77.1 ± 7.3	RC/1	50.2 ± 4.2	-26.9 ± 8.4
(SnMe$_3$)$_2$, g		IMR/40		-29.8
(SnEt$_3$)$_2$, l	-217.4 ± 8.6	SB/1	62.8 ± 4.2	-154.6 ± 9.5
(SnPh$_3$)$_2$, c	661.4 ± 15.6	SB/1	188.3 ± 4.2	849.7 ± 16.2
Ph$_3$SnSnMe$_3$, g		IMR/40		360 ± 38
(SnEt$_3$)$_2$O, l	-427.0 ± 12.7	SB/1	52.3 ± 2.1	-374.7 ± 12.9
(SnPh$_3$)$_2$O, c	415.9 ± 10.0	SB/42	196.6 ± 8.4	612.5 ± 13.0
(SnMe$_3$)$_2$NMe, l	-133.9 ± 4.2	RC/2	50.2	-83.7 ± 8.4
(Bu$_3$SnN)$_2$C, l	-359.8 ± 4.2	SB/2		
(SnMe$_3$)$_3$N, c	-125.5 ± 8.4	RC/2	50.2	-62.8 ± 12.6
Me$_3$SnSiMe$_3$, g		IMR/9		-207.9
Me$_3$SnGeMe$_3$, g		IMR/9		-165.9

[a] IMR = ion–molecule reactions; RB = rotating-bomb combustion calorimetry; RC = reaction calorimetry; SB = static-bomb combustion calorimetry.
[b] Enthalpies of vaporization or sublimation from Reference 1 or 2, except when indicated otherwise.
[c] The value relies on the proton affinity value of NH$_3$, taken as 854 kJ mol^{-1}.
[d] Values from M. H. Abraham and R. J. Irving, J. Chem. Thermodyn., **12**, 539 (1980).

we analyze the remaining two gaseous-phase values, omitting the enthalpy of formation for tetraethyl tin, the statistical constants are $\alpha = -18.33$ and $\beta = 74.00$. The methylene increment, α, although within the range of variation for other homologous series, is very low and is comparable to the lowest of which we know[43]. Using the statistical constants above, the calculated enthalpies of formation for tetraethyl tin are (kJ mol^{-1}): -72.64 (g); -126.38 (l). The calculated and experimental values for gaseous- and liquid-phase tetraethyl tin differ by essentially identical amounts, showing that the measured enthalpy of vaporization is probably accurate.

The extremely low slope calculated from equation 1 for the gaseous tetraalkyl tin compounds affects the calculation for the methyl deviation[19] of tetramethyl tin from its homologous series linear relationship, and its magnitude (-13.23 per methyl group) seems large compared to its Pauling electronegativity (1.96). When the methyl deviations of tetramethyl tin and methyl iodide, both heteroatoms from the fifth row of the periodic table, are plotted versus the heteroatom electronegativity, neither point falls on the line containing the third- and fourth-row elements and the slope of the new line is not parallel to the line shown in Figure 1. It is noted that both the tin and the iodine methyl deviations were calculated using enthalpies of formation from just two other members of their respective homologus series, neither of which is within the desirable range of $n_c \geqslant 4$ per alkyl substituent.

If the conditions of equations 3–5 hold, the measured enthalpies of formation for the tin-containing compounds should fit the linear relationship of equation 7:

$$\Delta H_f^\circ(\text{SnR}_4, \text{ g}) = (a')\Delta H_f^\circ(\text{RH, g}) + (b') \qquad (7)$$

A plot of $\Delta H_f^\circ(SnR_4, g)$ against $\Delta H_f^\circ(RH, g)$, where R is alkyl, is presented in Figure 3 where the enthalpy-of-formation values used are both experimentally determined (Table 3) and calculated. The calculated values are for tetramethyl and tetraethyl tin and methane. As stated above, the enthalpy-of-formation values for tetraethyl tin may be incorrect. That the calculated value for tetraethyl tin results in a better linear fit with tetrapropyl and tetrabutyl tin is further confirmation of this supposition. And, as discussed for the alkyl germaniums, the methyl deviations of methane and tetramethyl tin are too different for their measured values to fit a linear relationship such as equation 7.

In Figure 3 the value for $Sn(i\text{-}Pr)_4$ falls above the line indicating that, compared to four n-propyl substituents on tin, four isopropyl substituents are destabilizing. Although repulsive interactions between the secondary alkyl groups are possibly responsible, the largeness of tin mitigates this and suggests electronic interactions such as those found for organolithiums[17]. Whichever interactions are responsible, they should obviously be smaller, for example, in the compound $Me_3Sn(i\text{-}Pr)$. Indeed one finds that the enthalpy of formation of this molecule is fit by the regression line obtained with the points for $SnMe_4$, Me_3SnEt and Me_3SnBz (equation 8):

$$\Delta H_f^\circ(Me_3SnR, \ g) = (0.8322 \pm 0.0092)\Delta H_f^\circ(RH, \ g) + (40.91 \pm 0.65) \qquad (8)$$

Although equation 8 is statistically sound, the three values used to define the correlation may not be entirely consistent. Whereas the difference $\Delta H_f^\circ(SnMe_4, \ g) - \Delta H_f^\circ(Me_3SnEt, \ g)$ almost matches the difference $\Delta H_f^\circ(MeH, \ g) - \Delta H_f^\circ(EtH, \ g)$, in keeping with an ideal unit slope, the difference $\Delta H_f^\circ(SnMe_4, \ g) - \Delta H_f^\circ(Me_3SnBz, \ g)$, $-103 \ kJ\,mol^{-1}$, is higher than the $-124 \ kJ\,mol^{-1}$ calculated for $\Delta H_f^\circ(MeH, \ g) - \Delta H_f^\circ(BzH, \ g)$. In other words, the data for $SnMe_4$ and Me_3SnEt would imply $\Delta H_f^\circ(Me_3SnBz, \ g) \approx 105 \ kJ\,mol^{-1}$, i.e. $22 \ kJ\,mol^{-1}$ higher than the value in Table 3.

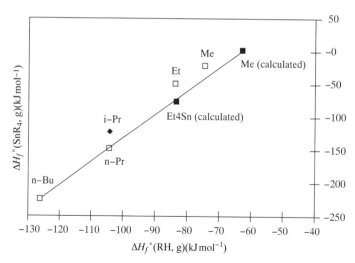

FIGURE 3. Enthalpies of formation of gaseous tin tetra-alkyls versus the enthalpies of formation of the corresponding hydrocarbons. The point Me (cal'c.) is for the calculated values of $\Delta H_f^\circ(CH_4)$ and $\Delta H_f^\circ(SnMe_4)$. The point Et4Sn (cal'c.) is for the calculated value of $\Delta H_f^\circ(SnEt_4)$ and the measured value of $\Delta H_f^\circ(EtH)$. See text

Despite the previous remarks, organotin compounds are one of the few families of organometallic substances for which the thermochemical data justify the calculation of bond enthalpy terms. Some term values (consistent with the terms recommended by Pilcher and Skinner for hydrocarbons)[27] are given in Table 4. A few words on this selection are, however, appropriate.

The value tabulated for the term $E(Sn-H)$, 252.60 kJ mol^{-1}, relies on the enthalpy of formation of SnH$_4$. This result is very close to those obtained from Me$_3$SnH (252.96 kJ mol^{-1}) and Et$_3$SnH (252.29 kJ mol^{-1}), but rather different from the values derived from Me$_2$SnH$_2$ (244.27 kJ mol^{-1}) and Et$_2$SnH$_2$ (258.73 kJ mol^{-1}), suggesting that the overall set of enthalpies of formation is not consistent and that the experimental results for the dihydrides may be inaccurate.

The term $E(C-H)_s^{Sn}$ (401.4 kJ mol^{-1}) represents the average of the values derived from the experimentally measured enthalpies of formation of gaseous SnEt$_4$, SnPr$_4$ and SnBu$_4$ (400.23, 402.51 and 401.49 kJ mol^{-1}, respectively). It can be used to predict, for example, ΔH_f°(Me$_3$SnBz, g) = 106 kJ mol^{-1}. This result is in keeping with the above discussion on the slope of equation 8 and it may indicate that the experimental result in Table 3 is too low. However, the discrepancy may also imply that one or more of the bond terms used in the calculation [e.g. $E(Sn-C) = 217.27$ kJ mol^{-1} and $E(C-H)_s^{Sn} = 401.4$ kJ mol^{-1}] are not appropriate. In other words, and referring to equation 8, it is possible that the experimental value of ΔH_f° (Me$_3$SnBz, g) is correct and that the molecule should not be included in the same family[20a,b] as SnMe$_4$ and Me$_3$SnEt.

The value selected for the term $E(Sn-C_b)$, 233.29 kJ mol^{-1}, calculated from the enthalpy of formation of Me$_3$SnPh, is about 15 kJ mol^{-1} higher than the result derived from the tetraphenyl compound (218.46 kJ mol^{-1}), which indicates repulsive interactions between the phenyl groups in SnPh$_4$. If $E(Sn-C_b) = 233.29$ kJ mol^{-1} is used, together with other appropriate terms in Table 4 and elsewhere[24], to calculate the enthalpy of atomization of the five-membered metallacycle compound Ph$_2$Sn(CH$_2$)$_4$, one obtains 15419 kJ mol^{-1}. The experimental value (15390 kJ mol^{-1}) is 29 kJ mol^{-1} lower, meaning that the compound is less stable than predicted on the basis of the Laidler scheme and therefore indicating that either steric hindrance between the phenyl groups, strain in the metallacycle or both are responsible for the destabilization. A large contribution from the metallacycle strain is, however, unlikely, after the discussion for the analogous germanium molecule and also having in mind that for cyclopentane the strain is 27 kJ mol^{-1}. Some interaction between the phenyl rings has therefore to be admitted and, indeed, if one uses the Laidler term derived from SnPh$_4$ (218.46 kJ mol^{-1}), the calculated enthalpy of atomization of Ph$_2$Sn(CH$_2$)$_4$ becomes 15389 kJ mol^{-1}, almost identical

TABLE 4.   Laidler terms for some organotin compounds[a]

Term	Value (kJ mol^{-1})	Source[b]
$E(Sn-H)$	252.60	SnH$_4$
$E(Sn-C)$	217.27	SnMe$_4$
$E(C-H)_s^{Sn}$	401.4	SnEt$_4$, SnPr$_4$, SnBu$_4$
$E(C-H)_t^{Sn}$	381.12	Me$_3$SnPr-i
$E(Sn-C_b)$	233.29	Me$_3$SnPh
$E(Sn-C_d)$	229.39	Me$_3$SnCHCH$_2$
$E(Sn-Sn)$	147.02	(SnMe$_3$)$_2$

[a] The values must be used together with the terms for hydrocarbons given in Reference 24.
[b] See text for a more detailed explanation of the selection.

to the experimental result. This conclusion seems to hold for the six-membered metalla-cycle compound $Ph_2Sn(CH_2)_5$. The enthalpy of atomization 16563 kJ mol^{-1}, calculated with $E(Sn-C_b) = 218.46$ kJ mol^{-1}, is close to the experimental value, 16556 kJ mol^{-1}.

The value assigned to the term $E(Sn-C_d)$, 229.39 kJ mol^{-1}, which was calculated from the enthalpy of formation of $Me_3SnCHCH_2$, differs significantly from those relying on the gaseous enthalpies of formation for $Et_3SnCHCH_2$ (207.42), $Sn(CHCH_2)_4$ (248.46) and $Ph_3SnCHCH_2$ (240.34). It must be noted that the last value was calculated by considering $E(Sn-C_b) = 218.46$ kJ mol^{-1}, on the assumption of some degree of steric interactions involving the phenyl groups. These interactions are probably smaller than in $SnPh_4$, so that the correct value for $E(Sn-C_b)$ in $Ph_3SnCHCH_2$ must be higher than 218.46, but of course lower than the value in Table 4, 233.29 kJ mol^{-1}. So, while the discrepancy between the terms obtained from $Me_3SnCHCH_2$ and $Ph_3SnCHCH_2$ was expected, the same cannot be said with regard to $E(Sn-C_d)$ calculated from $Et_3SnCHCH_2$. A simple application of the reasoning used to explain the type of correlations using equation 2, namely that $\Delta H_f^\circ(Et_3SnCHCH_2, g) - \Delta H_f^\circ(Me_3SnCHCH_2, g) \approx 3[\Delta H_f^\circ(EtH, g) - \Delta H_f^\circ(MeH, g)] = -28.2$ kJ mol^{-1} leads to $\Delta H_f^\circ(Et_3SnCHCH_2, g) \approx 64$ kJ mol^{-1}, i.e. 24 kJ mol^{-1} lower than the experimental result in Table 3. Note that this rule works well for other cases, e.g. $\Delta H_f^\circ(Et_3SnH, g) - \Delta H_f^\circ(Me_3SnH, g) = -25.1\pm9.4$ kJ mol^{-1} (from Table 3), and therefore there is reason to be suspicious about the inconsistency between the literature values for $\Delta H_f^\circ(Me_3SnCHCH_2, g)$ and $\Delta H_f^\circ(Et_3SnCHCH_2, g)$. As seen in Table 4, the former value has been selected in the present survey.

Another example illustrating the apparent problems with the enthalpy-of-formation data for some tin compounds concerns the term $E(Sn-Sn)$. In order to avoid the complication related to the term $E(Sn-C_b)$, the bond enthalpy $E(Sn-Sn)$ in Table 4 was calculated from $\Delta H_f^\circ[(SnMe_3)_2, g]$. As a confirmation, one can then derive the value for the same term from $\Delta H_f^\circ[(SnEt_3)_2, g]$—they should be similar. Yet, this calculation yields $E(Sn-Sn) = 223.18$ kJ mol^{-1}, a value which looks unreasonably high and questions the experimental value of $\Delta H_f^\circ[(SnEt_3)_2, g]$ in Table 3.

The above analysis reveals that some of the thermochemical data for organotin compounds may not be as accurate as one could hope. Although the information is in general of much better quality than in the case of germanium and lead analogues, we believe that some values in Table 3 should be redetermined. Other examples could have been used to illustrate this point (see also the next section), but once again we wish to resist the temptation of recommending data that in some cases conflict with the available experimental results. By a judicious use of the Laidler terms in Table 4 and/or correlations similar to those in equation 2, it is rather simple to assess other values from Table 3 and predict new data.

B. Bond Dissociation Enthalpies

To our knowledge, the experimental measurements of bond dissociation enthalpies in organotin compounds are limited to the values of $D(Me_3Sn-Me) = 297 \pm 17$ kJ mol^{-1} and $D(Et_3Sn-Et) = 264 \pm 17$ kJ mol^{-1}, as recommended in McMillen and Golden's review[32d], and $D(Bu_3Sn-H) = 308.4 \pm 8.4$ kJ mol^{-1}, obtained by photoacoustic calorimetry (in isooctane)[44]. The first two values, together with the experimental enthalpies of formation of gaseous $SnMe_4$ and $SnEt_4$ (Table 3) and the enthalpies of formation of methyl and ethyl radicals[33], lead to $\Delta H_f^\circ(SnMe_3, g) = 130 \pm 17$ kJ mol^{-1} and $\Delta H_f^\circ(SnEt_3, g) = 100 \pm 18$ kJ mol^{-1}. These results are handles to derive other bond dissociation enthalpies by using the enthalpies of formation of parent compounds in

Table 3. A sample of those data, namely values of $D(\text{Sn}-\text{H})$ and $D(\text{Sn}-\text{C})$, are presented in Table 5. Some gaps in this table (or new entries) can be filled with help of the methods described above. For example, it is reasonable to assume that $D(\text{Et}_3\text{Sn}-\text{Pr-}i)$ will be about 9 kJ mol^{-1} lower than $D(\text{Et}_3\text{Sn}-\text{Et})$ — such is the difference $D(\text{Et}-\text{H})-D(i\text{-Pr}-\text{H})$; note also that $D(\text{Me}_3\text{Sn}-\text{Et})-D(\text{Me}_3\text{Sn}-\text{Pr-}i) = 12$ kJ mol^{-1} is quite close to the prediction. Incidentally, this simple method can also be applied to assess some data. Suppose, for instance, that the value of $D(\text{Et}_3\text{Sn}-\text{Me})$ is to be derived. The enthalpy of formation of gaseous Et$_3$SnMe is not available, but can easily be estimated as -36 kJ mol^{-1}, either by adding the difference $\Delta H_f^\circ(\text{MeH, g}) - \Delta H_f^\circ(\text{EtH, g}) = 9.4$ kJ mol^{-1} to $\Delta H_f^\circ(\text{SnEt}_4, \text{g})$, or using the Laidler scheme. That value affords $D(\text{Et}_3\text{Sn}-\text{Me}) = 283$ kJ mol^{-1}, which looks sensible. By contrast, if the experimental value $\Delta H_f^\circ(\text{Et}_3\text{SnMe}, \text{l})$ in Table 3 were considered, together with $\Delta H_v(\text{Et}_3\text{SnMe}) \approx 46$ kJ mol^{-1} [45], a totally unreasonable value of $D(\text{Et}_3\text{Sn}-\text{Me})$, 427 kJ mol^{-1}, would be obtained.

Despite the large uncertainties assigned to the enthalpies of formation of SnMe$_3$ and SnEt$_3$, their values look fairly consistent: it can be expected, for example, that the Sn$-$H bond dissociation enthalpies in compounds R$_3$SnH (R = linear alkyl) are not very sensitive to R. The values in Table 5 are indeed rather similar and they overlap with the photoacoustic result for $D(\text{Bu}_3\text{Sn}-\text{H})$, 308.4 ± 8.4 kJ mol^{-1} (note that the error bars assigned to the enthalpies of formation of SnMe$_3$ and SnEt$_3$ were not considered when calculating the uncertainties of data in Table 5). As in the cases of SnMe$_3$ and SnEt$_3$, the enthalpy of formation of SnBu$_3$ radical, evaluated as -36 kJ mol^{-1} by using the experimental value of Sn$-$H bond dissociation enthalpy and $\Delta H_f^\circ(\text{Bu}_3\text{SnH}, \text{g}) = -126$ kJ mol^{-1} (estimated with the methods described above), is a handle to calculate other bond dissociation enthalpies, using data in Table 3.

Tributyltin hydride is a well known reducing agent[46], in keeping with the low value of Sn$-$H bond dissociation enthalpy. This weakness of Sn$-$H bond has been used by Rathke

TABLE 5.   Bond dissociation enthalpies for some organotin compounds[a] (kJ mol^{-1})

L	$D(\text{Me}_3\text{Sn}-\text{L})^b$	$D(\text{Et}_3\text{Sn}-\text{L})^c$
H	322.7 ± 4.2	317.7 ± 8.4
Me	297 ± 17	$(283 \pm 5)^d$
Et	278.3 ± 4.9	264 ± 17
i-Pr	266.6 ± 5.2	
t-Bu	248.2 ± 6.5	
Ph	346.7 ± 9.5	
Bz	247.0 ± 8.3	
CH$_2$CH	333.1 ± 15.6	306.8 ± 11.6
Me$_3$Sn	286.5 ± 8.4	

[a] The enthalpies of formation of organic radicals were taken from Reference 33, except in the cases of i-Pr and t-Bu, which were quoted from P. W. Seakins, M. J. Pilling, J. T. Niiranen, D. Gutman, and L. N. Krasnoperov, *J. Phys. Chem.*, **96**, 9847 (1992).
[b] The values rely on $\Delta H_f^\circ(\text{Me}_3\text{Sn}, \text{g}) = 129.8 \pm 17.1$ kJ mol^{-1}, which was calculated from $D(\text{Me}_3\text{Sn}-\text{Me})$. The uncertainties assigned to other $D(\text{Me}_3\text{Sn}-\text{L})$ do not include the uncertainty in $\Delta H_f^\circ(\text{Me}_3\text{Sn}, \text{g})$.
[c] The values rely on $\Delta H_f^\circ(\text{Et}_3\text{Sn}, \text{g}) = 99.7 \pm 17.6$ kJ mol^{-1}, which was calculated from $D(\text{Et}_3\text{Sn}-\text{Et})$. The uncertainties assigned to other $D(\text{Et}_3\text{Sn}-\text{L})$ do not include the uncertainty in $\Delta H_f^\circ(\text{Et}_3\text{Sn}, \text{g})$.
[d] Estimated. See text.

and coworkers to probe the energetics of carbon dioxide activation by measuring the equilibrium constant of reaction 9 in tetrahydrofuran at several temperatures $(115-175\,^{\circ}\text{C})^{47}$:

$$Bu_3SnH \text{ (soln)} + CO_2 \text{ (g)} \longrightarrow Bu_3SnOC(O)H \text{ (soln)} \tag{9}$$

A van't Hoff plot led to $\Delta H_r(9) = -76.6 \pm 0.8 \text{ kJ mol}^{-1}$ and $\Delta S_r(9) = -84.5 \pm 0.8 \text{ J/(mol K)}$.

III. LEAD COMPOUNDS

A. Enthalpies of Formation

There is general agreement that static-bomb combustion calorimetry is inherently unsatisfactory to determine enthalpies of formation of organolead compounds[2,3]. Unfortunately, as shown in Table 6 only three substances have been studied by the rotating-bomb method. The experimentally measured enthalpies of formation of the remaining compounds in Table 6 were determined by reaction-solution calorimetry and all rely on $\Delta H_f^{\circ}(\text{PbPh}_4, \text{c})$.

There are not enough data to perform a linear analysis according to equation 1. Only one enthalpy-of-formation value is needed, however, to calculate a methyl deviation[19]. So employing $\Delta H_f^{\circ}(\text{PbEt}_4, \text{g})$ results in a methyl deviation per alkyl group for tetramethyl lead of -14.01, similar to that obtained for tetramethyl tin. We are not surprised, then, that the difference $\Delta H_f^{\circ}(\text{PbMe}_4, \text{g}) - \Delta H_f^{\circ}(\text{PbEt}_4, \text{g}) = 26.4 \text{ kJ mol}^{-1}$ is nearly the same as the difference $\Delta H_f^{\circ}(\text{SnMe}_4, \text{g}) - \Delta H_f^{\circ}(\text{SnEt}_4, \text{g}) = 25.0 \text{ kJ mol}^{-1}$. Applying the type of analysis in equation 2 to the halogen compounds, we find differences of: $\Delta H_f^{\circ}(\text{HI}, \text{g}) - \Delta H_f^{\circ}(\text{HBr}, \text{g}) = 62.8 \text{ kJ mol}^{-1}$[148], $\Delta H_f^{\circ}(\text{Ph}_3\text{PbI}, \text{g}) - \Delta H_f^{\circ}(\text{Ph}_3\text{PbBr}, \text{g}) = 50.7 \text{ kJ mol}^{-1}$ and $1/2[\Delta H_f^{\circ}(\text{Ph}_2\text{PbI}_2, \text{g}) - \Delta H_f^{\circ}(\text{Ph}_2\text{PbBr}_2, \text{g})] = 56.5 \text{ kJ mol}^{-1}$. [Note that the error bar of ΔH_f° (PbPh_4, c), which is the source of the large uncertainties assigned to the enthalpies of formation of the halogen compounds in Table 6, cancels when the differences are calculated, so that the correct uncertainty of the last two differences is ca 10 kJ mol^{-1}.] The previous analysis suggests that the following Laidler terms derived from the experimental data in Table 6 are fairly reliable[24]: $E(\text{Pb}-\text{C}) = 151.67 \text{ kJ mol}^{-1}$ and $E(\text{C}-\text{H})_s^{\text{Pb}} = 400.40 \text{ kJ mol}^{-1}$. Yet we hesitate to

TABLE 6.   Standard enthalpies of formation of organolead compounds (kJ mol^{-1})

Compound	ΔH_f° (l/c)	Method/Ref.^{a}	ΔH_v° or $\Delta H_s^{\circ b}$	ΔH_f° (g)
Me$_2$Pb=CH$_2$, g		IMR/7		246^c
PbMe$_4$, l	98.1 ± 4.4	RB/1	38.0 ± 0.4^d	136.1 ± 4.4
Me$_3$Pb(t-Bu), g		IMR/9		29.0
PbEt$_4$, l	53.1 ± 5.0	RB/1	56.6 ± 1.0^d	109.7 ± 5.1
PbPh$_4$, c	515.3 ± 15.4	RB/1	194.6 ± 6.3	709.9 ± 16.7
Ph$_3$PbBr, c	271.5 ± 18.0	RC/1	134.7 ± 3.3	406.2 ± 18.3
Ph$_2$PbBr$_2$, c	36.0 ± 17.6	RC/1	141.8 ± 0.8	177.8 ± 17.6
Ph$_3$PbI, c	326.8 ± 15.5	RC/1	130.1 ± 0.4	456.9 ± 15.5
Ph$_2$PbI$_2$, c	152.7 ± 15.5	RC/1	138.0 ± 4.2	290.7 ± 16.1
(Me$_3$Pb)$_2$, g		IMR/9		162.0

^{a}IMR = ion–molecule reactions; RB = rotating-bomb combustion calorimetry; RC = reaction calorimetry.
^{b}Enthalpies of vaporization or sublimation from Reference 1, except when indicated otherwise.
^{c}The value relies on the proton affinity value of NH$_3$, taken as 854 kJ mol^{-1}.
^{d}Values from M. H. Abraham and R. J. Irving, *J. Chem. Thermodyn.*, **12**, 539 (1980).

extend this scheme to the organolead halides. While bond enthalpy analysis in its simplest form would predict thermoneutrality for the gas-phase reactions 10:

$$Ph_4Pb + Ph_2PbX_2 \longrightarrow 2Ph_3PbX \ (X = Br, I) \qquad (10)$$

use of the values in Table 6 result in descrepancies of ca 40 kJ mol^{-1}, similar to the analysis of Reference 49.

B. Bond Dissociation Enthalpies

The only experimental values of bond dissociation enthalpies that have been reported in the literature are given in McMillen and Golden's review[32d]: D(Me$_3$Pb$-$Me) = 238 ± 17 kJ mol^{-1} and D(Et$_3$Pb$-$Et) = 230±17 kJ mol^{-1}. These values and the data in Table 6 imply ΔH_f°(PbMe$_3$, g) = 227 ± 18 kJ mol^{-1} and ΔH_f°(PbEt$_3$, g) = 221 ± 18 kJ mol^{-1}.

IV. FINAL COMMENTS

It is somewhat disappointing to realize that the thermochemistry of germanium, tin and lead organometallic compounds is still at the level achieved ten years ago, in contrast to the considerable recent efforts to probe the energetics of the silicon analogues. The data analysis in the previous sections shows that many key values are either missing or require experimental confirmation. To a certain extent, an overall discussion of the thermochemical data for Ge, Sn and Pb is therefore hindered by the probable inaccuracies and the uncertainties that affect those values.

The main purpose of these final comments is to show a few general trends in the thermochemistry of Group 14 organometallic compounds, helped by some (hopefully) reliable values. And one of the trends is revealed by a rather usual plot[1,2], in which the *mean* bond dissociation enthalpies of the species MR$_4$ (i.e. one-fourth of the enthalpy required to break all the M$-$R bonds) are represented as a function of the enthalpy of formation of M in the gaseous state. As observed in Figure 4, for R = H and Me, $\bar{D}$(M$-$H) and $\bar{D}$(M$-$Me) increase with the enthalpy of formation (or sublimation) of M. It is noted, on the other hand, that the differences $\bar{D}$(M$-$H) $-$ $\bar{D}$(M$-$Me) vary from 47.7 kJ mol^{-1}

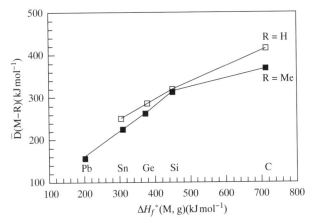

FIGURE 4. Mean M$-$R bond dissociation enthalpies in MR$_4$ compounds (M = element from Group 14; R = H or Me) versus the enthalpies of formation of gaseous M

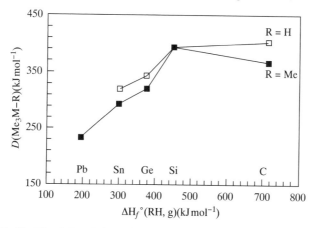

FIGURE 5. Me_3M-R bond dissociation enthalpies (M = element from Group 14; R = H or Me) versus the enthalpies of formation of gaseous M

for M = C to only 4.2 kJ mol^{-1} for M = Si. The upper limit of the range found for carbon can be attributed to the low value of $\bar{D}(C-Me)$ in the sterically congested neopentane molecule. The fact that $\bar{D}(Si-H)$ and $\bar{D}(Si-Me)$ are almost identical (4.2±3.0 kJ mol^{-1}) is interesting, but also surprising. Although no explanation can be offered, it is apparent from Figure 4 that the trend of $\bar{D}(M-H)$ is smoother than $\bar{D}(M-Me)$, which could mean that the experimental value of $\Delta H_f^\circ(SiMe_4, g)$ is affected by a large error. This is, however, a rather unlikely supposition. Recall that in order to obtain $\bar{D}(Si-H) - \bar{D}(Si-Me) \approx 20$ kJ mol^{-1}, the value of $\Delta H_f^\circ(SiMe_4, g)$ would have to increase by ca 60 kJ mol^{-1}. We believe, therefore, that the small difference observed in Figure 4 is genuine.

The plot shown in Figure 5 is similar to the one in Figure 4, except that it now involves the bond dissociation enthalpies $D(Me_3M-H)$ and $D(Me_3M-Me)$. Again it is noted that the largest difference $D(Me_3M-H) - D(Me_3M-Me)$, 37.2 ± 1.2 kJ mol^{-1}, is for carbon, and the smallest (1.3 ± 5.2 KJ mol^{-1})[37] is for silicon. Although the value of $\Delta H_f^0(SiMe_3, g)$ is surrounded by some controversy[28,37,50], this will affect equally both $D(Me_3Si-H)$ and $D(Me_3Si-Me)$, so that we can trust the value for the difference, assuming that the enthalpies of formation of $SiMe_4$ and Me_3SiH are reliable. With regard to the difference $D(Me_3Ge-H) - D(Me_3Ge-Me)$, its value is unsettled. The experimental results in Table 2 are affected by large uncertainties and, as discussed above, none of them seems consistent with the enthalpy of formation of GeMe$_3$ radical derived from photoacoustic calorimetry results and an estimate of $\Delta H_f^0(Me_3GeH, g)$ made with the Laidler scheme. In Section I.B a value of $D(Me_3Ge-Me) = 323$ KJ mol^{-1} was recommended, which implies $D(Me_3Ge-H) - D(Me_3Ge-Me) = 23$ KJ mol^{-1}. Finally, the enthalpies of formation of Me_3SnH and $SnMe_4$ seem well established and lead to $D(Me_3Sn-H) - D(Me_3Sn-Me) = 25.7 \pm 4.6$ KJ mol^{-1}. In conclusion, an apparent singularity for silicon is also suggested by the trend in Figure 5. And again it remains to be explained.

The group 14 M—M bond enthalpy may also be derived as one-half the enthalpy of the atomization process[6]

$$M\ (s) \longrightarrow M\ (g) \qquad (11)$$

where M (s) is the tetrahedral/tetracoordinate allotrope of the element (there is no such allotrope of lead). Using diamond and white tin, the following values are obtained $(kJ\,mol^{-1})$: C, 357; Si, 228; Ge, 188; Sn, 151 in encouraging consonance to earlier discussed values. In that the final M (g) neglects hybridization and steric effects, any agreement is surprising. Only future research efforts to investigate the organometallic thermochemistry of the three heavier elements of Group 14 will settle this and the many other questions raised in the present survey.

V. ACKNOWLEDGMENTS

JFL thanks the Chemical Science and Technology Laboratory, National Institute of Standards and Technology, for partial support of his research. JAMS thanks Junta Nacional de Investigação Cientifica e Technológica, Portugal (Project PMCT/C/CEN/42/90) for financial support. A travel grant from The Luso-American Foundation for Development, Portugal, is also acknowledged.

VI. REFERENCES

1. G. Pilcher and H. A. Skinner, in *The Chemistry of the Metal–Carbon Bond*, (Eds. F. R. Hartley and S. Patai), Wiley, New York, 1982.
2. V. I. Tel'noi and I. B. Rabinovich, *Russ. Chem. Rev.*, **49**, 603 (1980).
3. G. Pilcher, in *Energetics of Organometallic Species* (Ed. J. A. Martinho Simões), Kluwer, Dordrecht, 1992.
4. G. Pilcher, personal communication.
5. J. B. Pedley and J. Rylance, *Sussex–N.P.L. Computer Analysed Thermochemical Data*, University of Sussex, Brighton, 1977.
6. D. D. Wagman, W. H. Evans, V. B. Parker, R. H. Schumm, I. Halow, S. M. Bailey, K. L. Churney and R. L. Nuttall, *J. Phys. Chem. Ref. Data*, **11**, Suppl. No. 2 (1982).
7. W. J. Pietro and W. J. Hehre, *J. Am. Chem. Soc.*, **104**, 4329 (1982).
8. L. H. Long and C. I. Pulford, *J. Chem. Soc., Faraday Trans. 2*, **82**, 567 (1986).
9. M. F. Lappert, J. B. Pedley, J. Simpson and T. R. Spalding, *J. Organometal. Chem.*, **29**, 195 (1971).
10. N. K. Lebedev, E. G. Kiparisova, B. V. Lebedev, A. M. Sladkov and N. A. Vasneva, *Proc. Acad. Sci. USSR, Chem. Ser.*, 374 (1980).
11. N. K. Lebedev, B. V. Lebedev, E. P. Kiparisova, A. M. Sladkov and N. A. Vasneva, *Doklady Phys. Chem.*, **246**, 548 (1975).
12. A. S. Carson, J. Dyson, P. G. Laye and J. A. Spencer, *J. Chem. Thermodyn.*, **20**, 1423 (1988).
13. A. S. Carson, E. H. Jamea, P. G. Laye and J. A. Spencer, *J. Chem. Thermodyn.*, **20**, 1223 (1988).
14. (a) E. J. Prosen, W. H. Johnson and F. D. Rossini, *J. Res. Natl. Bur. Stand.*, **37**, 51 (1946).
    (b) J. D. Cox and G. Pilcher, *Thermochemistry of Organic and Organometallic Compounds*, Academic Press, London, 1970.
    (c) P. Sellers, G. Stridh and S. Sunner, *J. Chem. Eng. Data*, **23**, 250 (1978).
    (d) J. F. Liebman, K. S. K. Crawford and S. W. Slayden, in *Supplement S: The Chemistry of Sulphur-containing Functional Groups* (Eds. S. Patai and Z. Rappoport), Wiley, Chichester, 1993.
    (e) S. W. Slayden, J. F. Liebman and W. G. Mallard, 'Thermochemistry of halogenated organic compounds', in *Supplement D: The Chemistry of Halides, Pseudo-halides and Azides*, Vol. 2 (Eds. S. Patai and Z. Rappoport), Wiley, Chichester, 1995.
15. In the least-squares analyses of equation 1, the individual enthalpies of formation were weighted inversely as the squares of the uncertainty intervals.
16. J. S. Chickos, D. G. Hesse, J. F. Liebman and S. Y. Panshin, *J. Org. Chem.*, **53**, 3424 (1988).
17. (a) J. F. Liebman, J. A. Martinho Simões and S. W. Slayden, 'Aspects of the thermochemistry of organolithium compounds', in *Lithium Chemistry: A Theoretical and Experimental Overview* (Eds. A. -M. Sapse and P. von R. Schleyer), Wiley, New York, in press.
    (b) J. F. Liebman, J. A. Martinho Simões and S. W. Slayden, *Structural Chemistry*, **6**, 65 (1995).
18. R. L. Montgomery and F. D. Rossini, *J. Chem. Thermodyn.*, **10**, 471 (1978).

19. In Reference 17b, we followed the general method of Reference 18 by expressing the experimental standard enthalpy of formation of each member of a homologous functional group series $\{CH_3-(CH_2)_m-Z\}$ as $\Delta H_f^\circ = (\alpha' \cdot n_m) + \beta'$, where α' is the 'universal' methylene increment slope $(-20.6 \, kJ \, mol^{-1}$, determined from the correlation with carbon number in the n-alkane series) and n_m is the number of methylene groups in the molecule. The term β' is thus a calculated enthalpy-of-formation value for CH_3-Z. We then defined the methyl deviation as $\delta(CH_3-Z) = \Delta H_f^\circ (CH_3 Z_{\text{experimental}}) - \Delta H_f^\circ (CH_3 Z_{\text{calculated}})$, where $(CH_3 Z_{\text{calculated}})$ is the mean value of β' for a series as calculated for each individual member bonded to a given Z. For series containing more than one alkyl group bonded to a heteroatom, consideration was limited to identically substituted heteroatoms and the methyl deviation was corrected for the number of alkyl groups. The methyl deviation per methyl group for tetramethyl germanium is $-5.45 \, kJ \, mol^{-1}$.

20. (a) A. R. Dias, J. A. Martinho Simões, C. Teixeira, C. Airoldi and A. P. Chagas, *J. Organometal. Chem.*, **335**, 71 (1987).

    (b) A. R. Dias, J. A. Martinho Simões, C. Teixeira, C. Airoldi and A. P. Chagas, *J. Organometal. Chem.*, **361**, 319 (1989).

    (c) A. R. Dias, J. A. Martinho Simões, C. Teixeira, C. Airoldi and A. P. Chagas, *Polyhedron*, **10**, 1433 (1991).

    (d) D. Griller, J. A. Martinho Simões and D. D. M. Wayner, in *Sulfur-Centered Reactive Intermediates in Chemistry and Biology* (Eds. C. Chatgilialoglu and K. -D. Asmus), Plenum, New York, 1991.

    (e) J. A. Martinho Simões, in *Energetics of Organometallic Species* (Ed. J. A. Martinho Simões), NATO ASI Series, Kluwer, Dordrecht, 1992.

    (f) J. P. Leal, A. Pires de Matos and J. A. Martinho Simões, *J. Organometal. Chem.*, **403**, 1 (1991).

    (g) J. P. Leal and J. A. Martinho Simões, *Organometallics*, **12**, 1442 (1993).

    (h) H. P. Diogo, J. A. Simoni, M. E. Minas da Piedade, A. R. Dias and J. A. Martinho Simões, *J. Am. Chem. Soc.*, **115**, 2764 (1993).

    (i) J. P. Leal and J. A. Martinho Simões, *J. Organometal. Chem.*, **460**, 131 (1993).

    (j) J. F. Liebman, J. A. Martinho Simões and S. W. Slayden, 'Thermochemistry of organo-arsenic, antimony and bismuth compounds', in *The Chemistry of Organic Arsenic, Antimony and Bismuth Compounds* (Ed. S. Patai), Wiley, Chichester, 1994.

21. Fluorine-containing species are notable exceptions. See References 14e and 17b.

22. J. B. Pedley, R. D. Naylor and S. P. Kirby, *Thermochemical Data of Organic Compounds*, Chapman and Hall, London, 1986.

23. Enthalpies of formation for the n-alkanes were taken from Reference 22. The constants from equation 1 for $n_c = 4-12$, 16, 18 are: $\alpha = -20.63 \pm 0.06$, $\beta = -43.20 \pm 0.46$, $r^2 = 0.99994$. The measured enthalpy of formation for methane is $-74.4 \, kJ \, mol^{-1}$.

24. The following auxiliary data $(kJ \, mol^{-1})$ were quoted from Reference 1: $E(C-C) = 358.46$, $E(C-H)_p = 411.26$, $E(C-H)_s = 407.40$, $E(C-H)_t = 404.30$, $E(C=C) = 556.50$, $E(C_d-H)_2 = 424.20$, $E(C_d-H)_1 = 421.41$, $E(C_d-C) = 378.05$, $E(C-C_b) = 372.81$, $E(C_b-H) = 421.41$, $E(C_b-H_b) = 499.44$, $\Delta H_f^\circ (Ge, g) = 377$, $\Delta H_f^\circ (Sn, g) = 301.2$, $\Delta H_f^\circ (Pb, g) = 195.20$, $\Delta H_f^\circ (C, g) = 716.67$ and $\Delta H_f^\circ (H, g) = 218.00$.

25. We thank Dr A. Carson (University of Leeds, UK) and Dr G. Pilcher (University of Manchester, UK) for helpful discussions on this subject.

26. K. B. Clark and D. Griller, *Organometallics*, **10**, 746 (1991).

27. R. Walsh, in *The Chemistry of Organic Silicon Compounds* (Eds. S. Patai and Z. Rappoport), Wiley, Chichester, 1988.

28. There is recent experimental evidence that the alkylation of silanes actually increases $D(Si-H)$, namely $D(Me_3Si-H)$ is about 14 $kJ \, mol^{-1}$ higher than $D(H_3Si-H)$. See A. Goumri, W. -J. Yuan and P. Marshall, *J. Am. Chem. Soc.*, **115**, 2539 (1993).

29. J. M. Kanabus-Kaminska, J. A. Hawari, D. Griller and C. Chatgilialoglu, *J. Am. Chem. Soc.*, **109**, 5267 (1987).

30. A. R. Dias, H. P. Diogo, D. Griller, M. E. Minas da Piedade and J. A. Martinho Simões, in *Bonding Energetics in Organometallic Compounds* (Ed. T. J. Marks), *ACS Symp. Series* No. 428, 1990.

31. J. M. Kanabus-Kaminska, B. C. Gilbert and D. Griller, *J. Am. Chem. Soc.*, **111**, 3311 (1989).

32. (a) A. M. Doncaster and R. Walsh, *J. Phys. Chem.*, **83**, 578 (1979).

    (b) B. Ruscic, M. Schwarz and J. Berkowitz, *J. Chem. Phys.*, **92**, 1865 (1990).

    (c) P. N. Noble and R. Walsh, *Int. J. Chem. Kinet.*, **15**, 547 (1983).

(d) D. F. McMillen and D. M. Golden, *Ann. Rev. Phys. Chem.*, **33**, 493 (1982).

(e) G. P. Smith and R. Patrick, *Int. J. Chem. Kinet.*, **15**, 167 (1983).

(f) P. N. Noble and R. Walsh, *Int. J. Chem. Kinet.*, **15**, 561 (1983).

33. Unless indicated otherwise, the enthalpies of formation of alkyl and aryl radicals were taken from J. A. Martinho Simões and J. L. Beauchamp, *Chem. Rev.*, **90**, 629 (1990).

34. $E(Ge-H) = 289.55$ kJ mol^{-1} was derived as ΔH°_{atom} (GeH$_4$, g)/4 (the enthalpy of formation was quoted from Table 1).

35. An identical procedure was applied for estimating ΔH°_f (Me$_3$SiH, g) $= -166.3$ kJ mol^{-1}, and ΔH°_f (Me$_3$SnH, g) $= 26.3$ kJ mol^{-1}, and these results are in very good agreement with the experimental values, -163.4 ± 4.0 kJ mol^{-1} (Reference 27) and 25.1 ± 4.2 (Reference 2), respectively.

36. See Reference 20e and references cited therein.

37. G. Pilcher, M. L. P. Leitão, Y. Meng-Yan and R. Walsh, *J. Chem. Soc., Faraday Trans.*, **87**, 841 (1991).

38. W. P. Neumann and K. -D. Schultz, *J. Chem. Soc., Chem. Commun.*, 43 (1982).

39. A. S. Carson, E. H. Jamea, P. G. Laye and J. A. Spencer, *J. Chem. Thermodyn.*, **20**, 923 (1988).

40. D. B. Chambers and F. Glocking, *Inorg. Chim. Acta*, **4**, 150 (1970).

41. A. S. Carson, P. G. Laye and J. A. Spencer, *J. Chem. Thermodyn.*, **17**, 277 (1985).

42. A. S. Carson, J. Franklin, P. G. Laye and H. Morris, *J. Chem. Thermodyn.*, **7**, 763 (1975).

43. From Reference 14d, the gaseous α values are -18.19 for di-n-alkyl sulfites and -18.10 for di-n-alkyl sulfates. The α values were calculated from the diethyl, dipropyl and dibutyl substituted species in each of the homologous series.

44. T. J. Burkey, M. Majewski and D. Griller, *J. Am. Chem. Soc.*, **108**, 2218 (1986).

45. As noted by the enthalpy-of-vaporization data in Table 3 for the compounds SnMe$_4$, Me$_3$SnEt and Me$_2$SnEt$_2$, replacing a methyl group by an ethyl group increases ΔH_v by *ca* 5 kJ mol^{-1}. See Reference 16.

46. See W. P. Newmann, *Synthesis*, 665 (1987) and references cited therein.

47. R. J. Klingler, I. Bloom and J. W. Rathke, *Organometallics*, **4**, 1893 (1985).

48. J. D. Cox, D. D. Wagman and V. A. Medvedev (Eds.), *CODATA Key Values for Thermodynamics*, Hemisphere, New York, 1989.

49. R. S. Butler, A. S. Carson, P. G. Laye and W. V. Steele, *J. Chem. Thermodyn.*, **8**, 1153 (1976).

50. J. A. Seetula, Y. Feng, D. Gutman, P. W. Seakins and M. J. Pilling, *J. Phys. Chem.*, **95**, 1658 (1991).

CHAPTER 5

ESR of organogermanium, organotin and organolead radicals

JIM ILEY

Physical Organic Chemistry Research Group, Chemistry Department, The Open University, Milton Keynes, MK7 6AA, UK
Fax: (+44)1908-653-744; e-mail: J.N.ILEY@OPEN.AC.UK

I. INTRODUCTION

Various radicals of Group IV elements have been studied, largely because of an interest in the effect that going down a group in the periodic table has upon the types of radical formed and the differences in their structures. Among the radicals that have been observed are the neutral radicals $R_3M^{\bullet}$ and $R_5M^{\bullet}$, the radical cations $R_4M^{+\bullet}$ and $R_3MMR_3^{+\bullet}$, and the radical anions $R_4M^{-\bullet}$ and $R_3MMR_3^{-\bullet}$. As well as these heavy-atom centred radicals, attention has also been paid to carbon-centred radicals which have the heavy atom α or β to the radical centre. While these will be described here, more attention is given to the radicals in which the unpaired electron is directly associated with the germanium, tin or lead atom.

The chemistry of organic germanium, tin and lead compounds
Edited by S. Patai © 1995 John Wiley & Sons Ltd

II. THE NEUTRAL $R_3M^\bullet$ AND $R_5M^\bullet$ RADICALS (M = Ge, Sn OR Pb)

A. $R_3M^\bullet$

1. Direct detection

The radicals $R_3M^\bullet$ have been generated by a variety of means. The most common solution phase method involves hydrogen atom abstraction from R_3MH using, for example, $Bu^tO^\bullet$ (equation 1)[3,8,11,12]. In the solid phase, γ-irradiation of R_4M generates, amongst other radicals, $R_3M^\bullet$[1,4,6,7]. Other methods that have been employed are homolytic substitution at[5], or thermolysis of[10,11,19], R_3MMR_3 (equation 3), the reaction of R_3MCl with sodium metal (equation 4)[24], the photolysis of R_3MLi (equation 5), or the photolysis of the divalent species R_2M[21] or $(R_2N)_2M$[17,18] (equation 6).

$$R_3MH \xrightarrow{\ Bu^tO^\bullet\ } R_3M^\bullet \tag{1}$$

$$R_4M \xrightarrow{\ \gamma\ } R_3M^\bullet \tag{2}$$

$$R_3MMR_3 \quad \begin{matrix} \xrightarrow{\ \Delta\ } 2R_3M^\bullet \\ \xrightarrow{\ Bu^tO^\bullet\ } R_3M^\bullet + R_3MOBu^t \end{matrix} \tag{3}$$

$$R_3MCl + Na \longrightarrow R_3M^\bullet + NaCl \tag{4}$$

$$R_3MLi \xrightarrow{\ h\nu\ } R_3M^\bullet \tag{5}$$

$$R_2M \text{ or } (R_2N)_2M \xrightarrow{\ h\nu\ } R_3M^\bullet \text{ or } (R_2N)_3M^\bullet \tag{6}$$

As evidenced by the ESR spectral data collected together in Table 1, the most commonly studied radicals of this kind are those with tin and germanium centres; the corresponding lead centred radicals have been the subject of only a few investigations.

The parent radicals $H_3M^\bullet$ have been observed for M = Ge and Sn[1,2]. These radicals display coupling to all three protons and to the Ge or Sn atom. Since anisotropy in the a_{Ge} and a_{Sn} hyperfine coupling has not yet been reported, the structure of these radicals has to be inferred indirectly. The isotropic coupling to the M atom arises from spin density in the 4s orbital of germanium or the 5s orbital of tin. Coupling of a single electron to a pure 4s Ge orbital would give rise to hyperfine splitting of 535G, and to a pure 5s Sn orbital 15672G. Comparison of these values with those in Table 1 reveals that for $H_3Ge^\bullet$ the unpaired spin density in the 4s orbital is $75/535 = 0.14$, while for $H_3Sn^\bullet$ the unpaired spin density in the 5s orbital is $380/15400 = 0.025$. The large s contribution to the Ge 4s orbital containing the unpaired spin indicates that the radical must be pyramidal, unlike the radical of the first Group IV element, $H_3C^\bullet$, which is planar. An estimate of the H—Ge—H bond angle θ can be obtained from the relationship

$$\cos(\pi - \theta) = s_b/(1 - s_b)$$

where s_b is the contribution of the s orbital to the three Ge—H bonds[1]. Thus, $s_b = (1 - 0.14)/3 = 0.29$, and $\theta = 114°$ (actually the unpaired spin density requires a correction for spin polarization in the three bonding orbitals, but the correction has little affect on the conclusions arrived at). The tetrahedral bond angle is $109°$, and the trigonal angle $120°$. It follows that the structure of $H_3Ge^\bullet$ is significantly pyramidal, and the out-of-plane angle of the germanium atom is $14°$. A similar analysis for $H_3Sn^\bullet$ gives $\theta = 118°$, which would suggest that this radical is almost planar but, as discussed below, for other tin radicals the a_{Sn} hyperfine coupling is much larger, consistent with a tetrahedral structure for these too.

TABLE 1. ESR spectral data for $R_3M^\bullet$ radicals

Radical	M = Ge				M = Sn				M = Pb		
	g	a_H (G)	a_{Ge} (G)	Ref.	g	a_H (G)	a_{Sn} (G)	Ref.	g	a_{Pb} (G)	Ref.
$H_3M^\bullet$	2.0092[a]; 2.0062[b]	13.91[a] (3H); 14.56[b] (3H)	75	1,2	2.003[a]; 2.025[b]	26	380	1			
$Me_3M^\bullet$	2.0104	5.5 (9H)	84.7	3,4	2.017	3.1	1983 (^{119}Sn); 1899 (^{117}Sn); 1959[a] (^{119}Sn); 1350[b] (^{119}Sn)	4,5	1.917[a]; 2.113[b]	3040[a]; 1793[b]	7
$Et_3M^\bullet$	2.0089	4.75 (6H); 0.56 (9H)		8	2.000[a]; 2.027[b]		1950[a] (^{119}Sn); 1350[b] (^{119}Sn)	6	1.904[a]; 2.091[b]	2625[a]; 1400[b]	7
$(PhMe_2CCH_2)_3M^\bullet$	2.0096	5.10 (6H)		9	ca 2.0		1350[b] (^{119}Sn)	6			
$(Me_3CCH_2)_3M^\bullet$	2.0107	5.14 (6H)		12	2.0150	3.1 (6H)	1380[b] (^{119}Sn); 1325 (^{117}Sn)	10,11			
$(Me_3CCH_2)_2MeM^\bullet$	2.0106	5.54 (3H); 6.68 (2H); 3.67 (2H)		12	2.0170	3.4 (6H)		10			
$Ph_3M^\bullet$	2.0054	0.93 (ortho); 0.46 (meta); 0.93 (para)	109[a]; 71.5[b]	8,23	2.007; 2.002; 1.988		2335[a] (^{119}Sn); 1632[b] (^{119}Sn)	13			
$(2,4,6\text{-}Me_3C_6H_2)_3M^\bullet$	2.0066	0.70 (o-Me); 0.70 (m-H); 0.70 (p-Me)	68.4	14,15	2.0073			11			
$(2,4,6\text{-}Pr^i_3C_6H_2)_3M^\bullet$					2.0078			10,11			
$\{(Me_3Si)_2CH\}_3M^\bullet$	2.0078	3.8 (3H)	92	16	1.995[a]; 2.016[b]	2.1 (3H)	1678 (^{119}Sn); 1602 (^{117}Sn); 2110[a] (^{119}Sn); 1390[b] (^{119}Sn)	19			
$\{(Me_3Si)_2N\}_3M^\bullet$	1.9991	10.6 (a_N)	171	16	2.0094	10.9 (a_N)	1776 (^{119}Sn); 1698 (^{117}Sn)	16,17, 21			
$\{(Bu^tMe_3Si)N\}_3M^\bullet$	1.9998	12.9 (a_N)	173	18	1.9912	12.7 (a_N)	3426 (^{119}Sn); 3176 (^{117}Sn)	16			

[a] Parallel component.
[b] Perpendicular component.

The $Me_3Ge^•$ radical displays a similar a_{Ge} hyperfine coupling to $H_3Ge^•$, indicative of a pyramidal structure for this radical too. The small value of the proton hyperfine coupling a_H as compared with that for the $Me_3C^•$ radical (22.7G) is consistent with this structure. The corresponding $Me_3Sn^•$ radical exhibits an even smaller value for the proton hyperfine coupling. However, unlike $H_3Sn^•$, there is considerable hyperfine coupling to the tin atom consistent with significant spin density in the Sn 5s orbital. Using the relationships

$$a_{iso} = (a_\| + 2a_\perp)/3 \text{ and } a_{aniso} = (a_\| - a_\perp)/3$$

the parallel and perpendicular components of the ^{119}Sn coupling allow the isotropic and anisotropic coupling to be evaluated. The isotropic and anisotropic couplings, 1550G and 200G respectively, indicate that the unpaired spin density resides in an orbital with 10% 5s and 78% 5p character (the anisotropic coupling of an electron in a pure Sn 5p orbital is 261G). Thus *ca* 90% of the spin density resides on the tin atom. The radical therefore has a pyramidal structure, though it is somewhat flattened from sp^3 hybridization: the s content of an sp^3 orbital is 25% whereas here it is 11%. The $Me_3Pb^•$ radical does not display proton hyperfine coupling. The isotropic coupling to the ^{207}Pb nucleus, 2209G, indicates an unpaired spin density in the Pb 6s orbital of 22% if one assumes the isotropic coupling of a single electron in a pure Pb 6s orbital to be 9990G[7]. Such a value, if correct, would indicate a pyramidal structure for the radical. However, two observations relating to the $Me_3Pb^•$ radical remain unexplained[7]. The first is the low value for the $g_\|$ component, the second is the large value of the anisotropic hyperfine coupling to ^{207}Pb. The data in Table 1 give a value of 416G which, when compared with the theoretical coupling of an electron in a Pb 6p orbital of 175G, suggests an unpaired spin density of 237%, which of course cannot be correct[7].

From these data, and the similarity of the data for the other radicals contained in Table 1, it therefore appears that, unlike carbon-centred radicals, the tricoordinate trialkyl radicals of Group IV elements have the tetrahedral structure **1**.

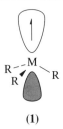

(1)

Certain sterically bulky trialkyl-germyl and stannyl radicals exhibit a non-equivalence of the CH protons α to the radical centre[9,11,12,20], either through an alternating linewidth effect or through the existence of two different proton hyperfine coupling constants. Thus, whereas coupling to six equivalent protons is observed at temperatures $> 40\,^\circ$C for $(PhMe_2CCH_2)_3Ge^•$, at $-60\,^\circ$C a quartet coupling of 5.65G to only three protons is observed. This can be accommodated by a radical that involves restricted rotation about the C—Ge bond and that has a preferred conformation in which one of the protons is almost orthogonal to the orbital containing the unpaired electron. Using the relationship

$$a_H = B \cos^2 \theta$$

where θ is the dihedral angle between the C—H bond and the unpaired electron, it is possible to calculate a value for B of 11G (assuming that the hyperfine coupling to the protons in $Me_3Ge^•$ arises from freely rotating methyl groups for which $\langle \cos^2 \theta \rangle = \frac{1}{2}$). The coupling of 5.65G for the $(PhMe_2CCH_2)_3Ge^•$ radical therefore corresponds to an angle $\theta = \pm 44°$. The dihedral angle of the remaining proton is thus $76°$ or $164°$, of which only the former is consistent with a coupling of less than 1G. Consequently, the preferred conformation of the radical is **2** or its equivalent enantiomeric conformation. However, the two distinct hyperfine couplings observed for $(Me_3CCH_2)_2MeGe^\bullet$ imply a conformation like that of **3** (assuming that interaction of the proton with the smaller lobe of the singly occupied orbital is roughly a quarter that of the larger lobe).

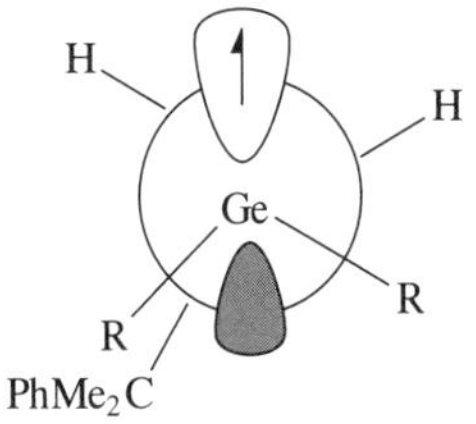

| (2) | (3) |

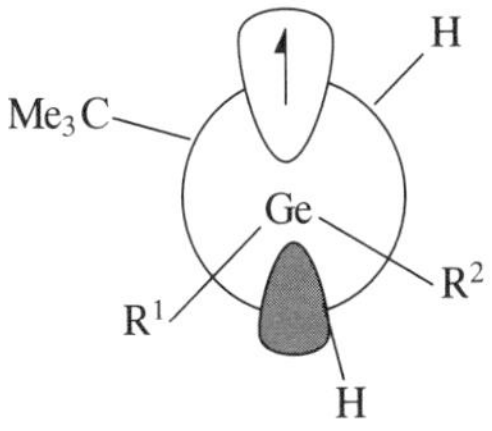

While the stannyl radical $(Me_3CCH_2)_3Sn^\bullet$ exhibits free rotation about the $C-Sn$ bond[9], $(PhMe_2CCH_2)_3Sn^\bullet$ exhibits an alternating linewidth effect consistent with a preferred conformation[9,20]. However, the absence of an assigned hyperfine coupling at lower temperatures precludes the assignment of the preferred orientation. The more heavily substituted phenyl radical $(Ph_3CCH_2)_3Sn^\bullet$ displays two different proton hyperfine couplings, 6.0G and 0.8G[9]. The proton hyperfine coupling for $Me_3Sn^\bullet$, 3.1G, enables a value for B of 6.2G to be calculated for the tin-centred radicals. Together with the observed hyperfine couplings, this identifies the preferred conformation of $(Ph_3CCH_2)_3Sn^\bullet$ as **4**, in which the dihedral angle of the proton with the largest coupling is $-10°$.

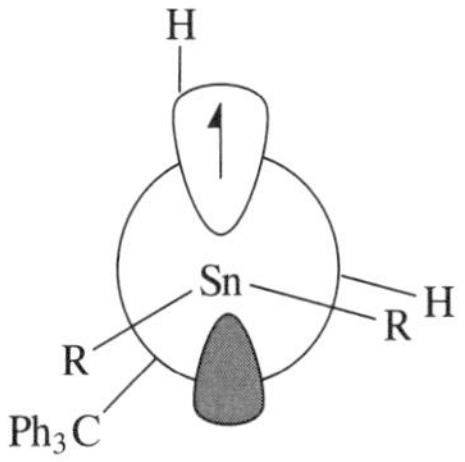

(4)

The triaryl radicals of germanium exhibit lower g values than their trialkyl counterparts. In part, this arises from increased delocalization of the unpaired spin density onto the aryl rings (and the $Ar_3Ge^\bullet$ radicals do show hyperfine coupling to the ring protons). For example, spin densities for the radicals $Ph_mMe_{3-m}Ge^\bullet$, calculated by the Hückel method (Table 2), reveal that there is a linear correlation between the g value of the radical and

TABLE 2. Relationship between calculated spin densities and the g and a_H values of $Ar_3Ge^\bullet$ radicals

Radical	g value	Spin density	a_H (ring protons)
$Ph_3Ge^\bullet$	2.0054	0.82	0.93 (o- and p-)
			0.46 (m-)
$Ph_2MeGe^\bullet$	2.0070	0.86	0.97 (o- and p-)
			0.49 (m-)
$PhMe_2Ge^\bullet$	2.0086	0.91	1.2 (o- and p-)
			0.6 (m-)

the spin density at the germanium atom[8,14]. At least 82% of the unpaired spin resides at the germanium atom, as compared with the corresponding triphenylmethyl radical in which only 42% of the spin density resides at the central carbon atom.

For the series $Ph_m Me_{3-m} Ge^{\bullet}$, the ring proton hyperfine couplings increase as m decreases (Table 2), consistent with an increase in the unpaired spin density at the ring carbon atoms. Hückel molecular orbital calculations bear this out[8,14]. Overall, however, the small values of the ring proton hyperfine couplings reveal that there is only a small Ge 4p–C 2p overlap in these radicals.

The isotropic germanium hyperfine coupling appears to be smaller for the triaryl radicals as compared with the trialkyl radicals. Since the hyperfine coupling to the germanium atom should increase markedly with increased s-character of the orbital containing the unpaired electron, this observation would imply that the aryl-substituted radicals are somewhat more planar than their alkyl-substituted analogues.

Like $Ar_3 Ge^{\bullet}$, $Ar_3 Sn^{\bullet}$ radicals display lower g values than their alkyl counterparts. However, for the tin-centred radicals this appears to be due to more subtle differences in the pyramidal structure, rather than due to delocalization of the unpaired spin density. Whereas for $Ar_3 Ge^{\bullet}$ the hyperfine coupling to the central germanium atom is smaller for the aryl radicals, for $Ar_3 Sn^{\bullet}$ the tin hyperfine coupling is larger than for the alkyl radicals. A comparison of the isotropic values of a_{Sn} (which of course is related to the unpaired spin density in the Sn 5s orbital) for $Ph_3 Sn^{\bullet}$ and $Me_3 Sn^{\bullet}$ (Table 1) reveals that there is an approximately 20% greater contribution of the 5s orbital in the former radical. This implies that, rather than being more planar than $Me_3 Sn^{\bullet}$ as would be expected from delocalization, $Ph_3 Sn^{\bullet}$ is in fact more pyramidal. This can be attributed to the higher electronegativity of Ph as compared with Me. This fits with the generally accepted tenet that the greater the difference in electronegativity between the central atom and the groups bonded to it, the greater will be the pyramidal nature of the radical. Significantly, sterically bulky triaryltin radicals have g values and a_{Sn} hyperfine coupling constants lying between those of $Me_3 Sn^{\bullet}$ than those of $Ph_3 Sn^{\bullet}$. This is because the bulky radicals adopt a slightly more planar structure in order to relieve unfavourable steric interactions[19]. Indeed, there is a linear relationship between the isotropic g value of the radical and the out-of-plane angle, ϕ, subtended by the tin atom, suggesting that the extent of the pyramidal nature of the radical alone determines g[19].

Also in contrast to $Ar_3 Ge^{\bullet}$, the corresponding $Ar_3 Sn^{\bullet}$ radicals do not display hyperfine coupling to the aryl ring protons. Hyperfine coupling in $Me_3 Ge^{\bullet}$ is about twice that in $Me_3 Sn^{\bullet}$, so by comparison to the aryl protons in $Ar_3 Ge^{\bullet}$ it might be expected that coupling to the aryl protons in $Ar_3 Sn^{\bullet}$ should be no more than 0.5G, which is considerably less than the linewidth of the signals, ca 2G[11,22]. The even smaller overlap expected between Sn 5p–C 2p as compared with Ge 4p–C 2p would account for the lack of any observable coupling for the tin-centred radicals.

The above differences in behaviour between the germanium- and tin-centred radicals can therefore be ascribed to (a) the greater difference in electronegativity between tin and carbon as opposed to germanium and carbon, which manifests itself in the greater influence of the pyramidal nature of the tin radicals, and (b) the poorer overlap of the Sn 5p–C 2p as compared with Ge 4p–C 2p orbitals.

2. Spin trapping

Aliphatic and aromatic nitro compounds react with all three $R_3 M^{\bullet}$ radicals to generate intermediate nitroxyl radicals of general structure $R_3 M-O-N(O^{\bullet})-R'$. For the tin series, such radicals are implicated in the denitration of nitroalkanes[25]. The persistence of these radicals decreases with the nature of R' in the order Me (minutes) $<$ Et $<$ Bu^t (hours)[28].

Attempts to trap tin-centred radicals directly using the more conventional spin trap nitroso-durene, (2,3,5,6-tetramethylnitrosobenzene), have been unsuccessful[29]. The ESR spectral parameters for some representative examples of $R_3M-O-N(O^\bullet)-R'$ radicals are detailed in Table 3. These reveal that the nitrogen hyperfine coupling, a_N, is largely unaffected by the nature of the heteroatom M, and that for M = Sn, Pb coupling to the metal nucleus is observable. The nitrogen hyperfine coupling is more sensitive to the nature of the R' group: for alkyl groups a_N is large and reflects a pyramidal structure at the nitrogen atom; for aryl groups a_N is much smaller, reflecting a flatter geometry at nitrogen and spin delo-calization onto the aryl ring. The latter effect results in hyperfine coupling with the aryl ring protons. The sterically bulky aryl system $2,4,6$-$Bu_3{}^tC_6H_2$ displays an intermediate value for the nitrogen hyperfine coupling, which may be attributed to a diminution of spin delocalization due to rotation of the aryl ring out of the plane of the orbital containing the unpaired electron.

1,2-Dicarbonyl compounds also act as excellent spin traps for neutral $R_3M^\bullet$ radicals, forming adducts of structure **5**, **6** or **7** depending upon the dicarbonyl compound and the nature of the ligand bound to the metal centre. Of course, for cyclic dicarbonyl compounds such as *ortho* quinones the *trans* structure **7** is not accessible because of geometric constraints. Data for these radicals are contained in Table 4.

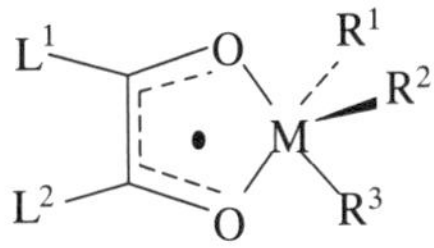

(5)

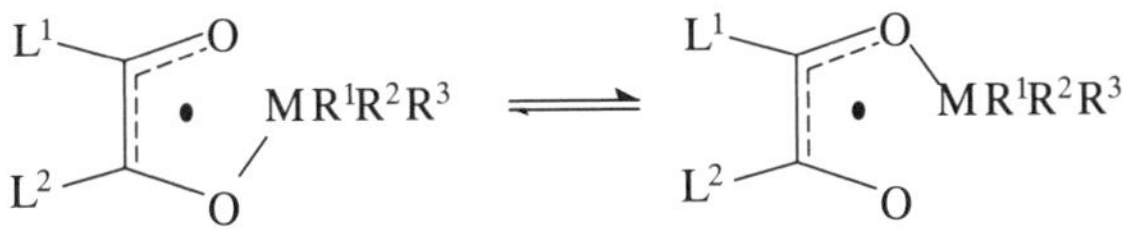

(6)

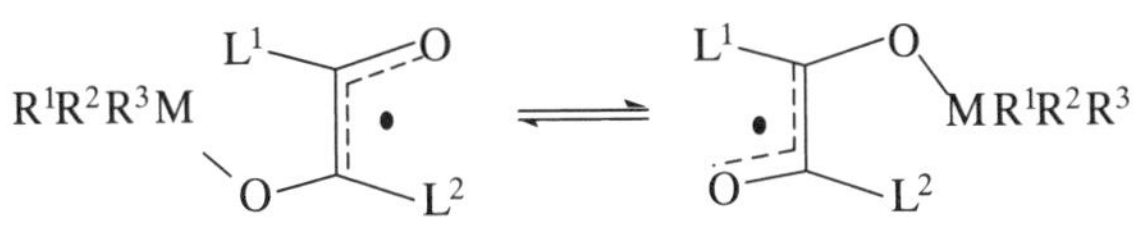

(7)

The pentacoordinate structure **5** is observed when the central atom M acts as a Lewis acid by virtue of it carrying at least two strongly electronegative atoms such as Cl. In such structures coupling is only observed to the chlorine ligand that occupies the apical position, and the L^1 and L^2 groups are not identical. At temperatures above $0\,^\circ C$, spectral changes reveal that the pentacoordinate structure becomes fluxional so that coupling to two chlorine atoms is observed and the L^1 and L^2 groups become equivalent[30]. As the number of electronegative ligands attached to the M atom decreases (and consequently its Lewis acidity) the structure of the radical favours the tetracoordinate species **6** or **7**. A lack of coupling to the chlorine atom and, for the biacetyl-trapped radicals, a larger difference in the hyperfine coupling constants for the L^1 and L^2 groups argues against the pentacoordinate structure **5**. When one of the R ligands is Cl, the radical displays

TABLE 3. ESR spectral parameters for nitroxyl radicals formed from R₃M• and nitro compounds

Radical	M = Ge			M = Sn					M = Pb			
	a_N (G)	a_H (G)	Ref.	g	a_N (G)	a_{Sn} (G)	a_H (G)	Ref.	g	a_N	a_{Pb}	Ref.
Ph₃MO–N(O•)–Buᵗ				2.0048	29.0	3.0		26	2.0037	28.75	7.25	26
Me₃MO–N(O•)–Buᵗ	30.3		27		30.1	4.78		27	2.0052	28.0	6.0	26
Bu₃MO–N(O•)–Buᵗ				2.00527	28.77	5.05	0.19 (9H)	25, 28				
Me₃MO–N(O•)–Me	29.6	9.48 (3H)	27	2.00511	28.06	2.69	10.14 (3H)	25, 28				
Bu₃MO–N(O•)–Me												
Me₃MO–N(O•)–Ph	14.9	3.41 (*ortho*-H) 1.16 (*meta*-H) 3.41 (*para*-H)	27		14.10		3.36 (*ortho*-H) 1.14 (*meta*-H) 3.72 (*para*-H)	27				
Me₃MO–N(O•)–(2,4,6-Bu₃ᵗC₆H₂)	22.45	0.89 (*meta*-H)	27		21.30	4.17	0.97 (*meta*-H)	27				

TABLE 4. ESR spectral data for the radicals formed between 1,2-dicarbonyl compounds and $R^1R^2R^3M^\bullet$

1,2-Di-carbonyl	R^1, R^2, R^3 ligands	M = Ge			M = Sn						M = Pb				
		Type	a_H (G)	Ref.	Type	g	a_H (G)	a_{Sn} (G)	a_{Cl} (G)	Ref.	Type	g	a_H (G)	a_{Pb} (G)	Ref.
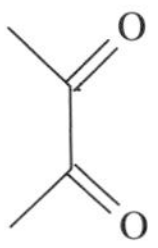 (biacetyl)	$R^1 = R^2$ $= R^3 = Cl$				A	2.0030	9.98 (3H) 9.10 (3H)		0.88 (1Cl)	30					
	$R^1 = Bu$, $R^2 = R^3$ $= Cl$				A	2.0028	9.98 (3H) 9.10 (3H)		0.88 (1Cl)	30					
	$R^1 = R^2$ $= Bu$, $R^3 = Cl$				B	2.0039	10.8 (3H) 7.40 (3H)			30					
	$R^1 = R^2$ $= R^3 = Bu$				B	2.0040	8.54 (6H)	10.9		30	B	2.0038	8.0 (6H)		26
					C	2.0045	7.3 (6H)				C	2.0044	6.0 (6H)		
	$R^1 = R^2$ $= R^3 = Me$	B or C	14.3 (3H) 2.0 (3H)	32											
	$R^1 = R^2$ $= R^3 = Et$				B	2.0039	8.9 (6H)	9.0		26	B	2.0036	8.3 (6H)	6.0	26
					C	2.0050	8.0 (6H)				C	2.0042	7.1 (6H)		
	$R^1 = R^2$ $= R^3 = Ph$														
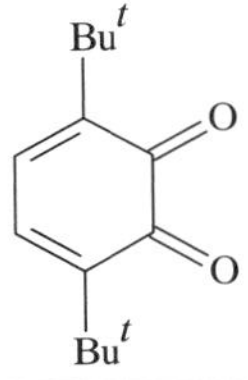 Bu^t (3,5-di-tert-butyl-o-benzoquinone) Bu^t	$R^1 = R^2$ $= R^3 = Cl$				A	2.0033	4.0 (2H)	10.0	0.63 (1Cl)	31					

(continued overleaf)

TABLE 4. (*continued*)

1,2-Di-carbonyl	R^1, R^2, R^3 ligands	M = Ge Type	M = Ge a_H (G)	M = Ge Ref.	M = Sn Type	M = Sn g	M = Sn a_H (G)	M = Sn a_{Sn} (G)	M = Sn a_{Cl} (G)	M = Sn Ref.	M = Pb Type	M = Pb g	M = Pb a_H (G)	M = Pb a_{Pb} (G)	M = Pb Ref.
	$R^1 = Bu$, $R^2 = R^3 = Cl$				A		4.4 (2H)		0.60 (1Cl)	31					
	$R^1 = Ph$, $R^2 = R^3 = Cl$										B	2.0028	4.3 (1H) 2.1 (1H)	4.8	31
	$R^1 = R^2 = Bu$, $R^3 = Cl$				B	2.0037	4.9 (1H) 2.6 (1H)	25.2		31					
	$R^1 = R^2 = R^3 = Me$				B		3.6 (2H)	13.0		31	B	2.0044	3.4 (2H)	6.3	26, 31
	$R^1 = R^2 = R^3 = Bu$				B	2.0036	3.6 (2H)	13.6 (^{119}Sn) 12.8 (^{117}Sn)		28, 31					
	$R^1 = R^2 = R^3 = Ph$				B	2.0030	3.7 (2H) 5.2 (1H) 2.1 (1H)	12.0		26, 31, 34	B	2.0038	3.0 (2H)	3.0	26, 31
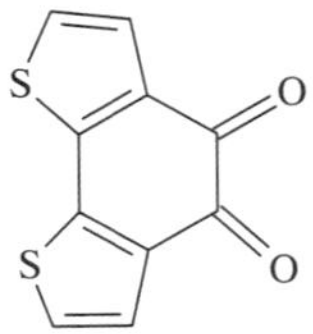 (fused dithieno diketone structure)	$R^1 = R^2 = R^3 = Ph$	B	1.17 (1H) −0.52 (1H) 0.52 (1H) 0.36 (1H)	33	B	2.00394	0.91 (2H) 0.25 (2H)	8.0 (^{119}Sn) 7.6 (^{117}Sn)		33					

coupling to two different L^1 and L^2 groups. For the radical derived from 3,6-di-*tert*-butyl-1,2-benzoquinone, these different couplings become identical at higher temperatures, an indication that the four-coordinate species **6** becomes fluxional. The two different four-coordinate species **6** and **7** have been distinguished for Sn and Pb when the R^1, R^2 and R^3 ligands are alkyl or aryl. In such cases two radicals are observed, one that displays an alternating linewidth effect with temperature and one that displays no such linewidth effect. The *cis* radical **6** is expected to be the more rapidly fluxional of the two and is therefore assigned to the radical that shows no linewidth effect. The *trans* radical **7** is then assigned to the system that exhibits fluxionality on a time scale commensurate with the ESR experiment.

The radical observed for M = Ge probably has the *trans* structure **7** given the rather large difference in the hyperfine coupling to the two methyl groups. It is of interest to note that this radical appears to be somewhat less fluxional than the similar Sn or Pb radicals.

B. $R_5M^\bullet$

Very few radicals of this structure have been examined. In the germanium series only the parent $H_5Ge^\bullet$ has been reported[35], and in the tin series $Me_5Sn^\bullet$ and $Ph_5Sn^{\bullet}$[6,13]. The germanium radical was generated by reacting GeH_4 with H atoms in a xenon matrix, while the tin radicals were generated by X- or γ-irradiation of the parent R_4Sn compounds. The data for these radicals are contained in Table 5. Those for $H_5Ge^\bullet$ reveal that the unpaired spin couples to two different types of proton: one set containing three equivalent protons, the second containing two equivalent protons. This can be accommodated by a radical that has trigonal bipyramidal symmetry, such as **8**.

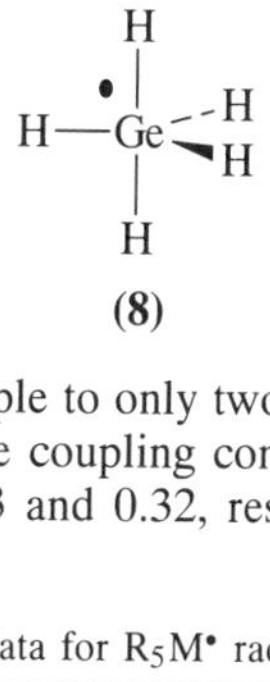

$$H-\overset{\displaystyle H}{\underset{\displaystyle H}{\overset{\bullet}{Ge}}}\overset{\textstyle H}{\underset{\textstyle H}{\diagdown}}$$

(8)

The $Me_5Sn^\bullet$ radical appears to couple to only two of the methyl groups. Moreover, the isotropic and anisotropic tin hyperfine coupling constants indicate that the Sn 5s and 5p orbital contributions are roughly 0.03 and 0.32, respectively (Table 6). Thus, compared

TABLE 5. ESR spectral data for $R_5M^\bullet$ radicals

Radical	g	a_H (G)	a_M (G)	Ref.
$H_5Ge^\bullet$	1.9965^a	13.5^a (3H)		35
	2.0169^b	15.9^b (3H)		
		6.66^a (2H)		
		7.85^b (2H)		
$Me_5Sn^\bullet$		18 (7H)	650^a	6
			400^b	
$Ph_5Sn^\bullet$	1.994		1258^a	13
	2.007		921^b	
	2.021			

^{a} Parallel component.
^{b} Perpendicular component.

TABLE 6.   Tin spin densities for the radicals $R_5Sn^\bullet$ and $R_3Sn^\bullet$

Radical	a_{iso} (G)	a_{aniso} (G)	5s	5p	5s + 5p
$Me_5Sn^\bullet$	483	83	0.03	0.32	0.35
$Me_3Sn^\bullet$	1550	200	0.10	0.77	0.87
$Ph_5Sn^\bullet$	1033	112	0.07	0.43	0.50
$Ph_3Sn^\bullet$	1866	234	0.10	0.90	1.00

with $Me_3Sn^\bullet$ the unpaired spin density at the tin atom is markedly reduced. The same is also true of $Ph_5Sn^\bullet$. These radicals also have D_{3h} symmetry in which the two apical groups share two electrons in a three-centre bonding orbital involving the tin $5p_z$ orbital, while the unpaired electron resides in a non-bonding orbital significantly localized on the ligands.

III. THE RADICAL CATIONS $R_4M^{+\bullet}$ AND $R_3MMR_3^{+\bullet}$

Exposures of dilute solutions of the parent compounds to ^{60}Co γ-irradiation in $CFCl_3^{36-40}$, $SiCl_4{}^{41}$ or $SnCl_4{}^{41}$ matrices at 77 K give rise to the corresponding radical cations. Radicals derived from germanium, tin and lead have been successfully generated by these methods. The radical cations $R_4M^{+\bullet}$ fragment readily according to the equation

$$R_3M \overset{+\bullet}{\_} R \longrightarrow R_3M^+ + R^\bullet$$

The ESR spectral data for selected radicals of this type are collected together in Table 7. These reveal that significant structural reorganization takes place upon ionization. For $Me_4Ge^{+\bullet}$, coupling of the unpaired electron with two different pairs of methyl groups is evident. Thus, distortion from the tetrahedral T_d symmetry of the parent compound has taken place and D_{2d} and D_{4h} are clearly ruled out. The most appropriate geometry for this radical cation is the trigonal bipyramidal structure **9** in which the unpaired electron occupies the $2a_1$ molecular orbital[36].

(9)

In contrast, the corresponding tin and lead radical cations, $Me_4Sn^{+\bullet}$ and $Me_4Pb^{+\bullet}$ respectively, exhibit coupling to only one unique methyl group. Other tetraalkyltin radical cations behave similarly, where coupling is observed to only one of the alkyl groups (for the radical cation containing the Bu^t group the hyperfine coupling arises from the *tert*-butyl hydrogen atoms). The positive hole is therefore believed to be localized in one of the C—M bonds. For the tin radical cation, the parallel and perpendicular components of the tin hyperfine coupling (the tin isotopes ^{119}Sn and ^{117}Sn both have $I = \frac{1}{2}$ and a natural abundance of 8.68% and 7.67%, respectively) enable an estimate of the spin density at the tin atom to be made. The isotropic and anisotropic components of the tin coupling arise from the spin density in the tin 5s and 5p orbitals, respectively. Using the data in Table 7 values for a_{iso} and a_{aniso} of 122G and 44G can be calculated. Since the isotropic coupling of an electron in a pure tin 5s orbital is 15672G and the anisotropic coupling in a pure tin 5p orbital is $261G^{44}$, it follows that the spin density at the tin atom resides in

TABLE 7. ESR spectral parameters for $R_4M^{+\bullet}$ and $R_3MMR_3^{+\bullet}$ radical cations

Radical	M = Ge			M = Sn					M = Pb		
	g	a_H (G)	Ref.	g	a_H (G)	a_{Sn} (G)	a_C (G)	Ref.	g	a_H (G)	Ref.
$Me_4M^{+\bullet}$	2.0196^a	14.7 (6H) 4.2 (6H)	36	2.029^a 1.999^b 2.0444^c	13.7 (3H)	210^b 78^c	120^b 53^c	37, 38, 39	2.111	14.7 (3H)	38
	2.0165^a	12.0 (6H) 4.0 (6H)	43	2.0194^a	13.2 (6H)	150^b 133^c		41			
$Me_3Bu^tM^{+\bullet}$											
$Me_3HM^{+\bullet}$				2.046	7.6 (9H)	88^c		38			
				1.988^a 1.960^b 2.027^c		2420^b (^{119}Sn) 2310^b (^{117}Sn) 1670^c (^{119}Sn) 1600^c (^{117}Sn)		39			
$Me_2H_2M^{+\bullet}$				1.972^b 2.046^c	68 (1H)	2700^b (^{119}Sn) 2580^b (^{117}Sn) 1090^c (^{119}Sn) 1090^c (^{117}Sn)		39			
				1.975^b 2.026^c	85 (2H)	3220^b (^{119}Sn) 3080^b (^{117}Sn) 2380^c (^{119}Sn) 2270^c (^{117}Sn)		39			
$MeH_3M^{+\bullet}$				1.976^b 2.027^c	85 (2H)	3150^b (^{119}Sn) 3010^b (^{117}Sn) 2380^c (^{119}Sn) 2270^c (^{117}Sn)		39			

(continued overleaf)

TABLE 7. (*continued*)

Radical	M = Ge			M = Sn					M = Pb		
	g	a_H (G)	Ref.	g	a_H (G)	a_{Sn} (G)	a_C (G)	Ref.	g	a_H (G)	Ref.
$H_4M^{+\bullet}$				1.984^b 2.016^c	175 (2H)	3650^b (^{119}Sn) 3490^b (^{117}Sn) 3180^c (^{119}Sn) 3040^c (^{117}Sn)		39			
				1.991^b 2.020^c	85 (1H)	3100^b (^{119}Sn) 2960^b (^{117}Sn) 2370^c (^{119}Sn) 2270^c (^{117}Sn)		39			
$Bu_4M^{+\bullet}$				1.999^b 2.047^c	14 (2H)	200^b 78^c		40			
$(Me_3M)_2^{+\bullet}$	2.0302^a 2.0031^b 2.0441^c	5.25 (18H)^a 5.39 (18H)^b 5.18 (18H)^c	42	2.110^c	3.4 (18H)	238^b 100^c		38 43			
				2.011	8.0 (3H)			41			

^{a} Isotropic value.
^{b} Parallel component.
^{c} Perpendicular component.

an orbital comprised of 0.8% 5s and 17.6% 5p. That is, the orbital has 96% p-character, indicating that the tin atom is essentially planar. As the spin density at the tin atom is only 18%, the remainder must reside with the unique methyl group. The ^{13}C hyperfine coupling constants are consistent with this analysis. Using values of 1130G and 33G for the isotropic and anisotropic coupling to the carbon 2s and 2p orbitals respectively[45], the spin density at the methyl carbon atom is 6.7% in the 2s orbital and 67.7% in the 2p orbital. Thus, the orbital on carbon has 10% s-character suggesting that, while it is close to becoming planar, it still retains some pyramidal structure. The proton hyperfine coupling in a 'free' methyl radical is 22.5G; for a methyl radical bearing 75% of the spin density, as is the case here, the coupling would therefore be ca 16.5G. So the value of 13.7 observed for $Me_4Sn^{+\cdot}$ supports the concept of an almost planar though slightly pyramidal methyl group. Therefore, the $Me_4Sn^{+\cdot}$ and $Me_4Pb^{+\cdot}$ radical cations are best represented by the structure 10, which has C_{3v} symmetry.

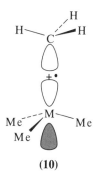

(10)

An alternative report of the $Me_4Sn^{+\cdot}$ radical cation describes hyperfine coupling to two equivalent methyl groups and tin hyperfine coupling which is much less anisotropic[41]. Accordingly, only 3% of the spin density is located at the tin centre, which is approximately sp^2-hybridized. The structure of the radical cation must have C_{2v} symmetry, i.e. similar to that for $Me_4Ge^{+\cdot}$, 9. The discrepancy between these two different descriptions of the same radical cation may well lie with the different matrices used; the species with C_{2v} symmetry is formed in a $SiCl_4$ matrix, whereas the species with C_{3v} symmetry is formed in a $CFCl_3$ matrix. The smaller molecular size of the latter should enable it to accommodate the greater reorganization required to adopt C_{3v} symmetry.

As the alkyl groups are replaced by the sterically less demanding proton, changes in the structure of the radical cations are observed[41]. For $SnH_4^{+\cdot}$ two radicals are observed; both have large tin hyperfine couplings (indicating significantly greater 5s participation than in $Me_4Sn^{+\cdot}$) but one exhibits coupling to only one proton whereas the other couples to two protons. The C_{3v} species has structure 11, which is similar to 10 except that the tin centre is associated with 80% of the spin density and is essentially sp^3-hybridized. The C_{2v} species has a structure similar to 9. The C_{2v} structure converts to the C_{3v} structure with time or on annealing, making the latter the thermodynamically more stable.

The proton hyperfine coupling indicates that the $MeSnH_3^{+\cdot}$ radical cation has the C_{2v} structure. In contrast, the $Me_2SnH_2^{+\cdot}$ radical cation gives rise to two structures: one which has C_{2v} symmetry, and a second which has either C_{3v} symmetry (as in 11) or C_{2v} symmetry (as in 9 in which only one of the equatorial groups is occupied by an H atom). In the absence of proton hyperfine coupling, the structure of the $Me_3SnH^{+\cdot}$ radical cation is uncertain and could have either symmetry, though the tin hyperfine couplings indicate that the majority of the spin density is associated with the tin atom and that there

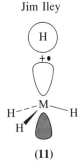

(11)

is a substantial contribution, *ca* 12%, from the tin 5s orbital. Thus it would appear that the greatest structural change in these radical cations occurs when there are four alkyl substituents. In such cases the contribution of the tin atom 5s orbital almost completely disappears rendering the tin system planar. This has been interpreted in terms of a greater stretching of an Sn—C bond relative to an Sn—H bond[39].

The radical cations of hexamethyldigermane and hexamethyldistannane both exhibit proton hyperfine coupling to all eighteen hydrogen atoms. These results suggest a symmetrical structure for both species, which can be accommodated if the unpaired electron occupies a bonding orbital between the two metal atoms, i.e. the electron lost is from the metal–metal σ bond. For hexamethyldistannane, the anisotropy of the tin hyperfine coupling constants enables the 5s and 5p contributions to be calculated as 1% 5s and 18% 5p. These values are very similar to those calculated for $Me_4Sn^{+\bullet}$, revealing that, as in **12**, the tin orbitals containing the unpaired electron have 95% p-character making the Me_3Sn units almost planar.

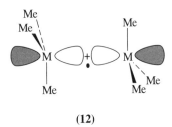

(12)

IV. THE RADICAL ANIONS $R_4M^{-\bullet}$ AND $R_3MMR_3^{-\bullet}$

Compared to the radical cations of Group IV elements, radical anions are somewhat less well studied. In general, they are formed through an electron capture process during the high energy irradiation of the parent tetravalent compounds[6,7,46]. However, certain aryl-substituted radical anions have been generated by alkali metal reduction[47,48]. The ESR spectral data for some of these radicals are contained in Table 8.

The structure of the $Me_4Sn^{-\bullet}$ and $Me_3ClSn^{-\bullet}$ radicals is almost certainly the trigonal bipyramidal structure **13**, in which the unpaired electron occupies the s/p hybrid orbital in the equatorial plane. Using the data in Table 8, together with the isotropic and anisotropic coupling constants for the tin 5s and 5p orbitals, it is possible to calculate unpaired spin densities at the tin atom in these radicals of 0.12 (5s) and 0.55 (5p) for $Me_4Sn^{-\bullet}$ and 0.15 (5s) and 0.53 (5p) for $Me_3ClSn^{-\bullet}$. The spin density at the tin atom is thus *ca* 0.65, somewhat smaller than the corresponding value for the $R_3Sn^{\bullet}$ radicals (*ca* 0.9–1.0) but larger than that for $R_5Sn^{\bullet}$ (*ca* 0.5). This reflects the transfer of spin density away from the central tin atom to the ligands as the number of ligands increases.

TABLE 8. ESR spectral data for the radical anions $R_4M^{-\bullet}$ and $R_3MMR_3^{-\bullet}$

Radical	M = Ge				M = Sn				M = Pb			
	g	a_X (G)	a_{Ge} (G)	Ref.	g	a_X (G)	a_{Sn} (G)	Ref.	g	a_X (G)	a_{Pb} (G)	Ref.
$Me_4M^{-\bullet}$					ca 2.0		2101[a] 1672[b]	6	1.926[a] 2.087[b]		3692[a] 2331[b]	7
$Me_3ClM^{-\bullet}$						30 (Cl)	2696[a] 2280[b]	6	ca 2.00		2218	7
$Ph_3ClM^{-\bullet}$									1.96[a] 2.00[b]	38[a] (Cl) 13[b] (Cl)	2165	7
$Ph_3BrM^{-\bullet}$									1.95[a] 1.98[b]	201[a] (Br) 117[b] (Br)	2440[a] 2370[b]	7
$(Ph_3MO)Ph_3M^{-\bullet}$	2.0051 2.0002 1.9993		138[a] 105[b]	46								
$Ph_3MMPh_3^{-\bullet}$					2.006		1730[a] 1462[b]	6	1.936[a] 1.954[b]		3737[a] 3046[b]	7
$Me_3MC_6H_4MMe_3^{-\bullet}$		1.88 (4H)		47								
$(1\text{-}C_{10}H_7)Me_3M^{-\bullet}$		21.5[c]	1.98	48		21.5[c]	37.1	48				
$(2\text{-}C_{10}H_7)Me_3M^{-\bullet}$		24.1[c]	0.99	48		24.04[c]	17.7	48				

[a] Parallel component.
[b] Perpendicular component.
[c] Sum of the proton hyperfine couplings.

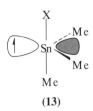

(13)

The isotropic and anisotropic germanium hyperfine couplings in the radical anion $Ph_3GeOGePh_3^{-\bullet}$ imply that the unpaired electron occupies an orbital that has contributions of 0.13 Ge 4s and 0.65 Ge 4p. Thus, the unpaired electron is significantly localized on one of the germanium atoms. The 4s/4p ratio of 5 is of sufficiently similar magnitude to those mentioned above for the tin radicals (4.6 and 3.5, respectively) to assume that this radical also has a trigonal bipyramidal structure at the germanium atom.

Though it is possible that the lead species have similar structures, calculation of the lead 6s and 6p spin densities for $Me_4Pb^{-\bullet}$ (using values for the hyperfine coupling of an electron in pure 6s and 6p orbitals of 29086G and 232G, respectively[44]) reveals that the contribution of the 6s orbital is 0.1 whereas the contribution of the 6p orbital is 1.96. The latter is clearly far too high. For $Ph_3ClPb^{-\bullet}$ a structure intermediate between the trigonal bipyramidal structure **13** and the tetrahedral structure **14** is favoured[7]. This is because the spectrum of this radical gives different a_{Cl} values from the high-field ^{207}Pb than from the central component, suggesting that at least two of the principle components of the ^{207}Pb and ^{35}Cl hyperfine tensors are well separated.

$$Ph\diagdown \atop Ph\text{--}\diagup \atop Ph} \overset{\bullet}{M}\text{---}Cl$$

(14)

Interestingly, unlike the $Ph_3GeOGePh_3^{-\bullet}$ and $Ph_3XPb^{-\bullet}$ radical anions, which demonstrate that the radical centre is largely associated with the heavy atom, the aryl-substituted radical anions of germanium and tin that are formed by alkali metal reduction appear to have the unpaired spin density associated with the aryl rings (see the last three rows in Table 8). Why this discrepancy exists is unclear. Nevertheless, the very small values for the germanium and tin hyperfine coupling constants in the 1- and 2-(trimethylgermyl/stannyl)naphthalenes is clear evidence that there is little unpaired electron spin density at the germanium or tin atoms. Moreover, a comparison of the sum of the aryl ring proton hyperfine coupling constants with the corresponding sum (26.9G) for the naphthalene radical anion (in which the unpaired electron is entirely associated with the aryl ring) reveals that in these germanium- and tin-substituted naphthalene radical anions the unpaired electron is 80–90% associated with the aryl ring[48]. Hückel molecular orbital calculations for the 1,4-bis(trimethylgermyl)benzene radical anion identifies each germanium as carrying 0.25 of the unpaired spin density; of the remaining 0.5, 0.3 is shared by the two *ipso* carbon atoms and 0.2 by the remaining four ring carbon atoms[47].

The tin hyperfine coupling constants for the hexaphenylditin radical anion enable values for the occupancy of the unpaired electron in the tin 5s and 5p orbitals of 0.10 and 0.34, respectively, to be calculated. This would imply that the electron occupies a tin sp^3 hybrid orbital. Since the radical contains two tin atoms, the total spin density residing with the tin centres is *ca* 0.9. There is very little spin density residing with the ligands, and the most likely structure for the radical is one in which the unpaired electron occupies an Sn—Sn σ^* antibonding orbital, as in **15**.

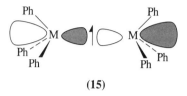

(15)

The hexaphenyldilead radical anion almost certainly has a similar structure since the g values are all less than 2.0023. Calculation of the unpaired spin population in the lead 6s and 6p orbitals leads to values of 0.11 and 0.99. Once again the spin population is too large, especially for the 6p orbital. Nevertheless, the calculations do show that the spin density is probably entirely associated with the lead atoms.

V. THE RADICALS $R_3MCH_2^\bullet$ AND $R_3MCH_2CH_2^\bullet$

There has been significant interest in carbon-centred radicals with a Group IV heavy atom positioned α or β to the radical centre. This stems from a desire to identify the nature of the interaction between the heavy atom and the unpaired spin density. It is well known, for example, that allylsilanes undergo radical addition to the double bond readily, suggesting that a silicon atom positioned β to the new radical centre has a stabilizing effect[49]. Radicals of the type discussed here have been generated by a variety of procedures, including: hydrogen atom abstraction using, for example, $Bu^tO^{\bullet[50-52]}$, addition of the appropriate $R_3M^\bullet$ radical to an alkene[50,51,53,54], halogen atom abstraction from the corresponding halomethyl compound[55] and γ-irradiation of the parent compounds[6,56,57,59].

$$R_3MCH_3 \text{ or } R_3MCH_2CH_3 \xrightarrow{Bu^tO^\bullet} R_3MCH_2^\bullet \text{ or } R_3MCH_2CH_2^\bullet$$

$$R_3M^\bullet + H_2C{=}CH_2 \longrightarrow R_3M{-}CH_2{-}\dot{C}H_2$$

$$R_3M^\bullet + R_3MCH_2X \longrightarrow R_3MCH_2^\bullet + R_3MX$$

$$R_3MCH_3 \text{ or } R_3MCH_2CH_3 \xrightarrow{\gamma} R_3MCH_2^\bullet \text{ or } R_3MCH_2CH_2^\bullet$$

Data for the most simple α- and β-substituted radicals in this series are contained in Table 9. Substituted radicals, whether on the α- or β-carbon atoms or on the Group IV atom, have largely similar values.

The data in Table 9 reveal that, for $R_3MCH_2^\bullet$, there is a definite decrease in the isotropic g values on increasing the atomic number of the Group IV element: 2.0023 (Ge), 2.0000 (Sn), 1.9968 (Pb). These α-radicals are planar with isotropic hyperfine couplings to the CH_2 protons of approximately 21.5G no matter which heavy atom is present. This coupling compares favourably with that for the $Me^\bullet$ radical, 23G, suggesting that ca 95% of the spin density is localized on the carbon atom. A more direct comparison with the neopentyl radical, $Me_3CCH_2^\bullet$, which has a proton hyperfine coupling of 21.8G, would suggest that the spin is entirely localized on the α-carbon atom. The constancy of the magnitude of the proton hyperfine couplings implies that there is little delocalization of the unpaired spin onto the heavy atom. For $Me_3SnCH_2^\bullet$ the anisotropy of the tin hyperfine coupling constants allows for the calculation of the unpaired spin density in the tin 5s and 5p orbitals. The isotropic coupling of 140G corresponds to a 5s occupancy of 0.9%, and the anisotropic coupling of 12G corresponds to a 5p occupancy of 4.6%, the overall

TABLE 9. ESR spectral data for the $R_3MCH_2^{\bullet}$ and $R_3MCH_2CH_2^{\bullet}$ radicals

Radical	M = Ge				M = Sn				M = Pb			
	g	a_H (G)	a_{Ge} (G)	Ref.	g	a_H (G)	a_{Sn} (G)	Ref.	g	a_H (G)	a_{Pb} (G)	Ref.
$Me_3MCH_2^{\bullet}$	2.0029 2.0020 2.0020	27^a 18.5^b		58	2.003 2.001 1.996	26.7^a 18.7^b	164^a (^{119}Sn) 128^b (^{119}Sn) 156^a (^{117}Sn)	6,56	2.0029 1.9938 1.9938	27.5^a 18.7^b	158^a	56
$Me_3MCH_2CH_2^{\bullet}$	2.00255	20.7 (α-H) 16.57 (β-H) 0.14 (Me)	$ca\ 30^c$	54, 57	2.00205	19.69 (α-H) 15.84 (β-H) 0.15 (Me)	488.9 (^{119}Sn) 467.7 (^{117}Sn)	54				

^{a} Parallel component.
^{b} Perpendicular component.
^{c} For $Et_3GeCH_2CH_2^{\bullet}$.

occupancy of the tin orbitals being consistent with that expected from consideration of the proton hyperfine coupling. Similar calculation for the $Me_3PbCH_2^{\bullet}$ yields values for the occupancy of the lead 6s and 6p orbitals of 0.6% and 4.3%, respectively. Again, the constancy in these values suggests that there is little π-delocalization in these radicals, and interaction of the unpaired spin with the heavy atom is attributed to a spin polarization mechanism through the M−C bond[4].

The β-substituted radicals, $R_3MCH_2CH_2^{\bullet}$, display coupling to both the α- and β-protons, as well as to the methyl groups on the heavy atom. Coupling to the α-protons is somewhat smaller and more variable than in the corresponding α-substituted radicals. This implies an interaction between the heavy atom and the carbon bearing the unpaired spin. Even so, comparison of the hyperfine coupling of the α-protons with those of the ethyl radical (21.8G) implies that for the germyl-substituted radical at least 90% of the unpaired spin resides on the carbon atom, while for the tin radical the value is 95%. More revealing is the magnitude of the coupling to the β-protons. This is much smaller than that expected for free rotation of the methylene group about the C−C bond (for example, the β-proton hyperfine coupling in the propyl radical is 30.33G). The radical must therefore adopt a preferred conformation. Using the relationship

$$a_H = B \cos^2 \theta$$

and taking a value for B from the propyl radical of 60G, the observed hyperfine couplings of ca 16G correspond to a value of $\theta = 120°$. Thus the preferred conformation of these radicals is **16**, in which the C−M bond eclipses the 2p orbital containing the unpaired spin. Not surprisingly, the hyperfine coupling to the β-protons increases with temperature as would be expected for increased rotational mobility.

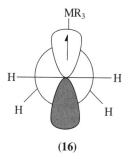

(16)

Consistent with this interpretation is the observation that the hyperfine coupling to the heavy atom is much larger than in the corresponding $R_3MCH_2^{\bullet}$ radicals. Moreover, for the tin-containing radical the hyperfine coupling is essentially isotropic[57] which, together with its magnitude, rules out p−d orbital overlap as the mechanism of spin delocalization. This isotropic coupling corresponds to a tin 5s contribution of ca 2.5%. Hyperconjugation of the unpaired electron into the sp^3 orbital of the heavy atom is a much more likely explanation. Stabilization of these Group IV heavy atom β-substituted radicals by the mechanism just described is probably the reason such radicals can be formed by hydrogen atom abstraction at the primary site in the parent compounds. Hydrogen atom abstraction from such sites in the equivalent carbon radicals is not observed. Interestingly, carbon-centred radicals with two heavy atom groups positioned β to the radical centre, i.e. $(R_3MCH_2)_2CH^{\bullet}$, have been found to possess both heavy atom groups positioned so that the C−M bonds eclipse the p-orbital containing the unpaired electron. Whether these groups lie *syn* or *anti* has not been established, but they are thought to be *anti* for steric reasons[60].

Finally, similar stabilization has been observed for the homoallylic analogues **17**, where the preferred conformation is one in which the C—M bond eclipses the π-system containing the unpaired spin[53]. Clearly, this mechanism of spin delocalization is of general nature.

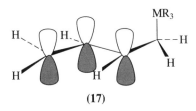

(17)

VI. REFERENCES

1. G. S. Jackel and W. Gordy, *Phys. Rev.*, **176**, 443 (1968).
2. K. Nakamura, T. Takayanagi, M. Okamoto, K. Shimokoshi and S. Sato, *Chem. Phys. Lett.*, **164**, 593 (1989).
3. S. W. Bennett, C Eaborn, A Hudson, H. A. Hussain and R. A. Jackson, *J. Organometal. Chem.*, **16**, P36 (1969).
4. R. V. Lloyd and M. T. Rogers, *J. Am. Chem. Soc.*, **95**, 2459 (1973).
5. G. B. Watts and K. U. Ingold, *J. Am. Chem. Soc.*, **94**, 491 (1972).
6. S. A. Fieldhouse, A. R. Lyons, H. C. Starkie and M. C. R. Symons, *J. Chem. Soc., Dalton Trans.*, 1966 (1974).
7. R. J. Booth, S. A. Fieldhouse, H. C. Starke and M. C. R. Symons, *J. Chem. Soc., Dalton Trans.*, 1506 (1976).
8. H. Sakurai, K. Mochida and M. Kira, *J. Am. Chem. Soc.*, **97**, 929 (1975).
9. M. Lehnig, W. P. Neumann and E. Wallis, *J. Organometal. Chem.*, **333**, 17 (1987).
10. A. F. El-Farargy, M. Lehnig and W. P. Neumann, *Chem. Ber.*, **115**, 2783 (1982).
11. M. Lehnig, H. U. Buschhaus, W. P. Neumann and T. Apoussidis, *Bull. Soc. Chim. Belg.*, **89**, 907 (1980).
12. K. Mochida, *Bull. Chem. Soc. Jpn.*, **57**, 796 (1984).
13. T. Berclaz and M. Geoffroy, *Radiat. Phys. Chem.*, **12**, 91 (1978).
14. H. Sakurai, K. Mochida and M. Kira, *J. Organometal. Chem.*, **124**, 235 (1977).
15. M. J. S. Gynane, M. F. Lappert, P. Rivière and M. Rivière-Baudet, *J. Organometal. Chem.*, **142**, C9 (1977).
16. J. D. Cotton, C. S. Cundy, D. H. Harris, A. Hudson, M. F. Lappert and P. W. Lednor, *Chem. Commun.*, 651 (1974).
17. P. J. Davidson, A. Hudson, M. F. Lappert and P. W. Lednor, *Chem. Commun.*, 829 (1973).
18. M. J. S. Gynane, D. H. Harris, M. F. Lappert, P. P. Power, P. Rivière and M. Rivière-Baudet, *J. Chem. Soc., Dalton Trans.*, 2004 (1977).
19. M. Lehnig, T. Apoussidis and W. P. Neumann, *Chem. Phys. Lett.*, **100**, 189 (1983).
20. H. U. Buschhaus, M. Lehnig and W. P. Neumann, *Chem. Commun.*, 129 (1977).
21. M. Westerhausen and T. Hildenbrand, *J. Organometal. Chem.*, **411**, 1 (1991).
22. M. Lehnig and K. Dören, *J. Organometal. Chem.*, **210**, 331 (1981).
23. M. Geoffroy, L. Ginet and E. A. C. Lucken, *Chem. Phys. Lett.*, **38**, 321 (1976).
24. J. E. Bennett and J. A. Howard, *Chem. Phys. Lett.*, **15**, 322 (1972).
25. H. -G. Korth, R. Sustmann, J. Dupuis and B. Giese, *Chem. Ber.*, **120**, 1197 (1987).
26. A. G. Davies, J. A. -A. Hawari, C. Gaffney and P. G. Harrison, *J. Chem. Soc., Perkin Trans. 2*, 631 (1982).
27. K. Reuter and W. P. Neumann, *Tetrahedron Lett.*, 5235 (1978).
28. A. G. Davies and J. A. -A. Hawari, *J. Organometal. Chem.*, **201**, 221 (1980).
29. D. Rehorek and E. Janzen, *J. Organometal. Chem.*, **268**, 135 (1984).
30. P. J. Barker, A. G. Davies, J. A. -A. Hawari and M. -W. Tse, *J. Chem. Soc., Perkin Trans. 2*, 1488 (1980).
31. A. G. Davies and J. A. -A. Hawari, *J. Organometal. Chem.*, **251**, 53 (1983).
32. J. Cooper, A. Hudson and R. A. Jackson, *J. Chem. Soc., Perkin Trans. 2*, 1933 (1973).

33. A. Alberti, A. Hudson and G. F. Pedulli, *J. Chem. Soc., Faraday Trans. 2*, **76**, 948 (1980).
34. A. Y. Brezgunov, A. A. Dubinskii, O. G. Poluektov, A. I. Prokof'ev, S. D. Chemerisov and Y. S. Lebedev, *Zhur. Strukt. Khim.*, **33**, 69 (1992).
35. K. Nakamura, N. Masai, M. Okamoto, S. Sato and K. Shimokoshi, *J. Chem. Phys.*, **86**, 4949 (1987).
36. B. W. Walther and F. Williams, *J. Chem. Soc., Chem. Commun.*, 270 (1982).
37. M. C. R. Symons, *J. Chem. Soc., Chem. Commun.*, 869 (1982).
38. B. W. Walther, F. Williams, W. Lau and J. Kochi, *Organometallics*, **2**, 688 (1983).
39. A. Hasegawa, S. Kaminaka, T. Wakabayashi, M. Hayashi, M. C. R. Symons and J. Rideout, *J. Chem. Soc., Dalton Trans. 2*, 1667 (1984).
40. E. Butcher, C. J. Rhodes, M. Standing, R. S. Davidson and R. Bowser, *J. Chem. Soc., Perkin Trans. 2*, 1469 (1992).
41. L. Bonazzola, J. P. Michault and J. Roncin, *New J. Chem.*, **16**, 489 (1992).
42. J. T. Wang and F. Williams, *J. Chem. Soc., Chem. Commun.*, 666 (1981).
43. M. C. R. Symons, *J. Chem. Soc., Chem. Commun.*, 1251 (1981).
44. J. R. Morton and K. F. Preston, *J. Magn. Reson.*, **30**, 577 (1978).
45. P. W. Atkins and M. C. R. Symons, *The Structure of Inorganic Radicals*, Elsevier, Amsterdam, 1967.
46. M. Geoffroy and L. Ginet, *Helv. Chim. Acta*, **59**, 2536 (1976).
47. A. A. Allred and L. W. Bush, *J. Am. Chem. Soc.*, **90**, 3352 (1968).
48. H. Bock, W. Kaim, and H. Tesmann, *Z. Natarforsch.*, **33B**, 1223 (1978).
49. H. Sakurai, A. Hosomi and M. Kumada, *J. Org. Chem.*, **34**, 1764 (1969).
50. P. J. Krusic and J. K. Kochi, *J. Am. Chem. Soc.*, **91**, 6161 (1969).
51. P. J. Krusic and J. K. Kochi, *J. Am. Chem. Soc.*, **93**, 846 (1971).
52. A. Hudson and H. A. Hussain, *J. Chem. Soc. (B)*, 792 (1969).
53. T. Kawamura, P. Meakin and J. K. Kochi, *J. Am. Chem. Soc.*, **94**, 8065 (1972).
54. T. Kawamura and J. K. Kochi, *J. Am. Chem. Soc.*, **94**, 648 (1972).
55. H. Sakurai and K. Mochida, *J. Organometal. Chem.*, **154**, 353 (1978).
56. A. R. Lyons, G. W. Nielson and M. C. R. Symons, *J. Chem. Soc., Faraday Trans. 2*, **68**, 807 (1972).
57. A. R. Lyons and M. C. R. Symons, *J. Chem. Soc., Faraday Trans. 2*, **68**, 622 (1972).
58. J. H. Mackey and D. E. Wood, *Mol. Phys.*, **18**, 783 (1970).
59. K. Höppner and G. Lassmann, *Z. Naturforsch.*, **23A**, 1758 (1968).
60. M. Kira, M. Akiyama and H. Sakurai, *J. Organometal. Chem.*, **271**, 23 (1984).

CHAPTER 6

Photoelectron spectroscopy (PES) of organometallic compounds with C–M (M = Ge, Sn, Pb) bonds

CARLA CAULETTI and STEFANO STRANGES

Dipartimento di Chimica, Università di Roma 'La Sapienza', Piazzale Aldo Moro, 5, 00185 Rome, Italy
Fax: (+39) 6 490 631; e-mail: CAULETTI@SCI.UNIROMA1.IT

The chemistry of organic germanium, tin and lead compounds
Edited by S. Patai   © 1995 John Wiley & Sons Ltd

I. INTRODUCTION

A. Generalities and Scope

An understanding of the properties and behaviour of a chemical compound requires knowledge not only of the three-dimensional arrangement of atoms in the molecule, but also information on the electronic structure, i.e. electron distribution, bonding character and stability of the bonds. This is of fundamental importance also for an understanding of the chemical and physical properties of new systems. In this subject, photoelectron spectroscopy (PES) proved to be a very powerful technique, which involves the application of the photoelectric effect to the study of photoionization processes occurring in matter interacting with electromagnetic radiation. In particular, when the matter is in the gas phase and the light is of short wavelength, i.e. in the far ultraviolet (UV), the entities involved in the photoionization are free molecules and the electrons ejected ('photoelectrons') belong to occupied molecular orbitals (MOs). This chapter will deal with UV gas-phase photoelectron spectroscopy (abbreviated to PES throughout the chapter) studies on organometallic compounds of germanium, tin and lead, containing carbon–metal bonds. The analysis of the literature data will be done with the purpose of emphasizing regularities that could be observed in bonding characters along series of analogous compounds, possibly related to similarities or differences in chemical behaviour. In such an analysis, we shall sometimes refer to silicon- and carbon-containing analogues of the compounds investigated, for comparison.

B. Fundamental Aspects of PES

The fundamental process of the photoemission is the ejection of the photoelectrons by a material irradiated with photons of sufficient energy $(h\nu)$, the excess kinetic energy (KE) of the emitted electrons being related to the ionization energy (IE) through the equation

$$h\nu = IE + KE \qquad (1)$$

The latter provides a means of determining IE by measuring KE of photoelectrons generated by a radiation of known $h\nu$. When the ionizing source has an energy in the region of the soft X-rays (up to a few thousand electron volts) both valence and core electrons can be ionized, but the primary information concerns the latter ones. In this case the technique is usually named X-ray photoelectron spectroscopy (XPS) or, with the historic acronym, ESCA (electron spectroscopy for chemical applications). When $h\nu$ is in the field of the medium or high energies, such as those produced in an inert gas discharge (typically 21.22 eV in He I 2p $\longrightarrow$ 1s or 40.81 eV in He II 2p $\longrightarrow$ 1s), only the valence electrons can be ionized and the technique is named UV photoelectron spectroscopy (UPS or UV PES). If the target is in the gas phase, the information obtained concerns the electronic structure of free molecules, as already mentioned in the previous section.

Several excellent books and reviews contain an extensive description of the principles and the chemical applications of PES[1-23], including investigations on organic[21,22] and organometallic[14,16,18-20] compounds.

We shall describe in this section only the most important aspects of PES, to make the reader able to better understand the meaning of the information that can be extracted from the analysis of the photoelectron spectrum of a free molecule.

1. PES and molecular orbitals

The photoelectron spectrum of a molecular species M consists of a diagram reporting the number of photoelectrons ejected with a given KE (corresponding to a given IE, following equation 1) versus this KE (or IE). Strictly speaking, the IEs measured for the various filled orbitals of the molecule M represent the differences between the energy levels of the different final ionic states M^+ and the ground state of the neutral M. A correlation between the energy sequence of the ionic levels and the sequence of the MOs of the neutral ground state can be found by applying an approximation, known as Koopmans' theorem[24]:

$$IE_i = -\varepsilon_i \tag{2}$$

stating that the IE corresponding to the removal of an electron from the ith MO (IE_i) is equal to the negative of the self-consistent field (SCF) orbital energy ε_i.

Koopmans' theorem does not obviously apply to open-shell molecules, and implies some approximations, first of all neglect of the relaxation phenomena following the photoionization, and of the difference in electron-correlation effects in the molecule and in the different states of the molecular ion. However, equation 2 mainly offers a good way of evaluating the energy sequence of the MOs, allowing correct qualitative and semiquantitative comparisons, mainly within series of related molecules, where the approximations should affect in a similar way the correlation IE_i/ε_i. In this framework, we could say that the sequence of the bands of a PE spectrum is a good representation of the electronic structure of the molecule, namely of the sequence of the occupied molecular orbitals. Such one-to-one correspondence does obviously fail in some cases, in addition to the already-mentioned case of open-shell systems, where a multiplicity of ionic states may arise from the ionization of one or more of the open shells. For instance, the ionization of one electron with simultaneous excitation of a second electron to an unoccupied excited orbital may occur [shake-up or configuration interaction (CI) processes]. Furthermore, effects such as spin–orbit coupling or the Jahn–Teller effect may remove the degeneracy in the molecular ion. In all these cases, we will observe more ionization peaks in the spectrum than MOs in the neutral molecule.

2. Intensity of the bands in a photoelectron spectrum

The relative intensities of the bands, i.e. the band–area ratios, are very meaningful for the interpretation of a PE spectrum since they are proportional to the relative probabilities of ionization. The absolute value of the area of a spectral band depends, among other factors to be discussed shortly, also on the density of the target, which is quite difficult to measure, so that usually the spectral intensities are given in arbitrary units. For the purpose of the analysis of the electronic structure of a molecule, the intensity ratio between the different bands is sufficient to give valuable indications.

The relative probabilities of ionization out of the various occupied MOs, named relative partial cross-sections (σ), are determined by several factors, the most important ones being the following:

(i) σ_i for ionization out of the ith MO is proportional to the number of equivalent electrons occupying this MO (i.e. to the degeneracy of the level) in the case of closed-shell molecules, or to the statistical weight of the ionic state produced, in the case of open-shell systems.

(ii) σ_i depends on the character of the ionized MO and on the energy of the ionizing radiation. This dependence follows different trends for the various orbitals. For instance, σ_i of an orbital with at least one node in the radial part of the wave function presents a characteristic behaviour as a function of photon energy, with a minimum for a certain value of this energy, which is characteristic of the orbital itself. It is therefore particularly interesting to study the variation of the photoelectron cross section by using ionizing radiation of different wavelength. The most recent developments of the technique, extending the sources from the simple laboratory ones to the very sophisticated synchrotron or storage ring radiation, allow one to investigate the σ trend continuously over a wide range of photon energy. A deeper description of this aspect of the photoelectron spectroscopy investigation can be found in References 25 and 26. Some empirical rules can be extracted from the numerous investigations carried out mainly with He I and He II sources. For instance, σ for atomic orbitals increases going down a group of the Periodic Table under He I radiation, whilst usually decreasing under He II. In fact, the cross section of p orbitals of chlorine, bromine and sulphur decreases significantly on passing from He I to He II, whilst carbon 2p atomic orbitals have approximately the same cross section. In the case (of most interest for us) of molecular orbitals, a model, known as the Gelius model[27], correlates the cross section of the jth MO (σ_j) with that of the contributing atomic orbitals (AOs):

$$\sigma_j = \sum_{A,i} P_{ji_A} \sigma_i^A \qquad (3)$$

where the summation extends over atomic orbitals Φ_i^A, on the different centres A; σ_i^A are the atomic cross sections and P_{ji_A} are factors describing the occupancy of the MO.

3. Assignment of a photoelectron spectrum

The main criteria to assign photoelectron spectra are as follows:

(i) Analysis of the vibrational structure of the bands, whenever present. This is particularly useful for small molecules, whose PE spectra often present resolved vibrational features, due to the fact that the removal of an electron causes changes in the geometry on passing from the neutral molecule to the molecular ion whenever the ionized orbital is of bonding or antibonding character. On the other hand, the ionization of lone-pair electrons leaves the geometry unchanged, therefore giving rise to sharp and narrow bands, without vibrational structure.

(ii) Observation of variations in band–intensity ratios as a function of photon energy. This criterion is an application of what was discussed in Section I.B.2 above. In particular, by applying equation 3 one can identify the predominant AO character in the MO associated with a particular photoelectron band. For instance, the ionization of a MO with a significant contribution of chlorine 3p orbitals gives rise to a band which decreases in intensity on switching from He I to He II ionizing radiation.

(iii) Comparison of the spectra of series of related compounds. Trends in IEs along series of analogous compounds are often an assignment criterion of very significant chemical meaning. In fact, the variations and/or similarities observed can be placed in relation to various factors, such as polarization effects, size and steric hindrance of substituents, presence or absence of conjugative interactions, and so on.

(iv) Comparison of the experimental data with theoretical calculations. The improvement of theoretical models and the development of fast computers have made this criterion perhaps the most powerful tool to assign the PE spectra, even of large molecules. The theoretical calculations of interest for us may be performed at various levels of sophistication, and can be roughly divided in two categories:

(a) calculations of ionization energies, i.e. the difference in energy between the various ionic states and the molecular ground state, taking into account all or most of the effects following the photoionization, including relaxation, correlation and relativistic effect;

(b) calculations of the eigenvalues of the molecular orbitals and their composition in the ground state of the molecule. In this case, the comparison between experimental and theoretical data is meaningful only in the framework of Koopmans' theorem (equation 2). These calculations are of different degrees of sophistication, ranging from the exact Hartree–Fock self-consistent-field (HF-SCF) to *ab initio* SCF, to semiempirical, to empirical ones.

C. Chemical Applications of PES

Applications of PES in chemistry can be roughly divided into two major categories: (a) analysis of unimolecular reactions; (b) insight into molecular electronic structure and bonding. In this section we will place emphasis on the latter point, selecting some of the most interesting applications in the field of organometallic chemistry.

1. Role of the d orbitals in chemical bonding

The contribution of the d orbital, both of metal and non-metal elements, to the bonding, for instance through a $d\pi$–$p\pi$ interaction, has been a controversial matter for a long time, and can receive from PES important help in clarification. We will discuss some examples on this point in the following sections.

2. Bonded and non-bonded interactions

With the designations bonded and non-bonded interactions, one refers to interactions between orbitals centred at atoms which are bonded directly to each other or, respectively, through one or more groups. Among the simplest bonded interactions the π–π conjugations are of particular importance, for instance those between the π system of an aromatic ring and a π-type orbital of a substituent, leading to the splitting of originally degenerate MOs. Also σ–σ and σ–π conjugation between adjacent bonds may be relevant in the bonding character. PES is particularly suitable to study this kind of interactions.

Typical non-bonded interactions are the 'through-space' and 'through-bond' ones. If, for instance, in a molecule there are two equivalent lone-pair orbitals, these may interact either directly, 'through space', or indirectly, 'through bond', i.e. through the MOs of groups linked to both the atoms carrying the lone pair. The mechanism of these interactions is schematized in Figure 1.

A type of non-bonded interaction which is often very significant in organometallic compounds is the so-called 'hyperconjugation', namely a σ–π or $d\pi$–$p\pi$ conjugation between the π systems of the organic moiety and $\sigma_{M–C}$ orbitals. We will find several examples of these kinds of interaction studied by PES in the following sections.

3. Molecular geometry and conformation

PES has been of great help in elucidating and observing geometrical implications on the electronic structure of molecules. Changes in bond length or in the angle between bonds on passing from the molecular ground state to the ionic states, and the most stable molecular conformation in the gas phase could often be established by this technique.

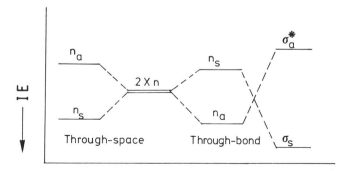

FIGURE 1. Schematic diagram showing the effect of 'through-space' and 'through-bond' interactions on the ionization energies of lone-pair orbitals

II. TETRAHEDRAL HYDRIDES

It is useful, before entering a discussion of the subject of the present review, namely that dealing with molecules containing M$-$C bonds, to describe the electronic structure as resulting from theoretical and PES studies of the simplest derivatives of elements, i.e. the tetrahedral hydrides MH_4 (M = Ge, Sn). The four M$-$H σ bonding orbitals transform in the T_d point group as $t_2 + a_1$ irreducible representations. We should therefore expect in the PE spectrum, in the framework of the Koopmans' approximation, two bands, that at lower IE being related to the t_2 orbital and approximately three times more intense than the one at higher IE, related to the a_1 orbital. Actually, the Jahn$-$Teller distortion in the MH_4^+ ion does affect the spectra. Theoretical studies on GeH_4^{+28} and SnH_4^{+29} molecular ions indicated an equilibrium structure after the Jahn$-$Teller distortion of C_s symmetry for the former ion and of C_{2v} symmetry for the latter one. In both cases there is a removal of degeneracy of the t_2 orbital towards three non-degenerate levels leading to the appearance of structures, if not of well-resolved peaks, in the first band of the PE spectra of GeH_4^{30-32} and of SnH_4^{32}. The theoretical investigations took into account also the spin$-$orbit splitting whose effect should not be negligible especially in the heavier SnH_4 molecule, but no experimental evidence of this effect could be observed, probably due to a not excellent resolution. Cradock[31] managed to detect, in the spectrum of GeH_4, two bands, at 26.9 and 27.4 eV, associated with the inner-valence 3d levels of Ge, split into t_2 and e components and ionized by the He II radiation.

III. ALKYL AND HALOALKYL DERIVATIVES

A. Tetraalkyls

After the pioneering work by Evans and coworkers[33], in which the PE spectra of the tetramethyl derivatives of all the elements of the XIV group were extensively analysed, showing again the effect of the Jahn$-$Taller distorsion upon ionization of the t_2 level (HOMO) of main M$-$C bonding character, several investigations on both symmetrical and unsymmetrical alkyls[34$-$41] appeared, which draw attention to particular aspects, such as the role played by the metal d-orbitals (both filled and empty) in the bonding, trends in the energy and nature of the radical cation states of series of related molecules (with different alkyl groups or different central metal M), also through free-energy relationships between ionization energies and chemico-physical parameters such as the Taft σ^* constant, the variation of the photoionization cross section of the various orbitals as a function of photon energy, using both laboratory sources, namely He I and He II, and synchrotron radiation as ionizing radiation. In most of these investigations a comparison between the experimental data and the results of calculations at various levels of sophistication is essential for the interpretation of the spectra and the discussion of the resulting chemical implications. Some theoretical papers on semiempirical MNDO calculations on tetraalkyls of Group XIV elements, along with other derivatives, appeared quite recently[42,43], but the method which proved, in our opinion, the most suitable for compounds containing germanium, tin and lead is that of the pseudopotentials, which are introduced to deal with all core electrons. Several different schemes have been proposed for the construction of pseudopotentials[44], some of them including relativistic and core-polarization effects, very significant for heavy atoms. The subject of the participation of metal d orbitals was treated quite extensively. All the authors agree that this participation is actually negligible, if any. Boschi and coworkers[35] observed that the splitting between the two t_1 and e CH levels in Me_4M (M = C, Si, Ge, Sn, Pb) drops monotonically from neopentane to tetramethylplumbane, consistently with the decreasing through-space interaction in the Me_4 tetrahedron following the increasing size of the central atom. A contribution of the d orbitals would tend to increase the splitting in the opposite sense. For tetramethylstannane,

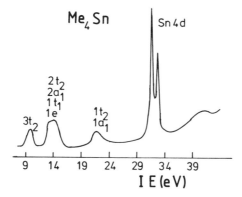

FIGURE 2. The photoelectron spectrum of Me_4Sn for $h\nu = 49$ eV. Reproduced by permission of Elsevier Science from Reference 41

this conclusion is confirmed by the shape and IE of the two maxima of the band observed for the $4d_{5/2}$ and $4d_{3/2}$ Sn^{-1} states in the PE spectrum obtained with synchrotron radiation (see Figure 2). The IE values of 31.5 and 32.6 eV, slightly different from those obtained for atomic tin (31.9 and 33.2 eV)[45], and the narrow profile of the two peaks indicate that the 4d orbitals retain their atomic character in the molecule. Also Beltram and coworkers[38], who studied a series of group XIV alkyls and alkyl hydrides, among which are the compounds Et_4M (M = Si, Ge, Sn), confine the contribution of d orbitals of M to a hyperconjugative interaction, not affecting the three highest occupied molecular orbitals.

Wong and coworkers[37] studied a series of both symmetrical and unsymmetrical tetraalkyltin compounds with different alkyl substituents, focusing their attention on the effect of these on the ionization energy of the highest occupied MOs. It is useful to recall the type of molecular orbitals deriving from the triply degenerate t_2 HOMO in the symmetrical R_4M (local symmetry T_d) upon substitution of one or more R:

$$R_3MX(\text{local symmetry } C_{3v}) : e + a_1$$

$$R_2MX_2(\text{local symmetry } C_{2v}) : b_2 + a_1 + b_1$$

$$RMX_3(\text{local symmetry } C_{3v}) : a_1 + e$$

Figure 3 reproduces a correlation diagram for the triply degenerate t_2 MO of Me_4Sn as a result of successive substitution of ethyl for methyl substituents. The dashed line is drawn through weighted average ionization energies (IEs) for each $Me_{4-n}Et_nSn$. These IEs correlate well with Taft $\Sigma\sigma^*$ parameters. The electronic effect of alkyl substituent R on the splitting of the e and a_1 levels in a series of $RSnMe_3$ compounds is shown in Figure 4, where this splitting is reported in a diagram versus the Taft σ^* values. A linear correlation is found which passes through the origin. It follows that for a series of this kind the first IE from HOMO can be simply related to that of Me_4Sn (9.70 eV) from the Taft σ^* value of R, through the equation

$$IE_1(RSnMe_3) = 9.70 + 4.3\sigma^*.$$

The study of the photoionization cross section as a function of photon energy for the different orbitals of Me_4Sn, which can be a powerful tool for the assignment of the spectra and the analysis of the contribution of the various atomic orbitals to the molecular orbitals, has been carried out by the authors of References 11 and 12 by using He I and He II as ionizing source, and of Reference 13 by using synchrotron radiation. Bertoncello

Carla Cauletti and Stefano Stranges

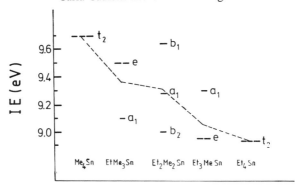

FIGURE 3. Correlation diagram for the t_2 MO of Me_4Sn as a result of successive substitution of ethyl for methyl ligands. Reprinted with permission from Reference 37. Copyright (1979) American Chemical Society

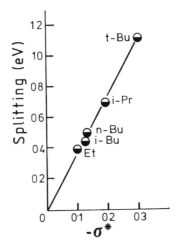

FIGURE 4. Electronic effects of alkyl ligands (R) measured by Taft σ^* values on the splitting between the e and a_1 MOs in a series of alkyltrimethyltin $RSnMe_3$ compounds. Note the extrapolation through the origin. Reprinted with permission from Reference 37. Copyright (1979) American Chemical Society

and coworkers[40] assumed a 20% contribution of tin 5p orbital and a 70% of carbon 2p to the t_2 σ_{Sn-C} bonding MO, on the ground of pseudopotential calculations and of the increase of the intensity ratio between the first and second band under He II radiation. Since previous results by the same research group on $Me_3Sn-SnMe_3$[46] indicated that the photoionization cross section of Sn 5p decreases with respect to C 2p and increases with respect to H 1s orbital, the authors ascribed the variation in band intensity ratio in the spectrum of Me_4Sn to a dominant C 2p contribution to the HOMO. On the other hand, Bancroft and coworkers[39], who assumed that the Sn 5p cross section increases on going from He I to He II spectra, attributed to the Sn–C orbitals a 30–40% metal p character. The investigation with synchrotron radiation in the photon energy range 21–70 eV by Novak and coworkers[41] confirms the dominant carbon 2p character of these orbitals, on the ground of the slow decrease in cross section with increasing photon energy. On the

TABLE 1. Experimental ionization energies (eV) for bivalent metal alkyls MR_2 [M = Ge, Sn, Pb; R = $CH(SiMe_3)_2$] and first atomic ionization energies (eV) of M. Reproduced by permission of the Royal Society of Chemistry from Reference 48

Alkyls	MO^a		M	Atomic first IE^b
	$a_1 M$	$b_2(M-C\ \sigma_{asym})$		
GeR_2	7.75	8.87	Ge	7.5
SnR_2	7.42	8.33	Sn	7.0
PbR_2	7.25	7.98	Pb	6.8

^{a} Symmetry symbols refer to C_{2v} local symmetry of MC_2 frameworks.
^{b} Calculated from the weighted average of the p^2 terms of the atom and of the p^1 terms of the ion data from Reference 49.

other hand, the calculated subshell cross sections of Yeh and Lindau[47] predict for the C 2p orbital a fall from 6.1 Mb at 21 eV photon energy to 1.9 Mb at 41 eV, while the corresponding change for the Sn 5p subshell is from 1.17 to 0.04 Mb.

B. Dialkyls

Very little has been done regarding PES of dialkyl derivatives of Group XIV elements. Harris and coworkers[48] investigated the He I spectra of the monomeric, presumed-bent in gas phase, compounds MR_2 [M = Ge, Sn, Pb; R = $CH(SiMe_3)_2$], along with some bivalent amides and halides. Table 1 reports the ionization energies of the first band with the assignment. It contains also, for comparison purposes, the first IE of atomic Ge, Sn and Pb, calculated from the weighted average of the p^2 terms of the atom and of the p^1 terms of the ion taken from the literature[49]. It appears clearly that the HOMO in all the three molecules has a substantial metal 'lone pair' np character, although it can mix with both metal ns and carbon 2p orbitals. This mixing should produce an opposite effect on the IE values. The comparison of the observed IEs in the molecules with the calculated atomic IEs hints to a balance of these effects, since the deviation of the 'molecular' IEs from the 'atomic' one is invariably less than 3%. The second IE corresponds to ionization from the b_2 C−M−C antisymmetric bonding MO.

C. Alkyl Hydrides and Haloalkyl Derivatives

Several PES studies of compounds of general formula $R_x MX_{4-x}$ (M = Si, Ge, Sn, Pb; X = H, halogen, x = 1−4) have been carried out[38,40,50−61] with the main aim of studying the influence of the X substituent(s) on the metal−carbon bond(s). We have already analysed the effect of the substitution of one or more R on the symmetry of the highest occupied MOs, and consequently on the photoelectron spectra. A nice example of this in a series of alkylgermahydrides is shown in Figure 5, from Reference 38, where this effect appears clearly. The triply degenerate t_2 band in the spectrum of Et_4Ge (1) (broadened by the Jahn−Teller effect) is split into the e and a component in Et_3GeH (2), the former level being of σ_{Ge-C} character and the latter one of σ_{Ge-H} character; into three components (b_2, a_1 and b_1) in Et_2GeH_2 (3), where the b_2 and b_1 orbitals are localized respectively on the σ_{Ge-C} and σ_{Ge-H} bonds, while the a_1 orbital is delocalized through the whole H_2GeC_2 skeleton; and finally, again into two components, a_1 and e in order of increasing IE, in $EtGeH_3$ (4), the former one being related to the only σ_{Ge-C} bond. It is noteworthy that the σ_{Ge-C} orbitals lie always higher in energy (lower in ionization energy) with respect to the σ_{Ge-H}, and that the fourth σ orbital is too deep in energy to be ionized by the He I radiation.

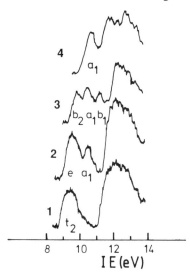

FIGURE 5. The He I photoelectron spectra of compounds **1**–**4** (**1** = Et$_4$Ge; **2** = Et$_3$GeH; **3** = Et$_2$GeH$_2$; **4** = EtGeH$_3$). Reproduced by permission of Elsevier Science from Reference 38

The introduction of one or more halogen atoms in an alkyl derivative strongly complicates the electronic structure and the bonding situation of the molecule, due to the possible interactions between the σ_{M-C} and σ_{M-X} (X = halogen) orbitals. The electrons localized on the halogens (halogen lone pairs) usually do not interact with the bonding orbitals. The analysis of the PE spectra of these compounds is rarely straightforward, and controversies on some assignments have appeared in the literature. The joint use of the comparison between He I and He II spectra and of continuously improving computational methods to calculate the eigenvalues of the various MOs is of great help in solving ambiguities and uncertainties. For these molecules, the observation of the changes in band intensity ratios between He I and He II spectra is particularly useful. For instance, the photoionization cross-section of chlorine 3p atomic orbitals is lower under He II than that of C 2p orbitals[62], so that a decrease in intensity of a band on passing from He I to He II ionizing radiation is a strong indication of a significant participation of Cl 3p orbitals to the MO associated with that particular band. Such a decrease is obviously dramatic for the ionization of the chlorine lone pairs.

An example of a series of alkyltin halides is that of the general formula Me$_n$SnCl$_{4-n}$ (n = 0–4), re-examined completely by Bertoncello and coworkers[40] on the basis of pseudopotential calculations, after several studies by different authors[54–56,58]. The He I and He II spectra of Me$_2$SnCl$_2$ are shown in Figure 6, from which the effect of decrease in intensity of bands A, B and C with respect to band D, and of band B with respect to bands A and C on passing from He I to He II is very clear. Following Reference 40 (see Table 2), band A, at 10.43 eV, associated with the HOMO of b_1 symmetry, is a σ_{Sn-C} bonding MO with some contribution from chlorine 3p orbitals; band B, at 11.31 eV, arises from the ionization of three orbitals ($a_1 + b_2 + a_2$) of the chlorine lone pairs, and band C, at 12.16 eV, again from three orbitals ($b_1 + a_1 + b_2$) of main σ_{Sn-Cl} bonding character. The broad band D, peaking at 14.30 eV, is associated with four CH$_3$-based MOs ($b_1 + a_2 + a_1 + b_2$), the partially resolved band X with a MO(a_1) of mixed Sn(5s)−C and Sn(5s)−Cl bonding character. The assignment is analogous for the other

TABLE 2.   Experimental ionization energies and pseudopotential *ab initio* results for $Me_2SnCl_2{}^a$. Reprinted with permission from Reference 40. Copyright (1986) American Chemical Society

$$
\begin{array}{c}
\text{Cl} \\
| \\
\text{Cl}\overset{(.68)}{—}\text{Sn}\overset{(.82)}{—}\text{C}\overset{(.76)}{—}\text{H}\quad \text{H}+.15 \\
\text{-.34}\quad |+.95-.61\quad \diagdown \text{H} \\
\text{CH}_3
\end{array}
$$

| | IE (eV) | | %pop | | | | | | |
| | | | Sn | | | | | | |
MO	calcd[b]	exptl[c]	s	p	d	2 Cl	2 C	6 H	Dominant character
$4b_1$ (HOMO)	10.25	10.43 (A)	13	2	29		51	5	Sn−C bonding MO
$6a_1$	11.12 ⎫	1	6	3	70	18		2 ⎫	
$4b_2$	11.12 ⎬ 11.31 (B)		1	1	98			⎬	Cl lone pairs
$2a_2$	11.37 ⎭		1		98			1 ⎭	
$3b_1$	12.06 ⎫		15		65	16		4 ⎫	
$5a_1$	12.21 ⎬ 12.16 (C)		14	1	68	13		4 ⎬	Sn−Cl bonding MOs
$3b_2$	12.38 ⎭		13	2	79	3		3 ⎭	
$2b_1$	14.33 ⎫		2		1	54		43 ⎫	
$1a_2$	14.33 ⎬ 14.30 (D)			1		55		44 ⎬	CH3-based MOs
$4a_1$	14.59 ⎪		3		1	54		42 ⎪	
$2b_2$	14.64 ⎭		4		4	51		41 ⎭	
$3a_1$	15.90	15.93 (D′)	40			33	20	7	Sn(5s)−C, Sn(5s)−Cl

[a] Gross atomic charges and overlap populations (in parentheses) are in electrons.
[b] Values rescaled by nine tenths; energies in eV.
[c] Experimental values from Reference 57.

alkyltin chlorides in which the HOMO is always of predominant Sn−C bonding character and a mixing between σ_{Sn-C} and σ bonds is present especially in a quite deep a_1 MO, with significant contribution (*ca* 40%) from the Sn 5s atomic orbital. The broad band E, with maximum at 22.16 eV in Figure 6, is assigned by the authors of Reference 57 to the ionization of inner-valence C 2s-based orbitals. The doublet F and F′ relates to the $^2D_{5/2}$ and $^2D_{3/2}$ final ionic states produced by ionization of Sn 4d subshell, affected by spin−orbit splitting. They are produced not by the primary He IIα radiation, so that the true IE values are not in the 25–26, but in the 32–34 eV range, precisely 32.59 ($^2D_{5/2}$) and 33.61 ($^2D_{3/2}$) eV. The primary doublet is obscured by more intense He Iα signals. It is highly desirable that these molecules be investigated also with synchrotron radiation, to better explore the inner-valence region, as has already been done by some authors for $SnMe_4{}^{41}$, $SnCl_2{}^{63,64}$ and $SnCl_4{}^{64}$. With the data presently available one can say that the half-widths of the Sn $4d^{-1}$ bands found in methyltin halides (*ca* 0.35 eV) are lower than in the tin halides (for instance, *ca* 0.45 eV in $SnCl_2$ following Reference 65) indicating that the presence, in the crystal-field expansion, of a $C_2{}^°$ electric-field gradient term is of minor importance.

In Reference 40 the analysis of the PE spectra of the pseudohalide derivative Me_3SnNCS is also reported. In this molecule, in which the group Sn−N−C−S is practically linear as observed in the solid state structure[66], the HOMO (6e, ionizing at 8.73 eV) is a NCS π orbital, mainly localized on sulphur, while the following MOs (5e, ionizing at 10.45 eV, and $7a_1$, ionizing at *ca* 12.5 eV) are of predominant σ_{Sn-C} bonding character. The latter MO presents, following the pseudopotential calculation, some σ_{C-S}

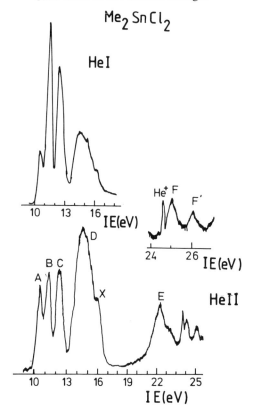

FIGURE 6. The photoelectron spectra of Me_2SnCl_2. Close-up of the 4d region of the He II spectrum. Reproduced by permission of The Royal Society of Chemistry from Reference 57

and σ_{N-C} antibonding contribution, whilst it does not carry any Sn—N bonding character. It ionizes approximately at the same energy as the 4e $\pi_{N=C}$ bonding MO, giving rise in the spectrum to a quite broad double band. The large separation between 4e and 5e MOs (1.94 eV) does not allow any mixing between them, i.e. between the $\pi_{N=C}$ MO of the NCS moiety and the σ_{Sn-C} MO of the Me_3Sn moiety, at variance with the case of Me_3SnCl.

Drake and coworkers studied a series of germanium derivatives[54−56,58] of formula Me_3GeX, Me_2GeX_2 and $MeGeX_3$ (X = H, F, Cl, Br). The analysis of the He I and He II spectra was supported by CNDO/2 calculations. Figure 7 shows the PE spectra of the compounds of the series Me_2GeX_2 (X = H: **5**, X = F: **6**, X = Cl: **7**), and Table 3 reports the experimental IEs and the results of the calculations. Following the assignment proposed, the HOMO (IE = 10.65 eV) is mainly localized on the Ge—C bonds in **5** and **6**, with a not negligible contribution from H 1s and F 2p orbitals. In **7** the HOMO seems to present a significant antibonding contribution from chlorine 3p orbitals, which overrides the inductive stabilizing effect expected with respect to the HOMO of **5** (IE = 10.74 eV). It may be of interest to compare this assignment with that proposed in Reference 57 for the analogues molecule Me_2SnCl_2. In the tin derivative the calculation (actually of a different kind, using pseudopotentials, as already mentioned) suggests for the HOMO a lower, though not negligible, contribution of the chlorine orbitals, and therefore a dominant σ_{Sn-C}

TABLE 3. Ionization energies (eV), CNDO/2 calculated eigenvalues (eV), orbital symmetries and parentage for Me_2GeH_2, Me_2GeF_2 and Me_2GeCl_2

	5				6				7		
MO	Calcd	Exptl	Parentage	MO	Calcd	Exptl	Parentage	MO	Calcd	Exptl	Parentage
$3b_1$	15.46	10.74	Ge(p)0.32 C(p)0.52 H(s)0.15	$4b_1$	15.17	10.45	Ge(p)0.22 F(p)0.19 C(p)0.47 H(s)0.10	$4b_1$	13.87	10.65	Ge(p)0.12 Cl(p)0.52 C(p)0.29 H(s)0.04
$4a_1$	15.51	11.04	Ge(p)0.32 H(^{1}s)0.24 C(p)0.30 H(s)0.15	$6a_1$	16.33	12.24	Ge(p)0.15 F(p)0.50 C(p)0.26 H(s)0.09	$6a_1$	14.53	11.20	Ge(p)0.08 Cl(p)0.75 C(p)0.13 H(s)0.04
$3b_2$	15.29	11.50	Ge(p)0.27 H(^{1}s)0.45 C(p)0.11 H(s)0.17	$4b_2$	17.00	13.16	Ge(p)0.10 F(p)0.75 C(p)0.06 H(s)0.08	$4b_2$	14.84	11.47	Ge(p)0.05 Cl(p)0.92
				$2a_2$	19.56	13.76	F(p)0.92 C(p)0.04	$2a_2$	15.65	12.40	Cl(p)0.99 Cl(s)0.03
				$3b_2$	19.59	13.88	F(p)0.80 C(p)0.10 H(s)0.10	$3b_2$	16.70	12.75	Ge(p)0.11 Cl(p)0.63 C(p)0.09 H(s)0.12
				$5a_1$	20.18	14.30	Ge(p)0.03 F(p)0.75 C(p)0.10 H(s)0.09	$5a_1$	17.62	13.50	Ge(p)0.14 Cl(s)0.03 Cl(p)0.58 C(p)0.14 H(s)0.10
				$3b_1$	19.77	14.56	F(p)0.35 C(p)0.33 H(s)0.32	$3b_1$	17.80	13.90	Ge(p)0.14 Cl(p)0.37 C(p)0.28 H(s)0.20
$1a_2$	19.98	13.60	C(p)0.50 H(s)0.50	$1a_2$	20.66	14.80	F(p)0.08 C(p)0.47 H(s)0.45	$1a_2$	20.36	14.78	C(p)0.51 H(s)0.43
$2b_1$	20.14	13.80	Ge(p)0.06 C(p)0.50 H(s)0.44	$2b_1$	21.95	15.38	Ge(p)0.12 F(p)0.43 C(p)0.25 H(s)0.19	$2b_1$	20.92	15.20	Ge(p)0.11 Cl(p)0.08 C(p)0.44 H(s)0.34
$3a_1$	21.66	14.30	Ge(s)0.09 Ge(p)0.05 C(p)0.48 H(s)0.36	$4a_1$	22.64	15.80	Ge(p)0.09 F(p)0.12 C(p)0.41 H(s)0.34	$4a_1$	21.63	15.47	Ge(s)0.16 Cl(s)0.26 Cl(p)0.20 C(p)0.22 H(s)0.15
$2b_2$	22.62	14.90	Ge(p)0.20 H(^{1}s)0.10 C(p)0.39 H(s)0.32	$2b_2$	22.73	16.80	Ge(p)0.11 F(s)0.03 F(p)0.22 C(p)0.35 H(s)0.28	$2b_2$	21.80	16.35	Ge(p)0.05 Cl(s)0.20 Cl(p)0.09 C(p)0.35 H(s)0.30
$2a_1$	23.08	16.90	Ge(s)0.37 Ge(p)0.07 H(^{1}s)0.23 C(p)0.20 H(s)0.12	$3b_1$	22.84	17.52	Ge(s)0.36 F(s)0.04 F(p)0.29 C(p)0.22 H(s)0.08	$3a_1$	22.33	17.00	Ge(s)0.132 Ge(p)0.06 Cl(p)0.03 C(p)0.45 H(s)0.31

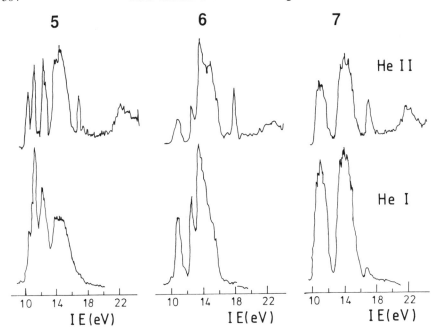

FIGURE 7. The photoelectron spectra of compounds 5–7 (5 = Me$_2$GeH$_2$; 6 = Me$_2$GeF$_2$; 7 = Me$_2$GeCl$_2$). Reproduced by permission of the National Research Council of Canada from Reference 56

bonding character. Figure 8 shows a correlation of the ionization energies across the series MeGeCl$_3$, Me$_2$GeCl$_2$ and Me$_3$GeCl. The main points of interest are a steady destabilization of the highest a_1 orbital (chlorine lone pair) along the series (observed also in the analogous fluorides), likely due to increasing electron-donor ability of the alkyl-metal moiety, and the opposite stabilization of the $4a_1$ MO, a delocalized $\sigma_{C-Ge-Cl}$ orbital. This stabilization was ascribed, following the indications of the CNDO/2 calculations, to an increasing participation of the Ge 4s orbital. An interesting aspect of the electronic structure of the series Me$_3$GeX (X = F, Cl, Br) is the similarity in the values of their first IE: 10.49 eV (F), 10.50 eV (Cl) and 10.0 eV (Br), although the related MOs, all of e symmetry, have a different parentage. In Me$_3$GeF and in Me$_3$GeCl the HOMO is in fact of main σ_{Ge-C} bonding character, whilst in Me$_3$GeBr it is practically a pure lone-pair bromine orbital.

IV. UNSATURATED CARBON DERIVATIVES

A. Aromatic Derivatives

In this section the PES investigations on compounds containing at least one aromatic group are described. Let us start from the most recent study, on the analysis of He I and He II spectra of the molecules of the MPh$_4$ series [M = C (8), Si (9), Ge (10), Sn (11), Pb (12)][67]. The authors assumed for these molecules a S_4 symmetry in the gas phase, on the basis of the only known gas-phase molecular structure, that of SnPh$_4$[68]. In molecules of the MPh$_4$ series, when viewed along any of the four bonds between the central atom and the phenyl rings, the three remaining rings form a propeller-like arrangement. The

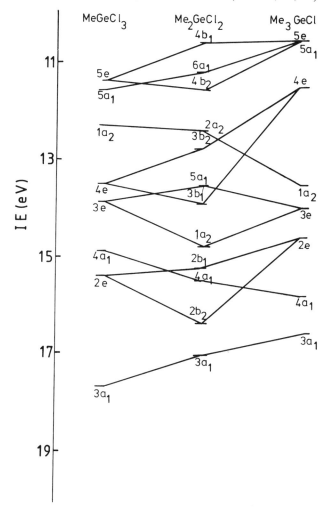

FIGURE 8. Correlation diagram of the ionization energies across the series MeGeCl$_3$, Me$_2$GeCl$_2$ and Me$_3$GeCl. Reproduced by permission of the National Research Council of Canada from Reference 56

molecules are, however, flexible due to allowed rotation of benzene rings around the C–M bonds. Figure 9 shows the PE spectra of Group XIV tetraphenyls and of benzene. The spectra, of good quality and resolution, of all the compounds are very similar to each other, and are closely related to the spectrum of benzene[69]. This suggests that the MPh$_4$ spectra might be interpreted in terms of weak interactions between four benzene molecules. The authors proposed, also on the ground of Extended Hückel MO calculations, that the M–Ph bonding is formed mainly by interaction of one from each of the $3e_{2g}$, $3e_{1g}$ benzene orbitals with metal p orbitals. On this basis (see Table 4), band B is assigned to ionization of orbitals formed from the benzene $3e_{2g}$ orbitals, which in MPh$_4$, of S_4 symmetry, give rise to one b and two e orbitals. This band moves to progressively lower IE

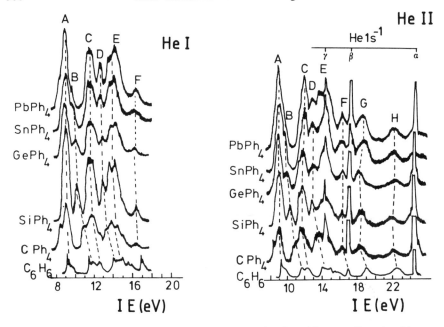

FIGURE 9. The photoelectron spectra of Group XIV tetraphenyls and benzene. Reproduced by permission of Elsevier Science from Reference 67

with the increasing atomic mass of the central atom, analogously to the equivalent band in MH_4 spectra[6], confirming the bonding M-Ph character of the related MOs. The interaction of the metal p orbitals with the benzene $3e_{1u}$ and $3a_{1g}$ combinations is expected to be less important. However, band D, progressively separating from band E with increasing central atom mass, and band F were assigned to ionization of MOs with some σ_{M-Ph} character, with the main contribution respectively from $3e_{1u}$ and $3a_{1g}$ benzene levels. No bands associated with bonding MOs involving central atom s orbitals were observed. The four sets of weakly interacting $1e_{1g}$ π orbitals (HOMO) of benzene give rise to band A, the remaining five e_{2g} weakly interacting with the $4a_{2u}$ orbitals to band C, the five unperturbed $3e_{1u}$ and orbitals derived from weakly interacting combinations of $1b_{2u}$ and $2b_{1u}$ orbitals to band E.

The He I PE spectra of some triphenyl derivatives of Group XIV elements were investigated by Distefano and coworkers[70]: $HMPh_3$ (M = C, Si, Ge, Sn), $BrGePh_3$, $ClSnPh_3$. They found a similar trend of the IEs of orbitals of M−Ph bonding character as in MPh_4, i.e. a decreasing IE with increasing size of the central metal (IE = 11.0 eV in $HCPh_3$, 10.35 eV in $HSiPh_3$, 10.11 eV in $HGePh_3$, 9.6 eV in $HSnPh_3$), and a sizeable interaction among the π orbitals of the aromatic rings only for the C, but not for the Si, Ge and Sn derivatives.

One of the most discussed subjects in the analysis of the electronic structure of unsaturated carbon derivatives of the XIV Group elements is that of the hyperconjugation, i.e. of the role played in the bonding by the conjugation between the π orbitals of the unsaturated moieties with σ-type orbitals (where π and σ denomination refers to the original character of the interacting orbitals) or by the $p_\pi-d_\pi$ conjugation. A theoretical treatment of hyperconjugation and its role in determining trends in properties, such as ionization energies, in Group XIV compounds was done by Pitt[71].

TABLE 4. Ionization energies (eV) and assignments for group XIV tetraphenyls. Reproduced by permission of Elsevier Science from Reference 67

Band	A	B	C	D	E	F	G	H
Assignment[a]	$2x(a+b+2xe)$	$b+2xe$	$2x(2xe+a)+b$	$b+2xe$	$3x(a+b+2xe)+a$	$a+b+2xe$	$2x(a+b+2xe)$	$2x(a+b+2xe)$
8	8.41, 8.97	10.93	11.68	13.22	13.82, 14.52	16.50	18.5	22.1
9	8.96	10.17	11.63	12.89	13.72, 14.38	16.26	18.4	21.9
10	8.95	10.03	11.75	12.85	13.77, 14.26	16.17	18.4	22.4
11	9.04	9.77	11.55	12.74	13.99	16.01	18.5	21.9
12	8.95	8.37	11.55	12.74	13.68, 14.14	16.2	18.4	22.0

[a] **8** = CPh_4; **9** = $SiPh_4$; **10** = $GePh_4$; **11** = $SnPh_4$; **12** = $PbPh_4$.

TABLE 5. Inductive and hyperconjugative parameters for Me_3M in Me_3MCH_2Ph and changes in C—M orbital energy due to hyperconjugative interactions. Reproduced by permission of Elsevier Science from Reference 72

	M = C	M = Si	M = Ge	M = Sn
$\sigma W'$ (eV)$^{-1}$	0.04	0.91	0.82	1.13
β_{CM}^2 (eV)$^{-2}$	0.21	0.50	0.68	1.05
δE_{CM} (eV)$^{-1}$	−0.09	−0.15	−0.17	−0.22

Some PES investigations, aimed at elucidating this aspect of the bonding in aromatic derivatives, were performed for various molecules[72−76], including naphthalene derivatives.

Bischof and coworkers[72] analysed the He I spectra of the compounds of the two series Me_3MCH_2Ph and Me_3MPh (M = C, Si, Ge, Sn) in terms of the inductive and hyperconjugative effect of the substituent on the orbital energy of a π MO. The change in the orbital energy (E_μ) of a π MO Ψ_μ due to the inductive effect of a substituent at position i is given by

$$\delta E_\mu^I = q_{\mu_i} \delta W_i$$

where q_{μ_i} is the orbital density of Ψ_μ at position i, and δW_i is the change in effective valence state ionization energy of carbon atom i, while the change due to the hyperconjugative effect is given by

$$\delta E_\mu^H = \beta^2 q_{\mu_i} \Delta E^{-1}$$

where β is the effective resonance integral between the 2p AO of carbon atom i and the substituent, and δE is the difference in energy between Ψ_μ and the orbital of the substituent involved in hyperconjugation. In the compounds of the series Me_3MCH_2Ph one of the M—C p-type MOs, lying along the benzyl—M bond, can hyperconjugate with the π orbitals of the benzene ring, while the other two p-type M—C MOs remain degenerate. The ionizations of the M—C bonding orbitals therefore give rise in the PE spectra to a single peak and to a doublet (broadened by the Jahn−Teller effect). By comparison between the spectra of the benzyl derivates with that of toluene, the inductive and hyperconjugative parameters reported in Table 5 were found. It is evident that the inductive contribution is almost constant along the series, apart from a near-zero value for M = C, and follows the Allred−Rochow[77] ordering. The hyperconjugative contribution rises with the increasing size of M. In the series M_3MPh the benzene π MOs may be also perturbed by a direct $p_\pi-d_\pi$ conjugative interaction with d AOs of M. On the basis of the experimental results and of considerations analogous to those mentioned for the other series, the authors gave an estimation of this effect, which was found quite large and decreasing along the series Si > Ge > Sn.

A $p_\pi-d_\pi$ effect was found also in the silicon derivative in the series p-$ClC_6H_4MMe_3$ (M = C, Si, Ge, Sn)[74]. In this case the hyperconjugation minimizes the difference in orbital energy $\pi_2-\pi_3$, where π_3 is the phenyl orbital affected by the conjugation, which is stabilizing for the chlorine 3p orbitals, destabilizing for the M empty d orbitals.

B. Heteroaromatic Derivatives

This section deals with a few PES studies on molecules in which an element of the Group XIV is the heteroatom in a heteroaromatic compound[78−80].

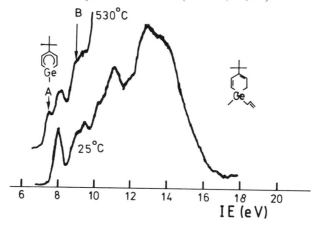

FIGURE 10. The He I photoelectron spectrum of 1-allyl-4-*t*-butyl-1-germacyclohexadiene (**13a**) at 25 and 530 °C. Bands A and B arise from ionization of **14a** (See text in Section IV.B for structures **13** and **14**). Reproduced by permission of VCH Verlagsgesellschaft mbH from Reference 79

Märkl and coworkers[78,79] performed variable-temperature photoelectron spectroscopy (VT PES) experiments on germabenzene derivatives. Precisely, the allylcyclohexadiene **13** yielded by gas-phase pyrolysis (450–550 °C, *ca* 0.05 mbar) the germabenzene **14**:

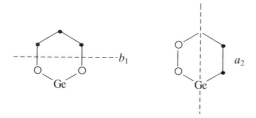

(**13**)   (**14**)

(a) R^1 = Me, R^2 = CMe₃; (b) R^1 = Me, R^2 = Et; (c)R^1 = Me, R^2 = CHMe₂

Figure 10 shows the PE spectra of **13a** at 25 °C and of the product of its pyrolysis at 530 °C. Only partial conversions can be achieved in these reactions. However, bands referring to the germabenzene derivative could be identified: bands A and B, at 7.40 and 8.90 eV, respectively, were ascribed to the ionization of **14a** π MOs, precisely the b_1 and a_2 orbitals.

The measured IEs for all the germabenzenes were compared and found to be in good agreement with those of substituted silabenzene derivatives, found in the literature[81].

C. Alkenyl and Alkynyl Derivatives

In this section the PES investigations on compounds containing C=C and/or C≡C groups, bonded directly or through saturated carbon to germanium, tin or lead, are described[36,73,82-96]. As for the aromatic derivatives, most of these discussions deal with the existence and the extent of $\sigma-\pi$ and/or $d_\pi-p_\pi$ hyperconjugation in different series of molecules.

Schweig's research group carried out several investigations on this subject[73,82,83] ranging from allyl and vinyl[73,83] to cyclopentene[82] derivatives. Figure 11 shows, along with the photoelectron spectra, a correlation diagram of the highest occupied MOs of the series tetraethylgermane, triethylvinylgermane, triethylallylgermane and of the series tetra-n-butylstannane, tri-n-butylvinylstannane, tri-n-butylallylstannane. It is clear that in the allyl derivatives of both germanium and tin, the splitting between the HOMO, of main M−C character, and the MO of substantial πC=C character is larger that in the vinyl derivatives, indicating a more important $\sigma-\pi$ hyperconjugation in the allylgermane and allylstannane[83]. On the basis of these experimental findings the authors suggested for the latter compounds the following conformation:

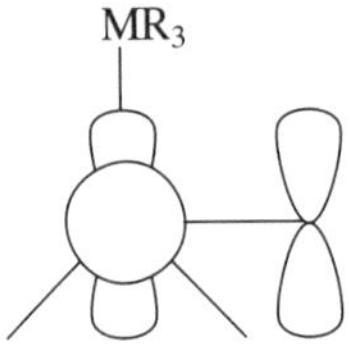

M = Ge, R = Et; M = Sn, R = Bu

the only one allowing a significant hyperconjugation, contrary to the conformation:

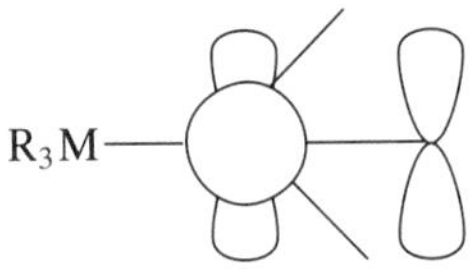

The same research group demonstrated, (by comparing the photoelectron spectrum with that of related silicon and carbon derivatives and with MNDO/2 calculations) that in 1,1-dimethyl-1-germa-3-cyclopentene, such as in the silicon analogue, the $d_\pi-p_\pi$ transannular antibonding hyperconjugation, previously postulated by other authors[84,85] to account for the planarity of the latter molecule, is not at all significant[82].

The comparison between the $\sigma-\sigma$ and the $\sigma-\pi$ conjugation in cyclopropylcarbinyltrimethyltin

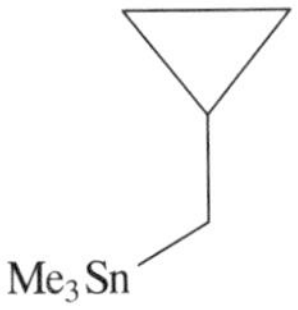

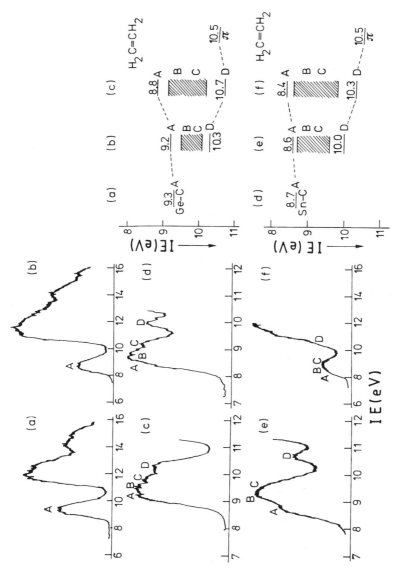

FIGURE 11. The He I photoelectron spectra and correlation diagrams of the highest MOs of tetraethylgermane (a), triethylvinylgermane (b), triethylallyl-germane (c), tetra-*n*-butylstannane (d), tri-*n*-butylvinylstannane (e) and tri-*n*-butylallylstannane (f). The shaded areas correspond to not-well-resolved features in the photoelectron spectra. Reproduced by permission of Elsevier Science from Reference 83

and allyltrimethyltin

$$\begin{array}{c} CH{=}CH_2 \\ \diagup \\ Me_3SnCH_2 \end{array}$$

was the main aim of Brown and coworkers in their PES investigation[86]. The σ–σ conjugation is the delocalization of the Sn–C σ bond into an adjacent C–C σ bond (of the cyclopropyl or of the allyl moiety in this case). The authors found that the energy of interaction of cyclopropane orbitals with the Sn–C σ bond (1.6 eV) may be compared with the σ–π interaction in allyltrimethyltin (2.2 eV).

A series of allyltin compounds, of formula $R_{3-n}Cl_nSnCH_2CH{=}CH_2$ (R = Me, n-Bu; $n = 0$–2) and $Ph_3SnCH_2CH{=}CH_2$ was investigated by Cauletti and coworkers[89], again with the aim of determining the extent of the σ–π hyperconjugation and its dependence on the dihedral angle ϑ between the C(1)–C(2)–C(3) plane and the C(3)–Sn bond (see Figure 12). This angle can vary because of the (in principle) free rotation around the C(2)–C(3) bond, and any σ–π conjugation will be at a maximum for $\vartheta = 90°$ and absent for $\vartheta = 0°$. The conformation of the free molecule will therefore be the result of the energy balance between electron repulsion due to steric hindrance of the substituents on the tin atom and the stabilization due to the σ–π conjugation. The assignment of the photoelectron spectra was supported by a fragment analysis based on LCBO (Linear Combination of Bond Orbitals) calculations[98]. The He I spectrum and the correlation diagram for the molecule $Me_3SnCH_2CH{=}CH_2$ is displayed in Figure 13. The diagram was built on the basis of the LCBO model and taking into account the inductive effects. It appears clearly that a σ/π hyperconjugation does exist, and that its extent is rather important. The energy separation between the two resulting σ/π MOs, one antibonding and one bonding, was calculated by the LCBO method as

$$\Delta E = [(\Delta E°)^2 + 4\beta^2]^{1/2}$$

where $\Delta E°$ is the energy difference between the two interacting levels [$\pi_{C=C}$ and $a'(\sigma_{Sn-C})$] and β is the resonance integral. It follows that

$$\beta = [(\Delta E)^2 - \Delta E°^2]^{1/2}/2$$

The σ–π resonance integral for $Me_3SnCH_2CH{=}CH_2$ was found to be 1.08 eV, a value very close to that in other allyl and benzyl compounds of Group XIV elements, already mentioned[73].

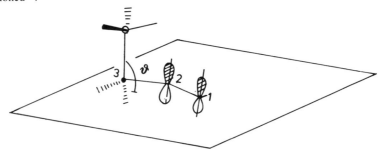

FIGURE 12. Definition of the dihedral angle ϑ for a tin–allyl compound. Reproduced by permission of Elsevier Science from Reference 89

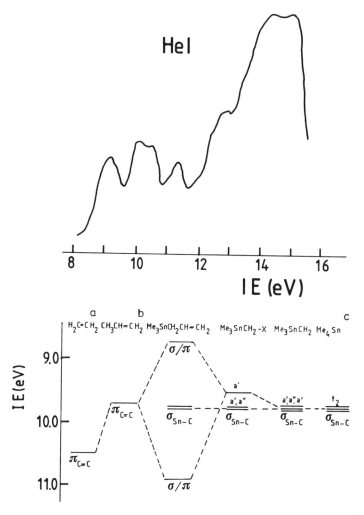

FIGURE 13. The photoelectron spectrum and correlation diagram of Me₃SnCH₂CH=CH₂ **(14)**. Reproduced by permission of Elsevier Science from Reference 89; (a) IE from Reference 99; (b) IE from Reference 100; (c) IE from Reference 33

Table 6 reports the resonance integrals for all the compounds of the series with the exception of $Ph_3SnCH_2CH=CH_2$ **(15)**. Some points are of interest: (i) On going from the methyl to the butyl derivatives, the extent of the hyperconjugation, constantly present, decreases, probably due to a deviation of the angle ϑ from 90° caused by the greater bulk of the butyl groups. (ii) Within each subseries the extent of the $\sigma-\pi$ conjugation falls upon substitution of the alkyl groups by chlorine atoms, probably because of a participation of 3p atomic orbitals of chlorine in the function involved in the hyperconjugation. The spectra suggest the presence of a significant $\sigma-\pi$ interaction also in the triphenyl derivative **15**, where this interaction is restricted to the allyltin moiety, without any contribution from the π orbitals of the phenyl rings.

TABLE 6. Resonance integrals for σ/π interaction $(\beta(\sigma/\pi))$ in the series $R_{3-n}Cl_nSnCH_2CH=CH_2$. Reproduced by permission of Elsevier Science from Reference 89

Compound	E° of interacting levels (eV)		ΔE° (eV)	E (eV)		ΔE (eV)	β (eV)
	$\sigma(Sn-C)$	$\pi(C=C)$		$(\sigma/\pi)^+$	$(\sigma/\pi)^-$		
$Me_3SnCH_2CH=CH_2$	-9.50	-9.70	0.2	-10.87	-8.70	2.17	1.08
$Me_2ClSnCH_2CH=CH_2$	-9.83	-9.70	0.13	-10.65	-8.95	1.70	0.85
$MeCl_2SnCH_2CH=CH_2$	-10.30	-9.70	0.60	-10.69	-9.33	1.36	0.61
$Bu_3SnCH_2CH=CH_2$	-8.95	-9.70	0.75	-10.30	-8.40	1.90	0.87
$Bu_2ClSnCH_2CH=CH_2$	-8.95	-9.70	0.75	-10.25	-8.63	1.62	0.72
$BuCl_2SnCH_2CH=CH_2$	-9.53	-9.70	0.17	-10.42	-9.24	1.18	0.58

An interesting example of PES study of molecules in which an alkyltin moiety is bonded to a conjugate system of two C=C bonds through a CH_2 group is described in Reference 95. The compounds studied by PES supported by pseudopotential calculations and compared with NMR data were those defined in Figure 14. To determine the equilibrium conformation of these molecules in the gas phase is a difficult task, due to the free rotation around the $C(1)-C(2)$ σ-bond. Contributions to the total energy are given by two effects of opposite sign: repulsive interaction between methyl groups and the hydrogens bonded at $C(2)$ and $C(3)$ atoms or $\pi_{C=C}$ electrons, that increases the energy, and $\sigma-\pi$ mixing between the σ_{Sn-CH_2} and the $\pi_{C=C}$ bond, that decreases the energy. To find the value of minimum energy, the authors performed several calculations on the molecule **16**, varying the dihedral angle ϑ between 0° and 180°, while keeping fixed the dihedral angles Φ at 60°, 180° and 300° (values giving the lowest energy value). They found a minimum in the total energy curve for $\vartheta = 108.75^\circ$, the barrier to the rotation being maximum, 6.81 kcal mol^{-1}. The energy of the 0° conformer is higher than that of the 180° one, in both cases the σ/π hyperconjugation being symmetry-forbidden. It is noteworthy that the minimum energy conformation is not that with $\vartheta = 90^\circ$, as expected in view of a possibly better σ/π conjugation. The He I and He II spectra, along with the PSHONDO calculations performed for the most stable conformer, indicated a fairly strong $\sigma-\pi$ conjugation, following the scheme of the correlation diagram shown in Figure 15, which was assumed valid also for **17**, given the strong similarity of the photoelectron spectra. The analysis of the electronic structure of these molecules correlates well with the ^{13}C NMR chemical shifts and ^{13}C$-^{119}$Sn couplings, some of which were taken from previous literature data[101].

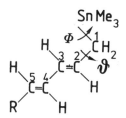

FIGURE 14. *trans-E* isomer of the trimethylstannylpentadiene (R = H, **16**) and the trimethylstannyl-hexadiene (R = Me, **17**). Reproduced by permission of Freund Publishing House, Ltd., London, from Reference 95

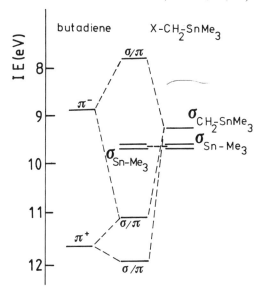

FIGURE 15. σ/π Hyperconjugation between one σ_{Sn-CH_2} bond and two split $\pi_{C=C}$ bonds in trialkyl-stannyldienes. Reproduced by permission of Freund Publishing House, Ltd., London from Reference 95

The He I and He II photoelectron spectra of tetravinylstannane $Sn(CH=CH_2)_4$ were studied by Novak and coworkers[87]. They assumed a rather weak interaction between the vinyl groups, and between them and the central tin atom.

Alkynyltin and alkynyllead derivatives were extensively studied by PES, together with NMR and pseudopotential calculations[88,92,93,97]. For the sake of simplicity we will divide the compounds investigated into two classes:

(a) Linear acetylides and diacetylides, of the general structures:

(I) $R_3MC{\equiv}CR'$

$$M = Sn \begin{cases} R = Et, R' = H \\ R = Me, R' = C{<}^{CH_3}_{CH_2} \\ R = R' = Me \end{cases}$$

$$M = Pb \begin{cases} R = Me, R' = SiMe_3 \\ R = R' = Me \end{cases}$$

(II) $Me_3MC{\equiv}CMMe_3$, M = Sn, Pb

(III) $Me_3MC{\equiv}CC{\equiv}CMMe_3$, M = Si, Sn; R = Me

A representative spectrum, i.e. that of $Me_3SnC{\equiv}CSnMe_3$ (18), is displayed in Figure 16, along with the correlation diagram, built starting from the experimentally known IEs of Me_3SnH^{38} and acetylene[102], taking into account the reciprocal inductive effects of the various fragments (represented in Figure 16 by the arrow). It appears clearly that bands A and C are related to MOs (e') of mixed $\sigma-\pi$ character, whilst

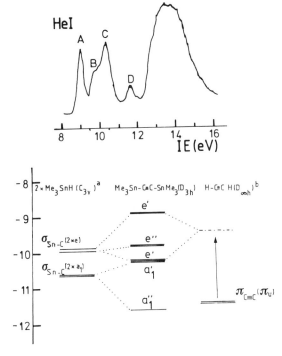

FIGURE 16. The photoelectron spectrum and correlation diagram of Me$_3$SnC≡CSnMe$_3$ (**18**): (a) IE values taken from Reference 38; (b) IE taken from Reference 102. Reprinted with permission from Reference 88. Copyright (1985) American Chemical Society

band B arises from ionization of an orbital (e'') substantially localized on the σ_{Sn-CH_3} bonds. The pseudopotential calculation was in agreement with this scheme. It is evident that the presence and the extent of the $\sigma-\pi$ conjugation in this kind of molecule depends critically upon the nature of the substituents R and R', which may induce shifts, sometimes quite large, of the interacting starting energy levels in the correlation diagrams. For instance, in the class **I**, the $\sigma-\pi$ interaction is not very important in **19** and negligible in Me$_3$SnC≡CMe (**20**), whilst being quite important in the diacetylide Me$_3$SnC≡CC≡CSnMe$_3$ (**21**). The nature of the HOMO is therefore different in the various molecules, being of mixed $\sigma-\pi$ character in **18, 20** and in Et$_3$SnC≡CH (**22**), whilst being of predominant $\pi_{C≡C}$ character in **19** and **20**.

$$Me_3SnC≡CC\overset{\displaystyle CH_3}{\underset{\displaystyle CH_2}{<}}$$

(**19**)

The bonding situation is quite similar in the lead acetylides, where the hyperconjugation is significant in Me$_3$PbC≡CSiMe$_3$ (**23**) and in Me$_3$PbC≡CPbMe$_3$ (**24**), whilst being negligible in Me$_3$PbC≡CMe (**25**), as in the tin analogue. The similarity of the chemical bonding in **20** and **25** was confirmed by NMR measurements[103]. On the whole, the results of these investigations are consistent with the expectation of a higher $\sigma-\pi$ interaction in tin and lead acetylides with respect to alkyl-, silyl- and germyl-substituted analogues, also studied extensively by PES[102,104–110]. However, a search for regularities in IE trends along series of acetylides of Group XIV elements, such as that done in Reference 111, is not very meaningful. In fact, in the alkyl and silyl acetylides the mechanism of the $\sigma-\pi$ conjugation is relatively simple, the orbital energy of the σ orbitals being always lower than that of the $\pi_{C\equiv C}$ orbitals, and the effect of the conjugation being therefore always destabilizing for the latter orbitals. The situation is more doubtful in the case of germyl derivatives and very complicated in the tin and lead acetylides, as just discussed.

(b) Tetrahedral and pseudotetrahedral acetylides of the structures:

$$M(C\equiv CR)_4 \ (\mathbf{26})$$

$$M = Sn; \ CMe_3, \ SiMe_3$$

$$M = Pb; \ R = Me, \ SiMe_3$$

and

$$Me_2Pb(C\equiv CMe)_2$$

In these molecules both experimental and theoretical results indicate that the $\sigma-\pi$ interaction is practically non-existent.

D. Acetylgermane

We place in this section a study by Ramsey and coworkers[112], dealing with the molecule $Me_3GeC(O)Me$, whose photoelectron spectrum, compared with that of the silicon analogue, with UV absorption spectroscopy measurements and with the results of CNDO/2 calculations, allows one to ascertain the presence of a strong mixing between the localized oxygen lone pair and metal–carbon bond. The authors suggested that the HOMO is a σ-type antibonding combination of O lone pair and Ge—C localized orbitals and that the longest-wavelength singlet–singlet transition must therefore be regarded as $\sigma \rightarrow \pi^*$.

V. COMPOUNDS CONTAINING METAL–METAL BONDS

This section will deal with PES investigations on molecules where direct bonds between Group XIV metals (equal or different) are present[46,113–116]. Szepes and coworkers[113] studied the He I spectra of molecules of the general structure $Me_3MM'Me_3$ (M = M' = C, Si, Ge, Sn; M = Sn, M' = Si; M = Sn, M' = Ge), which were assigned with the aid of CNDO/2 calculations. They found that the HOMO is in all cases highly localized at the metal–metal bond. Table 7 reports the observed IEs for the outermost MOs. The authors

TABLE 7. Ionization energies (eV) for $Me_3MM'Me_3$ derivatives. Reproduced by permission of Elsevier Science from Reference 113

M—M'	M—M'(a_1)	$C_3M(e)$		$C_3M'(e)$	$C_3M(a_1)$
Ge—Ge	8.6	9.7	10.2		
Sn—Sn	8.20	9.2	9.6		
Sn—Si	8.32		9.42	10.1	11.65
Sn—Ge	8.33		9.5	10.15	11.63

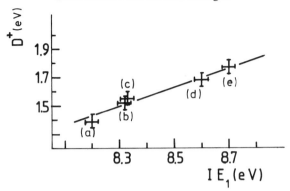

FIGURE 17. Plot of the ionic bond dissociation energy ($D^+ = \text{AP} - \text{IE}$) for the formation of the $[\text{Me}_3\text{M}]^+$ ion from $\text{Me}_3\text{MM}'\text{Me}_3$ vs IE_I for $\text{Me}_3\text{SnSnMe}_3$ (a), $\text{Me}_3\text{SiSnMe}_3$ (b), $\text{Me}_3\text{GeSnMe}_3$ (c), $\text{Me}_3\text{GeGeMe}_3$ (d) and $\text{Me}_3\text{SiSiMe}_3$ (e). The appropriate appearance potentials (APs) and ionization energies (IEs) are taken from Reference 117. Reproduced by permission of Elsevier Science from Reference 113

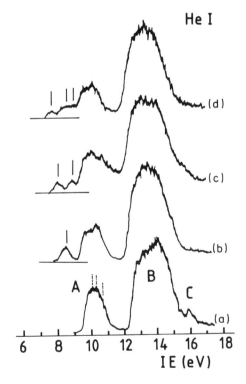

FIGURE 18. The photoelectron spectra of Me_4Ge (a), $\text{Me}_3\text{GeGeMe}_3$ (b), $\text{Me}(\text{Me}_2\text{Ge})_3\text{Me}$ (c) and $\text{Me}(\text{Me}_2\text{Ge})_4\text{Me}$ (d). Reproduced by permission of The Chemical Society of Japan from Reference 116

made a plot of the ionization energy of the HOMO versus the ionic bond dissociation energy taken from the literature[117], finding a satisfactory linear relationship, as shown in Figure 17. This good correlation is explained by a gradual weakening of the M−M' bond upon ionization of the HOMO going down the group.

Mochida and coworkers investigated series of peralkylated catenates of silicon, germanium and tin[115,116]. In particular, they studied some permethylated linear polygermanes, $Me(Me_2Ge)_nMe$ ($n = 2$–5), with the aim of investigating the electron-donor ability of the germanium–germanium σ bonds, i.e. the ability to transfer an electron to an electrophile acting as an electron acceptor. Figure 18 shows the He I spectra of the $Me(Me_2Ge)_nMe$ ($n = 2$–4) together with the spectrum of Me_4Ge. One can see that on going from Me_4Ge to $Me_3GeGeMe_3$ a band at low IE appears. This band is split into two and three bands, respectively, on going to $Me(Me_2Ge)_3Me$ and $Me(Me_2Ge)_4Me$. These bands were therefore ascribed to ionization out of orbitals of Ge−Ge σ bonding character.

VI. COMPLEX COMPOUNDS

With the designation 'complex compounds' we mean germanium, tin and lead derivatives in which the metal atoms are bonded to quite large groups, giving rise to complex molecules, although they can not always be classified as classical 'complexes'. We will further divide this section into subsections, trying to identify classes of compounds with some analogies.

A. Nitrogen-containing Compounds

The first IE of a series of substituted diazoalkanes of formula $Me_3MC(N)_2CO_2Et$ (M = Si, Ge, Sn, Pb), Me_3MCHN_2 (M = Si, Ge) and $(Me_3M)_2CN_2$ (M = Si, Ge, Sn, Pb) were measured by He I PES[118]. The electronic structure of these molecules was also investigated by Mössbauer, NMR and IR spectroscopy. The HOMO, of π_{CNN} character, is increasingly destabilized by substitution of hydrogen in diazomethane CH_2N_2 for organometallic groups at the α-C-atom. The measured IE values correspondingly decrease along the series H > Si = Ge > Sn > Pb.

Andreocci and coworkers[119] studied a series of four-membered cyclic amines, of formula

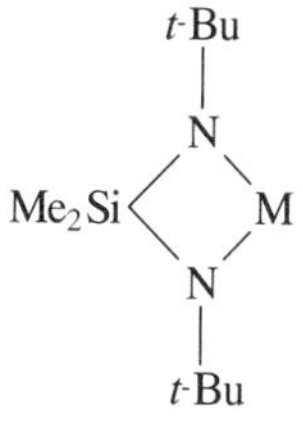

$$Me_2Si\underset{N}{\overset{N}{\diagup}}M \quad M = Sn\ (27),\ Pb\ (28) \quad \text{and} \quad Me_2M\underset{N}{\overset{N}{\diagup}}SnMe_2 \quad M = Si\ (29),\ Sn\ (30)$$

with t-Bu groups on the nitrogen atoms.

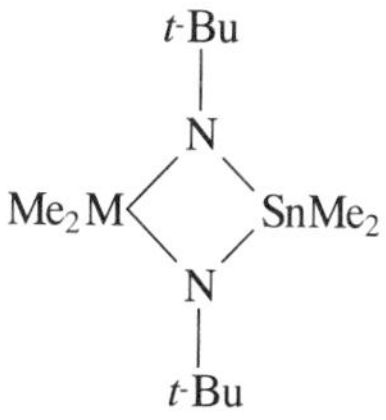

The main aim was, on the one hand, to ascertain the presence of a 'through-space' interaction between the two nitrogen lone pairs and the role played by the Si−C and Sn−C bonds in the mechanism of this interaction, and on the other hand, to evaluate the effect of the change in tin oxidation state ($+2$ in **27**, $+4$ in **30**) on the electronic structure of these molecules. The He I and He II spectra were interpreted by comparison with literature PES data on the non-cyclic molecules[48]:

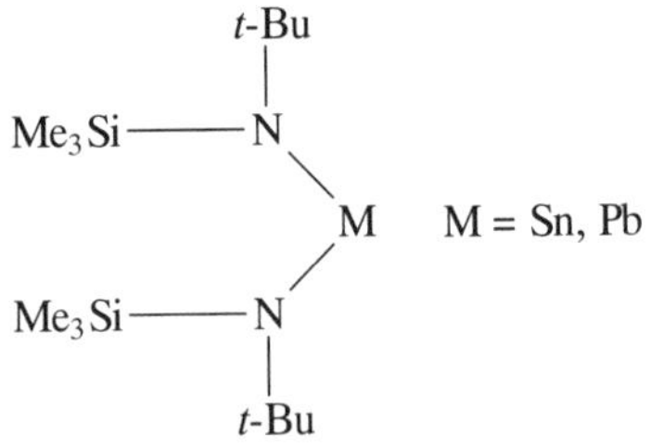

$$
\begin{array}{c}
t\text{-Bu} \\
|\\
Me_3Si\text{---}N \\
\diagdown \\
M \qquad M = Sn, Pb \\
\diagup \\
Me_3Si\text{---}N \\
|\\
t\text{-Bu}
\end{array}
$$

as well as with the results of pseudopotential calculations performed on model molecules containing methyl in place of the t-butyl substituents, and with NMR data. The cyclic compounds were assumed to be monomeric species in the gas phase, with a planar geometry of the ring. Table 8 reports the IE values measured for the various photoelectron bands with the assignment. Some points emerge clearly:

(i) The 'through-space' interaction between the nitrogen lone-pair orbitals, perpendicular to the ring, is significantly stronger in **27** and **28** than in **29** and **30**, this resulting from the larger splitting between the antisymmetric (n_N^{asym}) and the symmetric (n_N^{sym})

TABLE 8.   Ionization energies and assignments for cyclic amines 27–30. Reproduced by permission of Verlag der Zeitschrift für Naturforschung from Reference 119

Compounds	Band label[a]	IE (eV)	Assignment
27	$3a_2$	7.40	n_N^{asym}
	$9a_1$	8.09	n_{Sn}
	$5b_2$	8.40	n_N^{sym}
	$6b_1$	9.50	σ_{Sn-N}
	$4b_2$	10.5[b]	σ_{Si-C}
28	$3a_2$	7.12	n_N^{asym}
	$9a_1$	7.89	n_{Pb}
	$5b_2$	8.11	n_N^{sym}
	$6b_1$	9.03	σ_{Pb-N}
	$4b_2$	10.2[b]	σ_{Si-C}
29	$4a_2$	7.40	n_N^{asym}
	$8b_2$	7.78	n_N^{sym}
	$11a_1$	9.28	$\sigma_{Sn-N} + \sigma_{SiN}$
	$7b_1$	9.28	$\sigma_{Sn-N} + \sigma_{SiN}$
	$7b_2$	9.85[b]	$\sigma_{Sn-C} + \sigma_{SiC}$
	$6b_2$	10.49[b]	$\sigma_{Sn-C} + \sigma_{SiC}$
30	$4a_2$	7.11	n_N^{asym}
	$8b_2$	7.34	n_N^{sym}
	$11a_1$	8.83	σ_{Sn-N}
	$7b_1$	8.83	σ_{Sn-N}
	$7b_2$	9.48	σ_{Sn-C}
	$6b_2$	10.00	σ_{Sn-C}

[a] MO labels of dimethyl compounds are used.
[b] Shoulder.

combination of these orbitals in the M(II) derivatives. This effect can be explained, from theoretical calculations, by the fact that in **29** and **30** the n_N^{asym} orbital ($8b_2$) may combine, and actually does, with the $6b_2$ MO, of σ_{M-C} character.

(ii) The replacing of tin with lead on passing from **27** to **28** leads to a systematic lowering of the IEs due to the stronger metal character of Pb compared to Sn. The calculations suggest a mixed s–p character of the n_M non-bonding orbitals, as already found in the non-cyclic amines.

(iii) The IE values of the Si-containing compound **29** are systematically higher than the corresponding ones of **30**, due to the higher electronic charge localized on the nitrogen atoms. This can be explained by both the stronger metal character of Sn compared to Si, and the missing d_π(Si) ⟵⟶ p_π(N) interaction in **30**.

B. Sulphur- and selenium-containing Compounds

In this section we will describe some PES investigations on Group XIV element derivatives in which one or more sulphur or selenium atoms are bonded directly or indirectly to the metal[120–132]. The aspects investigated preferentially for these molecules were the electronic and conformational situation, the role played in the bonding by the chalcogen lone-pair orbitals and possible hyperconjugative interaction (n–σ or p_π–d_π).

The electronic interactions between the MMe$_3$ substituents and the sulphur nπ orbital were analysed[121] on the basis of the semilocalized orbitals approximation in two series of the structures S(MMe$_3$)$_2$ and MeSMMe$_3$ (M = C, Si, Ge, Sn, Pb).

Three electronic effects are relevant in both series:

(i) the reciprocal electrostatic effects of the various groups,
(ii) the n–σ interaction between the sulphur lone pair and σ_{M-C} orbitals,
(iii) the p_π–d_π interaction between the sulphur lone pair and the d orbitals of M.

In all the compounds the HOMO was found to be of main sulphur lone-pair character. Figure 19 shows the variation of the orbital energy along the two series. On the basis of several considerations, the authors proposed that the most important interaction when M = Si is the p_π–d_π one, while the +I effect of the PbMe$_3$ group predominates when M = Pb, and when M = Ge, Sn there is no evidence for the predominance of one of the three electronic effects.

Conformational and electronic structure considerations are the main subjects treated in studies on dithiolanes and related open-chain species[123,124,129] of the structures **I** and **II**.

In these molecules two kinds of interactions are present: a 'through-space' interaction between the two sulphur lone-pair orbitals, giving rise to an antibonding (n_-) and a

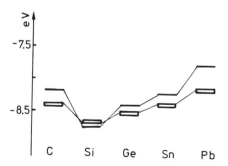

FIGURE 19. Comparison between the energies of the sulphur nπ MO: ▭, S(MMe$_3$)$_2$, MeSMMe$_3$; ▬. Reproduced by permission of Elsevier Science from Reference 121

R—S—CH₂ and Me—S—R structures

M = C, Si, Sn; R = Me   M = C, Si, Sn;
M = Si, Sn, Pb; R = Ph

(I)   (II)

bonding (n_+) MO, and the interaction between the lone-pair orbitals and the orbital of

suitable energy and symmetry of the R₂M moiety. The extent of such interactions depends

of course critically on the molecular geometry. The conformation of the dithiolanes (series **I**), according to Dreiding models, must contain a non-planar cycle. Therefore, due to the low symmetry, both n_- and n_+ combinations can interact with other orbitals. On the other hand, following CNDO/2 calculations performed by Bernardi and coworkers[123], the most stable conformer for the series **II** of the open-chain compounds is that with a planar Me₂M moiety, and therefore only the n_+ sulphur combination can interact with the rest of the molecule.

Figure 20 shows the photoelectron spectra of **31** and of $Me_2Sn(SMe)_2$ (**32**). The assignment of the spectrum of **31** is not straightforward, and controversies can be found in

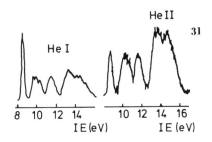

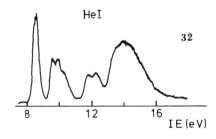

FIGURE 20. The photoelectron spectra of compounds **31** and **32**. Reproduced by permission of Elsevier Science from References 123 and 129

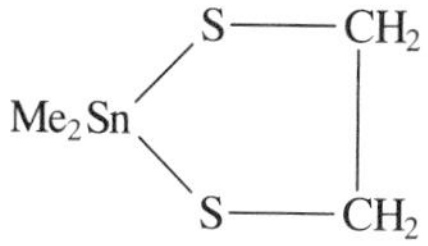

$$
\begin{array}{c}
\text{Me}_2\text{Sn}
\overset{\displaystyle \text{S}\text{——}\text{CH}_2}{\underset{\displaystyle \text{S}\text{——}\text{CH}_2}{\Big\langle}}
\end{array}
$$

(31)

literature. The dramatic decrease in intensity of the first band on passing from He I to He II ionizing radiation supports the assignment to the ionization of both n_+ and n_- sulphur lone-pair combination, practically unsplit, at variance with the carbon and silicon analogues, indicating a very small, if any, 'through-space' interaction. The following three peaks, giving rise to the second broad band, were attributed to ionization of the $\sigma^+_{\text{Sn–S}}$ and $\sigma^-_{\text{Sn–S}}$ and $\pi(\text{C–Sn–C})$ orbitals, the ordering in IE not being determined with certainty. It is interesting to observe the peaks deriving from the Sn 4d pseudo-valence orbitals, at 34.08 $(^2D_{5/2})$ and 35.07 $(^2D_{3/2})$ eV. These IE values are slightly lower than in SnCl$_4$ (34.91 and 36.05 eV)[133] and higher than in Me$_2$SnCl$_2$ (32.59 and 33.61 eV)[133], suggesting that the bidentate ligand acts as an electronegative group. The spectrum of 32 is fairly similar, although the first band is broader (Bernardi and coworkers[123] could identify a shoulder in the low-IE side), indicating a small interaction between the sulphur lone pairs. The corresponding band in the carbon and silicon derivatives is split into two components, indicating a different mechanism of interactions. For instance, the hyperconjugative effects, stabilizing n^- and destabilizing n^+, are probably of the same order of magnitude when M = C and M = Si, whilst being practically absent when M = Sn. Furthermore, d-orbital participation is significant only in the case of the silicon derivative, while in the tin derivative 32 mainly the inductive effect operates.

Modelli and coworkers[126] studied by PES and ETS (electron transmission spectroscopy) some silicon and tin derivatives of thiophene and furan, with the aim of following the energy gap between the HOMO and the LUMO as a function of the substituents. In particular they investigated the following tin derivatives:

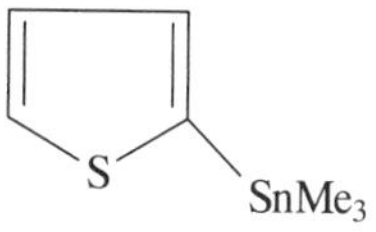

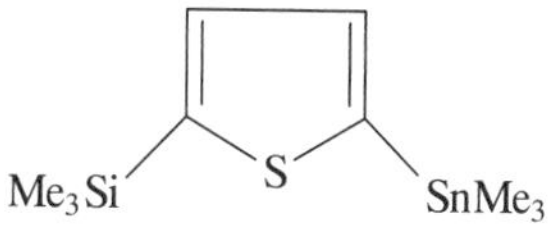

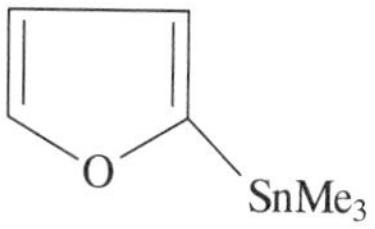

The experimental results indicated that the filled and empty π MOs of thiophene and furan are perturbed in opposite directions by the MR$_3$ substituents, causing a reduction of their energy gap. Furthermore, the empty orbitals of the substituents do not stabilize significantly the filled ring π orbitals, whilst mixing significantly with the unoccupied π^* orbitals.

A spectroscopic detection of germathiones and silathiones, previously characterized only by chemical trapping, was described elsewhere[127,128]. Dimethylgermathione (33) and dimethylsilathione were generated from the corresponding trimers (Me$_2$MS)$_3$[M = Ge(34), Si] by pyrolysis (250–300°C, 5×10^{-2} mbar).

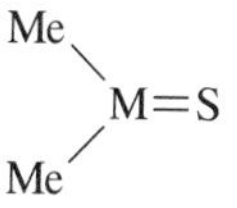

$$
\begin{array}{c}
\text{Me} \\
\diagdown \\
\text{M}{=}\text{S} \quad (\text{M = Ge (33), Si}) \\
\diagup \\
\text{Me}
\end{array}
$$

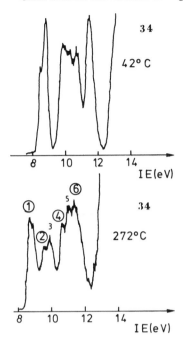

FIGURE 21. The He I photoelectron spectra of compound **34** (M = Ge) at 42 and 272 °C. Reproduced by permission of Elsevier Science from Reference 127

Figure 21 shows the photoelectron spectra of **34** (M = Ge) at 42 and 272 °C. The evident change in the shape of the spectrum clearly indicates the decomposition of the trimer. Band 1 was attributed to ionization of the sulphur lone pair of the monomeric species **33** (M = Ge), band 2 is related to the π Ge=S bond, bands 4 and 6 to the σ Ge−S and σ Ge−C bonds respectively. Bands 3 and 5 were assigned to the ionization of a dimeric species. The assignment was supported by pseudopotential calculations. Also, the photoelectron spectrum of the dimeric species (n-Bu$_2$GeS)$_2$ was detected.

A joint study of cyclic compounds and smaller open units, which can be considered in some way precursors of the former ones, was carried out by Cauletti and coworkers[132] on the following systems:

$$
\begin{array}{ccc}
\text{Me}_3\text{Sn} & & \text{Me}\diagdown\quad\diagup\text{Me} \\
\diagdown & & \text{Sn} \\
& \text{X} \quad\text{and}\quad & \text{Me}\diagdown\text{X}\diagup\quad\diagdown\text{X}\diagup\text{Me} \quad (\text{X = S, Se}) \\
\diagup & & \text{Me}\diagup\text{Sn}\diagdown\quad\diagup\text{Sn}\diagdown\text{Me} \\
\text{Me}_3\text{Sn} & & \text{Me}\quad\text{X}\quad\text{Me}
\end{array}
$$

The main aim was to verify the changes in the tin−chalcogen bonding on passing from open to cyclic molecules, also in light of previous experience on some Group XIV thiospiranes[134]. The PES results were analysed on the basis of pseudopotential calculations, and compared with NMR data, namely $\delta(^{119}S)$ and $\delta(^{77}Se)$. Both the experimental and theoretical results indicated that in both the monomeric and trimeric species the HOMO has a predominant chalcogen p lone-pair character, though with a not negligible contribution of orbitals of the CH_3 groups via hyperconjugative interaction. Such an interaction is allowed by the twisted-boat conformation assumed by the cyclic molecules (C_2 symmetry). The expected 'through-space' interaction between the three p lone pairs in the trimers was proven to be quite weak, given that only one PES band in the IE region between 8 and 9 eV, typical of the chalcogen lone-pair ionizations, could be observed and assigned to ionization of the three quasi-degenerate lone-pair orbitals. It is interesting to observe that the spectra of Si and Ge sulphur analogues just discussed[127,128] are quite similar, though showing an evident shoulder on the low-energy side of the first band. This hints at a similar geometry for these molecules, with a slightly greater interaction between the sulphur lone pairs in the Si and Ge derivatives. On the other hand, the analogous molecule with C atoms, 1,3,5-trithiane, shows a doublet of bands of intensity ratio 1:2 in the PE spectrum[135], indicating quite important sulphur lone-pair interaction, probably due to the different molecular geometry (chair conformation, C_{3v} symmetry). Both in the monomeric and in the trimeric molecules, extensive delocalization of σ-electron density was suggested by the calculations and the NMR results, which showed a remarkable deshielding of the ^{119}Sn nuclei, attributed not only to the mentioned high degree of delocalization of σ-electron density but also to the presence of electronegative ligands.

A nice example of conformational studies in vapour phase by PES is represented by the investigation described in Reference 131 on trimethylphenylthiostannane (35). In this molecule the dihedral angle ϑ (see Figure 22) between the plane of the phenyl ring and the C–S–Sn plane determines the extent of the conjugation between the π system of the phenyl group, on the one hand, and the sulphur lone pair and/or the σ_{Sn-S} bond, on the other, the maximum interaction being with the sulphur lone pair for $\vartheta = 0°$ and with the σ_{Sn-S} orbital for $\vartheta = 90°$. The He I and He II spectra of 35 are shown in Figure 23. The band A, dramatically decreasing in intensity on passing from He I to He II ionizing radiation, clearly related to the sulphur 3p lone pair, falls at 8.39 eV, an IE value higher than that of the corresponding band in HSPh (8.28 eV[122]) and MeSPh (8.07 eV[136]), despite the greater electron-donor ability of the Me_3Sn group relative to H and Me. In the latter two compounds a significant n–π interaction between the sulphur lone pair and the π system of the phenyl ring is present, destabilizing the HOMO, differently than in 35, where the HOMO is essentially of S 3p lone-pair character. The following system of the three bands B, C and D, at 8.63, 9.20 and 9.79 eV, respectively, is typical of conjugative interaction of a π orbital of the phenyl ring (e_{1g} in benzene) in a monosubstituted benzene. In this case bands B and D represent the antibonding and bonding combinations, respectively, of the π phenyl orbital with the σ_{Sn-S} bond, while band C is related to the unperturbed component of the phenyl π orbital. This is a clear

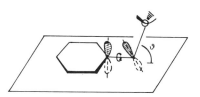

FIGURE 22. Definition of the dihedral angle ϑ in trimethylphenylthiostannane (35). Reproduced by permission of Elsevier Science from Reference 131

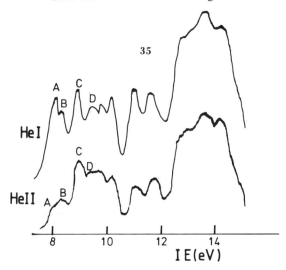

FIGURE 23. The photoelectron spectra of compound **35**. Reproduced by permission of Elsevier Science from Reference 131

indication of a molecular conformation with ϑ around 90° allowing an important $\sigma-\pi$ conjugation, causing the splitting of 1.16 eV between bands B and D.

It is interesting to observe that the σ_{Sn-S} bond in Me_3SnSMe, which is not involved in any conjugative interaction, ionizes at 9.55 eV[120], i.e. at a much higher IE than that of the first band (B) of the triplet in the spectrum of **35**.

The electronic structure of some alkyltin derivatives with sulphur-containing ligands was studied by Cauletti and coworkers[125], by means of PES supported by CNDO/2 calculations. The compounds investigated were trialkyltindialkyldithiocarbamates: $R_3SnS_2CNR_2$ (R = Me, Et), dialkyltin bisdialkyldithiocarbamates: $R_2Sn(S_2CNR_2')_2$ (R = R' = Me; R = n-Bu, R' = Me), trimethyltindimethyldithiophosphinate: $Me_3SnS_2PMe_2$ and dialkyltin bisdiethyldithiophosphinates: $R_2Sn(S_2PEt_2)_2$ (R = Me, n-Bu).

The analysis of the experimental and theoretical results on these complex molecules leads to some generalizations about the sequence of MOs in such compounds. The highest energy levels in both dithiocarbamato and dithiophosphinato complexes are π-type orbitals of predominant S 3p character; the following orbitals are of σ_{Sn-S} and σ_{Sn-C} type. There are indications, originating from comparison between He I and He II spectra, that in the dithiocarbamates both ligand sulphur atoms are involved in the coordinative bonding with Sn.

C. Acetylacetonate Derivatives

This section deals with PES studies on alkyl– and chloro–tin complexes, with acetylacetonate and related ions as ligands[53,137]. Cauletti and coworkers[53] investigated the following compounds:

$R_3Sn(acac)$ (R = Me, Et; acac = acetylacetonate)

$R_2Sn(acac)_2$ (R = Me, n-Bu; acac = acetylacetonate)

$R_2Sn(Tacac)_2$ (R = Me; Tacac = trifluoroacetylacetonate)

along with simpler molecules [Me_3SnOMe, R_2SnCl_2 (R = Me, n-Bu)] for comparison.

Fragalà and coworkers[137] analysed the He I and He II spectra of β-diketonato complexes of formula Me$_2$Sn(pd)$_2$ (**36**) and Cl$_2$Sn(pd)$_2$ (**37**) (pd = pentane-2,4-dionate), having a pseudo-octahedral *trans* or *cis* configuration, with the aim both of elucidating the stereochemistry of such complexes and of gaining more direct insight into the bonding of the alkyl– or halo–tin moiety with the bidentate ligand. The interaction between the orbitals of the ligands (π_3, n$_−$ and n$_+$ for each chelate ring, following the notation of Evans and coworkers[138]) and the Sn orbitals depends critically on the molecular conformation. The molecule of **36** is known to have in both solution and the solid state a *trans*-'pseudo-octahedral' geometry, whilst that of **37** has a *cis* structure, as shown in Figure 24. The authors assumed the same geometry in the vapour phase. Correspondingly, the photoelectron spectra of the two compounds look quite different, as appears clearly

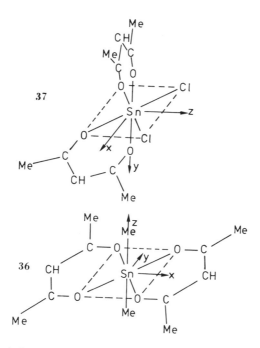

FIGURE 24. Geometrical arrangement of **36** (D_{2h} symmetry) and **37** (C_2 symmetry). Reproduced by permission of The Royal Society of Chemistry from Reference 137

TABLE 9.   Ionization energies and assignments for Me$_2$Sn(pd)$_2$. Reproduced by permission of the Royal Society of Chemistry from Reference 137

Band labelling	IE (eV)	Assignment
A	8.35	$b_{1u}(\pi_3)$
B	8.93	$b_{2g}(\pi_3)$
C	9.16	$b_{1g}(n_−)$
D	9.63	$b_{2u}(n_−)$
E	10.16	$b_{1u}[\sigma(Sn−C)]$
F	10.51	$a_g(n_+)$

from Figure 25. On the basis of a localized bond model and of comparison of the band intensity ratios in the He I and He II spectra, the assignments reported in Tables 9 and 10 were carried out. Some interesting information can be extracted from the results obtained:
    (i) In the *trans* complex **36** the MOs localized on the equatorial ligands are profoundly involved in bonding with the 5 p_y and 5 p_x orbitals of Sn; furthermore, some

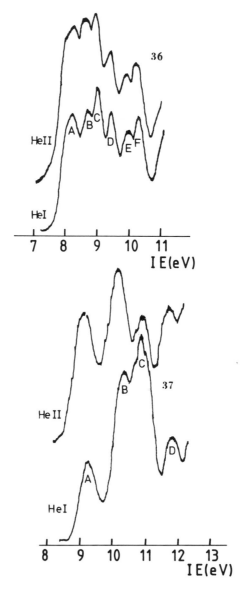

FIGURE 25. The photoelectron spectra of compounds **36** and **37**. Reproduced by permission of The Royal Society of Chemistry from Reference 137

TABLE 10.   Ionization energies and assignments for $Cl_2Sn(pd)_2$.
Reproduced by permission of the Royal Society of Chemistry from
Reference 137

Band labelling	IE (eV)	Assignment
A	9.10	$a + b(\pi_3)$
B	10.20	$a + b(n_-) + a[\sigma(Sn-Cl)]$
	(10.57)	
C	10.74	Cl lone pairs
	(10.86)	
D	11.64	$a + b(n_-) + a[\sigma(Sn-Cl)]$

Sn 5s electron density is transferred to equatorial bonding via the $a_g[\sigma(Sn-C)]-a_g(n_+)$ interaction.

(ii) In the *cis* complex **37** significant interaction between MOs of the ligands and those of Sn does not exist, the observed IEs of the highest occupied MOs being only affected by the potential field due to the tin atom, experienced by the ligand.

D. Cyclopentadienyl Derivatives

In this section PES studies of sandwich compounds of Group XIV elements with cyclopentadienyl and related ligands[139-141] will be described. Cradock and coworkers[139] studied the Sn(II) and Pb(II) cyclopentadienyl (Cp) complexes, $SnCp_2$ (**38**) and $PbCp_2$ (**39**), along with other heavy-metal derivatives, namely $HgCp_2$, TlCp and InCp. $SnCp_2$ and $PbCp_2$ are of considerable interest since they are the most stable organic derivatives of these metals in the oxidation state +2. They have centrally bonded metal atoms, i.e. a pentahapto coordination implying a delocalized π system extending over all the five

carbon atoms of the ring, as in the cyclopentadienide anion, $C_5H_5^-$:

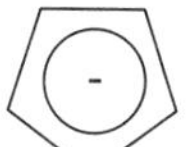

In the Group XIV compounds the two rings are not parallel, due to the influence of the metal lone pair, giving rise to a bent-sandwich molecular structure[142] (C_{2v} symmetry), different from that in other stable cyclopentadienyl derivatives, such as $Fe(Cp)_2$, where the two rings are parallel (D_{5h} symmetry in the eclipsed conformation and D_{5d} symmetry in the staggered one).

Baxter and coworkers[140] also studied **38** and **39**, together with the pentamethylated analogues, $SnCp'_2$ (**40**) and $PbCp'_2$ (**41**) ($Cp' = Me_5C_5$). The authors analysed the PE spectra with the help of Xα-SW calculations on the molecule of **38**, assuming the molecular geometry shown in Figure 26. The computed IE values of **38** and the experimental IE values of compounds **38–41** are reported in Table 11. It is clear that the two highest occupied MOs ($6a_2$ and $9b_2$, very close in energy) are localized on the π system of the two rings, while the bonding ring Sn MOs are the $11a_1$, $6b_1$, $10a_1$, $8b_2$ and $9a_1$. The calculation indicated that the least bonding orbital is $11a_1$ in which the primary interaction occurs between the ring π MOs and the Sn $5p_z$ AO, with significant participation also of the Sn 5s AO. In $6b_1$, the bonding interaction is with the Sn $5p_x$. The strongest bonding occurs in the $8b_2$ MO. The $9a_1$ MO arises from interaction of the Sn 5s AO with an a_1 combination of ring π MOs. The Sn lone pair is the $10a_1$ orbitals. The other compounds gave rise to similar PE spectra, whose assignment followed the same lines as discussed for **38**. The observation that the averages of IE values for the $6a_2$ and $9b_2$ MOs are practically identical for the couples **38, 39** and **40, 41** is consistent with the localization of these MOs on the rings and the consequent independence of their IE of the central

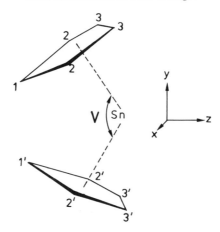

FIGURE 26. The geometry of SnCp$_2$ (**40**), omitting the hydrogen atoms; $V = 146°$, $d_{Sn-C} = 2.70$ Å, $d_{CC} = 1.42$ Å, $d_{CH} = 1.14$ Å. The molecule has C_{2v} symmetry and the equivalent carbon atoms have identical numbers. The Sn, C$_1$ and C$_1'$ atoms lie in the yz plane. Reprinted with permission from Reference 140. Copyright (1982) American Chemical Society

TABLE 11. Calculated ionization energies for SnCp$_2$ (**38**) and experimental ionization energies for SnCp$_2$ (**38**), PbCp$_2$ (**39**), SnCp$_2'$ (**40**) and PbCp$_2'$ (**41**). Reprinted with permission from Reference 140. Copyright (1982) American Chemical Society

	Calculated		Experimental IE values (eV)			
Assignment	MO^a	IE(eV)	**38**	**39**	**40**	**41**
Ring π	$6a_2$ ⎱	6.60	7.57	7.55	6.60	6.33
	$9b_2$ ⎰	6.61	7.91	7.85	6.60	6.88
Ring-Sn	$11a_1$ ⎱	7.31	8.85	8.54	7.64	7.38
Bonding MO	$6b_1$ ⎰	7.64	8.85	8.88	7.64	7.38
Sn lone pair	$10a_1$	8.74	9.58	10.10	8.40	8.93
Ring-Sn	$8b_2$	11.25	10.5	10.6	9.4	9.38
Ring σ	$5a_2$ ⎱	13.34 ⎫				
	$5b_1$ ⎰	13.35 ⎪	11.2	12.0	10.2	10.0
Ring-Sn	$9a_1$	13.41 ⎬ to	to	to	to	
Ring σ	$8a_1$ ⎱	13.46 ⎪ 14.0	14.5	16.0	16.0	
	$7b_2$ ⎰	13.41 ⎭				

^{a}The orbital labelling employed is that of the Xα-SW calculation on **38** and it has been used for the other compounds.

atom. The IE values of the metal 'lone-pair' MOs are larger for **40** and **41** than for **38** and **39**, indicating a greater participation for these MOs of s AO of the metal in the lead than in the tin derivatives.

Bruno and coworkers[141] investigated some Cp$'$ derivatives of Ge and Sn by He I and He II PES, namely GeCp$_2'$ (**42**), SnCp$_2'$ (**40**) and GeCp$'$Cl. The authors, on the basis of the band-intensity ratio variations on passing from the He I to the He II spectra (see Figure 27), make a different analysis of the electronic structure of **40** than the authors of Reference 140. They observed that the bands A and B present comparable total half-widths, and do not show any significant metal dependence. Furthermore, on the whole, the low-IE spectral region is similar for both **42** and **40** to that for highly symmetrical cyclopentadienyl derivatives, such as MgCp$_2$[143] and MgCp$_2'$[144], where the rings are

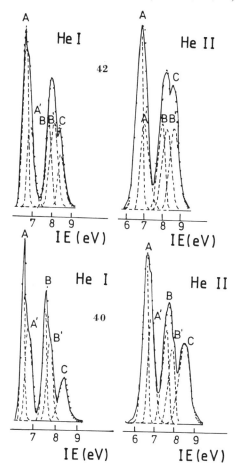

FIGURE 27. The photoelectron spectra of compounds **40** and **42**. The solid line refers to the experimental data; the dashed line shows the Gaussian components; the points refer to the convoluted Gaussian components. Reproduced by permission of Elsevier Science from Reference 141

parallel. This is consistent with electron diffraction data for **42** in the vapour phase[145], indicating an equilibrium conformation with almost parallel ring ligands. In conclusion, the authors of this investigation propose that the four highest occupied MOs in both **42** and **40**, corresponding to the spectral doublet A, B, are of main ring π character, though with some admixture of metal p orbitals, while the following orbital, corresponding to band C, is related to the metal s lone-pair orbital. Also, the high IE associated with the latter orbital (8.46 eV in **42**, 8.39 eV in **40**) is indicative of a pronounced 'inertness' of the metal lone pair, and therefore of almost ionic metal–ligand bonding character.

VII. CARBONYL DERIVATIVES

This section contains a description of a PES study on an interesting class of complexes, of formula $L_3MCo(CO)_4$[146] (M = Si, Ge, Sn, Pb; L = Cl, Br, Me). The chemistry of

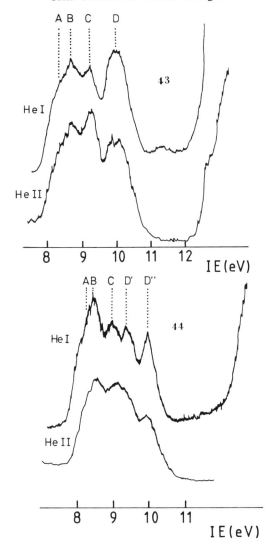

FIGURE 28. Low-IE region of the photoelectron spectra of compounds **43** and **44**. Reproduced with permission from Reference 146

complexes with main group metal–cobalt bonds attracted a great deal of attention[147] also due to the catalytic activity of octacarbonyldicobalt in reactions involving organosilicon hydrides. A strong intermetallic bonding (M–Co) was suggested on the basis of measurements and calculations of bond dissociation energies from appearance potential data and mass spectroscopy[148]. The low-IE region of the He I and He II spectra of $Me_3SnCo(CO)_4$ (**43**) and $Me_3PbCo(CO)_4$ (**44**) is shown in Figure 28, while a qualitative MO diagram is reproduced in Figure 29. From this scheme and the observation of band-intensity ratio variations on switching from the He I to the He II ionizing radiation, the authors suggested

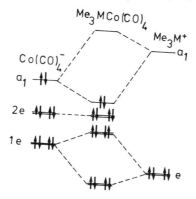

FIGURE 29. Qualitative MO diagram of the interaction between Me_3M^+ and $Co(CO_4)^-$ fragments. Reproduced with permission from Reference 146

that the HOMO, corresponding to band A, has a strong metal–metal character. The following two MOs, corresponding to bands B and C, are derived by the $2e$ and $1e$ levels of the $Co(CO_4)$ moiety, mainly localized on the cobalt 3d orbitals, although some interaction occurs with the $e(\sigma M–C)$ orbital of the Me_3M^+ fragment. The fourth MO, associated with band D, is of predominant M–C character. Band D is well split in **44** into two components, D' and D'', with a separation of 0.55 eV, due to the spin–orbit effect, larger for Pb than for Sn (in **43** band D is broad, but not clearly split), supporting the view of an important participation of metal p AOs to this MO.

As for the halogen-substituted derivatives, $Cl_3MCo(CO)_4$ and $Br_3MCo(CO)_4$, some indication of multiple bonding was found, namely a δ donative bonding from cobalt to M.

No bands due to Sn(4d) or Pb(5d) ionizations could be observed in the complexes of these metals.

VIII. REFERENCES

1. D. W. Turner, C. Baker, A. D. Baker and C. R. Brundle, *Molecular Photoelectron Spectroscopy*, Wiley, New York, 1970.
2. J. H. D. Eland, *Photoelectron Spectroscopy*, 2nd ed., Butterworths, London, 1974.
3. J. W. Rabalais, *Principles of Ultraviolet Photoelectron Spectroscopy*, Wiley, New York, 1977.
4. R. E. Ballard, *Photoelectron Spectroscopy and Molecular Orbital Theory*, Adam Hilger, Bristol, 1978.
5. J. Berkowitz, *Photoabsorption, Photoionization, and Photoelectron Spectroscopy*, Academic Press, New York, 1979.
6. A. D. Baker, *Acc. Chem. Res.*, **3**, 17 (1970).
7. A. Hamnet and A. F. Orchard, in *Electronic Structure and Magnetism of Inorganic Compounds*, Vol. 1, Chemical Society, London, 1972, p. 1.
8. A. D. Baker, C. R. Brundle and M. Thompson, *Chem. Soc. Rev.*, **3**, 355 (1972).
9. S. Evans and A. F. Orchard, in *Electronic Structure and Magnetism of Inorganic Compounds*, Vol. 2, Chemical Society, London, 1973, p. 1.
10. A. Hamnet and A. F. Orchard, in *Electronic Structure and Magnetism of Inorganic Compounds*, Vol. 3, Chemical Society, London, 1974, p. 218.
11. A. F. Orchard, in *Electronic States of Inorganic Compounds: New Experimental Techniques* (Ed. P. Day), Reidel, Dordrecht, 1975, p. 267.
12. M. Thompson, P. A. Hewitt and D. S. Wooliscroft, in *Handbook of X-Ray and Ultraviolet Photoelectron Spectroscopy* (Ed. D. Briggs), Heyden, London, 1977, p. 341.
13. C. Furlani and C. Cauletti, *Structure and Bonding*, Vol. 35, Springer-Verlag, Berlin, 1978, p. 119.

14. A. H. Cowley, *Prog. Inorg. Chem.*, **26**, 45 (1979).
15. R. G. Egdell and A. W. Potts, in *Electronic Structure and Magnetism of Inorganic Compounds*, Vol. 2, Chemical Society, London, 1980, p. 1.
16. J. C. Green, *Structure and Bonding*, Vol. 43, Springer-Verlag, Berlin, 1981, p. 37.
17. C. Cauletti and C. Furlani, *Comments Inorg. Chem.*, **5**, 29 (1985).
18. J. C. Maire, *J. Organomet. Chem.*, **218**, 45 (1985).
19. H. Van Dam and A. Oskam, *Transition Metal Chemistry (N.Y.)*, **9**, 125 (1985).
20. D. L. Lichtenberger, G. E. Kellogg and L. S. K. Pang, *Acc. Chem. Res., Symp. Ser.*, **357**, 265 (1987).
21. R. S. Brown and F. P. Torgensen, in *Electron Spectroscopy: Theory, Techniques and Applications*, Vol. 5, (Eds. C. R. Brundle and A. D. Baker), Academic Press, London, 1984, p. 1.
22. C. Cauletti and G. Distefano, in *The Chemistry of Organic Selenium and Tellurium Compounds*, Vol. 2, (Ed. S. Patai), Wiley, Chichester, 1987, p. 1.
23. H. Bock and B. Solouki, in *The Chemistry of Organic Silicon Compounds* (Eds. S. Patai and Z. Rappoport), Wiley, Chichester, 1989, p. 555.
24. T. Koopmans, *Physica*, **1**, 104 (1934).
25. M. O. Krause, in *Synchrotron Radiation Research* (Eds. H. Winick and S. Doniach), Plenum Press, New York, 1980, p. 101.
26. G. Margaritondo, *Introduction to Synchrotron Radiation*, Oxford University Press, New York, Oxford, 1988, p. 139.
27. U. Gelius and K. Siegbahn, *Faraday Discuss. Chem. Soc.*, **54**, 257 (1972).
28. R. C. Binning, Jr. and L. A. Curtis, *J. Chem. Phys.*, **92**, 3688 (1990).
29. J. Fernandez, J. Arria and A. Dargelos, *Chem. Phys.*, **94**, 397 (1985).
30. B. P. Pullen, T. A. Carlson, W. E. Moddeman, G. K. Schweitzer, W. E. Bull and F. A. Grimm, *J. Chem. Phys.*, **53**, 768 (1970).
31. S. Cradock, *Chem. Phys. Lett.*, 291 (1971).
32. A. W. Potts and W. C. Price, *Proc. R. Soc. London, Ser. A*, **326**, 165 (1972).
33. S. Evans, J. C. Green, P. J. Joachim, A. F. Orchard, D. W. Turner and J. P. Maier, *J. Chem. Soc., Faraday Trans. 2*, 905 (1972).
34. A. E. Jonas, G. K. Schweitzer, F. A. Grimm and T. A. Carlson, *J. Electron Spectrosc. Relat. Phenom.*, **1**, 29 (1972/3).
35. R. Boschi, M. F. Lappert, J. B. Pedley, W. Schmidt and B. T. Wilkins, *J. Organomet. Chem.*, **50**, 69 (1973).
36. A. Hosomi and T. G. Traylor, *J. Am. Chem. Soc.*, **97**, 3682 (1975).
37. C. L. Wong, K. Mochida, A. Gin, M. A. Weiner and J. K. Kochi, *J. Org. Chem.*, **44**, 3979 (1979).
38. G. Beltram, T. P. Fehlner, K. Mochida and J. K. Kochi, *J. Electron Spectrosc. Relat. Phenom.*, **18**, 153 (1980).
39. G. M. Bancroft, E. Pellach and J. S. Tse, *Inorg. Chem.*, **21**, 2950 (1982).
40. R. Bertoncello, J. P. Daudey, G. Granozzi and U. Russo, *Organometallics*, **5**, 1866 (1986).
41. I. Novak, J. M. Benson, A. Svensson and A. W. Potts, *Chem. Phys. Lett.*, **135**, 471 (1987).
42. M. J. S. Dewar, G. L. Grady and J. J. P. Stewart, *J. Am. Chem. Soc.*, **106**, 6771 (1984).
43. M. J. S. Dewar, G. L. Grady and E. F. Healy, *Organometallics*, **6**, 186 (1987).
44. G. Igel-Mann, H. Stoll and H. Preuss, *Mol. Phys.*, **65**, 1321 (1988) and references cited therein.
45. P. Gerard, M. O. Krause and T. A. Carlson, *Z. Physik D*, **2**, 123 (1986).
46. G. Granozzi, R. Bertoncello, E. Tondello and D. Ajò, *J. Electron Spectrosc. Relat. Phenom.*, **36**, 207 (1985).
47. J. J. Yeh and I. Lindau, *At. Data Nucl. Data Tables*, **32**, 1 (1985).
48. D. H. Harris, M. F. Lappert, J. B. Pedley and G. J. Sharp, *J. Chem. Soc., Dalton Trans.*, 945 (1976).
49. C. E. Moore, Atomic Energy Levels, US Dept. of Commerce, National Bureau of Standards, 1952.
50. E. A. V. Ebsworth, *Kem. Kozl.*, **42**, 1 (1974). CHEMABS 082(02)009776
51. A. Flamini, E. Semprini, F. Stefani, S. Sorriso and G. Cardaci, *J. Chem. Soc., Dalton Trans.*, 731 (1976).
52. G. C. Stocco and A. F. Orchard, *Chim. Ind. (Milan)*, **59**, 122 (1977)
53. C. Cauletti, C. Furlani, and M. N. Piancastelli, *J. Organomet. Chem.*, **149**, 289 (1978).
54. J. E. Drake, B. M. Glavincevski and K. Gorzelska, *J. Electron Spectrosc. Relat. Phenom.*, **16**, 331 (1979)

55. J. E. Drake, B. M. Glavincevski and K. Gorzelska, *J. Electron Spectrosc. Relat. Phenom.*, **17**, 73 (1979).
56. J. E. Drake, B. M. Glavincevski and K. Gorzelska, *Can. J. Chem.*, **57**, 2278 (1979).
57. I. Fragalà, E. Ciliberto, R. G. Egdell and G. Granozzi, *J. Chem. Soc., Dalton Trans.*, 145 (1980).
58. J. E. Drake and K. Gorzelska, *J. Electron Spectrosc. Relat. Phenom.*, **21**, 365 (1981).
59. K. G. Gorzelska, *Diss. Abstr. Int.*, **44**, 810B (1983).
60. J. Langó, L. Szepes, P. Császár and G. Innorta, *J. Organomet. Chem.*, **269**, 133 (1984).
61. G. Lespes, J. Fernandez and A. Dargelos, *Chem. Phys.*, **115**, 453 (1987).
62. R. G. Egdell, A. F. Orchard, D. R. Lloyd and N. V. Richardson, *J. Electron Spectrosc. Relat. Phenom.*, **12**, 415 (1977).
63. S. Stranges, M. Y. Adam, C. Cauletti, M. de Simone, C. Furlani, M. N. Piancastelli, P. Decleva and A. Lisini, *J. Chem. Phys.*, **97**, 4764 (1992).
64. S. Stranges, M. Y. Adam, M. de Simone, P. Decleva, A. Lisini, C. Cauletti, M. N. Piancastelli and C. Furlani, *J. Chem. Phys.*, in press.
65. A. W. Potts and M. L. Lyus, *J. Electron Spectrosc. Relat. Phenom.*, **13**, 327 (1978).
66. R. A. Forder and G. M. Sheldrick, *J. Organomet. Chem.*, **21**, 115 (1970).
67. I. Novak and A. W. Potts, *J. Organomet. Chem.*, **262**, 17 (1984).
68. A. V. Belyakov, L. S. Khaikin, L. V. Vilkov, E. T. Bogoradovski and V. S. Zavgorodnii, *J. Mol. Struct.*, **72**, 233 (1981).
69. K. Kimura, S. Katsumata, Y. Achiba, T. Yamazaki and S. Iwata, in *Handbook of He(I) Photoelectron Spectra of Fundamental Organic Molecules*, Japan Scientific Society Press, Tokyo, 1981, p. 188 and references cited therein.
70. G. Distefano, S. Pignataro, L. Szepes and J. Borossay, *J. Organomet. Chem.*, **104**, 173 (1976).
71. C. G. Pitt, *J. Organomet. Chem.*, **61**, 49 (1973).
72. P. K. Bischof, M. J. S. Dewar, D. W. Goodman and T. B. Jones, *J. Organomet. Chem.*, **82**, 89 (1974).
73. A. Schweig, U. Weidner and G. Manuel, *J. Organomet. Chem.*, **67**, C4 (1974).
74. Y. Limouzin and J. C. Maire, *J. Organomet. Chem.*, **92**, 169 (1975).
75. H. Bock, W. Kaim and H. Tesmann, *Z. Naturforsch.*, **33b**, 1223 (1978).
76. W. Kaim, H. Tesmann and H. Bock, *Chem. Ber.*, **113**, 3221 (1980).
77. A. L. Allred and E. G. Rochow, *J. Inorg. Nucl. Chem.*, **5**, 264 (1958).
78. G. Märkl, D. Rudnick, R. Schulz and A. Schweig, *Angew. Chem.*, **94**, 211 (1982); *Angew. Chem., Int. Ed. Engl.*, **21**, 221 (1982).
79. G. Märkl, D. Rudnick, R. Schulz and A. Schweig, *Angew. Chem. Suppl.*, 523 (1982).
80. G. V. Lopukhova, V. A. Godik, A. N. Rodionov and D. N. Shigorin, *Zh. Fiz. Khim.*, **62**, 199 (1988).
81. H. Bock, R. A. Bowling, B. Solouki, T. J. Barton and G. T. Burns, *J. Am. Chem. Soc.*, **102**, 429 (1980).
82. A. Schweig, U. Weidner and G. Manuel, *Angew. Chem.*, **84**, 899 (1972); *Angew. Chem., Int. Ed. Engl.*, **11**, 837 (1972).
83. A. Schweig, U. Weidner and G. Manuel, *J. Organomet. Chem.*, **54**, 145 (1973).
84. J. Laane, *J. Chem. Phys.*, **50**, 776 (1969).
85. J. F. Blanke, T. H. Chao and J. Laane, *J. Mol. Spectrosc.*, **38**, 483 (1971).
86. R. S. Brown, D. F. Eaton, A. Hosami, T. G. Traylor and J. M. Wright, *J. Organomet. Chem.*, **66**, 249 (1974).
87. I. Novak, T. Cvitas and L. Klasinc, *J. Organomet. Chem.*, **220**, 145 (1981).
88. C. Cauletti, C. Furlani, G. Granozzi, A. Sebald and B. Wrackmeyer, *Organometallics*, **4**, 290 (1985).
89. C. Cauletti, C. Furlani, F. Grandinetti, and D. Marton, *J. Organomet. Chem.*, **315**, 287 (1986).
90. V. N. Baidin, I. V. Shchirina-Eingorn, M. M. Timoshenko, Yu. V. Chizhov, I. I. Kritskaya and D. N. Kravtsov, *Metallorg. Khim.*, **1**, 1310 (1988).
91. C. Cauletti, C. Furlani and A. Sebald, *Gazz. Chim. Ital.*, **118**, 1 (1988).
92. M. V. Andreocci, M. Bossa, C. Cauletti, S. Stranges, B. Wrackmeyer and K. Horchler, *Inorg. Chim. Acta*, **162**, 83 (1989).
93. M. V. Andreocci, M. Bossa, C. Cauletti, S. Stranges, B. Wrackmeyer and K. Horchler, *Phys. Scr.*, **41**, 800 (1990).
94. C. Cauletti and C. Furlani, *Atti Accad. Naz. Lincei, Sc. Fis. e Nat., Sez. 9*, **2**, 191 (1991).
95. E. Anagnostopoulos, M. V. Andreocci, C. Cauletti, S. Stranges, A. Paz-Sandoval and N. Zuniga Villareal, *Main Group Metal Chem.*, **14**, 233 (1991).

96. M. V. Andreocci, M. Bossa, C. Cauletti, S. Stranges, K. Horchler and B. Wrackmeyer, *J. Mol. Struct. (Theochem)*, **254**, 171 (1992).
97. M. V. Andreocci, M. Bossa, C. Cauletti, S. Stranges, K. Horchler and B. Wrackmeyer, to appear.
98. A. Modelli and G. Distefano, *Z. Naturforsch., A*, **36**, 344 (1981) and references cited therein.
99. C. R. Brundle, M. B. Robin and N. A. Kuebler, *J. Am. Chem. Soc.*, **94**, 1466 (1972).
100. M. J. S. Dewar and S. D. Worley, *J. Chem. Phys.*, **50**, 654 (1969).
101. M. Jones and W. Kitching, *Aust. J. Chem.*, **37**, 1863 (1984).
102. G. Bieri, F. Burger, E. Heilbonner and J. P. Maier, *Helv. Chim. Acta*, **60**, 2213 (1977).
103. B. Wrackmeyer, *J. Magn. Reson.*, **42**, 287 (1981).
104. H. Bock and H. Seidl, *J. Chem. Soc. (B)*, 1158 (1968).
105. H. Bock and H. Alt, *Chem. Ber.*, **103**, 1784 (1970).
106. F. Brogli, E. Heilbronner, V. Hournung and E. Kloster-Jensen, *Helv. Chim. Acta*, **56**, 2171 (1973).
107. W. Ensslin, H. Bock and C. J. Becker, *J. Am. Chem. Soc.*, **96**, 2757 (1974).
108. F. Brogli, E. Heilbronner, J. Wirz, E. Kloster-Jensen, R. G. Bergman, K. P. C. Vollhardt and A. J. Ashe III, *Helv. Chim. Acta*, **58**, 2620 (1975).
109. P. Carlier, J. E. Dubois, P. Masclet and G. Mouvier, *J. Electron Spectrosc. Relat. Phenom.*, **7**, 55 (1975).
110. G. Bieri, E. Heilbronner, E. Kloster-Jensen and J. P. Maier, *Phys. Scr.*, **16**, 202 (1977).
111. M. I. MacLean and R. E. Sacher, *J. Organomet. Chem.*, **74**, 197 (1974).
112. B. G. Ramsey, A. Brook, A. R. Bassindale and H. Bock, *J. Organomet. Chem.*, **74**, C41 (1974).
113. L. Szepes, T. Korányi, G. Náray-Szabó, A. Modelli and G. Distefano, *J. Organomet. Chem.*, **217**, 35 (1981).
114. H. Sakurai, M. Ichinose, M. Kira and T. G. Traylor, *Chem. Lett.*, 1383 (1984).
115. K. Mochida, S. D. Worley and J. K. Kochi, *Bull. Chem. Soc. Jpn.*, **58**, 3389 (1985).
116. K. Mochida, S. Masuda and Y. Harada, *Chem. Lett.*, 2281 (1992).
117. M. F. Lappert, J. B. Pedley, J. Simpson and T. R. Spalding, *J. Organomet. Chem.*, **29**, 195 (1971).
118. A. Fadini, E. Glozbach, P. Krommes and J. Lorberth, *J. Organomet. Chem.*, **149**, 297 (1978).
119. M. V. Andreocci, C. Cauletti, S. Stranges, B. Wrackmeyer and C. Stader, *Z. Naturforsch.*, **46b**, 39 (1991).
120. D. C. Frost, F. Herring, A. Katrib, C. McDowell and R. McLean, *J. Phys. Chem.*, **76**, 1030 (1972).
121. G. Distefano, A. Ricci, F. P. Colonna and D. Pietropaolo, *J. Organomet. Chem.*, **78**, 93 (1974).
122. G. Distefano, S. Pignataro, A. Ricci, F. P. Colonna and D. Pietropaolo, *Ann. Chim.*, **64**, 153 (1974).
123. F. Bernardi, G. Distefano, A. Modelli, D. Pietropaolo and A. Ricci, *J. Organomet. Chem.*, **128**, 331 (1977).
124. M. A. Delmas and J. C. Maire, *J. Organomet. Chem.*, **161**, 13 (1978).
125. C. Cauletti, G. Nicotra and M. N. Piancastelli, *J. Organomet. Chem.*, **190**, 147 (1980).
126. A. Modelli, G. Distefano, D. Jones and G. Seconi, *J. Electron Spectrosc. Relat. Phenom.*, **31**, 63 (1983).
127. C. Guimon, G. Pfister-Guillouzo, H. Lavayssiere, G. Dousse, J. Barrau and J. Satge, *J. Organomet. Chem.*, **249**, C17 (1983).
128. G. Pfister-Guillouzo and C. Guimon, *Phosphorus Sulfur*, **23**, 197 (1985).
129. E. Andoni, C. Cauletti and C. Furlani, *Inorg. Chim. Acta*, **76**, L35 (1983).
130. E. Andoni, *Bul. Shkencave Nat.*, **37**, 67 (1983). CHEMABS 102(26)228858
131. C. Cauletti, F. Grandinetti, A. Sebald and B. Wrackmeyer, *Inorg. Chim. Acta*, **117**, L37 (1986).
132. C. Cauletti, F. Grandinetti, G. Granozzi, A. Sebald and B. Wrackmeyer, *Organometallics*, **7**, 262 (1988).
133. R. G. Egdell, I. L. Fragalà and A. F. Orchard, *J. Electron Spectrosc. Relat. Phenom.*, **17**, 267 (1979).
134. E. Andoni, M. Bossa, C. Cauletti, C. Furlani and A. Palma, *J. Organomet. Chem.*, **244**, 343 (1983).
135. D. A. Sweigart and D. W. Turner, *J. Am. Chem. Soc.*, **94**, 5599 (1972).
136. H. Bock, G. Wagner and J. Kroner, *Tetrahedron Lett.*, 3713 (1971).
137. I. Fragalà, E. Ciliberto, P. Finocchiaro and A. Recca, *J. Chem. Soc., Dalton Trans.*, 240 (1979).
138. S. Evans, A. Hamnett, A. F. Orchard and D. R. Lloyd, *Faraday Discuss. Chem. Soc.*, **54**, 227 (1972).

139. S. Cradock and W. Duncan, *J. Chem. Soc., Faraday Trans. 2*, **74**, 194 (1978).
140. S. G. Baxter, A. H. Cowley, J. G. Lasch, M. Lattman, W. P. Sharum and C. A. Stewart, *J. Am. Chem. Soc.*, **104**, 4064 (1982).
141. G. Bruno, E. Ciliberto, I. L. Fragalà and P. Jutzi, *J. Organomet. Chem.*, **289**, 263 (1985).
142. A. Almenningen, A. Haarland and T. Motzfeld, *J. Organomet. Chem.*, **7**, 97 (1967).
143. S. Evans, M. L. H. Green, B. Jewit, A. F. Orchard and C. F. Pygall, *J. Chem. Soc., Faraday Trans. 2*, **68**, 1847 (1972).
144. C. Cauletti, J. C. Green, M. R. Kelly, P. Powell, J. van Tilborg, J. Robbins and J. Smart, *J. Electron Spectrosc. Relat. Phenom.*, **19**, 327 (1980).
145. L. Fernholt, A. Haaland, P. Jutzi, X. F. Kohl and R. Seip, *Acta Chem. Scand., Ser. A*, **38**, 211 (1984).
146. J. A. Louwen, R. R. Andréa, D. J. Stufkens and A. Oskam, *Z. Naturforsch.*, **38b**, 194 (1983).
147. R. A. Burnham and S. R. Stobart, *J. Chem. Soc., Dalton Trans.*, 1489 (1977).
148. R. A. Burnham and S. R. Stobart, *J. Organomet. Chem.*, **86**, C45 (1975).

CHAPTER 7

Analytical aspects of organo-germanium compounds

JACOB ZABICKY and SARINA GRINBERG

Institutes for Applied Research, Ben-Gurion University of the Negev, Beer-Sheva 84110, Israel
Fax: +(972)-7-271612;   e-mail: zabicky@bgumail.bgu.ac.il

The chemistry of organic germanium, tin and lead compounds
Edited by S. Patai   © 1995 John Wiley & Sons Ltd

ABBREVIATIONS
The following abbreviations are used in Chapters 7–9 dealing with analytical aspects of organometallic compounds containing group 14 elements:

AAS	atomic absorption spectrometry
AED	atomic emission detector
AES	atomic emission spectrometry
AFD	atomic fluorescense detector
ASV	anodic stripping voltametry
CP-MAS-NMR	cross polarization-MAS-NMR
CVD	chemical vapour deposition
DCP-AES	direct current plasma AES
DPASV	differential pulse ASV
ECD	electron capture detector
EIMS	electron impact MS
ENDOR	electron–nucleus double resonance
EPAAS	electrostatic precipitation-AAS
ESCA	electron spectroscopy for chemical analysis
ESR	electron spin resonance
EXAFS	extended X-ray absorption fine structure
ETAAS	electrothermal AAS
FABMS	field absorption MS
FDMS	field desorption MS
FIA	flow injection analysis
FID	flame ionization detector
FPD	flame photometric detector
FTIR	Fourier-transform infra-red
GC	gas chromatography
GC-...	GC combined with detectors of various types
GFAAS	graphite furnace AAS
HREELS	high resolution electron-energy loss spectroscopy
ICP-...	inductively coupled plasma combined with detectors of various types
IDMS	isotope dilution MS
INAA	instrumental neutron activation analysis
INS	inelastic neutron scattering
ISS	ion scattering spectrometry
ITD	ion trap detector
LAMMA	laser microprobe mass analysis
LC	liquid chromatography
LEAF	laser-excited atomic fluorescence
LEAF-ETA	LEAF electrothermal atomisation
LEI	laser-enhanced ionisation
LOD	limit(s) of detection
MAS-NMR	magic-angle spinning NMR
MS	mass spectrometry
NMR	nuclear magnetic resonance
NQR	nuclear quadrupole resonance
PC-FIA	packed column FIA
PID	photoionization detector
PVC	poly(vinyl chloride)
RNAA	radiochemical neutron activation analysis
RSD	relative standard deviation

SCE	supercritical extraction
SCF-MO	self-consistent field molecular orbital
SCF-MS	self-consistent field multiple scattering
SIMS	secondary ion MS
SNR	signal-to-noise ratio
TEL	tetraethyllead
TLC	thin layer chromatography
TML	tetramethyllead
TPD	temperature-programmed desorption
UPS	UV photoemission spectroscopy
UVV	UV-visible
VUV	vacuum-UV
XPS	X-ray photoelectron spectroscopy
XRD	X-ray diffraction
XRF	X-ray fluorescence

I. INTRODUCTION

A. Remarks on the Analysis of Group 14 Organometallics

1. General

During the organization of the literature sources gathered on the analytical aspects pertaining to organometallic compounds containing Ge, Sn and Pb, it was realized that the subject needs to be treated in three separate chapters for various reasons: The profusion of the material, the scant number of sources dealing with two or more elements of the group, and predominantly, the driving force underlying analytical research for each one of the elements. Thus, at present environmental pollution is by far the predominant theme with organometallics containing Pb and to a lesser extent Sn, due to their technological importance and their impact on biological systems. The main underlying theme with organotin compounds is structural analysis, due to their continuously increasing importance as intermediates in organic synthesis. Organogermanium compounds have not yet reached the technological importance of the other two groups.

2. Elemental analysis

Two subjects of fundamental importance in elemental analysis of organometallic compounds are sample preparation and end analysis. In many published articles emphasis is placed on one or the other when trying to find optimal analytical conditions for a certain type of sample. For example, an article dealing with an improved method for trace concentrations of tin may be based on the use of some additives to the usual digestion reagents; once the tin present in the sample is converted to an adequate form, measurement can proceed either as described in the article or by alternative methods compatible with the nature of the digested sample. No separation of the subjects is attempted in this section, and the reader should judge by himself, on perusal of the cited literature and many additional sources, what combination of sample preparation and finishing methods best fits the problem on hand. In the following sections two trends are reviewed for elemental analysis: the classical macro or semi-micro methods and the modern approach, heavily based on expensive instruments, allowing faster and more accurate determinations, and extremely low detection limits. It should be pointed out that the modern trend is much influenced by the stringent demands of environmental and occupational protection agencies.

Some of the older methods of analysis of the metallic element are mentioned, not only for their historical value, but also because they may serve as the basis for the development of new methods requiring simpler instrumentation than those in vogue today. Modern devices used in the analytical laboratory, with prices in the range of tens if not hundreds of thousands $US, and requiring highly trained maintenance and operating personnel would usually be unacceptable for on-line automatic control in industrial plants.

Organic chemists frequently make use of analytical instrumentation operating with extremely small amounts of sample, for example GC and most of the spectroscopic methods, but they rarely require analyses at trace or ultratrace concentration levels. The demands of environmental and occupational protection drove analytical sensitivity to concentrations in the subnanogram per litre level for elements such as lead. However, instrumental sensitivity and accuracy are not sufficient. To illustrate the labours involved in a reliable analysis of Pb in that concentration range, one should imagine a speck of lead the size of a bacterium distributed uniformly in one litre volume. Another speck of lead of that size could have fallen in the sample, carried by the air or shaken off from the dress of the analyst. To avoid such contingencies, reliable analyses involve painstaking procedures for sample collection and preparation, besides adequate handling of the sensitive instrumentation required for the actual determination. This makes such analyses extremely expensive. A comprehensive review of trace analysis of the elements appeared recently, including underlying theory, sampling and instrumental methods[1a].

3. Speciation

Trace level analyses are frequently required for forensic, clinical and toxicological applications and for better understanding of the fate of individual pollutants in the environment, where not only the element is determined but the organic species are also identified and individualy quantified. These analyses can be performed with or without actual separation of the individual species.

4. Structural analysis

Classifying a compound as organometallic does not confer on it sufficient 'functionality', in the sense established for ordinary organic compounds in *The Chemistry of Functional Groups* series of books. Thus, a certain untoward filling of crowding and lack of unity may be developed when trying to discuss the organometallic compounds of the present volume. In fact, several tomes could have been compiled for the chemistry of organotin compounds. The structural features of organometallic compounds of group 14 give rise to various types of spectroscopic properties discussed in detail elsewhere. In the three 'analytical chapters' examples published in the recent literature are shown with scant comment, that may serve as leading references for analytical problems involving similar structures. For example, NMR spectroscopy of organic compounds appears in the literature with a profusion of values for chemical shifts, coupling constants and solvent effects, and a steadily growing pool of information on MAS-NMR (magic-angle spinning NMR). Such information is not discussed in the analytical chapters.

Coordination numbers higher than 4 for C are hardly encountered in ordinary organic chemistry. A familiar example for a coordination 5 is the trigonal bipyramidal transition state involved in a Walden inversion. However, with organometallic compounds containing Ge, Sn and Pb coordination numbers of 5 and 6 are frequent, and even 7 is occasionally encountered, especially in the solid state. Such structural features are analysed by NMR and Moessbauer spectroscopies, and most frequently by single-crystal XRD (X-ray diffraction). Innumerable solid organometallic compounds containing Ge, Sn and Pb atoms have been analysed by XRD crystallography alone or in addition to other spectroscopic methods for structural assessment. Of special interest are direct observation of coordination numbers of the metal atoms with surrounding ligands and their spacial arrangement, shapes of cyclic groups in the molecule and abnormal bond lengths and angles.

TABLE 1. Industrial organogermanium compounds[a]

Compound and CAS registry number	NIOSH/OSHA RTECS[b]
Allyltriethylgermane [1793-90-4]	LY 5360000
Butyltrichlorogermane [4872-26-8]	LY 4940000
Chlorogermane	WH 6790000
Chlorotributylgermane [2117-36-4]	LY 5230000
Dibutyldichlorogermane [4593-81-1]	LY 5020000
Dichlorogermane	
Digermane	
Ethyltriiodogermane [4916-38-5]	LY 5100000
Germane [7782-65-2]	LY 4900000
Isopropyltriiodogermane [21342-26-7]	LY 5125000
Methyltriiodogermane [1111-91-7]	LY 5150000
Propyltriethylgermane [994-43-4]	LY 5390000

[a] See also Reference 5.
[b] *Registry of Toxic Effects of Chemical Substances* of National Institute for Occupational Safety and Health/Occupational Safety and Health Administration.

5. Derivatization

Derivatization of organic compounds has been traditionally used in organic analysis as additional evidence for structural features, to simplify analytical procedures, to improve the sensitivity or accuracy of the analysis, etc. It is worthwhile recalling briefly the requirements for a good derivatizing scheme that were summarized elsewhere in the *Functional Group* series[1b], because such schemes will be an important part of the analytical chapters.

i. Functional selectivity. Only the functional group of interest reacts in a predetermined way, while the rest of the molecule remains untouched.

ii. Analytical compatibility. The properties to be measured should be enhanced in the derivative.

iii. Stability. Derivatives should be stable under ordinary laboratory conditions and those involved in the analytical process.

iv. Ease of handling. The derivatizing process should be easy to perform within a reasonable time.

B. Remarks on the Analysis of Organogermanium Compounds

Among the organometallic compounds of group 14 elements, those of germanium are technologically the least important. They have, however, a certain potential in medical applications and in high-tech fields, such as electronics, optics and radiation detectors, where elemental germanium itself is important. Germanes, for example, can be used for deposition of germanium thin films. A complex continental–marine environmental correlation for methylated germanium pollutants has been described[2]. Table 1 lists organogermanium compounds that have found industrial application with references to occupational protection protocols, where analytical methods for the particular compound can be found. Analysis of organogermanium compounds has been reviewed[3,4].

II. ELEMENTAL ANALYSIS

A. Determination in Organic Samples

Germanium in organometallic compounds can be determined by a modification of the combustion tube method. The sample is mixed with a large excess of chromic oxide, the 'organic' elements are burnt in the tube and are absorbed downstream, whereas the non-volatile germanium dioxide residue is determined by the weight difference of the combustion tube[6]. Ordinary tube combustion of organogermanium compounds can lead to erratic C-H analyses. This was overcome by carrying out the combustion very slowly[7]. Volatile organogermanium compounds can be carried by the oxygen stream into the combustion tube, where they leave a deposite of germanium dioxide, to be weighed at the end of the process[8].

Organogermanium compounds can be mineralized by wet oxidative digestion for 4 h at $70°C$, in aqueous potassium persulphate, at pH 12. After dilution to an adequate concentration germanium can be determined by ICP-AES (inductively coupled plasma atomic emission spectrometry)[9].

B. Trace Analysis

1. Atomic absorption and emission spectroscopies

A method combining FIA (flow injection analysis), hydride generation and ICP-AES allowed 150 analyses per hour; LOD (limits of detection) 0.4 $\mu g/L$ at SNR (signal-to-noise ratio) 3, for 1 mL aqueous sample, RSD (relative standard deviation) 3-4%. Interference from transition metals was reduced with EDTA[10]. Daily intake of high doses of germanium compounds for 12-16 months resulted in high Ge concentrations in hair and nails, as determined by ICP-AES. Corresponding analyses were under the LOD of the method for individuals exposed to normal Ge levels[11].

The chemical reactions undergone by Ge in the graphite furnace during analysis by GFAAS (graphite furnace AAS) were elucidated by AAS (atomic absorption spectrometry), XRD, electron microscopy and molecular absorption spectroscopy. The Na_2GeO_3 deposited during the drying stage is reduced by C to elementary Ge. Volatile GeO formed at temperatures over 1100 K leads to analytical losses. Excess NaOH enhances the Ge absorbance, due to GeO reduction by Na at temperatures over 1500 K. Treatment of the furnace with carbide-forming elements also enhances the Ge absorbance[12].

Two methods were examined for digestion of biological samples prior to trace element analysis. In the first one a nitric acid–hydrogen peroxide–hydrofluoric acid mixture was used in an open system, and in the second one nitric acid in a closed Teflon bomb. The latter method was superior for Ge determination, however, germanium was lost whenever hydrogen fluoride had to be added for disolving silicious material. End analysis by ICP-AES was used for Ge concentrations in the $\mu g/g$ range[13].

A method based on hydride generation and DCP-AES (direct current plasma AES) was proposed with claims of reduced sensitivity to instrumental parameters. Addition of L-cystine or L-cysteine reduces interference from transition metals; interference of other hydride-forming elements is negligible; LOD 20 ng Ge/L[14]. The presence of easily ionized elements for signal enhancement is an important feature of DCP-AES. Alkaline and alkaline earth elements improve the performance of this method with hydride generation for group 14 elements. The RSD showed by 2 μg Ge/L was 4.7% in water and 4.1% in 1 M KCl[15]. Use of ammonium peroxodisulphate achieved simultaneous signal enhancement and suppression of interference by Al, As, Cu, Hg, Pb and Sn at the 1 g/L level. Cd, Co, Ni and Zn at this level interferred with the determination[16].

2. Spectrophotometric methods

Germanium solutions in the presence of phenylfluorone (1) and lauryldimethylammonium bromide yield coloured compounds, λ_{max} 508 nm, $\varepsilon 1.60 \times 10^5$. This allows germanium determination at concentrations of 8–80 $\mu g/L$. Only Sb, Sn, Cr(VI) and Mo interfere with the analysis[17]. Precipitation and filtration of the Ge(IV)–1 complex formed in the presence of zephiramine chloride (2), followed by spectrophotometric analysis of the filter, was applied to determination ppb levels of Ge in groundwater and spring water[18]. Trace analysis of Ge in water was carried out spectrophotometrically after extraction at pH 3.3–4.0, with heptanol containing 0.1 M of benzilic acid (HL, 3) and 5×10^{-5} M of phenylfluorone (HL', 1). A complex $Ge(OH)_2L_2$ is formed and subsequently transformed into $Ge(OH)_2L'_2$, with λ_{max}510–515 nm, ε 120,000; LOD 2 μg Ge/L of aqueous phase, RSD 0.5–0.7% for 0.5–5.0 μg Ge[19]. A combination of FIA with phenylfluorone complex formation was also reported[20]. A complex is formed between Ge ions and 3,7-dihydroxyflavone (4) in 4 M phosphoric acid solution that can be measured by its fluorescence at 444 nm, using the Hg line at 405 nm for excitation; RSD 0.71% for 0.44 μg Ge. A preliminary extraction of $GeCl_4$ from a concentrated HCl solution will avoid interference by Sb(III), Sn(IV) and In(III)[21]. An analogous method consists of forming a complex with mandelic acid (5) and malachite green (6) which is extracted into chlorobenzene and measured at λ_{max}628 nm, ε 133,000. Interference by Fe, Ti, Sn(IV), Mo and Sb(III) is eliminated by chelating agents[22].

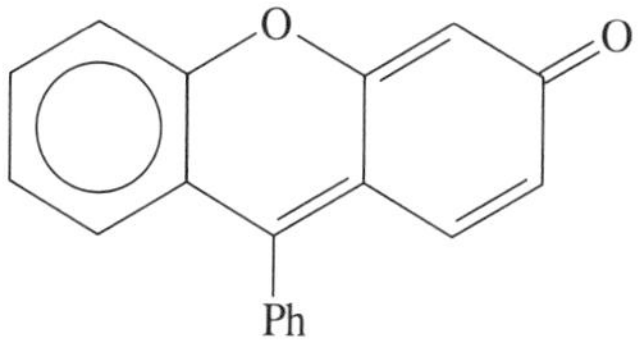

Ph

[n-$C_{14}H_{29}Me_2(PhCH_2)N^+$]Cl$^-$ $Ph_2C(OH)CO_2H$

(1)   (2)   (3)

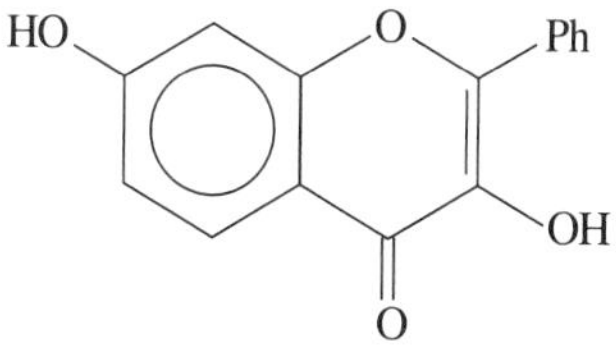

HO ... O ... Ph ... OH ... O $PhCH(OH)CO_2H$ Me_2N ... N^+Me_2 ... Ph

(4)   (5)   (6)

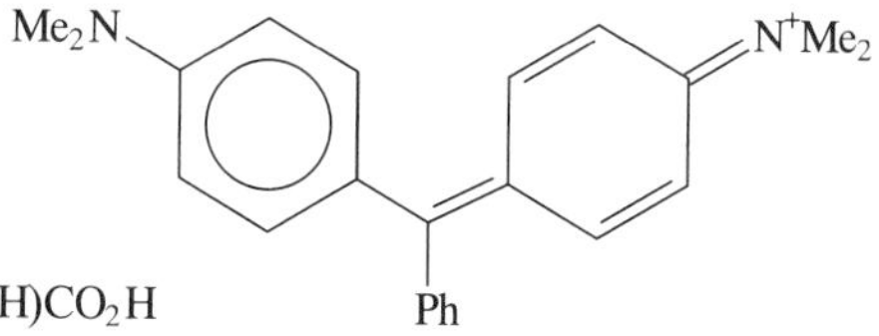

A selective method for determination of Ge(IV) consists in reacting the sample with iodide in the presence of sulphuric acid. The tetraiodogermane(IV) formed is extracted into chloroform and a complex is formed with reagent 7, $\varepsilon = 15,000$ at $\lambda_{max} = 395$ nm; LOD 0.1 μg Ge/L[23].

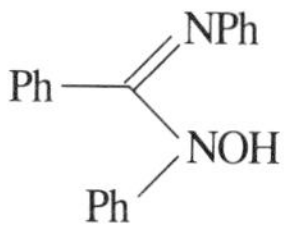

Ph — NPh / NOH, Ph

(7)

Germanium in samples dissolved in water or dioxane or dispersed into borax disks can be determined by XRF (X-ray fluorescence), using K_α radiation and arsenic as internal standard[24].

3. Electrochemical methods

A plasticized ion-selective electrode was developed for the complexes of Sb(III) and Ge(IV) with 3,5-dinitrocatechol (8)[25].

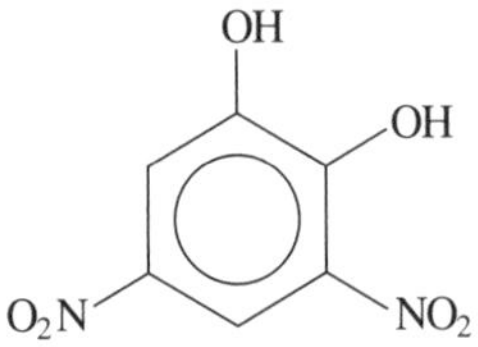

OH

OH

O_2N NO_2

(8)

A comparative study was carried out for various methods for trace analysis of germanium in biological samples: germanium tetrachloride extraction with carbon tetrachloride followed by electrochemical determination based on Ge accumulation in a hanging Hg drop electrode of Ge(IV)–diol complexes with phenolic reagents (e.g. pyrogallol) followed by cathodic stripping voltametry, LOD 0.1 μg/L, 13% RSD; spectrophotometric determination of the phenylfluorone (1) complex, LOD 5 μg Ge/L, 6% RSD; ETAAS (electrothermal AAS) using alkaline samples in the presence of an oxidant and palladium nitrate–magnesium nitrate as matrix modifiers, LOD 20 μg Ge/L, 8% RSD[26,27].

III. TRACE ANALYSIS ALLOWING SPECIATION

A. Chromatographic Methods

The behaviour of alkylgermanium compounds is similar to that of alkylsilicon compounds[28,29]. Equation 1 is an empirical correlation between the retention time, t_R, relative to that of mesitylene taken as 100, and the number of normal alkyl groups of tetraalkygermanes[30]. Linear correlations were also found for straight-chain compounds of formula $Ge_mSi_nH_{2(m+n+1)}$, $m, n = 0, 1, \ldots$, when taking $\log_{10} t_R$ as function of the number of heteroatoms[31].

$$\log_{10} t_R = 0.14 + 0.14 n_{Me} + 0.45 n_{Et} + 0.69 n_{n\text{-Pr}} + 0.93 n_{n\text{-Bu}} \qquad (1)$$

Organosilicon and organogermanium compounds were separated at 330–350 K on a Cromaton column coated with squalene. Improved quantitative determination was achieved by accumulation of preliminary decomposition products of the organometallic compounds in a graphite atomizer, followed by ETAAS[32].

Permethylated oligogermanes (9) and various photolysis degradation products (10) were scavenged with carbon tetrachloride and other agents and determined by GC-MS[33].

Me(GeMe$_2$)$_n$Me   Me(GeMe$_2$)$_n$Cl

(n = 3–6, 9)   (n = 1–4, 10)

B. Miscellaneous Methods

A method was devised for the precise measurement of the flow of dense vapours dissolved in an inert carrier gas, and was demonstrated with a stream of tetraethylgermane vapour[34].

Organogermanium chorides $R_nGeCl_{(4-n)}$ (n = 1–3) are strong and hard Lewis acids. When these compounds are extracted into alkaline aqueous solutions they become hydrolysed to the corresponding hydroxides. On acidification of these aqueous solutions with hydrochloric acid Ge$-$Cl bonds are formed to various extents, depending on the acid concentration and the nature of the organometallic ion. The chlorides can be extracted into an organic solvent. The series of extraction constants K_{ex}, depicted in equation 2 for various organogermanium halides, reflects the coordination ability of the central germanium atom with chlorine atoms and the hydrophobic properties of the alkyl groups. Such behaviour allows separation of individual components of organogermanium halides by liquid–liquid extraction[9]. Inorganic germanium and organogermanium compounds extracted into aqueous solutions can be determined by ICP-AES methods either in alkaline or acid media. Detection limits may be as low as 3.1×10^{-8} M. Trialkylgermanium compounds showed higher sensitivity than inorganic or other organogermanium compounds, and should be determined separately[35].

$$Et_3GeCl > Me_3GeCl > Et_2GeCl_2 > Me_2GeCl_2 > PhGeCl_3$$

$$> EtGeCl_3 > MeGeCl_3 > GeCl_4 \tag{2}$$

IV. STRUCTURAL ANALYSIS

The aryl groups depicted in formula **11** appear throughout Sections IV and V.

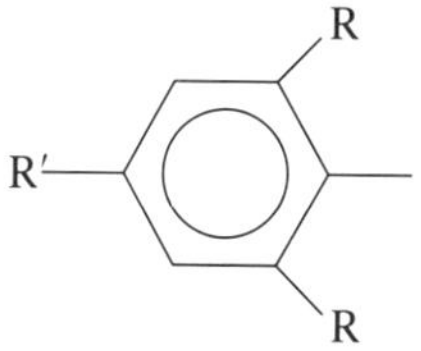

$Ar^1 : R = R' = i\text{-}Pr$
$Ar^2 : R = R' = (Me_3Si)_2CH$
$Ar^3 : R = R' = Me$
$Ar^4 : R = Et, R' = H$
$Ar^5 : R = R' = t\text{-}Bu$

(11)

A. Vibrational and Rotational Spectra

The IR vibration frequencies of compounds MH$_4$ and MeMH$_3$ were calculated *ab initio* for the metallic elements of group 14, including M = Ge, and were compared with experimental data from various sources[36]. Many constants pertaining to rotational and vibrational spectra of compounds MH$_3$X (M = C, . . . , Sn; X = F, . . . , I) were calculated *ab initio*, showing good agreement with experimental values[37]. The microwave spectra of 18 isotopic species of methylfluorogermane (MeGeH$_2$F) were interpreted taking into account the GeH$_2$ angle and the tilt angle of the methyl group[38].

The stretching band of germane (GeH$_4$) in the ν_{Ge-H} 2000 cm^{-1} region was recorded in a high resolution IR spectrometer and analysed. The rotational constants and Coriolis parameters were in accord with those obtained for the band in the 3000 cm^{-1} region. A C_{3v} symmetric top geometry was assigned to the molecule[39–41]. The IR band at ν_{Ge-H} 2040 cm^{-1} served to investigate the kinetics and mechanism of CVD (chemical vapour deposition) of a thin germanium layer, by thermolysis of trimethylgermane at 420–670 K[42] and 673–873 K[43]. Fourier transform microwave spectroscopy was applied to determine the tensorial spin–rotation and spin–spin interaction constants of germane[44].

The vibrational frequencies ν_s and ν_{as} of the Ge$-$Y$-$Ge in hexasubstituted digermyl chalcogenides (**12**) were studied. The electronegative effect of CF$_3$ groups displaces the bands to higher frequencies[45].

$$R_3Ge-Y-GeR_3$$

(12, R = H, Me, CF$_3$; Y = O, S, Se, Te)

Laser Raman and IR spectra of H_3SiGeH_3 were analysed on the basis of fundamental, overtone and combination frequencies, assuming a C_{3v} point group symmetry, and the frequencies were asigned to the various vibrational modes[46]. The structure of gaseous H_3GeSiH_3 was determined by analysis of the electron diffraction intensities and rotational constants. The Ge—Si bond length obtained from this analysis (0.2364 nm) is 0.004 to 0.007 nm shorter than values calculated *ab initio*[47].

The molecular structure of cyclopropylgermane (13) was investigated, based on the rotational spectra of 41 isotopomers obtained by combining the isotopes ^{70}Ge, ^{72}Ge, ^{74}Ge, ^{76}Ge, ^{13}C and ^2H attached to Ge[48]. The barrier to internal rotation of the germyl group in 13 was determined from its IR spectrum and compared to that of cyclopropylsilane[49]. The IR and Raman vibrational spectra of cyclobutylgermane (14) were assigned[50] and the rotational spectra of three isotopic species of equatorial and axial conformers were identified and assigned for the ground state, including the rotational constants for all six species[51].

(13) 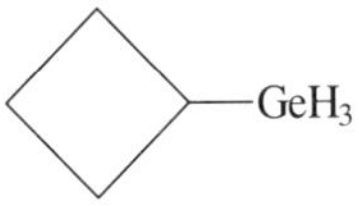   **(14)**

Tetraalkynylgermanes [Ge(C≡CR)$_4$, R = Me, Ph] were characterized spectroscopically IR($\nu_{C≡C}$ = 2180 cm^{-1}), MS and ^{13}C and ^{73}Ge NMR spectroscopy[52].

The structure of the stable germanethione 31 (see reaction 14 in Section V) was characterized by FT-Raman spectrum ($\nu_{Ge=S}$ 521 cm^{-1}, in good agreement with 518 cm^{-1} reported for $Me_2Ge=S$), UVV spectrum (λ_{max} = 450 nm, ε100, Ge=S $n-\pi^*$ transition) and FABMS (field absorption). See also Tables 2 and 3[53].

Properties of the carbonyl group of ferrocenyl germyl ketones (15) were studied by IR, NMR and XRD. The carbonyl basicity was assessed based on hydrogen bonding with phenol in carbon tetrachloride solution (see also Tables 2 and 3). The spectra were compared with those of the analogous Si compounds and those of acetyl- and benzoylgermanes[54].

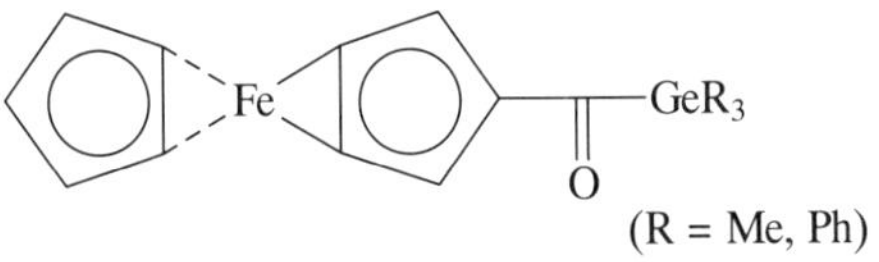

(R = Me, Ph)

(15)

The Raman, mid- and far-IR spectra of liquid and solid CH_3GeD_2NCO and CD_3GeH_2NCO were assigned and applied for the conformational analysis of these isocyanates[55a].

The kinetics of the autooxidation of trimethylgermane (Me_3GeH) vapours, in the range 493–533 K, leading to bis(trimethylgermyl) oxide ($Me_3Ge-O-GeMe_3$) was followed by FTIR[55b].

The nature of the surface of organogermanium films, obtained by magnetically enhanced rf-plasma deposition from tetraethylgermane, was examined by ESCA (electron spectroscopy for chemical analysis) and FTIR methods[56].

B. NMR Spectroscopy

See introductory remarks in Section I.A.4. Information on NMR spectra of organogermanium compounds appears in Table 2. Most sources usually include chemical shift values, δ, and coupling constants, J, however, in many cases other features are also discussed. A general discussion of ^{73}Ge NMR appeared in the literature[57].

C. X-ray Crystallography

See introductory remarks in Section I.A.4. Table 3 summarizes examples of crytallographic analysis published in the recent literature, with brief comments about structural features of organogermanium compounds.

D. Miscellaneous Methods

The molecular structure of gaseous cyclopropylgermane (**13**) was investigated, based on electron diffraction analysis and theoretical calculations. A 2–3.4° tilt of the GeH$_3$ group towards the ring was found, attributed to a hyperconjugative interaction between the group and the ring[94] (see also Section IV.A). The structure of gaseous tetrasilylgermane was determined by electron diffraction analysis, showing T_d symmetry, with almost freely rotating silyl groups[61] (see also Table 2). The symmetry of gaseous tetraphenylgermane is debatable on the basis of its electron diffraction spectrum[95]. The same method was used for investigating the molecular structure of gaseous trimethylgermyl formate (HCO$_2$GeMe$_3$). The Ge$-$O$-$C=O fragment is not planar and the H$-$C=O plane is nearly perpendicular to the Ge$-$O$-$C plane[96].

The conformational analysis of compounds RGeBr$_3$ (R = BrCH$_2$, Br$_2$CH, BrCH=CH, CH$_2$=CHCH$_2$, CH$_2$=CBrCH$_2$, CH$_2$=C(SiMe$_3$) and i-PrO$_2$CCH$_2$) was studied by ^{79}Br and ^{81}Br NQR (nuclear quadrupole resonance) spectroscopy[97].

Using Auger spectroscopy and XPS (X-ray photoelectron spectroscopy), it was concluded that CVD of germane on the Si(100)2×1 surface at 110 K proceeds by chemisorption, according to the mechanism depicted in reaction 3[98]. Digermane (Ge$_2$H$_6$), on the other hand, undergoes molecular adsorption without scission on Si(100)2×1 at 110 K or on Ge(111) at 120 K. At 150 K digermane breaks up and a surface covered with chemisorbed GeH$_3$ groups is generated[99,100]. Homoepitaxial growth of thin germanium films on the Ge(100) surface, by digermane chemisorption followed by H$_2$ elimination was observed by differential reflectance, before and after chemisorption. The reflectance can be measured spectroscopically for structure characterization, or at a fixed wavelength for rate determination. The IR frequency of the Ge$-$H bond does not change on chemisorption. Various mechanisms for H$_2$ elimination were discussed[101]. Heteroepitaxial growth of germanium with digermane was also investigated[102]. Epitaxial growth of thin germanium layers by chemisorption of digermane on deuterated Si(100)(2×1)-D surfaces, using UV radiation, was studied by means of TPD (temperature-programmed desorption) and Auger electron spectroscopy[103]. The CVD mechanism of germanium layers on the Si(100) surface using diethylgermane (Et$_2$GeH$_2$) was studied by a combination of Auger electron spectroscopy, TPD and HREELS (high resolution electron-energy loss spectroscopy). Chemisorbed Et$_2$GeH and H species were reported. The desorption peak temperature is about 693 K for ethylene and 780 K for hydrogen[104,105]. CVD of ultrafine germane layers on Si(100)7×7 surfaces by chemisorption of diethylgermane was investigated using FTIR and laser-induced thermal desorption. Upon adsorption of diethylgermane the surface shows Si$-$Et, Si$-$H, and Ge$-$H species, but probably no Ge$-$Et species. Si$-$Et moieties undergo H β-elimination, as shown in reaction 4, induced

TABLE 2.   Structural determination of organogermanium compounds by NMR spectroscopy

Compounds	Comments	Refs.
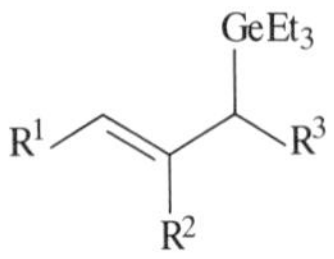 GeEt$_3$, R^1—CH=C(R^2)—CH(R^3)GeEt$_3$	^{1}H and ^{13}C NMR. Allyltrimethygermanes of E-configuration.	58
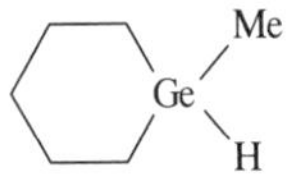 Cyclohexyl-Ge(Me)(H)	^{13}C and ^{73}Ge NMR. Spectroscopic data and molecular mechanics calculations indicate a preferred axial over equatorial configuration (3:2 ratio) for the Me group.	59
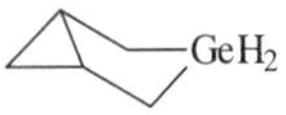 (cyclohexyl)GeH$_2$	^{13}C and ^{73}Ge NMR. Chair conformation as determined from spectroscopy, in accordance with molecular structure calculations, similarly to the analogous carbocyclic compound. Introduction of bulky substituents results in a preferred boat conformation.	60
$(SiH_3)_n GeH_{4-n}$, $n = 2-4$, $(SiH_3)_n GeH_{3-n} CH_3$, $n = 1-3$	^{1}H and ^{29}Si NMR. The ^{1}H NMR quadruplet bands of the silyl groups are further split into deciplets.	61
$R_3 SnCl$, R = 9-triptycyl	^{1}H and ^{13}C NMR. The three triptycyl groups have no rotation about the C-Sn axis due to mechanical interlocking.	62
$(CHF_2)_m (CF_3)_n SnGeMe_3$ $m + n = 3$	^{1}H, ^{13}C, ^{19}F and ^{119}Sn NMR.	63
Organogermanium compounds in reaction **A** (see end of table)	^{13}C and ^{73}Ge NMR.	64
16–18 (see end of table)	^{1}H, ^{13}C, ^{73}Ge and ^{119}Sn NMR. The molecular structure of these compounds, bearing the bulky Si(SiMe$_3$)$_3$ and Ge(SiMe$_3$)$_3$ groups, was analysed, showing the steric effects. NMR spectra point to a straight line tetragermane backbone for compound **16**.[a]	65
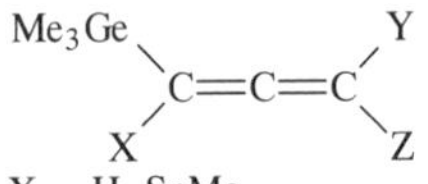 $Me_3 Ge$ X / C=C=C / Y Z, X = H, SnMe$_3$ Y = H, SiMe$_3$, GeMe$_3$, SnMe$_3$, SEt Z = SnMe$_3$.	^{13}C, ^{29}Si and ^{119}Sn NMR.	66
$Me_{4-n} Ge(CH=CH_2)_n$, $n = 1-4$	^{1}H, ^{13}C and ^{73}Ge NMR. Good correlation for chemical shifts with those of Si analogues. Some discrepancy with MNDO calculations.	67
$Me_{4-n} Ge(C\equiv CH)_n$, $n = 1-4$	^{1}H, ^{13}C and ^{73}Ge NMR. Increasing the number of ethynyl groups decreases the electron density at the Ge atom, displacing the signal upfield. The following correlation was found for the chemical shifts of analogous Ge and Sn compounds of this series: $\delta_{73Ge} = 0.481\delta_{119Sn} - 0.842$.	68

TABLE 2.   (*continued*)

Compounds	Comments	Refs.
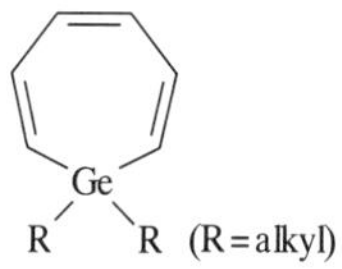 Ge R R (R=alkyl)	^{1}H and ^{13}C NMR aided by spectral stimulation of ^{1}H NMR.	69
Cp$^{\#}$GeCp$^{\#}$	^{13}C NMR, IR, Raman, MS and powder XRD. A Ge(II) compound. Parallel planes staggered conformation proposed for decaphenylgermanocene.[c]	70,71
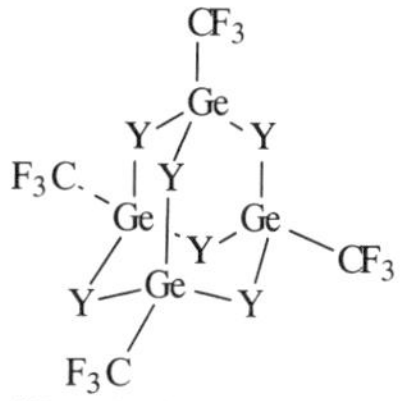 (Y = S, Se)	^{13}C and ^{19}F NMR and MS. Stable adamantane-like structure.	72
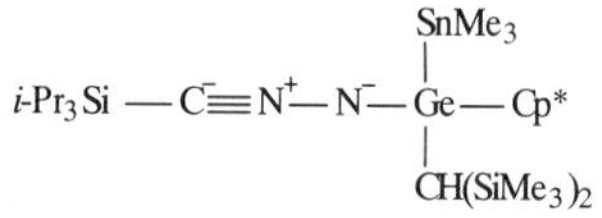	^{1}H, ^{13}C and ^{119}Sn NMR.[d]	73
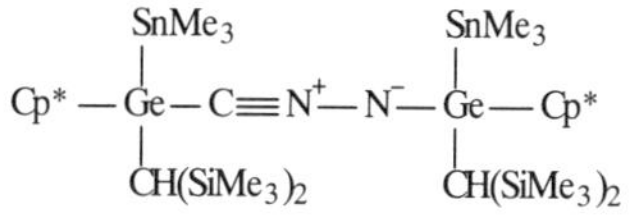	^{1}H, ^{13}C and ^{119}Sn NMR. Two diastereoisomers and two different SnMe$_3$ signals.[d]	73
(Me$_3$Si)$_3$MM(SiMe$_3$)$_3$ (M = Ge, Sn; **19**)	^{1}H and ^{13}C NMR. The analogue with M = Pb and Sn could not be synthesized.[a]	74
(**31**) (see reaction 14 in Section V)	^{1}H and ^{13}C NMR, FT-Raman, UVV and FABMS. A germanethione stabilized by bulky aryl substituents.[a]	53
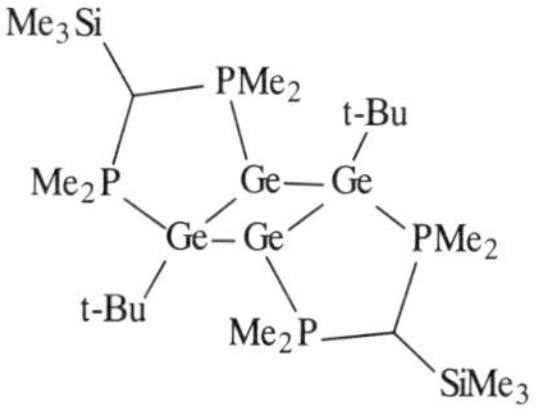	^{1}H, ^{13}C and ^{31}P NMR and XRD. This yllide is the first example synthesized of a homocyclic organonogermanium compound containing alternating 4- and 3-coordinated Ge atoms.	75

(*continued overleaf*)

TABLE 2. (*continued*)

Compounds	Comments	Refs.
$R_{3-n}H_nGeML_m$ $R_2Ge\overset{\displaystyle H}{\underset{\displaystyle |}{}}\overset{\displaystyle ML_m}{\underset{\displaystyle |}{}}GeR_2$ $R = Et, Ph, Ar^3; n = 1, 2;$ $ML_m = FeCp(CO)_2, WCp(CO)_3$	^{1}H and ^{13}C NMR, IR. Spectroscopic data correlate with a negative charge on the Ge atom coordinated to the M atom.[b]	76
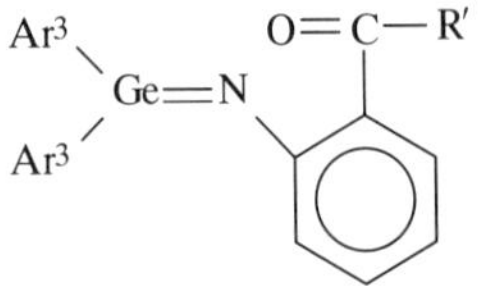 $\underset{Ar^3}{\overset{Ar^3}{>}}Ge=N$—(aryl)—$O=C$—$R'$ $R' = NMe_2, OMe$	^{1}H and ^{13}C NMR. These are rare examples of thermally stable monomeric germaimines. At room temperature the mesityl and the N-Me groups are not equivalent. At 55 °C the split signals coalesce. This is possibly due to interaction between Ge and NMe_2. Stability is adduced to intramolecular Ge−O coordination, electron withdrawal by the C=O group and bulkiness of the mesityl groups.[b]	77

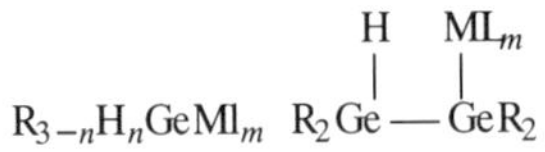

[a] See also crystallographic analysis in Table 3.
[b] See formula **11** for the aryl groups.
[c] Cp[#] = pentaphenylcyclopentadienyl anion.
[d] Cp* = pentamethylcyclopentadienyl anion.

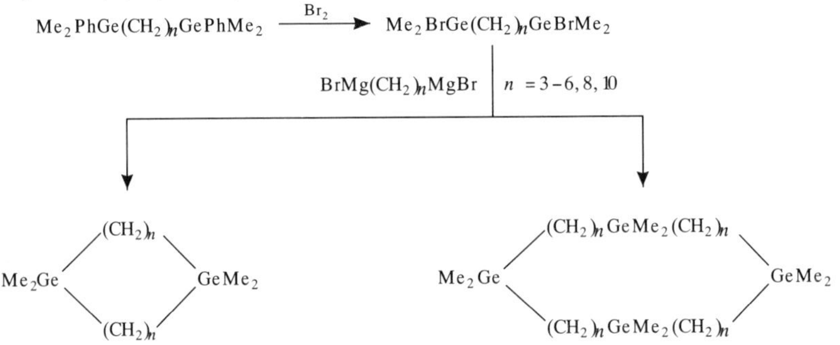

$$Me_2PhGe(CH_2)_nGePhMe_2 \xrightarrow{Br_2} Me_2BrGe(CH_2)_nGeBrMe_2$$

$$BrMg(CH_2)_nMgBr \qquad n = 3-6, 8, 10$$

(A)

$$Me_2Ge\overset{(CH_2)_n}{\underset{(CH_2)_n}{<}}GeMe_2$$

$$Me_2Ge\overset{(CH_2)_nGeMe_2(CH_2)_n}{\underset{(CH_2)_nGeMe_2(CH_2)_n}{<}}GeMe_2$$

$(Me_3Si)_3GeGeCl_2GeCl_2Ge(SiMe_3)_3$

(16)

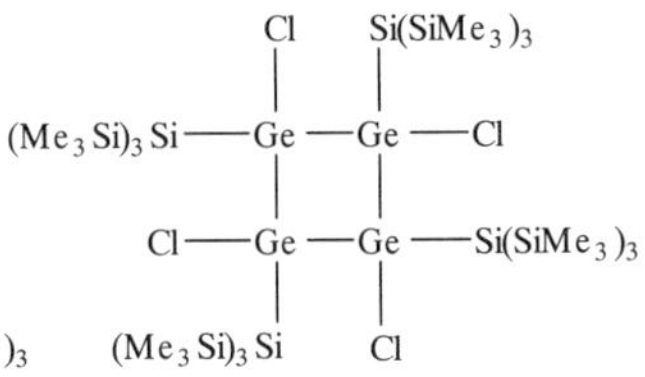

$$\begin{array}{ccc} & Cl & Si(SiMe_3)_3 \\ & | & | \\ (Me_3Si)_3Si\text{—}Ge\text{—}Ge\text{—}Cl \\ & & \\ Cl\text{—}Ge\text{—}Ge\text{—}Si(SiMe_3)_3 \\ & | & | \\ (Me_3Si)_3Si & Cl \end{array}$$

(17)

$Cl_2Sn[Ge(SiMe_3)_3]_2$

(18)

TABLE 3. Structural determination of organogermanium compounds by XRD crystallographic methods

Compounds	Comments	Refs.
$(Me_3Si)_3MM(SiMe_3)_3$ (M = Ge, Sn; **19**)	Only the crystalline analogue with M = Sn could be analysed. See also Table 2.	74
17, 18 (see end of Table 2)	The cyclotetragermane ring of compound **17** has an average torsion angle of 17.2°. Large distortions in the coordination angles of Sn are observed in compound **18**: Ge—Sn—Ge 142.1°, Cl—Sn—Cl 98.8°. See also Table 2.	65

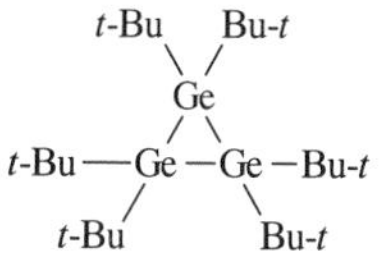

t-Bu   Bu-t
\ /
Ge
/ \
t-Bu—Ge —Ge —Bu-t
/ \
t-Bu   Bu-t

Longest Ge—Ge and Ge—C bonds measured to the date of publication. — 78

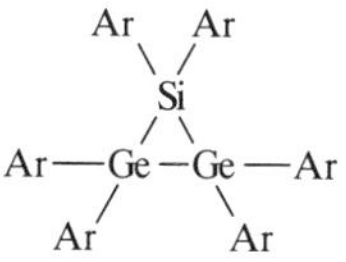

Ar   Ar
\ /
Si
/ \
Ar—Ge —Ge —Ar
/ \
Ar   Ar   (**20**)

Stable highly substituted siladigermiranes (**20**) and cyclotrigermanes (**21**). — 79

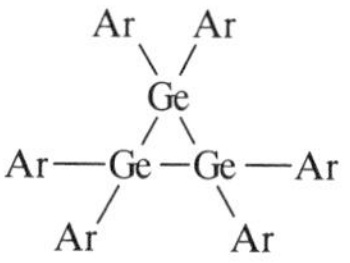

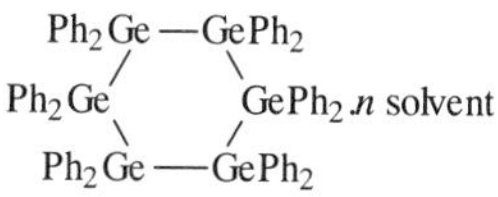

Ar   Ar
\ /
Ge
/ \
Ar—Ge —Ge —Ar
/ \
Ar   Ar   (**21**)

Ph_2Ge —$GePh_2$
/ \
Ph_2Ge $GePh_2$ .n solvent
\ /
Ph_2Ge —$GePh_2$

Chair conformation. Molecular packing in crystal changes with number of molecules and nature of solvate. — 80

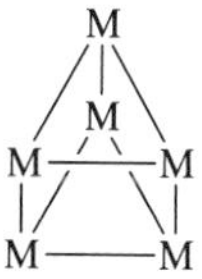

M
/ | \
/ M \
M—/—\—M
| / \ |
M———M
(M = GeCH(SiMe_3)_2)

D_{3h} molecular symmetry. All Ge—Ge bonds considerably longer than in other polygermanes. Also MS; ^1H, ^{13}C and ^{29}Si NMR; λ_{max} 280 nm, ε 32 200 (hexane); exhibits thermochromism. — 81

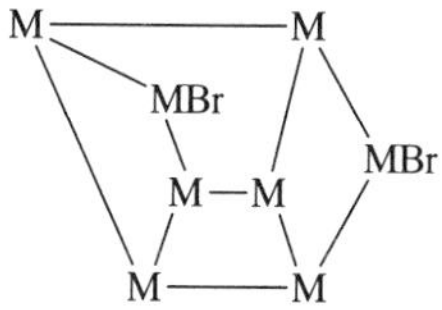

M———M
\ MBr \
\ \ MBr
\ M—M /
\ / \ /
M———M
(M = GeBu-t)

C_2 molecular symmetry. Stable in air, moisture. Good thermal stability. — 82

(continued overleaf)

TABLE 3.   (*continued*)

Compounds	Comments	Refs.
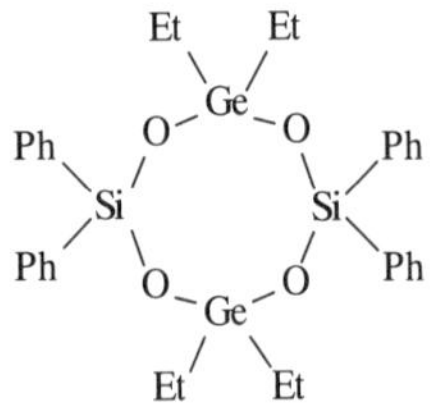	Puckered eight-membered ring. Ethyl groups in disordered arrangement.	83
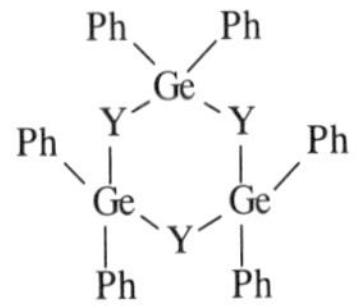 (Y = O, S)	The six-membered heterocyclic ring has twisted boat conformation for Y = S and is nearly planar for Y = O.	84
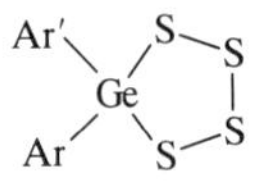 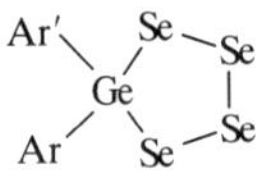	Tetrachalcogenagermolanes.	85, 86
Ph₃GeOH	Crystallizes with eight independent molecules in the asymmetric unit, arranged in two groups of four molecules. At the core of each group lie four oxygen atoms in a flattened tetrahedral arrangement.	87
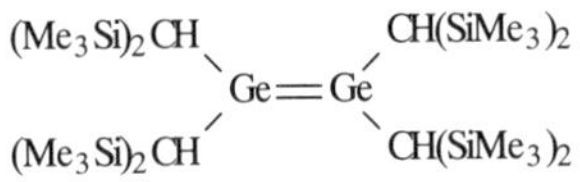	The two Ge atoms and the four C atoms attached to them form a centrosymmetric group, however, these atoms are not coplanar. The fold angle of the R–Ge–R plane with the Ge–Ge line is 32°. The analogous Sn compound was also investigated.	88
31 (see reaction 14 in Section V)	The Ge=S distance, 0.2049 nm, was the shortest one known to the date of publication. The planes of the aryl groups lie almost perpendicular to each other. See also Table 2.	53
[Li([12]crown-4)₂]⁺[Ge(SiMe₃)₃]⁻	The Si–Ge–Si angle (101.6°) is slightly lower than the Si–Si–Si angle of analogous Si compounds.	89
22 (see end of table)	One doubly bidentate Lewis base 2,2′-bipyrimidine is coordinated with one germanium atom, while in the case of the analogous lead compound coordination takes place with two lead atoms.	90
O₃(GeCH₂CH₂CO₂H)₂	The basic structure of carboxyethylgermanium sesquioxide, a low-toxicity γ-interferon inducing agent, consists of a 12-membered ring containing six Ge tetrahedra bridged by oxygen atoms.	91

TABLE 3. (*continued*)

Compounds	Comments	Refs.
34 (see reaction 21 in Section V)	Twisted digermacyclobutene ring because of steric interaction between *N*-*t*-Bu groups. In the analogous Sn compound this ring is planar.	92

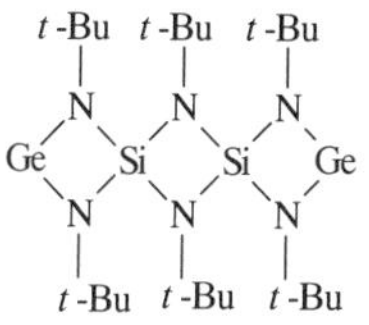
t-Bu *t*-Bu *t*-Bu
  | | |
  N   N   N
 Ge   Si   Si   Ge
  N   N   N
  | | |
t-Bu *t*-Bu *t*-Bu
 	
(23)
	A Ge(II) compound. This dispiro compound was prepared by the same method used for the pentacyclo compound **83** of Table 8 in Chapter 8 using analogous Sn reagents.	93

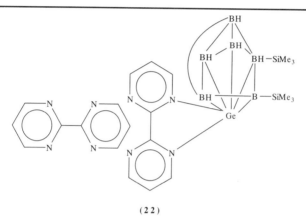

(22)

by the laser radiation[106]. Diethylgermane is claimed to undergo the chemisorption shown in reaction 5 on Si(111)7×7 surfaces. Ethylene becomes desorbed at 700 K and hydrogen at 800 K[107]. The chemisorption of $GeMe_4$ on Si(100) surface at 110 K was studied by Auger and UV-photoelectron spectroscopies. The TPD species characterized by MS were $GeMe_4$, $GeMe_3$, CH_3 and H_2[108].

$$GeH_4(g) \longrightarrow GeH_3(ad) + H(ad) \tag{3}$$

$$Si-CH_2-CH_3(ad) \longrightarrow Si-H(ad) + CH_2=CH_2(ad) \tag{4}$$

$$Et_2GeH_2(g) \longrightarrow Ge-CH_2-CH_3(ad) \longrightarrow Ge-H(ad) + CH_2=CH_2(ad) \tag{5}$$

Dimethylgermylene is a very reactive intermediate that may be detected by UVV spectrophotometry. It undergoes dimerization as shown in reaction 6 or can be scavenged by various reagents[109].

$$2Me_2Ge: \longrightarrow Me_2Ge=GeMe_2 \tag{6}$$

$$\lambda_{max}450 \text{ nm} \qquad \lambda_{max}350 \text{ nm}$$

Compounds **24** and **25**, containing highly methylated aryl groups, yield the corresponding arene radical cation on treatment with $AlCl_3$ in CH_2Cl_2 solution, as demonstrated by

ESR and ENDOR (electron-nucleus double resonance) spectroscopies[110].

$$Ar_3MMAr_3 \ (M = Ge, Sn)$$

(24)

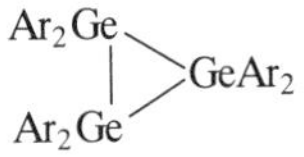

$$\begin{array}{c} Ar_2Ge \\ | \quad \rangle GeAr_2 \\ Ar_2Ge \end{array}$$

(25)

The photoelectronic properties of poly(dihexylgermane) were investigated by photoluminescence spectroscopy, after one- and two-photon absorption. The spectra were compared with those of the analogous poly(dihexylsilane)[111].

Diorganylgermylenes, the analogues of diorganylcarbenes, were characterized by their UVV spectra in hydrocarbon matrices at 77 K, λ_{max} 420–558 nm. In the presence of p-electron substrates they form adducts with a bathochromic shift[112]. The structure of dimethylgermylene trapped in a hydrocarbon matrix at 77 K was analysed by IR and UVV spectroscopy[113].

The first direct characterization of the structure of a sterically hindered diarylgermylene in solution was claimed for compound 26 (Ar = Ar^5 in formula 11), using EXAFS (extended X-ray absorption fine structure), UVV (λ_{max} 420 nm) and ^1H and ^{13}C NMR[114]. The UVV spectra of diarylgermylenes (26), their dimerization products (27) and the complexes formed with compounds containing p-electrons (28) were studied in solution[115,116].

$$Ar_2Ge(II) \qquad Ar_2Ge{=}GeAr_2 \qquad Ar_2Ge \cdots XR_n$$
(26)   (27)   (28, X = N, O, S, Cl, $n = 0$–2)

V. DERIVATIZATION

Trialkylgermyl chlorides undergo the metathesis shown in reaction 7 with silver dimesylamide[117].

$$R_3GeCl + AgN(SO_2Me)_2 \longrightarrow R_3GeN(SO_2Me)_2 + AgCl \qquad (7)$$

Bulky substituents confer stability on organogermanium compounds. Scheme 1 shows possible derivatization reactions for some substitued germanes and digermanes. Compounds d and e are obtained from reactions a_4 and b_3 as mixtures of diasteroisomers, however, reactions a_5, b_4 and b_5 are more stereospecific. Compounds b–i in the scheme were characterized by ^1H NMR spectra[118].

Trimethylgermyl substituents in aromatic compounds are easily removed with halogen, yielding the corresponding aryl halide, as shown in reaction 8[119]. Another example of phenyl group displacement was carried out with bromine as in reaction 2 (see end of Table 2)[64].

$$ArGeMe_3 + X_2 \xrightarrow[X_2 = Cl_2, Br_2, ICl]{} ArX \qquad (8)$$

Alkylaryldichlorogermanes produce substituted cyclogermanes. In the case shown in reaction 9, three of the four possible configurations of the tetragermane product 29 are formed, but none is all-cis. It is possible to carry out an electrophilic displacement of the phenyl groups to form the 1,2,3,4-tetrachloro derivative 30, which is all-trans, the configuration of 29 notwithstanding[120].

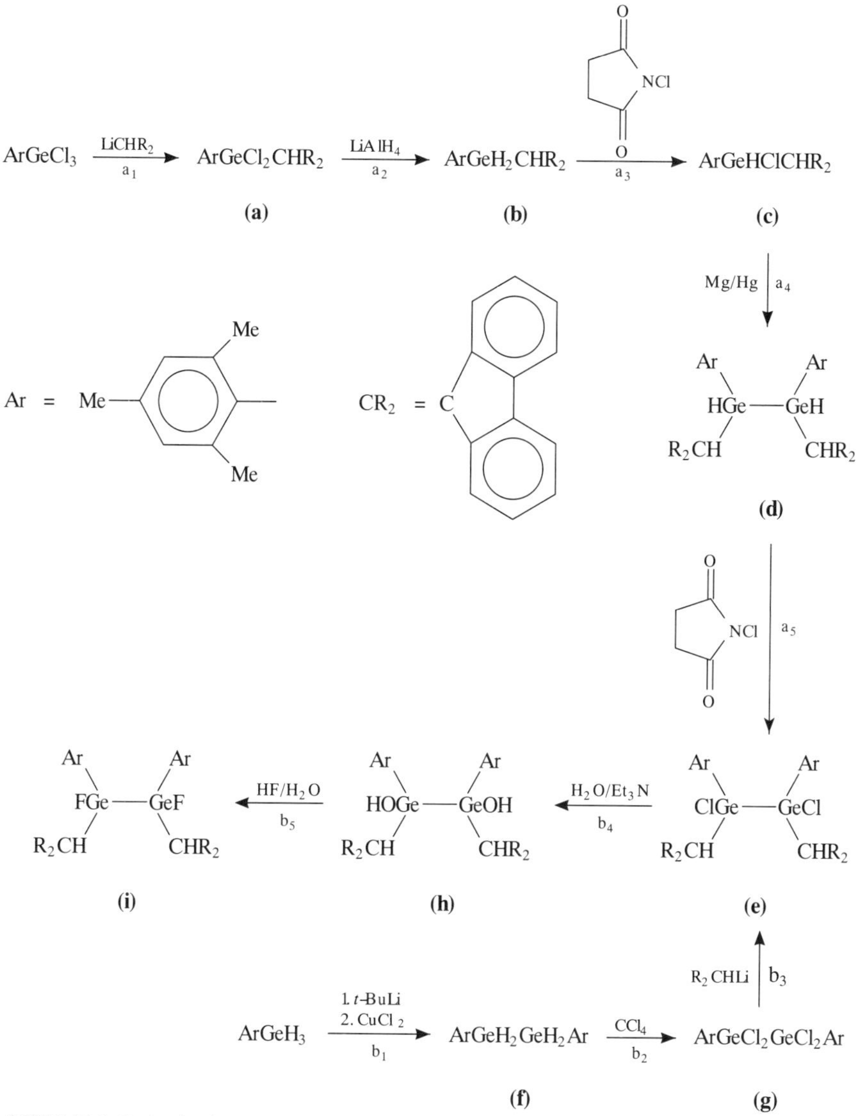

SCHEME 1. Derivatization reactions for germanes and digermanes stabilized by bulky substituent groups[118].

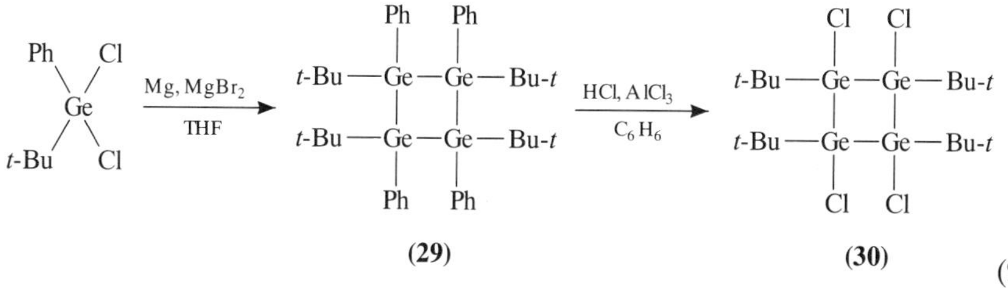

Certain heterylgermanes undergo transmetallation reactions with organolithium compounds. This is shown in reaction 10, as well as the effectiveness of the various organolithium compounds[121].

$$Het_2GeMe_2 \xrightarrow{\text{RLi}} HetGeRMe_2 + R_2GeMe_2$$

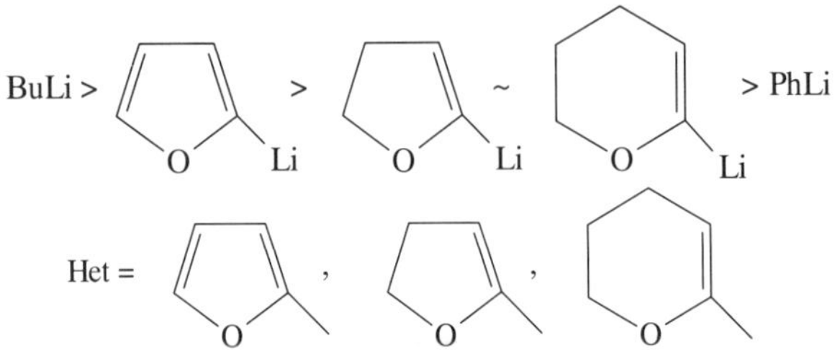

$$BuLi > \quad \bigg\langle \!\!\! \bigcirc \!\!\! \text{—Li} \quad > \quad \bigg\langle \!\!\! \bigcirc \!\!\! \text{—Li} \quad \sim \quad \bigg\langle \!\!\! \bigcirc \!\!\! \text{—Li} \quad > PhLi$$

$$Het = \quad , \quad , \quad \tag{10}$$

Alkoxygermanes and (alkylthio)germanes can be determined after acetylation with acetic anhydride according to reaction 11, hydrolysis of the germyl acetate and titration of the resulting acetic acid. The acetate and thioacetate esters produced in reaction 11 are stable under the conditions of the subsequent hydrolysis[122]. An alternative method is based on formation of germyl bromides in an anhydrous medium, as shown in reaction 12, and titrating the excess acid. Due to the different stability of the organometallic bromides, the determination can be carried out in the presence of alkoxysilanes[123].

$$R'_{4-n}Ge(XR)_n + Ac_2O \xrightarrow{X = O, S} R'_{4-n}Ge(OAc)_n + nAcXR \tag{11}$$

$$R'_{4-n}Ge(XR)_n + nHBr \xrightarrow[\substack{X=O,S \\ R,R'=Me,i\text{-}Pr,Bu,Ph}]{} R'_{4-n}GeBr_n + nRXH \tag{12}$$

Tetraalkynylgermanes react with boranes to yield spirobigermoles, as shown in reaction 13. Treatment with acetic acid displaces the boryl groups[52].

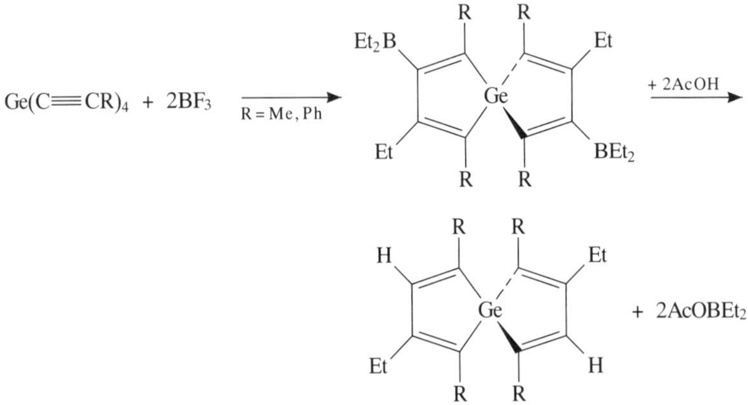

$$Ge(C{\equiv}CR)_4 + 2BF_3 \xrightarrow{R = Me, Ph}$$

$$+ 2AcOH \longrightarrow$$

$$+ 2AcOBEt_2 \tag{13}$$

Simple dialkylgermanethiones are stable only at very low temperatures. A stable diarylgermanethione (**31**) was synthesized with stongly sterically hindered aryl groups **11**, as shown in reaction 14. The diarylgermanethione **31** is a solid crystalline compound, stable at room temperature. However, it undergoes addition reactions across the Ge=S double bond with water and organic compounds, as shown in reactions 15–18[53].

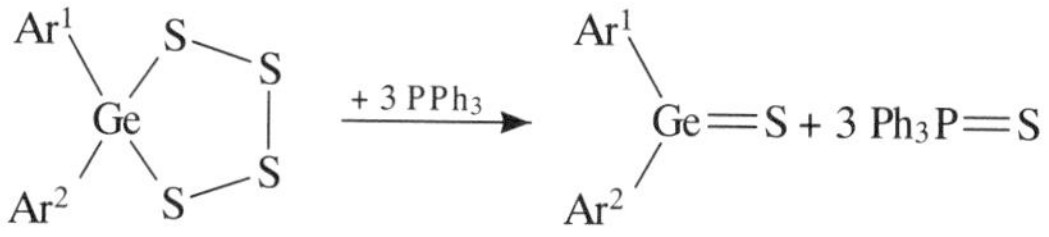

$$\begin{array}{c}\text{Ar}^1 \\ \diagdown \\ \text{Ge} \\ \diagup \\ \text{Ar}^2 \end{array}\!\!\begin{array}{c}\text{S}\!-\!\text{S} \\ \diagdown \\ \text{S}\!-\!\text{S} \end{array} \xrightarrow{+3\,PPh_3} \begin{array}{c}\text{Ar}^1 \\ \diagdown \\ \text{Ge}\!=\!\text{S} + 3\,Ph_3P\!=\!S \\ \diagup \\ \text{Ar}^2 \end{array}$$ (14)

(31)

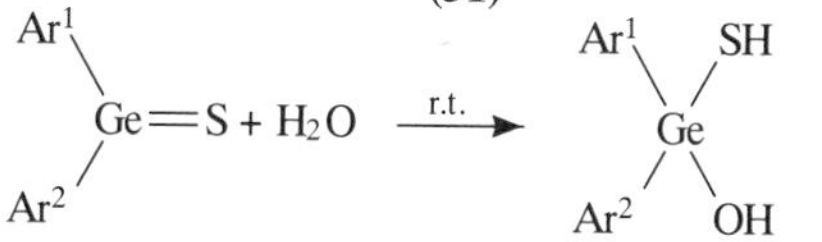

$$\begin{array}{c}\text{Ar}^1 \\ \diagdown \\ \text{Ge}\!=\!\text{S} + H_2O \\ \diagup \\ \text{Ar}^2 \end{array} \xrightarrow{r.t.} \begin{array}{c}\text{Ar}^1 \quad \text{SH} \\ \diagdown \diagup \\ \text{Ge} \\ \diagup \diagdown \\ \text{Ar}^2 \quad \text{OH} \end{array}$$ (15)

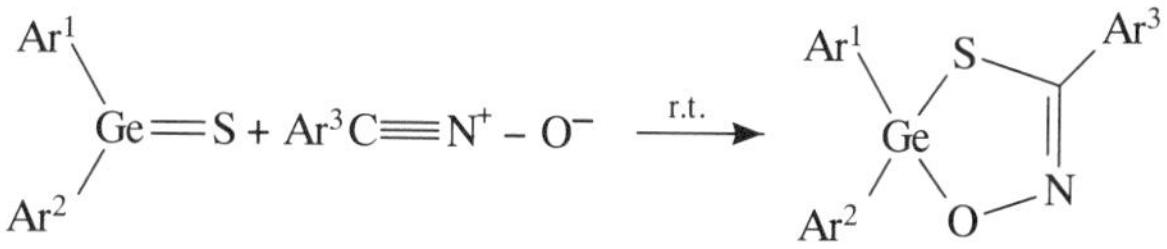

$$\begin{array}{c}\text{Ar}^1 \\ \diagdown \\ \text{Ge}\!=\!\text{S} + Ar^3C\!\equiv\!N^+ \!-\! O^- \\ \diagup \\ \text{Ar}^2 \end{array} \xrightarrow{r.t.} \begin{array}{c}\text{Ar}^1 \quad \text{S}\!\!\diagup\!\!\text{Ar}^3 \\ \diagdown \diagup \quad \diagdown \\ \text{Ge} \qquad \text{C} \\ \diagup \diagdown \quad \diagup \\ \text{Ar}^2 \quad \text{O}\!-\!N \end{array}$$ (16)

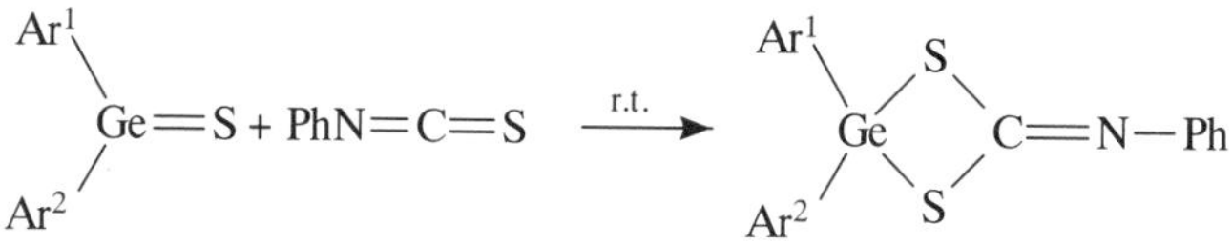

$$\begin{array}{c}\text{Ar}^1 \\ \diagdown \\ \text{Ge}\!=\!\text{S} + PhN\!=\!C\!=\!S \\ \diagup \\ \text{Ar}^2 \end{array} \xrightarrow{r.t.} \begin{array}{c}\text{Ar}^1 \quad \text{S} \\ \diagdown \diagup \diagdown \\ \text{Ge} \qquad \text{C}\!=\!N\!-\!Ph \\ \diagup \diagdown \diagup \\ \text{Ar}^2 \quad \text{S} \end{array}$$ (17)

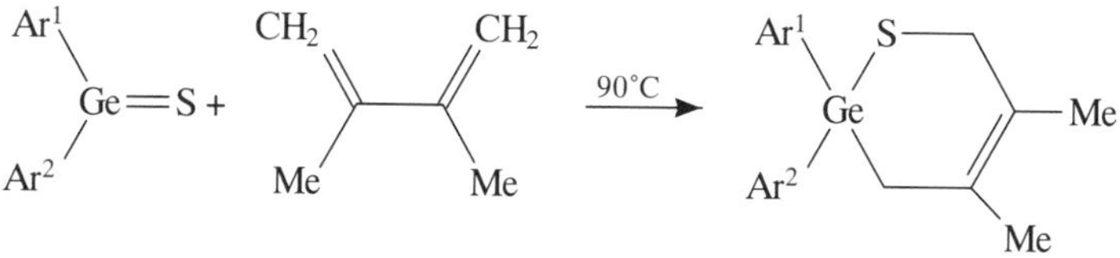

$$\begin{array}{c}\text{Ar}^1 \\ \diagdown \\ \text{Ge}\!=\!\text{S} + \\ \diagup \\ \text{Ar}^2 \end{array}\begin{array}{c}\text{CH}_2 \quad \text{CH}_2 \\ \diagdown\!\!\diagdown \quad \diagup\!\!\diagup \\ \text{Me} \quad \text{Me} \end{array} \xrightarrow{90^\circ C} \begin{array}{c}\text{Ar}^1 \quad \text{S} \\ \diagdown \diagup \diagdown \\ \text{Ge} \qquad \!\!-\!\text{Me} \\ \diagup \diagdown \diagup \\ \text{Ar}^2 \qquad \text{Me} \end{array}$$ (18)

Germaphosphenes undergo 1,2-addition with phenylacetylene, as shown in reaction 19[124].

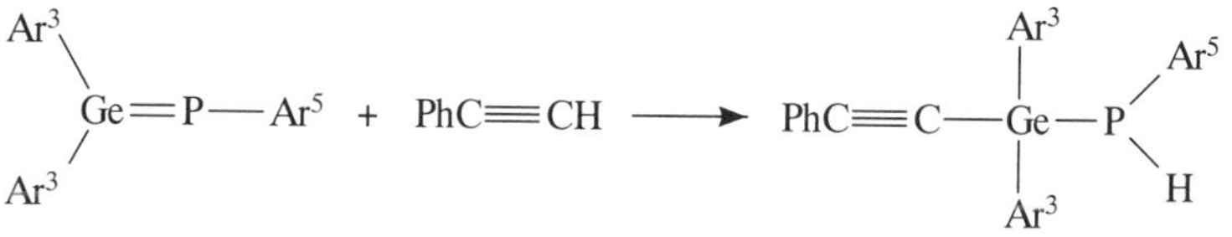

$$\begin{array}{c}\text{Ar}^3 \\ \diagdown \\ \text{Ge}\!=\!P\!-\!Ar^5 + PhC\!\equiv\!CH \\ \diagup \\ \text{Ar}^3 \end{array} \longrightarrow \begin{array}{c}\qquad \text{Ar}^3 \quad \text{Ar}^5 \\ \qquad | \quad \diagup \\ PhC\!\equiv\!C\!-\!\text{Ge}\!-\!P \\ \qquad | \quad \diagdown \\ \qquad \text{Ar}^3 \quad \text{H} \end{array}$$ (19)

Digermenes undergo [2 + 2] cycloaddition with phenylacetylene, as shown in reaction 20[125,126].

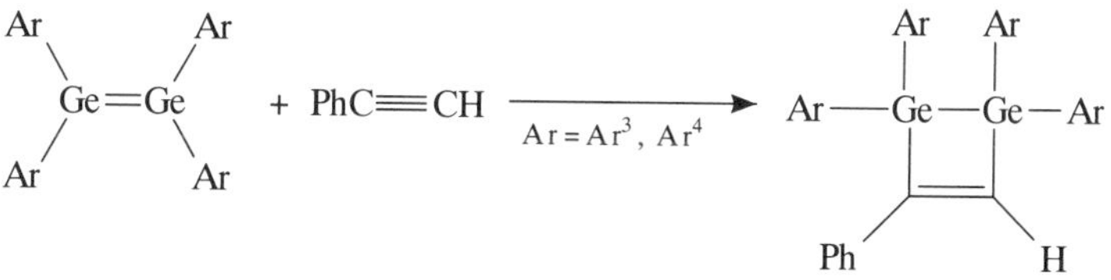

$$\begin{array}{c}\text{Ar} \qquad \text{Ar} \\ \diagdown \qquad \diagup \\ \text{Ge}\!=\!\text{Ge} + PhC\!\equiv\!CH \\ \diagup \qquad \diagdown \\ \text{Ar} \qquad \text{Ar} \end{array} \xrightarrow[Ar=Ar^3,\,Ar^4]{} \begin{array}{c}\text{Ar} \quad \text{Ar} \\ | \quad\ | \\ \text{Ar}\!-\!\text{Ge}\!-\!\text{Ge}\!-\!\text{Ar} \\ |\qquad\quad | \\ \quad\diagdown\!\!=\!\!\diagup \\ Ph \qquad H \end{array}$$ (20)

Organogermanium(II) compound **32** reacts with organic azides to yield stable dispiro compounds **33**[127]. This derivatizing reaction may be applicable to different types of organogermanium(II) compounds. Another effective derivatizing possibility is treatment with acetylenic compounds, as shown in reaction 21 (see Table 3)[92].

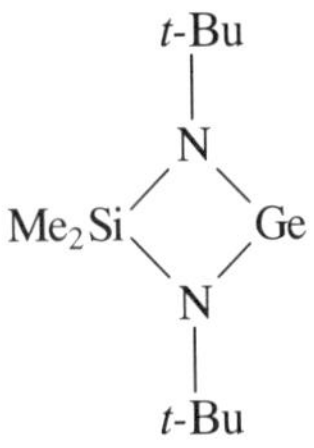

(32)

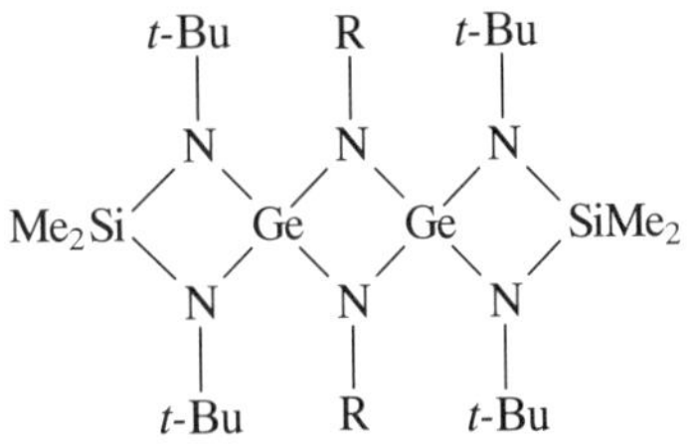

(33)   R = Ar, ArCO, ArSO$_2$

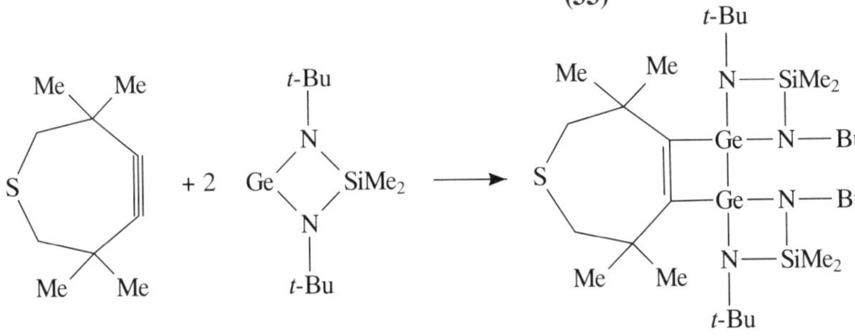

(34)

VI. REFERENCES

1. (a) Z. B. Alfassi (ed.), *Determination of Trace Elements*, VCH, Weinheim, Germany, 1994.
    (b) J. Zabicky, in *The Chemistry of Sulphenic Acids and their Derivatives* (Ed. S. Patai), Wiley-Interscience, Chichester, 1990, pp. 83ff.
2. B. L. Lewis, M. O. Andreae and P. N. Froelich, *Mar. Chem.*, **27**, 179 (1989); *Chem. Abstr.*, **112**:62200.
3. T. R. Crompton, *Chemical Analysis of Organometallic Compounds*. Volume 3. *Elements of Group IVB*, Academic Press, London, 1974, pp. 1–12.
4. M. Anke and M. Glei, in *Handbook on Metals in Clinical and Analytical Chemistry*, (Eds. H. G. Seiler, A. Sigel and H. Sigel), Dekker, New York, 1994, pp. 381–386.
5. N. I. Sax, *Dangerous Properties of Industrial Materials*, 6th ed., Van Nostrand Reinhold, New York, 1984.
6. T. Arány-Halmos and A. Schneer Erdey, *Magy. Kem. Lab.*, **20**, 164 (1965); *Chem. Abstr.*, **63**:6303a.
7. H. Pieters and W. J. Buis, *Microchem. J.*, **8**, 383 (1964).
8. M. P. Brown and G. W. A. Fowles, *Anal. Chem.*, **30**, 1689 (1958).
9. Y. Sohrin, *Anal. Chem.*, **63**, 811 (1991).
10. F. Nakata, H. Sunahara, H. Fujimoto, M. Yamamoto and T. Kumamaru, *J. Anal. At. Spectrom.*, **3**, 579 (1988).
11. H. Morita, S. Shimomura, K. Okagawa, S. Saito, C. Sakigawa and H. Sato, *Sci. Total Environ.*, **58**, 237 (1986).
12. A. Kolb, G. Mullervogt, W. Stossel and W. Wendl, *Spectrochim. Acta Part B, At. Spectr.*, **42**, 951 (1987).
13. Y. Narusawa and I. Matsubara, *Nippon Kagaku Kaishi*, 195 (1994).
14. I. D. Brindle, X. C. Lee and X. F. Li, *J. Anal. At. Spectrom.*, **4**, 227 (1989).

7. Analytical aspects of organogermanium compounds 361

15. I. D. Brindle and X. C. Le, *Anal. Chem.*, **61**, 1175 (1989).
16. I. D. Brindle and C. M. C. Ponzoni, *Analyst (London)*, **112**, 1547 (1987).
17. K. Kania, *Chem. Anal. (Warsaw)*, **36**, 911 (1991); *Chem. Abstr.*, **117**:219256.
18. I. Nakatsuka, K. Takahashi, K. Ohzeki and R. Ishida, *Analyst (London)*, **114**, 1473 (1989).
19. A. K. Charikov, M. N. Ptushkina and E. A. Klochkova, *Zh. Anal. Khim.*, **41**, 1596 (1986); *Chem. Abstr.*, **106**:43038.
20. K. Shimada, M. Nakajima, H. Wakabayashi and S. Yamato, *Chem. Pharm. Bull.*, **37**, 1095 (1989); *Chem. Abstr.*, **111**:247061.
21. A. Murata, N. Sugiyama and T. Suzuki, *Bunseki Kagaku*, **36**, 27 (1987); *Chem. Abstr.*, **106**:130845.
22. S. Sato and H. Tanaka, *Talanta*, **36**, 391 (1989).
23. N. Nashine and R. K. Mishra, *Anal. Chim. Acta*, **285**, 365 (1994).
24. M. Schülunz and A. Köster-Pflugmacher, *Z. Anal. Chem.*, **232**, 93 (1967).
25. A. A. Abrutis, *Zh. Anal. Chim.*, **47**, 347 (1992); *Chem. Abstr.*, **117**:183835c.
26. C. Schleich and G. Henze, *Fresenius Z. Anal. Chem.*, **338**, 140 (1990).
27. C. Schleich and G. Henze, *Fresenius Z. Anal. Chem.*, **338**, 145 (1990).
28. A. D. Snegova, L. K. Markov and V. A. Ponomarenko, *Zh. Anal. Khim.*, **19**, 610 (1964); *Chem. Abstr.*, **61**:6402f.
29. G. Garzo, J. Fekete and M. Blazso, *Acta Chim. Akad. Sci. Hung.*, **51**, 359 (1967); *Chem. Abstr.*, **67**:17613h.
30. J. A. Semlyen, G. R. Walker, R. E. Blofeld and C. S. G. Phillips, *J. Chem. Soc.*, 4948 (1964).
31. C. S. G. Phillips and P. L. Timms, *Anal. Chem.*, **35**, 505 (1963).
32. V. T. Demarin, V. A. Ktylov, E. A. Nokolaev, N. K. Rudnevskii and L. V. Sklemina, *J. Anal. Chem. USSR*, **42**, 239 (1987); *Chem. Abstr.*, **106**:226838.
33. K. Mochida, H. Chiba and M. Okano, *Chem. Lett.*, 109 (1991).
34. M. Gazicki, H. Schalko, P. Svasek, F. Olcaytug and F. Kohl, *J. Vac. Sci. Technol. A, Vac. Surf. Films*, **10**, 51 (1992).
35. Y. Sohrin, *Anal. Chim. Acta*, **247**, 1 (1991).
36. T. A. Hein, W. Thiel and T. J. Lee, *J. Phys. Chem.*, **97**, 4381 (1993).
37. W. Schneider and W. Thiel, *Chem. Phys.*, **159**, 49 (1992).
38. M. Hayashi, S. Kaminaka, M. Fujitake and S. Miyazaki, *J. Mol. Spectrosc.*, **135**, 289 (1989).
39. Q. S. Zhu, B. A. Thrush and A. G. Cobiette, *Chem. Phys. Lett.*, **150**, 181 (1989).
40. Q. S. Zhu and B. A. Thrush, *J. Chem. Phys.*, **92**, 2691 (1990).
41. Q. S. Zhu, H. B. Qian and B. A. Thrush, *Chem. Phys. Lett.*, **186**, 436 (1991).
42. V. V. Azatyan, R. G. Aivazyan, N. M. Pavlov and T. A. Sinelnikova, *Kinet. Catal.*, **34**, 518 (1993).
43. P. G. Harrison, J. McManus and D. M. Podesta, *J. Chem. Soc., Chem. Commun.*, 291 (1992).
44. W. Stahl and H. Dreizler, *Z. Naturforsch., A: Phys. Sci.*, **42**, 1402 (1987).
45. A. Haas and H. J. Kutsch, *Chem. Ber.*, **121**, 803 (1988).
46. S. Mohan, A. R. Prabakaran and F. Payami, *J. Raman Spectrosc.*, **20**, 119 (1989).
47. H. Oberhammer, T. Lobreyer and W. Sundermeyer, *J. Mol. Struct.*, **323**, 125 (1994).
48. K. J. Epple and H. D. Rudolph, *J. Mol. Spectr.*, **152**, 355 (1992).
49. M. B. Kelly, J. Laane and M. Dakkouri, *J. Mol. Spectrosc.*, **137**, 82 (1989).
50. J. R. Durig, T. S. Little, T. J. Geyer and M. Dakkouri, *J. Phys. Chem.*, **93**, 6296 (1989).
51. J. R. Durig, T. J. Geyer, P. Groner and M. Dakkouri, *Chem. Phys.*, **125**, 299 (1988).
52. R. Koster, G. Seidel, I. Klopp, C. Kruger, G. Kehr, J. Suss and B. Wrackmeyer, *Chem. Ber.*, **126**, 1385 (1993).
53. N. Tokitoh, T. Matsumoto, K. Manmaru and R. Okazaki, *J. Am. Chem. Soc.*, **115**, 8855 (1993).
54. H. K. Sharma, F. Cervantes-Lee and K. H. Pannell, *J. Organomet. Chem.*, **409**, 321 (1991).
55. (a) J. R. Durig and G. M. Attia, *Spectrochim. Acta, Part A*, **44A**, 517 (1988).
    (b) P. G. Harrison and D. M. Podesta, *Organometallics*, **13**, 1569 (1994).
56. M. Gazicki, J. Schalko, F. Olcaytug, M. Ebel, H. Ebel, J. Wernisch and H. Yasuda, *J. Vac. Sci. Technol. A, Vac. Surf. Films*, **12**, 345 (1994).
57. E. Liepins, I. Zicmane and E. Lukevics, *J. Organomet. Chem.*, **341**, 315 (1988).
58. J. Yamaguchi, Y. Tamada and T. Takeda, *Bull. Chem. Soc. Jpn.*, **66**, 607 (1993).
59. Y. Takeuchi, M. Shimoda, K. Tanaka, S. Tomoda, K. Ogawa and H. Suzuki, *J. Chem. Soc., Perkin Trans. 2*, 7 (1988).
60. Y. Takeuchi, I. Zicmane, G. Manuel and R. Boukherroub, *Bull. Chem. Soc. Jpn.*, **66**, 1732 (1993).

61. T. Lobreyer, H. Oberhammer and W. Sundermayer, *Angew. Chem., Int. Ed. Engl.*, **32**, 586 (1993).
62. J. M. Chance, J. H. Geiger and K. Mislow, *J. Am. Chem. Soc.*, **111**, 2326 (1989).
63. R. Eujen, N. Jahn and U. Thyrmann, *J. Organomet. Chem.*, **465**, 153 (1994).
64. S. Aoyagi, K. Tanaka, I. Zicmane and Y. Taekuchi, *J. Chem. Soc., Perkin Trans. 2*, 2217 (1992).
65. S. P. Mallela and R. A. Geanangel, *Inorg. Chem.*, **33**, 1115 (1994).
66. E. Liepins, I. Birgele, E. Lukevics, E. T. Bogoradovsky and V. S. Zavgorodny, *J. Organomet. Chem.*, **402**, 43 (1991).
67. Y. Takeuchi, H. Inagaki, K. Tanaka and S. Yoshimura, *Magn. Reson. Chem.*, **27**, 72 (1989).
68. E. Liepins, M. V. Petrova, E. T. Bogoradovsky and V. S. Zavgorodny, *J. Organomet. Chem.*, **410**, 287 (1991).
69. Y. Nakadaira, R. Sato and H. Sakurai, *J. Organomet. Chem.*, **441**, 411 (1992).
70. C. Janiak, H. Schumann, C. Stader, B. Wrackmeyer and J. J. Zuckermann, *Chem. Ber.*, **121**, 1745 (1988).
71. M. J. Heeg, R. H. Herber, C. Janiak, J. J. Zuckermann, H. Schumann and W. F. Manders, *J. Organomet. Chem.*, **346**, 321 (1988).
72. A. Haas, C. Kruger and H. J. Kutsch, *Chem. Ber.*, **120**, 1045 (1987).
73. C. Leue, R. Reau, B. Neumann, H. G. Stammler, P. Jutzi and G. Bertrand, *Organometallics*, **13**, 436 (1994).
74. S. P. Mallela and R. A. Geanangel, *Inorg. Chem.*, **32**, 5623 (1993).
75. H. H. Karsch, G. Baumgartner and S. Gamper, *J. Organomet. Chem.*, **462**, C3 (1993).
76. A. Castel, P. Riviere and J. Satge, *J. Organomet. Chem.*, **462**, 97 (1993).
77. M. Rivierebaudet, A. Khallaayoun and J. Satge, *J. Organomet. Chem.*, **462**, 89 (1993).
78. M. Weidenbruch, F. T. Grimm, M. Herrndorf, A. Schafer, K. Peters and H. G. Vornschnering, *J. Organomet. Chem.*, **341**, 335 (1988).
79. K. M. Baines, J. A. Cooke, N. C. Payne and J. J. Vittal, *Organometallics*, **11**, 1408 (1992).
80. M. Goto, S. Tokura and K. Mochida, *Nippon Kagaku Kaishi*, 202 (1994).
81. A. Sakiguchi, C. Kabuto and H. Sakurai, *Angew. Chem., Int. Ed. Engl.*, **28**, 55 (1989).
82. M. Weidenbruch, F. T. Grimm, S. Pohl and W. Saak, *Angew. Chem., Int. Ed. Engl.*, **28**, 198 (1989).
83. M. Akkurt, T. R. Kok, P. Faleschini, L. Randaccio, H. Puff and W. Schuh, *J. Organomet. Chem.*, **470**, 59 (1994).
84. K. Tanaka, H. Yuge, K. Ogawa, S. Aoyagi and Y. Takeuchi, *Heterocycles*, **37**, 1599 (1994).
85. N. Tokitoh, H. Suzuki, T. Matsumoto, Y. Matsuhashi, R. Okazaki and M. Goto, *J. Am. Chem. Soc.*, **113**, 7047 (1991).
86. N. Tokitoh, T. Matsumoto and R. Okazaki, *Tetrahedron Lett.*, **33**, 2531 (1992).
87. G. Ferguson, J. F. Gallagher, D. Murphy, T. R. Spalding, C. Glidewell and H. D. Holden, *Acta Crystallogr., Sect. C, Cryst. Struct. Commun.*, **48**, 1228 (1992).
88. D. E. Goldberg, P. B. Hitchcock, M. F. Lappert, K. M. Thomas, A. J. Thorne, T. Fjeldberg, A. Haaland and B. E. R. Schilling, *J. Chem. Soc., Dalton Trans.*, 2387 (1986).
89. A. Heine and D. Stalke, *Angew. Chem., Int. Ed. Engl.*, **33**, 113 (1994).
90. N. S. Hosmane, K. J. Lu, U. Siriwardane and M. S. Shet, *Organometallics*, **9**, 2798 (1990).
91. Z. X. Xie, S. Z. Hu, Z. H. Chen, S. A. Li and W. P. Shi, *Acta Crystallogr., Sect. C, Cryst. Struct. Commun.*, **49**, 1154 (1993).
92. A. Krebs, A. Jacobsenbauer, E. Haupt, M. Veith and V. Huch, *Angew. Chem., Int. Ed. Engl.*, **28**, 603 (1989).
93. M. Veith and R. Lisowsky, *Angew. Chem., Int. Ed. Engl.*, **27**, 1087 (1988).
94. M. Dakkouri, *J. Am. Chem. Soc.*, **113**, 7109 (1991).
95. E. Csavari, I. F. Shishkov, B. Rozsondai and I. Hargittai, *J. Mol. Struct.*, **239**, 291 (1990).
96. K. B. Borisenko, M. V. Popik, S. V. Ponomarev, L. V. Vilkov, A. V. Golubinskii and T. V. Timofeeva, *J. Mol. Struct.*, **321**, 245 (1994).
97. V. P. Feshin, P. A. Nikitin, E. V. Feshina, T. K. Gar and N. A. Viktorov, *Zh. Obshch. Khim.*, **58**, 111 (1988); *Chem. Abstr.*, **110**:24016.
98. D. A. Klug, W. Du and C. M. Greenlief, *J. Vac. Sci. Technol. A, Vac. Surf. Films*, **11**, 2067 (1993).
99. D. A. Klug, W. Du and C. M. Greenlief, *Chem. Phys. Lett.*, **197**, 352 (1992).
100. G. Q. Lu and J. E. Crowell, *J. Phys. Chem.*, **98**, 3415 (1993).
101. G. Eres and J. W. Sharp, *J. Vac. Sci. Technol. A, Vac. Surf. Films*, **11**, 2463 (1993).
102. J. W. Sharp and G. Eres, *J. Cryst. Growth*, **125**, 553 (1992).

103. C. Isobe, H. C. Cho and J. E. Crowell, *Surf. Sci.*, **295**, 117 (1993).
104. A. Mahajan, B. K. Kellerman, N. M. Russell, S. Banerjee, A. Campion, J. G. Ekerdt, A. Tash, J. M. White and D. J. Bonser, *J. Vac. Sci. Technol. A, Vac. Surf. Films*, **12**, 2265 (1994).
105. W. Du, L. A Keeling and C. M. Greenlief, *J. Vac. Sci. Technol. A, Vac. Surf. Films*, **12**, 2281 (1994).
106. P. A. Coon, M. L. Wise, Z. H. Walker and S. M. George, *Surf. Sci.*, **291**, 337 (1993).
107. P. A. Coon, M. L. Wise, Z. H. Walker, S. M. George and D. A. Roberts, *Appl. Phys. Lett.*, **60**, 2002 (1992).
108. C. M. Greenlief and D. A. Klug, *J. Phys. Chem.*, **96**, 5424 (1992).
109. K. Mochida, N. Kanno, R. Kato, M. Kotani, S. Yamauchi, M. Wasaka and H. Hayashi, *J. Organomet. Chem.*, **415**, 191 (1991).
110. M. Lehnig, T. Reiche and S. Reiss, *Tetrahedron Lett.*, **33**, 4149 (1992).
111. H. Tachibana. Y. Kawabata, A. Yamaguchi, Y. Moritomo, S. Koshihara and Y. Tokura, *Phys. Rev. B.*, **45**, 8752 (1992).
112. W. Ando, H. Itoh and T. Tsumuraya, *Organometallics*, **8**, 2759 (1989).
113. W. P. Neumann, *Chem. Rev.*, **91**, 311 (1991).
114. K. Mochida, A. Fujii, K. Tohji, N. Tsuchiya and Y. Udagawa, *Organometallics*, **6**, 1811 (1987).
115. W. Ando, H. Itoh, T. Tsumuraya and H. Yoshida, *Organometallics*, **7**, 1880 (1988).
116. S. Konieczny, S. J. Jacobs, J. K. B. Wilkins and P. P. Gaspar, *J. Organomet. Chem.*, **341**, C17 (1988).
117. A. Blaschette, T. Hamann, A. Michalides and P. G. Jones, *J. Organomet. Chem.*, **456**, 49 (1993).
118. M. A. Chaubonderedempt, J. Escudie and C. Couret, *J. Organomet. Chem.*, **467**, 37 (1994).
119. S. M. Moerlein, *J. Org. Chem.*, **52**, 664 (1987).
120. A. Sekiguchi, T. Yatabe, H. Naito, C. Kabuto and H. Sakurai, *Chem. Lett.*, 1697 (1992).
121. V. Gevorgyan, L. Borisova and E. Lukevics, *J. Organomet. Chem.*, **441**, 381 (1992).
122. J. S. Fritz and G. H. Schlenk, *Anal. Chem.*, **31**, 1801 (1959).
123. J. A. Magnuson and E. W. Knaub, *Anal. Chem.*, **37**, 1607 (1965).
124. J. Escudié, C. Couret, M. Andrianarison and J. Sagté, *J. Am. Chem. Soc.*, **109**, 386 (1987).
125. S. A. Batcheller and S. Masamune, *Tetrahedron Lett.*, **29**, 3383 (1988).
126. W. Ando and T. Tsumuraya, *J. Chem. Soc., Chem. Commun.*, 770 (1989).
127. B. Klein and W. P. Neumann, *J. Organomet. Chem.*, **465**, 119 (1994).

CHAPTER **8**

Analytical aspects of organotin compounds

JACOB ZABICKY and SARINA GRINBERG

Institutes for Applied Research, Ben-Gurion University of the Negev, Beer-Sheva 84110, Israel
Fax: (+972)-7-271612; e-mail: zabicky@bgumail.bgu.ac.il

The chemistry of organic germanium, tin and lead compounds
Edited by S. Patai © 1995 John Wiley & Sons Ltd

The abbreviations used in this chapter appear after the list of contents of the chapter on organogermanium compounds, Chapter 7, page 339.

I. INTRODUCTION

Among the organometallic compounds containing elements of group 14 treated in the present volume, those of tin have the widest technological applications[1]. The variety of these compounds can be appreciated from the list presented in Table 1. Tetraorganotins are frequently the starting material for the preparation of organotins with the Sn atom bonded to fewer organic radicals. Triorganotins found application as industrial and agricultural biocides, wood preservatives and marine antifoulants. Diorganotins are used as poly(vinyl chloride) stabilizers, catalysts in the manufacture of plastic materials and other applications. Monoorganotins are synergists in poly(vinyl chloride) stabilization. Functionalized organotin compounds are also intermediates in organic synthesis research; especially important are compounds with a trialkylstannyl group attached to a carbon–carbon double

TABLE 1.   Industrial organotin compounds[a]

Compound and CAS registry number	NIOSH/OSHA RTECS[b]
Allyltriphenylstannane [76-63-1]	WH 6705000
(2-Biphenyloxy)tri-n-butylstannane [3644-37-9]	WH 6711000
Bis(acetoxydibutylstannyl) oxide [5967-09-9]	JN 8740000
Bis(butanoyloxy)dibutylstannane [28660-63-1]	WH 7070000
Bis(butoxymaleyloxy)dibutylstannane [15546-16-4]	WH 6712000
Bis(butoxymaleyloxy)dioctylstannane [29572-02-8]	WH 6714000
Bis(butylthio)dimethylstannane [1000-40-4]	WH 6715200
Bis(p-chlorophenylthio)dimethylstannane [55216-04-1]	WH 6714300
Bis(decanoyloxy)di-n-butylstannane [3465-75-6]	WH 6715310
Bis(dibutyldithiocarbamato)dibenzylstannane [64653-03-8]	WH 6715330
Bis(dibutyldithiocarbamato)dimethylstannane [64653-05-0]	WH 6714370
Bis(1,3-dichloro-1,1,3,3-tetraethyldistannoxane)	JN 8735500
Bis(dodecanoyloxy)dibutylstannane [77-58-7]	WH 7000000
Bis(dodecanoyloxy)dioctylstannane [3648-18-8]	WH 7562000
Bis(2-ethylhexyloxycarbonylmethylthio)dibutylstannane [10584-98-2]	WH 7125000
Bis(2-ethylhexyloxy)dibutylstannane [2781-10-4]	WH 6714500
Bis(2-ethylhexyloxy)(maleyloxy)dibutylstannane	WH 6717000

TABLE 1. *(continued)*

Compound and CAS registry number	NIOSH/OSHA RTECS[b]
Bis(formyloxy)dibutylstannane [7392-96-3]	WH 7125000
Bis(hexanoyloxy)dibutylstannane [19704-60-0]	WH 6718000
Bis(isooctyloxycarbonylmethylthio)dibutylstannane [25168-24-5]	WH 6719000
Bis(isooctyloxycarbonylmethylthio)dimethylstannane [26636-01-1]	WH 6721000
Bis(isooctyloxycarbonylmethylthio)dioctylstannane [26401-97-8]	WH 6723000
Bis(isooctyloxymaleyloxy)dioctylstannane [3356-89-9]	WH 6727000
Bis(methoxymaleyloxy)dibutylstannane [15546-11-9]	WH 6729000
Bis(methoxymaleyloxy)dioctylstannane [60494-19-1]	WH 6730000
Bis(octadecanoyloxy)dibutylstannane [13323-62-1]	WH 6733300
Bis(octanoyloxy)dibutylstannane [13323-62-1]	WH 6731000
Bis(octanoyloxy)diethylstannane [2641-56-7]	
Bis(pentanoyloxy)dibutylstannane [3465-74-5]	WH 7133000
Bis(p-phenoxyphenyl)diphenylstannane [17601-12-6]	WH 6733500
Bis(phenylthio)dimethylstannane [4848-63-9]	WH 6733700
Bis(phenylthio)diphenylstannane [1103-05-5]	WH 6733800
Bis(propanoyloxy)dibutylstannane [3465-73-4]	WH 7135000
Bis(tetradecanoyloxy)dibutylstannane [28660-67-5]	WH 6733850
Bis(tributylstannyl)cyclopentadienyliron [12291-11-1]	LK 0725000
Bis(tributylstannyl) itaconate [25711-26-6]	WH 8585250
Bis(tributylstannyl) oxide [56-35-9]	JN 8750000
Bis(triethylstannyl)acetylene	
Bis(triethylstannyl) sulphate [57-52-3]	XQ 7175000
Bis(trifluoroacetoxy)dibutylstannane [52112-09-1]	WH 6734000
Bis(trimethylhexyl)dichlorostannane [64011-34-8]	WH 7230000
Bis(triphenylstannyl) sulphate [3021-41-8]	XQ 7380000
Bis(triphenylstannyl) sulphide [77-80-5]	JN 8850000
Bis(tris(β,β-dimethylphenethyl)stannyl) oxide [13356-08-6]	JN 8750000
Bromotributylstannane [1461-23-0]	WH 6735000
Bromotriethylstannane [2767-54-6]	WH6740000
Bromotripentylstannane [3091-18-7]	WH6750000
Bromotripropylstannane [2767-61-5]	WH6760000
Butylstannoic acid [2273-43-0]	WH 6770000
Butyltrichlorostannane [1118-46-3]	WH 6780000
Butyltris(isooctyloxycarbonylmethylthio)stannane [25852-70-4]	WQ 4150000
Chlorotribenzylstannane [3151-41-5]	WH 6800000
Chlorotributylstannane [1461-22-9]	WH 6820000
Chlorotributylstannane complex [56573-85-4]	WH 6820000
Chlorotriethylstannane [994-31-0]	WH 6850000
Chlorotriisobutylstannane [7342-38-3]	WH 6845000
Cyanatotributylstannane [4027-17-2]	WH 687100
Diacetoxydibutylstannane [1067-33-0]	WH 6880000
Diallyldibromostannane [17381-88-3]	WH 6881000
Dibromodibutylstannane [996-08-7]	WH 6882000
Dibromodiphenylstannane [4713-59-1]	WH 6883100
Dibutyldichlorostannane [683-18-1]	WH 7100000
Dibutyldifluorostannane [563-25-7]	WH 7130000
Dibutyldiiodostannane [2865-19-2]	WH 7128000
2,2-Dibutyl-1,3-dioxa-2-stanna-7,9-dithiacyclododecan-4,12-dione	JG 7880000
2,2-Dibutyl-1,3-dioxa-2-stanna-7-thiacyclodecan-4,10-dione	JH 4780000
N,N-Dibutyldithiocarbamic acid tribytylstannyl ester [67057-34-5]	WH 7138000
Dibutyltin maleate [15535-69-0]	WH7175000

(continued overleaf)

TABLE 1. (continued)

Compound and CAS registry number	NIOSH/OSHA RTECS[b]
Dibutylstannathione [4253-22-9]	WH 7195000
Dibutylstannone [818-08-6]	WH 7175000
Dibutyltin dinonylmaleate [59239-37-8]	WH 7540000
Dichlorobis(2-ethylhexyl)stannane [25430-97-1]	
Dichlorodiethylstannane [866-55-7]	WH 7200000
Dichlorodihexylstannane [2767-41-1]	WH 7220000
Dichlorodimethylstannane [753-73-1]	WH 7245000
Dichlorodioctylstannane [3542-36-7]	WH 7247000
Dichlorodipentylstannane [1118-42-9]	WH 7250000
Dichlorodiphenylstannane [1135-99-5]	WH 7253000
Dichlorodiphenylstannane dipyridine complex [25868-47-7]	WH 7254000
Dichlorodipropylstannane [867-36-7]	WH 7255000
(2,4-Dichlorophenoxy)tributylstannane [39637-16-6]	WH 7260000
Diethyldiiodostannane [2767-55-7]	WH 7270000
Diethylphenylstannyl acetate [64036-46-0]	WH 5700000
Difluorodimethylstannane [3582-17-0]	WH 7285000
Diisobutylstannone [67947-30-6]	WH 7310000
Diisopentyloxystannane [63979-62-4]	WH 7350000
Diisopropyloxystannane [23668-76-0]	WH 7525000
O,O-Diisopropyl S-tricyclohexylstannyl phosphorodithionate [49538-98-9]	WH 7400000
Dimethylstannone [2273-45-2]	WH 7526500
2,2-Dioctyl-1,3-dioxa-2-stanna-7-thiadecan-4,10-dione	
2,2-Dioctyl-1,3-dioxa-2-stannepin-4,7-dione [16091-18-2]	JH 4745000
Dioctylstannathione [3572-47-2]	WH 7690000
Dioctylstannone [870-08-6]	WH 7620000
Dioctyltin mercaptide	XQ 2975000
Dioctyltin β-mercaptopropionate [3033-29-2]	RP 4400000
Dipentylstannone [2273-46-3]	WH7700000
Diphenylstannane [1011-95-6]	WH 7875000
Diphenylstannone polymer [31671-16-6]	WH 8100000
Dipropylstannone [7664-98-4]	WH8225000
Fluorotributylstannane [1983-10-4]	WH 8275000
Formyloxytribenzylstannane [17977-68-3]	WH 8277000
Glycolyloxytributylstannane	
Hexabutyldistannoxane [56-35-9][c]	
Hexahydro-2,4,6-trioxo-1,3,5-tris(tributylstannyl)-s-triazine [752-58-9]	WH 8880000
Hexakis(2-methyl-2-phenylpropyl)distannoxane [13356-08-6][c]	
Hexamethyldistannane [661-69-8]	WH 8280000
Hydroxytributylstannane [1067-97-6]	WH 8310000
Hydroxytricyclohexylstannane [13121-70-5][c]	WH 8750000
Hydroxytrimethylstannane [56-24-6]	WH 8400000
Hydroxytriphenylstannane [76-87-9]	WH 8575000
Iodotributylstannane [7342-47-4]	WH 8580000
Iodotrimethylstannane [811-73-4]	WH 8581000
Iodotriphenylstannane [894-09-7]	WH 8582000
Iodotripropylstannane [7342-45-2]	WH 8583000
(3-Methacrylyloxypropyl)tributylstannane [2155-70-6]	WH 8585200
Methyltrichlorostannane [993-16-8]	WH 8585500
Methyltris(2-ethylhexyloxycarbonylmethylthio)stannane [57583-34-3]	WH 8586000
Octyltrichlorostannane [3091-25-6]	WH 8590000
Octyltris(2-ethylhexyloxycarbonylmethylthio)stannane [27107-89-7]	WH 8595000
Phenoxytriethylstannane [1529-30-2]	WH 8597000
Salicyloxytributylstannane [4342-30-7]	WH 8600000
Tetrabutylstannane [1461-25-2]	WH 8605000

TABLE 1. (*continued*)

Compound and CAS registry number	NIOSH/OSHA RTECS[b]
Tetraethylstannane [597-64-8]	WH 8625000
Tetraethynylstannane	
Tetraisopropylstannane [2429-42-0]	WH 8628000
Tetrakis(4-phenoxyphenyl)stannane [17068-17-6]	WH 8629000
(2-(2,2,3,3-Tetramethylbutylthio)acetoxy)tributylstannane [73927-97-6]	WH 8635000
Tetramethylstannane [594-27-4]	WH 8630000
Tributylstannane [688-73-3]	WH 8675000
Tributylstannanecarbonitrile [2179-92-2]	WH 6792000
Tributylstannyl 4-acetamidobenzoate [2857-03-6]	WH 5670000
Tributylstannyl acetate [56-36-0]	WH 5775000
Tributylstannyl chloroacetate [5847-52-9]	WH 6795000
Tributylstannyl 4-chlorobutyrate [33550-22-0]	WH 6797000
Tributylstannyl 4,4-dimethyloctanoate [28801-69-9]	WH 8588000
Tributylstannyl 2-ethylhexanoate [5035-67-6]	WH 8255000
Tributylstannyl iodoacetate [73927-91-0]	WH 8576000
Tributylstannyl 2-iodobenzoate [73927-93-2]	WH 8578000
Tributylstannyl 4-iodobenzoate [73940-88-2]	WH 8578200
Tributylstannyl 3-iodopropionate [73927-95-4]	WH 8579200
Tributylstannyl isocyanate [681-99-2]	WH 8538500
Tributylstannyl laurate [3090-36-6]	WH 8584000
Tributylstannyl linoleate [24124-25-2]	WH 8585000
Tributylstannyl methacrylate [2155-70-6]	WH 8692000
Tributylstannyl methanesulphonate [13302-06-2]	WH 8585300
Tributylstannyl nonanoate [4027-14-9]	WH 8589000
Tributylstannyl oleate [3090-35-5]	WH 8700000
Tributylstannyl 2-(2,4,5-trichlorophenoxy)propionate [73940-89-3]	WH 8709000
Tributylstannyl undecanoate [69226-47-7]	WH 8710000
Tributyl(2,4,5-trichlorophenoxy)stannane [73927-98-7]	WH 8707000
Triethylstannyl trifluoroacetate [429-30-1]	WH 8810000
Triisopropylstannyl acetate [19464-55-2]	WH 6300000
Trimethylstannyl acetate [1118-14-5]	WH 6475000
Trimethylstannyl isothiocyanate [15597-43-0]	WH 8583600
Trimethylstannyl sulphate [63869-87-4]	XQ 7225000
Trimethylstannyl thiocyanate [4638-25-9]	WH 8637000
Triphenylstannyl acetate [668-34-8][c]	
Triphenylstannyl benzoate [910-06-5]	WH 6710500
Triphenylstannyl hydroperoxide	
Triphenylstannyl levulinate [23292-85-5]	WH 8596200
Triphenylstannyl methanesulphonate [13302-08-4]	WH 8585400
Triphenylstannyl propiolate [57410-20-2]	WH 5702000
Triphenylstannyl thiocyanate [7224-23-9]	WH 8900000
1-(Triphenylstannyl)-1*H*-1,2,4-triazole [974-29-8]	WH 8638000
Tripropylstannyl acetate [3267-78-5]	WH 6700000
Tripropylstannyl iodoacetate [73927-92-1]	WH 8577000
Tripropylstannyl isothiocyanate [31709-32-7]	WH 8583700
Tripropylstannyl trichloroacetate	WH 8715000
Tris(dibutylbis(2-hydroxyethylthio)stannane) *O*-ester with bis(boric acid) [34333-07-8]	XK 4860000

[a] See also Reference 3.
[b] *Registry of Toxic Effects of Chemical Substances* of National Institute for Occupational Safety and Health/Occupational Safety and Health Administration.
[c] See also Reference 4.

bond. All this commercial and research activity has brough about development of analytical methods for structural characterization and quality control.

Elsewhere in *The Chemistry of Functional Groups* series appears a brief discussion on the stages in the lifetime of chemicals[2]. Organotin compounds are usually very toxic and they constitute a potential source of harmful pollution with both acute and long-term effects. Increasing concern with environmental and occupational issues has also contributed to the development of analytical methods. Table 1 lists organotin compounds that have found industrial application with references to occupational protection protocols where analytical methods for the particular compound can be found.

The environmental impact of tin is appreciable, as it is one of the three most enriched metals — only lead and tellurium precede — in the atmospheric particular matter, as compared with the abundance of the element in the earth crust (2.2 ppm). Tin releases to the environment can be methylated by aquatic organisms, yielding organometallic species of toxicity comparable to that of methylated mercury[5].

Analysis of organotin compounds has been reviewed[6,7].

II. ELEMENTAL ANALYSIS

A. General

See introductory comments in Sections I.A.1–2 of Chapter 7.

B. Determination in Organic Samples

The methods described under this heading are adequate for pure or concentrated organotin compounds. They are of historical interest and should be compared with modern ones as for detection limits, simplicity of procedure and outlays. The main difficulty involved in the determination of tin by combustion or wet digestion methods is the tendency of this element to yield mixtures of Sn(II) and Sn(IV) compounds. This requires further work-up to oxidize or reduce the element to a uniform valence. Thus, for example, oxygen flask combustion yields a mixture of stannous and stannic oxides, that can be dissolved with the aid of chromous sulphate and oxidized by air to the stannic and chromic state; the stannic ions can be reduced to stannous ions with sodium hypophosphite and titrated with standard potassium iodate[8]. Volatile organotin compounds can be vaporized with an oxygen stream and passed though a weighed silica combustion tube. The residue obtained in the tube is further ignited to constant weight, until totally converted to stannic oxide[9]. Wet digestion of organotin samples in sulphonitric mixture is to be followed by strong heating with a burner for 2–3 h and ashing[10]. After wet digestion and reduction to stannous ions, these can be determined by complexometric methods using EDTA[11,12].

C. Trace Analysis of Elemental Tin

1. General

In the present section attention is paid to total tin analysis without regard to speciation, which is dealt with in Section III.

2. Atomic absorption and emission spectroscopies

A continuous hydride generator was proposed, based on $NaBH_4$ reduction of tin compounds, in the presence of a small amount of L-cysteine, nitrogen stripping and measurement of SnH_4 by AAS (atomic absorption spectrometry)[13]. Treatment with Br_2, followed by hydride generation and ETAAS (electrothermal AAS) in a quartz tube was proposed for non-volatile organotin compounds; LOD (limits of detection) 1.5 µg Sn/L, RSD 3% for 4 µg Sn[14]. Sediments or biological tissues were digested in a mixture of

HNO_3, HF and $HClO_4$, evaporated and redissolved; 20 μL aliquots were introduced into a hydride generator. Stannanes were collected in tubes coated with pyrolytic graphite at 800 °C. End analysis was by GFAAS (graphite furnace AAS), heating the furnace up to 2700 °C, with Zeeman effect background correction; LOD 360 ng/g sediment, RSD 5%[15].

After digesting marine biological material and converting to chlorides, total Sn can be determined by AAS using ammonium dihydrogen phosphate as matrix modifier and Zeeman background correction. In a separate run, tributyltin can be extracted with n-hexane, converted to chlorides and determined in the same way. Absolute LOD 30 ng for both analyses, 7.8% RSD for samples containing 200 ng Sn[16]. Tin in blood can be determined by GFAAS using Ni, phosphoric acid and ascorbic acid as matrix modifiers; LOD 2.5 μg/L for 10 μL samples (25 pg Sn)[17]. A simplified method for biological samples consists of wet ashing and GFAAS after adding ascorbic acid; LOD 0.02 μg Sn/g or 2 μg/L, with linear calibration up to 1 mg/L[18].

An optimization study of GFAAS parameters for Sn determination in environmental and geological samples involved ashing temperature and noble metals as matrix modifiers. GFAAS and ICP-MS (inductively coupled plasma-MS) gave comparable results[19]. Treatment of the graphite furnace with La achieved an 11-fold signal enhancement for GFAAS determination of tin[20].

Trace amounts of Sn could be determined by DCP-AES (direct current plasma atomic emission spectrometry) after hydride generation. Interference by transition metal elements was eliminated on addition of L-cystine, however, interference by Au(III), Pd(II) and Pt(IV) remained significant; LOD 20 ng Sn/L[21]. The presence of easily ionized elements for signal enhancement is an important feature of DCP-AES. Alkaline and alkaline earth elements improve the performance of this method with hydride generation for group 14 elements. The RSD showed by 2 μg Sn/L was 4.7% in water and 4.1% in 1 M KCl. A slight increase of background is observed in the case of Sn[22].

Up to 30% hydrogen can be added to the carrier gas stream of a helium microwave plasma torch, in the determination of arsenic, bismuth and tin. The argon torch accepts up to 20% hydrogen; LOD about 2.5 μg Sn/L, with linear dynamic range over 3 orders of magnitude[23].

Determination of the various elements collected in the cryogenic trap can be done by fractional volatilization followed by AAS[24].

Inorganic stannous or stannic ions were extracted from urine after addition of hydrochloric acid, using n-hexane-benzene solvent containing 0.05% of tropolone (**1**). Treatment with the corresponding Grignard reagent yielded tetrapentylstannane, that was determined by GC-FPD (GC-flame photometric detector); absolute LOD 3 pg Sn with ca 80% recovery from rat urine[25].

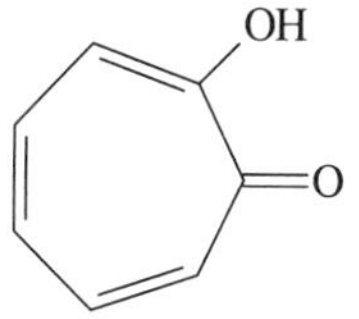

(1)

The spectrum of tin emitted from a triggered spark source in the far UV region (17.5–200 nm) has been analysed[26]. The emission lines in this region may be useful for development of new analytical methods.

3. Spectrophotometric methods

The complex of Sn(IV) ions and pyrocatechol violet (**2**) in a flow system is concentrated on Sephadex QAE A-25 gel and subsequently determined by visible spectrophotometry at 576 nm. The linear range of the method is 2–40 μg/L with LOD 0.3 μg/L[27a].

An alternative to AAS for the end analysis of stannane generated by hydrogenation, could be collection in permanganate solution and spectrophotometric determination with phenylfluorone (**3**). This was applied to submicrogram Sn/L concentrations in fresh and marine waters[27b]. Determination by hydride generation-AAS was found to be about 20 times more sensitive than by spectrophotometry of the phenylfluorone (**3**) complex[28].

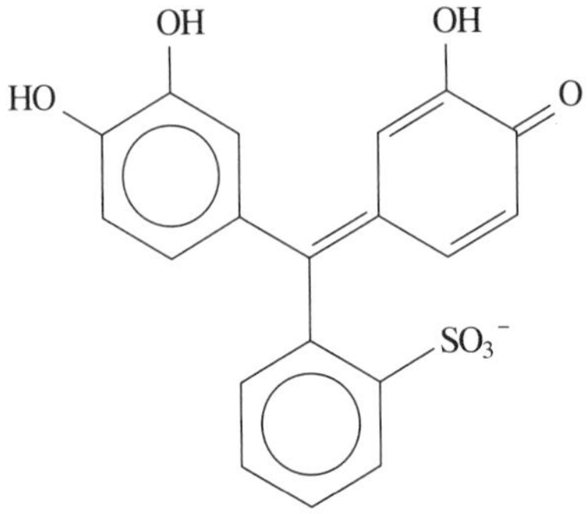

(2) 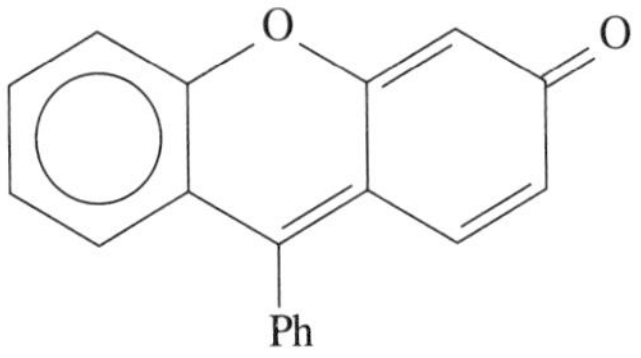   (3)

4. Electrochemical methods

The conditions for reliable cyclic voltametry determination of trace Sn concentrations in sea water were investigated. All organotin compounds should be converted to Sn(II) by UV-photolysis; adsorption on mercury drop in the presence of 40 μM of tropolone (**1**); cyclic voltametry stripping shows two cathodic peaks, corresponding to the two-step process Sn(IV) $\rightarrow$ Sn(II) $\rightarrow$ Sn(0)[29]. A complex of Sn ions with catechol can be accumulated in a glassy carbon mercury film electrode, followed by stripping voltametry measurement in the cathodic direction, at pH 4.2–4.7. Interference occurs when Cu, Cd and Cr are present; LOD 0.5 μg/L for 300 s accumulation[30].

5. Nuclear activation

The neutron activation reaction ^{116}Sn (n, γ) ^{117m}Sn is recommended for determination of tin in biological samples, because of the convenient half life-time of 13.6 days of the radioactive product. Some interferences that have to be considered and removed are decay of ^{47}Ca and ^{199}Au. Interference by ^{122}Te (n, γ) ^{123m}Te is possible but it is considered to be marginal because of the low tellurium concentrations in biological samples; LOD 16.5 ng/g and 4.9 ng/g with Compton supression, for 20,000 s counting time in a sample of ca 0.05 g[31]. For activation analysis of tin in blood-serum a process was investigated consisting of dry ashing, digestion with sulphuric acid and ammonium iodide, extraction of SnI$_4$ with toluene and measurement. The preferred decay process is ^{113}Sn $(t_{1/2} = 115.09$ d$) \rightarrow ^{113m}$In $(t_{1/2} = 1.658$ h; γ391.7 keV); LOD 0.5 μg Sn/L. Except for very low tin concentrations, measurement of ^{119m}Sn $(t_{1/2} = 13.61$ d; γ158.5 keV) is also convenient[32]. Comparison between AAS and neutron activation analysis, in both INAA (instrumental neutron activation analysis) and RNAA (radiochemical neutron activation analysis) modes, gave similar results for the tin content in human tissues[33].

6. Miscellaneous methods

Determination of Sn by wave-length dispersive XRF requires special corrections for sample transparency and background; LOD 3.5–4 ppm[34].

An ultasensitive simultaneous multi-element method of determination for As, Se, Sb and Sn in aqueous solution, consists of hydride generation, collection in a cryogenic trap and end analysis by GC-PID (photoionization detector); LOD ca 1 ng Sn/L for a 28 mL sample. No drying or CO_2 scrubbing is necessary before the cold trap[35].

Instead of AES the molecular emission of SnO can be stimulated in an oxycavity placed in a H_2/N_2 flame, and measured at 408 nm. The recommended sample preparation consists of hydride generation and concentration by cold-trap collection; LOD 80 μg Sn(II)/L in a 1 mL sample[36].

The analytical response of inorganic and organic tin compounds of formula $R_n SnX_{n-4}$ was studied for direct hydride generation and measurement with a non-dispersive AFD (atomic fluorescense detector). Tributyltin and phenyltin compounds gave unsatisfactory results. This was corrected by warming the sample with a dilute Br_2–HNO_3 solution[37].

III. TRACE ANALYSIS ALLOWING SPECIATION

A. General

See the introductory comments in Sections I.A.1 and I.A.2 of Chapter 7. This section is complementary to Section II.C above, dealing with trace analysis of tin, however, here attention is paid to the various organotin compounds present in the sample and not only to the overall tin content. It should be pointed out that innumerable examples appear in the literature, showing variations on procedural details required for a particular problem. The present account, although selective to a certain point, does not pretend to be critical on the subject.

Glass and polycarbonate containers were found to be suitable for seawater samples containing dibutyltin and tributyltin derivatives. Polyethylene containers were unsuitable[38]. Polycarbonate containers performed better than glass and Perspex for storing environmental samples containing tributyltin compounds[39]. Storing environmental waters for 2–3 months at $-20\,^\circ$C in polycarbonate containers showed no significant loss of analyte[40].

The biodegradation and photolysis paths of butyltin compounds, and especially tributyltin, in natural waters was investigated[41].

Butyltin and phenyltin compounds in samples of environmental, industrial and biological origin can be determined by the following general scheme[42–48]:

(a) Optional conversion of the organometallic species to organometallic chlorides with hydrochloric acid.

(b) Extraction of the chlorides or dithiocarbamate complexes with an organic solvent (e.g. ether containing 0.1% of tropolone (1), pentane, hexane, benzene or dichloromethane).

(c) Derivatizing with a Grignard reagent or sodium borohydride to generate volatile compounds — alkylated or hydrogenated derivatives, possibly preceded by a cleanup step. In the case of stannyl chlorides, some methods do not proceed with further derivatization. Derivatization of organometallic species with sodium tetraethylborate (NaBEt$_4$) has been reviewed[49].

(d) End analysis that can possibly include a cleanup or pre-concentration step, and a separation step of the various organometallic species, and measurement with a variety of detectors. Not all speciation analyses require an actual separation of the analytes.

B. Chromatographic and Other Phase Separation Methods

The most frequently applied end analysis methods are GC-FPD, GC-AAS and GC-MS. The sensitivity of GC-FPD to Sn was studied for the sulphur mode at 393 nm and the tin mode at 610 nm. The latter was found to be better for butyltin and phenyltin compounds[50]. End analysis of the hexyl Grignard derivatives gave similar results for GC-FPD and GC-MS[51]. Dibutylstannane and tributylstannane, obtained by $NaBH_4$ hydrogenation of their corresponding precursors, could be determined in paints and at 10 ng Sn/L levels in seawater by GC-FPD[52-55]. A method for Sn, Pb and Hg organometallics was evaluated by the US Environmental Protection Agency; it consists of measuring the pentylated Grignard derivatives by capillary GC-AED (GC-atomic emission detector); LOD 1–2.5 μg/L for 0.5 μL on-column injection, with good linearity in the 2.5–2500 μg/L range[56]. Samples of marine food products were hydrolysed enzymatically, extracted, derivatized and determined by GC-AAS; LOD: 0.8 and 0.7 ng Sn/g for tetramethylstannane and tetraethylstannane, respectively[57]. Tributyltin and triphenyltin compounds in industrial waste water were converted to the chlorides, extracted with hexane, derivatized by propylation and measured by GC-FPD or GC-MS, with good recovery; LOD 10 ng Bu_3SnCl/L, RSD 2.2–6.8%; LOD 30 ng Ph_3SnCl/L, RSD 1.8–7.3%[58]. Mono- and dioctyltin species in water, cell suspensions, urine and feces were extracted with ether, butylated and determined by GC-MS; LOD 5 μg monooctyl tin/L and 50 μg dioctyltin/L, with mean coefficient of variation of ca 15%[59]. After solvent extraction of the organometallic chorides and conversion to hydrides, these can be cleaned on a SEP-PAK silica cartridge, eluted with hexane, concentrated by stripping with nitrogen and determined by GC-FPD, using a DB-I wide-bore column; LOD 20–25 ng/L of organitin chloride in a 400 mL sample[60]. In the case of sediments, inorganic sulphur compounds are also extracted by the organic solvent and have to be removed. LOD for FPD at SNR 10: 3 ng/L for water and 0.5 μg/kg for sediments and biological samples[44] and lower[47].

When using DB-1 capillary columns, it is necessary to pretreat the column with an AcOEt solution of HBr, to obtain sharp peaks of organotin halides[61]. Conditioning with HBr was also applied to OV-225/Uniport HP columns, for the finishing steps of the analysis of tributyltin and triphenyltin traces in fish and shellfish[62]. GC-ITD (GC-ion trap detector) allows easy detection of organometallics yielding ions with rich isotopic patterns, as is the case of organotin compounds; LOD 5 ng/L for SNR 3, in natural waters containing compounds of formula $Bu_{4-n}SnHex_n$, $n = 1–3$[63]. A similar method was developed for butyltin species in sewage and sludges[64], and butyltin and phenyltin species in aquatic environments[44]. A method for butyltin and phenyltin products in biological and sediment samples, in the ng/g range consists of converting the organotin residues to chlorides, extracting with EtOAc, generating the hydrides and performing the end analysis by GC-ECD (GC-electron capture detector), using a silica gel column; LOD 10–20 ng/g for $BuSnCl_3$, 0.5–1 ng/g for Bu_2SnCl_2, 1–2 ng/g for Bu_3SnCl and Ph_2SnCl_2, and 2.5–5 ng/g for $PhSnCl_3$ and Ph_3SnCl[65]. Trace concentrations of chlorotributyltin in fresh or marine water are extracted into a solid phase which can be shipped to the analytical laboratory or preserved until analysis without deterioration. Elution is carried out with AcOEt, followed by splitless injection and measurement by GC-ECD. Strict exclusion of water from the GC column is required to avoid partial decomposition of the analyte. Recoveries at the 0.1 μg/L level are 94–101%[66]. LC using LEI (laser-enhanced ionization) as selective detector for Sn allows determination of organotin species in sediments; LOD 3 µg/L[67]. Extraction of sediments with acidic MeOH was followed by hydride formation and determination by GC-AAS with quartz tube[68].

The handling of environmental waters and sediments for organotin speciation was discussed[69]. The concentration of organotin chlorides in animal tissues and urine can be determined by solvent extraction, pentylation and GC analysis[70,71]. On-line hydride generation was attempted, by modifying the top of a GC column with NaBH$_4$[72]. Di- and tributyltin in animal tissue were determined by GC-FPD of an extract. The chromatograph had a hydride formation reactor at the injection port, a wide-bore capillary column and a 600 nm filter; absolute LOD 6 pg Sn[73]. Five extraction methods of tributyltin and triphenyltin from fish tissue before GC determination were examined. The best results were obtained for alkali hydrolysis followed by extraction with a 1:1 hexane–AcOEt mixture[74].

A method for tributyltin in sediments consists of extraction with anhydrous acetic acid, hydride generation, cold trapping and end analysis by GC-AAS using a quartz furnace[75]. Reduction with NaBH$_4$ followed by solvent extraction, concentration and GC-FPD was proposed for simulaneous determination of di- and tributyltin residues in sea water; LOD 10 ng/L for 1 L sample, with 87.1–98.4% of Sn recovery[76].

A sensitive determination of organotin compounds in sediments is based on separation of the chlorides $R_n SnCl_{4-n}$, $n = 1-3$, R = Me, n-Bu, Ph, by GC-FPD or GC-ECD using a DB-608 open tubular column with HCl doping of the carrier gas; LOD 30 ng Sn/g of sediment[77]. A modification of this method uses GC-FPD with DB-1 capillary column and a 611.5 nm filter. The column requires special pretreatment with an HBr/EtOAc solution[78].

A method for sample preparation allows determination of total tin and tributyltin ions in biological materials. End analysis by ETAAS, using a tungstate-treated graphite tube, allows LOD for tributyltin Sn of 0.4 ng/g[79]. An alternative method for sea water uses *in situ* concentration of Sn hydrides on a zirconium-coated graphite tube, followed by ETAAS; absolute LOD 20 and 14 pg for tributyltin ion and total Sn, respectively, with corresponding RSD of 5.6 and 3.4%[80].

C$_{18}$ liquid–solid extraction discs were used to separate di- and tributyltin compounds dissolved in aqueous matrices. After derivatizing *in situ* with a Grignard reagent the discs were subjected to a combination of static and dynamic SCE (supercritical extraction) procedures with carbon dioxide, with efficiency ranging from 92 to 102%. LOD 6–9 ng/L for tributyltin in 1 L of seawater, using capillary GC coupled to a single tin-selective FPD[81]. Reversed-phase flash liquid chromatography on C$_{18}$ columns with CH$_2$Cl$_2$–Mecn mixtures as eluting solvent, was proposed as a preparative method for separation of organotin compounds for its effectiveness and the inertness of the column. The method can be adapted for analysis of mixtures with ΔR_f as low as 0.05[82].

Dialkyltin dichlorides are emitted from hot PVC plastics. Airborne particles of these compounds can be determined by collecting in glass fibre filters, dissolving with hexane, converting to the hydride and separating by GC on a capillary column with FID (flame ionization detector) measurement; LOD *ca* 20 µg Sn/L of solution or *ca* 0.3 µg Sn/m^3 of air, for 120 L samples, about 50 times better than with ECD[83]. Dialkyltin compounds in PVC were extracted with THF. Their complexes with 8-hydroxyquinoline (**4**) were determined by reverse-phase HPLC-FPD at 380 nm[84]. A solution of sodium dodecyl sulphate, a micellar medium, served as mobile phase in the reverse-phase HPLC separation of organotin compounds, using ICP-MS for end analysis[85].

The detection limits of FPD can be improved by an approximate factor of 30 by using quartz surface-induced luminescence, at *ca* 390 nm, instead of the usual tin hydride

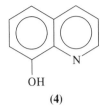

OH

(4)

gas-phase luminescence; absolute LOD *ca* 0.3 pg Sn for SnPr$_4$ and 2–3 pg for pentylated derivatives of butyltin ions[86].

Tributyltin can be determined in tissues and sediments by a method consisting of extraction with CH$_2$Cl$_2$, washing the extract with NaOH, re-extracting the washings, hydride generation and end analysis by GFAAS with Zeeman background correction; LOD 0.0025 μg tributyltin/g tissue with good Sn recovery[87a].

Tributyltin oxide and its metabolites were determined in urine after conversion to chlorides with HCl, extraction with ether containing tropolone (**1**), conversion to hydrides with NaBH$_4$, extraction with hexane and GC-AAS end-analysis; LOD 1 μg/L for BuSnH$_3$ and Bu$_2$SnH$_2$, and 2 μg/L for Bu$_3$SnH[87b].

A method for simultaneous speciation of trace levels of inorganic and organometallic tin in animal tissues consists of homogenizing the tissue, introducing a sample into a hydride generator, collecting the hydrides in a cryogenic trap packed with GC material, evaporating each hydride into a He stream by slowly rising the temperature of the trap, and performing the end analysis by AAS; LOD 11–25 ng Sn/g, for 0.1 g of tissue, depending on the tin species[88–90]. A similar method was applied to environmental water samples[91]. Dibutyltin and tributyltin species in sea water are determined by hydride generation, cryogenic trapping and end analysis by ETAAS; LOD 2 ng Sn/L for both species[92]. Application of flameless AAS to hydride derivatives was proposed, by passing the liberated hydrides into an evacuated, heated quartz tube[93].

Ion chromatography using a weak cation-exchange column with direct conductivity detection was applied in the determination of aqueous tributyltin ions, with short elution times; LOD 0.01 ppm, without preconcentration or derivatization. No organotin remains adsorbed on the column[94].

The SCE-GC combination began to be explored for trace analysis of organotin compounds and other pollutants. Its main advantage is that SCE accelerates the extraction-preconcentration operations involved in the analytical process[95].

Separation of organotin compounds by TLC and end analysis by ETAAS was also reported[96]. A sensitive method for determining the traces of organotin fungicides (for example Ph$_3$SnR, R = CH$_2$CO$_2$H, OH, Cl), is based on TLC of an extract using a suspension of *Curvularia lunata* spores as developing reagent[97].

C. Electroanalytical Methods

The electrochemical behaviour of trace concentrations of triphenylstannyl acetate, using a mercury film-glassy carbon electrode, was investigated by various measuring methods: cyclic voltametry, differential pulse voltametry and controlled-potential electrolysis. Determination by DPASV (differential pulse ASV) of water and fish samples has LOD 2.5 nM[98].

A mechanism was proposed for electrochemical reduction by ASV (anodic stripping voltametry) of aqueous tributylstannyl cations using mercury electrodes, in the presence of nitrate and chloride ions. It consists of three main processes, taking place at different potentials, as shown in reactions $1-3^{99}$.

$$Bu_3Sn^+(bulk) \longrightarrow Bu_3Sn^+(adsorbed) \overset{+ e^-}{\rightleftharpoons} Bu_3Sn^\circ(adsorbed) \quad (ca \;-0.8 \text{ V}) \quad (1)$$

$$Bu_3Sn^\circ(adsorbed) \rightleftharpoons Bu_3Sn^\circ(desorbed) \quad (ca \;-1.3 \text{ V}) \tag{2}$$

$$Bu_3Sn^+ (adsorbed) \overset{conc. \; Cl^-}{\rightleftharpoons} Bu_3Sn^+ (desorbed) \quad (ca \;-0.06 \text{ V}) \tag{3}$$

A combination of HPLC and amperometric detection was proposed for determination of tributylstannyl oxide in antifouling paint. The detector is of the static hanging Hg drop type in a flow cell, the solvent is CH_2Cl_2/THF, containing tetrabutylammonium perchlorate as supporting electrolyte. The oxidation mechanism depicted in reactions 4 and 5 was proposed[100].

$$2Bu_3SnOSnBu_3 + 2Hg \rightleftharpoons [Bu_3SnHgSnBu_3]^{2+} + Hg(OSnBu_3)_2 + 2e^- \quad (4)$$

$$Hg(OSnBu_3)_2 + 2Hg \rightleftharpoons [Bu_3SnHgSnBu_3]^{2+} + 2HgO + 2e^- \quad (5)$$

Accurate speciation of tributylstannyl ions in the presence of other degradation products was carried out by a.c. polarography, directly on the organic extract, without derivatization[101]. The degradation of tributyltins in aqueous solution was studied by differential pulse polarography[102].

D. Miscellaneous Methods

Determination of trace levels of tributyltin residues in sediments can be accomplished by solvent extraction, dilution and FIA (flow injection analysis) into the ionspray of a tandem mass spectrometer, using the $179/291 \; m/z$ pair; LOD 0.2 μg Sn/g[103].

The biopolymeric phenolic material derived from spent bleach liquour was caracterized by derivatizing with bis(tributyltin) oxide followed by ^{119}Sn NMR[104].

IV. STRUCTURAL ANALYSIS

A. General

Stuctural analysis of ordinary organic compounds is treated in the series *The Chemistry of Functional Groups* according to the particular functional group of each volume. In the case of organotin compounds all functional groups have to be dealt with together, making it practically impossible to give anything more than a superficial account of the most important problems. More detailed discussions of stuctural characterization by spectroscopic methods appear in other chapters of this book. The short exposition presented here gives leading references to application of spectroscopic methods to various functional groups involving tin heteroatoms and various problems arising in their structural characterization. In many of the examples presented the same multiply substituted aryl groups recur, and are represented by the same symbols throughout the rest of the chapter, as shown in formula 5.

B. Rotational and Vibrational Spectra

See introductory remarks in Section I.A.4 of Chapter 7. The IR vibration frequencies of compounds MH_4 and $MeMH_3$ were calculated *ab initio* for the metallic elements of group 14, including M = Sn, and compared with experimental data from various sources[105]. Many parameters pertaining to rotational and vibrational spectra of compounds

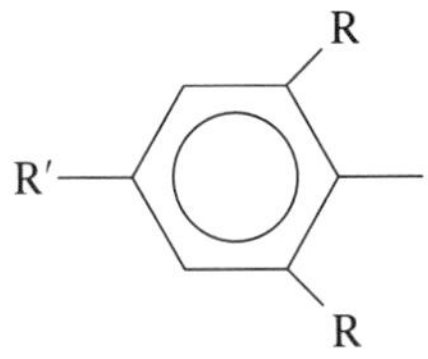

R

$$Ar^1 : R = R' = i\text{-}Pr$$
$$Ar^2 : R = R' = t\text{-}Bu$$
$$Ar^3 : R = R' = (Me_3Si)_2CH$$
$$Ar^4 : R = R' = Me$$
$$Ar^5 : R = R' = CF_3$$
$$Ar^6 : R = Et, R' = H$$

R'

R

(5)

MH_3X ($M = C,...,Sn$; $X = F,...,I$) were calculated *ab initio*, showing good agreement with experimental values. Spectral features of as yet unknown compounds such as SnH_3F were predicted[106].

Whether SnH_4 is a symmetric or spherical top was discussed based on the rotational spectrum of the sixth stretching vibrational overtone, obtained by photoacoustic spectroscopy[107]. The IR band at ν_{Sn-H} 1844 cm^{-1} served to investigate the kinetics and mechanism of chemical vapour deposition of a thin tin layer, by thermolysis of trimethylstannane at 378–503 K[108].

Surface phenomena taking place between dibutylstannane, tributylstannane or tetraalkylstannanes and partially hydroxylated or dehydroxylated silica were investigated using IR, ^{13}C and ^{119}Sn NMR and ^{119}Sn MAS-NMR (magic-angle spinning NMR) spectroscopies. Physisorption takes place at room temperature. Hydrogen bonding is noted between the H ligand of the terminal methyl groups of the alkyl ligands and the surface silanol groups of hydrated silica. At 100 °C, the silanol IR band at 3747 cm^{-1} decreases and an additional band appears at 3697 cm^{-1}. At 200 °C Si–O–SnR$_3$ groups are formed on the surface and RH is evolved. In the case of Bu_3SnH, H_2 is given off[109,110] (see also Table 2).

The structure of chlorotrimethylstannane in various solvents was studied by far-IR spectroscopy. The Sn–Cl stretching frequency was linearly correlated with the solvent donor number. The Cl–Sn–C angles of the solvated compound were estimated from the ratio of absorption intensities of the symmetric and asymmetric Sn–C stretching bands. These angles were correlated to the solvent donor number, and varied from tetrahedral to trigonal bipyramidal as the donor number increased. Alcohols acted both as donors in O-Sn bonds and acceptors in OH···Cl bonds[111,112]. The Sn–C and Sn–Cl stretching frequencies of triphenylstannyl chloride were studied[113].

The normal vibrations, symmetry and force constants of compounds **6** and **7** were deduced from their IR and Raman spectra, in the solid state and solution. The symmetry is D_{3d}, with staggered methyl groups. The SnCCSn chain is linear, and the C≡C bond longer than in ordinary acetylenic compounds[114].

$$Me_3Sn-C\equiv C-SnMe_3 \qquad (CD_3)_3Sn-C\equiv C-Sn(CD_3)_3$$
$$\textbf{(6)} \qquad\qquad\qquad \textbf{(7)}$$

The structure of compounds **8**, containing trifluoromethyl and halo groups in many combinations, was analysed by ^{13}C, ^{19}F and ^{119}Sn NMR, IR in the gas phase and Raman spectroscopy in the liquid phase[115].

$$(CF_3)_n SnX_{4-n}, \; X = F, \, Cl, \, Br, \, I; \; n = 1-4$$
$$\textbf{(8)}$$

Tetraalkynylstannanes [$Sn(C\equiv CPh)_4$] were characterized spectroscopically by ^{13}C NMR, MS and IR ($\nu_{C\equiv C}$ 2152 cm^{-1})[116].

The methyl groups of difluorodimethylstannane (Me_2SnF_2) are in disordered arrangement as rotors, as evidenced by IR, Raman and ^1H NMR spectroscopies. However, a phase transition involving such rotors was recorded at room temperature when the pressure increased to ca 40 kbar[117] (see also Table 5).

Tunnel spitting and libration lines of high resolution INS (inelastic neutron scattering) spectra of $(Me_3Sn)_2SO_4\cdot2D_2O$, at low temperatures, pointed to the presence of two types of methyl groups. These types do not correspond to those inferred by XRD (X-ray diffraction) crystallography of $(Me_3Sn)_2SO_4\cdot2H_2O$ (see Table 8). Reasons for such discrepancy are discussed[118]. Mixtures of tetramethylstannanes, of which 1–4 of the methyl groups were CD_3 groups, showed shifting and splitting of tunnelling INS lines, in the crystalline state at low temperatures. This is explained on the basis of intra- and intermolecular interaction of rotating methyl groups[119]. Mixtures of tetramethylstannane and tetramethylplumbane were studied at 2 K by high resolution INS, revealing rotational tunnelling transitions for the methyl groups[120].

Organotin(IV) compounds are Lewis acids. The crystalline structure of Lewis acid–base adducts between compounds $R_iSnCl_{4-i}(i = 0,...,3)$ and Ni complexes with Schiff bases was determined by XRD, ^{119}Sn Mössbauer and IR spectroscopy. Adducts are formed in 1:1, 1:2 and 2:1 ratios[121].

The ν_{OH} stretching frequency of trimesityltin hydroxide (Ar^4_3SnOH) shows a sharp band at 3629 cm^{-1}, pointing to a monomeric non-associated structure. This is in contrast with the broad band of trimethyltin hydroxide[122] (see also Table 8).

The structure of a series of dihalodimethystannane adducts with bidentate N,N'-dimethyl-2,2'-biimidazole (DMBIm) was analysed using conductivity measurements, ^1H NMR and IR spectroscopies. In one case, [(Me_2SnBr_2). DMBIm], XRD crystallographic analysis was also performed (see compound **9** in Table 8). Adducts with 1:1 and 2:1 stoichiometric ratios were non-conducting. Cordination numbers at the Sn atom are found from IR spectra to be 5 for adducts [($Me_2SnX_2)_2$ DMBIm] and 6 for adducts [($Me_2SnX_3)_2$ DMBIm]$^{2-}$ ($NEt_4)_2^+$ (see also Table 5)[123].

Bis(stannyl) dicarboxylates **10** and **11** showed usually tetracoordinated Sn atoms by IR and Mössbauer spectroscopy. See also Tables 5 and 8[124].

$$R'\diagdown\diagup CO_2SnR_3 \qquad \diagup CO_2SnR_3$$
$$\diagup\diagdown CO_2SnR_3 \qquad \diagdown CO_2SnR_3$$

R = Bu, Cy, Ph

R' = H, Me, Ph

(10)   **(11)**

The mobility of the franmework of azastanntrane **12** (see Table 5) was analysed by FTIR and NMR spectroscopy[125,126]. The IR and Raman frequencies of the N–Sn coordination bond in compound **13** are correlated with the Sn–X bond lengths of the various X substituents[127].

The IR and Raman spectra of the thiophosphoric acid ester **14** are consistent with a polymeric structure including tetra- and pentacoordinated tin atoms[128].

Stannylenes, the analogues of carbenes, are stabilized by sterically hindered groups such as mesitylene. Dimethylstannylene, a less stable species, was trapped in an argon matrix at 5 K and analysed by IR spectroscopy[129].

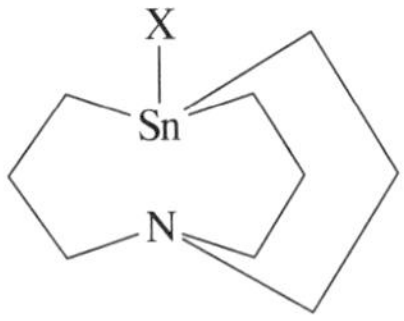

$$X$$
$$\overset{|}{Sn}$$
(bicyclic structure with Sn and N)

(X = Hal, Me)   SP(OSnMe$_3$)$_3$

(13)   (14)

The nature of the species obtained by chemisorption of organotin compounds on various reactive surfaces was investigated by FTIR. Alkoxystannanes bind effectively to reactive silica and alumina surfaces using the residual OH groups and yielding the corresponding alcohol. The nature of the surface activity varies from acidic for silica to basic for γ-alumina. Equilibrium 6 illustrates the chemisorption behaviour of a tributylalkoxystannane in the latter case. Monomeric (15, 16), dimeric (17) and trimeric (18) adsorbed species were postulated[130].

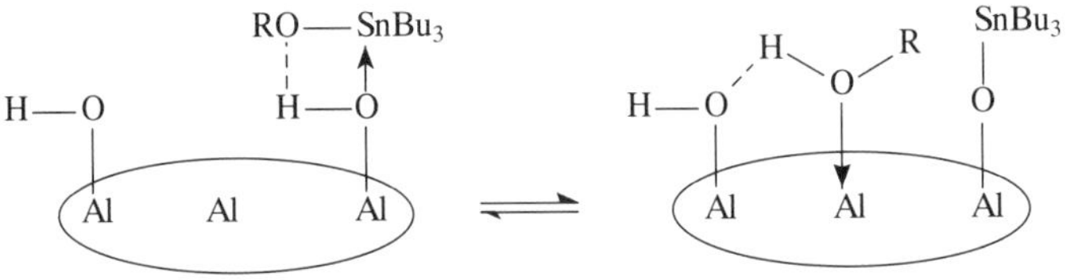

$$RO\!-\!SnBu_3$$

$$H\!-\!O \qquad H\!-\!O$$

(structures over Al Al Al) $\rightleftharpoons$ (structures over Al Al Al)   (6)

(15)

Bu$_3$Sn   SnBu$_3$ (structures 16, 17, 18 over Si Si / Al Al / Si Si)

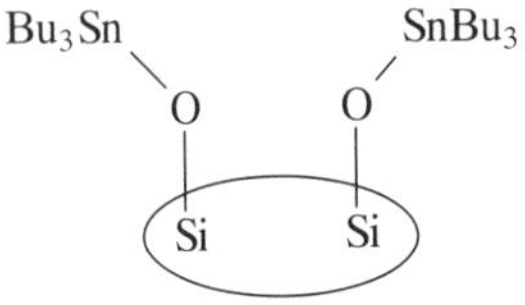

(16) 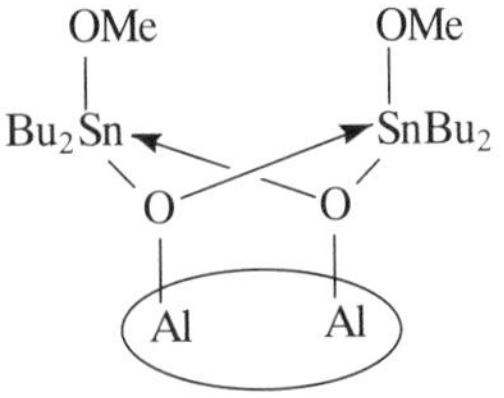   (17) 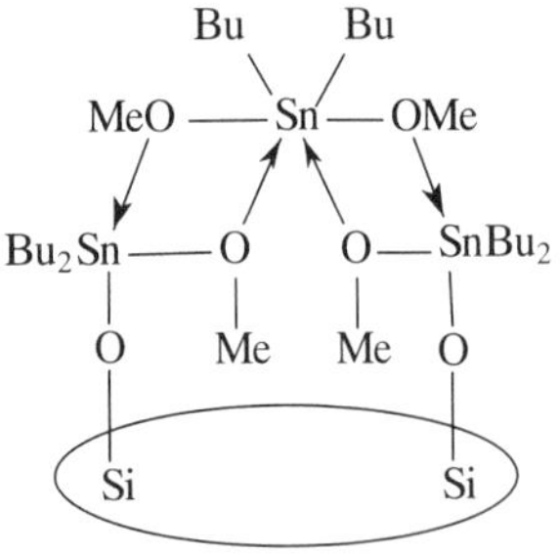   (18)

C. NMR and Mössbauer spectroscopies

See introductory remarks in Section I.A.2 of Chapter 7. The literature sources reporting NMR spectra usually include chemical shift values, δ, and coupling constants, J, and the reader interested in these values is directed to those sources. In the present chapter structural analysis based on NMR and Mössbauer spectra is distributed in various tables, according to some relevant functional feature. Compounds with mixed or undefined features appear in one of these tables, according to the nearest analogy.

Alkyl- and cycloalkylstannanes and functional derivatives with substituents attached to C atoms appear in Table 2.

TABLE 2. Structural determination of stannaalkanes, stannacycloalkanes and their derivatives by NMR and Mössbauer spectroscopic methods

Compounds	Comments	References
$SnMe_4$	^{119}Sn Mössbauer. The frozen solid compound shows a slight band broadenening due to one of the $Sn-C$ bonds becoming slightly shorter and inducing a small quadrupole splitting (see Table 8). This explanation does not seem to apply to the much wider band-broadening observed for the compound isolated in a low-temperature argon matrix.	131
SnD_nH_{4-n}, $SnD_nH_{3-n}{}^+$, $n = 0, \ldots$	^1H and ^{119}Sn NMR. Deuterium isotope effects on shielding constants and spin–spin coupling constants.	132
Bu_2SnH_2, Bu_3SnH and Bu_4Sn	^{13}C and ^{119}Sn NMR and ^{119}Sn MAS-NMR of compounds chemisorbed on silica gel. Peaks appearing at δ 28.5, 26.6, 12.6 and 7.8 ppm can be assigned to β-, γ-, δ- and α-carbon atoms of butyl groups, respectively (see also Section IV.B).	109, 110
Bu_3Sn^+	^{119}Sn NMR. The ion is produced in solution on hydrogen abstraction from Bu_3SnH by Ph_3C^+ ions. No significant interaction between the tin cation and the counterion is observed.	133
$Me(OCH_2CH_2)_3CH_2SnHBu_2$ (19)	^1H, ^{13}C and ^{119}Sn NMR. See discussion on reaction 26 in Section V.A.7.	134a
$(CF_3)_nSnH_{4-n}$, $n = 1-3$	^1H, ^{13}C, ^{19}F and ^{119}Sn NMR, vibrational spectra and physical properties. Unstable compounds, prepared at $-40\,^\circ C$.	134b
$(CHF_2)_m(CF_3)_nSnMe_{4-m-n}$ $m + n = 1-4$; $(CHF_2)_m(CF_3)_nSnGeMe_3$ $m + n = 3$	^1H, ^{13}C, ^{19}F and ^{119}Sn NMR and ^{119}Sn CP-MAS-NMR^c. Linear correlations between various coupling constants were found.	135
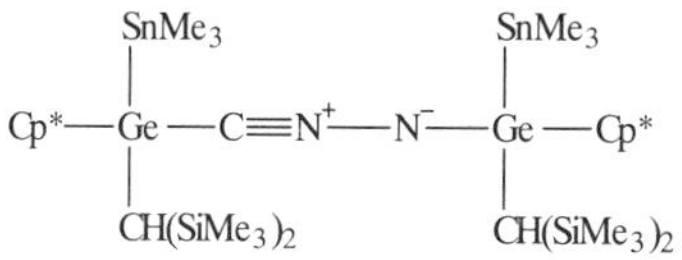 $i\text{-}Pr_3Si-C\!\equiv\!N^+-N^--Ge-Cp^*$ with $SnMe_3$ (top) and $CH(SiMe_3)_2$ (bottom) substituents on Ge (20)	^1H, ^{13}C and ^{119}Sn NMR. Two diastereoisomers and two different $SnMe_3$ signals were found for compound 21^a.	136
$Cp^*-Ge-C\!\equiv\!N^+-N^--Ge-Cp^*$ with $SnMe_3$ (top, both) and $CH(SiMe_3)_2$ (bottom, both) substituents (21)		
$(Me_3Si)_3MM(SiMe_3)_3$ $M = Ge, Sn$ (22)	^1H and ^{13}C NMR. The analogue with $M = Pb$ could not be synthesized^a.	137
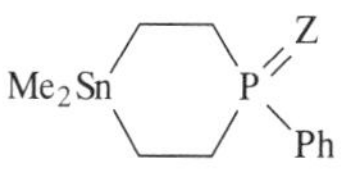 Me_2Sn ring with P bearing Z and Ph; $Z = O, S, Se,$ lone pair (23)	^1H, ^{31}P and ^{119}Sn NMR^a.	138

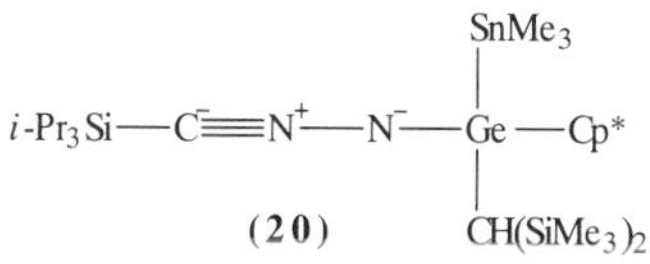

(continued overleaf)

TABLE 2.   (*continued*)

Compounds	Comments	References
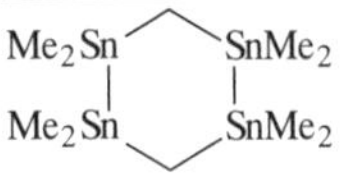 Me$_2$Sn $\diagup$ SnMe$_2$ Me$_2$Sn $\diagdown$ SnMe$_2$	^{119}Sn NMR. The two-path coupling constants $^{2+4}J(^{119}$Sn^{119}Sn) and $^{3+3}J(^{119}$Sn^{119}Sn) were assigned and interpreted.	139
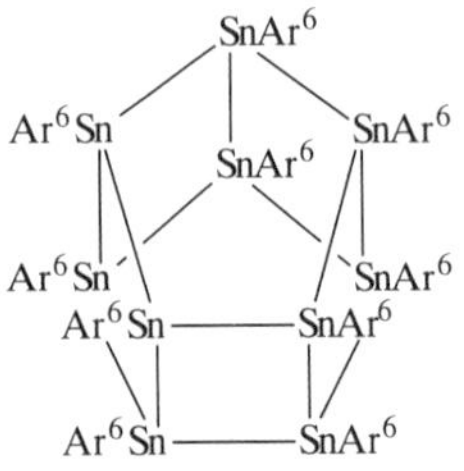 (24)	^{1}H, ^{13}C and ^{119}Sn NMR, UVV. Temperature-dependent spectra due to restricted rotation of Sn—C axis^{a,b,c}.	140
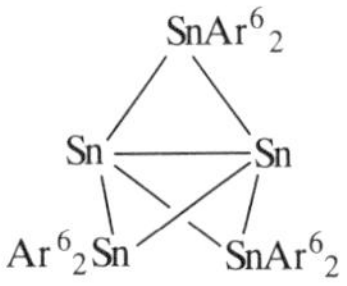 SnAr$^6_2$ Sn———Sn Ar$^6_2$Sn $\diagdown$ SnAr$^6_2$	^{119}Sn NMR, UVV. NMR supports the propellane structure^{b,c}.	141

---

^{a} See also Table 8 for X-ray crystallographic analysis.
^{b} See formula 5 in Section IV.A for aryl groups.
^{c} CP-MAS-NMR = cross-polarization-MAS-NMR, UVV = UV-visible spectroscopy.

TABLE 3.   Structural determination of open-chain and cyclic alkenylstannanes, alkynylstannanes and their derivatives by NMR and Mössbauer spectroscopic methods

Compounds	Comments	References
(CH$_2$=CH)$_2$SnH$_2$ vs Bu$_2$SnH$_2$	^{119}Sn NMR. The following linear correlation was found for the chemical shifts of ^{119}Sn of divinylstannane (δ_{Vi}) and dibutylstannane (δ_{Bu}), in non-coordinating and coordinating solvents. Four-, five- and six-coordinated Sn species could be distingushed according to the δ values: $\delta_{Vi} = (1.03 \pm 0.04)\delta_{Bu} - (155.00 \pm 8.77)$	142
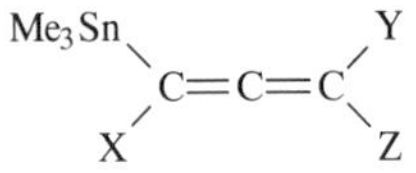 Me$_3$Sn $\diagdown$ $\diagup$ Y C=C=C X$\diagup$ $\diagdown$ Z	^{13}C, ^{29}Si and ^{119}Sn NMR. X = Z = H, SiMe$_3$, GeMe$_3$, SnMe$_3$, SEt; Y = H, SnMe$_3$.	143
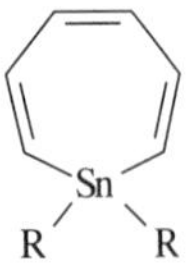 (25)	^{1}H and ^{13}C NMR, aided by spectral simulation of ^{1}H NMR.	144

TABLE 3. (*continued*)

Compounds	Comments	References
Alkynylstannanes	The sign of the following coupling constants was determined: $^1J(^{119}Sn^{13}C\equiv)$, $^2J(^{119}SnC\equiv^{13}C)$, $^3J(^{119}SnC\equiv C^{13}C)$, $^3J(^{119}SnC\equiv C^1H)$ and $^4J(^{119}SnC\equiv CC^1H)$.	145
$Me_{3-n}t\text{-}Bu_nSnC\equiv CX$, $n = 0-3$, X = various	^{119}Sn NMR. Correlation between isotope shifts due to replacement of ^{12}C by ^{13}C.	146
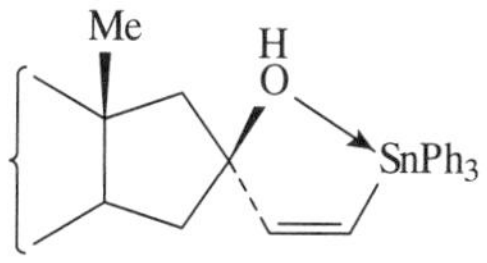	^1H, ^{13}C and ^{119}Sn 1D and 2D NMR. The structure of steroidal compounds bearing a hydroxy group and a β-stannylvinyl group at the 17 position was assigned based on NMR spectra. Coordination between oxygen and tin atoms in solution was observed.	147

Stannanes bearing aliphatic groups with unsaturation in the carbon chain are grouped in Table 3, including both open-chain and cyclic vinyl-, allyl- and alkynylstannanes.

In Table 4 appear stannanes with aromatic groups attached to the Sn atom.

Table 5 includes compounds with electronegative atoms singly bonded to Sn, such as halo, hydroxy, alkoxy, acyloxy, amino and their analogues. Acyloxy and analogous groups may appear as bidentate functions due to additional coordinative bonds established with a second oxygen atom, but this takes place with expansion of the coordination number of the tin atom to values higher than 4.

Organotin compounds with Sn atoms bearing a double bond appear in Table 6. These compounds are usually unstable at room temperature or when exposed to air. Attachment of sterically crowded groups to the tin atom greatly contributes to the stabilization of these compounds.

Organotin compounds with the tin atom in oxidation state 2 appear in Table 7.

D. X-ray Crystallography

See introductory remarks in Section I.A.2 of Chapter 7. Table 8 summarizes examples published in the recent literature, with brief comments about structural features of the compounds.

E. Miscellaneous Methods

Tetramethylstannane is a good precursor for glow discharge[252,253] and plasma deposition[254] of polymeric organotin films. Tin oxide films are deposited when plasma[254] or ArF eximer laser[255] are employed with tetramethylstannane in the presence of oxygen. Organotin polymeric films were analysed by various techniques: Auger electron spectroscopy, XPS (X-ray photoelectron spectroscopy), SIMS (secondary ion MS), ISS (ion scattering spectrometry) and XRD[252]. In one work the carbon content of the organometallic film was determined by combustion in an automatic C analyser and by ESCA (electron spectroscopy for chemical analysis) and the tin content by AAS[253]. In the case of tin oxide deposition the Sn−C to Sn−O ratio could be determined by AES[254,255] and the oxidation state of tin by UPS (UV photoemission spectroscopy) induced by synchrotron radiation[255].

TABLE 4. Structural determination of organometallic compounds containing tin atoms attached to aromatic rings by NMR and Mössbauer spectroscopic methods

Compounds	Comments	References
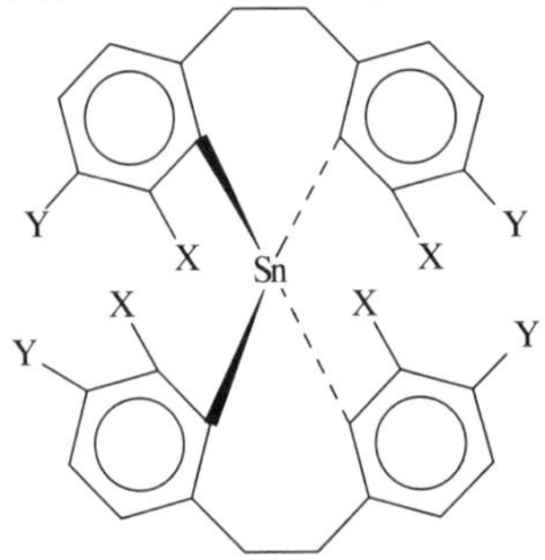 (26) X = Y = H (27) X–Y = CH=CHCH=CH (28) X = Me, Y = H	Temperature dependence and simulation of ^{1}H NMR were used for determining the rates and activation energies of conformational transformations. Bridged tetraarylstannanes **26–28** can adopt either D_2 or S_4 symmetry. From calculations it was shown that the achiral form S_4 has a higher energy level. It was proposed that transitions between one D_2 form to its antipodal proceed through the S_4 form[a].	148
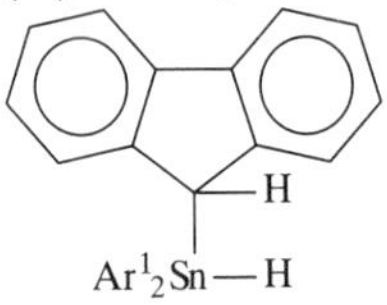 Ar$^1_2$Sn—H (29)	^{1}H, ^{13}C and ^{119}Sn NMR, IR and MS. Derived from compound **66** (see Table 6) by treatment with LiAlH$_4$[b].	149
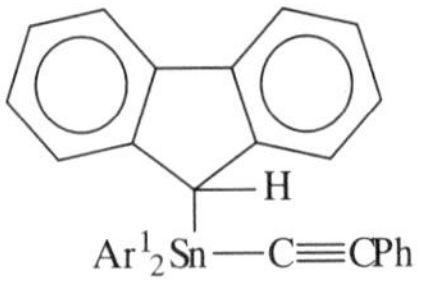 Ar$^1_2$Sn—C≡CPh (30)	^{1}H, ^{13}C and ^{119}Sn NMR, IR and MS. Derived from compound **66** (see Table 6) by treatment with the Brønsted acid H–C≡CPh. See also compounds **40**, **41** (Table 5) and **80** (Table 8)[b].	149
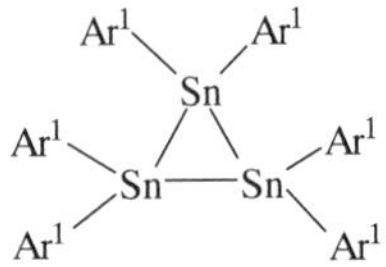 (31)	^{1}H, ^{13}C and ^{119}Sn NMR. **31** is in thermal equilibrium in solution with distannene **69** (Table 6), as shown by NMR and UV spectroscopies.	150
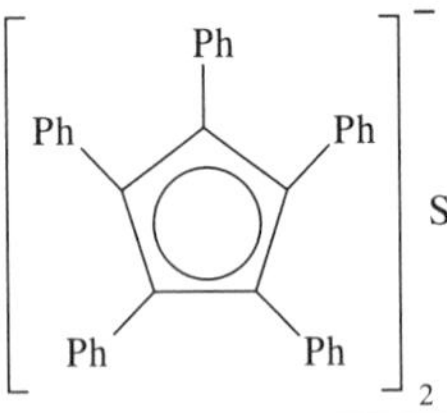	^{13}C NMR, ^{119}Sn CP-MAS-NMR and Mössbauer, IR, Raman, powder XRD. Parallel cyclopentadienyl anions with staggered conformation.	151, 152

[a] See also Table 8 for X-ray crystallographic analysis.
[b] See formula **5** in Section IV.A for aryl groups.

TABLE 5. Structural determination of organometallic compounds containing tin atoms attached to single-bonded electronegative atoms, by NMR and Mössbauer spectroscopic methods

Compounds	Comments	References
Me(OCH$_2$CH$_2$)$_3$CH$_2$SnBu$_2$Cl (32)	^{1}H, ^{13}C and ^{119}Sn NMR. See discussion on reaction 26 in Section V.A.7.	134
Me$_2$SnF$_2$	^{1}H NMR. A phase transition involving the methyl groups takes place at 70.4 K at ordinary pressure (see also Section IV.B).	117
[Me$_2$SnF$_4$]$^{2-}$[NH$_4$]$^+{}_2$	Mössbauer and IR, Raman spectroscopy. The methyl groups are *trans* to each other. Normal N−H···F hydrogen bonds[b].	153
Me$_3$SnCl (33)	^{119}Sn Mössbauer. The frozen solid compound is polymeric, while the compound isolated in a low-temperature argon matrix is monomeric as in the gas phase. Frozen solutions in polar solvents show significant solvent solute interactions. Compare to behaviour of compound 79 in Table 8.	154
MeSnCl$_3$, Me$_2$SnCl$_2$, Me$_3$SnCl	^{13}C and ^{119}Sn NMR. The molecular structure and preferred orientation of the organometallic compounds dissolved in liquid crystals was examined. Strong effects of Lewis acid–base interactions were observed.	155
Bz$_3$SnX X = various groups	^{13}C and ^{119}Sn NMR. A study in coordinating and non-coordinating solvents shows that $\delta(^{119}$Sn) and $^1J(^{119}$Sn, ^{13}C) depend on the coordination number and stereochemistry of the central Sn atom. Results indicate interaction between the polarized Sn−C σ-bond and the π-system of the benzyl groups.	156
R$_3$SnCl, R = 9-triptycyl	^{1}H and ^{13}C NMR. The three triptycyl groups have no rotation about the C−Sn axis due to mechanical interlocking.	157
(Bu$_3$Sn)$_2$O	^{119}Sn Mössbauer. The degradation of this antifouling agent in a neoprene rubber matrix takes place by progressive dealkylation and hydrolysis during service in a marine environment.	158
R$_2$SnCl$_2$ + F$^-$ RSnCl$_3$ + F$^-$ (R = Ph, Me, n-Bu, t-Bu)	^{19}F and ^{119}Sn NMR. Intermolecular and intramolecular dynamics of a CH$_2$Cl$_2$ solution of chlorinated organotin compounds and fluoride ions lead to multiple replacement reactions. Complex ions such as [Ph$_2$SnCl$_{3-n}$F$_n$]$^-$, $n = 1$–3 are formed, that are stereochemically rigid at $-100\,^\circ$C and fluxional at $-80\,^\circ$C.	159
Me$_2$SnCl$_2$ coordination complexes with nitrosoanilines	^{13}C, ^{15}N and ^{119}Sn NMR and Mössbauer. Two types of complex can be distingushed. Coordination enhances the quinonoid contribution in the ligand structure.	160
[SnX$_4$L$_2$]$^{2-}$ (X = Cl, Br)	^{119}Sn Mössbauer. For *trans* octahedral complex ions the Sn−X bond length (y, Å) is empirically correlated with the partial quadrupole splitting (x) associated with the L ligand (an organic radical) by the following equation: $y = a(4x) + b$. The correlation can yield information on the Sn−L bond length and the sign of the quadrupole splitting.	161, 162

(*continued overleaf*)

TABLE 5. (*continued*)

Compounds	Comments	References

X	a	b
Cl	-0.044 ± 0.002	2.420 ± 0.003
Br	-0.048 ± 0.002	2.589 ± 0.003

Compounds	Comments	References
R_3SnF (R = Me, n-Bu, t-Bu, Ph); $Ar_3{}^4SnF$ (**33′**)	^{119}Sn NMR, spin relaxation methods; ^{119}Sn CP-MAS-NMR. n-Bu$_3$SnF exists in polymeric form in solution. $Ar^4{}_3SnF$ is monomeric in the solid state, all the rest are chain polymeric with F atoms equally shared by Sn atoms[a].	163, 164
$[Ph_3SnX_2]^-$, $[Ph_3SnF]Z$, $[Ph_3SnX_3]^{2-}$, $[Ph_3SnX_2]^-Z$ (X = F, Cl; Z = HMPA, DMSO)	^{19}F and ^{119}Sn NMR. Fluorine exchange occurs between 4- and 5-coordinated tin complexes, possibly through chlorine and fluorine-bridged species.	165
R_3SnX R = Bu, Ph; X = ClO$_4$, BF$_4$	^{13}C, ^{31}P, ^{37}Cl and ^{119}Sn NMR. Solvents induce changes in the ^{119}Sn and ^{37}Cl line widths and ^{119}Sn–^{13}C scalar couplings. Spectra indicate equilibrium in solution between tetrahedral and trigonal bipyramidal coordination states of Sn. Compounds showed ^{119}Sn–^{31}P coupling when dissolved in totally deuterated hexamethylphosphortriamide $[((CD_3)_2N)_3PO]$ solvent.	166
$(Me_3Sn)_3N$	^{1}H, ^{15}N and ^{119}Sn NMR. The value and sign of several coupling constants were determined: $^1J\,(^{119}Sn^{15}N)$, $^3J\,(^{15}N^1H)$, $^2J\,(^{119}SnC^1H)$.	167

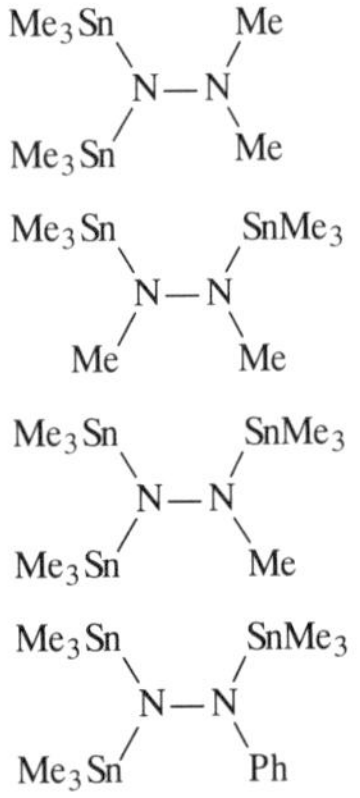

Me$_3$Sn   Me
   \　　　／
    N—N
   ／　　　\
Me$_3$Sn   Me

Me$_3$Sn   SnMe$_3$
   \　　　／
    N—N
   ／　　　\
Me　　　Me

Me$_3$Sn   SnMe$_3$
   \　　　／
    N—N
   ／　　　\
Me$_3$Sn   Me

Me$_3$Sn   SnMe$_3$
   \　　　／
    N—N
   ／　　　\
Me$_3$Sn   Ph

^{15}N and ^{119}Sn NMR. A correlation was proposed between the chemical shifts of ^{15}N in the hydrazines and the chemical shifts of ^{15}N in the corresponding amines, obtained by replacing an N atom by a CH group. This is helpful to support the non-trivial assignments of ^{15}N resonances in the hydrazines. Geminal coupling constants $^2J\,(^{119}SnN^{15}N)$ were observed for the first time. Their relative magnitude is related to the probability of the Sn–N bond having *cis* orientation with the lone electron pair on the ^{15}N atom in the $^{119}SnN^{15}N$ fragment.   168

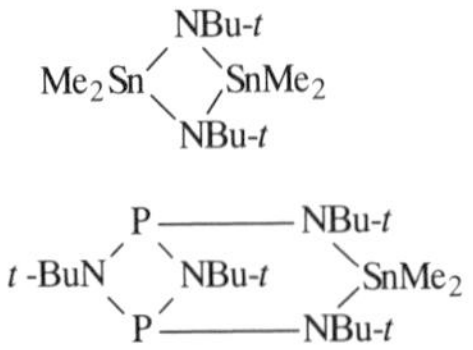

      NBu-*t*
Me$_2$Sn〈　　〉SnMe$_2$
      NBu-*t*

^{1}H, ^{13}C, ^{15}N, ^{31}P and ^{119}Sn NMR, XRD and MS. The SnMe$_2$ ring segment of **34** can be replaced by BPh and PPh moieties.   169

    P————NBu-*t*
t-BuN〈　\NBu-*t*　\SnMe$_2$
    P————NBu-*t*

(**34**)

TABLE 5.   (*continued*)

Compounds	Comments	References
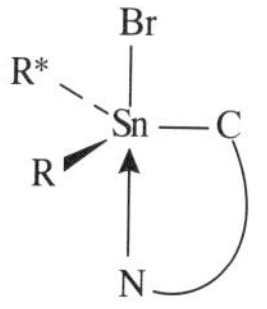 **(35)**	^{1}H, ^{13}C, ^{117}Sn and ^{119}Sn NMR and polarimetry. Coordination of the tin atom to the amino group in the γ-position stabilizes molecules of general structure **35**. The chiral group R* of compound **36** (see below this table), for example, induces chirality at the Sn centre. A slow racemization of the Sn centre takes place in solution, especially when the γ-amino group is attached to a flexible substituent. See also next entry[b].	170
37–39 X = Ph, I, Br, Me$_2$NCS$_2$ (see end of table)	^{119}Sn NMR and Mössbauer spectroscopy. The Sn atom has coordination 5 and the structure is monomeric whenever intramolecular N $\rightarrow$ Sn coordination can be established, as in compounds **37** and **38**. Compounds **39** have polymeric character, conferred by intermolecular N $\rightarrow$ Sn coordination[b].	171
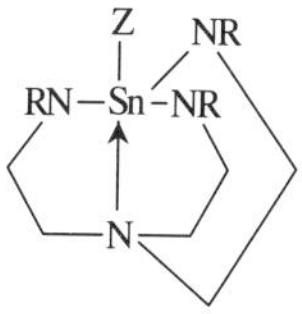 R = H, Me; Z = NH$_2$, F, Cl, Br, I, 1/2(C≡C), PhC≡C **(12)**	^{1}H NMR and FTIR were used for assessment of the mobility of the framework. ^{119}Sn CP-MAS-NMR showed strong transannular Sn $\rightarrow$ N coordination; in the case of Z = Cl, coupling is observed with quadrupolar ^{37}Cl[b].	125, 126
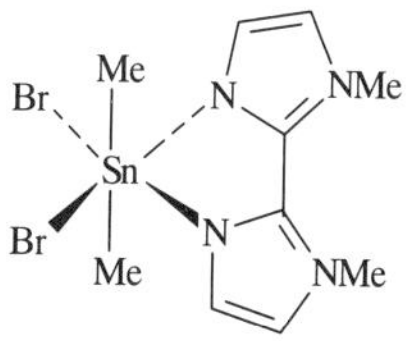 **(9)**	^{1}H NMR. Spectra show that the compound **9** and similar ones undergo extensive dissociation in CDCl$_3$. See also Section IV.B.	123
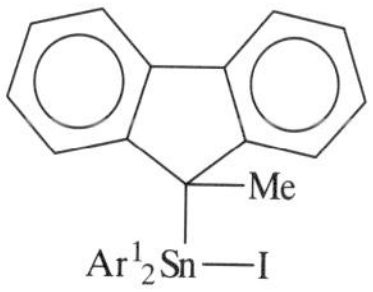 Ar$^1_2$Sn—I **(40)**	^{1}H, ^{13}C and ^{119}Sn NMR, IR and MS. Derived from compound **66** (see Table 6) by treatment with MeI[a].	147
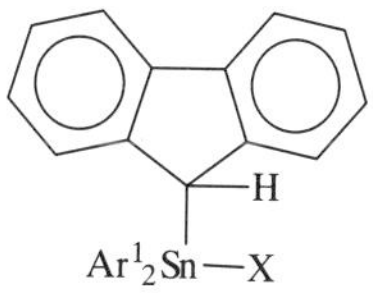 Ar$^1_2$Sn—X **(41)**	^{1}H, ^{13}C and ^{119}Sn NMR, IR and MS. Derived from compound **66** (see Table 6) by treatment with Brønstead acids HX, such as HOH, MeOH, HCl and H-NHPh[a].	147

(*continued overleaf*)

TABLE 5. (*continued*)

Compounds	Comments	References
Bu$_3$SnSR R = alkyl, aryl, heteryl	^{119}Sn NMR. The ^{119}Sn chemical shift moves upfield (-0.06 ppm/K) on increasing the temperature from -20 to $50\,^\circ$C.	172
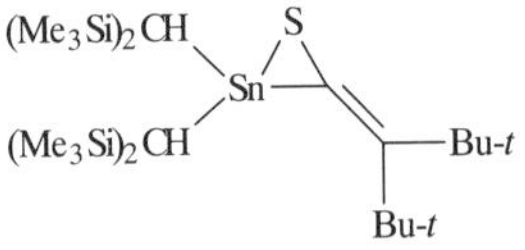 (**42**)	^{1}H, ^{13}C and ^{29}Si NMR. Formed by scavenging a solution of organotin(II) compound **72** (see Table 7) with a thioketene[b].	173
10, **11** (see Section IV.B)	^{119}Sn Mössbauer spectroscopy. Usually tetracoordinated Sn, however, pentacoordination may also occur[b].	124
44 (see reaction 7 below this table)	^{1}H, ^{13}C and ^{29}Si NMR. Formed on slow disproportionation of compounds **43**, as shown in reaction 7.	174
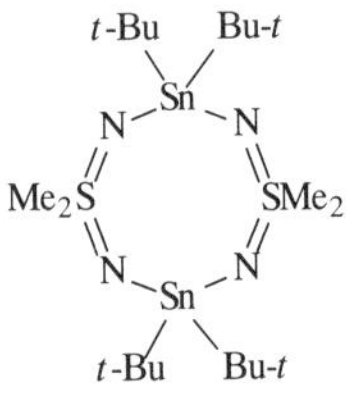 (**45**)	^{1}H NMR, MS and cryoscopy.	175
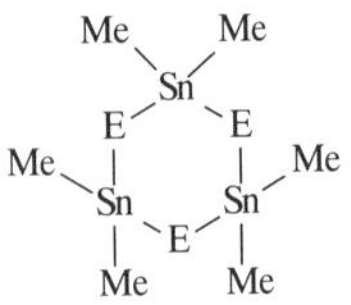 (E = S, Se, Te)	^{13}C, ^{77}Se, ^{119}Sn and ^{125}Te MAS-NMR. Molecular symmetry and conformation was deduced for the various crystalline forms of the dimethyltin chalcogenides.	176
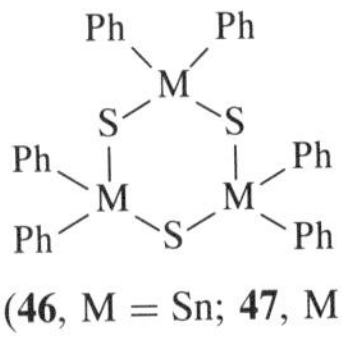 (**46**, M = Sn; **47**, M = Pb)	^{119}Sn and ^{207}Pb NMR and FDMS[c]. Ring segment exchange reaction in solutions containing mixtures of compounds **46** and **47**, leading to equilibrium mixtures of compounds containing Sn—S—Pb moieties.	177
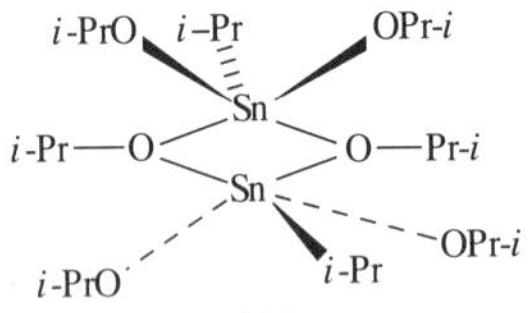 (**48**)	^{13}C and ^{119}Sn NMR and CP-MAS-NMR. Isopropyltriisopropoxystannane has monomeric structure in solution and the associated structure **48** in the solid phase.	178a

TABLE 5.   (*continued*)

Compounds	Comments	References
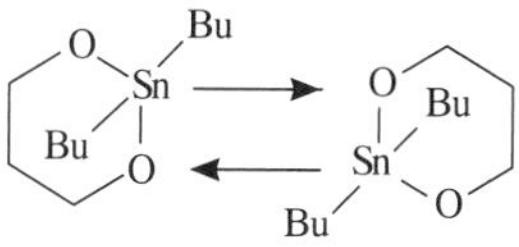	^{119}Sn NMR and CP-MAS-NMR, XRD. The solid has polymeric structure. In solution the dimer is in equilibrium with the trimer, etc. Collinear Bu−Sn−Bu configuration. Analogous stannaacetals with saccharides were also investigated.	178b–d
$R'_3SnOSiR_2OSiR_2OSnR'_3$ (R = Me, Ph; R' = Ph, Bu, $c\text{-}C_6H_{11}$) (49)	^{1}H, ^{13}C, ^{29}Si and ^{119}Sn NMR, ^{119}Sn Mössbauer and XRD.	179
Ph$_3$Sn−O$_2$CCH$_2$Cl (50)	Variable-temperature ^{119}Sn Mössbauer. Polymeric chain with *trans*-pentacoordinated Sn. The compound is isostructural with triphenylstannyl acetate (see compounds **81** in Table 8).	180
51, 52 (see below this table)	^{13}C and ^{119}Sn CP-MAS-NMR. 51 has ^{119}Sn bands at δ 94.3 and 68.6 ppm. The low-frequency resonance is more temperature dependent than the higher one. Tentative assignments were made for the propeller-like $2\pi/3$ reorientations of the SnMe$_3$ groups. Spectroscopic and crystallographic data on this compound can be also interpreted as belonging to the ionic structure **52**^{b}.	181, 182
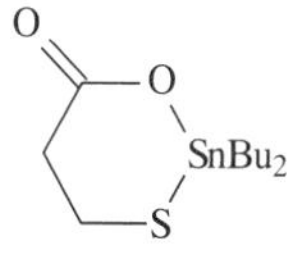 (53)	^{13}C and ^{119}Sn NMR. Used as PVC stabilizer. Monomeric or oligomeric form in solution depending on concentration and temperature. Spectra point to pentacoordinated tin due to association^b.	183
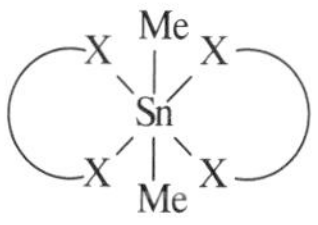 Chelates 54–56 (see below this table)	^{1}H, ^{13}C and ^{119}Sn NMR in solution and solid state. Five-membered ring chelates **54** and **55** do not conform to the correlations found by the authors for the Me−Sn−Me angle and the coupling constants^1J (^{119}Sn, ^{13}C) and ^2J(^{119}Sn, ^{1}H). Special solvent effects observed on molecular structure^b.	184
Stannyl diphenythiophosphonates 57–59 (see below this table)	^{1}H NMR, ^{119}Sn Mössbauer and IR. Cystalline organotin diphenylmonothiophophinates (**57**) have ligands attached to Sn through O atoms, however, certain Sn−S interaction appears to be present. The monomeric structure **58** was proposed for $n = 2$ and the polymeric structure **59** for $n = 3$.	185
60–62 (Y = S, Se, Te) (see below this table)	^{1}H, ^{13}C, ^{119}Sn and ^{125}Te NMR^{a,b}.	186, 187

(*continued overleaf*)

TABLE 5. (*continued*)

Compounds	Comments	References

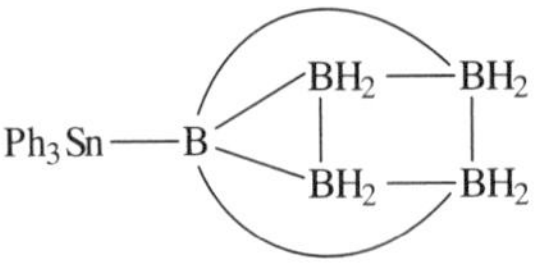

^{11}B and ^{119}Sn NMR.

188

$R_2Sn(O_2CCH_2Cl)_2$

^{1}H, ^{13}C and ^{29}Si NMR. R are sterically hindered groups, e.g. neopentyl, neohexyl, cyclohexyl, hexamethylene (**63**^{b}), SiMe$_3$.

189

DL-AcNHCHRSnBu$_3$

^{13}C and ^{29}Si NMR. Chemical shift and coupling constants depend on R, coordination number of Sn and solvent.

190

^{a} See formula 5 in Section IV.A for aryl groups.
^{b} See also Table 8 for X-ray crystallographic analysis.
^{c} FDMS = field desorption MS.

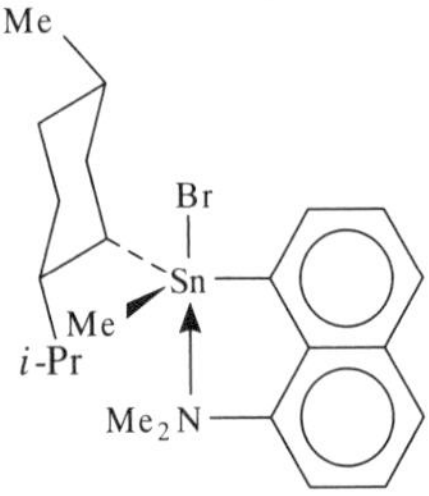

(36) 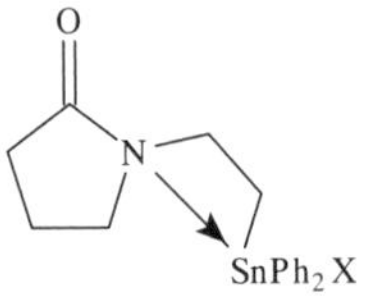   (37) 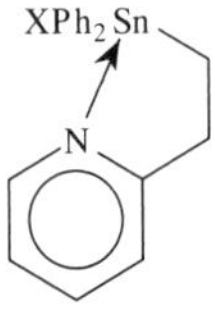   (38) 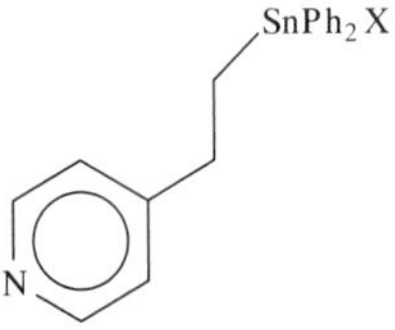   (39)

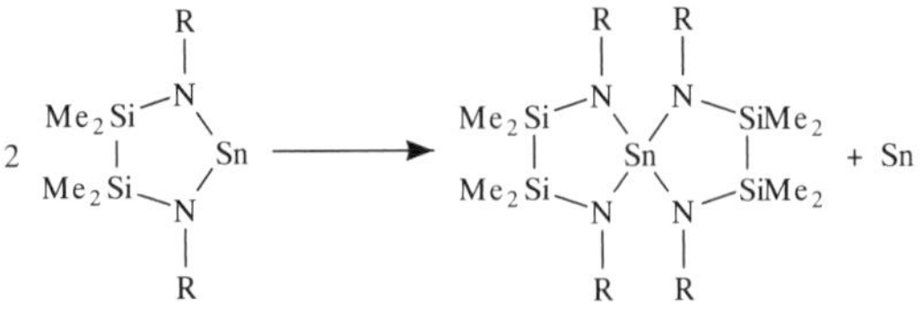

(43)   (44)   (7)

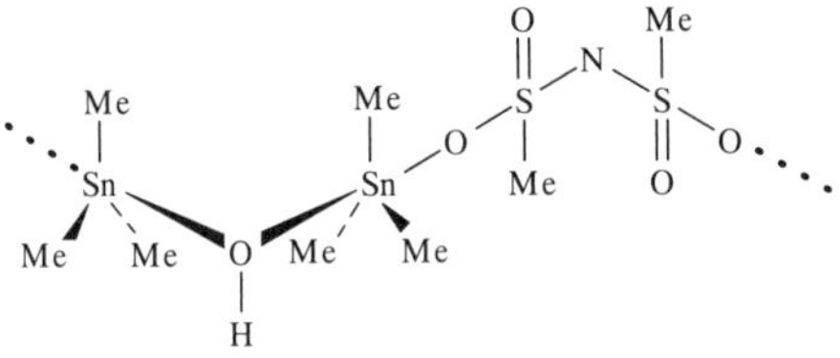

$[(Me_3Sn)_2OH]^+[(MeSO_2)_2N]^-$

(51)   (52)

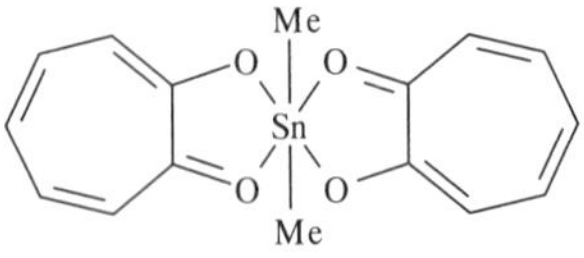

(54) 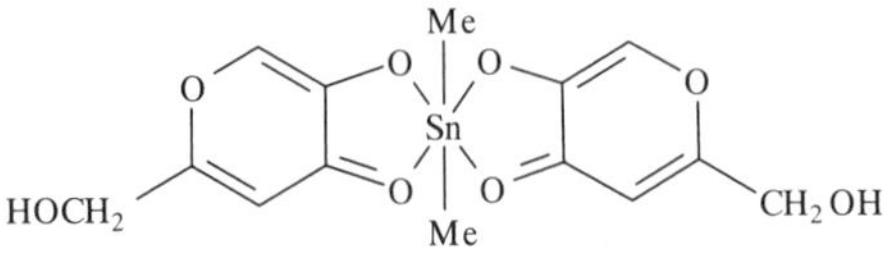   (55)

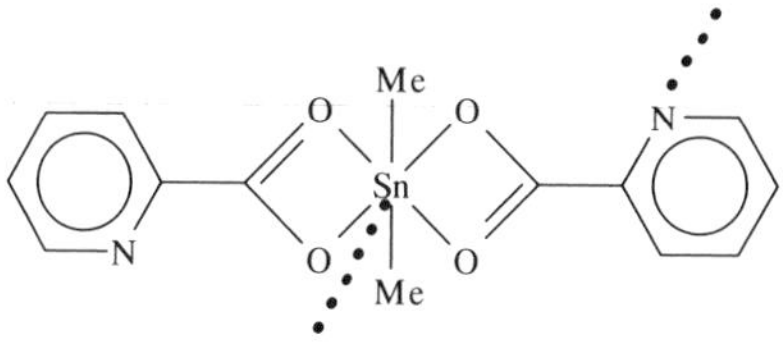

(56)

R$_n$ Sn[OP(S)Ph$_2$]$_{4-n}$
R = Me, Bu, Bz, Ph, Cy
n = 2, 3

(57)   (58) 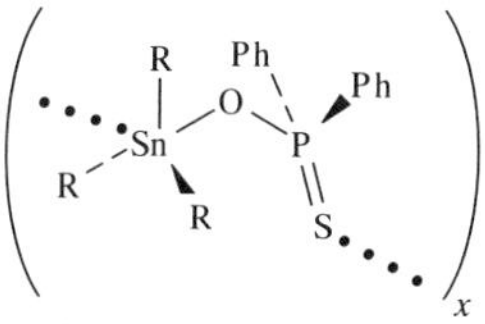   (59)

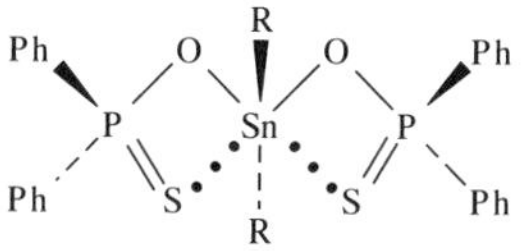

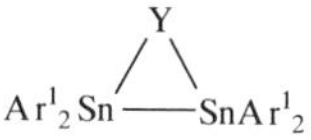

(60) 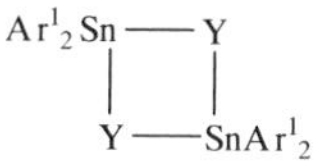   (61) 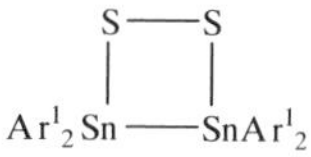   (62)

TABLE 6. Structural determination of organometallic compounds containing tin atoms supporting double bonds by NMR and Mössbauer spectroscopic methods

Compounds	Comments	References
64 (see below this table)	^{1}H, ^{11}B, ^{13}C and ^{119}Sn NMR. Mesomeric structure, as depicted in reaction 8; $\delta_{119_{Sn}}$ 835 ppm (compare compounds 64–70)^a.	191, 192
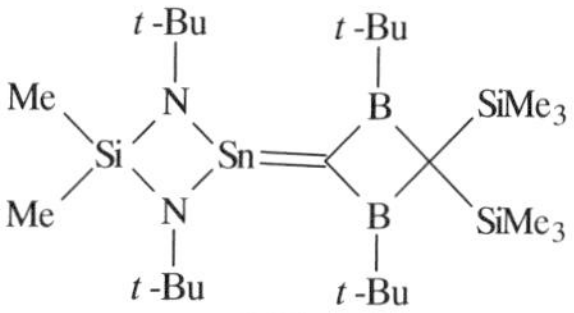	^{11}B, ^{13}C and ^{119}Sn NMR. Mesomeric structure, analogous to 64; $\delta_{119_{Sn}}$ 647 ppm (compare compounds 64–70).	191, 192
(65)		
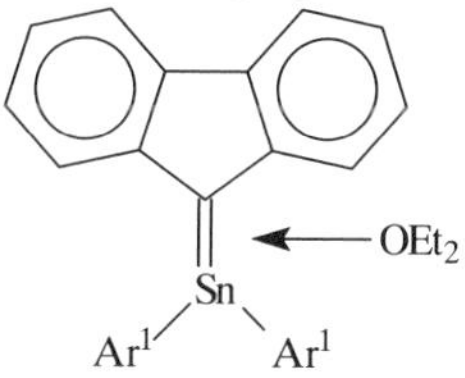 (66)	^{1}H, ^{13}C and ^{119}Sn NMR. $\delta_{119_{Sn}}$ 288 ppm (compare compounds 64–70). The upfield displacement relative to other examples in this table is attributed to shielding, due to complex formation with diethyl ether. See also compounds 29, 30 (Table 4), 40, 41 (Table 5) and 80 (Table 8)^b.	193
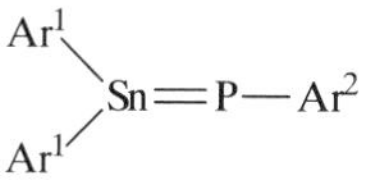 Ar^1\ Sn=P—Ar^2 Ar^1/ (67)	^{1}H, ^{13}C, ^{31}P and ^{119}Sn NMR. $\delta_{119_{Sn}}$ 499.5 ppm (compare compounds 64–70)^b.	194

(continued overleaf)

TABLE 6. (*continued*)

Compounds	Comments	References
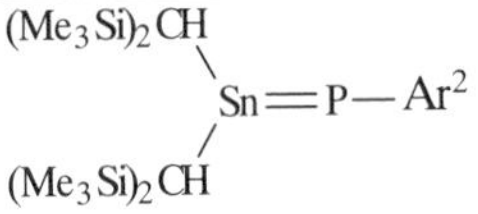 (Me$_3$Si)$_2$CH 　　　＼ 　　　　Sn＝P—Ar^2 　　　／ (Me$_3$Si)$_2$CH **(68)**	^{1}H, ^{31}P and ^{119}Sn NMR. $\delta_{119_{Sn}}$ 658.3 ppm (compare compounds **64–70**)^b.	195
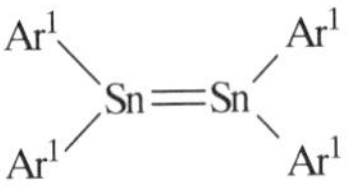 Ar^1＼　　　　／Ar^1 　　＼Sn＝Sn／ Ar^1／　　　　＼Ar^1 **(69)**	^{1}H, ^{119}Sn and ^{125}Te NMR. $\delta_{119_{Sn}}$ 427.3 ppm (compare compounds **64–70**). Does not dissociate in solution as does compound **70**, but is in thermal equilibrium with cyclotristannane **31** (see Table 4), as shown by NMR and UVV spectroscopies. **69** undergoes Te insertions as shown in reaction 9 (see below this table)^b.	150, 196
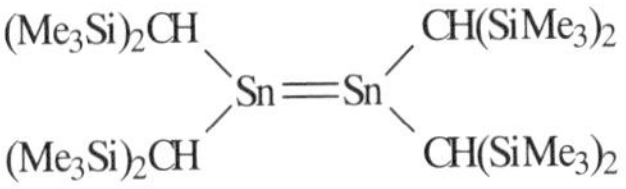 (Me$_3$Si)$_2$CH＼　　　／CH(SiMe$_3$)$_2$ 　　　　　＼Sn＝Sn／ (Me$_3$Si)$_2$CH＼　　　＼CH(SiMe$_3$)$_2$ **(70)**	^{13}C, ^{117}Sn and ^{119}Sn CP-MAS-NMR. $\delta_{119_{Sn}}$ 725 ppm (compare compounds **64–70**). Dissociates in solution into compound **72** (see Table 7). At low temperature (*ca* 220 K) ^{13}C–^{117}Sn and ^{13}C–^{119}Sn couplings appear as 3 lines that coalesce at higher temperatures due to conformational equilibrium^a.	173, 197a, 197b
Bu$_2$Sn=O complexes	^{119}Sn Mössbauer spectroscopy. Complexes of dibutyltin(IV) oxide with carbohydrates show Sn atoms coordinated to surrounding atoms in tetrahedral, trigonal bipyramidal and octahedral configuration. These types have several variants, depending on the alignment of the coordinated groups. Many of these complexes show simultaneous presence of two coordination types.	198

^{a} See also Table 8 for X-ray crystallographic analysis.
^{b} See formula **5** in Section IV.A for aryl groups.

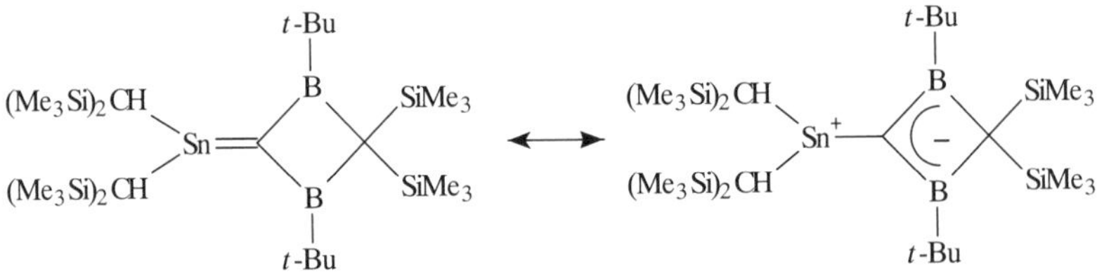

(8)

(64)

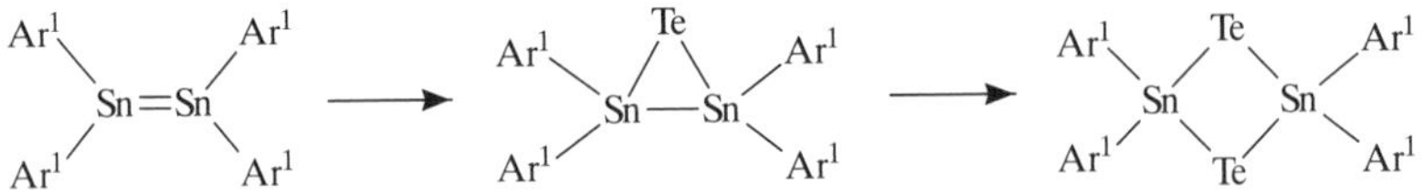

(9)

TABLE 7.   Structural determination of organometallic compounds containing tin(II) atoms by NMR and Mössbauer spectroscopy methods

Compounds	Comments	References
71 (see reaction 11 in Section IV.E)	^{119}Sn NMR. $\delta_{119_{Sn}}$ 2208 ppm points to divalent organotin compound.	199
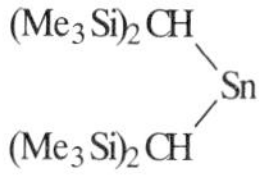 (Me$_3$Si)$_2$CH 　　　＼Sn (Me$_3$Si)$_2$CH＾ (72)	^{1}H NMR and ^{119}Sn Mössbauer spectroscopy. Formed in solution by dissociation equilibrium of distannene 70 (Table 6). See also compound 42 in Table 5.	173, 200
43 and 73 (see reaction 7 below Table 5 and below this table)	^{1}H, ^{13}C, ^{29}Si and ^{119}Sn NMR. Stable at low temperature in the absence of oxygen. Compounds 43 slowly undergo the disproportionation shown in reaction 7. The dimeric form 73 of 43 exists in solution.	174
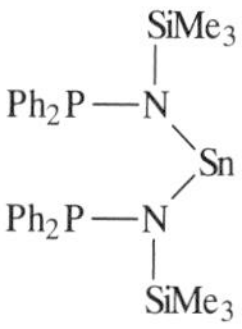 SiMe$_3$ 　　| Ph$_2$P—N 　　　　＼Sn Ph$_2$P—N＾ 　　| SiMe$_3$ (74)	^{1}H, ^{13}C, ^{29}Si and ^{31}P NMR.	201
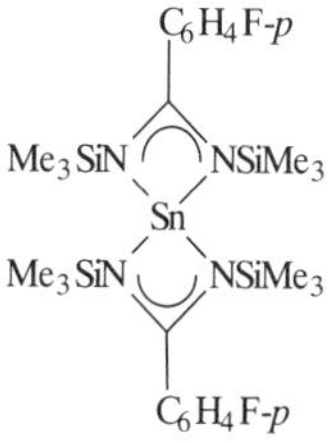 C$_6$H$_4$F-p 　| Me$_3$SiN　NSiMe$_3$ 　　Sn Me$_3$SiN　NSiMe$_3$ 　| C$_6$H$_4$F-p (75)	^{1}H, ^{13}C and ^{29}Si NMR.	201
76–78 (see below this table)	^{1}H, ^{13}C and ^{29}Si NMR. Sn(II) compounds stabilized by chelate formation.	202
Cp$^#$SnCp$^#$, Cp$^#$SnCp	^{13}C and ^{119}Sn NMR, CP-MAS-NMR, Mössbauer, IR, Raman, MS and powder XRD. A Sn(II) compound. Parallel staggered conformation proposed for decaphenylstannocene while the cyclopentadienyl groups are not parallel in pentaphenylstannocene^a.	151, 152
i-Pr 　| 　N Me$_2$Si　M (M = Sn, Pb) 　N 　| i-Pr	^{1}H, ^{13}C, ^{15}N, ^{29}Si, ^{119}Sn and ^{207}Pb NMR. For M = Sn at room temperature the structure is dimeric and fluxional. A mixture of Sn and Pb analogues gives the mixed dimer. For many analogous Sn(II) and Pb(II) compounds the following correlation holds: $\delta_{207_{Pb}} = 3.30\delta_{119_{Sn}} + 2336$	39, 203

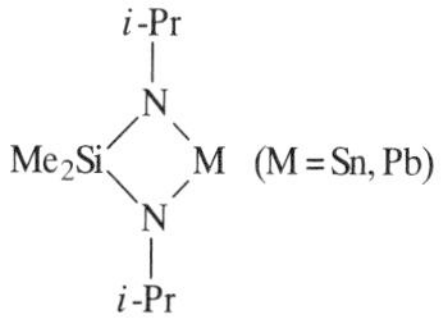

^{a} Cp = cyclopentadienyl anion; Cp$^#$ = pentaphenylcyclopentadienyl anion.

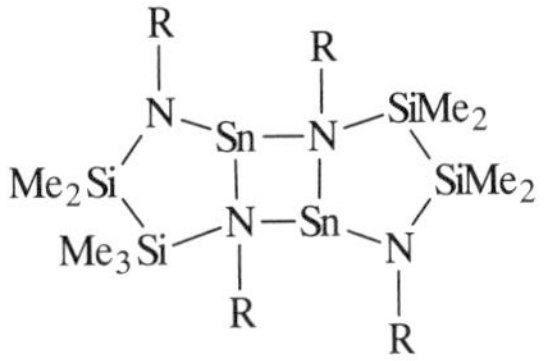

(73)

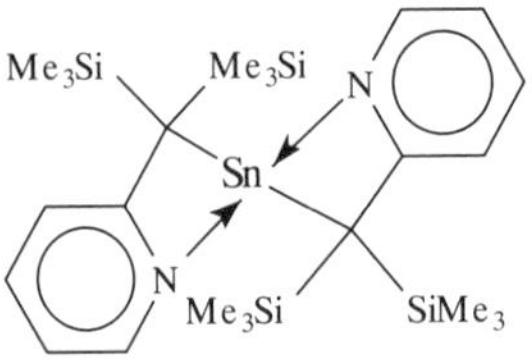

(76) 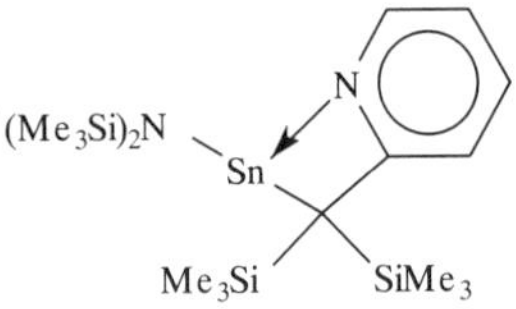   (77) 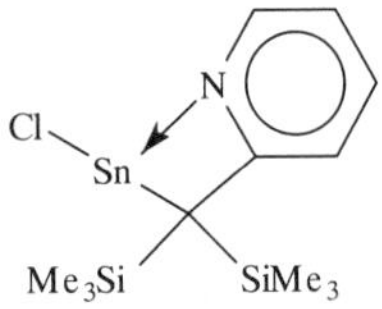   (78)

TABLE 8. Structural determination of organotin compounds by X-ray crystallography

Compounds	Comments	References
$SnMe_4$	SnI_4-like structure. Tetrahedral molecule with 3-fold axis of symmetry along a slightly shorter $Sn-C$ bond. The different $Sn-C$ bond lengths are in accord with NMR, INS and Mössbauer data (see Table 2).	131, 204
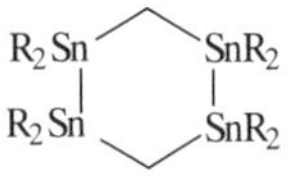 (R = Me, Ph)	Octaphenyl-substituted compound has chair conformation. Octamethyl-substituted compound has twisted boat conformation with slightly different $Sn-Sn$ bond lengths, corresponding to observed differences in $^{119}Sn-^{119}Sn$ coupling constants.	205a, b
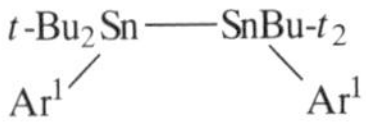	Eclipsed conformation with syn-oriented aryl groups. The $Sn-Sn$ bond is very long, 0.3034 nm. ^1H NMR points to restricted rotation about $Sn-C$ and $Sn-Sn$ bonds[a].	206
$(Me_3Si)_3MM(SiMe_3)_3$ (M = Ge, Sn; 22) (see Table 2)	Only the crystalline analogue with M = Sn could be analysed.	137
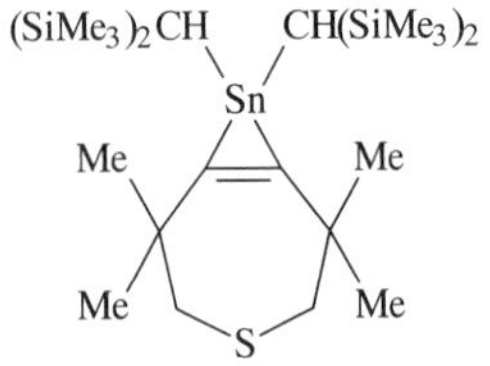	First example of a 3-stannacyclopropene. The $Sn-C_{sp^2}$ bond lengths, 0.2135 nm, are comparable to $Sn - C_{sp^2}$ bond lengths found in acyclic compounds. Also Raman of the solid at $-60\,^\circ C$ and IR spectra.	207
$Cl_2Sn[Ge(SiMe_3)_3]_2$	Bulky groups cause largely distorted coordination angles of Sn: $Ge-Sn-Ge$ 142.1°, $Cl-Sn-Cl$ 98.8°.	208

TABLE 8. *(continued)*

Compounds	Comments	References
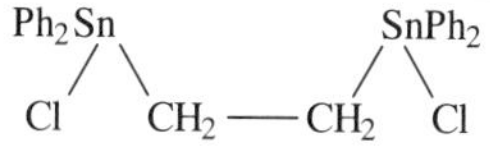 Ph$_2$Sn⟍ ⟋SnPh$_2$ Cl ⟍CH$_2$—CH$_2$⟋ Cl	Complex with a chloride salt. The organotin compound is a powerful bidentate Lewis acid, binding the chloride ion.	209a
Ph$_3$SnCl (**79**)	Loosely packed unassociated molecules. The Sn atoms have slightly distorted tetrahedral configuration. The same structure is prevalent also at 110 K. The behaviour of solid Me$_3$SnCl is different (see **33** in Table 5).	209b
Ar$^4{}_3$SnX (X = F, OH)	Both compounds are monomeric molecules with tetrahedral configuration at the Sn atom. The Sn—X distances in the crystal can be taken as the 'normal' bond lengths (Sn—F 0.1961 nm (av.) and Sn—O 0.1999 nm). See also Section IV.B^a.	122
28 (see Table 4)	The structure of this compound is very close to the chiral form D_2.	148
37, X = Me$_2$NCS$_2$ (see end of Table 5)	Coordination of Sn atom is 5. Structure is monomeric due to intramolecular N—Sn coordination.	171
(Me$_3$Sn)$_2$SeO$_4$·2H$_2$O	The two trimethylstannyl groups are crystallographically equivalent, however, the three methyl groups are rotationally and vibrationally different, as shown by INS spectroscopy at low temperature.	210
42 (see Table 5)	The C—Sn—S angle in this stannathiacyclopropane system is small (45.7°).	173

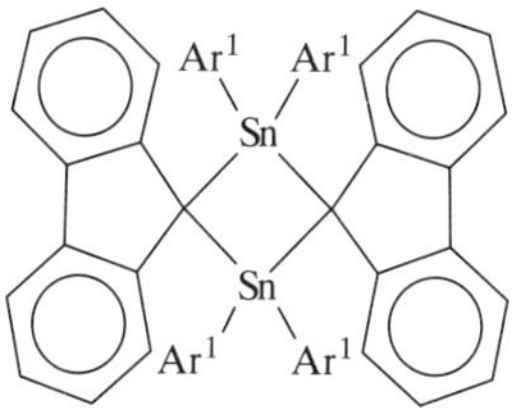

The structure drawing (**80**) with fluorene rings and two Sn atoms, Ar^1 substituents:

Produced by slow dimerization of compound **66** (see Table 6) in solution. The 4-membered ring is planar and perpendicular to the fluorene substituents^a. — 149

(**80**)

| 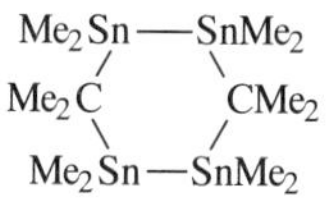 Me$_2$Sn—SnMe$_2$
 Me$_2$C⟍ ⟋CMe$_2$
 Me$_2$Sn—SnMe$_2$ | Chair conformation, independent molecules. | 211 |

| 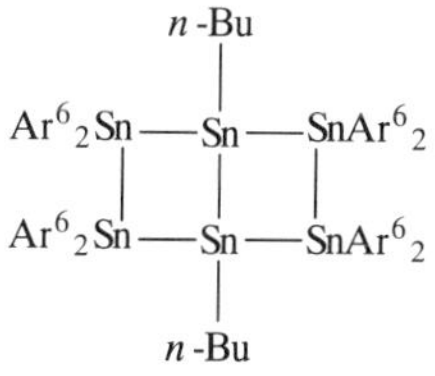 *n*-Bu
 |
 Ar$^6{}_2$Sn—Sn—SnAr$^6{}_2$
 | | |
 Ar$^6{}_2$Sn—Sn—SnAr$^6{}_2$
 |
 n-Bu | No plane or axis of symmetry. The dihedral angle between the *cis*-fused rings is 131.9°; the 4-membered rings are puckered, with angles in the 87.2–92.9° range, similar to other tetrastannacyclobutane compounds. Also MS and NMR spectra^a. | 212 |

(continued overleaf)

TABLE 8. (*continued*)

Compounds	Comments	References
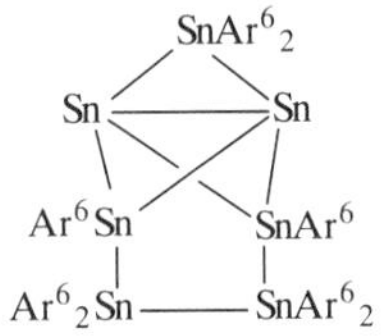 $SnAr^6{}_2$ Sn — Sn Ar^6Sn $SnAr^6$ $Ar^6{}_2Sn$ — $SnAr^6{}_2$	A perstanna(1.1.1)propellane polycondensed structure[a].	213
23 (see Table 2)	Polymeric structure; distorted trigonal bipyramidal configuration at Sn.	138
24 (see Table 2)	Normal bond distances and angles. A few other pergerma- and perstannaprismanes are known.	140
$(Me_3Sn)_2SO_4 \cdot 2H_2O$	Trigonal bispyramidal configuration of Sn, with *trans* O atoms of sulphate and water. Two types of Me groups are distinguished, however, these are different from those implied by INS (see Section IV.B).	118
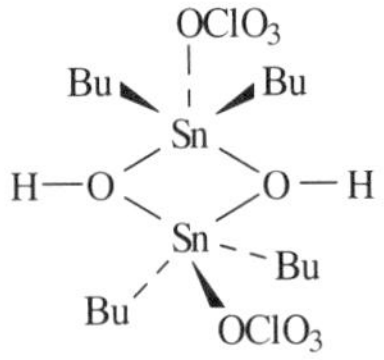 $OClO_3$ Bu Bu Sn H—O O—H Sn Bu Bu $OClO_3$	Dibutylhydroxystannyl perchlorate has dimeric structure with pentacoordinated Sn. Similar molecules containing carboxylate groups were also examined.	214, 215
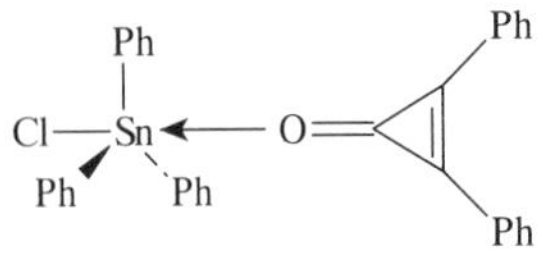 Ph Ph Cl—Sn ← O = Ph Ph Ph	Adduct has distorted trigonal bipyramidal structure at the Sn atom with clorine atoms in *trans*-positions.	216
$Ph_3SnO\text{-}i\text{-}Bu \cdot i\text{-}BuOH$	Analysis shows H-bonded chains with coplanar Sn links to phenyl groups in a distorted trigonal bipyramidal configuration.	217
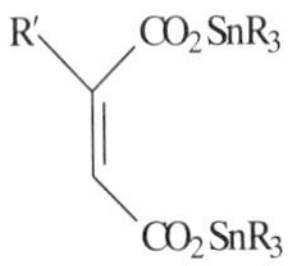 R' CO_2SnR_3 CO_2SnR_3 (**10**, $R' = Ph$)	Usually tetrahedral Sn coordination. In some cases one Sn is tretracoordinated and the other pentacoordinated. See also Section IV.B and Table 5.	124
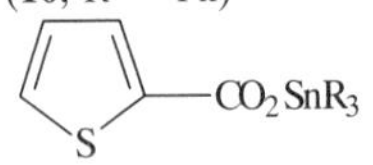 $—CO_2SnR_3$ S (R = Me, Ph)	The triphenylstannyl ester is monomeric with tetrahedral coordination. The trimethylstannyl ester is polymeric with pentacoordinated Sn in distorted *trans*-trigonal bipyramidal configuration.	218,219
$PhCH(OH)CO_2SnMe_3$	Tetrahedral Sn coordination, with H-bonding between the OH and CO_2 groups of neighbouring molecules	220

TABLE 8. (*continued*)

Compounds	Comments	References
RCO_2SnR_3' ($R = H$, Me; $R' = Ph$, $c\text{-}C_6H_{11}$) (**81**)	Polymeric trigonal bypiramidal Sn structure with *trans* $O-Sn-O$ binding. Also variable-temperature ^{119}Sn Mössbauer spectroscopy (see compound **50** in Table 5). These compounds are important pesticides.	221
$CO_2Sn(c\text{-}C_6H_{11})_3$ | $CO_2Sn(c\text{-}C_6H_{11})_3$	Two types of pentacoordinated Sn atoms: one with *cis*$-O-Sn-O$ binding, involving one oxalato species and the other one with *trans*$-O-Sn-O$ binding, involving two different oxalato species.	222a
$AcOSnPh_2SnPh_2OAc$	Polymeric crystalline structure. Monomer units linked by two isobidentate acetate moieties, with pentacoordinated Sn atoms	222b
$[Me_2Sn(H_2O)_4]^{2+}$	The cation has octahedral coordination with *trans*$-O-Sn-O$ configuration.	223
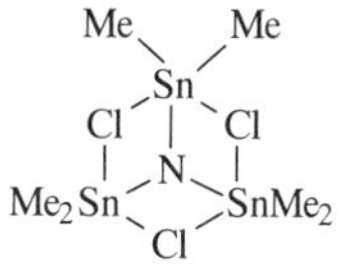	Planar configuration at N atom. Bridged Cl atoms have lower potential energy than unbridged ones.	224
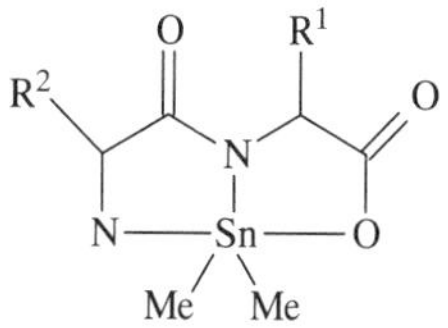	The crystalline structure of the coordination products of dimethyltin oxide with peptides, yielding potential antitumor agents, is being elucidated by various groups using X-ray crystallography and other methods. The adjacent formula shows the basic structure of the complex with a dipeptide.	225
12, $Z = NH_2$, $PhC\equiv C$ (see Table 5)	Strong transannular Sn$-$N coordination; transannular Sn$-$N bond is longer than the other three.	125, 126
35, **36** (see Table 5)	Distorted trigonal bipyramidal configuration, with N and Br atoms in *trans*-positions. Only one diastereoisomer of **36** appears in the solid state.	170
51, **52** and similar structures (see below Table 5)	Hydrogen bonding between OH groups of one polymer string and N atoms of a neighbouring string confer a ladder-type character to the structure. The tin atoms in **51** have a slightly distorted trigonal bipyramidal coordination, with two oxygen atoms lying on the axis. Spectroscopic and crystallographic data on compound **51** can be also interpreted as belonging to the ionic structure **52**.	181,182,226
$[Me_2SnF_4]^{2-}[NH_4]^+_2$	Two crystallographically independent centrosymmetric anions but only one kind of ammonium cations. Nearly linear $N-H\cdots F$ hydrogen bonds produce tridimensional network. Average Sn$-$F distance is 0.2127 nm, differing from the sum of covalent radii. See also Table 5.	153

(continued overleaf)

TABLE 8. (*continued*)

Compounds	Comments	References
$[(Ph_3Sn)_3 \cdot (C_2O_4)_2]^-$ Me_4N^+	An unusual structure showing pentacoordinated Sn atoms with both *cis*- and *trans* $-O-Sn-O$ configurations.	227
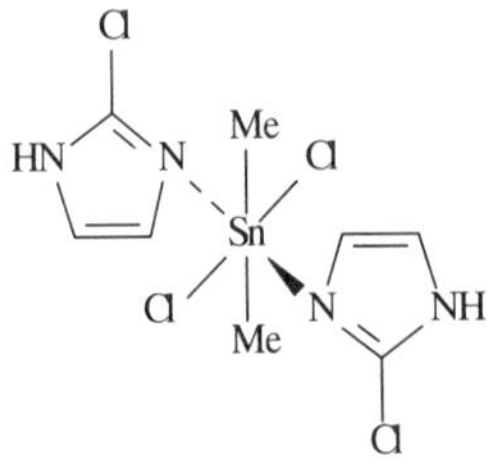	Octahedral coordination at Sn. The structure is made more compact by a network of $NH \cdots Cl-Sn$ hydrogen bonds with neighbouring molecules.	228
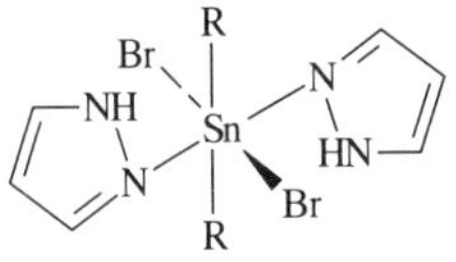	Octahedral coordination at Sn, with R = cyclohexyl. $NH \cdots Br$ hydrogen bonds are formed, as attested by IR spectra and some of the foreshortened N–Hal distances.	229
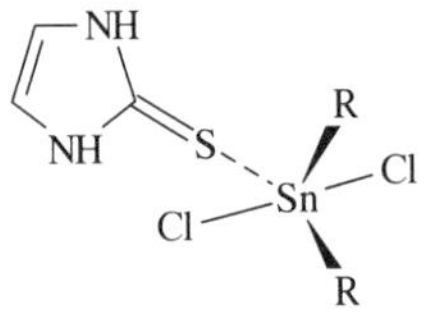	Distorted trigonal bipyramidal, with R = *n*-Bu. $NH \cdots Cl$ hydrogen bonds are formed, as attested by IR spectra and some of the foreshortened N–Hal distances. Other similar structures are known, such as R = Ph, which also yields an all-*trans* complex with two molecules of 1-methyl-2(3H)-imidazolinethione, that has octahedral configuration.	229, 230
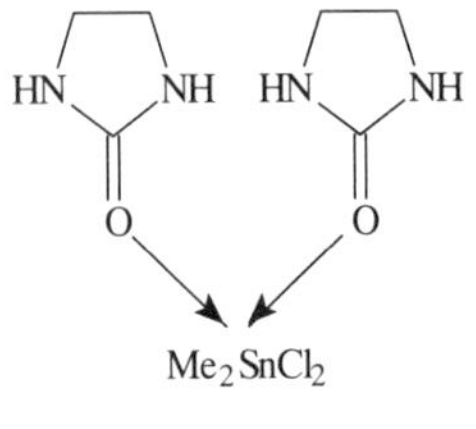 Me_2SnCl_2	Octahedral coordination at Sn. Cl–Sn–O bonds are almost collinear.	231
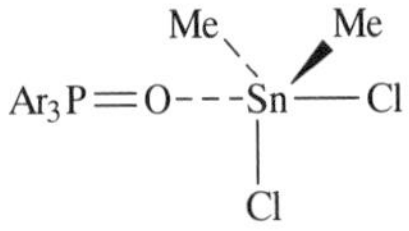	Pentacoordinated Sn with distorted trigonal bipyramidal configuration.	232
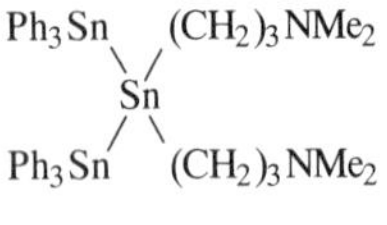	Tetrahedral coordination of all Sn atoms. No intra- or intermolecular interactions between Sn and N atoms are present.	233
53 (see Table 5)	Unit cell consisting of two independent cyclic hexamers of compound **53**, with carboxylic groups forming bridges between Sn atoms.	183

TABLE 8. (*continued*)

Compounds	Comments	References
Ar^1-Sn with Ar^3, CO_2Me, S, CO_2Me (thiastannete ring structure)	First 1,2-thiastannete stable at room temperature ever synthesized. Puckered four-membered ring[a].	234
t-Bu_3Sn-E-$Sn$$t$-$Bu_3$ (E = S, Se, Te)	The repulsion between the bulky substituents causes unusually large anisotropic thermal motion and increased Sn–E–Sn angles: S 134.2°, Se 127.4°, Te 122.3°	235
Ar^3, Ar, Sn, Y ring structure	Ar = Ph, Ar^1, Ar^4; Y = S, Se[a].	236
Me_2Sn with Te-$SnMe_2$, Te, Te-$SnMe_2$	Two crystalline forms, both with twisted boat conformations.	237
R, R, M=M, R, R structure	See formula **70** below Table 6. Centrosymmetric distribution of substituents on the M=M moiety. The fold angle of the R–M–R plane with the M–M line is 32° for M = Ge and 41° for M = Sn. Sn–Sn distance is 0.277 nm.	197
Ar^5, Ar^5, Sn=Sn, Ar^5, Ar^5 structure	Centrosymmetric distribution of substituents on the M=M moiety. The fold angle of the Ar–Sn–Ar plane with the Sn–Sn line is 46°. The Sn–Sn distance, 0.364 nm, points to a very weak bond, as compared to the preceding entry[a].	238, 239
64 (see below Table 6)	The angle between the plane determined by B–C–B and the C–Sn line is 16°. Sn is slightly pyramidal but the C atom in the cycle is significantly so.	191, 192
Ar^5, Ar^5, Sn=C, N–Ar^4, 153.9°, 175° structure	This stannaketeneimine has very bent geometry at the central carbon atom[a].	240
82 (see reactions 37 and 44 in Sections V.A.10 and V.A.12)	Sn and the three N atoms are coplanar, with the two Sn–N single bonds slightly shorter than most Sn–N single bonds.	241
Ar^4, Ar^3, Sn, Y, Sn, Ar^4, Ar^3 structure	Y = S, Se[a]. Ring is planar for *trans* configuration. Fold angles for *cis* configuration:	242

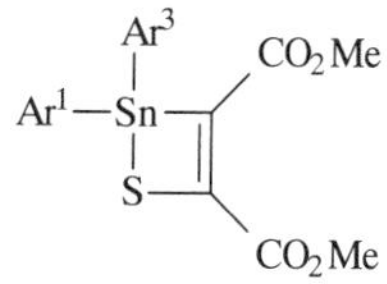

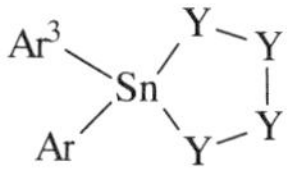

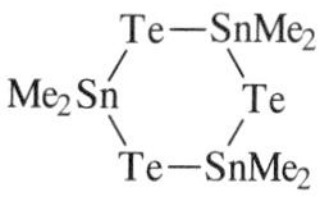

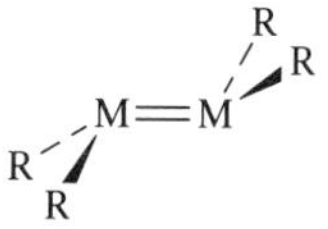

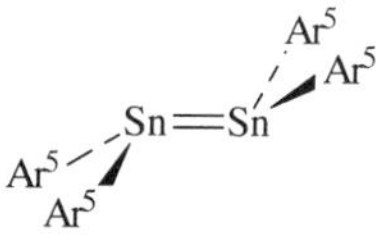

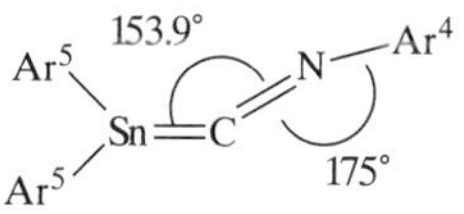

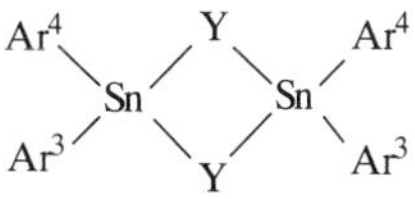

Y	Y–Y axis	Sn–Sn axis
S	39.8°	40.9°
Se	43.6°	41.3°

(*continued overleaf*)

TABLE 8. *(continued)*

Compounds	Comments	References
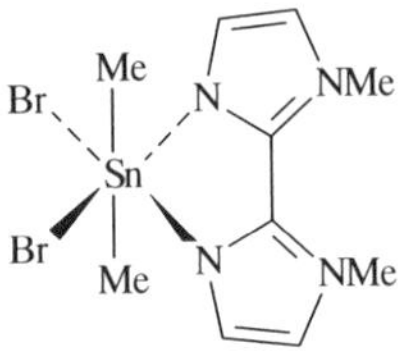 **(9)**	Discrete molecules with octahedral coordination and Me groups in *trans*-configuration.	123
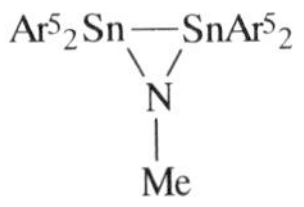 Ar^5_2Sn—$SnAr^5_2$ with N—Me bridge	Shorter Sn—Sn bond than in Sn=Sn compounds. Planar trigonal configuration at N atom[a].	243
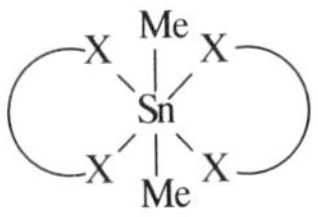 Chelates **54–56** (see below Table 5)	Distorted *cis* dimethyl octahedral geometry. Compound **56** is polymeric, with coordination 7 for the Sn atom.	184
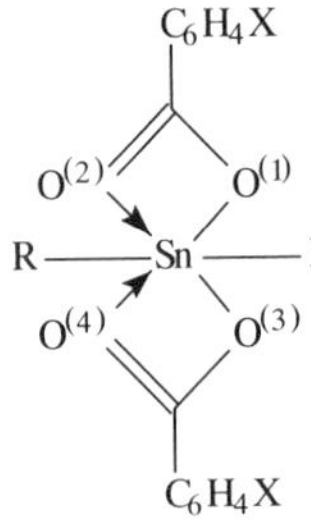 C_6H_4X structure with $O^{(1)}$, $O^{(2)}$, $O^{(3)}$, $O^{(4)}$, R groups	R = Me, Bu. Compounds of the series with X = *p*-Me, *p*-NH$_2$, *p*-Br are monomeric. Compounds of the series with X groups in *ortho*-positions show dimeric structure and have heptacoordinated Sn atoms. The great variation of bond strength in these compounds can be appreciated from the bond lengths in the dimeric case with R = Bu and X = *o*-Cl: Sn—$O^{(1)}$ 0.2122 nm; Sn—$O^{(2)}$ 0.2574 nm; Sn—$O^{(3)}$ 0.2104 nm; Sn—$O^{(4)}$ 0.2612 nm and Sn—$O^{(4')}$ 0.3881 nm (weak Sn—O binding to a carboxyl group of an adjacent molecule).	244
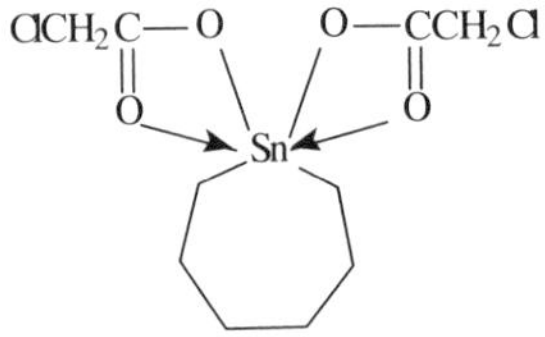 $ClCH_2C$—O O—CCH_2Cl structure **(63)**	Hexacoordinated Sn atom with two bidentate acyloxy groups. See also Table 5.	189
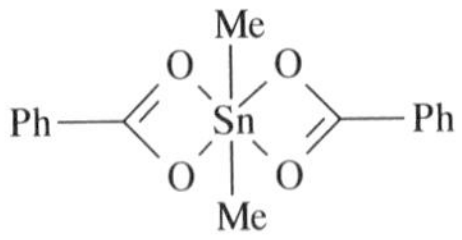 Ph—Sn—Ph with Me and O groups	Three Sn—O distances: *ca* 0.21 nm, *ca* 0.25 nm and *ca* 0.29 nm. The author favours a monomeric structure over a dimeric one, which would imply heptacoordination for the Sn atoms.	245

TABLE 8. (*continued*)

Compounds	Comments	References
$RSnX_nY_{3-n}$ R = Ph, Me; X = Cl, Br; Y = Et_2N-CS_2; $EtO-CS_2$; $(EtO)_2PS_2$; $n = 0-3$.	In the particular case of $PhSn(S_2C-NEt_2)_2$ the configuration about Sn is a distorted trigonal bipyramidal. As in the preceding entries, the dithiocarboxylate group acts as a bidentate ligand, showing two distinct Sn–S distances: 0.2487 and 0.2794 nm.	246
62 (see below Table 5)	Folded 4-membered ring, dihedral angle at Sn–Sn 22°. The >Sn–Sn< bond arrangement to aryl groups is nearly coplanar[a].	187
$[Me_3SnOPPh_2O]_4$	Inorganic 16-member $Sn_4P_4O_8$ cyclic skeleton.	247
20, 21 (see Table 2)	Compound **21** has two diastereoisomers and the two $SnMe_3$ groups are different.	136
105 (see reaction 43 in Section V.A.12)	Planar distannacyclobutene ring. The C=C bond is shorter than expected and the C–Sn bond longer. This may explain the slow dissociation this compound undergoes in solution. In the analogous Ge compound this ring is twisted.	248
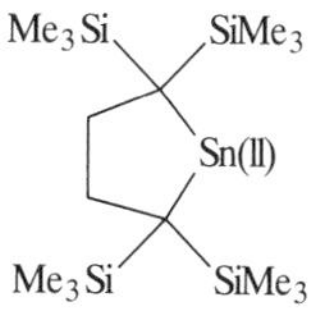	First case encountered of a stannylene compound that is monomeric in the solid state; shortest Sn-to-Sn distance is 714 pm. $[(Me_3Si)_2CH]_2Sn(II)$, for example, is dimeric as solid and monomeric as vapour or in solution.	249
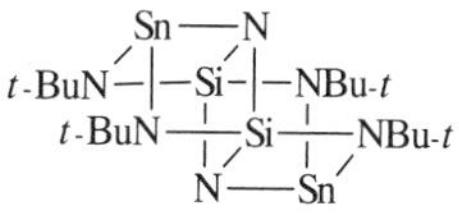	Sn(II) compound with tri-coordinated Sn atoms. This pentacyclo compound was prepared by the same method used for the dispiro compound **23** of Table 3 in Chapter 7 using analogous Ge reagents.	250
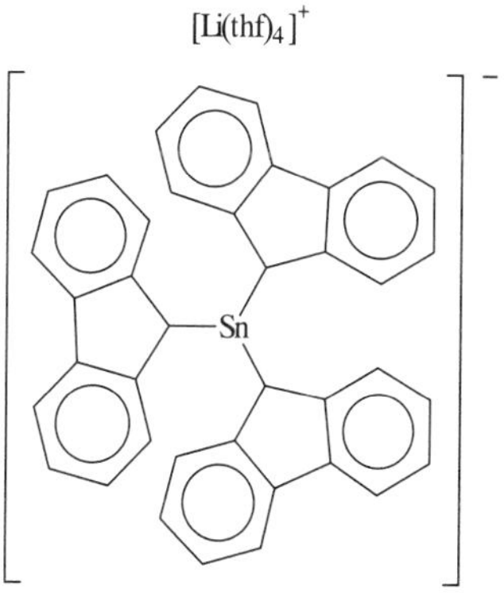	Sn(II) compound. The Sn atom and the bonds to fluorenyl groups lie in a distorted pyramidal configuration. See also following entry.	251

(*continued overleaf*)

TABLE 8. (continued)

Compounds	Comments	References
$[Mg(thf)_6]^{2+}$ 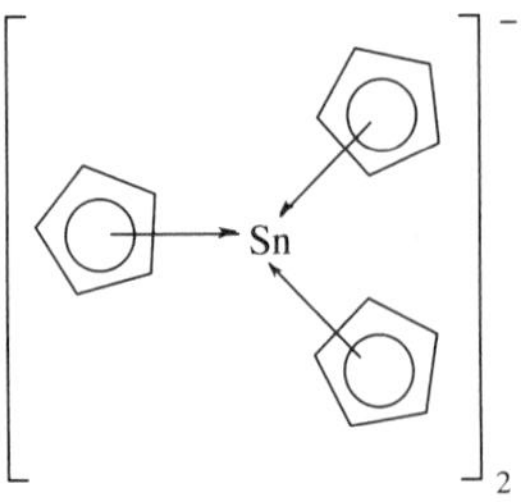	Sn(II) compound. The Sn atom and the bonds to cyclopentadienyl anions are coplanar. See also preceding entry.	251

^{a} See formula 5 in Section IV.A for aryl groups.

The electronic structure of dichorodiphenylstannane was calculated by the SCF-MS molecular orbital method and compared to that of dichlorodiphenylplumbane[256].

The solvent effects on charge-transfer spectra between $Me_3Sn-NCS$ and I_2 was investigated. Onsager's theory of dielectrics was used to estimate the stabilisation energy of excited states[257].

Racemic mixtures of compounds of the type 26–28 (see Table 4) could be resolved by HPLC using a chiral static phase, and the configuration was determined by circular dichroism[258].

The molecular structure of compound 84 in the gaseous phase was analysed by electron diffraction. It has chair conformation with C_s symmetry. The axial methyl groups are twisted outwards to relieve steric repulsion[259]. The symmetry of gaseous tetraphenylgermane is debatable on the basis of its electron diffraction spectrum[260]. Hexamethyldistannane ($Me_3SnSnMe_3$) has D_3 symmetry in the gas phase, as shown by electron diffraction[261].

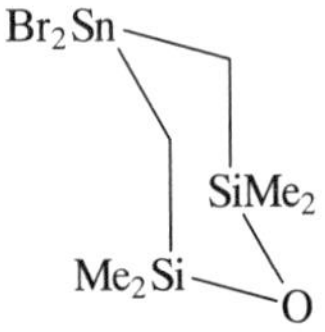

(84)

The structure of free radical 85 produced by dissociation of a stannyl benzopinacolate, was investigated by ESR spectroscopy and confirmed by ESR simulation. Free radical 85 is a good source for stannyl free radicals, as shown in reaction 10[262].

$$\underset{\substack{| \quad\quad\quad | \\ R_3SnO \quad OSnR_3}}{Ph_2C-CPh_2} \rightleftharpoons 2Ph_2\overset{\cdot}{C}-OSnR_3 \longrightarrow 2Ph_2C=O + 2R_3Sn^{\cdot} \quad (10)$$

(85)

Compounds 86, containing highly methylated aryl groups, yield the corresponding arene radical cation on treatment with $AlCl_3$ in CH_2Cl_2 solution, as demonstrated by ESR and

ENDOR (electron-nucleus double resonance) spectroscopies[263].

$$Ar_3MMAr_3 \ (M = Ge, Sn)$$
$$(86)$$

Compound **87** was the first case found of a stannanethione stable in solution. It was synthesized via the organotin(II) intermediate **71**; reaction 11 was followed by UVV spectrophotometry. When the less crowded Ar^4 was used instead of substituent Ar^1 (see formula **5**), the stannathione underwent dimerization to **88**[199]. See also Table 7.

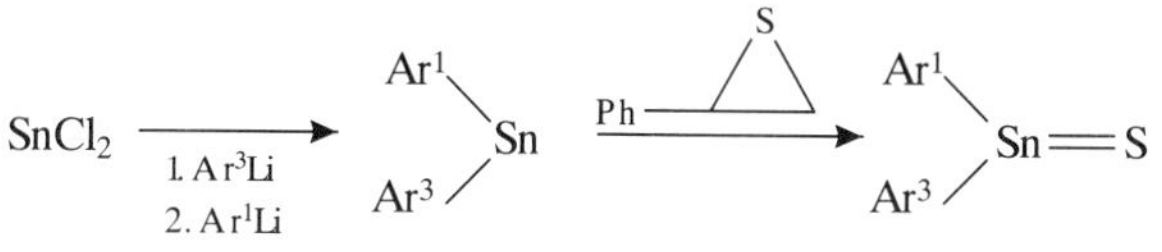

$$(71) \ \lambda_{max} \ 561 \ nm \qquad (87) \ \lambda_{max} \ 473 \ nm$$

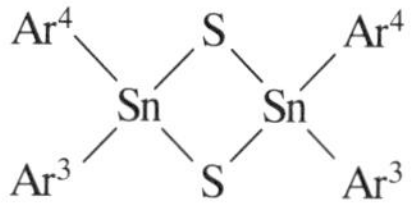

(88)

The UPS ($h\nu = 8-17$ eV), VUV (vacuum-UV, $h\nu = 6.20-11.28$ eV) and UVV spectra of methylstannane ($MeSnH_3$) were correlated with various ionization potentials and orbital transitions. The low Sn—C bond energy was confirmed[264]. UPS of tetramethylstannane in the $h\nu$ range $21-70$ eV were determined using synchrotron radiation at the magic angle θ_m, to simplify various considerations with respect to the radiation. The bands for $h\nu = 49$ eV were analysed[265].

A determination of fluoride in water, involving an organotin compound as analytical aid, consists of neutron activation of the sample, followed by precipitation and masking reactions of interfering nuclides, extraction of fluoride ions with a 3×10^{-2} M solution of chlorotrimethylstannane in toluene and measuring the activated fluorine emission in organic solution. Detection limit is *ca* 1 μg F/L[266]. Dichlorodiphenylstannane is a chromogenic reagent for detection and determination of flavones and flavonols on TLC plates[267].

V. DERIVATIZATION

See introductory remarks in Section I.A.5 of Chapter 7. Grouping organotin compounds according to a roughly defined functionality allows one in many cases to depict some common ways of behaviour. One should be cautious, however, not to stretch analogies too far, as the chemistry of organotin compounds needs still much exploration.

The traditional way of producing derivatives for analytical purposes aims at preserving, as much as possible, the important structural features of the analyte. In the case of organotin compounds this means that the tin moieties of the compound should appear in the derivative, as occurs in the methods described in Section V.A. Nevertheless, many organotin compounds are important intermediates in the synthesis of organic compounds devoid of metallic atoms. Such procedures can afford good derivatizing schemes, as described in Section V.B.

A. Organotin Derivatives

1. Stannanes containing Sn–H bonds

Stannane can be prepared and conserved at low temperatures. Its behaviour in the presence of various reagents is summarized in Scheme 1. It should be noted that reagents such as isopropylamine and acetic acid bring about decomposition of stannane, without themselves being apparently affected[268]. Some of these reactions can be developed into detection or determination methods for this compound.

Reagents catalysing decomposition

$$SnH_4 \xrightarrow{i\text{-PrNH}_2} Sn + 2H_2$$

$$SnH_4 \xrightarrow{AcOH} Sn + 2H_2$$

Halogenating reagents

$$SnH_4 \xrightarrow{BF_3 \cdot Et_2O} SnF_4$$

$$SnH_4 + PhCH_2Cl \longrightarrow SnCl_4 + PhCH_3 + Ph_nCH_{4-n}$$

Reductions by SnH₄

$$SnH_4 + Me_2CO \longrightarrow Sn + Me_2CHOH$$

$$SnH_4 + PhCH{=}O \longrightarrow Sn + PhCHOH$$

$$SnH_4 + PhNO_2 \longrightarrow Sn + PhNH_2$$

$$SnH_4 + 4CH_2{=}CHCN \longrightarrow Sn(CH_2CH_2CN)_4$$

SCHEME 1. Reactions of stannane[268]

Trialkylstannanes undergo palladium-catalysed regioselective addition to 1,3-butadienes, as shown in reaction 12[269].

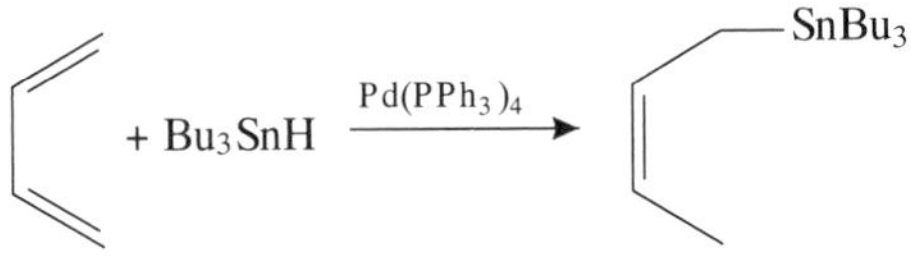

$$(12)$$

2. Arylstannanes

Phenyl groups attached to tin can be replaced aided by mercury(II) chloride, as shown in reaction 13[270].

$$(RPh_2Sn)_2CR'_2 + 4HgCl_2 \longrightarrow (RCl_2Sn)_2CR'_2 + 4PhHgCl \qquad (13)$$

$$R = Ph,\ CH_2SiMe_3$$

$$R' = H,\ Me$$

3. Compounds with multiple group 14 heteroatoms

Typical derivatives of this type of compounds are obtained by addition to carbon–carbon double and triple bonds, usually accompanied by metal–metal bond scission. The highly crowded cyclotristannane **89** reacts readily with phenylacetylene to yield a distannacyclobutene **31** (see Table 4), as shown in reaction 14. This points to intermediate **69** (Table 6) as possible dissociation product of **89** before undergoing (2 + 2) cycloaddition[271].

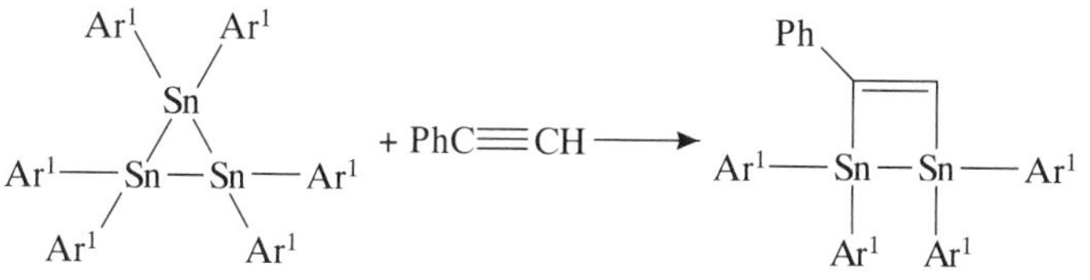

$$
\begin{array}{c}
Ar^1 \diagdown \quad \diagup Ar^1 \\
Sn \\
Ar^1 \diagdown \quad \diagdown \\
Ar^1 - Sn - Sn - Ar^1 \\
\diagup \quad \diagdown \\
Ar^1 \qquad Ar^1
\end{array}
\quad + \; PhC \equiv CH \longrightarrow
\begin{array}{c}
Ph \diagdown \\
\square \\
Ar^1 - Sn - Sn - Ar^1 \\
| \quad | \\
Ar^1 \quad Ar^1
\end{array}
\qquad (14)
$$

(89)   **(31)**

Silylstannanes undergo regiospecific addition to alkenes and alkynes with scission of the Sn–Si bond, as illustrated in reactions 15 and 16[272].

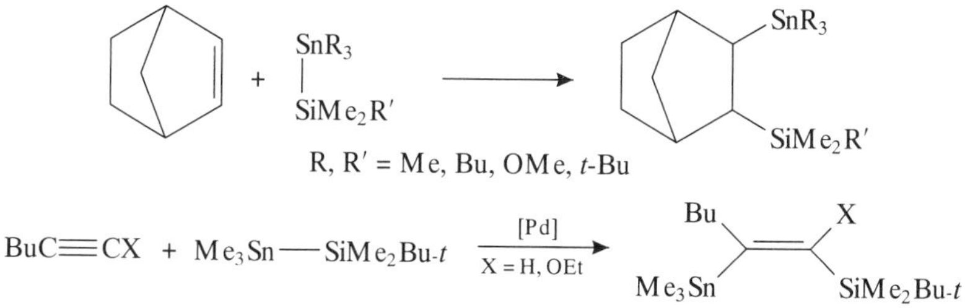

$$
\text{(norbornadiene)} + \begin{array}{c} SnR_3 \\ | \\ SiMe_2R' \end{array} \longrightarrow \text{(norbornane with } SnR_3 \text{ and } SiMe_2R') \qquad (15)
$$

$$
R, R' = Me, Bu, OMe, t\text{-}Bu
$$

$$
BuC \equiv CX \; + \; Me_3Sn - SiMe_2Bu\text{-}t \xrightarrow[\;X = H, OEt\;]{[Pd]}
\begin{array}{c}
Bu \diagdown \qquad \diagup X \\
\diagup \qquad \diagdown \\
Me_3Sn \qquad SiMe_2Bu\text{-}t
\end{array}
\qquad (16)
$$

4. Stannanes bearing reactive functional groups on alkyl groups

Trialkylstannyl groups—e.g. tributylstannyl—resist reaction under conditions that affect other parts of organic molecules. This property may be useful for mechanistic and structural studies.

Tetraalkylstannanes with an α-halo substituent undergo α, β-elimination in the presence of 1,8-diazabicyclo[5.4.0]undec-7-ene (DBU, **90**), yielding vinylstannanes with predominant E-configuration, as shown in reaction 17[273].

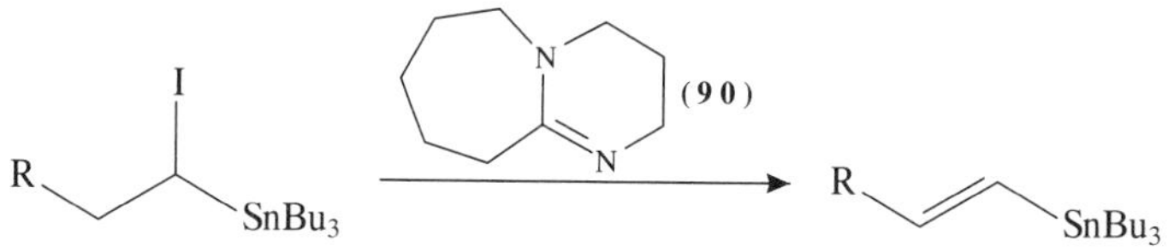

$$
\begin{array}{c}
I \\
R \diagup \diagdown \diagup \\
SnBu_3
\end{array}
\xrightarrow{\;(90)\;}
\begin{array}{c}
R \diagup \diagdown \diagup \\
SnBu_3
\end{array}
\qquad (17)
$$

Protected enol ethers yield the corresponding aldehydes, as shown in reaction 18[274a].

$$
Bu_3Sn \diagdown \diagup \diagdown _{OR} \xrightarrow{\;Bu_4NF \cdot OEt_2\;} Bu_3Sn \diagdown \diagup \diagdown _{CH=O} \qquad (18)
$$

α, β-Unsaturated ketone **91** undergoes a Michael-type addition with compound **92**, followed by an intramolecular carbonyl–methylene condensation leading to compound **93**.

The latter product can be reduced to an alcohol (**94**, X = H) and subsquently acetylated
(**94**, X = Ac)[274b].

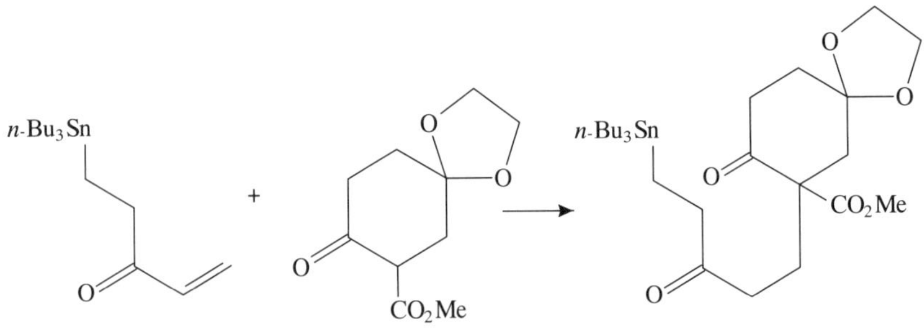

(**91**)   (**92**)

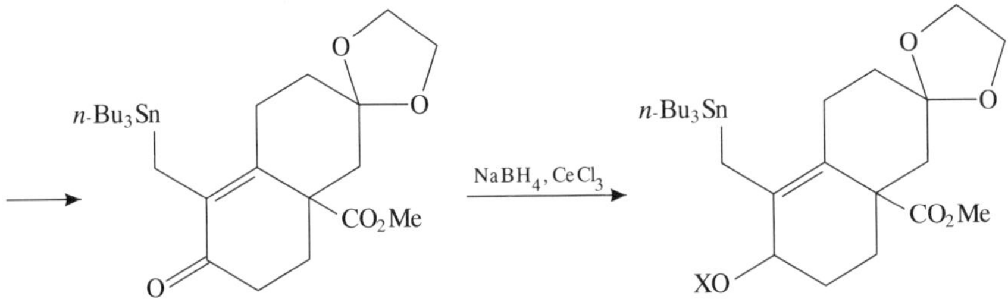

(**93**)   (**94**)

5. Alkenylstannanes

The most typical derivatizing process for vinylstannanes would be cross-coupling, as
shown in various examples in Section V.B. However, in certain cases other reactions may
afford convenient derivatives preserving the stanyl moiety.

Vinylstannanes with hydroxy groups in the allylic position undergo enantioselective
and diasteroselective hydrogenation in the presence of rhodium catalysts, as illustrated in
reaction 19[275]. Vinylstannanes can be converted into β-stannylacrylic esters in a two-step
synthesis, as shown in reaction 20[276].

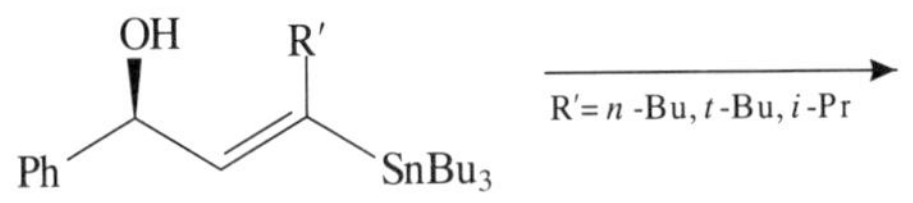

(19)

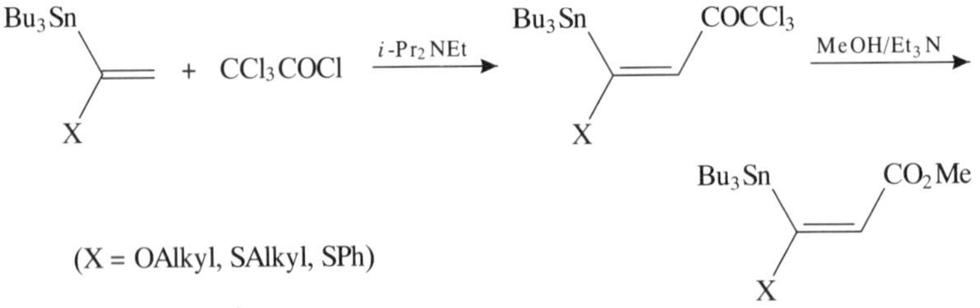

(X = OAlkyl, SAlkyl, SPh)

(20)

Trialkyl(2-borylvinyl)stannanes react with sulphur bis(trimethylsilylimide) to yield heterocyclic ylide compounds, as shown in reaction 21[277].

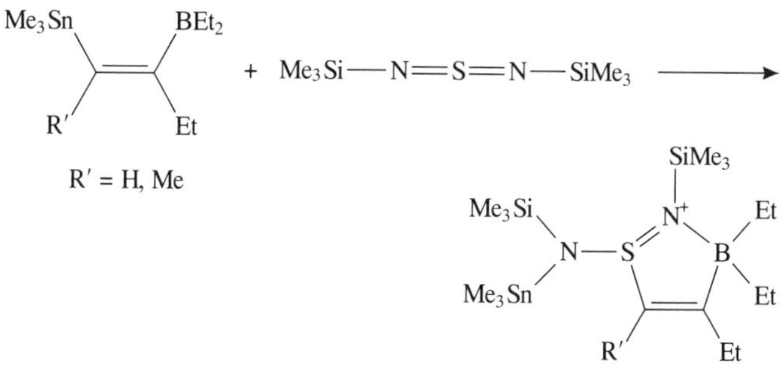

$$
\begin{array}{c}
\text{Me}_3\text{Sn} \quad\quad \text{BEt}_2 \\
\diagdown \quad \diagup \\
\diagup\diagdown\diagup\diagdown \\
\text{R}' \quad\quad \text{Et}
\end{array}
\; + \; \text{Me}_3\text{Si}\!-\!\text{N}\!=\!\text{S}\!=\!\text{N}\!-\!\text{SiMe}_3 \longrightarrow
$$

R' = H, Me

(21)

6. Alkynylstannanes

In many respects the reactions of alkynylstannanes are similar to those of vinylstannanes.

Alkynylstannanes yield Z-vinylstannanes stereoselectively, when treated with zirconocene hydrochoride. This can be easily followed by other substitution processes, such α-iodination of the vinylstannane, as shown in reaction 22[278].

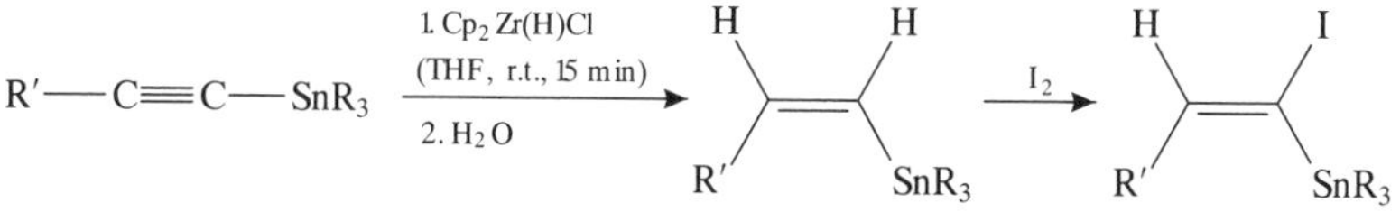

$$
\text{R}'\!-\!\text{C}\!\equiv\!\text{C}\!-\!\text{SnR}_3 \;\xrightarrow[\substack{2.\,\text{H}_2\text{O}}]{\substack{1.\,\text{Cp}_2\,\text{Zr(H)Cl} \\ (\text{THF, r.t., 15 min})}}
$$

(22)

Dialkynylstannanes were synthesized and characterized by their NMR spectra. These compounds react with boranes yielding 1-stanna-4-bora-2,5-cyclohexadienes (97), stannoles (98), as shown in reaction 24, and stannolines (99), as shown in reaction 25. When R' ≠ H it is possible to isolate at low temperature the tautomeric intermediates 95 and 96 of reaction 23, one of which is a π-stabilised stannyl cation complex[279]. Application of the same borane derivatization scheme to tetraalkynylstannanes leads to formation of 1-stannaspiranes where the rings are combinations of 97, 98, and 99 structures[116,280].

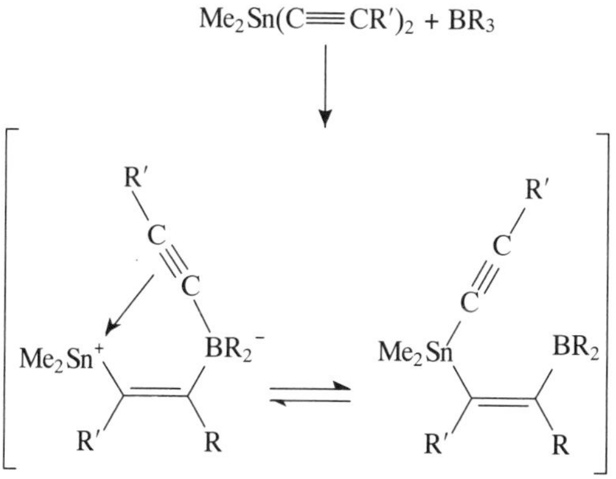

$$
\text{Me}_2\text{Sn(C}\!\equiv\!\text{CR}')_2 + \text{BR}_3
$$

(23)

(95)   (96)

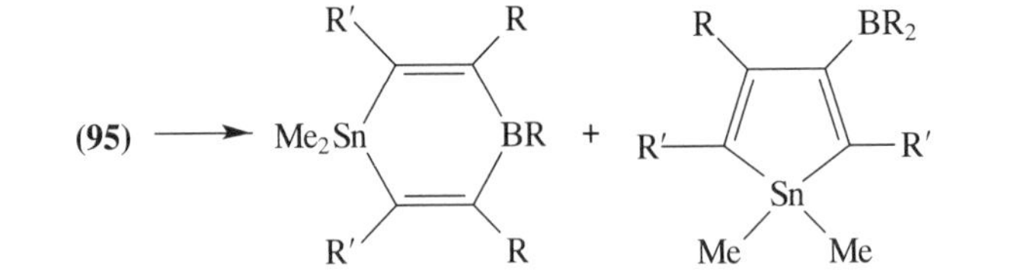

$$(95) \longrightarrow \text{Me}_2\text{Sn} \underset{R'}{\overset{R'}{\diagdown}} \text{BR} + \underset{\text{Me}}{\overset{R}{\diagdown}} \text{Sn} \underset{\text{Me}}{\overset{BR_2}{\diagdown}} R' \qquad (24)$$

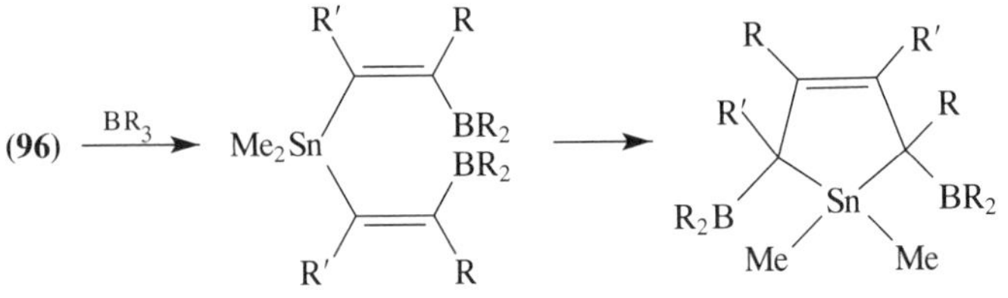

(97)   (98)

$$(96) \xrightarrow{\text{BR}_3} \text{Me}_2\text{Sn} \underset{R'}{\overset{R'}{\diagdown}} \underset{BR_2}{\overset{BR_2}{\diagdown}} \longrightarrow \underset{R_2B}{\overset{R}{\diagdown}} \text{Sn} \underset{\text{Me}}{\overset{R}{\diagdown}} BR_2 \qquad (25)$$

(99)

7. Stannyl halides

Organotin halides are versatile reagents for which many derivatizing schemes can be devised. Trace analysis of organotin componds in environmental samples often involve halostannanes, as was discussed in Section III.

The effect of pK_a on the stability constant of the 1:1 monodentate complexes of Me_3SnX with amino acids in aqueous solutions was studied by potentiometric methods[281].

Dialkyldichlorostannanes undergo reduction to dialkylstannanes (**100**), as shown in reaction 26. Product **100** is a good dehalogenation reagent for more complex stannyl chlorides, such as compound **32** (see Table 5) yielding compound **19** (see Table 2)[134].

$$\text{Bu}_2\text{SnCl}_2 \xrightarrow{\text{LiAlH}_4} \text{Bu}_2\text{SnH}_2 \qquad (26)$$
$$\text{(100)}$$

Trialkylstannyl chlorides undergo dealkylation in the presence of iodine chloride, as shown in reaction 27. The alkyl iodide product reacts further very slowly in the case of primary alkyl groups, however, reaction 28 proceeds readily for R = i-Pr. The mechanism involves a charge transfer complex that can be detected in the reaction mixture. The compounds involved in the process can be analysed by GC, NMR and UVV spectroscopy[282].

$$\text{R}_3\text{SnCl} + \text{ICl} \longrightarrow \text{R}_2\text{SnCl}_2 + \text{RI} \qquad (27)$$

$$\text{R} = \text{Me, Et, } i\text{-Pr, } n\text{-Bu}$$

$$\text{RI} + \text{ICl} \longrightarrow \text{RCl} + \text{I}_2 \qquad (28)$$

Trialkylfluorostannanes are insoluble polymeric solids (see entry **33′** in Table 5). They convert slowly into the corresponding chloride or bromide according to reaction 29[283].

$$\text{R}_3\text{SnF} + \text{NaX} \xrightarrow{\text{THF}} \text{R}_3\text{SnX} \qquad (29)$$

$$\text{X} = \text{Cl, Br}$$

In the particular case of bis(dichloroorganostannyl)methane structures, obtained from the *gem*-distannyl compounds shown in reaction 13, treatment with sodium chalcogenides will yield the adamantane analogues shown in reaction 30^{270}.

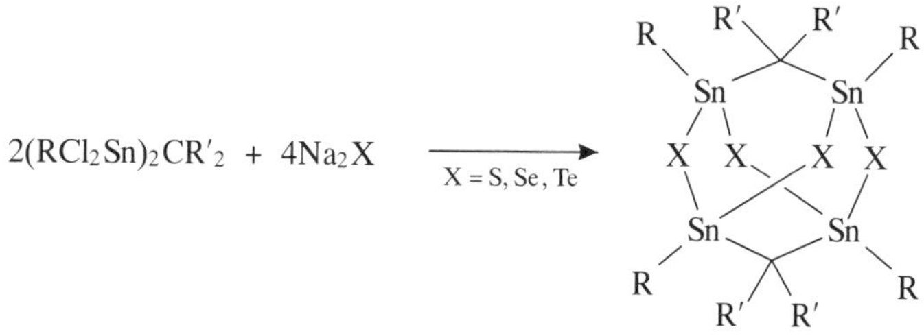

$$2(RCl_2Sn)_2CR'_2 + 4Na_2X \xrightarrow[X = S, Se, Te]{} \quad (30)$$

8. Organotin alkoxides, amines and analogous compounds

A method for determination of volatile thiols consists of preparing the corresponding tributyltin mercaptides according to reaction 31. After concentrating the mercaptides they are hydrolysed with aqueous hydrochloric acid, and the salted-out thiols are determined by GC-FID. This was applied to analysis of thiols in cigarette smoke[284].

$$RSH + Bu_3SnOH \longrightarrow Bu_3SnSR + H_2O \qquad (31)$$

Alkoxytrialkylstannanes react with alkynylcyclobutenones (**101**) to yield stannylated quinones[285]. The stannyl group can be displaced by aryl or hetaryl groups, as shown in reaction 32^{286a}.

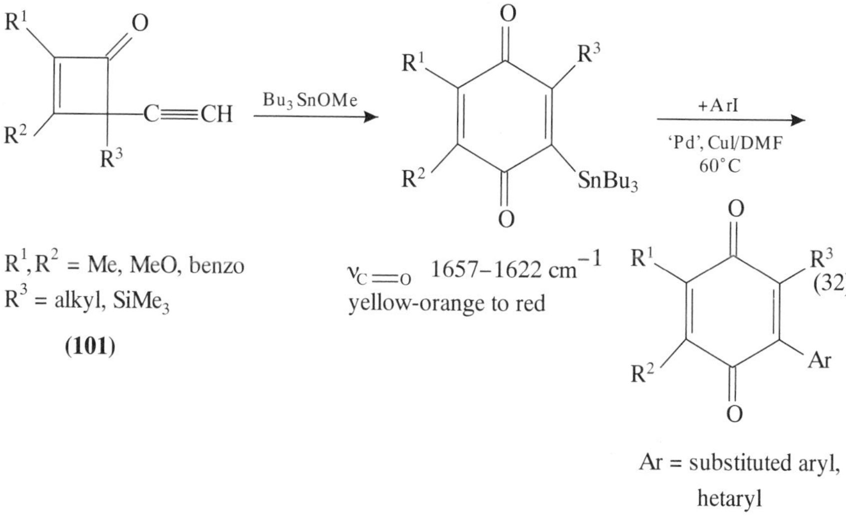

$R^1, R^2 = $ Me, MeO, benzo
$R^3 = $ alkyl, SiMe$_3$

(101)

$\nu_{C=O}$ 1657–1622 cm^{-1}
yellow-orange to red

$$(32)$$

Ar = substituted aryl,
hetaryl

Dialkoxydialkyl stannanes cross-couple with derivatives of chiral tartaric acid, preserving the original chirality, as shown in reaction 32'. The dialkylstannylene acetals produced in this reaction are useful reagents for synthesis of chiral compounds[286b].

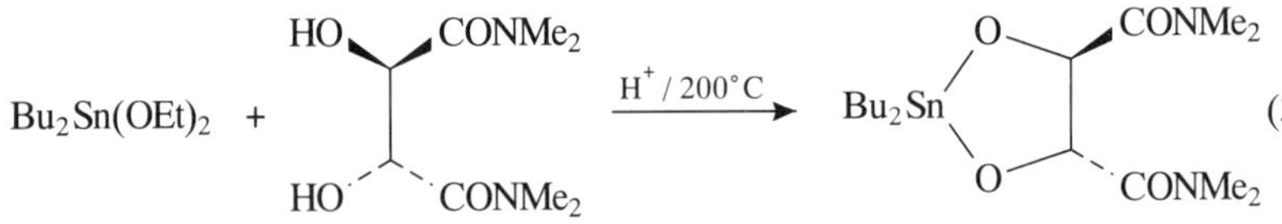

$$\text{Bu}_2\text{Sn(OEt)}_2 \; + \quad \underset{\text{HO}}{\overset{\text{HO}}{\bigwedge}} \quad \xrightarrow{\;\text{H}^+\,/\,200^\circ\text{C}\;} \quad \text{Bu}_2\text{Sn} \qquad (32')$$

Bis(stannyl) tellurides yield aryl stannyl sulphides, as shown in reaction 33[287].

$$\text{Ph}_3\text{Sn}-\text{Te}-\text{SnPh}_3 + \text{ArS}-\text{SAr} \longrightarrow 2\text{Ph}_3\text{Sn}-\text{S}-\text{Ar} + \text{Te} \qquad (33)$$

Reaction 34[288] served as model for the displacement of the group $Z = \text{NMe}_2$ in compounds **12** (see Table 5) by a variety of substituents, such as halogeno, and alkynyl groups[125,126].

$$\text{R}_3\text{SnNR}_2' + \text{HA} \longrightarrow \text{R}_3\text{SnA} + \text{HNR}_2' \qquad (34)$$

9. Organotin oxides

Di-t-butyltin oxide has a trimeric structure (**102**). Arylboronic acids can condense into trimeric structures and can displace tin sectors from **102**, as shown in reaction 35. The structure of these compounds was elucidated by X-ray crystallography[289].

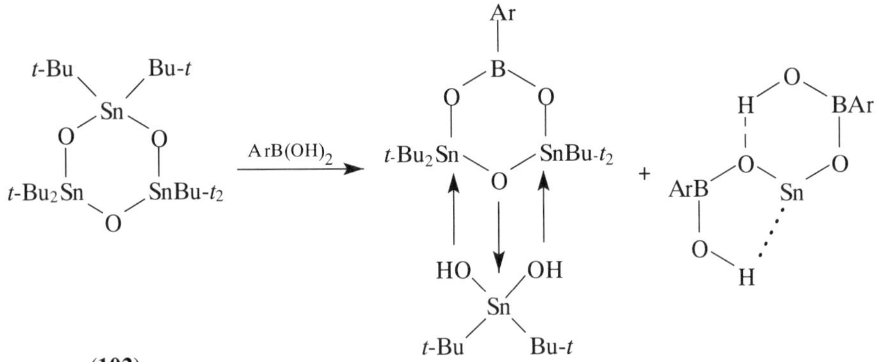

(102)

$$(35)$$

Dialkyltin oxides yield acetal adducts with 2,3,4-tri-O-benzylpyranoses possessing the proper configuration at $C_{(1)}$ and $C_{(4)}$, as is the case of glucose, mannose and galactose, with OH at $C_{(1)}$ and CH_2OH at $C_{(4)}$ *cis* to each other. Reaction 36 illustrates the process for tribenzilated galactopyranose[290].

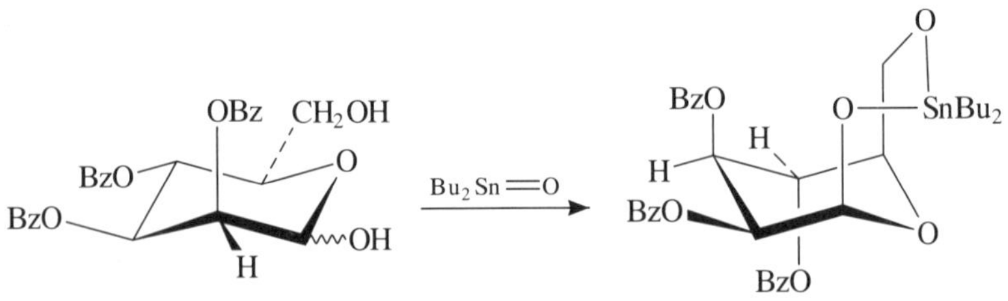

$$(36)$$

10. Organotin compounds containing double-bonded tin atoms

Only a few stable organometallic compounds containing double-bonded tin are known. Stannaimines are very reactive and can be scavenged in various ways. The sterically crowded groups of compound **82** confer to it some stability. Compound **82** decomposes slowly in solution below 0 °C, according to reaction 37. See also Table 8 and reaction 44[241].

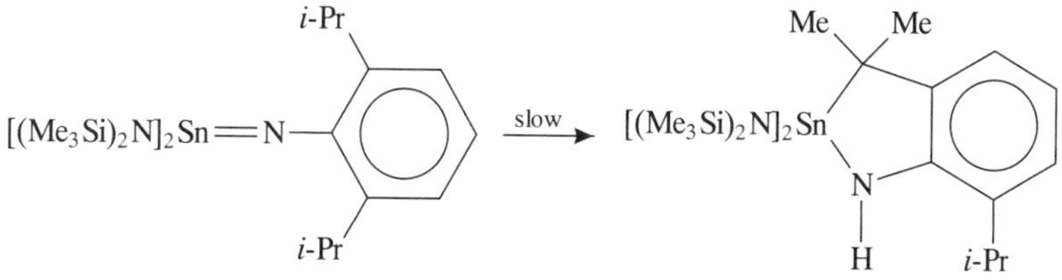

(37)

(82)

Stannanethione **87** is stable in solution. It undegoes the [2 + 2] cycloaddition reaction 38[199].

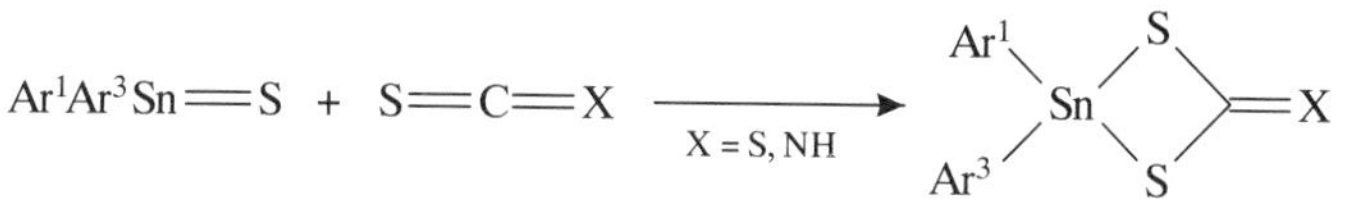

(38)

(87)

Hydrogenation reactions with LiAlH$_4$ and addition of acidic compounds XH, afford good derivatizing and scavenging schemes for compounds like **66** (see Table 6), which are very reactive[149]. For example, compounds **29, 30** (Table 4), **40** and **41** (Table 5) are derived in such a manner.

11. Miscellaneous functional groups

Reaction 39 shows the condensation of acylstannanes with primary amines. The only isomer produced was tentatively ascribed the Z-configuration[291]. Acylstannanes undergo Wittig-type processes with organophosphorus compounds to yield vinyl stannanes, as illustrated in reactions 40 and 41. In the latter reaction products are preferably of Z-configuration[292].

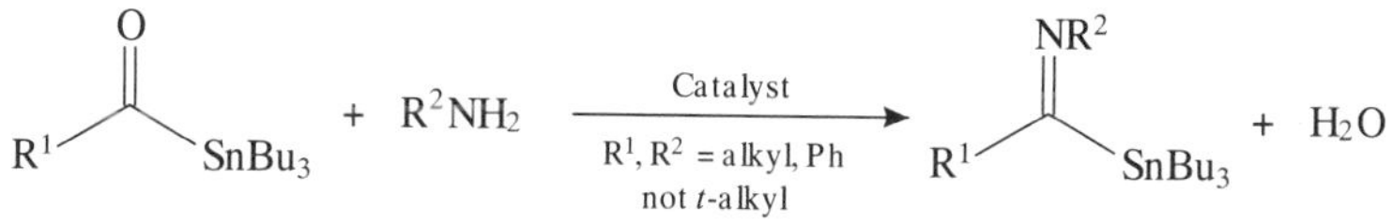

(39)

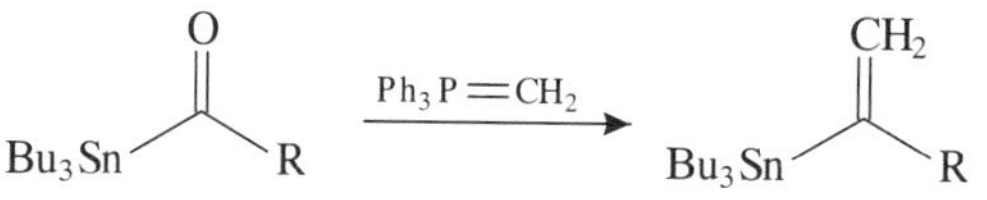

(40)

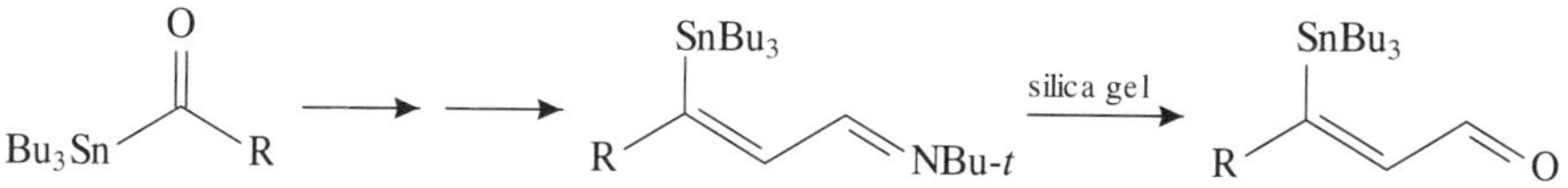

(41)

Reaction 42 shows the thermal rearrangement of bis(stannyl)diazomethanes into carbodiimides, which can be further derivatized in many ways[293].

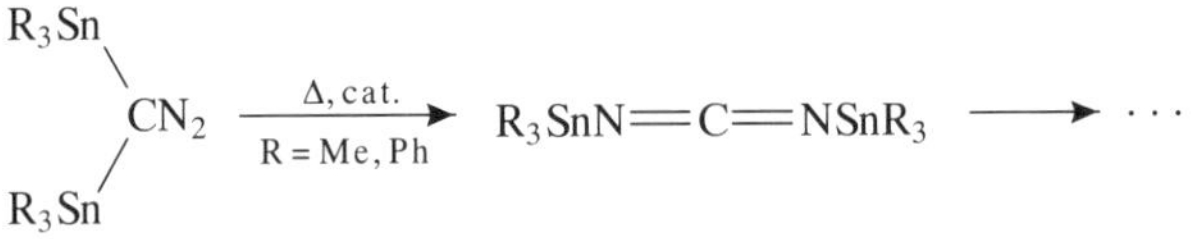

$$R_3SnN=C=NSnR_3 \longrightarrow \cdots \quad (42)$$

with the reaction scheme showing the reactant with R_3Sn groups and CN_2, conditions Δ, cat., $R = Me, Ph$.

12. Organotin(II) compounds

Organotin(II) compound **103** reacts with organic azides to yield stable dispiro compounds **104**[294]. This derivatizing reaction may be applicable to different types of organotin(II) compounds. Another effective derivatizing possibility is with acetylenic compounds, as shown in reaction 43 (see Table 8)[248].

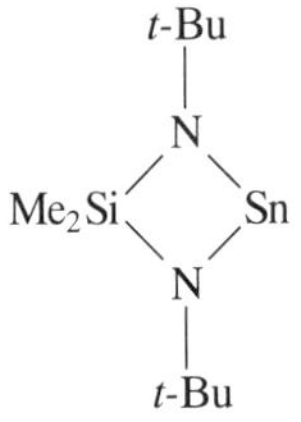

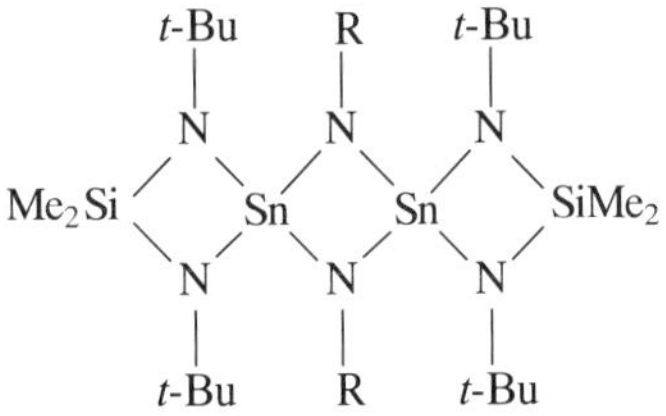

R = Ar, ArCO, ArSO$_2$

(103)   **(104)**

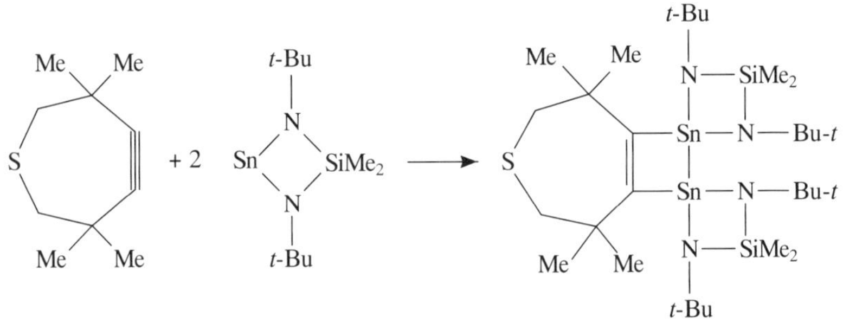

(105)

With sterically crowded substituents it is possible to obtain stannaimines, as shown in reaction 44. If R is a 2,6-diisopropylphenyl group, then compound **82** of reaction 37 is produced[241].

$$[(Me_3Si)_2N]_2Sn + N_3R \xrightarrow{-30\,°C,\ hexane} [(Me_3Si)_2N]_2Sn=NR + N_2 \quad (44)$$

B. Derivatives Containing No Tin

1. Allyltin compounds

Allylstannanes undergo diasterospecific additions to chiral α-alkoxyaldehydes, as shown in reaction 45[295]. Stereospecific additions to aldehydes are attained in the presence of chiral catalysts[296].

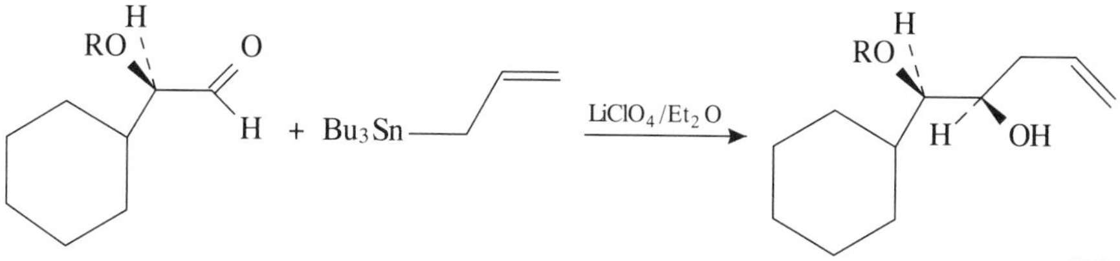

(45)

Pentadienyltrimethylstannanes undergo regioselective conjugate additions to aldehydes, catalysed by Lewis acids. The dominant product obtained depends on the catalyst used, as shown in reaction 46. In the case of titanium tetrachloride catalysis the reaction is also stereoselective and only one diasteroisomer is obtained[297]. Reaction with chiral aldehydes leads to asymmetric induction with similar organotin compounds[298].

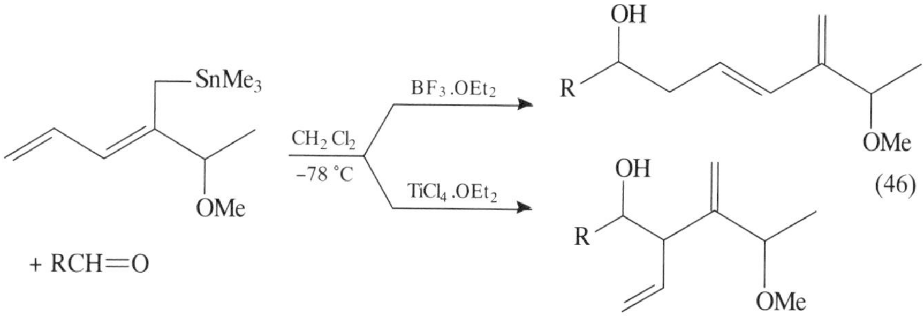

(46)

Tetraalkylstannanes with double bonds at allylic or more remote positions undergo cross-coupling with allylsilanes to yield dienes, as illustrated in reaction 47[299].

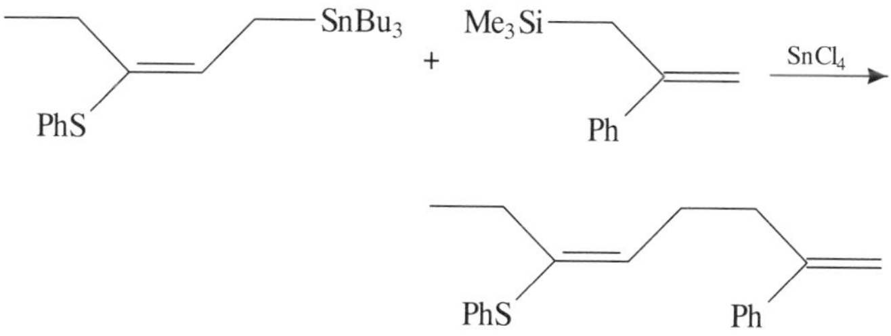

(47)

2. Vinyltin and alkynyltin compounds

An important derivatizing scheme of vinyltin compounds is based on cross-coupling reactions with reagents bearing good leaving groups. Such reactions are often regiospecific or stereospecific, however, in many cases a specific palladium catalyst is required. Various examples of the versatility of vinyltin compounds follow.

Trialkylvinylstannanes react with 1-bromoalkynes, as shown in reaction 48[300].

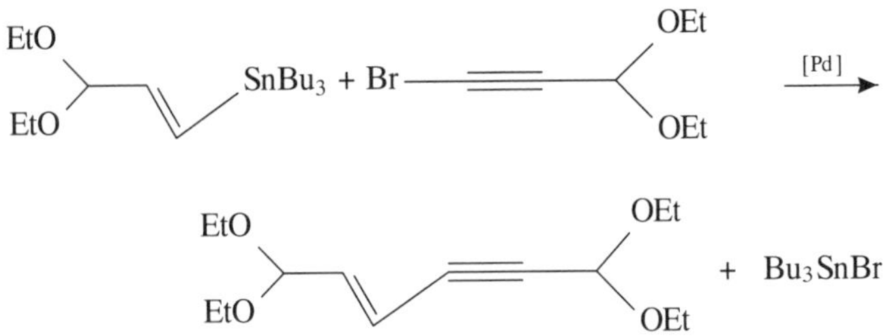

$$
\begin{array}{c}
\text{EtO}\diagdown \\
\text{EtO}\diagup
\end{array}
\diagdown\text{SnBu}_3 + \text{Br} \textemdash \textemdash \text{\equiv}\textemdash
\begin{array}{c}
\diagup\text{OEt} \\
\diagdown\text{OEt}
\end{array}
\xrightarrow{\text{[Pd]}}
$$

(48)

$$
\begin{array}{c}
\text{EtO}\diagdown \\
\text{EtO}\diagup
\end{array}
\diagdown\diagup\text{\equiv}\diagup
\begin{array}{c}
\diagup\text{OEt} \\
\diagdown\text{OEt}
\end{array}
\quad + \quad \text{Bu}_3\text{SnBr}
$$

The Z- or E-configuration of the vinyl groups of linchpins (**106, 107**) is preserved when the stannyl groups are displaced by triflate esters, as shown in reactions 49 and 50[301].

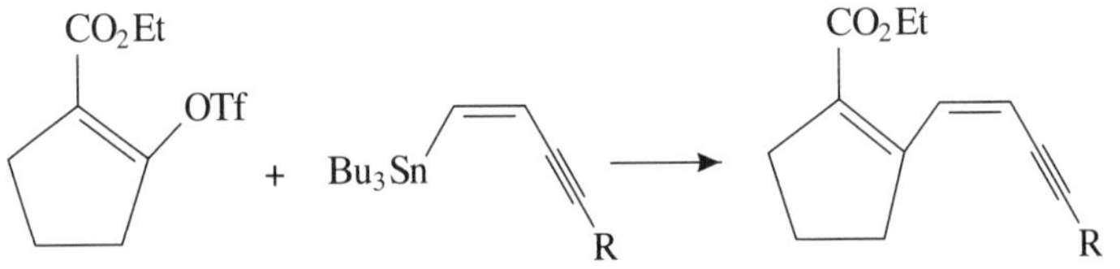

$$
\begin{array}{c}
\text{CO}_2\text{Et} \\
\diagup\text{OTf}
\end{array}
\quad + \quad \text{Bu}_3\text{Sn}\diagup\diagdown_R
\longrightarrow
\begin{array}{c}
\text{CO}_2\text{Et} \\
\end{array}_R
$$

(49)

(106)

$$
\begin{array}{c}
\text{CO}_2\text{Et} \\
\diagup\text{OTf}
\end{array}
\quad + \quad
\text{Bu}_3\text{Sn}\diagup\diagdown_R
\longrightarrow
\begin{array}{c}
\text{CO}_2\text{Et}
\end{array}^R
$$

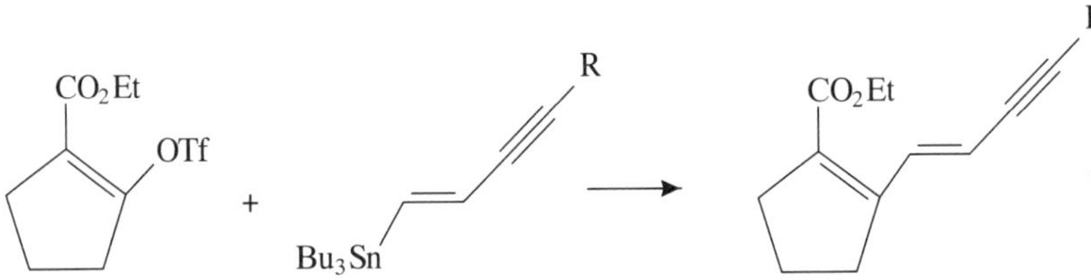

(50)

(107)

α, β-Unsaturated esters stannylated in the α-position can undergo stannyl displacement by various groups, as shown in reactions 51 and 52[302a,b]. Vinylstannanes possessing a remote carbonyl or tosylate group yield cyclic compounds in the presence of butyllithium, as illustrated in reactions 53–55[303].

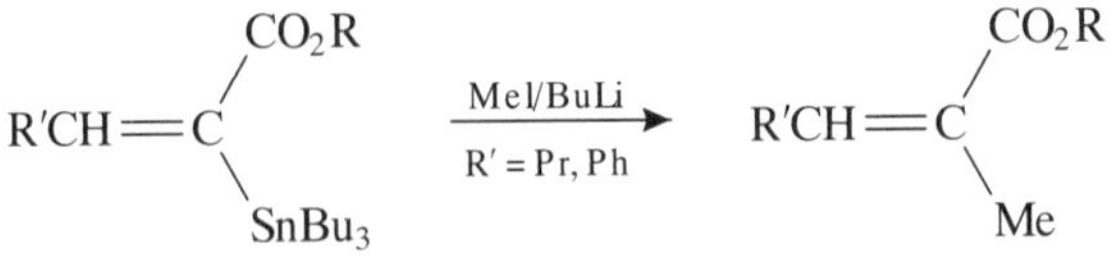

$$
\text{R}'\text{CH}\text{=}\text{C}
\begin{array}{c}
\diagup\text{CO}_2\text{R} \\
\diagdown\text{SnBu}_3
\end{array}
\xrightarrow[\text{R}' = \text{Pr, Ph}]{\text{MeI/BuLi}}
\text{R}'\text{CH}\text{=}\text{C}
\begin{array}{c}
\diagup\text{CO}_2\text{R} \\
\diagdown\text{Me}
\end{array}
$$

(51)

$$
\text{R}'\text{CH}\text{=}\text{C}
\begin{array}{c}
\diagup\text{CO}_2\text{R} \\
\diagdown\text{SnBu}_3
\end{array}
\xrightarrow[\text{R}' = \text{Ph}]{\text{CH}_2\text{=CHCH}_2\,\text{Cl/Pd(0)}}
\text{R}'\text{CH}\text{=}\text{C}
\begin{array}{c}
\diagup\text{CO}_2\text{R} \\
\diagdown\text{CH}_2\text{CH}\text{=CH}_2
\end{array}
$$

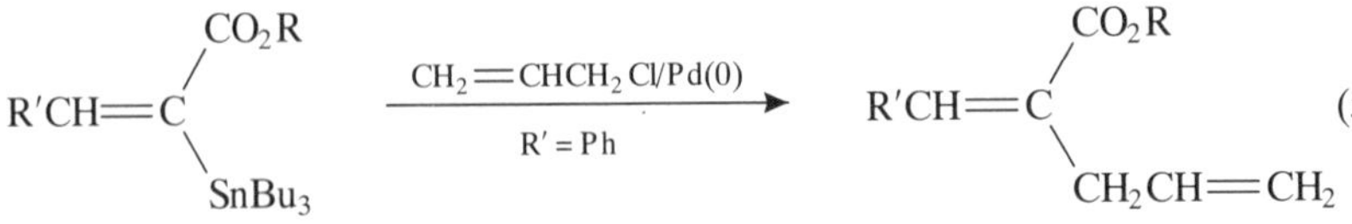

(52)

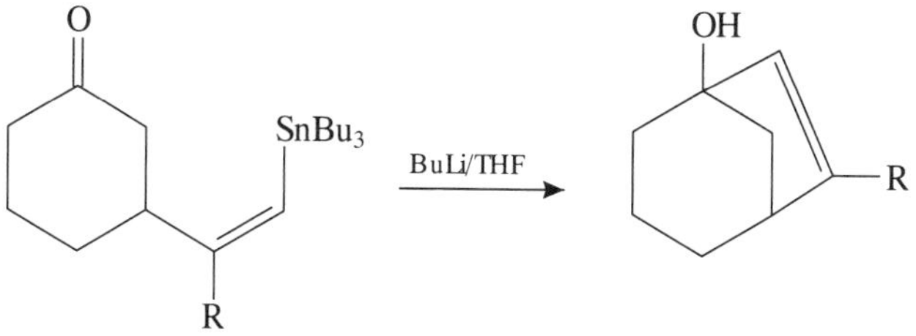

(53)

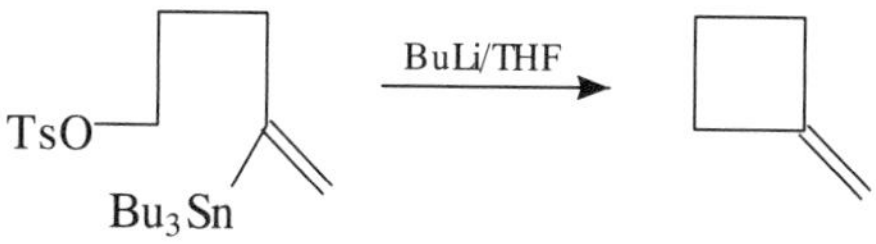

(54)

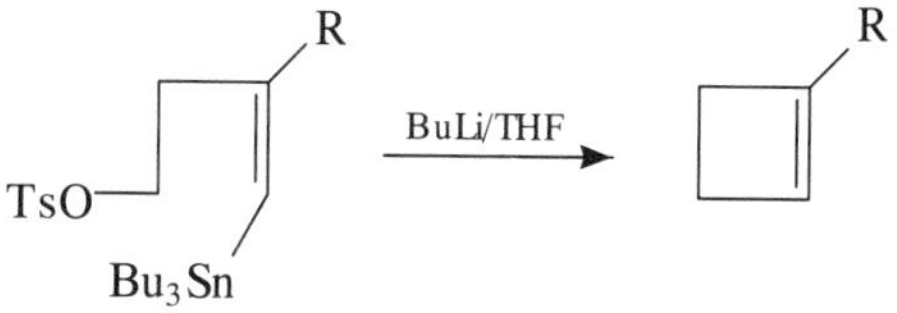

(55)

Cross-coupling with an aryl iodide is also possible, as shown in reaction 56[304]. Similarly, bis(trialkylstannyl)acetylene yields diarylacetylenes[305].

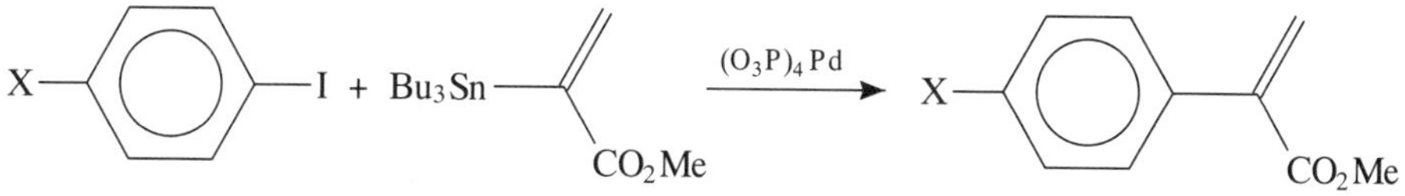

$X = $ H, Me, Br, MeO, O_2N, F_3C, Ac, AcO

(56)

(E)-α,β-Unsaturated carboxylic acids are obtained by derivatizing chlorotributylstannane, as depicted in reaction 57[306].

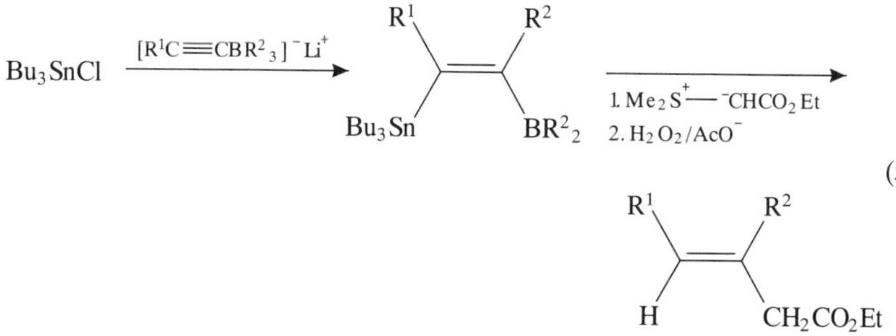

(57)

Trialkylvinylstannanes undergo cross-coupling reactions with acyl chlorides, as shown by reactions 58[307] and 59[308]. These acylations can be conducted under reductive conditions to saturate the carbon–carbon double bond, as illustrated in reaction 60[309]. Also, trialkylcyclopropylstannanes yield similar derivatives, as shown in reactions 61[308] and 62[310]. In the particular case of reaction 62, involving geminal sulphoxy and stannyl groups, an ester is formed as the result of a Pummerer-type rearrangement[310].

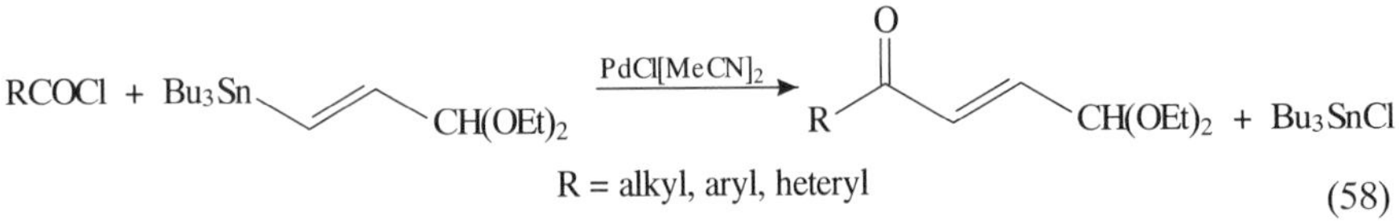

$$\text{RCOCl} + \text{Bu}_3\text{Sn}\diagup\diagdown\text{CH(OEt)}_2 \xrightarrow{\text{PdCl[MeCN]}_2} \text{R}\diagdown\text{C(O)}\diagup\diagdown\text{CH(OEt)}_2 + \text{Bu}_3\text{SnCl} \qquad (58)$$

R = alkyl, aryl, heteryl

$$\begin{array}{c}\text{SnBu}_3\\ \diagup\diagdown\\ \text{SO}_2\text{Ph}\end{array} \xrightarrow{\text{RCOCl}} \begin{array}{c}\text{COR}\\ \diagup\diagdown\\ \text{SO}_2\text{Ph}\end{array} + \text{Bu}_3\text{SnCl} \qquad (59)$$

$$\text{Bu}_3\text{Sn}\diagup\diagdown\text{SnBu}_3 \xrightarrow{\text{RCOCl/[H, Pd]}} \text{R}\diagdown\diagup\diagdown\text{R} \qquad (60)$$

$$\begin{array}{c}\text{SO}_2\text{Ph}\\ \triangleright\!\!<\\ \text{SnBu}_3\end{array} \xrightarrow{\text{RCOCl}} \begin{array}{c}\text{SO}_2\text{Ph}\\ \triangleright\!\!<\\ \text{COR}\end{array} + \text{Bu}_3\text{SnCl} \qquad (61)$$

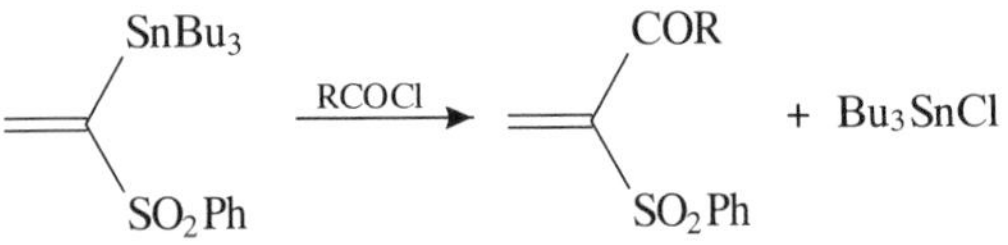

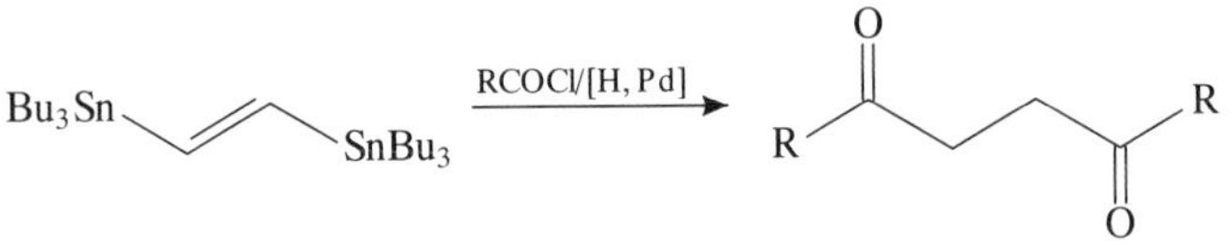

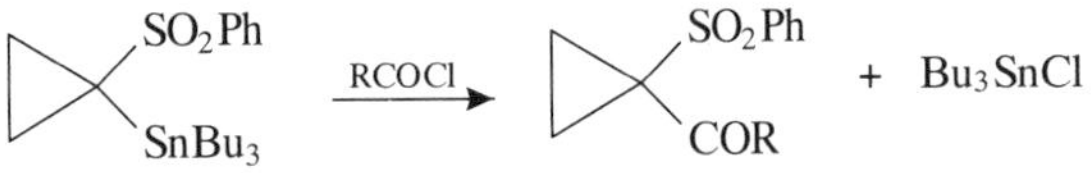

$$\begin{array}{c}\text{SOPh}\\ \triangleright\!\!<\\ \text{SnBu}_3\end{array} \xrightarrow{\text{RCOCl}} \begin{array}{c}\text{SPh}\\ \triangleright\!\!<\\ \text{O}_2\text{CR}\end{array} + \text{Bu}_3\text{SnCl} \qquad (62)$$

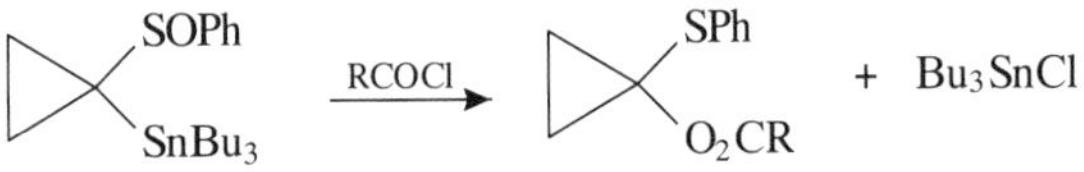

Vinyl fluorides are conveniently prepared from the corresponding vinylstannanes by electrophilic fluorination with reagent **108**, as shown, for example, in reaction 63. Reagent **108** is more easily handled than usual fluorinating agents, such as SF$_4$, F$_2$, CsSO$_4$F, or XeF$_2$/AgPF$_6$[311].

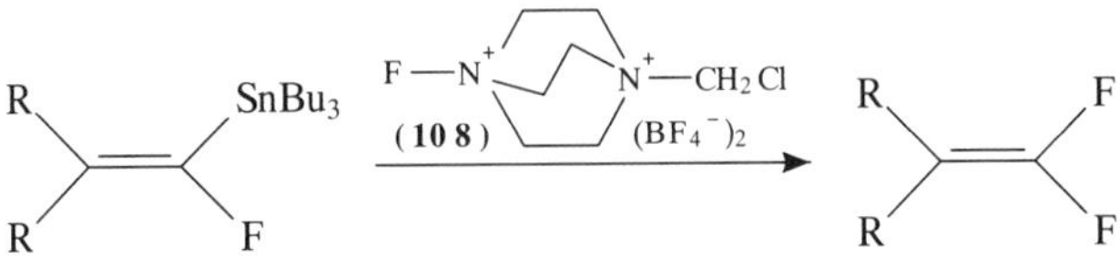

$$\begin{array}{c}\text{R}\\ \diagup\!\!=\!\!\diagdown\\ \text{R}'\quad\text{F}\end{array}\!\!\text{SnBu}_3 \xrightarrow[(108)\quad(BF_4^-)_2]{F-N^+\cdots N^+-CH_2Cl} \begin{array}{c}\text{R}\quad\text{F}\\ \diagup\!\!=\!\!\diagdown\\ \text{R}'\quad\text{F}\end{array} \qquad (63)$$

Vinyl ethers stannylated in the α-position undergo palladium-catalysed coupling with aryl hlides, as shown, for example, in reaction 64[312] and 65[313a].

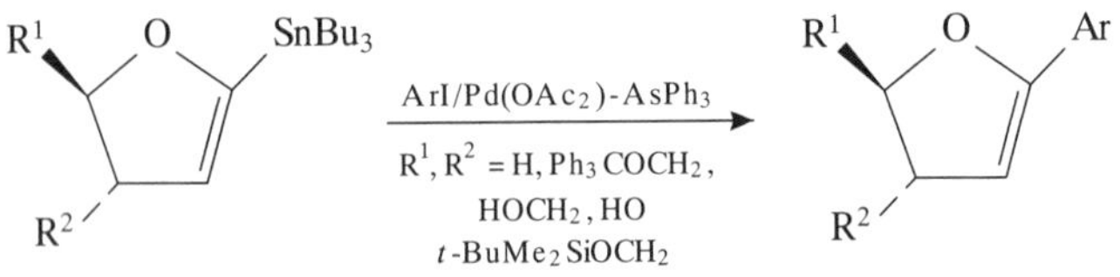

$$\begin{array}{c}\text{R}^1\quad\text{O}\quad\text{SnBu}_3\\ \end{array} \xrightarrow[\substack{R^1, R^2 = H, Ph_3COCH_2,\\ HOCH_2, HO\\ t\text{-BuMe}_2SiOCH_2}]{\text{ArI/Pd(OAc}_2)\text{-AsPh}_3} \begin{array}{c}\text{R}^1\quad\text{O}\quad\text{Ar}\\ \text{R}^2 \end{array} \qquad (64)$$

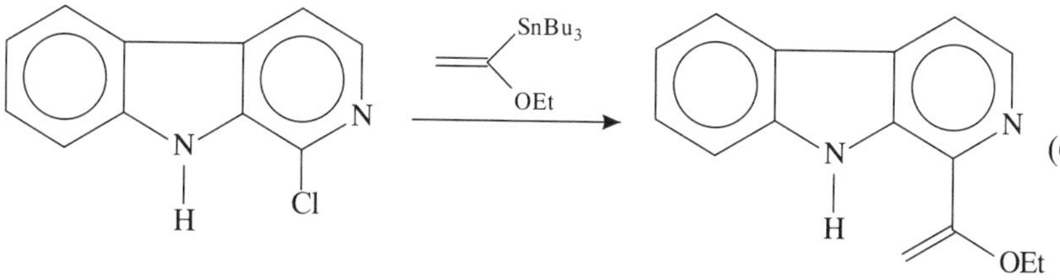

(65)

Vinylstannane compounds containing a remote vinyl halide moiety undergo a stereospecific internal coupling reaction, catalysed by cuprous iodide, leading to conjugated dienes, as illustated, for example, in reaction 66[313b].

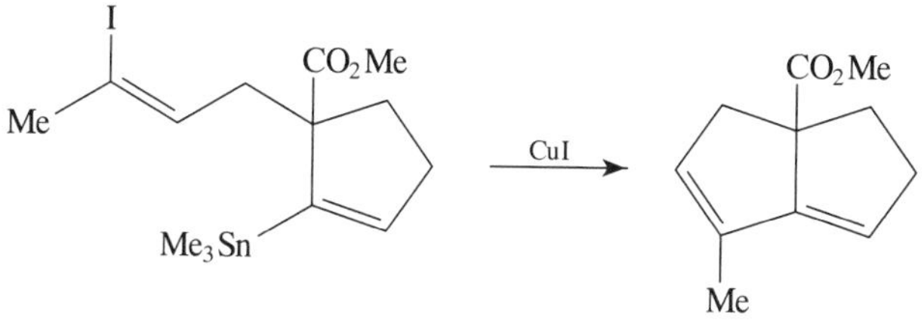

(66)

Stannyl groups attached to quinones can be displaced by aryl, heteryl and other groups, as shown in reactions 32[286] and 67[314].

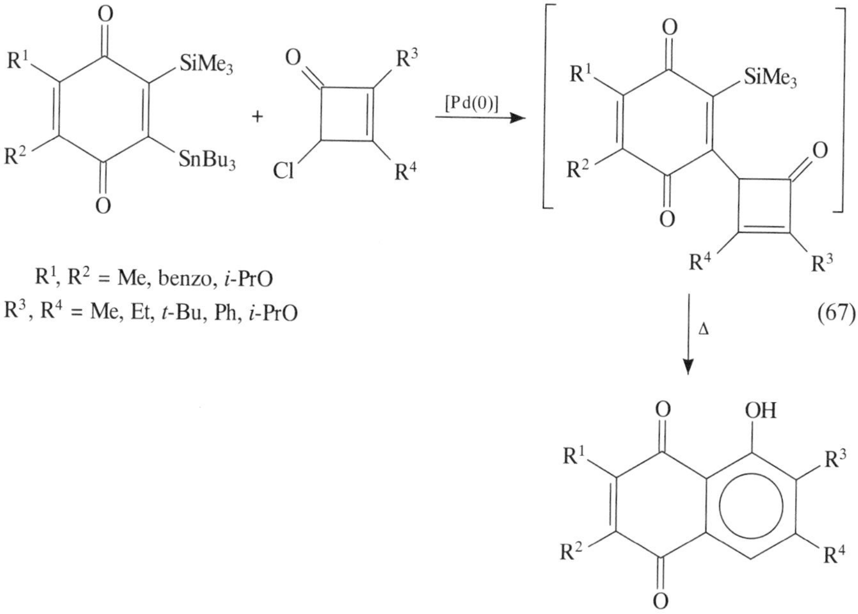

R^1, R^2 = Me, benzo, i-PrO
R^3, R^4 = Me, Et, t-Bu, Ph, i-PrO

(67)

3. Aryltin compounds

A great variety of aromatic compounds can be derived from aryltrialkylstannanes by electrophilic *ipso*-substitution with trialkylstannyl acting as leaving group. For example, the following reagents can provide the electrophile participating in the process: RCOCl in the presence of $AlCl_3$, NOCl, CNCl, $ArSO_2Cl$, Br_2, SO_2 and SO_3[315]. Cross-coupling with a vinyl triflate appears in reaction 68[316].

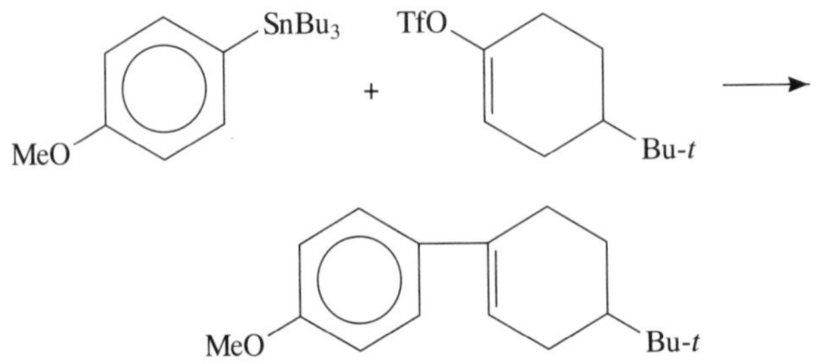

$$
\begin{array}{c}
\text{(68)}
\end{array}
$$

Aromatic iodination takes place by displacement of a trialkylstanyl group with a sodium iodide solution, in the presence of an oxidizing agent such as chloramine T, iodogen or peracetic acid. Iomazenil labeled with ^{123}I, a physiological tracer, was thus derived from its direct precursor **109**, using labelled sodium iodide[317].

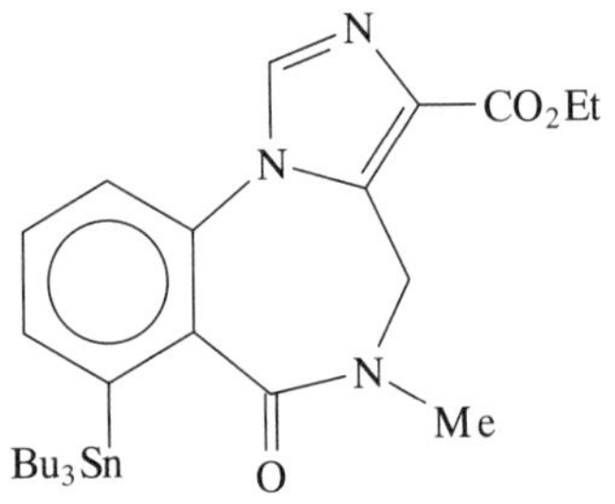

(109)

Aromatic fluorination can be carried out by a regiospecific destannylation process shown in reaction 69. This is an effective method for producing fluorinated *m*-tyrosine and other radiopharmaceuticals, as shown in reaction 70. The process can be applied for radiolabelling with ^{18}F, denoted as F* in these reactions, and the products used as radioactive tracers for clinical and fundamental investigations[318–321].

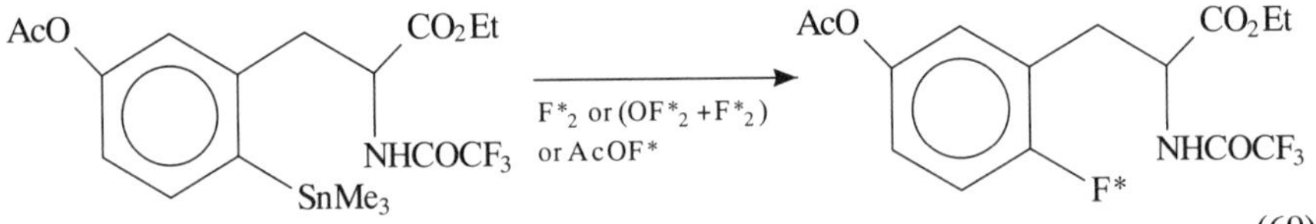

$$
\begin{array}{c}
\text{(69)}
\end{array}
$$

$$
\begin{array}{c}
\text{(70)}
\end{array}
$$

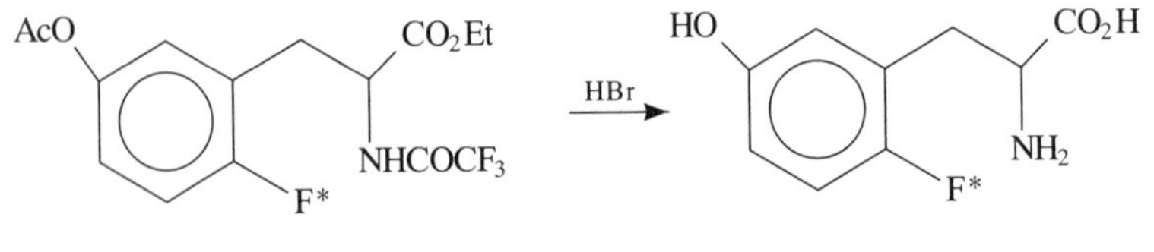

4. Organotin alkoxylates and carboxylates

In α,β-unsaturated stannyl esters the acyl moiety is displaced by mercury(II) salts without affecting carbon–carbon double bonds, as shown in reaction 71[322].

$$RCH{=}CHCO_2SnBu_3 + HgCl_2 \longrightarrow RCH{=}CHCO_2H + BuHgCl \qquad (71)$$

5. Trialkyl- and tetraalkylstannanes

Stannanes bearing hydrogen on the tin atom are good reducing agents. Thus, for example, tributylstannane can remove α-bromo or α-alkoxy groups from glycine derivatives[323]. Polymeric reagents containing Sn–H groups have been proposed as reagents for organic synthesis[315].

Tetraalkylstannanes with a β-acyloxy group undergo highly stereospecific elimination to yield Z- or E-unsaturated products, depending on whether the organotin compound has *erythro*- or *threo*-configuration, as depicted in reactions 72 and 73, respectively[324].

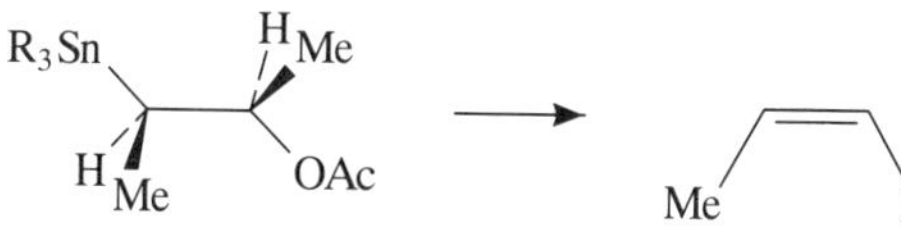
R3Sn
H
Me
H
Me
OAc
Me
Me

(72)

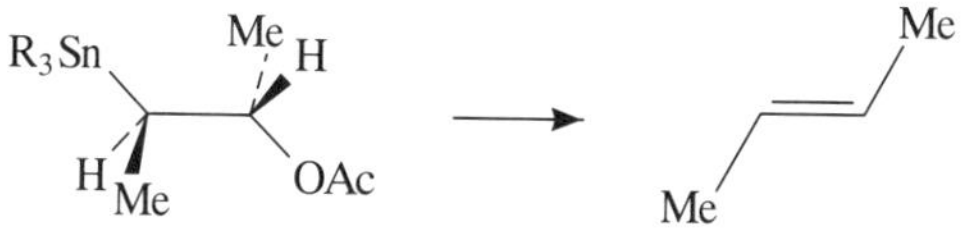
R3Sn
Me
H
H
Me
OAc
Me
Me

(73)

Tetraalkylstannanes undergo transmetallation reactions leading to reactive intermediates that may be combined with electrophilic substrates, as shown, for example, in reaction 74[325].

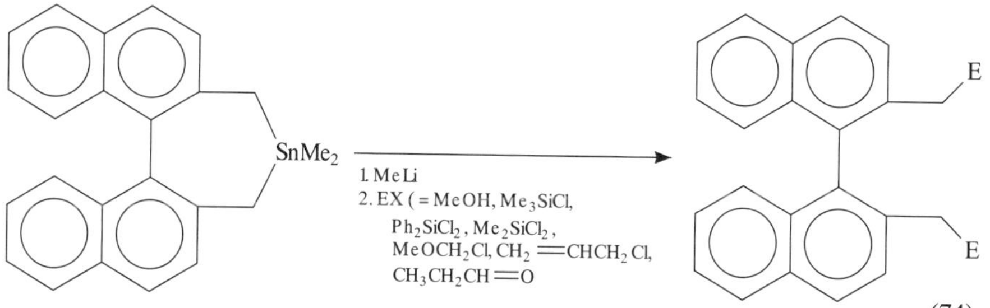
SnMe2
1. MeLi
2. EX (= MeOH, Me3SiCl,
Ph2SiCl2, Me2SiCl2,
MeOCH2Cl, CH2=CHCH2Cl,
CH3CH2CH=O
E
E

(74)

6. Organotin compounds containing miscellaneous functions

α-Aminotetraalkylstannanes afford stable carbanions for reaction with electrophiles if the molecule bears an N-(2-methoxyethyl) group. The process takes place preserving the configuration at the α-position. This is illustrated in reaction 75, where butyllithium promotes formation of a carbanion by displacement of the $SnBu_3$ group; carbon dioxide reacts with the carbanion and the adduct is reduced to a primary alcohol by lithium aluminium hydride[326]. Chiral N-acylated α-aminotetraalkylstannanes can undergo similar stereospecific transmetallation reactions, followed by quenching with an electrophile, as shown, for example, in reaction 76[327].

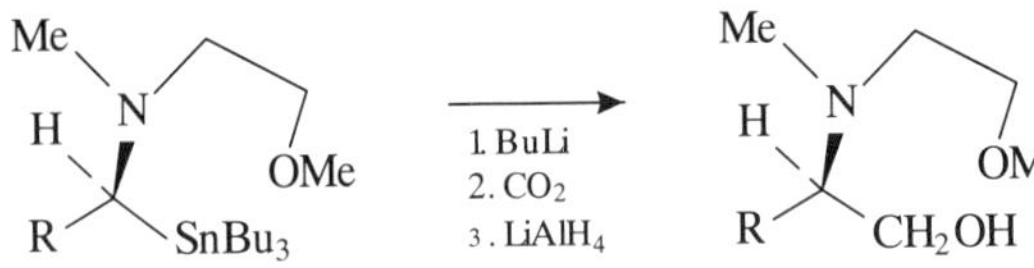

$$(75)$$

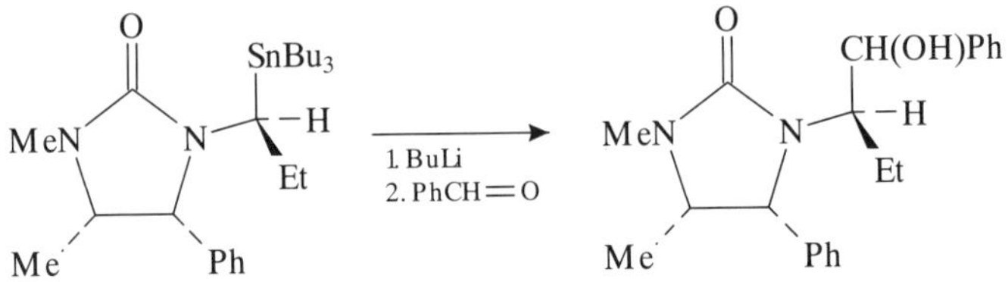

$$(76)$$

(only one diasteroisomer)

Acetal ethers containing the trialkylstannyl group undergo the interesting rearrangement 77, catalysed by Ti(IV) compounds[328].

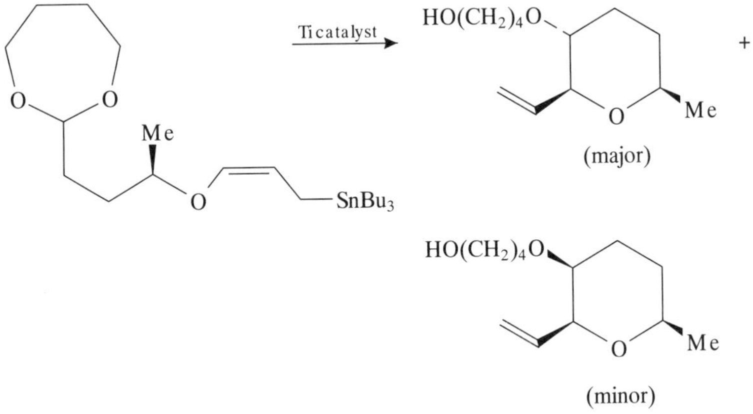

$$(77)$$

VI. REFERENCES

1. M. H. Gitlitz and M. K. Moran, in *Kirk-Othmer Encyclopedia of Chemical Technology*, 3rd edn., Vol. 23, Wiley, New York, 1983, pp. 42–77.
2. J. Zabicky, in *Supplement C: The Chemistry of Triple-Bonded Functional Groups* (Ed. S. Patai), Vol. 2, Wiley, Chichester, 1994, pp. 192–230.
3. N. I. Sax, *Dangerous Properties of Industrial Materials*, 6th edn., Van Nostrand Reinhold, New York, 1984.
4. C. R. Worthing (Ed.), *The Pesticide Manual*, British Crop Protection Council, 1991.
5. J. T. Byrd and M. O. Andreae, *Science*, **218**, 565 (1982).
6. T. R. Crompton, *Chemical Analysis of Organometallic Compounds*. Volume 3. *Elements of Group IVB*, Academic Press, London, 1974, pp. 13–98.
7. J. P. Anger and J. P. Curtes, in *Handbook on Metals in Clinical and Analytical Chemistry* (Eds. H. G. Seiler, A. Sigel and H. Sigel), Dekker, New York, 1994, pp. 613–626.
8. R. Reverchon, *Chim. Anal. (Paris)*, **47**, 70 (1965).
9. M. P. Brown and G. W. A. Fowles, *Anal. Chem.*, **30**, 1689 (1958).
10. J. G. A. Luijten and G. J. M. van der Kerk, in *Investigations in the Field of Organotin Chemistry*, Tin Research Institute, Greenford, U.K., 1966, p. 84. Cited in Reference 6.

11. R. Geyer and H. J. Seidlitz, *Z. Chem.*, **4**, 468 (1964).
12. V. Chromy and J. Vrestál, *Chem. Listy*, **60**, 1537 (1966).
13. X. C. Lee, W. R. Cullen, K. J. Reimer and I. D. Brindle, *Anal. Chim. Acta*, **258**, 307 (1992).
14. J. M. Rabadán, J. Galbán, J. C. Vidal and J. Aznárez, *J. Anal. At. Spectrom.*, **5**, 45 (1990).
15. R. E. Sturgeon, S. S. Berman and S. N. Willie, *Anal. Chem.*, **59**, 2441 (1987).
16. J. C. McKie, *Anal. Chim. Acta*, **197**, 303 (1987).
17. M. Chiba, A. Shinohara and Y. Inaba, *Microchem. J.*, **49**, 275 (1994).
18. T. Itami, M. Ema, H. Amano and H. Kawasaki, *J. Anal. Toxicol.*, **15**, 119 (1991).
19. A. Brezezinska-Paudyn and J. C. Van Loon, *Fresenius Z. Anal. Chem.*, **331**, 707 (1988).
20. W. Q. Qi, S. Q. Lin, S. Chen and S. Kaoru, *Bunseki Kagaku*, **38**, 228 (1989); *Chem. Abstr.*, 112:29953.
21. I. D. Brindle and X. C. Le, *Analyst (London)*, **113**, 1377 (1988).
22. I. D. Brindle and X. C. Le, *Anal. Chem.*, **61**, 1175 (1989).
23. R. Pereiro, M. Wu, J. A. C. Broekaert and G. M. Hieftje, *Spectrochim. Acta B*, **49**, 59 (1994).
24. J. Ohyama, *Bunseki Kagaku*, **38**, T119 (1989); *Chem. Abstr.*, 111:159867.
25. S. Ohhira and H. Matsui, *J. Chromatogr., Biomed. Appl.*, **662**, 173 (1993).
26. G. J. Vanhetof and Y. N. Yoshi, *Phys. Scripta*, **48**, 714 (1993).
27. (a) L. F. Capitán-Vallvey, M. C. Valencia and G. Mirón, *Anal. Chim. Acta*, **289**, 365 (1994).
    (b) T. Kiriyama and R. Kuroda, *Mikrochim. Acta*, **1**, 261 (1991).
28. H. Narasaki and E. Y. Kimura, *Bunseki Kagaku*, **38**, T95 (1989); *Chem. Abstr.*, 111:76595.
29. C. M. G. Van der Berg, S. H. Khan and J. P. Riley, *Anal. Chim. Acta*, **222**, 43 (1989).
30. S. B. O. Adeloju and F. Pablo, *Anal. Chim. Acta*, **270**, 143 (1992).
31. M. Rosbach, *J. Radioanal. Nucl. Chem., Articles*, **169**, 239 (1993).
32. J. Versieck and L. Vanballenberghe, *Anal. Chem.*, **63**, 1143 (1991).
33. M. Chiba, V. Iyengar, R. R. Greenberg and T. Gills, *Sci. Total Environ.*, **148**, 39 (1994).
34. R. A. Couture, *X-Ray Spectrom.*, **22**, 92 (1993).
35. S. H. Vien and R. C. Fry, *Anal. Chem.*, **60**, 465 (1988).
36. I. Z. Alzamil and A. Townshend, *Anal. Chim. Acta* **209**, 275 (1988).
37. A. D'Ulivo, *Talanta* **35**, 499 (1988).
38. K. Takahashi, *Nippon Kagaku Kaishi* 893 (1990); *Chem. Abstr.*, 113:120538.
39. R. J. Carter, N. J. Turoczy and A. M. Bond, *Environ. Sci. Technol.*, **23**, 615 (1989)
40. A. O. Valkirs, P. F. Seligman, G. J. Olson, F. E. Brinckman, C. L. Matthias and J. M. Bellama, *Analyst (London)* **112**, 17 (1987).
41. N. Watanabe, S. Sakai and H. Takatsuki, *Water Sci. Technol.*, **25**, 117 (1992).
42. T. Tsuda, H. Nakanishi, S. Aoki and J. Takebayashi, *J. Chromatogr.*, **387**, 361 (1987).
43. C. A. Krone, D. W. Brown, D. G. Burrows, R. G. Bogar, S. L. Chan and U. Varanasi, *Mar. Environ. Res.*, **27**, 1 (1989); *Chem. Abstr.*, 111:210186.
44. H. Harino, M. Fukushima and M. Tanaka, *Anal. Chim. Acta*, **264**, 91 (1992).
45. J. L. Gómez-Ariza, E. Morales and M. Ruiz-Benítez, *Appl. Organomet. Chem.*, **6**, 279 (1992).
46. W. M. R. Dirkx and F. C. Adams, *Mikrochim. Acta*, **109**, 79 (1992).
47. I. Tolosa, J. Dachs and J. M. Bayona, *Mikrochim. Acta*, **109**, 87 (1992).
48. J. L. Gómez-Ariza, E. Morales and M. Ruiz-Benítez, *Analyst (London)*, **117**, 641 (1992).
49. S. Rapsomanikis, *Analyst (London)*, **119**, 1429 (1994).
50. S. Ohhira and H. Matsui, *J. Chromatogr.*, **566**, 207 (1991).
51. J. Greaves and M. A. Unger, *Biomed. Environ. Mass Spectrom.*, **15**, 565 (1988).
52. K. Takahashi and Y. Ohyagi, *J. Oil Colour Chem. Assoc.*, **73**, 493 (1990).
53. K. Takahashi, *Nippon Kagaku Kaishi*, 380 (1990); *Chem. Abstr.*, 113:118198.
54. K. Takahashi, *J. Oil Colour Chem. Assoc.*, **74** 331 (1991).
55. K. Takahashi, *Nippon Kagaku Kaishi*, 367 (1991); *Chem. Abstr.*, 115:35292.
56. Y. Liu, V. López-Ávila, M. Alcaraz and W. F. Beckert, *J. High Resolut. Chromatog.*, **17**, 527 (1994).
57. D. S. Forsyth and C. Cleroux, *Talanta*, **38** 951 (1991).
58. Y. Hattori, H. Yamamoto, K. Nagai, K. Nonaka, H. Hashimoto, S. Nakamura, M. Nakamoto, K. Annen, S. Sakamori, H. Shiraishi and M. Morita, *Bunseki Kagaku* **40**, 25 (1991); *Chem. Abstr.*, 114:128691.
59. K. Ruthenberg and C. Madetzki, *Chromatographia*, **26**, 251 (1988).
60. M. Nagase, *Anal. Sci.*, **6**, 851 (1990).
61. S. Hashimoto and A. Otsuki, *J. High Resolut. Chromatogr.*, **14**, 397 (1991).

62. M. Takeuchi, K. Mitsuishi, H. Yamanobe, Y. Watanabe and M. Doguchi, *Bunseki Kagaku*, **38**, 522 (1989); *Chem. Abstr.*, **111**:213455.
63. S. Reader and E. Pelletier, *Anal. Chim. Acta*, **262**, 307 (1992).
64. Y. K. Chau, S. Z. Zhang and R. J. Maguire, *Analyst (London)*, **117**, 1161 (1992).
65. T. Tsuda, S. Aoki, H. Nakanishi and J. Takebayashi, *J. Chromatogr.*, **387**, 361 (1987).
66. O. Evans, B. J. Jacobs and A. L. Cohen, *Analyst (London)* **116**, 15 (1991)
67. K. S. Epler, T. C. Ohaver, G. C. Turk and W. A. MacCrehan, *Anal. Chem.*, **60**, 2062 (1988).
68. J. J. Cooney, A. T. Kronick, G. J. Olson, W. R. Blair and F. E. Brinckman, *Chemosphere*, **17**, 1795 (1988).
69. L. Schebek, M. O. Andreae and H. J. Tobschall, *Int. J. Environ. Anal. Chem.*, **45**, 257 (1991).
70. S. Ohhira and H. Matsui, *J. Anal. Toxicol.*, **16**, 375 (1992).
71. S. Ohhira and H. Matsui, *J. Chromatogr., Biomed. Appl.*, **662**, 173 (1993).
72. S. Clark, J. Ashby and P. J. Craig, *Analyst (London)*, **112**, 1781 (1987).
73. J. J. Sullivan, J. D. Torkelson, M. M. Wekell, T. A. Hollingworth, W. L. Saxton, G. A. Miller, K. W. Panaro and A. D. Uhler, *Anal. Chem.*, **60**, 626 (1988).
74. H. Kurosaki, H. Yokohama and K. Ozaki, *Bunseki Kagaku*, **40**, T65 (1991); *Chem. Abstr.*, **114**:246026.
75. M. Astruc, R. Pinel and A. Astruc, *Mikrochim. Acta*, **109**, 73 (1992).
76. K. Takahashi, *Nippon Kagaku Kaishi*, 367 (1991); *Chem. Abstr.*, **115**, 35292 (1991).
77. K. W. M. Siu, P. S. Maxwell and S. S. Berman, *J. Cromatogr.*, **475**, 373 (1989).
78. S. Hashimoto and A. Otsuki, *J. High Resolut. Chromatogr.*, **14**, 397 (1991).
79. F. Y. Pang, Y. L. Ng, S. M. Phang and S. L. Tong, *Int. J. Environ. Anal. Chem.*, **53**, 53 (1993).
80. Z. M. Ni, H. B. Hang, A. Li, B. He and F. Z. Xu, *J. Anal. At. Spectrom.*, **6**, 385 (1991).
81. R. Alzaga and J. M. Bayona, *J. Chromatogr. A*, **655**, 51 (1993).
82. V. Farina, *J. Org. Chem.*, **56**, 4985 (1991).
83. S. Vainiotalo and L. Hayri, *J. Chromatogr.*, **523**, 273 (1990).
84. I. L. Row, Y. L. Liu and C. W. Whang, *J. Chin. Chem. Soc. (Taipei)*, **37**, 203 (1990); *Chem. Abstr.*, **113**:213195.
85. H. Suyani, D. Heitkemper, J. Creed and J. Caruso, *Appl. Spectrosc.*, **43**, 962 (1989).
86. G. B. Jiang, P. S. Maxwell, K. W. M. Siu, V. T. Luong and S. S. Berman, *Anal. Chem.*, **63**, 1506 (1991).
87. (a) M. D. Stephenson and D. R. Smith, *Anal. Chem.*, **60**, 696 (1988).
     (b) D. Dyne, B. S. Chana, N. J. Smith and J. Cocker, *Anal. Chim. Acta*, **246**, 351 (1991).
88. O. F. X. Donard, S. Rapsomanikis and J. H. Weber, *Anal. Chem.*, **58**, 772 (1986).
89. L. Randall, O. F. X. Donard and J. H. Weber, *Anal. Chim. Acta*, **184**, 197 (1986).
90. J. S. Han and J. H. Weber, *Anal. Chem.*, **60**, 316 (1988).
91. J. Ohyama, *Bunseki Kagaku*, **38**, T119 (1989); *Chem. Abstr.*, **111**:159867.
92. P. W. Balls, *Anal. Chim. Acta*, **197**, 309 (1987).
93. B. Zhang, K. Tao and J. Feng, *Spectrochim. Acta, Part B*, **44B**, 247 (1989).
94. S. S. Lindsay and J. J. Pesek, *J. Liq. Chromatogr.*, **12**, 2367 (1989).
95. S. B. Hawthorne, D. J. Miller and M. S. Krieger, *J. High Resolut. Chromatogr.*, **12**, 714 (1989).
96. M. Astruc, A. Astruc and R. Pinel, *Mikrochim. Acta*, **109**, 83 (1992).
97. M. Adinarayana, U. S. Singh and T. S. Dwivedi, *J. Chromatogr.*, **435**, 210 (1988).
98. C. B. Pascual and V. A. V. Beckett, *Anal. Chim. Acta*, **224**, 97 (1989).
99. A. M. Bond, N. J. Turoczy and R. J. Carter, *J. Electroanal. Chem.*, **365**, 125 (1994).
100. A. M. Bond and N. M. McLachlan, *Anal. Chim. Acta*, **204**, 151 (1988).
101. M. Ochsenkuhn-Petropoulou, G. Poulea and G. Parissakis, *Mikrochim. Acta*, **109**, 93 (1992).
102. L. P. Pettinato and L. R. Sherman, *Microchem. J.*, **47**, 96 (1993).
103. K. W. M. Siu, G. J. Gardner and S. S. Berman, *Anal. Chem.*, **61**, 2320 (1989).
104. E. Kolehmainen, J. Paasivirta, R. Kauppinen, T. Otollinen, S. Kasa and R. Herzschuh, *Int. J. Environ. Anal. Chem.*, **43**, 19 (1991).
105. T. A. Hein, W. Thiel and T. J. Lee, *J. Phys. Chem.*, **97**, 4381 (1993).
106. W. Schneider and W. Thiel, *Chem. Phys.*, **159**, 49 (1992).
107. M. Halonen and X. W. Zhan, *J. Chem. Phys.*, **101**, 950 (1994).
108. P. G. Harrison, J. McManus and D. M. Podesta, *J. Chem. Soc., Chem. Commun.*, 291 (1992).
109. C. Nedez, A. Theolier, F. Lefebvre, A. Choplin, J. M. Basset and J. F. Joly, *J. Am. Chem. Soc.*, **115**, 722 (1993).
110. C. Nedez, A. Choplin, F. Lefebvre, J. M. Basset and E. Benazzi, *Inorg. Chem.*, **33**, 1099 (1994).

111. W. Linert, V. Gutmann and A. Sotriffer, *Vib. Spectrosc.*, **1**, 199 (1990); *Chem. Abstr.*, **114**, 130409.

112. W. Linert, A. Sotriffer and V. Gutmann, *J. Coord. Chem.*, **22**, 21 (1990); *Chem. Abstr.*, **115**, 8941.

113. D. Tudela and J. M. Calleja, *Spectrochim. Acta A, Mol. Spectr.*, **49**, 1023 (1993).

114. A. V. Belyakov, V. S. Nikitin and M. V. Polyakova, *Zh. Fiz. Khim.*, **63**, 652 (1989).

115. R. Eujen and U. Thurmann, *J. Organomet. Chem.*, **433**, 63 (1992).

116. R. Koster, G. Seidel, I. Klopp, C. Kruger, G. Kehr, J. Suss, and B. Wrackmeyer, *Chem. Ber.*, **126**, 1385 (1993).

117. D. M. Adams and J. Haines, *J. Chem. Soc., Faraday Trans.*, **88**, 3587 (1992).

118. D. Zhang, L. Walz, M. Prager and A Weiss, *Ber. Bunsenges. Phys. Chem.*, **91**, 1283 (1987).

119. Z. Da, M. Prager and A. Weiss, *J. Chem. Phys.*, **94**, 1765 (1991).

120. M. Prager, D. Zhang and A. Weiss, *Physica B*, **180**, 671 (1992).

121. D. Cunningham, J. F. Gallagher, T. Higgins, P. McArdle, J. McGinley and M. Ogara, *J. Chem. Soc., Dalton Trans.*, 2183 (1993).

122. H. Reuter and H. Puff, *J. Organomet. Chem.*, **379**, 223 (1990).

123. C. López, A. S. González, M. E. García, J. S. Casas, J. Sordo, R. Graziani and U. Casellato, *J. Organomet. Chem.*, **434**, 261 (1992).

124. A. Samuel-Lewis, P. J. Smith, J. H. Aupers, D. Hampson and D. C. Povey, *J. Organomet. Chem.*, **437**, 131 (1992).

125. W. Plass and J. G. Verkade, *Inorg. Chem.*, **32**, 5145 (1993).

126. W. Plass and J. G. Verkade, *Inorg. Chem.*, **32**, 5153 (1993).

127. K. Schenzel, A. Kolbe and P. Reich, *Monatsh. Chem.*, **121**, 615 (1990).

128. A. F. Shihada, *Z. Naturforsch. B*, **48**, 1781 (1993).

129. W. P. Neumann, *Chem. Rev.*, **91**, 311 (1991).

130. D. Ballivettkatchenko, J. H. Z. Dossantos and M. Malisova, *Langmuir*, **9**, 3513 (1993).

131. C. Obayashi, H. Sato and T. Tominaga, *J. Radioanal. Nuc. Chem., Letters*, **127**, 75 (1988).

132. K. L. Leighton and R. E. Wasylishen, *Can. J. Chem.*, **65**, 1469 (1987).

133. M. Kira, T. Oyamada and H. Sakurai, *J. Organomet. Chem.*, **471**, C4 (1994).

134. (a) F. Ferkous, D. Messadi, B. Dejeso, M. Degueilcastaing and B. Maillard, *J. Organomet. Chem.*, **420**, 315 (1991).

    (b) R. Eujen, N. Jahn and U. Thurmann, *J. Organomet. Chem.*, **434**, 159 (1992).

135. R. Eujen, N. Jahn and U. Thurmann, *J. Organomet. Chem.*, **465**, 153 (1994).

136. C. Leue, R. Reau, B. Neumann, H. G. Stammler, P. Jutzi and G. Bertrand, *Organometallics*, **13**, 436 (1994).

137. S. P. Mallela and R. A. Geanangel, *Inorg. Chem.*, **32**, 5623 (1993).

138. H. Weichmann and J. Meunierpiret, *Organometallics*, **12**, 4097 (1993).

139. T. N. Mitchell, R. Faust, B. Fabisch and R. Wickenkamp, *Magn. Reson. Chem.*, **28**, 82 (1990).

140. L. R. Sita and I. Kinoshita, *J. Am. Chem. Soc.*, **113**, 1856 (1991).

141. L. R. Sita and R. D. Bickerstaff, *J. Am. Chem. Soc.*, **111**, 6454 (1989).

142. J. Holecek, K. Handlir, M. Nadvornik, S. M. Teleb and A. Lycka, *J. Organomet. Chem.*, **339**, 61 (1988).

143. E. Liepins, I. Birgele, E. Lukevics, E. T. Bogoradovsky and V. S. Zavgorodny, *J. Organomet. Chem.*, **402**, 43 (1991).

144. Y. Nakadaira, R. Sato and H. Sakurai, *J. Organomet. Chem.*, **441**, 411 (1992).

145. B. Wrackmeyer, K. H. Von Locquenghien, E. Kupce and A. Sebald, *Magn. Reson. Chem.*, **31**, 45 (1993).

146. E. Liepins, I. Birgele, E. Lukevics, E. T. Bogoradovsky and V. S. Zavgorodny, *J. Organomet. Chem.*, **390**, 139 (1990).

147. F. Kayser, M. Biesemans, H. Pan, M. Gielen and R. Willem, *J. Chem. Soc., Perkin Trans. 2*, 297 (1994).

148. K. Okada, M. Minami, H. Inokawa and M. Oda, *Tetrahedron Lett.*, **34**, 567 (1993).

149. G. Anselme, J. -P. Declercq, H. Dubourg, H. Ranaivonjatovo, J. Escudié and C. Couret, *J. Organomet. Chem.*, **458**, 49 (1993).

150. A. Schäfer, M. Weidenbruch, W. Saak, S. Pohl and H. Marsmann, *Angew. Chem., Int. Ed. Engl.*, **30**, 834 (1991).

151. C. Janiak, H. Schumann, C. Stader, B. Wrackmeyer and J. J. Zuckermann, *Chem. Ber.*, **121**, 1745 (1988).

152. M. J. Heeg, R. H. Herber, C. Janiak, J. J. Zuckermann, H. Schumann and W. F. Manders, *J. Organomet. Chem.*, **346**, 321 (1988).
153. D. Tudela, *J. Organomet. Chem.*, **471**, 63 (1994).
154. C. Obayashi, H. Sato and T. Tominaga, *J. Radioanal. Nuc. Chem., Letters*, **137**, 87 (1989).
155. H. Fujiwara and Y. Sasaki, *J. Phys. Chem.*, **91**, 481 (1987).
156. A. Lycka, J. Holecek, J. Jirman and A. Kolonicny, *J. Organomet. Chem.*, **333**, 305 (1987).
157. J. M. Chance, J. H. Geiger and K. Mislow, *J. Am. Chem. Soc.*, **111**, 2326 (1989).
158. D. W. Allen, S. Bailey, J. S. Brooks and B. F. Taylor, *Chem. Ind. (London)*, **23**, 762 (1988).
159. D. Dakternieks and H. J. Zhu, *Organometallics*, **11**, 3820 (1992).
160. M. Cameron, B. G. Gowenlock, R. V. Parish and G. Vasapollo, *J. Organomet. Chem.*, **465**, 161 (1994).
161. D. Tudela, M. A. Khan and J. J. Zuckerman, *J. Chem. Soc., Chem. Commun.*, 558, (1989).
162. D. Tudela and M. A. Khan, *J. Chem. Soc., Dalton Trans.*, 1003 (1991).
163. K. M. Lo, V. G. K. Das, W. H. Yip and T. C. W. Mak, *J. Organomet. Chem.*, **408**, 167 (1991).
164. Y. W. Kim, A. Labouriau, C. M. Taylor, W. L. Earl and L. G. Werbelow, *J. Phys. Chem.*, **98**, 4919 (1994).
165. M. H. Jang and A. F. Janzen, *J. Fluorine Chem.*, **66**, 129 (1994).
166. U. Edlund, M. Arshadi and D. Johnels, *J. Organomet. Chem.*, **456**, 57 (1993).
167. B. Wrackmeyer and H. Zhou, *Magn. Reson. Chem.*, **28**, 1066 (1990).
168. B. Wrackmeyer, T. Gasparis-Ebeling and H. Nöth, *Z. Naturforsch., B: Chem. Sci.*, **44**, 653 (1989).
169. G. Linti, H. Nöth, E. Scheider and W. Storch, *Chem. Ber.*, **126**, 619 (1993).
170. H. Schumann, B. C. Wassermann and J. Pickardt, *Organometallics*, **12**, 3051 (1993).
171. M. F. Mahon, K. C. Molloy and P. C. Waterfield, *Organometallics*, **12**, 769 (1993).
172. F. Thunecke and D. Schulze, *Z. Chem.*, **30**, 444 (1990).
173. T. Ohtaki, Y. Kabe and W. Ando, *Organometallics*, **12**, 4 (1993).
174. B. Wrackmeyer, C. Stader, K. Horchler and H. Zhou, *Inorg. Chim. Acta*, **176**, 205 (1990).
175. D. Hansgen, H. Salz, S. Rheindorf and C. Scrage, *J. Organomet. Chem.*, **443**, 61 (1993).
176. I. D. Gay, C. H. W. Jones and R. D. Sharma, *J. Magn. Reson.*, **84**, 501 (1989).
177. B. M. Schmidt and M. Drager, *J. Organomet. Chem.*, **399**, 63 (1990).
178. (a) H. Reuter and D. Schroder, *J. Organomet. Chem.*, **455**, 83 (1993).
   (b) T. B. Grindley, R. Thangarasa, P. K. Bakshi and T. S. Cameron, *Can. J. Chem.*, **70**, 197 (1992).
   (c) T. B. Grindley, R. E. Wasylishen, R. Thangarasa, W. P. Power and R. D. Curtis, *Can.J. Chem.*, **70**, 205 (1992).
   (d) T. S. Cameron, P. K. Bakshi, R. Thangarasa and T. B. Grindley, *Can.J. Chem.*, **70**, 1623 (1992).
179. B. J. Brisdon, M. F. Mahon, K. C. Molloy and P. J. Schofield, *J. Organomet. Chem.*, **465**, 145 (1994).
180. S. W. Ng, K. L. Chin, C. Wei, V. G. K. Das and R. J. Butcher, *J. Organomet. Chem.*, **376**, 277 (1989).
181. J. Kummerlen, I. Lange, W. Milius, A. Sebald and A. Blaschette, *Organometallics*, **12**, 3541 (1993).
182. A. Blaschette, E. Wieland, P. G. Jones and I. Hippel, *J. Organomet. Chem.*, **445**, 55 (1993).
183. T. P. Lockhart, *Organometallics*, **7**, 1438 (1988).
184. T. P. Lockhart and F. Davidson, *Organometallics*, **6**, 2471 (1987).
185. C. Silvestru, I. Haiduc, F. Caruso, M. Rossi, B. Mahieu and M. Gielen, *J. Organomet. Chem.*, **448**, 47 (1993).
186. A. Schäfer, M. Weidenbruch, W. Saak, S. Pohl and H. Marsmann, *Angew. Chem., Int. Ed. Engl.*, **30**, 834 (1991).
187. A. Schäfer, M. Weidenbruch, W. Saak, S. Pohl and H. Marsmann, *Angew. Chem., Int. Ed. Engl.*, **30**, 962 (1991).
188. L. Barton and D. K. Srivastava, *J. Chem. Soc., Dalton Trans.*, 1327 (1992).
189. X. Q. Kong, T. B. Grindley, P. K. Bakshi and T. S. Cameron, *Organometallics*, **12**, 4881 (1993).
190. J. Klein, F. Thunecke and R. Borsdorf, *Monatsh. Chem.*, **123**, 801 (1992).
191. H. Meyer, G. Baum, W. Massa, S. Berger and A. Berndt, *Angew. Chem., Int. Ed. Engl.*, **26**, 546 (1987).
192. A. Berndt, H. Meyer, G. Baum, W. Massa and S. Berger, *Pure Appl. Chem.*, **59**, 1011 (1987).
193. G. Anselme, H. Ranaivonjatovo, J. Escudié, C. Couret and J. Satgé, *Organometallics*, **11**, 2748 (1992).

194. H. Ranaivonjatovo, J. Escudié, C. Couret and J. Satgé, *J. Chem. Soc., Chem. Commun.*, 1047 (1992).

195. C. Couret, J. Escudié, J. Satgé, A. Raharinirina and J. D. Andriamizaka, *J. Am. Chem. Soc.*, **107**, 8280 (1985).

196. S. Matsamune and L. R. Sita, *J. Am. Chem. Soc.*, **107**, 6390 (1985).

197. (a) K. W. Zilm, G. A. Lawless, R. M. Merrill, J. M. Millar and G. G. Webb, *J. Am. Chem. Soc.*, **109**, 7236 (1987).

(b) D. E. Goldberg, P. B. Hitchcock, M. F. Lappert, K. M. Thomas, A. J. Thorne, T. Fjeldberg, A. Haaland and B. E. R. Schilling, *J. Chem. Soc., Dalton Trans.*, 2387 (1986).

198. K. Burger, L. Nagy, N. Buzas, A. Vertes and H. Mehner, *J. Chem. Soc., Dalton Trans.*, 2499 (1993).

199. N. Tokitoh, M. Saito and R. Okazaki, *J. Am. Chem. Soc.*, **115**, 2065 (1993).

200. P. J. Davidson, D. H. Harris and M. F. Lappert, *J. Chem. Soc., Dalton Trans.*, 2268 (1976).

201. U. Kilimann, M. Noltemeyer and F. T. Edelmann, *J. Organomet. Chem.*, **443**, 35 (1993).

202. B. S. Jolly, M. F. Lappert, L. M. Engelhardt, A. H. White and C. L. Raston, *J. Chem. Soc., Dalton Trans.*, 2653 (1993).

203. B. Wrackmeyer, K. Horchler, H. Zhou and M. Veith, *Z. Naturforsch. Sect. B, J. Chem. Sci.*, **44**, 288 (1989).

204. B. Krebs, G. Henkel and M. Dartmann, *Acta Cryst., Sect. C, Cryst. Struct. Commun.*, **45**, 1010 (1989).

205. (a) J. Munier-Piret, M. Van Meerssche, M. Gielen and K. Jurkschat, *J. Organomet. Chem.*, **252**, 289 (1983).

(b) H. Preut and T. N. Mitchell, *Acta Cryst., Sect. C, Cryst. Struct. Commun.*, **45**, 35 (1989).

206. M. Weidenbruch, J. Schlaefke, K. Peters and H. G. Vonschnering, *J. Organomet. Chem.*, **414**, 319 (1991).

207. L. R. Sita and R. D. Bickerstaff, *J. Am. Chem. Soc.*, **110**, 5208 (1988).

208. S. P. Mallela and R. A. Geanangel, *Inorg. Chem.*, **33**, 1115 (1994).

209. (a) K. Jurkschat, F. Hesselbarth, M. Dargatz and J. Lehmann, *J. Organomet. Chem.*, **388**, 259 (1990).

(b) J. S. Tse, E. J. Gabe and F. L. Lee, *Acta Crystallogr., Sect. C, Cryst. Struct. Commun.*, **42**, 1876 (1986).

210. D. Zhang, M. Prager, S. Q. Dou and A. Weiss, *Z. Naturforsch., A: Phys. Sci.*, **44**, 151 (1989).

211. H. Preut and T. N. Mitchell, *Acta Cryst., Sect. C, Cryst. Struct. Commun.*, **47**, 951 (1991).

212. L. R. Sita and R. D. Bickerstaff, *J. Am. Chem. Soc.*, **111**, 3769 (1989).

213. L. R. Sita and I. Kinoshita, *J. Am. Chem. Soc.*, **114**, 7024 (1992).

214. J. B. Lambert, B. Kuhlmann and C. L. Stern, *Acta Cryst., Sect. C, Cryst. Struct. Commun.*, **49**, 887 (1993).

215. V. B. Mokal, V. K. Jain and E. R. T. Tiekink, *J. Organomet. Chem.*, **431**, 283 (1992).

216. S. W. Ng and V. G. K. Das, *J. Cryst. Spect. Res.*, **23**, 929 (1993).

217. H. Reuter and D. Schroeder, *Acta Crystallogr., Sect. C, Cryst. Struct. Commun.*, **49**, 954 (1993).

218. S. W. Ng, V. G. K. Das, F. van Meuss, J. D. Schagen and L. H. Savers, *Acta Cryst., Sect. C, Cryst. Struct. Commun.*, **45**, 568 (1990).

219. G. K. Sandhu, S. P. Verma and E. R. T. Tiekink, *J. Organomet. Chem.*, **393**, 195 (1990).

220. T. V. Sizova, N. S. Yashina, S. V. Petrosyan, A. V. Yatsenko, V. V. Chernishev and L. A. Aslanov, *J. Organomet. Chem.*, **453**, 171 (1993).

221. K. C. Molloy, I. W. Nowell and K. Quill, *J. Chem. Soc., Dalton Trans.*, 101 (1987).

222. (a) S. W. Ng, V. G. K. Das, S. L. Li and T. C. W. Mac, *J. Organomet. Chem.*, **467**, 47 (1994).

(b) S. Adams, M. Drager and B. Matiasch, *J. Organomet. Chem.*, **326**, 173 (1987).

223. I. Hippel, P. G. Jones and A. Blaschette, *J. Organomet. Chem.*, **448**, 63 (1993).

224. C. Kober, J. Kroner and W. Storch, *Angew. Chem., Int. Ed. Engl.*, **32**, 1608 (1993).

225. G. Stocco, G. Guli and G. Valle, *Acta Crystallogr., Sect. C, Cryst. Struct. Commun.*, **48**, 2116 (1992).

226. E. Kello, V. Vrabel, V. Rattay, J. Sivy and J. Kozisek, *Acta Cryst., Sect. C, Cryst. Struct. Commun.*, **48**, 51 (1992).

227. S. W. Ng and V. G. K. Das, *J. Organomet. Chem.*, **456**, 175 (1993).

228. U. Casellato, R. Graziani and A. S. Gonzalez, *Acta Crystallogr., Sect. C, Cryst. Struct. Commun.*, **48**, 2125 (1992).

229. G. Bandoli, A. Dolmella, V. Perruzo and G. Plazzogna, *J. Organomet. Chem.*, **452**, 75 (1993).

230. E. García Martínez, A. Sánchez Gonzalez, J. S. Casas, J. Sordo, G. Valle and U. Russo, *J. Organomet. Chem.*, **453**, 47 (1993).
231. P. Tavridou, U. Russo, G. Valle and D. Kovalademertzi, *J. Organomet. Chem.*, **460**, C16 (1993).
232. M. Wada, T. Fujii, S. Iijima, S. Hayase, T. Erabi and G. E. Matsubayashi, *J. Organomet. Chem.*, **445**, 65 (1993).
233. D. Schollmeyer, H. Hartung, C. Mreftaniklaus and K. Jurkschat, *Acta Cryst., Sect. C, Cryst. Struct. Commun.*, **47**, 2365 (1991).
234. N. Tokitoh, Y. Matsuhashi and R. Okazaki, *J. Chem. Soc., Chem. Commun.*, 407 (1993).
235. R. J. Batchelor, F. W. B. Einstein, C. H. W. Jones and R. D. Sharma, *Inorg. Chem.*, **27**, 4636 (1988).
236. Y. Matsuhashi, N. Tokitoh and R. Okazaki, *Organometallics*, **12**, 1351 (1993).
237. R. J. Batchelor, F. W. B. Einstein and C. H. W. Jones, *Acta Crystallogr., Sect. C, Cryst. Struct. Commun.*, **45**, 1813 (1989).
238. H. Grützmacher, H. Pritzkow and F. T. Edelmann, *Organometallics*, **10**, 23 (1991).
239. U. Lay, H. Pritzkow and H. Grützmacher, *J. Chem. Soc., Chem. Commun.*, 1260 (1992).
240. H. Grützmacher, S. Freitag, R. Herbstirmer and G. S. Sheldrick, *Angew. Chem., Int. Ed. Engl.*, **31**, 437 (1992).
241. G. Ossing, A. Meller, S. Freitag and R. Herbstirmer, *J. Chem. Soc., Chem. Commun.*, 497 (1993).
242. N. Tokitoh, Y. Matsuhashi, M. Goto and R. Okazaki, *Chem. Lett.*, 1595 (1992).
243. H. Grützmacher and H. Pritzkow, *Angew. Chem., Int. Ed. Engl.*, **30**, 1017 (1991).
244. S. P. Narula, S. K. Bharadwaj, Y. Sharda, R. O. Day, L. Howe and R. R. Holmes, *Organometallics*, **11**, 2206 (1992).
245. E. R. T. Tiekink, *J. Organomet. Chem.*, **408**, 323 (1991).
246. D. Dakternieks, H. J. Zhu, D. Masi and C. Mealli, *Inorg. Chim. Acta*, **211**, 155 (1993).
247. M. G. Newton, I. Haiduc, R. B. King and C. Silvestru, *J. Chem. Soc., Chem. Commun.*, 1229 (1993).
248. A. Krebs, A. Jacobsenbauer, E. Haupt, M. Veith and V. Huch, *Angew. Chem., Int. Ed. Engl.*, **28**, 603 (1989).
249. M. Kira, R. Yauchibara, R. Hirano, C. Kabuto and H. Sakurai, *J. Am. Chem. Soc.*, **113**, 7785 (1991).
250. M. Veith and R. Lisowsky, *Angew. Chem., Int. Ed. Engl.*, **27**, 1087 (1988).
251. A. J. Edwards, M. A. Paver, P. R. Raitby, C. A. Russell, D. Stalke, A. Steiner and D. S. Wright, *J. Chem. Soc., Dalton Trans.*, 1465 (1993).
252. E. Kny, L. L. Levenson, W. J. James and R. A. Auerback, *Thin Solid Films*, **85**, 23 (1981).
253. C. Oehr and H. Suhr, *Thin Solid Films*, **155**, 65 (1987).
254. R. K. Sadhir, W. J. James and R. A. Auerback, *Thin Solid Films*, **97**, 23 (1982).
255. R. Larciprete, E. Borsella, P. Depadova, M. Magiantini, P. Perfetti and M. Fanfoni, *J. Vacuum Sci. Tech. A, Vacuum Surf. Films*, **11**, 336 (1993).
256. E. M. Berksoy and M. A. Whitehead, *J. Organomet. Chem.*, **410**, 293 (1991).
257. P. Verbiest, L. Verdonck and G. P. Van der Kelen, *Spectrochim. Acta A, Mol. Spectr.*, **49**, 405 (1993).
258. K. Okada, M. Minami and M. Oda, *Chem. Lett.*, 1999 (1993).
259. A. V. Belyakov, A. V. Golubinskii, L. V. Vilkov, V. I. Shiryaev, E. M. Styopina, E. A. Kovalyova and V. S. Nikitin, *Russian Chem. Bull.*, **42**, 346 (1993).
260. E. Csavari, I. F. Shishkov, B. Rozsondai and I. Hargittai, *J. Mol. Struct.*, **239**, 291 (1990).
261. A. Haaland, A. Hammel, H. Thomassen and H. V. Volden, *Z. Naturforsch., B: Chem. Sci.*, **45**, 1143 (1990).
262. M. J. Tomaszewski and J. Warketin, *J. Chem. Soc., Chem. Commun.*, 1407 (1993).
263. M. Lehnig, T. Reiche and S. Reiss, *Tetrahedron Lett.*, **33**, 4149 (1992).
264. G. Lespes, J. Fernandez and A. Dargelos, *Chem. Phys.*, **115**, 453 (1987).
265. I. Novak, J. M. Benson, A. W. Potts and A. Svensson, *Chem. Phys. Lett.*, **135**, 471 (1987).
266. V. P. Kolotov and E. A. Arafa, *J. Radioanal. Nuc. Chem., Articles* **172**, 357 (1993).
267. A. Hiermann and F. Bucar, *J. Chromatogr.*, **675**, 276 (1994).
268. G. H. Reifenberg and W. J. Considine, *Organometallics*, **12**, 3015 (1993).
269. H. Miyake and K. Yamamura, *Chem. Lett.*, 507 (1992).
270. D. Dakternieks, K. Jurkschat, H. Wu and E. R. T. Tiekink, *Organometallics*, **12**, 2788 (1993).
271. M. Weidenbruch, A. Schäfer, H. Kilian, S. Pohl, W. Saak and H. Marsmann, *Chem. Ber.*, **125**, 563 (1992).

272. Y. Obora, Y. Tsuji, M. Asayama and T. Kawamura, *Organometallics*, **12**, 4697 (1993)
273. J. M. Chong and S. B. Park, *J. Org. Chem.*, **58**, 523 (1993).
274. (a) V. Gevorgyan and Y. Yamamoto, *J. Chem. Soc., Chem. Commun.*, 59 (1994).
     (b) S. K. Kim and P. L. Fuchs, *J. Am. Chem. Soc.*, **115**, 5934 (1993).
275. M. Lautens, C. H. Zhang and C. M. Cruden, *Angew. Chem., Int. Ed. Engl.*, **31**, 232 (1992).
276. C. Boot, H. Imanieh, P. Quayle and S. Y. Lu, *Tetrahedron Lett.*, **33**, 413 (1992).
277. B. Wrackmeyer, K. Wagner and R. Boese, *Chem. Ber.*, **126**, 595 (1993).
278. B. H. Lipschutz, R. Keil and J. C. Barton, *Tetrahedron Lett.*, **33**, 5861 (1992).
279. B. Wrackmeyer, S. Kundler and R. Boese, *Chem. Ber.*, **126**, 1361 (1993).
280. B. Wrackmeyer, G. Kehr and R. Boese, *Chem. Ber.*, **125**, 643 (1992).
281. M. M. Shoukry, *J. Inorg. Biochem.*, **48**, 271 (1992).
282. P. Verbiest, L. Verdonck and G. P. Vanderkelen, *Int. J. Chem. Kinet.*, **25**, 107 (1993).
283. T. N. Mitchell, K. Kwetkat and B. Godry, *Organometallics*, **10**, 1633 (1991).
284. M. Wronski, *J. Chromatogr.*, **555**, 306 (1991).
285. L. S. Liebeskind and B. S. Foster, *J. Am. Chem. Soc.*, **112**, 8612 (1990).
286. (a) L. S. Liebeskind and S. W. Riesinger, *J. Org. Chem.*, **58**, 408 (1993).
     (b) J. L. Parrain, J. C. Cintrat and J. P. Quintard, *J. Organomet. Chem.*, **437**, C19 (1992).
287. C. J. Li and D. N. Harpp, *Tetrahedron Lett.*, **33**, 7293 (1992).
288. K. Jones and M. F. Lappert, *J. Organomet. Chem.*, **3**, 295 (1965).
289. P. Brown, M. F. Mahon and K. C. Molloy, *J. Chem. Soc., Dalton Trans.*, 3503 (1992).
290. S. Kopper and A. Brandenburg, *Justus Liebigs Ann. Chem.*, 933 (1992).
291. H. Ahlbrecht and V. Baumann, *Synthesis, Stuttgart*, 981 (1993).
292. J. B. Verlhac, H. A. Kwon, and M. Pereyre, *J. Chem. Soc., Perkin Trans. 1*, 1367 (1993).
293. G. Veneziani, R. Reau and G. Bertrand, *Organometallics*, **12**, 4289 (1993).
294. B. Klein and W. P. Neumann, *J. Organomet. Chem.*, **465**, 119 (1994).
295. K. H. Henry, P. A. Grieco and C. T. Jagoe, *Tetrahedron Lett.*, **33**, 1817 (1992).
296. G. E. Keck, D. Krishnamurty and M. C. Grier, *J. Org. Chem.*, **58**, 6543 (1993).
297. Y. Nishigaichi, M. Fujimoto, K. Nakayama, A. Takuwa, K. Hamada and T. Fujiwara, *Chem. Lett.*, 2339 (1992).
298. A. H. McNeil and E. J. Thomas, *Tetrahedron Lett.*, **33**, 1369 (1992).
299. T. Takeda, Y. Takagi, H. Takano and T. Fujiwara, *Tetrahedron Lett.*, **33**, 5381 (1992).
300. I. Beaudet, J. L. Parrain and J. P. Quintard, *Tetrahedron Lett.*, **33**, 3647 (1992).
301. B. H. Lipshutz and M. Alami, *Tetrahedron Lett.*, **34**, 1433 (1993).
302. (a) C. Acuña and A. J. Zapata, *Synth. Commun.*, **18**, 1125 (1988).
     (b) A. J. Zapata, C. Fortoul, and C. Acuña, *J. Organomet. Chem.*, **448**, 69 (1993).
303. A. Barbero, P. Cuadrado, A. M. González, F. J. Pulido, R. Rubio and I. Fleming, *Tetrahedron Lett.*, **33**, 5841 (1992).
304. J. I. Levin, *Tetrahedron Lett.*, **34**, 6211 (1993).
305. C. H. Cummins, *Tetrahedron Lett.*, **35**, 857 (1994).
306. M. Z. Deng, N. S. Li and Y. Z. Huang, *J. Org. Chem.*, **58**, 1949 (1993).
307. J. L. Parrain, I. Beaudet, A. Duchene, S. Watrelot and J. P. Quintard, *Tetrahedron Lett.*, **34**, 5445 (1993).
308. M. Pohmakotr and S. Khosavanna, *Tetrahedron*, **49**, 6483 (1993).
309. M. Pérez, A. M. Castaño and A. M. Echavarrén, *J. Org. Chem.*, **57**, 5047 (1992).
310. M. Pohmakotr, S. Sithikanchanakul and S. Khosavanna, *Tetrahedron*, **49**, 6651 (1993).
311. D. P. Matthews, S. C. Miller, E. T. Jarvi, J. S. Sabol and J. R. McCarthy, *Tetrahedron Lett.*, **34**, 3057 (1993).
312. H. C. Zhang, M. Brakta and G. D. Daves, *Tetrahedron Lett.*, **34**, 1571 (1993).
313. (a) F. Bracher and D. Hildebrand, *Justus Liebigs Ann. Chem.*, 837 (1993).
     (b) E. Piers and T. Wong, *J. Org. Chem.*, **58**, 3609 (1993).
314. J. P. Edwards, D. J. Krysan and L. S. Liebeskind, *J. Am. Chem. Soc.*, **115**, 9868 (1993).
315. W. P. Neumann, *J. Organomet. Chem.*, **437**, 23 (1992).
316. V. Farina, B. Krishnan, D. R. Marshall and G. P. Roth, *J. Org. Chem.*, **58**, 5434 (1993).
317. Y. Zea-Ponce, R. M. Baldwin, S. S. Zoghbi and R. B. Innis, *Appl. Radiat. Isot.*, **45**, 63 (1944).
318. M. Namavarri, A. Bishop, N. Satyamurthy, G. Bida and J. R. Barrio, *Appl. Rad. Isotop.*, **43**, 989 (1992).
319. M. Namavarri, N. Satyamurthy, M. E. Phelps and J. R. Barrio, *Appl. Rad. Isotop.*, **44**, 527 (1993).
320. M. J. Adam, J. M. Lu and S. Jivan, *J. Labelled Compd. Radiopharm.*, **34**, 565 (1994).

321. M. J. Adam, J. Lu and S. Jivan, *J. Nuc. Med.*, **35**, P251 (1994).
322. C. Deb and B. Basu, *J. Organomet. Chem.*, **443**, C24 (1993).
323. C. J. Easton and S. C. Peters, *Tetrahedron Lett.*, **33**, 5581 (1992).
324. B. Jousseaume, N. Noiret, M. Pereyre, J. M. Frances and M. Petraud, *Organometallics*, **11**, 3910 (1992).
325. J. M. Chong, G. K. MacDonald, S, B, Park and S. H. Wilkinson, *J. Org. Chem.*, **58**, 1266 (1993).
326. A. F. Burchat, J. M. Chong and S. B. Park, *Tetrahedron Lett.*, **34**, 51 (1993).
327. W. H. Pearson, A. C. Lindbeck and J. W. Kampf, *J. Am. Chem. Soc.*, **115**, 2622 (1993).
328. J. Yamada, T. Asano, I. Kadota and Y. Yamamoto, *J. Org. Chem.*, **55**, 6066 (1990).

CHAPTER 9

Analytical aspects of organolead compounds

JACOB ZABICKY and SARINA GRINBERG

Institutes for Applied Research, Ben-Gurion University of the Negev, Beer-Sheva 84110, Israel
Fax: (+972)-7-271612; e-mail: zabicky@bgumail.bgu.ac.il

The chemistry of organic germanium, tin and lead compounds
Edited by S. Patai   © 1995 John Wiley & Sons Ltd

The abbreviations used in this chapter appear after the list of contents of Chapter 7, p.339, on organogermanium compounds.

I. INTRODUCTION

Among all organometallics in the market two organolead compounds have been for many years economically the most important, namely TML (tetramethyllead) and TEL (tetraethyllead), due to their antiknock properties when used as additives to automobile fuels[1]. However, the preponderance of these two compounds is receding, because of the ban that has been imposed on their use in many countries. Some minor applications of organolead compounds are as antifouling agents for marine paints, lubricant additives and biocides[2].

Elsewhere in *The Chemistry of Functional Groups* series appears a brief discussion on the stages in the lifetime of chemicals[3]. In the case of organolead compounds the metal atoms are always a potential source of harmful pollution with both acute and long-term effects. Increasing concern with environmental and occupational usues has also contributed to development of new analytical methods. Table 1 lists organolead compounds that have found industrial application with references to occupational protection protocols where analytical methods for the particular compound can be found.

See also Section I.A of Chapter 7. Analysis of organolead compounds has been reviewed[1,5,6].

TABLE 1. Industrial organolead compounds[a]

Compound and CAS registry number	NIOSH/OSHA RTECS[b]
Chlorotriethylplumbane [1067-14-7]	TP 4025000
Chlorotripropylplumbane [1520-71-4]	TP 4375000
Dibutyllead diacetate [2587-84-0]	TP 4400000
Diethyllead diacetate [15773-47-4]	OG 0470000
Dihexyllead diacetate [18279-21-5]	OG 0580000
Dimethyllead diacetate [20917-34-4]	TP 4450000
Dipentyllead diacetate [18279-20-4]	OG 0750000
(3-Methylbut-3-en-1-ynyl)triethylplumbane	
Tetraethylplumbane (TEL) [78-00-2]	TP 4550000
Tetramethylplumbane (TML) [75-74-1]	TP 4725000
Tetrapropylplumbane [3440-75-3]	TP 4900000
Tetravinylplumbane	
Tributylplumbyl acetate	
Triethylplumbyl fluoroacetate [562-95-8]	AH 9670000
Triethylplumbyl furoate [73928-18-4]	TP 4505000
Triethylplumbyl oleate [63916-98-3]	OG 6125000
Triethylplumbyl phenylacetate [73928-21-9]	TP 4515000
Triethylplumbyl phosphate [56267-87-9]	OG 6200000
Triphenylplumbyl acetate [1162-06-7]	TP 3922200
Triphenylplumbyl hexafluorosilicate [27679-98-7]	OG 6475000
Tripropylplumbane [6618-03-7]	TP 5160000

[a] See also Reference 4.
[b] *Registry of Toxic Effects of Chemical Substances* of National Institute for Occupational Safety and Health/Occupational Safety and Health Administration.

II. ELEMENTAL ANALYSIS

A. Lead in Organometallic Compounds

After digestion with hot fuming sulphonitric mixture, the solution is neutralized and brought up to pH 10–10.5 and titrated with EDTA against eriochrome black T indicator (**1**)[7]. The presence of lead does not seem to affect CH determination by the usual combustion methods[8]. Ionic halogen can be titrated directly by placing the sample in ethanol or acetone solution. If titration is precluded by the nature of the organometallic compound, sodium peroxide digestion in a bomb is recommended. It is also convenient to reduce the digestion product with hydrazine to avoid loss of free halogen, especially when this procedure is applied to compounds containing bromine or iodine[7].

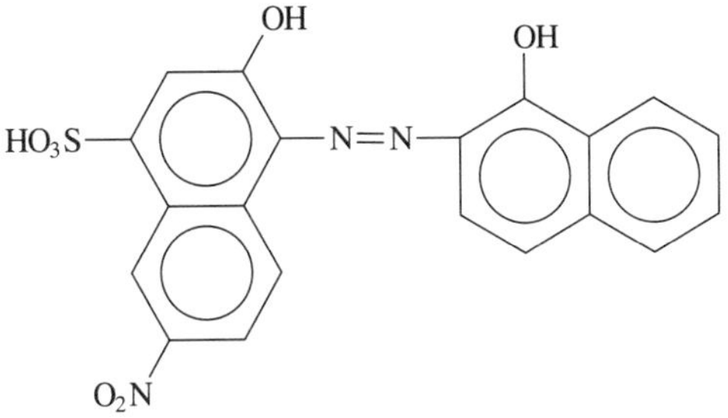

(**1**)

Table 2 presents ASTM standards for the determination of the lead content stemming from TEL and TEM additives in petroleum-derived products.

B. Trace Analysis of Lead

1. General

Analytical problems involving pollution by organolead compounds may be of various types: Determination of total lead, speciation of the original polluter and its degradation products, and identification of the individual source of pollution. Techniques have been developed for dealing with concentration levels of lead compounds involved in such problems, that have receded down to sub-ppb levels, while the sample size has also shrunk, sometimes down to a few milligrams or microlitres. These trends are in pace with the ever-sharpening demands of the authorities regarding environmental quality and occupational safety. Given today's lead pollution levels in the environment, the latter demands impose application of intricate procedures to ensure analytical reliability.

Various strategies can be applied for trace analysis of lead. The details of sample preparation are usually dictated by the nature of the matrix, and involve destruction of the original matrix, solvent extraction of the lead compounds as such or in the form of specific complexes, preconcentration and further chemical transformations of the sample. However, many methods disregard one or more of these operations. End analysis procedures take profit of specific spectral or electrochemical properties, concentration levels of lead and interferring elements in the sample, required precision, available equipment and personal tastes of the analyst. The methods described in the following sections are the results of studies focused on specific aspects of the analytical problem. The information was organized arbitrarily according to the instrumentation serving for the end analysis, however, the analyst should feel free to choose details from various sources that fit his particular case, as long as they are compatible with each other.

TEL and TML additives to fuel are released to the environment by evaporation during vehicle refueling, after spillages, or in exhaust gases together with unburnt fuel. They

TABLE 2.   ASTM standards for lead content in petroleum-derived products[9]

Standard No.	Vol. in Ref. 9	Range of application	Remarks
D 2787	24	*ca* 10 ppm	For gas turbine fuel. Digestion with conc. HCl to PbCl$_2$; spectrophotometric determination of dithizone (**2**) complex.
D 1368	23	< 1 mg Pb/L	Organometallic lead mineralized by treatment with bromine and steam; spectrophotometric determination of dithizone (**2**) complex.
D 2547	24	0.05–1.1 g Pb/L	Mineralization of organic lead, precipitation of PbCrO$_4$, redissolution in HCl/NaCl solution, addition of KI and iodometric titration of liberated I$_2$.
D 1262	23	> 0.1%	For new and used greases. Wet ashing with H$_2$SO$_4$, HNO$_3$ and H$_2$O$_2$; gravimetric determination by electrolytic deposition of PbO$_2$ on Pt anode.
D 3229	25	2.5–12.5 mg Pb/L	XRF measurements at 0.1175 nm for the L-α_1 line of Pb, 0.1144 nm for the L-α_1 line of Bi as internal standard and 0.1194 nm for the background.
D 2579	24	0.26–1.35 g Pb/L	See standard D 3229 above.
D 3116	25	0.25–25 mg Pb/L	PbR$_4$ is converted to ionic organolead by ICl and then to inorganic lead ions; spectrophotometric determination of dithizone (**2**) complex.
D 3341	25	0.025–1.3 g Pb/L	PbR$_4$ converted to ionic organolead by ICl and then to inorganic lead; EDTA titration.
D 3237	25	2.5–25 mg Pb/L	PbR$_4$ converted to R$_3$PbI by I$_2$; determination by AAS.
D 1949	23		TML and TEL in gasoline are separated by distillation to a cut-off temperature; TML appears in the distillate and TEL remains in the boiling residue; determination of both organometallics by one of the methods described above.

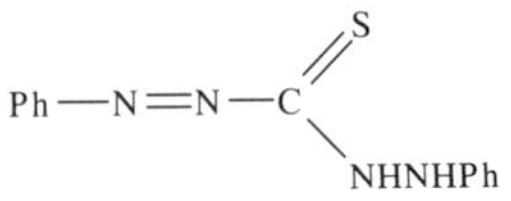

$$\text{Ph}-\text{N}=\text{N}-\text{C}\begin{array}{c}\diagup\!\!\!\!^S\\ \diagdown\\ \text{NHNHPh}\end{array}$$

(**2**)

decompose slowly in the environment into organometallic ions (R$_3$Pb$^+$ and R$_2$Pb^{2+}), and ultimately into inorganic forms.

Residues of commercial organolead compounds and their degradation products are the subject of multiple analytical methods intended for detection and determination in general cases or in specific matrices. Despite the progressively expanding ban on the use of TEL and TML in many countries, a continuous interest remains regarding their long-term effects on living organisms and the environment.

Various reference materials have been described, to help improving the reliability of trace elemental analysis of lead and other heavy elements, for clinical and environmental applications. Such materials include blood[10,11], diets, feces, air filters, dust[11], foodstuffs[12] and biological tissues[13].

The significance of the variability in analytical results obtained for Pb and other metallic elements in street dust was investigated. Coefficients of variation larger than 20% for Pb were taken as reflecting variability of concentration between samples. An important factor affecting variability of heavy metal content was particle size distribution. A decrease of Pb content was observed for dust fractions with nominal diameter decreasing from 1000 μm down to about 150 μm, when Pb concentrations rose with decreasing diameter[14]. Wiping and vacuum collection were compared as sampling methods of residential dust for Pb determination. Analytic results were usually much higher for the wiping method[15]. An automated apparatus for the determination of Pb in air particulates was described[16].

Three methods for trace metal preconcentration were examined: liquid–liquid extraction aided by a chelating agent, concentration on a synthetic chelating resin and reductive precipitation with NaBH₄. The latter method gave 1000-fold preconcentration factors with total recovery of Pb and other elements[17]. Preconcentration of nanogram amounts of lead can be carried out with a resin incorporating quinolin-8-ol (**3**)[18]. Enhancement factors of 50–100 can be achieved by such preconcentration procedures followed by determination in a FIA (flow injection analysis) system; limits of detection are a few μg Pb/L[19].

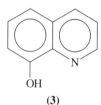

(3)

Tin(IV) antimonate is an inorganic cation exchanger with high selectivity for lead ions[20]. This may be useful for trace analyses that are sensitive to lead interference.

2. Atomic absorption and emission spectroscopies

a. Flame AAS. After combustion of the sample in an oxygen flask, lead and cadmiun were determined by flame AAS[21]. After elution with acid, lead determination can be carried out by AAS[22]. Lead and other trace metals were determined in soil samples by AAS after digestion with aqua regia; LOD 25 μg Pb/L[23]. Flame AAS in a setup that included FIA, a nebuliser interface and computer evaluation of the signal showed LOD of 0.06 μM Pb in blood, with RSD of 5% at normal levels. The peripheral equipment brought about a 12-fold improvement in the performance of a 10-year-old instrument[24]. A design was proposed for a FIA system for Pb trace analysis by flame AAS, enabling eluent conservation and a 50-fold improvement of detection limits[25].

The behaviour of various organolead compounds in gasoline in a hydride generator was investigated. No hydride is formed derived from TML or TEL. These compounds are carried to the silica tube of an AAS; LOD 17 ng Pb/g in a 20 mL sample, RSD 2.31%[26]. The efficiency of lead hydride generation for AAS determination can be improved if treatment is carried out in solutions containing the hydrogenating reagent and various additives: NaBH₄ with lactic acid and potassium dichromate in the ppm range[27]; KBH₄ in the presence of oxalic acid and ammonium cerium(III) nitrate in the μg/L range[28]; NaBH₄ with nitroso-R salt (**4**), RSD 3.8% for 20 μg Pb/L[29].

b. Electrothermal AAS. Biological and environmental samples containing trace amounts of lead and other pollutants were preconcentrated selectively using an ammonium diethyldithiophosphate-citrate complexing agent in a FIA system. After elution, determination was carried out by GFAAS; LOD 0.04 μg Pb/L; RSD 2% for 3 μg Pb/L[30].

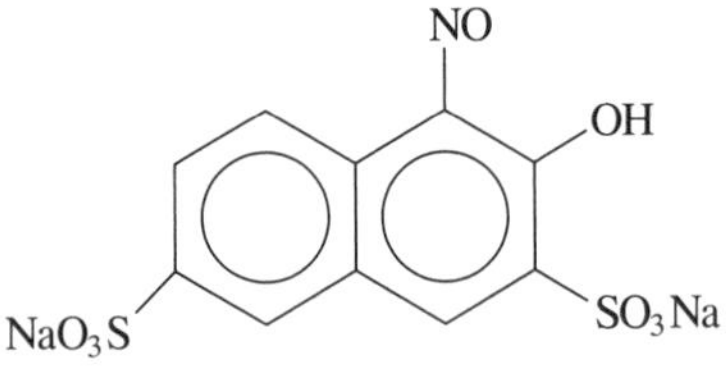

NO

OH

Nao$_3$S   SO$_3$Na

(4)

A combination of hydride generation with GFAAS was proposed for signal enhancement, instead of flame AAS. It was found that older tubes had better sensitivity than new ones[31]. GFAAS is commonly used for determination of low lead concentrations in water. The method suffers from matrix interferences and matrix modifiers can be added for their abatement. The graphite furnace itself may be problematic, because it reduces the oxides of interferring metals present in the matrix. Use of tungsten foil ribbon as heat transducer, in an instrument furnished with Zeeman background correction, and nickel(II)-ammonium tartrate as matrix modifier was evaluated at 30 μg/L lead levels[32]. A method for water samples and biological tissues consists of ethylation of organolead ions with NaBEt$_4$ followed by *in situ* concentration and ETAAS with Zeeman background correction; LOD 0.18 and 0.21 ng/L of Pb for Me$_3$Pb$^+$ and Me$_2$Pb^{2+}, respectively, for a 50 mL sample, RSD 4% at 100 ng/L[33]. Another preconcentration procedure consists of forming iodinated complex ions, extracting with diisobutyl ketone in the presence of zephiramine (**5**), followed by ETAAS; LOD 1 μg Pb/L (SNR 3), RSD 2.7% at 10 μg/L[34]. Lead and other metals in water are determined after extraction with MeCCl$_3$ containing ammonium pyrrolidinedithiocarbamate (**6**) as chelating agent, followed by ETAAS; LOD 0.7 μg Pb/L[35].

CS$_2$NH$_4$

N

n-C$_{14}$H$_{29}$Me$_2$(PhCH$_2$)N$^+$ H

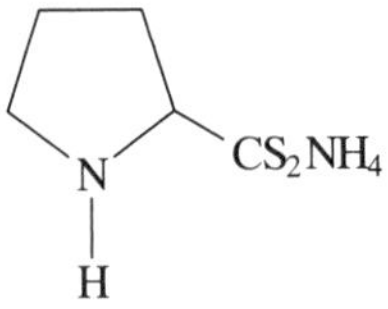

(5)   **(6)**

After a complex digestion process for slurries of marine sediments, the determination of Pb was carried out by ETAAS, using Pd and magnesium nitrate as chemical modifiers; LOD 0.22 μg/L, RSD 5% at 400 μg Pb/g[36].

The total Pb contained in alkyllead vapours present in the atmosphere was determined by filtering the inorganic Pb particulates, adsorbing the organometallic vapours on activated carbon, dissolving the adsorbed alkyllead compounds in hot HNO$_3$ and ETAAS; LOD 2 ng Pb/m^3 for 1 m^3 sample; RSD 9.5%[37]. Most methods for determination of lead concentrations in air involve sample collection for 0.5–24 h. A near real time method for direct, *in situ* determination was devised, based on EPAAS (electrostatic precipitation-AAS) from an electrothermal vaporizer into a tungsten rod, followed by injection of the rod into an electrothermal atomizer and AAS measurement. Background levels and detection limits were in the sub-ng/m^3 levels. Disturbed air showed higher lead content[38,39].

GFAAS end analysis of Pb in blood was critically discussed in a review, including methods of introducing the sample into the furnace, matrix interferences and ways of improving the precision of the method, such as use of a stabilized temperature platform[40]. An evaluation program was carried out for quality control materials and proficiency for long-term Pb in blood analysis using microsampling and Zeeman AAS techniques[41]. A comparison study for direct Pb determination in blood between flame AAS and ETAAS with L'vov platform showed that the latter method is more precise in the 35–400 μg Pb/L concentration range[42]. GFAAS determination of Pb in blood, using graphite tubes coated with pyrolitic graphite, had LOD 1.2 μg Pb/L, injecting 10 μL aliquots of a 1:10 solution of blood in Triton X-100 and an ammonium dihydrogen phosphate-magnesium nitrate matrix modifier; mean relative error of 1%[43]. Samples of avian blood were prepared using a diluent and a matrix modifier, and were determined using GFAAS with L'vov platform and Zeeman effect background correction; LOD as low as 4 ng/g[44].

A comparative study was carried out of ashing and atomization techniques in the GFAAS method for direct determination of Pb in bovine liver[45]. Bismuth nitrate was proposed as matrix modifier for ETAAS determination of Pb in animal tissues; LOD 0.9 μg/L[46].

A rapid digestion of vegetable tissue with nitric-perchloric acid mixture was followed by ETAAS using ammonium hydrogen phosphate as matrix modifier. Recovery of Pb was complete at the ppm level with RSD < 4%[47].

After nitric acid digestion As and Pb are determined in wine by ETAAS, using small amounts of Pd and magnesium nitrate as chemical modifiers; LOD 5.5 μg Pb/L for 8 ml sample, which is 20–30 times smaller than the maximum allowed concentrations[48]. Direct determination of lead in wine can be carried out by GFAAS with L'vov pyrolytic graphite platform and Zeeman effect background correction, after a 10-fold dilution with 1% nitric acid using ammonium dihydrogen phosphate as matrix modifier; LOD (3 σ) 4 μg/L, quantation limit (10 σ) 14 μg/L, short-term RSD 2.1%, long-term RSD 7.4%[49]. It was shown that lead in wine probably originates from atmospheric contamination[50].

A simple and fast method for determination of Pb in cereal-derived products consists of making a slurry in 20% ethanol and determining by ETAAS using phosphate as chemical modifier; LOD 8 ng Pb/g, RSD 4.5–14%[51]. Pd was found to be better than ammonium dihydrogen phosphate as matrix modifier for ETAAS determination of Pb in food slurries[52]. The slurry method for sample preparation is also applicable to plant material[53]. A broad interlaboratory study was carried out for the direct determination of lead in fatty materials by GFAAS with or without platform[54].

c. Tracing Pb sources and other applications of MS detectors. Of the four stable isotopes of lead three stem from the following radioisotope decay reactions: $^{238}U \rightarrow {}^{206}Pb$, $^{235}U \rightarrow {}^{207}Pb$ and $^{232}Th \rightarrow {}^{208}Pb$. Thus, every batch of lead has its own isotopic composition, depending on the original content of radioactive elements in the rock. Determination of lead and lead isotope ratios by the ICP-MS (inductively coupled plasma-MS) method has become routine[55–57]. One source claims an isotopic LOD of 2 ng/L[58]. Application of the isotope dilution technique to ICP-MS analysis of Pb traces in food was examined. Repeated scanning in the 203 to 210 *m/e* range gave better precision (RSD 0.5%) than repeated measurements at each mass unit (RSD 2.5%)[59].

An excellent example of the painstaking procedures necessary for determination of contaminations at the ng/g level was shown in a work dealing with the IDMS (isotope dilution MS) technique with ^{208}Pb and isotope systematics. This was applied for tracing the origin of lead in canned pineapple juice samples, based on the isotopic composition of possible individual lead sources. The ICP-MS method could also have been applied for the same purpose[60]. Lead was found to attach preferentially to certain high molecular weight fractions of human and rat plasma and red cell hemolysate. This study was performed by applying ICP-MS analytical techniques[61].

A combination of FIA with ICP-MS was applied for direct determination of lead in aqueous samples. Calibration of the ^{208}Pb signal by addition of standard was found to be better than external calibration and isotope dilution. In the 50 μg Pb/L range RSD was 1% for wine and 3.5% for urine[62].

d. Direct coupled plasma. Lead content in the frying oil of food vendors showed dependence on the nearby traffic intensity. Oil samples were emulsified in water, adding 1 mg Pb/L, the lead was extracted with EDTA solution and determined by DCP-AES. The lead addition was made to enter the linear response range of the instrument[63]. The presence of easily ionized elements for signal enhancement is an important feature of DCP-AES. Alkaline and alkaline earth elements improved the performance of hydride generation followed by DCP-AES for group 14 elements. RSD 6% at 20 μg Pb/L in water and 4.6% in 1 M CsCl[64].

e. Laser-excited atomic fluorescence. LEAF-ETA (laser-excited atomic fluorescence electrothermal atomization) was used for determination of sub-pg/g Pb concentrations in Antarctic ice, without preconcentration or chemical treatment, using 20 μL samples. Results compared well with those of IDMS[65]. LAEF using a copper vapour laser was developed for determination of lead in the low ng/L range of concentrations. This allows direct determination of polluted water sources without preconcentration[66]. The lowest LOD for Pb in seawater to the date of publication of the reference, 3 fg (absolute) or 1 ng/L, was achieved by a specially modified method, using LEAF[67]. Three different laser systems were evaluated for LEAF-ETA determination of Pb. The direct line fluorescence at 405.78 nm had LOD in the subfemtogram range[68].

3. Electrochemical methods

Extracts containing the diethyl dithiocarbamate (**7**) complexes of Cd, Tl, In and Pb were determined by ASV (anodic stripping voltammetry), using a mercury hanging drop electrode; LOD 10^{-8} M Pb[69]. The effects of solvent and pH buffer composition on the results of potentiometric stripping analysis of Pb, Cd and Cu were investigated[70]. After digestion, microsamples containing traces of lead can be determined by ASV, using a mercury ultramicroelectrode of 10 μm diameter. For example, RSD 4.1% was found for 5 μL samples containing 0.1 μM Pb^{2+} concentrations[71]. A method taking about 4 min per analysis is the scan staircase voltammetry modification of the ASV method, by which the Pb plated on a rotating mercury film electrode is exposed to potential step increases of 10 mV every 64 μs, finishing the Pb stripping operation in only 4 ms. Electrode rotation needs not be stopped and oxygen does not affect measurements; LOD 0.1 μg Pb/L[72]. Purging out dissolved oxygen in the exchange solution at the stripping step is important for Pb determination by DPASV at the ppb level. Sensitivity is greater in the presence of nitrate than chloride ions[73].

$$Et_2NCS_2{}^-Na^+$$

(**7**)

A method for Pb and other metals in seawater consists of an adsorption step of the complex with 8-hydroxyquinoline (**3**) followed by differential cathodic stripping voltammetry; LOD 0.03 nM Pb, after 10 min adsorption[74].

Determination of lead in 70 μL samples of whole blood can be carried out in a few minutes by a procedure including treatment with a matrix-modifying solution containing hydrochloric acid, Hg(II) ions, Triton X-100 and Bi(III) as internal standard. After deposition of lead amalgam on a glassy carbon electrode by a pulsed potential cycle, analysis is finished by stripping potentiometry; LOD 3.5 ng/L, RSD *ca* 10%[75]. An automated method was devised for stripping potentiometry determination of Pb(II) in whole blood with Cd(II) serving as internal standard, consisting of treating 0.2–0.4 ml samples with HCl in a computerized FIA system[76,77]. A three-in-one electrode assembly was devised for stripping potentiometry determination of Cd(II) and Pb(II) at the low ng/L level, using medium exchange in batch mode, and multiple stripping in a hanging stripping medium drop. The best medium is an acetic acid–ammonium acetate solution; LOD 0.5 ng Pb/L, RSD 5.4% at 20 ng Pb/L[78].

Low-cost, mercury-free, gold-coated, disposable electrodes were proposed to replace the usual mercury electrodes used in potentiometric stripping analysis, for determination of trace concentrations of metals in environmental, clinical and industrial samples. The electrodes are mass produced by the screen-printing method and do not cause environmental pollution on disposal. Determination of lead in 5–20 μg/L concentrations showed good precision[79]. Simultaneous ASV determination of Pb(II) and Cu(II) was studied after deposition of the metals in atomic form on a gold film electrode[80]. A method was proposed for electrode preparation for the FIA-ASV combination, where mercury(II) acetate is incorporated into a Nafion perflourosulphonate that coats a glassy carbon electrode. On applying a negative potential a 3-dimensional thin mercury film is formed in the coating, that is stable under flowing conditions[81]. In a variation of this type of coated electrode, dicyclohexano-18-crown-6 (**8**) was incorporated to the Nafion; LOD 5×10^{-10} M for 3 min electrodeposition; RSD 8% at 2×10^{-9} M level[82a]. Chemically modified electrodes with complexing capabilities similar to free crown ether or cryptand ligands can be prepared from a carbon paste incorporating these type of reagents[82b]. Chitosan, a derivative of chitin, was used to modify glassy carbon electrodes, for preconcentration and determination of trace concentrations of Pb(II) in water, achieving a 14-fold improvement of the electrode sensitivity[83]. Band electrodes, fabricated from ultrathin carbonized polyacrylonitrile film and coated with mercury, can be used for ASV determination of Pb in sub-microlitre samples. Application to blood analysis was studied[84].

(8)

A method for trace analysis of Pb, alone or simultaneously with Cu and Cd, consists of potentiostatic electrodeposition of the analytes on a porous electrode, from a flowing sample. After stopping the flow and removing oxygen from the system, the deposit is measured by chronopotentiometry at constant current. The process can be repeated and the signals accumulated to improve the SNR[85]. An interference noted in the Pb and Cd determination in fruit juice by the AOAC ASV method was attibuted to the presence of Sn(IV); it was suppressed by increasing the tartaric acid concentration in the supporting electrolyte[86].

A method for electroanalytical determination of lead traces in gasoline is based on an apparatus consisting of a glassy carbon electrode, a platinum electrode and a standard calomel electrode. The determination of a sample placed in an acid ethanolic solution containing mercuric ions, consists of three steps: A mercury film is formed on the carbon electrode at potentials of -0.50 V or lower, lead amalgam is formed by pulses of -1.50 to -1.20 V, and the measurement is carried out by stripping at -0.90 to -0.30 V. The cell is diffusion controlled[87]. Lead in gasoline can be determined by polarography of an aqueous emulsion. Lead becomes preferentially adsorbed on the mercury drop[88]. A method claimed to be a good alternative to AAS for Pb in gasoline at ppb levels is ASV of the aqueous extract obtained after treatment with ICl reagent[89].

Direct determination of lead in urine by DPASV was carried out without pretreatment of the sample, after adding fume silica to remove organic components that interfere with the method[90]. Samples of foodstuffs (0.3 g) can be digested in special 10 ml quartz vessels to be used for stripping voltametry; LOD 4 ng Pb/g[91]. The stability constants of cadmium and lead with oxalate and sulphate ions were determined by differential pulse polarography and direct current polarography at 1×10^{-7} to 1×10^{-4} M concentrations[92]. DPASV and ETAAS gave close, though significantly different, results in the analysis of Pb in deciduous teeth[93]. A computerized system was proposed for ASV determination of lead in wines and vinegars[94].

Wet oxidation of samples of biological origin with a mixture of nitric acid and perchloric acid is a common procedure that may be problematic in certain cases. An alternative procedure for sample preparation is irradiation with a strong UV source. Acidified samples of hair (0.1 g) containing about 2 μg Pb and other trace elements, were irradiated for 3 h with a 500 W source; a buffer (pH 5.5) and a small amount of catechol violet (9) were added, and the complex of the dye with the trace elements was determined polarographically; RSD ca 5% for Pb[95].

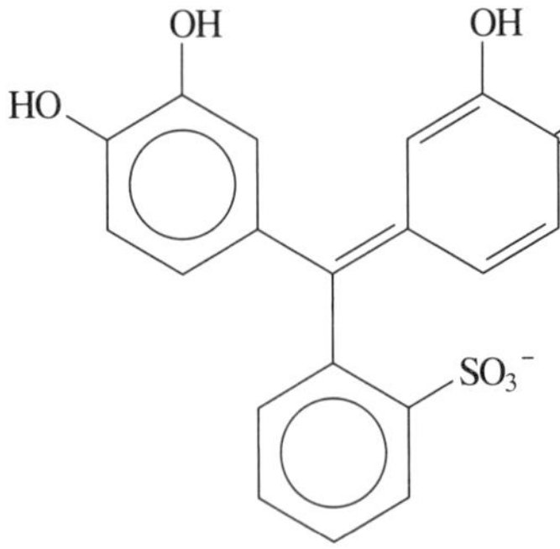

(9)

Selectivity for lead ion is induced by incorporating N,N,N',N'-tetrabutyl-3,6-dioxaoctanediamide (10) to the membrane of a graphite solid state sensor[96].

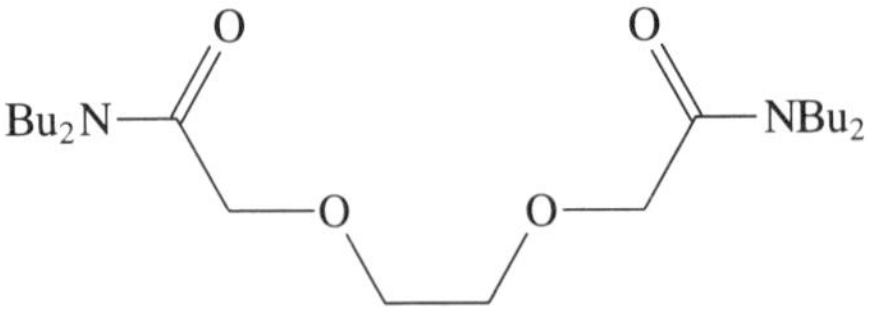

(10)

4. Spectrophotometric methods

a. Complexes in solution. The complex formed by lead ions with catechol violet (**9**), in the presence of cationic surfactants, can be determined spectrophotometrically (ε 4.20–5.06 × 10^4 L/mol cm)[97]. Preconcentration in a Cholex-100 column, extraction of lead ions from an acidic medium into chloroform with crown ether **8**, addition of dithizone (**2**) and absorbance measurement showed good selectivity for lead; LOD 5 mg Pb/L[98].

Determination of Pb(II) ion by classical or reversed FIA consists of a preconcentration step either on columns packed with a chelating sorbent (PC-FIA) or on a mercury film, followed by spectrophotometric determination of the complex with 4-(2-pyridylazo)resorcinol (**11**, λ_{max} 518 nm) in borate buffer solution; RSD 3–6% at 0.01–1 μM. End analysis by ASV was also applied[99].

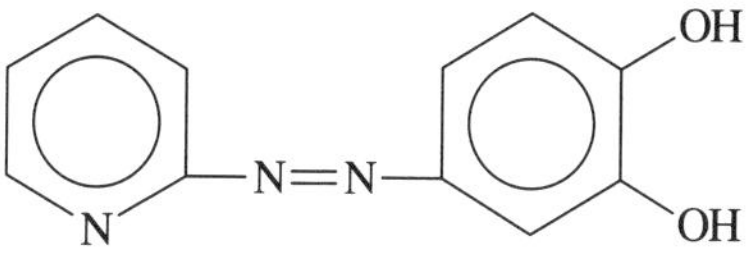

(11)

Hexaoxacycloazochrom (**12**) is a selective, sensitive (ε 1.5 × 10^5/M cm) complexing reagent for Pb ions; LOD 2 μg Pb/L in water[100].

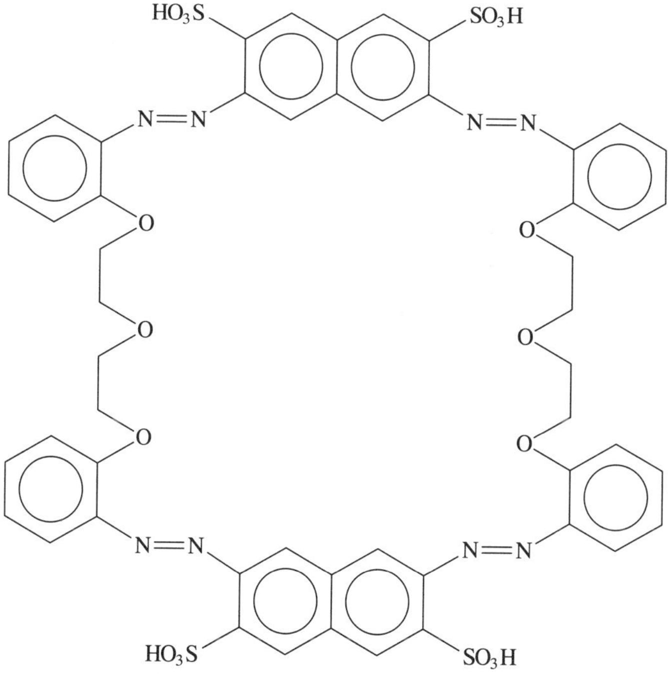

(12)

b. Optical sensors. Bulk optodes are optical sensors based on a membrane incorporating an ion-selective chelating agent—the ionophore—and a proton-selective dye—the chromoionophore, allowing determination of specific ions by contact with the sample solution. Lead ion-selective optodes, with subnanomolar detection limits and with good selectivity over alkali and alkaline earth cations, were developed based on a PVC membrane containing ionophore ETH5435 (**13**) and chromoionophore ETH5418 (**14**), or pairs of similar compounds. Measurements are spectrophotometric and depend on the pH of the solution. For example, in the particular case of an optode based on compounds **13** and **14**, LOD for lead ions are 3.2×10^{-12} M at pH 5.68 and 3.2×10^{-10} at pH 4.70[101].

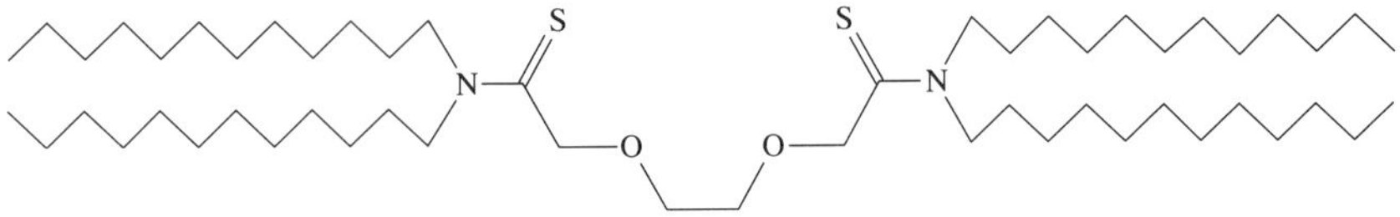

(13)

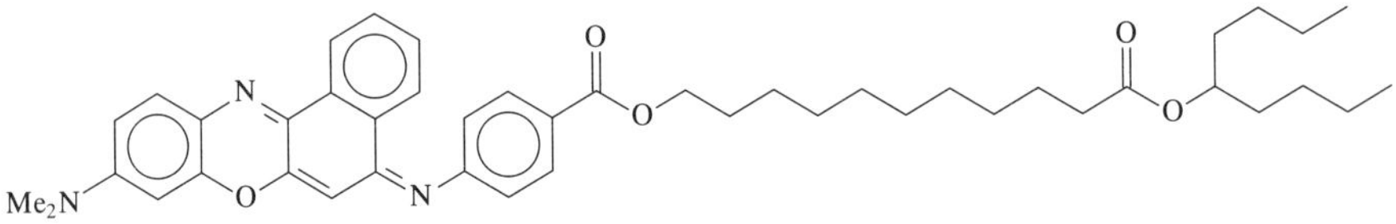

(14)

5. Nuclear activation

Activation analysis of low concentrations of Pb can be performed using 23 MeV protons to yield three bismuth radionuclides with half-lifetimes ($t_{1/2}$) in the range of 0.5–15.3 days. The following nuclear reactions take place: ^{204,206}Pb(p, xn)^{204}Bi, $x = 1$, 3; ^{206,207}Pb(p, xn)^{205}Bi, $x = 2$, 3; and ^{206,207,208}Pb(p, xn)^{206}Bi, $x = 1$, 2, 3. The plausible reaction ^{209}Pb(p, tn)^{206}Bi is less effective by a factor of 2×10^{-4}. The Bi ions produced are separated from the matrix by ion exchange chromatography before measurement; LOD 10 ng/g of environmental samples[102–104]. Another convenient activation is the α-particle reaction ^{208}Pb(^{4}He, 2n)^{210}Po, where the product is an α-emitter with $t_{1/2}$ 138.4 day[105].

Methods were described for diminishing the systematic errors in Pb activation analysis, stemming from the variable isotopic composition of Pb. Results with ICP-AES and ICP-MS were taken as standards for comparison[106].

A substoichiometric isotope dilution method, using ^{203}Pb (a γ-emitter with $t_{1/2}$ 52 h), was proposed for Pb determination in biological or soil samples. The method was applied after plasma ashing of biological samples or digestion of soil samples and a series of extractions with organic solvents containing dithizone (**2**) and re-extractions with aqueous acid. Absolute LOD 1 μg Pb for 0.1–0.3 g of biological samples; RSD 10%[107].

6. Miscellaneous methods

A GC-MS method for determining the isotope composition of lead in blood and urine samples is based on preparation of Pb(C$_6$H$_4$F-p)$_4$ using the corresponding Grignard reagent[108].

The feasibility was explored of measuring the lead content in bones of live patients by XRF, using a ^{99m}Tc-labelled radiopharmaceutical as internal standard. The method requires improvement of the detection capabilities of extant instrumentation, since the patient could be exposed to dangerously high doses of radiation[109]. A calibration method was described for the XRF determination of Pb in dust, using the Pb L_β line; LOD 0.7 μg Pb/m^3 air[110].

The origin of lead present in individual calcite particles could be ascribed by the LAMMA (laser microprobe mass analysis) technique. At low laser irradiances, the desorption mode, information is gathered on metallic species adsorbed on the surface of the particle. At high irradiances the particle is evaporated, revealing the components that coprecipitated with calcite[111].

The binding state of Pb in aquatic bryophites depends on the bryophite species, as determined by XPS. The binding energies were found to be lower than that of PbSO$_4$[112].

III. TRACE ANALYSIS ALLOWING SPECIATION

A. General

Organolead compounds in the environment were reviewed[113]. Results from a simulation study of R$_4$Pb degradation in the environment showed that R$_3$Pb$^+$ probably goes directly into the inorganic form and not through stepwise degradation[114].

The analytical strategies for trace concentrations of organolead compounds are quite similar to those mentioned for total lead in Section II.B.1. However, some important details are different. During the sample preparation stage one has to avoid destruction of the organolead species if they have to be separated from the matrix. The methods for end analysis should include either a separation step, usually by chromatography, or a detection method capable of distinguishing between the various organolead species present in the mixture, usually by UV-visible spectrophotometric or electrochemical methods.

A method was described for production of trace levels of TML for calibration purposes[115].

B. Chromatographic Methods

Ionic alkyllead compounds derived from tetraalkylplumbanes form complexes with sodium diethyldithiocarbamate (7), that can be extracted into pentane. Inorganic lead is complexed with EDTA and determined separately. The organic phase is evaporated and the haloorganoplumbanes are n-butylated with the corresponding Grignard reagent in THF, and the analysis is finished by GC-AAS; LOD in the ng Pb/L level for 0.5 L samples of water[116,117]. The same general procedure was applied to analysis of potable water and soil samples at the ultra-trace level[118] and of blood with LOD 3 μg/L[119]. A method was evaluated for environmental monitoring of alkyllead residues in grass and tree leaves, that began with the preparation of an extract of the biological material in 25% aqueous ammonia, and was followed up as described before[116,120].

Methods for speciation of alkyllead compounds in environmental samples such as air, wet atmosphere deposition and dust were summarized. The steps consist of extraction and derivatizing, and end analysis by GC-AAS[121]. A procedure for determination of the various organolead ionic species present in the atmosphere in trace amounts consists of scrubbing a sample of air in water, extracting with n-hexane, derivatizing by butylation or propylation, and end analysis by GC-ETAAS. No interference from inorganic Pb or alkylplumbanes occurred; LOD is sub-ng Pb/m^3 of air[122]. The use of a large cryogenic trap allowed GC-AAS analysis in samples $\leqslant$ 50 m^3 for 1–10 pg Pb/m^3 concentrations[123].

LOD in the range of 0.02 to 0.1 pg of lead, depending on the volatility of the species, were found after derivatizing with propyl or butyl Grignard reagents, GC separation and measurement by microwave-induced plasma AES[124].

Very high sensitivity was claimed after separation of the Grignard-ethylated derivatives by HPLC and volatilizing into a quartz tube AAS. The linear dynamic range of the response at optimum sensitivity was very limited[125]. A modification of the HPLC-quartz tube AAS method included a thermospray micro-atomizer before the quartz tube. Such an arrangement allows speciation of organolead compounds in water and sediments by liquid extraction of these matrices, without derivatization, at sub-ng Pb/g levels[126].

Determination of organolead metabolites of tetraalkyllead in urine can be carried out after solid-phase enrichment and end analysis using reversed-phase HPLC with chemical reaction detector and by LC-MS (thermospray[127]). The chemical derivation consists of conversion to the dialkyllead form, as shown in reaction 1, followed by complex formation with 4-(2-pyridylazo)resorcinol (11) and spectrophotometic measurement at 515 nm[128].

$$
\begin{array}{c}
R_4Pb + I_2 \\
\\
R_3Pb^+ + I_2
\end{array}
\searrow\!\!\!\!\!\nearrow \quad R_2Pb^{++}
\tag{1}
$$

A new method for Sn, Pb and Hg organometallics is now under evaluation by the US Environmental Protection Agency. It consists of measuring the pentylated Grignard derivatives by a capillary GC-atomic emission detector; LOD $1-2.5$ μg/L for 0.5 μL on column injection, with good linearity in the $2.5-2500$ μg/L range[129].

Organolead and organoselenium compounds were separated satisfactorily by high-performance capillary electrophoresis, using β-cyclodextrin-modified micellar electrokinetic chromatography with on-column UVV detector set at 210 nm[130].

C. Spectrophotometric Methods

Dithizone (2) associates with ionic organolead and inorganic lead compounds to yield complexes with distinct UVV spectra, that can be used for determination of mixtures of TEL and TML degradation products, as shown in reaction 2. Determination of the tetraorganolead compound in many procedures is by difference, involving separate determination of total lead and ionic lead forms in the sample. Tetraorganolead compounds convert to ionic forms by the action of iodine, as shown in reaction 3. Finishing analysis can be carried out spectrophotometrically by the dithizone method[131,132].

$$
Pb^{2+} + R_2Pb^{2+} + R_3Pb^+ \xrightarrow[\text{(2)}]{\text{PhN=NCSNHNHPh}} PbDz_2 + R_2PbDz_2 + R_3PbDz
\tag{2}
$$

$$
R_4Pb + I_2 \longrightarrow R_3PbI + RI
\tag{3}
$$

Analysis by GC of the various organolead species present in gasoline requires special detectors because of the profusion of species with retention times near those of the organometallic compounds. An old determination method consisted of scrubbing the separated species in iodine solution, followed by spectrophotometric determination of the complex with dithizone (2)[133,134] (see also Table 2).

D. Electrochemical Methods

The oxidation processes shown in reactions 4 and 5 are the basis for differential pulse polarography/cyclic voltametry determination of dialkyllead and trialkyllead species[135].

$$
2R_3PbX + 2Hg \rightleftharpoons [R_3Pb-Hg-PbR_3]^{2+} + HgX_2 + 2e^-
\tag{4}
$$

$$2R_2PbX_2 + 2Hg \rightleftharpoons [R_2XPb-Hg-PbR_2X]^{2+} + HgX_2 + 2e^- \qquad (5)$$

An alternative to the chromatographic procedures consists of extracting the diethyldithiocarbamate complexes into hexane, re-extracting the organolead ions into acidified water and determining lead by DPASV; LOD 0.5 ng Pb/L. The mechanism of the electrochemical process involved was discussed[136]. The stability constants of the complexes of triethylplumbyl and trimethylplumbyl ions with diethyldithiocarbamate (7) were measured by electrochemical techniques[137]. Relatively large amounts of inorganic Pb interfere with the determination of alkyllead species by DPASV. It is recommended to co-precipitate Pb(II) ions as sulphate with $BaSO_4$ before proceeding with the electrochemical determination. The precipitate need not be removed; LOD 2×10^{-10} M Pb[138]. Hexaethyldiplumbane content in TEL can be measured by various electroanalytical methods[139-141].

E. Miscellaneous Methods

A combination of FIA and iodination of tetraalkyllead in emulsions, followed by AAS Pb determination, was proposed. The different behaviour shown by TEL and TML can be used for speciation of these compounds[142].

IV. STRUCTURAL ANALYSIS

A. Vibrational and Rotational Spectra

See introductory remarks in Section I.A.4 of Chapter 7. The IR vibration frequencies of compounds MH_4 and $MeMH_3$ were calculated *ab initio* for the metallic elements of group IVb and compared with experimental data from various sources. No experimental data exist for PbH_4 or $MePbH_3$, however, these were estimated from a correction made on the values of SnH_4 and $MeSnH_3$, respectively. Estimated strong bands may be helpful for identification of compounds in the gas phase; for example, at ca 650 and 1830 cm^{-1} for PbH_4 and at ca 660 and 1790 cm^{-1} for $MePbH_3$[143].

Compounds of formula $(CF_3)_nPbR_{4-n}$ and $(CF_3)_nPbR_{3-n}X$ (**15**, $n = 0-4$, R = Me, Et, X = hal) were characterized spectroscopically by IR, Raman and MS. See also Table 3[144,145].

Tetramethylplumbane mixtures with tetramethylstannane were studied at 2 K by high resolution INS (inelastic neutron scattering), revealing rotational tunnelling transitions for the methyl groups[146].

B. Nuclear Magnetic Resonance Spectroscopy

See introductory remarks in Section I.A.4 of Chapter 7. Information on NMR spectra of organolead compounds appears in Table 3. Most sources usually include chemical shift values, δ, and coupling constants, J; however, in many cases other features are also discussed.

C. X-ray Crystallography

See introductory remarks in Section I.A.2 of Chapter 7. Table 4 summarizes examples published in the recent literature, with brief comments about structural features of organolead compounds.

TABLE 3. Structural determination of organolead compounds by NMR spectroscopy

Compounds	Comments	References
$(CF_3)_n PbR_{4-n}$, $(CF_3)_n PbR_{3-n}X$ (**15**, $n = 0-4$, R = Me, Et, X = hal)	^1H, ^{13}C, ^{19}F and ^{207}Pb NMR. See reaction 8 in Section V.	144, 145
$R_3Pb-PbR_3$ (R = c-C_6H_{11}, Ph, methylated Ph)	^{207}Pb CP-MAS-NMR.[a] Conformational analysis of solid hexasubstituted diplumbanes.	147
Alkynylplumbanes	Sign of the following coupling constants has been determined: $^1J(^{207}Pb^{13}C\equiv)$, $^2J(^{207}PbC\equiv^{13}C)$, $^3J(^{207}PbC\equiv^{13}C)$, $^3J(^{207}PbC\equiv C^1H)$ and $^4J(^{207}PbC\equiv CC^1H)$.	148
16–21 (see below this table)	^1H, ^{13}C, ^{29}Si and ^{207}Pb NMR. Compounds **16–20** are intermediates in the formation process of plumbole **21** from a plumbylacetylene with the aid of trialkylboron, according to reaction 6.	149
$Cp^{\#}PbCp^{\#}$	^{13}C and ^{207}Pb NMR, CP-MAS-NMR, IR, Raman, MS and powder XRD.[a] A Pb(II) compound. Parallel staggered conformation proposed for decaphenylplumbocene.[b]	150, 151a
Mes_3PbCOR, R = Me, Ph	^1H, ^{13}C and ^{207}Pb NMR and XRD. First stable acylplumbanes ever synthesized.	151b
22, 23 (see below this table)	^1H, ^{13}C, ^{29}Si and ^{31}P NMR and XRD.	152
$(Me_3Pb)_3N$	^1H, ^{15}N and ^{207}Pb NMR. The value and sign of coupling constants were determined: $^1J(^{207}Pb^{15}N)$ the largest value for this constant published so far, $^3J(^{15}N^1H)$.	153
$Me_2Si\begin{array}{c}i\text{-Pr}\\|\\N\\\diagdown\\ \diagup\\N\\|\\i\text{-Pr}\end{array}M$ (M = Sn, Pb)	^1H, ^{13}C, ^{29}Si, ^{119}Sn and ^{207}Pb NMR. For M = Pb at room temperature an equilibrium between the monomer and dimer is established. At low temperature the structure resembles that of the M = Sn dimer. A mixture of Sn and Pb analogues gives the mixed dimer. For many analogous Sn(II) and Pb(II) compounds the following correlation holds: $\delta_{207Pb} = 3.30\delta_{119Sn} + 2336$.	154, 155
24, 25 (see below this table)	^1H, ^{13}C, ^{29}Si and ^{207}Pb NMR. The diaminoplumbylene **24** exists as dimer **25**.	156
$\begin{array}{c}Ph\diagdown\diagup Ph\\M\\Ph-S\diagdown\diagup S-Ph\\|\quad\quad|\\M\diagdown\diagup M\\Ph\diagup S\diagdown Ph\end{array}$ (**26**, M = Sn; **27**, M = Pb)	A ring segment exchange reaction was studied by NMR in solutions containing mixtures of **26** and **27**, leading to equilibrium mixtures of compounds containing Sn–S–Pb moieties.	157

[a] CP-MAS-NMR = cross polarization–magic angle spinning–NMR, XRD = X-ray diffraction.

$$Me_3PbC\equiv CSiMe_3 + BR_3 \rightarrow \textbf{16} + \textbf{17} \qquad (6)$$

[b] $Cp^{\#}$ = pentaphenylcyclopentadienyl anion.

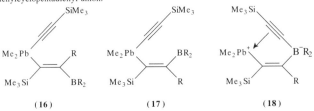

(16)   (17)   (18)

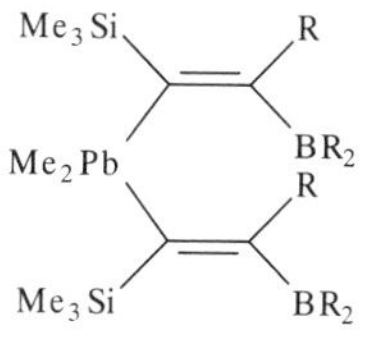

$$Me_3Si \quad R \quad BR_2$$

(19) 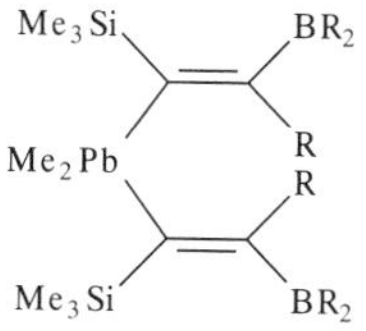   **(20)** 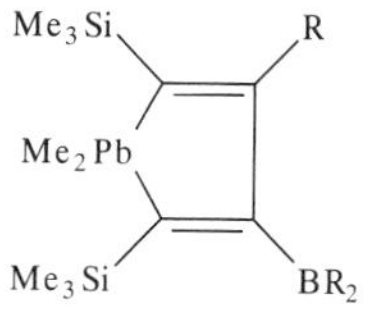   **(21)**

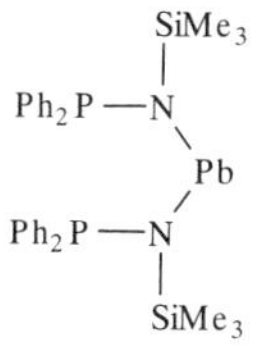

(22) 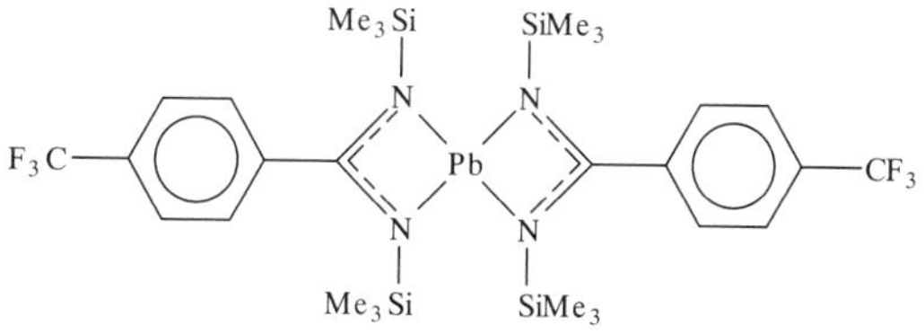   **(23)**

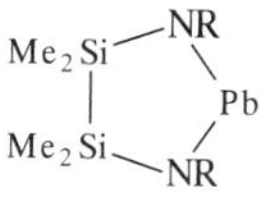

(24) 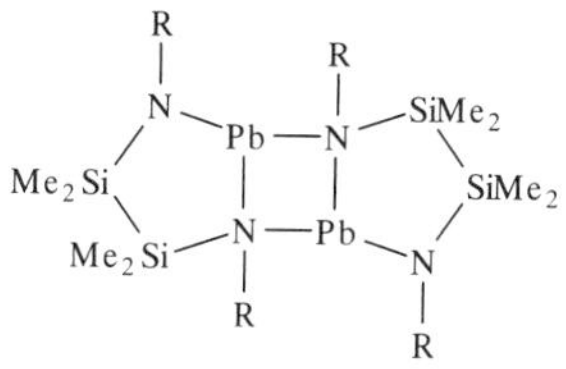   **(25)**

TABLE 4. Structural determination of organolead compounds by X-ray crystallography

Compounds	Comments	References
o-XC$_6$H$_4$Pb(OAc)$_3$ (X = Me, Cl)	o-Substituted phenyllead triacetate has the lead atom heptacoordinated to three chelating acetate groups and the aryl group, in a deformed pentagonal bipyramidal configuration.	158
29 (see reaction 17 in Section V)	This compound was the first stable diaryllead(II) synthesized. It remains stable to its melting temperature at 58 °C and it decomposes slowly in solution. The shortest Pb—Pb distance is 0.7316 nm which points to a monomeric structure. The closeness of four F atoms to Pb confers stability to the crystal.	159
28 (see below this table)	Plumbacarborane complexes with bidentate Lewis bases. Pb is heptacoordinated.	160a,b

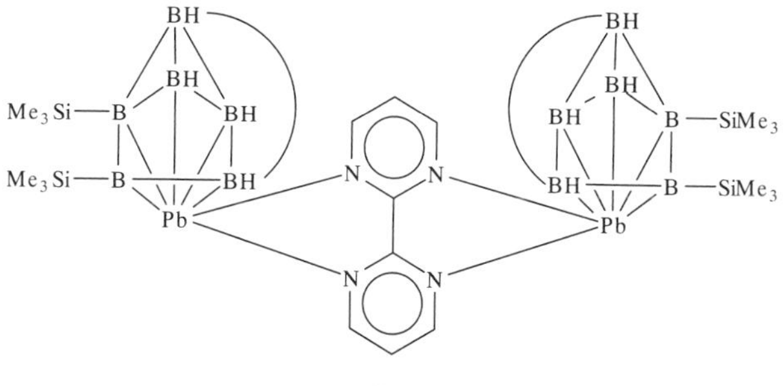

(28)

D. Miscellaneous Methods

The electronic structure of dichlorodiphenylplumbane was calculated by the SCF-MS (self-consistent field multiple scattering) molecular orbital method and compared to that of dichlorodiphenylstannane. The results suggest that one has to look for ^{35}Cl NQR (nuclear quadrupole resonance) frequencies of dichlorodiphenylplumbane in the 5–6 MHz region[161a].

The mass spectra of tetramethylplumbane and hexamethyldiplumbane were discussed based on SCF-MO calculations of the various radicals and ions that are formed in the ionization chamber[161b].

V. DERIVATIZATION

Acetal ethers containing the trialkylplumbyl group undergo the interesting rearrangement 7, catalysed by Ti(IV) compounds. The analogous tin compounds fail to give this reaction[162].

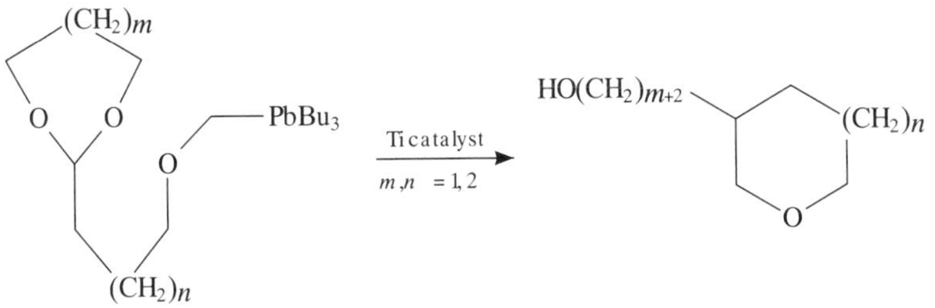

$$
\begin{array}{c}
\text{(CH}_2)_m \\[-2pt]
\displaystyle O \qquad O \\[-2pt]
\qquad O \!-\!\!-\text{PbBu}_3 \\[-2pt]
\text{(CH}_2)_n
\end{array}
\quad
\xrightarrow[m,n\,=\,1,2]{\text{Ti catalyst}}
\quad
\begin{array}{c}
\text{HO(CH}_2)_{m+2} \\
\text{(CH}_2)_n \\
O
\end{array}
\qquad (7)
$$

The alkyl groups of PbR_4 (R = Me, Et) can be replaced stepwise by halogen atoms, and these may be further substituted by other groups, as illustrated in reaction 8 for the first of three similar CF_3 group introductions[144].

$$PbR_4 + X_2 \xrightarrow{-RX} PbR_3X \xrightarrow{+(CF_3)_2Cd\cdot D} (CF_3)PbR_3 \qquad (8)$$
$$\textbf{(15)}$$

Chlorotriphenylplumbane[163,164] and dichlorodiphenylplumbane[165] react with sodium sulphide to yield bis(triphenylplumbyl) sulphide, as shown in reactions 9 and 10, respectively.

$$2Ph_3PbCl + Na_2S \longrightarrow (Ph_3Pb)_2S + 2NaCl \qquad (9)$$

$$3Ph_2PbCl_2 + 3Na_2S \longrightarrow (Ph_3Pb)_2S + PbS + S + 3NaCl \qquad (10)$$

Aryllead triacetates undergo aromatic nucleophilic displacement of the lead triacetate moiety by various reagents, including iodide, azide, and 'soft' carbanions, as shown in reaction 11[166,167]. Boron trifluoride-dimethyl ether complex leads to the formation of aryl fluorides (reaction 12), however, ortho-substituted aryl groups afford poor yields[168]. The derivative of 2-(phenylthio)benzofuran-3(2H)-one shown in reaction 13 affords good yields of the aryl derivative, after heating for a few hours at 60 °C, in dry pyridine, in the presence of N,N,N',N'-tetramethylguanidine[169]. Amines do not follow this path of nucleophilic substitution[166]. Azoles and their benzo derivatives and the anions of amides and imides become arylated by aryllead triacetates, in the presence of catalytic amounts of copper(II) acetate, as illustrated in reactions 14 and 15[170,171].

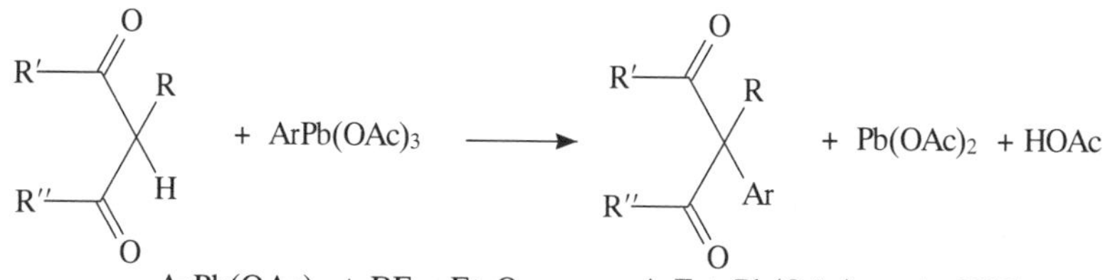

$$R' - C(=O) - CR(H) - C(=O) - R'' + ArPb(OAc)_3 \longrightarrow R' - C(=O) - CR(Ar) - C(=O) - R'' + Pb(OAc)_2 + HOAc \quad (11)$$

$$ArPb(OAc)_3 + BF_3 \cdot Et_2O \longrightarrow ArF + Pb(OAc)_2 + AcOBF_2 \quad (12)$$

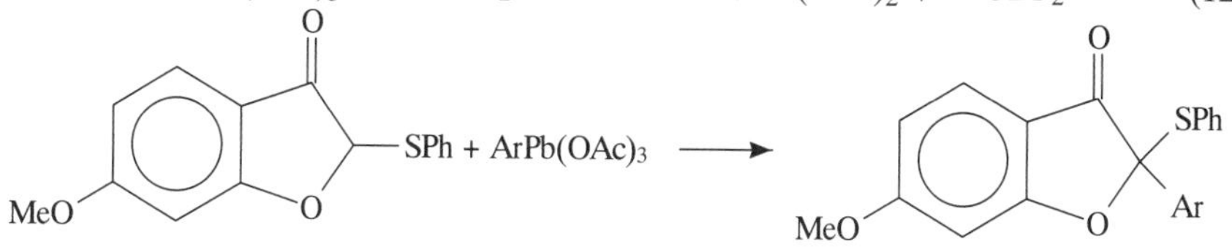

(13)

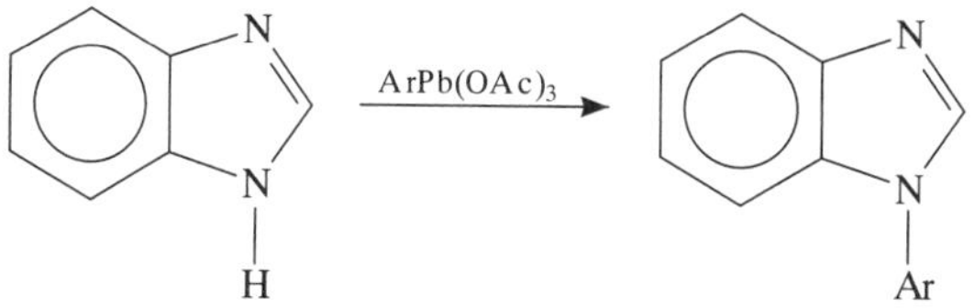

(14)

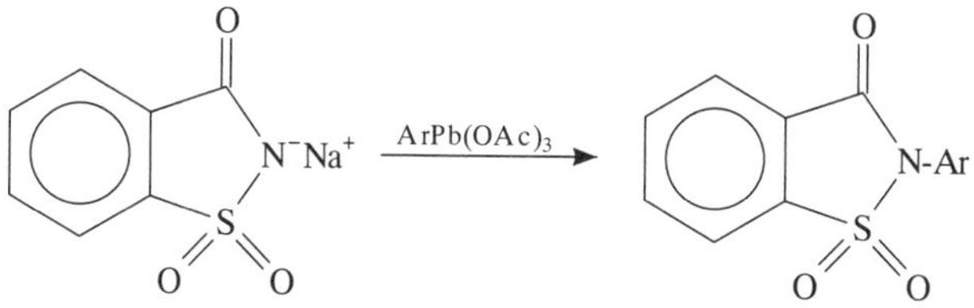

(15)

Trialkyllead chlorides undergo the metathesis shown in reaction 16, with silver dimesylamide[172].

$$R_3PbCl + AgN(SO_2Me)_2 \longrightarrow R_3PbN(SO_2Me)_2 + AgCl \quad (16)$$

Reaction 17 takes place with quantitative formation of a precipitate. See also Table 4[159].

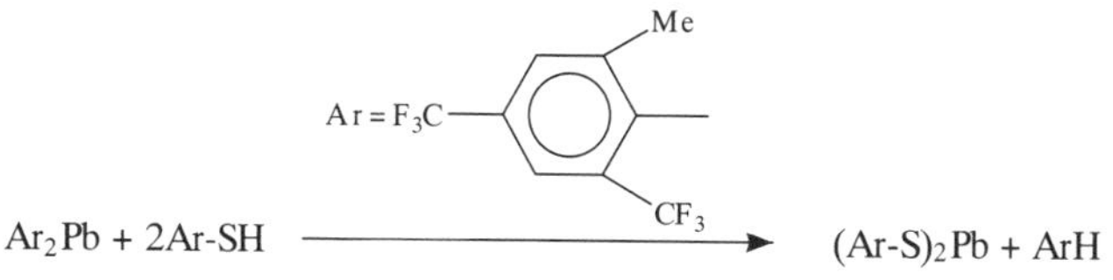

$$Ar_2Pb + 2Ar\text{-}SH \longrightarrow (Ar\text{-}S)_2Pb + ArH \quad (17)$$

(29)

VI. REFERENCES

1. W. B. McCormack, R. Moore and C. A. Sandy, in *Kirk-Othmer Encyclopedia of Chemical Technology*, 3rd edn., Vol. 14, Wiley, New York, 1981, pp. 180–195.
2. D. S. Carr, in *Ullmann's Encyclopedia of Industrial Chemistry*, 5th edn. (Eds. B. Elvers, S. Hawkins and G. Schulz), VCH, Weinheim, 1990, pp. 249 ff.

3. J. Zabicky, in *Supplement C: The Chemistry of Triple-bonded Functional Groups* (Ed. S. Patai), Vol. 2, Wiley, Chichester, 1994, pp. 192–230.

4. N. I. Sax, *Dangerous Properties of Industrial Materials*, 6th edn., Van Nostrand Reinhold, New York, 1984.

5. T. R. Crompton, *Chemical Analysis of Organometallic Compounds*. Volume 3. *Elements of Group IVB*, Academic Press, London, 1974, pp. 99–189.

6. J. M. Christensen and J. Kristiansen, in *Handbook on Metals in Clinical and Analytical Chemistry* (Eds. H. G. Seiler, A. Sigel and H. Sigel), Dekker, New York, 1994, pp. 425–440.

7. L. C. Willemsens and G. J. M. Van der Kerk, 1965, cited in Reference 5; *Chem. Abstr.*, **64**:8235b.

8. D. Colaitis and M. Lesbre, *Bull. Soc. Chim. France*, 1069 (1952).

9. *Annual Book of ASTM Standards*, American Society for Standards and Materials, Philadelphia, U.S.A., 1976.

10. R. A. Braithwaite and A. J. Girling, *Fresenius Z. Anal. Chem.*, **332**, 704 (1988).

11. M. Vahter and L. Friberg, *Fresenius Z. Anal. Chem.*, **332**, 726 (1988).

12. H. Schauenburg and P. Weigert, *Fresenius Z. Anal. Chem.*, **338**, 449 (1990).

13. K. R. Sperling, *Fresenius Z. Anal. Chem.*, **332**, 565 (1988).

14. J. E. Ferguson, *Can. J. Chem.*, **65**, 1002 (1987).

15. M. R. Farfel, P. S. J. Lees, C. A. Rohde, B. S. Lim, D. Bannon and J. J. Chisolm, *Environ. Res.*, **65**, 291 (1994).

16. G. Torsi and G. Bergamini, *Ann. Chim. (Rome)*, **79**, 45 (1989); *Chem. Abstr.*, **111**:44517.

17. E. Mentasti, A. Nicolotti, V. Porta and C. Sarzanini, *Analyst (London)*, **114**, 1113 (1989).

18. R. Purohit and S. Devi, *Talanta*, **38**, 753 (1991).

19. H. L. Lancaster, G. D. Marshall, E. R. Gonzalo, J. Ruzicka and G. D. Christian, *Analyst (London)*, **119**, 1459 (1994).

20. K. G. Varshney and U. Gupta, *Bull. Chem. Soc. Jpn.*, **63**, 1515 (1990).

21. Y. Horiba, S. Yamanaka and Y. Yamamoto, *Nippon Kagaku Kaishi*, 115 (1988); *Chem. Abstr.*, **108**:113281.

22. R. Purohit and S. Devi, *Anal. Chim. Acta*, **259**, 53 (1992).

23. S. Gucer and M Demir, *Anal. Chim. Acta*, **196**, 277 (1987).

24. O. Nygren, C. A. Nilsson and A. Gustavsson, *Analyst (London)*, **113**, 591 (1988).

25. S. R. Bysouth, J. F. Tyson and P. B. Stockwell, *J. Autom. Chem.*, **11**, 36 (1989); *Chem. Abstr.*, **112**:68761.

26. C. Nerin, J. Cacho and S. Olavide, *Anal. Chem.*, **59**, 1918 (1987).

27. J. Y. Madrid, J. Meseguer, M. Bonilla and C. Camara, *Anal. Chim. Acta*, **237**, 181 (1990).

28. J. X. Li, Y. M. Liu and T. Z. Lin, *Anal. Chim. Acta*, **231**, 151 (1990).

29. S. Z. Zhang, H. B. Han and Z. M. Ni, *Anal. Chim. Acta*, **221**, 85 (1989).

30. R. L. Ma, W. VanMol and F. Adams, *Anal. Chim. Acta*, **293**, 251 (1994).

31. I. Aroza, M. Bonilla and Y. Madrid, *J. Anal. At. Spectrom.*, **4**, 163 (1989).

32. I. Sekerka and J. F. Lechner, *Anal. Chim. Acta*, **254**, 99 (1991).

33. R. E. Sturgeon, S. N. Willie and S. S. Berman, *Anal. Chem.*, **61**, 1867 (1989).

34. T. Kumamary, Y. Okamoto, S. Hara, H. Matsuo and M. Kiboku, *Anal. Chim. Acta*, **218**, 173 (1989).

35. S. C. Apte and A. M. Gunn, *Anal. Chim. Acta*, **193**, 147 (1987).

36. P. Bermejo-Barrera, C. Barciela-Alonso, M. Aboal-Somoza and A. Bermejo-Barrera, *J. Anal. At. Spectrom.*, **9**, 469 (1994).

37. O. Royset and Y. Thomassen, *Anal. Chim. Acta*, **188**, 247 (1986).

38. J. Sneddon, *Appl. Spectrosc.*, **44**, 1562 (1990).

39. J. Sneddon, *Anal. Chim. Acta*, **245**, 203 (1991).

40. K. S. Subramanian, *Sci. Total Environ.*, **89**, 237 (1989).

41. S. T. Wang, S. Pizzolato and F. Peter, *Sci. Total Environ.*, **71**, 37 (1988).

42. C. G. Bruhn, M. O. Jerardino, P. C. Maturana, G. M. Navarrete and J. M. Piwonka, *Anal. Chim. Acta*, **198**, 113 (1987).

43. J. A. Navarro, V. A. Grandillo, O. E. Parra and R. A. Romero, *J. Anal. At. Spectrom.*, **4**, 401 (1989).

44. D. J. Pain, C. Metayer and J. C. Amiard, *Int. J. Environ. Anal. Chem.*, **53**, 29 (1993).

45. C. L. Chakrabarti, R. Karwowska, B. R. Hollebone and P. M. Johnson, *Spectrochim. Acta, Part B*, **42B**, 1217 (1987).

46. R. Chakraborty, S. S. Bhattacharyya and A. K. Das, *Bull. Chem. Soc. Jpn.*, **66**, 2233 (1993).

47. R. Puchades, A. Maquieira and M. Planta, *Analyst (London)*, **114**, 1397 (1989).
48. S. N. F. Bruno, R. C. Campos and A. J. Curtius, *J. Anal. At. Spectrom.*, **9**, 341 (1994).
49. W. R. Mindak, *J. Assoc. Off. Anal. Chem.*, **77**, 1023 (1994).
50. P. L. Teissedre, R. Lobinski, M. T. Cabanis, J. Szpunarlobinska, J. C. Cabanis and F. C. Adams, *Sci. Total Environ.*, **153**, 247 (1994).
51. P. Viñas, N. Campillo, I. López García and M. H. Córdoba, *Fresenius Z. Anal. Chem.*, **349**, 306 (1994).
52. S. Lynch and D. Littlejohn, *J. Anal. At. Spectrom.*, **4**, 157 (1989).
53. P. Viñas, N. Campillo and I. López García, *Food Chem.*, **50**, 317 (1994).
54. D. Firestone, *J. Assoc. Off. Anal. Chem.*, **77**, 951 (1994).
55. H. T. Delves and M. J. Campbell, *J. Anal. At. Spectrom.*, **3**, 343 (1988).
56. M. J. Campbell and H. T. Delves, *J. Anal. At. Spectrom.*, **4**, 235 (1989).
57. M. Viczian, A. Lasztity and R. M. Barnes, *J. Anal. At. Spectrom.*, **5**, 293 (1990).
58. T. A. Hinners, E. M. Heithmar, J. M. Henshaw and T. M. Spittler, *Anal. Chem.*, **59**, 2658 (1987).
59. H. M. Crews, R. C. Massey, D. J. McWeeny, J. R. Dean and L. Ebdon, *J. Res. Natl. Bur. Stand.*, **93**, 464 (1988).
60. S. J. Lang and K. J. R. Rosman, *Anal. Chim. Acta*, **235**, 367 (1990).
61. B. Gercken and R. M. Barnes, *Anal. Chem.*, **63**, 283 (1991).
62. J. Goossens, L. Moens and R. Dams, *Anal. Chim. Acta*, **293**, 171 (1994).
63. H. Ibrahim, *J. Am. Oil. Chem. Soc.*, **68**, 678 (1991).
64. I. D. Brindle and X. C. Le, *Anal. Chem.*, **61**, 1175 (1989).
65. M. A. Bolshov, C. F. Boutron and A. V. Zybin, *Anal. Chem.*, **61**, 1758 (1989).
66. V. Cheam, J. Lechner, I. Sekerka, R. Desrosiers, J. Nriagu and G. Lawson, *Anal. Chim. Acta*, **269**, 129 (1992).
67. V. Cheam, J. Lechner, I. Sekerka and R. Desrosiers, *J. Anal. At. Spectrom.*, **9**, 315 (1994).
68. J. A. Vera, M. B. Leong, N. Omenetto, B. W. Smith, B. Womack and J. D. Winefordner, *Spectrochim. Acta, Part B*, **44B**, 939 (1989).
69. J. Labuda and M. Vanickova, *Anal. Chim. Acta*, **208**, 219 (1988).
70. C. Labar and L. Lamberts, *Electrochim. Acta*, **33**, 1405 (1988).
71. J. X. Peng and W. R. Jin, *Anal. Chim. Acta*, **264**, 213 (1992).
72. B. Svensmark, *Anal. Chim. Acta*, **197**, 239 (1987).
73. A. R. Fernando and J. A. Plambeck, *Anal. Chem.*, **61**, 2609 (1989).
74. C. M. G. Van den Berg, *J. Electroanal. Chem. Interfac. Electrochem.*, **215**, 111 (1986).
75. D. Jagner, L. Renman and Y. D. Wang, *Electroanalysis*, **6**, 285 (1994).
76. L. Almestrand, D. Jagner and L. Renman, *Anal. Chim. Acta*, **193**, 71 (1987).
77. L. Almestrand, M. Betti, C. Hua, D. Jagner and L. Renman, *Anal. Chim. Acta*, **209**, 339 (1988).
78. D. Jagner, E. Sahlin and L. Renman, *Talanta*, **41**, 515 (1994).
79. J. Wang and B. Trian, *Anal. Chem.*, **65**, 1529 (1993).
80. E. P. Gil and P. Ostapczuk, *Anal. Chim. Acta*, **293**, 55 (1994).
81. R. R. Dalangin and H. Gunasingham, *Anal. Chim. Acta*, **291**, 81 (1994).
82. (a) S. Dong and Y. Wang, *Talanta*, **35**, 822 (1988).
    (b) S. V. Prabhu, R. P. Baldwin and L. Kryger, *Electroanalysis*, **1**, 13 (1989).
83. J. R. Xu and B. Liu, *Analyst (London)*, **119**, 1599 (1994).
84. J. Y. Wang, X. Rongrong, T. Baomin, J. Y. Wang, C. L. Renschler and C. A. White, *Anal. Chim. Acta*, **293**, 43 (1994).
85. E. Beinrohr, P. Csemi, A. Manova and J. Dzurov, *Fresenius Z. Anal. Chem.*, **349**, 625 (1994).
86. J. J. Specchio, *J. Assoc. Off. Anal. Chem.*, **71**, 857 (1988).
87. D. Jagner, L. Renman and Y. D. Wang, *Anal. Chim. Acta*, **267**, 165 (1992).
88. J. L. Guinon and R. Grima, *Analyst (London)*, **113**, 613 (1988).
89. P. Laukkanen, *Fresenius Z. Anal. Chem.*, **349**, 693 (1994).
90. J. L. Stauber and T. M. Florence, *Anal. Chim. Acta*, **237**, 177 (1990).
91. B. Ogorevc, V. Hudnik and A. Krasna, *Anal. Chim. Acta*, **196**, 183 (1987).
92. L. Nyholm and G. Wikmark, *Anal. Chim. Acta*, **223**, 429 (1989).
93. N. Ivicic and M. Blanusa, *Frezenius Z. Anal. Chem.*, **330**, 643 (1988).
94. S. Mannino, G. Fregapane and M. Bianco, *Electroanalysis*, **1**, 177 (1989).
95. C. F. Liu and K. Jiao, *Anal. Chim. Acta*, **238**, 367 (1990).
96. A. Borraccino, L. Camoanella, M. P. Sammartino, M. Tommasseti and M. Battilotti, *Sensors Actuators B—Chem.*, **7**, 535 (1992).

97. M. Jarosz and A. Swietlow, *Microchem. J.*, **37**, 322 (1988).
98. E. A. Novikov, L. K. Shpigun and Y. A. Zolotov, *Anal. Chim. Acta*, **230**, 157 (1990).
99. V. Kuban and R. Bulawa, *Collect. Czech. Chem. Commun.*, **54**, 2674 (1989).
100. T. V. Petrova, T. G. Dzherayan, A. V. Sultanov and S. B. Savvin, *Zh. Anal. Khim.*, **43**, 2221 (1988); *Chem. Abstr.*, **111**:208275.
101. M. Lerchi, E. Bakker, B. Rusterholz and W. Simon, *Anal. Chem.*, **64**, 1534 (1992).
102. G. Wauters, J. Hoste, K. Strijckmans and C. Vandecasteele, *J. Radioanal. Nucl. Chem., Articles*, **112**, 23 (1987).
103. G. Wauters, J. Hoste and C. Vandecasteele, *J. Radioanal. Nucl. Chem., Articles*, **110**, 477 (1987).
104. K. Strijckmans, G. Wauters, S. Vanwinckel, J. Dewaele and R. Dams, *J. Radioanal. Nucl. Chem., Articles*, **131**, 11 (1989).
105. M. E. Vargas, J. D. Batchelor and E. A. Schweikert, *J. Radioanal. Nucl. Chem., Articles*, **119**, 81 (1987).
106. G. Wauters, C. Vandecasteele and K. Strijckmans, *J. Radioanal. Nucl. Chem., Articles*, **134**, 221 (1989).
107. Z. L. Xue, B. R. Bao and Y. D. Chen, *J. Radioanal. Nucl. Chem., Lett.*, **119**, 513 (1987).
108. S. K. Agarwal, M. Kinter and D. A. Herold, *Clin. Chem.*, **40**, 1494 (1994).
109. P. J. Mountford, S. Green, D. A. Bradley, A. D. Lewis and W. D. Morgan, *Phys. Med. Biol.*, **39**, 773 (1994).
110. J. C. Galloo and R. Guillermo, *Analusis*, **17**, 576 (1989); *Chem. Abstr.*, **112**:164101.
111. L. C. Wouters, R. E. Van Grieken, R. W. Linton and C. F. Bauer, *Anal. Chem.*, **60**, 2218 (1988).
112. M. Soma, H. Seyama and K. Satake, *Talanta*, **35**, 68 (1988).
113. A. W. P. Jarvie, *J. Total Environ.*, **73**, 121 (1988).
114. R. J. A. Van Cleuvenbergen, W. Dirkx, P. Quevauviller and F. Adams, *J. Environ. Anal. Chem.*, **47**, 21 (1992).
115. P. R. Fielden and G. M. Greenway, *Anal. Chem.*, **61**, 1993 (1989).
116. D. Chakraborti, W. R. A. De Jonghe, W. E. Van Mol, R. J. A. Van Cleuvenbergen and F. C. Adams, *Anal. Chem.*, **56**, 2692 (1984).
117. R. J. A. Van Cleuvenbergen and F. Adams, *Environ. Sci. Technol.*, **26**, 1354 (1992).
118. D. Chakraborti, W. Dirkx, R. J. A. Van Cleuvenbergen and F. C. Adams, *Sci. Total Environ.*, **84**, 249 (1989).
119. O. Nygren and C. A. Nilsson, *J. Anal. At. Spectrom.*, **2**, 805 (1987).
120. R. J. A. Van Cleuvenbergen, D. Chakraborti and F. C. Adams, *Anal. Chim. Acta*, **228**, 77 (1990).
121. W. M. R. Dirkx, R. J. A. Van Cleuvenbergen and F. C. Adams, *Mikrochim. Acta*, **109**, 133 (1992).
122. C. N. Hewitt, R. M. Harrison and M. Radojevic, *Anal. Chim. Acta*, **188**, 229 (1986).
123. C. N. Hewitt and P. J. Metcalfe, *Sci. Total Environ.*, **84**, 211 (1989).
124. R. Lobinski and F. C. Adams, *Anal. Chim. Acta*, **262**, 285 (1992).
125. J. S. Blais and W. D. Marshall, *J. Anal. At. Spectrom.*, **4**, 641 (1989).
126. J. S. Blais and W. D. Marshall, *J. Anal. At. Spectrom.*, **4**, 6271 (1989).
127. C. R. Blackley and M. L. Vestal, *Anal. Chem.*, **55**, 750 (1983).
128. M. Blaszkewicz, G. Baumhoer, B. Neidhart, R. Ohlendorf and M. Linscheid, *J. Chromatogr.*, **439**, 109 (1988).
129. Y. Liu, V. López-Ávila, M. Alcaraz and W. F. Beckert, *J. High Resolut. Chromatogr.*, **17**, 527 (1994).
130. C. L. Ng, H. K. Lee and S. F. Y. Li, *J. Chromatogr. A*, **652**, 547 (1993)
131. S. R. Henderson and L. J. Snyder, *Anal. Chem.*, **31**, 2113 (1959).
132. S. R. Henderson and L. J. Snyder, *Anal. Chem.*, **33**, 1172 (1961).
133. W. W. Parker, G. Z. Smith and R. L. Hudson, *Anal. Chem.*, **33**, 1170 (1961).
134. W. W. Parker and R. L. Hudson, *Anal. Chem.*, **35**, 1334 (1963).
135. A. M. Bond and N. M. McLachlan, *J. Electroanal. Chem. Interfac. Electrochem.*, **218**, 197 (1987).
136. N. Mikac and M. Branica, *Anal. Chim. Acta*, **264**, 249 (1992).
137. N. Mikac and M. Branica, *Electroanalysis*, **6**, 37 (1994).
138. N. Mikac and M. Branica, *Anal. Chim. Acta*, **212**, 349 (1988).
139. J. E. De Vries, A. Lauw-Zecha and A. Pellecer, *Anal. Chem.*, **31**, 1995 (1959).
140. L. N. Vertyulina and I. A. Korschunov, *Khim. Nauka Prom.*, **4**, 136 (1959); *Chem. Abstr.*, **53**:12096c.

141. G. Tagliavini, *Anal. Chim. Acta*, **34**, 24 (1966).
142. R. Borja, M. de la Guardia, A. Salvador, J. L. Burguera and M. Burguera, *Fresenius Z. Anal. Chem.*, **338**, 9 (1990).
143. T. A. Hein, W. Thiel and T. J. Lee, *J. Phys. Chem.*, **97**, 4381 (1993).
144. R. Eujen and A. Patorra, *J. Organomet. Chem.*, **438**, 57 (1992).
145. R. Eujen and A. Patorra, *J. Organomet. Chem.*, **438**, C1 (1992).
146. M. Prager, D. Zhang and A. Weiss, *Physica B*, **180**, 671 (1992).
147. A. Sebald and R. K. Harris, *Organometallics*, **9**, 2096 (1990).
148. B. Wrackmeyer, K. H. Von Locquenghien, E. Kupce and A. Sebald, *Magn. Reson. Chem.*, **31**, 45 (1993).
149. B. Wrackmeyer and K. Horchler, *J. Organomet. Chem.*, **399**, 1 (1990).
150. C. Janiak, H. Schumann, C. Stader, B. Wrackmeyer and J. J. Zuckermann, *Chem. Ber.*, **121**, 1745 (1988).
151. (a) M. J. Heeg, R. H. Herber, C. Janiak, J. J. Zuckermann, H. Schumann and W. F. Manders, *J. Organomet. Chem.*, **346**, 321 (1988).
(b) R. Villazana, H. Sharma, F. Cervantes-Lee and K. H. Pannell, *Organometallics*, **12**, 4278 (1993).
152. U. Kilimann, M. Noltemeyer and F. T. Edelmann, *J. Organomet. Chem.*, **443**, 35 (1993).
153. B. Wrackmeyer and H. Zhou, *Magn. Reson. Chem.*, **28**, 1066 (1990).
154. B. Wrackmeyer, K. Horchler, H. Zhou and M. Veith, *Z. Naturforsch. Sect. B, J. Chem. Sci.*, **44**, 288 (1989).
155. B. Wrackmeyer, C. Stader and K. Horchler, *J. Magn. Reson.*, **83**, 601 (1989).
156. B. Wrackmeyer, C. Stader, K. Horchler and H. Zhou, *Inorg. Chim. Acta*, **176**, 205 (1990).
157. B. M. Schmidt and M. Drager, *J. Organomet. Chem.*, **399**, 63 (1990).
158. F. Huber, H. Preut, D. Scholz and M. Schurmann, *J. Organomet. Chem.*, **441**, 227 (1992).
159. S. Brooker, J. K. Buijink and F. T. Edelmann, *Organometallics*, **10**, 25 (1991).
160. (a) N. S. Hosmane, K. J. Lu, H. Zhu, U. Siriwardane, M. S. Shet and J. A. Maguire, *Organometallics*, **9**, 808 (1990).
(b) N. S. Hosmane, K. J. Lu, U. Siriwardane and M. S. Shet, *Organometallics*, **9**, 2798 (1990).
161. (a) E. M. Berksoy and M. A. Whitehead, *J. Organomet. Chem.*, **710**, 293 (1991).
(b) C. Glidewell, *J. Organomet. Chem.*, **398**, 271 (1990).
162. J. Yamada, T. Asano, I. Kadota and Y. Yamamoto, *J. Org. Chem.*, **55**, 6066 (1990).
163. D. P. Thompson and P. Boudjouk, *J. Org. Chem.*, **53**, 2104 (1988).
164. J. H. So and P. Boudjouk, *Synthesis*, **53**, 2104 (1988).
165. S. R. Bahr and P. Boudjouk, *Inorg. Chem.*, **31**, 4015 (1992).
166. J. Morgan and J. T. Pinhey, *J. Chem. Soc., Perkin Trans. 1*, 1673 (1993).
167. J. Morgan, I. Buys, T. W. Hambley and J. T. Pinhey, *J. Chem. Soc., Perkin Trans. 1*, 1677 (1993).
168. G. Demeio, J. Morgan and J. T. Pinhey, *Tetrahedron*, **49**, 8129 (1993).
169. D. M. X. Donnelly, J. M. Kielty, A. Cormons and J. P. Finet, *J. Chem. Soc., Perkin Trans. 1*, 2069 (1993).
170. P. López-Alvarado, C. Avendaño and J. C. Menéndez, *Tetrahedron Lett.*, **33**, 659 (1992).
171. P. López-Alvarado, C. Avendaño and J. C. Menéndez, *Tetrahedron Lett.*, **33**, 6875 (1992).
172. A. Blaschette, T. Hamann, A. Michalides and P. G. Jones, *J. Organomet. Chem.*, **456**, 49 (1993).

CHAPTER 10

Synthesis of M(IV) organometallic compounds, where M = Ge, Sn, Pb

JOHN M. TSANGARIS

Department of Chemistry, University of Ioannina, GR-45100 Ioannina, Greece
Fax: 30-651-44831

RUDOLPH WILLEM and MARCEL GIELEN

Free University of Brussels, Pleinlaan 2, B-1050 Brussels, Belgium
Fax: 322-629-3281; e-mail: mgielen@vnet3.vub.ac.be

The chemistry of organic germanium, tin and lead compounds
Edited by S. Patai © 1995 John Wiley & Sons Ltd

I. ORGANOGERMANIUM(IV), ORGANOTIN(IV) AND ORGANOLEAD(IV) COMPOUNDS CONTAINING AT LEAST ONE M−C σ- OR π-BOND AND/OR M−M BOND

A. Organogermanes

The oldest method for generating a Ge−Ge bond is the Würtz reaction from a triorganogermanium halide and sodium[1]:

$$2 \; R_3GeCl + 2 \; Na \longrightarrow R_3Ge-GeR_3 + 2 \; NaCl$$

where R = alkyl, aryl, cyclohexyl. The compounds Me_6Ge_2[2,3], $(iso\text{-}Pr)_6Ge_2$[4], Ph_6Ge_2[5] were prepared in this way.

The reactions can be carried out with or without solvent, using mostly sodium. Potassium and lithium can also be used in special circumstances. Phenylcyclopentyldigermane **1**[6] and decamethylcyclopentagermane **2**[7,8] are among other compounds prepared by this method.

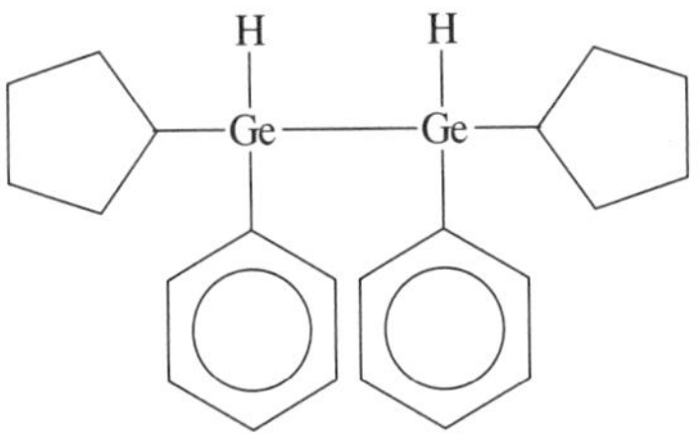

        (1) 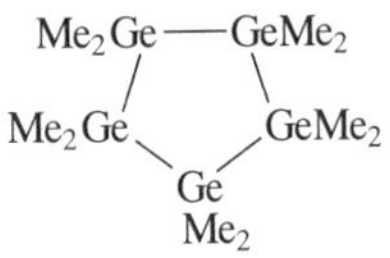   **(2)**

For the preparation of **2** dimethylgermanium dichloride was used in a reaction with Li in THF.

By applying the Würtz reaction to dialkyl and diaryl germanium dihalides, both linear and cyclic oligomers[9] were obtained; for example, Ph_2GeCl_2, when heated with sodium in xylene, gives octaphenylcyclotetragermane[10]. Usually Me_2GeCl_2 in Li/THF gives mixtures of cyclic oligomers $(Me_2Ge)_n$ where $n = 4-6$ and a mixture of insoluble linear polymers containing up to fifty Me_2Ge units[7]. Another method used extensively for the preparation of organogermanium compounds is the Grignard reaction. An excess of Grignard reagent reacting with $GeCl_4$ or $GeCl_2$ produces tetraorganogermanes together with hexaorganodigermanes, with yields ranging from 7 to 10%[11]. It is noteworthy that an excess of magnesium added to R_3GeX compounds reacting with Grignard reagents provides satisfactory yields in the digermane compounds, which suggests that a germylmagnesium halide is a precursor to the digermane[12].

$$2 \; R_3GeX + Mg \xrightarrow{\text{ether}} [R_3GeMgX] + R_3GeX \longrightarrow R_6Ge_2 + MgX_2$$

In this way $(iso\text{-}Pr)_6Ge_2$ can be prepared together with the by-products $(iso\text{-}Pr)_4Ge$, $(iso\text{-}Pr)_8Ge_4$ and a polymer $(iso\text{-}Pr_2Ge)_n$[13].

The preparation of some dimethylvinylhalogermanes with Grignard reagents as synthetic tools was reported[27]:

$$2 \; Me_2GeBr_2 + 4 \; H_2C{=}CHMgBr \xrightarrow{\text{THF}} Me_2Ge(CH{=}CH_2)_2 + [Me_2(H_2C{=}CH)Ge]_2$$
$$\phantom{2 \; Me_2GeBr_2 + 4 \; H_2C{=}CHMgBr \xrightarrow{\text{THF}} \;\;} 32\% \phantom{Me_2Ge(CH{=}CH_2)_2 +} 14\%$$

The dimethyldivinylgermane can be easily converted to dimethylvinyliodogermane $Me_2IGeCH{=}CH_2$ by I_2 in $CHCl_3$.

The direct synthesis can also be used in the case of bromide:

$$Me_2GeBr_2 + H_2C=CHMgBr \xrightarrow[\text{reflux}]{\text{THF}} Me_2BrGeCH=CH_2 \qquad 35\%$$

A 90% yield in Me_4Ge is reported[27]:

$$GeCl_4 + 4\ MeMgBr \xrightarrow{\text{THF}} Me_4Ge + 4\ MgBrCl$$

Other diorganogermanes also prepared by Grignard reactions were: $Me_2EtGeGeMe_2Et$[2], $Me_3GeGeEt_3$[2,14], $MePh_2GeGeMePh_2$[11], $(CH_2=CH)_6Ge_2$[15], Pr_6Ge_2[16], n-Bu_6Ge_2[16-18], (β-styryl)$_6Ge_2$[19], (o-tolyl)$_6Ge_2$ and (p-tolyl)$_6Ge_2$[13].

In Grignard reactions involving $GeCl_4$, complications usually arise because of the possible generation of intermediates like XR_2GeMgX. Such intermediates can explain the formation of cyclotetragermanes[20].

Finally, if a mixture of Grignard reagents is used for the reaction with $GeCl_4$, mixed hexaalkyldigermanes are obtained, such as $Me_3GeGeMePr_2$, $Me_2PrGeGeMe_2Pr$ from MeMgI and PrMgI or $Me_3GeGe(iso$-$Pr)_2Me$, $Me_2(iso$-$Pr)GeGeMe_2(iso$-$Pr)$ from MeMgI and (iso-Pr)MgI.

Alkyllithium compounds are only rarely used for the preparation or organogermanes or -digermanes like, e.g. Et_4Ge or Et_6Ge_2 by reaction with $GeCl_4$ or $GeBr_4$[2,21].

Organolithium compounds can also be used for the preparation of tetraneopentylgermane[30], 3:

$$GeCl_4 + 4\ Me_3CCH_2Li \xrightarrow{n\text{-hexane}} (Me_3CCH_2)_4Ge + (Me_3CCH_2)_3GeCl$$
$$(3)$$

Using alkyllithium compounds, some organometallic bis(p-hydroxyphenyl) derivatives 4 of Ge, Sn and Pb can be obtained in yields of 55 to 80% (Scheme 1).

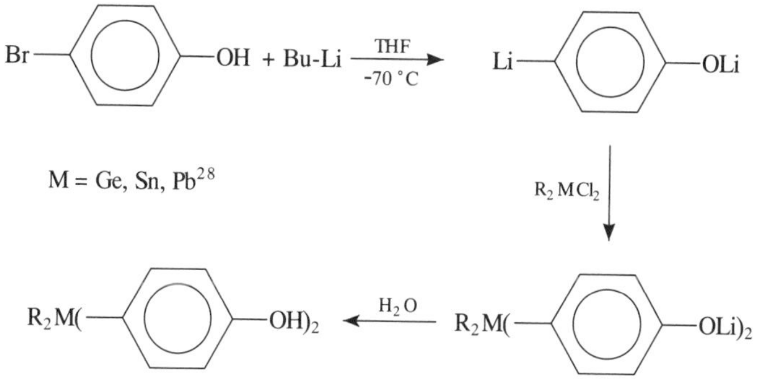

SCHEME 1

By the same procedure, p-disubstituted organometallic derivatives of diphenyl-N-methylamine 5 were synthesized in 48 to 66% yield[54] (Scheme 2).

Tetra-1-adamantylgermane was prepared[29] in 16% yield by refluxing adamantyl bromide with $GeCl_4$ and Na in cyclohexane:

$$GeCl_4 + 4\ C_{10}H_{15}Br \xrightarrow[\text{cyclohexane}]{\text{Na}} (C_{10}H_{15})_4Ge$$

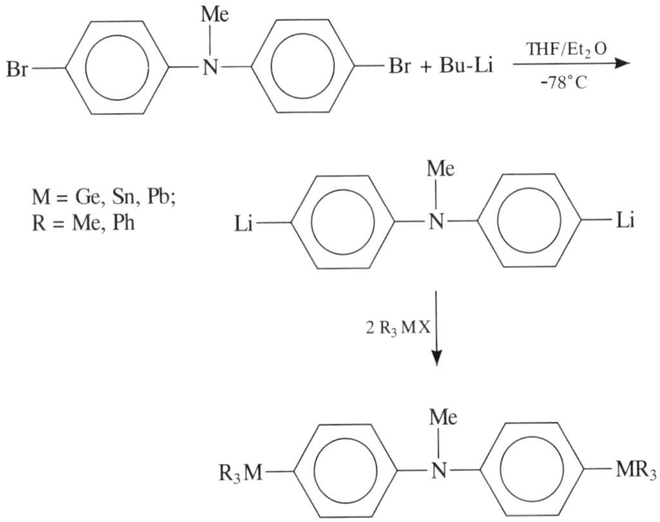

$$Br—\langle\bigcirc\rangle—\underset{\underset{Me}{|}}{N}—\langle\bigcirc\rangle—Br + Bu\text{-}Li \xrightarrow[-78°C]{THF/Et_2O}$$

M = Ge, Sn, Pb;
R = Me, Ph

$$Li—\langle\bigcirc\rangle—\underset{\underset{Me}{|}}{N}—\langle\bigcirc\rangle—Li$$

$$\Big\downarrow 2\,R_3MX$$

$$R_3M—\langle\bigcirc\rangle—\underset{\underset{Me}{|}}{N}—\langle\bigcirc\rangle—MR_3$$

(5)

SCHEME 2

Alkylaluminum compounds react with $GeCl_4$ and generate either tetraalkylmonogermanes or di- and polygermanes, as well as sometimes alkylgermanium polymers[22]. Higher alkyl groups favor the production of tetraalkylgermanium compounds.

A very interesting and relatively convenient way to synthesize organogermanium compounds involves triorganogermyl metallic salts: $R_3Ge^{(-)}M^+$. These salts can be obtained from triorganogermyl halides, e.g. R_3GeCl and the alkali metal in HMPT[17]. The triorganogermyl metallic salt prepared reacts with R_3GeCl. Symmetric or mixed hexaorganodigermanes can be prepared in this way[23]:

$$R_3GeCl \xrightarrow{K,\ (Me_2N)_3PO} R_3Ge^{(-)}K^+ \xrightarrow{R_3'GeCl} R_3Ge\text{—}GeR_3'$$

In the case of the latter reaction between $R_3Ge^{(-)}M^{(+)}$ and $R_3'GeCl$, the symmetric hexaorganodigermanes are obtained as side-products[24]:

$$R_3'GeCl \xrightarrow{R_3Ge^{(-)}K^+} R_3GeGeR_3' + R_3GeGeR_3 + R_3'GeGeR_3'$$

The existence of side-products can be explained by considering that the primary product is the mixed compound that undergoes subsequently a nucleophilic attack of $R_3Ge^{(-)}$ at the Ge—Ge bond, giving rise to $^{(-)}GeR_3$ or $^{(-)}GeR_3'$ species leading eventually to the side-products[14,24].

Cyclic organogermanes are obtained from variants of the above-mentioned methods[11,26]. The easily obtained cyclic germanes are converted into linear organopolygermanes (Scheme 3).

Branched chain organogermanes can also be prepared[5]:

$$Ph_3GeLi \xrightarrow{GeI_2} (Ph_3Ge)_3GeLi \xrightarrow{H_2O} (Ph_3Ge)_3GeH \xrightarrow{CH_3I} (Ph_3Ge)_3GeCH_3$$

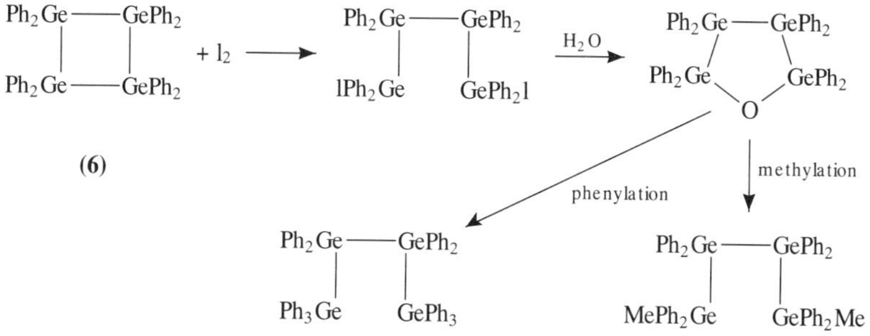

$$\text{(Scheme with germanium structures)}$$

Ph$_2$Ge—GePh$_2$ / Ph$_2$Ge—GePh$_2$ (square, compound (6)) + I$_2$ ⟶ Ph$_2$Ge—GePh$_2$ / IPh$_2$Ge   GePh$_2$I $\xrightarrow{H_2O}$ Ph$_2$Ge—GePh$_2$ / Ph$_2$Ge   GePh$_2$ (with O bridge)

(6)

phenylation ↙   ↓ methylation

Ph$_2$Ge—GePh$_2$ / Ph$_3$Ge   GePh$_3$

Ph$_2$Ge—GePh$_2$ / MePh$_2$Ge   GePh$_2$Me

SCHEME 3

A quite simple preparation of octaphenylcyclotetragermane **6** can be accomplished by a reaction of diphenylgermane with diethylmercury[11,25]:

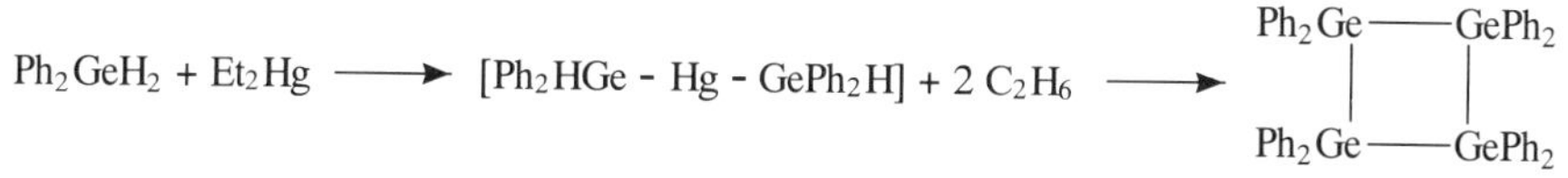

$$Ph_2GeH_2 + Et_2Hg \longrightarrow [Ph_2HGe\text{-}Hg\text{-}GePh_2H] + 2\,C_2H_6 \longrightarrow \begin{array}{c} Ph_2Ge\text{—}GePh_2 \\ | \quad\quad | \\ Ph_2Ge\text{—}GePh_2 \end{array}$$

(6)

The obtained unstable bis(hydrodiphenylgermyl)mercury polymerizes and decomposes to **6**. Ph$_3$GeH reacts with PhLi and produces a mixture of Ph$_6$Ge$_2$ and Ph$_4$Ge[26].

An important method of alkylation or arylation comparable to the Würtz reaction uses germanium tetraiodide[31]:

$$GeI_4 + 4\,RI + Zn \xrightarrow[160-200\,°C]{\text{inert atmosphere}} GeR_4 \quad (R = Bu\ or\ Ph)$$

An interesting preparation leading to digermanes and monogermanes consists of allowing Et$_3$GeLi to react with trityl chloride[32] (Scheme 4). The mechanism of this reaction might involve a one-electron transfer from triethylgermyllithium to trityl chloride.

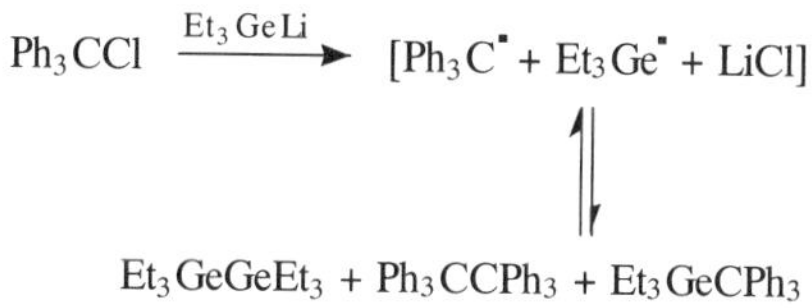

$$Ph_3CCl \xrightarrow{Et_3GeLi} [Ph_3C^{\cdot} + Et_3Ge^{\cdot} + LiCl]$$

$$\updownarrow$$

$$Et_3GeGeEt_3 + Ph_3CCPh_3 + Et_3GeCPh_3$$

SCHEME 4

A very original synthesis providing germyl-substituted ferrocenes **7** involves organolithium compounds[33,34] (Scheme 5).

It is possible to introduce germanium-containing moieties into carboranes by several methods:

For small nido-carboranes according to the following reaction[35]:

$$Me_3GeX + MC_2Me_2B_4H_5 \xrightarrow{Et_2O} Me_3GeC_2Me_2B_4H_5 + MX$$

$$M = Na,\ Li;\ X = Cl,\ Br$$

A B–Ge–B derivative is first obtained which thereafter isomerizes to the carborane.

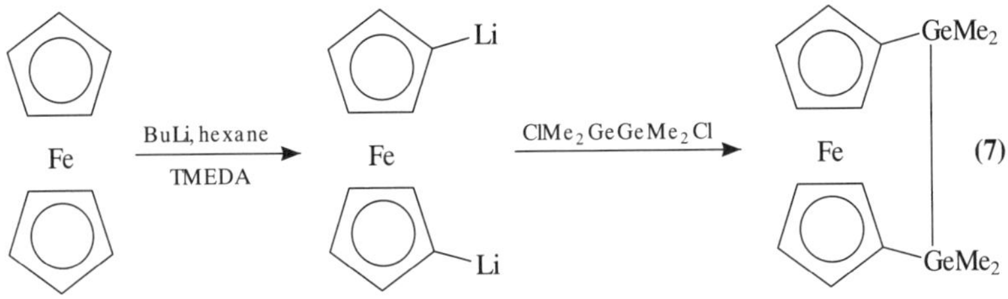

SCHEME 5

For germyl carboranes[36]:

$$C_2B_4H_8 + NaH \xrightarrow{\text{THF}} Na^+[C_2B_4H_7]^{(-)} + H_2 \xrightarrow{\text{Me}_3\text{GeCl}} Me_3GeC_2B_4H_7$$

For germa- and stanna-decarboranes:

$$B_{10}H_{14} + NaH \xrightarrow{\text{THF}} Na^+[B_{10}H_{13}]^{(-)} + H_2 \xrightarrow[\substack{M = \text{Ge, Sn;} \\ X = \text{Cl, Br}}]{\text{Me}_3\text{MX}} Me_2MB_{10}H_{12}$$

with yields of 5 to 18%[37].

An important synthesis method of organogermanes is the hydrogermylation of olefinic and acetylenic derivatives in the presence of a catalyst. Examples of such reactions are listed below[38]:

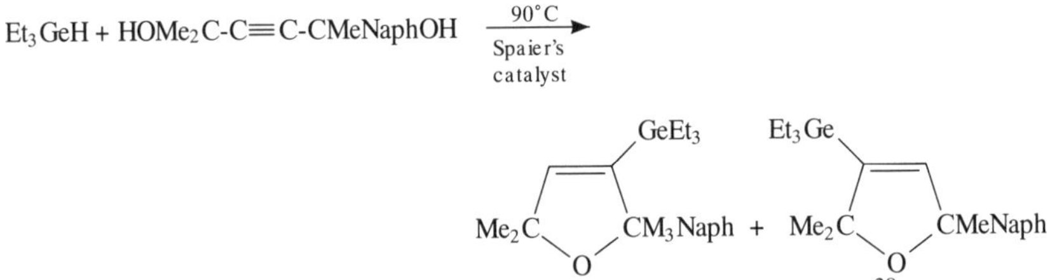

$$Et_3GeH + HOMe_2C\text{-}C\equiv C\text{-}CMeNaphOH \xrightarrow[\substack{\text{Spaier's} \\ \text{catalyst}}]{90°C}$$

$$R_3GeH + HC\equiv COBu \longrightarrow R_3GeCH=CHOBu \quad (R = \text{any alkyl})^{39}$$

$$R_3GeH + H_2C=C=CHCH_3 \xrightarrow[\text{AIBN}]{H_2PtCl_6} R_3GeCH_2CH=CHCH_3 + R_3GeCH(CH_3)CH=CH_2$$

The product distribution obtained shows that the hydride adds at the central sp carbon atom of the allene[40].

The relative amounts of isomers obtained depends on the catalyst and its concentration, on the organogermane used and on the ratio of reactant concentrations[41].

It is noteworthy that several of the triorganogermylethylenes synthesized as described above provide new starting materials for the preparation of other interesting germanium compounds. Germaacetylides were also used as useful synthons. The Diels–Alder reaction applied to germaacetylides and α-pyrone[42] leads to aromatic germanium compounds:

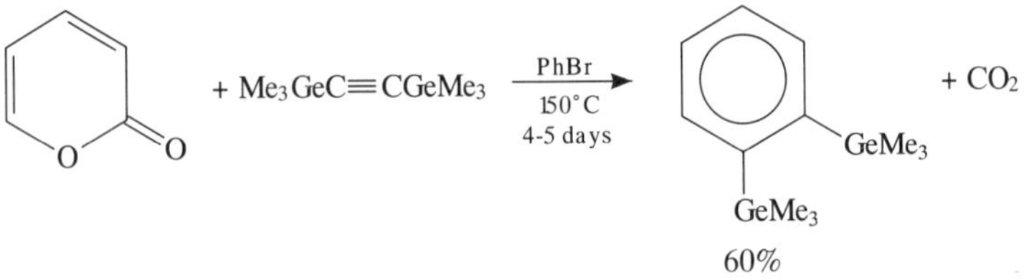

60%

1,3-Alkenynes containing germanium add to dienes[43]:

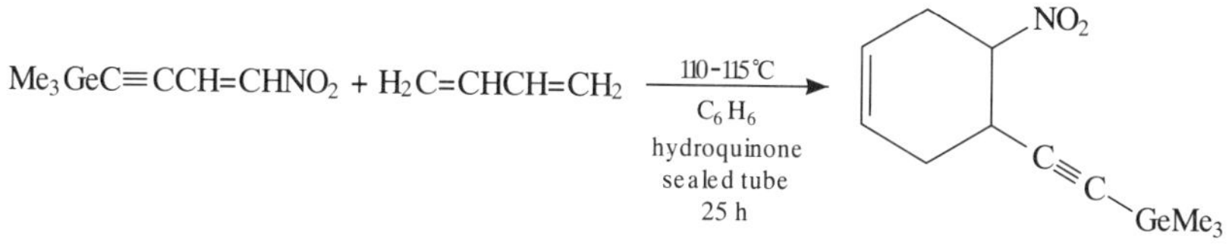
Me3GeC≡CCH=CHNO2 + H2C=CHCH=CH2
110-115°C
C6H6
hydroquinone
sealed tube
25 h
NO2
C≡C
GeMe3

Cycloadditions of germa-substituted ethylenes and acetylenes to dienes provide cyclic germa compounds (Scheme 6)[44].

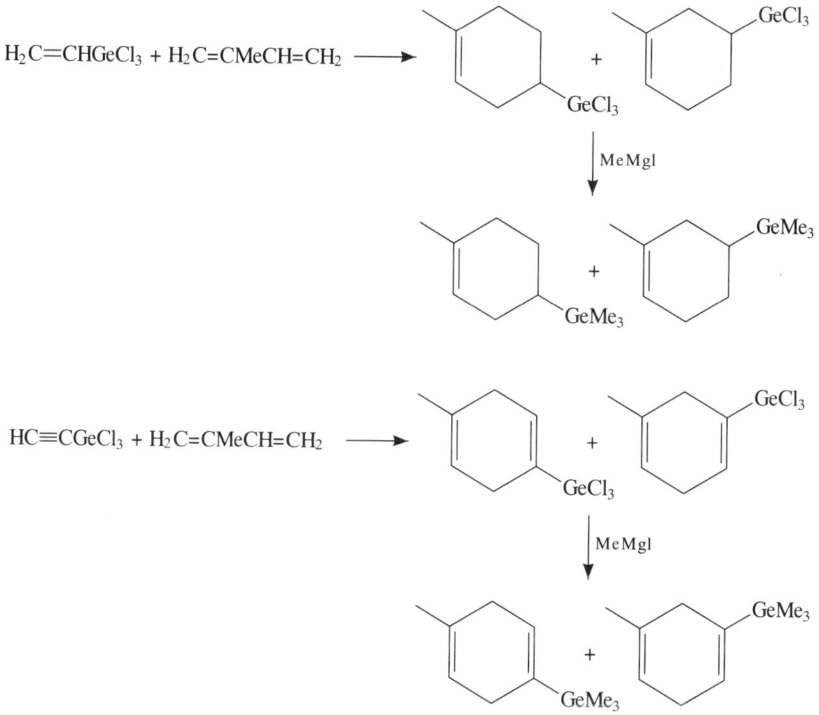
H2C=CHGeCl3 + H2C=CMeCH=CH2
GeCl3
+
MeMgl
GeMe3
HC≡CGeCl3 + H2C=CMeCH=CH2

SCHEME 6

A remarkable example of cycloaddition of organometallic dienophiles with 1,1-dimethyl-2,5-diphenylsilacyclopentadiene[45] is shown in Scheme 7.

Some other reactions of hydrogermylation of allenes (and acetylenes) are given below:

$$\text{RCH=C=CHR} + \text{Ph}_3\text{GeH} \longrightarrow \text{RCH=CHCHRGePh}_3$$

The catalyst used, $Pd(PPh_3)_4Et_3B$, induces a radical addition of Ph_3GeH[57]. The reaction can also be carried out with Ph_3SnH.

The allylic germanes and stannanes obtained are useful synthons. The hydrogermylation of acetylenes can also be performed with dimethyl(2-thiophenyl)germane[58]. This compound can be prepared quantitatively by the Grignard reaction of Cl_2GeMe_2 with 2-bromothiophene (Scheme 8).

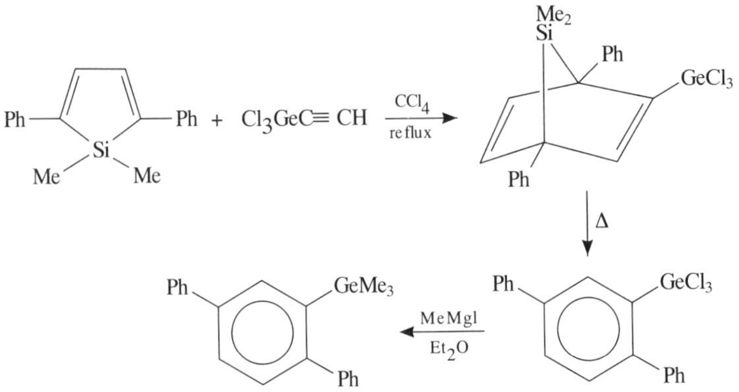
Ph
Ph
Si
Me
Me
Cl3GeC≡CH
CCl4
reflux
Me2
Si
Ph
GeCl3
Ph
Δ
Ph
GeCl3
Ph
MeMgI
Et2O
Ph
GeMe3
Ph

SCHEME 7

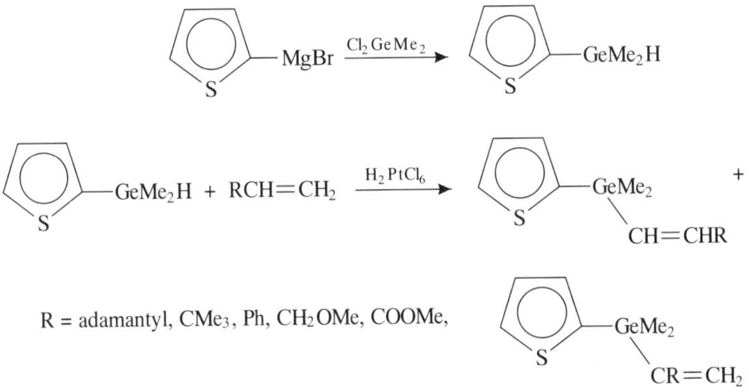
S
MgBr
Cl2GeMe2
GeMe2H
GeMe2H + RCH=CH2
H2PtCl6
GeMe2
CH=CHR
R = adamantyl, CMe3, Ph, CH2OMe, COOMe,
GeMe2
CR=CH2

SCHEME 8

Applications of the synthetic methods already mentioned for organogermanes are found in various preparation routes for germacyclanes. Some of these procedures are presented below. Organohydridohalogermanes react analogously in hydrogermylation reactions with 1,4- or 1,5-dienes in the presence of $H_2PtCl_6 \cdot 6\ H_2O$ in *i*-PrOH solution as a catalyst using toluene as a solvent for the whole reaction to provide a germacyclane, e.g.

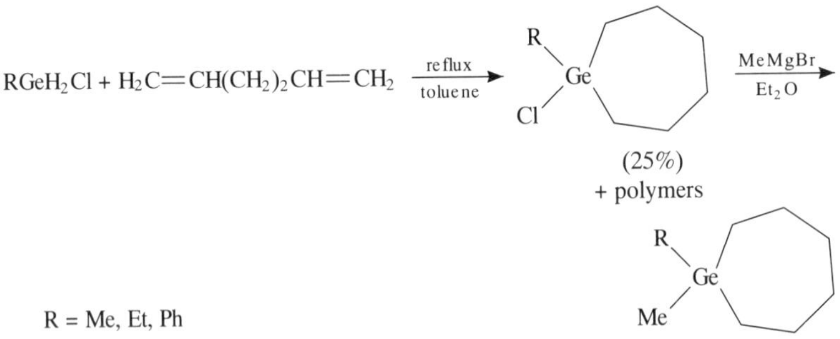
RGeH2Cl + H2C=CH(CH2)2CH=CH2
reflux
toluene
R
Ge
Cl
(25%)
+ polymers
MeMgBr
Et2O
R
Ge
Me
R = Me, Et, Ph

The method can also be utilized for the preparation of the corresponding germacyclohexane using $CH_2=CH-CH_2-CH=CH_2$ as starting material[46,47].

2,2'-Dilithiobibenzyl **8** is used in a reaction with diorganogermanium dichloride for the synthesis of 10,11-dihydro-5H-dibenzo[b,f]germepins[48], **9** (Scheme 9).

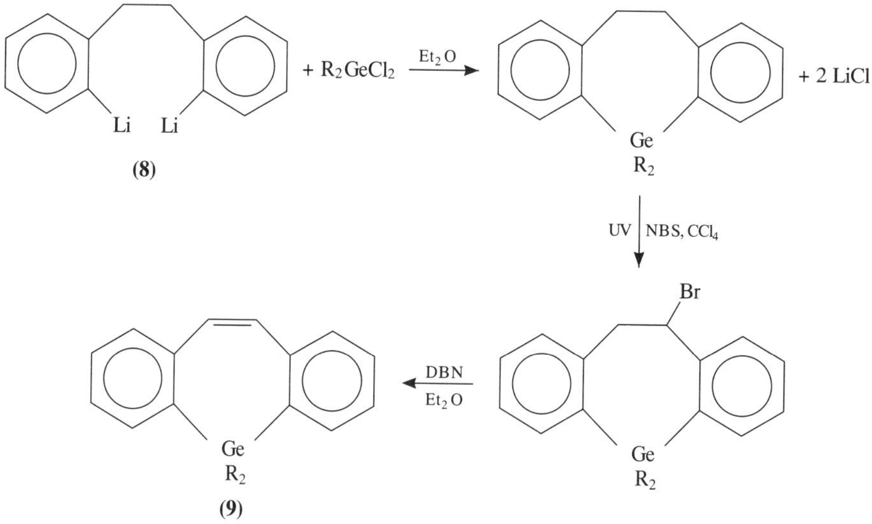

$$ \text{(8)} \quad + R_2GeCl_2 \xrightarrow{Et_2O} \quad \text{Ge } R_2 \quad + 2\,LiCl $$

UV | NBS, CCl₄

$$ \text{(9)} \xleftarrow[Et_2O]{DBN} \quad \text{Br} \quad \text{Ge } R_2 $$

SCHEME 9

A convenient method to prepare digermacyclenes is the reaction of GeI_2 with alkynes[49].

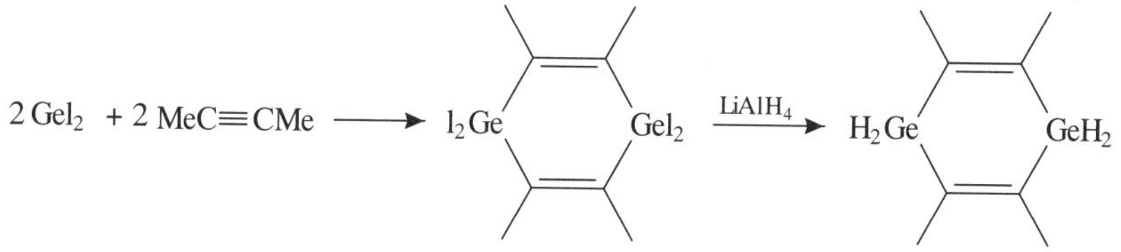

$$ 2\,GeI_2 + 2\,MeC{\equiv}CMe \longrightarrow I_2Ge{\cdots}GeI_2 \xrightarrow{LiAlH_4} H_2Ge{\cdots}GeH_2 $$

Reaction of dimethylphenylgermyllithium, **10**, with phenyldihydrochlorogermane, **11**, generates 1,2-diphenyl-1,1-dimethyldigermane, **12**:

$$ \underset{\textbf{(10)}}{PhMe_2GeLi} + \underset{\textbf{(11)}}{PhH_2GeCl} \longrightarrow \underset{\textbf{(12)}}{PhMe_2GeGePhH_2} + LiCl $$

Upon reaction of **12** with dibutylmercury in benzene, followed by immediate addition of 2,3-dimethylbutadiene, an interesting cyclogermadiene is produced[50] (Scheme 10). $Et_2Ge=CH_2$ was not isolated, but was trapped by the butadiene.

The reaction of a germacyclohexadiene, **13**, with perfluoro-2-butyne, **14**, followed by thermolysis of the obtained product, yields a digermacyclobutane[51] (Scheme 11). The existence of a germanium-carbon pπ-pπ double bond in the intermediate complex is likely. The intermediate was not isolated as such, but as its dimer. Compounds containing a carbon-germanium double bond were prepared by Satgé and coworkers[59], like the fluorenylidenedimethylgermanium, where stabilization arises from change transfer in the aromatic system.

Compound **15** was dehydrofluorinated by Me_3CLi in Et_2O to **16**, obtained as an etherate complex. After thermolysis and reduction of **16** by $LiAlH_4$, **17**, stabilized by the aromatic fluorene rings, is generated.

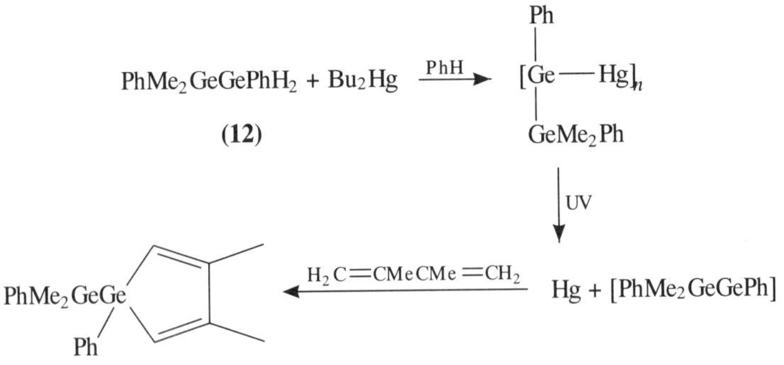

$$PhMe_2GeGePhH_2 + Bu_2Hg \xrightarrow{\text{PhH}} \begin{matrix} Ph \\ | \\ [Ge-Hg]_n \\ | \\ GeMe_2Ph \end{matrix}$$

(12)

$$\xleftarrow{H_2C=CMeCMe=CH_2} Hg + [PhMe_2GeGePh]$$

(with the bicyclic product drawn: PhMe$_2$GeGe—Ph ring)

$\downarrow$ UV

SCHEME 10

(13) + F$_3$CC≡CCF$_3$ **(14)** $\xrightarrow[\substack{70-80°C \\ 5\,h}]{\text{sealed tube}}$ 80-90%

$\downarrow \substack{N_2 \\ \text{flow}}$ 450°C

$$Et_2Ge\text{—}\square\text{—}GeEt_2 \xleftarrow{} [Et_2Ge=CH_2] +$$

35%

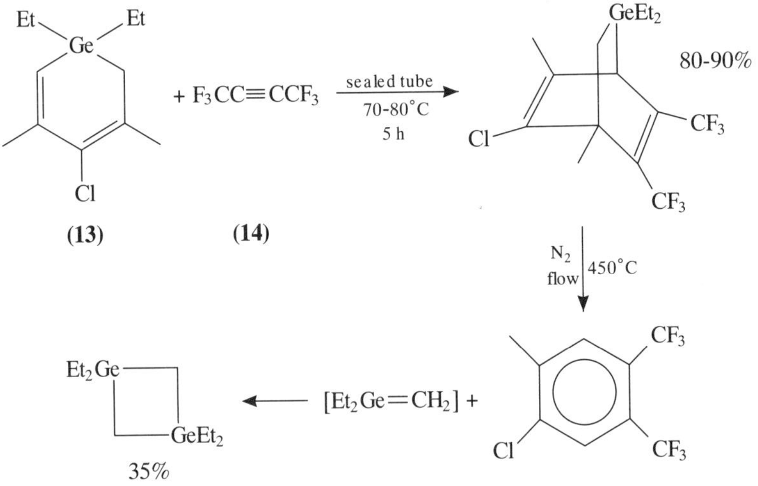

SCHEME 11

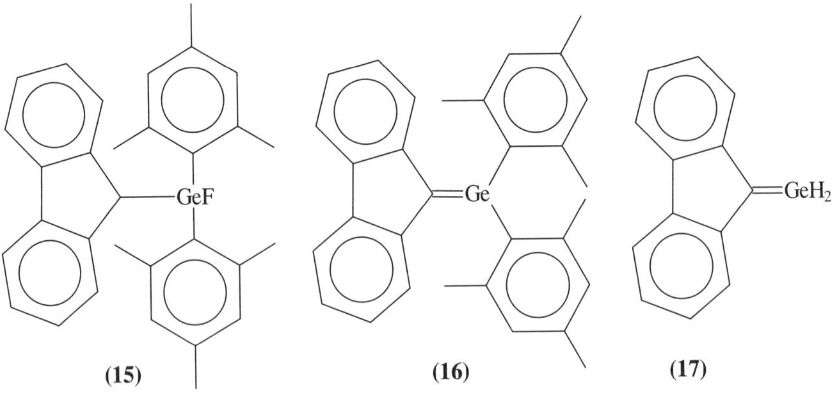

(15)   **(16)**   **(17)**

By the same method, two other stable fluorenylidenegermanes were prepared[60], one of them, **18**, possessing the bulky bisyl groups, [(Me$_3$Si)$_2$CH], and the other, **19**, one bisyl and one mesityl.

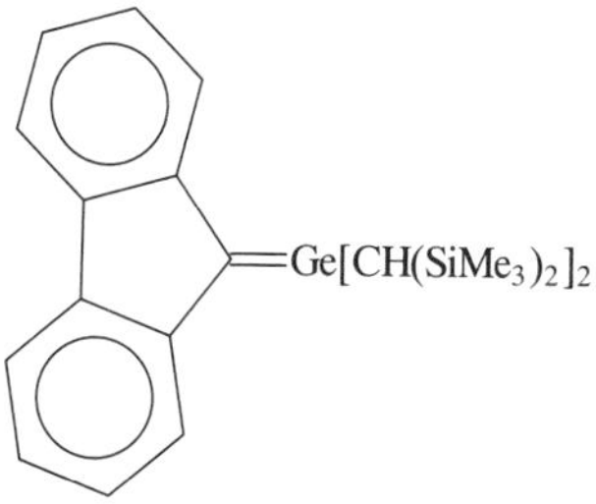

(18) 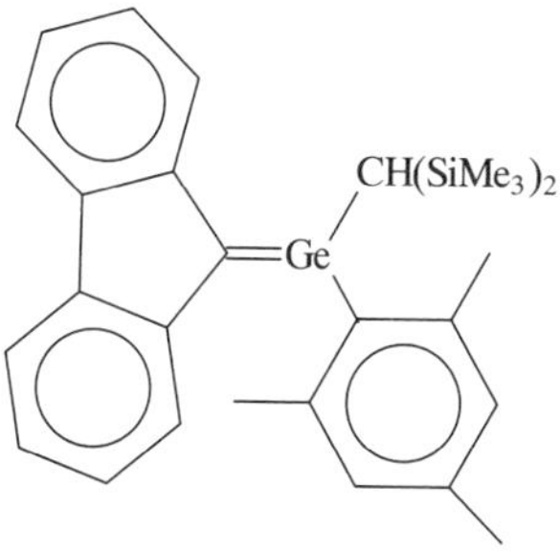   **(19)**

From the germene **20**[61], where R = mesityl, the first germapyrazoline **21** was prepared by a [2+3]cycloaddition of diazomethane to the Ge=C bond. Compound **21** gives, via a germirane intermediate, 9-methylenefluorene and dimesitylgermylene R$_2$Ge after thermolysis or photochemical decomposition. The latter can be trapped by 2,3-dimethylbutadiene to give germacyclopentene **22**.

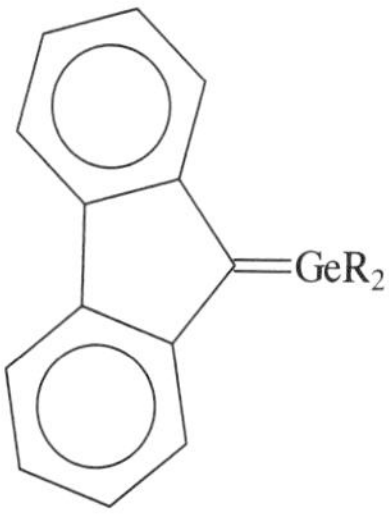

(20) 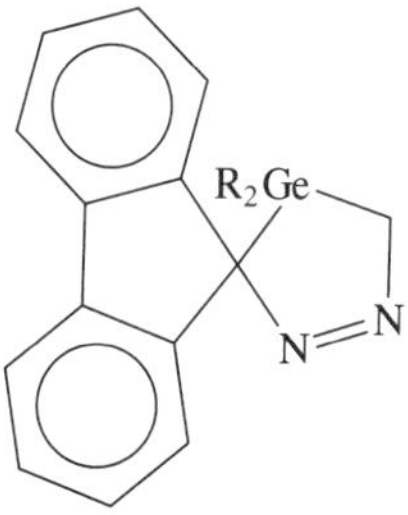   **(21)** 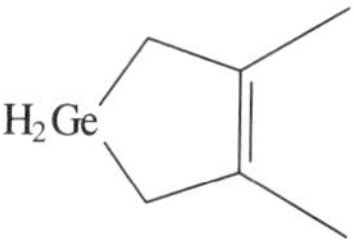   **(22)**

The organodichlorodigermanes (see Section III.A) and organodihydridodigermanes (see Section II.A) are good starting materials to synthesize 1,2-digermacyclenes[52] and 1,2-digermacyclanes[53]:

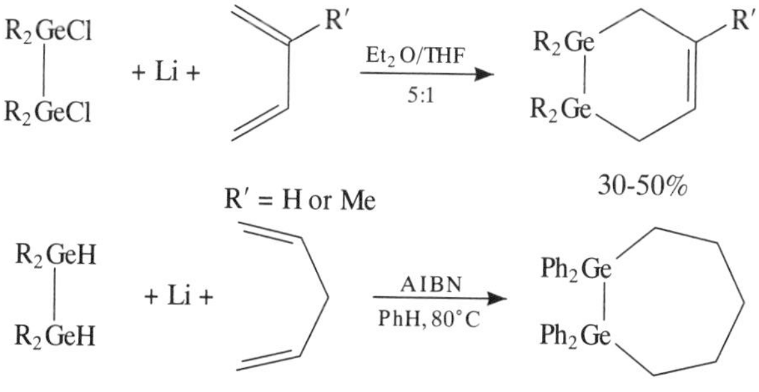

R' = H or Me   30-50%

Germacyclohexanes can be prepared by the method of Mazerolles[55] (Scheme 12). For the preparation of analogous compounds with R' = Me, Ph, the reactions in Scheme 13 can be used[56]:

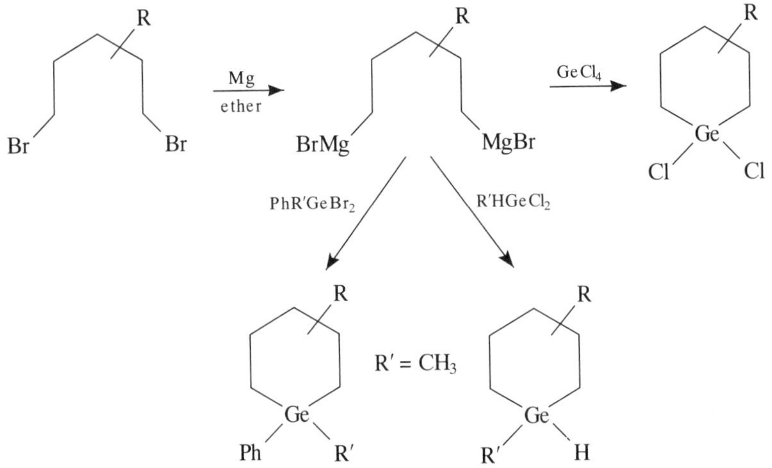

SCHEME 12

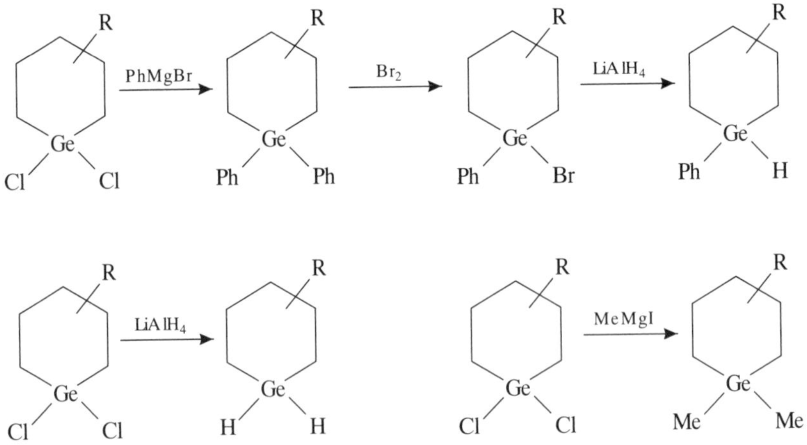

SCHEME 13

Besides 1,2- and 1,4-digermacyclohexenes and 1,2-digermacyclanes, interesting 1,2-digermacyclobutenes can be prepared as well[62].

Tetrakis(2,6-diethylphenyl)digermene **23** undergoes smooth cycloaddition with PhC≡CPh to yield a four-membered 1,2-digermacyclobutene:

A three-membered ring can also be prepared by reaction of diazomethane with **23**:

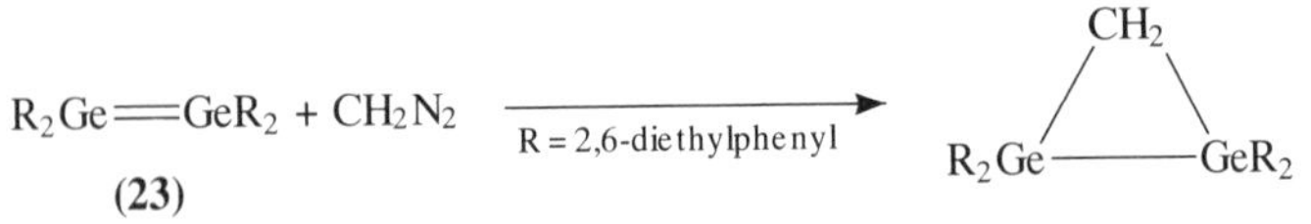

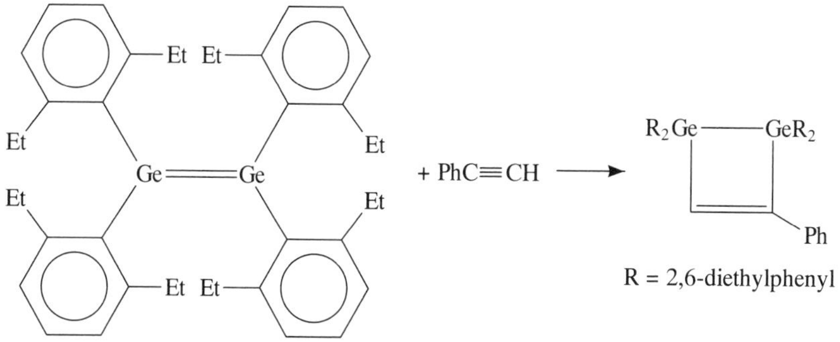

(23)

The germacyclopropene **24** reacting with the germabornadiene **25** gives the air stable 1,2-digermacyclobutene **26**, $\Delta^{1,7}$-2,2,6,6,8,8,9,9-octamethyl-4-thia-8,9-digerma-bicyclo[5,2,0]nonene[63].

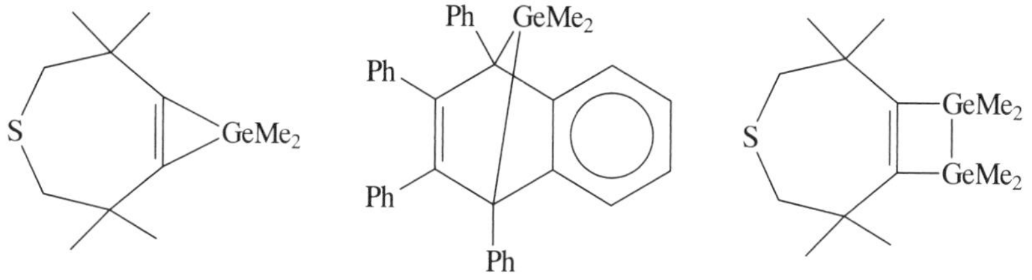

(24)   (25)   (26)

Three-membered rings containing germanium as the digermirene **27** can be expanded after reaction with acetylenic hydrocarbons[64]:

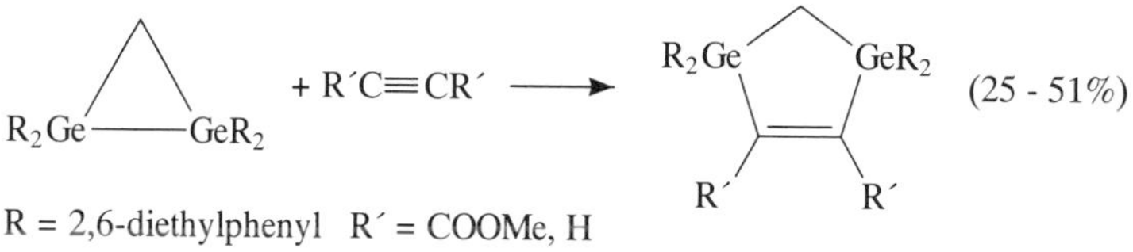

R = 2,6-diethylphenyl   R' = COOMe, H

(27)

The reaction is carried out in benzene in the presence of $Pd(PPh_3)_4$, $PdCl_2(PPh_3)_2$, $PdCl_2(PhCN)_2$ or $NiCl_2(PPh_3)_2$ as a catalyst.

A characteristic application of the Grignard reaction enables ring closure to prepare bicyclic monogermanes. Thus the bis-Grignard reagent derived from *cis*-1,3-(bromomethyl)cyclopentane **28** reacts with R_2GeCl_2 (R = Ph, Me) to give 3-germabicyclo[3.1.1]octanes **29**:

In the case where R = Ph, a bromine cleavage of one phenyl group of **29** allows, by subsequent nucleophilic displacement of the resulting bromide, the preparation of several other alkyl or phenyl-substituted derivatives of germanium[65].

Some bicyclogermanes are not unimportant since they provide a source of germylene radicals, GeR_2, which can be trapped by suitable reactants, such as 1,3-dienes.

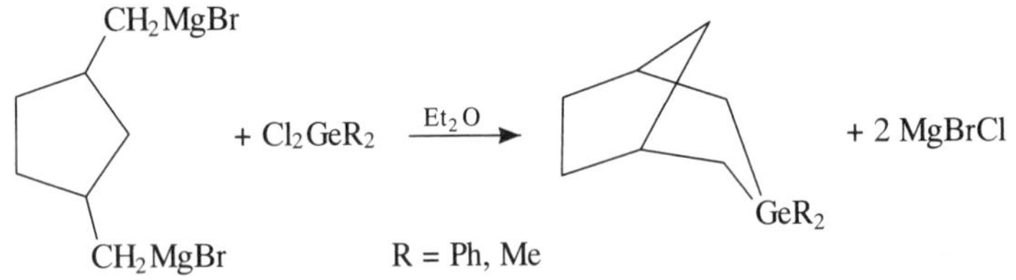

(28)   (29)

Dimethylgermylenes can be produced from 7-germanorbornadiene **30**. When **30** is reacted in benzene solution at $70\,^\circ$C with olefins of the type RR$'$C=CH$_2$, where R = 2-furyl, 2-thiophenyl and R$'$ = H, Me, germacyclopentanes **31** are obtained in 50 to 80% yield[66].

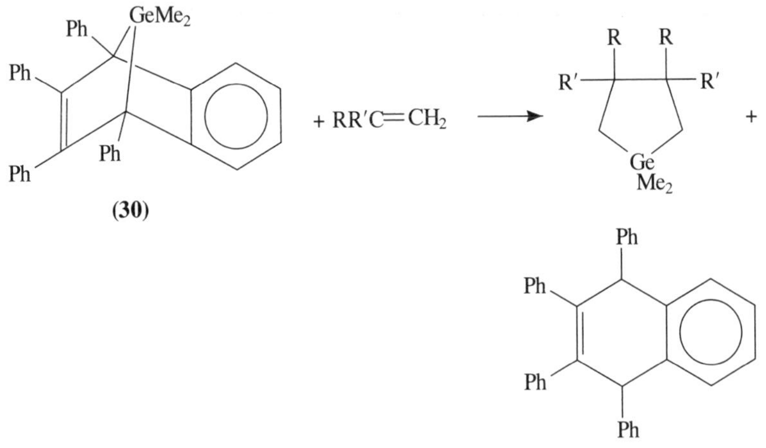

(30)

(31)

3,4-Dimethylcyclopentene-1-germylene, **33**[67], is found among various pyrolysis products of 2,3,7,8-tetramethyl-5-germa-spiro[4,4]nona-2,7-diene, **32**.

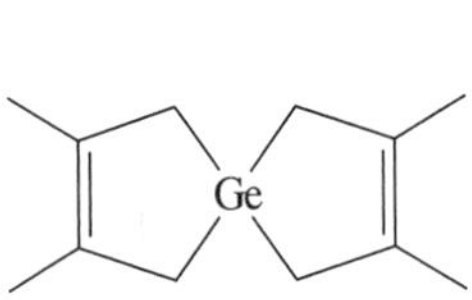

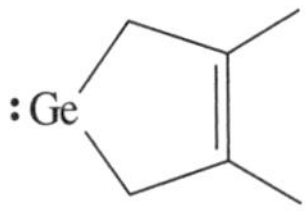

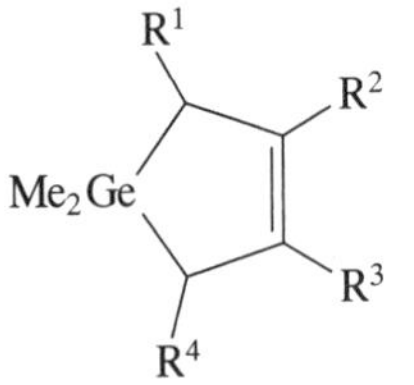

(32)   (33)   (34)

Upon reaction of Me$_2$GeCl$_2$ with Li in the presence of substituted 1,3-dienes of the type R^{1}CH=CR^{2}CR^{3}C=CHR^4 in THF, promoted by ultrasounds, germacyclopent-3-enes of the general type **34** are formed in yields of 15 to 54%.

The gas-phase pyrolysis of 1,1,3,4-tetramethyl-1-germacyclopent-3-ene leads to the extrusion of dimethylgermylene Me$_2$Ge that could be trapped by 1,3-dienes[68].

A dialkylgermylene is isolated after reaction of $Me_5C_5Ge-CH(SiMe_3)_2$ with $LiC(SiMe_3)_3$, which affords $(Me_3Si)_3C-Ge-CH(SiMe_3)_2$ as a monomer, stable at room temperature in the solid state. This contrasts with Lappert's[142] germylene $[(Me_3Si)_2CH]_2Ge$, which exhibits a dimeric structure in which the presence of a Ge=Ge double bond is evident[69] (see Section I.B for an analogous case with tin).

Bis(dimethylgermyl)alkanes of the type **35**, with $n = 1$ and 2, can be cyclized by ironpentacarbonyl, $[Ru(CO)_4]^{(-)}$ or rutheniumpentacarbonyl under UV irradiation giving digermairon heterocycles **36**:

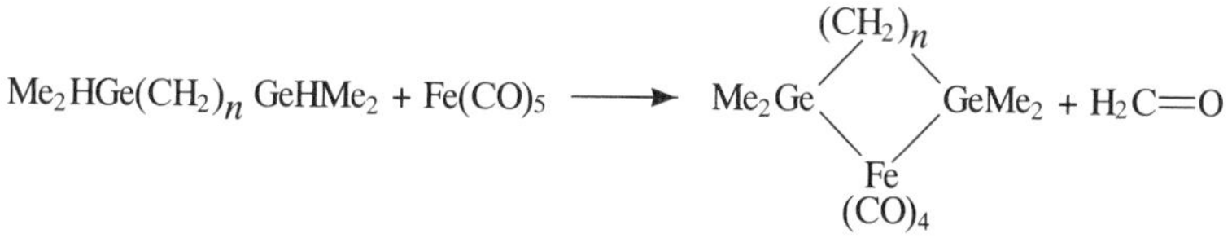

$$Me_2HGe(CH_2)_n GeHMe_2 + Fe(CO)_5 \longrightarrow$$

(35)   **(36)**

Compound **36**, with $n = 1$, reacts with oxygen or sulfur to give dioxa- or dithia-digermolanes **37**. These unstable compounds easily lose oxygen or sulfur and generate, respectively, oxa- or thiadigermetanes **38**. Compound **38** can be easily decomposed into Me_2GeY and $Me_2Ge=CH_2$[70].

(37) Y = O, S   **(38)**

Germanium can also be incorporated into large spiro heteropolyacetylenic cycles **39**[71] (Scheme 14).

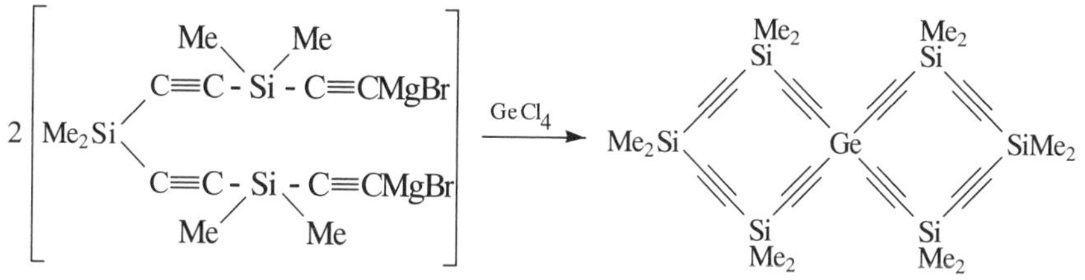

(39)

SCHEME 14

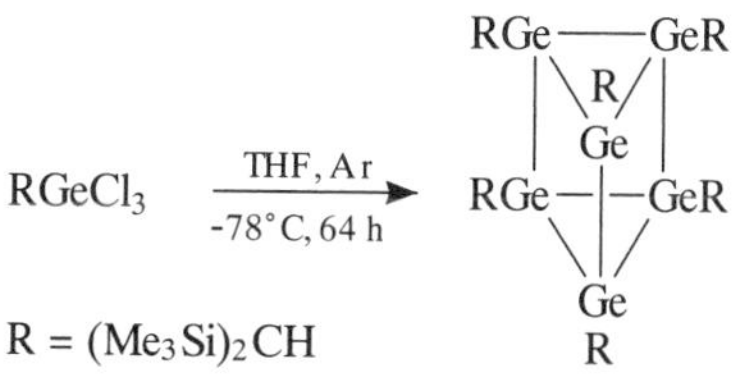

$R = (Me_3Si)_2CH$

(39a)

SCHEME 15

Like in the case of $[RM]_{2n}$ prismanes where $M = Sn$ (see Section I.B), a germanium-containing [3]prismane, stable up to $200\,^{\circ}C$, $[RGe]_6$, with $R = [(CH_3)_3Si]_2CH$, **39a**, is prepared[186] from bis(trimethylsilyl)methyltrichlorogermane. One of the interesting cyclogermanes is 1-methyl-1-germaadamantane **39b**, which can be prepared by a Grignard synthesis (Scheme 16)[187].

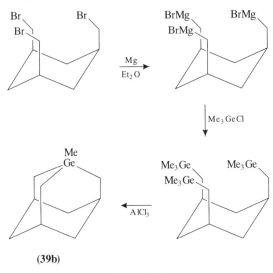

(39b)

SCHEME 16

B. Organostannanes

The main method for the preparation of different types of organostannanes is the ancient Würtz reaction between triorganotin halides or diorganotindihalides. Reactions proceed with or without solvent. Usual solvents are benzene, toluene, xylene, diethyl ether and ethanol. Occasionally liquid ammonia is also used. The preferred metals for this reaction are sodium and lithium. In the case of triorganotin halides, the reaction leads to symmetric hexaorganoditin compounds[72]:

$$2\ R_3SnX + 2\ Na \longrightarrow R_3SnSnR_3 + 2\ NaX$$

Unsymmetric organostannanes can be prepared by the reaction of triorganotin sodium compounds in liquid ammonia with a triorganotin halide[73,74]:

$$R_3SnNa + R'_3SnI \longrightarrow R_3SnSnR'_3 + 2\ NaI$$

Diorganotin dihalides react differently, since they produce polytin disodium compounds[75–77], which consequently can be treated with triorganotin halides:

$$3\ Me_2SnBr_2 + 8Na \longrightarrow \underset{NaSn\underset{Me_2}{\overset{Me_2}{\diagdown}}Sn\underset{Me_2}{\diagup}SnNa}{} + 6\ NaBr$$

$$\underset{NaSn\underset{Me_2}{\overset{Me_2}{\diagdown}}Sn\underset{Me_2}{\diagup}SnNa}{} \xrightarrow{Me_3SnBr} \underset{Me_3Sn\underset{Me_2}{\overset{Me_2}{\diagdown}}Sn\underset{Me_2}{\diagup}Sn\diagdown SnMe_3}{}$$

The dodecaorganopentatin compounds were prepared using this procedure[75].

Analogously Ph_3SnNa reacts in monoglyme with Ph_2SnCl_2 and yields octaphenyltritin, Ph_8Sn_3[78]. Furthermore, hexaphenylditin was prepared by treating Ph_3SnCl with potassium graphite KC_8[79]. Polystannacycloalkanes like dodecamethylhexastannacyclohexane are prepared by the reaction of Me_2SnCl_2 with sodium in liquid ammonia[80].
Medium-sized polystannacycloalkanes of the type **40** can be prepared by similar methods from 1,ω-bis(bromodimethylstannyl)alkanes[86]:

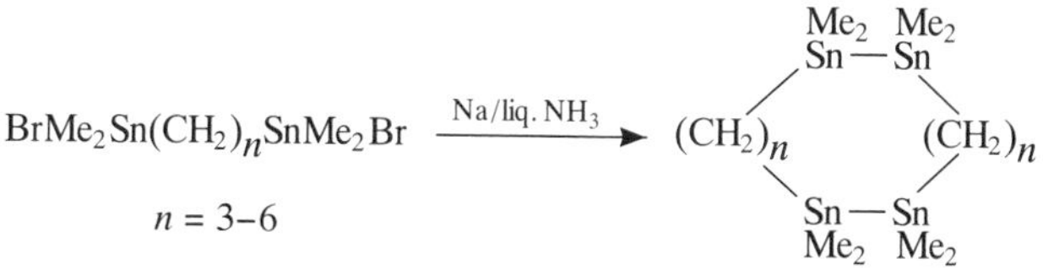

$$BrMe_2Sn(CH_2)_nSnMe_2Br \xrightarrow{Na/liq.\ NH_3} $$

$$n = 3\text{-}6$$

(**40**)

Analogous reactions can be achieved with 1,1- or 1,2-bis(bromodimethylstannyl)-1-alkenes.
Distannacycloalkanes of the type **42** can be prepared analogously from 1,ω-dichloroalkanes and triorganotinsodium compounds[124]:

$$Cl(CH_2)_nCl + 2\ Me_3SnNa \longrightarrow Me_3Sn(CH_2)_nSnMe_3 \xrightarrow{MCl} ClMe_2Sn(CH_2)_nSnMe_2Cl$$

$$n = 1\ \text{or}\ 3\ \text{to}\ 6 \qquad\qquad\qquad\qquad\qquad\qquad\qquad\qquad (\textbf{41})$$

$$M = HgCl, SnClMe_2, SnCl_3$$

$$\textbf{41} + BrMg(CH_2)_m MgBr \longrightarrow$$

$$m = 1\text{-}4$$

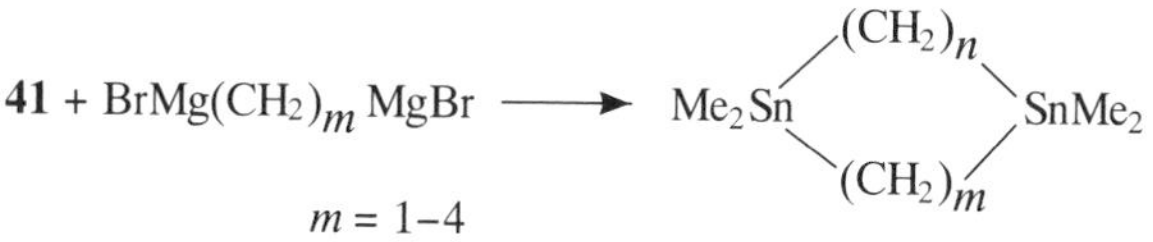

(**42**)

The Würtz reaction can be used to incorporate some functionality in an alkyl chain bound to tin in a stannane[111], e.g. by reaction of Ph_3SnCH_2I or $(ICH_2)_4Sn$ respectively with RSNa (where RH = benzothiazole, benzoxazole, 1-methylimidazole, pyrimidine). Compounds of the type Ph_3SnCH_2SR or $(RSCH_2)_4Sn$ were obtained.
Hexamethylditin can be prepared by reaction of dimethyltindihydride and sodium in liquid ammonia followed by the action of MeI[81]:

$$Me_2SnH_2 \xrightarrow{Na/liq.\ NH_3} NaMe_2SnSnMe_2Na \xrightarrow{MeI} Me_3SnSnMe_3$$

Triorganostannyllithium compounds R_3SnLi can be obtained by direct reaction of R_3SnCl compounds with lithium in tetrahydrofuran[82]. Me_3SnLi in THF yields the complex $(Me_3Sn)_3SnLi\cdot 3THF$[83] after solvent evaporation.
On the other hand, if $SnCl_2$ is reacted with Me_3SnLi in THF, hexamethylditin Me_6Sn_2 is generated, together with $(Me_3Sn)_3SnLi\cdot 3THF$. $(Me_3Sn)_3SnMe$ can be obtained in low yield by treating $(Me_3Sn)_3SnLi$ in THF with MeBr. Through the same route, Ph_3SnLi treated with $SnCl_2$ leads to $(Ph_3Sn)_3SnLi$[84] which, after methylation by CH_3I, yields $(Ph_3Sn)_3SnMe$.
The reaction of triorganostannyllithium compounds with $SnCl_2$, $SnCl_4$ or with triorganotin or diorganotin compounds has wide applications.

A stannacyclobutene[85] can be prepared from trichloromesityltin: the reaction of mesitylithium (MesLi) with $SnCl_4$ in hexane gives 40% of $MesSnCl_3$ which, treated with two equivalents of 2,4,6-tri-*t*-butylphenyllithum in hexane–petroleum ether, yields 52% of **43**. Subsequent conversion of **43** by Me_3CLi in hexane–petroleum ether generates the stannacyclobutene **44**.

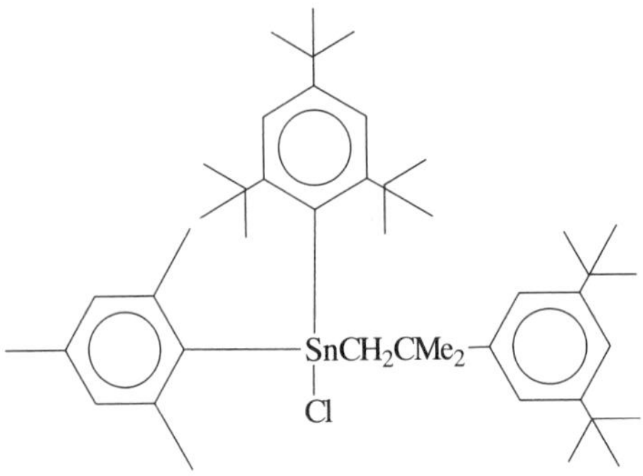
SnCH2CMe2
Cl

(43)

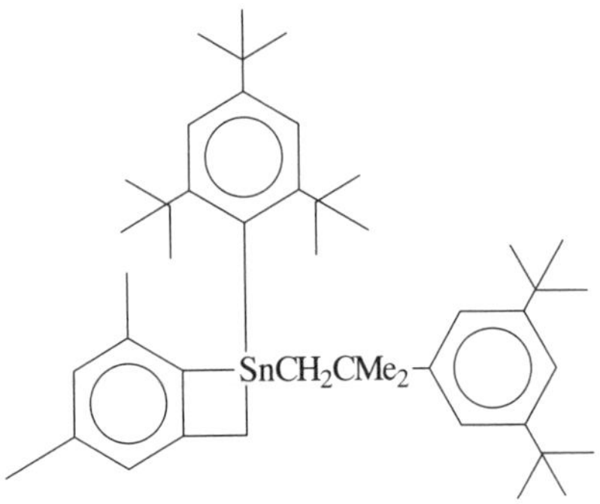
SnCH2CMe2

(44)

On the other hand, a 1,3-distannacyclobutane **45** was prepared by an interesting sequence of reactions[121] (Scheme 17).

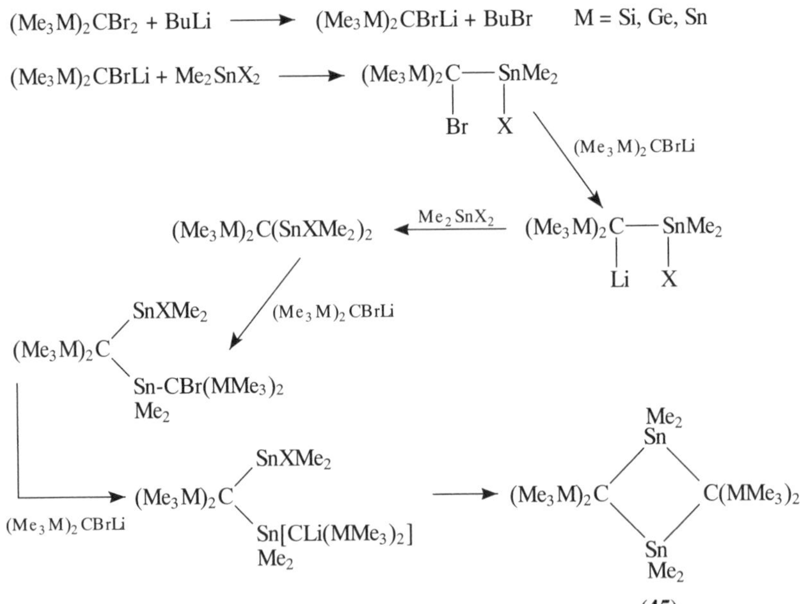
(Me3M)2CBr2 + BuLi
(Me3M)2CBrLi + BuBr
M = Si, Ge, Sn
(Me3M)2CBrLi + Me2SnX2
(Me3M)2C—SnMe2
Br X
(Me3M)2CBrLi
(Me3M)2C(SnXMe2)2
Me2SnX2
(Me3M)2C—SnMe2
Li X
SnXMe2
(Me3M)2C
Sn-CBr(MMe3)2
Me2
(Me3M)2CBrLi
SnXMe2
(Me3M)2C
Sn[CLi(MMe3)2]
Me2
(Me3M)2CBrLi
Me2
Sn
(Me3M)2C
C(MMe3)2
Sn
Me2

(45)

SCHEME 17

The organolithium reagents are so efficient for the generation of Sn–C bonds that a double Sn=C bond, identified as such, is obtained by the action of *t*-butyl lithium[87] on chloro- or fluorostannanes, yielding, for instance, the bis(2,4,6-triisopropylphenyl)(fluorenyl)stannene diethyl ether adduct **46** (see Section I.A for an analogous situation with germanium).

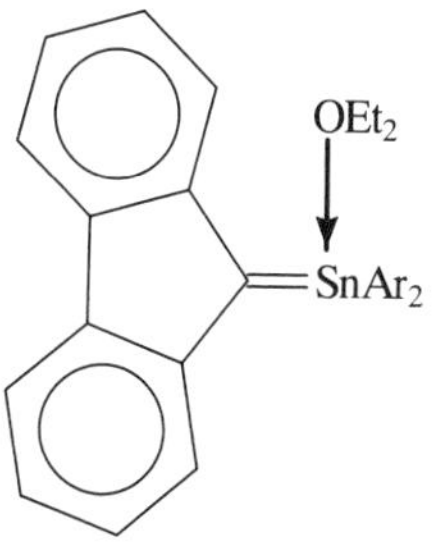

OEt$_2$

=SnAr$_2$ Ar = 2,4,6-triisopropylphenyl

(46)

The chloro- or fluorostannanes used were of the type R$_2'$XSn-CHR$_2$, where R$'$ = bis(trimethylsilyl)methyl or 2,4,6-triisopropylphenyl, R$_2$C is the fluorenylidene moiety and X = Cl or F. Compounds of this type are prepared from R$_2'$SnX$_2$ and LiCHR$_2$.

An aryllithium reagent, prepared from 1-bromo-2,6-diethylbenzene and butyllithium, reacts with SnCl$_2$ in THF at 0 °C to produce hexakis(2,6-diethylphenyl)tristannacyclopropane, **47**, in 50–55% yield[88a]. If the reaction above is performed at 0 °C in Et$_2$O, the 1-butyl-2,2,3,3,4,5,5,6,6-nonakis(2,6-diethylphenyl)hexastannabicyclo[2,2,0]hexane, **48**, can be isolated in 1.5% yield. This compound turns out to be the first example of a polycyclic polystannane[88b].

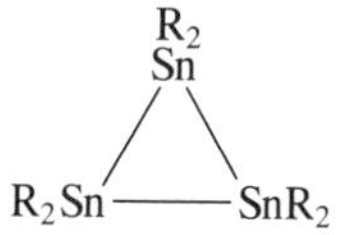

R$_2$Sn

/   \

R$_2$Sn———SnR$_2$

R = 2,6-diethylphenyl

(47)

R
Sn
/   \
R$_2$Sn   Sn   SnR$_2$
| Bu |
R$_2$Sn   SnR$_2$

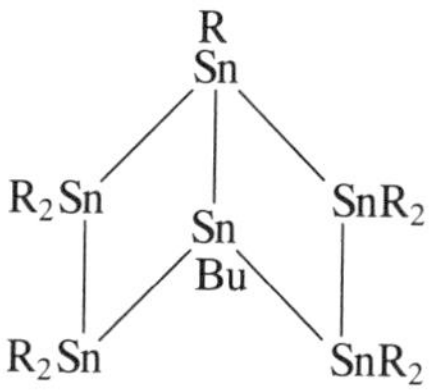

R = 2,6-diethylphenyl

(48)

Thermolysis of compound **47** in the presence of benzophenone or naphthalene at 200–220 °C provides two products, 2,2,4,4,5,5-hexakis(2,6-diethylphenyl)pentastanna-[1,1,1]propellane, **49**, and octakis(2,6-diethylphenyl)octastannacubane[89], **50**, as well as traces of R$_3$SnSnR$_3$.

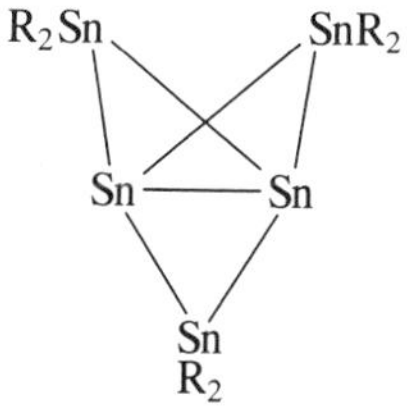

R$_2$Sn   SnR$_2$

Sn———Sn

Sn
R$_2$

(49)

R Sn——— Sn R
R Sn——— Sn R
R Sn——— Sn R
Sn——— Sn
R   R

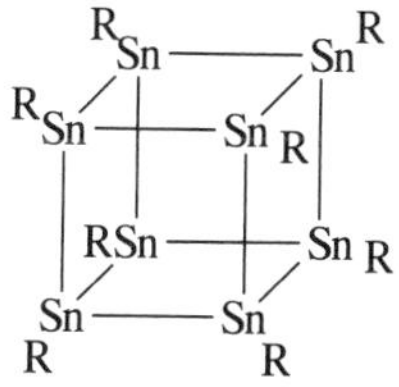

(50)

Decakis(2,6-diethylphenyl)decastanna[5]prismane, **51**, can be prepared[90] in an analogous way, together with by-products. The perstanna[n]prismanes are probably products generated by the thermal bimolecular disproportionation of the highly reactive diorganostannylenes R_2Sn: produced during the course of the pyrolysis:

$$2\ R_2Sn: \Longrightarrow R\overset{\bullet}{S}n: + R_3\overset{\bullet}{S}n$$

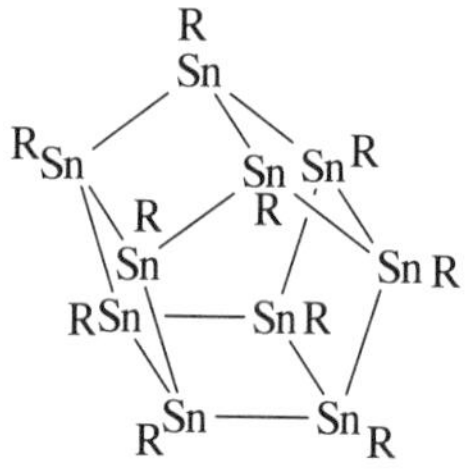

(51) 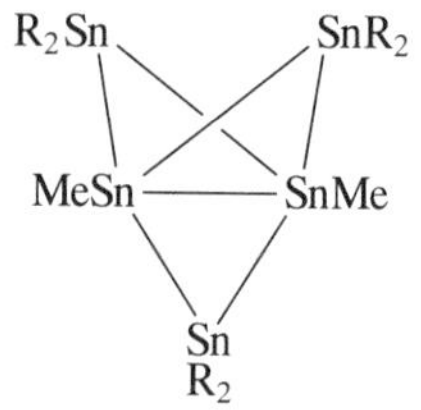   (52)

Successive addition of methyllithium and methyl iodide to pentastannapropellane in pentane gives the bicyclo compound **52**[91] in two steps.

Another use of organolithium compounds to produce cyclostannanes is the cyclization of the 2,2'-dilithiobiphenyl compounds of type **53** by Bu_2SnCl_2[92].

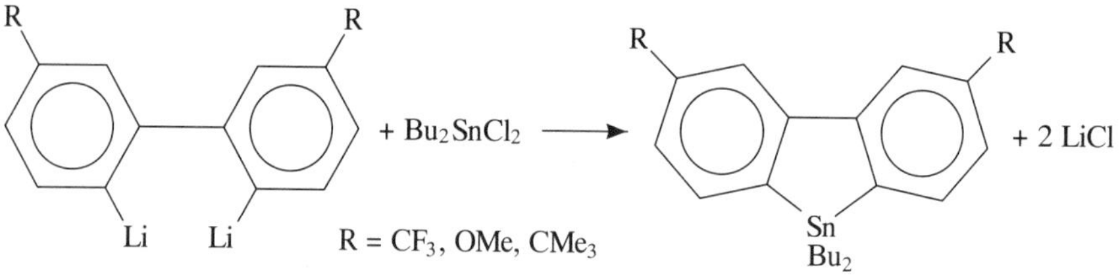

$+\ Bu_2SnCl_2 \longrightarrow$ $+\ 2\ LiCl$

$R = CF_3,\ OMe,\ CMe_3$

(53)

Stannylation of furan can be performed by Bu_3SnCl[93], as illustrated by the following 5-substituted lithium derivatives:

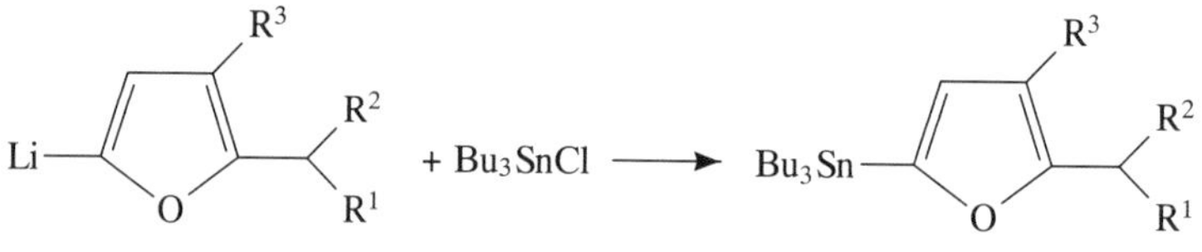

$+\ Bu_3SnCl \longrightarrow$

(Trialkylstannyl)lithium reagents are efficient compounds to stannylate large organic molecules like triptycene derivatives. (Trimethylstannyl)lithium in THF at $0\,°C$ in inert atmosphere, reacting with 1-bromotriptycene, **54**, gives 1-trimethylstannyltriptycene[135], **55**:

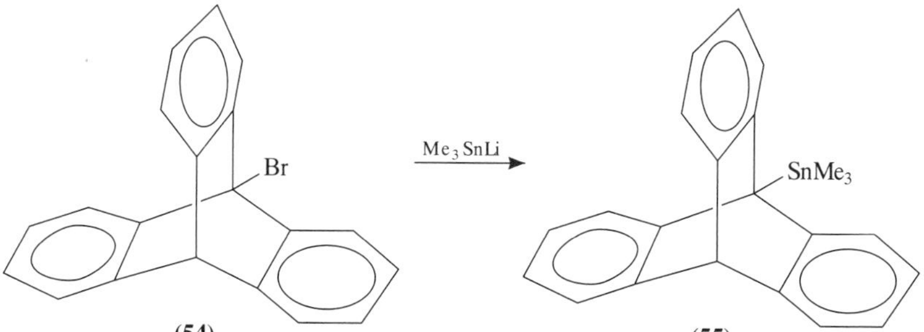

(54)   **(55)**

The same reagents give also a general access to acylstannanes[136a]:

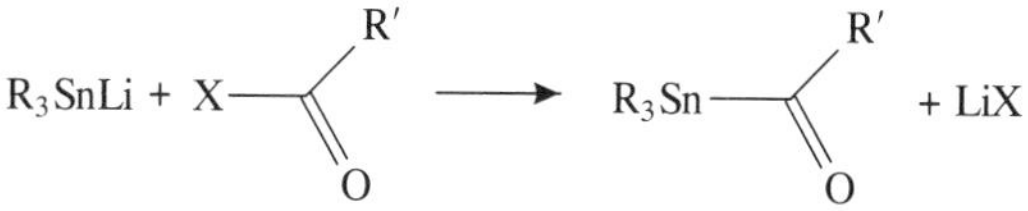

R = Me, Bu; X = Cl, OEt, SPh; R′ = Ph, Pr, Ph-3-t-Bu, 2-furyl, 2-thienyl

In particular, triorganostannyl(vinyl)ketones of the type **56** can be obtained[136b].

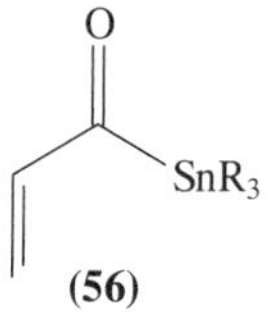

(56)

Using the olefination reactions mediated by $CrCl_2$[137], Hodgson prepared E-alkenylstannanes of the type **57**[138]:

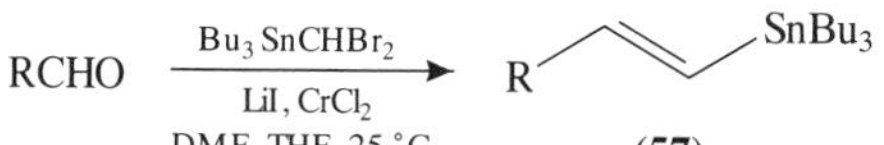

(57)

(R = C_8H_{17}, Ph, HC≡CH−CH_2−CH_2−)

$Bu_3SnCHBr_2$ was prepared from CH_2Br_2, Bu_3SnCl and lithium diisopropylamide, LDA.

The Grignard reaction can be used to prepare hexaorganoditin compounds from $SnCl_2$[72]. Reaction of diorganotin compounds like t-Bu_2SnCl_2 with Grignard reagents like t-BuMgCl in refluxing THF produces a cyclic tetratin compound[94].

Optically active (−)-menthyltin(IV) derivatives, in which tin is directly bound to a chiral carbon atom, can be synthesized easily by the reaction of the chiral Grignard reagent (−)-menthylMgCl **58** with Me_3SnCl yielding (−)-menthyltrimethyltin **59**[95]:

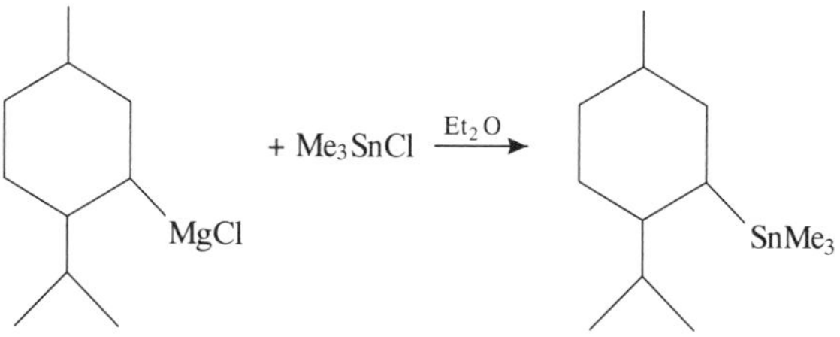

(58)   **(59)**

t-Butyl-(8-N,N-dimethylaminonaphthyl)($-$)menthyltin hydride, in which the tin atom is a chiral center, can be prepared analogously.

The first optically active organotin, where the metal atom is the only chiral center, was synthesized[132] using tetra-p-anisyltin as starting material. A racemic triorganotin hydride was first prepared by the sequence of reactions

$$An_4Sn \xrightarrow{SnCl_4} An_2SnCl_2 \xrightarrow{MeMgI} An_2MeSnCl \xrightarrow{NpMgBr} An_2SnMeNaph \xrightarrow[PhCl/MeOH]{HCl, 30°C}$$

$$AnMeNaphSnCl \xrightarrow{LiAlH_4} AnMeNaphSnH$$

Upon addition of ($-$)methylacrylate to the hydride, two diastereomers were obtained, as evidenced by NMR, and separated. Racemic tetraorganotin derivatives can be obtained following analogous routes using alternatively Grignard reactions and bromodemethylations[133,134]. Bromodemethylations must necessarily be carried out in a polar solvent like methanol.

$$Me_4Sn \xrightarrow[MeOH]{Br_2} Me_3SnBr \xrightarrow{c\text{-}C_6H_{11}MgBr} c\text{-}C_6H_{11}SnMe_3 \xrightarrow[MeOH]{Br_2} c\text{-}C_6H_{11}SnMe_2Br \xrightarrow[99\%]{i\text{-}PrMgBr}$$

$$c\text{-}C_6H_{11}SnMe_2Pr\text{-}i \xrightarrow[MeOH]{Br_2} (c\text{-}C_6H_{11})i\text{-}PrSnMeBr \xrightarrow{RMgBr}$$

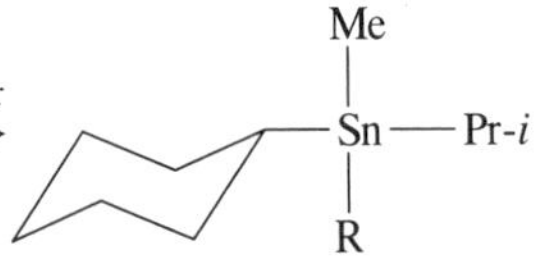

That Grignard reagents are widely used for the preparation of various types of organotin compounds is further illustrated by 1-trimethylstannyl-2-phenylethene, isolated[120] in 63% yield, besides 1,4-diphenylbutadiene (28%) as a side-product.

$$PhCH=CH-Br \xrightarrow[Et_2O/THF]{Mg} PhCH=CHMgBr \xrightarrow{BrSnMe_3} PhCH=CHSnMe_3$$

Expansion of tin-containing macrocycles can likewise be performed by Grignard reagents[139] (see, e.g., Scheme 18).

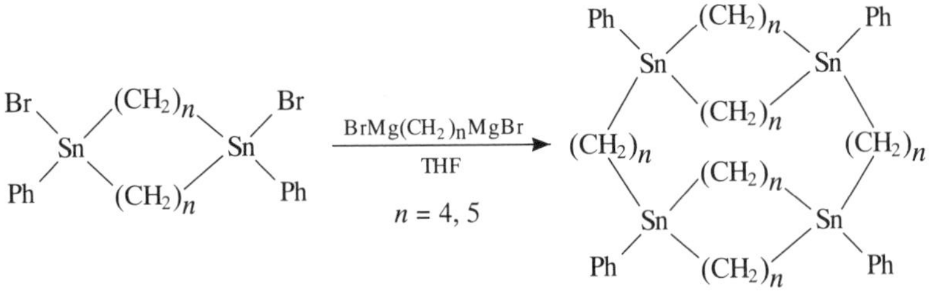

SCHEME 18

Tributyltin isocyanate **61** treated by phenylmagnesium bromide gives tributylphenyltin **62** in 100% yield. Reagent **61** is prepared[140] by reaction of Bu_3SnH with triethylammonium methylcarboxysulfamoyl hydroxide **60**:

$$\text{Bu}_3\text{SnH} \xrightarrow[\text{C}_6\text{H}_6]{[\text{Et}_3\overset{+}{\text{N}}\text{SO}_2\text{NHCO}_2\text{Me}]\text{OH}^{(-)}\ (\mathbf{60})} \underset{(\mathbf{61})}{\text{Bu}_3\text{Sn}-\text{NCO}} \xrightarrow[\text{Et}_2\text{O}]{\text{PhMgBr}} \underset{(\mathbf{62})}{\text{Bu}_3\text{SnPh}}$$

Another important route to obtain organostannanes is the hydrogenolytic cleavage of tin–nitrogen bonds[96,97]. This reaction obeys the general scheme

$$\text{R}_3\text{SnH} + \text{R}'_3\text{SnNEt}_2 \longrightarrow \text{R}_3\text{Sn}^{(-)}\text{R}'_3\text{Sn}\overset{+}{\text{N}}\text{HEt}_2 \longrightarrow \text{R}_3\text{SnSnR}'_3 + \text{HNEt}_2$$

A polar mechanism is proposed, where the hydrogen atom of the triorganotin hydride attacks in a rate-determining step the nitrogen atom of the triorganostannyl diethylamine. The method can also be applied to the generation of Ge–Sn bonds[119]. The intermediate product **63** can react with $\text{C}_6\text{H}_5\text{C}\equiv\text{CH}$, $(\text{C}_6\text{H}_5)_3\text{GeH}$ or $(\text{C}_6\text{H}_5)_3\text{SnH}$.

$$\text{Et}_2\text{Sn}(\text{NEt}_2)_2 + \text{Ph}_3\text{GeH} \longrightarrow \underset{(\mathbf{63})}{\text{Ph}_3\text{GeSnEt}_2(\text{NEt}_2)} \xrightarrow{\text{PhC}\equiv\text{CH}} \text{Ph}_3\text{GeSnEt}_2\text{C}\equiv\text{CPh}$$

$$\downarrow \begin{array}{l} \text{Ph}_3\text{MH} \\ \text{M = Ge, Sn} \end{array}$$

$$\text{Ph}_3\text{GeSnEt}_2\text{MPh}_3 + \text{HNEt}_2$$

Diorganotin dihydrides[98] and triorganotin monohydrides[99] are used in the presence of amines to prepare linear or cyclic polytin compounds[100].

In the case of triorganotin hydrides, the reaction with R_3SnX (where X = OAc, Cl, I, CN) requires the presence of triethylamine as a catalyst[99].

$$i\text{-Bu}_3\text{SnI} + \text{Ph}_3\text{SnH} \xrightarrow[68\%]{\text{PhH, Et}_3\text{N}} i\text{-Bu}_3\text{SnSnPh}_3$$

or

$$\text{Et}_3\text{SnCN} + \text{Ph}_3\text{SnH} \xrightarrow{\text{Et}_2\text{O, Et}_3\text{N}} \text{Et}_3\text{SnSnPh}_3$$

Me_2SnCl_2 or Ph_2SnCl_2, treated with LiAlH_4, in the presence of Et_2O and Et_3N, provide the cyclic hexamers $[\text{Me}_2\text{Sn}]_6$ or $[\text{Ph}_2\text{Sn}]_6$[98].

Also R_2SnH_2 treated with pyridine containing traces of R_2SnCl_2 (R = Me or Ph) gives as intermediate $\text{H-}[\text{R}_2\text{Sn}]_6\text{-H}$, which loses H_2 yielding the cyclic hexamer[100].

Upon reaction of diorganotin dihydrides with carboxylic acids, compounds of the type $[(\text{AcO})\text{R}_2\text{Sn}]_2$ are obtained which can be consequently converted into $(\text{HR}_2\text{Sn})_2$[101] (Scheme 19). Analogous reactions can be performed using the dialkyltin(II) polymers $(\text{R}_2\text{Sn})_n$ with benzoic acid, giving rise to 1,2-dibenzoates[102]:

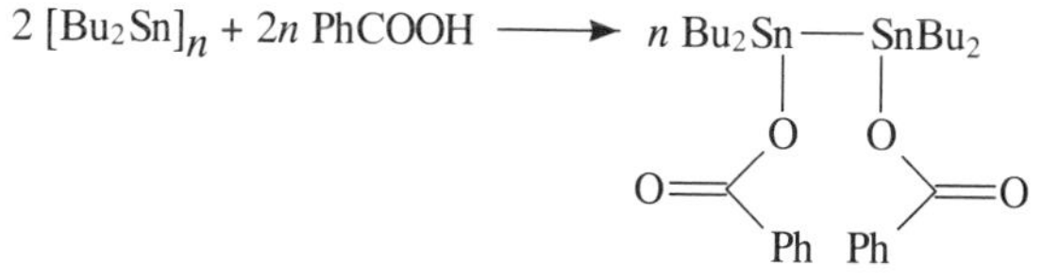

$$2\,[\text{Bu}_2\text{Sn}]_n + 2n\,\text{PhCOOH} \longrightarrow n\,\text{Bu}_2\text{Sn}-\text{SnBu}_2$$

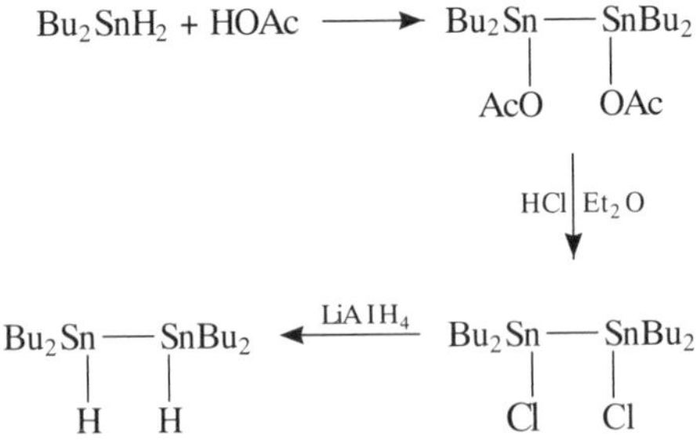

$$Bu_2SnH_2 + HOAc \longrightarrow Bu_2Sn \overset{|}{\underset{AcO}{}} \!\!\!\!\! - \!\!\!\!\! \overset{|}{\underset{OAc}{}} SnBu_2$$

$$\downarrow HCl \, | \, Et_2O$$

$$Bu_2Sn \overset{|}{\underset{H}{}} \!\!\!\!\! - \!\!\!\!\! \overset{|}{\underset{H}{}} SnBu_2 \quad \xleftarrow{\;LiAlH_4\;} \quad Bu_2Sn \overset{|}{\underset{Cl}{}} \!\!\!\!\! - \!\!\!\!\! \overset{|}{\underset{Cl}{}} SnBu_2$$

SCHEME 19

In the last years the monomeric dialkyltin(II) moiety R_2Sn, where $R = CH(SiMe_3)_2$, became more extensively involved in the preparation of diorganostannylenes. This compound is prepared by the following reaction[103a,b]:

$$2\, Li[CH(SiMe_3)_2] + SnCl_2 \xrightarrow[-20\,^\circ C]{Et_2O} Sn[CH(SiMe_3)_2]_2 + 2LiCl$$
$$(64)$$

Actually, **64** is known to be dimeric in the solid state but monomeric in dilute solution or in the gas phase. The first monomeric dialkyl- and diarylstannylenes are 2-pyridylbis[(trimethylsilyl)methyl]-substituted stannylenes and bis[2,4,6-tris(trifluoromethyl)phenyl]stannylene; it should be stressed, however, that the coordination number around Sn in the solid state is not 2 in these compounds. The first actual monomer with coordination number 2 in the solid state was found to be 2,2,5,5-tetrakis(trimethylsilyl)cyclopentane-1-stannylene, **65**, prepared by the following reaction[141]:

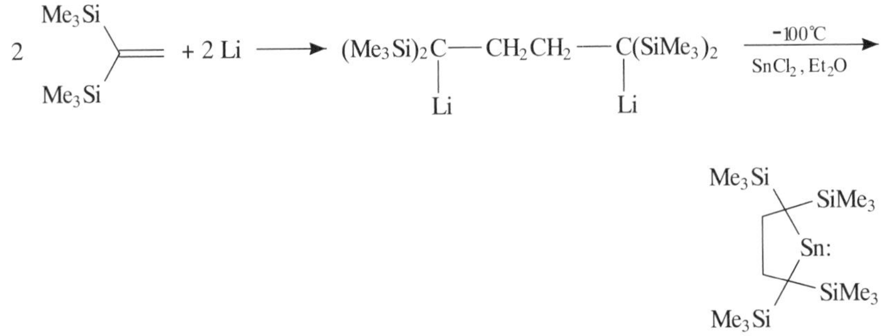

$$2 \; \underset{Me_3Si}{\overset{Me_3Si}{>}} \!\!\!\!= \quad + \; 2\,Li \longrightarrow (Me_3Si)_2C \!\!-\!\! CH_2CH_2 \!\!-\!\! C(SiMe_3)_2 \xrightarrow[SnCl_2,\,Et_2O]{-100\,^\circ C}$$
$$\underset{Li}{} \qquad \underset{Li}{}$$

(65)

Compound **64**, after reaction with 2,3-dimethylbutadiene, produces a cyclic stannylene **66** in an oxidative addition step[104].

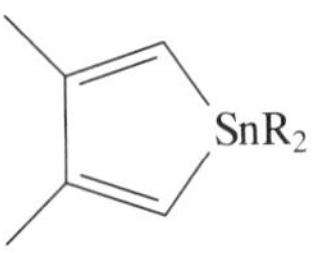

(66)

A series of compounds of the type $[(Me_3Si)_2CH]_2M$: have been prepared with M = Ge, Sn, Pb[142]. They are used to prepare various compounds by oxidative additions, such as

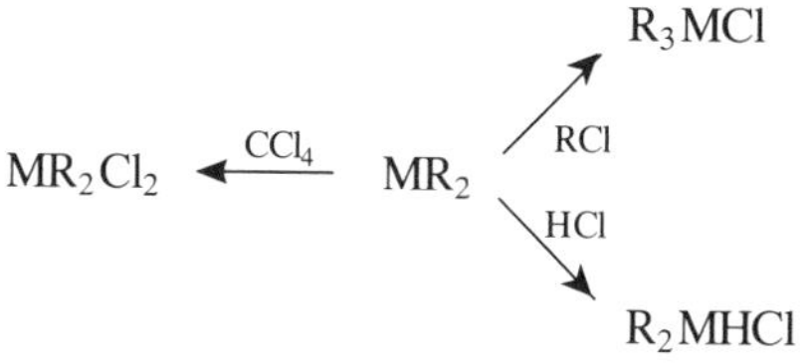

$$MR_2Cl_2 \xleftarrow{CCl_4} MR_2 \begin{array}{c} \nearrow R_3MCl \\ \xleftarrow{RCl} \\ \searrow \\ R_2MHCl \end{array}$$

with the pathways labelled R_3MCl, RCl, HCl, and R_2MHCl

Starting from the stable Sn(II) stannocene, bis-cyclopentadienyltin(II), $(C_5H_5)_2Sn^{105a}$, an interesting organotin compound **67** could be prepared, being the first molecule containing both Sn(II)−C and Sn(IV)−C bonds[105b]:

$$(C_5H_5)_2Sn + 2\,Me_3SnNEt_2 \xrightarrow{PhH} (Me_3SnC_5H_4)_2Sn + 2\,NHEt_2$$
$$(67)$$

Among the metallocenes, essentially only Sn(II) stannocenes in which Sn(II) is sandwiched between two cyclopentadienyl monoanions display a bent geometry[105c]. The first tin(IV) stannocene, where Sn(IV) is sandwiched between organic or organometallic moieties, a bis(carborane)tin(IV) compound having also a bent geometry, has been prepared by the following reaction[143a]:

$$1\text{-Sn(II)-}[2\text{-(SiMe}_3)\text{-3-Me-2,3-C}_2\text{B}_4\text{H}_4] \xrightarrow[\text{TiCl}_4,\ 27\,^\circ\text{C},\ 73\%]{\text{THF/C}_6\text{H}_6}$$
$$1,1'\text{-Sn(IV)}[2\text{-(SiMe}_3)\text{-3-Me-2,3-C}_2\text{B}_4\text{H}_4]_2$$

The compound $1\text{-Sn(II)}[2\text{-(SiMe}_3)\text{-3-Me-2,3-C}_2\text{B}_4\text{H}_4]$, **68**, is a half-sandwiched *closo* compound having a bare tin atom occupying one of the vertices of a pentagonal bipyramid. It can be prepared by a reaction between the dianion $2\text{-(SiMe}_3)\text{-3-Me-2,3-C}_2\text{B}_4\text{H}_4]^{(2-)}$ and $SnCl_4$ after a reductive insertion of the tin atom into the carborane cage[143b]. On the other hand, the compound $1,1'\text{-commo-Sn(IV)}[2\text{-(SiMe}_3)\text{-3-Me-2,3-C}_2\text{B}_4\text{H}_4]_2$ acquires a bent geometry, having as a π-complex Sn(IV) sandwiched between two $[2\text{-(SiMe}_3)\text{-3-}$ Me-2,3-$C_2B_4H_4]$ carborane moieties.

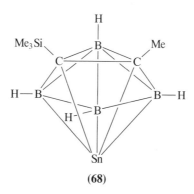

(68)

Polyatomic anions of tin can be prepared in solution, using alloys of tin with alkali metals which are remarkably soluble in liquid ammonia. The obtained colored solutions contain the cluster anion $[Sn_9]^{4-}$, **68a**[106]. Upon treatment with a crown ether in ethylenediamine, crystalline compounds $[Na(crypt)^+]_4[Sn_9]^{4-}$ could be prepared where crypt $= N(CH_2CH_2OCH_2CH_2OCH_2CH_2)_3N$:

$$9\ Sn + 4\ Na + 4\ crypt \xrightarrow[\text{ethylenediamine}]{\text{liq. } NH_3} [Na(crypt)^+]_4[Sn_9]^{4-}$$

The cluster $[Sn_9]^{4-}$ has a monocapped square antiprismatic geometry. On the other hand, the cluster $[Sn_5]^{2-}$, **68b**, has a trigonal bipyramidal geometry, as shown by X-ray crystallography[107].

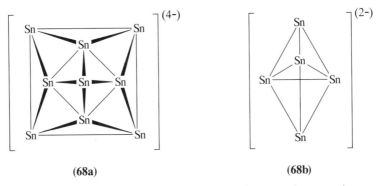

(68a)   (68b)

In the following we describe some special synthesis routes for preparing organotin compounds. Acyltrimethyltin compounds, $RC(O)SnMe_3$, are easily prepared[108] by treating RCOCl (R = Me, Et, Ph, p-Tol, PhCH=CH, PhC≡C, Cl_2CH) with $Me_3SnSnMe_3$ in THF containing $(Ph_3P)_4Pd$ or $(Ph_3P)_2Pd(CH_2Ph)Cl$.

Propenylstannanes are prepared by stannyl cupration of allenes[109]:

$$H_2C=C=CH_2 \xrightarrow{\text{stannylcupration}} $$

while allyltributyltin can be prepared by a transmetallation reaction from allylmagnesium bromide[110]:

$$CH_2=CHCH_2MgBr + Bu_3SnOH \longrightarrow CH_2=CHCH_2SnBu_3 + MgOHBr$$

Tetraalkyltin compounds are prepared by treating R_3Al with $Sn(OAc)_4$ (R = C_{1-18}) in THF[111].

Cyclic stannanes can be generated by reaction of stannylenes with alkynes[112]. For example, bis[bis(trimethylsilyl)methyl]tin reacts with cyclooctyne to provide $\Delta^{1,18}$-9,10-(distanna-9,9,10,10-tetrakis[bis(trimethylsilyl)methyl])bicyclo[6,2,0]-decene, **69**. This reaction is a typical oxidative addition on stannylenes.

Synthesis of polystannacycloalkane macrocycles was accomplished[113] by cyclocondensation of $Me_2Sn(CH_2CH_2CH_2CH_2Cl)_2$ with $NaMe_2Sn(CH_2)_4SnMe_2Na$ in THF and in liquid ammonia to give 1,1,6,6,11,11-hexamethyl-1,6,11-tristannacyclopentadecane, **70**.

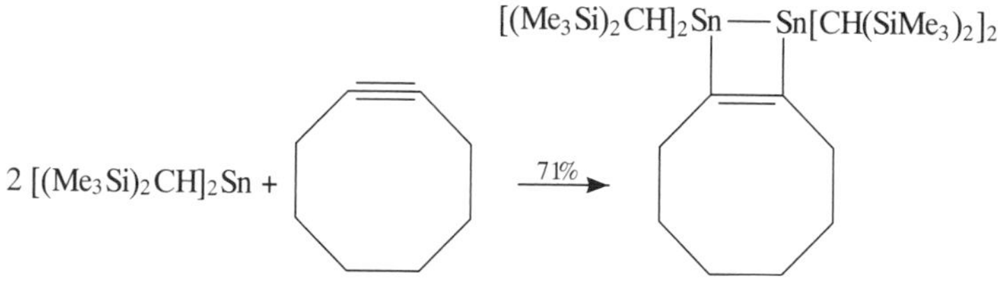

$$[(Me_3Si)_2CH]_2Sn \longrightarrow Sn[CH(SiMe_3)_2]_2$$

$$2\ [(Me_3Si)_2CH]_2Sn\ + \quad\xrightarrow{\ 71\%\ }$$

(69)

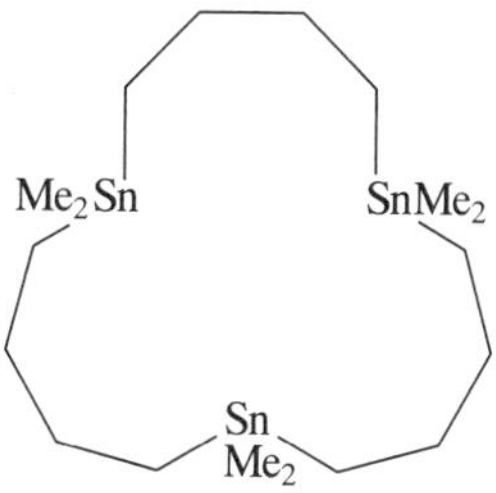

(70)

A similar method of synthesis of large stannacycloalkanes involves the reaction of $Me_2HSn-(CH_2)_3-SnHMe_2$ with Na, followed by $[Cl(CH_2)_3]_2SnMe_2$ in liquid ammonia providing tristannacyclododecane, **71**, in 35% yield[144].

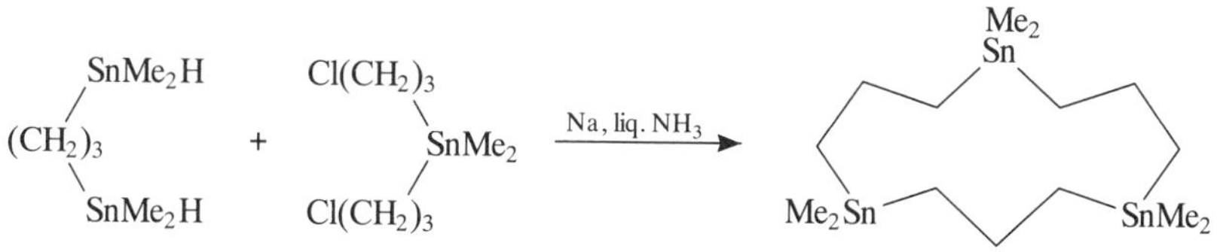

$$
\begin{array}{cc}
SnMe_2H & Cl(CH_2)_3 \\
(CH_2)_3 & + \qquad SnMe_2 \\
SnMe_2H & Cl(CH_2)_3
\end{array}
\xrightarrow{Na,\ liq.\ NH_3}
$$

(71)

For compounds with one or more functionalities, like for example $CH_2=CH(COOEt)NHAc$, the Sn—C bond can be introduced by hydrostannylation with R_3SnH (R = Ph, c-Hex, Bu, Me), providing $R_3SnCH_2CH(COOEt)NHAc$ compounds in good yields (40–80%). Upon halogenation of the phenyl derivative of this class of compounds, $Ph_2ClSnCH_2CH(COOEt)NHAc$ is obtained, whereas hydrolysis with NaOH followed by acidification yielded 75% of the compound[114]

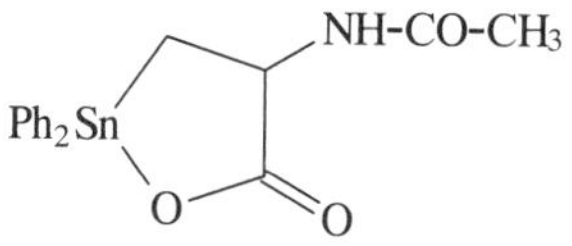

Analogous compounds can be prepared from $ClCH_2CH(COOEt)NHAc$ after reaction with $SnPh_3SnK$ in THF.

Another aspect of the previous procedure is the stannylcupration of N-protected propargylamines[115]. Thus $RNHCH_2C{\equiv}CH$ (R = t-butoxycarbamoyl) reacts with $Bu_3SnCu(Bu)(CN)Li_2$ under very mild conditions and provides $Bu_3SnCH=CHCH_2NHR$ in excellent yields.

Spiro stannyl complexes can be prepared[116] from tetraalkynyltin compounds $Sn(C\equiv CR)_4$ (R = Me, Et, n-Pr, i-Pr, n-Bu) upon reaction with BEt_3. This synthesis route involves a π-coordinated diorganotin compound, **72**, as an intermediate which, upon heating in toluene, gives the spiro compound **73**.

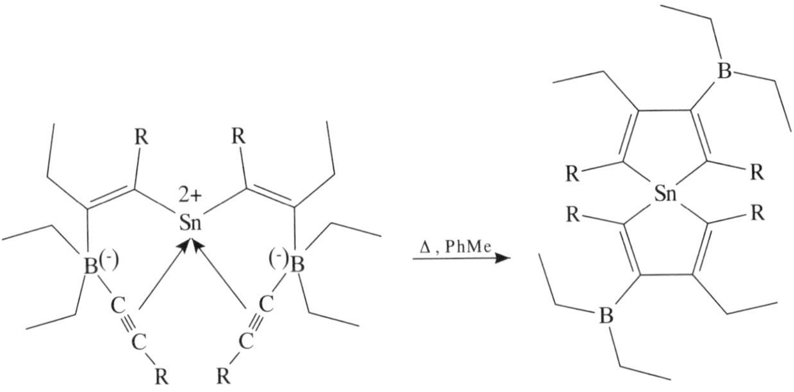

$$\xrightarrow{\Delta,\,PhMe}$$

(72)   (73)

Spirostannanes can also be prepared by organoboration of tetrakis(trimethylsilyl-ethynyl)tin with acyclic trialkylboranes BR_3 to give the stannaspiro[4,4]nona-1,3,6,8-tetraenes[117] **74** (R = ethyl, neopentyl). A stannacyclopentadiene was prepared by this type of organoboration of alkynylstannanes[123]. Treating bis(trimethylsilylethynyl)dimethyltin with a trialkylboron leads to the substituted stannacyclopentadiene **75**.

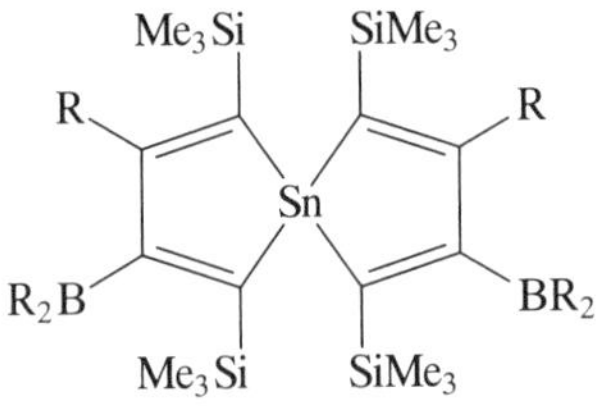

(74)

$$Me_2Sn(C\equiv CSiMe_3)_2 + BR_3 \longrightarrow$$

R = Me, Et, i-Pr, i-Bu   (75)

A Grignard reaction enables the preparation of hexakis(2,4,6-triisopropylphenyl)tris-tannacyclopropane, **76**, obtained from 2,4,6-triisopropylphenylmagnesium bromide and

$SnCl_2$. Cleavage in toluene of the obtained hexaalkyltristannacyclopropane affords $RSn{\equiv}SnR$ and $R_2Sn{=}SnR_2$. Subsequent reaction of $R_2Sn{=}SnR_2$ with phenylacetylene provides the 1,2-distannacyclobutene **77**[118].

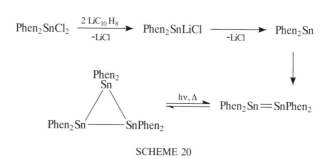

R = 2,4,6-$(Me_2CH)_3C_6H_2$

(76)   (77)

Similar reactions were performed earlier photochemically from **76**, which can be converted into tetrakis(2,4,6-triisopropylphenyl)distannene.

A fast equilibrium between these two compounds is established at room temperature or above[188]. An analogous reaction[189] between di-9-phenanthryltin dichloride, $Phen_2SnCl_2$ and lithium naphthalide at $-78\,^\circ C$ yields hexa-9-phenanthryltristannacyclopropane via a stannylene (Scheme 20).

SCHEME 20

Using alkyllithium compounds, a distannane with a very long Sn–Sn bond, 3.03Å, and a *syn*-peripheral conformation[122] was prepared. Thus, reaction of di-*t*-butyl(chloro)2,4,6-triisopropylphenyl)tin, **78**, with *t*-butyllithium gives 1,1,2,2-tetra-*t*-butyl-1,2-(2,4,6-triisopropylphenyl)ditin, **79**, which displays a restricted rotation about the Sn–C bond as well as the Sn–Sn bond, even at high temperatures.

$$t\text{-}Bu_2ArSnCl + t\text{-}BuLi \longrightarrow$$

Ar = 2,4,6-i-$Pr_3C_6H_2$

(78)   (79)

An elegant method of generating distannanes[145] starts from CH_2I_2 and lithium 4,4'-di-t-butyldiphenyl (YLi_2) in the presence of $MgBr_2$. In a first stage, 1,2-ethylenedimagnesium halide is generated. Stannylation of the latter yields 10 to 40% of $Me_3Sn(CH_2)_n SnMe_3$ (Scheme 21).

$$2\ CH_2I_2 + YLi_2 + 2\ MgBr_2 \xrightarrow{Et_2O} BrMgCH_2CH_2MgBr + 2\ LiBr + YI_2$$

$$\downarrow Me_3SnX$$

$$Me_3SnCH_2CH_2SnMe_3$$

$$(X = I, Br)$$

SCHEME 21

From the easily obtainable triorganostannyl esters, stannylation of benzene rings can be performed[146] using Na or Mg at temperatures lower than 150 °C.

$2\text{-}BrC_6H_4COOSnBu_3$ was added to Mg in THF at 40 °C and yielded, after acid hydrolysis, 45% of the corresponding acid:

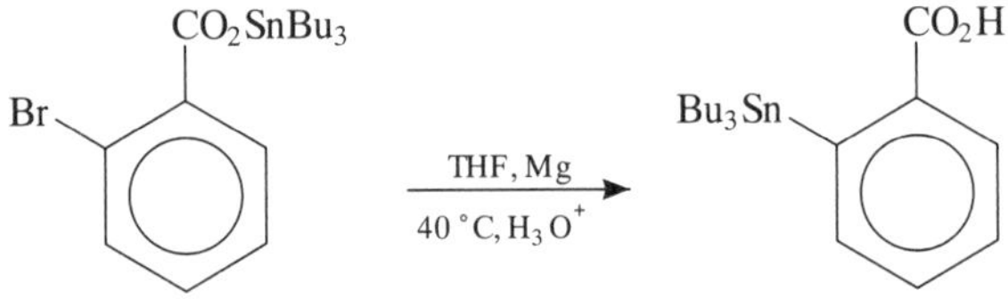

Acetylenic compounds and acetylenic organotin compounds are widely used in the synthesis of many organostannane derivatives of importance in further synthetic procedures.

Stannaalkenes can be prepared from stannalkynes[125] using $Pd(dba)_2/P(OEt)_3$ as catalyst (dba = dibenzylideneacetone):

$$Me_3SnC{\equiv}CR \quad + Me_3SnSnMe_3 \xrightarrow{Pd(dba)_2/P(OEt)_3} (Me_3Sn)_2C{=}CR(SnMe_3)$$
$$R = H\ or\ COOEt$$

This type of addition is very convenient to prepare 1,2-distannyl-1-alkenes[126]:

$$Me_3SnSnMe_3 + RC{\equiv}CH \xrightarrow{Pd(PPh_3)_4} R(Me_3Sn)C{=}C(SnMe_3)H$$
$$(Z\text{-}\ and\ E\text{-}isomers)$$

In addition, 2,3-bis(trialkylstannyl)-1-alkenes can be produced upon the proper addition to an allene:

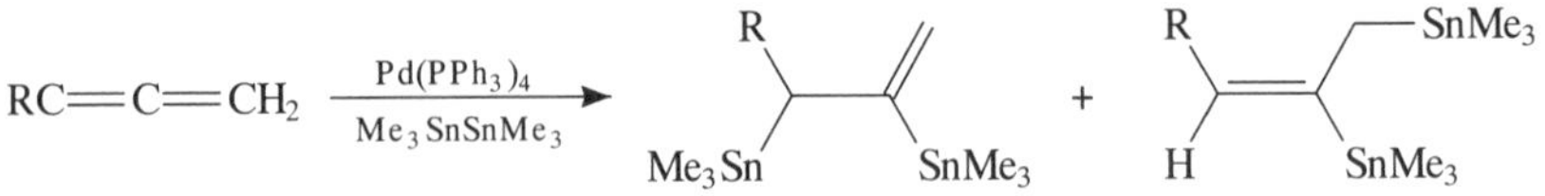

A very elegant method to prepare distannacyclopentenes consists of adding tetrastannacyclohexanes[155] to 1-alkynes and dimethylallene[127] with ring contraction to five-membered rings:

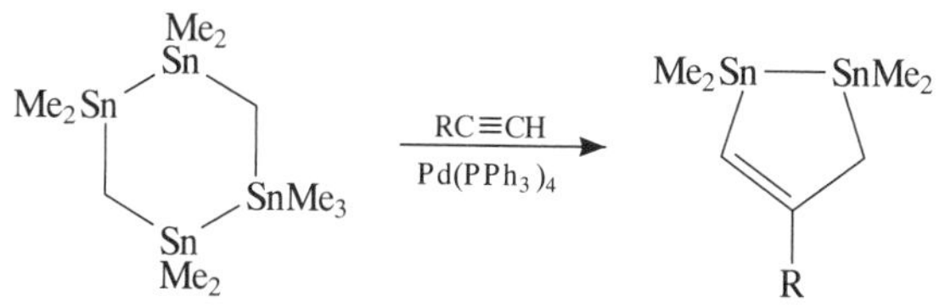

The method can be applied to larger tetrastannacycloalkanes[126]:

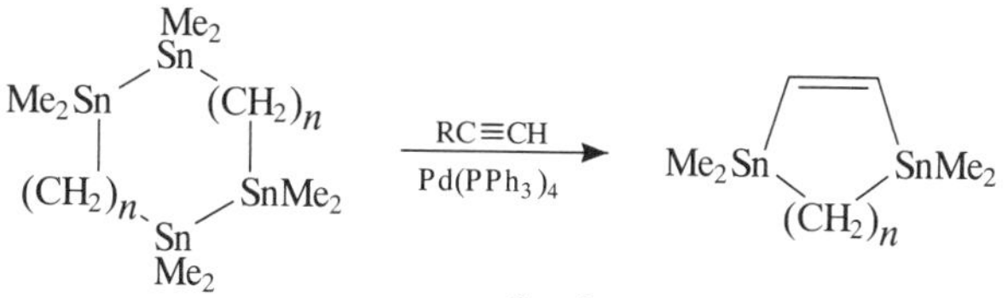

$$n = 3 - 6$$

The use of acetylenic organotin compounds for the preparation of ring derivatives is quite common and general[147].

By regiospecific [4+2] cycloadditions of functional alkenyltin derivatives with dienes such as 1,3-butadiene, 2,3-dimethyl-1,3-butadiene or 1-substituted 1,3-butadienes, polyfunctional cyclic vinyltin compounds **80** are obtained.

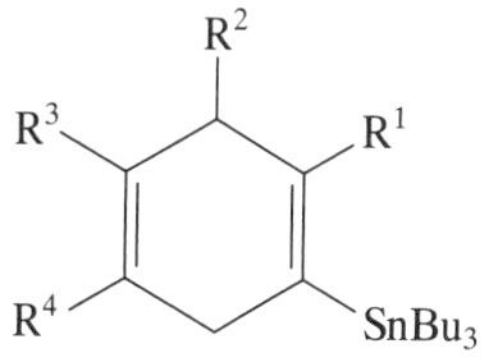

R^1 = COMe, CO_2Me, CN, $CONMe_2$
R^2 = H, Me, Et
R^3, R^4 = cyclopentadienyl

(80)

Hydrostannation refers to the addition of triorganotin hydride to different types of alkenes or alkynes, in the presence of catalysts, making possible the preparation of various types of stannylalkanes or stannylalkenes[130].

Hydrostannation, applied to alkynes with Pd as catalyst, produces vinylstannanes[129,131] as mixtures of two regioisomers, a and b:

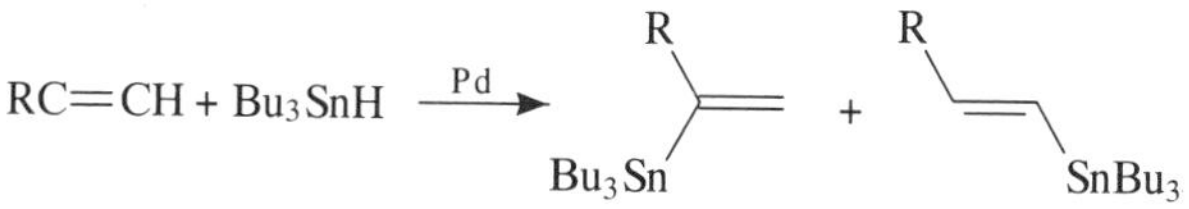

Conjugated diynes can react analogously and give regio- and steroselective products; for instance,

$$RC{\equiv}CC{\equiv}CR + Bu_3SnH \xrightarrow{Pd} RC{\equiv}CC(SnBu_3){=}CRH$$

Bromoalkynes reacting with two equivalents of tributyltin hydride exclusively provide the E-isomer:

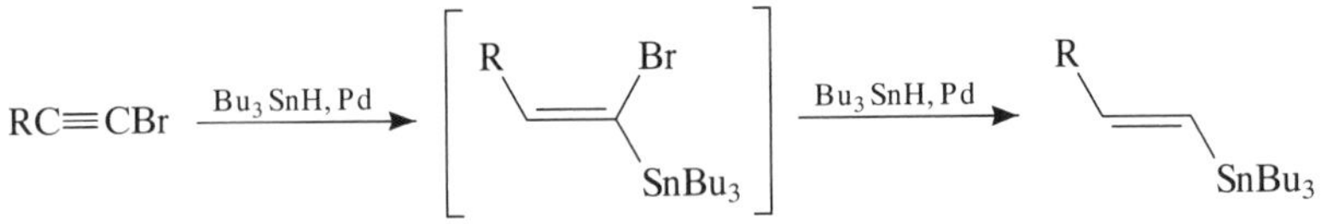

$$RC{\equiv}CBr \xrightarrow{Bu_3SnH, Pd} \left[\underset{SnBu_3}{\overset{R \qquad Br}{\diagdown \diagup}} \right] \xrightarrow{Bu_3SnH, Pd} \underset{SnBu_3}{\overset{R}{\diagdown \diagup}}$$

Using triethylborane, hydrostannation of alkynes can also be performed to achieve the addition of the triphenylstannyl group to the terminal acetylenic carbon. The reaction is regioselective, giving a mixture of E- and Z-1-triphenylstannyl-1-alkenes:

$$RC{\equiv}CH + Ph_3SnH \xrightarrow{Et_3B} RCH{=}CHSnPh_3$$

The mechanism of the addition involves radicals and the reaction can be applied to cyclizations[148].

An efficient method to introduce the trimethylstannyl group to α, β-acetylenic esters $RC{\equiv}CCOOR'$, where R = alkyl, or functionalized alkyl, and R' = Me, Et, involves the trimethylstannylcopper(I) reagent [Me$_3$SnCuSPh]Li. The Me$_3$Sn moiety of this reagent adds to triple bonds, affording a mixture of E- and Z-isomers of trimethylstannyl-α, β-unsaturated esters[149], $Me_3SnC(R){=}CHCOOR'$.

Besides the addition of distannanes to allenes where both Z- and E-isomers are generated, the same type of reaction occurs between $RC{\equiv}CH$ and hexaalkyldistannanes but, in this case, only the Z-isomer is formed.

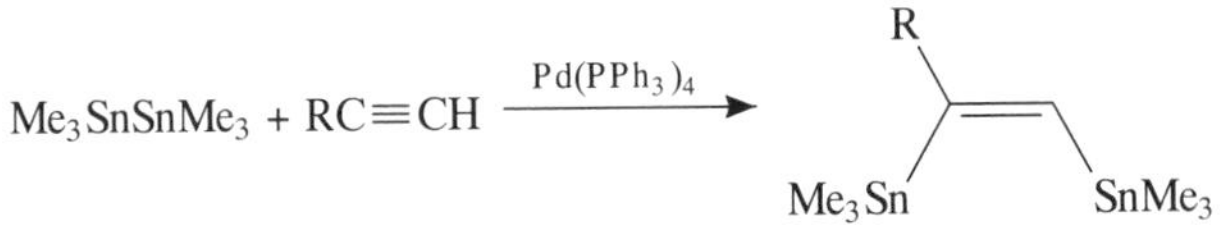

$$Me_3SnSnMe_3 + RC{\equiv}CH \xrightarrow{Pd(PPh_3)_4} \underset{Me_3Sn \qquad\quad SnMe_3}{\overset{R}{\diagdown \diagup}}$$

R = MeOCH$_2$, Me$_2$NCH$_2$, Ph, Me(HO)CH, Me$_2$(HO)C

Halodestannylation reactions of these distannylalkenes were described by Mitchell[126], yielding mainly Z-halostannylalkene compounds of the type **81**:

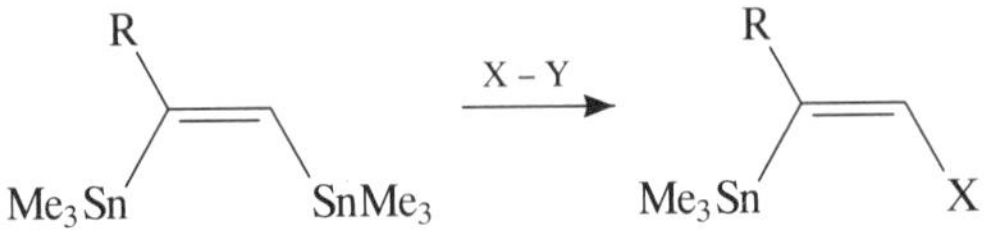

$$\underset{Me_3Sn \qquad\quad SnMe_3}{\overset{R}{\diagdown \diagup}} \xrightarrow{X-Y} \underset{Me_3Sn \qquad\quad X}{\overset{R}{\diagdown \diagup}}$$

(81)

R = MeOCH$_2$, Me$_2$NCH$_2$, Ph, Me(HO)CH, Me$_2$(HO)C
X-Y = Br$_2$, I$_2$, NBS

In contrast, the E-isomer fails to accomplish the same reaction and only in the case where R is MeOCH$_2$ and X-Y = Br$_2$ does a bromodemethylation take place;

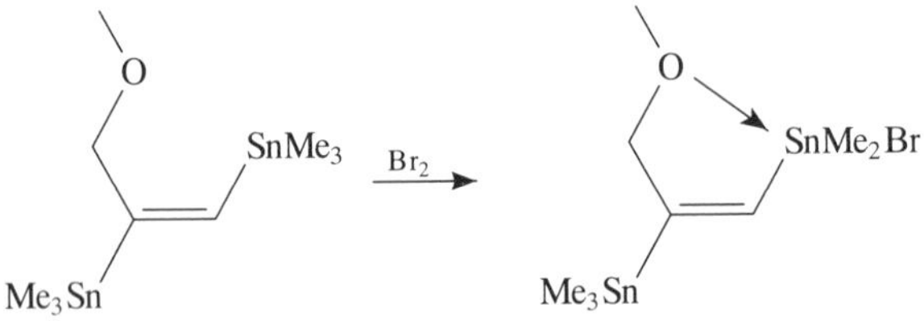

An efficient and easily performed preparation of *Z*-vinyl-stannanes has been reported[150]. The method consists of the hydrozirconation of stannylacetylenes by dicyclopentadienylzirconium chlorohydride, $Cp_2Zr(H)Cl$:

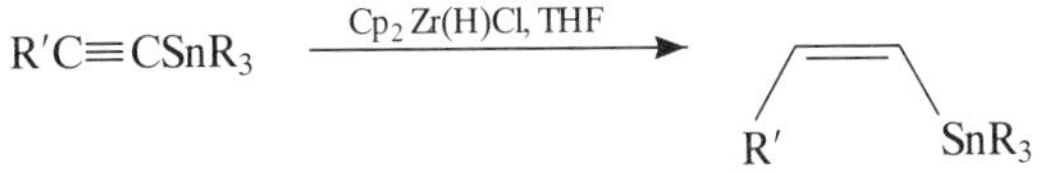
R′C≡CSnR3
Cp2 Zr(H)Cl, THF
R′
SnR3

It is noteworthy that hydrostannation of alkenes and conjugated dienes[128] can be performed either at high pressures, or under intense UV irradiation, or by using initiators like AIBN, as in the case of the hydrostannation of phenylcyclohexene **82**:

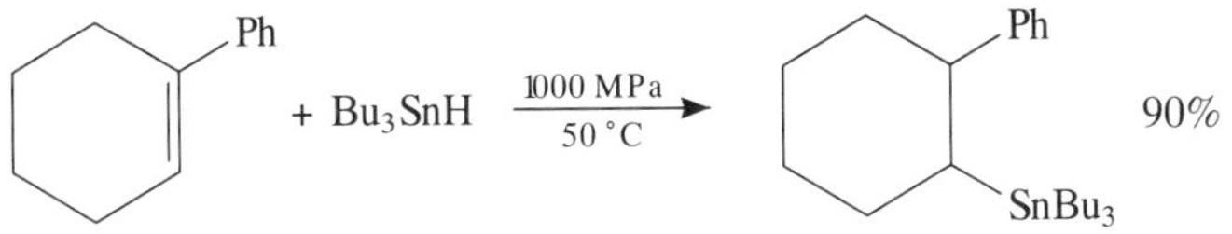
Ph
+ Bu3SnH
1000 MPa
50 °C
Ph
SnBu3
90%

(82)

of 2,3-dimethyl-1,3-butadiene **83**:

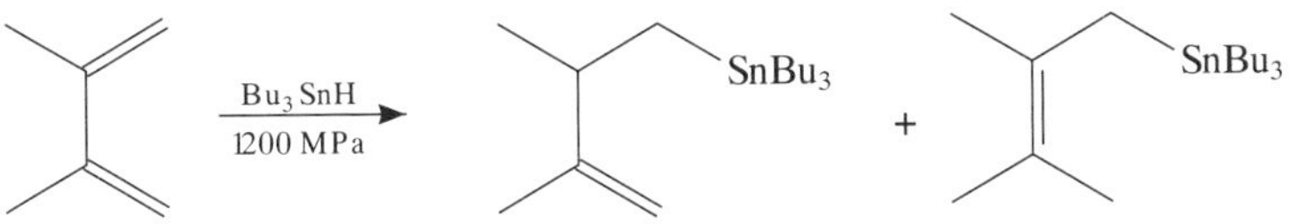
Bu3 SnH
1200 MPa
SnBu3
+
SnBu3

(83) 10% 90%

and of 1,3-menthadiene **84**:

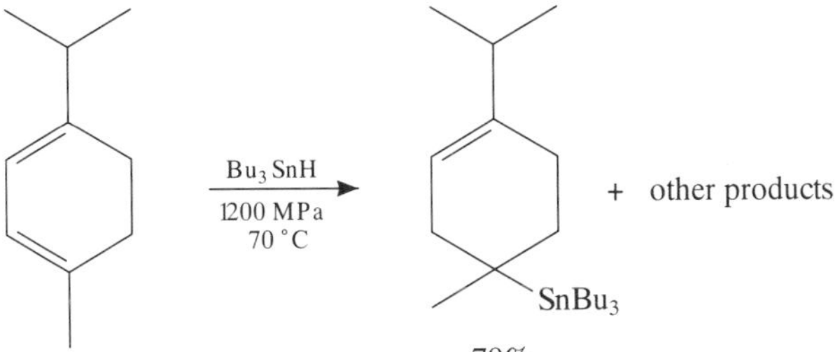
Bu3 SnH
1200 MPa
70 °C
+ other products
SnBu3
70%

(84)

Using tin–nitrogen compounds, further stannylation of already stannylated allenes can be accomplished giving stannylacetylenes[151]:

$$H_2C{=}C{=}CHSnRR'_2 \xrightarrow[\Delta,\ R''=Me,\ Et]{RR'_2Sn{-}NR''_2} RR'_2SnCH_2C{\equiv}CSnRR'_2\ \ 75\% + (RR'_2Sn)_2C{=}C{=}CH_2\ \ 20\%$$

$R = R' = Me, Et$

$R = t\text{-}Bu, R' = Me$

If Br_2 or I_2 in CCl_4 is reacted with $R_3SnCH_2C{\equiv}CSnR_3$, a 1-halo-1-triorganostannylallene, $CH_2{=}C{=}C(X)\text{-}SnR_3$ (X = Br, I), is obtained. The iodo derivative isomerizes into $R_3Sn-C\equiv C-CH_2I$. On the other hand, stannylation of stannylalkynes leads to polystannylallenes[152]:

$$Me_3SnC\equiv CCH_2R \xrightarrow[\Delta]{Me_3Sn-NEt_2} (Me_3Sn)_2C=C=CHR + (Me_3Sn)_2C=C=CR(SnMe_3)$$

Two similar reactions produce other polystannylallenes[153,154]:

$$\begin{array}{c} R_3SnCH=C=CH_2 \\ R=Me, Et \end{array} \xrightarrow[\substack{210-230°C \\ inert\ atm}]{R_3Sn-NEt_2} (R_3Sn)_2C=C=C(SnR_3)_2$$

$$Me_3SnC\equiv CCH_2X \xrightarrow[\Delta]{2\ Me_3Sn-NEt_2} (Me_3Sn)_2C=C=C(SnMe_3)X$$

or, depending on the molar ratio of the reactants,

$$Me_3SnC\equiv CCH_2X \xrightarrow[\Delta]{1\ Me_3Sn-NEt_2} (Me_3Sn)_2C=C=CHX$$

$$X = SEt, SiMe_3, GeMe_3, SnMe_3$$

C. Organoplumbanes

For the preparation of organoplumbanes only lead(II) compounds were used, mainly $PbCl_2$. The use of $PbCl_4$ or K_2PbCl_6 is excluded because these compounds are strong oxidizing agents and are easily transformed into Pb(II) by the organic or organometallic compounds used or generated during the synthesis. Industrial preparations of lead-organometallics include the well known synthesis of tetraalkyllead compounds achieved by alkylation of lead–alkali metal alloys by alkyl halides, the electrolysis of tetraethyla-luminate or -borate, or of Grignard reagents with lead anodes.

The main synthetic routes for the preparation of organoplumbanes are the alkylation of lead(II) chloride by organo-magnesium, -lithium, -aluminum and -boron compounds[156].

Tetraalkyllead compounds can also be used as starting materials for the synthesis of functionalized and specific organoplumbanes. In the latter case, cleavage of one or more of the attached alkyl groups is achieved so as to introduce subsequently new groups, different from the original ones.

Characteristic procedures for this purpose are[157−159]:

$$C_6F_5MgBr \xrightarrow[Et_2O]{Ph_3PbCl} C_6F_5PbPh_3$$

$$C_6F_5MgBr \xrightarrow[Et_2O]{PbCl_2} (C_6F_5)_4Pb$$

$$C_5H_5Na \xrightarrow[PhH,\ N_2]{R_3PbCl} C_5H_5PbR_3 \qquad R = Me,\ Et$$

$$Me_3CCH_2MgCl \xrightarrow[Et_2O,\ -10°C]{PbCl_2} (Me_3CCH_2)_3Pb-Pb(CH_2CMe_3)_3$$

A useful reagent for the preparation of a number of organoplumbanes is triphenylplumbyllithium, Ph_3PbLi. This compound is prepared as follows:

$$\text{PhLi} \xrightarrow[-10\,^\circ\text{C}]{\text{PbCl}_2,\ \text{Et}_2\text{O}} \text{Ph}_3\text{PbLi}$$

The corresponding Me_3PbLi is prepared differently[179]:

$$\text{Me}_3\text{PbBr} \xrightarrow[-78\,^\circ\text{C}]{\text{Li, THF}} \text{Me}_3\text{PbLi}$$

Reaction of Ph_3PbLi with H_2O_2 at $0\,^\circ\text{C}$ generates the red compound $(\text{Ph}_3\text{Pb})_4\text{Pb}$. The use of Ph_3PbLi as plumbation agent is illustrated by the two following reactions[160,161]:

$$\text{Ph}_3\text{PbLi} \xrightarrow[\text{THF, } -60\,^\circ\text{C}]{\text{CCl}_4} (\text{Ph}_3\text{Pb})_4\text{C}$$

$$\text{Ph}_3\text{PbLi} \xrightarrow[\text{THF, } -60\,^\circ\text{C}]{\text{PhCH}_2\text{Cl}} \text{Ph}_3\text{PbCH}_2\text{Ph}$$

Organolead compounds with functionalized alkyl or aryl groups directly bound to the lead atom are useful synthetic tools in organolead chemistry.

Organolead compounds with a carboxylic acid function bound to the alkyl group can be prepared by addition of Ph_3PbOH to ketenes[162]:

$$\text{Ph}_3\text{PbOH} + \text{CH}_2{=}\text{C}{=}\text{O} \longrightarrow \text{Ph}_3\text{PbCH}_2\text{COOH}$$

Upon reaction with organic halides, trialkylplumbylsodium[163] or triphenylplumbyllithium[164] are converted into useful organoplumbanes with additional functionalities, such as

$$\text{Ph}_3\text{PbNa} \xrightarrow{\text{Cl(CH}_2)_2\text{NEt}_2} \text{Ph}_3\text{Pb(CH}_2)_2\text{NEt}_2$$

$$\text{Ph}_3\text{PbLi} \xrightarrow[n = 3-5]{\text{Br(CH}_2)_n\text{Br}} \text{Ph}_3\text{Pb(CH}_2)_n\text{Br}$$

The above type of reaction can also be used to synthesize bis-triorganoplumbylalkanes[163,164]:

$$\text{Ph}_3\text{PbLi} \xrightarrow{\text{CH}_2\text{Br}_2} (\text{Ph}_3\text{Pb})_2\text{CH}_2$$

$$\text{Ph}_3\text{PbLi} \xrightarrow[n = 3-5]{\text{Br(CH}_2)_n\text{Br}} \text{Ph}_3\text{Pb(CH}_2)_n\text{PbPh}_3$$

The reaction of Ph_3PbLi with various heterocycles leads to interesting functionalized organolead compounds[165], e.g.

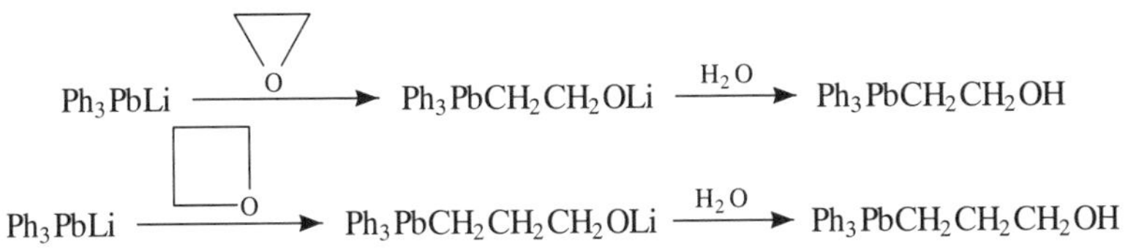

$$\text{Ph}_3\text{PbLi} \xrightarrow{\triangle\text{O}} \text{Ph}_3\text{PbCH}_2\text{CH}_2\text{OLi} \xrightarrow{\text{H}_2\text{O}} \text{Ph}_3\text{PbCH}_2\text{CH}_2\text{OH}$$

$$\text{Ph}_3\text{PbLi} \xrightarrow{\square\text{O}} \text{Ph}_3\text{PbCH}_2\text{CH}_2\text{CH}_2\text{OLi} \xrightarrow{\text{H}_2\text{O}} \text{Ph}_3\text{PbCH}_2\text{CH}_2\text{CH}_2\text{OH}$$

$$Ph_3PbLi \xrightarrow{\quad} Ph_3PbCH_2CH_2COOLi \xrightarrow{MeOH} Ph_3PbCH_2CH_2COOMe$$

Alkenyl- or alkynyllead compounds can be prepared by specific methods, some of which can be generalized.

The vinyllead compounds can be prepared using Grignard reagents:

$$R_2PbCl_2 \xrightarrow[Et_2O]{H_2C=CHMgBr} R_2Pb(CH=CH_2)_2$$

$$R = Et[166], Ph[167]$$

$$Ph_3PbCl \xrightarrow[Et_2O]{H_2C=CHMgBr} Ph_3PbCH=CH_2 [168,169]$$

Stable p-allylphenyllead compounds can be prepared from p-allyphenylmagnesium bromide and phenyllead chlorides of the general type $Ph_{4-n}PbCl_n$ ($n = 1-3$)[171,172]:

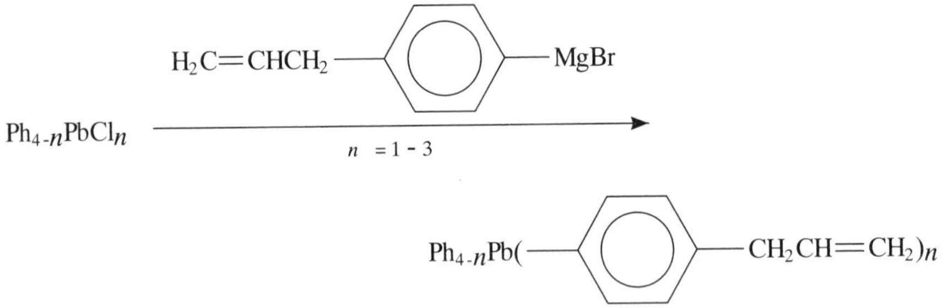

$$Ph_{4-n}PbCl_n \xrightarrow{H_2C=CHCH_2-\bigcirc-MgBr}_{\quad n\ =1\ -\ 3\quad} Ph_{4-n}Pb(-\bigcirc-CH_2CH=CH_2)_n$$

Allyllead compounds, which are very unstable, can be prepared from tetraethyllead by reaction with 2-methyl-3-chloropropene[170]:

$$Et_4Pb \xrightarrow{CH_2=C(CH_3)CH_2Cl} Et_3PbCH_2C(CH_3)=CH_2$$

Besides the early preparation by Gilman[173] of alkynyllead compounds using sodium acetylides and triorganolead halides, and the following modification of the method[174]:

$$R'_3PbX \xrightarrow{MC\equiv CR} R'_3PbC\equiv CR$$

$$R' = alkyl; R = alkyl, aryl, H$$

several alkynyllead compounds were prepared using alkynyl-metal compounds including alkynyl Grignard reagents[175]:

$$Ph_2PbCl_2 \xrightarrow{Li-C\equiv C-Ph} Ph_2Pb(C\equiv CPh)_2$$

$$Na_2PbCl_6 \xrightarrow{\text{Na}-C\equiv C-Ph} Ph(C\equiv CPh)_4$$

$$Ph_3PbCl \xrightarrow{\text{ClMg}-C\equiv C-Ph} Ph_3PbC\equiv CR$$

Trimethyllead hydroxides or methoxides, sometimes used in direct reactions with acetylenic derivatives for the preparation of alkynylplumbanes, are very suitable reagents[176]:

$$Me_3PbOMe \xrightarrow[\text{PhH, 80}^\circ C]{\text{H}-C\equiv C-H} Me_3PbC\equiv CPbMe_3$$

Alkynyllead compounds were also prepared by the decarboxylation of triphenylplumbyl esters of alkyne carboxylic acids[176]:

$$Ph_3PbCl \xrightarrow{\text{PhC}\equiv CCOONa} PhC\equiv CCOOPbPh_3 \xrightarrow[112^\circ C]{-CO_2} PhC\equiv CPbPh_3$$

The catenation ability of lead being considerably lower than that of germanium or tin, plumbanes with Pb—Pb bonds are quite difficult to prepare, even though a number of such compounds have been synthesized. As already pointed out, the main method for the preparation of this type of compound is the reaction of an appropriate Grignard reagent with $PbCl_2$[159], e.g.

$$PhMe_2CCH_2MgI \xrightarrow[\text{Et}_2O, -10^\circ C]{PbCl_2} [(PhMe_2CCH_2)Pb]_2$$

Another preparative route leading to diplumbanes is the reaction of triphenylplumbyl lithium with triphenyllead chloride[177]:

$$Ph_3PbLi + Ph_3PbCl \xrightarrow{\text{Et}_2O} Ph_3PbPbPh_3$$

A very satisfactory method to prepare R_6Pb_2 compounds is the following[178]:

$$R_3PbOMe + B_2H_6 \xrightarrow{\text{Et}_2O} R_3PbBH_4 \xrightarrow{\text{MeOH}} R_3PbH + B(OMe)_3$$
$$\downarrow$$
$$R_3PbPbR_3$$

By organoboration of bis(alkynyl)plumbanes **85**, the compound **86**, stabilized by a π-bonding interaction, is prepared[180]. (Scheme 22a). The cation **86** is unstable and, at room temperature, decomposes rapidly to rearrange into a 1,4-plumbabora-2,5-cyclohexadiene, **87**.

If **86** is reacted with an excess of BR_3, 2,5-bis(dialkylboryl)-3-plumbolene compounds, **88**, are obtained.

The compound **86** in which R = *i*-Pr and R′ = R″ = Me[181] decomposes and rearranges intramolecularly to product **89** which, upon reaction with methanol, gives **90**. In the same way[182], the reaction of bis(trimethylsilylethynyl)dimethyllead **91** with trialkylboranes gives 3-(dialkylboryl)-4-alkyl-2,5-bis(trimethylsilyl)-1,1-dimethylplumbole **93** via the intermediate **92** (Scheme 22b):

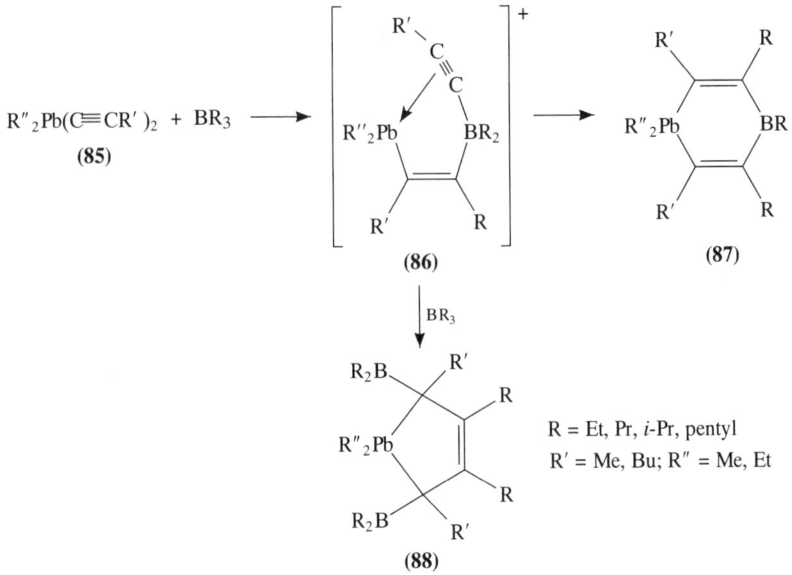

$R''_2Pb(C\equiv CR')_2$ + BR_3 ⟶

(85)

(86)

⟶

(87)

R_2B, R'

(88)

R = Et, Pr, *i*-Pr, pentyl
R' = Me, Bu; R'' = Me, Et

SCHEME 22a

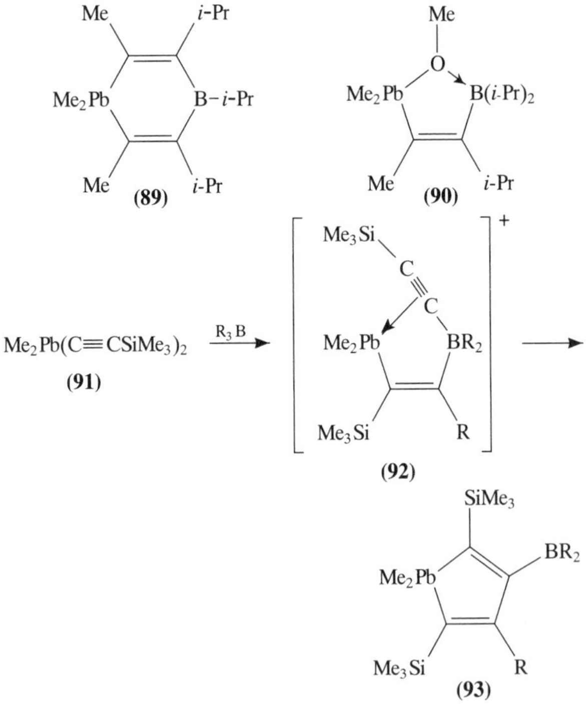

(89)

(90)

$Me_2Pb(C\equiv CSiMe_3)_2$ $\xrightarrow{R_3B}$

(91)

(92)

(93)

SCHEME 22b

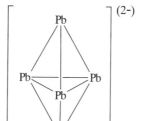

(94)

A crystalline compound $[Na(crypt)]^+_2Pb_5]^{2-}$ has been prepared[183] in which $[Pb_5]^{2-}$, **94**, is a trigonal bipyramidal cluster the structure of which has been characterized by X-ray diffraction, with Pb–Pb distances of 3.00–3.22 Å [crypt = $N(CH_2CH_2OCH_2CH_2OCH_2CH_2)_3N$]. See Section I.B for analogous tin compounds.

This cluster is prepared from the reaction of an alloy of Pb with an alkali metal, dissolved in liquid ammonia with the crown ether, crypt, dissolved in ethylenediamine.

Hydroplumbation of unsaturated compounds is more difficult than hydrogermylation and hydrostannylation. Actually, olefins are unreactive towards $HPbBu_3$, except when they are activated. Under such conditions they are successfully subjected to hydroplumbation in ether at $0\,^\circ C$[184]:

$$Bu_3PbH \xrightarrow{H_2C=CH-\Sigma} Bu_3PBCH_2CH_2\text{-}\Sigma \qquad \Sigma = CN, CO_2Me, Ph$$

Activated acetylenes are definitely more reactive towards hydroplumbation than alkenes[185]:

$$Bu_3PbH \xrightarrow[-50\,^\circ C]{MeO_2CC\equiv CCO_2Me} Bu_3Pb(MeO_2C)C=CH(CO_2Me)$$

$$Bu_3PbH \xrightarrow[-70\,^\circ C]{HC\equiv CCN} Bu_3PbCH=CHCN \qquad 91\% \textit{ cis } \text{and } 9\% \textit{ trans}$$

Allenes bearing Me_3Pb groups are prepared[247] from alkylboranes containing organometallic substituents $[Et_2BC(Et)=C(R)MMe_3$, with M = Sn, R = H, Me; M = Pb, R = CH_3, $PbMe_3$], after reaction with alkynyllead compounds of the type $Me_3PbC\equiv CR'$ (R' = Me, CMe_3, $PbMe_3$):

$$Et_2BC(Et)=C(R)MMe_3 + Me_3PbC\equiv CR' \longrightarrow$$

$$Me_3PbCR'=C=C(Et)C(BEt_2)(R)(MMe_3)$$

In the case R' = Me, 3-borolenes are prepared:

Plumbanes with particular or bulky substituents can be prepared by several special procedures. For example, tetra-p-nitrophenyllead can be prepared by a disproportionation reaction, induced by hydrazine, of Ar_2PbX_2, where $Ar = p\text{-}NO_2C_6H_4$, $X = Cl$, Br, I[190]:

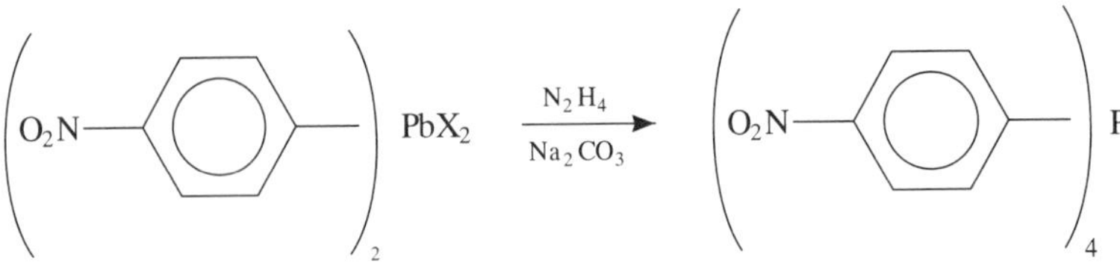

$$\left(O_2N-\!\!\!\left\langle \bigcirc \right\rangle\!\!\!- \right)_2 PbX_2 \xrightarrow[Na_2CO_3]{N_2H_4} \left(O_2N-\!\!\!\left\langle \bigcirc \right\rangle\!\!\!- \right)_4 Pb$$

Transition metal carbonyl derivatives are easily prepared[191]:

$$Ph_2PbCl_2 \xrightarrow{NaMn(CO_5)} Ph_2Pb[Mn(CO)_5]_2$$

Boron-containing derivatives are obtained by the sequence of the following reactions[192]:

$$Ph_3PbCl + C[B(OMe)_2]_4 \xrightarrow[MeOLi]{BuLi\ or} Ph_3PbC[B(OMe)_2]_3 \xrightarrow[30\%]{THF} (Ph_3Pb)_2C[B(OMe)_2]_2$$

The reaction can also take place with Ph_3SnCl.
The monolead compound disproportionates extremely rapidly which makes its isolation almost impossible.
Alkyllead-substituted organosilicon compounds can be prepared as follows[193]:

$$RCl \xrightarrow[PbCl_2]{Mg} R_3PbMgCl \xrightarrow{ClCH_2SiMe_3} R_3PbCH_2SiMe_3$$
$$R = Me,\ Et$$

[Tris(trimethylsilyl)methyl]trimethyllead, $[(Me_3Si)_3C]PbMe_3$, was prepared analogously. It can be mono- or dihalogenated without any cleavage of the $Pb-C(SiMe_3)_3$ bond, to yield $[(Me_3Si)_3C]PbMe_2X$ or $[(Me_3Si)_3C]PbMeX_2$ ($X = Cl$, Br)[194].
The first compound which was proven to contain a direct silicon–lead bond has been synthesized as follows[195]:

$$Mg(SiMe_3)_2 \xrightarrow[Et_2O,\ -78\,^\circ C]{PbCl_2} Pb(SiMe_3)_4$$
$$\textbf{(94a)}$$

Tetrakis(trimethylsilyl)lead **94a** can be isolated as pale yellow crystals moderately sensitive to light, soluble in hydrocarbons and ethers, decomposing above $80\,^\circ C$.
Cycloaddition reactions are used to prepare organolead compounds with functionalized substituents. (Triphenylplumbyl)pyrazole, **95**, can be prepared from triorganoplumbylalkynes by 1,3-dipolar cycloadditions with diazomethane[196]:

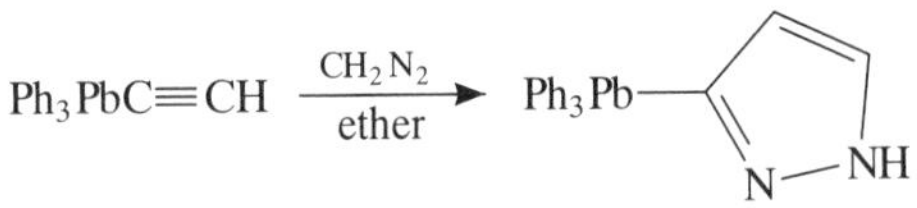

$$Ph_3PbC\!\equiv\!CH \xrightarrow[ether]{CH_2N_2} Ph_3Pb-\!\!\left[\!\!\begin{array}{c}\diagup\diagdown \\ N-\!NH\end{array}\!\!\right]$$

$$\textbf{(95)}$$

Trimethyllead diazoacetic acid ethyl ester reacts with activated alkenes or acetylenes to provide organolead pyrazoles or pyrazolines[197]:

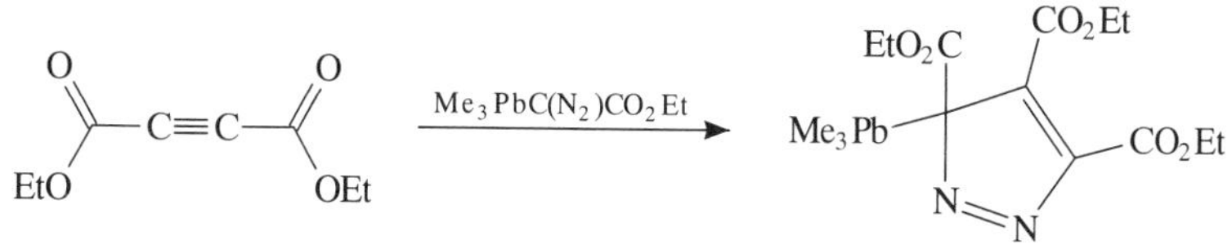
O
O
C≡C
EtO
OEt
$Me_3PbC(N_2)CO_2Et$
EtO_2C
CO_2Et
Me_3Pb
CO_2Et
N
N

The diazoacetic acid ethyl ester is prepared by the following reaction[197]:

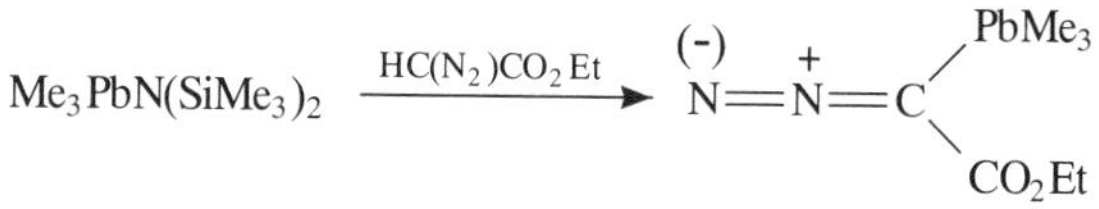
$Me_3PbN(SiMe_3)_2$
$HC(N_2)CO_2Et$
(-)
+
N=N=C
$PbMe_3$
CO_2Et

Organolead tetrazoles, for instance **96**, have been prepared from Ph_3PbN_3, CS_2 and RC_6H_4NCS (R = H, *p*-Br, OMe, *p*-Me)[372]:

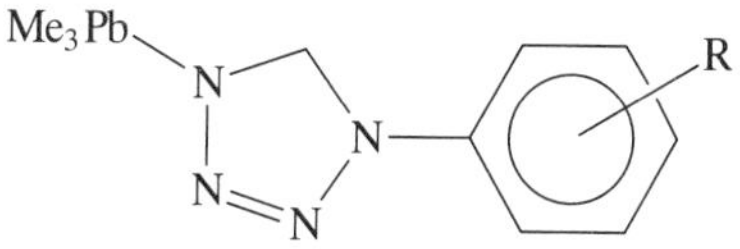
Me_3Pb
N
N
N
N
R

(96)

Ph_3PbLi reacts with Ph_3SnCl or Ph_3GeCl and gives rise to $Ph_3PbSnPh_3$ or $Ph_3PbGePh_3$, respectively, but there is no reaction with Ph_3SiCl. When Ph_3SiLi reacts with Ph_3PbCl, Ph_6Pb_2 and Ph_6Si_2 are formed[198].

There is increasing interest in the preparation of mixed alkyllead compounds or alkylplumbanes involving functional groups.

For example, trialkylcyclopentadienyllead compounds, $R_3Pb(C_5H_5)$, and their dialkyldicyclopentadienyl analogs, $R_2Pb(C_5H_5)_2$ (R = Me, Et), can be prepared from the corresponding methyl or ethyl chloroplumbanes in absolute benzene at room temperature, after reaction with cyclopentadienylsodium under nitrogen[199].

A stereodivergent synthesis of derivatives of 1,2-diols with an S_E2-retention pathway can be performed via α-alkoxyorganolead compounds.

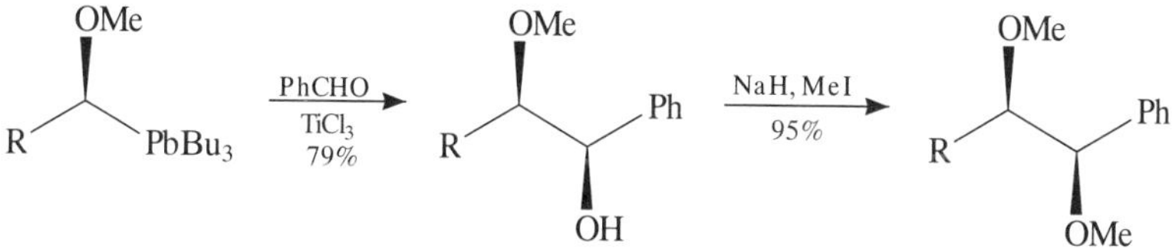
OMe
R
$PbBu_3$
PhCHO
$TiCl_3$
79%
OMe
R
Ph
OH
NaH, MeI
95%
OMe
R
Ph
OMe

These compounds can also be prepared from the corresponding α-methoxy-stannanes[200]:

$$\underset{R = Ph,\ C_7H_{15}}{R(MeO)CHSnBu_3} \xrightarrow[-78\,^{\circ}C]{BuLi,\ Bu_3PbBr} R(MeO)CHPbBu_3$$

It is worth noting the synthesis of α-alkoxyalkyltributyllead compounds by reaction of tributylplumbyllithium with α-chloroethers[377]:

$$R'OCH_2Cl \xrightarrow[THF]{Bu_3PbLi} R'OCH_2PbBu_3$$

$$R' = CH_3,\ CH_2Ph,\ CH_2CH{=}CH_2$$

This procedure is more convenient than the conventional method, which involves a trans-metalation reaction from an α-alkoxyalkyltrialkyltin(IV).

Bu$_3$PbLi adds to enones and generates γ-oxoorganolead compounds:

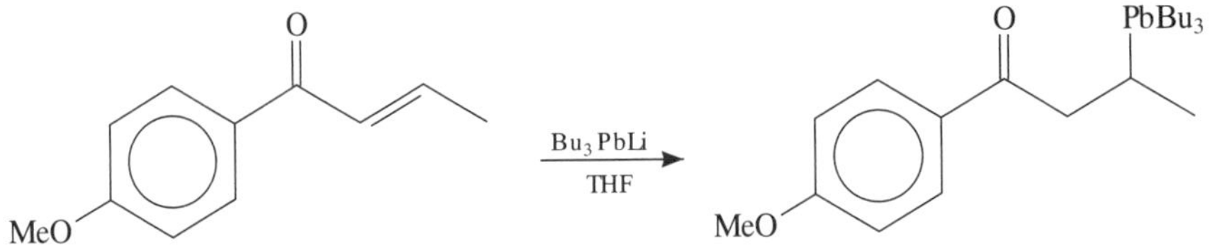

Mixed alkyl plumbanes can be prepared via trialkylplumbylmagnesium chlorides[378] by the sequence of reactions

$$PbX_2 \xrightarrow[THF, 5\,^\circ C]{RMgCl} R_2Pb$$

$$R_2Pb \xrightarrow[THF]{RMgCl} R_3PbMgCl$$

$$R_3PbMgCl \xrightarrow[THF]{R'Cl} R_3PbR'$$

Analogous reactions were also used for the preparation of silyl derivatives, through reaction with (chloromethyl)trimethylsilane:

$$Me_3PbMgCl \xrightarrow{Me_3SiCH_2Cl} Me_3PbCH_2SiMe_3$$

which had previously been prepared by the reaction[379]

$$Me_3SiCH_2MgCl \xrightarrow{Me_3PbCl} Me_3PbCH_2SiMe_3$$

A similar method was also proposed by Willemsens and van der Kerk[380] for the preparation at hexaryldileads in high yields:

$$PbCl_2 \xrightarrow{ArMgX} Ar_3PbMgX$$

$$2Ar_3PbMgX + XCH_2CH_2X \longrightarrow Ar_6Pb_2 + 2MgX_2 + CH_2{=}CH_2$$

$$Ar = Ph,\ o\text{-},\ m\text{-},\ p\text{-} \ Tol, p\text{-An} \ 1\text{-},\ 2\text{-Naph}$$

The synthesis of tetrakis[4-(1-phenylvinyl)phenyl]lead **97** can be performed as follows[381]:

$$PbCl_2 \xrightarrow{ArMgBr} Ar_3PbMgBr$$

$$Ar_3PbMgBr + ArMgBr + BrCH_2CH_2Br \longrightarrow Ar_4Pb + 2MgBr_2 + CH_2{=}CH_2$$

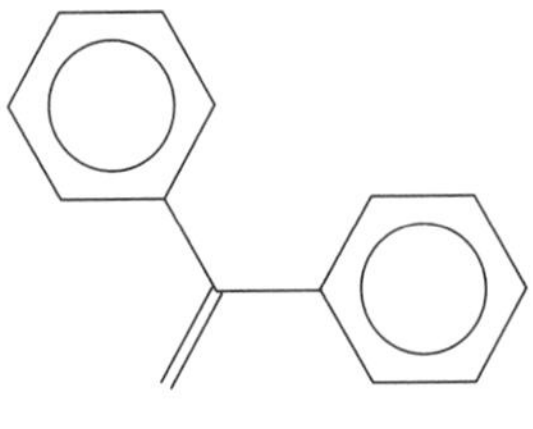

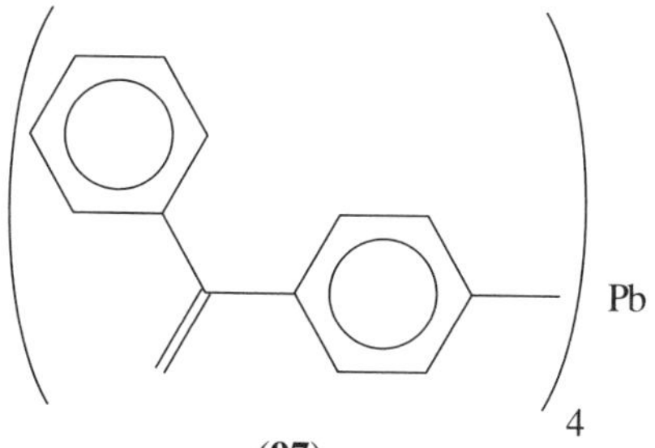

(97)

This compound **97** can be easily polymerized.

Some useful mixed ethylmethyllead compounds $PbMe_{4-n}Et_n$ $(1 \leqslant n \leqslant 3)$ can be prepared by exchanging alkyl groups between tetraalkylead and either trialkylaluminum compounds or dialkylaluminum chlorides. Exchange occurs between $PbMe_4$ and Al_2Et_6, $Al_2Et_4Cl_2$, $Al_2Et_3Cl_3$ or $Al_2Et_2Cl_4$ as well as between $PbEt_4$, Al_2Me_6 or $Al_2Me_4Cl_2$, $Al_2Me_3Cl_3$ or $Al_2Me_2Cl_4$[364]. It is worth pointing out the importance of PbR_4 for alkylations[365], as well as the new type of alkylating reagent $PbR_4 + TiCl_4$[366].

Insertion of thienyl or furyl groups into alkyllead compounds can be carried out using organolithium reagents. Thus, organolead(IV) tricarboxylates can be prepared as follows via tin(IV) precursors:

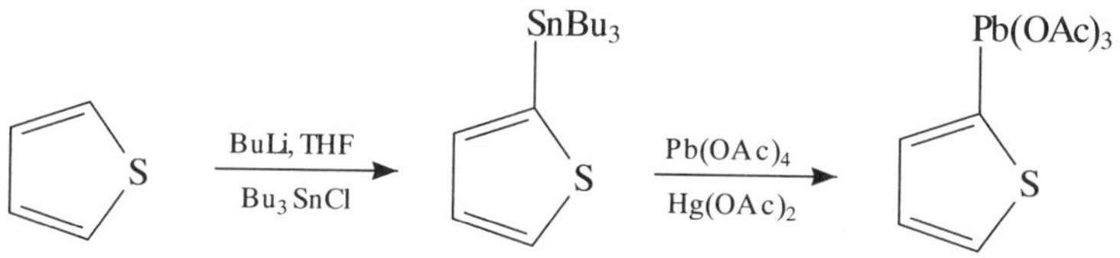

The special reagent used for the replacement of Sn(IV) by Pb(IV) is lead tetraacetate in the presence of mercury diacetate acting as a catalyst[367a]. The corresponding 2-furyl compounds can be prepared analogously:

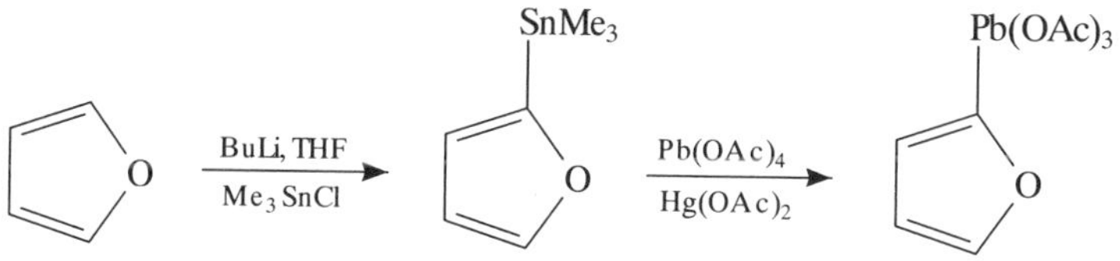

Applying this method, 2-thienyllead triacetate, 2-thienyllead tribenzoate, 3-thienyllead triacetate, 2-furyllead triacetate and 3-furyllead triacetate were prepared[367a]. All these compounds are useful reagents in heterocyclic chemistry[367b,367c].

Lead tetraacetate can also be used for the preparation of aryllead(IV) triacetates by Hg(II)-catalyzed arylation of $Pb(OAc)_4$ from $ArSnBu_3$[373]:

$$ArSnBu_3 \xrightarrow[\text{CHCl}_3,\ 70-80\%]{Pb(OAc)_4,\ Hg(OAc)_2} ArPb(OAc)_3$$

Thiophene derivatives like 3-(trimethylplumbyl)thiophene, **98**, can be synthesized by the reaction of Me_3PbX or Me_2PbX_2 (X = Cl, Br, I), with 3-lithiothiophene or 2,3-dilithiothiophene[368]:

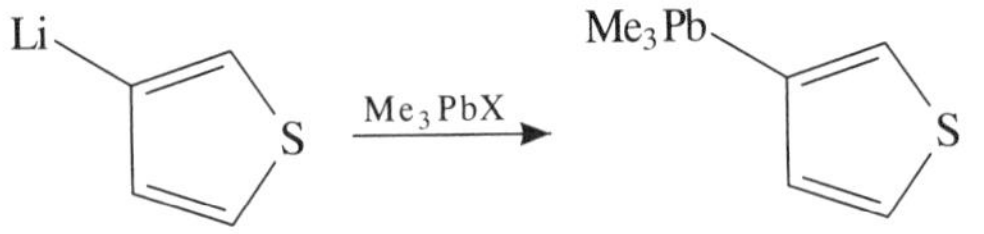

(98)

These thiophenyllead derivatives can be easily electropolymerized.

The dianion of 2,3-dimethylbutadiene reacts with Me_3PbCl to yield 2,3-bis(trimethyl-plumbylmethyl)butadiene in 30% yield[382]; metalation of 2,3-dimethyl-1,3-butadiene to generate the dianion can be carried out by t-BuOLi:

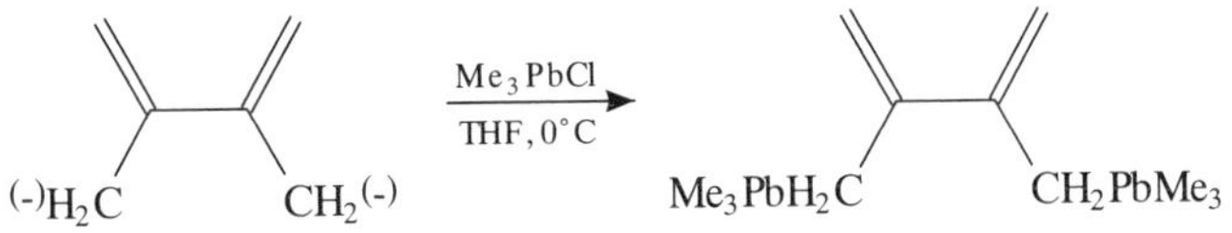

Unsymmetrical dilead derivatives have been synthesized from R_3PbCl and R'_3PbLi at $-60\,°C$ in THF although migration of R or R' does occur. However, the crystal structure of $Pb_2Ph_3(p$-Tol$)_3$ exhibits two independent molecules of $Ph(p$-Tol$)_2PbPb(p$-Tol$)Ph_2$[369] in the crystal lattice. On the other hand, (tricyclohexylstannyl)tricyclohexyllead $(C_6H_{11})_3Sn-Pb(C_6H_{11})_3$ was prepared from $(C_6H_{11})_3PbLi$ and $(C_6H_{11})_3SnCl$ in THF at $-50\,°C$[370]. The mixed Sn/Pb compound decomposes at room temperature.

Organolead compounds containing a lead-transition metal bond are also known. When a lead(II) or lead(IV) compound reacts with $Co_2(CO)_6L_2$ complexes (L = $tert$-phosphine, -arsine or a phosphite), the blue, air-stable $Pb[Co(CO)_3L]_4$ derivatives are obtained[371].

The same compounds are also obtained after reaction of lead(II) acetate with $Na[Co(CO)_3L]$. Other organolead compounds with a lead transition-metal bond have been prepared. $[Et_4N][CpNb[PbR_3](CO)_3]$ (where R = Et, Ph) was synthesized by the reaction of $Na[CpNb(H)(CO)_3]$ with R_3PbCl followed by treatment with $[Et_4N]Cl$[374]. $\{[(\eta^5-C_5H_5)_2HM]_2Pb\}(OAc)_2$, where M = Mo or W, was generated by reaction of the corresponding hydride $[MH_2(\eta^5-C_5H_5)_2]$ with Me_3PbOAc at $20\,°C$ according to the following reaction[375]:

$$MH_2(\eta^5\text{-}C_5H_5)_2 \xrightarrow{Me_3PbOAc} \{[(\eta^5\text{-}C_5H_5)MH]_2Pb\}(OAc)_2$$

Analogous compounds possessing lead–vanadium bonds were also prepared[376]:

$$Na(diglyme)_2V(CO)_6 \xrightarrow[\text{Ph}_3\text{PbCl, Et}_4\text{N}^+\text{Cl}^-]{\text{Na, liq. NH}_3} [Et_4N][V(PbPh_3)(CO)_5]$$

or

$$[Et_4N]\,[V(\eta^5\text{-}C_5H_5)(H)(CO)_3] \xrightarrow{\text{Ph}_3\text{PbCl}} [Et_4N][V(\eta^5\text{-}C_5H_5)(PbPh_3)(CO)_3]$$

A simpler compound, $Ph_3PbV(CO)_6$, involving a V–Pb bond, is obtained from $Ph_3Pb(C_5H_5)$ by reaction with $V(CO)_6$[412].

1,1-Diorgano-1-plumbacycloalkanes or -alkadienes, like 1,1-diethylplumbole, are prepared by the following reactions[383,384]:

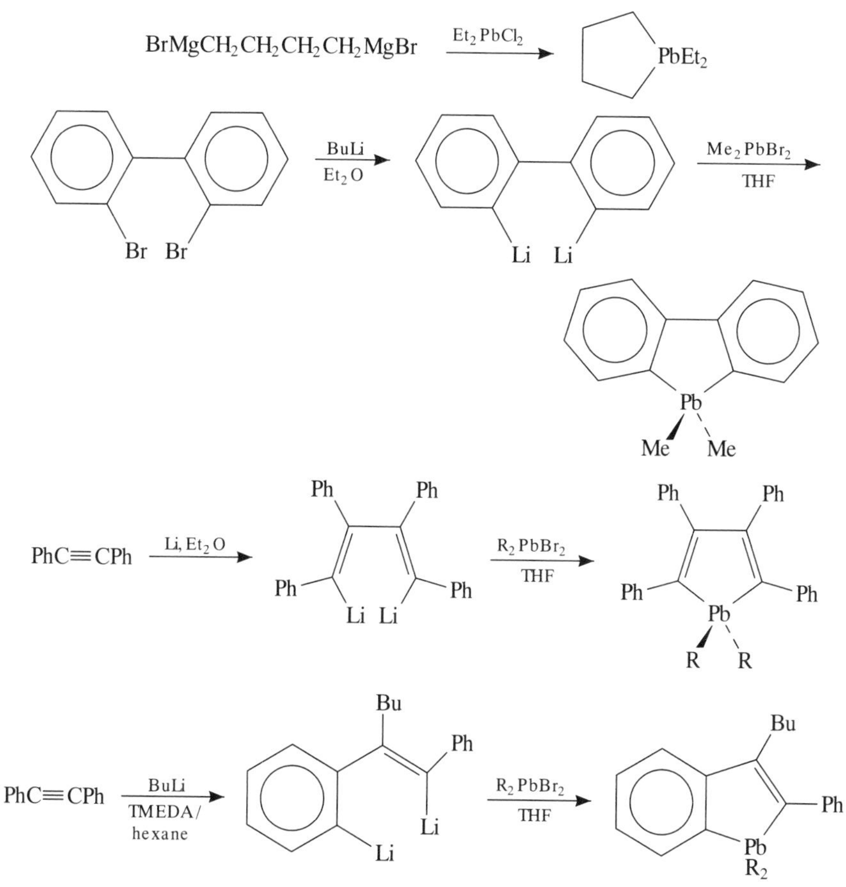

Addition of $PbCl_2$ to the above-mentioned dilithiated derivatives generates spiroplumbanes (**99**, **99a** and **99b**)[385]. The preparation of spiroplumbanes can also be performed

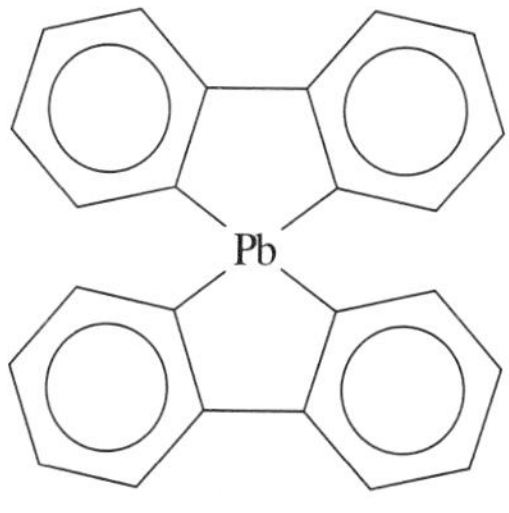

(**99**) 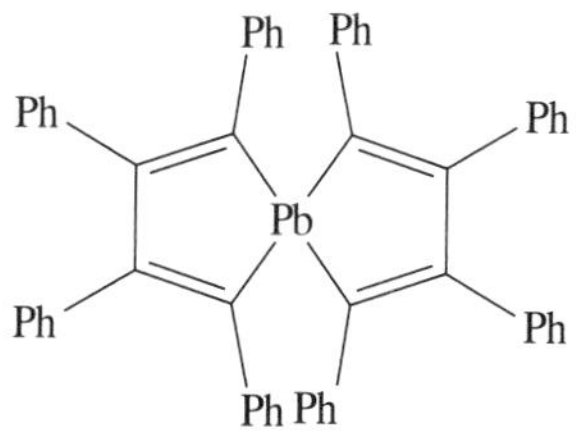   (**99a**) 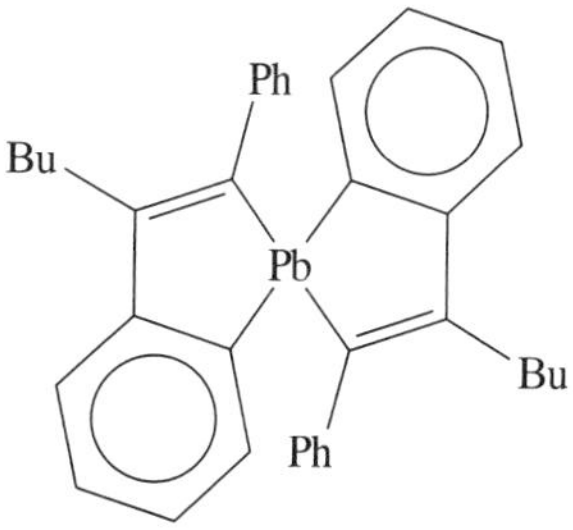   (**99b**)

by the reaction of the di-Grignard reagent of 1,4-dichloro- or -dibromobutane with $PbCl_2$[386] or $KPbCl_6$[387] in Et_2O:

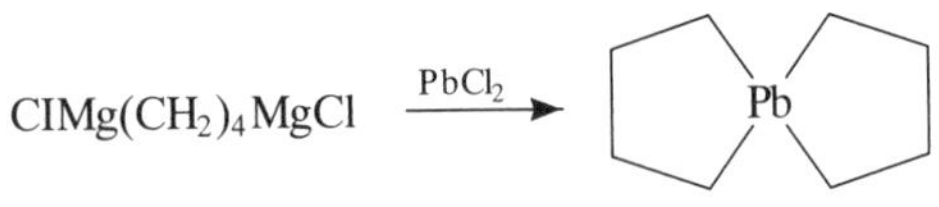

$$ClMg(CH_2)_4MgCl \xrightarrow{PbCl_2}$$

(99c)

This 5-plumba-spiro[4,4]nonane, **99c**, is prepared in 9% yield.

Related plumbacyclopentanes and plumbacyclohexanes can be prepared analogously[387]. Polymeric organoplumbanes can be prepared from bis(p-vinylphenyl)diphenyllead, (p-CH_2=$CHC_6H_4)_2PbPh_2$, by reaction with Ph_2SnH_2[402a]:

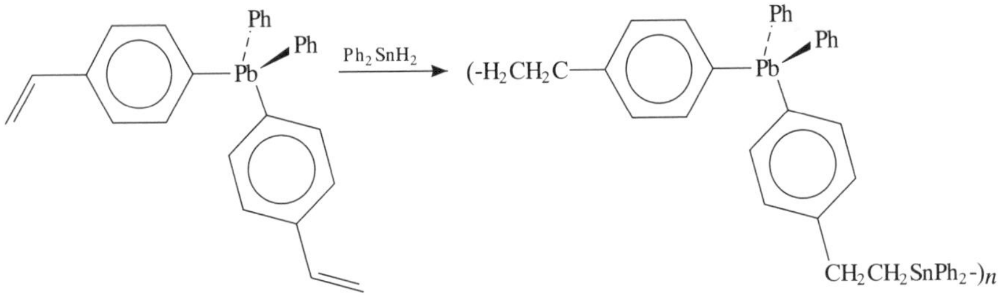

Diphenyldistyrenyllead is prepared[402b] by the following reaction:

$$Ph_2PbCl_2 \xrightarrow[Et_2O]{p\text{-}H_2C=CHC_6H_4\text{-}MgCl} (p\text{-}H_2C=CHC_6H_4)_2PbPh_2$$

II. ORGANOMETALLIC HYDRIDES OF GERMANIUM(IV), TIN(IV) AND LEAD(IV)

A. Organogermanium Hydrides

Apart from the classic preparation of simple germanes Ge_nH_{2n+2} (where n = 2–5) from the hydrolysis of Mg_2Ge[201], the following solvent-free reaction[202] gives satisfactory yields of digermanes:

$$KGeH_3 + GeH_3Cl \longrightarrow Ge_2H_6 + KCl$$

$KGeH_3$, which is a very useful germanium hydride salt in various synthetic routes, is usually prepared by the reaction[203a] of GeH_4 with potassium in dimethoxyethane. It can also be made by the reaction[203b]

$$Ge_2H_6 + KH \longrightarrow KGeH_3 + GeH_4$$

The synthesis of hydrides of the types $RGeH_3$, $RR'GeH_2$, $RR'R''GeH$, RGe_2H_5, 1,1-$RR'Ge_2H_4$, 1,2-$RR'Ge_2H_4$, 1,1,1-$RR'R''Ge_2H_3$, 1,1,2-$RR'R''Ge_2H_3$, 1,1,2,2-$RR'R''R'''Ge_2H_2$, etc., where some, all or none of the organic groups are identical, is of essential importance in organogermanium chemistry[206,207].

The preparation of the first three types is quite difficult, especially when simple alkyl groups, such as $MeGeH_3$, Et_2GeH_2, Et_3GeH or a phenyl one, Ph_3GeH[203], are involved.

In contrast, tris(pentafluorophenylgermane), **100**, is easily obtained[204]:

$$(C_6F_5)_3GeBr + Et_3GeH \xrightarrow[\text{no solvent}]{\text{1 h reflux}} \underset{\textbf{(100)}}{(C_6F_5)_3GeH} + Et_3GeBr$$

or

$$(C_6F_5)_3GeBr \xrightarrow[\text{Et}_2O, \text{ toluene}]{\text{LiAlH}_4} (C_6F_5)_3GeH$$

Compounds like $RGeH_3$, R_2GeH_2 can be prepared quite easily[205a] when $R = CH_2SiH_3$:

$$Cl_3SiCH_2Cl \xrightarrow[\text{Bu}_2O]{\text{LiAlH}_4} H_3SiCH_2Cl \xrightarrow{\text{NaGeH}_3} \underset{\textbf{(101)} \ 35\%}{H_3SiCH_2GeH_3} \xrightarrow[\text{HCl}]{\text{AlCl}_3} (H_3SiCH_2)_2GeH_2 + GeH_4$$

1-Germa-3-silapropane, **101**, is quite useful in organogermanium chemistry[205b],

$$H_3GeCH_2SiH_3 \xrightarrow[\text{AlCl}_3]{\text{HCl}} H_3GeCH_2SiH_2Cl \xrightarrow[\text{AlCl}_3]{\text{HCl}} H_3GeCH_2SiHCl_2 \xrightarrow{\text{H}_2\text{O}} (H_3GeCH_2SiH_2)_2O$$

the latter conversion being achieved at room temperature.

Organodigermanes can be obtained by coupling reactions[208]

$$PhGeH_2Cl \xrightarrow{\text{Mg/Hg}} PhH_2GeGeH_2Ph$$

$$PhGeH_2Cl + PhGeCl_2H \xrightarrow{\text{Mg/Hg}} PhH_2GeGeH_2Ph + PhH_2GeGeHPhGePhH_2$$

Ph_2GeHCl undergoes a Würtz reaction using Mg in THF to produce a dihydridotetraorganodigermane[209]:

$$2\ Ph_2GeHCl + Mg \xrightarrow{\text{THF}} Ph_2HGeGePh_2H$$

After partial dearylation of the product obtained, using HBr, stereogenic germanium atoms are generated giving the d, l and *meso* forms of 1,2-diphenyl-1,2-dibromo-digermane, **102**.

$$PhHBrGeGePhHBr$$
$$\textbf{(102)}$$

Trichlorogermane $HGeCl_3$ is a very useful compound in synthetic organogermanium chemistry[210,211]. Its preparation can be performed by several methods. One of them involves the reduction of $GeCl_4$ using mono- or diorganosilicon hydrides[212]:

$$GeCl_4 + R_2SiH_2 \xrightarrow[\text{Et}_2O, \ 4-10 \text{ h, } Et_3NH^+Cl^{(-)}]{\text{room temperature}} HGeCl_3 \cdot 2Et_2O + R_2SiHCl$$

$$(R = Me, \ Et, \ Pr, \ Ph)$$

Quantitative yields of $HGeCl_3 \cdot Et_3N$ are obtained if the above reaction is carried out in the presence of an equivalent amount of triethylamine[213]:

$$GeCl_4 + RR'SiHCl + Et_3N \xrightarrow[\text{hydrocarbon}]{0\,^\circ C} HGeCl_3 \cdot Et_3N + RR'SiCl_2$$

$$(R = \text{alkyl, aryl, H; } R' = \text{Cl, alkyl, aryl})$$

1,1,3,3-Tetramethyldisiloxane (TMDS) is a very convenient reagent to prepare $HGeX_3$ (where X = F, Cl, Br, I), as adducts, from tetrahalogermanes[214]:

$$(CH_3)_2HSi-O-SiH(CH_3)_2 + GeX_4 \xrightarrow[5-7\,h]{Et_2O} HGeX_3 \cdot 2Et_2O + (CH_3)_2XSi-O-SiH(CH_3)_2$$

Digermane hydrides are obtained by the following methods[215]. Reaction of R_2GeH_2 (where R = Ph or mesityl), with Me_3CLi, produces R_2GeHLi with 80 to 95% yields:

$$R_2GeH_2 + Me_3CLi \xrightarrow{Et_2O} R_2HGeLi$$
$$\textbf{(103)}$$

The latter, diarylhydrogermyllithium **103**, reacts with $PhGeClH_2$, and 1,1-diaryl-2-phenyl-digermane **104** is produced:

$$R_2HGeLi + PhGeClH_2 \xrightarrow{35\%} R_2HGeGePhH_2$$
$$\textbf{(103)} \qquad\qquad\qquad \textbf{(104)}$$

Reaction of **103** with $MgBr_2$ in ether produces a germanium-containing Grignard reagent with a Ge–Mg bond:

$$R_2HGeLi + MgBr_2 \xrightarrow{Et_2O} R_2HGeMgBr + LiBr$$

The latter reacts further in ether with PhH_2GeCl resulting in 1,1,2-triaryldigermanium hydrides:

$$R_2HGeMgBr + PhH_2GeCl \xrightarrow{Et_2O} R_2HGeGeH_2Ph + MgClBr$$

with R = Ph, mesityl (yield: 56% and 67%, respectively).

Germanium hydrides (as well as silicon hydrides) can be obtained in high yields by reduction of the corresponding halides or alkoxides by $LiAlH_4$ in the presence of phase-transfer catalysts[244].

B. Organotin Hydrides

Organotin hydrides $R_n SnH_{4-n}$, where R represents an alkyl or aryl group, are generally prepared quite easily by reduction of the corresponding organotin halides using either $LiAlH_4$ or $NaBH_4$ in ether or dioxane:

$$MeSnCl_3 \xrightarrow{LiAlH_4,\ Et_2O} MeSnH_3$$

$$Me_2SnCl_2 \xrightarrow{LiAlH_4,\ Et_2O} Me_2SnH_2$$

$$R_3SnCl \xrightarrow{LiAlH_4,\ Et_2O} R_3SnH$$

R = Me[216], Ph[217,218], o-[219], m-[219], p-tolyl[220], p-fluorophenyl[220], mesityl[220], o-biphenyl-yl[219].

When an organotin oxide or alkoxide is the starting material, reduction is performed with $LiAlH_4$[221] or organosilicon hydrides like methylhydropolysiloxane $(MeHSiO)_n$[222], triphenylsilane or 1,1,3,3-tetraphenyl disiloxane[223]. Diborane can likewise be used for this kind of reduction[227]:

$$R_3SnOMe \xrightarrow{B_2H_6} R_3SnH$$

Several less general methods for preparing organotin hydrides are also used, though often with less satisfactory yields. However, reduction of N,N-diethylaminostannanes by di-n-butylaluminum hydride or diborane occurs with high yields[224]:

$$R_{4-n}Sn(NEt_2)_n \xrightarrow[\text{or } B_2H_6]{Bu_2AlH} R_{4-n}SnH_n$$

Reduction by R_2AlH of organotinchlorides can also be carried out under mild conditions[236]:

$$R_{4-n}SnCl_n \xrightarrow{R_2AlH} R_{4-n}SnH_n$$
$$n = 1, 2 \text{ or } 3$$

A new organotin hydride with a polar tail, di-n-butyl(4,7,10-trioxaundecyl)stannane, useful for various syntheses in organic chemistry, is prepared by the following sequence of reaction[237] (Scheme 23). Bu_2HSnCl is prepared in ether from $LiAlH_4$ and Bu_2SnCl_2 at low temperatures.

$$CH_3(OCH_2CH_2)_2OH \xrightarrow[BrCH_2CH=CH_2]{KOH, THF} CH_3(OCH_2CH_2)_2OCH_2CH=CH_2$$

$$\downarrow Bu_2SnHCl$$

$$CH_3(OCH_2CH_2)_2O(CH_2)_3SnBu_2H \xleftarrow[Et_2O]{LiAlH_4} CH_3(OCH_2CH_2)_2O(CH_2)_3SnBu_2Cl$$

SCHEME 23

Acid hydrolysis of triphenyltin lithium or bis(triphenyltin) magnesium provides triphenyltin hydride[225,226]:

$$Ph_3SnCl \xrightarrow[THF]{Mg} (Ph_3Sn)_2Mg \xrightarrow{H_3O^+} Ph_3SnH$$

$$Ph_3SnCl \xrightarrow[THF]{Li} Ph_3SnLi \xrightarrow{H_3O^+} Ph_3SnH$$

As for organogermanium hydrides, a simple organotin hydride, like tri-n-butyltin hydride, can be used for the reduction of structurally more complicated diorganotin dichlorides[228]:

$$2 Bu_3SnH + R_2SnCl_2 \longrightarrow R_2SnH_2 + 2 Bu_3SnCl$$

A convenient preparation of tri-n-butyltin hydride is the hydrogenation of $(Bu_3Sn)_2O$ with $NaBH_4$ in EtOH, giving the hydride in 95% yield after 30 minutes reaction[229].

Mixed trimethylstannylcuprates, prepared from R_3SnH (R = Me, Bu) and $Bu_2Cu(CN)Li_2$ in THF by transmetalation reactions, enable easy introduction of the R_3Sn

moiety into a number of organic compounds[230]:

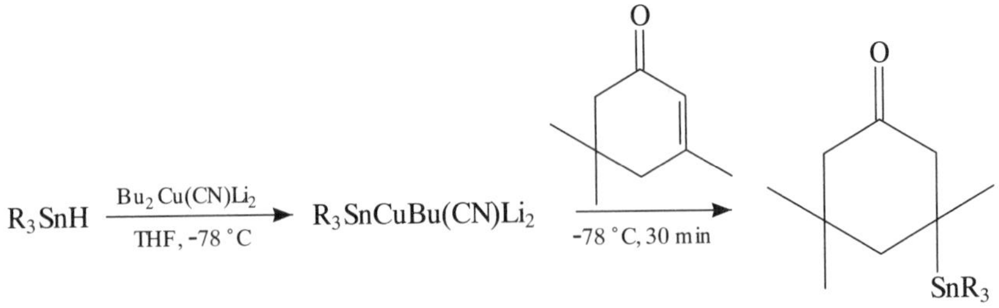

$$R_3SnH \xrightarrow[\text{THF},-78\,°C]{Bu_2Cu(CN)Li_2} R_3SnCuBu(CN)Li_2 \xrightarrow{-78\,°C,\,30\,min}$$

(83%)

Mixed alkyltin hydrides are relatively difficult to prepare. Nevertheless, a convenient synthesis starts from the reaction of $SnBr_4$ or SnI_4 with 2,4,6-tri-t-butylphenyllithium **105**:

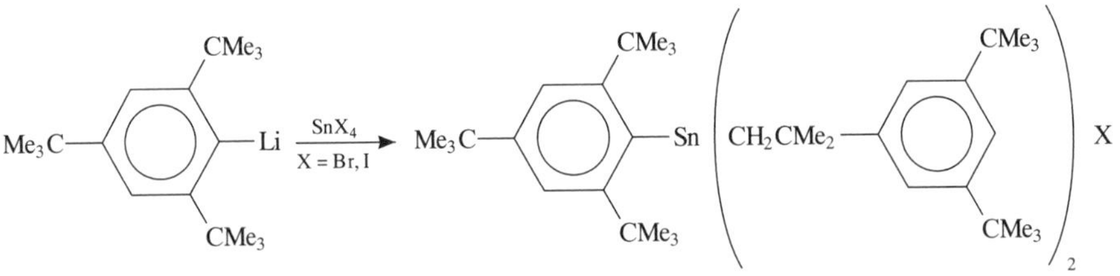

$$Me_3C-\underset{CMe_3}{\overset{CMe_3}{\bigcirc}}-Li \xrightarrow[X=Br,I]{SnX_4} Me_3C-\underset{CMe_3}{\overset{CMe_3}{\bigcirc}}-Sn\left(CH_2CMe_2-\underset{CMe_3}{\overset{CMe_3}{\bigcirc}}\right)_2 X$$

(105)   **(105a)**

For steric reasons, an aryldialkyltin halide ArR_2SnX is produced (X = Br, I) after transfer and isomerization of the 2,4,6-tri-t-butylphenyl group. Reaction of this mixed aryldialkyltin halide with t-butyllithium then provides the desired mixed triorganotin hydride ArR_2SnH[231]. The structures of **105a** and **105b** were confirmed by X-ray diffraction [R = 2-methyl-2-(3,5,-di-t-butylphenyl)propyl].

Alkylhalotin hydrides are prepared by reaction of an alkyltin hydride with either an alkyltin halide with the same alkyl groups[232] or with hydrogen halides[233], e.g.:

$$Bu_2SnCl_2 + Bu_2SnH_2 \rightleftharpoons Bu_2SnHCl$$

$$Bu_2SnH_2 \xrightarrow{HCl} Bu_2SnHCl$$

Both reactions are conducted at room temperature. The first reaction is an equilibrium which is usually shifted to the right.

Tetraorganodihydrodistannanes of the type $R_2HSnSnHR_2$ are prepared by reduction of the corresponding tetraorganodihalodistannanes[101].

Pentaorganodistannanes $R_3SnSnR'_2H$ are obtained from the sequence of reactions:

$$R_3SnH \xrightarrow{PhN=C=O} R_3SnNPh(CH=O)$$
(106)

$$R_3SnNPh(CH=O) \xrightarrow{R'_2SnH_2} R_3SnSnR'_2H$$

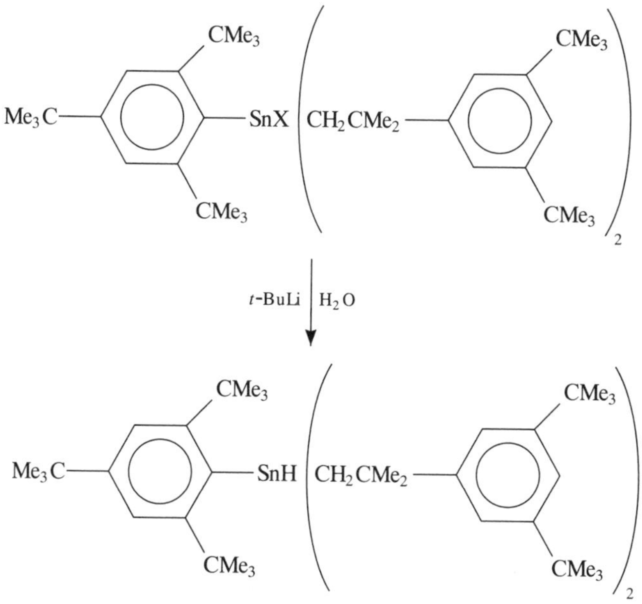

(105b)

The preparation of trialkyl(N-phenyl-formamido)tin(IV) compounds of the type **106** from phenylisocyanates is followed by a reduction using dialkyl or diaryltin dihydrides[234].

An application of the reduction method of organotin halides by simple organotin hydrides is the preparation of 1,ω-bis(hydridodimethylstannyl)alkanes[235]:

$$BrMe_2Sn(CH_2)_n SnMe_2Br \xrightarrow[50-60\,°C]{Bu_3SnH} HMe_2Sn(CH_2)_n SnMe_2H$$

$$n = 1-6; \text{ for } n = 1 \text{ yield } 80\%$$

η^6-1-Mesitylene chromiumtricarbonyl, $(\eta^6$-1,3,5-Me$_3$C$_6$H$_3)$Cr(CO)$_3$, **107**, reacts photochemically with Ph$_3$SnH giving a hydridostannyl complex of peculiar structure, **108**, as evidenced by X-ray diffraction analysis[238].

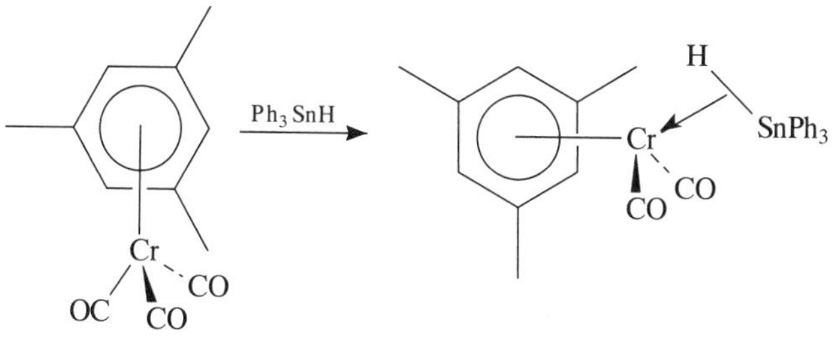

(107)   **(108)**

The bond Sn—H is coordinated in η^2 fashion.

Analogous complexes of the type $(CO)_n(dppe)M(H)SnR_3$ can be prepared by thermal reactions of $(CO)_4(R_3P)W(THF)$ or $(CO)_3(dppe)ML$ with R_3SnH (dppe = $Ph_2PCH_2CH_2PPh_2$; M = Cr, Mo, W; L = THF, dioxane; R = Ph, Me).

The complexes $(CO)_3(dppe)M(H)SnPh_3$ decompose in benzene at room temperature to hexaphenyldistannane, $Ph_3SnSnPh_3$, which is a method of Sn—Sn bond generation. $[(n\text{-Bu})_3PCo(CO)_3]_3SnH$ is an unusually stable organotin hydride. It is prepared from $Na[Co(CO)_3P(n\text{-Bu})_3]$ and tin(II) sulfate in aqueous diglyme. Attempts to prepare the corresponding organolead hydride, $[(n\text{-Bu})_3PCo(CO)_3]_3PbH$, were unsuccessful[246].

C. Organolead Hydrides

Organolead hydrides are prepared from the reduction of organolead halides, usually the methyl and ethyl derivatives, by KBH_4 or from the reduction of trialkyllead methoxides R_3PbOCH_3 (R = Me, Et, n-Pr, n-Bu) with B_2H_6.

Reductions of alkyllead halides with $LiAlH_4$ are among the most successful ones and are, despite complications in some cases[241], easy to perform. R_3PbCl can easily be reduced in liquid ammonia by KBH_4[239,241] through an R_3PbBH_4 intermediate that is converted into the trialkyllead hydride by NH_3:

$$R_3PbCl \xrightarrow[\substack{2\,NH_3,\,-5°C \\ (R = Me,\,Et)}]{1\,KBH_4} R_3PbH$$

It is noteworthy to add that potassium borohydride and trimethyllead chloride, upon reaction in liquid ammonia, first produce trimethyllead borohydride which, on distillation, gives $H_3B \cdot NH_3$, ammonia and trimethylplumbane[245]:

$$Me_3PBH_4 \cdot x\,NH_3 \longrightarrow Me_3PbH + (x\text{-}1)NH_3 + NH_3BH_3$$

Trimethylplumbane consequently reacts with ammonia giving the unstable green ammonium trimethylplumbate, which decomposes into $PbMe_2$, subsequently converted into the red pentamethyldiplumbate:

$$Me_3PbH + NH_3 \longrightarrow NH_4[PbMe_3]$$

$$NH_4[PbMe_3] \longrightarrow PbMe_2 + MeH + NH_3$$

$$NH_4[PbMe_3] + PbMe_2 \longrightarrow NH_4[Pb_2Me_5]$$

With a special distillation apparatus[239,241] Me_3PbH can be isolated at $-35°C$.

Reduction of trialkyllead chloride by B_2H_6 yields the same intermediate, which is in turn converted into the trialkyllead hydride by methanol[240]:

$$R_3PbCl \xrightarrow[\substack{2\,MeOH,\,-78°C \\ R = Me,\,Et}]{1\,B_2H_6} R_3PbH$$

In the case of reductions by $LiAlH_4$ the reaction conditions depend on the type of alkyllead halide used, R_3PbX, R_2PbX_2 or $RPbX_3$:

$$R_3PbCl \xrightarrow[\text{liq. } NH_3]{LiAlH_4} R_3PbH^{241,242}$$

$$R_2PbCl_2 \xrightarrow[Et_2O]{LiAlH_4} R_2PbH_2{}^{243}$$

$$R = Me, Et, n\text{-}Pr, i\text{-}Pr, n\text{-}Bu, \text{cyclohexyl}$$

The solvents used are liquid NH_3, ether, THF and mixtures of THF or ether with diglyme.

III. ORGANOMETALLIC HALIDES OF GERMANIUM(IV), TIN(IV) AND LEAD(IV)

A. Organogermanium Halides

The preparation of compounds of the general types R_3GeCl, R_2GeCl_2 and $RGeCl_3$ is of significant importance in the organometallic chemistry of germanium, since they are frequently used as starting materials for the synthesis of more complicated organogermanium compounds.

The action of alkyl halides on elemental germanium in the presence of a copper/gallium catalyst leads to the simultaneous preparation of several or all of the above-mentioned organogermanium halides[248]:

$$MeCl \xrightarrow[\substack{Cu,\ Ga \\ 400\,°C,\ 5\ h}]{Ge} \underset{85\%}{Me_3GeCl} + \underset{10\%}{Me_2GeCl_2} + \underset{5\%}{MeGeCl_3}$$

Analogously[249]

$$C_6H_5Br \xrightarrow[Cu]{Ge} C_6H_5GeBr_3 + (C_6H_5)_2GeBr_2$$

Ionizing radiation promotes the alkylation in the following reaction[250]:

$$RBr \xrightarrow[220\,°C,\ 13\ h]{Ge,\ \text{sealed tube}} RGeBr_3$$

$$R = n\text{-}C_4H_9,\ C_6H_{13},\ C_7H_{15},\ C_6H_5$$

The yield varies from 48 to 55%.

The method of alkylation of germanium tetrahalides in the presence of catalysts[251] is more effective:

$$GeX_4 \xrightarrow[140\,°C,\ 5\ h]{RX/Cu} RGeX_3$$

$$R = n\text{-}Bu,\ C_7H_{15};\ X = Br,\ I;\ \text{yield: } 45\text{-}59\%$$

$$GeBr_4 \xrightarrow[\substack{\text{sealed tube} \\ 200\,°C,\ 16\ h}]{PhBr,\ Cu} \underset{51\%}{PhGeBr_3} \xrightarrow{PhI} \underset{29\%}{Ph_2GeBr_2}$$

or[252]

$$GeI_4 \xrightarrow[\substack{5\ h,\ \text{inert atm} }]{RI,\ 180\text{-}200\,°C,\ Zn} RGeI_3$$
$$R = n\text{-}C_4H_9,\ n\text{-}C_6H_{13},\ n\text{-}C_8H_{17}$$
$$\text{yield: } 51\text{-}65\%$$

A less usual but useful route to aryltrichlorogermanes involves the reaction of germanium tetrachloride with aryltrichlorosilanes(**109**)[253]:

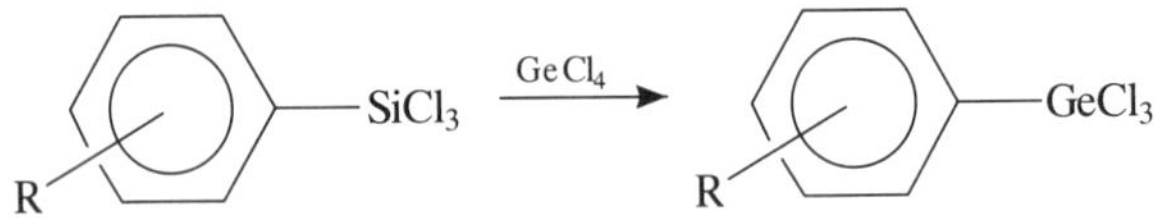

(109)

Insertion of germanium into the α, ω-dihaloalkanes and chloromethyl-silanes provides useful germanium halogen derivatives[254]:

$$X(CH_2)_n X \xrightarrow[370-380\,°C]{Ge/Cu} X(CH_2)_n GeX_3 + X_3Ge(CH_2)_n GeX_3$$

$$X = Cl, Br; \; n = 1-5$$

$$XMe_2SiCH_2Cl \xrightarrow[370-400\,°C]{Ge/Cu} \underset{5-8\%}{XMe_2SiCH_2GeCl_3} + \underset{26-47\%}{(XMe_2SiCH_2)_2GeCl_2}$$

$GeCl_2$ can be inserted into carbon–halogen bonds upon release from its dioxane complex *in situ*[255],

$$GeCl_2 \cdot O(CH_2CH_2)_2O \xrightarrow[reflux]{BuCl} BuGeCl_3$$

and[256]

$$GeCl_2 \cdot O(CH_2CH_2)_2O \xrightarrow[reflux]{RC_6H_4CH_2Cl} RC_6H_4CH_2GeCl_3$$

$$(R = p\text{-Me, H, } p\text{-Br, } p\text{-Cl, } m\text{-CF}_3, \text{ yield } 80-95\%)$$

$HGeCl_3$ can also be used to introduce the $GeCl_3$ moiety into organic compounds, for example by its addition to a double bond[257]:

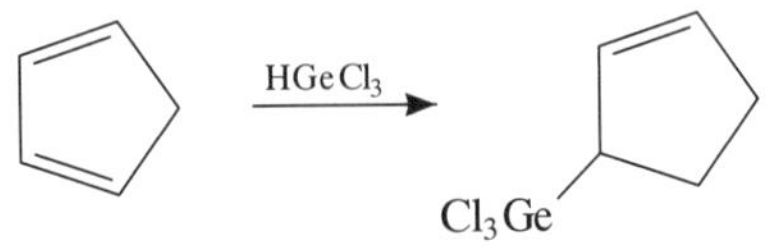

$HGeCl_3$ can also be prepared by reaction of elemental Ge with $HCl \cdot Et_2O$; actually the adduct $HGeCl_3 \cdot 2\, Et_2O$ is obtained[258].

When $HGeCl_3 \cdot 2\, Et_2O$ is reacted with acetylene[259] or propyne[260], a double germylation takes place:

$$RC\equiv CH \xrightarrow{HGeCl_3 \cdot 2\, Et_2O} (Cl_3Ge)RC=CH(GeCl_3)$$

In the absence of ether, the addition of $HGeCl_3$ proceeds in a different way[261,265]:

$$HC\equiv CH \xrightarrow{HGeCl_3} H_2C=CHGeCl_3 + Cl_3GeCH_2CH_2GeCl_3$$

$HGeCl_3$ or $RGeH_3$ is used in reactions with tetralkyltins or tin(IV) tetrachloride for the synthesis of alkyldichlorogermanes[262,263], which are difficult to obtain:

$$HGeCl_3 \xrightarrow{R_4Sn} RHGeCl_2$$

$$RGeH_3 \xrightarrow{SnCl_4} RH_2GeCl + RHGeCl_2$$

The so-called superacid[264a,b] $HGeCl_3$ reacts with aromatic compounds and generates longlived carbocations stabilized by the trichlorogermyl anion[264c]. The carbocations **110** are prepared from the reaction of $HGeCl_3$ with $1,3\text{-}(RO)_2C_6H_4$ at $-50\,°C$ in CD_2Cl_2. Allowing the reaction mixture to come to room temperature gives rise to 1,1,3,5-tetrakis(trichlorogermyl)-3-alkoxycyclohexane, **111**.

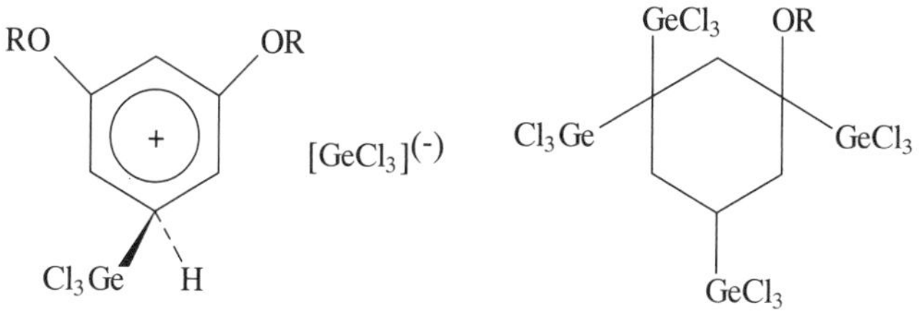

R = Me, Et

(110)   **(111)**

As the temperature rises, partial decomposition of **110** occurs giving rise to $GeCl_3^{(-)}$ moieties which attack **110** providing **111** in low yields. Compound **110** was observed by NMR. The structure of **111** was determined by X-ray diffraction.

Powdered germanium reacts with Cl_3SiCH_2Cl at $350\,°C$ and generates $Cl_3SiCH_2GeCl_3$ or $Cl_3SiCH_2-GeCl_2-CH_2SiCl_3$.

While, as noted earlier, $HGeCl_3$ reacts with acetylene to produce $Cl_3GeCH_2CH_2GeCl_3$, reaction with $CH_2=CHSiCl_3$ gives rise to $Cl_3SiCH_2CH_2GeCl_3$. In contrast, $HGeCl_3 \cdot NEt_3$ reacts with $HC\equiv CHCH_2Cl$ to yield $Cl_3GeCH_2C(GeCl_3)=CH_2$.

All the above chlorogermanes and chlorogermasilanes can be reduced by $LiAlH_4$ in tetraline in the presence of benzyltriethylammonium chloride as a catalyst, providing germanes or silagermanes[265].

Halodigermanes were first prepared by Bulten and Noltes[266] by an exchange reaction:

$$R_3GeGeR_3 \xrightarrow[275\,°C]{GeCl_4} R_2ClGeGeR_3 + R_2ClGeGeR_2Cl$$

Useful chlorogermanes can be synthesized by redistribution reactions[264]:

$$Et_3GeGeMe_2GeEt_3 \xrightarrow{SnCl_4} Et_3GeGeMeClGeEt_3 + Et_3GeGeMe_2GeEt_2Cl$$

and

$$R_4M \xrightarrow{M'Cl_4} R_3MCl + RM'Cl_3$$

(M = Ge, M' = Ge, Sn; M = Sn, M' = Ge)

Analogously[267]

$$Et_3GeGeEt_3 \xrightarrow{i\text{-PrBr}} Et_2BrGeGeEt_3$$

1,4-Diiodo-octaphenyltetragermane is prepared from octaphenylcyclotetragermane by reaction with iodine:

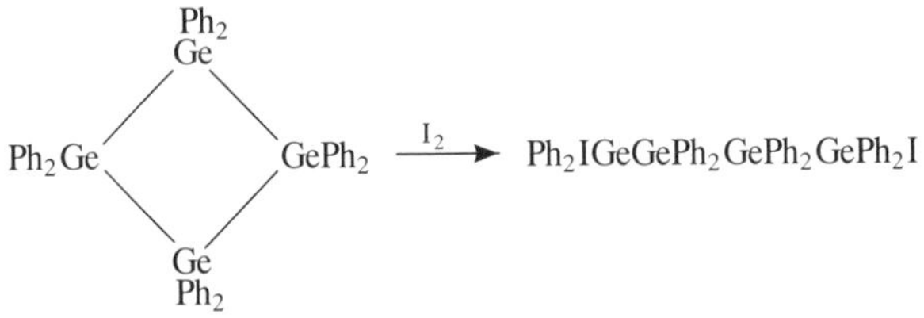

$$Ph_2IGeGePh_2GePh_2GePh_2I$$

The 1,1,2,2-tetrachlorodigermane **112**, $(OC)_3MnGeCl_2GeCl_2Mn(CO)_3$, in which each germanium atom is bound to a manganese atom, was synthesized[268] by reacting $GeCl_4$ with $NaMn(CO)_5$ [prepared from $Mn_2(CO)_{10}$ and NaHg in THF].

The same reaction yields, in addition, a mixture of by-products $Cl_3GeMn(CO)_5$ and $Cl_2Ge[Mn(CO)_5]_2$.

The associated bromine derivatives are prepared by a different procedure, involving bromodephenylation:

$$Ph_2GeBr_2 \xrightarrow[\text{THF}]{NaMn(CO)_5} BrPh_2Ge-Mn(CO)_5 \xrightarrow[72\,°C]{Br_2} Br_2PhGe-Mn(CO)_5$$

$$\xrightarrow{Br_2} Br_3Ge-Mn(CO)_5$$

From the reaction of dihalodigermanes with alkyl halides catalyzed by $Pd(PPh_3)_4$ an unexpected germylene insertion takes place[269]:

$$ClMe_2GeGeMe_2Cl + RX \xrightarrow{catalyst} RMe_2GeCl + RMe_2GeX$$

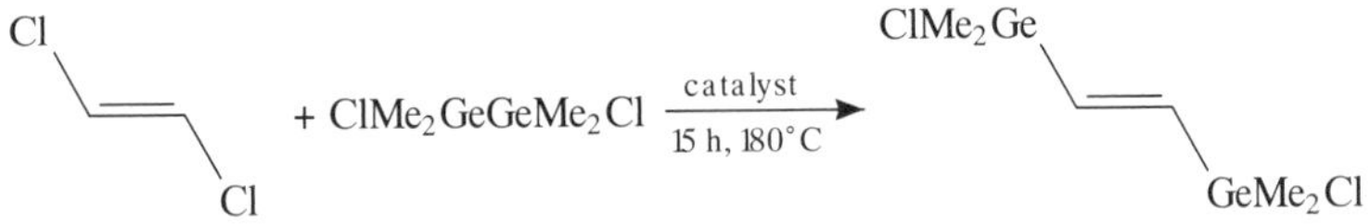

Useful syntheses[270] of chlorotrimethylgermane, chloropentamethyldigermane and 1,2-dichlorotetramethyldigermane are achieved by the following reactions:

$$Me_4Ge \xrightarrow{conc.\ H_2SO_4,\ NH_4Cl} Me_3GeCl\ (86\%)$$

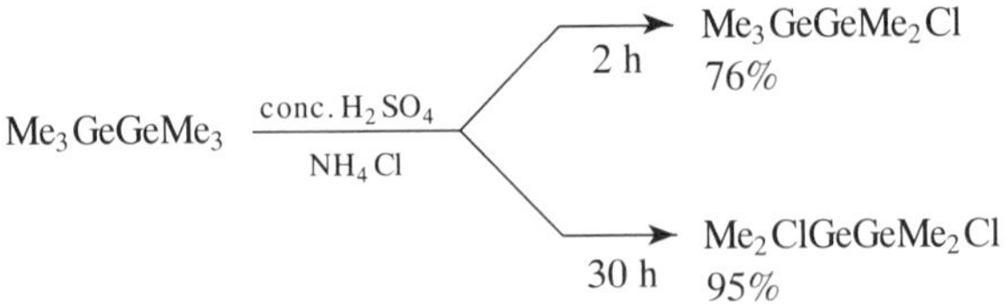

The above syntheses are demethylation reactions induced by concentrated H_2SO_4, followed by substitution of hydrogenosulphate for chloride from NH_4Cl.

Iodine, as in the case of germacyclanes, reacts smoothly with digermanes[271] or trigermanes[272] yielding at $-48\,^{\circ}C$ iododigermanes or iodotrigermane. The use of bromine leads to degradation of the hydrides:

$$Ge_2H_6 + I_2 \longrightarrow Ge_2H_5I + HI$$

$$Ge_3H_8 + I_2 \longrightarrow Ge_3H_7I + HI$$

The unstable di- or triiodogermane can be converted into the bromide or chloride after reaction with the corresponding silver halides[273]. Silver halides can also be used for the preparation of digermane monohalides[275]:

$$Ge_2H_6 \xrightarrow[X = Cl,\,Br]{AgX} Ge_2H_5X$$

An important and convenient method[274] to synthesize triphenylgermanium halides, useful for the preparation of polynuclear clusters, is the reaction of PhMgBr with GeCl$_4$ in toluene:

$$GeCl_4 \xrightarrow[reflux,\,72\%]{PhMgBr} Ph_4Ge$$

The Ph$_4$Ge obtained can be converted into Ph$_3$GeBr by reaction with bromine in 1,2-dibromoethane (82% yield).

The useful reagents monobromogermane and monochlorogermane can be prepared from BBr$_3$ resp. SO$_2$Cl$_2$ and GeH$_4$[276,277]:

$$BBr_3 + GeH_4 \longrightarrow H_3GeBr + HBr$$

$$SO_2Cl_2 + GeH_4 \longrightarrow H_3GeCl + SO_2 + HCl$$

Halomethylgermanes are prepared from diazomethane and GeCl$_4$ or GeBr$_4$:

$$CH_2N_2 + \underset{(X = Cl,Br)}{GeX_4} \longrightarrow X_3GeCH_2X + N_2$$

The associated trihydride H$_3$GeCH$_2$X is obtained after reduction with LiAlH$_4$.
In the case where X = Cl, exchange of halogen with KI leads to H$_3$GeCH$_2$I[278].
Mercuric chloride in THF can be used to introduce chlorine into germanium trihydrides[279] with a functional group in the aliphatic chain (Scheme 24). Finally, γ-hydroxygermylenes are produced.

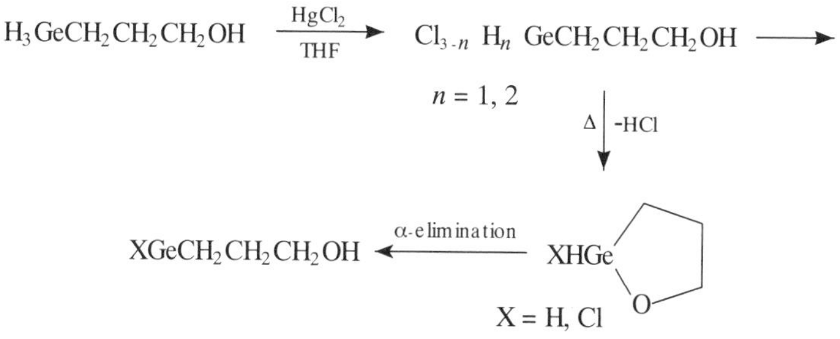

$$H_3GeCH_2CH_2CH_2OH \xrightarrow[THF]{HgCl_2} Cl_{3-n}\,H_n\,GeCH_2CH_2CH_2OH \longrightarrow$$

$$n = 1,\,2$$

$$\Delta \downarrow -HCl$$

$$XGeCH_2CH_2CH_2OH \xleftarrow{\alpha\text{-elimination}} XHGe \overset{}{\underset{O}{\diagup}}$$

$$X = H,\,Cl$$

SCHEME 24

Perfluoroalkylgermanium derivatives, like tetrakis(trifluoromethyl)germanium, can be prepared by fluorination of organogermanium compounds[280]:

$$(CH_3)_4Ge \xrightarrow{F_2} (CF_3)_4Ge$$

GeF_2 undergoes insertion into the Si–Co bond of $Ph_3SiCo(CO)_4$ and generates germanium fluorides containing a Si–Ge–Co moiety:

$$GeF_2 + Ph_3SiCo(CO)_4 \longrightarrow Ph_3SiGeF_2Co(CO)_4 \longrightarrow Ph_3SiF + FGeCo(CO)_4$$

Upon addition onto dienes, $FGeCo(CO)_4$, which is an active germylene, gives rise to a germacyclopentane containing a F–Ge–Co moiety[281]:

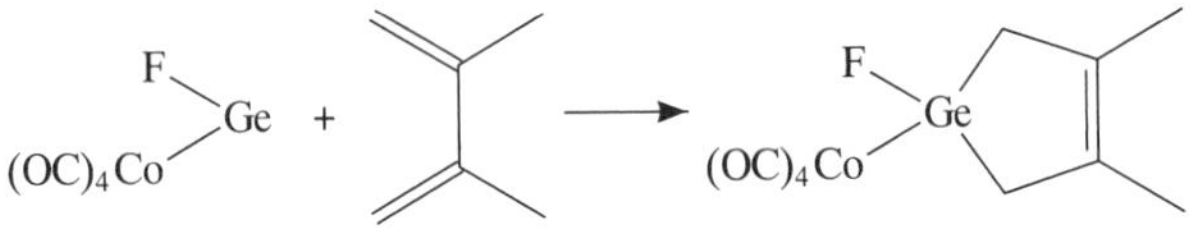

A 1,1,2,2-tetrachloro-1,2-digermacyclobutene, **113**, has been obtained[282] by the reaction of the dioxane complex of $GeCl_2$ with 3,3,6,6-tetramethyl-1-thiacyclohept-4-yne:

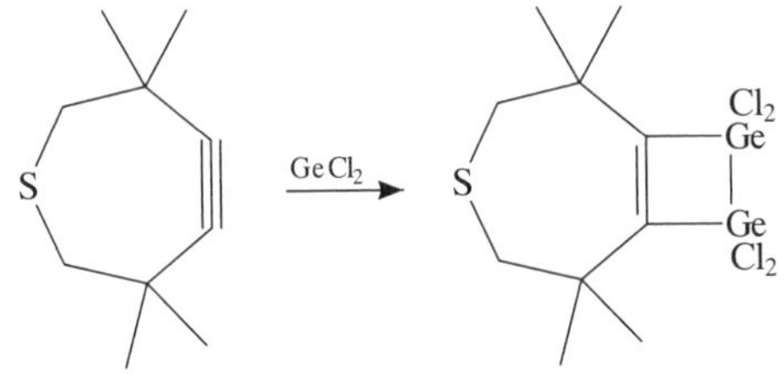

(113)

Germylenes can also be used for the preparation of several 1,2-dibromodigermanes or polygermanes[283]. A possible source of germylenes is the reaction:

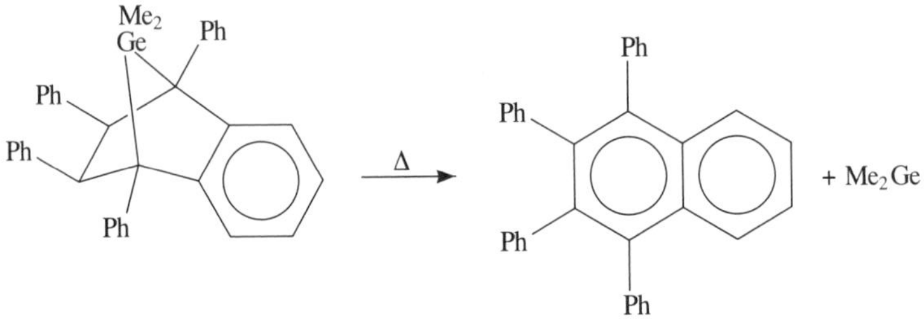

The produced germylene reacts with benzyl bromide to generate $(PhCH_2)Me_2GeBr$ and Me_2GeBr_2.

Other products can also be obtained since Me_2GeBr_2 is a germylene scavenger; it gives rise to bromine-containing oligogermanes ($n = 1, 2$):

$$Me_2GeBr_2 \xrightarrow{Me_2Ge} BrMe_2GeGeMe_2Br \xrightarrow{nMe_2Ge} BrMe_2Ge\text{-}(GeMe_2)_n\text{-}GeMe_2Br$$

From the last compound, cyclogermanes can be prepared.

Reaction of GeBr$_4$ with t-butyllithium gave a mixture of dibromo-di-t-butylgermane and 1,1,2,2-tetrabromo-1,2-di-t-butyl-digermane, **114**.

$$GeBr_4 + t\text{-}BuLi \longrightarrow (t\text{-}Bu)_2GeBr_2 + t\text{-}BuBr_2GeGeBr_2 - t\text{-}Bu$$
$$(20\%) \qquad\qquad \textbf{(114)} \quad (20\%)$$

Reaction of **114** with an excess of naphthalene lithium gave, besides unreacted material, a major product, **115**, 4,8-dibromoocta-t-butyl-tetracyclo[3.3.0.0^{2,7}.0^{3,6}]octagermane[284]:

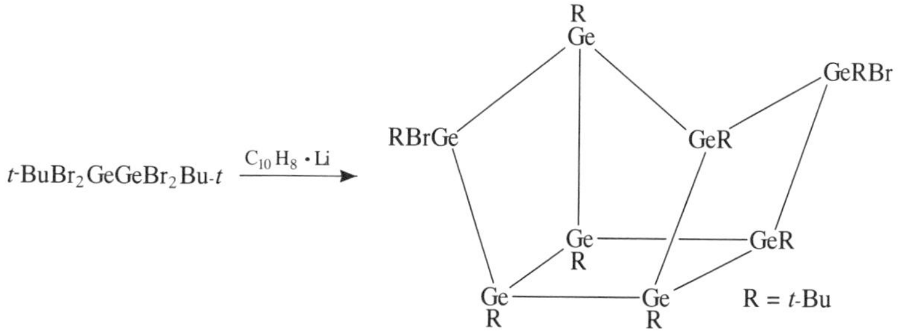

$$t\text{-}BuBr_2GeGeBr_2Bu\text{-}t \xrightarrow{\ C_{10}H_8 \cdot Li\ }$$

(115)

The structure of **115** was identified by X-ray analysis.

The dichloro analog of **115**, 4,8-dichloroocta-t-butyl-tetracyclo[3.3.0.0^{2,7}.0^{3,6}] octagermane **116**, was prepared[285] by the reduction of either 1,2-di-t-butyl-1,1,2,2-tetrachlorodigermane or t-butyltrichlorogermane with naphthalene lithium, together with 1,2,3,4,5-penta-t-butyl-1,2,3,4,5-pentachloropentagermacyclopentane, **117**:

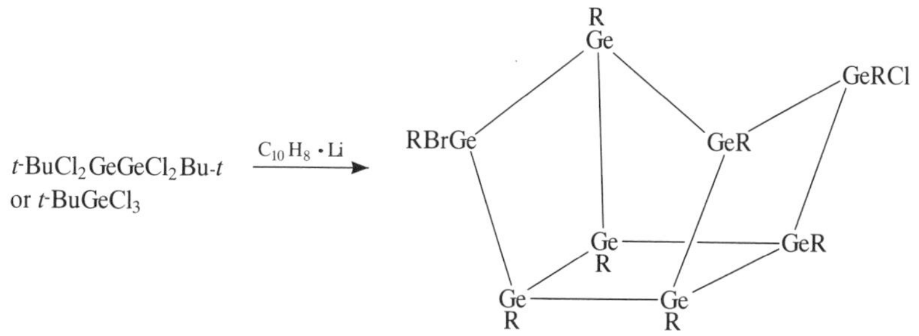

$$t\text{-}BuCl_2GeGeCl_2Bu\text{-}t \xrightarrow{\ C_{10}H_8 \cdot Li\ }$$
or t-BuGeCl$_3$

(116)

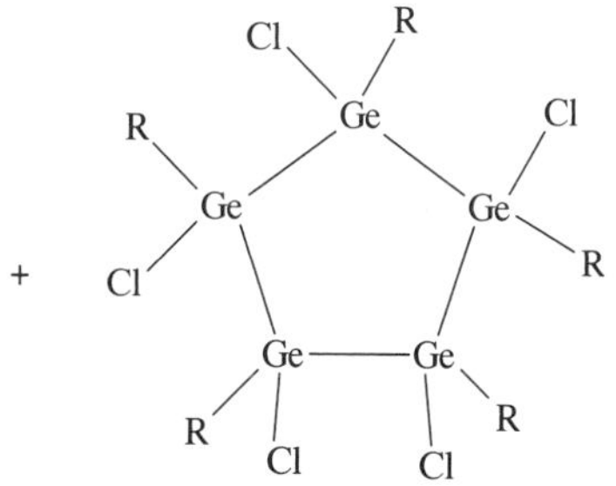

(117)

The insertions of different germylenes GeRR' (R = R' = F; R = F, R' = Co(CO)$_4$; R = R' = Cl; R = R' = Me) were studied. Germylenes like GeF$_2$ (which is used as its dioxane complex $GeF_2 \cdot C_4H_8O_2$) and GeRF (R = Ph, alkyl) can be inserted into various compounds, giving rise to organogermanium halides. Two examples follow[286,287]:

$$Ph_3SnSnMe_3 + GeF_2 \xrightarrow[15h]{130\,^\circ C} Ph_3SnGeF_2SnMe_3$$

$$EtX + RGeF \xrightarrow{120\,^\circ C} REtGeFX \; (R = Ph, \; F; \; X = Br, I)$$

In addition to being a fluorination agent for tetramethylgermane, bis(trifluoromethyl)mercury $(CF_3)_2Hg$ is used to react with $GeBr_4$, GeI_4 and $SnBr_4$ in sealed tubes, in order to generate trifluoromethylmetal halides of the type $(CF_3)_nMX_{4-n}$ for $n = 1-4$ when M = Ge and $n = 1, 2$ for M = Sn[288]. For example:

$$(CF_3)_2Hg \xrightarrow{GeBr_4} (CF_3)GeBr_3 + (CF_3)_2GeBr_2 + (CF_3)_3GeBr + (CF_3)_4Ge$$
$$\phantom{(CF_3)_2Hg \xrightarrow{GeBr_4}} 55\% \qquad\quad 22\% \qquad\quad 18\% \qquad\quad 4\%$$

Several types of $(CF_3)_3GeX$ compounds are prepared with X = H, I, Br, Cl, F, according to the following reactions:

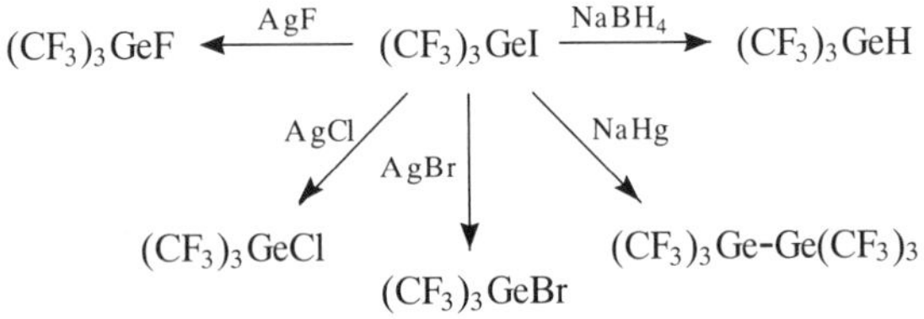

$$(CF_3)_3GeF \xleftarrow{AgF} (CF_3)_3GeI \xrightarrow{NaBH_4} (CF_3)_3GeH$$

AgCl / AgBr | NaHg

$$(CF_3)_3GeCl \qquad\qquad (CF_3)_3Ge\text{-}Ge(CF_3)_3$$
$$(CF_3)_3GeBr$$

By the same procedure, hydrogenation of all types of halides $(CF_3)_nGeX_{4-n}$ (X = halogen, $n = 1, 3$) with $NaBH_4$ leads to useful hydrides[289]:

$$(CF_3)_nGeX_{4-n} \xrightarrow[5-20\,^\circ C]{NaBH_4/H_3PO_4} (CF_3)_nGeH_{4-n}$$

$$n = 1-3, \; X = Cl, Br$$

Similarly, $(CF_3)_2GeHBr$ could be converted into the halides $(CF_3)_2GeHX$ (X = F, Cl, I) by treatment with AgX.

For the preparation of methylgermanium halides, tetramethylsilene can be used[290]. The catalytic amount of $AlCl_3$ or $AlBr_3$ used in this process enables the selective mono-, di-, tri- or tetra-methylation of $GeCl_4$.

The mechanism proposed involves the initial formation of Me_4Ge, regardless of the initial proportions of the reagents, followed by disproportionation reactions.

$SnCl_4$ can be used to introduce chlorine atoms into alkylgermanes via ligand redistribution reactions of the type $R_4Ge/SnCl_4$, $R_6Ge_2/SnCl_4$ or $R_8Ge_3/SnCl_4$. Using these reactions, especially of the first type in nitromethane, a mixed trialkylchlorogermane is prepared by the sequence of reactions[291] which involves Grignard reactions and dealkylations:

$$Me_3GeBu \xrightarrow[MeNO_2]{SnCl_4} Me_2BuGeCl \xrightarrow{PrMgBr} Me_2PrBuGe$$

$$\downarrow EtMgBr$$

$$EtPrBuGeCl \longleftarrow MeEtPrBuGe \longleftarrow MePrBuGeCl$$

B. Organotin Halides

The synthesis of organotin halides is of paramount importance for organotin chemistry, because these compounds are considered as the starting materials for the preparation of a great number of organotin substances.

The oldest method for the synthesis of the organotin halides is the reaction of alkyl or aryl halides with elemental tin. This reaction exhibits numerous variants and uses a wide variety of catalysts; however, yields are usually unsatisfactory and the method is no longer common nowadays.

There are two very efficient methods for the preparation of organotin halides. The first one[413] consists of the halo- or protodemetallation of tetraorganotin or hexaorganoditin compounds by a halogen or hydrogen halide, as well as by metal halides and alkyl or aryl halides. The other one[413,414] involves the alkylation or arylation of either tin(II) or tin(IV) halides, carried out by Grignard reagents or alkyl and aryl metals (mainly of Li, Al and Hg).

In addition to the above-mentioned methods, more specific preparations, described below, are also used.

Several patents improve the old 'direct' method[413,415]; benzyl or octyltin chlorides and bromides have been obtained by refluxing mixtures of tin granules and the appropriate organic halides, using the corresponding alcohol as solvent, in the presence of butylamine, cyclohexylamine or aniline and phenol[292]. On the other hand, heating a tin foil with butyl chloride, triphenylarsine and iodine for 32 hours at $170-180\,^{\circ}$C resulted in 95% conversion of tin into $BuSnCl_3$ (16%), Bu_2SnCl_2 (78%) and Bu_3SnCl (6%)[293]. Heating a mixture of tin leaves, magnesium, butyl iodide and butanol at $220-225\,^{\circ}$C under nitrogen results in the formation of Bu_2SnI_2 in good yield[294].

High yields are also obtained for the reaction of alkyl iodides with tin, producing dialkyltin diiodides, in the presence of catalysts, either nitrogen-containing heterocycles or quaternary ammonium compounds[295]. Actually, tetraalkylammonium salts were found to be the most effective catalysts, as they give quantitative conversion of tin in 24 h at $101\,^{\circ}$C in the presence of iodine[296]. The mechanism of this catalytic conversion is given in Scheme 25.

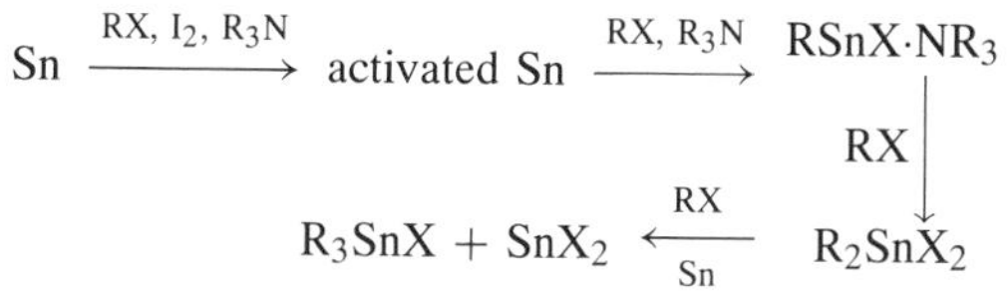

$$Sn \xrightarrow{RX,\ I_2,\ R_3N} \text{activated Sn} \xrightarrow{RX,\ R_3N} RSnX \cdot NR_3$$

$$\downarrow RX$$

$$R_3SnX + SnX_2 \xleftarrow[Sn]{RX} R_2SnX_2$$

SCHEME 25

Diorganotin dihalides R_2SnX_2 (X = Cl, Br, I, R = Bu, octyl, Ph) are prepared in excellent yields from RX and Sn powder using $(Bu_4N)^+I^-$, Me_3SiI, $(Ph_3MeAs)^+I^-$ in the presence or absence of iodine or inorganic iodides[297].

1,5-Crown-5 catalyzed the synthesis of functionalized dialkyldibromostannanes R_2SnBr_2 (R = CH_2CO_2Et, $CH_2CH=CH_2$) with yields of 51 and 65%, respectively, from RBr and tin powder[298].

Upon reaction of elemental tin with dimethyl itaconate in the presence of anhydrous HCl in ethylene glycol, dimethyl or diethyl ether, 92% of $MeO_2CCH_2CH(CO_2Me)CH_2SnCl_3$ is obtained as sole product[299].

Organotin compounds R_nSnCl_{4-n} (R = alkyl, n = 1-4) were prepared by reaction of tin powder with alkyl halides in the presence of $R_aN[(CH_2CH_2O)_mCH_2CH_2OR']_{3-a}$ (R = alkyl, aryl, arylalkyl, R' = H, alkyl, aryl, a = 0, 1, 2, m = 0-20). Iodine should be added as catalyst. Stirring for 3–6 h at 170–175 °C provides 84% of either R_3SnCl or R_2SnCl_2 (R = octyl)[300]. Diallyldibromostannane can be prepared in the absence of oxygen from allyl bromide and tin at room temperature[301].

Dimethyltin dichloride can be synthesized from methyl chloride, tin(II) chloride and molten tin. The reaction proceeds in high yields in $NaAlCl_4$ melt[359]. High yields in triorganotin halides are also obtained from tin and lower alkyl halides, provided an equimolar amount of halide is added to the reaction medium[360].

Elemental tin reacts with functional or simple alkyl halides in solution with hexamethylphosphotriamide (HMPT) and with CuI as a catalyst, to give very good yields of diorganotin dihalides complexed by HMPT. When these complexes are treated with KOH, the corresponding diorganotin oxides are generated. They are readily converted into free diorganotin halides by hydrogen halides[346]:

$$R_2SnX_2 \cdot 2HMPT \xrightarrow{\text{KOH}} (R_2SnO)_n \xrightarrow{\text{HBr}} R_2SnBr_2$$

Reaction of a Grignard reagent with these complexes provides mixed tetraorganotin derivatives:

$$Me_2SnBr_2 \cdot 2HMPT \xrightarrow{\text{PhMgBr}} Me_2SnPh_2$$

This is an easy and convenient preparation method of such compounds.

Halostannanes H_3SnX (X = Cl, Br, I) are formed by the action of hydrogen halides on stannane at room temperature[302,303].

Several halogenating agents like CCl_4 or CH_3OCH_2Cl were successfully used[305] for the preparation of organotin halides. $SnCl_4$, $SnBr_4$[306], $SOCl_2$[307], $(NH_4)_2PbCl_6$[308], Br_2[309], ICl and IBr[310] are equally suitable reagents.

Grignard reagents are used successfully for the alkylation or arylation of $SnCl_4$[311], in particular when they are applied for the introduction of large organic groups onto the tin moiety[312]. Grignard alkylation of $SnCl_4$ with RX (R = butyl, cyclohexyl, octyl; X = Cl, Br), in toluene or xylene in the presence of benzyldimethylamine gave R_2SnCl_2 in 58 to 80% yield[327].

It is noteworthy that R_3SnCl or R_2SnCl_2, treated with $R'MgCl$, gave R_3SnR' or $R_2SnR'_2$, respectively, and that the latter compounds, after reaction with $SnCl_4$, were converted into $R_2R'SnCl$ (R, R' = Me, Pr, Bu, cyclohexyl, n-hexyl, n-octyl)[328], dealkylation reactions occurring mostly with methyl as a leaving group.

A number of trifluoromethylphenyltin(IV) derivatives, $Bu_nSn[2,4-(CF_3)_2C_6H_3]_{4-n}$ ($n = 1–3$), have been obtained by the reaction of the corresponding butyltin chlorides and 2,4-bis(trifluoromethyl)phenylmagnesium bromide[313].

Organotin trifluorides $RSnF_3$ (R = alkyl, alkenyl or aryl) are easily prepared by treating corresponding organotin tricarboxylates with concentrated HF in benzene[314]:

$$RSn(O_2CCH_2CH=CH_2)_3 \xrightarrow[20-25\,°C]{HF,\ C_6H_6} RSnF_3$$

The synthesis of $Sn(CF_3)_4$ can be achieved by the reaction of SnI_4 with the trifluoromethyl radical in a radio-frequency discharge[315].

Usually, the trifluoromethyl-substituted stannanes of general type $(CF_3)_nSnH_{4-n}$ ($n = 1–3$) can be prepared from the corresponding halides with Bu_3SnH at $-40\,°C$[316,317]. $(CF_3)_2Hg$ or $(CF_3)_2Cd$ can be used for the preparation of $(CF_3)SnBr_3$, $(CF_3)_2SnBr_2$,

$(CF_3)_3SnBr$ or $(CF_3)_4Sn$, by reaction with $SnBr_4$. Mixtures of the above compounds are obtained[317]:

$$SnBr_4 \xrightarrow{(CF_3)_2Hg} \underset{55\%}{CF_3SnBr_3} + \underset{16\%}{(CF_3)_2SnBr_2} + \underset{8\%}{(CF_3)_3SnBr} + \underset{<0.5\%}{(CF_3)_4Sn}$$

Reaction of $(CF_3)_4Sn$ with HI or BI_3 provides $(CF_3)_3SnI$.

Dimethyltin(IV) difluorosulfates in liquid SbF_5 and dimethyltin(IV) fluorides in superacids like $HF-MF_5$ (M = Sb, Nb, Ta), as well as in $HF-SnF_4$, give rise to the dimethyltin(IV) fluorometalates $Me_2Sn(MF_6)_2$, $Me_2Sn(Sb_2F_{11})_2$ and $Me_2Sn(SnF_5)_2$[361].

Anhydrous $SnCl_2$ can be used with Grignard reagents in ether solution, providing R_2SnCl_2 in moderate yields (35–55%). The intermediate diorganotin(II) species are oxidized by sodium hypochlorite to the diorganotin oxides, which are converted into the diorganotin(IV) chlorides by dilute HCl[318].

SnX_2 (X = Cl, Br), the polymeric species $(SnR_2)_n$ and the stannylenes SnR_2 can be used for the preparation of organotin halides[319]:

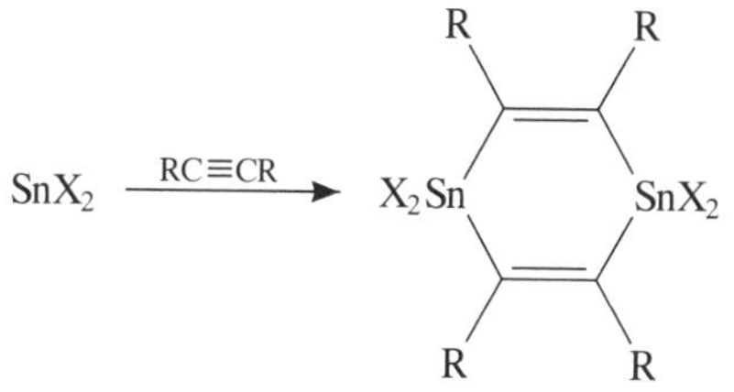

$$(Bu_2Sn)_n + nRX \longrightarrow nR_2SnX_2 \text{[320]}$$

$$X = Cl, Br, I$$

$$SnR_2 + HCl \longrightarrow R_2SnHCl$$

$$SnR_2 + MeI \longrightarrow R_2MeSnI$$

$SnCl_2$ reacts with triphenylethenyllithium $Li(CPh=CPh_2)$ in Et_2O-n-hexane–THF to yield an alkenyltin(II) derivative which, subsequently treated with butyl bromide, yields a mixed alkyltin halide[362]:

$$SnCl_2 \xrightarrow{LiCPh=CPh_2} Sn(CPh=CPh_2)_2 \xrightarrow{BuBr} BuSn(CPh=CPh_2)_2Br$$

Tin(II) halides can also react with lead(II) acetate to yield dihalotin(IV) diacetate[321]:

$$SnX_2 \xrightarrow{(CH_3CO_2)_2Pb} (CH_3CO_2)_2SnX_2$$

On the other hand, SnX_2 easily reacts with α,ω-dihaloalkanes affording ω-haloalkyltin trichlorides, provided R_3Sb is used as a catalyst[322]:

$$SnX_2 \xrightarrow[\substack{R_3Sb \\ m \geqslant 3}]{X(CH_2)_mX} X_3Sn(CH_2)_mX$$

These compounds are versatile starting materials for the synthesis of a variety of ω-functionally substituted organotin compounds: $R_{3-n}X_nSn(CH_2)_mY$ (R = alkyl, phenyl) with n = 0–3, X = Cl, Br, Y = Br, NMe_2, NEt_2, COOH, CH(OH)R, R_3Sn.

Haloalkyltin compounds were obtained by the following sequence of reactions[363]:

$$R_3SnH \xrightarrow{CH_2=CH(CH_2)_nOH} R_3Sn(CH_2)_{n+2}OH \xrightarrow{PPh_3/CCl_4} R_3Sn(CH_2)_{n+2}Cl$$

ω-Haloalkyltin compounds can be prepared[323] in a different way by treating organotin halides $R_{4-m}SnCl_m$ ($m = 1$, R = Me, Et, Ph; $m = 2$, R = Me) with sodium in liquid ammonia and by adding a solution of $X(CH_2)_nX$ (X = Cl, $n = 3, 4$; X = Br, $n = 3$; X = I, $n = 4$) in ether to that mixture. Organotin compounds of the type $R_{4-m}Sn[(CH_2)_nX]_m$ are formed in 28 to 72% yields.

Alkyltin trihalides with chain lengths of up to 18 carbon atoms can be prepared in the presence of trialkylantimony catalysts[340]:

$$SnCl_2 + OctBr \xrightarrow[20\ h,\ 130\ ^\circ C]{Et_3Sb} \underset{94\%}{OctSnCl_2Br}$$

An oxidative addition is likewise taking place[341] during the reaction of tin(II) acetylacetonate $Sn(acac)_2$ with alkyl iodides. The reaction is promoted by light:

$$Sn(acac)_2 \xrightarrow[R\ =\ Me,\ Pr]{RI} RSn(acac)_2I$$

The same type of addition occurs in the reaction of cyclopentadienyl tin(II) with MeI[341]:

$$(C_5H_5)_2Sn \xrightarrow[3\ h,\ light]{MeI,\ 5\ ^\circ C} Me(C_5H_5)_2SnI$$

Cyclopentadienyltin(IV) halides Cp_nSnX_{4-n} (X = Cl, Br, I) can be prepared by halogenation of Cp_2Sn[342]:

$$Cp_2Sn + X_2 \longrightarrow Cp_2SnX_2$$

$$2Cp_2Sn + 2HgX_2 \longrightarrow Cp_3SnX + CpSnX + Hg_2X_2$$

or by the reaction

$$Cp_2Sn + SnCl_4 \longrightarrow Cp_2SnCl_2 + SnCl_2$$

Arylcopper can be used as arylation agent to react with diorganotin dibromides, providing interesting aryl-substituted organotin bromides[324]:

$$ArCu + Ph_2SnBr_2 \longrightarrow ArPh_2SnBr + CuBr$$

$$(Ar = C_6H_4CH_2NMe_2, C_6H_4NMe_2, C_6H_4OMe)$$

Alkylation by alkyllithium or alkylcopper (as above) of dialkyltin dibromides can also provide chiral trialkyltin halides[324].

Reaction of $SnCl_4$ with RLi, where R is the supermesityl radical, $[1,3,5\text{-}(Me_3C)_3C_6H_2]$, gave 30% of **117A** $(Me_3C)_2C_6H_3CMe_2CH_2SnCl_2R$. The same reaction with an excess of $SnCl_4$ gave **117A** and the isomeric bis(supermesityl)tin dichloride R_2SnCl_2[325].

Diaryl- or dialkylmercury compounds can be used for the synthesis of mixed aryltin chlorides[326], or for alkylation of SnF_2[338] or SnX_4[339] (Scheme 26). Functional halostannanes can be prepared by a variety of methods.

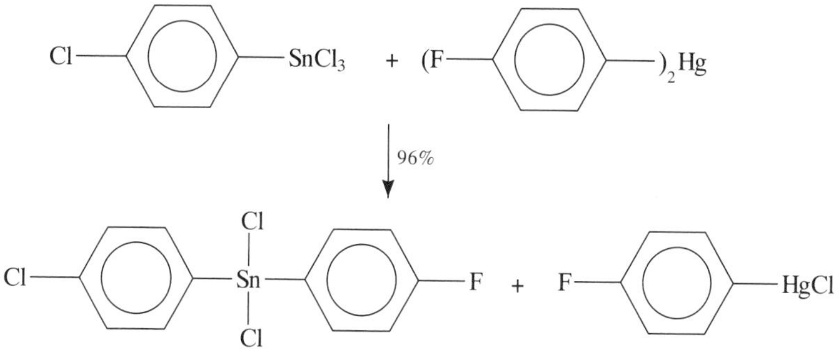

Cl—⟨C₆H₄⟩—SnCl₃   +   (F—⟨C₆H₄⟩—)₂Hg

↓ 96%

Cl—⟨C₆H₄⟩—Sn(Cl₃)—⟨C₆H₄⟩—F   +   F—⟨C₆H₄⟩—HgCl

$$4\,SnF_2 + R_2Hg \xrightarrow{\Delta} RSnF_3 + HgF_2 + 2\,Sn$$

R = Et, Pr, Bu, Pent, Hex

$$(PhCH_2)_2Hg + SnX_4 \xrightarrow{benzene} PhCH_2HgX + PhCH_2SnX_3$$

X = Cl, Br, I

SCHEME 26

The prepararation of α-iodoalkyltin iodides can be performed in THF by reaction of Me_2SnCl_2 with iodomethylzinc iodide, $IZnCH_2I$, in molar ratio 1:1[329]:

$$Me_2SnCl_2 \xrightarrow[THF]{IZnCH_2I} Me_2ISnCH_2I$$

The compound obtained can subsequently be converted into hexamethyl-1,3,5-tristannacyclohexane by Mg in THF in 22% yield:

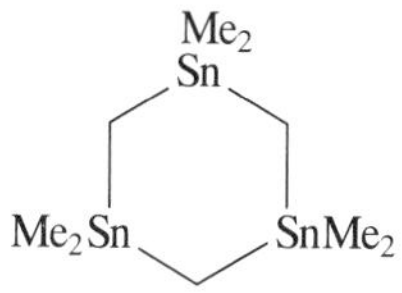

Me₂Sn / Me₂Sn—SnMe₂ (ring structure)

Other syntheses of functional organotin halides include that of β-chlorovinyltin compounds[330]:

$$(ClHC=CH)_2Hg + Bu_3SnH \longrightarrow Bu_3SnCH=CHCl$$
38% cis, 62% trans

and of diorganotin chloride alkoxides[331]:

$$Sn(OCH_3)_4CH_3COCl \longrightarrow Cl_2Sn(OCH_3)_2 \cdot CH_3COOCH_3$$

The reaction of alcohols with $SnCl_2$ in the presence of N-chlorosuccinimide affords alkoxytrichlorotin(IV) compounds or their disproportionation products, dialkoxydichlorotin(IV) and $SnCl_4$[332].

Reaction of tin(II) halides with halomethylsilanes at 140–180 °C affords $H_3SiCH_2SnX_3$[333a]. In general

$$RR'R''SiCH_2Cl + SnCl_2 \xrightarrow[\text{THF}]{140-200\,^\circ C} RR'R''SiCH_2SnCl_3$$

$$R, R', R'' = \text{alkyl, aryl or alkoxy}$$

Compounds of the type $Cl_n Me_{3-n} SiCH_2SnCl_3$ can also be prepared by using onium salts as catalysts[333b]:

$$Cl_n Me_{3-n} SiCH_2Cl + SnCl_2 \xrightarrow[140-180\,^\circ C]{R_4E^+X^-} Cl_n Me_{3-n} SiCH_2SnCl_3$$

$$E = N, P, As, Sb; X = Cl, Br; n = 1-3$$

Redistribution reactions are considered as quite important for the preparation of alkyltin halides having mainly identical alkyl groups. Actually, the redistribution reactions are halogenation reactions of organotin halides by SnX_4 ($X = F, Cl, Br, I$).

More precisely, depending on the molar ratio of the reactants, one of the following reactions can proceed:

$$(i) \quad 3\, R_4Sn + SnX_4 \longrightarrow 4\, R_3SnX$$

$$(ii) \quad R_4Sn + SnX_4 \longrightarrow 2\, R_2SnX_2$$

$$(iii) \quad R_4Sn + 3\, SnX_4 \longrightarrow 4\, RSnX_3$$

Since secondary reactions take place in these processes, they are actually redistributions either between products:

$$(iv) \quad R_3SnX + RSnX_3 \longrightarrow 2\, R_2SnX_2$$

or between products and reactants:

$$(v) \quad 2\, R_3SnX + SnX_4 \longrightarrow 3\, R_2SnX_2$$

$$(vi) \quad 2\, RSnX_3 + SnR_4 \longrightarrow 3\, R_2SnX_2$$

$$(vii) \quad R_2SnX_2 + SnR_4 \longrightarrow 2\, R_3SnX$$

$$(viii) \quad R_2SnX_2 + SnX_4 \longrightarrow 2\, RSnX_3$$

The reactions (v) to (viii) proceed when an excess of either SnX_4 or SnR_4 is used for the reactions (i) to (iii).

The first redistribution reactions were introduced early by Kocheshkov[334]. They remain still the subject of extended interest in the field of organometallic syntheses of Ge(IV), Sn(IV) and Pb(IV) compounds[291,335,336]. For example, the redistribution reactions occurring between allyltrialkyltin or crotyltributyltin compounds and $SnCl_4$ at $-50\,^\circ C$ were studied by NMR[335]:

$$CH_2{=}CHCH_2SnR_3 + SnCl_4 \xrightarrow{-50\,^\circ C} CH_2{=}CHCH_2SnCl_3 + R_3SnCl$$

$$MeCH{=}CHCH_2SnBu_3 + SnCl_4 \xrightarrow{-50\,^\circ C} CH_2{=}CHCHMeSnCl_3 + Bu_3SnCl$$

$$\Big\updownarrow$$

$$MeCH{=}CHCH_2{-}SnCl_3$$

Halogen exchange reactions were also described[336]:

$$i\text{-Pr}_2\text{SnBr}_2 + \text{SnCl}_4 \longrightarrow i\text{-Pr}_2\text{SnCl}_2 + \text{SnBr}_2\text{Cl}_2$$

Halogen exchange reactions can also be used for the preparation of mixed diorganotin dihalides of the type R_2SnXY[337] (X, Y = Cl, Br, I):

$$\text{R}_2\text{SnX}_2 + \text{R}_2\text{SnY}_2 \longrightarrow 2\text{R}_2\text{SnXY}$$

$$\text{R}_2\text{SnX}_2 + \text{Me}_3\text{SiY} \longrightarrow \text{R}_2\text{SnXY} + \text{Me}_3\text{SiX}$$

$$(\text{R} = \text{Me, Bu})$$

Y represents a halogen with a higher atomic mass than X.
This reaction is likewise applicable to diorganotin dihalides:

$$\text{Bu}_2\text{SnCl}_2 + 2\text{Et}_3\text{SiI} \longrightarrow \text{Bu}_2\text{SnI}_2 + 2\text{Et}_3\text{SiCl}$$

Reactions of K_2SnO_2 with $^{13}\text{CH}_3\text{I}$ afford ^{13}C-labeled trialkyltin iodides[343]:

$$3\ \text{K}_2\text{SnO}_2 + 9\ \text{CH}_3\text{I} \longrightarrow 3\ (\text{CH}_3)_3\text{SnI} + 2\ \text{KIO}_3 + 4\ \text{KI}$$

while organotin oxides reacting with acyl chlorides provide an easy route to organotin dichlorides[344]:

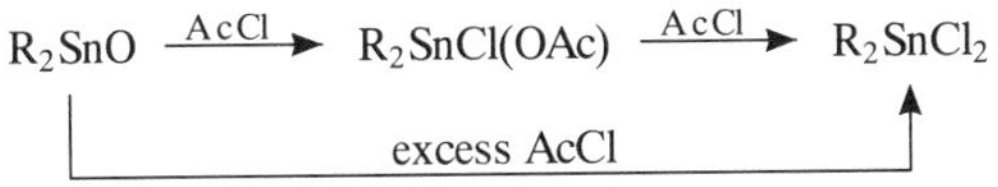

$$\text{R}_2\text{SnO} \xrightarrow{\text{AcCl}} \text{R}_2\text{SnCl(OAc)} \xrightarrow{\text{AcCl}} \text{R}_2\text{SnCl}_2$$

$$\text{excess AcCl}$$

$$(\text{R} = \text{Me, Et, Pr, Bu, Oct})$$

Monohalomonohydrotin compounds can be prepared in excellent yields by reactions of triphenyltin hydrides with dialkyl- or diphenyltin dihalides[345]:

$$\text{Ph}_3\text{SnH} + \text{Ph}_2\text{SnX}_2 \rightleftharpoons \text{Ph}_3\text{SnX} + \text{Ph}_2\text{SnHX}$$

$$\text{Ph}_3\text{SnH} + \text{R}_2\text{SnX}_2 \rightleftharpoons \text{Ph}_3\text{SnX} + \text{R}_2\text{SnHX}$$

$$\text{X} = \text{Cl, Br, I}$$

Hydrostannylation, as a counterpart to hydrogermylation for tin, using HSnCl_3, is not of common use in organotin chemistry. In contrast, R_3SnH is very commonly used (see Section I.B).

Nevertheless, β-carbonylalkyltin chlorides are obtained[347] in high yields by reactions of the carbonyl-activated alkenes $\text{RR}'\text{C=CR}''\text{-C(R}''')\text{=O}$ with HCl and SnCl_2, yielding compounds of the type **118**, while with tin metal and HCl compounds of the type **119** are obtained.

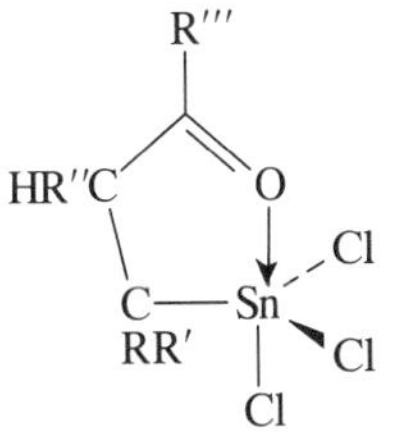

(118) 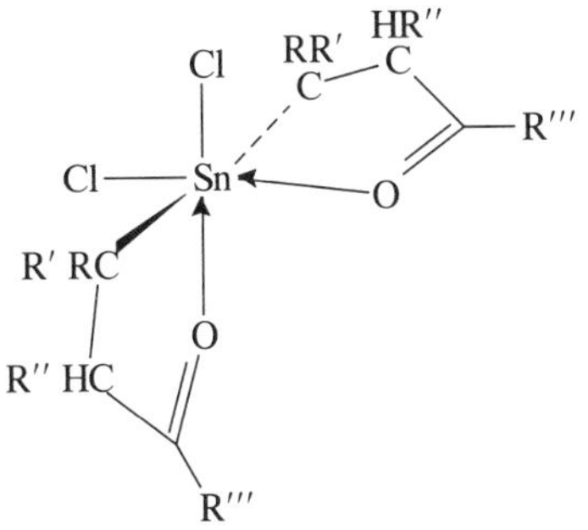   **(119)**

Although this synthesis is extremely simple, its mechanism is not yet clearly understood. It has been proposed that such reactions can be rationalized[348] in terms of 1,4-additions, where the initial attack by a proton occurs at the carbonyl oxygen, giving rise to a very stable carbocation that is not only allylic but also stabilized by the mesomeric donating oxygen. This step is followed by the nucleophilic addition of $SnCl_3^{(-)}$ and by a ketoenol tautomerism (Scheme 27).

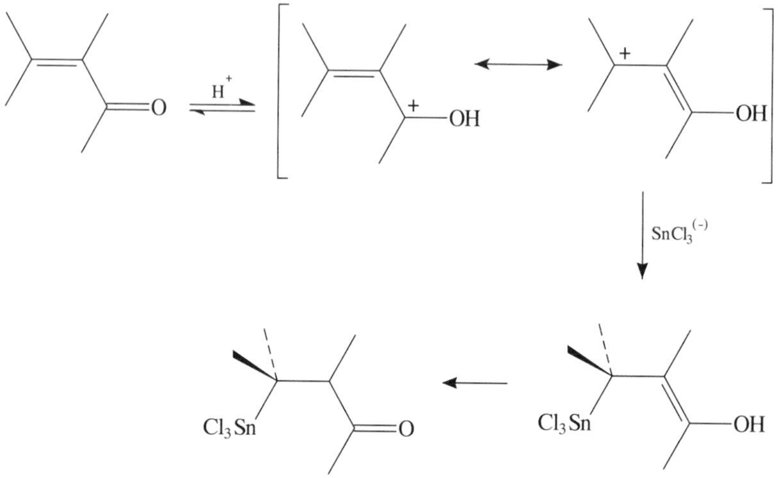

SCHEME 27

The above mechanism requires the existence in the reaction mixture of species like $HSnX_3$ or H_2SnX_4. These intermediates are extremely unstable but could be stabilized as etherates. From the phases $SnCl_2-HCl-Et_2O$ or $SnBr_2-HBr-Et_2O$, $HSnCl_3 \cdot 2Et_2O$ and $H_2SnBr_4 \cdot 3Et_2O$ could indeed be prepared[349]. The above systems can easily be produced as pale yellow oils, poorly soluble in ether, by dissolving elemental tin into a dry solution of gaseous HCl or HBr in absolute ether. They are smoking in air and are highly acidic.

Diorganotin dihalides react with aryllithium, arylcopper or arylgoldlithium providing triorganotin halides[350]:

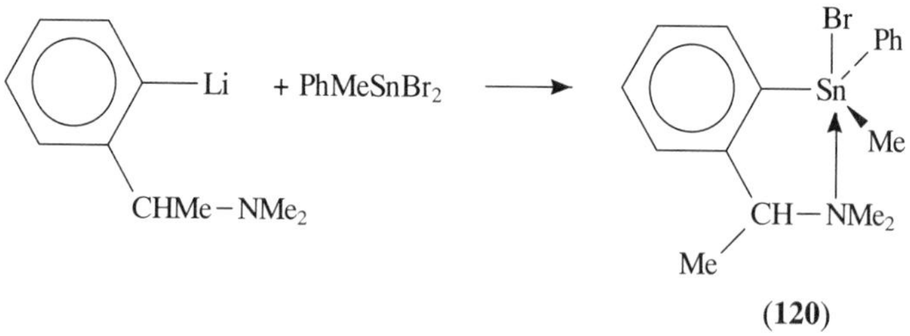

(120)

The compound 120, as well as 118, has a typical five-coordinate tin(IV) atom with a trigonal bipyramidal geometry. In addition, compound 120 is chiral.

Next, some special methods for preparing organotinhalides are presented. 1,2-Dichlorotetraalkyldistannanes ClR_2SnSnR_2Cl (R = Me, Et, Bu) have been prepared by the electrolysis of acetonitrile solutions of the appropriate dialkyltin dichloride using a mercury cathode[351]. α,ω-Dichlorodistannanes, useful for the preparation of large tin macrocycles, can be prepared by halodephenylation of the corresponding parent phenyl compounds, using HBr or HCl[352]:

$$Ph_3Sn(CH_2)_n SnPh_3 \xrightarrow{HBr} Ph_2BrSn(CH_2)_n SnPh_2Br$$

Extension of the chain can be achieved by a Grignard reaction:

$$Ph_2BrSn(CH_2)_n SnPh_2Br \xrightarrow{BrMg(CH_2)_n Cl} Cl(CH_2)_n (Ph_2)Sn(CH_2)_n Sn(Ph_2)(CH_2)_n Cl$$

These halides can be used for the preparation of large macrocycles as the reactions in Scheme 28 indicate. In contrast, ω-hydroxyalkyltetrafluorostannates are obtained from stannacycloalkanes by rapid oxidation via an intermediate peroxide, detected by polarographic analysis, followed by cleavage of the Sn—Ph bonds[353] (Scheme 29).

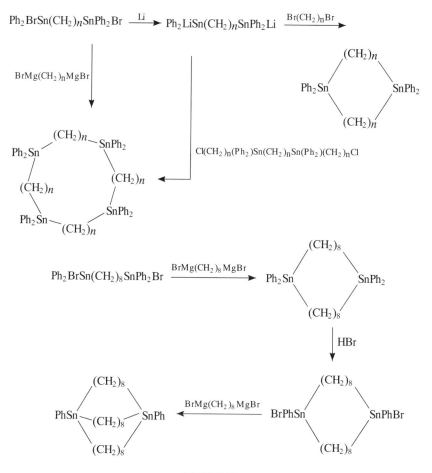

SCHEME 28

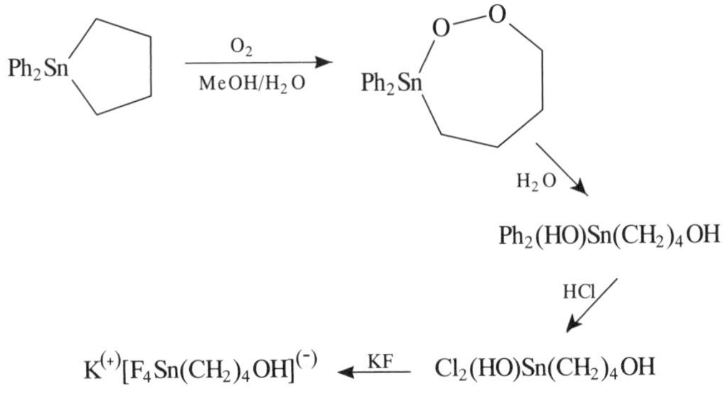

$$Ph_2Sn \overset{O_2}{\underset{MeOH/H_2O}{\longrightarrow}} Ph_2Sn$$

$$\downarrow H_2O$$

$$Ph_2(HO)Sn(CH_2)_4OH$$

$$\downarrow HCl$$

$$K^{(+)}[F_4Sn(CH_2)_4OH]^{(-)} \xleftarrow{KF} Cl_2(HO)Sn(CH_2)_4OH$$

SCHEME 29

Several organotin(IV) halides with transannular $Sn \leftarrow N$ interactions were synthesized according to the following reactions[354]:

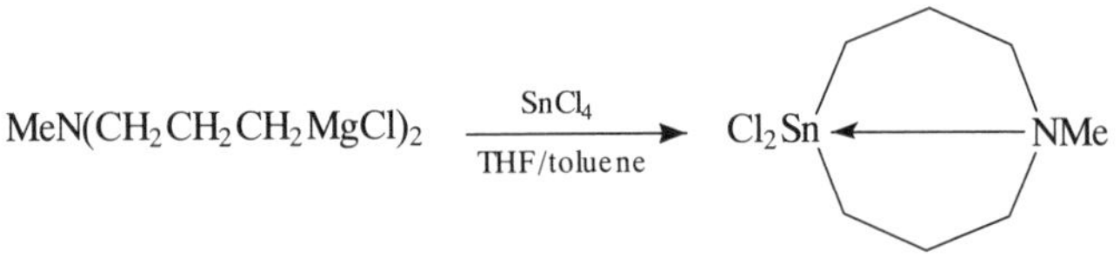

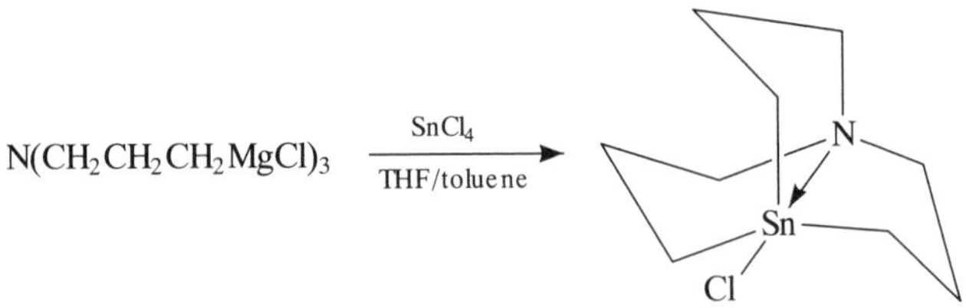

$$MeN(CH_2CH_2CH_2MgCl)_2 \xrightarrow[THF/toluene]{SnCl_4} Cl_2Sn \longleftarrow NMe$$

$$N(CH_2CH_2CH_2MgCl)_3 \xrightarrow[THF/toluene]{SnCl_4}$$

Trineophyltin chloride can be prepared analogously from $SnCl_4$ and the appropriate Grignard reagent[355]:

$$C_6H_5CMe_2CH_2MgCl \xrightarrow{SnCl_4} (C_6H_5CMe_2CH_2)_3SnCl$$

Mesitylmagnesium bromide RMgBr, another bulky Grignard reagent, reacts with $SnCl_4$ and gives a mixture of diorganotin halides: R_2SnCl_2, R_2SnBr_2 and $R_2SnClBr$. Reaction with $SnCl_4$ of the mixture obtained gives the full range of $RSnBr_3$, $RSnClBr_2$, $RSnCl_2Br$ and $RSnCl_3$.

When mesityllithium reacts with $SnCl_4$ in a 2:1 molar ratio, only R_2SnCl_2 is generated, in contrast to the more sterically demanding supermesityl lithium that, when reacting with $SnCl_4$ or $SnBr_4$, gives the interesting $R'_2XSnSnR'_2X$ compound[356]. Note that hydrogen chloride reacts with hexaphenylditin to yield the analogous derivative $Ph_2ClSnSnPh_2Cl$[304]. Mixed diorganotin dihalides R_2SnXY (XY = FCl, ClBr, ICl and IBr) have been obtained by reaction of Br_2, I_2 and HgI_2 with triorganotin halides R_3SnX (R = Ph, $PhCH_2$, p-MeC_6H_4 or cyclohexyl; X = F, Cl, Br) through facile cleavage of the Sn—C bond, like in the following reactions[357]:

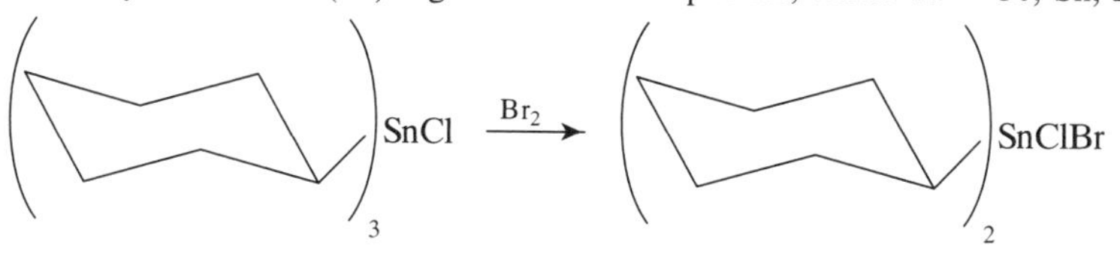

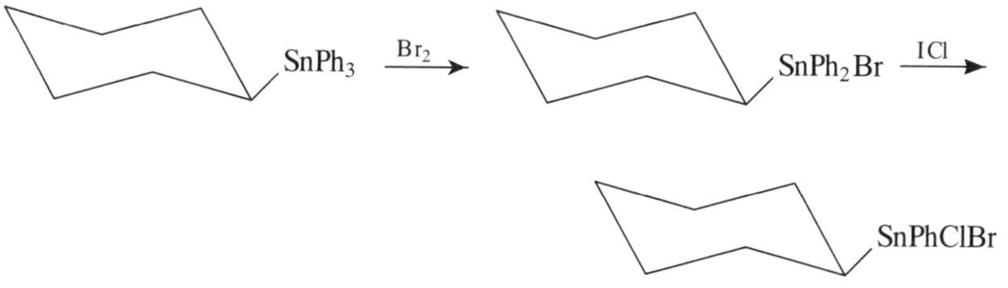

(121)

Note that the latter reaction gives the racemic phenylcyclohexyltin bromide chloride **121**. Triorganotin fluorides can be prepared by the use of new fluorinating systems. 18-Crown-6 or dibenzo-24-crown-8 can act as solid–liquid phase transfer catalysts for CsF. Trialkyltin mercaptides can be fluorodestannylated by CsF in the presence of crown ethers or alkyl bromides[358]:

$$CsF + Bu_3Sn-SCH_2Ph + CH_3(CH_2)_5Br \longrightarrow PhCH_2S(CH_2)_5CH_3 + Bu_3SnF + CsBr$$

C. Organolead Halides

Tri-n-butyllead chloride (Bu_3PbCl) can be easily prepared either by dealkylation of Bu_4Pb by HCl[388] or by chlorination of hexabutyldilead, $Bu_3PbPbBu_3$[389]:

$$Bu_4Pb \xrightarrow[\text{pet. ether}]{HCl, \, 0\,°C} Bu_3PbCl$$

$$(Bu_3Pb)_2 \xrightarrow[\text{ether, } N_2]{Cl_2, \, -60\,°C} Bu_3PbCl$$

When the halogenations of Bu_4Pb or Bu_6Pb_2 are carried out for a long time at $-10\,°C$ with an excess of chlorine gas, dibutyllead dichloride is generated[389]:

$$Bu_4Pb \xrightarrow[N_2]{Cl_2} Bu_2PbCl_2$$

$$(Bu_3Pb)_2 \xrightarrow[N_2]{Cl_2} Bu_2PbCl_2$$

Hexaneopentyldilead and hexaneophyldilead are converted into $(neopentyl)_3PbI$ and $(neophyl)_3PbI$, respectively[159], by iodination.

In contrast with the halogenations of hexabutyldilead and of other hexalkyldileads[390]:

$$R_3PbPbR_3 \xrightarrow[C_6H_6]{Br_2} R_3PbBr$$

those of hexaphenyldilead are very complex[156,177] because they may proceed in two ways simultaneously:

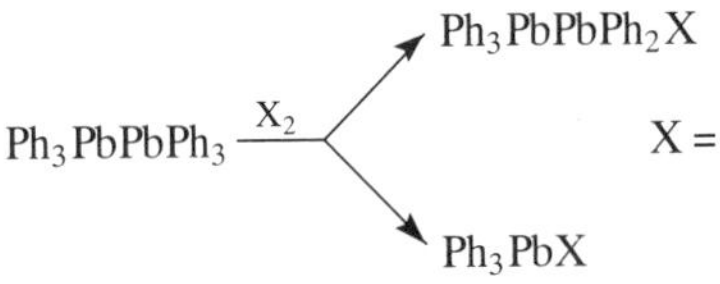

$$Ph_3PbPbPh_3 \xrightarrow{X_2} \begin{cases} Ph_3PbPbPh_2X \\ \\ Ph_3PbX \end{cases} \quad X = Cl, Br, I$$

Ar_2PbX_2 and Ar_3PbX can be prepared, respectively, from Ar_3PbX or Ar_6Pb_2, according to the following reactions[392]:

$$Ar_3PbX + HX \longrightarrow Ar_2PbX_2 + ArH$$

$$Ar_6Pb_2 + 3HX \longrightarrow Ar_3PbX + PbX_2 + 3ArH$$

Diaryllead dihalides can also be obtained from the following reaction:

$$Ph_2Pb(OAc)_2 + 2HX \longrightarrow Ph_2PbX_2 + 2AcOH \ (X = F, Cl, Br, I)^{393}$$

Insertion of PbI_2 into alkyl iodides in the presence of Me_3Sb as a catalyst provides monoalkylleadtriiodides[394]:

$$PbI_2 + RI \xrightarrow{Me_3Sb} RPbI_3$$

Plumbylalkenyl or plumbylvinyl halogen compounds are prepared by special methods.
   Triphenyl-3-butenyllead, prepared from $CH_2=CHCH_2CH_2MgBr$ and Ph_3PbCl, reacting with HCl, provides butenyldiphenyllead chloride[395]:

$$Ph_3Pb\text{-}CH_2CH_2CH=CH_2 + HCl \longrightarrow Ph_2PbCl(CH_2CH_2CH=CH_2) + C_6H_6$$

Tetravinyllead reacts with BCl_3[396], MX_3 (M = P, As, Sb)[397] or $M'X_n$ (M' = Al, Si, P)[166,398] (X = Cl, Br) giving tri- or divinyllead halides:

$$3(CH_2=CH)_4Pb + BCl_3 \xrightarrow{-70\,°C} (CH_2=CH)_3B + 3(CH_2=CH)_3PbCl$$

$$(CH_2=CH)_4Pb + 2MX_3 \xrightarrow{CCl_4} (CH_2=CH)_2PbX_2 + 2CH_2=CH-MX_2$$

Plumbyl halides with a lead–manganese bond have been prepared by the following dephenylation or demethylation reactions[191]:

$$Ph_2Pb[Mn(CO)_5]_2 \xrightarrow[-60\,°C]{X_2} X_2Pb[Mn(CO)_5]_2$$

$$Et_3PbCl \xrightarrow{NaMn(CO)_5} Et_3PbMn(CO)_5 \xrightarrow[25\,°C]{Cl_2} Et_2ClPbMn(CO)_5$$

When triaryl- or diaryllead halides react with Me_4NX or Et_4NX, arylhaloplumbates are produced, like $[Me_4N][Ph_3PbCl_2]$[399] or $[Et_4N][Ph_3PbXY]$ (X or Y = Cl, Br, I), $[Et_4N][Ph_6Pb_2X_2Y]$ (X or Y = Cl, Br), $[Et_4N][Ph_2PbX_3]$ (X = Cl, Br, I) and $[Me_4N]_2[Ph_2PbX_4]$ (X = Cl, Br)[393,400,401]. These plumbates can be obtained easily by mixing Ph_3PbCl or Ph_2PbCl_2 with Me_4NX or Et_4NX in acetonitrile, n-propanol or ethanol. Analogous organoplumbates can be prepared from R_4NX ($R = Me, Et$), Ph_4PX and Ph_4SbX and diaryllead dihalides, providing mixed halostannates of the types $[Ph_2PbXYZ]^{(-)}$ and $[Ph_2PbXYZ_2]^{(2-)}$ ($XY = ClBr, ICl, BrI, Z = Cl, Br, I$)[403].

Tetrakis(trifluoromethyl)lead (cf analogous compounds of germanium and tin in Sections III.A and III.B) has been prepared from R_4Pb ($R = Ph, p$-tolyl) and $(CF_3)_4Sn$ by a CF_3–aryl exchange. $(CF_3)_4Pb$ can be separated from the excess of $(CF_3)_4Sn$ by 1,10-phenanthroline which selectively complexes $(CF_3)_4Sn$, being a stronger Lewis acid[404]. Analogous mixed plumbanes like $(CF_3)_nPbR_{4-n}$ ($n = 1, 2$) can be prepared from $(CF_3)_nPbR_{3-n}X$ ($R = Me, Et, n = 0-2$) and $(CF_3)_2Cd$ by stepwise halide–CF_3 exchanges[405].

The compounds $(CH_3)_3Pb(CF_3)$, $(CH_3)_2Pb(CF_3)_2$ and $(CH_3)Pb(CF_3)_3$ have been prepared from hexafluoroethane and $(CH_3)_3PbX$ in a radiofrequency discharge tube in 0.2–1.9% yields[406].

$Me_3Pb(CF_3)$ can be prepared by the reaction[407]:

$$Me_4Pb + CF_3I \longrightarrow Me_3Pb(CF_3) + MeI$$

Similar compounds containing the CCl_3 or CBr_3 moiety were also prepared:

$$R_3PbOR' + HCX_3 \longrightarrow R_3PbCX_3 + R'OH \qquad (R = Ph, Et)[408]$$

$$Ph_3PbOMe + Br_3CCHO \longrightarrow [Br_3CCH(OMe)OPbPh_3] \longrightarrow Ph_3Pb(CBr_3)$$
$$+ HCOONa[409]$$

Phenylplumbyllithium reacting with chloroform or carbon tetrachloride provides organolead compounds containing C–Cl bonds[410]:

$$Ph_3PbLi + CHCl_3 \longrightarrow Ph_3PbCHCl_2 + LiCl$$

$$2Ph_3PbLi + CCl_4 \longrightarrow (Ph_3Pb)_2CCl_2 + 2LiCl$$

Diazomethane or diazoethane are used to prepare analogous ethyllead compounds[411]:

$$Et_3PbCl \xrightarrow{CH_2N_2} Et_3PbCH_2Cl$$

$$Et_3PbCl \xrightarrow{CH_3CHN_2} Et_3PbCHCl(CH_3)$$

$$Et_2PbCl_2 \xrightarrow{CH_2N_2} Et_2Pb(CH_2Cl)_2$$

Compounds of the type $R_3Pb(CH_2)_nBr$ can be prepared as follows[164]:

$$Ph_3PbLi + Br(CH_2)_nBr \longrightarrow Ph_3Pb(CH_2)_nBr + LiBr$$

Functional organolead compounds containing a C–Cl bond can also be prepared by special methods[165]:

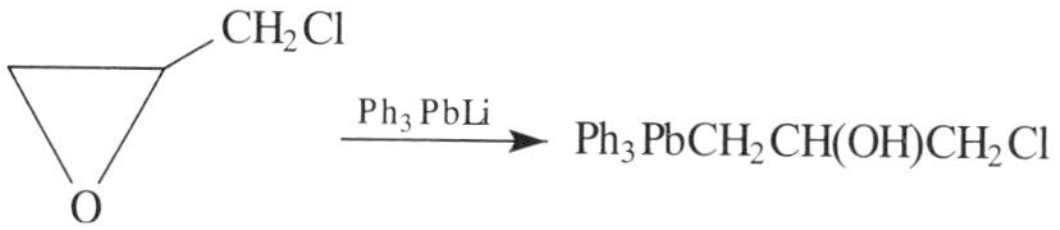

Functional organolead compounds containing C–F bonds can be prepared when R_3PbCl reacts with C_6F_5Br in the presence of $P(NEt_3)_3$ to give $R_3PbC_6F_5$.

526   J. M. Tsangaris, R. Willem and M. Gielen

The same reaction can be used for the preparation of Ge and Sn analogs because R_3MX (R = alkyl, M = Ge, Sn, X = Cl, Br) leads to such compounds[391].

IV. REFERENCES

1. J. M. Shackelford, H. de Schmertzing, C. H. Heutler and H. Podall, *J. Org. Chem.*, **28**, 1700 (1963).
2. J. A. Semlyen, G. R. Walker and C. S. G. Phillips, *J. Chem. Soc.*, 1197 (1965).
3. M. P. Brown and G. W. Fowles, *J. Chem. Soc.*, 2811 (1958).
4. A. Carrick and F. Glockling, *J. Chem. Soc. (A)*, 623 (1966).
5. F. Glockling and K. A. Hooton, *J. Chem. Soc.*, 1859 (1963).
6. P. Mazerolles and J. Dubac, *Compt. Rend.*, **257**, 1103 (1963).
7. O. M. Nefedov, G. Gargo, T. Szekely and V. I. Shirayaev, *Dokl. Akad. Nauk SSR*, **164**, 45 (1965) (Engl. transl.).
8. O. M. Nefedov and T. Székely, *Symposium on Organosilicon Chemistry*, Scientific Communications, Prague, 1965, p. 65.
9. O. M. Nefedov and M. N. Manakov, *Angew. Chem.*, **78**, 1039 (1966); *Angew Chem., Int. Ed. Engl.*, **5**, 1021 (1966).
10. W. P. Neumann and K. Kühlein, *Tetrahedron Lett.*, 1541 (1963).
11. W. P. Neumann and K. Kühlein, *Ann. Chem.*, **683**, 1 (1968).
12. F. Glockling and K. A. Hooton, *J. Chem. Soc.*, 3509 (1962).
13. H. H. Anderson, *J. Am. Chem. Soc.*, **75**, 814 (1953).
14. E. J. Bulten and J. G. Noltes, *J. Organomet. Chem.*, **11**, 19 (1968).
15. S. Cawly and S. S. Danyluk, *Can. J. Chem.*, **41**, 1850 (1963).
16. A. Koster-Pflugmacher and A. Hirsch, *Naturwissenschaften*, **54**, 645 (1967).
17. E. J. Bulten and J. G. Noltes, *Tetrahedron Lett.*, 4389 (1966).
18. E. W. Abel, R. P. Busch, C. R. Jenkins and T. Zobel, *Trans. Faraday Soc.*, **60**, 1214 (1964).
19. K. H. Bir and D. Kräft, *Z. Anorg. Allg. Chem.*, **311**, 235 (1961).
20. P. Mazerolles, *Bull. Soc. Chim. Fr.*, 1911 (1961).
21. R. Schwarz and M. Schmeisser, *Chem. Ber.*, **69**, 579 (1939).
22. F. Glockling and J. R. C. Light, *J. Chem. Soc. (A)*, 623 (1967).
23. C. Tamborski, F. E. Ford, W. L. Lehn, G. J. Moore and E. J. Soloski, *J. Org. Chem.*, **27**, 619 (1962).
24. E. J. Bulten and J. G. Noltes, *Tetrahedron Lett.* 1443 (1967); *Recl. Trav. Chim. Pays-Bas*, **91**, 1042 (1972).
25. W. P. Neumann, *Angew. Chem.*, **75**, 679 (1963).
26. O. H. Johnson and D. M. Harris, *J. Am. Chem. Soc.*, **72**, 5564 (1950).
27. R. C. Job and M. D. Curtis, *Inorg. Nucl. Chem. Lett.*, **8**, 251 (1972).
28. W. E. Davidson, B. R. Laliberte, C. M. Goddard and M. C. Henry, *J. Organomet. Chem.*, **36**, 283 (1972).
29. R. M. C. Roberts, *J. Organomet. Chem.*, **63**, 159 (1973).
30. P. J. Davidson, M. F. Lappert and R. Pearse, *J. Organomet. Chem.*, **57**, 269 (1973).
31. N. V. Fomine, N. I. Sheverdina, E. I. Dobrova, I. V. Sosnina and K. A. Kocheshkov, *Dokl. Akad. Nauk SSSR*, **210**, 621 (1973); Engl. transl., 439.
32. V. T. Bychkov, N. S. Vyajankin, G. A. Abakumov, O. V. Linzina and G. A. Razuvaev, *Dokl. Akad. Nauk SSSR*, **202**, 593 (1972); *Chem. Abstr.*, **76**, 139726d (1972).
33. K. Yamamoto and M. Kumada, *J. Organomet. Chem.*, **35**, 297 (1972).
34. M. Kumada, T. Kondo, K. Mimura, M. Ishikawa, K. Yamamoto, S. Ikeda and M. Kondo, *J. Organomet. Chem.*, **43**, 293 (1972).
35. C. G. Savory and M. G. H. Wallbridge, *J. Chem. Soc., Dalton Trans.*, 918 (1972).
36. M. L. Thomson and R. N. Grimes, *Inorg. Chem.*, **11**, 1925 (1972).
37. R. E. Loffredo and A. D. Norman, *J. Am. Chem. Soc.*, **93**, 5587 (1971).
38. I. M. Gverdtsiteli and M. O. Charturiya, *Zh. Obshch. Khim.*, **42**, 1773 (1972); Engl. transl., 1760.
39. M. A. Kazankova, T. I. Zverkova and I. F. Lutsenko, *Dokl. Vses. Konf. Khim. Atsetilena*, **173** (1972); through *Chem. Abstr.*, **79**, 42625n (1973).
40. M. Massol, Y. Cabadi and J. Satgé, *Bull. Soc. Chim. Fr.*, 3235 (1971).
41. R. J. P. Corriu and J. J. E. Moreau, *J. Organomet. Chem.*, **40**, 73 (1972).
42. D. Seyferth and D. L. White, *J. Organomet. Chem.*, **34**, 119 (1972).

10. Synthesis of M(IV) organometallic compounds, where M = Ge, Sn, Pb   527

43. K. B. Rall, A. I. Vil'davskaya and A. A. Petrov, *Zh. Obshch. Khim.*, **42**, 1598 (1972); *Chem. Abstr.*, **77**, 101804k (1972).
44. P. Mazerolles, A. Laporterie and J. Dubac, *Compt Rend.*, **257**, 387 (1972).
45. A. Laporterie, J. Dubac, P. Mazerolles and M. Lesbre, *Tetrahedron Lett.*, 4653 (1971).
46. K. I. Kobrakov, T. I. Chernysheva, N. S. Nametkin and A. Ya. Sideridu, *Dokl. Akad. Nauk SSSR*, **196**, 100 (1971); *Chem. Abstr.*, **75**, 6042k (1971).
47. K. I. Kobrakov, T. I. Chernysheva and N. S. Nametkin, *Dokl. Akad. Nauk SSSR*, **198**, 1340 (1971); *Chem. Abstr.*, **75**, 129908e (1971).
48. J. Y. Corey, M. Dueber and M. Malaidza, *J. Organomet. Chem.*, **36**, 49 (1972).
49. J. V. Scribeli and M. D. Curtis, *J. Organomet. Chem.*, **40**, 317 (1972).
50. P. Rivière, J. Satgé and D. Soula, *J. Organomet. Chem.*, **63**, 167 (1973).
51. T. J. Barton, E. A. Kline and P. M. Carvey, *J. Am. Chem. Soc.*, **95**, 3078 (1973).
52. P. Mazerolles, M. Joanny and G. Tourrou, *J. Organomet. Chem.*, **60**, 63 (1973).
53. P. Rivière, J. Stagé and D. Soula, *Compt. Rend.*, **277**, 893 (1973).
54. B. R. Laliberte and S. A. Leone, *J. Organomet. Chem.*, **37**, 209 (1972).
55. P. Mazerolles, *Bull. Soc. Chim. Fr.*, 1907 (1962).
56. Y. Takeuchi, *Main Group Met. Chem.*, **12**, 305 (1989).
57. Y. Yashifumi, K. Oshimara and T. Utimoto, *Bull. Chem. Soc. Jpn.*, **61**, 2693 (1988).
58. E. Lukevics, R. Ya. Sturkovich, O. A. Pudova, *Zh. Obshch Khim.*, **58**, 815 (1988); *Chem. Abstr.*, **110**, 75677x (1989).
59. M. Lazraq, J. Escudié, C. Couret, J. Satgé, M. Drayer and R. Damnel, *Angew. Chem.*, **100**, 885 (1988).
60. M. Lazraq, C. Couret, J. Escudié, J. Satgé and M. Soufiaoui, *Polyhedron*, **10**, 1153 (1991).
61. M. P. Egorov, S. P. Kolesnikov, O. M. Nefedov and A. Krebs, *J. Organomet. Chem.*, **375**, 65 (1988).
62. S. A. Batcheller and S. Masamune, *Tetrahedron Lett.*, 3383 (1988).
63. S. P. Kolesnikov, M. P. Egorov, A. M. Gal'minas and O. M. Nefedov, *Metalloorg. Khim.*, **2**, 1356 (1989); *Chem. Abstr.*, **113**, 24074e (1990).
64. T. Tsumuraya and W. Arido, *Organometallics*, **8**, 2286 (1989).
65. A. G. Sommese, S. E. Cremer, J. A. Campbell and M. R. Thomson, *Organometallics*, **9**, 1784 (1990).
66. S. P. Kolesnikov, M. P. Egorov, A. M. Gal'minas and O. M. Nefedov, *Metalloorg. Khim.*, **2**, 1351 (1989); *Chem. Abstr.*, **113**, 40865u (1990).
67. P. Mazerolles, V. W. Khebashesku, S. E. Boganov and O. M. Nefedov, *Izv. Akad. Nauk SSSR, Ser. Khim.*, **6**, 1428 (1989); *Chem. Abstr.*, **112**, 118994m (1990).
68. D. Lei and P. P. Gaspar, *Polyhedron*, **10**, 1221 (1991).
69. P. Jutzi, A. Becker, H. G. Stammler and B. Neumann, *Organometallics*, **10**, 1647 (1991).
70. J. Barrau, N. Ben Hamida, A. Agrebi and J. Satgé, *Organometallics*, **8**, 1585 (1989); J. Barrau, N. Ben Hamida and J. Satgé, *J. Organomet. Chem.*, **282**, 315 (1985).
71. M. G. Voronkov, O. G. Yarosh, L. V. Zhilitskaya, A. I. Albanov and V. Yu. Vitkovskii, *Metalloorg. Khim.*, **4**, 462 (1991); *Chem. Abstr.*, **115**, 29446d (1991).
72. G. Bähr and R. Gelins, *Chem. Ber.*, **91**, 825 (1958).
73. C. A. Kraus and W. V. Sessions, *J. Am. Chem. Soc.*, **47**, 2361 (1925).
74. C. A. Kraus and R. H. Bulland, *J. Am. Chem. Soc.*, **48**, 2131 (1926).
75. C. A. Kraus and W. N. Greer, *J. Am. Chem. Soc.*, **47**, 2568 (1925).
76. C. A. Kraus and A. M. Neal, *J. Am. Chem. Soc.*, **51**, 2403 (1929).
77. T. Harada, *Bull. Chem. Soc. Jpn.*, **15**, 81 (1940).
78. W. P. Neumann, K. Konig and G. Burkhardt, *Ann. Chem.*, **677**, 18 (1964).
79. F. Glockling and D. Kingston, *Chem. Ind.*, 1037 (1961).
80. T. L. Brown and G. L. Morgan, *Inorg. Chem.*, **2**, 736 (1963).
81. S. F. A. Kettle, *J. Chem. Soc.*, 2936 (1959).
82. C. Tamborski, F. E. Ford and E. J. Soloski, *J. Org. Chem.*, **28**, 181 (1963).
83. W. L. Wells and T. L. Brown, *J. Organomet. Chem.*, **11**, 271 (1968).
84. H. Gilman and F. K. Cartledge, *J. Organomet. Chem.*, **5**, 48 (1966).
85. M. Weidenbruch, K. Schaeters, J. Schlaefke, K. Peters and H. -G. Von Schering, *J. Organomet. Chem.*, **415**, 343 (1991).
86. T. N. Mitchell, D. Rutschow and B. Vieler, *Main Group Met. Chem.*, **13**, 89 (1990).

87. G. Anselme, C. Couret, J. Escudié, S. Richelme and J. Satgé, *J. Organomet. Chem.*, **418**, 321 (1991); G. Anselme, H. Ranaivonjetovo, J. Escudié, C. Couret and J. Satgé, *Organometallics*, **11**, 2348 (1992).

88. (a) S. Mosamune, L. R. Sita and D. J. Williams, *J. Am. Chem. Soc.*, **105**, 630 (1983). (b) L. R. Sita and R. D. Bickenstaff, *J. Am. Chem. Soc.*, **111**, 6454 (1989).

89. L. R. Sita and I. Kinoshita, *Organometallics*, **9**, 2865 (1990).

90. L. R. Sita and I. Kinoshita, *J. Am. Chem. Soc.*, **113**, 1856 (1991).

91. L. R. Sita and I. Kinoshita, *J. Am. Chem. Soc.*, **113**, 5070 (1991).

92. A. H. Brune, R. Hohenadel, G. Schmidtberg and U. Ziegler, *J. Organomet Chem.*, **402**, 171 (1991).

93. M. Yamamoto, H. Munatata, K. Kishikawa, S. Kohmoto and K. Yamada, *Bull. Chem. Soc. Jpn.*, **63**, 2366 (1992).

94. W. V. Farrar and H. A. Skinner, *J. Organomet. Chem.*, **1**, 434 (1964).

95. H. Schumann, B. C. Wassermann and E. F. Hahn, *Organometallics*, **11**, 2803 (1992).

96. K. Jones and M. F. Lappert, *Organomet. Chem. Rev.*, **1**, 67 (1966); H. M. J. C. Creemers and J. G. Noltes, *Recl. Trav. Chim. Pays-Bas*, **84**, 382 (1965); R. Sommer, W. P. Neumann and B. Schneider, *Tetrahedron Lett.*, 3875 (1964).

97. H. M. J. C. Creemers, F. Verbeek and J. G. Noltes, *J. Organomet. Chem.*, **8**, 469 (1967); H. M. J. C. Creemers, J. G. Noltes and G. J. M. Van der Kerk, *Recl. Trav. Chim. Pays-Bas*, **83**, 1284 (1964).

98. W. P. Neumann and K. Koning, *Angew. Chem.*, **76**, 891 (1964); *Angew Chem., Int. Ed. Engl.*, **3**, 751 (1964).

99. W. P. Neumann, B. Schneider and R. Sommer, *Ann. Chem.*, **692**, 12 (1968).

100. W. P. Neumann and K. Koning, *Ann. Chem.*, **677**, 1 (1964).

101. A. K. Sawyer and H. G. Kuivila, *J. Am. Chem. Soc.*, **85**, 1010 (1963).

102. N. S. Vyazankin and V. T. Bychkov, *Zh. Obshch. Khim.*, **35**, 685 (1965); *Chem. Abstr.*, **66**, 46460p (1967).

103. (a) P. J. Davidson and M. F. Lappert, *J. Chem. Soc., Chem. Commun.*, 317 (1973). (b) P. J. Davidson, D. H. Harris and M. F. Lappert, *J. Chem. Soc., Dalton Trans.*, 2268 (1976).

104. J. D. Cotton, P. J. Davidson and M. F. Lappert, *J. Chem. Soc., Dalton Trans.*, 2275 (1976).

105. (a) E. O. Fischer and H. Grubert, *Z. Naturforsch.*, **B11**, 423 (1956). (b) E. J. Bulten and H. A. Budding, *J. Organomet. Chem.*, **157**, 66 (1978). (c) J. L. Atwood, W. E. Hunter, A. H. Cowley, R. A. Jones and C. A. Stewart, *J. Chem. Soc., Chem. Commun.*, 925 (1981).

106. J. Tehan, B. L. Barnett and J. L. Dye, *J. Am. Chem. Soc.*, **96**, 7203 (1974).

107. P. A. Edwards and J. D. Corbett, *Inorg. Chem.*, **16**, 903 (1977); *J. Am. Chem. Soc.*, **99**, 3313 (1977).

108. T. N. Mitchell and K. Kwetkat, *Synthesis*, **11**, 1001 (1990).

109. A. Barbero, P. Cuadrado, I. Fleming, A. Gonzalez and F. J. Pulido, *J. Chem. Soc., Chem. Commun.*, 1030 (1990).

110. G. N. Halligan and L. C. Blaszezak, *Org. Synth.*, **68**, 104 (1990).

111. E. Ruf, German Pat. DE 4,006,043, 18 Apr. 1991; *Chem. Abstr.*, **115**, 49984f (1991).

112. L. R. Sita, I. Kinoshita and S. P. Lee, *Organometallics*, **9**, 1664 (1990).

113. K. Jurkschat, A. Rühlmann and A. Tzschach, *J. Organomet. Chem.*, **381**, 653 (1990).

114. K. Doelling, H. Hartung, D. Schollmayer and H. Weichmann, *Z. Anorg. Allg. Chem.*, **600**, 153 (1991).

115. L. Copella, A. Degl'Innocenti, A. Mordini, G. Reginato, A. Ricci and G. Seconi, *Synthesis*, **12**, 1201 (1991).

116. B. Wrackmeyer, G. Kehr and R. Boese, *Angew. Chem.*, **103**, 1374 (1991); *Angew. Chem., Int. Ed. Engl.*, **30**, 1370 (1991).

117. B. Wrackmeyer, G. Kehr and R. Boese, *Chem. Ber.*, **125**, 643 (1992).

118. M. Weidenbruch, A. Schaefer, H. Kilian, S. Pohl, W. Saak and H. Marsmann, *Chem. Ber.*, **125**, 563 (1992).

119. H. M. J. Creemers and J. G. Noltes, *J. Organomet. Chem.*, **7**, 227 (1967).

120. D. Seyferth, L. G. Vaughen and R. Suzuki, *J. Organomet. Chem.*, **1**, 437 (1964).

121. D. Seyferth and J. L. Lefferts, *J. Am. Chem. Soc.*, **96**, 6237 (1974).

122. M. Weidenbruch, J. Schlaefke, K. Peters and H. G. von Schnering, *J. Organomet. Chem.*, **414**, 319 (1991).

123. B. Wrackmeyer, *J. Organomet. Chem.*, **364**, 331 (1989).
124. D. Farah, K. Swami and H. G. Kuivila, *J. Organomet. Chem.*, **429**, 311 (1992).
125. T. N. Mitchell and B. Kowell, *J. Organomet. Chem.*, **437**, 127 (1992).
126. T. N. Mitchell, *Main Group Met. Chem.*, **12**, 425 (1989).
127. H. Killing and T. N. Mitchell, *Organometallics*, **3**, 1927 (1984).
128. A. Rahm, *Main Group Met. Chem.*, **12**, 277 (1989).
129. H. X: Zhang, F. Guibé and G. Balavoine, *Tetrahedron Lett.*, 619, 623, 3874 (1988).
130. M. Pereyre, J. P. Quintard and A. Rahm, *Tin in Organic Synthesis*, Butterworths, London, 1986.
131. F. Guibé, *Main Group Met. Chem.*, **12**, 437 (1988).
132. H. Mokhtar-Jamaï and M. Gielen, *Bull. Soc. Chim. Fr.*, 32 (1972).
133. S. Boué, M. Gielen, J. Nasielski, J. Lieutenant and R. Spielmann, *Bull. Soc. Chim. Belg.*, **78**, 135 (1969).
134. M. Gielen, B. De Poorter, M. T. Sciot and J. Topart, *Bull. Soc. Chim. Belg.*, **82**, 27 (1973).
135. W. Adcock and V. Sankarlyer, *Magn. Reson. Chem.*, **29**, 381 (1991).
136. (a) A. Capperucci, A. Degl'Innocenti, C. Faggi, G. Reginato, A. Ricci, P. Dembeck and G. Seconi, *J. Org. Chem.*, **54**, 2966 (1989).
    (b) A. Ricci, A. Degl'Innocenti, A. Capperucci, G. Reginato and A. Mordini, *Tetrahedron Lett.*, 1899 (1991).
137. P. Cintas, *Synthesis*, **13**, 248 (1992).
138. D. M. Hodgson, *Tetrahedron Lett.*, 5603 (1992).
139. M. T. Blanda and M. Newcomb, *Tetrahedron Lett.*, 350 (1989).
140. M. Ratier, D. Khatmi, G. Duboudin and T. D. Minch, *Synth. Commun.*, **19**, 1929 (1989).
141. M. Kira, R. Yauchibara, R. Hirano, C. Kabuto and H. Sakurai, *J. Am. Chem. Soc.*, **113**, 7785 (1991).
142. M. F. Lappert, *Rev. Si, Ge, Sn, Pb Cmpds.*, **9**, 129 (1986).
143. (a) L. Jia, H. Zhang and N. S. Hosmane, *Organometallics*, **11**, 2957 (1992).
    (b) R. W. Chapman, J. G. Kester, K. Folting, W. E. Streib and J. L. Todd, *Inorg. Chem.*, **31**, 979 (1992).
144. K. Jurkschat, H. G. Kuivila, S. Lin and J. A. Zubieta, *Organometallics*, **8**, 2755 (1989).
145. N. J. R. Van Eikema Hommes, F. Bichelkaupt and G. W. Klumpp, *Recl. Trav. Chim. Pays-Bas* **107**, 393 (1988).
146. E. Nietschmann, A Tzzach, J. Heinicke, U. Pape, U. Thuist, H. D. Pfeiffer, D. Pat. (East), 268, 699, 11 Feb. 1988; *Chem. Abstr.*, **112**, P77545v (1990).
147. B. Jousseaume and P. Villeneuve, *Tetrahedron*, **45**, 1145 (1989).
148. K. Nozaki, K. Oshima and K. Utimoto, *Tetrahedron*, **45**, 923 (1989).
149. E. Piers, M. J. Chong and H. E. Morton, *Tetrahedron*, **45**, 363 (1989).
150. B. H. Lishutz, R. Kell and J. C. Barton, *Tetrahedron Lett.*, 5861 (1992).
151. E. T. Bogoradovski, V. S. Zavgorodnii and A. A. Petrov, *Zh. Obshch. Khim.*, **58**, 455 (1988); *Chem. Abstr.*, **110**, 8326w (1989).
152. E. T. Bogoradovski, V. S. Zavgorodnii and A. A. Petrov, *Zh. Obshch. Khim.*, **58**, 2797 (1988); *Chem. Abstr.*, **110**, 192958f (1989).
153. E. T. Bogoradovski, V. S. Zavgorodnii and A. A. Petrov, USSR Patent SU 1,558, 916, 23 Apr. 1990; *Chem. Abstr.*, **113**, P115571f (1990).
154. E. T. Bogoradovski, V. S. Zavgorodnii and A. A. Petrov, *Zh. Obshch. Khim.*, **60**, 871 (1990); *Chem. Abstr.*, **113**, 152638d (1990).
155. (a) K. Jurkschat and M. Gielen, *J. Organomet. Chem.*, **69**, 236 (1982).
    (b) T. N. Mitchell, B. Fabisch, R. Winkenkamp, H. G. Kuivila and T. J. Karol, *Rev. Si, Ge, Sn and Pb Cmpds.*, **9**, 57 (1986).
156. L. C. Willemsens, *Organolead Chemistry*, International Lead Zinc Research Organization, Inc., New York, 1964.
157. D. E. Fenton and A. G. Massey, *J. Inorg. Nucl. Chem.*, **27**, 329 (1965).
158. H. P. Fritz and K. E. Schwarzhaus, *Chem. Ber.*, **97**, 1390 (1964); *J. Organomet. Chem.*, **1**, 297 (1964).
159. H. Zimmer and O. A. Homberg, *J. Org. Chem.*, **31**, 947 (1966).
160. L. C. Willemsens and G. J. M. van der Kerk, *Recl. Trav. Chim. Pays-Bas*, **84**, 43 (1965).
161. L. C. Willemsens and G. J. M. van der Kerk, *J. Organomet. Chem.*, **2**, 271 (1964).
162. L. C. Willemsens and G. J. M. van der Kerk, *J. Organomet. Chem.*, **4**, 241 (1965).
163. H. Gilman and L. Summers, *J. Am. Chem. Soc.*, **74**, 5924 (1952).

164. H. Gorth and M. C. Henry, *J. Organomet. Chem.*, **9**, 117 (1967).
165. L. C. Willemsens and G. J. M. van der Kerk, *J. Organomet. Chem.*, **4**, 34 (1965).
166. B. Bartocha, C. M. Douglas and M. Y. Gray, *Z. Naturforsch.*, **14b**, 809 (1959).
167. A. K. Holliday and R. E. Pendlebury, *J. Organomet. Chem.*, **7**, 281 (1967).
168. M. C. Henry and J. G. Noltes, *J. Am. Chem. Soc.*, **82**, 558 (1960).
169. D. Seyferth and M. A. Weiner, *J. Am. Chem. Soc.*, **84**, 361 (1962).
170. F. Glockling and D. Kingston, *J. Chem. Soc.*, 3001 (1959).
171. E. A. Puchinyan and Z. M. Manulkin, *Dokl. Akad. Nauk USSR*, **19**, 47 (1962); *Chem. Abstr.*, **57**, 13788c (1962).
172. E. A. Puchinyan and Z. M. Manulkin, *Tr. Tashkentsk. Farmetsent. Inst.*, **3**, 2127 (1962); *Chem. Abstr.*, **61**, 677 (1964).
173. M. R. McCorde and H. Gilman, *Proc. Iowa Acad. Sci.*, **45**, 133 (1938); *Chem. Abstr.*, **33**, 7728 (1939).
174. J. C. Masson and P. Cadiot, *Bull. Soc. Chim. Fr.*, 3518 (1965).
175. *Part V, Lead Chemistry*, ILZRO Res. Dig. No. 13, International Lead Zinc Research Organization, Inc., New York, 1967
176. A. G. Davies and R. J. Puddephatt, *J. Chem. Soc.*, 2663 (1967).
177. G. J. M. van der Kerk, *Ind. Eng. Chem.*, **58**, 29 (1966).
178. E. Amberger and R. Hönigschmid-Grossich, *Chem. Ber.*, **99**, 1973 (1966).
179. B. Wrackmeyer and K. Horchler, *Z. Naturforsch. B*, **44**, 1195 (1989).
180. B. Wrackmeyer and K. Horchler, *Z. Naturforsch. B*, **45**, 437 (1990).
181. B. Wrackmeyer, K. Horchler and R. Boese, *Angew. Chem.*, **101**, 1563 (1989).
182. B. Wrackmeyer and K. Horchler, *J. Organomet. Chem.*, **399**, 1 (1990).
183. J. D. Corlett and P. A. Edwards, *J. Chem. Soc., Chem. Commun.*, 984 (1975).
184. W. P. Neumann and K. Kühlein, *Angew. Chem.*, **77**, 808 (1965).
185. A. J. Leusink and G. J. M. van der Kerk, *Recl. Trav. Chim. Pays-Bas*, **84**, 1617 (1965).
186. A. Sekiguchi, C. Kabuto and H. Sakurai, *Angew. Chem. Int. Ed. Engl.*, **28**, 55 (1989).
187. P. Boudjouk and C. A. Kapfer, *J. Organomet. Chem.*, **296**, 339 (1985).
188. S. Matsamune and L. R. Sita, *J. Am. Chem. Soc.*, **107**, 6390 (1985).
189. J. Fu and W. P. Neumann, *J. Organomet. Chem.*, **272**, 65 (1984).
190. E. Kunze and F. Huber, *J. Organomet. Chem.*, **63**, 287 (1973).
191. H. -J. Haupt, W. Schubert and F. Huber, *J. Organomet. Chem.*, **54**, 231 (1973).
192. D. S. Matteson and G. L. Larson, *J. Organomet. Chem.*, **57**, 225 (1973).
193. K. C. Williams and S. E. Cook, U. S. Patent 3,775,454; *Chem. Abstr.*, **80**, 37289d (1974).
194. F. Glockling and N. M. N. Cowde, *Inorg. Chim. Acta*, **50**, 149 (1982).
195. L. Rösch and U. Starke, *Angew. Chem., Int. Ed. Engl.*, **22**, 557 (1983).
196. G. Guillerm, A. L'. Honoré, L. Veniard, G. Pourcelot and J. Benarm, *Bull. Soc. Chim. Fr.*, 2739 (1973).
197. R. Grüning and J. Lorbeth, *J. Organomet. Chem.*, **78**, 221 (1974).
198. N. Kleiner and M. Dräger, *J. Organomet. Chem.*, **279**, 151 (1974).
199. H. P. Fritz and K. E. Schwarzhaus, *J. Organomet. Chem.*, **1**, 297 (1964); *Chem. Ber.*, **97**, 1390 (1964).
200. J. Yamada, H. Abe and Y. Yamamoto, *J. Am. Chem. Soc.*, **112**, 6118 (1990).
201. E. Amberger, *Angew. Chem.*, **71**, 372 (1959).
202. W. R. Bornhorst and M. A. Ring, *Inorg. Chem.*, **7**, 1009 (1968).
203. (a) G. Thirase, E. Weiss, H. J. Hennig and H. Leahert, *Z. Allg. Anorg. Chem.*, **417**, 221 (1975).
     (b) C. Eaborn and I. D. Jenkins, *J. Chem. Soc., Chem. Commun.*, 780 (1972).
204. M. N. Bochkarev, L. P. Maiorova and N. S. Vyazankin, *J. Organomet. Chem.*, **55**, 89 (1973).
205. (a) C. H. Van Dyke, E. M. Kifer and G. A. Gibbon, *Inorg. Chem.*, **11**, 408 (1972).
     (b) G. A. Gibbon, E. W. Kifer and C. H. Van Dyke, *Inorg. Nucl. Chem. Lett.*, **6**, 617 (1970).
206. K. M. Mackay, R. D. George, R. Robinson and R. Watt, *J. Chem. Soc. (A)*, 1920 (1968).
207. K. M. Mackay and R. D. Watt, *J. Organomet. Chem.*, **14**, 123 (1968).
208. P. Rivière and J. Stagé, *Helv. Chim. Acta*, **55**, 1164 (1972).
209. F. Feher and P. Plichta, *Inorg. Chem.*, **10**, 609 (1971).
210. S. P. Kolesnikov, *Main Group Met. Chem.*, **12**, 305 (1989).
211. V. F. Mironov, *Main Group Met. Chem.*, **12**, 355 (1989).
212. N. S. Nametkin, D. V. Kuz'min, T. I. Chermysheva, V. K. Korolev and N. A. Leptukhina, *Dokl. Akad. Nauk SSSR*, **201**, 1116 (1971); *Chem. Abstr.*, **77**, 5581q (1972).

213. N. S. Nametkin, V. K. Korolev and O. V. Kuz'min, *Dokl. Akad. Nauk SSSR*, **205**, 1111 (1972); Engl. transl., 660.
214. V. F. Mironov and T. K. Gar, *Zh. Obshch. Khim.*, **45**, 103 (1975); *Chem. Abstr.* **82**, 98082u (1975).
215. A. Castel, P. Rivière, J. Stagé and Y. Hooniko, *J. Organomet. Chem.*, **243**, C1 (1988).
216. A. F. Finholt, A. C. Bond, Jr, K. E. Wilzbach and H. J. Schlesinger, *J. Am. Chem. Soc.*, **69**, 2692 (1947).
217. G. Wittig, F. J. Meyer and G. Langer, *Justus Liebigs Ann. Chem.*, **571**, 167 (1951).
218. H. Gilman and J. Eisch, *J. Org. Chem.*, **20**, 763 (1955).
219. A. Stern and E. I. Becker, *J. Org. Chem.*, **29**, 3221 (1964).
220. D. H. Lorenz, P. Shapiro, A. Stern and E. I. Becker, *J. Org. Chem.*, **28**, 2332 (1963).
221. W. J. Considine and J. J. Ventura, *Chem. Ind. (London)*, 1683 (1962).
222. K. Hoi and S. Kumano, *Kogyo kagaku Zasshi*, **70**, 82 (1963); *Chem. Abstr.*, **67**, 11556v (1967).
223. K. Hayashi, J. Syoda and I. Shihara, *J. Organomet. Chem.*, **10**, 81 (1967).
224. M. R. Kula, J. Lorbeth and E. Amberger, *Chem. Ber.*, **97**, 2087 (1964).
225. C. Tamborski and E. J. Soloski, *J. Am. Chem. Soc.*, **83**, 3734 (1961).
226. C. Tamborski, F. E. Ford and E. J. Soloski, *J. Org. Chem.*, **28**, 181 (1963).
227. E. Amberger and M. R. Kula, *Chem. Ber.*, **96**, 2560 (1963).
228. A. K. Sawyer, J. E. Brown and G. S. May, *J. Organomet. Chem.*, **11**, 192 (1968).
229. J. Szammer and L. Otvos, *Chem. Ind. (London)*, **23**, 764 (1988).
230. H. B. Lipshutz and C. D. Reuter, *Tetrahedron Lett.*, 4617 (1989).
231. M. Weidenbruck, K. Schäfers, S. Pohl, W. Saak, K. Peters and H. G. von Schnering, *Z. Anorg. Allg. Chem.*, **570**, 75 (1989).
232. A. K. Sawyer and H. G. Kuivila, *Chem. Ind. (London)*, 260 (1961).
233. A. K. Sawyer, J. E. Brown and E. L. Hanson, *J. Organomet. Chem.*, **3**, 464 (1965).
234. H. M. J. C. Creemers and J. G. Noltes, *Recl. Trav. Chim. Pays-Bas*, **84**, 382 (1965).
235. T. N. Mitchell and B. S. Brank, *Organometallics*, **10**, 936 (1991).
236. W. P. Neumann and H. Niermann, *Justus Liebigs Ann. Chem.*, **653**, 164 (1962).
237. F. Ferkus, D. Messadi, B. De Jeso, M. Degueil-Castaing and B. Maillard, *J. Organomet. Chem.*, **420**, 315 (1991).
238. H. Piane, U. Kirchgessner and U. Schubert, *Chem. Ber.*, **124**, 743 (1991).
239. R. Duffy, J. Feeney and A. K. Holliday, *J. Chem. Soc.*, 1144 (1962).
240. E. Amberger and R. Höningschmid-Grossich, *Chem. Ber.*, **99**, 1673 (1966).
241. E. Amberger, *Angew. Chem.*, **72**, 494 (1960).
242. W. E. Becker and S. E. Cook, *J. Am. Chem. Soc.*, **82**, 6264 (1960).
243. W. P. Neuman and K. Kühlein, *Angew. Chem.*, **77**, 808 (1965).
244. V. N. Gevorgyan, L. M. Igmatovich and E. Lukevics, *J. Organomet. Chem.*, **284**, C31 (1985).
245. R. Duffy and A. K. Holiday, *J. Chem. Soc.*, 1679 (1961).
246. P. Hackett and A. R. Monning, *J. Organomet. Chem.*, **66**, C17 (1974).
247. B. Wrackmeyer and K. Horchler von Locquenhien, *Z. Naturforsch. B*, **46**, 1207 (1991).
248. G. Ya. Zueva, N. V. Luk'yankina and V. A. Ponomarenko, *Izv. Akad. Nauk SSSR, Ser. Khim.*, **20**, 2777 (1971); *Chem. Abstr.*, **76**, 140974q (1972).
249. M. Weidenbruch and N. Wessal, *Chem. Ber.*, **105**, 173 (1972).
250. V. V. Pozdeev, N. V. Fomina, N. I. Sheverdina and K. A. Kocheshkov, *Izv. Akad. Nauk SSSR, Ser. Khim.*, **21**, 2051 (1972); *Chem. Abstr.*, **78**, 43634n (1963).
251. N. V. Fomina, N. I. Sheverdina and K. A. Kocheshkov, *Dokl. Akad. Nauk SSSR*, **201**, 1128 (1971); *Chem. Abstr.*, **76**, 72616x (1972).
252. N. V. Fomina, N. I. Sheverdina, E. I. Dobrova, I. V. Sosnina and K. A. Kocheshkov, *Dokl. Akad. Nauk SSSR*, **210**, 621 (1973); *Chem. Abstr.*, **79**, 42630s (1973).
253. E. A. Chernyshev, M. E. Kurek and A. N. Polivanov, USSR Pat. 316,693 (1970); *Chem. Abstr.*, **76**, 72616x (1972).
254. V. F. Mironov, T. K. Gar, A. A. Buyakov, V. M. Slobodina and T. P. Guntsadze, *Zh. Obshch. Khim.*, **42**, 2010 (1972); *Chem. Abstr.*, **78**, 29940c (1973).
255. S. P. Kolesnikov, B. L. Perl'mutter and O. M. Nefedov, *Dokl. Akad Nauk SSSR*, **196**, 594 (1971); *Chem. Abstr.*, **79**, 92339z (1973).
256. O. M. Nefedov, S. P. Kolesnikov, B. L. Perl'mutter and A. I. Ioffe, *Dokl. Akad. Nauk SSSR*, **211**, 110 (1973); *Chem. Abstr.*, **79**, 91425n (1973).
257. V. F. Mironov, L. N. Kalinina, E. M. Berliner and T. K. Gar, *Zh. Obshch. Khim.*, **40**, 2597 (1970); *Chem. Abstr.*, **75**, 49255z (1971).

258. T. K. Gar, E. M. Berliner, A. V. Kisin and A. V. Mironov, *Zh. Obshch. Khim.*, **40**, 2601 (1970); *Chem. Abstr.*, **75**, 5143g (1971).

259. V. F. Mironov and T. K. Gar, *Izv. Akad. Nauk SSSR, Ser. Khim.*, 1515 (1964); *Chem. Abstr.*, **64**, 14213f (1966).

260. T. K. Gar, V. M. Nosova, A. V. Kisin and V. F. Mironov, *Zh. Obshch. Khim.*, **48**, 838 (1978); *Chem. Abstr.*, **89**, 43651t (1978).

261. V. F. Mironov, L. N. Kalinina and T. K. Gar, *Zh. Obshch. Khim.*, **39**, 2486 (1969); *Chem. Abstr.*, **72**, 79180b (1970).

262. V. F. Mironov and A. L. Kravchenko, *Zh. Obshch. Khim.*, **34**, 1356 (1964); *Chem. Abstr.*, **61**, 677d (1964).

263. T. K. Gar, N. Yu Khromova, S. N. Tandura, V. M. Nosova, A. V. Kisin and V. F. Mironov, *Zh. Obshch. Khim.*, **52**, 622 (1982); *Chem. Abstr.*, **97**, 72464e (1982).

264. (a) G. A. Olah, G. K. S. Prakash and Y. Sommer, *Science*, **13**, 206 (1979).
     (b) V. B. Kazanskii, O. M. Nefedov, A. A. Pankov, V. Yu. Boronkov, S. P. Kolesnikov and I. V. Lyudkovskaya, *Izv. Acad. Nauk. USSR, Ser. Khim.*, 698 (1983); *Chem. Abstr.*, **98**, 190684v (1983).
     (c) S. P. Kolesnikov, S. L. Povarov, A. I. Lutsenko, D. S. Yufit, Yu.T. Struchkov and O. M. Nefedov, *Izv. Akad. Nauk SSSR, Ser. Khim.*, 895 (1990); *Chem. Abstr.*, **113**, 97726d (1990).

265. H. Schmidbaur and J. Rott, *Z. Naturforsch. B*, **45**, 961 (1990).

266. E. J. Bulten and J. G. Noltes, *Tetrahedron Lett.*, 3471 (1966).

267. N. S. Vyazankin, E. N. Gladyshev, S. P. Korneva and G. A. Razuvaev, *Zh. Obshch. Khim.*, **34**, 1465 (1965); *Chem. Abstr.*, **61**, 5680f (1964).

268. A. N. Nesmeyanov, K. N. Anisimov, N. E. Kolobava and A. B. Nutonova, *Dokl. Akad. Nauk SSSR*, **176**, 844 (1967); Engl. transl. 876.

269. N. P. Redy, T. Hayashi and M. Tamka, *Chem. Lett.*, 677 (1991).

270. J. Barrau, G. Rima, M. El Amine and J. Satgé, *Synth. React. Inorg. Met. Org. Chem.*, **18**, 21 (1988).

271. K. M. Mackay and P. J. Roebuck, *J. Chem. Soc.*, 1195 (1964).

272. K. M. Mackay and P. Robinson, *J. Chem. Soc.*, 5121 (1965).

273. K. M. Mackay, P. Robinson, A. G. MacDiarmid and E. J. Spanier, *J. Inorg. Nucl. Chem.*, **28**, 1377 (1966).

274. Z. Xu, *Huaxne Shiji*, **13**, 254 (1991); *Chem. Abstr.*, **115**, 208119w (1991).

275. E. J. Bulten and W. Drenth, *J. Organomet. Chem.*, **61**, 179 (1973).

276. J. W. Anderson and J. E. Drake, *Synth. React. Inorg. Met. Org. Chem.*, **1**, 155 (1971).

277. J. W. Anderson, G. K. Barker, J. E. Drake and R. T. Hemmings, *Can. J. Chem.*, **50**, 1607 (1972).

278. J. M. Bellama and C. J. McCormick, *Inorg. Nucl. Chem. Lett.*, **7**, 533 (1971).

279. J. Barrau, J. Satgé and M. Massol, *Helv. Chim. Acta*, **56**, 1638 (1973).

280. E. K. Liu and R. J. Lagow, *J. Chem. Soc.*, 450 (1977).

281. A. Castel, P. Rivière, J. Satgé, J. J. E. Moreau and R. J. P. Corriu, *Organometallics*, **2**, 1498 (1983).

282. A. A. Espenbetov, Yu.T. Struchkov, S. P. Kolesnikov and O. M. Nefedov, *J. Organomet. Chem.*, **275**, 33 (1984).

283. J. Köcher and W. P. Neumann, *J. Am. Chem. Soc.*, **106**, 3861 (1984).

284. M. Weidenbruch, F. T. Grimm, S. Pohl and W. Saak, *Angew. Chem., Int. Ed. Engl.* **28**, 198 (1989).

285. A. Sekiguchi, H. Naito, H. Naneki, K. Ebata, C. Kabuto and H. Sakurai, *J. Organomet. Chem.*, **368**, C1 (1989).

286. P. Rivière, J. Satgé and A. Boy, *J. Organomet. Chem.*, **96**, 25 (1975).

287. C. F. Shaw and A. L. Allsed, *Organomet. Chem. Rev.*, **A5**, 59 (1970).

288. R. J. Lagow, R. E. Eujen, L. L. Gerchman and J. A. Morrison, *J. Am. Chem. Soc.*, **100**, 1722 (1978).

289. R. Eujen, F. E. Laufs and E. Petrauskas, *J. Organomet. Chem.*, **299**, 29 (1986).

290. M. Bordeau, S. M. Djanel and J. Dunoguès, *Organometallics*, **4**, 1087 (1985).

291. E. J. Bulten and W. Drenth, *J. Organomet. Chem.*, **61**, 179 (1973).

292. J. Safar, J. Tomiska, J. Sule and E. Cahelova, Czech Patent, 148,776, 15/5/73; *Chem. Abstr.*, **79**, 137295a (1973).

293. T. Onuma, T. Inone, B. Uno and Y. Kawai, Japan Patent, 72 41,337 19/10/72; *Chem. Abstr.*, **78**, 29199d (1973).

294. H. Yamaguchi and T. Akabare, Japan Patent, 73 05,724 24/1/73; *Chem. Abstr.*, **78**, 136412d (1973).
295. T. Katsumura, R. Suzuki and T. Mastunaga, Japan Patent, 76 32,971 22/8/72; *Chem. Abstr.*, **78**, 4360a (1973).
296. H. Matschiner, R. Voigtländer and A. Tzschach, *J. Organomet. Chem.*, **70**, 387 (1974).
297. O. G. Chee and V. G. Kumar Das, *Appl. Organomet. Chem.*, **2**, 109 (1988).
298. S. A. Kotlyar, G. V. Dimitrishchuk, R. L. Savranskaya and N. G. Luck'yanenko, *Zh. Obshch. Khim.*, **58**, 1443 (1988); *Chem. Abstr.*, **110**, 192972c (1989).
299. O. K. Sang Jung and S. Y. Sohn, *Bull. Korean Chem. Soc.*, **12**, 256 (1991).
300. R. Malguzzi, M. Sandri and M. Rosenthal, Faming Zhuanli Shenqing, Gonkai Shuomingshu CN 105267; *Chem. Abstr.*, **116**, 194594d (1992).
301. V. V. Pozdeev and V. E. Gelfan, *J. Gen. Chem. USSR*, **43**, 1196 (1973); *Chem. Abstr.*, **79**, 66510d (1973).
302. J. R. Webster, M. M. Millard and W. L. Jolly, *Inorg. Chem.*, **10**, 879 (1971).
303. J. M. Bellama and R. A. Gsell, *Inorg. Nucl. Chem. Lett.*, **7**, 365 (1971).
304. A. K. Sawyer and G. Belletete, *Synth. React. Inorg. Met.-Org. Chem.*, **3**, 301 (1973).
305. P. Rivière and J. Satgé, *Helv. Chim. Acta*, **55**, 1164 (1972).
306. L. Mel'nichenko, N. Zemlyanskii and K. Kocheshkov, *Izv. Akad. Nauk SSSR, Ser. Khim.*, 184, (1972); *Chem. Abstr.*, **77**, 19756n (1972).
307. R. Paul, K. Soniand and S. Naruba, *J. Organomet. Chem.*, **40**, 355 (1976).
308. B. Pant and W. Davidsohn, *J. Organomet. Chem.*, **39**, 295 (1972).
309. T. Geisler, C. Cooper and A. Norman, *Inorg. Chem.*, **11**, 1710 (1972).
310. A. Falarini, R. A. N. McLean and N. Wabidia, *J. Organomet. Chem.*, **73**, 59 (1974).
311. R. K. Ingham, S. D. Rosenberg and H. Gilman, *Chem. Rev.*, **60**, 459 (1960).
312. W. T. Reichle, *Inorg. Chem.*, **5**, 87 (1966).
313. M. Barnard, P. J. Smith and R. E. White, *J. Organomet. Chem.*, **77**, 189 (1974).
314. V. Shiryaev, L. Makhalikina, F. Huzmina, V. Krylov, V. Osipov and V. F. Mironov, *Zh. Obshch. Khim.*, **43**, 2232 (1973); *Chem. Abstr.*, **80** 48116w (1974); V. F. Mironov, T. T. Kuz'mina, L. V. Makhalikina and V. I. Shirayev, USSR Patent, 374,320 30/3/73; *Chem. Abstr.*, **79**, 53577f (1973).
315. R. A. Jacob and R. L. Lagow, *J. Chem. Soc., Chem. Commun.*, 104 (1993).
316. R. Eujen, N. Jahn and U. Thurmann, *J. Organomet. Chem.*, **434**, 159 (1992).
317. R. Eujen and U. Thurmann, *J. Organomet. Chem.*, **433**, 63 (1992).
318. C. Gopinathan, S. K. Pandit, S. Gopinathan, J. R. Unni and P. A. Awasarkar, *Ind. J. Chem.*, **11**, 605 (1973).
319. P. G. Harrison, *Inorg. Nucl. Chem. Lett.*, **8**, 555 (1972).
320. S. Kozima, K. Kobayashi and M. Savowisi, *Bull. Chem. Soc. Jpn.*, **49**, 2837 (1976).
321. P. F. R. Ewings and P. G. Harrison, *Inorg. Chim. Acta*, **18**, 165 (1976).
322. E. J. Bulten, H. F. M. Grunter and H. F. Martens, *J. Organomet. Chem.*, **117**, 329 (1976).
323. H. Werchmann and B. Beusch, *Z. Chem.*, **29**, 184 (1989).
324. G. van Koten and J. G. Noltes, *J. Am. Chem. Soc.*, **98**, 5393 (1976); G. Van Koten, C. A. Schoor and J. G. Noltes, *J. Organomet. Chem.*, **99**, 157 (1975).
325. M. Weidenbruch, K. Schaefers, S. Pohl, W. Saak, K. Peters and H. G. Von Schering, *J. Organomet. Chem.*, **346**, 171 (1988).
326. S. I. Pombrik, L. S. Golovchenko and D. N. Kravtsov, *Metalloorg. Khim.*, **2**, 1198 (1989); *Chem. Abstr.*, **112**, 235447w (1990).
327. J. Kizlik and V. Rattay, *Ropa Uhlie*, **32**, 341 (1990); *Chem. Abstr.*, **114**, 185647e (1991).
328. G. H. Reifenberg and H. M. Gitzlitz, German Patent 2,218,211 9/12/72; *Chem. Abstr.*, **78**, 43716r (1973).
329. D. Seyferth, S. Andrews and R. Lambert, *J. Organomet. Chem.*, **37**, 69 (1972).
330. A. Nesmeyanov and A. Borisov, *Izv. Akad. Nauk SSSR, Ser. Khim.*, 1667 (1974); *Chem. Abstr.*, **81**, 105651p (1974).
331. N. Yoshino, Y. Kondo and T. Yoshino, *Synth. React. Inorg. Met.-Org. Chem.*, **3**, 397 (1973).
332. M. Masaki, K. Fukui, I. Uchida and H. Yasuno, *Bull. Chem. Soc. Jpn.*, **48**, 2310 (1975).
333. (a) V. F. Mironov, V. Shirayev, V. V. Yankov and N. Gladchenko USSR Patent 396,340 29/8/73; *Chem. Abstr.*, **80**, 15075d (1974).
   (b) V. F. Mironov, V. I. Shirayev, V. V. Yankov, A. E. Gladchenko and A. D. Naumov, *J. Gen. Chem. USSR*, **44**, 776 (1974); *Chem. Abstr.*, **81**, 49766r (1974).

334. K. A. Kocheshkov, *Chem. Ber.*, **61**, 1659 (1928).
335. Y. Naruta, Y. Nishigaichi and K. Maruyama, *Tetrahedron*, **45**, 1067 (1989).
336. V. Bhushan, K. L. Gupta and G. C. Saxena, *Synth. React. Met.-Org. Chem.*, **20**, 363 (1990).
337. D. Armitage and A. Tarassoli, *Inorg. Chem.*, **14**, 1210 (1975).
338. M. Molnar, *Bull. Sci. Cons. Acad. Sci. Arts. RSF Yougoslavie Sect. A*, **19**, 65 (1974); *Chem. Abstr.*, **81**, 49767s (1974).
339. V. S. Petrosyan, S. G. Sakharov and O. A. Reutov, *Izv. Akad. Nauk SSSR, Ser. Khim.*, 743 (1974); *Chem. Abstr.*, **81**, 13620z (1974).
340. E. J. Bulten, *J. Organomet. Chem.*, **97**, 167 (1975).
341. K. D. Bos, E. J. Bulten and J. G. Noltes, *J. Organomet. Chem.*, **99**, 373 (1975).
342. U. Schraer, H. J. Albert and W. P. Neumann, *J. Organomet. Chem.*, **102**, 291 (1975).
343. J. D. Kennedy, *J. Labelled Compd.*, **11**, 285 (1975).
344. S. Kohana, *J. Organomet. Chem.*, **99**, 644 (1975).
345. A. K. Sawyer, J. E. Brown, G. S. May, A. K. Sawyer Jr. R. E. Schofield and W. E. Sprague, *J. Organomet. Chem.*, **124**, 13 (1977).
346. P. Fostein and J. C. Pommier, *J. Organomet. Chem.*, **114**, 67 (1976).
347. R. E. Hutton, J. W. Burley and V. Oakes, *J. Organomet. Chem.*, **156**, 369 (1978).
348. E. J. Bulten and J. W. vander Hurk, *J. Organomet. Chem.*, **162**, 161 (1978).
349. A. G. Galinos and I. M. Tsangaris, *Angew. Chem.*, **70**, 24 (1958).
350. G. van Koten, J. T. B. H. Jastrzebski, J. G. Noltes, W. M. G. F. Pontenagel, J. Kroon and A. L. Spek, *J. Am. Chem. Soc.*, **100**, 5021 (1978).
351. M. Devaud, M. Engele, C. Feasson and J. L. Lecat, *J. Organomet. Chem.*, **281**, 181 (1985).
352. Y. Azuma and M. Newcomb, *Organometallics*, **3**, 1917 (1984); M. Newcomb, M. T. Blonda, Y. Azuma and T. J. Delord, *J. Chem. Soc., Dalton Trans.*, 1159 (1984).
353. M. Devaud and P. Leponsez, *J. Chem. Res. (S)*, 100 (1982).
354. K. Jurkschat and A. Tzschach, *J. Organomet. Chem.*, **272**, C13 (1984).
355. K. J. Tacke, M. Link, H. Joppien and L. Ernst, *Z. Naturforsch. B*, **41**, 1123 (1986).
356. P. Brown, M. F. Mahon and K. C. Molloy, *J. Organomet. Chem.*, **435**, 265 (1992).
357. P. Raj, *Ind. J. Chem.*, **17A**, 616 (1979).
358. M. Gingras and W. N. Harpp, *Tetrahedron Lett.*, 4669 (1988).
359. A. Von Ruchr, W. Sunderneyer and W. Towel, *Z. Anorg. Allg. Chem.*, **499**, 75 (1983).
360. F. S. Holland, *Appl. Organomet. Chem.*, **1**, 449 (1987).
361. S. P. Mallela, S. Yap, J. R. Sams and F. Aubke, *Rev. Chim. Miner.*, **23**, 572 (1986).
362. C. J. Cardin, D. J. Cardin, J. M. Kelly, R. J. Norten, A. Roy, B. J. Hathaway and T. J. King, *J. Chem. Soc., Dalton Trans.*, 671 (1983).
363. M. Gielen and J. Topart, *Bull. Soc. Chim. Belges*, **80**, 655 (1971).
364. K. Jaworski and J. Przybylowicz, *Bull. Pol. Acad. Sci. Chem.*, **39**, 479 (1991); *Chem. Abstr.*, **117**, 191957j (1992).
365. Y. Yamamoto and J. Yamada, *J. Am. Chem. Soc.*, **109**, 4395 (1987).
366. Y. Yamamoto, J. Yamada and A. Tetsuye, *Tetrahedron*, **48**, 5587 (1992).
367. (a) J. T. Pinhey and E. G. Roche, *J. Chem. Soc., Perkin Trans.*, 2415 (1988).
      (b) H. C. Bell, J. R. Kalman, J. T. Pinhey and S Sternhell, *Tetrahedron Lett.*, 853 (1974).
      (c) D. de Vos, W. A. A. van Barnefeld, D. C. van Beelen, H. O. van der Kooi, J. Wolters and A. van der Gen, *Recl. Trav. Chim. Pays-Bas*, **94**, 97 (1973).
368. S. Ritter and R. E. Noftle, *Chem. Mater.*, **4**, 872 (1992).
369. N. Kleiner and M. Dräger, *J. Organomet. Chem.*, **288**, 295 (1985).
370. N. Kleiner and M. Dräger, *Z. Naturforsch., Teil 5*, **40**, 477 (1985).
371. P. Hackett and A. R. Manning, *Polyhedron*, **1**, 45 (1982).
372. S. N. Bhattacharya, A. K. Saxena and R. Prem, *Indian J. Chem. Sect. A*, **21**, 141 (1982).
373. R. P. Kozyrod and J. T. Pinhey, *Tetrahedron Lett.*, 1301 (1983).
374. I. Pforr, F. Näumann and D. Rehder, *J. Organomet. Chem.*, **258**, 189 (1983).
375. M. M. Kubicki, R. Kergoat, W. -E. Guerchais and P. L'Harridon, *J. Chem. Soc., Dalton Trans.*, 1791 (1984).
376. R. Tolay and D. Rehder, *J. Organomet. Chem.*, **262**, 25 (1984).
377. I. Suzuki, T. Furuta and Y. Yamamoto, *J. Organomet. Chem.*, **443**, C6 (1993).
378. K. C. Williams, *J. Organomet. Chem.*, **22**, 141 (1970).
379. H. Schmidbaur, *Chem. Ber.*, **97**, 270 (1964).
380. L. C. Willemsens, G. J. M. Van der Kerk, *J. Organomet. Chem.*, **21**, 123 (1970).

381. G. Beinert and J. Herz, *Makromol. Chem.*, **181**, 59 (1980).
382. R. B. Bates, B. Gordon III, T. K. Highsmith and J. J. White, *J. Org. Chem.*, **49**, 2981 (1984).
383. E. C. Juenge and S. Gray, *J. Organomet. Chem.*, **10**, 465 (1967).
384. D. C. Van Beelen, J. Wolters and A. van der Gen, *J. Organomet. Chem.*, **145**, 359 (1978).
385. D. C. Van Beelen, J. Wolters and A. van der Gen, *Recl. Trav. Chim. Pays-Bas*, **98**, 437 (1979).
386. K. C. Williams, *J. Organomet. Chem.*, **19**, 210 (1969).
387. E. C. Juenge and H. E. Jack, *J. Organomet. Chem.*, **21**, 359 (1970).
388. B. C. Saunders and G. J. Stacey, *J. Chem. Soc.*, 919 (1949).
389. L. C. Willemsens, *Investigations in the Field of Organolead Chemistry*, International Lead Zinc Research Organization, Inc., New York, 1965, p. 111.
390. G. Singh, *J. Org. Chem.*, **31**, 949 (1966).
391. V. V. Bardin, L. S. Pressman, L. N. Rogoza and G. G. Furin, *J. Fluorine Chem.*, **53**, 213 (1991).
392. H. J. Emeléus and P. R. Evans, *J. Chem. Soc.*, 511 (1964).
393. E. Kunze and F. Huber, *J. Organomet. Chem.*, **57**, 345 (1973).
394. G. Ghobert and M. Devaud, *J. Organomet. Chem.*, **153**, C23 (1978).
395. H. Gillman, E. Towne and H. L. Jones, *J. Am. Chem. Soc.*, **55**, 4698 (1933).
396. A. K. Holliday and R. E. Pendlebury, *3rd Int. Symp. Organomet. Chem.*, München, 1967, Abstr. p. 160.
397. L. Maier, *Tetrahedron Lett.*, 1 (1949).
398. B. Bartocha, U.S. Patent 3.100.917 (1963); *Chem. Abstr.*, **60**, 551 (1964).
399. K. Hills and M. C. Henry, *J. Organomet. Chem.*, **3**, 159 (1965).
400. I. Warf, R. Cuenca, E. Besso and M. Onyczchuk, *J. Organomet. Chem.*, **277**, 245 (1984).
401. F. Huber and E. Schönafinger, *Angew. Chem., Int. Ed. Engl.*, **7**, 72 (1968).
402. (a) J. G. Noltes and G. J. M. van der Kerk, *Recl. Trav. Chim. Pay-Bas*, **80**, 623 (1961).
    (b) J. G. Noltes, H. A. Budding and G. J. M. van der Kerk, *Recl. Trav. Chim. Pay-Bas*, **79**, 400 (1960).
403. S. K. Misra, K. Singhal, F. S. Siddiqui, A. Ranjan and P. Raj, *Ind. J. Phys. Natl. Sci.*, **3A**, 16 (1983).
404. R. Eugen and A. Patorra, *J. Organomet. Chem.*, **438**, C1 (1992).
405. R. Eugen and A. Patorra, *J. Organomet. Chem.*, **438**, 57 (1992).
406. M. A. Guera, R. L. Armonstrong, W. J. Bailey and R. J. Lagow, *J. Organomet. Chem.*, **254**, 53 (1983).
407. H. D. Kaesz, J. R. Phillipsand and F. G. A. Stone, *J. Am. Chem. Soc.*, **82**, 6228 (1960).
408. A. G. Davies and R. J. Puddephatt, *J. Chem. Soc.*, 2663 (1967).
409. A. G. Davies and R. J. Puddephatt, *J. Organomet. Chem.*, **5**, 590 (1966).
410. L. C. Willemsens and G. J. M. van der Kerk, *Investigations in the Field of Organometallic Chemistry*, International Lead Zinc Research Organization, Inc., New York, 1965, p. 62.
411. A. Ya. Yakubovich, E. N. Merkulova, S. R. Makarov and G. I. Gavrilov, *Zh. Obshch. Khim.*, **22**, 2060 (1952); *Chem. Abstr.*, **47**, 9257 (1953).
412. A. Dush, M. Hoch and D. Rehder, *Chimia*, **42**, 179 (1988).
413. A. K. Sawyer, *Organotin Compounds*, Vol. 1, Marcel Dekker, New York, 1971, pp. 83–85.
414. G. E. Coates, M. L. Green and K. Wade, in *Organometallic Compounds*, Vol. 1, *Main Group Elements*, (Eds. G. E. Coates and K. Wade), Methuen, London, 1967, p. 432.
415. R. C. Poller, *Chemistry of Organotin Compounds*, Logos Press, London, 1970, p. 55.

CHAPTER **11**

Acidity, complexing, basicity and H-bonding of organic germanium, tin and lead compounds: experimental and computational results[*]

AXEL SCHULZ and THOMAS M. KLAPÖTKE[**]

Institut für Anorganische und Analytische Chemie, Technische Universität Berlin, Strasse des 17. Juni 135, D-10623 Berlin, Germany
Fax: +49 30 31426468;   e-mail: tmk@wap0205.chem.tu-berlin.de

---

[*] In this chapter, full lines are used both for covalent chemical bonds as well as for partial bonds and for coordination.
[**] Author to whom correspondence should be addressed.

The chemistry of organic germanium, tin and lead compounds
Edited by S. Patai   © 1995 John Wiley & Sons Ltd

ABBREVIATIONS

The following abbreviations are used in addition to the well known abbreviations which
are listed in each volume.

ATM	absolute true minimum	Naph	naphthyl (naphthalenid)
BDE	bond dissociation energy	PA	proton affinity
B.E.	bond energy	Pz	pyrazolyl ring
CI	configuration interaction	SET	single electron transfer
Dep	2,6-diethylphenyl	Tb	2,4,6-tris[bis(trimethylsilyl)methyl]-phenyl
Dip	2,6-diisopropyl	Tip	2,4,6-triisopropylphenyl
DME	dimethyl ether	TM	true minimum
ECP	effective core potential	TMTAA	dibenzotetramethyltetraaza[14]-annulene
Mes	mesityl	TS	transition state

I. OUTLINE

The aim of this review is to focus on the hydrogen bonding, the acidity and basicity,
complexing as well as some aspects of computational chemistry concerning the organo-
element chemistry of germanium, tin and lead. This chapter is not exhaustive in scope, but
rather consists of surveys of the most recent decade of work in this still developing area.
This chapter emphasizes the synthesis, reactions and molecular structures of the class of

compounds outlined above (less attention is paid to mechanism, spectroscopic properties and applications which can be found in other specialized chapters). Especially the single-crystal X-ray diffraction technique has elucidated many novel and unusual structures of molecules and of the solid state in general. Not unexpectedly, certain organo-element compounds present problems concerning their classification as n-coordinated species since it is sometimes difficult to distinguish between a weak long-range interaction in the solid state and the fact that two atoms can be forced a little bit closer together by crystal lattice effects.

Since organo-element chemistry is the discipline dealing with compounds containing at least one direct element–carbon bond in this chapter, we discuss germanium, tin and lead species in which at least one organic group is attached through carbon. Classical species containing E–C σ-covalent bonds (E = Ge, Sn, Pb) as well as π-complexes involving dative bonds will also be considered (for definition compare also the old definition by N. V. Sidgwick from 1927, quoted in ref. 1, p. 1082).

The electronic structure and physical properties of any molecule can in principle be determined by quantum-mechanical calculations. However, only in the last 20 years, with the availability and aid of computers, has it become possible to solve the necessary equations without recourse to rough approximations and dubious simplifications[2]. Computational chemistry is now an established part of the chemist's armoury. It can be used as an analytical tool in the same sense that an NMR spectrometer or X-ray diffractometer can be used to rationalize the structure of a known molecule. Its true place, however, is a predictive one. Therefore, it is of special interest to predict molecular structures and physical properties and compare these values with experimentally obtained data. Moreover, quantum-mechanical computations are a very powerful tool in order to elucidate and understand intrinsic bond properties of individual species.

II. INTRODUCTION

Considering the chemical reactivity and group trends of germanium, tin and lead it can be stated that germanium is somewhat more reactive and more electropositive than silicon. Alkyl halides react with heated Ge (as with Si) to give the corresponding organogermanium halides. Tin, however, is notably more reactive and electropositive than Ge and Pb powder is pyrophoric whereas the reactivity of the metal is usually greatly diminished

TABLE 1. Physical properties of Group 14 elements Ge, Sn and Pb

	$_{32}$Ge	$_{50}$Sn	$_{82}$Pb
Electron configuration	[Ar]3d^{10}4s^{2}4p^2	[Kr]4d^{10}5s^{2}5p^2	[Xe]4f^{14}5d^{10}6s^{2}6p^2
Atomic weight (g/mol)	72.61	118.71	207.20
Electronegativity:			
Pauling	2.01	1.96	2.33
Allred–Rochow	2.02	1.72	1.55
Sanderson	2.31	2.02	2.0
Ionization potential[3]	(1) 7.899	7.344	7.416
(eV)	(2) 15.934	14.632	15.032
	(3) 34.220	30.502	31.937
	(4) 45.710	40.734	42.32
Relative electron density	17.4	17.8	15.3
B.E.(E–E) (kcal mol^{-1})	45	36	—
B.E.(E–C) (kcal mol^{-1})	61	54	31
B.E.(E–H) (kcal mol^{-1})	69	60	49
B.E.(E–Cl) (kcal mol^{-1})	82	77	58
Covalent bond radius (Å)	1.225	1.405	(1.750)
van der Waals radius (Å)	2.10	2.17	2.02

by the formation of a thin, coherent protective oxide layer. The steady trend towards increasing stability of M^{II} rather than M^{IV} compounds in general in the sequence Ge, Sn, Pb is an example of the 'inert-pair effect' which is well established for the heavier main group metals. However, with respect to the scope of this review it has to be emphasized that the organometallic chemistry of Sn and Pb which is almost entirely confirmed to the M^{IV} state is a notable exception of this trend.

There is a steady decrease in the E−E bond strength (E = Ge, Sn, Pb). In general, with the exception of E−H bonds (E = Ge, Sn, Pb), the strength of other E−X bonds (X = Cl, C) diminishes less noticeably, though the absence of Ge analogues of silicone polymers speaks for the lower stability of the Ge−O−Ge linkage.

Table 1 summarizes some physical properties of the elements Ge, Sn and Pb[1].

III. H-BONDING

A. Introduction

The valence electrons of germanium, tin and lead have the ns^2np^2 ($n = 4, 5, 6$) configuration in the ground state. Tetravalent organoelement compounds of Ge, Sn and Pb can conveniently be described in terms of sp^3 hybridization at the central metal atom resulting in tetrahedrally coordinated species. One possibility to classify tetravalent organoelement compounds of Ge, Sn and Pb is the specification by the number of covalently bound organic substituents: R_4E, R_3EX, R_2EX_2 and REX_3, where R may be any organic group directly coordinated through an E−C bond and X is a univalent group or atom with a E−X bond (X not C). If one X is represented by hydrogen (X = H) the polarity of the E−H bond (E = Ge, Sn, Pb) depends on the type of the group R and the other X. For instance, the reversal of the generally low polarity of the germanium–hydrogen bond in the sequence of hydrogermanes has often been observed[4]:

$$R_3Ge^{\delta+} - H^{\delta-} \quad R'_nX_{3-n}Ge-H \quad X_3Ge^{\delta-} - H^{\delta+}$$

$$R = \text{alkyl}; \; R' = \text{alkyl or aryl}; \; X = \text{univalent group}$$

The polarity of the E−C and E−H bonds increases descendingly in the group from germanium to lead (cf Table 1).

B. Reactions

1. Germanium

The most stable of the germanium hydrides, R_nGeH_{4-n} ($n = 0,1,2,3$), are the triorganogermanium hydrides, which are prepared by the reduction of the corresponding halides with lithium alanate, or with amalgamated zinc and hydrochloric acid[5−7].

The lower polarity of the Ge−H bond compared with the Si−H bond in the corresponding silicon hydrides explains the differences in the reactivity of organogermanium hydrides vs organosilicon hydrides. The germanium hydrides are more stable towards alkalis and R_3GeH compounds are not attacked by KOH while analogous silanes release hydrogen to form silanolates. The organogermanium hydrides also differ from silanes in their reactions with organolithium reagents where they react like triphenylmethane (equations 1−3):

$$\text{Ph}_3\text{SiH} + \text{RLi} \longrightarrow \text{Ph}_3\text{SiR} + \text{LiH} \tag{1}$$

$$\text{Ph}_3\text{GeH} + \text{RLi} \longrightarrow \text{Ph}_3\text{GeLi} + \text{RH} \tag{2}$$

$$\text{Ph}_3\text{CH} + \text{RLi} \longrightarrow \text{Ph}_3\text{CLi} + \text{RH} \tag{3}$$

a. Alkyl- and arylgermanes. Homolytic bond dissociation energies of organometallic compounds, which are generally derived from the heats of formation of metal-centred radicals, are rather rare since these materials are not suitable for conventional gas-phase techniques. One possibility to determine metal–hydrogen bond dissociation energies is the laser-induced photoacoustic calorimetry[8,9]. With this technique the Ge–H bond dissociation (BDE) energies of several alkyl- and aryl-substituted germanium hydrides have been determined. According to equations 4 and 5 the bond dissociation energies of the Ge–H bond can be obtained by combination of both equations and applying literature values for the heats of formation of *t*-BuOH, *t*-BuOOBu-*t* and H· (equation 6).

$$t\text{-BuOOBu-}t \longrightarrow 2t\text{-BuO·} \tag{4}$$

$$2t\text{-BuO·} + 2R_3GeH \longrightarrow 2t\text{-BuOH} + 2R_3Ge· \tag{5}$$

$$\Delta H_{4,5} = 2\Delta H_f(t\text{-BuOH}) + 2\Delta H_f(R·) - \Delta H_f(t\text{-BuOOBu-}t) - 2\Delta H_f(R_3GeH) \tag{6}$$

Compared with GeH_4, the bond dissociation energies of these hydrides are not affected by alkyl substitution and are in the range of $81.6-82.6\ kcal\,mol^{-1}$. Aryl substitution results in a slightly weakened Ge–H bond ($79.2-80.2\ kcal\,mol^{-1}$). In the cases of phenyl-, diphenyl- and triphenylgermane[10], the Ge–H bond dissociation energies have been found to be identical within experimental error (Table 2). The Ge–H BDE value for GeH_4 was determined by Walsh and coworkers to be $82.7\ kcal\,mol^{-1}$[11,12] while a value of $85.5\ kcal\,mol^{-1}$ was reported on the basis of photoionization mass spectrometric techniques[13]. For Me_3GeH the Ge–H BDE value was experimentally determined to be equal to $81.6\ kcal\,mol^{-1}$[14]. Results from a recent theoretical study place the Ge–H bond dissociation energy at $84.8\ kcal\,mol^{-1}$[15].

Compared with tri-*n*-butyltin hydride which has been widely employed in organic chemistry and in kinetic studies of rearrangements of carbon-centred radicals, the corresponding germanium compound, tri-*n*-butylgermanium hydride, is not so common in organic synthesis but it may offer an alternative since it is less reactive than tin hydride[16,17]. A detailed kinetic study of hydrogen abstraction from tri-*n*-butylgermanium hydride by primary and secondary alkyl radicals has been reported[18]. Reactions as shown in Figure 1 were investigated kinetically and compared with those of tri-*n*-butyltin hydride[19].

It was found that the germanium hydride is a notably less reactive hydrogen donor than the analogous tin hydride, presumably because Ge–H bonds are *ca* $8-10\ kcal\,mol^{-1}$ stronger than comparable Sn–H bonds[20] and there is larger activation energy required for the attack upon the germane (Table 3). Therefore, the reaction with germane is less exothermic.

TABLE 2. Values of $\Delta H_{(4,5;eq.\ 6)}$ for the reaction between di-*tert*-butoxy radicals and organogermanes and bond dissociation energies, BDE(Ge–H)

Germane	$\Delta H_{(4,5;eq.\ 6)}$ (kcal mol^{-1})	BDE(Ge–H) (kcal mol^{-1})
GeH_4		82.7 ± 2.4[11,12]
Me_3GeH	-7.0 ± 0.5	81.6 ± 0.5
Et_3GeH	-5.8 ± 0.8	82.3 ± 0.6
Bu_3GeH	-5.0 ± 0.8	82.6 ± 0.6
$PhGeH_3$	-11.8 ± 1.0	79.2 ± 0.7
Ph_2GeH_2	-11.2 ± 1.0	79.5 ± 0.7
Ph_3GeH	-9.9 ± 1.4	80.2 ± 0.8

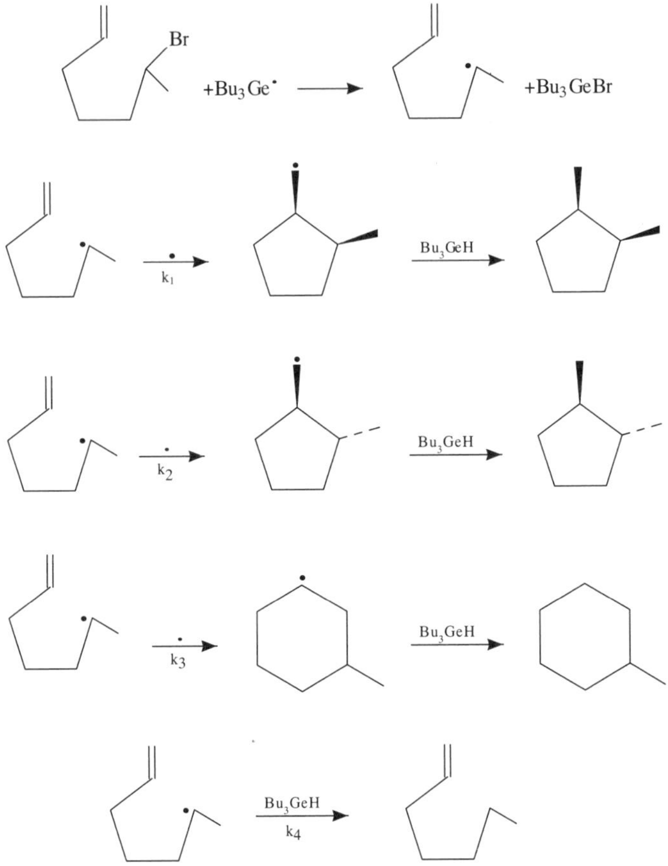

FIGURE 1. Reaction of 1-methyl-5-hexenyl with n-Bu$_3$GeH. Reprinted with permission from Reference 19. Copyright (1987) American Chemical Society

TABLE 3. Activation energies for the reaction of alkyl radicals R· with Bu$_3$GeH and Bu$_3$SnH

Alkyl radical	Hydride	E (kcal mol^{-1})	R·
Primary	Bu$_3$GeH	4.70 ± 0.62	5-hexenyl
Primary	Bu$_3$SnH	3.69 ± 0.32	combined data for ethyl and n-butyl radical
Secondary	Bu$_3$GeH	5.52 ± 0.35	1-methyl-5-hexenyl
Secondary	Bu$_3$SnH	3.47 ± 0.49	isopropyl

The thermal stability of germanes decreases in the order: saturated tetraalkylgermanes > vinylgermanes > allylgermanes > diallygermacyclopent-3enes > 5-germaspiro [4.4] nona-2,7-dienes. Compared with the corresponding organosilicon compounds, the Ge derivatives are less stable and decompose at lower temperatures[21]. Thermolysis mechanisms as deduced from gas-phase studies are consistent with the presence of germylenes as intermediates.

Thus, the mechanism of the thermolysis of tri- and tetraethylgermane is probably a radical process according to equations 7–13. In case of the tetraethylgermane the reaction starts with the homolytic cleavage of one germanium carbon bond (equation 7)[22]. While the ethyl radical decomposes to give ethylene and hydrogen, the germyl radical can recombine with atomic hydrogen radical to form triethylgermane (equations 8 and 9).

$$(C_2H_5)_4Ge \longrightarrow (C_2H_5)_3Ge\cdot + C_2H_5\cdot \qquad (7)$$

$$C_2H_5\cdot \longrightarrow C_2H_4 + H\cdot \qquad (8)$$

$$(C_2H_5)_3Ge\cdot + H\cdot \longrightarrow (C_2H_5)_3GeH \qquad (9)$$

The next steps of the decomposition are the same as for the triethylgermane, giving the same thermolysis products (equations 10–13). The energy of the germanium–hydrogen bond is greater than that of the germanium–carbon bond (Ge–H 82.7 kcal mol^{-1}, see above; Ge–C 76.2 kcal mol^{-1}); the reaction proceeds by successive homolytic cleavages of ethyl radicals, finally leading to the products ethylene, hydrogen and elemental germanium.

$$(C_2H_5)_3GeH \longrightarrow C_2H_4 + (C_2H_5)_2GeH_2 \qquad (10)$$

$$(C_2H_5)_2GeH_2 \longrightarrow C_2H_4 + C_2H_5GeH_3 \qquad (11)$$

$$C_2H_5GeH_3 \longrightarrow C_2H_4 + GeH_4 \qquad (12)$$

$$GeH_4 \longrightarrow Ge + 2H_2 \qquad (13)$$

Hydrogen abstraction in α and β positions of Et$_4$E (E = C, Si, Ge, Sn) in a reaction with *tert*-butoxyl radicals has been reported[23]. It was pointed out that the activation energy was greater at the β position than at the α position, and followed the order Sn>Ge>Si as illustrated by the relative reactivities (Figure 2). The activating influence of the Et$_3$E (E = Si, Ge, Sn) is clear-cut and unequivocal: in each case and in both positions there is an increase in reactivity compared with Et$_4$C, and there is a monotonic increase from silicon to tin, although the differences between silicon and germanium are relatively small (post-transition metal effect).

Interaction of alkadienylidenecarbenes, R$_2$C=C=C=C, with group 14 hydrides results in the formation of Si, Ge and Sn functionalized butatrienes R$_2$C=C=C=CHER$_3$ (Figure 3)[24]. The reaction is general for alkyl- and aryl-substituted carbenes with isolated

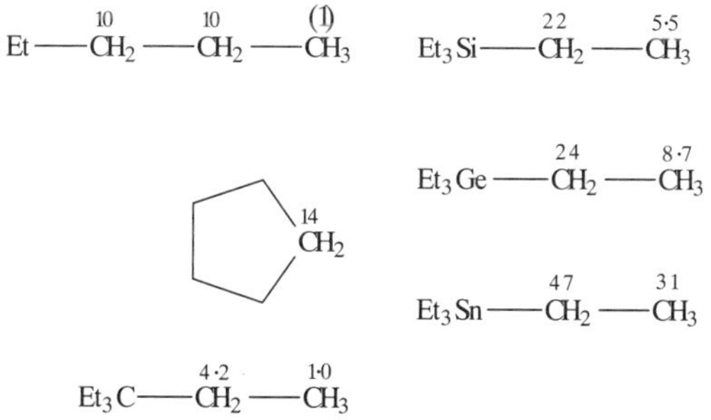

FIGURE 2. Relative reactivities toward the *tert*-butoxyl radical of individual positions (per hydrogen atom) relative to a hydrogen atom in a methyl group in pentane. Reprinted with permission from Reference 23. Copyright (1985) American Chemical Society

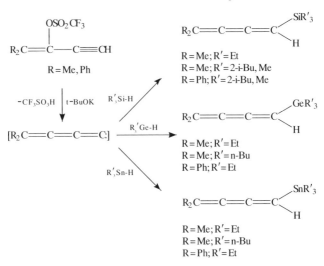

FIGURE 3. Reaction behaviour of alkadienylidenecarbenes, $R_2C=C=C=C$, with group 14 hydrides. Reprinted with permission from Reference 24. Copyright (1981) American Chemical Society

yields of 26–88%. Although the parent systems are too unstable to isolate, IR evidence clearly indicates their formation. All so-prepared cumulenes are stable, isolable but moderately oxygen-sensitive compounds that rearrange or polymerize upon prolonged standing at room temperature.

Branched-chain polyorganogermanes were prepared by the reaction of triphenylgermyl-lithium with germanium diiodide, which led to the formation of tris(triphenylgermyl)-germyllithium. The hydrolysis of tris(triphenylgermyl)-germyllithium led to tris(triphenyl-germyl)german (equation 14)[25]. An analogous behaviour was observed in the reaction of tris(triphenylgermyl)-germyllithium with methyl iodide (equation 15).

$$3Ph_3GeLi + GeI_2 \xrightarrow{-2LiI} [Ph_3Ge]_3GeLi \xrightarrow{H_2O} [Ph_3Ge]_3GeH + LiOH \quad (14)$$

$$[Ph_3Ge]_3GeLi + CH_3I \longrightarrow [Ph_3Ge]_3GeCH_3 + LiI \quad (15)$$

Alkylation or arylation of the germanium–hydrogen bond of trihalogermanes usually leads to organohalogermane species (equation 16)[26].

$$Cl_3GeH \cdot NEt_3 + RI \longrightarrow Et_3N \cdot IH + RGeCl_3 \quad (16)$$

The reaction of GeH_4 with sodium silylsilanide in diglyme yielded the new sodium silyl-germanides $NaGeH_n(SiH_3)_{3-n}$ ($n = 2,1,0$) with the evolution of hydrogen and SiH_4 (Figure 4)[27,28]. In the next step the reaction with p-toluene sulphonic acid methyl ester (p-TosCH$_3$) led to methylsilylgermanes (**a**–**c**)[29] isolated by preparative gas chromatography, whereas in the reaction with nonafluorobutanesulphonic acid silyl ester the corresponding silylgermanes (**A**–**C**) were obtained. The structure of $Ge(SiH_3)_4$ was determined by electron diffractometry in the gaseous state. The electron diffraction study revealed the expected T_d symmetry with nearly free rotation along the Ge–Si bonds[30].

b. Germylamines and germylimines. Trigermylamine is obtained by treatment of germyl chloride with ammonia according to reaction (17).

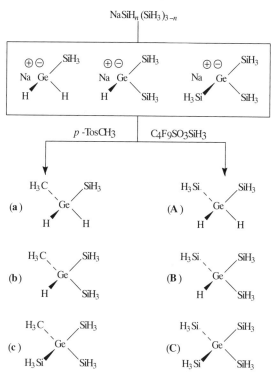

FIGURE 4. Reactions of sodium silylgermanides, $NaGeH_n(SiH_3)_{3-n}$ ($n = 0,1,2$). Reproduced from Reference 28 by permission of Verlag der Zeitschrift für Naturforschung

$$3GeH_3Cl + 4NH_3 \longrightarrow 3NH_4Cl + (GeH_3)_3N \qquad (17)$$

The molecular structure of trigermylamine has been determined by electron diffraction[31] indicating a planar skeleton of the heavy atoms and a $r_g(Ge-N) = 1.838$ Å. Trigermylamine is an extremely unstable compound and decomposes predominantly as indicated in equation (18).

$$x\,(GeH_3)_3N \longrightarrow 3(GeH_2)_x + xNH_3 \qquad (18)$$

The reaction of germane with trifluoronitrosomethane has recently been reported. The first example of a stable N-trifluoromethylgermaimine, $CF_3N{=}GeH_2$, was prepared according to equation (19)[32]:

$$CF_3NO + GeH_4 \longrightarrow CF_3N{=}GeH_2 + H_2O \qquad (19)$$

At room temperature the N-trifluoromethylgermaimine is a colourless gas that undergoes polymerization on standing for two days at room temperature. It was assumed that the stability of the Ge=N double bond is caused by the highly electronegative trifluoromethyl group or, as in the following example, by the bis(trifluoromethyl)-nitroxy group. The Ge=N double bond in $CF_3N{=}GeH_2$ was chemically confirmed by its reaction with hydrogen iodide, resulting in the addition product as illustrated in equation 20.

$$CF_3N{=}GeH_2 + HI \longrightarrow CF_3NHGeH_2I \qquad (20)$$

CF_3NHGeH_2I is a volatile pale yellow liquid. Another reaction confirming the presence of the double bond and the GeH_2 moiety is the reaction with bis(trifluoromethyl)nitroxyl which is a powerful hydrogen abstractor. According to the molar ratio of either 1:2 or 1:4 two bis(trifluoromethyl)nitroxy derivatives were isolated as shown by equations 21 and 22.

$$CF_3N{=}GeH_2 + 2(CF_3)_2NO \longrightarrow CF_3N{=}GeHON(CF_3)_2 + (CF_3)_2NOH \quad (21)$$
$$\textbf{(A)}$$

$$CF_3N{=}GeH_2 + 4(CF_3)_2NO \longrightarrow CF_3N{=}Ge[ON(CF_3)_2]_2 + 2(CF_3)_2NOH \quad (22)$$
$$\textbf{(B)}$$

Both compounds **A** and **B** are stable at room temperature. Cleavage of the $Ge-O$ bond in both bis(trifluoromethyl)nitroxy derivatives **A** and **B** by hydrogen chloride led to the corresponding chloro derivatives **C** and **D** (equations 23 and 24). **C** and **D** are stable at $-20\,^{\circ}C$ but, on standing for two days, both compounds polymerized.

$$CF_3N{=}GeHON(CF_3)_2 + HCl \longrightarrow CF_3N{=}GeHCl + (CF_3)_2NOH \quad (23)$$
$$\textbf{(C)}$$

$$CF_3N{=}Ge[ON(CF_3)_2]_2 + 2HCl \longrightarrow CF_3N{=}GeCl_2 + 2(CF_3)_2NOH \quad (24)$$
$$\textbf{(D)}$$

c. Alkyl- and arylhydridogermanium metal compounds. The organogermylalkali metal compounds R_3GeM (M = Li, Na, K or Cs) have been very useful reagents in organometallic synthesis. The synthesis has undergone a continuous development over the last decade and in the literature a wide range of preparative methods for these compounds has been published[33,34].

The hydridogermyl metals represent an interesting class of compounds because of the presence of two reactive positions at the germanium atom: the $Ge-metal$ and the $Ge-H$ bond. In the case of organohydridogermyl alkali metal compounds the germyl anion determines the reaction resulting in nucleophilic or SET-type reactions, while in the case of organohydridogermyl transition metal complexes the activity of the $Ge-H$ bond can predominate[35].

Hydrogermolysis leads to arylgermyllithium compounds; their stability depends on the group R and the solvent (equation 25)[36,37]. If a large excess of ^tBuLi is used, the corresponding diaryldilithiogermane is formed (equation 26). Whereas the reaction of two equivalents of *tert*-BuLi with asymmetrically substituted dimesityldiphenylgermane led to the formation of the corresponding monolithiated product, the less kinetically stabilized $Ph_4Ge_2H_2$ reacts with *tert*-BuLi to give polygermanes and Li hydride (equation 27). Treatment of $Ph_4Ge_2H_2$ with an excess of *tert*-BuLi, however, afforded the aryldigermyl dilithium compound according to equation 28.

$$R_nGeH_{4-n} \xrightarrow{\ ^tBuLi/THF\ } R_nH_{3-n}GeLi \quad (25)$$

$$R = Ph, Mes;\ n = 1, 2, 3$$

$$2R_2GeH_2 \xrightarrow{\ ^tBuLi/THF\ } R_2HGeLi + R_2GeLi_2 \quad (26)$$
$$\quad\quad R = Ph \quad\quad\quad\quad\quad\ 41\% \quad\ 59\%$$
$$\quad\quad R = Mes \quad\quad\quad\quad\ 28\% \quad\ 71\%$$

$$Ph_2HGeGeHPh_2 \xrightarrow{\ 1^tBuLi\ } \{Ph_2HGeGeLiPh_2\} \xrightarrow{\ -LiH\ } 1/n\ (Ph_2Ge)_n \quad (27)$$

$$Ph_2HGeGeHPh_2 \xrightarrow{\ 2^tBuLi\ } Ph_2LiGeGeLiPh_2 \quad (28)$$

Tetrakis(2,6-diisopropylphenyl)digermene is synthesized directly from bis(2,6-diisopropylphenyl)dichlorogermane by reductive coupling with lithium naphthalenide. Surprisingly, the treatment of the tetrakis(2,6-diisopropylphenyl)digermen **A** with excess of lithium naphthalenide resulted in a cleavage of the Ge−C forming the corresponding digermenyllithium compound **B** (germanium analogue of vinyl lithium) (equations 29a,b)[38]. The crystalline product is very air-sensitive and reacts with methanol to the digerman **C**.

$$Dip_2Ge=GeDip_2 \xrightarrow{\text{LiNaph/DME}} (Dip)_2Ge=Ge(Dip)(Li \cdot DME) + DipLi \quad (29a)$$
$$\textbf{(A)} \qquad\qquad\qquad\qquad \textbf{(B)}$$

$$\textbf{B} \xrightarrow{\text{MeOH}} (Dip)_2HGe-GeH(Dip)(OMe) + LiOMe \quad (29b)$$
$$\textbf{(C)}$$

Dip = 2,6-diisoprophylphenyl

In contrast to the observed reduction of **A** to **B**, treatment of tetrakis(2,6-diethylphenyl)digermen **D** with lithium naphthalenide under the same conditions led to products which were converted upon methanolysis to the corresponding diarylgermane **E**, tetraaryldigermane **F** and polymers (equation 30). Tetraaryldigermenes undergo smooth cycloadditions with diazomethane, phenyl azide, selenium, phenylacetylene and acetone to yield the corresponding three- or four-membered addition products[39].

$$(Dep)_2Ge=Ge(Dep)_2 \xrightarrow{\text{1. LiNaph/2. MeOH}} (Dep)_2GeH_2$$
$$\textbf{(D)} \qquad\qquad\qquad\qquad \textbf{(E)}$$

$$+ (Dep)_2HGe-GeH(Dep)_2 + \textbf{P} \quad (30)$$
$$\textbf{(F)}$$

Dep = 2,6-diethylphenyl; **P** = polymers

It has been established that triethylgermylcaesium metallates react with toluene in an equilibrium reaction forming triethylgermane (equation 31) if toluene is used as the solvent. The pK_a value of triethylgermane was obtained to be 33.3[40]. While the value for the t-butyldiethylgermane (33.1) does not differ much from that of triethylgermane, the pK_a value for phenyldiethylgermane is only 30.5, probably due to electron interaction of the aromatic substituent with the germanium atom.

$$Et_3GeCs + PhCH_3 \rightleftharpoons PhCH_2Cs + Et_3GeH \quad (31)$$

The replacement of an alkyl or phenyl substituent by an organo-silicon or -germanium group results in considerable changes in the properties of the germanium hydride. This was observed in the metallation reaction with alkalitriethylgermyl derivatives. The reaction of $R_3SiGeEt_2H$ with Et_3GeCs in benzene yielded $Et_3GeGeEt_2H$ while the analogous reaction with Et_3GeLi in HMPA led to $Me_3SiGeEt_3$ and $LiGeEt_2H$ (equations 32 and 33).

$$Me_3SiGeEt_2H + Et_3GeM \xrightarrow{M=Cs} Me_3SiCs + Et_3GeGeEt_2H \quad (32)$$

$$Me_3SiGeEt_2H + Et_3GeM \xrightarrow{M=Li} Me_3SiGeEt_3 + LiGeEt_2H \quad (33)$$

When Et_2GeHCl was treated with Et_3GeLi, a mixture of three germanium compounds was obtained according to equation 34.

$$Et_2GeHCl + Et_3GeLi \xrightarrow{\text{hexane/water}} Et_3GeGeEt_2H + Et_3GeH + Et_3GeGeEt_3 \quad (34)$$
$$82\% \qquad\qquad 11\% \qquad\qquad 7\%$$

Germanium–carbon bonds can easily be formed in the reaction of organohydrido germanium lithium compounds with organic halides as shown in equation 35[41].

$$PhH_2GeLi + ClCH_2OCH_3 \xrightarrow{-LiCl} PhH_2GeCH_2OCH_3 \qquad (35)$$

Ge–metal bonds can be built in analogy as described for Ge–C bonds by the reaction of organolithium compounds with metal halides. With *trans*-dichlorobis(triethylphosphine)platinum(II), new germyl transition metal complexes were synthesized (equation 36)[41].

$$2Ph_2LiGeGeLiPh_2 + 2trans\text{-}[PtCl_2(PEt_3)_2] \xrightarrow{-4\,LiCl} (Et_3P_2)Pt\overline{-Ge(Ph_2)-Ge(Ph_2)}$$

$$+ (Et_3P)_2Pt\overline{-Ge(Ph_2)-O-Ge(Ph_2)} + \ldots \qquad (36)$$

Interaction of the hydridogermyl complex $(\eta^5-C_5H_5)Re(NO)(PPh_3)-(GePh_2H)$ with *n*-BuLi in THF led quantitatively via an instable lithiocyclopentadienyl complex, $(\eta^5-C_5H_4Li)Re(NO)(PPh_3)(GePh_2H)$, to the rhenium centred anion $Li^+[(\eta^5-C_5H_4GePh_2H)Re(NO)(PPh_3)]^-$ (Figure 5)[42]. Adding $HBF_4\cdot OEt_2$ the germylcyclopentadienyl complex $(\eta^5-C_5H_4GePh_2H)Re(NO)(PPh_3)(H)$ was formed.

Some other organogermyl-, organohalogermyl- and organopolygermylmercury derivatives have been prepared by hydrogermolysis of dialkylmercury compounds (equations 37 and 38)[43,44].

$$2R_nX_{3-n}GeH + Bu_2Hg \xrightarrow{-2BuH} (R_nX_{3-n}Ge)_2Hg \qquad (37)$$

R = aryl, alkyl, CF_3, SiMe_3; X = halogen; $n = 1, 2, 3$

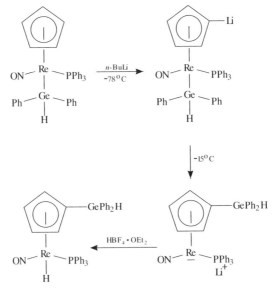

FIGURE 5. Syntheses of hydride complexes from germyl complexes. Reprinted with permission from Reference 42. Copyright (1991) American Chemical Society

$$2R_nGeH_{4-n} + (4-n)Bu_2Hg \xrightarrow{-2(4-n)BuH} (R_nGe)_2Hg_{4-n} \qquad (38)$$

$$R = \text{aryl, halogen; } n = 1, 2, 3$$

d. Cyclic compounds. Germacyclohexanes can be prepared by coupling the bis-Grignard reagent of 1,5-dibromopentane with tetrachlorogermane to yield 1,1-dichloro-1-germacyclohexane, followed by treatment with LAH[45]. Further treatment of the 1,1-dichloro-1-germacyclohexane with methylmagnesium iodide led to the formation of dimethylgermacyclohexane (Figure 6)[46].

Coupling of the bis-Grignard reagent of 1,5-dibromopentane with dibromo(methyl)-phenylgermane yielded 1-methyl-1phenyl-1-germacyclohexane. The phenyl group of 1-methyl-1-phenyl-1-germacyclohexane can be replaced by a bromine atom upon treatment with bromine, obtaining 1-bromo-1methyl-1-germacyclohexane. Further, replacement of the Br atom by hydrogen is possible in the reaction with LiAlH$_4$[47].

Both electrophilic and nucleophilic reagents are able to cleave the ring in cyclic germanium–carbon compounds due to the ring strain and the polarizability of the germanium–carbon bond, as illustrated in Figure 7[48,49].

2. Tin

Tin belongs to the long period elements from Rb to Xe and is a main group element because the 4d shell is filled with electrons. Since the valence electrons are $5s^2p^2$, tin occurs in two valences. Whereas valence 2 is formally always positive, valence 4 has amphoteric properties possessing the formal oxidation states $+4$ or -4, according to the covalently bound substituents and to the reaction partner.

a. Alkyl- and aryl stannanes. The reaction behaviour of stannane towards organic functional groups was the subject of a recent investigation[50]. This study has illustrated that stannane is able to react with some but not all organic functional groups (as organotin hydrides do). The addition of stannane to the double bond of acrylonitrile resulted in the formation of tetrakis(2-cyanoethyl)tin. In contrast to the high yields for the same reaction with organotin hydrides only low yields (5–35%) depending on the reaction conditions could be obtained, especially since the instability of stannane limits the reaction temperature to room temperature. However, temperatures of 70–100 °C are required for the addition of organotin hydrides to acrylonitrile in the absence of suitable catalysts. The stannane SnH$_4$ was shown to react with carboxylic acids to give hydrogen and the acetate Sn(OOCCH$_3$)$_4$ as a minor product (4.4%) by way of an unstable intermediate {SnH$_3$OAc} decomposing into tin metal, hydrogen and carboxylic acid (equations 39a,b) (cf organotin hydrides react with carboxylic acids yielding organotin carboxylates).

$$2SnH_4 + 2HOAc \longrightarrow 2\{SnH_3OAc\} + 2H_2 \qquad (39a)$$

$$2\{SnH_3OAc\} \longrightarrow 2Sn + 2H_2 + 2HOAc \qquad (39b)$$

The reaction of tri-n-butyltin hydride with acyl halides led to the formation of the corresponding aldehyde and ester[51]. Both these were obtained by free-radical chain processes involving acyl radicals as intermediates (equations 40–42). In contrast, Me$_3$CCOCl reacts with n-Bu$_3$SnH at temperatures as low as 60 °C forming the corresponding aldehyde by a facile, non-radical process[52]. It was suggested that the aldehyde was formed either in a concerted bimolecular reaction or via an unstable α-chloroalkoxytin, which easily converts into pivalaldehyde (equation 43).

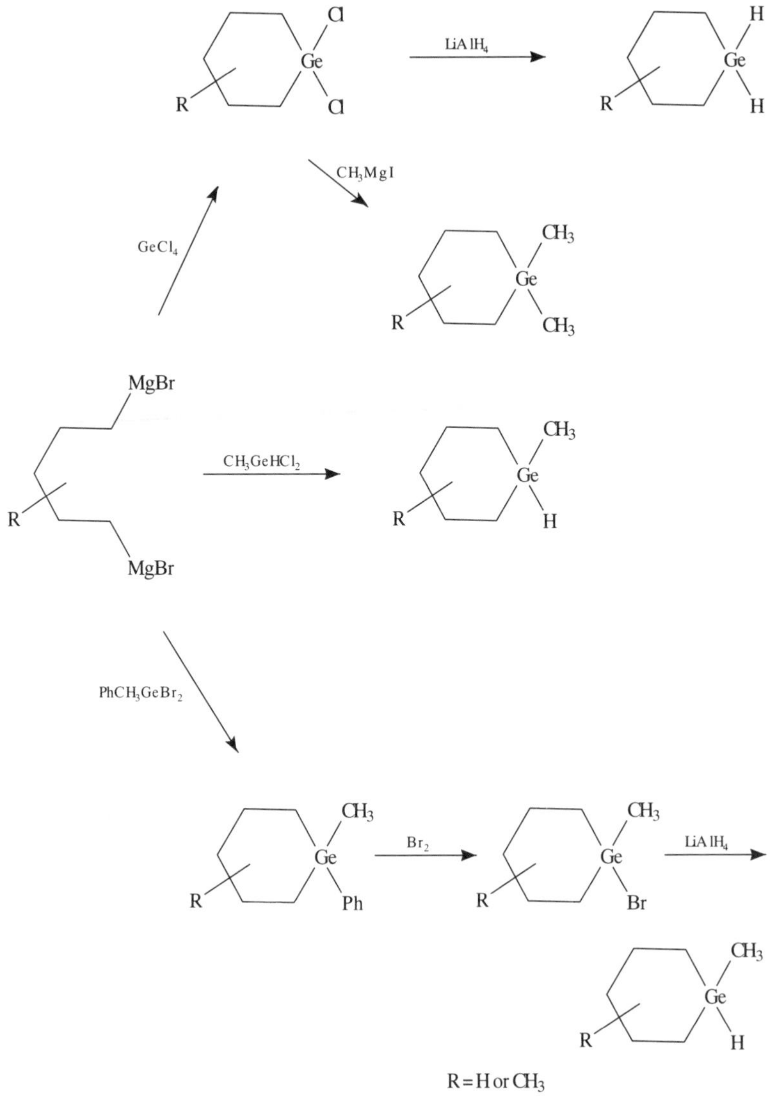

R = H or CH$_3$

FIGURE 6. Synthetic route to methylgermacyclohexane derivatives. Reproduced from Reference 46 by permission of The Royal Society of Chemistry

$$3RCOCl + 3n\text{-}Bu_3SnH \longrightarrow RCHO + RC(O)OCH_2R + 3\ nBu_3SnCl \quad (40)$$

$$RCOCl + n\text{-}Bu_3Sn\cdot \longrightarrow RC\cdot(=O) + n\text{-}Bu_3SnCl \quad (41)$$

$$RC\cdot(=O) + n\text{-}Bu_3SnH \longrightarrow RCHO + n\text{-}Bu_3Sn\cdot \quad (42)$$

$$n\text{-}Bu_3SnH + Me_3CCOCl \longrightarrow \{n\text{-}Bu_3SnOCH(Cl)CMe_3\} \longrightarrow$$

$$n\text{-}Bu_3SnCl + Me_3CCHO \quad (43)$$

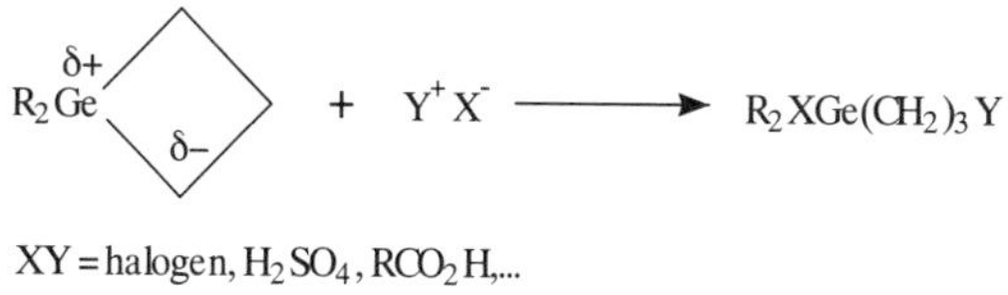

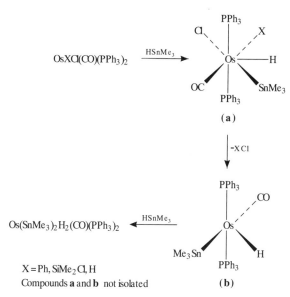

$$R_2Ge \overset{\delta+}{\underset{\delta-}{\diamond}} + Y^+X^- \longrightarrow R_2XGe(CH_2)_3Y$$

$XY = \text{halogen}, H_2SO_4, RCO_2H,...$

FIGURE 7. Ring cleavage of cyclic germanium–carbon compounds

$$OsXCl(CO)(PPh_3)_2 \xrightarrow{HSnMe_3}$$

(a)

$$\downarrow -XCl$$

$$Os(SnMe_3)_2H_2(CO)(PPh_3)_2 \xleftarrow{HSnMe_3}$$

X = Ph, SiMe$_2$Cl, H
Compounds **a** and **b** not isolated

(b)

FIGURE 8. Postulated mechanism for the formation of Os(SnMe$_3$)$_2$H$_2$(CO)(PPh$_3$)$_2$. Reproduced from Reference 53 by permission of Elsevier Sequoia S.A.

Treatment of either Os(Ph)X(CO)(PPh$_3$)$_3$ (X = Ph, SiMe$_2$Cl, H) or Os(SiMe$_2$Cl)Cl-(CO)(PPh$_3$)$_2$ with two equivalents of HSnMe$_3$ led to the formation of Os(SnMe$_3$)$_2$H$_2$(CO)-(PPh$_3$)$_2$[53]. Reactions with less than two equivalents of HSnMe$_3$ afforded the same product, but in reduced yield. In case of OsHCl(CO)(PPh$_3$)$_3$ three equivalents of the three methyltin hydrides were required to obtain Os(SnMe$_3$)$_2$H$_2$(CO)(PPh$_3$)$_2$, because the HCl generated in the first reductive step of the reaction reacts rapidly with trimethyltin hydride, giving trimethyltin chloride and hydrogen. The postulated mechanism for the reaction of the osmium complexes with the tin hydrides is illustrated in Figure 8. In the first step, the HSnMe$_3$ oxidatively adds to the osmium(II) complex forming an intermediate (**a**), which reductively eliminates HCl in the next step yielding Os(SnMe$_3$)$_2$H$_2$(CO)(PPh$_3$)$_2$.

Similar reactions of the other group 14 hydrides with ruthenium(0) and osmium(0) complexes have also been described.

Reactions of triorganotin hydrides with amines resulted in the formation of the corresponding ditin species according to equation 44[54−56].

$$2(C_6H_5)_3SnH \xrightarrow{n\text{-}C_6H_{13}NH_2} (C_6H_5)_6Sn_2 + H_2 \qquad (44)$$

This reaction has been studied, and it was assumed that the hydride attacks electrophilically on nitrogen in a polar reaction forming a tin–tin bond. This means that the amine behaves as a catalyst. No formation of hydrocarbons or ammonia was observed in this reaction.

3. Lead

Because of the lower metal–carbon and metal–hydrogen bond strength, organolead hydrides are particularly unstable species and represent the least stable of those of the group 14 elements. Triorganolead hydrides are obtained at low temperatures by reduction of the halides with $LiAlH_4$ (equation 45), but they decompose at $0\,°C$.

$$4R_3PbH \xrightarrow{\ T \geqslant 0\,°C\ } 3PbR_4 + Pb + 2H_2 \qquad (45)$$

The hydrides reduce organic halides and add to olefins, acetylenes and isocyanates (hydroplumbation)[5].

IV. COMPLEXING, ACIDITY AND BASICITY

A. Introduction

The concept of second- or outer-sphere coordination, orginally introduced by Werner[57], has played a major role in the subsequent development of the theory of bonding in metal complexes and has recently re-emerged as a means of describing higher-order bonding interactions in complexes with crown ether ligands and in systems involving supramolecular or host–guest interactions[58]. In molecular compounds, the preference for inner- over outer-sphere coordination may be expected to be dependent primarily on (i) the size of the central atom, (ii) the symmetries and energies of the available unoccupied orbitals, (iii) the electronegativity differences and (iv) special structural features of the ligating groups. Accordingly, with certain metals and ligands, complexes with unusual coordination numbers and geometries were obtained, but because of the manifold nature of the metal–ligand interactions, predictions as to the behaviour of a given metal or ligand are not generally possible.

We divided the Section VI.B into two parts according to the two formal oxidation states occurring in group 14 organometal chemistry.

A milestone in the development of isolable molecular compounds of germanium in low oxidation states (in particular Ge II) was the synthesis of $Ge[HC(SiMe_3)_2]_2$ by Lappert, Hitchcock amd coworkers[102]. Only a few additional examples of stable organogermanium(II) compounds, mostly containing substituted cyclopentadienyl ligands[59], have been synthesized since then. However, after the first synthesis of a germanium(II)amide derivative, quite a number of similar compounds with bulky N-substituents have been obtained[60]. Moreover, organogermanium compounds with coordination numbers higher than four have been reported.[61,62].

Tin has been observed in both valences +2 and ±4 in which the −4 state represents the Xe configuration in which formally four electrons have been accepted ($5s^25p^6$). The sp^3 hybridization to form tetrahedral covalent compounds can be discussed if one of the 5s electrons is promoted into the 5p orbital. When 5d orbitals are added to the hybridization, tin(IV) compounds with higher coordination numbers (of 5, 6, 7, 8) are formed. In general, tin(IV) is known to form compounds or complexes in which it adopts the coordination numbers 4 and 6, although compounds with numbers 2, 3, 5, 7 and 8 are

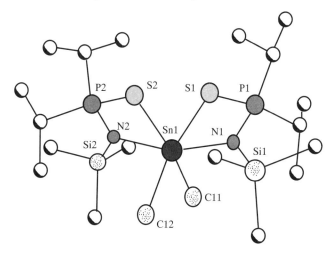

FIGURE 9. Molecular structure of $[(i\text{-Pr})_2P(S)NSiMe_3]_2SnCl_2$. Reproduced from Reference 64 by permission of Barth Verlagsgesellschaft mbH

also known[3]. Higher coordination numbers are favoured in compounds with electronegative substituents. However, hexa- and heptacoordinated organotin derivatives containing covalent R_2Sn and RSn groups have also been described. While derivatives of R_3Sn may still be pentacoordinated, no hexacoordinated compounds of this type are known[63].

Spirocycles have also been described. For example, the four-membered spirocycle $[(i\text{-Pr})_2P(S)NSiMe_3]_2SnCl_2$ recently reported by Roesky and coworkers was obtained from $(i\text{-Pr})_2P(S)N(SiMe_3)_2$ and $SnCl_4$ in 2:1 ratio under elimination of trimethylsilylchlorid (equation 46; Figure 9)[64].

$$Sn Cl_4 + 2(Me_3Si)_2N-P(S)(i\text{-Pr})_2 \xrightarrow{-2Me_3SiCl} [S-P(i\text{-Pr}_2)-N(SiMe_3)]_2SnCl_2 \quad (46)$$

This compound was characterized by X-ray studies. Two $Sn-N-P-S$ rings join at tin to form the spirocycle. In this complex, the two rings are arranged in cis-position as illustrated in Figure 9.

Lead occurs in complex compounds mostly in four and six coordination.

A large number of complexes containing heavier group 14 element–transition metal bonds is known[65–67]. General methods for the preparation of these complexes have been reviewed comprehensively[68–72].

B. Complexing

1. Complexing of germanium, tin and lead in the oxidation state +2

In the unstable +2 oxidation state, the valence electron configuration of the group 14 elements corresponds to a completely filled ns orbital. However, the lone pair located at the central metal has often steric influence on the structure.

There are some examples of macrocyclic complexes of germanium, tin and lead reported in the recent literature. Several crown ethers[73,75], tetraaza macrocycles[76] [for instance dibenzotetramethyltetraaza[14]annulene (TMTAA)], cyclic polyamines (polyazacycloalkanes)[77–80] or, as already mentioned above, poly(pyrazolyl)borate were

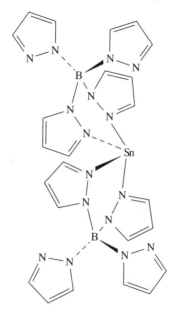

FIGURE 10. Molecular structure of [B(Pz)$_4$]$_2$ Sn(Pz = pyrazolyl ring). Reprinted with permission from Reference 82. Copyright (1993) American Chemical Society

used as ligands. These ligands are suitable for the preparation of kinetically highly stable complexes including organometallic compounds of the main-group metals. Some examples of recent research are described below.

There are only a few examples of complexes of tin(II) and lead(II)[81,82]. Figure 10 shows [B(Pz)$_4$]$_2$Sn where the tin atom is four-coordinated in the solid state, adopting a pseudo-trigonal bipyramidal structure with the lone pair of the tin(II) in the equatorial position. The analogous lead complex has also been described[83].

Proceeding from Pb(SCN)$_2^{84}$, Pb(CCl$_3$CO$_2$)$_2^{85}$ and Pb(NO$_3$)$_2^{86,87}$ some Pb(II) crown ether complexes were synthesized. [PbCl(18-crown-6)][SbCl$_6$] was obtained in low yields besides [Pb(CH$_3$)$_2$(18-crown-6)][SbCl$_6$] by reaction of Pb(CH$_3$)$_2$Cl$_2$ and antimony pentachloride in the presence of 18-crown-6 in acetonitrile. The complex cation possesses almost C$_{6v}$ symmetry according to the fact that the lead atom in the cation is surrounded in nearly hexagonal planar coordination of the six oxygen atoms of the crown ether while the Cl atom is arranged axially (Figure 11)[88]. Similar cations of tin ([SnCl(18-crown-6)]) were described some years ago[89,90]. It was assumed that the antimony pentachloride acts as a chlorinating agent as shown in equation 47.

$$Pb(CH_3)_2Cl_2 + 2SbCl_5 + 18\text{-crown-6} \longrightarrow [PbCl(18\text{-crown-6})][SbCl_6]$$

$$+ SbCl_3 + 2CH_3Cl \quad (47)$$

Furthermore, the structures of [Pb(18-crown-6)(CH$_3$CN)$_3$][SbCl$_6$]$_2$ and [Pb(15-crown-5)$_2$][SbCl$_6$]$_2$ were determined. The structure determination of [Pb(18-crown-6)(CH$_3$CN)$_3$][SbCl$_6$]$_2$ is restricted by disorder problems. Both compounds were formed by the reaction of PbCl$_2$, antimony pentachloride and the corresponding crown ether. The

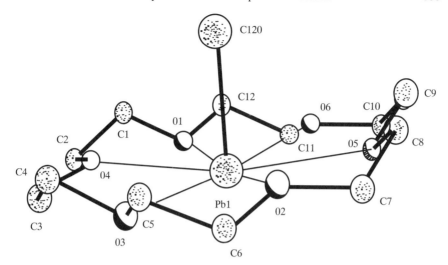

FIGURE 11. Molecular structure of the [PbCl(18-crown-6)]$^+$ cation in [PbCl(18-crown-6)][SbCl$_6$]. Reprinted from Reference 88 by permission of Barth Verlagsgesellschaft mbH

lead atom in the [Pb(18- crown-6)(CH$_3$CN)$_3$]$^{2+}$ cation is nona-coordinated via the six oxygen atoms of the crown ether and the N-atoms of the three acetonitrile molecules (Figure 12). No sterical action of the free electron lone pair at the lead atom could be observed.

In the cation [Pb(15-crown-5)$_2$]$^{2+}$ the lead atom is sandwich-like coordinated by the ten oxygen atoms of the two crown ether molecules. (Figure 13).

In the cation [Pb$_2$L$_5$]$^{4+}$ (Figures 14, and 15) the Pb^{2+} ion is coordinated by four nitrogen atoms of the macrocyclic ligand (mean Pb$-$N distance 2.5 Å), which describe a mean plane with the metal ion located 1.363(2) Å above[91]. The lead ion shows further interaction with another nitrogen atom N(1) and one oxygen atom O(1) of a perchlorate anion at a very long distance [O(1)$-$Pb 2.98(7) Å and N(1)$-$Pb 3.04(6) Å]. In addition, there are two further, even weaker interactions between the lead cation and two oxygen atoms of another perchlorate anion [Pb$-$O(2') 3.38(4) and Pb$-$O(3') 3.41(6) Å]. Taking into account all these interactions the arrangement of the eight donor atoms around the lead ion is rather asymmetric, leaving a free sphere which is supposed to be occupied by the lone pair of electrons at the Pb^{2+} centre. The presence of a 'gap' in the coordination sphere of the cation is one of the structural features ascribable to the stereochemical activity of the lone pair[92,93].

The first example of a square pyramidal Ge(II) complex with macrocyclic ligands is the red crystalline compound Ge(TMTAA), which can be prepared by treating the dilithium salt of TMTAA with an equimolar amount of GeCl$_2$ · dioxane in Et$_2$O at $-78\,^\circ$C[94]. In that compound the Ge atom is symmetrically coordinated to the four N-atoms of the macrocycle, which means that the GeN$_4$ subunit is square pyramidal, and the Ge atom is situated 0.909(4) Å above the N$_4$ plane (Figure 16).

The Ge(TMTAA) complex and the well known Sn(TMTAA) complex undergo facile oxidative addition reactions and reverse ylide formation with MeI and C$_6$F$_5$I because of the reactive M(II) (M = Sn, Ge) lone pair of electrons. In case of the oxidation with MeI it was assumed that, in solution, an ionic–covalent equilibrium exists (equation 48)[95].

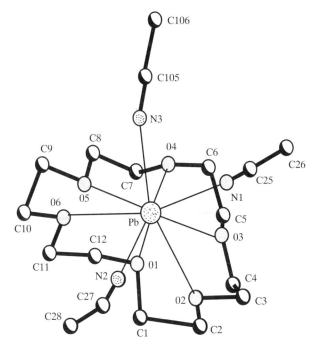

FIGURE 12. Structure of the cation $[Pb(18\text{-crown-}6)(CH_3CN)_3]^{2+}$. Reprinted from Reference 88 by permission of Barth Verlagsgesellschaft mbH

$$Ge(TMTAA)(Me)(I) \underset{}{\overset{THF}{\rightleftharpoons}} [Ge(TMTAA)(Me)]^+[I]^- \qquad (48)$$

The coordination sphere of M(TMTAA)(Me)(I) (M = Ge, Sn) can be regarded as a distorted trigonal prism in which the Me and the I ligands adopt a mutual *cis* arrangement (Figure 17).

In recent years rings and cages with Ge_4, Ge_6 and Ge_8 cores have become accessible[96] as well as Ge−Si heterocycles[97]. Quite recently a novel tricycle with a planar tetragermanium(I) four-membered ring was published[98]. Nucleophilic attack of Li^{t}Bu on $Ge_2[C(PMe_2)_2(SiMe_3)]_2$ led to the first germanium homocycle, $Ge_4[C(PMe_2)_2(SiMe_3)]_2(^tBu)_2$, containing alternating three- and four-coordinated germanium atoms with an average +1 oxidation state (Figure 18). In this complex $Ge_4[C(PMe_2)_2(SiMe_3)]_2(^tBu)_2$ the tricycle is formed by two diphosphinomethanide Ge−Ge bridging ligands on the opposite side of the planar four-membered rhombohedral germanium homocycle.

2. Complexing of germanium, tin and lead in the oxidation state +4

a. Germanium. i. Tricoordinated complexes. The germanethiones (RR′Ge=S) deserve much interest as these species belong to the class of multiply bonded compounds containing $np\pi-np\pi$ bonds ($n \geq 3$)[99]. The synthesis and crystal structure of the first kinetically stable, multiply bonded organoheteroatom diarylgermanethione was

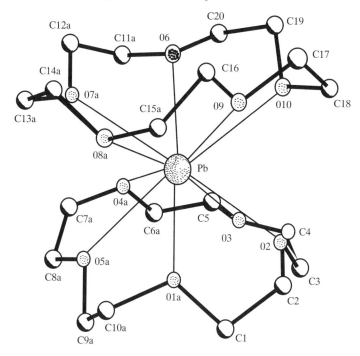

FIGURE 13. Structure of the sandwich-like cation $[Pb(15\text{-crown-}5)_2]^{2+}$. Reprinted from Reference 88 by permission of Barth Verlagsgesellschaft mbH

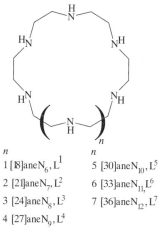

n		n	
1	[18]aneN$_6$, L^1	5	[30]aneN$_{10}$, L^5
2	[21]aneN$_7$, L^2	6	[33]aneN$_{11}$, L^6
3	[24]aneN$_8$, L^3	7	[36]aneN$_{12}$, L^7
4	[27]aneN$_9$, L^4		

FIGURE 14. Macrocyclic ligands of the $[3n]$aneN$_n$ type ($n = 6,7,8,\ldots, 12$). Reproduced from Reference 91 by permission of The Royal Society of Chemistry

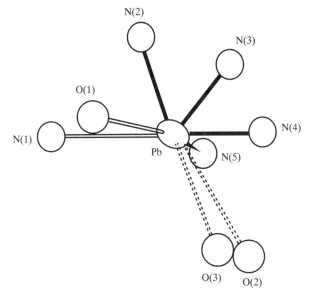

FIGURE 15. ORTEP representation of the coordination site of lead(II) in $[Pb_2L_5][ClO_4]$ showing the gap occupied by the lone pair (cf Figure 14). Reproduced from Reference 91 by permission of The Royal Society of Chemistry

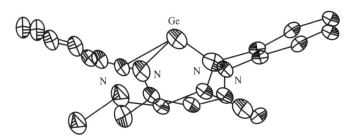

FIGURE 16. Molecular structure of Ge(TMTAA)(TMTAA = dibenzotetramethyltetraaza[14]annulene). Reprinted with permission from Reference 94. Copyright (1992) American Chemical Society

recently reported[100]. Treatment of tetrathiagermolane with triphenylphosphane causes desulphurization of the five-membered GeS_4 ring (equation 49)

$$(Tb)(Tip)\overline{GeSSSS} + 3Ph_3P \longrightarrow (Tb)(Tip)Ge=S + 3Ph_3PS \qquad (49)$$

The molecular structure of (Tb)(Tip)Ge=S was determined by X-ray investigation, which indicates that (Tb)(Tip)Ge=S exists as a monomer in the solid state (Figure 19). The sum of the bond angles around the germanium atom (359.4°) is nearly equal to 360° which indicates a trigonal-planar geometry. The value for the Ge=S bond distance (2.049 Å) is distinctly shorter than for a Ge−S single bond (2.21 Å[101]).

Treatment of MgClR(OEt$_2$) [R = CH(SiMe$_3$)$_2$] with GeCl$_2$ · dioxane in OEt$_2$ yielded the bright yellow crystalline Ge_2R_4 possessing a *trans*-fold C_{2h} framework with a dihedral

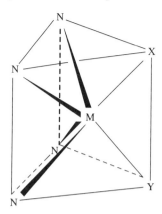

FIGURE 17. Coordination sphere of M(TMTAA)(Me)(I) complexes (M = Ge, Sn; X, Y = Me, I). Reprinted from Reference 95 with kind permission from Elsevier Science Ltd

angle of $32°^{102}$. The sum of angles at Ge is 348.5°, indicating a germanium coordination between a pyramidal and a planar structure. The Ge−Ge bond is slightly shorter (4%) than that in elemental germanium (Figure 20). The chemical properties of crystalline Ge_2R_4 are those of monomer GeR_2 behaving as a Lewis base or a substrate for valence expansion, to yield a tetravalent Ge adduct. Ge_2R_4 resembles Sn_2R_4, i.e. the M−M bond for both is exceedingly labile.

Another example of digermenes, Ge_2R_4, is represented by the (Z)-1,2-bis(2,6-diisopropylphenyl)-1,2-dimesityldigermene isolated by the group of Masamune[103]. Treatment of dichloro(2,6-diisopropylphenyl)mesitylgermane with two equivalents of lithium naphthalenide led to the formation of both the Z- and the E-isomer of 1,2-bis(2,6-diisopropylphenyl)-1,2-dimesityldigermene. According to spectroscopic data for both species a double bond can be postulated.[104–106].

The structure of the Z-isomer was determined (Figure 21) indicating considerable pyramidal distortion at the germanium, reflected in the large dihedral angle (36°) at these atoms. Kinetic investigations (^{1}H-NMR) of the $Z-E$ isomerization in C_6D_6 indicated that the equilibrium is shifted to the Z-isomer [(E)/(Z) values are 0.490 at 40.1 °C and 0.368 at 17.0 °C].

ii. Tetracoordinated complexes. Anomalous stability and spectroscopic properties have been observed in compounds containing the group $R_3GeR-C-X$ (X = halogen, NR_2, O-, S-, ...). This can be explained by intramolecular interaction between the germanium atom and the halogen (α-effect). The reason for this coordination interaction is the transition of s-electrons of the halogen atom participating in the C−X bond to vacant d-orbitals at the germanium centre. Owing to the α-effect a thermal induced decomposition results in an elimination reaction (e.g. Figure 22)[107,108].

Many spirocyclic compounds of germanium are known[109].

Treating hexafluorocumyl alcohol with n-BuLi/TMEDA and reacting the generated dianion of the alcohol solution of $GeCl_4$ in hexane at −78 °C[110] (followed by decomplexation of the TMEDA complex which can be accomplished by methylation with MeOTf) yielded the 8-Ge-4 spirogermanacycle (Figure 23)[111]. The X-ray structure showed the expected deviations due to the longer bonds to germanium. Interestingly, the increase is greater for the Ge−O bonds (average Δd 0.133 Å) than for the Ge−C bonds (average Δd 0.063 Å), and the resulting distortion from an ideal

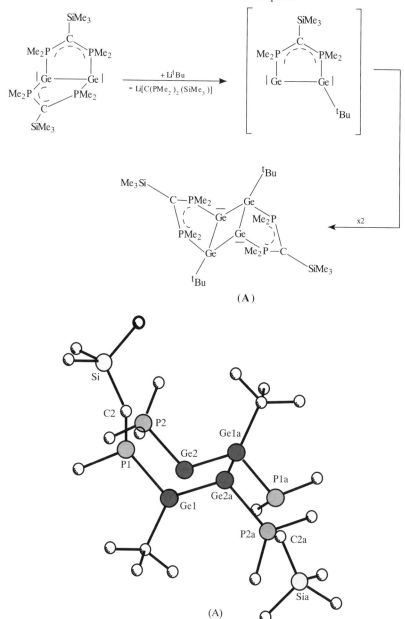

(A)

FIGURE 18. Synthesis and molecular structure of $Ge_4[C(PMe_2)_2(SiMe_3)]_2(^tBu)_2$ (A). Reproduced from Reference 98 by permission of Elsevier Sequoia S.A.

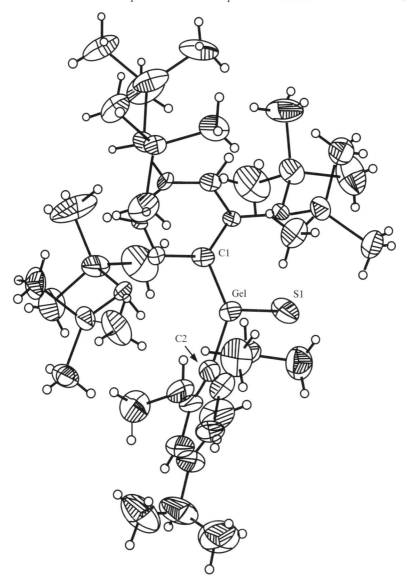

FIGURE 19. Molecular structure of Tb(Tip)Ge=S, where Tb = 2,4,6-tris[bis(trimethylsilyl)methyl]-phenyl, Tip = 2,4,6-triisopropylphenyl. Reprinted with permission from Reference 100. Copyright (1993) American Chemical Society

tetrahedral geometry is apparent [< O1−Ge−O11 110.4° (2), < O1−Ge−C8 91.6° (2), < O11−Ge−C18 91.4° (2), < O1−Ge−C18 112.7° (2), < O11−Ge−C8 112.8° (2), < C8−Ge−C18 137.7° (3)] (Figure 24). The distortion of the internal O−Ge−C angles leads to an enhanced Lewis acidity, which was illustrated by the promotion of pericyclic reactions involving activation of aldehyde carbonyl groups.

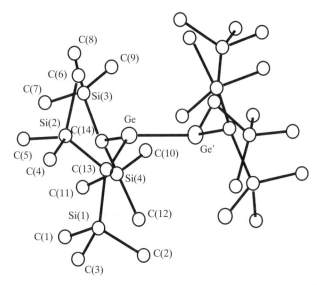

FIGURE 20. Molecular structure of Ge_2R_4 [R = $CH(SiMe_3)_2$]. Reproduced from Reference 102 by permission of The Royal Society of Chemistry

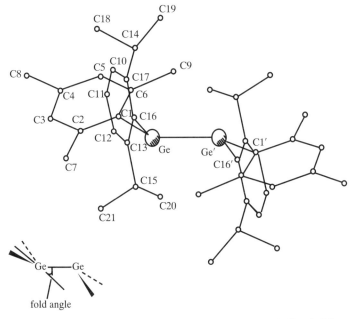

FIGURE 21. Molecular structure of (Z)-1,2-bis(2,6-diisopropylphenyl)-1,2-dimesityldigermene. Reprinted with permission from Reference 103. Copyright (1990) American Chemical Society

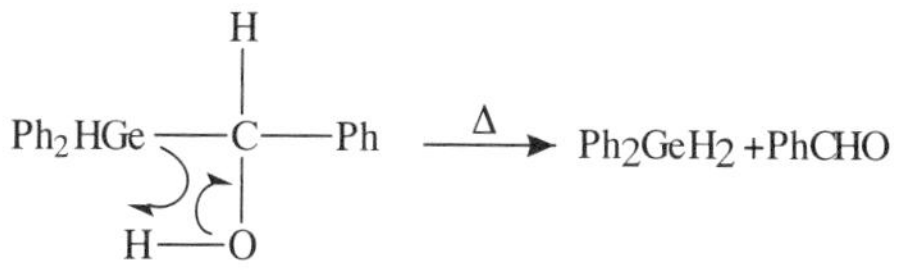

FIGURE 22. Thermal decomposition of $R_2HGe-C-X$ compounds (X = halogen, NR_2, O-, S-, ...)

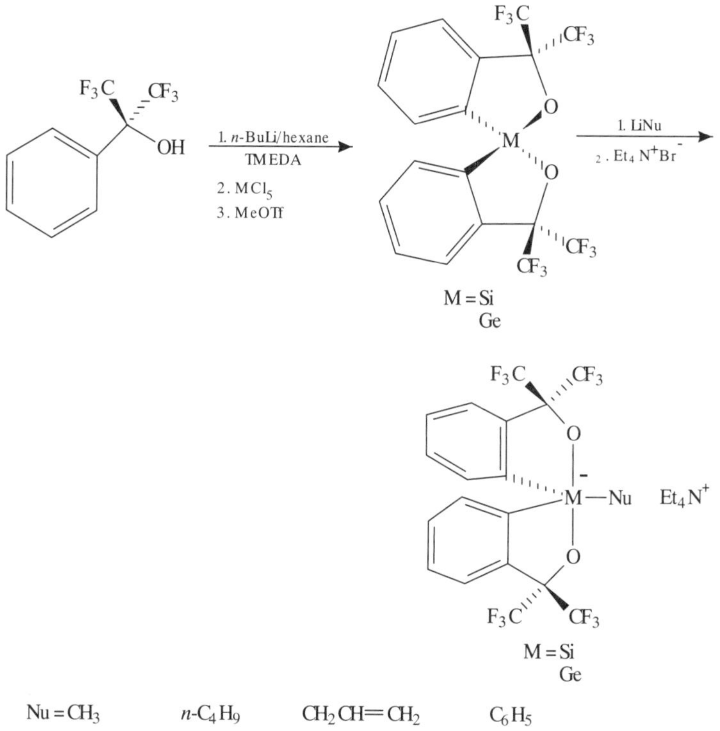

M = Si
Ge

M = Si
Ge

Nu = CH_3 n-C_4H_9 $CH_2CH=CH_2$ C_6H_5

FIGURE 23. Preparation of 8-M-4 spirocyclic compounds of Si and Ge(M = Si, Ge). Reprinted with permission from Reference 111. Copyright (1990) American Chemical Society

The 8-Ge-4 spirogermanacycle reacts with nucleophiles such as methyl- or n-butyllithium forming adducts containing the corresponding 10-Ge-5 anionic complexes (Figure 23). In case of the n-butyl adduct, crystals could be isolated and the structure was determined by X-ray methods. The anionic complex is primarily trigonal bipyramidal with an O−Ge−O angle of 173.8° (2) and average O−Ge−C angles of 83.5° (Figure 25).

Digermenes undergo oxygenation though several discrete pathways to provide the corresponding 1,2-digermadioxetane (A), 1,3-cyclodigermoxane (B), and digermoxirane derivatives (Figure 26)[112,113].

Compound A was obtained from a reaction of tetrakis(2,6-diethylphenyl)digermene with an excess of DMSO in toluene. The molecular structures of both A and B were determined by single-crystal X-ray analyses (Figure 27).

The Ge−Ge* bond in B is slightly shorter than a Ge−Ge single bond (2.45 Å) while the Ge−O bond lengths in A and B are somewhat longer. These deviations may be due to

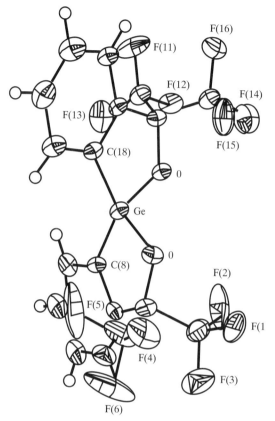

FIGURE 24. Molecular structure of the 8-Ge-4 spirogermanacycle prepared according to Figure 23. Reprinted with permission from Reference 111. Copyright (1990) American Chemical Society

the constraint in the strained ring system. The Ge_2O_2 moiety in **B** deviates significantly from planarity according to a $Ge-O-O-Ge^*$ torsion angle of 19.5° while in **A** the deviation from planarity is not really significant. In **A** the dihedral angles between the two $Ge-Ge^*-O$ and between the two $O-Ge-O^*$ planes are 8.8(2)° and 8.4(1)°, respectively. In **C** the $Ge-O$ bond length (1.857 Å) is long enough to accomodate the two germanium atoms with a $Ge-Ge$ distance of 2.617 Å , well beyond a $Ge-Ge$ single bond length. This justifies the conclusion that the cyclodigermoxane ring is made up of four equivalent localized $Ge-O$ bonds.

iii. Pentacoordinated complexes. The pentacoordinate state is not very common in germanium chemistry. Most of such pentacoordinate species contain chelating ligands and possess geometries between trigonal bipyramidal[114] and square pyramidal[115,116] or derive from internal coordination of a pendant heteroatom (mostly via N, O or S)[117-122]. Examples for structures of pentacoordinated complexes are shown in Figure 28[123]. Pentacoordinated germanium complexes form a series of structures showing progressive distortions from a trigonal bipyramid to a square- or rectangular-pyramidal geometry. A requirement for the formation of a square-pyramidal geometry seems to be the presence of two unsaturated five-membered rings[124].

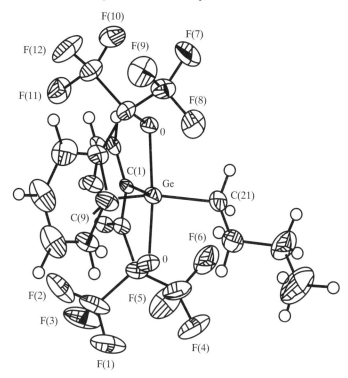

FIGURE 25. Molecular structure of the 8-Ge-4 spirogermanacycle *n*-butyl adduct, prepared according to Figure 23. Reprinted with permssion from Reference 111. Copyright (1990) American Chemical Society

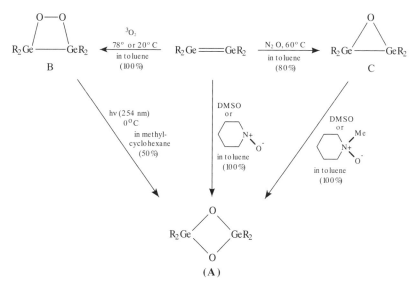

FIGURE 26. Oxygenation reactions of digermenes R = Dep or Dip

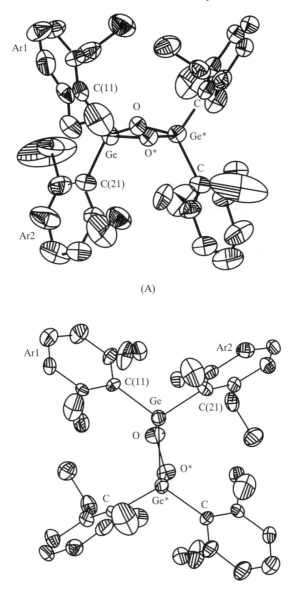

(A)

(B)

FIGURE 27. Molecular structure of the oxygenation products **A** and **B** of tetrakis(2,6-diisopropylphenyl)digermene (cf. Figure 26). Reprinted with permission from Reference 113. Copyright (1989) American Chemical Society

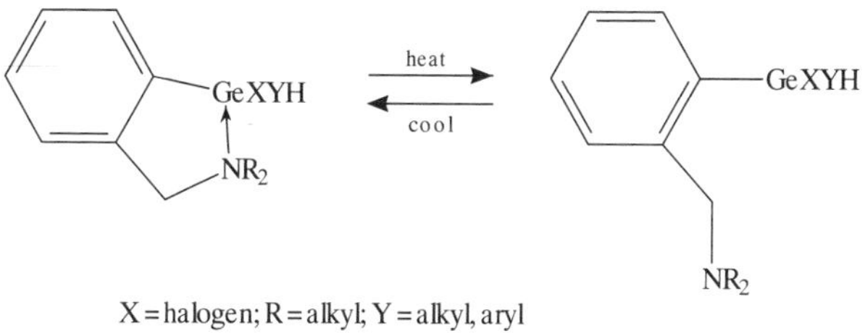

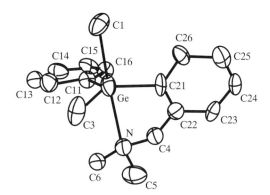

X = halogen; R = alkyl; Y = alkyl, aryl

FIGURE 28. Pentacoordinated Ge complexes

FIGURE 29. Molecular structure of the chlorogermane o-$(Me_2NCH_2)C_6H_4GeMePhCl$. Reproduced from Reference 123 by permission of Elsevier Sequoia S.A.

The crystal structure of o-$(Me_2NCH_2)C_6H_4GeMePhCl$ was determined by a single crystal X-ray diffration study. In this species germanium exists in a distorted trigonal bipyramidal coordination with a Ge−N bond distance of 2.479(11) Å or 2.508 Å (11), respectively (two slightly different conformations) (Figure 29).

Another example for a five-coordinated germanium complex (10-Ge-5 anionic complex) has already been described above (see Section IV.B.2.a.ii; Figure 23). An unusual rectangular pyramidal germanium(IV) anion was synthesized according to the reaction shown in Figure 30. The anion was characterized by ^{31}P and ^{1}H NMR studies[125].

Another interesting example of a five-coordinated anionic germanium complex is triethylammonium bis(1,2-benzenediolato)phenyl-germanate(IV), $[(C_6H_4O_2)_2GePh][Et_3NH]$[126], which was prepared by the reaction of phenylgermanium trichloride with catechol and triethylamine. The anionic spirocyclic germanate $[(C_6H_4O_2)_2GePh][Et_3NH]$ as well as the CH_3CN containing compound have a pseudo-2-fold axis coincident with the Ge−C bond while the Ge atom is surrounded by an apical equatorial-positioned phenyl group of a rectangular pyramid with four basal oxygen atoms (Figure 31).

In the case of the acetonitril coordinated germanium complex the Ge−O bonds to the oxygen atoms that are involved in the hydrogen bonding are elongated relative to the other two [1.874(5) Å, ax. and 1.870(5) Å, eq. to 1.858(5) Å, ax. and 1.838(5) Å, eq.]. Therefore, the complex geometry is distorted toward a rectangular pyramid. This shift from a trigonal bipyramid toward a more rectangular pyramid as a result of the hydrogen bonding can be explained by removal of electron density from the affected Ge−O bonds.

A nice example for an almost ideal trigonal bipyramidal structure was observed in the stable complex anion $[(CF_3)_3GeF_2]^-$[127], which was obtained from a reaction

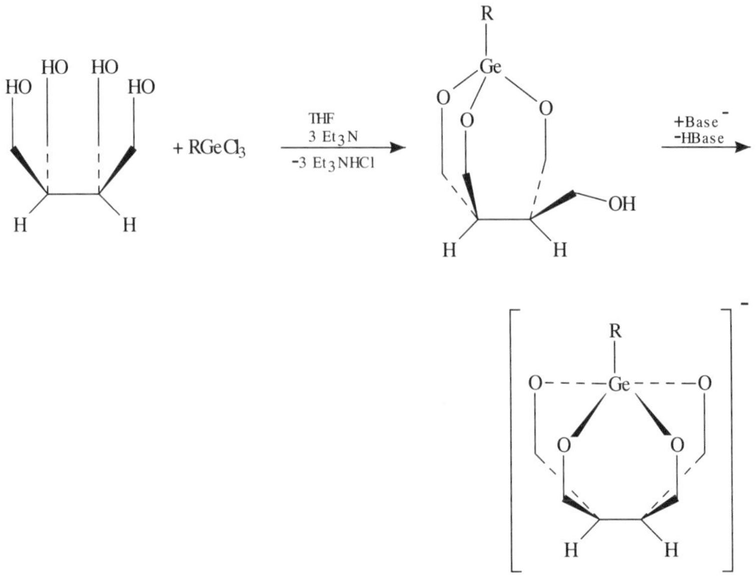

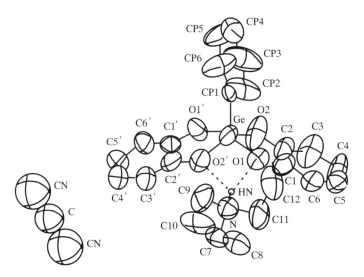

FIGURE 30. Preparation of the square-pyramidal Ge(IV) anion $RGe[O_4(CH_2)_4C_2H_2]$ (R = Me, Ph)

FIGURE 31. Molecular structure of $[(C_6H_4O_2)_2GePh][Et_3NH]\cdot\frac{1}{6}$ CH_3CN. Reprinted with permission from Reference 126. Copyright (1986) American Chemical Society

of $(CF_3)_3GeX$ (X = halogen) with F^- in an aqueous system. The germanium atom is coordinated in a trigonal bipyramidal fashion [< C1−Ge−C2 118.4° (2), < C2−Ge−C2′ 123.2° (3), < F1−Ge−F2 177.4(2)°], with the two fluoride ligands in axial positions and the three trifluoromethyl groups in equatorial positions. All three C-atoms lie within 0.02 Å in the trigonal plane (Figure 32). Depending on the orientation

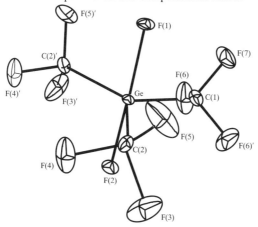

FIGURE 32. Molecular structure of the $[(CF_3)_3GeF_2]^-$ anion in $[NMe_4][(CF_3)_3GeF_2]$. Reproduced from Reference 127 by permission of Elsevier Sequoia S.A.

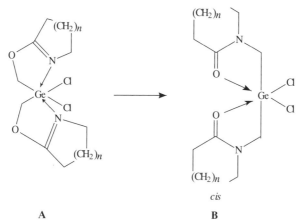

FIGURE 33. Preparation of (O→Ge)-chelate bis-(lactamo-N-methyl)-cis-dichlorogermanes. Reproduced from Reference 129 by permission of Elsevier sequoia S.A.

of the CF_3 groups, the highest possible symmetry for the $[(CF_3)_3GeF_2]^-$ is not D_{3h} but either C_{3h} or C_{3v}.

For further examples of pentacoordinated Ge complexes of the type $GeF_4(CH_2CH=CH_2)^-$ see also Section IV.B.2.b.iii.

iv. Hexacoordinated complexes. Compared with tetra- or pentacoordinated germanium complexes, so far hexacoordinated germanium species have received little attention. Hexacoordination can mostly be found in the literature dealing with intermediates of reactions with pentacoordinated germanium complexes.

Treatment of bis-(chloromethyl)dichlorogermane with N-trimethylsilyllactams in a 1:2 ratio afforded a thermodynamically instable hexacoordinated germanium complex (Figure 33,**A**) that isomerized with increasing temperature, yielding the (O−Ge)-chelate bis(lactamo-N-methyl)dichlorogermane (Figure 33,**B**)[128]. Recently, the reactivity of complex B towards trimethylsilyl triflate was investigated[129]. The reaction of B with trimethylsilyltriflate led to the abstraction of only one chlorine atom by the triflate group and an inversion of the configuration at the germanium atom forming bis-(lactamo-N-methyl)-*trans*(trifluoromethylsulphonyloxy)-chlorogermanes (Figure 34,C). According to the determined X-ray structures, the Ge atom has an octahedral coordination strongly distorted towards a capped trigonal bipyramidal configuration.

It was assumed that in sufficiently polar solvents, the corresponding germanium-triflate species are to a high degree dissociated into triflate anions and Ge-containing ions with pentacoordinated Ge atoms.

v. Germanium transition metal complexes. Compounds of germanium as well as silicon, tin and lead with delocalized p-electron systems have been of great interest in the last few years because of their assumed new chemical and electrical properties[130–132]. Recently, the synthesis (Figure 35) and X-ray structure (Figure 36) of a stable η^5-germacyclopentadienyl Ru-complex [(η^5-C$_5$Me$_5$)-Ru{η^5-C$_4$Me$_4$GeSi(SiMe$_3$)$_3$}] was reported (Figure 35)[133]. In 1 (Figure 35) the folded germanium heterocycle coordinates in an η^4-fashion to the ruthenium centre (Figure 37). The germanium is situated outside (0.97 Å) the plane formed by the four carbon atoms and the Ru–Ge bond distance of 2.985(2) Å indicates weak interaction between both atoms. Due to the electron density transfer from the Ru atom into the σ^*-orbital of the Ge–Cl2 bond in 1, this bond is significantly prolonged [cf d Ge–Cl2 2.246(2) Å, Ge–Cl3 2.151(3)]. In contrast to the η^4-Ru complex 1 in the η^5-germa-cyclopentadienyl–Ru complex 4, the five-membered Ge-heterocycle is planar (Ge is 0.02 Å outside the C4 plane, sum of angles at Ge is 358.1°). The Ru atom in 4 is surrounded by both planar five-membered rings (dihedral angle 8.6°).

The osmium complex Os[Ge(*p*-tolyl)$_3$]H(CO)$_2$(PPh$_3$)$_2$ was obtained by the reaction of Os(CO)$_2$(PPh$_3$)$_3$ with HGe(*p*-tolyl)$_3$ under irradiation (see Section III.B.2, tin hydrides). The single-crystal X-ray structure determination showed a highly distorted octahedral coordinated osmium central atom. The tetracoordinated germanium ligands and the hydride occupy the equatorial plane (Figure 38). Although the hydride was not localized in the

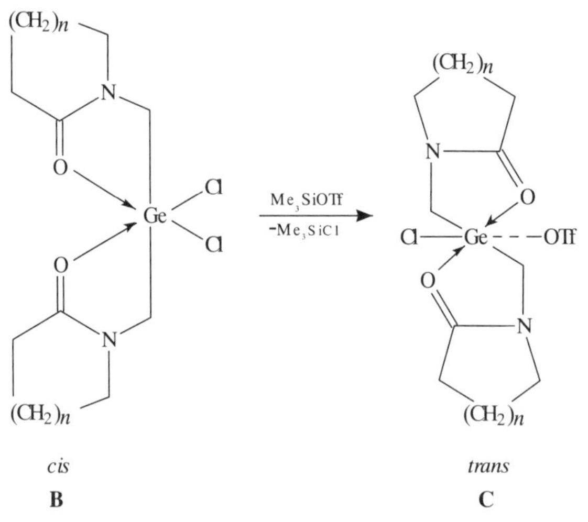

cis
B

trans
C

FIGURE 34. *cis–trans* Isomerization of (O→ Ge)-chelate bis-(lactamo-N-methyl)dichlorogermanes. Reproduced from Reference 129 by permission of Elsevier Sequoia S.A.

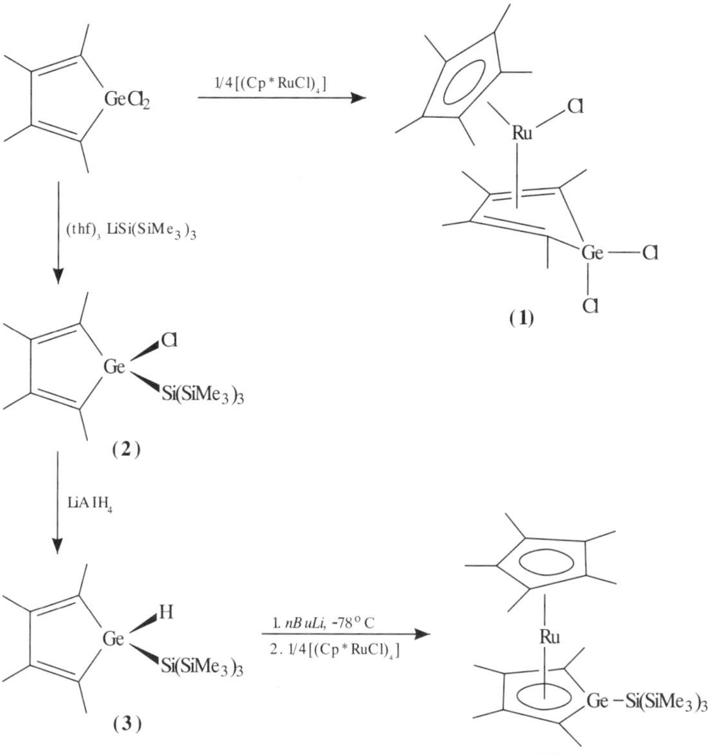

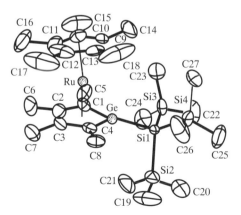

FIGURE 35. Preparation of the first stable germacyclopentadienyl complex. Reproduced from Reference 133 by permission VCH Verlagsgesellschaft mbH

FIGURE 36. Molecular structure of $[(\eta^5\text{-}C_5Me_5)\text{-}Ru\{\eta^5\text{-}C_4Me_4GeSi(SiMe_3)_3\}]$ (**4**). Reproduced from Reference 133 by permission of VCH Verlagsgesellschaft mbH

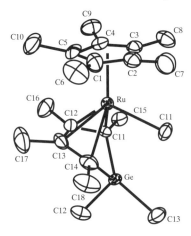

FIGURE 37. Molecular structure of $(\eta^5\text{-}C_5Me_5)(\eta^4\text{-}C_4Me_4GeCl_2)RuCl$ (**1**). Reproduced from Reference 133 by permission of VCH Verglagsgesellschaft mbH

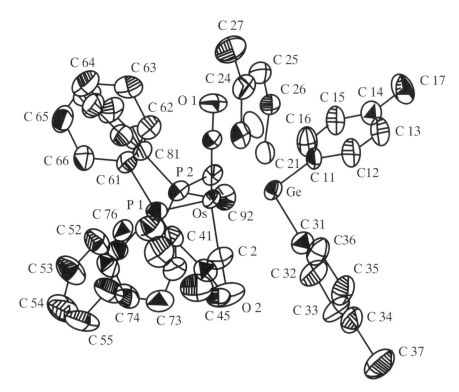

FIGURE 38. Molecular structure of $Os[Ge(p\text{-tolyl})_3]H(CO)_2(PPh_3)_2$. Reproduced from Reference 53 by permission of Elsevier Sequoia S.A.

structure determination, on the basis of spectroscopic data it was assumed that it is coordinated in the equatorial plane between the phosphorus and the germanium atom. The Os−Ge bond length is 2.5600(3) Å, representing the first experimentally determined Os−Ge bond length. The value of 2.55 Å compares nicely with the sum of covalent radii of Os and Ge[53,134].

b. Tin. i. Tricoordinated complexes. Reaction of borandiyl-boriran [135,136] with stannandiyl led to the formation of a stable stannaethene (Figure 39)[137] which was crystallized from pentane.

The tin as well as the carbon atom C2 are tricoordinated with a bonding distance between tin and carbon of 2.025(4) Å, but the distances of the tin atom to the tetracoordinated C atoms are 2.152(5) Å and 2.172(4) Å (Figure 40).

The tin and the carbon atoms are pyramidally distorted. The experimentally determined value of $d(Sn−C) = 2.025$ Å compares well with calculated bond data for $H_2C=SnH_2$[138].

ii. Tetracoordinated complexes. Addition of $SnCl_4$ to a mixture of lithium and Me_3SiCl in THF led to the formation of hexakis(trimethylsilyl)distannane, $(Me_3Si)_3Sn−Sn(SiMe_3)_3$, and $(Me_3Si)_4Sn$[139]. The mechanism of the formation of the distannane is not apparent. The tin is coordinated in the molecule in an essentially tetrahedral fashion (Figure 41) with the Si−Sn−Si angles slightly compressed (average 107.5°) and the Sn−Sn−Si angles slightly expanded (average 111.5°). According to the criteria of Bock, Meuret and Ruppert[140] for the presence of overcrowding in $(Me_3Si)_3E−E(SiMe_3)_3$ structures, $(Me_3Si)_3Sn−Sn(SiMe_3)_3$ with a Sn−Sn distance of 2.789(1) Å (<3.33 Å) would be expected to show evidence for steric interaction. The absence of appreciable Sn−Sn bond lengthening, in contrast to the enlarged Si−Si distance in the analogous silicon compound [141,142], indicates that the magnitude of steric strain is smaller in the distannane.

iii. Pentacoordinated complexes. The structures of tris(stannyl)amines could not be determined exactly. On the one hand, IR spectroscopic investigations indicate a pyramidal

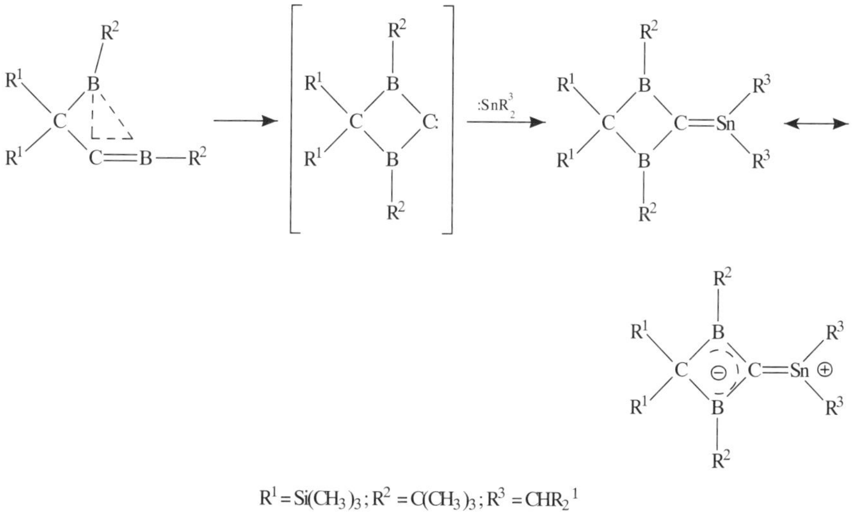

$R^1 = Si(CH_3)_3; R^2 = C(CH_3)_3; R^3 = CHR_2{}^1$

FIGURE 39. Preparation of stannaethene complexes. Reproduced from Reference 137 by permission of VCH Verlagsgesellschaft mbH

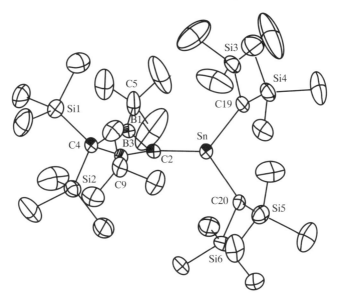

FIGURE 40. Molecular structure of $R^1_2C(BR^2)_2C=SnR^3_2$ [R^1 = SiMe$_3$, R^2 = CMe$_3$, R^3 = CH(SiMe$_3$)$_2$]. Reprinted with permission from Reference 139. Copyright (1993) American Chemical Society

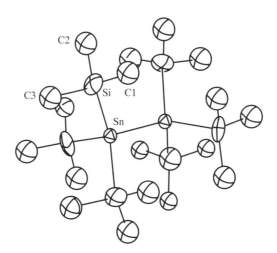

FIGURE 41. Molecular structure of (Me$_3$Si)$_3$Sn$-$Sn(SiMe$_3$)$_3$

Sn$_3$N frame[143] but, on the other hand, electron diffraction investigations speak for trigonal planar arrangement of the tin atoms surrounding the nitrogen centre[144] in agreement with the known structure of N(SiR$_3$)$_3$[145]. Treatment of tris(trimethylstannyl)amine with dimethyltin dichloride causes the exchange of the (CH$_3$)$_3$Sn groups yielding the functionalized tris(chlorodimethylstannyl)amine (equation 50)[146].

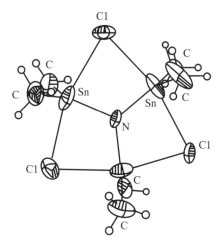

FIGURE 42. Molecular structure of $[ClMe_2Sn]_3N$. Reproduced from Reference 146 by permission of VCH Verlagsgesellschaft mbH

$$(Me_3Sn)_3N + 3\ Me_2SnCl_2 \xrightarrow{CH_2Cl_2} [ClMe_2Sn]_3N + 3\ Me_3SnCl \qquad (50)$$

The crystal structure of $[ClMe_2Sn]_3N$ indicates a trigonal planar coordinated nitrogen centre with an angle sum at nitrogen of $360°$ (Figure 42). The pentacoordination of the tin atoms is caused by essentially symmetrical $Sn-Cl-Sn$ bridges. The $Sn-N$ bonding distance was found to be very short (1.99 Å) compared to the sum of covalent radii (2.03 Å).

Semiempirical AM1 calculations for $[ClH_2Sn]_3N$ as a model compound for $[ClMe_2Sn]_3N$ agree well with the experimental data and indicate that the bridging structure compared with the unbridged structure is favoured by $104.5\ kcal\,mol^{-1}$ (Figure 43). The lone pair of electrons at the trigonal planar nitrogen atom is localized in a p-type orbital. The calculation could not confirm whether a $Sn-N$ double bond can be discussed.

The dibutyldichloro[1-methyl-2(3H)-imidazolinethione-S]tin(IV), $[Sn(C_4H_9)_2Cl_2 \cdot C_4H_6N_2S]$, consists of a distorted trigonal-bipyramidal coordinated tin atom, with C(5), C(9) and S in the trigonal plane and the chlorine atoms in the axial positions (Figure 44)[147]. The Cl(2)$-$Sn$-$S angles deviate substantially from $90°$ with $76.1°$ and $77°$ in A and B. The asymmetric unit comprises two molecules: A and B. The molecules are held together by two effective (and linear) hydrogen bonds.

Quite recently, it was established that the formation of the pentacoordinated germanium and tin compounds $EF_4(CH_2CH=CH_2)^-$ from $EF_3(CH_2CH=CH_2)$ (E = Ge, Sn) by the addition of F^- is exothermic[148]. The nucleophilicity of the allylic γ-carbon is much enhanced when the pentacoordinated $EF_4(CH_2CH=CH_2)^-$ is formed. These results are qualitatively similar to those found for the reaction of the corresponding silicon compound. The pentacoordinated Ge and Sn complexes have a significant Lewis acidity which allows them to form stable hexacoordinated intermediates in the course of the reaction.

iv. Hexacoordinated complexes. Transmetallation reactions of Schiff-base-type arylmercury compounds with metallic tin result in dichlorobisaryltin(IV) according to equation 51[149].

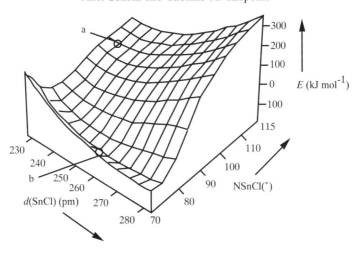

FIGURE 43. AM1 energy hypersurface of $[ClH_2Sn]_3N$. (a) with $Sn-Cl-Sn$ bridges and (b) without $Sn-Cl-Sn$ bridges. Reproduced from Reference 146 by permission of VCH Verlagsgesellschaft mbH

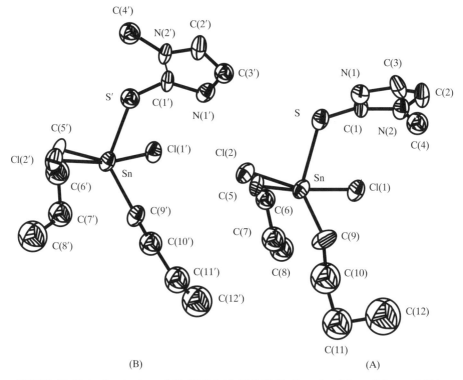

FIGURE 44. Molecular structure of $[Sn(C_4H_9)_2Cl_2 \cdot C_4H_6N_2S]$. Reproduced from Reference 147 by permission of Elsevier Sequoia S.A.

$$2 \text{ ArHgCl} + \text{Sn} \longrightarrow \text{Ar}_2\text{SnCl}_2 + 2 \text{ Hg} \tag{51}$$

$$\text{Ar} = (C_6H_5)-\text{CH}=\text{N}-(o-C_6H_4)$$

In the crystal structure of dichlorobis[2-(benzylideneamino)-5-tolyl]tin(IV) shown in Figure 45 the tin atom is covalently bound to two chlorine atoms and two carbon atoms of the N-phenyl rings and, in addition, there are intramolecular interactions between the nitrogen atoms of the N-phenyl rings with tin forming a four-membered ring system [cf Sn−N distances of 2.750(5) and 2.859(5) Å are significantly shorter than the sum of the van der Waals radii for Sn and N (3.6 Å)]. On account of the N−Sn intramolecular coordination, the Sn atom is six-coordinated in a very distorted octahedral fashion. Whereas two *cis* chlorine and two *cis* imino nitrogen atoms are in the equatorial plane, two carbon atoms of the *ortho*-metallated N-phenyl groups are considerably displaced from the *trans* axial positions [<C(1)−Sn−C(15) 139.3(2)°].

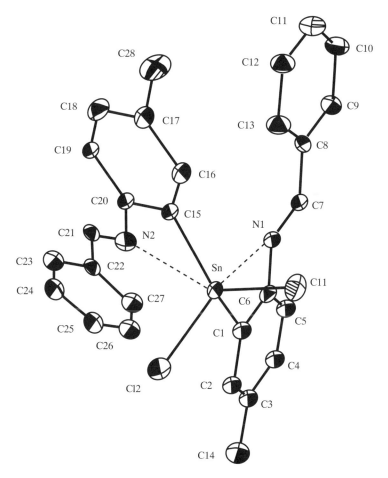

FIGURE 45. Molecular structure of dichlorobis[2-(benzylideneamino)-5-tolyl]tin(IV). Reproduced from Reference 149 by permission of Elsevier Sequoia S.A.

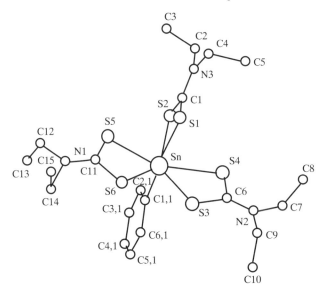

FIGURE 46. Molecular structure of PhSn(S$_2$CNEt$_2$)$_3$. Reproduced from Reference 150 by permission of Elsevier Sequoia S.A.

v. Heptacoordinated complexes. The tin atom in PhSn(S$_2$CNEt$_2$)$_3$ is coordinated in the form of a distorted pentagonal bipyramid with the phenyl group axial and two dithiocarbamate bridging groups axial-equatorial (Figure 46)[150]. Atom S(2) is 0.66 Å out of the plane made up by Sn, S(3), S(4), S(5) and S(6). The stepwise substitution of chlorine in RSnCl$_3$ (R = Me, Ph) by diethyldithiocarbamate leads to the series RSnCl$_2$(S$_2$CNEt$_2$), RSnCl(S$_2$CNEt$_2$)$_2$ and RSn(S$_2$CNEt$_2$)$_3$. In all cases the diethyldithiocarbamate behaves as a bidentate ligand and the tin atom changes its coordination sphere from four-coordinate in RSnCl$_3$ to five-coordinate in RSnCl$_2$(S$_2$CNEt$_2$), to six-coordinate in RSnCl(S$_2$CNEt$_2$)$_2$ and seven-coordinate in RSn(S$_2$CNEt$_2$)$_3$.

vi. Octacoordinated complexes. A nice example of an octacoordinated tin-*cis-β*-(methylthio)stilbene-*α*-thiol complex, Sn^{IV}[Ph(S)C=C(SCH$_3$)Ph]$_4$, has recently been published[151]. Unsaturated thiolthioethers may behave as monodentate, bidentate or *μ*-S-bridging ligands[152]. In mononuclear complexes of Ph(S)C=C(SCH$_3$)Ph, unsaturated thio-ether-thiol ligands may be monodentate with a non-interacting SCH$_3$ group or, as in the case of the Sn^{IV}[Ph(S)C=C(SCH$_3$)Ph]$_4$ complex, bidentate. That complex contains octacoordinated Sn(IV) with two sets of essentially tetrahedral Sn−S bonds (Figure 47). The coordination polyhedron of Sn^{IV}[Ph(S)C=C(SCH$_3$)Ph]$_4$ is that of a distorted cube. The S−Sn−S bond angles of the covalent Sn−S bonds are 107 and 114.5°, respectively, indicating a small deviation from ideal tetrahedral symmetry; the S−Sn−S angles of the coordinative Sn−S bonds of 80.1 and 125.9° are more distorted. The covalent Sn−S bonds have a length of 2.436(1) Å closely corresponding to the sum of the covalent radii of 2.44 Å, and are considered as equivalent tetrahedral Sn−S single bonds involving the 5s and 5p orbitals. The coordinative Sn−S(CH$_3$) bonds with a length of 3.599(2) Å are significantly shorter than the sum of the van der Waals radii (3.97 Å) which may be explained by involving the 6s and 6p orbitals of the tin atom.

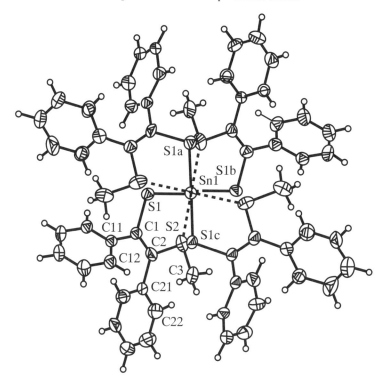

FIGURE 47. Molecular structure of $Sn^{IV}[Ph(S)C=C(SCH_3)Ph]_4$. Reproduced from Reference 151 by permission of VCH Verlagsgesellschaft mbH

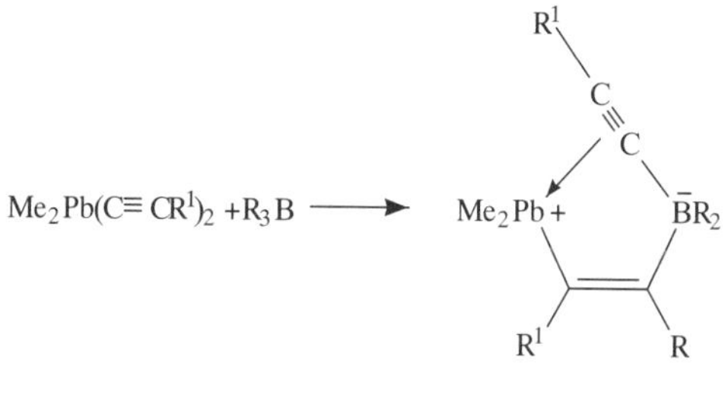

$$Me_2Pb(C\equiv CR^1)_2 + R_3B \longrightarrow Me_2Pb + \quad \bar{B}R_2$$

FIGURE 48. Preparation of triorganolead cations

c. Lead. i. Tetracoordinated complexes. The preparation and structural characterization of triorganolead cations have recently been reported (Figure 48)[153].

Reactions between bis(alkinyl)dimethylplumbanes and trialkylboranes resulted in the cleavage of the lead–alkinyl bond and the fast intramolecular transfer of the alkinyl group to the boron atom forming the alkinylborate anion and the triorganolead cation (structure **A**) (Figure 49)[154].

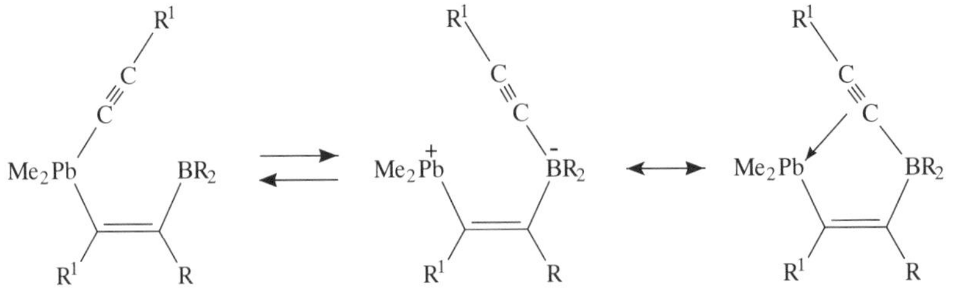

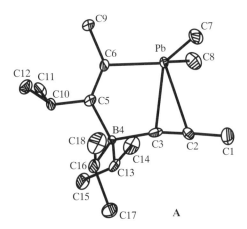

FIGURE 49. Reaction between bis(alkinyl)dimethylplumbans and trialkylborans

The cationic centre is stabilized by the asymmetric side-on-coordination of the C−C unit leading to a tetrahedral distorted coordination sphere at the lead atom(i.e. structure **B**). In the case R = i-Pr and R^1 = Me the compound was characterized by an X-ray investigation (Figure 50). The C−Pb−C angles have values between 81.3 and 123.7° [C6−Pb−C3 81.3(2)°, C6−Pb−C2 108.3(2)°, C6−Pb−C7 123.7(2)°, C6−Pb−C8 118.6(3)°, C7−Pb−C8 113.4(3)°] indicating the tetrahedral distortion (Figure 50). The electrophilic character of the Pb atom was shown in the interaction of **A** with pyridine; in this case the NMR resonance of the ^{207}Pb nucleus was shifted more than 200 ppm. Increasing the temperature, **A** rearranges in hexane in an intramolecular process forming compound **B** (Figure 51). In a similar reaction using tin instead of lead with BR₃, no product in analogy to compound **A** was observed. With an excess of triisopropylborane complex **A** was found to yield compound **C** (Figure 51)[155,156].

Recently, investigations of the reaction of MR₂Cl₂ (M = Ge, R = Me, Ph or Cl; M = Pb, R = Ph) with two equivalents of Li[Si(SiMe₃)₃] have shown that only symmetrical digermane and diplumbane products of the type [(Me₃Si)₃Si]R₂M−MR₂[Si(SiMe₃)₃] could be isolated[157,158]. An exception was found in the reaction of PbPh₂Cl₂ with two equivalents of Li[C(SiMe₃)₃] yielding the

FIGURE 50. Molecular structure of the alkinylborate anion−triorganolead cation complex according to structure **A** (Figure 49) with R = i-Pr and R^1 = Me. Reproduced from Reference 153 by permission of VCH Verlagsgesellschaft mbH

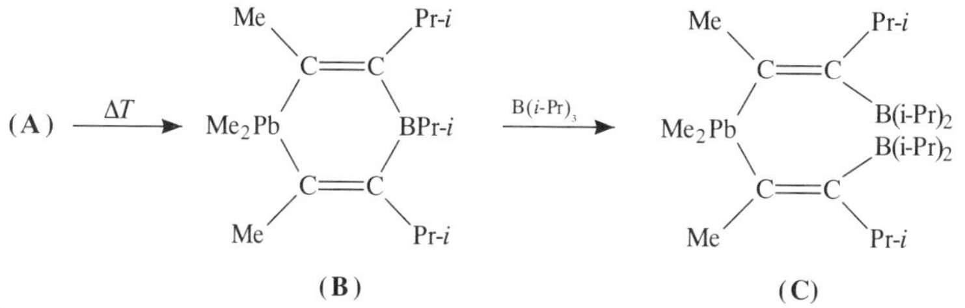

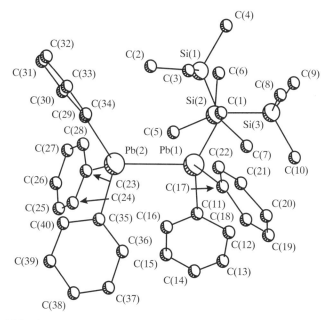

FIGURE 51. Thermal rearrangement of compound **A** (cf Figure 50) yielding complex **B** and reaction of **A** with an excess of $(i\text{-Pr})_3\text{B}$ to give compound **C**

FIGURE 52. Molecular structure of $Ph_3Pb-Pb[C(SiMe_3)_3Ph_2]$. Reproduced from Reference 159 by permission of The Royal Society of Chemistry

thermochromic compound $Ph_3Pb-Pb[C(SiMe_3)_3]Ph_2$ instead of the more symmetric $Pb[C(SiMe_3)_3]_2Ph_2$[159]. An X-ray crystal structure determination indicated a molecule consisting of a central pair of lead atoms, of which one is bound to three phenyl groups while the other coordinates to two phenyl groups and a $C(SiMe_3)_3$ ligand (Figure 52). The Pb—Pb bond distance is 2.908(1) Å, while Pb—C(1) is 2.280(10) Å and Pb—C(phenyl) is, on the average, 2.250 Å, which indicates that all distances to lead are slightly elongated. This distortion as well as the distortion which stems from the bond angles is consistent with the presence of considerable steric interaction between the $C(SiMe_3)_3$ group and the phenyl rings. The fact that the Pb—Pb and Pb—C(phenyl) distances are nearly as large as those in $[Pb(Si(SiMe_3)_3)]_2$ [where there are bulky $Si(SiMe_3)_3$ groups on both lead atoms] suggests that the steric influence of the $C(SiMe_3)_3$ group is greater than that of its silyl analogue, as would be expected from cone angle considerations[160].

The symmetrically substituted diplumbanes Pb_2R_6 are reasonably stable compounds and the X-ray crystal structures of $Pb_2(phenyl)_6$[161], $Pb_2(cyclohexyl)_6$[162] and $Pb_2(o$-tolyl$)_6$[163] have been determined. All these hexaorganyldiplumbanes were obtained by well established methods. The symmetrically substituted hexaaryldiplumbanes can be synthesized from $Ar_3PbMgBr$ by the reaction with (i) $PbCl_2$ and (ii) 1,2-dibromoethane[164].

The X-ray crystal structure of $Pb_2(o$-tolyl$)_6$ shows one centrosymmetric molecule per unit cell. Figure 53 clearly shows that there is no expansion of coordination of the lead atom in $Pb_2(o$-tolyl$)_6$. The bond distances of the tetrahedral coordinated lead atom to the carbon atoms are in the range of 2.242–2.249 Å ; the Pb–Pb distance was found to be 2.895(2) Å .

ii. Pentacoordinated complexes. Treating trimethyllead chloride with silver dimesyl-amide in acetonitrile or water led, in a metathetical reaction, to trimethyllead(IV) dime-sylamid (equation 52). The same reaction for germanium has also been reported[165].

$$Me_3PbCl + AgN(SO_2Me)_2 \longrightarrow Me_3PbN(SO_2Me)_2 + AgCl \qquad (52)$$

In the X-ray structure analysis of trimethyllead(IV)-dimesylamid the lead atom was found to be pentacoordinated, containing weak Pb–O [2.653(6) Å] interactions in the direction of the a-axis between the $Me_3Pb-N(SO_2Me)_2$ units (Figure 54) forming infinite parallel chains. Moreover, there are weak 1,4-interactions within one unit according to the 7% shorter distance than the sum of van der Waals radii of Pb (2.02 Å) and O (1.52 Å)[166]. The Pb atom has a distorted trigonal-bipyramidal arrangement ($< N–Pb–O$ 169.3°, average values: $< C–Pb–C$ 119.2°, $< C–Pb–N$ 95.5°, C–Pb–O 84.8°) in which the N-atom is trigonal-planar coordinated.

iii. Hexacoordinated complexes. Adding diphenyllead dichloride to a solution of O,O'-dibenzylphosphorodithioic acid and triethylamine diphenyllead bis(O,O'-dibenzyldithiophosphate), $Ph_2Pb[S_2P(OCH_2Ph)_2]_2$, was obtained (Figure 55)[167]. The lead in $Ph_2Pb[S_2P(OCH_2Ph)_2]_2$ is coordinated in a distorted octahedral fashion by two *trans* phenyl groups [angle C-Pb-C 165.0(9)°] and two bidentate dithiophosphate ligands,

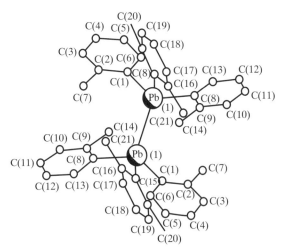

FIGURE 53. Molecular structure of $Pb_2(o$-tolyl$)_6$ illustrating the twisting of the o-methyl groups. Reprinted with permission from Reference 163. Copyright (1990) American Chemical Society

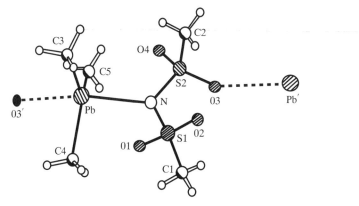

FIGURE 54. Molecular structure of $Me_3Pb-N(SO_2Me)_2$. Reproduced from Reference 165 by permission of Elsevier Sequoia S.A.

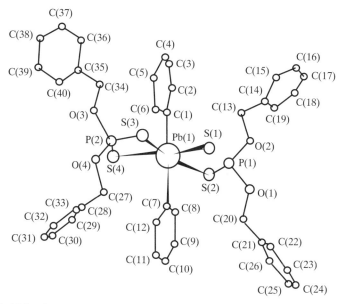

FIGURE 55. Molecular structure of diphenyllead bis(O,O'-dibenzyldithiophosphate). Reproduced from Reference 167 by permission of Elsevier Sequoia S.A.

each forming one short bond [2.723(6), 2.679(6) Å] and one long bond [2.940(7), 2.957(6) Å] to the lead atom. In the crystal lattice, adjacent pairs of molecules are linked by long [3.69(2) Å] lead···sulphur contacts to give overall an equatorial [PbS$_5$] arrangement.

iv. Lead transition metal complexes. Treatment of the anionic complex [Fe(CO)$_2$(dppe) {Si(OMe)$_3$}]$^-$ (**A**, Figure 56) (dppe = Ph$_2$PCH$_2$CH$_2$PPh$_2$) with Me$_3$PbCl in THF yielded a silyl plumbyl complex (Figure 56)[168]. Spectroscopic data indicate a *trans* arrangement of

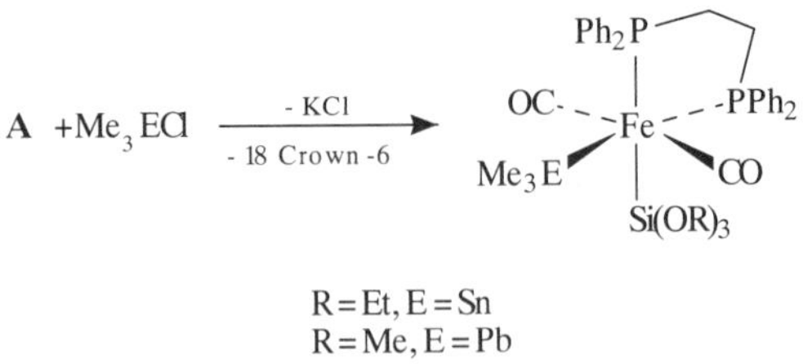

A +Me3ECl
- KCl
- 18 Crown -6
Ph2P
OC
Fe
PPh2
Me3E
CO
Si(OR)3
R=Et, E=Sn
R=Me, E=Pb

FIGURE 56. Preparation of silyl plumbyl and silyl stannyl complexes of iron. Reproduced from Reference 168 by permission of Elsevier Sequoia S.A.

B +Me3ECl
- KCl
- 18 Crown -6
Si(OEt)3
Ph2
H
P
ER3
Fe
P
CO
Ph2
H
E=Sn
Pb

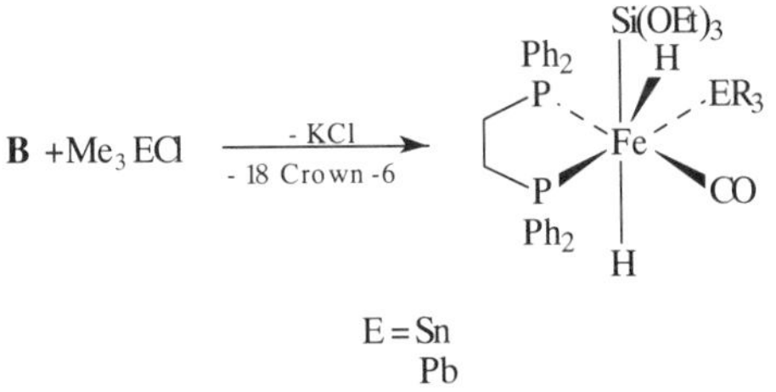

FIGURE 57. Preparation of the silyl-plumbyl-dihydrido complex $Fe(CO)(dppe)H_2(PbMe_3)[Si(OEt)_3]$. Reproduced from Reference 168 by permission of Elsevier Sequoia S.A.

Co1
C2
N2
C9
C1
C12
b
C15
N1
Pb
N6
C7
C8
Pb2
a
C13
C4
C10
C14
Pb
N5
Co2
C6
N6

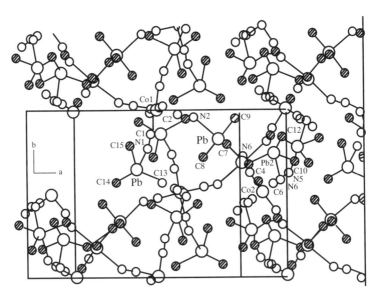

FIGURE 58. Elementary cell of $[(Me_3Pb)_3Co(CN)_6]_\infty$. Reprinted with permission from Reference 177. Copyright (1992) American Chemical Society

the two CO ligands as known for other bis(silyl) complexes [cf $Fe(CO)_2(dppe)(SiR_3)_2$[169]]. This means that PR_3 and ER_3 occupy *trans* positions.

Furthermore, the reaction of $[Fe(CO)(dppe)H_2\{Si(OEt)_3\}]^-$ (**B**, Figure 57) with Me_3PbCl led to the dppe-substituted silyl-plumbyl-dihydrido complex $Fe(CO)(dppe)H_2$ $(PbMe_3)[Si(OEt)_3]$ (Figure 57). The dihydride complex is less stable than the dicarbonyl complex and tends to fast decomposition at $-25\,°C$, especially in polar solvents.

A novel class of organometallic 3D polymers is represented by compounds of the type $[(R_3Sn^{IV})_3M^{III}(CN)_6]$ (R = alkyl or aryl and M = e.g. Fe or Co). In 1985 the X-ray diffraction study of $[(Me_3Sn^{IV})_3Co^{III}(CN)_6]$ indicated that the lattice is a 3D network of regularly interlinked, infinite $[-Co-C\equiv N-Sn-N\equiv C-]_\infty$ chains[170]. The availability of notably large cavities within these novel organometallic polymers can also be demonstrated chemically by the facile encapsulation of voluminous organic and organometallic guest cations G^{n+} into the negatively charged host lattice $[(Me_3Sn^{IV})_3Fe^{II}(CN)_6]^{-171-173}_\infty$. Up to now the compounds of this *class* have been modified in terms of the R_3Sn subunit[174,175] or Os and Ru were used as transition metals[176]. However, there are only few examples made up of R_3Pb fragments instead of R_3Sn units. In the literature the synthesis and X-ray crystal structure of $[(Me_3Pb)_3M(CN)_6]_\infty$ (M = Co, Fe) was reported[177]. The complex $[(Me_3Pb)_3Co(CN)_6]_\infty$, as the analogous tin compound, forms a three-dimensional coordination polymer (Figure 58). There are three crystallographically non-equivalent $[-Co-C\equiv N-Pb-N\equiv C-]_\infty$ chains in the lattice.

V. RELATIVISTIC EFFECTS

It was not until the 1970s that the full relevance of relativistic effects in heavy-element chemistry was discovered. However, for the sixth row (W ... Bi), relativistic effects are comparable to the usual shell-structure effects and therefore provide an explanation for many unusual properties also of lead chemistry[178-180]. The main effects on atomic orbitals are (i) the relativistic radial contraction and energetic stabilization of the s and p shells, (ii) the spin-orbit splitting and (iii) the relativistic radial expansion and energetic destabilization of the outer d and f shells. Relativistic effects increase, for all electrons, like Z^4 and, for valence shells, roughly like Z^2. For example, let us compare the radial 1s shrinkage for the elements Ge, Sn and Pb that are the subject of this chapter. The non-relativistic limit for an 1s electron is Z a.u. (cf. $c = 137$ a.u.). Thus the 1 s electron of Ge has v/c of $32/137 \approx 0.23$, whereas these values are 0.37 for tin and 0.60 for Pb [c is the finite speed of light and v is the average radial velocity of the electrons in the 1 s shell (v is roughly Z a.u.)]. As the relativistic mass increase is given by $m = m_0/[1-(v/c)^2]^{0.5}$ with the effective Bohr radius $a_0 = (4\pi\epsilon_0)/(\hbar^2/me^2)$ the radial 1 s shrinkage for germanium and tin is 3 or 8%, respectively, whereas it is as much as 20% for lead.

Many physical properties down group 14 (and not only there)[181] exhibit a saw-tooth behaviour, superimposed on the regular trend down a column. For example, the covalent radii increase from carbon (0.77 Å) to silicon/germanium (1.17/1.22 Å) to tin/lead (140/154 Å). It was proposed that the first anomaly (Ge) at row four is caused by the post-transition metal effect (d contraction), caused by an increase of the effective nuclear charge for the 4s electrons due to filling the first d shell (3d). A similar interpretation is possible for the second anomaly at row six (Pb). This effect is commonly called lanthanoid contraction due to the effect of filling the 4f shell. However, it is interesting to ask how much of this decrease (of the radius) does come from relativistic effects. A comparison of computed non-relativistic and relativistic bond-length contractions indicates that relativistic contractions of single-bond covalent radii substantially increase down the column. For lead, the relativistic contraction is already 0.11 Å (Table 4).

Axel Schulz and Thomas M. Klapötke

TABLE 4. Relativistic bond-length contractions

Element	Molecule	d_{nr} (Å)	d_r (Å)	d_{exp} (Å)	C (Å)[a]
H	H_2	0.73354	0.73352	0.74152	0.000017
Ge	GeH_4	1.596	1.586	1.527	0.01
Sn	SnH_4	1.736	1.717	1.701	0.02
Pb	PbH_4	1.890	1.782	1.754	0.11

[a] C, the relativistic contraction, is the difference between the non-relativistic and relativistic bond lengths.

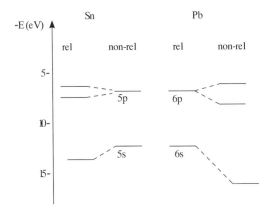

FIGURE 59. Relativistic and non-relativistic Hartree–Fock orbital energies for tin and lead

The term 'inert pair' is often used for the tendency of the $6s^2$ electron pair to remain formally unoxidized in the compounds of Pb(II) [and also in the case of Tl(I) and Bi(III) etc.]. As discussed above, this tendency can be related to relativity. Figure 59 shows the relativistic and non-relativistic valence orbital energies for Sn and Pb. The relativistic increase of the s–p gap leads to a $6s^2$ inert pair in the case of Pb. However, the situation is more complex if the local geometry at the heavy atom (Pb) is discussed. There are examples for both, stereochemically inactive and stereochemically active s^2 lone pairs.

It has been shown that in a relativistic treatment neither the orbital angular momentum l nor the spin angular momentum s of an electron are good quantum numbers, but the vector sum is:

$$\mathbf{j} = \mathbf{l} + \mathbf{s}$$

Thus for a p electron with $l = 1$ the two possible values are $j = \frac{1}{2}$ and $j = \frac{3}{2}$. The energetic splitting between these two j values is a relativistic effect and may rise up to a few electron volts for the valence electrons of the heaviest elements (cf. Pb, Table 5).

It is interesting to mention that $PbCl_2$ is stable whereas $PbCl_4$ is thermodynamically unstable and decomposes at room temperature to yield $PbCl_2$ and elemental Cl_2. On the other hand, organolead(IV) compounds have been prepared (and have been used extensively as fuel additives, such as $PbEt_4$) whereas species like $PbEt_2$ have yet to be prepared and are likely not possible, as they disproportionate to the more thermodynamically stable $PbEt_4$ and solid Pb. This tendency is also reflected in the reaction enthalpies of the methyl

TABLE 5. Energy terms for group 14 elements (eV)

Element E	Ionization energy	E_2 bond dissociation energy	Spin-orbit $^3P_2 - {}^3P_0$
C	11.26	6.1	0.005
Si	8.15	3.2	0.03
Ge	7.88	2.8	0.17
Sn	7.34	2.0	0.42
Pb	7.42	1.0	1.32

and fluorine substituted compounds (gas phase) (equations 53 and 54).

$$2\ Me_2Pb \longrightarrow Me_4Pb + Pb \quad \Delta H < 0 \tag{53}$$

$$2\ PbF_2 \longrightarrow PbF_4 + Pb \quad \Delta H > 0 \tag{54}$$

Recent *ab initio* calculations delineate the remarkable thermodynamic destabilization of lead(IV) compounds by electronegative substituents[182,183]. Based on population analyses of the molecular wave functions it was proposed that electronegative substituents increase the charge of the metal and increase the difference in the radial extensions of the 6s and 6p orbitals. By increasing the differences in the radial extensions of the s and p orbitals, 6th-row relativistic effects also contribute to a destabilization of the higher valence state.

VI. COMPUTATIONAL CHEMISTRY

A. Introduction

The molecular chemistry of the heavier main group elements has increased unexpectedly fast in the last decade[184]. For instance, the synthesis and characterization of compounds containing double bonds between heavier main group elements was one of the major goals in preparative chemistry in the years past. For a long time it was assumed that such double bonds would not be stable because of weak $p\pi$–$p\pi$ interaction, and it was not before the early 1980s when the first compounds containing kinetically stabilized P=P, Si=C, and Si=Si double bonds were reported. In 1984 the preparation and isolation of a complex containing the first stable Ge–Ge bond, with a bond order greater than one, was achieved[185,186].

Parallel to this progress in the experimental field, computational chemistry has developed step by step and the heavier elements of the main groups were also taken into consideration. This became possible through powerful *ab initio* programmes which were able to deal with the great number of electrons in heavier elements. Moreover, the improvement of pseudo-potential methods (see below) caused a situation in which calculations that include heavy elements (e.g. SnH_2) do not require more computation time than those for second-row species (e.g. CH_2[187]). Computational quantum-mechanical procedures like *ab initio* and semi-empirical methods[188–192] present an elegant way to get energy and structural data of educts and products, reactive intermediates and transition states and are therefore suitable to describe the whole reaction pathway on a theoretical basis. For the user of quantum-chemical *ab initio* calculations the programme packages GAUSSIAN 88[193], 90[194] and 92[195] of Pople and coworkers, CADPAC 4.1[196] etc. offer a maximum on flexibility and convenience.

Semi-empirical methods (e.g. CNDO, MNDO or PM3), which had been successfully applied to elements of the second period, failed for the heavier elements. This is probably due to the determination of the parameters for the d-atomic orbitals which turned out to be rather difficult. For this reason only *ab initio* calculations give satisfactory results if d-type atomic orbitals are present. *Ab initio* calculations are based on the single Slater determinant approximation of the wave function within the Hartree–Fock theory[197].

The difference between the Hartree–Fock energy and the exact solution of the Schrödinger equation (Figure 60), the so-called correlation energy, can be calculated approximately within the Hartree–Fock theory by the configuration interaction method (CI) or by a perturbation theoretical approach (Møller–Plesset perturbation calculation *n*th order, MPn). Within a CI calculation the wave function is composed of a linear combination of different Slater determinants. Excited-state Slater determinants are then generated by exciting electrons from the filled SCF orbitals to the virtual ones:

$$\Psi = \sum_{i}^{\infty} C_i \Psi_i^{SD}$$

This approach results in a matrix eigenvalue problem:

$$\underline{H}\vec{C} = E^{CI} S \vec{C}$$

and obeys the variation principle. If we use infinite Slater determinants, the exact energy value E^0(non-relativistic) (Figure 60) for the Schrödinger equation would be obtained.

In practice, it is not possible to use infinite Slater determinants. Usually CI programmes are written to permit single plus double excitations (CISD). In contrast to SCF(HF) and full CI, the CISD method is not size-consistent.

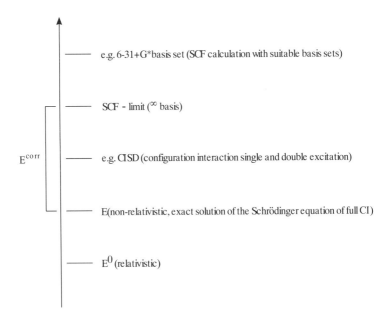

FIGURE 60. Schematic representation of the correlation energy

Within the SCF-CI method a fixed set of molecular orbitals is used. This means that during the calculation (leading to slow convergence) the individual molecular orbitals remain unchanged. A method where the linear expansion coefficients and the LCAO coefficients are optimized simultaneously is the multi-configuration SCF (MCSCF).

Even if electron correlation is taken completely into consideration (full CI; note this corresponds to the exact solution of the Schrödinger equation) the computed energy does not meet the true energy value for atoms or molecules. This remaining difference betwen computed (full CI) and true energy is due to relativistic effects. As the core or total relativistic energies go like Z^4 (the valence contribution roughly goes like Z^2) they become relevant especially for the heavy elements and for the sixth row (W...Pb...Bi); relativistic effects are comparable to the usual shell-structure effects. Due to the relativistic mass increase the effective Bohr radius will decrease for inner-shell electrons with large average speed (cf. Section V). The most widely used approach to include relativistic effects into quantum-mechanical computations is the pseudo-potential method[198-201]. The central idea of a pseudo-potential method is to omit the frozen inner shells, and the corresponding nodes in an atomic valence wave function, by considering instead the eigenvalue problem for a nodeless one-component pseudo-wave function. The pseudo-potential or effective core potential (ECP) corresponds to all interactions between the valence and the core electrons. Whereas relativistic SCF calculations for molecules containing heavy elements are extremely time-consuming, the ECP method is fast and can describe molecular properties almost as accurately as all-electron calculations. Furthermore, relativistic effects can easily be included either in a spin–orbit averaged way (one–component ECP) or by including spin–orbit coupling (two-component ECP).

B. π-Bonding in Group 14

In the last decade many theoretical studies concerning the structure of the model compound $H_2E=EH_2$ (E = Si, Ge, Sn, Pb) in the ground state were published. The geometries of group 14 double bonds have been much studied experimentally and theoretically because they effectively demonstrate breakdown of the 'classical double bond rule' (which assumes that elements possessing a principal quantum number greater than 2 should not form $p_\pi-p_\pi$ bonds). Moreover, these compounds also show a trend from planarity to *trans*-bent structures descending down the column. It is well known that the coordination geometry at the C=C double bond is planar (except where other geometric demands make it impossible [202,203]). However, both planar and *trans*-bent coordinated Si=Si double bonds have been synthesized with substituents of comparable steric demand. This indicates that for Si=Si bound systems the two geometrical forms are close in energy. For Ge=Ge and Sn=Sn, only *trans*-bent double bond structures have been observed. The angles α[204,205] (for definition see Figure 61) are usually greater for Sn=Sn compounds than for their Ge=Ge analogues[205].

Ab initio calculations reproduce the experimentally observed geometrical changes. Ethylene is planar (absolute true minimum; global minimum) at all theoretical levels. The computed geometry of disilene depends strongly on the basis functions and on electron

FIGURE 61. Definition of the folding angle α in $R_2E=ER_2$ systems (E = Si, Ge, Sn, Pb)

TABLE 6. Calculated structural parameters for $R_2E=ER_2$ trans-bent (E = Ge, Sn) compounds

E,R	$d(E=E)$ (Å)	$\alpha(^{deg})$	Method	Basis set	Reference
Ge,CH(SiMe$_3$)$_2$	2.347	32	X-ray	—	205
Ge, H	2.325	38.9	CI	ECP + DZP	211
Ge, H	2.272	36.2	RHF	DZ	212
Ge, H	2.3	40	RHF	cf Ref.	213
Ge, H	2.307	38.1	RHF	ECP + 21G	214
Ge, H	2.270	34.6	RHF	DZP	215
Ge, H	2.315	36.5	RHF	DZP	216
Sn,CH(SiMe$_3$)$_2$	2.768	41	X-ray	—	205
Sn, H	2.71	46	RHF	cf Ref.	213
Sn, H	2.687	48.2	RMP2		217
Sn, H	2.702	41.0	RHF	ECP+21G	214
Sn, H	2.712	48.9	RHF	DZP	215

correlation[206]. High-level ab initio calculations at the CI level indicate the absolute true minimum structure for the trans-bent arrangement[207]. In contrast to disilene, the geometries of digermene and distannene were found not to depend on the basis sets and the computational level employed (Table 6). In particular, RHF and UHF calculations give precisely the same results. This shows that digermene as well as distannene are stable singlets in the ground state. Thus, the diradical character hypothesis is not adequate to explain the non-planarity of these two molecules.[206]

These theoretical investigations predict for $H_2Ge=GeH_2$ a trans-bent structure with a relatively short Ge=Ge bond distance of 2.27 Å –2.33 Å (depending on the method used and the basis set) and a large angle α of 34–40°. For distannene, $H_2Sn=SnH_2$, the Sn−Sn bond distance in the trans-bent form was computed to be 2.7 Å and an even greater angle $\alpha(41-48°)$ was predicted. The structures of substituted germenes and stannenes were examined by X-ray investigations and in both species they show a distortion towards pyramidicity with an angle α which is greater for Sn ($\alpha = 41°$) than for Ge ($\alpha = 32°$)[213]. The Ge=Ge bond length of 2.35 Å is significantly shorter than a Ge−Ge single bond, e.g. 2.465 Å in $(GePh_2)_4$[208] or 2.463 Å in $(GePh_2)_6$[209], and is only slightly longer than the calculated distance in the model compound Ge_2H_4. In case of the tin species (2.77 Å) the bond length is slightly shorter than that of a Sn−Sn single bond, e.g. 2.810 Å in tetrahedral Sn_∞ (diamond structure)[210]. Considering the sizeable ligands in the E_2R_4 [E = Ge, Sn; R = CH(SiMe$_3$)$_2$] species, one can conclude that theory and experiment are in good agreement (Table 6).

The singlet potential surface for all group 14 $H_2E=EH_2$ species (E = C, Si, Ge, Sn and Pb) has been investigated by Trinquier by means of ab initio SCF+CI calculations using effective core potentials which include relativistic effects for tin and lead[216]. In all cases but carbon, the bridged structures were found to be true minima with the trans isomer being favoured over the cis by 2 kcal mol^{-1}. For carbon, a trans-bridged form was found to be a saddle point. The $H_2E=EH_2$ doubly bonded forms, planar or trans-bent, were found to be true minima in all cases except for lead, where it was only a saddle point. The most stable structures (absolute true minima) of the Si and Ge species are the trans-bent doubly bonded isomers, while the most stable structures of Sn_2H_4 and Pb_2H_4 are the trans-bridged forms (Tables 7 and 8). Besides, the system $H_2E=EH_2$ is expected to lose the direct E=E link to the benefit, for instance, of doubly bridged structures, as occurs in the tetrafluoro derivatives of disilene, digermene and distannene (in such polar structures, electrostatics could be a further stabilizing factor).

TABLE 7. Calculated relative energies at the CI level (in kcal mol^{-1})^{216} (ATM = absolute true minima, TM = true minima, SP = saddle point, CP2 = critical point of index 2)

		C_2H_4	Si_2H_4	Ge_2H_4	Sn_2H_4	Pb_2H_4
$2EH_2\,(^1A_1)$		192.0	53.7	35.9	33.2	28.7
H_3E-EH	C_s	79.1, SP	9.8, TM	2.4, TM	7.0, TM	17.5, TM
HE——EH C_{2v}, cis^b (bridged)	C_{2v}, cis^b	140.3, CP2	25.2, TM	11.6, TM	2.3, TM	2.0, TM
HE——EH C_{2h}, trans^b (bridged)	C_{2h}, trans^b	164.7, SP	22.5, TM	9.0, TM	0, ATM	0, ATM
Fig. 61	C_{2h}, trans-bent		ATM^a	0, ATM	9.1, TM	23.9, SP
$H_2E=EH_2$	D_{2h}, planar	0, ATM	0, TS^a	3.2, SP	18.5, SP	43.7, SP

^{a} At this level disilene was found to be planar, i.e. there is no stationary point corresponding to a trans-bent geometry. At a higher level the absolute true minima were found to be trans-bent (because of the very flat surface corresponding to this trans-wagging coordinate).
^{b} cis and trans refer to the terminal H atoms.

TABLE 8. Energy differences between the singlet and triplet states of EH$_2$ species (E = C, Si, Ge, Sn, Pb)

	CH_2	SiH_2	GeH_2	SnH_2	PbH_2
ΔE_{ST}(CI) kcal mol^{-1216}	-14.0	16.7	21.8	24.8	34.8

The trans-bent structure of Ge$_2$H$_4$, the two bridged structures and the germylgermylene structure were found to be real minima. The planar digermene form is the transition state for interconverting the two trans-bent forms.

Similar to the Ge$_2$H$_4$, in Sn$_2$H$_4$ the four main structures are true minima while planar distannene is a saddle point.

The two bridged forms of Pb$_2$H$_4$ and the plumbylplumbylene form were found to be true minima. Planar diplumbene is a saddle point with a single imaginary frequency corresponding to trans-bending, as for the Ge and Sn surface. Remarkably, the trans-bent form is no longer a local minimum, but a saddle point with a single imaginary frequency corresponding to C1 symmetry. That means no diplumbene should exist and only the bridged form is expected to be stable. The complexes E[(TeSi(SiMe$_3$)$_3$)$_3$]$_2$ (E = Sn, Pb) represent examples with bridging structures for tin and lead^{218}. In case of the tin species the Sn$_2$Te$_2$ core of the dimer exhibits a butterfly structure. Only the cis isomer was observed experimentally (Figure 62).

The geometry of the double bond is determined by the degree of orbital mixing in the R$_2$E=ER$_2$ species. Greater orbital mixing leads to pyramidalization or trans-bending of double bonds. The degree of mixing (and therefore the folding angle α) is determined by (i) the intrinsic $\pi-\sigma^*$ gap of the double bond (intrinsic because it is largely determined by its σ and π bond strength; stronger double bonds have larger $\pi-\sigma^*$ gaps) and (ii) electronegative substitution, which increases orbital mixing^{214}. Thus, for ethylene the intrinsic $\pi-\sigma^*$ gap is so large that no substituent can increase the orbital mixing

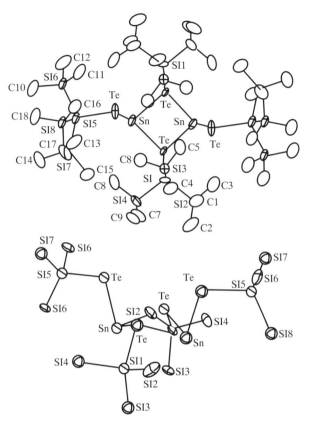

FIGURE 62. Molecular dimeric structure of Sn[(TeSi(SiMe$_3$)$_3$]$_2$. Reprinted with permission from Reference 218. Copyright (1993) American Chemical Society

sufficiently to a *trans*-bent C=C bond. On the other hand, the intrinsic $\pi-\sigma^*$ gaps of disilene, digermene and distannene are small enough to make orbital mixing possible. Consequently, the electron-withdrawing effect becomes important for the degree of orbital mixing.

In case of Ge and Sn the destabilization of the E−E $\sigma(a_g)$ bond in this *trans*-bent structure is compensated by stabilizing the highest occupied $\pi(b_u)$ orbital which interacts with the E−E σ^* orbital[213].

In conclusion, it has been established that the degree of orbital mixing, and therefore the folding angle α, depend on the energy difference between the $\pi(b_g)$ and the E−E σ^*-orbital which decreases with higher atom weights of E. That means that the stability of the trans-folded form relative to the planar double-bonded form in H$_2$E=EH$_2$ (E = C, Si, Ge, Sn; for Pb the *trans*-bent structure represents a transition state) as well as the angle α increase with increasing atomic number of E. This correlates with the increasing energy difference between the singlet and triplet state of the monomers[219]. Only for CH$_2$ is the triplet (^{3}B$_1$) the ground state[220,221]; SiH$_2$[222,223], GeH$_2$[224,225], SnH$_2$[223−226] and PbH$_2$[216] all have singlet ground states of ^{1}A$_1$ symmetry, with progressively increasing ^{1}A$_1$ to ^{3}B$_1$ energy gaps in the order of Si<Ge<Sn<Pb (Table 8).

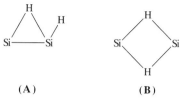

FIGURE 63. Bridged structures for Si_2H_2

FIGURE 64. Vinylidene-like structure of Ge_2H_2

The calculated dissociation energy for the gas-phase reaction

$$H_2E=EH_2 \longrightarrow 2\ EH_2$$

is $52-58\ \text{kcal mol}^{-1}$ [215] for E = Si, $30-45\ \text{kcal mol}^{-1}$ [213,215,211] for the germanium species and for tin $22-28\ \text{kcal mol}^{-1}$ [213,217]. Experimentally determined energies are not available. It is interesting that experimentally determined single E−E bond dissociation energies in molecules such as H_3GeGeH_3 and Me_3EEMe_3 (E = Ge, Sn) range between 48 and 72 kcal mol^{-1}. However, these experimentally obtained values are much higher than the computed ones for the H_2EEH_2 species of Ge and Sn.

Among small molecules, one of the most interesting structural discoveries in recent years was that of the monobridged equilibrium geometry of the Si_2H_2 molecule (Figure 63, **A**). The existence of such a structure (**A**) only 10.8 kcal mol^{-1} above the dibridged 'butterfly' global minimum (Figure 63, **B**) was proposed in 1990 by Colegrave. The microwave spectrum of the dibridged minimum structure of Si_2H_2 was observed and analysed[227,228]. The potential energy hyperface for the analogous germanium species has recently been published[229,230].

These studies predict three minima structures, the dibridged and monobridged isomers (Figure 63) and a vinylidene-like species (Figure 64). At the ZPVE (Zero-Point Vibrational Energies)-corrected TZP+f CCSD(T) level of theory, the monobridged and the vinylidene Ge_2H_2 structures are predicted to lie 8.9 and 11.0 kcal mol^{-1} above the dibridged global minimum (ATM).

C. The Heavier Congeners of Ethane: H_3X-YH_3 (X, Y = C, Si, Ge, Sn, Pb)

The rotational barriers along single bonds (exemplified by ethane) have been the focus of much experimental and theoretical effort[231−233]. The relationships between the central bond length and the rotational barriers are of particular interest. The barriers decrease but do not vanish at large X−Y separations (e.g. Pb_2H_6). Each of the lighter elements (X = C, Si, Ge) gives a correlation. Schleyer and coworkers published a detailed study on this subject[234]. They found that in most cases the X−Y distances obtained from all-electron 3-21(D,P) calculations agree well with the pseudo-potential results (Table 9). For lead compounds, the quasi-relativistic pseudo-potential calculation gives Pb−Y bonds substantially shorter than the all-electron results and it is obviously necessary to include

TABLE 9.   X–Y distances, R (Å), and rotational barriers, ΔE_{rot} (kcal mol)$^{-1}$, for $H_3X–YH_3$ molecules (X, Y = C, Si, Ge, Sn Pb)

		All-electron calculation			Pseudo-potential calculation			
X	Y	R_{st}^a	R_{ec}^a	ΔE_{rot}	R_{st}^a	R_{ec}^a	ΔE_{rot}	ΔE_{exp}^b
C	C	1.542	1.556	2.751	1.526	1.539	2.776	2.9
C	Si	1.883	1.893	1.422	1.883	1.893	1.388	1.7
C	Ge	1.990	1.999	1.104	1.996	2.004	0.986	1.24
C	Sn	2.188	2.193	0.498	2.178	2.184	0.520	*ca* 0.6
C	Pb	2.275	2.278	0.204	2.242	2.246	0.321	
Si	Si	2.342	2.355	0.949	2.355	2.364	0.823	1.22
Si	Ge	2.409	2.420	0.613	2.425	2.433	0.682	
Si	Sn	2.610	2.617	0.581	2.610	2.616	0.476	
Si	Pb	2.695	2.701	0.486	2.640	2.645	0.358	
Ge	Ge	2.499	2.513	0.664	2.499	2.506	0.528	
Ge	Sn	2.662	2.667	0.445	2.669	2.675	0.408	
Ge	Pb	2.741	2.745	0.395	2.705	2.709	0.315	
Sn	Sn	2.850	2.855	0.412	2.843	2.847	0.350	
Sn	Pb	2.928	2.930	0.309	2.869	2.873	0.286	
Pb	Pb	3.012	3.015	0.214	2.897	2.900	0.230	

$^a R_{st}$ and R_{ec} are X–Y distances for the staggered and eclipsed conformations,
^{b} Reference 234 and references cited therein.

relativistic effects for calculations of lead compounds (cf section V). Relativistic bond contractions for lead are well established[178–180,235].

The rotational barriers obtained from pseudo-potential and all-electron calculations generally agree within *ca* 0.15 kcal mol^{-1}, even when X, Y = Pb. Hence, relativistic effects do not appear to influence the barriers.

Within the NBO (Natural bond orbital) analysis non-covalent effects are taken into account by a second-order perturbation theory. This provides information concerning the interaction between the strictly localized, almost fully occupied bonding NBOs and the almost empty 'Rydberg-type' or anti-bonding NBOs. These interactions result in deviations from the ideal Lewis structure[236–238]. The energy contributions are relatively small. In case of the $H_3X–YH_3$ species, the non-covalent energy terms resulting from vicinal $\sigma HX \rightarrow \sigma^*HY$ (and $\sigma HY \rightarrow \sigma^*HX$) interactions are only contributions that favour the staggered conformation. Table 10 summarizes the second-order vicinal interactions for the symmetrical $H_3X–XH_3$ molecules. The column ΔE (st–ec) gives the net stabilization

TABLE 10.   Vicinal $\sigma HX \rightarrow \sigma^*HX$ delocalization for symmetrical $H_3X–XH_3$ molecules; analysis of the NBO-Fock matrix by second-order perturbation theory (energies in kcal mol^{-1})

X	Staggered $E\sigma\sigma^*$(total)	Eclipsed $E\sigma\sigma^*$(total)	ΔE (st–ec)	ΔE_{rot}
C	23.16	17.40	5.76	2.78
Si	8.22	7.32	0.90	0.82
Ge	7.08	6.54	0.52	0.53
Sn	5.88	5.58	0.3	0.35
Pb^a	8.46	8.22	0.24	0.23
Pb^b	4.32	4.08	0.24	0.31

^{a} Quasi-relativistic lead pseudo-potential.
^{b} Non-relativistic lead pseudo-potential.

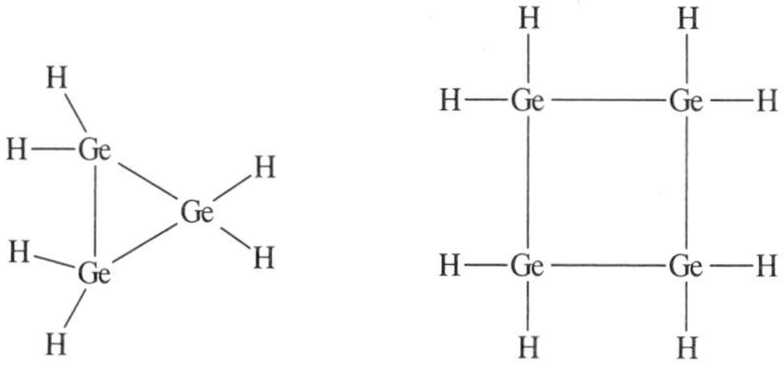

FIGURE 65. Structures of the cyclogermanes $(H_2Ge)_n$ $(n = 3, 4)$

obtained by subtracting the total vicinal delocalization for the eclipsed conformation from that for the staggered conformation. The correlation with the rotational barriers (ΔE_{rot}) is reasonable. The sum of the vicinal contributions decreases down the column from C to Pb. This decrease is probably due to poorer orbital overlap, since the orbital energy differences are smaller for the heavier and less electronegative elements. In conclusion, vicinal $\sigma HX \rightarrow \sigma^*HY$ and $\sigma HY \rightarrow \sigma^*HX$ interactions in the staggered conformations are responsible for the rotational barriers of the complete series. This is in agreement with a previous study of ethane. The lower barriers for the heavier species appear to be due to poorer orbital overlap and to increasingly smaller differences between anti-periplanar, syn-periplanar and $120°$ vicinal overlap.

D. Cyclotrigermane and Cyclotristannane

Since the synthesis of the first cyclotrigermane[239], the hexakis(2,6-dimethylphenyl)-cyclotrigermane, calculations have been carried out on that molecule using $(H_2Ge)_3$ as a model. Similar to the analogous Si compounds, in $(H_2Ge)_3$ a shorter $Ge-Ge$ distance of 2.448 Å was predicted for the four-membered ring $(H_2Ge)_4$ (2.462 Å). The energy due to the ring strain of 44.6 $kcal\,mol^{-1}$ for $(H_2Ge)_3$ is substantially greater than for $(H_2Ge)_4$ with 13.5 $kcal\,mol^{-1}$ (Figure 65)[240]. The $Sn-Sn$ bond distances in the corresponding tin cycles have computed values of 2.80 Å [for$(H_2Sn)_3$] and 2.81 Å [for $(H_2Sn)_4$].

VII. ACKNOWLEDGEMENTS

The authors wish to thank the Deutsche Forschungsgemeinschaft and the Fonds der Chemischen Industrie for continuous financial support.

VIII. REFERENCES

A. General Literature

Books

I. Omae (Ed.), *J. Organomet. Chem. Library*, **21** Elsevier, Amsterdam, 1989.

F. G. A. Stone and R. West (Eds.), *Advances in Organometallic Chemistry*, Volume 4, Academic Press, New York, 1966.

J. J. Zuckerman and A. P. Hagen (Eds.), *Inorganic Reactions and Methods*, Volume 9, VCH Verlagsgesellschaft, Weinheim, 1991.

E. G. Rochow and E. W. Abel, *The Chemistry of Germanium, Tin and Lead*, Pergamon, Oxford, 1975.

G. Bähr, H. O. Kalinowski and S. Pawlenko, in *Methoden der Organischen Chemie (Houben-Weyl), Met. Org. Verbindungen (Ge, Sn)*, Thieme, Stuttgart, 1978.

A superb comprehensive and up-to-date survey concerning the organic chemistry of tin is given in the multi-volume Gmelin series on Sn compounds.

Reviews and Book Chapters

M. Veith, *Angew. Chem.*, **99**, 1 (1987).
N. C. Norman, *Polyhedron*, **12**, 2431 (1993).
C. F. Shwa, III and A. L. Allred, *Organomet. Chem. Rev. A*, **5**, 96 (1970).
S. -W. Ng and J. J. Zuckerman, *Adv. Inorg. Chem.*, **29**, 297 (1985).
M. Veith, *Chem. Rev.*, **90**, 3 (1990).
J. Barrau, J. Escudié, and J. Satgé, *Chem. Rev.*, **90**, 283 (1990).
P. Jutzi, *Adv. Organomet. Chem.*, **26**, 217 (1986).
R. C. Poller, *J. Organomet. Chem.*, **239**, 189 (1982).
B. C. Pant, *J. Organomet. Chem.*, **66**, 321 (1974).
A. G. Davies and P. J. Smith, *Adv. Inorg. Chem.*, **23**, 1 (1980).
R. West, in *The Chemistry of Inorganic Ring Systems* (Ed. R. Steudel), Chap. 4, Elsevier, Amsterdam, 1992.
M. Weidenbruch, in *The Chemistry of Inorganic Ring Systems* (Ed. R. Steudel), Chap. 5, Elsevier, Amsterdam, 1992.
U. Klingebiel, S. Schütte and D. Schmidt-Bäse, in *The Chemistry of Inorganic Ring Systems* (Ed. R. Steudel), Chap. 6, Elsevier, Amsterdam, 1992.
A. Sekiguchi and H. Sakurai, in *The Chemistry of Inorganic Ring Systems* (Ed. R. Steudel), Chap. 7, Elsevier, Amsterdam, 1992.
M. Veith and S. Müller-Becker, in *The Chemistry of Inorganic Ring Systems* (Ed. R. Steudel), Chap. 8, Elsevier, Amsterdam, 1992.

B. Cited Literature

1. N. N. Greenwood and A. Earnshaw, *Chemistry of the Elements*, Pergamon, Oxford, 1984.
2. A. Hinchliffe, *Computational Quantum Chemistry*, Wiley, New York, 1988.
3. Iwao Omae, *J. Organomet. Chem. Library*, **21** (1989).
4. M. Lesbre, *Afinidad*, **29**, 171 (1972).
5. I. Haiduck and J. J. Zuckerman, *Basic Organometallic Chemistry*, Chap. 8, Walter de Gruyter, Berlin, New York, 1985.
6. H. G. Kuivila and O. F. Beumel Jr., *J. Am. Chem. Soc.*, **83**, 1246 (1961).
7. P. J. Stang and M. R. White, *J. Am. Chem. Soc.*, **103**, 5429 (1981).
8. C. Chatgilialoglu, D. Griller and M. Lesage, *J. Org. Chem.*, **53**, 3541 (1988).
9. M. Lesage, J. A. Martinho-Simoes and D. Griller, *J. Org. Chem.*, **55**, 5413 (1990).
10. K. B. Clark and D. Griller, *Organometallics*, **10**, 746 (1991).
11. P. N. Noble and R. Walsh, *Int. J. Chem. Kinet.*, **15**, 547 (1983).
12. M. J. Almond, A. M. Doncaster, P. N. Noble and R. Walsh, *J. Am. Chem. Soc.*, **104**, 4717 (1982).
13. B. Ruscic, M. Schwarz and J. Berkowitz, *J. Chem. Phys.*, **92**, 1865 (1990).
14. A. M. Doncaster and R. Walsh, *J. Phys. Chem.*, **83**, 578 (1979).
15. R. C. Binning Jr. and L. A. Curtiss, *J. Chem. Phys.*, **92**, 1860 (1990).
16. L. J. Johnston, J. Lusztyk, D. D. M. Wayner, A. N. Abeywickreyma, A. L. J. Beckwith, J. C. Scaiano and K. U. Ingold, *J. Am. Chem. Soc.*, **107**, 4594 (1985).
17. A. L. J. Beckwith, D. H. Roberts, C. H. Schiesser and A. Wallner, *Tetrahedron Lett.*, **26**, 3349 (1985).
18. J. Lusztyk, B. Maillard, D. A. Lindsay and K. U. Ingold., *J. Am. Chem. Soc.*, **105**, 3578 (1983).
19. J. Lusztyk, B. Maillard, S. Deycard, D. A. Lindsay and K. U. Ingold., *J. Org. Chem.*, **52**, 3509 (1987).
20. R. A. Jackson, *J. Organomet. Chem.*, **166**, 17 (1979).
21. P. Mazerrolles, R. Morancho and A. Reynes, *Rev. Silicon, Germanium, Tin, Lead Comp.*, **9**, 155 (1986).
22. T. A. Sladkova, O. P. Berezhanskaya, B. M. Zolotaray and G. A. Razuvaev, *Izv. Akad. Nauk SSSR, Otdel. Khim. Nauk*, **6**, 1316 (1978).
23. R. A. Jackson, K. U. Ingold, D. Griller and A. S. Nazaran, *J. Am. Chem. Soc.*, **107**, 208 (1985).
24. P. J. Stang and M. R. White, *J. Am. Chem. Soc.*, **103**, 5429 (1981).
25. F. Glockling and K. A. Hooton, *J. Chem. Soc. (London)*, 1849 (1963).

26. G. Bähr, H. O. Kalinowski and S. Pawlenko, in *Methoden der Organischen Chemie (Houben-Weyl)*, *Met. Org. Verbindungen (Ge, Sn)*, Thieme Verlag, Stuttgart, 1978.
27. T. Lobreyer, J. Oeler, W. Sundermeyer and H. Oberhammer, *Chem. Ber.*, in press.
28. F. Feher and M. Krancher, *Z. Naturforschung B*, **40**, 1301 (1985).
29. T. Lobreyer, J. Oeler and W. Sundermeyer, *Chem. Ber.*, **124**, 2405 (1991).
30. T. Lobreyer, H. Oberhammer and W. Sundermeyer, *Angew. Chem.*, **105**, 587 (1993).
31. C. Glidewell, D. W. H. Rankin and A. G. Robiette, *J. Chem. Soc. (A)*, 2936 (1970).
32. H. G. Ang and F. K. Lee, *J. Chem. Soc., Chem. Commun.*, 310 (1989).
33. P. Riviere, M. Riviere-Baudet and J. Satge, Germanium, in *Comprehensive Organometallic Chemistry* (Eds. G. Wilkinson, F. G. A. Stone and E. W. Abel), Vol. 2, Chap. 10, Pergamon Press, Oxford, 1982.
34. M. Draeger, L. Ross and D. Simon, *Rev. Silicon, Germanium, Tin, Lead Compd.*, **7**, 299 (1983).
35. A. Castel, P. Riviere and J. Satge, *J. Organomet. Chem.*, **462**, 97 (1993).
36. A. Castel, P. Riviere, J. Satge and Y. H. Ko, *Organometallics*, **9**, 205 (1990).
37. A. Castel, P. Riviere, J. Satge, Y. H. Ko and D. Desor, *J. Organomet. Chem.*, **397**, 7 (1990).
38. J. Park, S. A. Batcheller and S. Masamune, *J. Organomet. Chem.*, **367**, 39 (1989).
39. S. A. Batcheller and S. Masamune, *Tetrahedron Lett.*, **29**, 3383 (1988).
40. S. D. Pigarev, D. A. Bravo-Zhivotovskii, I. D. Kalikhman, N. S. Vyazankin and M. G. Voronkov, *J. Organomet. Chem.*, **369**, 29 (1989).
41. A. Castel, P. Riviere, J. Satge and D. Desor, *J. Organomet. Chem.*, **433**, 49 (1992).
42. K. Lee, A. M. Arif and J. A. Gladysz, *Organometallics*, **10**, 751 (1991).
43. N. L. Ermolaev, M. N. Bochkarev, G. A. Razuvaev, Y. K. Grishin and Y. A. Ustynyuk, *Zh. Obshch. Khim.*, **54**, 96 (1984).
44. D. A. Bravo-Zhivotovskii, I. S. Biltueva, O. A. Vyanzankina and N. S. Vyanzankin, *Izv. Akad. Nauk. SSSR, Ser. Khim.*, 1214 (1985).
45. P. Mazerolles, *Bull. Soc. Chim. Fr.*, **29**, 1907 (1962).
46. K. Mochida and K. Asami, *J. Organomet. Chem.*, **232**, 13 (1982).
47. Y. Takeuchi, M. Shimoda, K. Tanaka, S. Tomoda, K. Ogawa and H. Suzuki, *J. Chem. Soc., Perkin Trans. 2*, 7 (1988).
48. B. C. Pant, *J. Organomet. Chem.*, **66**, 321 (1974).
49. R. Damrauer, *Organomet. Chem. Rev. (A)*, **8**, 67 (1972).
50. G. H. Reifenberg and W. J. Considine, *Organometallics*, **12**, 3015 (1993).
51. E. J. Walsh and H. G. Kuivila, *J. Am. Chem. Soc.*, **88**, 576 (1966).
52. J. Lusztyk, E. Lusztyk, B. Maillard, L. Lunazzi and K. U. Ingold, *J. Am. Chem. Soc.*, **105**, 4475 (1983).
53. G. R. Clark, K. R. Flower, C. E. F. Rickard, W. R. Roper, D. M. Salter and L. J. Wright, *J. Organomet. Chem.*, **462**, 331 (1993).
54. A. Stern and E. I. Becker, *J. Org. Chem.*, **27**, 4052 (1962).
55. J. G. Noltes and G. J. M. v. d. Kerk, *Chem. Ind. (London)*, 294 (1959).
56. R. Sommers, W. P. Neumann and B. Schneider, *Tetrahedron Lett.*, 3875 (1964).
57. A. Werner, *Ber. Dtsch. Chem. Ges.*, **45**, 121 (1912).
58. J. F. Stoddard and R. Zarzycki, *Recl. Trav. Chim. Pays-Bas*, **107**, 515 (1988) and references cited therein.
59. P. Jutzi, E. Schlüter, M. B. Hursthouse, A. M. Arif and R. L. Short, *J. Organomet. Chem.*, **299**, 285 (1986).
60. M. Veith, *Angew. Chem.*, **100**, 1124 (1988).
61. M. Gielen and N. Sprecher, *Organomet. Chem. Rev.*, **1**, 455 (1976).
62. J. M. Dumas and M. Gomel, *Bull. Soc. Chim. Fr.*, 1885 (1974).
63. J. A. Zubieta and J. J. Zuckermann, *Prog. Inorg. Chem.*, **24**, 251 (1978).
64. T. Raubold, S. Freitag, R. Herbst-Irmer and H. W. Roesky, *Z. Anorg. Allg. Chem.*, **619**, 951 (1993).
65. K. M. Mackay, B. K. Nicholson, G. Wilkinson, F. G. A. Stone and E. W. Abel (Eds.), *Comprehensive Organometallic Chemistry*, Vol. 6, Pergamon, Oxford, 1982, p. 1043.
66. W. Petz, *Chem. Rev.*, **86**, 1019 (1986).
67. M. F. Lappert and R. S. Rowe, *Coord. Chem. Rev.*, **100**, 267 (1990).
68. M. L. H. Green, A. K. Hughes and P. Mountford, *J. Chem. Soc., Dalton Trans.*, 1407 (1991).
69. H. -G. Woo, W. P. Freeman and T. D. Tilley, *Organometallics*, **11**, 2198 (1992).
70. U. Schubert, S. Grubert, M. Schulz and S. Mock, *Organometallics*, **11**, 3163 (1992).

71. S. Zhang and T. L. Brown, *Organometallics*, **11**, 2122 (1992).
72. Y. Wakatsuki, H. Yamazaki, M. Nakano and Y. Yamamoto, *J. Chem. Soc., Chem. Commun.*, 703 (1991).
73. M. Schäfer and K. Dehnicke, *Z. Anorg. Allg. Chem.*, **615**, 127 (1992).
74. E. Hough, D. G. Nicholson and A. K. Vasudevan, *J. Chem. Soc., Dalton Trans.*, 427, (1987).
75. H. v. Arnim, K. Dehnicke, K. Maczek and D. Fenske, *Z. Naturforsch.*, **48b**, 1331 (1993).
76. F. A. Cotton and J. Czuchajowska, *J. Polyhedron*, **9**, 2553 (1990).
77. A. Bencini, A. Bianchi, P. Dapporto, E. Garcia-Espana, V. Marcelino, M. Micheloni, P. Paoletti and P. Paoli, *Inorg. Chem.*, **29**, 1176 (1990).
78. A. Bencini, A. Bianchi, M. Micheloni, P. Paoletti, P. Dapporto, P. Paoli and E. Garcia-Espana, *J. Inclusion Phenom.*, **12**, 291 (1992).
79. A. Bencini, A. Bianchi, P. Dapporto, E. Garcia-Espana, M. Micheloni, P. Paoletti and P. Paoli, *J. Chem. Soc., Chem. Commun.*, 1176 (1990).
80. A. Bencini, A. Bianchi, P. Dapporto, E. Garcia-Espana, M. Micheloni and P. Paoletti, *Inorg. Chem.*, **28**, 1188 (1989).
81. D. L. Reger, S. J. Knox, M. F. Huff, A. L. Rheingold and B. S. Haggerty, *Inorg. Chem.*, **30**, 1754 (1991).
82. D. L. Reger, M. F. Huff, S. J. Knox, R. J. Adams, D. C. Apperley and R. K. Harris, *Inorg. Chem.*, **32**, 4472 (1993).
83. D. L. Reger, M. F. Huff, A. Rheingold and B. S. Haggerty, *J. Am. Chem. Soc.*, **114**, 579 (1992).
84. H. J. Brügge, R. Fölsing, A. Köchel and W. Dreissig, *Polyhedron*, **4**, 1493 (1985).
85. S. T. Malinovskii, Yu. A. Simonov and A. Yu. Nazarenko, *Kristallografiya*, **35**, 1410 (1990).
86. N. I. Krasnova, Yu. A. Simonov, M. B. Korshunov and V. V. Yakshin, *Kristallografiya*, **32**, 499· (1987).
87. R. D. Rogers and A. H. Bond, *Inorg. Chim. Acta*, **192**, 163 (1992).
88. H. v. Amim, K. Dehnicke, K. Mackzek and D. Fenske, *Z. Anorg. Allg. Chem.*, **619**, 1704 (1993).
89. R. H. Herber and G. Carraquillo, *Inorg. Chem.*, **20**, 3693 (1981).
90. M. G. B. Drew and D. G. Nicholson, *J. Chem. Soc., Dalton. Trans.*, 1543 (1986).
91. A. Andres, A. Bencini, A. Carachalios, A. Bianchi, P. Dapporto, E. Garcia-Espana, P. Paoletti and P. Paoli, *J. Chem. Soc., Dalton Trans.*, 1993, 3507.
92. R. D. Hancock, M. S. Shaikjee, S. M. Dobson and J. C. A. Boeyens, *Inorg. Chim. Acta*, **134**, 229 (1988).
93. K. Byriel, K. R. Dunster, L. R. Gahan, C. H. L. Kennard, J. L. Jazzn, I. L. Swann and P. A. Duckworth, *Polyhedron*, **10**, 1205 (1992).
94. D. A. Atwood, V. A. Atwood, A. H. Cowley, J. L. Atwood and E. Roman, *Inorg. Chem.*, **31**, 3871 (1992).
95. D. A. Atwood, V. A. Atwood, A. H. Cowley and H. R. Gobran, *Polyhedron*, **12**, 2073 (1993).
96. A. Sekiguchi, T. Yatabe, H. Kamatani, C. Kabuto and H. Sakurai, *J. Am. Chem. Soc.*, **114**, 6260 (1992).
97. A. Heine and D. Stalke, *Angew. Chem.*, **106**, 121 (1994).
98. H. H. Karsch, G. Baumgartner and S. Gamper, *J. Organomet. Chem.*, **462**, C3 (1993).
99. J. Barrau, J. Escudie and J. Satge, *J. Chem. Rev.*, **90**, 283 (1990).
100. N. Tokitoh, T. Matsumoto, K. Manmaru and R. Okazaki, *J. Am. Chem. Soc.*, **115**, 8855 (1993).
101. (a) R. K. Chadha, J. E. Drake and A. B. Sarkar, *Inorg. Chem.*, **26**, 2885 (1987).
     (b) W. Ando, T. Kadowaki, Y. Kabe and M. Ishii, *Angew. Chem., Int. Ed. Engl.*, **31**, 59 (1992).
102. P. B. Hitchcock, M. F. Lappert, S. J. Miles and A. J. Thorne, *J. Chem. Soc., Chem. Commun.*, 480 (1984).
103. S. A. Batcheller, T. Tsumuraya, O. Tempkin, W. M. Davis and S. Masamune, *J. Am. Chem. Soc.*, **112**, 9394 (1990).
104. J. T. Snow, S. Murakami, S. Masamune and D. J. Williams, *Tetrahedron Lett.*, **25**, 4191 (1984).
105. S. Masamune, Y. Hanzawa and D. J. Masamune, *J. Am. Chem. Soc.*, **104**, 6136 (1982).
106. S. Masamune, in *Silicon Chemistry*, (Eds. E. R. Corey, J. Y. Corey and P. P. Gaspar), Chap. 25, Ellis Horwood, New York, 1988, p. 257.
107. J. Satge, M. Massol and P. Riviere, *J. Organomet. Chem.*, **56**, 1(1973).
108. P. Riviere, M. Riviere,-Boudet, S. Richelme, A. Castel and J. Satge, *J. Organomet. Chem.*, **168**, 43(1979).
109. *Gmelin Handbook of Inorganic Chemistry*, 8th ed., *Organogermanium Compounds*, Part 3, Springer, Berlin, 1990 pp. 342–348.

110. E. F. Perozzi, R. S. Michalak, G. D. Figuly, W. Stevenson, D. B. Dess, M. R. Ross and J. C. Martin, *J. Org. Chem.*, **46**, 1049 (1981).
111. S. E. Denmark, R. T. Jacobs, G. Dai-Ho and S. Wilson, *Organometallics*, **9**, 3015 (1990).
112. S. Masamune and S. A. Batcheller, *Tetrahedron Lett.*, **29**, 3383 (1988).
113. S. Masamune, S. A. Batcheller, J. Park and W. M. Davis, *J. Am. Chem. Soc.*, **111** 1888 (1989).
114. M. S. Bilton and M. Webster, *J. Chem. Soc., Dalton Trans.*, 722 (1972).
115. R. O. Day, J. M. Holmes, A. C. Sau and R. R. Holmes, *Inorg. Chem.*, **21**, 281 (1982).
116. R. R. Holmes, R. O. Day, A. C. Sau, C. A. Poutasse and J. M. Holmes, *Inorg. Chem.*, **24**, 193 (1985).
117. M. Dräger, *Z. Anorg.Allg. Chem.*, **423**, 53(1976).
118. M. Dräger, *Chem. Ber.*, **108**, 1723(1975).
119. M. S. Bilton and M. Webster, *J. Chem. Soc., Dalton Trans.*, 722 (1972).
120. L. O. Atovmyan, J. J. Bleidelis, A. A. Kemme and R. P. Shibaeva, *J. Struct. Chem. (Engl. Transl.)*, **11**, 295 (1970).
121. A. A. Kemme, J. J. Bleidelis, R. P. Shibaeva and L. O. Atovmyan, *J. Struct. Chem. (Engl. Transl.)*, **14**, 90 (1973).
122. S. N. Gurkova, A. I. Gusev, I. P. Segelman, N. V. Alekseev, T. K. Gar and N. V. Khromova, *J. Struct. Chem. (engl. Transl.)*, **22**, 461 (1981).
123. C. Breliere, F. Carre, R. J. P. Corriu, A. de Saxce, M. Poirier and G. Royo, *J. Organomet. Chem.*, **205**, C1(1981).
124. R. R. Holmes, *J. Am. Chem. Soc.*, **97**, 5379 (1975).
125. M. Ye and J. G. Verkade, *Inorg. Chem.* **32**, 2796 (1993).
126. R. R. Holmes, R. O. Day, A. C. Sau, C. A. Poutasse and J. M. Holmes, *Inorg. Chem.*, **25**, 607 (1986).
127. D. J. Brauer, J. Wilke and R. Eujen, *J. Organomet. Chem.*, **316**, 261 (1986).
128. E. P. Kramarova, G. I. Olenyeva, A. G. Shipov, Y. I. Baukov, A. O. Mozzhukhin, M. Y. Antipin and Y. T. Struchkov, *Metalloorg. Khim. (Russian J. Organomet. Chem.)*, **4**, 1016 (1991).
129. Y. Baukov, A. G. Shipov, L. S. Smirnova, E. P. Kramarova, S. Y. Bylikin, Y. E. Ovchinnikoy and Y. T. Struchkov, *J. Organomet. Chemistry*, **461**, 39 (1993).
130. P. Jutzi, M. Meyer, H. P. Reisenauer and G. Maier, *Chem. Ber.*, **122**, 1227 (1989).
131. G. Märkl and W. Schlosser, *Angew. Chem.*, **100**, 1009 (1988).
132. E. Colomer, R. J. P. Corriu and M. Lheureux, *Chem. Rev.*, **90**, 265 (1990).
133. W. P. Freeman, T. D. Tilley, A. L. Rheingold and R. L. Ostrander, *Angew. Chem.*, **105**, 1841 (1993).
134. L. Pauling, *The Nature of the Chemical Bond*, 2nd ed., Cornell University Press, Ithaca, New York, 1960.
135. R. Wehrmann, H. Klusik and A. Berndt, *Angew. Chem.*, **96**, 810 (1984).
136. P. H. M. Budzelaar, P. v. R. Schleyer and K. Krogh-Jespersen, *Angew. Chem.*, **96**, 809 (1984).
137. H. Meyer, G. Baum, W. Massa, S. Berger and A. Berndt, *Angew. Chem.*, **99**, 559 (1987).
138. K. D. Dobbs and W. J. Hehre, *Organometallics*, **5** 2057 (1986).
139. S. P. Mallela and R. A. Geanangel, *Inorg. Chem.*, **32**, 5623 (1993).
140. H. Bock, J. Meuret and K. Ruppert, *J. Organomet. Chem.*, **445**, 19 (1993).
141. S. P. Mallela, I. Bernal and R. A. Geanangel, *Inorg. Chem.*, **31**, 1626 (1992).
142. H. Bock, J. Meuret and K. Ruppert, *Angew. Chem., Int. Ed. Engl.*, **32**, 414 (1993).
143. R. E. Hesters and K. Jones, *J. Chem. Soc., Chem. Commun.*, 317 (1966).
144. L. S. Khaikin, A. V. Belyakov, A. V. Golubinskij, L. V. Vilkow, N. V. Girbasowa, E. T. Bogoradovskij and V. S. Zavgorodnij, *J. Mol. Struct.*, **66**, 191 (1980).
145. D. G. Anderson, D. W. H. Rankin, H. E. Robertson, G. Gundersen and R. Seip, *J. Chem. Soc., Dalton Trans.*, 161 (1990).
146. C. Kober, J. Kroner and W. Storch, *Angew. Chem.*, **105**, 1693(1993).
147. G. Bandoli, A. Dolmella, V. Peruzzo and G. Plazzogna, *J. Organomet. Chem.*, **452**, 47 (1993).
148. M. Hada, H. Nakatsuji, J. Ushio, M. Izawa and H. Yokono, *Organometallics*, **12**, 3398 (1993).
149. K. Ding, Y. Wu and Y. Wang, *J. Organomet. Chem.*, **463**, 77 (1993).
150. D. Daktemieks, H. Zhu, D. Masi and C. Mealli, *Inorg. Chim. Acta*, **211**, 155 (1993).
151. G. N. Schrauzer, R. K. Chadha, C. Zhang and H. K. Reddy, *Chem. Ber.*, **126**, 2367 (1993).
152. C. Zhang, R. K. Chadha, H. K. Reddy and G. N. Schrauzer, *Inorg. Chem.*, **30**, 3865 (1991).
153. B. Wrackmeyer, K. Horchler and R. Boese, *Angew. Chem.*, **101**, 1563 (1989).
154. B. Wrackmeyer, *J. Chem. Soc., Chem. Commun.*, 1624 (1988).

155. L. Killian and B. Wrackmeyer, *J. Organomet. Chem.*, **148**, 137 (1978).
156. S. Kerschl and B. Wrackmeyer, *Z. Naturforsch.*, **B39**, 1037 (1984).
157. S. P. Mallela and R. A. Geanangel, *Inorg. Chem.*, **30**, 1480 (1991).
158. S. P. Mallela and R. A. Geanangel, *Inorg. Chem.*, **32**, 602 (1993).
159. S. P. Mallela, J. Myrczek, I. Bernal and R. A. Geanangel, *J. Chem. Soc., Dalton Trans.*, 2891 (1993).
160. M. Aggarwal, M. A. Ghuman and R. A. Geanangel, *Main Group Met. Chem.*, **14**, 263 (1990).
161. H. Preut and F. Huber, *Z. Anorg. Allg. Chem.*, **419**, 92 (1976).
162. M. Kleiner and M. Dräger, *Z. Naturforsch.*, **40b**, 477 (1985).
163. A. Sebald and R. K. Harris, *Organometallics*, **9**, 2096 (1990).
164. L. C. Willemsens, *Investigations in the Field of Organolead Chemistry*, Schotanus en Jens, Utrecht, 1965.
165. A. Blaschette, T. Hamann, A. Michalides and P. G. Jones, *J. Organomet. Chem.*, **456**, 49 (1993).
166. A. Blaschette, P. G. Jones, A. Michalides and K. Linoh, *Z. Anorg. Allg. Chem.*, **619**, 392 (1993).
167. M. G. Begley, C. Gaffney, P. G. Harrison and A. Steel, *J. Organomet. Chem.*, **289**, 281 (1985).
168. U. Schubert, S. Gilbert and M. Knorr, *J. Organomet. Chem.*, **454**, 79 (1993).
169. M. Knorr, J. Müller and U. Schubert, *Chem. Ber.*, **120** 879 (1987).
170. K. Yünlü, N. Höck and R. D. Fischer, *Angew. Chem.*, **97**, 863 (1985).
171. P. Brandt, A. K. Brimah and R. D. Fischer, *Angew. Chem.*, **100**, 1578 (1988).
172. S. Eller, P. Brandt, A. Brimah, A. K. Schwarz and R. D. Fischer, *Angew. Chem.*, **101**, 1274 (1989).
173. S. Eller, M. Adam and R. D. Fischer, *Angew. Chem.*, **102**, 1157 (1990).
174. D. C. Apperley, N. A. Davies, R. K. Harries, A. K. Brimah, S. Eller and R. D. Fischer, *Organometallics*, **9**, 2672 (1990).
175. A. Bonardi, C. Carini, C. Pelizzi, G. Pelizzi, G. Predieri, P. Tarsconi, M. A. Zoruddu and K. C. Molloy, *J. Organomet. Chem.*, **401**, 283 (1991).
176. S. Eller and R. D. Fischer, *Inorg. Chem.*, **29**, 1289 (1990).
177. U. Behrens, A. K. Brimah, T. M. Soliman, R. D. Fischer, D. C. Apperley, N. A. Davies and R. K. Harris, *Organometallics*, **11**, 1718 (1992).
178. K. S. Pitzer, *Acc. Chem. Res.*, **12** 271 (1979).
179. P. Pyykkö and J. P. Desclaux, *Acc. Chem Res.*, **12**, 276 (1979).
180. P. Pyykkö, *Chem. Rev.*, **88**, 563 (1988).
181. K. C. H. Lange and T. M. Klapötke, in *The Chemistry of Organic Arsenic, Antimony and Bismuth Compounds* (Ed. S Patai), Wiley, New York, 1994.
182. M. Kaupp and P. v. R. Schleyer, *Angew. Chem.*, **104**, 1240 (1992).
183. M. Kaupp and P. v. R. Schleyer, *J. Am. Chem. Soc.*, **115**, 1061 (1993).
184. M. Veith, *Chem. Rev.*, **90**, 3 (1990).
185. J. Barrau, J. Escudie and J. Satge, *Chem. Rev.*, **90**, 283 (1990).
186. J. T. Snow, S. Murakami, S. Masumune and D. J. Williams, *Tetrahedron Lett.*, **25**, 4191 (1984).
187. W. Kutzelnigg, *Angew. Chem.*, **96**, 262(1984).
188. P. v. R. Schleyer, *J. Comp.-Aided Mol. Design*, **1**, 223(1988).
189. W. J. Hehre, L. Random, P. v. R. Schleyer and J. A. Pople, *Ab Initio Molecular Orbital Theory*, Wiley, New York, 1986.
190. T. Clark, *A Handbook of Computational Chemistry*, Wiley, New York, 1985
191. K. P. Lawley, *Ab Initio Methods in Quantum Chemistry*, Volumes I and II, Wiley, New York, 1987.
192. J. J. P. Stewart, *J. Comp.-Aided Mol. Design*, **4** 1(1990).
193. Gaussian 88, M. J. Frisch, M. Head-Gordon, H. B. Schlegel, K. Raghavachari, J. S. Binkley, C. Gonzalez, D. J. DeFrees, D. J. Fox, R. A. Whiteside, R. Seeger, C. F. Melius, J. Baker, R. L. Martin, L. R. Kahn, J. J. P. Stewart, E. M. Fleuder, S. Topiol and J. A. Pople, Gaussian, Inc., Pittsburgh, PA, 1988.
194. Gaussian 90, M. J. Frisch, M. Head-Gordon, G. W. Trucks, J. B. Foresman, H. B. Schlegel, K. Raghavachari, M. A. Robb, J. S. Binkley, C. Gonzalez, D. J. Fox, R. A. Whiteside, R. Seeger, C. F. Melius, J. Baker, R. L. Martin, L. R. Kahn, J. J. P. Stewart, S. Topiol and J. A. Pople, Gaussian, Inc., Pittsburgh, PA, 1990.
195. Gaussian 92, Revision B. M. J. Frisch, G. W. Trucks, M. Head-Gordon, P. M. W. Gill, M. W. Wong, J. B. Foresman, B. G. Johnson, H. B. Schlegel, M. A. Robb, E. S. Replogle, R. Gomperts, K. Andres, K. Raghavachari, J. S. Binkley, C. Gonzalez, R. L. Martin, D. J. Fox, D. J. DeFrees, J. Baker, J. J. P. Stewart and J. A. Pople, Gaussian, Inc., Pittsburgh, PA, 1992.

196. R. D. Amos and J. E. Rice, *CADPAC: The Cambridge Analytic Derivatives Package*, issue 4.1, Cambridge, 1990.
197. H. Primas and U. Müller-Herold, in *Elementare Quantenchemie*, Teubner, Stuttgart, 1990.
198. P. A. Christiansen, W. C. Ermler and K. S. Pitzer, *Ann. Rev. Phys. Chem.*, **36**, 407 (1985).
199. K. Balasubramanian and S. K. Pitzer, in *Ab Initio Methods in Quantum Chemistry* (Ed. K. P. Lawley), Vol. I, Wiley, New York, 1987, p. 287.
200. P. Durand and J. -P. Malrieu, in *Ab Initio Methods in Quantum Chemistry* (Ed. K. P. Lawley), Vol. I, Wiley, New York, 1987, p. 321.
201. P. Fuentealba, O. Reyes, H. Stoll and H. Preuss, *J. Chem. Phys.*, **87**, 5338 (1987).
202. A. Z. Khan and J. Sandstrom, *J. Am. Chem. Soc.*, **110**, 4843 (1988).
203. K. N. Houk, N. G. Rondan and F. K. Brown, *Isr. J. Chem.*, **23**, 3 (1983).
204. P. B. Hitchcock, M. F. Lappert, S. J. Miles and A. J. Thorne, *J. Chem. Soc, Chem. Commun.*, 480 (1984).
205. D. E. Goldberg, P. B. Hitchcock, M. F. Lappert, K. M. Thorne, T. Fjeldberg, A. Haaland and B. E. R. Schilling, *J. Chem. Soc., Dalton Trans.*, 2387 (1986).
206. H. Teramae, *J. Am. Chem. Soc.*, **109**, 4140 (1987).
207. G. Trinquier, *J. Am. Chem. Soc.*, **112**, 2130 (1990).
208. L. Ross and M. Dräger, *J. Organomet. Chem.*, **199**, 195 (1980).
209. M. Dräger and L. Ross, *Z. Anorg. Allg. Chem.*, **476**, 95 (1981).
210. J. C. Bailar, H. J. Emeleus, R. S. Nyholm and A. F. Trotman-Dickenson (Eds.) *Comprehensive Inorganic Chemistry* Vols. 1 and 2, Pergamon Press, Oxford, 1973.
211. G. Trinquier, J. -P. Malrieu and P. Riviere, *J. Am Chem. Soc.*, **104**, 4529 (1982).
212. S. Nagase and T. Kudo, *J. Mol. Struct.(Theochem)*, **103**, 35 (1983).
213. D. E. Goldberg, P. B. Hitchcock, M. F. Lappert, K. M. Thomas, A. J. Thorne, T. Fjeldberg, A. Haaland and B. E. R. Schilling, *J. Chem. Soc., Dalton Trans.*, 2387 (**1986**).
214. C. Liang and L. C. Allen, *J. Am. Chem Soc.*, **112**, 1039 (1990).
215. R. S. Grev, H. F. Schaefer and K. M. Baines, *J. Am. Chem. Soc.*, **112**, 9458 (1990).
216. G. Trinquier, *J. Am. Chem. Soc.*, **112**, 2130 (1990).
217. A. Marquez, G. G. Gonzalez and J. F. Sanz, *Chem. Phys.*, **138**, 99 (1989).
218. A. Seligson and J. Arnold, *J. Am. Chem. Soc.*, **115**, 8214 (1993).
219. G. Trinquier and J. P. Malrieu, *J. Am. Chem. Soc.*, **109**, 5303 (1987).
220. P. R. Bunker, P. Jensen, W. P. Kraemer and R. Beardsworth, *J. Chem. Phys.*, **85**, 3724 (1986).
221. H. F. Schaefer, *Science (Washington DC)*, **231**, 1100 (1986).
222. M. S. Gordon, *Chem.Phys. Lett.*, **114**, 348 (1985).
223. K. Balasubramanian and A. D. McLean, *J. Chem. Phys.*, **85**, 5117 (1986).
224. A. Selmani and D. R. Salahub, *J. Chem. Phys.*, **89**, 1529 (1988).
225. R. A. Phillips, R. J. Buenker, R. Beardsworth, P. R. Bunker, P. Jensen and W. P. Kraemer, *Chem. Phys. Lett.*, **118**, 60 (1985).
226. K. Balasubramanian, *J. Chem. Phys.*, **89**, 5731 (1988).
227. B. T. Colegrove and H. F. Schaefer, *J. Phys. Chem.*, **94**, 5593 (1990).
228. M. Bogey, H. Bolvin and C. Demuynck, *Phys. Rev. Lett.*, **66**, 413 (1991).
229. R. S. Grev, B. J. DeLeeuw and H. F. Schaefer, *Chem.Phys.Lett.*, **165**, 257 (1990).
230. Z. Palagyi, H. F. Schaefer and E. Kapuy, *J. Am. Chem. Soc.*, **115**, 6903 (1993).
231. R. M. Pitzer *Acc. Chem. Res.*, **16**, 207 (1983).
232. G. F. Musso and V. J. Magnasco, *J. Chem. Soc., Faraday Trans. 2*, **78**, 1609 (1982).
233. R. F. W. Bader, J. R. Cheeseman, K. E. Laidig, K. B. Wiberg and C. Breneman, *J. Am. Chem. Soc.*, **112**, 6530 (1990).
234. P. v. R. Schleyer, M. Kaupp, F. Hampel, M. Bremer and K. Mislow, *J. Am. Chem. Soc.*, **114**, 6791 (1992).
235. J. Almlöf and K. Faegri, *Theor. Chem. Acta*, **69**, 438 (1986).
236. A. E. Reed, R. Weinstock and F. Weinhold, *J. Chem. Phys.*, **83**, 735 (1985).
237. A. E. Reed and F. Weinhold, *J. Chem. Phys.*, **83**, 1736 (1985).
238. A. E. Reed L. A. Curtiss and F. Weinhold, *Chem. Rev.*, **88**, 899 (1988).
239. S. Masamune, Y. Hanzawa and D. J. Williams, *J. Am. Chem. Soc.*, **104**, 6137 (1982).
240. S. Nagase and M. Nakano, *J. Chem. Soc., Chem. Commun.*, 1077 (1988).

CHAPTER 12

Substituent effects of germanium, tin and lead groups

MARVIN CHARTON

Chemistry Department, School of Liberal Arts and Sciences, Pratt Institute, Brooklyn, New York 11205, USA
Fax: 718-722-7706; e-mail: M.CHARTON@ACNET.PRATT.EDU

The chemistry of organic germanium, tin and lead compounds
Edited by S. Patai © 1995 John Wiley & Sons Ltd

ABBREVIATIONS

Ak	alkyl
COMFA	comparative molecular field analysis
CR	diparametric equation with σ_c and σ_e as parameters
DF	degrees of freedom
DSP	dual substituent parameter
EA	electron acceptor
ED	electron donor
IMF	intermolecular force
LD	diparamatric equation with σ_l and σ_D as parameters
MCD	minimal conformational dependence
MLD	modified LD equation
MSI	minimal steric interaction
MYT	modified Yukawa–Tsuno
NCD	no conformational dependence
N_{SD}	number of standard deviations
NS	not significantly different from zero
Pn	phenylene
QSAR	quantitative structure–activity relationships
QSCR	quantitative structure–chemical property relationship
QSPR	quantitative structure–physical property relationship
QSRR	quantitative structure–reactivity relationship
SB	simple branching equation
SCD	strong conformational dependence
SPR	structure property relationships
SPQR	structure property quantitative relationships
SURS	Swain–Unger–Rosenquist–Swain
XB	expanded branching equation
YT	Yukawa–Tsuno

I. THE NATURE OF STRUCTURAL EFFECTS

A. Introduction

In this work we present models for the quantitative description of the structural effects of substituents whose first or second atom is silicon, germanium, tin or lead. Silicon has been included in this work because its behavior is analogous to that of the remaining elements of the group and there is much more information available for silicon containing substituents than there is for all of the other elements. There are only two types of substituent we shall consider here. They are:

1. Tetracoordinate substituents of the type $MZ^1Z^2Z^3$. M may be Si, Ge, Sn or Pb.
2. Tetracoordinate substituents of the type $CH_2MZ^1Z^2Z^3$.

The structural theory of organic chemistry was developed in the second half of the nineteenth century. With its inception arose the concept that chemical, physical and biological properties of all kinds must vary with structural change. The first structure–property relationships (SPR) reported were qualitative. As quantitative measurements of these properties accumulated attempts were made to develop quantitative models of the structural dependence of properties. We now consider these methods for the quantitative description of structural effects.

B. Structure–Property Quantitative Relationships (SPQR)

Quantitative descriptions of the structural dependence of properties are called structure–property quantitative relationships (SPQR). These relationships are classified according to the type of property:

1. Quantitative structure–chemical reactivity relationships (QSRR). Chemical reactivities involve the formation and/or cleavage of chemical bonds. Equilibrium constants, rate constants and oxidation–reduction potentials are typical examples of quantitative measures of chemical reactivity.

2. Quantitative structure–chemical property relationships (QSCR). Chemical properties involve the difference in intermolecular forces between an initial and a final state. Examples of quantitative measures of chemical properties are equilibrium constants for hydrogen bonding; partition coefficients; chromatographic properties such as capacity factors in high performance liquid chromatography, retention times in gas chromatography and R_F values in thin layer chromatography; melting and boiling points; solvent effects on equilibrium or rate constants; and solubilities.

3. Quantitative structure–physical property relationships (QSPR). There are two types of physical properties we must consider: ground state properties and properties which depend on the difference in energy between the ground state and an excited state. Examples of the former are bond lengths, bond angles and dipole moments. The latter include infrared, ultraviolet, nuclear magnetic resonance and other types of spectra, ionization potentials and electron affinities.

4. Quantitative structure–activity relationships (QSAR). These involve any type of property associated with biological activities. The bioactive substrates data range from pure enzymes through single celled organisms to large multicellular organisms. The data may be obtained in vitro or in vivo. The quantitative measures of bioactivity vary from rate and equilibrium constants for enzyme reactivity through binding to receptor sites to toxicities in large multicellular organisms.

1. The nature of SPQR

There are three different types of chemical species (molecules, ions, radicals, carbenes etc.) for which SPQR can be determined:

1. Species with the structure XGY where X is a variable substituent, Y an active site (an atom or group of atoms at which a measurable phenomenon takes place) and G is a skeletal group to which X and Y are bonded. In a data set G and Y are held constant and only X varies.

2. Species with the structure XY in which the variable substituent X is directly attached to the constant active site Y.

3. Species in which no distinction between substituent and active site is possible, the entire species is the active site and it varies. These species are designated X_Y.

The purpose of SPQR is to provide a quantitative description of the change in some measurable quantity Q with a corresponding change in the structure of the substituent X when all other pertinent variables such as the conditions of the measurement are held constant. Thus:

$$\left(\frac{\partial Q}{\partial X} \right)_{G,Y,T,P,Sv,I,\ldots} = Q_X \tag{1}$$

where G is the skeletal group, Y the active site, T the temperature, P the pressure, Sv the solvent, I the ionic strength, all of which are constant throughout the data set.

We assume that Q_X will be a linear function of some number of parameters which represent the effects of the structural variation of X. Then:

$$Q_X = a_1 p_{1X} + a_2 p_{2X} + a_3 p_{3X} + \cdots + a_0$$

$$= \sum_{i=1}^{n} a_i p_{iX} + a_0 \tag{2}$$

where the p_i are the parameters which account for the structural effect of X on Q. These parameters can be obtained in various ways:

1. From quantum chemical calculations.

2. From molecular mechanics calculations for steric effects.

3. From a reference set by definition. This method assumes that structural effects on the data set to be studied are a linear function of those which occur in the reference set.

4. From comparative molecular field analysis (COMFA).

5. In the case of steric parameters, from molecular geometry.

6. From topological methods.

Once suitable parameters are available the values of Q can be correlated with them by means of either simple linear regression analysis if the model requires only a single variable, or multiple linear regression analysis if it requires two or more variables. We consider here only those parameters which are defined directly or indirectly from suitable reference sets or, in the case of steric parameters, from molecular geometries.

2. The uses of SPQR

SPQR have three major uses: mechanistic, predictive and archival. Thus, they can be used to provide mechanistic information about chemical and enzymatic reactions, and the activity-determining step in bioactivities. They are useful in the prediction of chemical reactivities and properties, of physical properties and of biological activities. This has resulted in their wide use in the design of medicinal drugs and pesticides. In addition to the maximization of activity and minimization of side effects, desirable pharmaceutical properties such as improved solubility, longer shelf life and controlled release can be developed. They are also a major method in environmental science where they can be used to predict toxicities, biodegradabilities and other properties of environmental interest. Finally, SPQR provide a concise, efficient and convenient method for storing the results of experimental studies on the effect of structural changes upon properties.

C. The Types of Structural Effects

Structural effects are conveniently divided into three categories:
1. Electrical effects. These effects cause a variation in the electron density at the active site. They account for the ability of a substituent to stabilize or destabilize a cation, anion or radical.
2. Steric effects. These effects result from the repulsion between valence electrons in orbitals on atoms which are in close proximity but not bonded to each other.
3. Inter- and intramolecular force effects. These effects result from the interactions between the substituent and its immediate surroundings such as the medium, a surface or a receptor site. They also involve the effect of the substituent on the interactions of the skeletal group G and the active site Y with their surroundings.

Electrical effects are the major factor in chemical reactivities and physical properties. Intermolecular forces are usually the major factor in bioactivities. Either electrical effects or intermolecular forces may be the predominant factor in chemical properties. Steric effects only occur when the substituent and the active site are in close proximity to each other and even then rarely account for more than twenty percent of the overall substituent effect.

II. ELECTRICAL EFFECTS

A. Introduction

The earliest successful parametrization of electrical effects is that of Hammett[1-3]. Burkhardt reported the existence of QSRR at about the same time as Hammett but did not develop a general relationship[4]. Hammett defined the σ_m and σ_p constants using the ionization constants of 3- and 4-substituted benzoic acids in water at 25 °C as the reference set and hydrogen as the reference substituent to which all others are compared. For hydrogen the values of the σ_m and σ_p constants were defined as zero. Thus:

$$\sigma_X \equiv \log \frac{K_X}{K_H} \qquad (3)$$

These parameters were intended to apply to XGY systems in which the skeletal group is phenylene. Hammett found it necessary to define an additional set of parameters, σ_p^-, in order to account for substituent effects in systems with an active site that has a lone pair on the atom adjacent to the skeletal group. The reference set in this case was the ionization constants of 4-substituted phenols in water at 25 °C. Brown and his coworkers[5,6] later defined another set of constants, σ_p^+, to account for substituent effects in benzene derivatives with electronically deficient active sites. In this case the reference set was the rate constants for the solvolysis of 4-substituted cumyl chlorides in 90% aqueous acetone at 25 °C. Finally, Wepster and coworkers[7] and Taft[8] both independently proposed constants intended to represent substituent effects in benzene derivatives with minimal delocalized effect. Using the Taft notation these constants are written as σ_p^0. The reference systems were of the type 4-XPnCH$_2$Y as it was argued that the methylene group intervening between the phenylene (Pn) group and the active site acted as an insulator preventing conjugation between X and Y. These parameters differ in electronic demand. They are used in the Hammett equation which may be written in the form:

$$Q_X = \rho\sigma_X + h \qquad (4)$$

where Q_X is the value of the quantity of interest when the substituent is X, and σ_X is either σ_{mX}, σ_{pX}, σ_{pX}^0, σ_{pX}^+, or σ_{pX}^-; ρ and h are the slope and intercept of the line.

In using the Hammett equation it is necessary to make an *a priori* choice of parameters based on the location of the substituent and a knowledge of the electronic demand in the data set which is to be modelled. If such knowledge is unavailable it is necessary to correlate the data set with each different parameter. The parameter which gives the best fit is then assumed to be the proper choice and the electronic demand associated with it is that of the data set.

Taft and his coworkers[9-11] developed a diparametric model which separated the electrical effect into contributions from the 'inductive' and resonance effects. This separation depends on the difference in the extent of electron delocalization when a substituent is bonded to an sp^3-hybridized carbon atom in one reference system and to an sp^2-hybridized carbon atom in another. As the first case represents minimal delocalization and the second extensive delocalization, we have referred to the two effects as the localized and delocalized electrical effects. The diparametric model of electrical effects can be written in the form:

$$Q_X = L\sigma_{lX} + D\sigma_{DX} + h \qquad (5)$$

where σ_l and σ_D are the localized and delocalized electrical effect parameters, respectively. Taft and coworkers[11] suggested that four σ_D constants are required. They are σ_{RX}, $\sigma_{RX}{}^0$, $\sigma_{RX}{}^+$ and $\sigma_{RX}{}^-$, and they correspond to the σ_p constants described above. Charton noted that in cases of very large electron demand two additional σ_D constants were required, $\sigma_R{}^\oplus$ for highly electron-deficient (positive) active sites[12] and $\sigma_R{}^\ominus$ for active sites that are very electron-rich (negative)[13].

An alternative diparametric model was proposed by Yukawa and Tsuno[14] for use with electron-poor active sites. The equation was originally written as:

$$Q_X = \rho\sigma_X + \rho r(\sigma_X{}^+ - \sigma_X) \qquad (6)$$

A later version has the form[15]:

$$Q_X = \rho\sigma_X + \rho r(\sigma_X{}^+ - \sigma_X{}^0) \qquad (7)$$

A similar relationship:

$$Q_X = \rho\sigma_X + \rho r(\sigma_X{}^- - \sigma_X) \qquad (8)$$

has been proposed for electron-rich active sites[16]. We will refer to these relationships as the YT equations. They have the advantage that both *meta*- and *para*-substituted compounds may be included in the same data set on the assumption that ρ_m is equal to ρ_p. This assumption is usually a reasonable approximation but in some cases the difference between ρ_m and ρ_p ($\Delta\rho$) is significant. If the molecular geometry of the system of interest does not differ much from that of the benzoic acids, then $\Delta\rho$ is likely to be negligible.

Clearly there was a need for a more general model of electrical effects. Like the case of the Hammett equation the use of the LD equation for the description of chemical reactivities required either an *a priori* knowledge of the type of σ_D substituent constant required or a comparison of the results obtained using each of the available σ_D constants. The use of the YT equation has generally been restricted to electronically deficient active sites.

A triparametric model of the electrical effect has been introduced[17] that can account for the complete range of electrical effects on chemical reactivities of closed shell species (carbenium and carbanions), that is, reactions which do not involve radical intermediates. The basis of this model was the observation that the σ_D constants differ in their electronic demand. On the assumption that they are generally separated by an order of magnitude in this variable it is possible to assign to each σ_D type a corresponding value of the electronic demand, η. We find that the equation:

$$\sigma_{DX} = a_1\eta + a_0 = \sigma_e\eta + \sigma_d \tag{9}$$

is obeyed. The intercept of this linear relationship represents the intrinsic delocalized (resonance) effect, σ_{dX}; the slope represents the sensitivity of the X group to the electronic demand of the active site. On substituting equation 9 into the LD equation we obtain the triparametric LDR equation:

$$Q_X = L\sigma_{lX} + D\sigma_{dX} + R\sigma_{eX} + h \tag{10}$$

The σ_l values are identical to σ_I. The symbol was changed in order to be consistent with the other symbols used in the equation.

When the composition of the electrical effect, P_D, is held constant the LDR equation simplifies to the CR equation:

$$Q_X = C\sigma_{ldX} + R\sigma_{eX} + h \tag{11}$$

where σ_{ld} is a composite parameter. It is defined by the relationship:

$$\sigma_{ldX} = l\sigma_{lX} + d\sigma_{dX} \tag{12}$$

The difference between pure and composite parameters is that the former represent a single effect while the latter represent a mixture of two or more. The percent composition of these parameters is given by:

$$P_D = \frac{100d}{l+d} \tag{13}$$

If the constant value of P_D is written as k', then the σ_{ldX} parameter for a given value of k' is:

$$\sigma_{ldXk'} = \sigma_{lX} + \left[\frac{k'}{100} - k'\right]\sigma_{DX} \tag{14}$$

Writing:

$$k^* = \frac{k'}{100 - k'} \tag{15}$$

gives:

$$\sigma_{ldXk'} = \sigma_{lX} + k^*\sigma_{dX} \tag{16}$$

It has been shown that Yukawa–Tsuno equation for 4-substituted benzene derivatives is approximately equivalent to the CR equation[18,19]. This has led to the development of a modified Yukawa–Tsuno (MYT) equation which has the form:

$$Q_X = \rho\sigma_X + R\sigma_{eX} + h \tag{17}$$

with σ taking the value σ_m for 3-substituted benzene derivatives and σ_{50} for 4-substituted benzene derivatives, while σ_{eX} for substituents in the *meta* position is 0. The σ_{50} constants have k' equal to 50 and η equal to zero; they are therefore equal to the sum of the σ_l and σ_d values.

When the sensitivity to electronic demand is held constant, the LDR equation reverts to the LD equation (equation 5). The combination of 3- and 4-substituted benzene derivatives into a single data set can be done by using a modified form of the LD equation (the MLD equation):

$$Q_X = \rho'\sigma_X + D\sigma_{DX} + h \tag{18}$$

where σ is σ_m for 3-substituted and σ_l for 4-substituted while σ_D is 0 for 3-substituents. Again, the use of the MLD equation is restricted to systems for which $\Delta\rho$ is not significant.

When both the electronic demand and the composition of the electrical effect are held constant, a set of composite parameters having the form:

$$\sigma_{k'/kX} = l\sigma_{lX} + d\sigma_{dX} + r\sigma_{eX} \tag{19}$$

results in:

$$k' = P_D = \frac{100d}{(l+d)}; \quad k = \eta = \frac{r}{d} \tag{19a}$$

The Hammett substituent constants are special cases of these parameters.

The $\sigma_{k'/k}$ values describe the overall electrical effect of the X group. They are obtained from the expression:

$$\sigma_{k'/kX} = \sigma_{lX} + \left[\frac{P_D}{100 - P_D}\right](\sigma_{dX} + \eta\sigma_{eX}) \tag{20}$$

$$= \sigma_{lX} + k^*(\sigma_{dX} + k\sigma_{eX}) \tag{20a}$$

A plot of the $\sigma_{k'/kX}$ values for a group with P_D on the x axis, η on the y axis and $\sigma_{k'/k}$ on the z axis produces a surface that characterizes the electrical effect of the X group.

B. Estimation of Electrical Effect Parameters

It is frequently necessary to estimate values of electrical effect parameters for groups for which no measured values are available. This is generally important for substituents whose central atom is a main block element of the higher periods. Electrical effect parameters of substituents X whose structure can be written as MZ_n^i are known to be a function of the electrical effect of the Z^i when M is held constant[20-22]. It was later shown that when Z^i is held constant and M is allowed to vary, the substituent constant is a function of the Allred–Rochow electronegativity[23] of M, χ_M and the number of Z groups, n_Z[24-26]. When both Z and M vary, σ_l values for all groups of the type $X = MZ^1Z^2Z^3$, where Z may be either a group of atoms or a lone pair, are given by an equation of the form:

$$\sigma_{lX} = a\chi_m + L\Sigma\sigma_{lZ} + D\Sigma_{dZ} + R\Sigma\sigma_{eZ} + h \tag{21}$$

σ_{dX} and σ_{eX} for $MZ^1Z^2Z^3$ groups can be calculated from an equation of this type.

1. MZ¹Z²Z³ groups

Electrical effect substituent constants for group 14 elements other than carbon have been reviewed by Egorochkin and Razuvaev[27]. They report the values that are available in the literature for the σ_l, $\sigma_R°$, σ_m, σ_p and σ_p^+ constants. When we compare the results obtained for groups for which three or more values have been determined, we find that they are frequently in very poor agreement with each other. As the values of these substituent constants have been determined by a number of different methods and in a range of media, they are not directly comparable in many cases. A further factor is the existence of a steric component in many of the measured values. In order to establish this point, we have correlated the σ_l constants for $SiAk_3$, $Si(OAk)_3$, $Si(OAk)MePh$ and $Ge(OAk)_3$ groups (Ak = alkyl and $\sigma_R°$ for $SiAk_3$ obtained from C^{13} NMR spectroscopy with the equation:

$$\sigma_l = S\upsilon_{Ak} + h \tag{22}$$

σ_I values for Si(OAk)MePh groups were correlated with the simple branching equation in the form:

$$\sigma_{IX} = a_1 n_1 + a_0 \tag{23}$$

These methods of modelling steric effects will be discussed in detail in Section III of this work. With the exception of the correlation of σ_I for the Si(OAk)MePh groups with equation 22, all of the results were significant. The data sets were small but the overall trend seems clear; there is generally a dependence on steric effects for these substituent constants. We have therefore generally excluded the values determined by this method from the tables of substituent constants given in this work. Exceptions have been made in a few cases for groups for which values have not been determined by any other methods.

These results explain the deviation of these values from the general observation that substituent constants for groups of the type $M(ZAk)_n$ are usually constant within experimental error.

Values of σ_I for all substituents of the Group 14 elements, and for any other substituents of the type $MZ^1Z^2Z^3$ where Z can be a $p\pi$ or $d\pi$ bonded O atom, any other group of atoms or a lone pair can be estimated from the equation[19]:

$$\sigma_{IX} = 0.341 \ (\pm 0.0101)\Sigma\sigma_{IZ} + 0.128 \ (\pm 0.0194)\Sigma\sigma_{dZ} + 0.314 \ (\pm 0.0813)\Sigma\sigma_{eZ}$$

$$+ \ 0.0329 \ (\pm 0.0128)\chi + 0.348 \ (\pm 0.0182)n_{Od1} + 0.205 \ (\pm 0.0239)n_{Od2}$$

$$+ \ 0.296 \ (\pm 0.0143)n_{Op} + 0.149 \ (\pm 0.00707)n_{lp} - 0.0559 \ (\pm 0.0366) \tag{24}$$

where M is the atom of the group which is bonded to either the skeletal group or the active site, χ_M is the Allred–Rochow electronegativity[23], n_H is the number of hydrogen atoms attached to M, n_{Od1} and n_{Od2} the first and second oxygen atoms involved in $pd\pi$ bonding, n_{Op} the number of oxygen atoms involved in $pp\pi$ bonding and n_{lp} the number of lone pairs on M. On omitting the terms which are not involved in the calculation of parameter values for the Group 14 substituents, we have:

$$\sigma_{IX} = 0.341 \ (\pm 0.0101)\Sigma\sigma_{IZ} + 0.128 \ (\pm 0.0194)\Sigma\sigma_{dZ} + 0.314 \ (\pm 0.0813)\Sigma\sigma_{eZ}$$

$$+ \ 0.032 \ (\pm 0.0128)\chi - 0.0559 \ (\pm 0.0366) \tag{25}$$

Values calculated from equation 25 are reported in Table 1.

In order to develop estimation equations for the σ_d and σ_e of the Group 14 elements other than carbon, it is necessary to have a sufficient number of such values from other sources. To obtain these values we have made use of our work on modifying composite electrical effect substituent constants[28]. We must first rescale them by adding a large enough constant to make all of the values positive. It is sometimes useful to divide $\sigma^\bullet$ by a second constant to obtain values of a convenient size. They can then be modified by raising to an appropriate power. Thus:

$$\sigma^\bullet = \frac{(\sigma + c)^m}{c'} \tag{26}$$

where σ is the composite substituent constant, c and c' are constants, m the exponent and $\sigma^\bullet$ is the modified composite substituent constant. We have also shown that the η values obtained by modification of the σ_p constants with c equal to 2 are a linear function of the exponent m:

$$\eta = -0.439 \ (\pm 0.0347)m + 1.37 \ (\pm 0.0902) \tag{27}$$

TABLE 1. Values of σ_I, σ_d, and σ_e^a

X	σ_I	Ref.	σ_d	Ref.	σ_e	Ref.
Si						
SiCF$_3$	0.44	25	0.47	32	−0.023	33
Si(CN)$_3$	0.58	25	0.61	32	−0.020	33
SiMe$_2$H	−0.02f	27				
	−0.06	25	0.12	29	−0.044	30
SiMe$_3$	−0.11	21	0.13	21	−0.046	21
SiEt$_3$	−0.09	25	0.12	29	−0.051	30
	−0.11	21	0.13	21	−0.046	21
Si(NMe$_2$)$_3$	−0.04f	27				
	−0.30	25	0.33	29	−0.086	30
SiMe$_2$OMe	−0.05	25	0.19	29	−0.043	30
SiMe(OMe)$_2$	−0.01	25	0.17	29	−0.039	30
SI(OH)$_3$	0.10	25	0.12	32	−0.029	33
Si(OCF$_3$)$_3$	0.42	25	0.25	32	−0.021	33
Si(OMe)$_3$	0f	27				
	0.04	25	0.10	29	−0.031	30
Si(OEt)$_3$	−0.04f	27				
	0.01	25	0.08	29	−0.033	30
Si(OAc)$_3$	0.07c	27				
	0.29	25	0.24	32	−0.020	33
Si(O$_2$CCF$_3$)$_3$	0.76c	27				
Si(OPh)$_3$	0.14	25	0.23	32	−0.034	33
Si(OBz)$_3$	0.31	25	0.28	32	−0.021	33
Si(SCN)$_3$	0.48	25	0.44	32	−0.018	33
Si(SCF$_3$)$_3$	0.37	25	0.38	32	−0.023	33
Si(SMe)$_3$	0.15	25	0.13	32	−0.026	33
Si(SEt)$_3$	0.08c	27				
	0	25	0.27	32	−0.048	33
Si(SPh)$_3$	0.03	25	0.40	32	−0.055	33
Si(SeMe)$_3$	0	25	0.31	32	−0.051	33
Si(C$_2$H)$_3$	0.20	25	0.44	32	−0.043	33
SiVi$_3$	0.08c	27				
	−0.03	25	0.34	32	−0.055	33
SiMe$_2$Ph	0.02f	27				
	−0.07	25	0.17	29	−0.045	30
SiMePh$_2$	0.07f	27				
	−0.05	25	0.21	29	−0.039	30
Si(C$_6$F$_5$)$_3$	0.29	25	0.46	32	−0.036	33
SiPh$_3$	0.13f	27				
	−0.04	25	0.34	29	−0.062	30
			0.33	32	−0.055	33
SiMe$_2$OSiMe$_3$	−0.11	21				
SiMe(OSiMe$_3$)$_2$						
Si(OSiMe$_3$)$_3$						
SiMe$_2$Br	0.06	25	0.05	29	−0.026	30
SiMeBr$_2$	0.21	25	0.09	29	−0.035	30
SiBr$_3$	0.39f	27				
	0.37	25	**0.07**	29	−0.050	30
			0.30	32	**−0.018**	33
SiMeBrCl	0.25f	27				
SiMe$_2$Cl	0.11f	27				
	0.06	25	0.10	29	−0.030	30
SiMeCl$_2$	0.24f	27				
	0.21	25	0.23	29	−0.072	30

TABLE 1. (*continued*)

X	σ_I	Ref.	σ_d	Ref.	σ_e	Ref.
SiCl$_3$	0.39f	27				
	0.36	25	**0.09**	29	−0.059	30
			0.28	32	**−0.017**	33
SiMe$_2$F	0.08f	27				
	0.08	25	0.09	29	−0.029	30
SiMeF$_2$	0.24	25	−0.02	29	−0.015	30
SiF$_3$	0.42f	27				
	0.41	25	**0.15**	29	−0.090	30
			0.14	32	**−0.004**	33
SiI$_3$	0.28	25	0.35	32	−0.029	33
SiH$_3$	0.01f	27				
	0	25	0.11	29	−0.035	30
			0.13	32	−0.038	33
Si(SiMe$_3$)$_3$	−0.10	25	0.21	32	−0.052	33
Ge						
Ge(CF$_3$)$_3$	0.45	25	0.38	32	−0.023	33
Ge(CN)$_3$	0.59	25	0.52	32	−0.020	33
GeMe$_3$	−0.08	25	0.11	29	−0.050	30
GeEt$_3$	−0.08	25	0.11	29	−0.050	30
GeVi$_3$	−0.02	25	0.25	32	−0.055	33
Ge(C$_2$H)$_3$	0.21	25	0.35	32	−0.043	33
GePh$_2$Br	0.11	25	0.28	29	−0.065	30
GePh$_2$H	−0.01	25	0.19	29	−0.041	30
Ge(C$_6$F$_5$)$_3$	0.29	25	0.37	32	−0.036	33
GePh$_3$	−0.03	25	0.24	29	−0.055	30
Ge(SiMe$_3$)$_3$	−0.10	25	0.12	32	−0.052	33
Ge(NMe$_2$)$_3$	−0.30	25	0.18	32	−0.075	33
Ge(OH)$_3$	0.11	25	0.03	32	−0.029	33
Ge(OCF$_3$)$_3$	0.43	25	0.16	32	−0.010	33
Ge(OMe)$_3$	0.05	25	0.04	32	−0.035	33
Ge(OAc)$_3$	0.30	25	0.15	32	−0.020	33
Ge(OPh)$_3$	0.15	25	0.14	32	−0.034	33
Ge(OBz)$_3$	0.32	25	0.19	32	−0.021	33
Ge(SCF$_3$)$_3$	0.38	25	0.29	32	−0.023	33
Ge(SCN)$_3$	0.49	25	0.35	32	−0.018	33
Ge(SMe)$_3$	0.16	25	0.04	32	−0.026	33
Ge(SEt)$_3$	0.01	25	0.18	32	−0.048	33
Ge(SPh)$_3$	0.04	25	0.31	32	−0.055	33
Ge(SeMe)$_3$	0.01	25	0.22	32	−0.051	33
GeBr$_3$	0.59	27				
	0.37	25	0.55	29	−0.240	30
			0.21	32	**−0.018**	33
GeCl$_3$	0.63f	27				
	0.37	25	0.33	29	−0.15	30
			0.19	32	**−0.017**	33
GeF$_3$	0.74f	27				
	0.42	25	0.95	29	−0.046	30
			0.05	32	**−0.004**	33
GeI$_3$	0.29	25	0.26	32	−0.029	
GeH$_3$	0.01	25	0.02	29	−0.036	30
			0.04	32	−0.038	33
Sn						
Sn(CF$_3$)$_3$	0.44	25	0.48	32	−0.023	33
Sn(CN)$_3$	0.58	25	0.61	32	−0.020	33

(*continued overleaf*)

TABLE 1. (continued)

X	σ_I	Ref.	σ_d	Ref.	σ_e	Ref.
Sn						
SnMe$_3$	*0*	27				
	−0.09	25	0.12	29	−0.051	30
SnEt$_3$	0.02f	27				
	−0.09	25	0.12	29	−0.051	30
SnBu$_3$	−0.10	25	0.15	29	−0.051	30
SnPh$_2$Cl	0.09	25	0.35	32	−0.084	33
SnVi$_3$	−0.03	25	0.35	32	−0.055	33
Sn(C$_2$H)$_3$	0.20	25	0.46	32	−0.043	33
SnPh$_2$H	−0.02	25	0.25	32	−0.047	33
Sn(C$_6$F$_5$)$_3$	0.28	25	0.46	32	−0.036	33
SnPh$_3$	0.20f	27				
	−0.04	25	0.29	29	−0.054	30
			0.33	32	−0.055	33
Sn(Pn-F-4)$_3$	0.28f	27				
Sn(Pn-F-3)$_3$	0.33f	27				
Sn(SiMe$_3$)$_3$	−0.11	25	0.21	32	−0.052	33
Sn(NMe$_2$)$_3$	−0.30	25	0.27	32 ·	−0.075	33
Sn(OH)$_3$	0.10	25	0.12	32	−0.029	33
Sn(OCF$_3$)$_3$	0.42	25	0.26	32	−0.010	33
Sn(OMe)$_3$	0.04	25	0.13	32	−0.035	33
Sn(OAc)$_3$	0.29	25	0.25	32	−0.020	33
Sn(OPh)$_3$	0.14	25	0.24	32	−0.034	33
Sn(OBz)$_3$	0.31	25	0.28	32	−0.021	33
Sn(SCF$_3$)$_3$	0.37	25	0.38	32	−0.023	33
Sn(SCN)$_3$	0.48	25	0.45	32	−0.018	33
Sn(SMe)$_3$	0.15	25	0.14	32	−0.026	33
Sn(SEt)$_3$	0	25	0.28	32	−0.048	33
Sn(SPh)$_3$	0.03	25	0.41	32	−0.055	33
Sn(SeMe)$_3$	0	25	0.31	32	−0.051	33
SnBr$_3$	0.36	25	0.30	32	−0.018	33
SnCl$_3$	0.80f	27				
	0.36	25	**0.10**	29	−0.063	30
			0.29	32	**−0.017**	33
SnF$_3$	0.41	25	0.15	32	−0.004	33
SnI$_3$	0.28	25	0.36	32	−0.029	33
SnH$_3$	0	25	0.14	32	−0.038	33
Pb						
Pb(CF$_3$)	0.43	25	0.53	32	−0.023	33
Pb(CN)$_3$	0.57	25	0.67	32	−0.020	33
PbMe$_3$	−0.10	25	0.16	32	−0.044	33
PbEt$_3$	−0.10	25	0.18	32	−0.045	33
Pb(Vi)$_3$	−0.04	25	0.40	32	−0.055	33
Pb(C$_2$H)$_3$	0.19	25	0.50	32	−0.043	33
Pb(C$_6$F$_5$)$_3$	0.28	25	0.51	32	−0.036	33
PbPh$_3$	−0.04	25	0.28	29	−0.051	30
			0.39	32	−0.055	33
Pb(SiMe$_3$)$_3$	−0.11	25	0.27	32	−0.052	33
Pb(NMe$_2$)$_3$	−0.31	25	0.33	32	−0.075	33
Pb(OH)$_3$	0.09	25	0.18	32	−0.029	33
Pb(OCF$_3$)$_3$	0.41	25	0.32	32	−0.010	33
Pb(OMe)$_3$	0.03	25	0.19	32	−0.035	33
Pb(OAc)$_3$	0.29	25	0.30	32	−0.020	33
Pb(OPh)$_3$	0.13	25	0.29	32	−0.034	33
Pb(OBz)$_{3-}$	0.30	25	0.34	32	−0.021	33

TABLE 1. (*continued*)

X	σ_I	Ref.	σ_d	Ref.	σ_e	Ref.
Pb(SCF$_3$)$_3$	0.36	*25*	0.44	*32*	−0.023	*33*
Pb(SCN)$_3$	0.47	*25*	0.50	*32*	−0.018	*33*
Pb(SMe)$_3$	0.14	*25*	0.20	*32*	−0.026	*33*
Pb(SEt)$_3$	0.00	*25*	0.33	*32*	−0.048	*33*
Pb(SPh)$_3$	0.02	*25*	0.46	*32*	−0.055	*33*
Pb(SeMe)$_3$	0.00	*25*	0.37	*32*	−0.051	*33*
PbBr$_3$	0.36	*25*	0.36	*32*	−0.018	*33*
PbCl$_3$	0.36	*25*	0.34	*32*	−0.017	*33*
PbF$_3$	0.40	*25*	0.20	*32*	−0.004	*33*
PbI$_3$	0.27	*25*	0.41	*32*	−0.029	*33*
PbH$_3$	0.00	*25*	0.19	*32*	−0.038	*33*
C						
CH$_3$	0.00	*45*	−0.15	*46*	−0.029	*47*
CPh$_3$	0.09	*45*	−0.09	*46*	−0.028	*47*
C(NMe$_2$)$_3$	0.12	*45*	0.06	*46*	−0.056	*47*
C(OMe)$_3$	0.22	*45*	0.01	*46*	−0.041	*47*
C(OPh)$_3$	0.29	*45*	0.06	*46*	−0.032	*47*
CF$_3$	0.40	*45*	0.14	*46*	−0.020	*47*
CCl$_3$	0.35	*45*	0.10	*46*	−0.013	*47*
CBr$_3$	0.35	*45*	0.01	*46*	−0.013	*47*
CI$_3$	0.29	*45*	0.06	*46*	−0.013	*47*
C(OH)$_3$	0.26	*45*	0.03	*46*	−0.039	*47*
C(OAc)$_3$	0.28	*45*	0.05	*46*	−0.017	*47*
C(CF$_3$)$_3$	0.29	*45*	0.06	*46*	0.061	*47*
C(SMe)$_3$	0.22	*45*	0.08	*46*	−0.031	*47*
C(SEt)$_3$	0.19	*45*	−0.01	*46*	−0.034	*47*
CVi$_3$	0.08	*45*	−0.09	*46*	−0.026	*47*
C(C$_2$H)$_3$	0.21	*45*	0.00	*46*	−0.010	*47*
C(C$_6$F$_5$)$_3$	0.23	*45*	0.01	*46*	−0.003	*47*
C(SeMe)$_3$	0.20	*45*	−0.00	*46*	−0.033	*47*
C(SPh)$_3$	0.23	*45*	0.01	*46*	−0.028	*47*
C(SiMe$_3$)$_3$	−0.09	*45*	−0.21	*46*	−0.028	*47*
C(CN)$_3$	0.42	*45*	0.15	*46*	0.017	*47*
C(OCF$_3$)$_3$	0.38	*45*	0.12	*46*	−0.012	*47*
C(OBz)$_3$	0.32	*45*	0.08	*46*	−0.161	*47*
C(SCN)$_3$	0.41	*45*	0.14	*46*	0.000	*47*
C(SCF$_3$)$_3$	0.33	*45*	0.09	*46*	−0.007	*47*
CZ$_2$H						
CHAc$_2$	0.14	*45*	−0.04	*46*	−0.005	*47*
CHPh$_2$	0.06	*45*	−0.11	*46*	−0.028	*47*
CH(NMe$_2$)$_2$	0.08	*45*	−0.09	*46*	−0.047	*47*
CH(OMe)$_2$	0.14	*45*	−0.04	*46*	−0.037	*47*
CH(OPh)$_2$	0.19	*45*	−0.01	*46*	−0.031	*47*
CHF$_2$	0.26	*45*	0.04	*46*	−0.023	*47*
CHCl$_2$	0.23	*45*	0.02	*46*	−0.018	*47*
CHBr$_2$	0.23	*45*	0.02	*46*	−0.018	*47*
CHI$_2$	0.19	*45*	−0.01	*46*	−0.018	*47*
CH(OH)$_2$	0.17	*45*	−0.03	*46*	−0.035	*47*
CH(OAc)$_2$	0.18	*45*	−0.02	*46*	−0.021	*47*
CH(CF$_3$)$_2$	0.19	*45*	−0.01	*46*	−0.006	*47*
CH(SMe)$_2$	0.14	*45*	−0.04	*46*	−0.030	*47*
CH(SEt)$_2$	0.12	*45*	−0.06	*46*	−0.032	*47*
CHVi$_2$	0.05	*45*	−0.11	*46*	−0.027	*47*

(*continued overleaf*)

TABLE 1. (*continued*)

X	σ_I	Ref.	σ_d	Ref.	σ_e	Ref.
CZ₂H						
CH(C₂H)₂	0.14	*45*	−0.05	*46*	−0.16	*47*
CH(C₆F₅)₂	0.15	*45*	−0.04	*46*	−0.012	*47*
CH(SeMe)₂	0.14	*45*	−0.05	*46*	−0.032	*47*
CH(SPh)₂	0.15	*45*	−0.04	*46*	−0.028	*47*
CH(SiMe₃)₂	−0.06	*45*	−0.19	*46*	−0.029	*47*
CH(CN)₂	0.28	*45*	0.05	*46*	0.002	*47*
CH(OCF₃)₂	0.25	*45*	0.03	*46*	−0.018	*47*
CH(OBz)₂	0.21	*45*	0.00	*46*	−0.020	*47*
CH(SCN)₂	0.27	*45*	0.05	*46*	−0.009	*47*
CH(SCF₃)₂	0.22	*45*	0.01	*46*	−0.014	*47*
CH₂Z						
CH₂SiH₃	0.00	*45*	−0.15	*46*	−0.026	*47*
CH₂SiPh₃	−0.01	*45*	−0.16	*46*	−0.023	*47*
CH₂Si(OEt)₃	0.00	*45*	−0.15	*46*	−0.027	*47*
CH₂Si(OMe)₃	0.01	*45*	−0.14	*46*	−0.026	*47*
CH₂SiMe₃	−0.03	*45*	−0.17	*46*	−0.029	*47*
CH₂SiF₃	0.10	*45*	−0.08	*46*	−0.016	*47*
CH₂SiCl₃	0.08	*45*	−0.09	*46*	−0.019	*47*
CH₂SiEt₃	−0.03	*45*	−0.17	*46*	−0.029	*47*
CH₂CH₂SiMe₃	−0.01	*45*	−0.15	*46*	−0.034	*47*
CH₂CH₂SiEt₃	−0.01	*45*	−0.15	*46*	−0.034	*47*
CH₂Ge(OMe)₃	0.01	*45*	−0.14	*46*	−0.027	*47*
CH₂GeEt₃	−0.02	*45*	−0.16	*46*	−0.028	*47*
CH₂GeMe₃	−0.02	*45*	−0.06	*46*	−0.028	*47*
CH₂GeH₃	0.00	*45*	−0.15	*46*	−0.028	*47*
CH₂GePh₃	−0.01	*45*	−0.15	*46*	−0.024	*47*
CH₂GeF₃	0.10	*45*	−0.08	*46*	−0.018	*47*
CH₂GeCl₃	0.09	*45*	−0.08	*46*	−0.017	*47*
CH₂SnMe₃	−0.03	*45*	−0.16	*46*	−0.028	*47*
CH₂SnPh₃	−0.01	*45*	−0.16	*46*	−0.023	*47*
CH₂Sn(OMe)₃	0.01	*45*	−0.14	*46*	−0.025	*47*
CH₂SnCl₃	0.08	*45*	−0.09	*46*	−0.015	*47*
CH₂SnF₃	0.10	*45*	−0.08	*46*	−0.165	*47*
CH₂SnEt₃	−0.03	*45*	−0.16	*46*	−0.028	*47*
CH₂SnH₃	0.00	*45*	−0.15	*46*	−0.026	*47*
CH₂PbF₃	0.10	*45*	−0.08	*46*	−0.016	*47*
CH₂PbCl₃	0.08	*45*	−0.09	*46*	−0.139	*47*
CH₂PbPh₃	−0.01	*45*	−0.16	*46*	−0.024	*47*
CH₂Pb(OMe)₃	0.00	*45*	−0.14	*46*	−0.024	*47*
CH₂PbH₃	0.00	*45*	−0.15	*46*	−0.025	*47*

[a] Values in boldface are preferred. f indicates a value determined by F^{19} NMR spectroscopy, c indicates a value determined by C^{13} NMR spectroscopy. Numbers in italics in the columns headed Ref. refer to equations in the text used to estimate the values reported; numbers in ordinary typeface refer to references to the source from which the values were taken.

From this equation we may calculate the value of m for which η is zero to be 3.12. Values of $\sigma^{\bullet}$ for this exponent were calculated from equation 25 (with c' equal to 10) for 52 substituents. They were then correlated with the LDR equation in the form:

$$\sigma_X = l\sigma_{lX} + d\sigma_{dX} + r\sigma_{eX} + h \qquad (28)$$

to give:

$$\sigma^{\bullet} = 1.71 \ (\pm 0.0695)\sigma_{lX} + 1.50 \ (\pm 0.454)\sigma_{dX} + 0.847 \ (\pm 0.0293) \qquad (29)$$

Using σ_p values taken from Egorochkin and Razuvaev to calculate the appropriate $\sigma^{\blacklozenge}$ values and σ_I values calculated from equation 25, we have estimated σ_d values for Si, Ge, Sn and Pb groups. We then used the equation:

$$\sigma^{\blacklozenge} = 0.161\ (\pm 0.0121)\sigma_{IX} + 0.170\ (\pm 0.00822)\sigma_{dX}$$
$$+\ 0.238\ (\pm 0.0394)\sigma_{eX} + 0.701\ (\pm 0.00516) \tag{30}$$

obtained for $\sigma^{\blacklozenge}$ values calculated from σ_p values with $m = -0.5$ together with the values of σ_I and σ_d obtained above to estimate σ_e values for these groups. These σ_d and σ_e constants were then correlated with equation 28 in the form:

$$\sigma_X = l\,\Sigma\sigma_{IX} + d\,\Sigma\sigma_{dX} + r\,\Sigma\sigma_{eX} + h \tag{31}$$

to give the estimation equation equations:

$$\sigma_{dX} = 0.192\ (\pm 0.0226)\Sigma\sigma_{IZ} + 0.168\ (\pm 0.0247)\Sigma\sigma_{dZ}$$
$$-\ 0.512\ (\pm 0.0813)\Sigma\sigma_{eZ} - 0.320\ (\pm 0.0245)\chi_M + 0.690\ (\pm 0.0547) \tag{32}$$

$100R^2$, 92.79; $A100R^2$, 91.80; F, 67.55; S_{est}, 0.0355; S^0, 0.299; n, 26;

and:

$$\sigma_{eX} = 0.0162\ (\pm 0.00232)\Sigma\sigma_{IZ} + 0.0636\ (\pm 0.0144)\Sigma\sigma_{eZ} - 0.0376\ (\pm 0.00198) \tag{33}$$

$100R^2$, 77.27; $A100R^2$, 76.23; F, 35.69; S_{est}, 0.00571; S^0, 0.510; n, 24

respectively. The poor correlation observed for equation 32 is due to the small degree of variation in the σ_e values as a function of structure.

Values of σ_I, σ_d and σ_e for group 14 substituents are given in Table 1. Estimated values of σ_D parameters may be calculated from the equations[25]:

$$\sigma_{RX} = 0.934\sigma_{dX} + 0.308\sigma_{eX} - 0.0129 \tag{34}$$

$$\sigma_{RX}{}^+ = 1.05\sigma_{dX} + 2.14\sigma_{eX} - 0.0731 \tag{35}$$

$$\sigma_{RX}{}^- = 1.13\sigma_{dX} - 1.58\sigma_{eX} + 0.00272 \tag{36}$$

$$\sigma_{RX}{}^\oplus = 1.15\sigma_{dX} + 3.81\sigma_{eX} - 0.0262 \tag{37}$$

$$\sigma_{RX}{}^\ominus = 1.01\sigma_{dX} - 3.01\sigma_{eX} - 0.00491 \tag{38}$$

$$\sigma_{RX}{}^0 = 0.770\sigma_{dX} - 0.288\sigma_{eX} - 0.0394 \tag{39}$$

Values of these parameters are reported in Table 2. Estimated values of Hammett σ constants can be calculated from the relationships[25]:

$$\sigma_{mX} = 1.02\sigma_{IX} + 0.385\sigma_{dX} + 0.661\sigma_{eX} + 0.0152 \tag{40}$$

$$\sigma_{pX} = 1.02\sigma_{IX} + 0.989\sigma_{dX} + 0.837\sigma_{eX} + 0.0132 \tag{41}$$

$$\sigma_{pX}{}^0 = 1.06\sigma_{IX} + 0.796\sigma_{dX} + 0.278\sigma_{eX} - 0.00289 \tag{42}$$

$$\sigma_{pX}{}^+ = 1.10\sigma_{IX} + 0.610\sigma_{dX} + 2.76\sigma_{eX} + 0.0394 \tag{43}$$

$$\sigma_{pX}{}^- = 1.35\sigma_{IX} + 1.36\sigma_{dX} - 1.28\sigma_{eX} + 0.0176 \tag{44}$$

Table 3 presents values of the Hammett substituent constants.

TABLE 2. Values of $\sigma_D{}^a$

X	σ_R	$\sigma_R{}^+$	$\sigma_R{}^-$	$\sigma_R{}^\oplus$	$\sigma_R{}^\ominus$	$\sigma_R{}^0$
Si						
Si(CF$_3$)$_3$	0.42	0.37	0.57	0.43	0.54	0.33
Si(CN)$_3$	0.55	0.52	0.72	0.60	0.67	0.44
SiMe$_2$H	0.09	−0.04	0.21	−0.06	0.25	−0.70
SiMe$_3$						
SiEt$_3$						
Si(NMe$_2$)$_3$	−0.29	−0.52	−0.18	−0.62	−0.05	−0.23
SiMe$_2$OMe	0.15	0.03	0.28	0.03	0.32	0.12
SiMe(OMe)$_2$	0.13	0.02	0.26	0.02	0.28	0.10
Si(OH)$_3$	0.09	−0.01	0.18	0	0.20	0.06
Si(OCF$_3$)$_3$	0.22	0.17	0.30	0.22	0.28	0.16
Si(OMe)$_3$	0.1	−0.01	0.20	−0.01	0.23	0.07
Si(OEt)$_3$						
Si(OAc)$_3$	0.21	0.14	0.31	0.17	0.3	0.15
Si(O$_2$CCF$_3$)$_3$						
Si(OPh)$_3$	0.19	0.10	0.32	0.11	0.33	0.15
Si(OBz)$_3$	0.24	0.18	0.35	0.22	0.34	0.18
Si(SCN)$_3$	0.39	0.35	0.53	0.41	0.49	0.30
Si(SCF$_3$)$_3$	0.35	0.28	0.47	0.32	0.45	0.26
Si(SMe)$_3$	0.26	0.14	0.43	0.14	0.46	0.21
Si(SEt)$_3$	0.22	0.11	0.38	0.10	0.41	0.18
Si(SPh)$_3$	0.34	0.23	0.54	0.22	0.56	0.28
Si(SeMe)$_3$	0.26	0.14	0.43	0.14	0.46	0.21
Si(C$_2$H)$_3$	0.38	0.30	0.57	0.32	0.57	0.31
SiVi$_3$	0.29	0.17	0.47	0.16	0.50	0.24
SiMe$_2$Ph	0.13	0.01	0.27	0	0.30	0.10
SiMePh$_2$	0.17	0.05	0.31	0.05	0.34	0.14
Si(C$_6$F$_5$)$_3$	0.40	0.30	0.60	0.31	0.61	0.33
SiPh$_3$	0.28	0.16	0.46	0.14	0.49	0.23
SiMe$_2$OSiMe$_3$						
SiMe(OSiMe$_3$)$_2$						
Si(OSiMe$_3$)$_3$						
SiMe$_2$Br	0.03	−0.08	0.1	−0.07	0.13	0.01
SiMeBr$_2$	0.06	−0.05	0.16	−0.06	0.19	0.04
SiBr$_3$	0.26	0.20	0.37	0.25	0.35	0.20
SiMeBrCl						
SiMe$_2$Cl	0.07	−0.03	0.16	−0.26	0.19	0.05
SiMeCl$_2$	0.18	0.01	0.38	−0.04	0.44	0.16
SiCl$_3$	0.24	0.18	0.35	0.23	0.33	0.18
SiMe$_2$F	0.06	−0.04	0.15	−0.03	0.17	0.04
SiMeF$_2$	−0.04	−0.13	0.04	−0.11	0.02	−0.05
SiF$_3$	0.02	0.06	0.17	0.12	0.15	0.07
SiI$_3$	0.30	0.23	0.44	0.27	0.44	0.24
SiH$_3$	0.10	−0.02	0.21	−0.02	0.24	0.07
Si(SiMe$_3$)$_3$	0.17	0.04	0.32	0.02	0.36	0.14
Ge						
Ge(CF$_3$)$_3$	0.33	0.28	0.47	0.32	0.45	0.26
Ge(CN)$_3$	0.47	0.43	0.62	0.50	0.58	0.37
GeMe$_3$	0.07	−0.06	0.21	−0.09	0.26	0.06
GeEt$_3$	0.07	−0.06	0.21	−0.09	0.26	0.06
GeVi$_3$	0.20	0.07	0.37	0.05	0.41	0.17
Ge(C$_2$H)$_3$	0.30	0.20	0.47	0.21	0.48	0.24
GePh$_2$Br	0.23	0.08	0.42	0.05	0.47	0.20
GePh$_2$H	0.15	0.04	0.28	0.00	0.31	0.19

TABLE 2. (continued)

X	σ_R	σ_R^+	σ_R^-	$\sigma_R^\oplus$	$\sigma_R^\ominus$	σ_R^0
Ge(C$_6$F$_5$)$_3$	0.32	0.24	0.48	0.26	0.48	0.26
GePh$_3$	0.19	0.06	0.36	0.04	0.40	0.16
Ge(SiMe$_3$)$_3$	0.08	0.06	0.22	0.09	0.27	0.07
Ge(NMe$_2$)$_3$	0.13	−0.04	0.32	−0.10	0.40	0.12
Ge(OH)$_3$	0.01	−0.10	0.08	−0.10	0.11	−0.01
Ge(OCF$_3$)$_3$	0.13	0.07	0.20	0.12	0.19	0.09
Ge(OMe)$_3$	0.01	−0.11	0.10	−0.11	0.14	0.00
Ge(OAc)$_3$	0.12	0.04	0.20	0.07	0.21	0.08
Ge(OPh)$_3$	0.11	0.00	0.21	0.00	0.24	0.08
Ge(OBz)$_3$	0.16	0.08	0.25	0.11	0.25	0.11
Ge(SCF$_3$)$_3$	0.25	0.18	0.37	0.22	0.36	0.19
Ge(SCN)$_3$	0.31	0.26	0.43	0.31	0.40	0.24
Ge(SMe)$_3$	0.18	0.06	0.33	0.04	0.36	0.14
Ge(SEt)$_3$	0.14	0.01	0.28	0.00	0.32	0.11
Ge(SPh)$_3$	0.26	0.14	0.44	0.12	0.47	0.22
Ge(SeMe)$_3$	0.18	0.05	0.33	0.03	0.37	0.14
GeBr$_3$	0.18	0.11	0.27	0.15	0.26	0.13
GeCl$_3$	0.16	0.09	0.24	0.13	0.24	0.11
GeF$_3$	0.03	−0.03	0.07	0.02	0.06	0.00
GeI$_3$	0.22	0.14	0.34	0.16	0.34	0.17
GeH$_3$	0.01	−0.11	0.11	−0.12	0.15	0.00
Sn						
Sn(CF$_3$)$_3$	0.43	0.38	0.58	0.44	0.55	0.34
Sn(CN)$_3$	0.55	0.52	0.72	0.60	0.67	0.43
SnMe$_3$	0.08	−0.06	0.22	−0.08	0.27	0.07
SnEt$_3$	0.08	−0.06	0.22	−0.08	0.27	0.07
SnBu$_3$	0.11	−0.02	0.25	−0.05	0.30	0.09
SnPh$_2$Cl	0.29	0.11	0.53	0.06	0.60	0.25
SnVi$_3$	0.30	0.18	0.48	0.17	0.51	0.25
Sn(C$_2$H)$_3$	0.40	0.31	0.58	0.33	0.58	0.32
SnPh$_2$H	0.21	0.09	0.36	0.08	0.39	0.17
Sn(C$_6$F$_5$)$_3$	0.41	0.33	0.58	0.37	0.57	0.32
SnPh$_3$	0.28	0.16	0.46	0.14	0.49	0.23
Sn(Pn-F-4)$_3$						
Sn(Pn-F-3)$_3$						
Sn(SiMe$_3$)$_3$	0.17	0.04	0.32	0.02	0.36	0.14
Sn(NMe$_2$)$_3$	0.22	0.05	0.43	0.00	0.49	0.19
Sn(OH)$_3$	0.09	−0.01	0.18	0.00	0.20	0.06
Sn(OCF$_3$)$_3$	0.25	0.20	0.33	0.26	0.31	0.18
Sn(OMe)$_3$	0.09	−0.01	0.20	−0.01	0.23	0.07
Sn(OAc)$_3$	0.21	0.15	0.32	0.18	0.31	0.16
Sn(OPh)$_3$	0.20	0.11	0.33	0.12	0.34	0.16
Sn(OBz)$_3$	−0.28	−0.41	−0.28	−0.43	−0.22	−0.25
Sn(SCF$_3$)$_3$	0.34	0.28	0.47	0.32	0.45	0.26
Sn(SCN)$_3$	0.40	0.36	0.54	0.42	0.50	0.31
Sn(SMe)$_3$	0.27	0.16	0.44	0.16	0.46	0.22
Sn(SEt)$_3$	0.23	0.12	0.40	0.11	0.42	0.19
Sn(SPh)$_3$	0.35	0.24	0.55	0.24	0.58	0.29
Sn(SeMe)$_3$	0.26	0.14	0.43	0.14	0.46	0.21
SnBr$_3$	0.26	0.20	0.37	0.25	0.35	0.20
SnCl$_3$	0.25	0.20	0.36	0.24	0.34	0.19
SnF$_3$	0.13	0.08	0.18	0.13	0.16	0.08
SnI$_3$	0.31	0.24	0.46	0.28	0.45	0.25
SnH$_3$	0.11	−0.01	0.22	−0.01	0.25	0.08

(continued overleaf)

TABLE 2. (*continued*)

X	σ_R	σ_R^+	σ_R^-	$\sigma_R^\oplus$	$\sigma_R^\ominus$	σ_R^0
Pb						
$Pb(CF)_3$	0.48	0.43	0.64	0.50	0.60	0.38
$Pb(CN)_3$	0.61	0.59	0.79	0.67	0.72	0.48
$PbMe_3$	0.12	0.00	0.25	−0.01	0.29	0.01
$Pb(Vi)_3$	0.34	0.23	0.54	0.22	0.56	0.28
$Pb(C_2H)_3$	0.44	0.36	0.64	0.38	0.63	0.36
$Pb(C_6F_5)_3$	0.46	0.40	0.65	0.44	0.63	0.37
$PbPh_3$	0.33	0.22	0.53	0.21	0.56	0.28
$Pb(SiMe_3)_3$	0.22	0.01	0.39	0.09	0.42	0.18
$Pb(NMe_2)_3$	0.27	0.11	0.49	0.07	0.55	0.24
$Pb(OH)_3$	0.17	0.05	0.25	0.07	0.26	0.11
$Pb(OCF_3)_3$	0.28	0.24	0.38	0.30	0.35	0.21
$Pb(OMe)_3$	0.15	0.05	0.27	0.06	0.29	0.12
$Pb(OAc)_3$	0.26	0.20	0.37	0.24	0.36	0.20
$Pb(OPh)_3$	0.25	0.16	0.38	0.18	0.39	0.19
$Pb(OBz)_3$	0.30	0.24	0.42	0.28	0.40	0.23
$Pb(SCF_3)_3$	0.39	0.34	0.54	0.39	0.51	0.31
$Pb(SCN)_3$	0.45	0.41	0.60	0.48	0.55	0.35
$Pb(SMe)_3$	0.32	0.21	0.50	0.22	0.51	0.26
$Pb(SEt)_3$	0.28	0.17	0.45	0.17	0.47	0.23
$Pb(SPh)_3$	0.40	0.29	0.61	0.29	0.62	0.33
$Pb(SeMe)_3$	0.32	0.21	0.50	0.20	0.52	0.26
$PbBr_3$	0.32	0.27	0.44	0.32	0.41	0.24
$PbCl_3$	0.30	0.25	0.41	0.30	0.39	0.23
PbF_3	0.17	0.13	0.24	0.19	0.21	0.12
PbI_3	0.36	0.30	0.51	0.34	0.50	0.28
PbH_3	0.15	0.04	0.28	0.05	0.30	0.12
C						
CH_3						
CPh_3	−0.11	−0.23	−0.06	−0.24	−0.01	−0.10
$C(NMe_2)_3$	−0.09	−0.26	−0.02	−0.31	−0.10	−0.07
$C(OMe)_3$	−0.02	−0.15	−0.08	−0.17	0.13	−0.02
$C(OPh)_3$	0.03	−0.08	0.12	−0.08	0.15	0.02
CF_3	0.10	0.02	0.18	0.05	0.19	0.17
CCl_3	0.08	0.00	0.14	0.04	0.14	0.04
CBr_3	0.08	0.00	0.14	0.04	0.14	0.04
CI_3	0.04	−0.04	0.09	−0.01	0.09	0.01
$C(OH)_3$	0.00	−0.12	0.10	−0.14	0.14	0.00
$C(OAc)_3$	0.03	−0.06	0.09	−0.03	0.10	0.00
$C(CF_3)_3$	0.04	−0.01	0.07	0.05	0.05	−0.01
$C(SMe)_3$	−0.01	−0.13	0.06	−0.13	0.10	−0.02
$C(SEt)_3$	−0.03	−0.16	−0.04	−0.17	0.09	−0.04
CVi_3	−0.10	−0.22	−0.06	−0.23	−0.02	−0.10
$C(C_2H)_3$	−0.02	−0.09	0.02	−0.06	0.02	−0.04
$C(C_6F_5)_3$	−0.01	−0.12	0.06	−0.12	0.09	−0.02
$C(SeMe)_3$	−0.02	−0.14	0.06	−0.05	0.09	−0.03
$C(SPh)_3$	−0.01	−0.12	0.06	−0.12	0.09	−0.02
$C(SiMe_3)_3$	−0.22	−0.35	−0.19	−0.37	−0.13	−0.19
$C(CN)_3$	0.13	0.12	0.14	0.21	0.10	0.07
$C(OCF_3)_3$	0.10	0.03	0.16	0.07	0.15	0.06
$C(OBz)_3$	0.06	−0.02	0.12	0.00	0.12	0.03
$C(SCN)_3$	0.13	0.08	0.17	0.15	0.15	0.08
$C(SCF_3)_3$	0.07	0.02	0.11	0.07	0.09	0.03

TABLE 2. (*continued*)

X	σ_R	$\sigma_R{}^+$	$\sigma_R{}^-$	$\sigma_R{}^\oplus$	$\sigma_R{}^\ominus$	$\sigma_R{}^0$
CZ₂H						
CHAc₂	−0.05	0.13	0.04	0.09	0.03	0.07
CHPh₂	−0.12	−0.25	−0.08	−0.26	−0.03	−0.12
CH(NMe₂)₂	−0.11	−0.27	−0.02	−0.31	0.05	−0.10
CH(OMe)₂	−0.06	−0.19	0.02	−0.21	0.07	−0.06
CH(OPh)₂	−0.03	−0.15	0.04	−0.16	0.08	−0.04
CHF₂	0.02	−0.08	0.08	−0.07	0.10	0.00
CHCl₂	0.00	−0.09	0.05	−0.07	0.07	−0.02
CHBr₂	0.00	−0.09	0.05	−0.07	0.07	−0.02
CHI₂	−0.03	−0.12	0.02	−0.11	0.04	−0.04
CH(OH)₂	−0.05	−0.18	0.02	−0.19	0.07	−0.05
CH(OAc)₂	−0.04	−0.14	0.01	−0.13	0.04	−0.05
CH(CF₃)₂	−0.02	−0.09	0.00	−0.06	0.00	−0.05
CH(SMe)₂	−0.06	−0.18	0.00	−0.19	0.04	−0.06
CH(SEt)₂	−0.08	−0.20	−0.01	−0.22	−0.03	−0.08
CHVi₂	−0.12	−0.25	−0.08	−0.26	−0.04	−0.12
CH(C₂H)₂	−0.06	−0.16	−0.03	−0.14	−0.01	−0.07
CH(C₆F₅)₂	−0.05	−0.14	−0.02	−0.12	−0.01	−0.17
CH(SeMe)₂	−0.07	−0.19	0.00	−0.21	−0.04	−0.07
CH(SPh)₂	−0.06	−0.17	0.00	−0.18	0.04	−0.06
CH(SiMe₃)₂	−0.20	−0.34	−0.17	−0.36	−0.11	−0.18
CH(CN)₂	0.03	−0.02	0.06	0.04	0.04	0.00
CH(OCF₃)₂	0.10	0.03	0.16	0.07	0.15	0.06
CH(OBz)₂	0.06	−0.02	0.12	0.00	0.12	0.03
CH(SCN)₂	0.13	0.08	0.17	0.15	0.15	0.08
CH(SCF₃)₂	0.07	0.02	0.11	0.07	0.09	0.03
CH₂Z						
CH₂SiH₃	−0.16	−0.29	−0.12	−0.30	−0.08	−0.15
CH₂SiPh₃	−0.17	−0.29	−0.14	−0.30	−0.10	−0.16
CH₂(OEt)₃	−0.16	−0.29	−0.12	−0.30	−0.08	−0.15
CH₂Si(OMe)₃	−0.15	−0.28	−0.11	−0.29	−0.07	−0.14
CH₂SiMe₃	−0.18	−0.31	−0.14	−0.33	−0.09	−0.16
CH₂SiF₃	−0.09	−0.19	−0.06	−0.18	−0.04	−0.10
CH₂SiCl₃	−0.10	−0.21	−0.07	−0.20	−0.04	−0.10
CH₂SiEt₃	−0.18	−0.31	−0.14	−0.33	−0.09	−0.16
CH₂CH₂SiMe₃	−0.16	−0.30	−0.11	−0.33	−0.05	−0.14
CH₂CH₂SiEt₃	−0.16	−0.30	−0.11	−0.33	−0.05	−0.14
CH₂Ge(OMe)₃	−0.15	−0.28	−0.11	−0.29	−0.06	−0.14
CH₂GeEt₃	−0.17	−0.30	−0.13	−0.32	−0.08	−0.15
CH₂GeMe₃	−0.17	−0.30	−0.13	−0.32	−0.08	−0.16
CH₂GeH₃	−0.16	−0.29	−0.12	−0.30	−0.07	−0.15
CH₂GePh₃	−0.16	−0.28	−0.13	−0.29	−0.08	−0.15
CH₂GeF₃	−0.09	−0.20	−0.06	−0.19	−0.03	−0.10
CH₂GeCl₃	−0.09	−0.19	−0.06	−0.18	−0.03	−0.10
CH₂SnMe₃	−0.17	−0.30	−0.14	−0.31	−0.09	−0.16
CH₂SnPh₃	−0.17	−0.29	−0.14	−0.30	−0.10	−0.16
CH₂Sn(OMe)₃	−0.15	−0.27	−0.12	−0.28	−0.07	−0.14
CH₂SnCl₃	−0.09	−0.19	−0.06	−0.18	−0.04	−0.10
CH₂SnF₃	−0.09	−0.19	−0.06	−0.18	−0.04	−0.10
CH₂SnEt₃	−0.17	−0.30	−0.13	−0.32	−0.08	−0.15
CH₂SnH₃	−0.16	−0.29	−0.13	−0.30	−0.08	−0.15
CH₂PbF₃	−0.09	−0.19	−0.06	−0.18	−0.04	−0.10
CH₂PbCl₃	−0.10	−0.20	−0.08	−0.18	−0.05	−0.10
CH₂PbPh₃	−0.17	−0.30	−0.14	−0.30	−0.09	−0.16
CH₂Pb(OMe)₃	−0.15	−0.27	−0.12	−0.28	−0.07	−0.14
CH₂PbH₃	−0.16	−0.28	−0.13	−0.29	−0.08	−0.15

[a] The σ_D constants were calculated from equations 34 through 39.

TABLE 3.   Values of Hammett-type substituent constants[a]

X	σ_m	σ_p	σ_p^0	σ_P^+	σ_P^-
Si					
Si(CF$_3$)$_3$	0.63	0.91	0.83	0.75	1.28
Si(CN)$_3$	0.83	0.19	1.09	0.99	1.66
SiMe$_2$H	−0.03	0.03	0.12	−0.07	0.16
SiMe$_3$					
SiEt$_3$					
Si(NMe$_2$)$_3$	−0.44	−0.62	−0.56	−0.66	−0.66
SiMe$_2$OMe	−0.01	0.11	0.08	−0.02	0.26
SiMe(OMe)$_2$	−0.40	0.14	0.11	−0.02	0.28
SI(OH)$_3$	0.14	0.21	0.19	0.14	0.35
Si(OCF$_3$)$_3$	0.53	0.68	0.64	0.63	0.94
Si(OMe)$_3$	0.08	0.15	0.13	0.07	0.29
Si(OEt)$_3$					
Si(OAc)$_3$	0.39	0.53	0.49	0.45	0.76
Si(O$_2$CCF$_3$)$_3$	0.53	0.68	0.64	0.63	0.94
Si(OPh)$_3$	0.22	0.36	0.32	0.24	0.56
Si(OBz)$_3$	0.42	0.59	0.54	0.49	0.84
Si(SCN)$_3$	0.66	0.92	0.85	0.79	1.29
Si(SCF$_3$)$_3$	0.52	0.75	0.68	0.62	1.06
Si(SMe)$_3$	0.14	0.32	0.27	0.14	0.56
Si(SEt)$_3$	0.09	0.24	0.20	0.07	0.45
Si(SPh)$_3$	0.16	0.39	0.33	0.16	0.67
Si(SeMe)$_3$	0.10	0.28	0.23	0.09	0.50
Si(C$_2$H)$_3$	0.36	0.62	0.55	0.41	0.94
SiVi$_3$	0.08	0.27	0.22	0.06	0.51
SiMe$_2$Ph	−0.02	0.07	0.05	−0.06	0.21
SiMePh$_2$	0.02	0.13	0.10	−0.01	0.29
Si(C$_6$F$_5$)$_3$	0.45	0.72	0.66	0.50	1.10
SiPh$_3$	0.06	0.25	0.20	0.04	0.48
SiMe$_2$OSiMe$_3$					
SiMe(OSiMe$_3$)$_2$					
Si(OSiMe$_3$)$_3$					
SiMe$_2$Br	0.08	0.10	0.09	0.06	0.20
SiMeBr$_2$	0.24	0.29	0.28	0.23	0.47
SiBr$_3$	0.49	0.66	0.61	0.57	0.94
SiMeBrCl					
SiMe$_2$Cl	0.10	0.15	0.13	0.08	0.27
SiMeCl$_2$	0.27	0.40	0.38	0.21	0.71
SiCl$_3$	0.48	0.64	0.60	0.56	0.91
SiMe$_2$F	0.11	0.16	0.15	0.10	0.28
SiMeF$_2$	0.24	0.23	0.23	0.25	0.33
SiF$_3$	0.48	0.57	0.54	0.56	0.77
SiI$_3$	0.42	0.62	0.56	0.48	0.91
SiH$_3$	0.04	0.11	0.09	0.01	0.24
Si(SiMe$_3$)$_3$	−0.04	0.08	0.04	−0.09	0.24
Ge					
Ge(CF$_3$)$_3$	0.60	0.83	0.77	0.70	1.17
Ge(CN)$_3$	0.80	1.11	1.03	0.95	1.55
GeMe$_3$	−0.06	0	−0.01	−0.12	0.12
GeEt$_3$	−0.06	0	−0.01	0.12	0.12
GeVi$_3$	0.06	0.19	0.16	0.02	0.40
Ge(C$_2$H)$_3$	0.35	0.54	0.49	0.36	0.83
GePh$_2$Br	0.19	0.35	0.32	0.15	0.63
GePh$_2$H	0.05	0.16	0.13	0.03	0.32
Ge(C$_6$F$_5$)$_3$	0.43	0.64	0.59	0.48	0.96

TABLE 3. (continued)

X	σ_m	σ_p	σ_p^0	σ_P^+	σ_P^-
GePh$_3$	0.04	0.17	0.14	0	0.37
Ge(SiMe$_3$)$_3$	−0.08	−0.01	−0.03	−0.14	0.11
Ge(NMe$_2$)$_3$	−0.27	0.18	−0.20	−0.39	−0.05
Ge(OH)$_3$	0.12	0.13	0.13	0.10	0.24
Ge(OCF$_3$)$_3$	0.51	0.60	0.58	0.58	0.83
Ge(OMe)$_3$	0.06	0.07	0.07	0.02	0.18
Ge(OAc)$_3$	0.37	0.45	0.43	0.41	0.65
Ge(OPh)$_3$	0.20	0.28	0.26	0.20	0.45
Ge(OBz)$_3$	0.40	0.51	0.48	0.45	0.75
Ge(SCF$_3$)$_3$	0.50	0.67	0.62	0.57	0.95
Ge(SCN)$_3$	0.64	0.84	0.79	0.74	1.18
Ge(SMe)$_3$	0.12	0.24	0.21	0.10	0.45
Ge(SEt)$_3$	0.06	0.16	0.14	0.03	0.34
Ge(SPh)$_3$	0.14	0.32	0.27	0.12	0.56
Ge(SeMe)$_3$	0.08	0.20	0.17	0.04	0.40
GeBr$_3$	0.46	0.58	0.55	0.52	0.83
GeCl$_3$	0.46	0.56	0.54	0.52	0.80
GeF$_3$	0.46	0.49	0.48	0.52	0.66
GeI$_3$	0.39	0.54	0.50	0.44	0.80
GeH$_3$	0.02	0.03	0.03	−0.03	0.13
Sn					
Sn(CF$_3$)$_3$	0.63	0.92	0.84	0.75	1.29
Sn(CN)$_3$	0.83	1.19	0.09	0.99	1.66
SnMe$_3$	−0.06	0.00	−0.02	−0.13	0.12
SnEt$_3$	−0.06	0.00	−0.02	−0.12	0.12
SnBu$_3$	−0.06	0.02	0.00	−0.12	0.15
SnPh$_2$Cl	0.20	0.39	0.36	0.13	0.74
SnVi$_3$	0.08	0.28	0.23	0.07	0.52
Sn(C$_2$H)$_3$	0.36	0.63	0.56	0.42	0.96
SnPh$_2$H	0.06	0.20	0.16	0.04	0.39
Sn(C$_6$F$_5$)$_3$	0.45	0.72	0.65	0.53	1.07
SnPh$_3$	0.06	0.25	0.20	0.04	0.48
Sn(Pn-F-4)$_3$					
Sn(Pn-F-3)$_3$					
Sn(SiMe$_3$)$_3$	−0.05	0.06	0.03	−0.10	0.22
Sn(NMe$_2$)$_3$	−0.24	−0.09	−0.13	−0.33	0.08
Sn(OH)$_3$	0.14	0.21	0.19	0.14	0.35
Sn(OCF$_3$)$_3$	0.54	0.71	0.66	0.64	0.98
Sn(OMe)$_3$	0.08	0.15	0.13	0.07	0.29
Sn(OAc)$_3$	0.39	0.54	0.50	0.46	0.77
Sn(OPh)$_3$	0.23	0.36	0.33	0.25	0.58
Sn(OBz)$_3$	0.21	0.04	0.10	0.15	0.08
Sn(SCF$_3$)$_3$	0.52	0.75	0.68	0.61	1.06
Sn(SCN)$_3$	0.67	0.93	0.86	0.79	1.30
Sn(SMe)$_3$	0.15	0.33	0.28	0.15	0.57
Sn(SEt)$_3$	0.09	0.25	0.21	0.08	0.46
Sn(SPh)$_3$	0.17	0.40	0.34	0.17	0.69
Sn(SeMe)$_3$	0.10	0.28	0.23	0.09	0.50
SnBr$_3$	0.49	0.66	0.61	0.57	0.94
SnCl$_3$	0.48	0.65	0.60	0.56	0.92
SnF$_3$	0.49	0.58	0.55	0.57	0.78
SnI$_3$	0.42	0.63	0.57	0.49	0.92
SnH$_3$	0.04	0.12	0.10	0.02	0.26

(continued overleaf)

TABLE 3. (*continued*)

X	σ_m	σ_p	$\sigma_p{}^0$	$\sigma_P{}^+$	$\sigma_P{}^-$
Pb					
Pb(CF)$_3$	0.64	0.96	0.87	0.77	1.35
Pb(CN)$_3$	0.84	1.24	1.13	1.02	1.72
Pb(Me)$_3$	−0.05	0.03	0.01	−0.09	0.16
Pb(Vi)$_3$	0.09	0.32	0.26	0.09	0.58
Pb(C$_2$H)$_3$	0.37	0.67	0.58	0.44	1.01
Pb(C$_6$F$_5$)$_3$	0.48	0.78	0.70	0.56	1.15
PbPh$_3$	0.09	0.31	0.25	0.08	0.56
Pb(SiMe$_3$)$_3$	−0.03	0.12	0.08	−0.06	0.30
Pb(NMe$_2$)$_3$	−0.22	−0.04	−0.09	−0.31	0.14
Pb(OH)$_3$	0.16	0.26	0.23	0.17	0.42
Pb(OCF$_3$)$_3$	0.55	0.74	0.68	0.66	1.02
Pb(OMe)$_3$	0.10	0.20	0.17	0.09	0.36
Pb(OAc)$_3$	0.41	0.59	0.54	0.49	0.84
Pb(OPh)$_3$	0.24	0.40	0.36	0.27	0.63
Pb(OBz)$_3$	0.44	0.64	0.58	0.52	0.91
Pb(SCF$_3$)$_3$	0.54	0.80	0.72	0.64	1.13
Pb(SCN)$_3$	0.68	0.97	0.89	0.81	1.36
Pb(SMe)$_3$	0.16	0.37	0.31	0.17	0.62
Pb(SEt)$_3$	0.11	0.30	0.25	0.11	0.53
Pb(SPh)$_3$	0.18	0.44	0.37	0.19	0.74
Pb(SeMe)$_3$	0.12	0.34	0.28	0.12	0.59
PbBr$_3$	0.51	0.72	0.66	0.60	1.01
PbCl$_3$	0.50	0.70	0.64	0.60	0.99
PbF$_3$	0.50	0.62	0.58	0.59	0.84
PbI$_3$	0.43	0.67	0.60	0.51	0.98
PbH$_3$	0.06	0.17	0.14	0.05	0.32
C					
CH$_3$					
CPh$_3$	0.05	−0.01	0.01	0.01	0.05
C(NMe$_2$)$_3$	0.08	0.03	0.06	−0.02	0.17
C(OMe)$_3$	0.22	0.21	0.23	0.17	0.38
C(OPh)$_3$	0.31	0.34	0.34	0.30	0.53
CF$_3$	0.46	0.53	0.52	0.50	0.76
CCl$_3$	0.40	0.46	0.44	0.45	0.64
CBr$_3$	0.40	0.46	0.44	0.45	0.64
CI$_3$	0.33	0.36	0.35	0.36	0.51
C(OH)$_3$	0.27	0.28	0.29	0.24	0.46
C(OAc)$_3$	0.31	0.33	0.33	0.33	0.48
C(CF$_3$)$_3$	0.34	0.37	0.35	0.40	0.49
C(SMe)$_3$	0.22	0.22	0.23	0.20	0.37
C(SEt)$_3$	0.18	0.17	0.18	0.15	0.30
CVi$_3$	0.04	−0.02	0.00	0.00	0.04
C(C$_2$H)$_3$	0.22	0.22	0.22	0.24	0.31
C(C$_6$F$_5$)$_3$	0.24	0.23	0.24	0.22	0.38
C(SeMe)$_3$	0.20	0.19	0.20	0.17	0.33
C(SPh)$_3$	0.24	0.23	0.24	0.22	0.38
C(SiMe$_3$)$_3$	0.18	0.31	−0.27	−0.26	−0.35
C(CN)$_3$	0.51	0.60	0.57	0.64	0.77
C(OCF$_3$)$_3$	0.44	0.51	0.49	0.50	0.71
C(OBz)$_3$	0.36	0.40	0.40	0.40	0.58
C(SCN)$_3$	0.49	0.58	0.55	0.58	0.78
C(SCF$_3$)$_3$	0.39	0.44	0.42	0.45	0.59

TABLE 3. (*continued*)

X	σ_m	σ_p	σ_p^0	σ_P^+	σ_P^-
CZ₂H					
CHAc₂	0.14	0.11	0.11	0.16	0.16
CHPh₂	0.02	−0.06	−0.04	−0.04	−0.02
CH(NMe₂)₂	0.03	−0.03	0.00	−0.06	0.06
CH(OMe)₂	0.12	0.08	0.10	0.07	0.20
CH(OPh)₂	0.18	0.17	0.18	0.16	0.30
CHF₂	0.28	0.30	0.30	0.29	0.45
CHCl₂	0.25	0.25	0.25	0.26	0.38
CHBr₂	0.25	0.25	0.25	0.26	0.38
CHI₂	0.19	0.18	0.18	0.19	0.28
CH(OH)₂	0.15	0.13	0.14	0.11	0.25
CH(OAc)₂	0.18	0.16	0.17	0.17	0.26
CH(CF₃)₂	0.20	0.20	0.19	0.23	0.27
CH(SMe)₂	0.12	0.09	0.10	0.09	0.19
CH(SEt)₂	0.09	0.05	0.07	0.05	0.14
CHVi₂	0.01	−0.07	−0.04	−0.05	−0.03
CH(C₂H)₂	0.13	0.09	0.10	0.12	0.16
CH(C₆F₅)₂	0.14	0.12	0.12	0.15	0.18
CH(SeMe)₂	0.11	0.07	0.09	0.06	0.17
CH(SPh)₂	0.13	0.10	0.12	0.10	0.20
CH(SiMe₃)₂	−0.14	−0.26	−0.23	−0.22	−0.28
CH(CN)₂	0.32	0.35	0.33	0.38	0.46
CH(OCF₃)₂	0.27	0.28	0.28	0.29	0.42
CH(OBz)₂	0.22	0.21	0.21	0.22	0.33
CH(SCN)₂	0.30	0.33	0.32	0.34	0.46
CH(SCF₃)₂	0.23	0.24	0.23	0.25	0.35
CH₂Z					
CH₂SiH₃	−0.06	−0.16	−0.13	−0.13	−0.15
CH₂SiPh₃	−0.07	−0.17	−0.15	−0.13	−0.18
CH₂Si(OEt)₃	−0.06	−0.16	−0.13	−0.13	−0.15
CH₂Si(OMe)₃	−0.05	−0.14	−0.11	−0.11	−0.13
CH₂SiMe₃	−0.10	−0.21	−0.18	−0.18	−0.22
CH₂SiF₃	0.08	0.02	0.04	0.06	0.06
CH₂SiCl₃	0.06	0.00	0.02	0.03	0.04
CH₂SiEt₃	−0.10	−0.21	−0.18	−0.08	−0.22
CH₂CH₂SiMe₃	−0.08	−0.17	−0.14	−0.16	−0.16
CH₂CH₂SiEt₃	−0.08	−0.17	−0.14	−0.16	−0.16
CH₂Ge(OMe)₃	−0.05	−0.14	−0.11	−0.11	−0.12
CH₂GeEt₃	−0.08	−0.19	−0.16	−0.16	−0.19
CH₂GeMe₃	−0.08	−0.19	−0.16	−0.16	−0.19
CH₂GeH₃	−0.06	−0.16	−0.13	−0.13	−0.15
CH₂GePh₃	−0.07	−0.16	−0.14	−0.13	−0.17
CH₂GeF₃	0.08	0.02	0.03	0.05	0.07
CH₂GeCl₃	0.06	0.01	0.02	0.04	0.05
CH₂SnMe₃	−0.09	−0.20	−0.17	−0.16	−0.21
CH₂SnPh₃	−0.07	−0.18	−0.15	−0.13	−0.18
CH₂Sn(OMe)₃	−0.04	−0.14	−0.11	−0.10	−0.13
CH₂SnCl₃	0.07	0.01	0.02	0.05	0.05
CH₂SnF₃	0.08	0.02	0.03	0.06	0.06
CH₂SnEt₃	−0.10	−0.20	−0.17	−0.17	−0.20
CH₂SnH₃	−0.06	−0.16	−0.13	−0.12	−0.15
CH₂PbF₃	0.08	0.02	0.04	0.06	0.06
CH₂PbCl₃	0.06	0.00	0.02	0.04	0.03
CH₂PbPh₃	−0.07	−0.18	−0.15	−0.14	−0.18
CH₂Pb(OMe)₃	−0.06	−0.14	−0.12	−0.11	−0.14
CH₂PbH₃	−0.06	−0.16	−0.13	−0.12	−0.15

[a] The Hammett substituent constants were calculated from equations 40 through 44.

2. CH_{3-n} $(MZ^1Z^2Z^3)_n$ groups

Values of σ_I for these groups were calculated from the equation[29]:

$$\sigma_{ICZ^1Z^2Z^3} = 0.248\Sigma\sigma_{IZ} - 0.00398 \tag{45}$$

Values of σ_d were estimated from the equation:

$$\sigma_{dCZ^1Z^2Z^3} = 0.175\ (\pm0.00921)\Sigma\sigma_{IZ} - 0.149\ (\pm0.0525) \tag{46}$$

$100r^2$, 93.77; F, 361.2; S_{est}, 0.0204; S°, 0.260; n, 26

while for values of σ_e the estimation equation used was:

$$\sigma_{eCZ^1Z^2Z^3} = 0.0226\ (\pm0.00536)\Sigma\sigma_{IZ} + 0.0197\ (\pm0.00655)\Sigma\sigma_{dZ} - 0.0287\ (\pm0.00208) \tag{47}$$

$100R^2$, 48.51; $A100R^2$, 45.94; F, 8.952; S_{est}, 0.00726; S°, 0.772; n, 22; r_{Id}, 0.777.

The poor quality of this regression equation is due to a structural dependence that is marginal at best. The σ_e values used to obtain equation 47 are all in the range from -0.041 to -0.014, spanning 0.027 unit. F, l, d and h are significant at the 99.5, 99.9, 99.0 and 99.9 percent confidence levels, respectively. The value of r_{Id} is significant at the 99.9 percent confidence level.

C. Electrical Effects of Group 14 Substituents

1. Classification of substituent electrical effects

It is traditional to classify substituents as either electron acceptor (electron withdrawing, electron sink), EA; or electron donor (electron releasing, electron source), ED. There is a third category as well, however, that consists of groups whose electrical effect is not significantly different from zero (NS groups). Groups vary in the nature of their electrical effect to a greater or lesser extent depending on the electronic demand of the phenomenon being studied, the skeletal group, if any, to which they are bonded, and the experimental conditions. Very few groups are in the same category throughout the entire range of P_D and η normally encountered. We have observed earlier that a plot of the $\sigma_{k'/k,X}$ values for a group with X = P_D, Y = η and Z = $\sigma_{k'/k}$, produces a surface that characterizes the electrical effect of the X group. A matrix of these values can be obtained by calculating them for values of P_D in the range 10 to 90 in increments of 10 and values of η in the range -6 to 6 in increments of 1. The resulting 9 by 13 matrix has 117 values. We define $\sigma_{k'/k,X}$ values greater than 0.05 as EA $\sigma_{k'/k,X}$ values less than -0.05 as ED and $\sigma_{k'/k,X}$ values between 0.05 and -0.05 as NS. The variability of the electrical effect of a group can be quantitatively described by the percent of the matrix area in the P_D-η plane in which the group is in each category (P_{EA}, P_{ED} and P_0). Approximate measures of these quantities are given by the relationships:

$$P_{EA} = \frac{n_{EA}}{n_T}, \quad P_0 = \frac{n_{NS}}{n_T}, \quad P_{ED} = \frac{n_{ED}}{n_T} \tag{48}$$

where n_{EA}, n_{NS}, n_{ED}, and n_T are the number of EA, the number of NS, the number of ED and the total number of values in the matrix. Matrices for a number of substituents are given in Table 4; values of P_{EA}, P_{ED} and P_0 for many substituents are reported in Table 5. We may now classify groups into seven types (p.635):

TABLE 4. Electrical effect substituent matrices[a]

η	10	20	30	P_D 40	50	60	70	80	90
Si(OPh)₃									
−6	0.19	0.25	0.33	0.43	0.57	0.79	1.15	1.88	4.05
−5	0.18	0.24	0.31	0.41	0.54	0.74	1.07	1.74	3.74
−4	0.18	0.23	0.30	0.38	0.51	0.69	0.99	1.60	3.43
−3	0.18	0.22	0.28	0.36	0.47	0.64	0.91	1.47	3.13
−2	0.17	0.21	0.27	0.34	0.44	0.59	0.84	1.33	2.82
−1	0.17	0.21	0.25	0.32	0.40	0.54	0.76	1.20	2.52
0	0.17	0.20	0.24	0.29	0.37	0.49	0.68	1.06	2.21
1	0.16	0.19	0.22	0.27	0.34	0.43	0.60	0.92	1.90
2	0.16	0.18	0.21	0.25	0.30	0.38	0.52	0.79	1.60
3	0.15	0.17	0.19	0.23	0.27	0.33	0.44	0.65	1.29
4	0.15	0.16	0.18	0.20	0.23	0.28	0.36	0.52	0.99
5	0.15	0.16	0.17	0.18	0.20	0.23	0.28	0.38	0.68
6	0.14	0.15	0.15	0.16	0.17	0.18	0.20	0.24	0.37
SiF₃									
−6	0.43	0.45	0.48	0.52	0.57	0.66	0.79	1.07	1.89
−5	0.43	0.45	0.48	0.52	0.57	0.65	0.78	1.05	1.85
−4	0.43	0.45	0.48	0.51	0.57	0.64	0.77	1.03	1.81
−3	0.43	0.45	0.48	0.51	0.56	0.64	0.76	1.02	1.78
−2	0.43	0.45	0.47	0.51	0.56	0.63	0.76	1.00	1.74
−1	0.43	0.45	0.47	0.51	0.55	0.63	0.75	0.99	1.71
0	0.43	0.45	0.47	0.50	0.55	0.62	0.74	0.97	1.67
1	0.43	0.44	0.47	0.50	0.55	0.61	0.73	0.95	1.63
2	0.42	0.44	0.47	0.50	0.54	0.61	0.72	0.94	1.60
3	0.42	0.44	0.46	0.50	0.54	0.60	0.71	0.92	1.56
4	0.42	0.44	0.46	0.49	0.53	0.60	0.70	0.91	1.53
5	0.42	0.44	0.46	0.49	0.53	0.59	0.69	0.89	1.49
6	0.42	0.44	0.46	0.49	0.53	0.58	0.68	0.87	1.45
GeH₃									
−6	0.04	0.08	0.12	0.19	0.28	0.41	0.64	1.08	2.42
−5	0.04	0.07	0.11	0.16	0.24	0.35	0.55	0.93	2.08
−4	0.03	0.06	0.09	0.14	0.20	0.30	0.46	0.78	1.74
−3	0.03	0.05	0.08	0.11	0.16	0.24	0.37	0.63	1.40
−2	0.02	0.04	0.06	0.09	0.13	0.18	0.28	0.47	1.05
−1	0.02	0.03	0.04	0.06	0.09	0.13	0.19	0.32	0.71
0	0.01	0.02	0.03	0.04	0.05	0.07	0.10	0.17	0.37
1	0.01	0.01	0.01	0.01	0.01	0.01	0.01	0.02	0.03
2	0.01	0.00	−0.01	−0.01	−0.03	−0.04	−0.07	−0.13	−0.31
3	0.00	−0.01	−0.02	−0.04	−0.06	−0.10	−0.16	−0.29	−0.66
4	0.00	−0.02	−0.04	−0.06	−0.10	−0.16	−0.25	−0.44	−1.00
5	−0.01	−0.03	−0.05	−0.09	−0.14	−0.22	−0.34	−0.59	−1.34
6	−0.01	−0.04	−0.07	−0.12	−0.18	−0.27	−0.43	−0.74	−1.68
GePh₃									
−6	0.03	0.11	0.21	0.35	0.54	0.83	1.30	2.25	5.10
−5	0.03	0.10	0.19	0.31	0.49	0.74	1.17	2.03	4.60
−4	0.02	0.08	0.17	0.28	0.43	0.66	1.04	1.81	4.11
−3	0.01	0.07	0.14	0.24	0.37	0.58	0.91	1.59	3.61
−2	0.01	0.06	0.12	0.20	0.32	0.49	0.79	1.37	3.12
−1	0.00	0.04	0.10	0.17	0.26	0.41	0.66	1.15	2.63
0	0.00	0.03	0.07	0.13	0.21	0.33	0.53	0.93	2.13
1	−0.01	0.02	0.05	0.09	0.16	0.25	0.40	0.71	1.64
2	−0.02	0.00	0.03	0.06	0.10	0.16	0.27	0.49	1.14

(continued overleaf)

TABLE 4. (continued)

η	10	20	30	P_D 40	50	60	70	80	90
GePh$_3$									
3	−0.02	0.01	0.00	0.02	0.05	**0.08**	**0.15**	**0.27**	**0.65**
4	−0.03	−0.03	−0.02	−0.02	−0.01	−0.00	0.02	0.05	**0.15**
5	−0.03	−0.04	−0.05	−0.05	*−0.07*	*−0.08*	*−0.11*	*−0.17*	*−0.35*
6	−0.04	−0.05	−0.07	−0.09	*−0.12*	*−0.16*	*−0.24*	*−0.39*	*−0.84*
Ge(NMe$_2$)$_3$									
−6	*−0.23*	*−0.14*	*−0.03*	**0.12**	**0.33**	**0.65**	**1.17**	**2.22**	**5.37**
−5	*−0.24*	*−0.16*	*−0.06*	**0.07**	**0.26**	**0.53**	**0.99**	**1.92**	**4.69**
−4	*−0.25*	*−0.18*	*−0.09*	0.02	**0.18**	**0.42**	**0.82**	**1.62**	**4.02**
−3	*−0.26*	*−0.20*	*−0.13*	−0.03	**0.11**	**0.31**	**0.65**	**1.32**	**3.35**
−2	*−0.26*	*−0.22*	*−0.16*	−0.08	0.03	**0.19**	**0.47**	**1.02**	**2.67**
−1	*−0.27*	*−0.24*	*−0.19*	−0.13	−0.05	**0.08**	**0.29**	**0.72**	**2.00**
0	*−0.28*	*−0.26*	*−0.22*	−0.18	−0.12	−0.03	**0.12**	**0.42**	**1.32**
1	*−0.29*	*−0.27*	*−0.26*	−0.23	−0.20	−0.14	−0.06	**0.12**	**0.65**
2	*−0.30*	*−0.29*	*−0.29*	−0.28	−0.27	−0.26	−0.23	*−0.18*	−0.03
3	*−0.31*	*−0.31*	*−0.32*	*−0.33*	*−0.35*	*−0.37*	*−0.41*	*−0.48*	*−0.71*
4	*−0.31*	*−0.33*	*−0.35*	*−0.38*	*−0.42*	*−0.48*	*−0.58*	*−0.78*	*−1.38*
5	*−0.32*	*−0.35*	*−0.38*	*−0.43*	*−0.50*	*−0.59*	*−0.76*	*−1.08*	*−2.05*
6	*−0.33*	*−0.37*	*−0.42*	*−0.48*	*−0.57*	*−0.71*	*−0.93*	*−1.38*	*−2.73*
Ge(OPh)$_3$									
−6	**0.19**	**0.24**	**0.30**	**0.38**	**0.49**	**0.67**	**0.95**	**1.53**	**3.25**
−5	**0.18**	**0.23**	**0.28**	**0.36**	**0.46**	**0.62**	**0.87**	**1.39**	**2.94**
−4	**0.18**	**0.22**	**0.27**	**0.33**	**0.43**	**0.56**	**0.79**	**1.25**	**2.63**
−3	**0.18**	**0.21**	**0.25**	**0.31**	**0.39**	**0.51**	**0.71**	**1.12**	**2.33**
−2	**0.17**	**0.20**	**0.24**	**0.29**	**0.36**	**0.46**	**0.64**	**0.98**	**2.02**
−1	**0.17**	**0.19**	**0.22**	**0.27**	**0.32**	**0.41**	**0.56**	**0.85**	**1.72**
0	**0.17**	**0.19**	**0.21**	**0.24**	**0.29**	**0.36**	**0.48**	**0.71**	**1.41**
1	**0.16**	**0.18**	**0.20**	**0.22**	**0.26**	**0.31**	**0.40**	**0.57**	**1.10**
2	**0.16**	**0.17**	**0.18**	**0.20**	**0.22**	**0.26**	**0.32**	**0.44**	**0.80**
3	**0.15**	**0.16**	**0.17**	**0.18**	**0.19**	**0.21**	**0.24**	**0.30**	**0.49**
4	**0.15**	**0.15**	**0.15**	**0.15**	**0.15**	**0.16**	**0.16**	**0.17**	**0.19**
5	**0.15**	**0.14**	**0.14**	**0.13**	**0.12**	**0.10**	**0.08**	0.03	*−0.12*
6	**0.14**	**0.13**	**0.12**	**0.11**	**0.09**	0.05	0.00	*−0.11*	*−0.43*
GeF$_3$									
−6	**0.43**	**0.44**	**0.45**	**0.47**	**0.49**	**0.53**	**0.59**	**0.72**	**1.09**
−5	**0.43**	**0.44**	**0.45**	**0.47**	**0.49**	**0.53**	**0.58**	**0.70**	**1.05**
−4	**0.43**	**0.44**	**0.45**	**0.46**	**0.49**	**0.52**	**0.57**	**0.68**	**1.01**
−3	**0.43**	**0.44**	**0.45**	**0.46**	**0.48**	**0.51**	**0.56**	**0.67**	**0.98**
−2	**0.43**	**0.43**	**0.44**	**0.46**	**0.48**	**0.51**	**0.56**	**0.65**	**0.94**
−1	**0.43**	**0.43**	**0.44**	**0.46**	**0.47**	**0.50**	**0.55**	**0.64**	**0.91**
0	**0.43**	**0.43**	**0.44**	**0.45**	**0.47**	**0.50**	**0.54**	**0.62**	**0.87**
1	**0.43**	**0.43**	**0.44**	**0.45**	**0.47**	**0.49**	**0.53**	**0.60**	**0.83**
2	**0.42**	**0.43**	**0.44**	**0.45**	**0.46**	**0.48**	**0.52**	**0.59**	**0.80**
3	**0.42**	**0.43**	**0.44**	**0.45**	**0.46**	**0.48**	**0.51**	**0.57**	**0.76**
4	**0.42**	**0.43**	**0.43**	**0.44**	**0.45**	**0.47**	**0.50**	**0.56**	**0.73**
5	**0.42**	**0.43**	**0.43**	**0.44**	**0.45**	**0.46**	**0.49**	**0.54**	**0.69**
6	**0.42**	**0.43**	**0.43**	**0.44**	**0.45**	**0.46**	**0.48**	**0.52**	**0.65**
SnH$_3$									
−6	0.04	**0.09**	**0.16**	**0.25**	**0.37**	**0.55**	**0.86**	**1.47**	**3.31**
−5	0.04	**0.08**	**0.14**	**0.22**	**0.33**	**0.49**	**0.77**	**1.32**	**2.97**
−4	0.03	**0.07**	**0.13**	**0.19**	**0.29**	**0.44**	**0.68**	**1.17**	**2.63**
−3	0.03	**0.06**	**0.11**	**0.17**	**0.25**	**0.38**	**0.59**	**1.02**	**2.29**

TABLE 4. (continued)

η	10	20	30	P_D 40	50	60	70	80	90
−2	0.02	0.05	**0.09**	**0.14**	**0.22**	**0.32**	**0.50**	**0.86**	1.94
−1	0.02	0.04	**0.08**	**0.12**	**0.18**	**0.27**	**0.42**	**0.71**	1.60
0	0.02	0.04	**0.06**	**0.09**	**0.14**	**0.21**	**0.33**	**0.56**	1.26
1	0.01	0.03	0.04	**0.07**	**0.10**	**0.15**	**0.24**	**0.41**	0.92
2	0.01	0.02	0.03	0.04	**0.06**	**0.10**	**0.15**	**0.26**	0.58
3	0.00	0.01	0.01	0.02	0.03	0.04	**0.06**	**0.10**	0.23
4	0.00	0.00	−0.01	−0.01	−0.01	−0.02	−0.03	−0.05	−0.11
5	−0.01	−0.01	−0.02	−0.03	−0.05	−0.07	−0.12	−0.20	−0.45
6	−0.01	−0.02	−0.04	−0.06	−0.09	−0.13	−0.21	−0.35	−0.79
SnF$_3$									
−6	**0.43**	**0.45**	**0.48**	**0.53**	**0.58**	0.67	0.82	1.11	1.98
−5	**0.43**	**0.45**	**0.48**	**0.52**	**0.58**	0.66	0.81	1.09	1.94
−4	**0.43**	**0.45**	**0.48**	**0.52**	**0.58**	0.66	0.80	1.07	1.90
−3	**0.43**	**0.45**	**0.48**	**0.52**	**0.57**	0.65	0.79	1.06	1.87
−2	**0.43**	**0.45**	**0.48**	**0.52**	**0.57**	0.65	0.78	1.04	1.83
−1	**0.43**	**0.45**	**0.48**	**0.51**	**0.56**	0.64	0.77	1.03	1.80
0	**0.43**	**0.45**	**0.47**	**0.51**	**0.56**	0.64	0.76	1.01	1.76
1	**0.43**	**0.45**	**0.47**	**0.51**	**0.56**	0.63	0.75	0.99	1.72
2	**0.43**	**0.45**	**0.47**	**0.50**	**0.55**	0.62	0.74	0.98	1.69
3	**0.43**	**0.44**	**0.47**	**0.50**	**0.55**	0.62	0.73	0.96	1.65
4	**0.42**	**0.44**	**0.47**	**0.50**	**0.54**	0.61	0.72	0.95	1.62
5	**0.42**	**0.44**	**0.47**	**0.50**	**0.54**	0.61	0.71	0.93	1.58
6	**0.42**	**0.44**	**0.46**	**0.49**	**0.54**	0.60	0.70	0.91	1.54
PbH$_3$									
−6	0.05	**0.10**	**0.18**	**0.28**	**0.42**	0.63	0.98	1.67	3.76
−5	0.04	**0.10**	**0.16**	**0.25**	**0.38**	0.57	0.89	1.52	3.42
−4	0.04	**0.09**	**0.15**	**0.23**	**0.34**	0.51	0.80	1.37	3.08
−3	0.03	**0.08**	**0.13**	**0.20**	**0.30**	0.46	0.71	1.22	2.74
−2	0.03	**0.07**	**0.11**	**0.18**	**0.27**	0.40	0.62	1.06	2.39
−1	0.03	**0.06**	**0.10**	**0.15**	**0.23**	0.34	0.53	0.91	2.05
0	0.02	0.05	**0.08**	**0.13**	**0.19**	0.29	0.44	0.76	1.71
1	0.02	0.04	**0.07**	**0.10**	**0.15**	0.23	0.35	0.61	1.37
2	0.01	0.03	0.05	**0.08**	**0.11**	0.17	0.27	0.46	1.03
3	0.01	0.02	0.03	0.05	**0.08**	0.11	0.18	0.30	0.68
4	0.00	0.01	0.02	0.03	0.04	0.06	0.09	0.15	0.34
5	0.00	0.00	0.00	0.00	0.00	0.00	0.00	0.00	0.00
6	0.00	−0.01	−0.02	−0.03	−0.04	−0.06	−0.09	−0.15	−0.34
C(SiMe$_3$)$_3$									
−6	−0.09	0.10	−0.11	−0.12	−0.13	−0.15	−0.19	−0.26	−0.47
−5	−0.10	−0.11	−0.12	−0.14	−0.16	−0.19	−0.25	−0.37	−0.72
−4	−0.10	−0.11	−0.13	−0.16	−0.19	−0.24	−0.32	−0.48	−0.97
−3	−0.10	−0.12	−0.14	−0.17	−0.22	−0.28	−0.38	−0.59	−1.22
−2	−0.11	−0.13	−0.16	−0.19	−0.24	−0.32	−0.45	−0.71	−1.48
−1	−0.11	−0.14	−0.17	−0.21	−0.27	−0.36	−0.51	−0.82	−1.73
0	−0.11	−0.14	−0.18	−0.23	−0.30	−0.41	−0.58	−0.93	−1.98
1	−0.12	−0.15	−0.19	−0.25	−0.33	−0.45	−0.65	−1.04	−2.23
2	−0.12	−0.16	−0.20	−0.27	−0.36	−0.49	−0.71	−1.15	−2.48
3	−0.12	−0.16	−0.22	−0.29	−0.38	−0.53	−0.78	−1.27	−2.74
4	−0.13	−0.17	−0.23	−0.30	−0.41	−0.57	−0.84	−1.38	−2.99
5	−0.13	−0.18	−0.24	−0.32	−0.44	−0.62	−0.91	−1.49	−3.24
6	−0.13	−0.18	−0.25	−0.34	−0.47	−0.66	−0.97	−1.60	−3.49

(continued overleaf)

TABLE 4. (continued)

η	10	20	30	P_D 40	50	60	70	80	90
CH(SiMe₃)₂									
−6	−0.06	−0.06	−0.07	−0.07	−0.08	−0.08	−0.10	−0.12	−0.20
−5	−0.07	−0.07	−0.08	−0.09	−0.11	−0.13	−0.17	−0.24	−0.47
−4	−0.07	−0.08	−0.09	−0.11	−0.13	−0.17	−0.23	−0.36	−0.73
−3	−0.07	−0.09	−0.10	−0.13	−0.16	−0.21	−0.30	−0.47	−0.99
−2	−0.07	−0.09	−0.12	−0.15	−0.19	−0.26	−0.37	−0.59	−1.25
−1	−0.08	−0.10	−0.13	−0.17	−0.22	−0.30	−0.44	−0.70	−1.51
0	−0.08	−0.11	−0.14	−0.19	−0.25	−0.35	−0.50	−0.82	−1.77
1	−0.08	−0.11	−0.15	−0.21	−0.28	−0.39	−0.57	−0.94	−2.03
2	−0.09	−0.12	−0.17	−0.23	−0.31	−0.43	−0.64	−1.05	−2.29
3	−0.09	−0.13	−0.18	−0.24	−0.34	−0.48	−0.71	−1.17	−2.55
4	−0.09	−0.14	−0.19	−0.26	−0.37	−0.52	−0.77	−1.28	−2.81
5	−0.10	−0.14	−0.20	−0.28	−0.39	−0.56	−0.84	−1.40	−3.07
6	−0.10	−0.15	−0.22	−0.30	−0.42	−0.61	−0.91	−1.52	−3.34
CH₂SiH₃									
−6	0.00	0.00	0.01	0.01	0.01	0.02	0.03	0.05	0.11
−5	0.00	0.00	−0.01	−0.01	−0.02	−0.02	−0.04	−0.06	−0.14
−4	0.00	−0.01	−0.02	−0.03	−0.04	−0.06	−0.10	−0.17	−0.38
−3	−0.01	−0.02	−0.03	−0.05	−0.07	−0.10	−0.16	−0.28	−0.62
−2	−0.01	−0.02	−0.04	−0.06	−0.10	−0.14	−0.22	−0.38	−0.86
−1	−0.01	−0.03	−0.05	−0.08	−0.12	−0.18	−0.29	−0.49	−1.11
0	−0.02	−0.04	−0.06	−0.10	−0.15	−0.23	−0.35	−0.60	−1.35
1	−0.02	−0.04	−0.08	−0.12	−0.18	−0.27	−0.41	−0.71	−1.59
2	−0.02	−0.05	−0.09	−0.14	−0.20	−0.31	−0.48	−0.82	−1.84
3	−0.03	−0.06	−0.10	−0.15	−0.23	−0.35	−0.54	−0.92	−2.08
4	−0.03	−0.06	−0.11	−0.17	−0.26	−0.39	−0.60	−1.03	−2.32
5	−0.03	−0.07	−0.12	−0.19	−0.29	−0.43	−0.67	−1.14	−2.57
6	−0.03	−0.08	−0.13	−0.21	−0.31	−0.47	−0.73	−1.25	−2.81
CH₂SiPh₃									
−6	−0.01	−0.02	−0.02	−0.02	−0.03	−0.04	−0.06	−0.10	−0.21
−5	−0.01	−0.02	−0.03	−0.04	−0.05	−0.08	−0.11	−0.19	−0.41
−4	−0.02	−0.03	−0.04	−0.06	−0.08	−0.11	−0.17	−0.28	−0.62
−3	−0.02	−0.03	−0.05	−0.07	−0.10	−0.15	−0.22	−0.37	−0.83
−2	−0.02	−0.04	−0.06	−0.09	−0.12	−0.18	−0.28	−0.47	−1.04
−1	−0.03	−0.04	−0.07	−0.10	−0.15	−0.22	−0.33	−0.56	−1.24
0	−0.03	−0.05	−0.08	−0.12	−0.17	−0.25	−0.38	−0.65	−1.45
1	−0.03	−0.06	−0.09	−0.13	−0.19	−0.28	−0.44	−0.74	−1.66
2	−0.03	−0.06	−0.10	−0.15	−0.22	−0.32	−0.49	−0.83	−1.86
3	−0.04	−0.07	−0.11	−0.16	−0.24	−0.35	−0.54	−0.93	−2.07
4	−0.04	−0.07	−0.12	−0.18	−0.26	−0.39	−0.60	−1.02	−2.28
5	−0.04	−0.08	−0.13	−0.19	−0.29	−0.42	−0.65	−1.11	−2.49
6	−0.04	−0.08	−0.14	−0.21	−0.31	−0.46	−0.71	−1.20	−2.69
CH₂SiMe₃									
−6	−0.03	−0.03	−0.03	−0.03	−0.03	−0.02	−0.02	−0.01	−0.01
−5	−0.03	−0.04	−0.04	−0.05	−0.06	−0.07	−0.09	−0.13	−0.26
−4	−0.04	−0.04	−0.05	−0.07	−0.08	−0.11	−0.16	−0.25	−0.52
−3	−0.04	−0.05	−0.07	−0.09	−0.11	−0.15	−0.22	−0.36	−0.76
−2	−0.04	−0.06	−0.08	−0.10	−0.14	−0.20	−0.29	−0.48	−1.04
−1	−0.05	−0.07	−0.09	−0.12	−0.17	−0.24	−0.36	−0.59	−1.30
0	−0.05	−0.07	−0.10	−0.14	−0.20	−0.29	−0.43	−0.71	−1.56
1	−0.05	−0.08	−0.12	−0.16	−0.23	−0.33	−0.49	−0.83	−1.82
2	−0.06	−0.09	−0.13	−0.18	−0.26	−0.37	−0.56	−0.94	−2.08

TABLE 4.   (*continued*)

η	10	20	30	P_D 40	50	60	70	80	90
3	*−0.06*	*−0.09*	*−0.14*	*−0.20*	*−0.29*	*−0.42*	*−0.63*	−1.06	−2.34
4	*−0.06*	*−0.10*	*−0.15*	*−0.22*	*−0.32*	*−0.46*	*−0.70*	−1.17	−2.60
5	*−0.07*	*−0.11*	*−0.17*	*−0.24*	*−0.35*	*−0.50*	*−0.76*	−1.29	−2.87
6	*−0.07*	*−0.12*	*−0.18*	*−0.26*	*−0.37*	*−0.55*	*−0.83*	−1.41	−3.13
CH_2GeMe_3									
−6	*−0.02*	*−0.02*	*−0.02*	*−0.01*	*−0.01*	*−0.01*	0.00	0.01	0.05
−5	*−0.02*	*−0.02*	*−0.03*	*−0.03*	*−0.04*	*−0.05*	*−0.07*	*−0.10*	*−0.20*
−4	*−0.03*	*−0.03*	*−0.04*	*−0.05*	*−0.07*	*−0.09*	*−0.13*	*−0.21*	*−0.45*
−3	*−0.03*	*−0.04*	*−0.05*	*−0.07*	*−0.10*	*−0.13*	*−0.20*	*−0.32*	*−0.70*
−2	*−0.03*	*−0.05*	*−0.06*	*−0.09*	*−0.12*	*−0.18*	*−0.26*	*−0.44*	*−0.96*
−1	*−0.03*	*−0.05*	*−0.08*	*−0.11*	*−0.15*	*−0.22*	*−0.33*	*−0.55*	*−1.21*
0	*−0.04*	*−0.06*	*−0.09*	*−0.13*	*−0.18*	*−0.26*	*−0.39*	*−0.66*	*−1.46*
1	*−0.04*	*−0.07*	*−0.10*	*−0.15*	*−0.21*	*−0.30*	*−0.46*	*−0.77*	*−1.71*
2	*−0.04*	*−0.07*	*−0.11*	*−0.16*	*−0.24*	*−0.34*	*−0.52*	*−0.88*	*−1.96*
3	*−0.05*	*−0.08*	*−0.12*	*−0.18*	*−0.26*	*−0.39*	*−0.59*	*−1.00*	*−2.22*
4	*−0.05*	*−0.09*	*−0.14*	*−0.20*	*−0.29*	*−0.43*	*−0.65*	*−1.11*	*−2.47*
5	*−0.05*	*−0.10*	*−0.15*	*−0.22*	*−0.32*	*−0.47*	*−0.72*	*−1.22*	*−2.72*
6	*−0.06*	*−0.10*	*−0.16*	*−0.24*	*−0.35*	*−0.51*	*−0.79*	*−1.33*	*−2.97*

[a] Values in boldface are electron accepting, values in italics are electron donating and values in ordinary typeface show no significant electrical effect.

TABLE 5.   Values of P_{EA}, P_0 and P_{ED}

X	P_{EA}	P_0	P_{ED}
Si			
$Si(CF_3)_3$	100	0.00	0.00
$Si(CN)_3$	100	0.00	0.00
$SiMe_2H$	41.9	24.8	33.3
$SiMe_3$	36.8	14.5	48.7
$SiEt_3$	36.8	14.5	48.7
$Si(NMe_2)_3$	4.27	1.71	94.0
$SiMe_2OMe$	53.8	30.8	15.4
$SiMe(OMe)_2$	60.7	31.6	7.69
$Si(OH)_3$	93.2	4.27	2.56
$Si(OCF_3)_3$	100	0.00	0.00
$Si(OMe)_3$	74.4	18.8	6.84
$Si(OEt)_3$	50.4	35.0	14.5
$Si(OAc)_3$	100	0.00	0.00
$Si(OPh)_3$	100	0.00	0.00
$Si(OBz)_3$	100	0.00	0.00
$Si(SCN)_3$	100	0.00	0.00
$Si(SCF_3)_3$	100	0.00	0.00
$Si(SMe)_3$	96.6	3.42	0.00
$Si(SEt)_3$	76.9	21.4	1.71
$Si(SPh)_3$	96.6	3.42	0.00
$Si(SeMe)_3$	80.3	19.7	0.00
$Si(C_2H)_3$	100	0.00	0.00
$SiVi_3$	74.4	25.6	0.00
$SiMe_2Ph$	47.0	24.8	28.2
$SiMePh_2$	59.0	33.3	7.69

(*continued overleaf*)

TABLE 5. (*continued*)

X	P_{EA}	P_0	P_{ED}
Si			
$Si(C_6F_5)_3$	100	0.00	0.00
$SiPh_3$	70.9	29.1	0.00
$SiMe_2Br$	71.8	17.1	11.1
$SiMeBr_2$	92.3	2.56	5.13
$SiBr_3$	100	0.00	0.00
$SiMe_2Cl$	81.2	12.0	6.84
$SiMeCl_2$	90.6	2.56	6.84
$SiCl_3$	100	0.00	0.00
$SiMe_2F$	85.5	9.40	5.13
$SiMeF_2$	88.9	3.42	7.69
SiF_3	100	0.00	0.00
SiI_3	100	0.00	0.00
SiH_3	57.3	30.8	12.0
$Si(SiMe_3)_3$	47.9	17.1	35.0
Ge			
$Ge(CF_3)_3$	100	0.00	0.00
$Ge(CN)_3$	100	0.00	0.00
$GeMe_3$	38.5	17.9	43.6
$GeEt_3$	38.5	17.9	43.6
$GeVi_3$	84.6	11.1	4.27
$Ge(C_2H)_3$	65.0	24.8	10.3
$GePh_2Br$	100	0	0
$GePh_2H$	64.1	29.9	5.98
$Ge(C_6F_5)_3$	64.1	26.5	9.40
$GePh_3$	60.7	27.4	12.0
$Ge(SiMe_3)_3$	36.8	15.4	47.9
$Ge(NMe_2)_3$	29.9	5.98	64.1
$Ge(OH)_3$	78.6	9.40	12.0
$Ge(OCF_3)_3$	100	0.00	0.00
$Ge(OMe)_3$	61.5	21.4	17.1
$Ge(OAc)_3$	100	0.00	0.00
$Ge(OPh)_3$	95.7	1.71	2.96
$Ge(OBz)_3$	100	0.00	0.00
$Ge(SCF_3)_3$	100	0.00	0.00
$Ge(SCN)_3$	100	0.00	0.00
$Ge(SMe)_3$	70.9	22.2	6.84
$Ge(SEt)_3$	89.7	9.40	0.85
$Ge(SPh)_3$	100	0.00	0.00
$Ge(SeMe)_3$	100	0.00	0.00
$GeBr_3$	100	0.00	0.00
$GeCl_3$	100	0.00	0.00
GeF_3	100	0.00	0.00
GeI_3	100	0.00	0.00
GeH_3	41.0	35.0	23.9
Sn			
$Sn(CF_3)_3$	100	0.00	0.00
$Sn(CN)_3$	100	0.00	0.00
$SnMe_3$	38.5	17.9	43.6
$SnEt_3$	38.5	17.9	43.6
$SnBu_3$	41.0	15.4	43.6
$SnPh_2Cl$	88.9	4.27	6.84
$SnVi_3$	75.2	24.8	0.00
$Sn(C_2H)_3$	100	0.00	0.00

TABLE 5. (*continued*)

X	P_{EA}	P_0	P_{ED}
SnPh$_2$H	67.5	28.2	4.27
Sn(C$_6$F$_5$)$_3$	100	0.00	0.00
SnPh$_3$	70.9	29.1	0.00
Sn(SiMe$_3$)$_3$	45.3	16.2	38.5
Sn(NMe$_2$)$_3$	35.9	6.84	57.3
Sn(OH)$_3$	93.2	4.27	2.56
Sn(OCF$_3$)$_3$	100	0.00	0.00
Sn(OMe)$_3$	74.4	18.8	6.84
Sn(OAc)$_3$	100	0.00	0.00
Sn(OPh)$_3$	100	0.00	0.00
Sn(OBz)$_3$	100	0.00	0.00
Sn(SCF$_3$)$_3$	100	0.00	0.00
Sn(SCN)$_3$	100	0.00	0.00
Sn(SMe)$_3$	94.4	2.56	0.00
Sn(SEt)$_3$	77.8	21.4	0.85
Sn(SPh)$_3$	98.3	1.71	0.00
Sn(SeMe)$_3$	80.3	19.7	0.00
SnBr$_3$	100	0.00	0.00
SnCl$_3$	100	0.00	0.00
SnF$_3$	100	0.00	0.00
SnI$_3$	100	0.00	0.00
SnH$_3$	58.1	32.5	9.40
Pb			
Pb(CF)$_3$	100	0.00	0.00
Pb(CN)$_3$	100	0.00	0.00
PbMe$_3$	42.7	16.2	41.0
Pb(Vi)$_3$	78.6	21.4	0.00
Pb(C$_2$H)$_3$	100	0.00	0.00
Pb(C$_6$F$_5$)$_3$	100	0.00	0.00
PbPh$_3$	78.6	21.4	0.00
Pb(SiMe$_3$)$_3$	53.8	17.1	29.1
Pb(NMe$_2$)$_3$	40.2	6.84	53.0
Pb(OH)$_3$	100	0.00	0.00
Pb(OCF$_3$)$_3$	100	0.00	0.00
Pb(OMe)$_3$	83.8	14.5	1.71
Pb(OAc)$_3$	100	0.00	0.00
Pb(OPh)$_3$	100	0.00	0.00
Pb(OBz)$_3$	100	0.00	0.00
Pb(SCF3)$_3$	100	0.00	0.00
Pb(SCN)$_3$	100	0.00	0.00
Pb(SMe)$_3$	97.4	2.56	0.00
Pb(SEt)$_3$	85.5	14.5	0.00
Pb(SPh)$_3$	97.4	2.56	0.00
Pb(SeMe)$_3$	88.0	12.0	0.00
PbBr$_3$	100	0.00	0.00
PbCl$_3$	100	0.00	0.00
PbF$_3$	100	0.00	0.00
PbI$_3$	100	0.00	0.00
PbH$_3$	66.7	29.9	3.42
C			
CZ$_3$			
CH$_3$	3.42	31.6	65.0
CPh$_3$	46.2	21.4	32.5
C(NMe$_2$)$_3$	59.0	12.0	29.1

(*continued overleaf*)

TABLE 5. (continued)

X	P_{EA}	P_0	P_{ED}
CZ_3			
C(OMe)$_3$	82.9	3.42	13.7
C(OPh)$_3$	93.2	2.56	4.27
CF$_3$	100	0.00	0.00
CCl$_3$	100	0.00	0.00
CBr$_3$	100	0.00	0.00
CI$_3$	100	0.00	0.00
C(OH)$_3$	88.0	2.56	9.40
C(OAc)$_3$	98.3	0.85	0.85
C(CF$_3$)$_3$	100	0.00	0.00
C(SMe)$_3$	83.7	6.83	9.40
C(SEt)$_3$	78.6	6.84	14.5
CVi$_3$	42.7	23.1	34.2
C(C$_2$H)$_3$	93.2	3.42	3.42
C(C$_6$F$_5$)$_3$	87.2	3.42	9.40
C(SeMe)$_3$	82.1	5.13	12.8
C(SPh)$_3$	87.1	3.42	9.40
C(SiMe$_3$)$_3$	0.00	0.00	100
C(CN)$_3$	100	0.00	0.00
C(OCF$_3$)$_3$	100	0.00	0.00
C(OBz)$_3$	100	0.00	0.00
C(SCN)$_3$	100	0.00	0.00
C(SCF$_3$)$_3$	100	0.00	0.00
CZ_2H			
CHAc$_2$	73.5	13.7	12.8
CHPh$_2$	26.5	32.5	41.0
CH(NMe$_2$)$_2$	46.0	18.0	35.9
CH(OMe)$_2$	67.5	12.0	20.5
CH(OPh)$_2$	80.3	5.13	14.5
CHF$_2$	94.0	2.56	3.42
CHCl$_2$	93.2	2.56	4.27
CHBr$_2$	93.2	2.56	4.27
CHI$_2$	85.5	4.27	10.3
CH(OH)$_2$	74.4	6.84	18.8
CH(OAc)$_2$	81.2	6.84	12.0
CH(CF$_3$)$_2$	94.9	2.56	2.56
CH(SMe)$_2$	69.2	11.1	19.7
CH(SEt)$_2$	62.4	12.8	24.8
CHVi$_2$	15.4	41.0	43.6
CH(C$_2$H)$_2$	70.1	11.1	18.8
CH(C$_6$F$_5$)$_2$	76.1	9.40	14.5
CH(SeMe)$_2$	65.8	12.0	22.2
CH(SPh)$_2$	70.9	10.3	18.8
CH(SiMe$_3$)$_2$	0.00	0.00	100
CH(CN)$_2$	100	0.00	0.00
CH(OCF$_3$)$_2$	95.7	1.71	2.56
CH(OBz)$_2$	87.2	3.13	7.69
CH(SCN)$_2$	100	0.00	0.00
CH(SCF$_3$)$_2$	93.2	2.56	4.27
CH_2Z			
CH$_2$SiH$_3$	0.85	32.5	66.7
CH$_2$SiPh$_3$	0.00	23.9	76.1
CH$_2$Si(OEt)$_3$	0.85	32.5	66.7

TABLE 5. (continued)

X	P_{EA}	P_0	P_{ED}
$CH_2Si(OMe)_3$	1.71	36.8	61.5
CH_2SiMe_3	0.00	16.2	83.8
CH_2SiF_3	48.7	23.1	28.2
CH_2SiCl_3	41.9	26.5	31.6
CH_2SiEt_3	0.00	16.2	83.8
$CH_2CH_2SiMe_3$	5.13	30.8	64.1
$CH_2CH_2SiEt_3$	5.13	30.8	64.1
$CH_2Ge(OMe)_3$	2.56	37.6	59.8
CH_2GeEt_3	0.85	22.2	76.9
CH_2GeMe_3	0.85	22.2	76.9
CH_2GeH_3	1.71	32.5	65.8
CH_2GePh_3	0.00	27.4	72.6
CH_2GeF_3	49.6	21.4	29.1
CH_2GeCl_3	42.7	26.5	30.8
CH_2SnMe_3	0.00	12.8	87.2
CH_2SnPh_3	0.00	23.9	76.1
$CH_2Sn(OMe)_3$	1.71	35.0	63.2
CH_2SnCl_3	41.9	27.4	30.8
CH_2SnF_3	48.7	23.1	28.2
CH_2SnEt_3	0.00	17.9	82.1
CH_2SnH_3	0.85	32.5	66.7
CH_2PbF_3	48.7	23.1	28.2
CH_2PbCl_3	36.8	29.1	34.2
CH_2PbPh_3	0.00	25.6	74.4
$CH_2Pb(OMe)_3$	0.00	35.0	65.0
CH_2PbH_3	0.00	31.6	68.4

1. Entirely EA ($P_{EA} = 100$). Examples: CF_3, $PO(OMe)_2$, $POPh_2$.
2. Predominantly EA ($100 > P_{EA} \geqslant 75$). Examples: NO_2, HCO, CN.
3. Largely EA ($75 > P_{EA} \geqslant 50$). Examples: Cl, C_2Ph, OCN.
4. Ambielectronic ($50 > P_{EA}$ or P_{ED}). SH, CH_2Ph, $SiMe_3$.
5. Largely ED ($75 > P_{ED} \geqslant 50$). Examples: Me, OH, NH_2.
6. Predominantly ED ($100 > P_{ED} \geqslant 75$). Examples: $P = PMe$, $P = POMe$.
7. Entirely ED ($P_{ED} = 100$). Example: $P = PNMe_2$.
The values in italics are based on estimated substituent constants.

2. The nature of group 14 substituent electrical effects

The overall electrical effect of a substituent, as noted above, is a function of its σ_l, σ_d and σ_e values. It depends on the nature of the skeletal group G, the active site Y, the type of phenomenon studied, the medium and the reagent if any. These are the factors that control the values of P_D and η, which in turn determine the contributions of σ_l, σ_d and σ_e.

a. Tricoordinate substituents. Values of P_{EA}, P_{ED} and P_0 in Table 5 show that for Group 14 elements other than carbon the substituents are most often found in the categories of entirely, predominantly or largely electron acceptors. Exceptions are the $M(NMe_2)_3$ groups. For M equal to Ge, Sn and Pb these are predicted to be largely electron donor groups, while for M equal to Si the group is predicted to be a predominantly electron donor group. The $M(SiMe_3)_3$ groups with M equal to Ge, Sn and Pb are also exceptions.

The two former are predicted to be amphielectronic, the latter to be largely electron donor. It is of interest to compare the electrical effects of MZ_3 groups with M equal to C with those in which M is equal to Si, Ge, Sn or Pb. No difference is observed when Z is halogen, OCF_3, SCN, SCF_3 and CF_3, for all of which the electrical effect is entirely electron acceptor. Significant differences appear for Z equal to H, Ph, Vi and particularly for NMe_2.

b. CH_{3-n} $(MZ_3)_n$ groups. When M equals Si, Z equals Me and n is 2 or 3, the group exhibits an entirely electron donor effect; when n is 1 it falls into the predominantly electron donor category. The groups for which Z is Me, n is 1 and M is Ge or Sn also fall into this category. When Z is F, n is 1 and M is Si, Ge, Sn or Pb, we predict that the group will be amphielectronic. When Z is Ph and n is 1, the group is largely or predominantly electron donor.

III. STERIC EFFECTS

A. Introduction

The concept of steric effects was first introduced qualitatively by Kehrmann[30]. By the end of the next decade V. Meyer[31] and J. Sudborough[32] had accumulated kinetic results supporting the steric effect explanation of rate retardation in the estification of suitably substituted benzoic and acrylic acids. Early reviews of steric effects are given by Stewart[33], Wittig[34] and somewhat later by Wheland[35].

B. The Nature of Steric Effects

1. Primary steric effects

Primary steric effects are due to repulsions between electrons in valence orbitals on atoms which are not bonded to each other. They are believed to result from the interpenetration of occupied orbitals on one atom by electrons on the other resulting in a violation of the Pauli exclusion principle. *All steric interactions raise the energy of the system in which they occur.* In terms of their effect on chemical reactivity, they may either decrease or increase a rate or equilibrium constant depending on whether steric interactions are greater in the reactant or in the product (equilibria) or transition state (rate).

2. Secondary steric effects

These effects are due to the shielding of an active site from the attack of a reagent, from solvation, or both. They may also be due to a steric effect on the reacting conformation of a chemical species that determines its concentration.

3. Direct steric effects

These effects can occur when the active site at which a measurable phenomenon occurs is in close proximity to the substituent. Among the many systems exhibiting direct steric effects are *ortho*-substituted benzenes, **1**, *cis*-substituted ethylenes, **2**, and the *ortho*- (1,2-, 2,1- and 2,3-) and peri- (1,8-) substituted naphthalenes, **3, 4, 5** and **6**, respectively. Other examples are *cis*-1,2-disubstituted cyclopropanes, *cis*-2,3-disubstituted norbornanes and *cis*-2,3-disubstituted [2.2.2]-bicyclooctanes, **7, 8** and **9**, respectively.

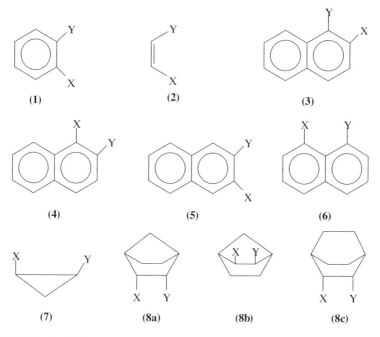

(1)   (2)   (3)

(4)   (5)   (6)

(7)   (8a)   (8b)   (8c)

4. Indirect steric effects

These effects are observed when the steric effect of the variable substituent is relayed by a constant substituent between it and the active site as in **10** where Y is the active site, Z is the constant substituent and X is the variable substituent. This is a buttressing effect.

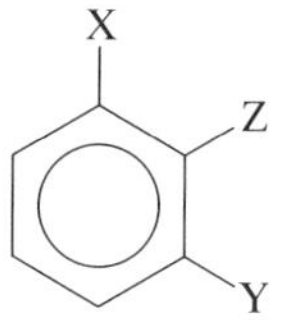

(10)

5. The directed nature of steric effects

Steric effects are vector quantities. This is easily shown by considering, for example, the pentyl and the 1,1-dimethylpropyl groups which have the same volume but a different steric effect. That of the former is less than half that of the latter. In order to account for this let us examine what happens when a nonsymmetric substituent is in contact with an active site. Consider, for example, the simple case of a spherical active site Y, in contact with a carbon substituent, $CZ^LZ^MZ^S$, where the superscripts, L, M and S represent the largest, the medium-sized and the smallest Z groups, respectively. There are three possible conformations of this system which are shown in topviews in Figure 1. As all steric interactions raise the energy of the system, the preferred conformation will be the one that results in the lowest energy increase. This is the conformation which presents the smallest face to the active site, conformation **C**. This is the basis of the minimum

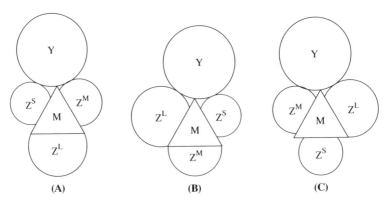

FIGURE 1. Possible conformations of a spherical active site adjacent to a substituent $MZ^{L}Z^{M}Z^{S}$ where L, M and S designate the largest, medium-sized and smallest groups, respectively. Conformation **A** has the lowest energy, conformation **C** the highest

steric interaction (MSI) principle which states: *a nonsymmetric substituent prefers that conformation which minimizes steric interactions.* The directed nature of steric effects permits us to draw a conclusion of vital importance: that the substituent volume is not an acceptable measure of its steric effect[36-38]. Although there are still some workers who are unable to comprehend this point, it is nevertheless true that group volumes are not appropriate as steric parameters. They are actually measures of group polarizability. In short, *steric effects are not directly related to bulk, polarizability is.*

C. The Monoparametric Model of Steric Effects

In the second edition of his book[33] Stewart proposed a parallel between the rate of esterification of 2-substituted benzoic acids and the molecular weights of the substituents, the nitro group strongly deviating from this relationship. The first actual attempt to define a set of steric parameters is due to Kindler[39]. It was unsuccessful; these parameters were later shown to be a function of electrical effects. The first successful parametrization of the steric effect is due to Taft[40], who defined the steric parameter E_S for aliphatic systems by the expression:

$$E_{SX} \equiv \delta \log \frac{k_X}{k_{Me}} \qquad (49)$$

where k_X and k_{Me} are the rate constants for the acid-catalyzed hydrolysis of the corresponding esters XCO_2Ak and $MeCO_2Ak$, respectively. The value of δ is taken as 1.000 for this purpose; $E_{So,X}$ parameters, intended to represent the steric effects of substituents in the *ortho* position of a benzene derivative, were defined for a few groups from the rates of acid-catalyzed hydrolysis of 2-substituted alkyl benzoates. These parameters are a mix of electrical and steric effects with the former predominating, and are therefore of no use as steric parameters.

The Taft $E_{S,X}$ values suffered from several deficiencies:

1. Their validity as measures of steric effects was unproven.

2. They were determined from average values of rate constants obtained under varying experimental conditions.

3. They were available only for derivatives of sp^3-hybridized carbon groups and for hydrogen.

4. The use of the methyl group as the reference substituent meant that they were not compatible with electrical effect substituent constants for which the reference substituent is hydrogen.

The first problem was resolved when it was shown that the E_s values for symmetric groups are a linear function of van der Waals radii[41]. The latter have long been held to be an effective measure of atomic size. The second and third problems were solved by Charton, who proposed the use of the van der Waals radius as a steric parameter[42] and developed a method for the calculation of group van der Waals radii for tetracoordinate symmetric top substituents MZ_3 such as the methyl and trifluoromethyl groups[43]. In later work the hydrogen atom was chosen as the reference substituent and the steric parameter υ was defined as:

$$\upsilon_X \equiv r_{VX} - r_{VH} = r_{VX} - 1.20 \tag{50}$$

where r_{VX} and r_{VH} are the van der Waals radii of the X and H groups in Angstrom units[44]. Expressing r_V in these units is preferable to the use of picometers because the coefficient of the steric parameter is then comparable in magnitude to the coefficients of the electrical effect parameters. Whenever possible, υ parameters are obtained directly from van der Waals radii or calculated from them. Recently, an equation has been derived which makes possible the calculation of υ values for nonsymmetric tetrahedral groups of the types $MZ_2^S Z^L$ and $MZ^S Z^M Z^L$ in which the Z groups are symmetric. These are considered to be primary values. For the greater number of substituents however, υ parameters must be calculated from the regression equations obtained for correlations of rate constants with primary values. The values obtained in this manner are considered to be secondary υ values. Available values of υ for Group 14 substituents, and for a number of other substituents as well, are reported in Table 6. All other measures of atomic size are a linear function of van der Waals radii. There is, therefore, no reason for preferring one measure of atomic size over another. As values of υ were developed for a wide range of substituent types with central atoms including oxygen, nitrogen, sulfur and phosphorus as well as carbon, these parameters provide the widest structural range of substituents for which a measure of the steric effect is available.

1. Steric classification of substituents

Substituents may be divided into three categories based on the degree of conformational dependence of their steric effects:

1. No conformational dependence (NCD). Groups of this type include monatomic substituents such as hydrogen and the halogens: cylindrical substituents such as the ethynyl and cyano groups, and tetracoordinate symmetric top substituents such as the methyl, trifluoromethyl and silyl groups.

2. Minimal conformational dependence (MCD). Among these groups are:

a. Nonsymmetric substituents with the structure $MH_n(lp)_{3-n}$, such as the hydroxyl and amino groups.

b. Nonsymmetric substituents with the structure $MZ_2^S Z^L$ where S stands for small and L for large.

3. Strong conformational dependence (SCD). These groups have the structures:

a. $MZ_2^L Z^S$ and $MZ^L Z^M Z^S$, where the superscript M indicates medium.

b. Planar π-bonded groups such as Ph and carboxy.

c. Quasi-planar π-bonded groups such as dimethylamino and cyclopropyl.

The steric parameter for NCD groups can be obtained directly from van der Waals radii or calculated from them. The values for SCD groups are often obtainable from van der Waals radii, although in some cases they must be derived as secondary values from regression equations obtained by correlating rate constants with known values of

TABLE 6. Values of υ and υ_i

X	υ	υ_1	υ_2	υ_3
Si				
Si(CF$_3$)$_3$			0.90	0.27
Si(CN)$_3$			0.40	0.40
SiMe$_2$H			0.52	0
SiMe$_3$	1.40	1.40	0.52	0
SiEt$_3$	1.40	1.40	0.52	0.52
SiMe$_2$OMe			0.52	0
SiMe(OMe)$_2$			0.52	0
SI(OH)$_3$			0.32	0
Si(OCF$_3$)$_3$			0.32	0.90
Si(OMe)$_3$			0.32	0.52
Si(OEt)$_3$			0.32	0.52
Si(OAc)$_3$			0.32	0.50
Si(O$_2$CCF$_3$)$_3$			0.32	0.50
Si(OPh)$_3$			0.32	0.57
Si(OBz)$_3$			0.32	0.50
Si(SCN)$_3$			0.60	0.40
Si(SCF$_3$)$_3$			0.60	0.90
Si(SMe)$_3$			0.60	0.52
Si(SEt)$_3$			0.60	0.52
Si(SPh)$_3$			0.60	0.57
Si(SeMe)$_3$			0.70	0.52
Si(C$_2$H)$_3$			0.58	0.58
SiVi$_3$			0.57	0.57
SiMe$_2$Ph		1.40	0.57	0.57
SiMePh$_2$			0.57	0.57
SiPh$_3$			0.57	0.57
SiMe$_2$OSiMe$_3$			0.52	0
SiMe(OSiMe$_3$)$_2$			0.52	0
Si(OSiMe$_3$)$_3$			0.32	1.40
SiMe$_2$Br		1.40	0.65	0
SiMeBr$_2$			0.65	0
SiBr$_3$	1.69	1.69	0.65	0
SiMeBrCl			0.65	0
SiMe$_2$Cl		1.40	0.55	0
SiMeCl$_2$			0.52	0
SiCl$_3$	1.50	1.50	0.55	0
SiMe$_2$F			0.52	0
SiMeF$_2$		1.01	0.52	0
SiF$_3$	1.01	1.01	0.27	0
SiI$_3$	1.93	1.93	0.78	0
SiH$_3$	0.70	0.70	0	0
Si(SiMe$_3$)$_3$			1.40	0.52
Ge				
Ge(CF$_3$)$_3$			0.90	0.27
Ge(CN)$_3$			0.40	0.40
GeMe$_3$	1.44	1.44	0.52	0
GeEt$_3$		1.44	0.52	0.52
GeVi$_3$			0.57	0.57
Ge(C$_2$H)$_3$			0.58	0.58
GePh$_2$Br			0.57	0.57
GePh$_2$H			0.57	0.57
GePh$_3$			0.57	0.57
Ge(SiMe$_3$)$_3$			1.40	0.52

TABLE 6. (*continued*)

X	v	v_1	v_2	v_3
Ge(OH)$_3$			0.32	0
Ge(OCF$_3$)$_3$			0.32	0.90
Ge(OMe)$_3$			0.32	0.52
Ge(OAc)$_3$			0.32	0.50
Ge(OPh)$_3$			0.32	0.57
Ge(OBz)$_3$			0.32	0.50
Ge(SCF$_3$)$_3$			0.60	0.90
Ge(SCN)$_3$			0.60	0.40
Ge(SMe)$_3$			0.60	0.52
Ge(SEt)$_3$			0.60	0.52
Ge(SPh)$_3$			0.60	0.57
Ge(SeMe)$_3$			0.70	0.52
GeBr$_3$	1.73	1.73	0.65	0
GeCl$_3$	1.53	1.53	0.55	0
GeF$_3$	1.05	1.05	0.27	0
GeI$_3$	1.96	1.96	0.78	0
GeH$_3$	0.72	0.72	0	0
Sn				
Sn(CF$_3$)$_3$			0.90	0.27
Sn(CN)$_3$			0.40	0.40
SnMe$_3$	1.55	1.55	0.52	0
SnEt$_3$		1.55	0.52	0.52
SnBu$_3$		1.55	0.52	0.52
SnPh$_2$Cl			0.57	0.57
SnVi$_3$			0.57	0.57
Sn(C$_2$H)$_3$			0.58	0.58
SnPh$_2$H			0.57	0.57
Sn(C$_6$F$_5$)$_3$				0.57
SnPh$_3$			0.57	0.57
Sn(Pn-F-4)$_3$			0.57	0.57
Sn(Pn-F-3)$_3$			0.57	0.57
Sn(SiMe$_3$)$_3$			1.40	0.52
Sn(OH)$_3$			0.32	0
Sn(OCF$_3$)$_3$			0.32	0.90
Sn(OMe)$_3$			0.32	0.52
Sn(OAc)$_3$			0.32	0.50
Sn(OPh)$_3$			0.32	0.57
Sn(OBz)$_3$			0.32	0.50
Sn(SCF$_3$)$_3$			0.60	0.90
Sn(SCN)$_3$			0.60	0.40
Sn(SMe)$_3$			0.60	0.52
Sn(SEt)$_3$			0.60	0.52
Sn(SPh)$_3$			0.60	0.57
Sn(SeMe)$_3$			0.70	0.52
SnBr$_3$			0.65	0
SnCl$_3$			0.55	0
SnF$_3$			0.27	0
SnI$_3$			0.78	0
SnH$_3$			0	0
Pb				
Pb(CF$_3$)$_3$			0.90	0.27
Pb(CN)$_3$			0.40	0.40
PbMe$_3$			0.52	0
Pb(Vi)$_3$			0.57	0.57

(*continued overleaf*)

TABLE 6.   (*continued*)

X	υ	υ_1	υ_2	υ_3
Pb				
Pb(C$_2$H)$_3$			0.58	0.58
PbPh$_3$			0.57	0.57
Pb(SiMe$_3$)$_3$			1.40	0.52
Pb(OH)$_3$			0.32	0
Pb(OCF$_3$)$_3$			0.32	0.90
Pb(OMe)$_3$			0.32	0.52
Pb(OAc)$_3$			0.32	0.50
Pb(OPh)$_3$			0.32	0.57
Pb(OBz)$_3$			0.32	0.50
Pb(SCF$_3$)$_3$			0.60	0.90
Pb(SCN)$_3$			0.60	0.40
Pb(SMe)$_3$			0.60	0.52
Pb(SEt)$_3$			0.60	0.52
Pb(SPh)$_3$			0.60	0.57
Pb(SeMe)$_3$			0.70	0.52
PbBr$_3$			0.65	0
PbCl$_3$			0.55	0
PbF$_3$			0.27	0
PbI$_3$			0.78	0
PbH$_3$			0	0
C				
CZ_3				
CH$_3$	0.52	0.52	0	0
C(OMe)$_3$		0.99	0.52	0
C(OPh)$_3$		0.99	0.32	0.57
CF$_3$	0.90	0.90	0.27	0
CCl$_3$	1.38	1.38	0.55	0
CBr$_3$	1.56	1.56	0.65	0
CI$_3$	1.79	1.79	0.78	0
C(OH)$_3$	0.99	0.99	0.32	0
C(OAc)$_3$		0.99	0.32	0.50
C(CF$_3$)$_3$			0.90	0.27
C(SMe)$_3$		1.46	0.52	0
C(SEt)$_3$		1.46	0.52	0.52
CVi$_3$			0.57	0.57
C(C$_2$H)$_3$	1.06	1.06	0.58	0.58
C(SeMe)$_3$			0.70	0.52
C(SPh)$_3$		1.46	0.57	0.57
C(SiMe$_3$)$_3$			1.40	0.52
C(CN)$_3$	0.90	0.90	0.40	0.40
C(OCF$_3$)$_3$		0.99	0.90	0.27
C(OBz)$_3$		0.99	0.50	0.57
C(SCN)$_3$			0.60	0.40
C(SCF$_3$)$_3$		1.46	0.90	0.27
CZ_2H				
CHAc$_2$			0.50	0.52
CH(NMe$_2$)$_2$			0.35	
CH(OMe)$_2$			0.32	0.52
CH(OPh)$_2$			0.32	0.57
CHF$_2$	0.68	0.68	0.27	0
CHCl$_2$	0.81	0.81	0.55	0
CHBr$_2$	0.89	0.89	0.65	0
CHI$_2$	0.97	0.97	0.78	0

TABLE 6. (*continued*)

X	υ	υ_1	υ_2	υ_3
$CH(OH)_2$			0.32	0
$CH(OAc)_2$			0.32	0.50
$CH(CF_3)_2$			0.90	0.27
$CH(SMe)_2$			0.60	0.52
$CH(SEt)_2$			0.60	0.52
$CHVi_2$			0.57	0.57
$CH(C_2H)_2$			0.58	0.58
$CH(C_6F_5)_2$				
$CH(SeMe)_2$			0.70	0.52
$CH(SPh)_2$			0.60	0.57
$CH(SiMe_3)_2$			1.40	0.52
$CH(CN)_2$			0.40	0.40
$CH(OCF_3)_2$			0.32	0.90
$CH(OBz)_2$			0.32	0.50
$CH(SCN)_2$			0.60	0.40
$CH(SCF_3)_2$			0.60	0.90
CH₂Z				
CH_2SiH_3		0.52	0.70	0
CH_2SiPh_3		0.52		0.57
$CH_2Si(OEt)_3$		0.52		0.32
$CH_2Si(OMe)_3$		0.52		0.32
CH_2SiMe_3		0.52	1.40	0.52
CH_2SiF_3		0.52	1.01	0.27
CH_2SiCl_3		0.52	1.50	0.55
CH_2SiEt_3		0.52	1.40	0.52
$CH_2CH_2SiMe_3$		0.52	0.52	1.40
$CH_2CH_2SiEt_3$		0.52	0.52	1.40
$CH_2Ge(OMe)_3$		0.52		0.32
CH_2GeEt_3		0.52	1.44	0.52
CH_2GeMe_3		0.52	1.44	0.52
CH_2GeH_3		0.52	0.72	0
CH_2GePh_3		0.52		0.57
CH_2GeF_3		0.52	1.05	0.27
CH_2GeCl_3		0.52	1.53	0.55
CH_2SnMe_3		0.52	1.55	0.52
CH_2SnPh_3		0.52		0.57
$CH_2Sn(OMe)_3$		0.52		0.32
CH_2SnCl_3		0.52		0.55
CH_2SnF_3		0.52		0.27
CH_2SnEt_3		0.52	1.55	0.52
CH_2SnH_3		0.52		0
CH_2PbF_3		0.52		0.27
CH_2PbCl_3		0.52		0.55
CH_2PbPh_3		0.52		0.57
$CH_2Pb(OMe)_3$		0.52		0.32
CH_2PbH_3		0.52		0

the steric parameter. Steric parameters for SCD groups of the nonsymmetric type are only obtainable from regression equations. In the case of planar π-bonded groups the maximum and minimum values of the steric parameter are available from the van der Waals radii. These groups are sufficiently common and important to require a more detailed discussion.

2. Planar π-bonded groups

These ($X_{p\pi}$) groups represent an especially difficult problem because their delocalized electrical effect depends on the steric effect when they are bonded to planar π-bonded skeletal groups, $G_{p\pi}$. An approach to the problem has been developed[45,46]. The σ_d and σ_e electrical effect parameters are a function of the dihedral angle formed by $X_{p\pi}$ and $G_{p\pi}$. The relationship generally used has the form:

$$P = P_0 \cos^2 \theta \tag{51}$$

where P is the property of interest, P_0 is its value when the dihedral angle is zero and θ is the dihedral angle. Thus:

$$\sigma_{dX,\theta} = \sigma_{dX,0} \cos^2 \theta \tag{52}$$

and:

$$\sigma_{eX,\theta} = \sigma_{eX,0} \cos^2 \theta \tag{53}$$

where $\sigma_{dX,0}$ and $\sigma_{eX,0}$ are the values of σ_d and σ_e when the substituent and skeletal group are coplanar ($\theta = 0$). The effective value of υ is given by the expression:

$$\upsilon = d \cos \theta + r_{vzs} - 1.20 \tag{54}$$

where Z^S is the smaller of the two Z groups attached to the central atom, M of the $X_{p\pi}$ group. There is no simple *a priori* way to determine θ. It could conceivably be estimated by molecular mechanics calculations, but there is some reason to believe that θ is a function of the medium. Alternatively, the $X_{p\pi}$ group can be included in the data set by means of an iteration procedure. The method requires an initial correlation of the data set with all $X_{p\pi}$ and other SCD groups excluded. This constitutes the basis set. The correlation equation used for this purpose is the LDRS equation in the form:

$$Q_X = L\sigma_{lX} + D\sigma_{dX} + R\sigma_{eX} + S\upsilon + h \tag{55}$$

The correlation is then repeated for each $X_{p\pi}$ group using υ values increasing incrementally by some convenient amount from the minimum, which represents the half-thickness of the group, to the maximum, which occurs when $X_{p\pi}$ is nearly perpendicular to $G_{p\pi}$. The proper value of θ is that which:

1. Results in the best fit of the data to the correlation equation.
2. Has the L, D, R, S and h values that are in best agreement with those of the basis set.

D. Multiparametric Models of Steric Effects

When the active site is itself large and nonsymmetric, or alternatively when the phenomenon studied is some form of bioactivity in which binding to a receptor is the key step, a simple monoparametric model of the steric effect will often be insufficient. It is then necessary to make use of a multiparametric model of steric effects. Five multiparametric models are available: that of Verloop[47], the simple branching model, the expanded branching model, the segmental model and the composite model. The Verloop model suffers from the fact that its parameters measure maximum and minimum distances perpendicular to the group axis. These maxima and minima may occur at any point in the group skeleton (the longest chain in the group). The steric effect, however, may be very large at one segment of the chain and negligible at others. If a data set is large, the likelihood that the maximum and minimum distances of all groups are located at the same segment and that it is this segment at which the steric effect is important is very small. The Verloop model will therefore not be discussed further. The composite model is a combination of the monoparametric υ model with the simple branching model. The method has proven useful in modelling amino acid, peptide and protein properties[48].

1. The branching equations

The simple branching model[44,46] for the steric effect is given by the expression:

$$S\psi = \sum_{i=1}^{m} a_i n_i + a_b n_b \tag{56}$$

where the a_i and a_b are coefficients, n_i is the number of branches attached to the i-th atom, and n_b is the number of bonds between the first and last atoms of the group skeleton. It follows that n_b is a measure of group length. Unfortunately, it is frequently highly collinear in group polarizability. For saturated cyclic substituents it is necessary to determine values of n_i from an appropriate regression equation. For planar π-bonded groups n_i is taken to be 1 for each atom in the group skeleton. For other groups n_i is obtained simply by counting branches. The model makes the assumption that all of the branches attached to a skeleton atom are equivalent. This is only a rough approximation. Distinguishing between branches results in an improved model, called the expanded branching equation:

$$S\psi = \sum_{i=1}^{m} \sum_{j=1}^{3} a_{ij} n_{ij} + a_b n_b \tag{57}$$

which allows for the difference in steric effect that results from the order of branching[44,46]. This difference is a natural result of the MSI principle. The first branch has the smallest steric effect, because a conformation in which it is rotated out of the way of the active site is possible. This rotation becomes more difficult with the second branch and impossible with the third. The problem with the expanded branching method is that it requires a large number of parameters. Rarely does one encounter a data set large enough to permit its use.

2. The segmental model

As both branching methods have problems associated with them, the segmental method[46] is often the simplest and most effective. In this model each atom of the group skeleton together with the atoms attached to it constitutes a segment of the substituent. Applying the MSI principle, the segment is considered to have that conformation which presents its smallest face to the active site. The segment is assigned the υ value of the group which it most resembles. Values of the segmental steric parameters υ_i, where i designates the segment number, are given in Table 6. Numbering starts from the first atom of the group skeleton, that is, the atom which is attached to the rest of the species. The expression for the steric effect using the segmental model is:

$$S\psi = \sum_{i=1}^{m} S_i \upsilon_i \tag{58}$$

When only steric effects are present:

$$Q_X = S\psi_X \tag{59}$$

In the general case electrical effects are also present and the general form of the LDRS equation:

$$Q_X = L\sigma_{DX} + D\sigma_{dX} + R\sigma_{eX} + S\psi_X + h \tag{60}$$

is required.

IV. INTERMOLECULAR FORCES

A. Introduction

Inter- and intramolecular forces (imf) are of vital importance in the quantitative description of structural effects on bioactivities and chemical properties. They can make a significant contribution to chemical reactivities and some physical properties as well. Types of intermolecular forces and their present parametrization are listed in Table 7.

B. Parametrization of Intermolecular Forces

1. Hydrogen bonding

For the description of hydrogen bonding two parameters are required, one to account for the hydrogen atom donating capacity and another to account for the hydrogen atom accepting capacity of a group. We have used for this purpose n_H, the number of OH and/or NH bonds in the substituent, and n_n, the number of lone pairs on oxygen and/or nitrogen atoms[49-51]. The use of these parameters is based on the argument that if one of the phases involved in the phenomenon studied includes a protonic solvent, particularly water, then all of the hydrogen bonds the substituent is capable of forming will indeed form. For such a system, hydrogen bond parameters defined from equilibria in highly dilute solution in an 'inert' solvent are unlikely to be a suitable model. The parametrization we have used accounts only for the number of hydrogen-donor and hydrogen-acceptor sites in a group. It does not take into account differences in hydrogen bond energy. A more sophisticated parametrization than that described above would be the use of the hydrogen bond energy for each type of hydrogen bond formed. Thus for each substituent the parameter E_{hbX} would be given by the equation:

$$E_{hbX} = \sum_{i=1}^{m} n_{hbi} E_{hbi} \tag{61}$$

TABLE 7. Intermolecular forces and the quantities upon which they depend[a]

Intermolecular force	Quantity
molecule–molecule	
Hydrogen bonding (hb)	E_{hb}, n_H, n_n
Dipole–dipole (dd)	dipole moment
Dipole–induced dipole (di)	dipole moment, polarizability
Induced dipole–induced dipole (ii)	polarizability
Charge transfer (ct)	ionization potential, electron affinity
ion–molecule	
Ion–dipole (Id)	ionic charge, dipole moment
Ion–induced dipole (Ii)	ionic charge, polarizability

[a] Abbreviations are in parentheses. The dd interactions are also known as Keesom interactions; di interactions are also known as Debye interactions; ii interactions are also known as London or dispersion interactions. Collectively, dd, di and ii interactions are known as van der Waals interactions. Charge transfer interactions are also known as donor–acceptor interactions.

where E_{hbX} is the hydrogen bonding parameter, E_{hbi} is the energy of the i-th type of hydrogen bond formed by the substituent X and n_{hbi} is the number of such hydrogen bonds. The validity of this parametrization is as yet untested. In any event, the parametrization we have used suffers from the fact that though it accounts for the number of hydrogen bonds formed, it does not differentiate between their energies and can therefore be only an approximation. Much remains to be done in properly parametrizing hydrogen bonding.

2. van der Waals interactions

These interactions (dd, di, ii) are a function of dipole moment and polarizability. It has been shown that the dipole moment cannot be replaced entirely by the use of electrical effect substituent constants as parameters. This is because the dipole moment has no sign. Either an overall electron donor group or an overall electron acceptor group may have the same value of μ. We have also shown that the bond moment rather than the molecular dipole moment is the parameter of choice. The dipole moments of MeX and PhX were taken as measures of the bond moments of substituents bonded to sp^3- and sp^2-hybridized carbon atoms, respectively, of a skeletal group. Application to substituents bonded to sp-hybridized carbon atoms should require a set of dipole moments for substituted ethynes. The polarizability parameter we have chosen, α, is given by the expression:

$$\alpha \equiv \frac{MR_X - MR_H}{100} = \frac{MR_X}{100} - 0.0103 \tag{62}$$

where MR_X and MR_H are the group molar refractivities of X and H, respectively[40-42]. The factor 1/100 is introduced to scale the α parameter so that its coefficients in the regression equation are roughly comparable to those obtained for the other parameters used. There are many other polarizability parameters including parachor, group molar volumes of various kinds, van der Waals volumes and accessible surface areas any of which will do as well, as they are all highly collinear in each other[43-45]. Proposing other polarizability parameters seems to be a popular occupation of many authors.

Values of α can be estimated by additivity from the values for fragments. They may also be estimated from group molar refractivities calculated from the equation:

$$MR_X = 0.320n_c + 0.682n_b - 0.0825n_n + 0.991 \tag{63}$$

where n_c, n_b, and n_n are the number of core, bonding and nonbonding electrons, respectively, in the group X^{36}.

3. Charge transfer interactions

These interactions can be roughly parametrized by the indicator variables n_A and n_D, where n_A takes the value 1 when the substituent is a charge transfer acceptor and 0 when it is not; n_D takes the value 1 when the substituent is a charge transfer donor and 0 when it is not. An alternative parametrization makes use of the first ionization potential of MeX (ip_{MeX}) as the electron donor parameter and the electron affinity of MeX as the electron acceptor parameter. We have generally found the indicator variables n_A and n_D to be sufficient. This parametrization accounts for charge transfer interactions directly involving the substituent. If the substituent is attached to a π-bonded skeletal group, then the skeletal group is capable of charge transfer interaction the extent of which is modified by the substituent. This is accounted for by the electrical effect parameters.

4. The intermolecular force (IMF) equation

We may now write a general relationship for the quantitative description of intermolecular forces:

$$Q_X = L\sigma_{lX} + D\sigma_{dX} + R\sigma_{eX} + A\alpha_X + H_1 n_{HX}$$

$$+ H_2 n_{nX} + Ii_X + B_{DX} n_{DX} + B_{AX} n_{AX} + S\psi_X + B^0 \qquad (64)$$

Values of the IMF parameters for Group 14 substituents are set forth in Table 8.

TABLE 8. Values of intermolecular force parameters[a]

X	μ_{MeX}	μ_{PhX}	α	n_H	n_n
Si					
$Si(CF_3)_3$	3.17	3.64	0.241	0	0
$Si(CN)_3$	4.16	6.08	0.260	0	0
$SiMe_2H$	0.08	0.30	0.193	0	0
$SiMe_3$	**0**	**0.31**	0.239	0	0
$SiEt_3$		**0.71**	0.380	0	0
$Si(NMe_2)_3$	2.08	2.90	0.536	0	0
$SiMe_2OMe$	0.11	0.67	0.261	0	1
$SiMe(OMe)_2$	0.27	0.83	0.283	0	2
$SI(OH)_3$	0.74	1.28	0.149	3	3
$Si(OCF_3)_3$	2.63	3.72	0.305	0	3
$Si(OMe)_3$	**1.61**	**1.62**	0.305	0	3
$Si(OEt)_3$	**1.72**	**1.68**	0.443	0	3
$Si(OAc)_3$	1.95	2.90	0.443	0	6
$Si(O_2CCF_3)_3$				0	6
$Si(OPh)_3$	**1.417**	1.94	0.902	0	3
$Si(OBz)_3$	2.13	3.17	1.022	0	6
$Si(SCN)_3$	3.32	4.81	0.473	0	0
$Si(SCF_3)_3$	2.63	3.92	0.485	0	0
$Si(SMe)_3$	**1.84**	1.64	0.485	0	0
$Si(SEt)_3$	0.52	1.25	0.623	0	0
$Si(SPh)_3$	0.94	1.92	1.100	0	3
$Si(SeMe)_3$	0.60	1.40	0.581	0	0
$Si(C_2H)_3$	1.88	3.11	0.356	0	3
$SiVi_3$	0.51	1.34	0.401	0	0
$SiMe_2Ph$	0.03	0.46	0.436	0	1
$SiMePh_2$	0.15	0.74	0.633	0	2
$Si(C_6F_5)_3$	2.38	3.63	0.791	0	0
$SiPh_3$	**0.28**	1.24	0.830	0	3
$SiMe_2OSiMe_3$			0.448	0	1
$SiMe(OSiMe_3)_2$			0.657	0	2
$Si(OSiMe_3)_3$			0.866	0	3
$SiMe_2Br$	0.39	0.78	0.272	0	0
$SiMeBr_2$	1.24	1.71	0.305	0	0
$SiBr_3$	**1.86**	**2.36**	0.338	0	0
$SiMeBrCl$			0.276	0	0
$SiMe_2Cl$	0.49	0.97	0.243	0	0
$SiMeCl_2$	1.52	2.06	0.247	0	0
$SiCl_3$	**1.91**	**2.43**	0.251	0	0
$SiMe_2F$	0.58	1.04	0.191	0	0
$SiMeF_2$	1.17	1.54	0.145	0	0
SiF_3	2.36	**2.80**	0.098	0	0

TABLE 8. (continued)

X	μ_{MeX}	μ_{PhX}	α	n_H	n_n
SiI$_3$	2.11	3.26	0.458	0	0
SiH$_3$	**0.7351**	0.72	0.101	0	0
Si(SiMe$_3$)$_3$	0.11	0.42	0.818	0	0
Ge					
Ge(CF$_3$)$_3$	3.04	4.36	0.336	0	0
Ge(CN)$_3$	4.04	5.74	0.275	0	0
GeMe$_3$	0.20	**0.623**	0.254	0	0
GeEt$_3$	0.20	0.37	0.395	0	0
GeVi$_3$	0.38	1.00	0.416	0	0
Ge(C$_2$H)$_3$	1.76	1.95	0.371	0	3
GePh$_2$Br	1.11	1.77	0.681	0	2
GePh$_2$H	0.31	0.90	0.602	0	2
Ge(C$_6$F$_5$)$_3$	2.21	2.35	0.806	0	0
GePh$_3$	0.31	0.91	0.845	0	3
Ge(SiMe$_3$)$_3$	0.29	0.03	0.833	0	0
Ge(NMe$_2$)$_3$	1.19	0.97	0.551	0	0
Ge(OH)$_3$	0.61	0.95	0.164	3	3
Ge(OCF$_3$)$_3$	2.50	3.39	0.320	0	3
Ge(OMe)$_3$	**1.91**	0.62	0.320	0	3
Ge(OAc)$_3$	1.82	2.57	0.458	0	6
Ge(OPh)$_3$	1.03	1.61	0.917	0	3
Ge(OBz)$_3$	2.00	2.84	1.037	0	6
Ge(SCF$_3$)$_3$	2.51	3.59	0.500	0	0
Ge(SCN)$_3$	3.19	4.48	0.488	0	0
Ge(SMe)$_3$	0.68	1.31	0.500	0	0
Ge(SEt)$_3$	0.40	0.92	0.638	0	0
Ge(SPh)$_3$	0.81	1.59	1.115	0	0
Ge(SeMe)$_3$	0.48	1.07	0.596	0	0
GeBr$_3$	2.30	3.22	0.353	0	0
GeCl$_3$	**2.63**	3.14	0.266	0	0
GeF$_3$	**3.8**	2.90	0.113	0	0
GeI$_3$	1.99	2.92	0.503	0	0
GeH$_3$	**0.644**	0.38	0.116	0	0
Sn					
Sn(CF$_3$)$_3$	3.19	4.73	0.289	0	0
Sn(CN)$_3$	4.16	6.08	0.328	0	0
SnMe$_3$	0.23	**0.51**	0.307	0	0
SnEt$_3$	0.23	0.41	0.448	0	0
SnBu$_3$	0.23	0.41	0.736	0	0
SnPh$_2$Cl	1.19	1.89	0.705	0	2
SnVi$_3$	0.53	1.38	0.469	0	0
Sn(C$_2$H)$_3$	1.90	3.15	0.424	0	3
SnPh$_2$H	0.38	1.06	0.655	0	2
Sn(C$_6$F$_5$)$_3$	2.33	3.68	0.859	0	0
SnPh$_3$	0.43	1.24	0.898	0	3
Sn(Pn-F-4)$_3$			0.895	0	3
Sn(Pn-F-3)$_3$			0.895	0	3
Sn(SiMe$_3$)$_3$	0.16	0.36	0.886	0	0
Sn(NMe$_2$)$_3$	**1.36**	0.58	0.604	0	0
Sn(OH)$_3$	0.74	1.28	0.217	3	3
Sn(OCF$_3$)$_3$			0.373	0	3
Sn(OMe)$_3$	0.45	0.95	0.373	0	3
Sn(OAc)$_3$	1.97	2.94	0.511	0	6

(continued overleaf)

TABLE 8. (*continued*)

X	μ_{MeX}	μ_{PhX}	α	n_H	n_n
Sn					
Sn(OPh)$_3$	2.43	3.44	0.970	0	3
Sn(OBz)$_3$	1.01	0.77	1.090	0	6
Sn(SCF$_3$)$_3$	2.63	3.92	0.553	0	0
Sn(SCN)$_3$	3.34	4.86	0.541	0	0
Sn(SMe)$_3$	0.83	1.68	0.553	0	0
Sn(SEt)$_3$	0.54	1.29	0.691	0	0
Sn(SPh)$_3$	0.96	1.97	1.168	0	0
Sn(SeMe)$_3$	0.60	1.40	0.649	0	0
SnBr$_3$	**3.20**	3.55	0.406	0	0
SnCl$_3$	**3.62**	**3.99**	0.319	0	0
SnF$_3$	2.38	3.28	0.166	0	0
SnI$_3$	**2.64**	3.30	0.556	0	0
SnH$_3$	**0.68**	0.76	0.169	0	0
Pb					
Pb(CF)$_3$	3.24	4.89	0.306	0	0
Pb(CN)$_3$	4.23	6.28	0.345	0	0
PbMe$_3$	0.21	0.26	0.324	0	0
Pb(Vi)$_3$	0.17	0.33	0.486	0	0
Pb(C$_2$H)$_3$	1.95	3.31	0.441	0	3
Pb(C$_6$F$_5$)$_3$	2.45	3.94	0.876	0	0
PbPh$_3$	0.56	1.50	0.915	0	3
Pb(SiMe$_3$)$_3$	0.04	0.62	0.906	0	0
Pb(NMe$_2$)$_3$	0.94	0.38	0.621	0	0
Pb(OH)$_3$	0.81	1.49	0.234	3	3
Pb(OCF$_3$)$_3$	2.72	3.99	0.390	0	3
Pb(OMe)$_3$	0.52	1.16	0.390	0	3
Pb(OAc)$_3$	2.07	3.16	0.528	0	6
Pb(OPh)$_3$	1.23	2.14	0.987	0	3
Pb(OBz)$_3$	2.20	3.38	1.107	0	6
Pb(SCF$_3$)$_3$	2.70	4.12	0.570	0	0
Pb(SCN)$_3$	3.38	5.02	0.558	0	0
Pb(SMe)$_3$	0.88	1.84	0.570	0	0
Pb(SEt)$_3$	0.64	1.51	0.708	0	0
Pb(SPh)$_3$	1.00	2.13	1.185	0	0
Pb(SeMe)$_3$	0.72	1.66	0.666	0	0
PbBr$_3$	2.54	3.81	0.423	0	0
PbCl$_3$	2.50	3.73	0.336	0	0
PbF$_3$	2.43	3.44	0.186	0	0
PbI$_3$	2.18	3.46	0.573	0	0
PbH$_3$	0.37	0.97	0.186	0	0
C					
CZ$_3$					
CH$_3$	**0**	**0.37**	0.046	0	0
CPh$_3$	0.27	0.33	0.775	0	0
C(NMe$_2$)$_3$	0.48	0.43	0.481	0	3
C(OMe)$_3$	1.13	1.38	0.250	0	6
C(OPh)$_3$	1.59	2.04	0.847	0	6
CF$_3$	2.29	3.03	0.040	0	0
CCl$_3$	1.97	2.67	0.191	0	0
CBr$_3$	1.97	2.67	0.283	0	0
CI$_3$	1.59	1.88	0.433	0	0
C(OH)$_3$	1.38	1.70	0.094	3	6
C(OAc)$_3$	1.52	2.05	0.386	0	9

TABLE 8. (continued)

X	μ_{MeX}	μ_{PhX}	α	n_H	n_n
$C(CF_3)_3$	1.59	2.27	0.166	0	0
$C(SMe)_3$	1.13	1.45	0.250	0	0
$C(SEt)_3$	0.94	1.18	0.568	0	0
CVi_3	0.22	0.29	0.346	0	0
$C(C_2H)_3$	1.06	1.50	0.301	0	3
$C(C_6F_5)_3$	1.18	1.53	0.736	0	0
$C(SeMe)_3$	1.01	1.28	0.526	0	0
$C(SPh)_3$	1.18	1.53	1.045	0	0
$C(SiMe_3)_3$	0.89	1.17	0.760	0	0
$C(CN)_3$	2.43	3.48	0.205	0	0
$C(OCF_3)_3$	2.17	2.93	0.250	0	3
$C(OBz)_3$	1.78	2.40	0.967	0	9
$C(SCN)_3$	2.38	3.31	0.418	0	0
$C(SCF_3)_3$	1.85	2.61	0.430	0	0
CZ_2H					
$CHAc_2$	0.62	0.98	0.250	0	4
$CHPh_2$	0.07	0.08	0.532	0	2
$CH(NMe_2)_2$	0.22	0.14	0.481	0	2
$CH(OMe)_2$	0.62	0.76	0.182	0	4
$CH(OPh)_2$	0.94	1.20	0.580	0	2
CHF_2	1.40	1.85	0.042	0	0
$CHCl_2$	1.20	1.64	0.144	0	0
$CHBr_2$	1.20	1.64	0.204	0	0
CHI_2	0.94	1.10	0.304	0	0
$CH(OH)_2$	0.80	0.98	0.078	2	4
$CH(OAc)_2$	0.87	1.17	0.274	0	6
$CH(CF_3)_2$	0.94	1.10	0.126	0	0
$CH(SMe)_2$	0.62	0.81	0.302	0	2
$CH(SEt)_2$	0.48	0.60	0.384	0	2
$CHVi_2$	0.02	0.03	0.246	0	0
$CH(C_2H)_2$	0.60	0.86	0.216	0	2
$CH(C_6F_5)_2$	0.67	0.99	0.506	0	0
$CH(SeMe)_2$	0.55	0.69	0.366	0	0
$CH(SPh)_2$	0.67	0.87	0.712	0	2
$CH(SiMe_3)_2$	0.70	0.93	0.523	0	0
$CH(CN)_2$	1.52	2.18	0.152	0	0
$CH(OCF_3)_2$	1.32	1.80	0.182	0	2
$CH(OBz)_2$	1.06	1.43	0.660	0	6
$CH(SCN)_2$	1.47	2.05	0.294	0	0
$CH(SCF_3)_2$	1.13	1.57	0.302	0	0
CH_2Z					
CH_2SiH_3	0.31	0.41	0.147	0	0
CH_2SiPh_3	0.38	0.48	0.876	0	3
$CH_2Si(OEt)_3$	0.31	0.41	0.489	0	3
$CH_2Si(OMe)_3$	0.24	0.31	0.351	0	3
CH_2SiMe_3	0.50	0.68	0.285	0	0
CH_2SiF_3	0.34	0.51	0.144	0	0
CH_2SiCl_3	0.27	0.39	0.297	0	0
CH_2SiEt_3	0.50	0.68	0.426	0	0
$CH_2CH_2SiMe_3$	0.36	0.52	0.331	0	0
$CH_2CH_2SiEt_3$	0.36	0.52	0.472	0	0
$CH_2Ge(OMe)_3$	0.24	0.31	0.366	0	3
CH_2GeEt_3	0.43	0.57	0.441	0	0

(continued overleaf)

TABLE 8. (*continued*)

X	μ_{MeX}	μ_{PhX}	α	n_H	n_n
CH_2Z					
CH_2GeMe_3	0.43	0.57	0.300	0	0
CH_2GeH_3	0.31	0.42	0.162	0	0
CH_2GePh_3	0.36	0.45	0.891	0	3
CH_2GeF_3	0.34	0.50	0.159	0	0
CH_2GeCl_3	0.29	0.45	0.312	0	0
CH_2SnMe_3	0.48	0.61	0.353	0	0
CH_2SnPh_3	0.38	0.48	0.944	0	3
$CH_2Sn(OMe)_3$	0.24	0.30	0.419	0	3
CH_2SnCl_3	0.29	0.46	0.365	0	0
CH_2SnF_3	0.34	0.51	0.212	0	0
CH_2SnEt_3	0.48	0.63	0.494	0	0
CH_2SnH_3	0.31	0.41	0.215	0	0
CH_2PbF_3	0.34	0.51	0.229	0	0
CH_2PbCl_3	0.27	0.43	0.382	0	0
CH_2PbPh_3	0.38	0.49	0.961	0	3
$CH_2Pb(OMe)_3$	0.29	0.35	0.436	0	3
CH_2PbH_3	0.31	0.40	0.232	0	0

[a] μ values in boldface are from A. L. McClellan, *Tables of Experimental Dipole Moments*, W. H. Freeman, San Francisco, 1963; A. L. McClellan, *Tables of Experimental Dipole Moments*, Vol. 2, Rahara Enterprises, El Cerrito, Cal., 1974. Other values were calculated from equations 83 and 84. α values were calculated assuming additivity.

V. APPLICATIONS

A. Introduction

We have applied the methods described above to a number of data sets involving chemical reactivities, chemical properties and physical properties. In several cases the data sets were chosen because they provided an opportunity to test the validity of some of the parameters estimated in this work. All of the data sets studied are reported in Table 9.

B. Chemical Reactivity

The correlation of pK_a values of 4-substituted benzoic acids in aqueous ethanol[52] has been carried out with a correlation equation[53,54] resulting from the addition of terms in the mole fraction of ethanol, ϕ_{EtOH}, and the polarizability of X, α_X, to the LDR equations. Thus:

$$pK_{aX} = L\sigma_{lX} + D\sigma_{dX} + R\sigma_{eX} + A\alpha_X + F\phi_X + h \tag{65}$$

The data used are given in Table 9. The best regression equation obtained was:

$$pK_{aX} = -1.51 \ (\pm 0.0873)\sigma_{lX} - 1.10 \ (\pm 0.0896)\sigma_{dX} + 3.59 \ (\pm 0.114)\phi_X + 4.78 \ (\pm 0.0524) \tag{66}$$

$100R^2$, 97.70; $A100R^2$, 97.57; F, 494.6; S_{est}, 0.133; S^0, 0.160; n, 39; P_D, 42.3 (± 3.98); η, 0; r_{ld}, 0.135; $r_{l\phi}$, 0.011; $r_{d\phi}$, 0.084.

We have calculated pK_a values for a number of Group 14 substituents. Values of pK_a calculated and observed, and of Δ, given by

$$\Delta = pK_{a_{obs}} - pK_{a_{calc}} \tag{67}$$

were reported in Table 9 for Group 14 substituents other than carbon.

TABLE 9.   Data used in correlations

1. pK_a, 4-XPnCO$_2$H, aq. EtOH, 25 °C[a].

Φ, 0.163: H, 5.16; Me, 5.37; OMe, 5.45; NO$_2$, 4.06; Br, 4.83; Cl, 4.83; F, 4.95; Φ, 0.236: H, 5.70; SiMe$_3$, 5.80; PO(OMe)$_2$, 4.90; Ac, 5.07; Me, 5.88; OMe, 6.03; Br, 5.35; Cl, 5.32; OCF$_3$, 5.19; CN, 4.70; SCF$_3$, 5.01; SF$_5$, 4.70; SMe, 5.74; CF$_3$, 4.99; OPh, 5.50; Φ, 0.371: H, 5.98; t-Bu, 6.16; CEt$_3$, 6.15; SiMe$_3$, 5.96; Φ, 0.481: H, 6.66; Me, 6.86; OMe, 7.08; NO$_2$, 5.41; Br, 6.19; Cl, 6.23; Φ, 0.696: H, 7.25; Me, 7.46; OMe, 7.59; NO$_2$, 5.93; Br, 6.74; Cl, 6.67; F, 6.98.

2. pK_a, XCO$_2$H, aq. EtOH, 25 °C[a].

Φ, 0.144: H, 4.18; Me, 5.36; Et, 5.59; CH$_2$Ph, 5.13; CHPh$_2$, 5.09; Ph, 5.10; Φ, 0.236: E-2-PhVn, 5.68; Vi, 5.46; E-2-MeVi, 5.91; CH$_2$Ph, 5.63; cHx^m, 6.49; C$_2$Ph, 3.58; Ph, 5.70; CCl$_3$, 1.98; CHCl$_2$, 2.44; Φ, 0.242: Me, 5.61; t-Bu, 6.41; CHPh$_2$, 5.77; SiMe$_3$, 6.60; Ph, 5.74; Φ, 0.281: H, 4.53; Me, 5.80; Et, 6.07; E-2-MeVi, 5.979; CH$_2$Ph, 5.62; CHPh$_2$, 5.64; c-Hx, 6.30; Ph, 5.76; Φ, 0.500: H, 5.25; Me, 6.55; Et, 6.85; CH$_2$Ph, 6.37; CHPh$_2$, 6.34; Φ, 0.553: Me, 6.57; Et, 7.19; E-2-MeVi, 6.88[b]. SiMe,$_3$, 7.74; CH$_2$Vi, 6.54[b].

3. ΔG_{acid}, (kcal mol)$^{-1}$, XCH$_2^-$ (g)[c].

CF$_3$SO$_2$, 339.8; Me$_2$P, 383.8; I, 379.4; SPh, 374.2; SiMe$_3$, 390.7; SMe, 386.0; SO$_2$Ph, 355.3; Vi, 384.1; Br, 285.8; SO$_2$Me, 358.2; Cl, 389.1; 2-Fur, 377.0; Ph, 373.7; OMe, 398.1; CN, 365.2; H, 408.8; SOMe, 366.4; Me, 412.2.

4. ΔG_{acid} (kcal mol)$^{-1}$, 4-XPnCH$_2^-$, (g)[c].

O$_2$N, 345.3; SO$_2$Ph, 352.1; SO$_2$Me, 352.1; NMe$_2$, 379.0; CHO, 352.6; B, 353.5; CN, 353.6; Ac, 354.9; CO$_2$Me, 355.4; Me, 374.8; SOMe, 359.3; CF$_3$, 359.8; H, 373.7; Cl, 366.2; F, 372.4.

5. k_{rel}, 4-XPnSiMe$_3$ + H$_3$O$^+$, HCl$_4$, aq. MeOH[d].

NMe$_2$, 3 × 10^7; OH, 10,700; OMe, 1510; Me, 32.2; Et, 19.5; i-Pr, 17.2; t-Bu, 15.6; SPh, 10.7; Ph, 355; SiMe$_3$, 2.5; H, 1; F, 0.75; Cl, 0.13; Br, 1.10.

6. k_{rel}, 3- or 4-XPnGeEt$_3$, HClO$_4$, aq. MeOH[e].

3-OMe, 0.51; Me, 1.78; F, 0.032; Cl, 0.019; Br, 0.019; CO$_2$H, 0.0177. 4-OMe, 39.2; Me, 12.4; Ph, 2.43; H, 1; Cl, 0.108; p-Br, 0.133; CO$_2$H, 0.0052.

7. k_{rel}, 3- or 4-XPnSn(cHx)$_3$, HClO$_4$, aq. EtOH[f].

3-X: OMe, 0.89; Me, 1.84; Cl, 0.039; 4-X: NMe$_2$, 20,000; OMe, 63; Me, 5.6; Et, 5.3; i-Pr, 6.95; t-Bu, 7.05; Ph, 1.77; H, 1; F, 0.62; Cl, 0.187; Br, 0.145; CO$_2$H, 0.030.

8. k_{rel}, 3- or 4-XPnSn(c-Hx)$_3$^{m} + I$_2$, CCl$_4$^{g}.

3-X: OMe, 22; Me, 42; Cl, 0.039; 4-X: OMe, 69; t-Bu, 13.9; i-Pr, 12.1; Et, 10.1; Me, 7.5; Ph, 2.9; H, 1; F, 0.22; Cl, 0.10; Br, 0.08; CO$_2$H, 0.0145.

9. 10^2K_2 (l/mol min), 3- or 4-XPnNMe$_2$ + MeI, MeAc, 25 °C[h].

3-X: OMe, 1.03; Me, 1.85; Cl, 0.16; Br, 0.14; 4-X: OMe, 9.0; Me, 3.4; H, 1.17; Cl, 0.39; Br, 0.37.

10. 10^2K_2 (l/mol s), 3- or 4-XPnCO$_2$H + Ph$_2$CN$_2$, EtOH, 35 °C[h,i].

3-X: OMe, 2.15; Me, 1.60; Br, 4.33; NO$_2$, 8.68; CPh$_3$, 1.85; 4-X: H, 1.78; Cl, 2.78; Br, 3.32; NO$_2$, 8.87; CPh$_3$, 2.01.

11. 10^2K_2 (l/mol s), 3- or 4-XPnCOCl + PhNH$_2$, PhH, 25 °C[h].

3-X: Me, 4.6; Br, 16.5; I, 15.4; 4-X: OMe, 1.49; Me, 3.1; H, 6.2; Cl, 9.3; Br, 10.1.

12. K, XCl + Me$_3$SiNH$_2$[j].

Si(CHCl$_2$)Me$_2$, 38.2; SiPh$_2$H, 27.6; SiPh$_2$Me, 16.09; SiPhViMe, 9.4; SiMe$_2$(CH$_2$Cl), 7.9; SiMe$_2$Ph, 3.5; SiMe$_2$H, 2.6; SiMe$_2$Vi, 1.7.

(continued overleaf)

TABLE 9. (*continued*)

13. IP (eV)[k].

H, 9.24; Me, 8.85; *t*-Bu, 8.40; OMe, 8.39; OH, 8.56; NH_2, 8.05; NMe_2, 7.37; Vi, 8.49; C_2H, 8.82; Ph, 8.39; F, 9.35; Cl, 9.10; Br, 8.99; I, 8.67; CHO, 9.71; Ac, 9.51; $CONH_2$, 9.45; Bz, 9.42; CN, 9.72; N_3, 8.72; SH, 8.47; SPh, 7.86; SMe, 8.07; $SiMe_3$, 8.22; SO_2Me, 9.74; SO_2Ph, 9.37; CF_3, 9.68; COF, 9.78; CO_2Me, 9.34; COCl, 9.54; *c*-Pr, 8.66.

14. (ΔU°_V) (kJ mol)$^{-1}$ at 298.15 K[l].

Me, 7.67; Et, 13.86; *t*-Bu, 19.94; Vi, 13.66; $ViCH_2$, 13.66; E-CH=CHMe, 19.51; $H_2C=C=CH$, 21.21; Ph, 35.58; CH_2Ph, 39.79; 2-Tpn, 36.43; 3-Tpn, 36.98; NO_2, 35.88; C_6F_5, 38.66; CH_2CN, 33.71; CCl_3, 30.14; $CONMe_2$, 47.76; $CHCl_2$, 28.30; CH_2Br, 25.78; CH_2I, 29.57; NMe_2, 19.72; CN, 30.92; OMe, 16.87; OPh, 44.43; CHO, 23.65; Ac, 28.79; COEt, 32.44; OAc, 30.02; CO_2Me, 30.02; OBz, 53.09; SMe, 25.18; SSMe, 35.42; SEt, 29.52, Vi, 13.66.

^{a}Ref. 52. ^{b}at 23 °C. ^{c}Ref. 55. ^{d}Ref. 57. ^{e}Ref. 58. ^{f}Ref. 59. ^{g}Ref. 60. ^{h}Ref. 61. ^{i}Ref. 62. ^{j}Ref. 63. ^{k}Refs. 64–72. ^{l}Ref. 73. ^{m}*c*-Hx = cyclohexyl.

pK_a values for XCO_2H in aqueous ethanol[52] were correlated with a form of equation 65 to which a term in υ_1, the first segmental steric parameter, was added to account for steric effects. This gave the correlation equation[53,54]:

$$pK_{aX} = L\sigma_{lX} + D\sigma_{dX} + R\sigma_{eX} + H\alpha_X + F\phi_X + S_1\upsilon_X + h \tag{68}$$

The best regression equation obtained was:

$$pK_{aX} = -10.0 \ (\pm 0.363)\sigma_{lX} - 2.57 \ (\pm 0.541)\sigma_{dX} - 13.5 \ (\pm 1.30)\sigma_{eX} - 1.14 \ (\pm 0.435)\alpha_X$$
$$+ 3.22 \ (\pm 0.267)\phi_X + 0.863 \ (\pm 0.163)\upsilon_{lX} + 3.73 \ (\pm 0.143) \tag{69}$$

$100R^2$, 97.52; $A100R^2$, 97.13; F, 202.9; S_{est}, 0.221; S^0, 0.174; n, 38; P_D, 20.4 (± 4.43); η, 5.24 ($\pm$?); r_{ld}, 0.030; r_{le}, 0.314; $r_{l\alpha}$, 0.231; $r_{l\phi}$, 0.269; $r_{l\upsilon}$, 0.219; r_{de}, 0.482; $r_{d\alpha}$, 0.164; $r_{d\phi}$, 0.024; $r_{d\upsilon}$, 0.323; $r_{e\alpha}$, 0.493; $r_{e\phi}$, 0.218; $r_{e\upsilon}$, 0.099; $r_{\alpha\phi}$, 0.171; $r_{\alpha\upsilon}$, 0.671; $r_{\phi\upsilon}$, 0.048.

Values of ΔG_{acid} for the gas phase protonation[55] of substituted methide ions, XCH_2^-, were correlated with the LDRA equation:

$$Q_X = L\sigma_{lX} + D\sigma_{dX} + R\sigma_{eX} + A\alpha_X + h \tag{70}$$

giving the regression equation:

$$\Delta G_{acid,X} = -54.4 \ (\pm 4.34)\sigma_{lX} - 41.2 \ (\pm 5.25)\sigma_{dX} + 119 \ (\pm 22.1)\sigma_{eX}$$
$$- 37.6 \ (\pm 13.8)\alpha_X + 405.9 \ (\pm 2.45) \tag{71}$$

$100R^2$, 96.11; $A100R^2$, 95.27; F, 80.20; S_{est}, 4.12; S^0, 0.232; n, 18; P_D, 43.1 (± 6.30); η, -2.89 (± 0.390); r_{ld}, 0.278; r_{le}, 0.234; $r_{l\alpha}$, 0.037; r_{de}, 0.460; $r_{d\alpha}$, 0.001; $r_{e\alpha}$, 0.565.

Values of Q_{obs}, Q_{calc}, and Δ for Group 14 substituents are given in Table 10. Values of ΔG_{acid} for the gas phase protonation[55] of 4-substituted phenylmethide ions, $4\text{-}XPnCH_2^-$, were also correlated with the LDRA equation giving the regression equation:

$$\Delta G_{acid,X} = -26.7 \ (\pm 1.98)\sigma_{lX} - 28.3 \ (\pm 1.50)\sigma_{dX} + 36.0 \ (\pm 6.43)\sigma_{eX} + 372.7 \ (\pm 0.993) \tag{72}$$

TABLE 10. Values of $Q_{X,\text{obs}}$, $Q_{X,\text{calc}}$ and Δ

X	$Q_{X,\text{obs}}$	$Q_{X,\text{calc}}$	Δ	N_{SD}	Q type	Parameters
MZ^{1}Z^{2}Z^3						
SiH$_3$	9.18	9.14	0.04	0.265	IP	$\sigma_l, \sigma_d, \sigma_e$
		9.16	0.02	0.132	IP	$\sigma_l, \sigma_d, \sigma_e$
	0.73	0.27	0.47	1.29	μ_{MeX}	σ_l, σ_d
	380.0	393.4	-13.4	3.25	ΔG acid	$\sigma_l, \sigma_d, \sigma_e, \alpha$
		392.2	-12.2	2.96	ΔG acid	
SiH$_2$Me	384.0	393.5	-9.5	2.31	ΔG acid	
SiEt$_3$	0.71	0.09	0.62	1.81	μ_{PhX}	$\sigma_l, \sigma_d, \sigma_e$
	5.95	6.12	-0.17	1.28	pK_a	σ_l, σ_d
SiHPh$_2$	9.13	8.62	0.51	3.38	IP	$\sigma_l, \sigma_d, \sigma_e, \alpha$
SiPh$_3$	0.28	0.47	-0.19	0.521	μ_{MeX}	σ_l, σ_d
		0.49	-0.21	0.575	μ_{MeX}	
	-0.54	-0.68	0.14	1.45	$\log k_{\text{rel}}$	σ_{c50}
		-0.65	0.11	1.14	$\log k_{\text{rel}}$	σ_{c50}
	0.303	0.538	0.235	5.06	$\log k_{\text{rel}}$	σ_{c50}
		0.530	0.227	4.89	$\log k_{\text{rel}}$	σ_{c50}
	0.820	0.918	-0.098	2.75	$\log k_{\text{rel}}$	σ_{c50}, σ_e
		0.932	-0.112	3.14	$\log k_{\text{rel}}$	
SiMe$_2$F	9.17	9.09	0.08	0.530	IP	$\sigma_l, \sigma_d, \sigma_e, \alpha$
SiF$_2$Me	9.55	9.19	0.36	2.38	IP	
SiF$_3$	10.23	9.39	0.84	5.56	IP	
		9.70	0.53	3.51	IP	
		9.71	0.52	3.44	IP	
	2.80	3.24	-0.54	1.58	μ_{PhX}	$\sigma_l, \sigma_d, \sigma_e$
	363.7	363.0	0.7	0.170	ΔG acid	$\sigma_l, \sigma_d, \sigma_e, \alpha$
		373.7	$-10.$	2.43	ΔG acid	$\sigma_l, \sigma_d, \sigma_e, \alpha$
		373.3	-9.6	2.33	ΔG acid	$\sigma_l, \sigma_d, \sigma_e, \alpha$
SiMe$_2$Cl	9.3	9.01	0.29	1.92	IP	$\sigma_l, \sigma_d, \sigma_e, \alpha$
SiMeCl$_2$	9.52	9.18	0.34	2.25	IP	$\sigma_l, \sigma_d, \sigma_e, \alpha$
SiCl$_3$	9.55	9.16	0.39	2.58	IP	$\sigma_l, \sigma_d, \sigma_e, \alpha$
		9.32	0.23	1.52	IP	$\sigma_l, \sigma_d, \sigma_e, \alpha$
		9.60	0.05	0.331	IP	$\sigma_l, \sigma_d, \sigma_e, \alpha$
	1.91	2.03	-0.12	0.329	μ_{MeX}	σ_l, σ_d
		2.41	-0.48	1.32	μ_{MeX}	
	2.40	3.48	-1.08	3.16	μ_{PhX}	$\sigma_l, \sigma_d, \sigma_e$
		2.37	0.03	0.0877	μ_{PhX}	
SiBr	1.86	2.45	-0.69	1.89	μ_{MeX}	σ_l, σ_d
		1.97	-0.11	0.301	μ_{MeX}	σ_l, σ_d
	2.36	3.55	-1.19	3.48	μ_{PhX}	$\sigma_l, \sigma_d, \sigma_e$
		2.30	0.06	0.175	μ_{PhX}	$\sigma_l, \sigma_d, \sigma_e$
SiMe$_2$OMe	9.34	8.97	0.37	2.45	IP	$\sigma_l, \sigma_d, \sigma_e, \alpha$
Si(OMe)$_3$	1.61	0.48	1.13	3.10	μ_{MeX}	
	1.62	0.95	0.95	2.78	μ_{PhX}	
Si(OEt)$_3$	1.72	0.22	1.46	4.71	μ_{MeX}	
	1.68	0.59	1.09	3.19	μ_{PhX}	
Si(OPh)$_3$	1.417	1.19	0.23	0.630	μ_{MeX}	
SiMe$_2$Ph	6.05	6.09	-0.04	0.181	$pK_{a_{XCO_2H}}$	$\sigma_l, \sigma_d, \sigma_e, \alpha,$ ϕ, υ
GeH$_3$	0.664	0.14	0.504	1.38	μ_{MeX}	
	3.5	4.31	0.8	ca 2.5	pK_a **(11)**	$\sigma_l, \sigma_d, \sigma_e$
GeMe$_3$	0.623	0.11	0.52	1.52	μ_{PhX}	
	8.98	8.80	0.18	1.19	IP	
	5.97	6.11	-0.14	1.05	pK_a	$\sigma_l, \sigma_d, \phi,$
	6.41	6.65	-0.24	1.09	pK_a	$\sigma_l, \sigma_d, \sigma_e, \alpha,$

(continued overleaf)

TABLE 10.   (*continued*)

X	$Q_{X,obs}$	$Q_{X,calc}$	Δ	N_{SD}	Q type	Parameters
						ϕ, υ
	7.43	7.66	−0.23	1.04	pK_a	ϕ, υ
GeEt$_3$	5.97	6.11	−0.14	1.05	pK_a	
GePh$_2$H	9.15	8.54	0.61	4.04	IP	
GeMe$_2$Ph	6.00	6.24	−0.24	1.09	pK_a	σ_l, σ_d, σ_e, α,
						ϕ, υ
GePh$_3$	−0.44	−0.57	0.13	1.35	log k_{rel}	σ_{c50}
	−0.43		0.01	0.104	log k_{rel}	
	0.348	0.504	0.156	3.36	log k_{rel}	
	0.461		0.113	2.50	log k_{rel}	
GeCl$_3$	2.63	2.28	0.35	0.959	μ_{MeX}	σ_l, σ_d
Ge(OMe)$_3$	1.91	0.35	1.56	4.27	μ_{MeX}	
Ge(OEt)$_3$	1.59	0.19	1.40	3.84	μ_{MeX}	
GeH$_3$	3.5	4.2	0.7	?	pK_a	σ_l, σ_d, σ_e
SnH$_3$	0.68	0.29	0.39	1.07	μ_{MeX}	
SnMe$_3$	0.51	0.09	0.42	1.15	μ_{MeX}	
	8.94	8.73	0.21	1.39	IP	
	5.98	6.12	−0.14	1.05	pK_a	σ_l, σ_d, ϕ,
SnEt$_3$	5.93	6.12	−0.19	1.43	pK_a	σ_l, σ_d, ϕ
SnCl$_3$	3.02	2.43	1.19	3.29	μ_{MeX}	
		2.05	1.57	4.30	μ_{MeX}	
	3.99	3.52	0.47	1.37	μ_{PhX}	
		2.38	1.71	5.00	μ_{PhX}	
SnBr$_3$	3.20	2.45	0.75	2.05	μ_{MeX}	
SnI$_3$	2.64	2.16	0.54	1.48	μ_{MeX}	
Sn(NMe$_2$)$_3$	(−)1.36	(−)0.98	−0.38	1.04	μ_{MeX}	
PbMe$_3$	8.82	8.71	0.12	0.795	IP	σ_l, σ_d, σ_e, α
PbEt$_3$	0.82	0.17	0.67	1.96	μ_{MeX}	
CZ^{1}Z^{2}Z^3						
CH$_2$SiMe$_3$	8.42	8.47	−0.05	0.331	IP	
	6.33	6.34	−0.01	0.0752	pK_a	
	6.08	5.86	0.22	1.65	pK_a	
CH(SiMe$_3$)$_2$	8.10	8.09	0.01	0.0662	IP	
	6.14	5.93	0.21	1.58	pK_a	
C(SiMe$_3$)$_3$	8.10	7.71	0.39	2.58	IP	
	6.05	5.99	0.06	0.462	pK_a	
CH$_2$GeMe$_3$	8.40	8.48	−0.08	0.530	IP	
	6.12^a	5.83	0.29	2.18	pK_a	

$100R^2$, 98.31; $A100R^2$, 98.03; F, 213.2; S_{est}, 1.51; S^0, 0.152; n, 15; P_D, 51.6 (±3.59); η, −1.27 (±0.217); r_{ld}, 0.173; r_{le}, 0.173; r_{de}, 0.086.

Kuznesoff and Jolly[56] have reported a pK_a value of 3.5 for the acid GeH$_3$CO$_2$H, **12**, in water. We have correlated the pK_a values of XCO$_2$H in water with the LDR equation to obtain the relationship:

$$pK_{aX} = -9.63 \ (\pm 0.696)\sigma_{lX} - 1.61 \ (\pm 0.799)\sigma_{dX} - 6.09 \ (\pm 1.78)\sigma_{eX} + 4.237 \ (\pm 0.129)$$
$$(73)$$

$100R^2$, 96.37; F, 159.1; S_{est}, 0.280; S^0, 0.211; n, 22; P_D, 14.3 (±7.24); η, 3.80 (±?); r_{ld}, 0.512; r_{le}, 0.159; r_{de}, 0.519.

Calculated and observed values of the pK_a of **11** (**11** = H$_3$GeCO$_2$H) are given in Table 10.

Rate constants for the protodetrimethylsilylation of 4-substituted phenyltrimethylsilanes were correlated with the LDRA equation giving the regression equation:

$$\log k_{\text{rel,X}} = -8.80 \ (\pm 1.11)\sigma_{\text{IX}} - 11.1 \ (\pm 1.11)\sigma_{\text{dX}} - 5.48 \ (\pm 2.62)\sigma_{\text{eX}} + 0.215 \ (\pm 0.246)$$

$100R^2$, 95.21; $A100R^2$, 94.34; F, 66.22; S_{est}, 0.555; S^0, 0.259; n, 14; P_{D}, 55.8 (± 7.11); η, 0.493 (± 0.231); r_{Id}, 0.672; r_{Ie}, 0.111; r_{de}, 0.357.

Rate constants for the protodetriethylgermylation of 3- or 4-substituted phenyltriethylgermanes and the protodetricyclohexylstannylation of 3- or 4-substituted phenyltricyclohexylstannanes were correlated with the modified Yukawa–Tsuno (MYT) equation (equation 17) to give the regression equations:

$$\log k_{\text{rel,X}} = -5.18 \ (\pm 0.165)\sigma_{\text{cX}} - 2.31 \ (\pm 1.03)\sigma_{\text{eX}} - 0.115 \ (\pm 0.0514) \tag{74}$$

$100R^2$, 99.17; $A100R^2$, 99.09; F, 595.3; S_{est}, 0.121; S^0, 0.104; n, 13; η, 0.446 (± 0.197); r_{ce}, 0.349

for the former reaction and:

$$\log k_{\text{rel,X}} = -4.42 \ (\pm 0.383)\sigma_{\text{cX}} - 7.00 \ (\pm 1.41)\sigma_{\text{eX}} + 0.0611 \ (\pm 0.0943) \tag{75}$$

$100R^2$, 96.44; $A100R^2$, 96.16; F, 162.5; S_{est}, 0.290; S^0, 0.211; n, 15; η, 1.58 (± 0.289); r_{ce}, 0.563

for the latter. Correlation of the relative rate constants for the iododetricyclohexylstannylation of 3- or 4-substituted phenyl tricyclohexylstannes with the MYT equation gave the regression equation:

$$\log k_{\text{rel,X}} = -4.90 \ (\pm 0.419)\sigma_{\text{cX}} - 4.04 \ (\pm 2.33)\sigma_{\text{eX}} + 0.171 \ (\pm 0.106) \tag{76}$$

$100R^2$, 93.40; $A100R^2$, 92.85; F, 77.80; S_{est}, 0.312; S^0, 0.290; n, 14; η, 0.825 (± 0.471); r_{ce}, 0.216.

We have also correlated rate constants for the reaction of 3- or 4-substituted N,N-dimethylanilines with methyl iodide, 3- or 4-substituted benzoic acids with diphenyldiazomethane and 3- or 4-substituted benzoyl chlorides with aniline with the MYT equation. The best regression equations obtained are:

$$\log k_{\text{X}} = -2.80 \ (\pm 0.155)\sigma_{\text{cX}} + 0.160 \ (\pm 0.0339) \tag{77}$$

$100r^2$, 97.89; F, 325.4; S_{est}, 0.0940; S^0, 0.165; n, 9.

$$\log k_{\text{X}} = 0.853 \ (\pm 0.0485)\sigma_{\text{cX}} + 0.282 \ (\pm 0.0187) \tag{78}$$

$100r^2$, 97.48; F, 309.0; S_{est}, 0.0464; S^0, 0.178; n, 10

and:

$$\log k_{\text{X}} = 1.29 \ (\pm 0.0825)\sigma_{\text{cX}} + 3.89 \ (\pm 0.824)\sigma_{\text{eX}} + 0.772 \ (\pm 0.0217) \tag{79}$$

$100R^2$, 99.29; $A100R^2$, 99.17; F, 350.1; S_{est}, 0.0357; S^0, 0.107; n, 8; η, 3.01 (± 0.607); r_{ce}, 0.686

respectively. Values of $\log k_{\text{calc}}$ and Δ for SiPh_3 and GePh_3 groups from these regression equations are reported in Table 10.

We have correlated equilibrium constants for the reaction:

$$ClSiZ^1Z^2Z^3 + Me_2NSiMe_3 \rightleftharpoons Me_2NSiZ^1Z^2Z^3 + ClSiMe_3$$

with the correlation equation:

$$\log K_X = L\sigma_{lX} + D\sigma_{dX} + A\alpha_X + h \tag{80}$$

A term in σ_{eX} was not included as the values of this parameter for the substituents in the data set were essentially constant. The best regression equation obtained was:

$$\log K_X = 11.5 \ (\pm 2.01)\sigma_{lX} + 5.42 \ (\pm 1.46)\sigma_{dX} + 0.329 \ (\pm 0.325) \tag{81}$$

$100R^2$, 91.47; $A100R^2$, 90.52; F, 42.89; S_{est}, 0.211; S^0, 0.342; n, 11; P_D, 32.1 (± 9.85); r_{ld}, 0.473.

This result supports the validity of the σ_l and σ_d constants for the substituents in the data set.

C. Chemical Properties (QSCR) and Physical Properties (QSPR)

Vertical ionization potentials of the π_s orbital in substituted benzenes were correlated with the LDRA equation (equation 70) to give the regression equation:

$$IP_X = 0.948 \ (\pm 0.150)\sigma_{lX} + 1.49 \ (\pm 0.0992)\sigma_{dX} + 3.78 \ (\pm 0.604)\sigma_{eX}$$
$$- 1.36 \ (\pm 0.390)\alpha + 9.25 \ (\pm 0.0751) \tag{82}$$

$100R^2$, 94.89; $A100R^2$, 94.30; F, 116.0; S_{est}, 0.151; S^0, 0.248; n, 30; P_D, 61.2 (± 6.06); η, -2.53 (± 0.367); r_{ld}, 0.098; r_{le}, 0.139; $r_{l\alpha}$, 0.119; r_{de}, 0.216; $r_{d\alpha}$, 0.115; $r_{e\alpha}$, 0.428.

Values of IP_{calc} and Δ calculated from the regression equation are reported in Table 10.

We have reported elsewhere[51] the results of the correlation of dipole moments for MeX and PhX where X is a symmetric substituent with the LDR equation. The regression equations obtained are for the MeX:

$$\mu_{MeX} = 5.11 \ (\pm 0.497)\sigma_{lX} + 1.99 \ (\pm 0.541)\sigma_{dX} + 0.0129 \ (\pm 0.205) \tag{83}$$

and for the PhX:

$$\mu_{PhX} = 5.47 \ (\pm 0.150)\sigma_{lX} + 4.30 \ (\pm 0.446)\sigma_{dX} + 6.94 \ (\pm 1.91)\sigma_{eX} + 0.420 \ (\pm 0.172) \tag{84}$$

Values of μ_{calc} and Δ for Group 14 substituents are reported in Table 10.

To illustrate the application of the imf and steric parameters we consider cohesive energies of MeX at 298.15 K taken from the compilation of Majer and Svoboda[73]. The data set (Table 9) contains no compounds capable of hydrogen bonding. We have therefore used the IMF equation in the form:

$$(\Delta U_V^0)_X = M\mu_{MeX} + A\alpha_X + S_1\upsilon_{1X} + S_2\upsilon_{2X} + h \tag{85}$$

as only dd, di and ii interactions should occur. Correlation of the data set with equation 85 gave the regression equation:

$$(\Delta U_V^0)_X = 5.60 \ (\pm 0.363)\mu_{MeX} + 131 \ (\pm 6.73)\alpha_X - 7.40 \ (\pm 1.87)\upsilon_{1X}$$
$$- 7.40 \ (\pm 3.87)\upsilon_{2X} + 8.25 \ (\pm 2.19) \tag{86}$$

$100R^2$, 95.44; $A100R^2$, 94.97; F, 114.7; S_{est}, 2.30; S^0, 0.232; n, 33; $r_{\mu\alpha}$, 0.218; $r_{\mu\upsilon1}$, 0.185; $r_{\mu\upsilon2}$, 0.110; $r_{\alpha\upsilon1}$, 0.143; $r_{\alpha\upsilon2}$, 0.461; $r_{\upsilon1\upsilon2}$, 0.124.

If the gas phase value of the dipole moment for $MeNO_2$, 3.56, is replaced by the liquid phase value, 4.39, the results are much improved. The regression equation is:

$$(\Delta U_V^0)_X = 5.45 \ (\pm 0.319)\mu_{MeX} + 132 \ (\pm 6.15)\alpha_X - 7.10 \ (\pm 1.71)\upsilon_{1X}$$

$$- 6.23 \ (\pm 3.52)\upsilon_{2X} + 7.41 \ (\pm 2.01) \tag{87}$$

$100R^2$, 96.22; $A100R^2$, 95.82; F, 177.9; S_{est}, 2.09; S^0, 0.211; n, 33; $r_{\mu\alpha}$, 0.237; $r_{\mu\upsilon1}$, 0.197; $r_{\mu\upsilon2}$, 0.136; $r_{\alpha\upsilon1}$, 0.143; $r_{\alpha\upsilon2}$, 0.461; $r_{\upsilon1\upsilon2}$, 0.124.

There is no significant difference between the coefficients of equations 86 and 87.

VI. THE VALIDITY OF ESTIMATED SUBSTITUENT CONSTANTS

The values calculated for various Group 14 substituents provide the only evidence we have for the validity of our parameter estimates. Table 10 presents the values of Q_{obs}, Q_{calc}, Δ, the number of the regression equation used to obtain the calculated value and the parameter types used in the calculation. The agreement between observed and calculated values is described in terms of the number of standard deviations, N_{SD}, defined as:

$$N_{SD} = \frac{\|\Delta\|}{S_{est}} \tag{88}$$

For N_{SD} less than or equal to 1, the agreement is considered excellent; greater than 1 and less than or equal to 2, it is considered good, greater than 2 and less than or equal to 3, it is considered fair; greater than 3, it is considered poor (unacceptable). Values of N_{SD} are also given in Table 9.

$MZ^1Z^2Z^3$ groups. The agreement between calculated and observed values for substituents in which Z groups are H, alkyl or aryl is generally good. We believe that electrical, steric and intramolecular force substituent constants for these substituents are probably reliable. The electrical effect substituent constants for groups in which Z is halogen or alkyl give mixed results. Generally, better agreement between calculated and observed Q values results from the use of the smaller values of σ_d and σ_e, an exception being the dipole moments of $SnHl_3$ (Hl = halogen). Preferred values of electrical effect substituent constants are given in Table 1 in boldface. The agreement between observed and calculated dipole moments is the only evidence available for groups in which Z is alkoxy or thiomethyl. The degree of agreement is generally unacceptable. This may well be due to the fact that the regression equations used were obtained for symmetric groups. The alkoxy and thiomethyl groups are probably nonsymmetric. It is interesting to note that good agreement is obtained for Z equal to phenoxy or dimethylamino.

$C(MZ^1Z^2Z^3)_nH_{3-n}$ groups. We have been able to make comparisons between calculated and observed values for the groups with M equal to Si and n equal to 1, 2 or 3; and for the group with M equal to Ge and n equal to 1. Agreement is generally good. There seems to be no need to involve any special capability for electron donation in these groups. If this is indeed the case, it will be necessary to modify the views generally held about the electrical effects of these groups.

We have been able to make comparisons for a number of groups with M equal to Si, and for some groups with M equal to Ge or Sn. Unfortunately, very little is available for

M equal to Pb, comparisons being limited to trialkylplumbyl groups. Obviously, much more experimental work is required before we arrive at a reliable overview of substituent effects of Group 14 elements other than carbon.

VII. APPENDIX I. GLOSSARY

This appendix is an updated and slightly modified version of one we have published elsewhere.

General

X   A variable substituent.
Y   An active site. The atom or group of atoms at which a measurable phenomenon occurs.
G   A skeletal group to which X and Y may be attached.
Parameter   An independent variable.
Pure parameter   A parameter which represents a single effect.
Composite parameter   A parameter which represents two or more effects.
Modified composite parameter   A composite parameter whose composition has been altered by some mathematical operation.
Monoparametric equation   A relationship in which the effect of structure on a property is represented by a single generally composite parameter. Examples are the Hammett and Taft equations.
Diparametric equation   A relationship in which the effect of structure on a property is represented by two parameters, one of which is generally composite. Examples discussed in this work include the LD, CR and MYT equations. Other examples are the Taft, Eherenson and Brownlee DSP (dual substituent parameter), Yukawa–Tsuno (YT) and the Swain, Unger, Rosenquist and Swain (SURS) equations. The DSP equation is a special case of the LDR equation with the intercept set equal to zero. It is inconvenient to use and has no advantages. The SURS equation uses composite parameters which are of poorer quality than those used with the LDR and DSP equations. The MYT equation has all the advantages of the YT equation and gives results which are easier to interpret.
Multiparametric equation   An equation which uses three or more parameters all of which may be either pure or composite.

Electrical effect parametrization

σ_l The localized (field) electrical effect parameter. It is identical to σ_I. Though other localized electrical effect parameters such as σ_I^q and σ_F have been proposed, there is no advantage to their use. The σ^* parameter has sometimes been used as a localized electrical effect parameter; such use is generally incorrect. The available evidence is strongly in favor of an electric field model for transmission of the effect.

σ_d The intrinsic delocalized (resonance) electrical effect parameter. It represents the delocalized electrical effect in a system with zero electronic demand.

σ_e The electronic demand sensitivity parameter. It adjusts the delocalized effect of a group to meet the electronic demand of the system.

σ_D A composite delocalized electrical effect parameter which is a function of σ_d and σ_e. Examples of σ_D constants are the σ_R^+ and σ_R^- constants. The $\sigma_{R,k}$ constants, where k designates the value of the electronic demand η, are also examples of σ_D constants.

σ_R A composite delocalized electrical effect parameter of the σ_D type with η equal to 0.380. It is derived from 4- substituted benzoic acid pK_a values.

$\sigma_R{}^0$ A composite delocalized electrical effect parameter of the σ_D type with η equal to -0.376. It is derived from 4- substituted phenylacetic acid pK_a values.

$\sigma_R{}^+$ A composite delocalized electrical effect parameter of the σ_D type with η equal to 2.04. It is derived from rate constants for the solvolysis of 4- substituted cumyl chlorides.

$\sigma_R{}^\oplus$ A composite delocalized electrical effect parameter of the σ_D type with η equal to 3.31. It is derived from ionization potentials of the lowest-energy π orbital in substituted benzenes.

$\sigma_R{}^\ominus$ A composite delocalized electrical effect parameter of the σ_D type with η equal to -2.98. It is derived from pK_a values of substituted nitriles.

$\sigma_R{}^-$ A composite delocalized electrical effect parameter of the σ_D type with η equal to -1.40. It is derived from pK_a values of substituted anilium ions.

$\sigma_{k'/k}$ A composite parameter which is a function of σ_l, σ_d, and σ_e. Its composition is determined by the values of k and k'. The Hammett σ_m and σ_p constants are of this type.

$\sigma_{Ck'}$ A composite constant that is a function of σ_l and σ_d; its composition is determined by the value of k'.

$\sigma^\blacklozenge$ An electrical effect modified composite parameter.

σ Any electrical effect parameter.

η The electronic demand of a system or of a composite electrical effect parameter that is a function of both σ_d and σ_e. It is represented in subscripts as k. It is a descriptor of the nature of the electrical effect. It is given by R/D, where R and D are the coefficients of σ_e and σ_d, respectively.

P_D The percent delocalized effect. It too is a descriptor of the nature of the electrical effect. It is represented in subscripts as k'.

LDR equation   A triparametric model of the electrical effect.

P_{EA} The percent of the $\sigma_{k'/k}$ values in a substituent matrix which exhibit an electron acceptor electrical effect.

P_{ED} The percent of the $\sigma_{k'/k}$ values in a substituent matrix which exhibit an electron donor electrical effect.

P_0 The percent of the $\sigma_{k'/k}$ values in a substituent matrix which do not exhibit a significant electrical effect.

Steric effect parametrization

r_V The van der Waals radius. A useful measure of group size. The internuclear distance of two nonbonded atoms in contact is equal to the sum of their van der Waals radii.

υ A composite steric parameter based on van der Waals radii. For groups whose steric effect is at most minimally dependent on conformation, it represents the steric effect due to the first atom of the longest chain in the group and the branches attached to that atom. The only alternative monoparametric method for describing steric effects is that of Taft which uses the E_S parameter. This was originally developed only for alkyl and substituted alkyl groups and for hydrogen. Hansch and Kutter[74] have estimated E_S values for other groups from the υ values using a method which, in many cases, disregards the MSI principle. It is best to avoid their use.

Simple branching equation (SB)   A topological method for describing steric effects which takes into account the order of branching by using as parameters n_i, the number of atoms other than H that are bonded to the i-th atoms of the substituent.

n_i The number of branches on the i-th atoms of a substituent. These are the steric parameters used in the SB equation.

Expanded branching equation (XB) A topological method for describing steric effects which takes into account the order of branching by using as parameters n_{ij}, the number of j-th branching atoms bonded to the i-th atoms of the substituent.

n_{ij} The number of j-th branches on the i-th atoms of a substituent. These are the steric parameters used in the XB model of steric effects.

n_b The number of bonds in the longest chain of a substituent. It is a steric parameter which serves as a measure of the length of a group along the group axis.

Segmental equation A steric effect model that separately parametrizes each segment of a substituent. It requires fewer parameters than the XB equation and is generally more effective than the SB equation.

v_i A steric parameter based on van der Waals radii that is a measure of the steric effect of the i-th segment of a substituent. The i-th segment consists of the i-th atom of the longest chain in the substituent and the groups attached to it. The MSI principle is assumed to apply and the segment is assigned the conformation that gives it the smallest possible steric effect.

MSI principle The principle of minimal steric interaction, which states that the preferred conformation of a group is that which results in the smallest possible steric effect.

Intermolecular force parametrization

α A polarizability parameter defined as the difference between the group molar refractivities for the group X and for H divided by 100. Many other polarizability parameters, such as the van der Waals volume, the group molar volume and the parachor, can be used in its place. All of these polarizability parameters are very highly linear in each other.

n_H A hydrogen-bonding parameter which represents the lone-pair acceptor (proton donor) capability of a group. It is defined as the number of OH and/or NH bonds in the group.

n_n A hydrogen-bonding parameter which represents the lone-pair donor (proton acceptor) capability of the group. It is defined as the number of lone pairs on O and/or N atoms in the group.

i A parameter which represents ion–dipole and ion–induced dipole interactions. It is defined as 1 for ionic groups and 0 for nonionic groups.

n_D A charge transfer donor parameter which takes the values 1 when the substituent can act as a charge transfer donor and 0 when it cannot.

n_A A charge transfer acceptor parameter which takes the values 1 when the substituent can act as a charge transfer acceptor and 0 when it cannot.

IMF equation A multiparametric equation which models phenomena that are a function of the difference in intermolecular forces between an initial and a final state.

Statistics

Correlation equation An equation with a data set is correlated by simple (one parameter) or multiple (two or more parameters) linear regression analysis.

Regression equation The equation obtained by the correlation of a data set with a correlation equation.

n The number of data points in a data set.

Degrees of freedom (DF) Defined as the number of data points (n), minus the number of parameters (N_p) plus 1 [$DF = n - (N_p + 1)$].

F statistic A statistic which is used as a measure of the goodness of fit of a data set to a correlation equation. The larger the value of F, the better the fit. Confidence levels can be assigned by comparing the F value calculated with the values in an F table for the N_p and DF values of the data set.

$100R^2$ A statistic which represents the percent of the variance of the data accounted for by the regression equation. It is a measure of the goodness of fit.

S_{est} The standard error of the estimate. It is a measure of the error to be expected in predicting a value of the dependent variable from the appropriate parameter values.

S^0 Defined as the ratio of S_{est} to the root-mean-square of the data. It is a measure of the goodness of fit. The smaller the value of S^0, the better the fit.

VIII. REFERENCES

1. L. P. Hammett, *J. Am. Chem. Soc.*, **59**, 96 (1937).
2. L. P. Hammett, *Trans. Faraday Soc.*, **34**, 156 (1938).
3. L. P. Hammett, *Physical Organic Chemistry*, 1st ed., McGraw-Hill, New York, 1940, pp. 184–228.
4. G. N. Burkhardt, *Nature*, **136**, 684 (1935).
5. H. C. Brown and Y. Okamoto, *J. Am. Chem. Soc.*, **79**, 1913 (1957).
6. L. M. Stock and H. C. Brown, *Adv. Phys. Org. Chem.*, **1**, 35 (1963).
7. H. van Bekkum, P. E. Verkade and B. M. Wepster, *Recl. Trav. Chim. Pays-Bas*, **78**, 815 (1959).
8. R. W. Taft, *J. Phys. Chem.*, **64**, 1805 (1960).
9. R. W. Taft, *J. Am. Chem. Soc.*, **79**, 1045 (1957).
10. R. W. Taft and I. C. Lewis, *J. Am. Chem. Soc.*, **80**, 2436 (1958).
11. S. Ehrenson, R. T. C. Brownlee and R. W. Taft, *Prog. Phys. Org. Chem.*, **10**, 1 (1973).
12. M. Charton, in *Molecular Structures and Energetics* 4 (Eds. A. Greenberg and J. F. Liebman), VCH Publ., Weinheim, 1987, pp. 261–317.
13. M. Charton, *Bull. Soc. Chim. Belg.*, **91**, 374 (1982).
14. Y. Yukawa and Y. Tsuno, *Bull. Chem. Soc. Jpn.*, **32**, 965, 971 (1959).
15. Y. Yukawa, Y. Tsuno and M. Sawada, *Bull. Chem. Soc. Jpn.*, **39**, 2274 (1966).
16. M. Yoshioka, M. Hamamoto and T. Kabota, *Bull. Chem. Soc. Jpn.*, **35**, 1723 (1962).
17. M. Charton, *Prog. Phys. Org. Chem.*, **16**, 287 (1987).
18. M. Charton and B. I. Charton, *Abstr. 10th Int. Conf. Phys. Org. Chem.*, Haifa, 1990, p. 24.
19. M. Charton, in *The Chemistry of Arsenic, Antimony and Bismuth* (Ed. S. Patai), Wiley, Chichester, 1994, pp. 367–439.
20. M. Charton, *J. Org. Chem.*, **28**, 3121 (1963).
21. M. Charton, *Prog. Phys. Org. Chem.*, **13**, 119 (1981).
22. M. Charton, in *The Chemistry of the Functional Groups, Supplement C, The Chemistry of Triple Bonded Groups* (Ed. S. Patai), Wiley, Chichester, 1983, pp. 269–323.
23. A. Allred and E. G. Rochow, *J. Inorg. Nucl. Chem.*, **5**, 264 (1958).
24. M. Charton, *J. Org. Chem.*, **49**, 1997 (1984).
25. M. Charton, in *The Chemistry of the Functional Groups. Supplement A, The Chemistry of Double Bonded Functional Groups*, Vol. 2, Part 1 (Eds. S. Patal and Z. Rappoport), Wiley, Chichester, 1989, pp. 239–298.
26. M. Charton, in *The Chemistry of Sulfenic Acids, Esters and Derivatives* (Ed. S. Patai), Wiley, Chichester, 1990, pp. 657–700.
27. A. N. Egorochkin and G. A. Razuvaev, *Uspekhi Khimii*, **56**, 1480 (1987).
28. M. Charton and B. I. Charton, *Abstr. VI-th Int. Conf. Correlation Anal. in Chem.*, Prague, 1994, p. O-1.
29. M. Charton, unpublished results.
30. F. Kehrmann, *Chem. Ber.*, **21**, 3315 (1888); **23**, 130 (1890); *J. prakt. chem.*, [2] **40**, 188, 257 (1889); [2] **42**, 134 (1890).
31. V. Meyer, *Chem. Ber.*, **27**, 510 (1894); **28**, 1254, 2773, 3197 (1895); V. Meyer and J. J. Sudborough, *Chem. Ber.*, **27**, 1580, 3146 (1894); V. Meyer and A. M. Kellas, *Z. phys. chem.*, 24, 219 (1897).
32. J. J. Sudborough and L. L. Lloyd, *Trans. Chem. Soc.*, **73**, 81 (1898); J. J. Sudborough and L. L. Lloyd, *Trans. Chem. Soc.*, **75**, 407 (1899).
33. A. W. Stewart, *Stereochemistry*, Longmans Green, London, 1907, pp. 314–443; 2nd ed., 1919, pp. 184–202.
34. G. Wittig, *Stereochemie*, Akademische Verlagsgesellschaft, Leipzig, 1930, pp. 333–361.

35. G. W. Wheland, *Advanced Organic Chemistry*, 3rd edn., Wiley, New York, 1960, pp. 498–504.
36. M. Charton and B. I. Charton, *J. Org. Chem.*, **44**, 2284 (1979).
37. M. Charton, *Top. Curr. Chem.*, **114**, 107 (1983).
38. M. Charton, in *Rational Approaches to the Synthesis of Pesticides* (Eds. P. S. Magee, J. J. Menn and G. K. Koan), American Chemical Society, Washington, D.C., 1984, pp. 247–278.
39. K. Kindler, *Ann. Chem.*, **464**, 278 (1928).
40. R. W. Taft, in *Steric Effects in Organic Chemistry* (Ed. M. S. Newman), Wiley, New York, 1956, pp. 556–675.
41. M. Charton, *J. Am. Chem. Soc.*, **91**, 615 (1969).
42. M. Charton, *Prog. Phys. Org. Chem.*, **8**, 235 (1971).
43. M. Charton, *Prog. Phys. Org. Chem.*, **10**, 81 (1973).
44. M. Charton, *Top. Curr. Chem.*, **114**, 57 (1983).
45. M. Charton, *J. Org. Chem.*, **48**, 1011 (1983); M. Charton, *J. Org. Chem.*, **48**, 1016 (1983).
46. M. Charton, *Stud. Org. Chem.*, **42**, 629 (1992).
47. A. Verloop, W. Hoogenstraaten and J. Tipker, *Drug Design*, **7**, 165 (1976).
48. M. Charton and B. I. Charton, *J. Theor. Biol.*, **99**, 629 (1982); M. Charton, *Prog. Phys. Org. Chem.*, **18**, 163 (1990).
49. M. Charton in *Trends in Medicinal Chemistry '88* (Eds. H. van der Goot, G. Domany, L. Pallos and H. Timmerman), Elsevier, Amsterdam, 1989, pp. 89–108.
50. M. Charton and B. I. Charton, *J. Phys. Org. Chem.*, **7**, 196 (1994).
51. M. Charton, *Abstr. 208th Mtg. Am. Chem. Soc.*, 1994, p. Agrochem.
52. V. A. Palm (Ed.), *Tables of Rate and Equilibrium Constants of Heterolytic Organic Reactions*, Vol. I, Moscow, 1975; Supplementary Vol. I, Tartu State University, Tartu, 1984.
53. M. Charton and B. I. Charton, *Abstr. 3rd Eur. Symp. Org. Reactivity*, Göteborg, 1991, p. 102; M. Charton, *Abstr. Fourth Kyushu International Symposium on Physical Organic Chemistry*, Fukuoka/Ube, 1991, pp. 42–48.
54. M. Charton and J. Shorter, *Abstr. 11th Int. Conf. Phys. Org. Chem.*, Ithaca, New York, 1992, p. 148; M. Charton and B. I. Charton, *Abstr. 11th Int. Conf. Phys. Org. Chem.*, Ithaca, New York, 1992, p. 149.
55. S. G. Lias, J. R. E. Bartmess, J. F. Liebman, J. L. Holmes, R. D. Levin and W. G. Mallard, *J. Phys. Chem. Ref. Data*, **17**, Suppl. 1 (1988).
56. D. M. Kuznesoff and W. J. Jolly, *Inorg. Chem.*, **7**, 2574 (1968).
57. C. Eaborn, *J. Chem. Soc.*, 4858 (1956); J. E. Baines and C. Eaborn, *J. Chem. Soc.*, 1436 (1956).
58. C. Eaborn and K. C. Pande, *J. Chem. Soc.*, 297, 5082 (1961).
59. C. Eaborn and J. A. Waters, *J. Chem. Soc.*, 542 (1961).
60. R. W. Bott, C. Eaborn and J. A. Waters, *J. Chem. Soc.*, 681 (1963).
61. R. A. Benkeser, C. E. DeBoer, R. E. Robinson and D. M. Sayre, *J. Am. Chem. Soc.*, **78**, 682 (1956).
62. R. A. Benkeser and R. G. Gosnell, *J. Org. Chem.*, **22**, 327 (1957).
63. R. H. Baney and R. J. Shindorf, *J. Organomet. Chem.*, **6**, 660 (1966).
64. J. F. Gal, S. Geribaldi, G. Pfister-Guillouzo and D. G. Morris, *J. Chem. Soc., Perkin Trans. Z*, 103 (1985).
65. M. H. Palmer, W. Moyes, M. Spiers and J. N. A. Ridyard, *J. Mol. Struct.*, **53**, 235 (1979).
66. E. J. McAlduff, B. M. Lynch and K. N. Houk, *Can. J. Chem.*, **56**, 495 (1978).
67. E. J. McAlduff and D. L. Bunbury, *J. Electron. Spectrosc. Rel. Phenom.*, **17**, 81 (1979).
68. T. Kobayashi, and S. Nagakora, *Bull. Chem. Soc. Jpn.*, **47**, 2563 (1974).
69. M. J. S. Dewar and S. D. Worley, *J. Chem. Phys.*, **50**, 654 (1969).
70. W. Kaim, H. Tesman and H. Bock, *Chem. Ber.*, **113**, 3221 (1980).
71. T. H. Gan, M. K. Livett and J. B. Peel, *J. Chem. Soc., Faraday Trans. Z*, **80**, 1281 (1984).
72. J. Bastide, J. P. Maier, and T. Kubota, *J. Electron. Spectrosc. Rel. Phenom.*, **9**, 307 (1976).
73. V. Majer and V. Svoboda, *Enthalpies of Vaporization of Organic Compounds. A Critical Review and Data Compilation*, Blackwell Scientific Publications, Oxford, 1985.
74. E. Kutter and C. Hansch, *J. Med. Chem.*, **12**, 647 (1969).

CHAPTER 13

The electrochemistry of alkyl compounds of germanium, tin and lead

MICHAEL MICHMAN

Department of Organic Chemistry, The Hebrew University of Jerusalem, 91904
Jerusalem, Israel.
Fax: + 972 2 585345; e-mail: michman@ums.huji.ac.il

The chemistry of organic germanium, tin and lead compounds
Edited by S. Patai © 1995 John Wiley & Sons Ltd

TERMS AND ABBREVIATIONS

Terminology follows the rules laid down by leading texts of electrochemistry, such as References 7, 8, 14 and 113.

Terms

Chemical reversibility. This is the common term for a reaction which can be run in two opposite directions. However, reversibility in connection with CV refers to the conditions of the CV experiment which is diffusion-controlled and dependent on scan rate. A reaction can be irreversible with respect to the time domain of the CV test yet still be chemically reversible.

Electrochemical and nonelectrochemical processes. A reaction is often designated by the letter E to mark it as an electrode reaction in contrast to C, a chemical (nonelectrodic) reaction. Reaction sequences can be marked accordingly as ECE, EEC, ECC, etc.

Abbreviations

AN	acetonitrile
ASV	anodic stripping voltammetry
CPE	constant potential electrolysis, controlled potential electrolysis
CV	cyclic voltammetry
DME	dropping mercury electrode
DPASV	differential pulsed anodic stripping voltammetry
DPP	differential pulsed polarography
$E_{1/2}$	half-wave potential
E_p	peak potential
ET	electron transfer
Fc/Fc^+	ferrocene reference electrode
gc	glassy carbon
LSV	linear sweep voltammetry

NHE	normal hydrogen electrode
NPP	normal pulse polarography
RDE	rotating disk electrode
SCE	standard calomel electrode
SET	single-electron transfer
SHE	standard hydrogen electrode
TBA(PF_6)	tetrabutylammonium hexafluorophosphate
TBAP	tetrabutylammonium perchlorate
TEAP	tetraethylammonium perchlorate

I. INTRODUCTION

It is interesting to consider what percentage of the available literature concerning the organometallic chemistry of Ge, Sn and Pb refers to electrochemical techniques. Our literature search for the period 1967–1993 (through June)[1] mentions electrochemistry for 240 out of 6500 organometallic references of Pb (3.7%), 121 out of 20,000 organometallic references of Ge (0.6%) and 528 out of 60,000 organometallic references of Sn (0.88%). Not all of these deal with actual C—M bonds and allowance must be made for some *noise* of irrelevant references. Specific authoritative texts make very scarce mention of electrochemical methods[2]. The Gmelin volumes concerning organolead compounds are exceptional as they provide extensive discussions of electrochemistry[3]. Gmelin volumes for organogermanium compounds have no specific treatment of electrochemistry and that for organotin compounds has very few entries[4].

As a synthetic method for alkylation of group 14 elements, there is no apparent advantage to electrochemistry over conventional methods. This is certainly true for the laboratory scale. Indeed, it is clear that many studies of the synthetic chemistry and reactivity of these compounds can be handled conveniently without resource to electrochemical methods. Still, the electrosynthesis of alkyllead compounds is among the largest electrochemical productions of organic or organometallic compounds, on the commercial scale. Peak interest in this was around 1970. Since that time, the declining market for lead compounds as gasoline additives is reflected in the decreasing number of papers and patents on their electrosynthesis. Tin compounds, on the other hand, do maintain a steady demand for a great variety of applications, and it has been stated that: 'Tin is unsurpassed by any other metal in the multiplicity of applications of its organometallic compounds'[5]. The extent of utilization of organogermanium compounds is much smaller[6a].

Electrochemical methods are of considerable importance in the analysis of group 14 compounds, especially in light of the increasing environmental concern. These have the advantage of being specific, sensitive, nondestructive and adaptable to on-line coupling with flow systems, and have been the objective of many recent studies. Other aspects of electrochemistry which attracted recent attention concern conductive polymers, special membrane preparations and sensor electrodes. Group 14 compounds and complexes have also served as models for a number of interesting mechanistic studies.

Many aspects of the electrochemistry of group 14 elements have been discussed in previous reviews, often in the more general context of the electrochemistry of organometallic compounds[7–12]. This chapter reviews the literature dealing with the electrochemistry of organic compounds of germanium, tin and lead, from 1967 through June 1993, with some citations through June 1994. Earlier literature is cited in the references given here. Though emphasis is given to compounds with carbon—metal bonds, our coverage also includes complexes and some organic compounds of these metals, which do not have an M—C bond. This is pertinent whenever data can only be evaluated by comparative projection. Note that we refrain from a classification of the material according to distinct oxidation states, nor do we discuss separately each of the metals.

Electrochemical reactions are often regarded as the direct experimental method to perform oxidations and reductions and to provide a straightforward means to follow electron transfer (ET) processes. Though this is basically true, it should be born in mind that the treatment of ET reactions is not confined to free energy relationships. Treatment based on the parameters comprising the Marcus theory has received considerable attention and is relevant to many mechanistic discussions[13]. This is particularly illustrated with the use of group 14 compounds as models for outer-sphere ET, as described below. It is also important to realize that electrode performance is very sensitive to experimental conditions which include the medium (solvent and electrolyte), ion strength, double-layer consistency, electrode surface and other parameters[14]. Potential readings, in particular, are dependent on arrangements and exact type of the reference electrode, which must be taken into account when considering results from different sources[15]. For example, the reference electrode Ag/AgCl tends to shift under various conditions, whereas SCE is well known for its consistency. Stress is therefore given to experimental detail when data are presented. Potential shifts are particularly evident in reductions on mercury[10], where secondary processes (e.g. adsorption) take place at the electrode.

It is generally held that in families of elements, properties change regularly with atomic number. This is clearly so for groups of elements at the beginning or end of periods but is much less the case for those positioned at the center of the periodic table. The chemical properties of the group 14 elements, in particular, show quite a number of intriguing irregularities[16]. Well known differences exist in the chemistry and properties of the hydrides of these elements. Furthermore, in the reduction of halides with zinc and hydrochloric acid, silicon differs from carbon, germanium reacts like carbon and tin resembles silicon:

$$R_3MX + Zn/HCl \longrightarrow R_3MH$$

M = C and Ge. No hydride forms with Si or Sn

Another illustration is given by the reactions of the corresponding triphenylhydrides with alkyllithium:

$$Ph_3MH + RLi \longrightarrow Ph_3MLi + RH \quad \text{for } M = C, Ge$$

$$Ph_3MH + RLi \longrightarrow Ph_3MR + LiH \quad \text{for } M = Si, Sn$$

These reductions are closely related to some of the electrodic reactions discussed below. It has also been noted that enthalpies of formation of many compounds of this group alternate in value along the series C, Si, Ge, Sn, Pb. Other changes are more regular, e.g. divalent compounds are more stable with the heavier congeners and very important with lead, whereas they are practically unknown for carbon and silicone. Obviously, these irregularities could be reflected in electrochemical experiments and one should therefore be very careful in judging trends of reactivity or irregularities, especially in complex compounds.

II. ELECTROSYNTHESIS OF ALKYLMETALLIC COMPOUNDS

A. Formation of the Carbon–Metal Bond

Group 14 metals can serve as sacrificial electrodes. Both anodic and cathodic reactions can be considered. Pb and Sn alkyls can be prepared by their use as a sacrificial metallic anode in a reaction with carbanions, for example in a *Grignard* reagent:

$$M + nRMgX \longrightarrow R_nM + \tfrac{1}{2}nMg + \tfrac{1}{2}nMgX_2$$

This is the basic reaction of the commercial *Nalco* process. On the other hand, reduction of alkyl halides at the cathode is often described in general terms as:

$$RX + e \longrightarrow [RX]^{\bullet -}$$

$$[RX]^{\bullet -} \longrightarrow R^{\bullet} + X^-$$

$$nR^{\bullet} + M \longrightarrow R_n M$$

The intriguing point is that the actual alkylation step may be the same at the anode and cathode, presumably by alkyl radicals which, in analogy to the *Paneth* reaction, alkylate the metal. The lifetime of the radical ion, reactivity of the radical ion or the radical towards the metal, stabilization of the radical by adsorption on the electrode surface, stabilization of each of the intermediates by solvation, their build-up in the double layer, the potential applied, all have an important contribution to the outcome. In certain cases the ET takes place catalytically, by a mediator or under the influence of surface effects[17]. It is therefore important to keep in mind the possible subtle differences between cases described below that otherwise appear similar.

III. ANODIC ALKYLATION

A. Alkyllead Compounds, The Nalco Process

Alkyllead compounds can be prepared by the oxidation of anionic alkyl groups on a lead anode, for example by oxidizing Grignard reagents. The process for tetraalkyllead is based on the electrolysis of a mixture of alkyl halide and the corresponding Grignard reagent over a lead anode and steel cathode. Tetraalkyllead forms at the anode while magnesium, deposited on the cathode, is consumed by the excess alkyl halide. In summary:

$$4RMgX + Pb \longrightarrow R_4Pb + 2MgX_2 + 2Mg$$

$$2Mg + 2RX \longrightarrow 2RMgX$$

The reaction is the electrochemical version of the well-known transmetallation with lead salts, and becomes significant when the lead anode oxidizes. The very high current yields (Ca 170–180%) imply that formation of R_4Pb by nonelectrodic reactions takes place as well:

$$4RMgX + 2PbCl_2 \longrightarrow R_4Pb + Pb + 2MgX_2 + 2MgCl_2$$

$$X = Br \text{ or } I, R = Me\text{- or } Et\text{-}$$

The concentration of the alkyl halide must be carefully controlled since excess could result in a Wurz-type side reaction,

$$RCl + RMgX \longrightarrow RR + MgCl_2$$

and shortage allows deposition of metallic magnesium which, besides impairing the cathode reaction, could short-circuit the cell. The Nalco reaction has been extensively reviewed[3,7,8,10,11]. It is not quite clear whether the oxidized metal attacks the Grignard reagent as in transmetallation or whether the radicals formed from oxidation of the anion attack the metal:

anode

$$RMgX - e \longrightarrow RMgX^{\bullet +} \longrightarrow R^{\bullet} + MgX^+$$

$$nR^{\bullet} + M \longrightarrow R_n M$$

cathode

$$2e + 2MgX^+ \longrightarrow Mg + MgX_2$$

Both routes are believed to take place. It should be kept in mind that the Grignard reagent is comprised of a complex equilibrium between several structures. Ethereal solutions of Grignard reagents are indeed conducting due to ions like RMg^+(sol) and $RMgX_2^-$(sol) and are electroactive at both the anode and cathode[7b]. The reaction mechanism is probably complicated and much remains to be clarified. The complementary step which is utilized to recycle magnesium is a cathodic alkylation of the type discussed below:

$$RX + Mg \longrightarrow RMgX$$

The Nalco process is the largest, commercially operated electroorganic process involving organometallics and, for a time, was second only to the hydrodimerization of acrylonitrile as an industrial electroorganic method. It was introduced in Freeport Texas in 1964 to produce 15,000 t/A of tetramethyllead or 18,000 t/A of tetraethyllead, and has been scaled-up later[18]. The Nalco process consists of the anodic electrooxidation of alkylmagnesium halides in ether in the presence of the corresponding RX, over a fixed bed of granular lead surrounded by steel cathodes and polypropylene separators. The cell is operated with an overall voltage of 15–30 V and current density kept at 1.5–3 $A\,dm^{-2}$. Selectivity of the conversion is around 95–99% and power demand is therefore not high (4–8 kWh/Kg product depending on exact cell consistency)[19,20]. The technical aspects are covered in detail by several reviews[21,3].

B. Patents Pertaining to the Nalco Process

A rich patent literature is available[3]; it provides little information on reaction kinetics and mechanisms but describes mainly the reaction as a production process. Some patents treat special aspects of the electrochemical process. These deal with reduction of RX, with the use of alkyl chloride in propylene carbonate using iodide as the catalyst[22] or with the use of alkyl bromides[23]. The anodic preparation by reaction of $KAlR_2X_2$ in the presence of RX with a Pb anode to yield R_4Pb is described[24]. An electrode has been developed to monitor the Grignard reagent concentration during the production of Me_4Pb[25]. The electrolytic apparatus for the preparation of organometallic compounds and technicalities of the cell structure have been discussed[26]. Active Grignard electrodes[27] and cells based on AC input are also described[28]. In a cell with granular metal electrodes, lead among other metals can be operated under AC, 110 V, 60 cps. The electrodes are separated by a fine mesh screen. Grignard reagents RMgX (R = Me, Et, Bu, Ph) are used in 1.5 M concentration, in THF and tetraethylene glycol diethyl ether. Among the products are homoleptic lead alkyls R_4Pb and Ph_4Pb. Cleaning-up procedures for the effluents of the electrolysis have received special attention[29], like treatment of bromides from the quaternary ammonium salt electrolyte used[30] or a method by which triorganolead salts of the type Me_3PbCl and Et_3PbCl in aqueous solutions are converted to Me_4Pb and Et_4Pb in 99% yield, using a carbon anode and a lead cathode, leaving the Pb content in the electrolyzed water at <2000 ppm, hence the water can be recycled in the industrial processes[31].

Among alkyllead compounds, those which were prepared on the largest scale are the homoleptic tetraethyllead and tetramethyllead and some heteroleptic R_4Pb with R being mixed Me- or Et- residues. The heaviest market for these compounds is as gasoline additives, being used as such since the early twenties and peaking in 1970–5. World consumption of these compounds as gasoline additives has declined by 66% in the years 1970–1986, from 365,200 to 124,000 t/A (excluding East Block markets)[32]. Other sources cite different numbers, like 700,000 t/A in 1977, falling to 500,000 t/A in 1980[33] and 400,000 t/A in 1977 worldwide, only 100,000 of which in the USA[34]. There are other numbers still[5b]. The reaction of lead/sodium alloy with an alkyl halide accounts for the largest fraction of these compounds on the market (80–90%). The remaining share is covered by the electrochemical Nalco process.

C. Mediated Anodic Transmetallation. Alkylation of Lead and Tin

In certain cases, a single-step electrochemical process can be derived out of a complex sequence of reactions. The electrolysis proposed by Ziegler[35] and Lehmkuhl[36], in which sodium tetramethylaluminate is electrolyzed between a lead anode and mercury in THF, is an example of complex reactions of very sensitive compounds which are translated into a simplified electrolytic procedure:

$$Pb + Hg + 4Na(Me_4Al) \longrightarrow Me_4Pb + 4Me_3Al + 4Na(Hg)_{amalgam}$$

$$Na(Hg)_{amalgam} + 4Me_3Al + 4MeCl \longrightarrow 4Na(Me_4Al) + NaCl + Hg$$

The overall electrolysis may be written up as:

$$4Na(Hg)_{amalgam} + 4MeCl + Pb \longrightarrow Me_4Pb + 4NaCl + Hg$$

This process was never put to commercial use.

Another study on the electrosynthesis of $(alkyl)_n M$ compounds (M = Ge, Pb, Sn; $n = 2, 4$) provides illustrative examples[37]. Sacrificial cathodes of Cd, Zn and Mg were used to produce the corresponding metal alkyls which are subsequently oxidized on sacrificial anodes of Ge, Sn and Pb. The cells are of very simple construction, with the proper metal electrodes. Diethylcadmium is utilized in this way for the manufacture of tetraethyllead from lead acetate and triethylaluminum in the following reaction sequence:

$$4Et_3Al + 6Pb(OAc)_2 \longrightarrow 3Pb + 4Al(OAc)_3 + 3Et_4Pb$$

$$Pb + 2EtI + Et_2Cd \longrightarrow Et_4Pb + CdI_2$$

Diethylcadmium is prepared in turn by the reaction of cadmium iodide with triethylaluminum.

These are well known nonelectrodic reactions[38]. The electrochemical processes are meant to take care of the large amount of elemental lead set free in transmetallation, and has been devised to confine all the reactions to a single-compartment cell process. The cell in the present example is undivided, with Cd and Pb electrodes in DMF or DMSO solutions containing TBAP and EtI(10%) with NaI(5%). A sacrificial cadmium cathode is oxidized to diethylcadmium by ethyl iodide or, less readily, with ethyl bromide.

$$EtI + 2e + Cd \longrightarrow Et_2Cd + 2I^-$$

Diethylcadmium can react at the lead or tin anode in several possible fashions:

1. *Heterogeneous route, at the electrode surface.* The lead anode is attacked and yields tetraethyllead as the main product. For this stage, several reaction routes are possible, e.g. diethylcadmium may be oxidized on the lead anode to produce ethyl radicals which, in turn, may oxidize metallic lead. Partially alkylated lead compounds thus formed are alkylated to tetraethyllead by ethyl iodide.

$$Et_2Cd - 2e \longrightarrow 2Et^\bullet + Cd^{2+}$$

$$2Et^\bullet + Pb \longrightarrow Et_2Pb$$

$$Et_2Pb + 2EtI \longrightarrow Et_4Pb + 2I^-$$

However, as stated above, lead can react directly with diethylcadmium and ethyl iodide.

2. *Homogeneous reaction, in solution*. Transalkylation occurs between diethylcadmium and lead ions from the oxidation of the anode:

$$2Et_2Cd + Pb^{4+} \longrightarrow Et_4Pb + 2Cd^{2+}$$

The reaction may start with Pb^{2+}, as in some well known transmetallations[9]:

$$4MeMgI + 2PbCl_2 \longrightarrow Me_4Pb + Pb + 2MgCl_2 + 2MgI_2$$

Neither reaction excludes the other, nor are these the only routes possible. The oxidation of alkylanions on a metallic anode is likely. Other processes are more clearly indicated when, instead of Cd, Zn is used as a cathode. In the latter case, zinc alkyls R_2Zn and RZnI are formed by cathodic reaction of RI. This allows an efficient preparation of R_4M where M = Pb, Sn and R = propyl, butyl or pentyl[39], in contrast to the cathodic reduction of alkyl halides which is practically limited to the methyl and ethyl groups.

The anodic processes with zinc compounds are similar to those suggested for electro-Grignard reactions:

$$4RZnI - 4e + M \longrightarrow R_4M + 4Zn^{2+} + 4I^-$$

$$2R_2Zn - 4e + M \longrightarrow R_4M + 2Zn^{2+}$$

The cathode materials may strongly affect the overall process. Often, consumption of the Zn cathode (determined by weight loss) significantly exceeds the value expected from a Faradaic process alone, and an additional nonelectrodic catalytic chain reaction is envisaged to allow for the discrepancy:

$$RI + e + Zn \longrightarrow RZn^{\bullet} + I^-$$

$$RZn^{\bullet} + RI \longrightarrow RZn^+ + I^- + R^{\bullet}$$

$$R^{\bullet} + Zn \longrightarrow RZn^{\bullet} \quad etc,$$

Evidence for such reaction is drawn from the enhancement of alkylation with bromides by added iodides. Bromides are relatively inactive in this system whereas iodides are very active as they also cause the catalytic reaction. Hence, while running electrolysis with EtBr yields little Et_4M (M = Pb, Sn), the addition of propyl iodide in both small and large amounts enhances the production of Et_4M[40]. Similar reactions have been proposed in other cases such as reductions on Hg[41], and their relative contribution depends on the alkyl group. For example, with 1-iodo-3-methylbutane this catalytic reaction is apparently absent. Another problem which is particularly related to the cathode material concerns the free metal ions. In the case of cadmium cathodes a major complication is formation of Cd^{2+} ions, which subsequently consume a certain amount of the reduction current at the expense of EtI. This is somewhat remedied by high concentrations of EtI, since the reduction potential of EtI is in any case negative in comparison with that of Cd^{2+}. Further inhibition of the reduction of Cd^{2+} is achieved by adding ethylenediamine as a complexing agent, but this does not improve the yield of the anodic process.

Tin compounds can be produced by the Nalco-type reactions[39] as well as by Cd and Zn mediated methods[42,43]. The use of organotin compounds in electrodically induced transmetallations has also been described in a study in which Grignard-type allylation of carbonyl compounds has been carried out by electrochemically recycled allyltin reagents[44].

D. Oxidation of Alkyl Halides

An interesting example is the electrolysis of alkyl halides on tin anodes and platinum cathodes which yields dialkyltin dihalides. The current yield is extremely high, *Ca* 5 g atoms/F, namely 10 times the charge equivalent. A mechanism proposed to account for the non-Faradaic reactions envisages the formation of alkyl radicals at the cathode and oxidation of halide anions with attack on tin[45]:

Catholyte reaction:

$$2RX + 2e \longrightarrow 2R^{\bullet} + 2X^{-}$$

$$2R^{\bullet} \longrightarrow RR$$

Anolyte reaction:

$$X^{-} - e \longrightarrow X^{\bullet}$$

$$Sn + X^{\bullet} \longrightarrow SnX^{\bullet}$$

$$SnX^{\bullet} + RX \longrightarrow RSnX + X^{\bullet}$$

$$RSnX^{\bullet} + RX \longrightarrow R_2SnX_2$$

The radical $X^{\bullet}$ carries on the non-Faradaic dissolution of the tin anode.

In several studies, electrosynthesis of tetraethyllead from EtBr on a Pb anode has been carried out in a two-phase system and empirical evaluation of reaction conditions was given[46,47].

IV. CATHODIC SYNTHESIS

A. Alkyllead Compounds by Reduction of Alkyl Halides

The classical large-scale method for preparation of tetraethyllead and tetramethyllead is by reaction of alkyl halide with sodium/lead alloy (composition Pb:Na 1/1)[38]. The product is isolated by steam distillation and yields are high:

$$4Pb/Na + 4EtCl \longrightarrow Et_4Pb + 4NaCl + 3Pb \quad 90\%$$

The major set-back of this method is the need to recycle large amounts of lead. The electrochemical processes described here and in the preceding section are meant to circumvent this difficulty.

The electroreduction of alkyl halides is known to yield transient radical-anions $[RX]^{\bullet -}$ and, subsequently, alkylmetal radicals. Final products isolated are R_nM, R_xMM_xR, RH and coupled RR^{48}. These reactions can be performed on Ge, Sn and Pb cathodes as well as on Hg and, as mentioned above, on Mg. They are in a sense the electrochemical version of the *Wurz* reaction, where a cathode rather than sodium provides the negative potential. The electrodic reactivity of alkyl halides has been reviewed recently[49,50]. The production of Et_4Pb from EtI by electrolysis at lead cathodes in alcohol was patented as early as 1925[51]. The yields were unsatisfactory in aqueous solutions, probably due to competitive hydrogen release. Results improved in aprotic media like AN[52] and propylene carbonate[53]. In both those solvents, Me_4Pb and Et_4Pb could be prepared in high yields with quaternary ammonium halides as electrolytes. Use of inorganic electrolytes gave mostly hydrocarbons and poor conversions to alkyllead in propylene carbonate[54], though under the same conditions in DMF a completely opposite trend in the influence of electrolytes was found and highest yields of R_4Pb were obtained[55] when sodium salts were used. Other observations, too, caused much confusion, such as the inability to extend the Et_4Pb synthesis to higher alkyl groups. In addition, electrode kinetics, reaction rates as expressed by the *Tafel* slope and values of diffusion currents showed irregularities and material balance was incomplete in many cases, i.e. when correlating consumption of charge, yield of R_4Pb and weight loss of Pb cathode. Many of these discrepancies can be explained by consideration of *non-Faradaic* reactions taking place on the electrode surface during the reduction, as in the case of ethyl iodide at a lead cathode. Indeed, prolonged electrolysis with rotating lead cathodes in a divided cell shows the formation of deactivating coatings on the electrode surface[56]. Kinetics at a clean electrode surface show the single-electron reduction as the rate-determining step:

$$EtI + e \longrightarrow I^- + Et^\bullet$$

Allowance must be made for surface adsorption, which will also explain the deactivation and morphological changes on the electrode surface:

$$EtI + e \longrightarrow I^- + Et^\bullet_{ads}$$

Yields of Et_4Pb are therefore constricted by adsorption and formation of passivating coatings, and rates are affected by side reactions of partially alkylated intermediates:

$$Et^\bullet_{ads} + Pb \longrightarrow Et_2Pb$$

Et_2Pb could undergo disproportionation, as well as react with the solvent or with EtI. Adventitious reactions are caused by the rather unstable $Et_3Pb^{52,56,57}$:

$$2Et_2Pb \longrightarrow Et_4Pb + Pb$$

$$Et_2Pb + EtI \longrightarrow Et_3PbI$$

$$Et_3PbI + e \longrightarrow Et_3Pb^\bullet + I^-$$

$$2Et_3Pb^\bullet \longrightarrow Et_6Pb_2$$

The high sensitivity to electrolyte, solvent, solubilities of intermediates and buildup of electrode coatings may seriously invalidate comparisons between different reaction conditions.

Experimental parameters of the electrosynthesis of tetraethyllead have been studied in recent years by Tomilov's group under a variety of conditions[58]. Improvement of the electrochemical synthesis of tetraethyllead under non-steady-state conditions was sought when a variable AC current regime was applied at several potential values. An increase in the current yield of tetraethyllead is observed while production of cathodic hydrogen falls off. A matrix formula for optimization of this procedure has been proposed[59]. Electrolysis of ethyl bromide was also tested on a cathode of drenched lead beads, in a set-up very similar to the Nalco electrolyzer. The cell was divided by an alund membrane, the catholyte was ethyl bromide with no other solvent, the anolyte was aqueous NaOH. Bu_4NBr was the electrolyte in both compartments and optimized conditions are described[60]. Many other optimization tests were made[61-63]. Efficiency of the various procedures was studied not only in terms of yield, but also in terms of selectivity and control of side products formation. An ubiquitously formed side product in the cathodic synthesis of tetraethyllead is the dimer hexaethyldilead $[Pb(CH_2CH_3)_3]_2$. The presence of an additional solvent in the aqueous solution has considerable effect on the ratio of the two products. Protic solvents such as aliphatic alcohols give a high proportion of dimer, whereas formation of tetraethyllead is favored in acetone and acetonitrile[64]. Results are summarized in Tables 1 and 2. There is, however, an optimal effect which depends on the ratio of water to co-solvent. This has been studied for acetone but not for other co-solvents which do not necessarily behave in the same manner. The values in the tables can therefore serve as qualitative indicators only. Detailed mechanistic studies of factors which may influence the extent of dimerization are cited in the following sections.

TABLE 1. Molar ratio (X) of tetraethyl-
lead/hexaethyldilead in the cathodic alkylation of lead[a]

Co-solvent	X
None	1.05
CH_3OH	1.03
C_2H_5OH	0.84
$(CH_3)_2CHOH$	1.28
$HOCH_2CH_2OH$	0.62
CH_3CN	2.94
CH_3COCH_3	3.8
$CH_3CH_2COCH_3$	2.5
THF	1.24
Dioxane	0.95

[a]Conditions: water 60 ml, solvent 40 ml,
Bu_4NBr 2 g, current density 0.01 $A\,cm^{-2}$. Data
from Reference 64.

TABLE 2. Molar ratio (X) of tetraethyl-
lead/hexaethyldilead in the cathodic alkylation of lead in various water-acetone
compositions[a]

% Acetone in water	X
0	1.05
20	3.5
40	3.9
60	3.04
80	2.38

[a]Conditions: Bu_4NBr 2 g, current density
0.01 $A\,cm^{-2}$. Data from Reference 64.

Electrochemical synthesis was utilized to prepare labeled compounds. Tetramethyllead
labeled with ^{14}C was prepared in a double compartment cell in DMF with $NaClO_4$, by
electrolyzing $^{14}CH_3I$ on lead electrodes. The method is reported as superior to transmetallation with methylmagnesium halide. It is also possible to incorporate lead isotopes.
$^{210}Pb^{2+}$ ions were deposited on a Cu foil and the latter was used as a sacrificial electrode in solutions of CH_3I. The yield of labeled tetramethyllead was 85%[65]. Synthesis of
^{210}Pb-labeled chlorotrimethylplumbane was also described[66].

B. Organotin Compounds by Cathodic Reaction

The electrolysis of alkyl halides on platinum cathode and tin anode has been mentioned
above. A completely different mechanism is associated with alkylation on tin cathodes.
Electroreduction of allyl bromide on tin electrodes yields tetraallylstannane (Ca 90%).
This is done in acetonitrile solutions with $LiClO_4$, Et_4NBr or Bu_4NBr as electrolyte and
followed by CV with Ag/AgBr reference. Yields decrease to 78% in DMF. The proposed
mechanism[67] in this case is:
Cathodic alkylation:

$$RBr + e \longrightarrow Br^- + [R]^\bullet_{adsorb}$$

$$[R]^{\bullet}_{adsorb} + \tfrac{1}{4}Sn \longrightarrow \tfrac{1}{4}R_4Sn$$

$$2[R]^{\bullet}_{adsorb} \longrightarrow RR$$

$$[R]^{\bullet}_{adsorb} + e \longrightarrow [R]^-$$

$$[R]^- + H \longrightarrow RH \ (R = allyl; H \ atoms \ from \ protic \ sources)$$

Electrochemical synthesis of tetraalkyl derivatives of tin and lead using alkyl sulfates as alkylating agents was described by Mengoli and coworkers[68]. The alkyl group can be functionalized. Thus reduction of cyanoethyl iodide on Sn will yield $Sn(CH_2CH_2CN)_4$[69]. Acetonitrile may react differently; its reduction on tin yields $(CH_3)_4Sn$ and cyanide anion, a process which is formally similar to the reaction mode of alkyl halides[70]. Electrolytic reduction of acrylonitrile on a tin cathode yields $Sn(CH_2CH_2CN)_4$ and $[Sn(CH_2CH_2CN)_3]_3$. The reaction is somewhat dependent on the pH of the solution. The highest yield is obtained at pH 8.5, whereas at pH lower than 4.0 no products form at all[71]. Like lead, the performance of tin cathodes may depend heavily on solvent and solution composition. Polarographic reduction of tin in DMSO, DMSO–H_2O and DMSO–AN is reversible, whereas in AN with little DMSO present, reduction is irreversible and up to six complexes of Sn(II) with DMSO are observed[72]. The synthesis of tetraethyltin by reduction of ethyl iodide on an electrode of Sn/Pd alloy is even more successful than reduction on a pure Sn electrode[73].

Linking together complementary anodic and cathodic reactions is considered as an energy-saving procedure when both are run in one divided cell on a given charge allocation. The cathodic synthesis of tetrakis(β-cyanoethyl)stannane was run in conjunction with anodic hypochlorination of allyl chloride by NaCl[74]. Another way of increasing efficiency is by incorporating homogeneously catalyzed reactions as a follow-up to the electrodic step. Several different reactions are brought together in a combined electrochemical process. Cinnamyltin alkyls prepared by reduction of cinnamyl halides and acetate couple with alkyl halides and acetate with C–C bond formation in the presence of palladium phosphine complexes. In the combined reaction, electrogenerated cinnamyltin alkyls couple with allyl halides and acetate under catalysis of Pd complexes. The first electrodic stage is:

$$PhCH=CHCH_2X + Bu_3SnCl + 2e \longrightarrow PhCH=CHCH_2SnBu_3$$

(X = Cl, yield 100%, X = acetate, yield 47%, in Et_4NOTs/DMF)

This is followed by catalytic coupling:

$$PhCH=CHCH_2SnBu_3 + PhCH=CHCH_2X \xrightarrow{\text{Pd catalyst}} PhCH=CHCH_2CH_2CH=CHPh$$

The passage of 1.0F/mol charge (constant current electrolysis) proved sufficient; the yield of coupling product was 92%. Lower yields were observed with higher charge. Several examples of such homo-couplings[75] were given.

V. ELECTRODIC REACTIONS OF GROUP 14 ALKYLMETALS

Reductions are also discussed under the section on analysis. The reduction on Hg has been extensively studied in connection with analytical applications (see Section VI), and is complicated by adsorption, transmetallation with mercury and reoxidations of transient products. Some disagreement as to the details is apparent in the primary literature[76]. Comparisons between different experimental settings should be made with critical appraisal.

A. Lead Compounds

The electrochemistry of dibutyllead diacetate[77] and triphenyllead acetate[76] was studied in detail by Fleet and Fouzder using polarography, DPP and CV (on GC electrodes). The solutions were made up of dibutyllead diacetate 1.084×10^{-4} M in acetate buffer (pH 7.0) containing 50% v/v ethanol. CV of dibutyllead diacetate showed three reduction waves at -0.25, -1.1 and -1.5 V (vs SCE); the first two were shown by coulometry to be single ET steps, the first being reversible. Polarography of the same solutions showed somewhat different values: -0.45 and -1.3 V for the first two $E_{1/2}$ values. The readings were -0.43 and -1.32 V by DPP. Cathodic shifts were noted as the pH was increased. In summary, reduction on GC by CV is proposed to take place as follows:

$$Bu_2Pb^{2+} \underset{-e}{\overset{+e}{\rightleftharpoons}} Bu_2Pb^{\bullet+} + e \longrightarrow Bu_2Pb: \longrightarrow (Bu_2Pb)_n \quad (polymer)$$

Side reactions are:

$$2Bu_2Pb^{\bullet+} \longrightarrow (Bu_2PbPbBu_2)^{2+}$$

$$2Bu_2Pb: \longrightarrow Bu_4Pb + Pb$$

whereas polarography using DME differs in the follow-up reaction of the radical cation:

$$Bu_2Pb^{2+} \underset{-e}{\overset{+e}{\rightleftharpoons}} Bu_2Pb^{\bullet+}$$

$$2Bu_2Pb^{\bullet+} \longrightarrow (Bu_2PbPbBu_2)^{2+} \longrightarrow Bu_4Pb + Pb^{2+} \quad (predominant)$$

$$Bu_2Pb: + 2H^+ + 2e \longrightarrow Bu_2PbH_2$$

$$Bu_2Pb: + Hg \longrightarrow Bu_2Hg + Pb$$

$$2Bu_2Pb: + Hg \longrightarrow Bu_4Hg + 2Pb$$

Polarography of Ph_3PbAc, 1.084×10^{-4} M in the same electrolyte solution gave, at pH 7.0, $E_{1/2}$ values of -0.425 and -1.075 V (SCE) which, too, show cathodic shifts with increasing pH. Unlike Bu_2PbAc_2, Ph_3PbAc undergoes a SET reduction to radicals, which adsorb on the DME and react with mercury:

$$Ph_3Pb^+ + e \longrightarrow Ph_3Pb^{\bullet}$$

$$Ph_3Pb^{\bullet} \longrightarrow Ph_3Pb^{\bullet}_{ads}$$

$$Ph_3Pb^{\bullet}_{ads} + Hg \longrightarrow PhHg^{\bullet} + Pb \quad (eventually \ Ph_2Hg)$$

CV on DME indicates the following reactions:

$$Ph_3Pb^+ \underset{-e}{\overset{+e}{\rightleftharpoons}} Ph_3Pb^{\bullet}$$

$$Pb^{2+} \underset{-2e}{\overset{+2e}{\rightleftharpoons}} Pb$$

$$PhHg^+ \underset{-e}{\overset{+e}{\rightleftharpoons}} PhHg^{\bullet}$$

The situation is different on GC where only one irreversible cathodic wave is observed at -1.6 V, associated with disintegration of the compound, and one anodic wave on the reverse scan, caused by stripping of the lead released by the reduction:

$$Ph_3Pb^+ + e \longrightarrow Ph_3Pb^\bullet$$

$$2Ph_3Pb^\bullet \longrightarrow Ph_6Pb_2$$

$$Ph_3Pb^\bullet \longrightarrow [3Ph^\bullet] + Pb$$

$$[3Ph^\bullet] + 3H^+ \longrightarrow 3PhH$$

$$Pb - 2e \longrightarrow Pb^{2+} \quad \text{Anodic stripping}$$

In all these reactions $Ph_3Pb^\bullet$ is considered as having a negligible lifetime. Triaryllead cations in dimethoxyethane were postulated earlier by Dessy and coworkers to involve both single- and double- ET reactions[78]:

$$Ph_3Pb^+ + e \xrightarrow{DME} Ph_3Pb^\bullet \xrightarrow{(+ Hg)} Ph_2Hg + Pb \quad \text{(first step)}$$

$$2Ph_3Pb^\bullet + 3Hg \longrightarrow 3Ph_2Hg + 3Hg$$

$$Ph_3Pb^+ + 2e \xrightarrow{DME} Ph_3Pb^- \quad \text{(at more negative potential)}$$

This was studied later by Kochkin and collaborators[79] and by Colliard and Devaud[80]. The latter studied the electrochemistry of di- and triphenyllead derivatives in water–alcohol and suggested the following reduction steps:

$$Ph_3Pb^+ + e \xrightarrow{H_2O/ROH} Ph_3Pb^\bullet$$

$$Ph_3Pb^\bullet + Hg \longrightarrow Ph_2Hg + Pb + PhH \quad -0.45 \text{ V} \quad \text{(first step)}$$

Reactions involved in these first stages include excessive reduction to benzene and Pb and are further complicated by reoxidation of the elemental Pb so produced, which would explain certain distortions in the polarographic wave at -0.45 V as caused by protic reactions:

$$Ph_3Pb^+ + 4e + 3H^+ \longrightarrow 3PhH_3 + Pb$$

Fleet and Fouzder have also observed these distortions when polarography was performed over DME. In fact, CV on GC of Ph_3PbAc, 1.084×10^{-4} M in acetate buffer pH 7.0 containing 50% v/v ethanol shows two well-defined peaks, at -1.7 V (a cathodic current) and -0.7 V (an oxidation current), and these are attributed to secondary reactions of the unstable triphenyl radical with imminent stripping of elemental lead:

$$Ph_3Pb^+ + e \longrightarrow Ph_3Pb^\bullet \longrightarrow Pb + 3Ph^\bullet$$

$$2Ph_3Pb^\bullet \longrightarrow Ph_6Pb_2 \text{ or } Ph_4Pb + Pb$$

$$3Ph^\bullet + 3H^+ \longrightarrow 3PhH$$

The oxidation current at -0.7 V is attributed to the reoxidation of the deposited lead:

$$Pb - 2e \longrightarrow Pb^{2+}$$

On DME, $Ph_3Pb^\bullet$ is adsorbed and transmetallation occurs in addition to its reactions above:

$$Ph_3Pb^\bullet(ads) + 3Hg \longrightarrow 3PhHg^\bullet + Pb$$

$$3PhHg^\bullet + 3e \longrightarrow 3PhHg^- \xrightarrow{+3H^+} 3PhH + Hg$$

$$2PhHg^\bullet \longrightarrow Ph_2Hg + Hg$$

Organomercurials formed in such processes are oxidized further and may inhibit the reduction of Pb(II) ions at similar potentials[81]:

$$PhHgCl + e \longrightarrow Cl^- + PhHg^\bullet \longrightarrow \tfrac{1}{2}Ph_2Hg + Hg \quad 0.07 \text{ V (SCE)}$$

$$PhHg^\bullet + e + H^+ \longrightarrow PhH + Hg \qquad\qquad -0.97 \text{ V (SCE)}$$

The electrochemical reduction of aryllead triacetates was studied by Chobert and Devaud[82], as a re-examination of some previous work[83] to detect the role of intermediates such as $[ArPb(OAc)_2]^\bullet$. The reductions were carried out by polarography in acetic acid or acidic alcohol solutions and show three diffusion controlled waves. The first step involves a single electron transfer to produce a radical anion which dimerizes, arylates the electrode or hydrolyzes to phenol:

$$ArPb(OAc)_3 + e \longrightarrow [ArPb(OAc)_2]^\bullet + OAc^-$$

$$2[ArPb(OAc)_2]^\bullet \longrightarrow [ArPb(OAc)_2]_2$$

$$[ArPb(OAc)_2]_2 + Hg \longrightarrow Ar_2Hg + 2Pb(OAc)_2$$

$$[ArPb(OAc)_2]^\bullet + OH^- \longrightarrow ArOH + Pb^{2+} + 2OAc^- + e$$

The subsequent steps involve, among several proposed reactions, arylation of the mercury cathode, and release of lead, but most importantly they indicate [ArPbH] as evidence for polymer formation by the proposed route:

$$ArPb(OAc)_3 + H^+ + 4e \longrightarrow [ArPbH] + 3AcO^-$$

$$m[ArPbH] \longrightarrow \frac{1}{m}[ArPbH]_m$$

$$2[ArPbH] + Hg \longrightarrow Ar_2HgPb_2 + H_2$$

$$[ArPbH] \longrightarrow ArH + Pb$$

Apart from lead, hydrogen, phenol and transmetallation products, the dimer and the polymer are considered the only stable reduction products of $ArPb(OAc)_3$.

Partially alkylated lead alkyls R_3PbCl and $R_2Pb(OAc)_2$, which are stabilized by having $(CH_3)_3SiCH_2^-$ as R, appear to be reduced on Hg (in CH_3OH) as follows:

$$R_3PbCl + e \longrightarrow Cl^- + R_3Pb^\bullet_{ads} \longrightarrow \tfrac{1}{2}(R_3Pb)_2Hg$$

The individual steps are associated with polarographic waves:

$$(R_3Pb)_2Hg \longrightarrow R_6Pb_2 + Hg$$

$$(R_3Pb)_2Hg \longrightarrow R_3PbOCH_3 + R_3PbH + Hg \quad (\text{in } CH_3OH)$$

$$(R_3Pb)_2Hg \longrightarrow 2R_2Pb + R_2Hg$$

$$(R_3Pb)_2Hg \longrightarrow 2R_3Pb^+ + Hg + 2e$$

Formation of R_6Pb_2 in this case is too slow to be observed. R_3PbH is unstable and decays in methanol:

$$R_3PbH + CH_3OH \longrightarrow 3RH + Pb(OCH_3)_2$$

Reduction of the acetate $\{(CH_3)_3SiCH_2\}_2Pb(OAc)_2$ yields a reactive plumbylene in the first step which reacts further as shown below:

$$R_2Pb^{2+} + 2e \longrightarrow R_2Pb \xrightarrow{Hg} Pb + 2RHg^\bullet$$

$$2RHg^\bullet \longrightarrow Hg + R_2Hg$$

$$R_2Pb + RHg^\bullet \longrightarrow R_3PbHg^\bullet \longrightarrow \tfrac{1}{2}(R_3Pb)_2Hg + \tfrac{1}{2}Hg$$

$$R_3PbHg^\bullet + RHg^\bullet \longrightarrow R_3PbHgR + Hg$$

The last two reactions are faster than the disproportionation of $RHg^\bullet$ and the solvolysis of plumbylene. R_3PbHgR reacts further by three routes:

a. Decomposition:
$$R_3PbHgR \longrightarrow R_4Pb + Hg$$

b. Solvolysis:
$$R_3PbHgR + CH_3OH \longrightarrow R_3PbH + RHgOCH_3$$

c. Oxidation:
$$R_3PbHgR \longrightarrow R_3Pb^+ + RHg^+ + 2e$$

The reactions above provide the interpretation of very detailed polarographic measurements. The oxidation products of aryllead derivatives on Hg electrodes were identified in other studies on Ph_4Pb, Ph_3PbCl, Ph_2PbCl_2, Ph_3PbOAc and $Ph(OAc)_3$. These oxidations also involve alkyl/aryl exchange[84].

B. Tin Compounds

A detailed survey of the electrochemistry of organotin compounds has been given by Dessy and coworkers in an early work[78]. Electroreduction of methylphenyltin dichloride and methylphenyltin dihydride (in methanol, LiCl) is complicated by several disproportionation and rearrangement reactions, which stem from intrinsic properties of the organotin compounds[85]. Bis(chloromethylphenyltin), the dimer formed by the reduction, is known to decompose even when isolated as a solid. This includes phenyl migration and disproportionation of the resultant chlorides:

$$PhMeClSn-SnClMePh \longrightarrow Ph_2MeSnCl + MeSnCl$$

$$MeSnCl \longrightarrow \tfrac{1}{3}[Me_3SnCl + SnCl_2 + Sn] + \tfrac{1}{2}[Me_2SnCl_2 + Sn]$$

$$PhMeClSn-SnClMePh \longrightarrow Ph_2Me_2Sn + SnCl_2$$

The reduction of R_3SnCH_2I on Hg has been studied in detail, where $R = CH_3$, C_2H_5 or Ph[86]. The reduction proceeds through the formation of the corresponding radical anion:

$$R_3SnCH_2I + e \longrightarrow [R_3SnCH_2I]^{\bullet-} \longrightarrow [R_3SnCH_2]^\bullet + I^-$$

$$[R_3SnCH_2]^\bullet + Hg \longrightarrow (R_3SnCH_2)_2Hg$$

$$[R_3SnCH_2]^\bullet + SH \longrightarrow R_3SnCH_3 + S^\bullet$$

$$S^\bullet + e \longrightarrow S^-$$

$$[R_3SnCH_2]^\bullet + e \longrightarrow [R_3SnCH_2]^-$$

$$[R_3SnCH_2]^- + SH \longrightarrow R_3SnCH_3 + S^-$$

$$[R_3SnCH_2I]^\bullet + [R_3SnCH_2]^\bullet \longrightarrow R_3SnCH_2I \longrightarrow [R_3SnCH_2]^-$$

$$SH = \text{protic solvent}$$

A complicated reaction pattern is also observed with dichlorotetraphenylditin[87]. The electrochemistry of this compound compound on Hg electrodes involves formation of intermediate SnHg compounds by reduction (see also Reference[88]). The polarogram of $Ph_2ClSn-SnClPh_2$ (in methanol/LiCl, on Hg) shows an anodic peak and two cathodic waves at -0.4, -0.55 and -1.35 (vs SCE). The oxidation involves between one and two electrons as determined by coulometry, and the proposed reactions are:

$$Ph_2ClSn-SnClPh_2 \longrightarrow Ph_3SnCl + \tfrac{1}{2}[Ph_2SnCl_2 + Sn^{2+}] + e$$

$$Ph_2ClSn-SnClPh_2 \longrightarrow 2Ph_2SnCl_2 + 2e$$

Reduction of the dimer appears to take place in two stages. In the first cathodic wave, a single-electron reduction generates a radical which decomposes in a chemical step in an ECE sequence:

$$Ph_2ClSn-SnClPh_2 + e \longrightarrow [Ph_2ClSn-SnClPh_2]^\bullet_{ads}$$

$$[[Ph_2ClSn-SnClPh_2]^\bullet_{ads} \longrightarrow \tfrac{1}{2}(Ph_2ClSn)_2Hg + [Ph_2Sn]_{ads}$$

$$\tfrac{1}{2}(Ph_2ClSn)_2Hg + Cl^- \longrightarrow Ph_2SnCl_2 + e + \tfrac{1}{2}Hg$$

$$[Ph_2Sn]_{ads} + 2Cl^- \longrightarrow Ph_2SnCl_2 + 2e$$

The oxidation in the last two reactions explains the anodic peak. Under the second cathodic wave an overall four-electron reduction yields a diphenyltin polymer and diphenyltin hydride:

$$Ph_2ClSn-SnClPh_2 + 4e \longrightarrow [Ph_2ClSn-SnClPh_2]^{2-} + 2Cl^-$$

$$[Ph_2ClSn-SnClPh_2]^{2-} \longrightarrow 2Ph^- + [PhSn]_2 \longrightarrow [Ph_2Sn] + Sn$$

$$[Ph_2ClSn-SnClPh_2]^{2-} \longrightarrow [Ph_2Sn] + [Ph_2Sn]^{2-}$$

$$[Ph_2ClSn-SnClPh_2] + [Ph_2Sn]^{2-} \longrightarrow [Ph_2Sn]_3 + [Ph_2Sn]_n$$

$$[Ph_2Sn]^{2-} + 2H^+ \longrightarrow Ph_2SnH_2$$

Similar mechanisms appear to apply for the reduction in methanol of R_3SnX and Ar_3SnX. Polarography over Hg characteristically involves three reduction waves; the first, for adsorption, is followed by two reduction stages[89-91]. The first two waves comprise a first step of the reaction in which bis(trialkylstannyl)mercury or bis(triarylstannyl)mercury is produced. For example:

$$Ph_3SnX + e \longrightarrow [Ph_3SnX-Hg]^\bullet_{ads}$$

$$Ph_3SnX + e \longrightarrow [Ph_3SnHg]^\bullet \longrightarrow (Ph_3Sn)_2Hg + Hg$$

The third wave consists of the second stage of reduction which involves stannylmercurate intermediates:

$$[Ph_3SnHg]^\bullet + e \longrightarrow [Ph_3SnHg]^-$$

$$[Ph_3SnHg]^- + Ph_3SnX \longrightarrow (Ph_3Sn)_2Hg + X^-$$

$$[Ph_3SnHg]^- + MeOH \longrightarrow MeO^- + Ph_3SnHgH \longrightarrow Ph_3SnH + Hg$$

The phenyl derivatives are more stable than the alkyl counterparts, but both disproportionate in solution and also react with methanol:

$$(Ph_3Sn)_2Hg \longrightarrow Ph_6Sn_2 + Hg$$

$$(Ph_3Sn)_2Hg + MeOH \longrightarrow Ph_3SnOMe + Ph_3Sn + Hg$$

$$(Ph_3Sn)_2Hg + 2MeOH \longrightarrow 2Ph_3SnOMe + Hg + H_2$$

These reductions are further utilized to alkylate organotin compounds. The stannylmercurate intermediates formed during electrolysis are reactive towards alkyl halides, mostly iodide and bromide. The reactivity pattern follows the order $R' = $ Me, Et, Bu, $>$ Ph. Alkylations are therefore observed when electrolysis is carried out in the presence of $R'X$:

$$(Ph_3Sn)_2Hg + R'X \longrightarrow Ph_3SnHgR' + Ph_3SnX$$

$$Ph_3SnHgR' \longrightarrow Ph_3SnR' + Hg$$

$$Ph_3SnHgR' + R'X \longrightarrow Ph_3SnR' + R'HgX$$

Bis(triphenylstannyl)mercury and alkyl(triphenylstannyl)mercury compounds were repeatedly observed as products from electrochemical reduction of triphenyltin chloride in the presence of alkyl halide[91] and earlier results are generally in agreement with this reduction scheme[92,93]. Dichlorotetraalkyldistannanes $R_2ClSnSnClR_2$ (R = Me, Et and Bu) were prepared by the electroreduction of the corresponding dialkyltin dichlorides on a mercury cathode. The distannanes were isolated in good yield by precipitation with acetate anions. They display a characteristic peak near -0.75 V/SCE.

The reduction on Hg cathodes of cyclic compounds in which tin is positioned in a strained ring has been studied. Cleavage of tin substituents without ring opening can be observed[94]:

$$Ph_2Sn-(CH_2)_n + 2Hg + 4OH^- \longrightarrow 2PhHgOH + (HO)_2Sn-(CH_2)_n + 4e$$

$$PhClSn-(CH_2)_n + 2Hg + 3OH^- \longrightarrow 2PhHgOH + (HO)_2Sn-(CH_2)_n + Cl^- + 2e$$

The results are significantly different when electrodes are inert and do not participate in transmetallation[95]. When reduction of Sn(IV) derivatives was carried out on RDE of glassy carbon or gold, in acetonitrile or DMF, a single wave was displayed by CV, and the curve conformed with the *Levitch* plot. The compounds examined were of the type R_nSnX_{4-n} where R = Me, Bu, Ph; $n = 2, 3$ and X = Cl, or compounds where R = Me, Ph; $n = 3$ and X = NO_3, N_3, NCS, NCO, OAc, OH.

C. Oxidation of Alkyltin Hydrides

Diphenyltin hydride Ph_2SnH_2 is stable in methanol. Polarography in MeOH/0.1 M LiCl shows two oxidation waves which are to some extent dependent on the electrolyte. $E_{1/2}$ values are -0.27 and -0.12 (V vs SCE). The oxidation is shown by LSV to follow the reaction[96]:

$$Ph_2SnH_2 + Cl^- \longrightarrow Ph_2SnHCl + H^+ + 2e \quad at \ -0.27 \ V$$

with a subsequent step:

$$Ph_2SnHCl + Cl^- \longrightarrow Ph_2SnCl_2 + H^+ + 2e \quad at \; -0.12 \; V$$

In MeOH/0.1 M NaClO$_4$, $E_{1/2}$ was found to be 0.1 and 0.4 (V vs SCE), which is a considerable positive shift. In this case the oxidation is shown by LSV to follow the reaction:

$$Ph_2SnH_2 \longrightarrow Ph_2SnH^+ + H^+ + 2e \; at \; 0.1 \; V$$

$$Ph_2SnH^+ \longrightarrow Ph_2Sn^{2+} + H^+ + 2e \; at \; 0.4 \; V$$

The protons released in these oxidations react with the hydrides in subsequent chemical steps:

$$Ph_2SnH_2 + HCl \longrightarrow Ph_2SnHCl + H_2$$

$$Ph_2SnHCl + HCl \longrightarrow 2PhH + SnCl_2$$

The mixed hydride chlorodiphenylstannane also forms by an exchange reaction with diphenyltin chloride and, in turn, converts rapidly to the dimer:

$$Ph_2SnH_2 + Ph_2SnCl_2 \longrightarrow 2Ph_2SnHCl$$

$$2Ph_2SnHCl \longrightarrow 2Ph_2ClSnSnClPh_2 + H_2$$

Owing to these nonelectrodic reactions, the apparent number of electron equivalents involved in the electrooxidation of Ph$_2$SnH$_2$ at -0.27 V is < 2. Reduction of the chlorohydride Ph$_2$SnHCl yields Ph$_2$SnH$_2$ as the main product:

$$Ph_2SnHCl + 2e \longrightarrow Ph_2SnH^- + Cl^-$$

$$Ph_2SnH^- + MeOH \longrightarrow Ph_2SnH_2 + MeO^-$$

However, Ph$_2$SnHCl is also consumed by two other side reactions. The first is precipitation of a polymer caused by basicity of MeO$^-$.

$$Ph_2SnHCl + OH^- \xrightarrow{(-H_2O)} Ph_2ClSn(Ph_2Sn)_n SnClPh + H_2$$

(The typical n is close to 4)

The second is a dimerization:

$$Ph_2SnHCl + Ph_2SnH^- \longrightarrow Ph_2HSnSnHPh_2 + Cl^-$$

Oxidation of methyltin compounds Me$_n$SnCl$_{4-n}$ at Hg electrodes in CH$_2$Cl$_2$ was studied by polarography and CPE. Oxidation of the mercury electrode is involved in these reactions and exchange of chloride and methyl groups is observed. The homoleptic Me$_4$Sn reacts irreversibly in CPE yielding the dimer Me$_3$SnSnMe$_3$ and alkylmercury compounds:

$$2Me_4Sn + 2Hg \longrightarrow Me_3SnSnMe_3 + 2MeHg^+ + 2e$$

The halides Me$_n$SnCl$_{4-n}$ have reversible CV responses and in CPE yield inorganic Sn(II), MeHgCl and unidentified compounds including unstable complexes of the formula Me$_{4-n}$Cl$_{n-1}$SnHgSnCl$_{n-1}$Me$_{4-n}^{88}$.

D. Reduction and Oxidation of Germanium Halides (see also Section XI)

Corriu and collaborators[97–99] studied the electroreduction of triorganohalosilanes and germanes. Polarography shows a single irreversible wave (a second probably resides outside the solvent window) for R_3SiX and two irreversible waves for R_3GeX. The reduction products are dimers R_3GeGeR_3 (from both reduction waves) and formation of R_3GeH at the potential corresponding to the second wave is particularly observed in protic solvents. Both reactions are considered a manifestation of the nucleophilic character of the anion R_3Ge^-:

$$R_3GeX + e \longrightarrow R_3GeX^{\bullet-}$$

$$R_3GeX^{\bullet-} + e \longrightarrow R_3Ge^- + X^-$$

$$R_3Ge^- + SolH \longrightarrow R_3GeH + Sol^-$$

$$R_3Ge^- + R_3GeX \longrightarrow R_3GeGeR_3 + X^-$$

Ph_3SnCl yields the radicals $Ph_3M^{\bullet}$ with greatest ease and reduction is actually proposed as an efficient route to dimers[100]. The first electron transfer is, however, reversible:

$$Ph_3SnCl + e \rightleftharpoons Ph_3SnCl^{\bullet-}$$

$$Ph_3SnCl^{\bullet-} \longrightarrow Ph_3SnCl^{\bullet} + Cl^-$$

$$Ph_3SnCl^{\bullet} \longrightarrow \tfrac{1}{2}Ph_3SnSnPh_3$$

Several values of the half-wave potential for the reduction first wave (vs g/AgCl) are given in Table 3. In a separate citation, the values are: R_3SnCl, $E_{1/2} = -0.9$ V; R_3GeCl, $E_{1/2} = -2.1$ V; R_3SiCl, $E_{1/2} = -2.4$ V; in this case vs SCE[101].

Generally ease of reduction of R_3MX is in declining order, with $M = Sn \gg Ge > Si$. The difference in first $E_{1/2}$ between Ph_3SnCl and Ph_3GeCl is extremely large, $+850$ mV, compared to $+200$ mV between Ph_3GeCl and Ph_3SiCl and, from the evidence of reversibility, $[Ph_3SnCl]^{\bullet-}$ is considerably more stable than the Ge and Si analogs. These are striking irregularities. The changes due to X are significant, around $100-200$ mV, though the shift from Ph_3SiCl to Ph_3SiBr is extraordinarily small, $+20$ mV, considering the other values. Electron affinities of the $Ge-X$ bonds are clearly higher than $Si-X$ (see also Reference 215). No such distinct differences are noted when $X =$ benzoyl[102]. The nature of the

TABLE 3. Half-wave potentials $E_{1/2}$, in polarography, of several halogermans in DMA or THF, 0.1 M TBAP vs Ag/AgI $(Bu_4NI$ sat$)^a$

	First $E_{1/2}$	Second $E_{1/2}$
Ph_3SnCl	-0.9^b	
Ph_3GeF	-1.85	-2.3
Ph_3GeCl	-1.75	-2.3
Ph_3GeBr	-1.60	-2.3
$(p\text{-}CF_3C_6H_4)_3GeCl$	-1.82	
$n\,Bu_3GeCl$	-1.50	
$Ph_2MeGeBr$	-1.35	
For comparison:		
$Ph_3GeGePh_3$	-2.30	
Ph_3SiF	-2.15	
Ph_3SiCl	-1.95	
Ph_3SiBr	-1.93	

^{a} Potential of this electrode is -0.44 V vs SCE.
^{b} Reference 100; other data from Reference 97.

ET process is not necessarily the same in analogous compounds of Si, Ge and Sn. The trends in oxidation potentials are also dicussed below in Sections VII to X.

Studies on the reduction of Ph_3GeX expose more problems related to interpretation of polarography. Whereas Corriu and coworkers stress the description by which the first ET creates a radical anion that reacts further (*vide infra*), Fleet and Fouzder[103] emphasize the formation of a radical in the first step:

$$Ph_3GeX + e \longrightarrow Ph_3Ge^{\bullet} + X^-$$

There are also differences in procedures which are nevertheless important to the mechanism. Corriu's group worked in aprotic solvents and noted the sensitivity to traces of water when the nucleophilic solvent catalyzes hydrolysis. The HX so formed is detected by polarography. Fleet and Fouzder, on the other hand, worked in acetate buffer and noted that in aqueous-organic media the product is strongly adsorbed on the electrode surface. Other studies, too, concur with the formation of the triphenylgermyl radical in the first ET step[104] and, moreover, propose the formation of a triphenylgermanium ion Ph_3Ge^+ by single electron oxidations of Ph_3GeCl, Ph_3GeBr and Ph_3GeI. Such oxidation is not observed on Pt but is detected on Hg electrode. Like other cases discussed in Section VI, oxidations take place during polarography on mercury. The evidence indicates formation of Ph_3Ge^+ in an ion pair with perchlorate, when the latter is present:

$$Ph_3Ge^+ + ClO_4^- \longrightarrow Ph_3GeClO_4$$

$$Ph_3Ge^+ + e \longrightarrow Ph_3Ge^{\bullet}$$

The electrochemical generation of the germyl anion has been the subject of a recent paper[105]. Evidence for its formation by SET reduction of Ph_3GeH on Pt in DMF with tetrabutylammonium tetrafluoroborate is based on ^{13}C NMR which shows a strong downfield shift of *Ca* 30 ppm for the *ipso* carbon of the anion. The anion tends to yield $Ph_3GeGePh_3$ above $20\,^{\circ}C$, but is also trapped by reactions at $-40\,^{\circ}C$ with O_2 and CH_3I:

$$Ph_3Ge^- + \tfrac{1}{2}O_2 \longrightarrow Ph_3GeO^-$$

$$Ph_3Ge^- + CH_3I \longrightarrow Ph_3GeCH_3 + I^-$$

E. Reduction and Oxidation of Benzoyl Derivatives

Only very scarce knowledge on this subject is available. Oxidation and reduction potentials of seven benzoylsilanes and three benzoylgermanes were measured[102]. The values are

TABLE 4. Oxidation and reduction of benzoyl-silanes and benzoylgermanes; peak potential values E_p, from CV in acetonitrile, TEAP 0.1 M on GC vs Ag/AgCl[102]

	red. E_p	ox. E_p
$C_6H_5COGeMe_3$	-1.93	1.68
$C_6H_5COSiMe_3$	-1.98	1.88
$C_6H_5COGePhMe_2$	-1.87	1.73
$C_6H_5COSiPhMe_2$	-1.80	2.02

dependent on substituents and most listed compounds were not substituted in a comparable way. The values for those compounds which do correspond are listed in Table 4.

Substitution of Si by Ge has no obvious effect on reduction but is clearly accompanied by easier oxidation. The oxidation reaction was not identified. In the silane series, values of reduction E_p are linear with Hammett σ constants for substituents on the aryl ring. No parallel test on germanes has been reported.

F. Halogenation

Tri-substituted germanes like Ph_3GeH and Ph_2MeGeH (and silanes) were halogenated by a cathodic process. This contrasts with the more familiar anodic halogenation. Electrolysis is conducted in AN on Pt electrodes with TBAP and Bu_4NBF_4, and the resulting products Ph_3GeCl, $Ph_2MeGeCl$, $(Ph_3Ge)_2O$, Ph_3GeF and Ph_2MeGeF appear to draw on the electrolyte for their source of chlorine, oxygen and fluorine, respectively. No clear mechanism has been outlined for the process[106].

VI. ANALYSIS

Basic aspects of the reactivity and analysis of lead, tin, germanium and mercury compounds have been dealt with in a previous review[107]. A thorough understanding of the mechanistic details is essential for a reliable analytical procedure and the development of electroanalytical techniques is closely related to the electrochemical properties of the M−C bonds and identification of the various polarographic steps[108,109]. A detailed review covers the polarographic studies concerned with reduction of group 14 elements on Hg, up to 1974[10].

A. Lead

The falling rate of production of alkyllead compounds in the last two decades is accompanied by increasing attention to their analysis and assay. Organolead is found in the environment, natural sweet-water, sea water, air, exhaust gases, sea fauna and gasoline. Tetraalkylleads are not easily reduced on Hg in aqueous solutions[110,111] and are relatively inactive in such conditions[107b]. Therefore, in early methods samples were converted to inorganic lead prior to analysis. Recently, methods for their straightforward polarographic analysis in organic media gained significance. The distinction between inorganic and organic lead is an important problem in environmental control. Therefore, the electrochemical reduction of the trimethyllead(IV) cation in sea water as distinct from inorganic lead has been studied[112]. Determinations by DPP, NPP, CV and ASV — in which the deposited lead is reoxidized — were compared[113]. The data for reduction of inorganic Pb(II) to Pb on Hg, vs Ag|AgCl(KCl) as reference, shows a two-electron process at −0.41 V (reversible), −0.72 V (DPP, NPP, CV) and −1.23 V (DPP). $(CH_3)_3Pb^+$ gives a single-electron reduction. The reaction yields in the first reduction stage $(CH_3)_3Pb^{\bullet}$ which is unstable, and lead which is absorbed on Hg as an amalgam, Pb(Hg):

$$(CH_3)_3Pb^+ + e \longrightarrow (CH_3)_3Pb^{\bullet} + Pb(Hg) + \text{other products}$$

A second reduction step observed in this analysis is attributed to the reduction potential for the methylmercury cation CH_3Hg^+:

$$CH_3Hg^+ + e \longrightarrow CH_3Hg_{ads} + e \longrightarrow CH_3Hg^-$$

Alkylmercury is therefore considered a product in these reactions:

$$(CH_3)_3Pb^\bullet + 3Hg \longrightarrow Pb + 3CH_3Hg_{ads}$$

and the second reduction is that of CH_3Hg_{ads}.

The first three methods, DPP, NPP and CV, clearly distinguish the reductions of $(CH_3)_3Pb^+$ and inorganic Pb^{2+} and can be applied for analysis of one in the presence of the other, but ASV may become inaccurate due to interference of CH_3Hg in reoxidation of deposited lead.

Conclusions from other studies regarding the reductions of $(CH_3)_3Pb^+$ are similar to those proposed earlier by Colombini and coworkers[111], who also studied the application of DPP for the electrochemical speciation and determination of organometallic species in natural waters[110] and the consecutive determination in natural waters of $(CH_3)_4Pb$, $(CH_3CH_2)_4Pb$, $(CH_3)_3PbI$, $(CH_3CH_2)_3PbI$, $(CH_3)_2PbI_2$, $(CH_3CH_2)_2PbI_2$ and elemental lead[114]. The method is based on selective extraction from water to an organic phase of tetramethyllead and tetraethyllead as well as inorganic lead, followed by DPP analysis.

Samples for determination of ionic alkyllead species in marine fauna were homogenized in the presence of salts and the alkyllead component was extracted with toluene and oxidized with HNO_3. Determination was by DPASV[115]. A method based on oxidation on Hg electrode has been described[116] for analysis of alkylleads in gasoline. Alkylation of Hg is involved, of course, but as an oxidation the method does not suffer from the background of atmospheric oxygen. The peak potentials E_p for oxidation of tetramethyllead and tetraethyllead on various cathodes are well resolved (Table 5).

Recently, trace analysis for $(CH_3)_4Pb$ and $(CH_3CH_2)_4Pb$ in the range of 0.1–30 mg/liter has been performed in a three-electrode cell by reducing the lead compounds by pulses at -1.2, -1.5, then stripping with an oxidizing current between -0.9 and -0.3 (all vs SCE). A carbon electrode was used which was freshly coated with mercury. The procedure is fast (ca 5 min) and applicable at low concentrations of tetraethyl- and tetramethyllead (ca 15 mg/liter), which is suitable for unleaded gasoline samples[117]. Electrochemical analysis with carbon and mercury-gold electrodes with detection limits of 310–340 ng was also reported[118]. Analysis of alkyllead derivatives (from gasoline) often suffers from interference by inorganic lead, which can be found in the natural environment in considerable concentrations. Methods were therefore developed for analysis by DPASV of alkylleads specifically, in the presence of large amounts of inorganic lead. In previous procedures, the latter is removed selectively by co-precipitation with $BaSO_4$ which was found superior to the more commonly used complexation by EDTA[119].

A polarographic DME detector was coupled to an HPLC system as well as a flow cell equipped with a glassy carbon anode. Gasoline samples with $(CH_3)_4Pb$ and $(CH_3CH_2)_4Pb$ in concentration ranges of 5×10^{-7} to 5×10^{-3} M were analyzed on a reverse-phase C18 column. Unlike DME, the GC electrode suffers much noise from constituents of the gasoline samples. The DME has no such problems and, being operable at much lower

TABLE 5.   Peak potentials E_p for oxidation of tetramethyllead and tetraethyllead on three different electrodes: in AN 0.1 M Et_4NClO_4 at $20\,^\circ C$, Ag|AgCl reference, calibrated at $+0.38$ V vs 10^{-3} ferrocene[116]

	Electrode		
	Pt	DME	GC
Tetramethyllead	+1.80	0.39	+1.1
Tetraethyllead	+1.26	+0.5	+1.65

potentials, is apparently the preferred electrode. Data were also collected for analysis in dichloromethane[120]. Liquid chromatography has been combined with the electrochemical and/or spectrophotometric detection and with the automated determination of several metals like lead, cadmium, mercury, cobalt, nickel and copper[121]. The electrochemical method for detection is based on the oxidation of the dithiocarbamate (dtc) complex, which is irreversible in dichloromethane (0.1 M Bu_4NClO_4) or in acetonitrile (0.1 M Et_4NClO_4):

$$3PbII(dtc)_2 \longrightarrow 2[PbIV(dtc)_3]^+ + Pb^{2+} + 4e$$

E_p^{ox} (vs Ag|AgCl in LiCl–acetone) $= +0.70$ V in dichloromethane and $+1.00$ V in acetonitrile as determined by CV at 200 mV s^{-1} on carbon. By controlling the detection potential, it proved possible to run simultaneous analysis of Pb, Cd, Hg and other metal ions in mixtures.

The oxidation mechanism of tetraethyllead and tetramethyllead on mercury cathodes has already been mentioned[120]. Studies under a short pulse on DME in dichloromethane indicate a single ET:

$$Et_4Pb + Hg \longrightarrow Et_3Pb^{\bullet} + EtHg^+ + e$$

$$2Et_3Pb^{\bullet} \longrightarrow Et_6Pb_2$$

The initial pathway is the same for tetramethyllead. However, under prolonged electrolysis additional reactions are observed with hexamethyldilead, namely:

$$Me_6Pb_2 + MeHg^+ + e \longrightarrow Me_5Pb_2^+ + Me_2Hg$$

Ph_4Pb, generally thought to be oxidized by a two-electron reaction, is also found to react by a SET under short pulses. However, $Ph_3P^{\bullet}$ is slow to dimerize and an alternative reaction occurs preferably:

$$2Ph_3Pb^{\bullet} + 2PhHg^+ \longrightarrow 2Ph_3Pb^+ + Ph_2Hg + Hg$$

Further reaction of Ph_3P^+ and Hg contributes the second ET for the overall two-electron oxidation. This was studied in detail for oxidations of tetraphenyllead, tetraethyllead and tetramethyllead at mercury electrodes in dichloromethane. The rationale of the mechanisms proposed above is based on the following observations[122].

Differential pulse polarograms for the oxidation of R_4Pb (R = Ph, Me, Et, Bu) in CH_2Cl_2 on DME all show an electrode response for oxidation. This DPP response is, however, spread over a wide potential range of 300 mV indicating that several processes are reflected by this wave. The DPP responses for several lead compounds are cited in Table 6.

The low-potential responses are detected on Hg. Solid electrodes like glassy carbon, gold or platinum show no response in this range. This implies that oxidation of the mercury electrode takes place as well. Indeed, the second DPP waves observed at $+0.68$ up to $+0.72$ V correspond to oxidation of mercury compounds such as Ph_2Hg. The

TABLE 6. DPP data for oxidation of lead compounds on DME[a]

Ph_4Pb	Bu_4Pb	Et_4Pb	Me_4Pb	Ph_3PbCl	Ph_2PbCl_2
+0.45,	+0.60	+0.57,	+0.41	+0.53	+0.52
+0.68	—	+0.72	—	—	—

[a] 5×10^{-4} in CH_2Cl_2 (TBAP 0.1 M), pulse amplitude 50 mV, drop time 0.5 s; reference is Ag|AgCl, 20 °C[122].

main response (first line in Table 6) is linear as long as concentration of lead compound remains within 10^{-6}–10^{-4} M. Above this, linearity is lost, implying absorption on Hg. Transmetallations from lead to mercury take place:

$$2R_4Pb + Hg \longrightarrow 2R_3Pb^+ + R_2Hg + 2e$$

$$2R_3Pb^+ + Hg \longrightarrow 2R_2Pb^{2+} + R_2Hg + 2e$$

Subsequent reactions of mercury compounds are observed.

To conclude this section, two exceptions to solution chemistry are noteworthy. In one, tetraethyllead has been collected from air onto activated carbon and the concentration in the solid adsorbent was determined by ASV[123]. In the other, a method for the potentiometric and CV measurement of solid samples, in solid, has been conceived by Bond and Scholz[124]. The insoluble compound is incorporated into a paraffin-impregnated graphite electrode. CV measurements were carried out of Hg and Pb diethyldithiocarbamate complexes. By running consecutive CV cycles it is possible to determine thermodynamic data, like stability constants. The Pb complex shows redox waves at -0.8 and -1.6 V (stripping) vs Ag|AgCl and the value $\log K = 17.7$ by this polarography compares well with $\log K = 18.3$ determined by extraction[125].

B. Tin

Organotin compounds are widely dispersed in the environment due to their many applications, e.g. as PVC stabilizers, in paints, in control of helminthic and protozoal infections in poultry, as anti-fouling agents. Their use on the hulls of ships is a cause of contamination of coastal waters[126]. Their determination in sea water by cathodic stripping voltammetry has often been done by their prior conversion to inorganic tin salts. Early literature on polarographic behavior of organotin compounds has been summarized[127].

Detection of trace amounts of dibutyltin dichloride by DPASV was studied by Kitamura's group[128]. The same method was also applied for the quantitative determination of tributyltin oxide in sea water[129]. The concentration of alkyl Pb and Sn in sea water has been determined by DPP and ASV[109]. With the same techniques it was possible to analyze inorganic Pb in the presence of dimethyl- and trimethyllead compounds. EDTA as complexing agent was used to avoid overlap of peak potentials of the inorganic lead ions and organic lead compounds. Trace analysis of all three proved practically possible[108]. In an earlier work the compounds Bu_2SnX_2 (X = chloride, laurate, maleate) in EtOH–H_2O solutions were studied methodically by pulse polarography, reverse pulse polarography, DPP and CV. CV for the three compounds shows clear reduction and strong reverse scans which reflect the reoxidation or stripping of the reduction products from the Hg cathode. Apart from the detailed redox data collected, use is made of the observation that products of reduction can be adsorbed, i.e. accumulated on the electrode surface. This explains the very strong anodic peak of the reverse scan, the stripping wave, which serves for determination of the tin compounds. The sensitivity is increased considerably by application of DPASV. A general scheme for the reduction of dibutyltin compounds of the type Bu_2SnX_2 (X = laureate, maleate) in EtOH–H_2O solutions has been proposed by Fleet and Fouzder[130]. The reduction takes place in three steps:

$$Bu_2Sn^{2+} \underset{-e}{\overset{+e}{\rightleftharpoons}} Bu_2Sn^{\circ+} \xrightarrow{+e} [Bu_2Sn\,:] \xrightarrow{e,\,2H_2O} Bu_2SnH_2 + [Bu_2SnSnBu_2]^{2+} + (Bu_2Sn)_n$$

Analysis by cathodic reduction of triphenyltin acetate in pesticide preparations[131] and in samples from marine sources[132] can reach detection limits of 10^{-8} M to 2.5×10^{-9} M, respectively. The reduction on mercury-film glassy carbon electrodes involves an initial adsorption peak at $E_p \sim -0.7$ V, reduction to triphenyltin radical at $E_p - 1.0$ V and further reduction to the anion at $E_p - 1.4$ V (all vs SCE); the values are pH-dependent:

$$Ph_3SnAc + e \longrightarrow [Ph_3Sn]^{\bullet}_{ads} + Ac^- \; E_p - 0.7 \text{ V}$$

$$Ph_3SnAc + e \longrightarrow [Ph_3Sn]^{\bullet} + Ac^- \; E_p - 1.0 \text{ V}$$

$$[Ph_3Sn]^{\bullet} + e \longrightarrow [Ph_3Sn]^- \; E_p - 1.4 \text{ V}$$

Complementary reaction:

$$2[Ph_3Sn]^{\bullet} \longrightarrow Ph_3SnSnPh_3 \quad \text{nonelectroactive}$$

$$[Ph_3Sn]^- + H^+ \longrightarrow Ph_3SnH^*$$

$$^*Ph_3SnH \text{ can be reoxidized at } -0.3 \text{ V.}$$

This is in accord with previous work[133].

VII. COMPLEXES OF GROUP 14 METALS

A. Anodic Synthesis

Synthesis with sacrificial electrodes is employed as a direct method in several other preparations of organometallic compounds and complexes. 3-Hydroxy-2-methyl-4-pyrone derivatives of Sn 1 (and of Zn, Cu, In and Cd as well) were prepared using the metal as an anode. The low oxidation state Sn(II) compound is obtained by direct electrolysis[134].

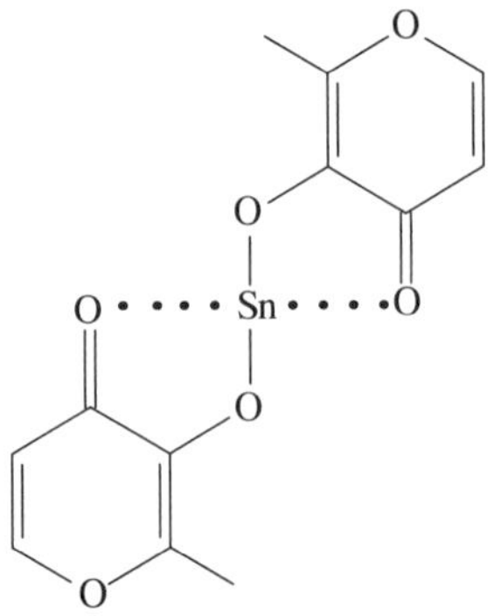

(1)

This direct electrochemical synthesis has proved efficient in the preparation of several other complexes, among these the tin derivatives of 3-hydroxy-2-phenylflavone (2) and 2-ethoxyphenol (3), respectively. The use of sacrificial electrodes proved very efficient; the produced complex precipitates during electrolysis and is easy to isolate[135].

Tin adducts of the type $Sn(O_2R)$ were obtained in the electrolysis of aromatic diols with tin as the sacrificial anode: $R(OH)_2 = 1, 2$-dihydroxybenzene (catechol), tetrabromo-cathechol, 2,3-dihydroxynaphthalene and 2,2'-dihydroxybiphenyl; yields, based on mass loss of the anode, range within 75–94 %[136].

The single-step electrosynthesis of O-ethylxanthato and N, N'-dimethyl dithiocarbamato complexes of Sn(IV) and Pb(II) was carried out by using Sn and Pb anodes in acetone

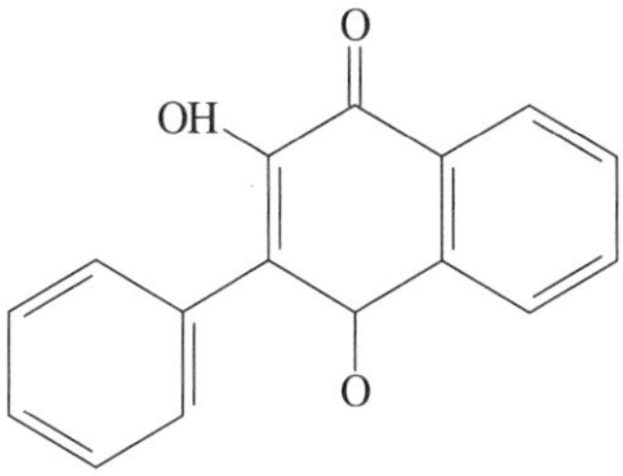

(2) 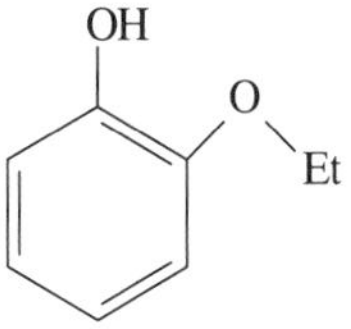   (3)

solutions of the dixanthogen or tetramethylthiuram disulfides. This was carried out in 0.1-liter cells with Pt as counterelectrode, with TBAP as electrolyte. An overall cell voltage of 50 V was applied yielding initial currents of 20–46 mA. Yields were 50–77 %. Potentials and current yields were not reported[137]. Lead complexes are important in lead production processes. Ethyl xanthate is used to collect lead in the flotation of Galena. Since redox processes are involved in complex formation, systematic electrochemical studies were undertaken. Lead xanthate can be oxidized as well as reduced:

$$Pb(EtOCS_2)_2 \longrightarrow Pb^{2+} + (EtOCS_2)_2 + 2e$$

$$Pb(EtOCS_2)_2 + 2e \longrightarrow Pb + 2EtOCS_2^{2-}$$

The potentiometry of both these reactions has been studied in detail[138].

Pb(II) oxalate was studied as well. Complex formation and stability constants were determined by polarography and CV. Results obtained on a solid Pt electrode (CV) confirm measurements obtained previously on DME[139].

Certain lead(II) complexes were shown to display reversible redox behavior on the CV scale. Examples are Pb(II) complexes of 2(o-hydroxyphenyliminomethyl)-pyrrole and 2(o-hydroxyphenyliminomethyl)-thiophene. Their stability constants were also determined by polarography[140]. Anodic exchange reactions of extracted metal chelates were carried out with 1-pyrrolidinecarbodithioic acid in isobutyl methyl ketone[141].

B. Complete Reduction of Complexes

Reduction of triphenyltin piperidyldithiocarbamate in acetone was shown by polarography and voltammetry to consist of two diffusion-controlled peaks and two peaks which seem to reflect adsorption[142]. Apparently, a dithiocarbamate group dissociates and triphenyltin radical forms by reduction. The latter partly dimerizes and partly reduces to triphenyltin anion.

A method for the preparation of dendritic Sn by the electrolysis of alkyltin complexes has been described. Bu_2SnBr_2 and Bu_3SnBr form as intermediates[143]. The method is aimed at recovery of elemental tin from residues of tin compounds and for reuse in the preparation of tin halides.

VIII. ELECTRODIC REACTIONS OF BIMETALLIC METALALKYLS AND ARYLS

A. Early Work

Works by Dessy and coworkers provide a comprehensive treatment of the electrochemistry of the metal–metal bonds in organometallic compounds. This includes the scission[144] and formation[145] of metal–metal bonds and also the reactivities and nucleophilicity of

the organometallic radicals and anions formed as intermediates in the wake of the electrochemical reaction. Bond cleavage of homo (M−M) and hetero (M−M′) dimetallic
molecules was assumed as taking place by either of two routes:
A single-electron reduction creating one anion and one radical:

$$M-M' + e \longrightarrow M :^- + M'^\bullet$$

A two-electron reduction creating two anions:

$$M-M' + 2e \longrightarrow M :^- + M' :^-$$

These reactions were carried out on mercury or platinum cathodes, with $Ag^+|Ag$ reference,
in dimethoxyethane with TBAP electrolyte. CPE was followed up by polarography.
Several findings are relevant to the subject at hand. Generally, it is expected that reduction
requires decreasing values of cathodic potentials for group 14 elements along the series
Si, Ge, Sn, Pb (first three entries in Table 7). This is in contrast to trends in transition
metal series like Cr, Mo and Fe, Ru. The alignment of potential values in heterodimetallic
compounds is therefore not unequivocal, as seen from the results cited in Table 7 (last
two entries). Polarography shows one-electron reductions for heterodimetallic compounds
of Ge and Sn, but more cathodic potentials and two-electron reductions for Pb compounds
of comparable structures.

An interesting remark is made by the authors considering these results. In both manganese and molybdenum complexes, $-SnMe_3$ and $-PbEt_3$ derivatives reduce at more
anodic potential, namely they are cleaved with greater ease, than the aryl counterparts. It is
proposed that electron-withdrawing substituents counterbalance the destabilizing effect of
nonbonding d electrons. (However, a stabilizing effect of d electrons is discussed below.)
A significant correlation of the reduction potentials to homodinuclear M−M bond strength
is pointed out. These are given as 42.2 for Si−Si, 37.6 for Ge−Ge and 34.2 (kcal/mol)
for Sn−Sn[146].

In an elegant extension, these results are utilized for the electrosynthesis of bimetallic
compounds[147]. In an example, hexaphenylditin is electrolyzed in the cathodic compartment of a divided cell. Polarography of Ph_6Sn_2 shows its reduction at $E_{1/2} = -2.9$ V on Pt
or Hg. CPE is carried out at -3.1 V until polarographic tests show the absence of a current
for reduction of Ph_6Sn_2 but a large anodic current for $[Ph_3Sn]^-$. Addition of $CpFe(CO)_2I$

TABLE 7.   Half-wave potential values for reduction of homo- and heterobimetallic compounds[144]

Compound	$E^a_{1/2}$	n^b	Products
$Ph_3 GeGePh_3$	−3.5	2	$2 [GePh_3]^-$
$Ph_3 SnSnPh_3$	−2.9	2	$2 [SnPh_3]^-$
$Ph_3 PbPbPh_3$	−2.0	2	$2 [PbPh_3]^-$
$Cp(CO)_3 MoSnPh_3$	−2.4	1	$[Cp(CO)_3 Mo]^-, [SnPh_3]^\bullet$
$Cp(CO)_2 MoSnMe_3$	−1.9	1	$[Cp(CO)_3 Mo]^-, [SnMe_3]^\bullet$
$Cp(CO)_3 MoPbPh_3$	−2.2	2	$[Cp(CO)_3 Mo]^-, [PbPh_3]^-$
$Cp(CO)_5 MnSnPh_3$	−2.5	1	$[Cp(CO)_5 Mn]^-, [SnPh_3]^\bullet$
$Cp(CO)_5 MnSnMe_3$	−1.9	1	$[Cp(CO)_5 Mn]^-, [SnMe_3]^\bullet$
$Cp(CO)_5 MnPbPh_3$	−2.1	2	$[Cp(CO)_5 Mn]^-, [PbPh_3]^-$
$Cp(CO)_5 MnPbEt_3$	−1.8	1	$[Cp(CO)_5 Mn]^-, [PbPh_3]^\bullet$
$Cp(CO)_2 FePbPh_3$	−2.1	2	$[Cp(CO)_2 Fe]^-, [PbPh_3]^-$
$Cp(CO)_2 FeSnPh_3$	−2.6	1	$[Cp(CO)_2 Fe]^-, [SnPh_3]^\bullet$

^{a} Vs $Ag^+|Ag$.
$^b n$ = number of electrons involved.

at this stage yields $CpFe(CO)_2SnPh_3$ identified by polarography $E_{1/2} = -2.61$ V and UV spectra. The synthesis can also be carried out the other way round, as shown below:

Sequence 1:

$$Ph_3SnSnPh_3 + 2e \longrightarrow 2Ph_3Sn^- \qquad - 3.1 \text{ V} \qquad\qquad \text{E}$$

$$Ph_3Sn^- + CpFe(CO)_2I \longrightarrow CpFe(CO)_2SnPh_3 + I^- \qquad \text{C}$$

Sequence 2:

$$[CpFe(CO)_2]_2 + 2e \longrightarrow 2[CpFe(CO)_2]^- \qquad - 2.4 \text{ V} \qquad \text{E}$$

$$[CpFe(CO)_2]^- + Ph_3SnCl \longrightarrow CpFe(CO)_2SnPh_{3+}Cl^- \qquad \text{C}$$

A Statement by Russel[148] is cited by Dessy in this connection, namely that: '...carbanions are capable of donating one electron to an acceptor especially when the central carbon atom of the anion is substituted by the heavier silicone, germanium or tin...':

$$R^- + S \longrightarrow R^\bullet + S^-$$

Hence, reduced metal alkyls of the group 14 tend to form radicals in the presence of a proper acceptor.

Whereas Dessy discussed the reduction of Ph_6Pb_2 where the anion Ph_3Pb^- and radicals $Ph_3Pb^\bullet$ form by ET, Doretti and Tagliavini find that a reversible oxidation, also involving the radical, occurs under certain conditions[149]:

$$Ph_6Pb_2 \underset{}{\overset{-e}{\rightleftharpoons}} 2Ph_3Pb^\bullet \rightleftharpoons 2Ph_3Pb^+ + 2e$$

This reaction is reversible on a platinized Pt electrode and radical formation is the key to reversibility. Nonelectrodic homolysis does not take place unless *platinum black* is present as catalyst. The evidence is based on considerable exchange between isotopic ^{210}Pb-labeled Ph_6Pb_2 and Ph_3PbNO_3 which only occurs in the presence of platinum black. A similar case has been shown for Ph_6Sn_2[150].

B. Recent Work

The electrochemistry of organotransition metal–tin compounds has been studied by Mann and coworkers[151], who also listed in detail the previous literature concerning these compounds. The experimental method is based on infrared spectroelectrochemical tests, using a flow-through, thin-layer cell of CaF_2, in addition to CV preparative electrolysis and double-potential step chronocoulometry. Compounds of the type M_nSnPh_{4-n} were examined where $M = Mn(CO)_5$, $CpMn(CO)_3$, $CpMo(CO)_3$ and $CpFe(CO)_2$. Both oxidations and reductions were studied. For example, oxidation of $Mn(CO)_5SnPh_3$ in CH_3CN takes place at 1.06 V (vs Ag/AgCl) and is irreversible on the CV time scale. It is, however, chemically reversible since the return CV curve shows a reduction peak for the $[Mn(CO)_5]^+$ ion and the resulting anion recombines with $SnPh_3(CH_3CN)_2^+$ to reproduce the starting material.

Oxidation in CH_3CN:

$$Mn(CO)_5SnPh_3 - 2e \longrightarrow [Mn(CO)_5](CH_3CN)^+ + SnPh_3(CH_3CN)_2^+ \qquad E_p = 1.4 \text{ V}$$

Reduction:

$$[Mn(CO)_5](CH_3CN)^+ + 2e \longrightarrow [Mn(CO)_5]^- \qquad E_p = -1.16 \text{ V}$$

Chemical step:

$$[Mn(CO)_5]^- + SnPh_3(CH_3CN)_2^+ \longrightarrow Mn(CO)_5SnPh_3$$

Survival of the ions is assisted by the solvent. Complexes with bridging tin atoms, and mixed complexes, are particularly interesting. Oxidation of the latter is followed by cleavage of either of the tin–metal bonds as in the following example:

$$[Mn(CO)_5]SnPh_2[CpFe(CO)_2] - e \longrightarrow [Mn(CO)_5]SnPh_2[CpFe(CO)_2]^+$$

Break up of Sn–Fe or Sn–Mn bonds is observed, possibly as a result of one more ET. The alternative fragments are found in nearly equimolecular amounts. It is uncertain whether this is due to similar cleavage rates or the results of follow-up equilibration. The results of this study can be considered as corresponding to the electrosynthesis of complexes described earlier in this section.

A recent study of the oxidation of several group-14 bimetallic (M–M) compounds shows that the shift in oxidation potentials clearly reflects the trend of decreasing E_p^{ox} with increasing atomic number, as expected for the group-14 line with the gradually increasing HOMO level[152]. The potential difference from M = C (no d electrons) to Si is much larger than from Si to Ge (Table 8). The oxidized products show low-intensity return currents on CV, indicating some extent of survival. The proposed products are the corresponding cations, however the nature of these oxidation products and subsequent intermediates is far from clarified and the line of compounds tested is not comprehensive enough. It seems clear though that the oxidation of Si–Si compounds requires higher potential than that of Ge–Ge derivatives, other substituents being equal, and that substitution of methyl by phenyl groups lowers the oxidation potential. It is inferred that interaction of the aromatic π electrons and available d orbitals raises the HOMO levels. These results stand in contrast to data on oxidation of silyl and germyl substituted pyrroles by the same authors[153].

Oxidation of organogermyl and organostannyl anions, as lithium compounds, shows opposite trends[154]. Irreversible oxidation voltammograms were measured in hexamethylphosphorous triamide (HMPA) solutions. The oxidation potentials are given in Table 9. These oxidation levels are correlated to the HOMO nonbonding orbital of the anions. The more anodic oxidation potential of Me$_3$SnLi as compared to Me$_3$GeLi is contrary to the trend of receding oxidation potential with increasing M observed with neutral bimetallic compounds. The higher oxidation potential of phenyl derivatives is also contrary to the observations made on the bimetal series[152] and has been attributed to steric hindrance. It appears that other factors like anion stabilization by π–d delocalization or

TABLE 8. Peak potentials for oxidation, E_p, of several group-14 alkyls in AN, TBAP on Pt, in V (vs Fc/Fc$^+$)[152]

Compound	E_p
Ph$_3$CH	1.81 ± 0.03
PhMe$_2$CH	1.89 ± 0.03
Ph$_3$SiSiMe$_3$	1.29 ± 0.03
(PhMe$_2$Si)$_2$	1.26 ± 0.03
(Me$_3$Si)$_2$	1.36 ± 0.03
Ph$_3$GeGeMe$_3$	1.16 ± 0.03
(Ph$_2$MeGe)$_2$	1.20 ± 0.03
(PhMe$_2$Ge)$_2$	1.24 ± 0.03
(Me$_3$Ge)$_2$	1.28 ± 0.03

TABLE 9. Oxidation potentials (E_p) of organogermyl- and organostannyllithium derivatives[a]

Anion	E_p (V)
Me₃SnLi	−0.10
Me₃GeLi	−0.90
Et₃GeLi	−0.88
n-Bu₃GeLi	−0.83
PhMe₂GeLi	−0.87
Ph₂MeGeLi	−0.49
Ph₃GeLi	−0.29

[a] E_p values are from CV at 100 mV s⁻¹ on Pt in HMPA with LiCl (0.1 M) at 25 °C vs SCE. Anion concentrations: 0.01–0.1 M[154].

anion strength at the electrode should not be excluded, and build-up of concentrations in the double layer should be much more dominant with the anions than with the neutral bimetallics. Differences in the structure of the double layer and the likelihood that the ET processes of anions and neutral bimetallics may not be comparable in terms of the Marcus theory must also be considered.

IX. COMPOUNDS WITH MACROCYCLIC LIGANDS

A. Porphyrin Complexes of Group 14 Elements (see also Sections XIV and XV)

Metalloporphyrins of Ge and Sn have received attention due — among other things — to their significant anticancer activity[155]. Their significance lies in the fine detail which these complexes provide on ET and redox reactions. Metalloporphyrins bearing silicon and germanium and containing carbon–metal σ-bonds have been reviewed[156]. Tin porphyrins with a carbon–metal bond have been analyzed by CV. The compounds are designated as (P)SnCH₃I in which P is the dianion [2,3,7,8,12,13,17,18-octaethylporphyrinato]²⁻ (oep), the dianion [5,10,15,20-tetra-p-tolylporphyrinato]²⁻ (tptp), or the dianions of either tetra-m-tolylporphyrin or of tetramesitylporphyrin. They are prepared by the addition of CH₃I to (P)Sn(II). The CV of (P)SnCH₃I is clearly distinguished from that of the starting material (P)Sn(II)[157] and the formation of (P)SnCH₃I on addition of CH₃I can therefore be followed by voltammetry. (P)SnCH₃I displays two reversible reduction waves which are attributed to the stepwise ET between the solvated cation and anion[158]:

$$[(P)SnCH_3(Sol)] \underset{-e}{\overset{+e}{\rightleftharpoons}} [(P)SnCH_3(Sol)]^{\bullet} \underset{-e}{\overset{+e}{\rightleftharpoons}} [(P)SnCH_3(Sol)]^{-}$$

It has been noted that alkylation with excess CH₃I yields (P)Sn(CH₃)₂ and (oep)Ge(CH₃)₂ when P = (oep). This is observed in CV with excess CH₃I, by the extra reduction potential at −1.5 V.

Carbenoid-bonded binuclear porphyrins of the type (DecP)SnFe(CO)₄ and (PalP)Sn Fe(CO)₄ can undergo two reversible single-electron reductions and one irreversible oxidation. (DecP) stands for 5-(4-N-decanoylaminophenyl)-10,15,20-triphenylporphyrin and (PalP) for 5-(4-N-palmitoylaminophenyl)-10,15,20-triphenylporphyrin. A two-electron oxidation causes cleavage of the Sn—Fe bond. The redox reactions can be clearly displayed by CV in pyridine, benzonitrile or methylene chloride. The precursor dichlorides have a somewhat different electrochemistry and show reversible oxidation as well as reduction on CV. (P)SnCl₂ is particularly interesting as its CV clearly shows a reversible oxidation of a Sn(IV) porphyrin, no such cases being previously known[159]. Two reversible reduction waves are also noted for several tin porphyrins (P)SnS and (P)SnSe, having the sulfide S^{2-} and selenide Se^{2-} groups as axial ligands to tin[160].

Voltammetric and thin-layer spectroelectrochemical measurements were taken of difluorogermanium(IV) porphyrins $GeF_2(P)$, where P is the dianion of (oep) or of 5,10,15,20-tetraphenylporphyrin (tpp) or of (tptp)[161]. These have one or two reversible waves for oxidation and for reduction with $E_{1/2}$ values as given in Table 10. Time-resolved and potential-resolved thin-layer spectra were taken on an optically transparent thin-layer Pt electrode. These two methods indicate that reversible single-electron transfers are involved and changes in the porphyrin Soret line (399 nm) and two Q bands (528 and 567 nm) imply that both the reduction and the oxidation ET reactions are centered on the ligand ring rather than on the metal atom. The conclusions are borne out by ESR measurements. The redox products are thus the corresponding porphyrin-based radical anion and radical cation.

The same techniques were used for the study of (P)Ge(Fc)$_2$ and (P)GePhFc, where Fc is a σ-bonded ferrocenyl group and P = oep or tpp. Reversible voltammograms were observed for (P)Ge(Fc)$_2$ and (P)GePhFc complexes (Table 11), single waves in case of oep and two complex waves in case of tpp[162]. Oxidation of the Fc component in tppGe(Fc)$_2$ and oepGe(Fc)$_2$ is reversible, and the potential difference between the first and second oxidation, 0.16–0.20, is in the range common to the analogous Fc$-$CH$_2-$Fc and Fc$-$Se$-$Fc[163]. A slight shift in the Soret line implies some charge delocalization on the porphyrin part. Oxidation of the porphyrin is accompanied by Ge$-$C bond cleavage, involving a Cp group, as evidenced from the substantial change in the Soret and Q bands. In oepGePhFc, Ge$-$C bond cleavage of a Ph group at > 1.2 V is implied. Reduction may include one or two reversible steps, to give a π-radical anion or a bianion, depending on the porphyrin. A comparison of these complexes with other (P)GeR$_2$ compounds[164] shows reduced molar absorptivities of the Soret lines (ca 50%) in the (P)GePhFc and

TABLE 10. $E_{1/2}$, in V (vs SCE), for oxidation and reduction of difluorogermanium(IV)porphyrins, GeF_2P, in CH_2Cl_2 and TBAP 0.1 M[161]

	$E_{1/2}$	oxidation (V)	$E_{1/2}$	reduction (V)
GeF$_2$oep	1.78	1.16	−1.25	−1.77
GeF$_2$tpp	1.72	1.36	−0.99	−1.43
GeF$_2$tptp	1.66	1.27	−0.99	−1.44

TABLE 11. $E_{1/2}$ in V (vs SCE), for oxidation and reduction of difluorogermanium(IV)(ferrocenyl)porphyrins, (P)Ge(Fc)$_2$ and (P)GePhFc, in benzonitrile and TBAP 0.1 M[162]

	Fc ligand[a]		P ligand[a]		P ligand[b]	
Ge(Fc)$_2$oep[c]	0.14	0.32	1.32	1.57	−1.59	
Ge(Fc)$_2$oep	0.17	0.34	1.30	—	−1.49	
GePhFcoep	0.22	—	1.13	—	−1.44	
Ge(Fc)$_2$tpp	0.27	0.43	1.40	—	−1.15	−1.71
GePhFctpp	0.29	—	1.24	1.51	−1.15	−1.70
GePh$_2$oep	—	—	0.88	1.39	−1.40	
GePh$_2$tpp	—	—	0.95	1.45	−1.10	−1.65

[a] Oxidation.
[b] Reduction.
[c] In CH_2Cl_2.

(P)GeFc$_2$ compounds, suggesting delocalization of electron density from the porphyrin to the ferrocenyl group. The conclusion is that the stability of Ge$-$C bonds is higher in the ferrocenyl complexes than in the alkyl complexes[165].

Iron$-$tin porphyrine complexes (P)Sn(II)Fe(CO)$_4$ and germanium$-$iron porphyrine complexes (P)Ge(II)Fe(CO)$_4$ were studied, where (P) $=$ the dianion of (oep), (tptp) or the m-tolyl analog (tmtp)[166] (see Reference 159). In these complexes Sn and Ge are assigned the low SnII and GeII oxidation states on the basis of IR and Mössbauer data. Two reversible reduction waves are observed for these complexes, reduction taking place on the porphyrin ligand. This is supported by potential-resolved thin-layer spectra. Potentials for the formation of the porphyrin π-radical anion are often linearly more cathodic, the higher the electronegativity of the central metal[167]. The results in the present set for Sn and Ge are in accord with this (Table 12). Electronegativity of GeII is taken as 2.01, and that of SnII as 1.96.

This trend has previously been observed when measurements were made in PhCN[168]. The reductions take place in discrete steps and do not cause dissociation of the complex:

$$(P)MFe(CO)_4 + e \longrightarrow (P)MFe(CO)_4{}^- + e \longrightarrow (P)MFe(CO)_4{}^{2-}$$

Oxidation, on the other hand, is irreversible and results in cleavage of the metal$-$metal bond. Sn compounds are oxidized at higher potentials than the corresponding Ge complexes. The products are $[(P)M(IV)]^{2+}$ (M $=$ Ge, Sn).

Another study by Neta and coworkers[169], on the redox chemistry of several metalloporphyrins in aqueous solution where metals were Zn, Pd, Ag, Cd, Cu, Sn and Pb, shows them to have well defined oxidation and reduction steps, yet the metals seem to exert merely an inductive effect on the porphyrin π level.

The redox chemistry of several Sn and Pb metalloporphyrins was also studied by Whitten and collaborators[170]. Oxidations that can involve the metal ion or the porphyrin ligand show complications related to the inability of the large divalent ions Sn(II) and Pb(II) to fit into the center of the porphyrin plane, whereas Sn(IV) and Pb(IV) are small enough to do so. The neutral lead(II) octaethylporphyrin (oep)Pb shows a three-step reversible redox behavior. It undergoes two reversible single-electron transfer oxidations and one reversible single-electron reduction, which are clearly displayed by CV. Comparisons of UV-vis spectra and ESR of (oep)Pb$^+$ with that of oxidized unattached porphyrin show a close fit, and imply that (oep)Pb$^+$ is the π-cation radical, namely the porphyrin is oxidized rather than the central lead atom. The lead atom is therefore considered to reside outside the ligand's plane.

$$(oep)Pb(II) \underset{+e}{\overset{-e}{\rightleftharpoons}} [(oep)Pb(II)]^+$$

TABLE 12.   Reduction potentials of porphyrin$-$iron carbonyl complexes of germanium and tin in CH$_2$Cl$_2$ with TBA(PF$_6$), 0.1 M, on Au electrodes vs SCE^a

	red. $E_{1/2}$ I	red. $E_{1/2}$ II	ox. E_p
(tptp)GeFe(CO)$_4$	$-1.02(-1.05)$	$-1.41(-1.45)$	0.58
(tmtp)GeFe(CO)$_4$	-1.02	-1.41	0.56
(oep)GeFe(CO)$_4$	-1.29	-1.75	0.42
(tptp)SnFe(CO)$_4$	-1.00	$-1.38(-1.39)$	0.69
(tmtp)SnFe(CO)$_4$	-0.97	-1.36	0.67
(oep)SnFe(CO)$_4$	-1.24	-1.70	0.50

^{a} Data from Reference 166 (values in parentheses from Reference 168).

The next oxidation step yields $[(oep)Pb(IV)]^{2+}$, which is easily demetallated by nucleophilic attack at the lead atom followed by protonation of the ligand:

$$[(oep)Pb(II)]^+ \underset{+e}{\overset{-e}{\rightleftharpoons}} [(oep)Pb(IV)]^{2+}$$

$$[(oep)Pb(IV)]^{2+} + H_2O \longrightarrow Pb(IV) + (oep)H_2$$

$$2H^+ + (oep)H_2 \longrightarrow (oep)H_4^{2+}$$

The dication $[(oep)Pb(IV)]^{2+}$ is unstable. Its absorption spectrum differs from that of doubly oxidized porphyrins of other metals like Zn or Mg, where the ligand is doubly oxidized. Spectrometry and voltammetry indicate that PbIV binds weakly to the unoxidized ligand (the Soret and adjacent lines are those of the unoxidized ligand). Another unique aspect of the lead complex is its instability, which stands in contrast to the stable Sn(IV) analog[171], implying the inability of PbIV to accommodate into the central cavity of the porphyrin plane.

B. Phthalocyanine Complexes of Group 14 Elements (see also Sections XIV and XV)

Redox properties of lead phthalocyanine in DMF (0.1 M TEAP) were studied in detail using CV, RDE and polarography[172]. On an RDE of Pt or Au, two oxidation steps, $E_{1/2} = +0.65$ and $+0.95$ V (SCE), are shown and three reversible reduction steps, $E_{1/2} = -0.75$, -1.04 and -1.92 V, are observed on a RDE of Au (SCE), whereas a fourth irreversible reduction is noted by Hg polarography only. The first oxidation step is reversible provided scanning by CV does not exceed the value of $+0.8$ V, since passivation of the anode probably occurs at more positive potentials. There is no evidence for PbIV formation. The three reduction steps are each a SET, but polarization at -0.8 to -1.0 V for extended time (10 min) indicates that a slow demetallation of the reduced complex takes place. The released Pb^{2+} is reduced to Pb^0. $E_{1/2}$ at $+0.65$ and at -0.75 V are characteristic of the phthalocyanine ligand itself and the data are indeed consistent with results for other metallophthalocyanines in which the (divalent) metal is not electroactive. It has been claimed[173] that the gap of $ca\,1.56$ V between the first oxidation step and the first reduction step of metallophthalocyanines with nonelectroreactive metals reflects the energy difference between the ligand centered HOMO and LUMO. In the present Pb complex this gap is only 1.40 V. This may be due to PbII being out of the ligand plane. PbII is given a diameter of 242 pm, which may be too large for incorporation into the coordination center. The first two half-wave potentials, -0.75 and -1.04 V, are cathodically shifted relative to those of the free ligand, -0.42 and -0.82 V, suggesting increased electron density, induced by PbII.

A recent review covers the redox chemistry of monomeric and oligomeric phthalocyanines in the form of monomers and stacks[174]. Of the group 14 elements, the electrochemical redox data described concerns mostly silicon derivatives and one germanium compound, m-oxobis(tetra-t-butyl) phthalocyanatogermanium[175].

C. Other Complexes

In the course of the synthesis and the study of several group 14 metalloles (silacyclopentadienes and germacyclopentadienes) their reduction potentials were recorded (vs SCE), with 0.1 M TBAP in dimethoxyethane on DME by polarography and by hanging mercury electrode voltammetry[176]. In $[(\eta^4\text{-germacyclopentadiene})Co(CO)_2]_2$

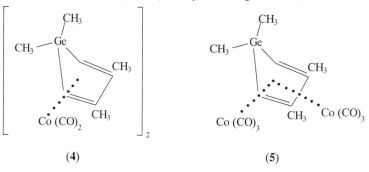

(4)   (5)

(4) and [(η^4-germacyclopentadiene)Co$_2$(CO)$_6$ (5), where in both the diene is 1,1,3,4-tetramethylgermole, reduction potential was -1.50 and -1.62, respectively. Ligand stability was retained.

X. GROUP 14 ELEMENTS IN CLUSTER COMPOUNDS

The redox chemistry of stannyl metalcarbonyl clusters has been studied in one case[177]. CV of the cluster PhSnCo$_3$(CO)$_{12}$ shows an irreversible reduction peak -0.74 V vs Ag wire, at 100 mV s^{-1}. Faster scans were not reported. It is concluded that the radical anion [PhSnCo$_3$(CO)$_{12}$]$^{\bullet-}$ is unstable, in contrast to the silicon cluster PhSiCo$_3$(CO)$_{11}$ which gives a stable radical anion (reported lifetime of 2.3 s) at -0.26 V (200 mV s^{-1}).

Formation of radical anions in the first step of reduction is more salient in the following cases[178]. Reduction by polarography of clusters such as η^5-CpFe(CO)$_2$X, where X = MPh$_3$, e.g. SiPh$_3$, GePh$_3$ and SnPh$_3$, yields stable radical anions. A reversible single-electron reduction is observed. The relative stability of the radical anions is associated with charge delocalization on the MPh$_3$ group, since with X being Cl, Br, I, GeCl$_3$ or SnCl$_3$ single-electron reduction results in bond cleavage at the Fe$-$X bond (see also Reference 179). It is also proposed that MPh$_3$ is a better electron donor (to Fe here, hence preventing bond fission) than MCl$_3$, on the basis of ionization potentials $E^1_{1/2}$, e.g. [η^5-CpFe(CO)$_2$]$^{\bullet}$ = 7.7 eV, [SnPh$_3$]$^{\bullet}$ = 6.29 eV and [SnCl$_3$]$^{\bullet}$ = 9 eV. Further reduction of the radical anion at $E^2_{1/2}$ (Table 13) causes Fe$-$M bond cleavage. It is hence surmised that reduction takes place at the σ-antibonding level:

$$\eta^5\text{-CpFe(CO)}_2\text{MPh}_3 + e \rightleftharpoons \eta^5\text{-Cp[Fe(CO)}_2\text{MPh}_3]^{\bullet-} \quad \text{at } E^1_{1/2}$$

$$[\eta^5\text{-CpFe(CO)}_2\text{MPh}_3]^{\bullet-} + e \longrightarrow [\eta^5\text{-CpFe(CO)}_2]^- + \text{MPh}_3^- \text{ or HMPh}_3\text{H}$$

TABLE 13.   First and second half-wave potentials, $E^1_{1/2}$ and $E^2_{1/2}$ in V (vs SCE), for reduction of η^5-CpFe(CO)$_2$MPh$_3$ in THF, TBAP 0.1 M[178]

M	$E^1_{1/2}$ (V)	$E^2_{1/2}$ (V)
Si	-1.95	-2.93
Ge	-1.90	-2.40
Sn	-1.86	-2.21

Whereas $SnPh_3^-$ is stable enough to be observed, $GePh_3^-$ and $SiPh_3^-$ react rapidly with proton donors (THF or water traces) to give $HMPh_3$. The anodic decrement in reduction potential from Si to Sn is observed for both steps and is related here to the increasing nucleophilicity of the anions $SnPh_3^- \ll GePh_3^- < SiPh_3^-$.

The radical anions η^5-Cp[Fe(CO)$_2$MPh$_3$]$^{\bullet-}$ of Si and Ge compounds are stable enough to be characterized by ESR (CPE in the ESR cavity) at 298 K. The Sn radical anion required cooling to 253 K, and even so some disproportionation into starting material and reduction products of the second stage is detected by polarography. Disproportionation is slow enough and does not interfere with distinction of the two reduction steps on an analytical scale, hence Fe(CO)$_2\eta^5$-CpMPh$_3$ shows reversible CV on Pt at 500 mV s^{-1}. However, CPE yields the products of the two-electron reduction directly.

It is therefore interesting to note that a series of capped tetrahedral clusters of the type **6** have their first reduction potential anodically shifted by an increment of 200–320 mV when the carbon cap (MeC− and PhC−) of the cluster is substituted by a Ge cap (MeGe− and PhCGe−)[180]. This potential has been associated with the LUMO level of the cluster and the HOMO of the resulting radical anion. Unlike cases such as radical anions of porphyrin complexes, where charge delocalization is centered at the macrocyclic ligand exclusively, no such separation of the group 14 element is in evidence in clusters like **6**, though they are in fact stabilized radical anions. Potential shifts are therefore as palpable in such clusters as in alkyl–M compounds. In another such example a special type of heterobimetallic (Pd/Pt–Ge/Sn) trinuclear cluster has been described which consists of a planar ring core, of alternating metal atoms M and M′ of the general formula M$_3$(CO)$_3$(μ_2-M′)$_3$ **(7)**, where M = Pd or Pt, M′ = Ge or Sn[181].

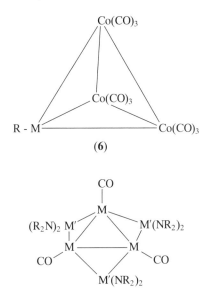

The compounds were analyzed by spectroscopy and CV and products of electrochemical reduction were further characterized by ESR. Crystal structures were obtained of [(M{μ_2-Sn(NR$_2$)$_2$}CO)]$_3$ where (M = Pd or Pt and R = −SiMe$_3$). These clusters show reversible single-electron reduction at (V): 1.36 and 1.16 when M = Pd and M′ = Ge and Sn, respectively, and at 1.13 and 1.04 when M = Pt and M′ = Ge and Sn, respectively (no reference electrode is specified). Sn compounds are reduced with greater ease. ESR implies that the radical anion of the Pt-based cluster is the more stable.

In examples studied by Wrighton and coworkers, the group 14 element is not part of a cluster network but still affects the HOMO of the ligand in the radical anion[182]. Charge-transfer properties of several complexes of the type $R_3EM(CO)_3L$, where R = Ph, Me; E = Sn or Ge; M = Mn or Re; L = 1,10-phenanthroline, 2,2'-bipyridine or 2,2'-biquinoline, were studied. The objective was to characterize the ground-state levels of oxidized and reduced forms. Among other results it was found that electroreduction at -1.40 and -1.42 V vs SCE was reversible, mainly charging the LUMO level of the L ligand, but oxidation between $+0.82$ to $+0.85$ V caused cleavage of the E−M σ-bond following the formation of a radical cation. The values given are for $Ph_3ERe(CO)_3$ (phen) (E = Sn and Ge, respectively). In this case the reduction potential is less negative and the oxidation potential is less positive when the atomic number of E is higher.

Several series of bimetallic complexes in which Ph_3Si-, Ph_3Ge- or Ph_3Sn- are bonded to transition metal carbonyl clusters were subjected to electrochemical reduction[101]. At the potentials applied, the intermetallic bond was severed leaving, as in the preceding case, the Ph_3M group intact, e.g.:

$$Ph_3MCo(CO)_3L + e \longrightarrow Ph_3M^\bullet + [Co(CO)_3L]^-$$

$$[Co(CO)_3L]^- + e \longrightarrow [Co(CO)_3]^- + L^{\bullet-}$$

$$Ph_3M^\bullet \longrightarrow \tfrac{1}{2}Ph_3MMPh_3$$

$$M = Si, Ge, Sn; L = PPh_3, P(OPh)_3 \text{ or } CO$$

The reduction potentials were determined by polarography, and by CV on a hanging Hg cathode, and were irreversible, though the presence of a reverse wave was noted. Evidently, here the group 14 elements are directly affected (Table 14). The cathodic shift is understandable when L in $Ph_3MCo(CO)_3L$ is changed from CO (an acceptor ligand) to the donors phosphine or phosphite. Anodic shift of the first reduction potential in the

TABLE 14. Reduction potentials $E_{1/2}$ (V vs SCE) of carbonyl complexes $Ph_3MCo(CO)_3L$, in THF, containing TBAP 0.1 M[101]

Compound	$E_{1/2}^a$
$Ph_3SiCo(CO)_4$	-1.70
$Ph_3GeCo(CO)_4$	-1.53
$Ph_3SnCo(CO)_4$	-1.00
$Ph_3SiCo(CO)_3PPh_3$	-1.87
$Ph_3GeCo(CO)_3PPh_3$	-1.78
$Ph_3SiCo(CO)_3P(OPh)_3$	-1.95
$Ph_3GeCo(CO)_3P(OPh)_3$	-1.85
$Co_2(CO)_8$	-0.30
$Cp(CO)_2FeSiPh_3$	-2.05^b
$Cp(CO)_2FeGePh_3$	-1.95^b
$Cp(CO)_2FeGeEt_3$	-2.15^b

^{a}Though on a CV scale, the curves are not reversible; anodic peaks are observed.
^{b}Data from Reference 100.

TABLE 15.   Redox peak potentials of several germanium and tin clusters, in THF, 0.1 M TBAP, working electrode graphite, counterelectrode Pt wire[183]

Compound	E_p (V vs SCE)[a]		
	reduction		oxidation
$(\mu_3\text{-Ge})[(\eta^5\text{-}C_5H_4CH_3)Mn(CO)_2]_3$	-1.38	-1.10	$+0.21$
$(\mu_3\text{-Sn})[(\eta^5\text{-}C_5H_4CH_3)Mn(CO)_2]_3$	-1.23	-1.02	$+0.22$
$(\mu\text{-Sn})[(\eta^5\text{-}C_5(CH_3)_5)Mn(CO)_2]_2$	-1.60	-1.48	$+0.16$
$(\mu\text{-Pb})[(\eta^5\text{-}C_5H_5)Mn(CO)_2]_2$	-1.37	-1.07	$+1.26$

[a] Values in second and third columns are irreversible.

line Si, Ge, Sn is as in the series $Ph_3MX^{97,100,101}$ (Section V). Here too the Ph_3M group is not cleaved and the reaction can be utilized for the preparation of dimers Ph_3MMPh_3.

Herrmann and coworkers[183] reported a series of Cp-manganese carbonyl complexes which bind Ge, Sn and Pb as central atoms linearly coordinated in clusters, to two Mn atoms in one series and trigonal-planar coordinated to three Mn atoms in another series **8** and **9**. The group 14 atoms are double-bonded to two Mn atoms in these compounds, or carry one double bond and two single bonds to three Mn atoms. Potentiometric measurements of these compounds show irreversible reductions and oxidation by CV. No products could be isolated from either reduction or oxidation. The exceptionally high oxidation potential of $(\mu\text{-Pb})[\eta^5\text{-}C_5H_5)Mn(CO)_2]_2$ as compared to the apparently similar Sn compound is noteworthy (Table 15).

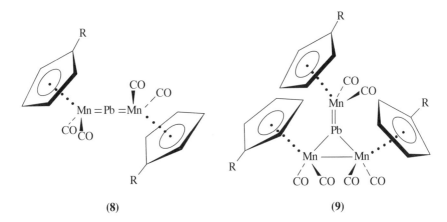

(8)   (9)

A. Oxidation States and Redox Chemistry

Dialkylstannyl units incorporated in transition-metal icosahedral cages such as $[Ni_{11}(SnR_2)(CO)_{18}^{-2}$ (R = n-Bu or Me) have been recently described[184]. This is a new family of icosahedral cages having the special feature of bearing an excess of 8 cluster valence electrons — a total of 158, over the predicted number of 150. This raises interesting bonding questions regarding the distribution and arrangement of the cluster valence orbitals. Among other tests, electrochemical measurements (technique not specified) are used to show that these dianions can be reversibly oxidized and reduced. Hence the effect of group 14 elements on results in complex and cluster molecules depends very much on their structural placement.

XI. ELECTROCATALYSIS

The term electrocatalysis is not unequivocal. The acceleration of an electrodic reaction can be achieved by the modification of the electrode surface, improving the *Tafel* kinetics, and would generally be regarded in terms of heterogeneous catalysis and surface phenomena[185]. It is sometimes used in terms of homogeneous catalysis and based on the idea that chains of reactions can be initiated by an electrodic step. In such a case, the initial electron transfer involves a reversible redox cycle using a compound capable of a reversible regeneration reaction. This reversible redox cycle is linked to subsequent reactions, and the acting compound is termed a mediator[186]. The method is known as *indirect electrolysis*. Its action can be followed by several competitive reactions. The success of a certain desired reaction route then depends on control of alternative pathways. The category of electrocatalysis may also include cases where a nonelectrodic reaction chain is initiated by an electrodic pulse. Such is the case below where a radical chain reaction is started by the radical $Ph_3Sn^\bullet$, electrogenerated from Ph_3SnH. A small promoting pulse is sufficient. This mimics the action of conventionally used initiators such as AIBN and peroxide. The ease of production of tin, germanium and lead alkyl radicals $R_3M^\bullet$ opens up interesting possibilities for their use as electroinitiators. Other combinations of catalysts in the subsequent reactions have been mentioned above (Section IIIc and d), and the use of palladium catalysts for coupling (see Section IV), the latter does not involve catalysis of the electrode reaction.

Indirect cathodic reduction of some triorganohalogermanes provides an example of one-electron redox catalysis[187]. The electrocatalytic reduction of Ph_3GeX to $Ph_3GeGePh_3$ is described, using pyrene, anthracene and 9,10-diphenylanthracene as mediators **A**. The first step is a reversible reduction of the mediator **A**. Peak currents for this reduction increase considerably when Ph_3GeX is added and the reduction of **A** becomes irreversible when the ratio Ph_3GeX/A exceeds 2. This is characteristic of a catalytic current reflecting the regeneration of **A**[17]. The rate of this regeneration is k_2. The process has a typical ECC sequence, clearly shown by CV:

$$A + e \rightleftharpoons A^{-\bullet}$$

$$A^{-\bullet} + R_3GeX \xrightarrow{(k_2/k_{-2})} R_3GeX^{-*} + A$$

$$R_3GeX^{-*} \xrightarrow{k_3} R_3Ge^* + X^-$$

$$2R_3Ge^* \longrightarrow R_3GeGeR_3$$

It is interesting to note that cleavage takes place at the GeX rather than the GeR bond. In this case, as with the direct reduction, the ET step (k_2) is rate-determining whereas k_3 is fast and, despite the fact that reduction potentials of R_3GeX are somewhat dependent on X, k_2 could be calculated (Table 16). It is noted, however, that tabulated half-wave reduction potentials of the mediators appear to be higher than those of R_3GeX (compare Table 16 with Table 3) though they appear very close on the CV trace.

Sn—Sn bond formation can be achieved by indirect electrolysis considering the relative ease of SnH-bond activation. Tributylin hydride is a known H atom donor. It is attacked by radicals like $\{Mn(CO)_3P(OPh)_3]_2\}^\bullet$, electrogenerated from the anion $\{Mn(CO)_3P(OPh)_3]_2\}^-$. The kinetics of hydrogen transfer and coupling of $Ph_3Sn^\bullet$ and $\{Mn(CO)_3P(OPh)_3]_2\}^\bullet$ was studied[188].

The propensity of organotin hydrides for SET reactions has been utilized to initiate radical chain reactions. Anodically promoted oxidation of Ph_3SnH to $[Ph_3Sn]^\bullet$ at 0.80 V (vs SCE) initiates the cyclization of several haloalkyne and haloalkene ethers as well as of some β-lactam derivatives. The catalytic cycle shown in Scheme 1 is based on

TABLE 16. Reduction of R_3GeX in the presence of mediators **A**. Half-wave reduction potentials of **A** (vs SCE), and calculated rates k_2, of regeneration of **A**[187].

R_3GeX, X =	A	$E_{1/2}$ (V)	k_2 (10^{-2} $M^{-1}s^{-1}$)
Br	anthracene	-2.00	2.1
Br	diphenylanthracene	-2.00	1.5
Cl	anthracene	-2.15	1.9
Cl	pyrene	-2.01	—

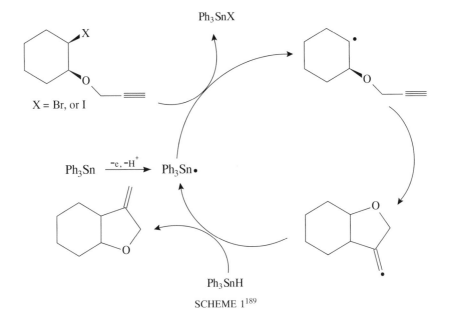

SCHEME 1[189]

experiments and control tests in a divided cell. The oxidation steps of Ph_3SnH have been analyzed by CV[189]:

$$Ph_3SnH \longrightarrow [Ph_3Sn]^{\bullet} + e \qquad +0.80 \text{ V}$$
$$[Ph_3Sn]^{\bullet} \longrightarrow [Ph_3Sn]^{+} + e \qquad +1.15 \text{ V}$$

The one-electron wave at 1.15 V is much reduced in intensity in the CV trace due to the fast consumption of $[Ph_3Sn]\cdot$ to produce $Ph_3SnSnPh_3$. The latter is recognized by its oxidation potential at 1.50 V. Ph_3SnH is the only source for radicals. CV of $Ph_3SnSnPh_3$ as well as of Ph_3SnCl shows only two-electron oxidations and reductions, hence no radical species should be expected from these compounds.

$$Ph_3SnSnPh_3 - 2e \longrightarrow 2[Ph_3Sn]^{+} \quad +1.50 \text{ V vs SCE}$$
$$Ph_3SnSnPh_3 + 2e \longrightarrow 2[Ph_3Sn]^{-} \quad -2.46 \text{ V vs SCE}$$
$$Ph_3SnCl + 2e \longrightarrow 2[Ph_3Sn]^{-} \quad -2.48 \text{ V vs SCE}$$

Indirect electrosynthesis of reactive formyl transition metal compounds involves an initial step of reduction of metal carbonyls to radicals followed by transfer of a hydrogen atom from trialkyltin hydrides[190]. Electroreduction of metal carbonyls yields products of dimerization and loss of CO from the radical anion. Electroreduction in the presence of R_3SnH yields the formylmetalcarbonyls:

$$Cr(CO)_6 + e \longrightarrow [Cr(CO)_6]^{\bullet-} \longrightarrow \tfrac{1}{2}Cr_2(CO)_{10}^{2-} + CO$$

In the presence of tin hydride the reactions are:

$$[Cr(CO)_6]^{\bullet-} + Bu_3SnH \longrightarrow Cr(CO)_5CHO^- + \text{tin products}$$

$$[Fe(CO)_5]^{\bullet-} + Bu_3SnH \longrightarrow Fe(CO)_4CHO^- + \text{tin products}$$

and

$$[Mn(CO)_4(Ph_3P)_2]^+ + e \longrightarrow [Mn(CO)_4(Ph_3P)_2]^\bullet$$

$$[Mn(CO)_4(Ph_3P)_2]^\bullet + Bu_3SnH \longrightarrow Mn(CO)_3(Ph_3P)_2CHO + \text{tin products}$$

In the absence of Bu_3SnH, further reduction occurs with loss of a phosphine group:

$$[Mn(CO)_4(Ph_3P)_2]^\bullet + e \longrightarrow [Mn(CO)_4(Ph_3P)]^- + Ph_3P$$

The last example is illustrated by CV. Under a scan rate of 500 mV s^{-1} the reduction peak of $Mn(CO)_3(Ph_3P)_2CHO$ at -2.3 V shows when CV of $[Mn(CO)_4(Ph_3P)_2]^+$ is carried on in the presence of Bu_3SnH. The hydrogen transfer from Bu_3SnH is confirmed by experiments with Bu_3SnD and is apparently rapid and in high yield. One equivalent of charge per metal carbonyl is required. Electrolysis was carried out on 5×10^{-2} M metalcarbonyl, in THF with 0.2 M of TBAP. Tin products were not identified.

XII. MECHANISTIC STUDIES OF ELECTRON TRANSFER, CONCERNING ORGANIC COMPOUNDS OF GROUP 14 ELEMENTS

Studies in which the electrochemical oxidation of alkyllead compounds is compared with nonelectrodic processes like oxidations by a given reagent or with energy enhancement tests like photoelectron spectroscopy, were carried out by Gardner and Kochi[191]. The model reaction is oxidative ET of tetraalkyllead, with hexachloroiridate(IV). Organolead compounds were chosen as models for alkyl-transfer reactions of organometallic compounds. The assumption is that transalkylation is basically caused by an electrophilic process, ET in this case, and hexachloroiridate(IV) was chosen as the oxidation agent, being known as a single-electron oxidant by inner-sphere as well as outer-sphere processes:

$$R_4Pb + Ir(IV)Cl_6^{2-} \longrightarrow R_4Pb^{\bullet+} + Ir(III)Cl_6^{3-}$$

$$R_4Pb^{\bullet+} \longrightarrow R^\bullet + R_3Pb^+$$

$$R^\bullet + Ir(IV)Cl_6^{2-} \longrightarrow RCl + Ir(III)Cl_5^{2-}$$

For instance, the reaction of Me_4Pb in acetic acid is:

$$Me_4Pb + 2IrCl_6^{2-} \longrightarrow Me_3PbOAc + MeCl + IrCl_6^{3-} + IrCl_5^{2-}$$

Oxidation potentials of R_4Pb were measured in AN with Pt electrodes vs Ag/AgCl, LiBF$_4$ as electrolyte. The value of n, denoting the number of electrons per molecule involved

TABLE 17. Oxidation potential of several tetraalkyllead derivatives under three current densities i/A (mA cm^{-2})[191][a]

R$_4$Pb	Potential V at a current density of:		
	$i/A = 1.0$	$i/A = 0.4$	$i/A = 0.0$[b]
Et$_4$Pb	1.67	1.58	1.52
Et$_3$MePb	1.75	1.67	1.62
Et$_2$Me$_2$Pb	1.84	1.74	1.68
EtMe$_3$Pb	2.01	1.88	1.80
Me$_4$Pb	2.13	2.00	1.90

[a] Determined in 0.002 M of R$_4$Pb, in AN, LiBF$_4$, 25 °C, ref Ag/AgCl.
[b] The value of $i/A = 0$ is by extrapolation from values determined in 0.1 mA cm^{-2} intervals between 0.3 and 1.2 mA cm^{-2}.

in the anodic reaction, was determined by thin-layer voltammetry. The actual values of the anodic oxidation potential are somewhat dependent on experimental parameters such as current density, since the reactions are irreversible. Even so, the trend from Et$_4$Pb to Me$_4$Pb is very clear as shown in Table 17. Provided that anodic oxidation of all the listed R$_4$Pb compounds takes place by the same mechanism, the listed values of oxidation potentials reflect the relative ease of removal of an electron.

$$R_4Pb - e \longrightarrow R_4Pb^{\bullet+} \text{anodic step} \qquad\qquad E$$

$$R_4Pb^{\bullet+} \longrightarrow R^{\bullet} + R_3Pb^{+} \quad \text{irreversible decomposition} \quad C$$

This is very elegantly borne out. A very good correlation is observed between the oxidation potentials of R$_4$Pb, the vertical ionization potentials from photoelectron spectroscopy and the reaction rates with Ir(IV)Cl$_6$$^{2-}$ which suggests that the rate-determining step is indeed the first step, that of ET (Reaction E). The linear shift with increasing number of methyl groups, from the homoleptic Et$_4$Pb to Me$_4$Pb, is of the same order versus each of the three parameters.

Klingler and Kochi have carried out measurement with the realization that ET should be evaluated in view of the two basic criteria of the Marcus theory, namely the standard potential and the reorganization energy factor. The preconditions for such experiments were, first, that the redox reaction under observation be an outer-sphere SET, polar reactions and other mechanisms excluded, and second, that the stereochemical structure of the substrate must be clearly defined. These authors have used several structural organometallic models for the comparison of heterogeneous (electrodic) and homogeneous ET processes. The structural identity of the ET substrate is essential for interpretations based on the Marcus theory. Tetraalkylstananes represented the tetrahedral complexes in models for irreversible oxidation[192]. It should in fact be noted that the determination by electrochemical tests of the standard potential of compounds such as R$_4$M (M = Ge, Sn, Pb) is not really straightforward, since the oxidation is irreversible. The first electrodic step of ET is much slower than subsequent reactions of the cation radicals:

$$R_4M - e \longrightarrow R_4M^{+\bullet} \longrightarrow \text{fast reactions}$$

The value of $E°$ was hence determined by the reaction of R$_4$M with Fe^{3+} complexes as outer-sphere SET oxidizers. Using five complexes with a range of different $E°$ values, from 1.15 to 1.42 V, the rate constants were determined[193]. This was followed up by Eberson who, by application of the Marcus theory, was able to determine from the $E°$ values (shown in Table 18) standard potentials and reorganization energies. Most compounds

TABLE 18.   Standard potentials E° (V vs NHE) for SET of R_4M in reactions with Fe^{3+} oxidizers[193]

Et_4Si	1.89
Et_4Ge	1.77
Et_4Sn	1.49
Et_4Pb	1.53^a
Et_2Me_2Sn	1.60
Et_2Me_2Pb	1.49
Pr_4Sn	1.52
$sec\text{-}Bu_4Sn$	1.41
$iso\text{-}Bu_4Sn$	1.42
$neo\text{-}Pen_4Sn$	1.52

^{a} Datum deviates.

indeed had very nearly the same reorganization energy value of 150 ± 15 kJ mol^{-1}[194]. An instructive discussion can be found in Eberson's book[13].

Criteria for evaluating ET reactions in group 14 homoleptic dimetal compounds were sought in a series of electrooxidations[195]. Values of E_p of oxidation to the radical cation were measured by CV at 100 mV s^{-1} on glassy carbon in acetonitrile with TBAP and Ag/AgCl reference. Here too, the ionization potentials were compared with gas-phase ionization energies from photoelectron spectra. The lowest energy band reflects oxidation at the dimetallic σ-level presuming Frank–Condon transitions. A good correlation between $E_{p\ oxid}$ and IP is shown. Anodic oxidation is irreversible.

$$R_3M-M'R_3 \longrightarrow [R_3M-M'R_3]^{\bullet+} + e$$

$$[R_3M-M'R_3]^{\bullet+} \longrightarrow R_3M^{\bullet} + M'R_3{}^{+}$$

$$M,M' = Si, Ge, Sn$$

The proposed subsequent reaction fits the fragmentation patterns observed in mass spectrometry where, even at 20 eV, group 14 centered radicals form in increasing order Si<Ge<Sn. It is believed that basic data of this kind can provide estimates of kinetic behavior of such reactions, where M−M bonds are cleaved by electrophiles and which depend on the ionization potentials of the former as well as the electron affinity of the latter.

XIII. EFFECT OF GROUP 14 ATOMS ON ENERGY LEVELS OF PROXIMAL BONDS

A. The β Effect

Substitution of carbon by another group 14 element may raise the HOMO level and consequently lower the oxidation potential of the resulting organometallic compound[196,197]. R_4Sn shows a low oxidation potential (Table 18) as compared to that of hydrocarbons, and organostannanes have been used as a source for radicals[198]. The effect is particularly manifest in ethers, where a secondary effect is attributed to the interaction of nonbonding p orbitals of oxygen with the C−Si and C−Sn σ-bond. This promotion is much larger with α-alkoxystannanes than with the silicon ethers. Results concerning α-silyl ethers and α-stannyl ethers are illustrative. Levels for HOMO energy as calculated by MNDO are shown in Table 19, where energy levels for the organometallic ethers are particularly high for Sn compounds. Direct measurements of the oxidation potentials of these compounds provide

Michael Michman

TABLE 19. HOMO level, energy values of several α-alkoxystannanes and silanes, with comparison to alkylmetals and to dimethyl ether[197]

Compound	eV^a
$CH_3OCH_2Si(CH_3)_3$	-10.2
$CH_3OCH_2Sn(CH_3)_3$	-9.8
$CH_3OCH[Si(CH_3)_3]Sn(CH_3)_3$	-9.5
$(CH_3)_4Si$	-11.3
$(CH_3)_4Sn$	-10.9
CH_3OCH_3	-11.0

^{a} Assuming a fixed angle of $90°$ in $Si-C-O-C$ and $Sn-C-O-C$.

experimental support. Values of oxidation potentials as determined by RDE voltammetry are given in Table 20 Electrochemical oxidation of ethers often requires high potentials, > 2.5 V vs Ag/AgCl, and all values for the listed ethers are significantly lower. The largest cathodic shift is clearly with alkoxystannanes.

Electrooxidation of α-alkoxystannanes on a preparative scale can therefore be carried out at lower potentials, with cleavage of the $C-Sn$ bond. Nucleophilic attack, e.g. by methanol, butanol or an amine on carbon at that α position, gives the product in high yields, 90–95% diether or 55% aminoether. A number of examples have been given:

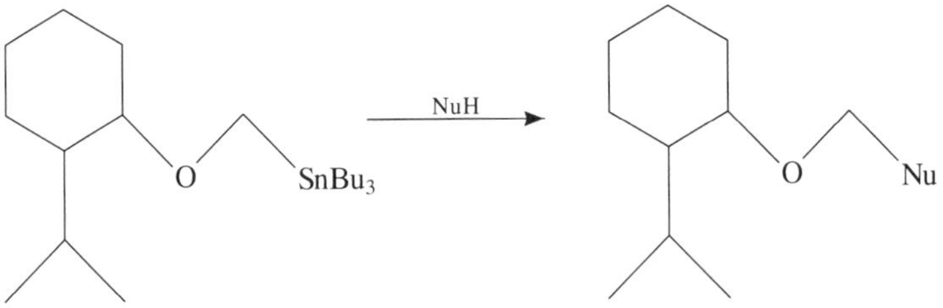

(10)   (11)

NuH = MeOH, BuOH, TsNHMe, HO—SiMe$_3$

The following mechanistic pathway seems to be the most plausible: A single ET is followed by the cleavage of a stannyl radical which is trapped by the solvent CH_2Cl_2, while the ene-oxonium residue is attacked by the nucleophile. Indeed, the electrolysis can be carried out with the nucleophile as solvent.

$$ROCH_2SnBu_3 - e \longrightarrow [ROCH_2SnBu_3]^{+\cdot} \longrightarrow [RO=CH_2]^+ + [Bu_3Sn]\cdot$$

$$[RO=CH_2]^+ + NuH \longrightarrow ROCH_2Nu + H^+$$

$$[Bu_3Sn]\cdot + CH_2Cl_2 \longrightarrow Bu_3SnCl$$

The electrolysis of unsaturated α-tributylstannyl ethers 12 yields cyclic compounds by a similar reaction, involving in this case carbon–carbon bond formation. Here, too, the evidence points to a facile ET with formation of an α-stannyl ether radical cation. This in turn cleaves to the tributylstannyl radical and an unsaturated ene-oxonium residue.

TABLE 20.   Oxidation potentials of α-alkoxystannanes and silane analogs, measured on glassy carbon RDE in 0.1 M $LiClO_4/CH_3CN$[197]

Compound[a]	$E_{1/2}$ (V vs Ag/AgCl)
$CH_3OCH_2Si(CH_3)_3$	1.85
$R'OCH_2Si(CH_3)_3$	1.83
$CH_3OCH_2SnR_3$	1.14
$R'OCH_2SnR_3$	1.09
$CH_3OCH[Si(CH_3)_3]_2$	1.54
$CH_3OCH(Si(CH_3)_3)SnR_3$	1.00
$(CH_3)_3SiCH_2OCH_2CH_2OCH_2Si(CH_3)_3$	~ 1.9[b]
$RSnCH_2OCH_2CH_2OCH_2SnR$	1.29
R_4Sn	1.52

[a] R = Bu, R' =

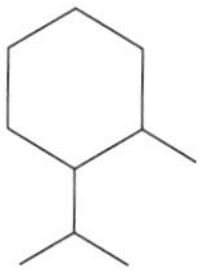

[b] Approximate.

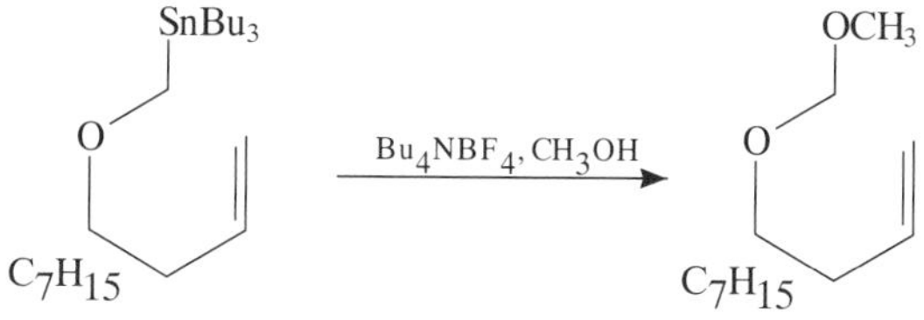

(12)   (13)

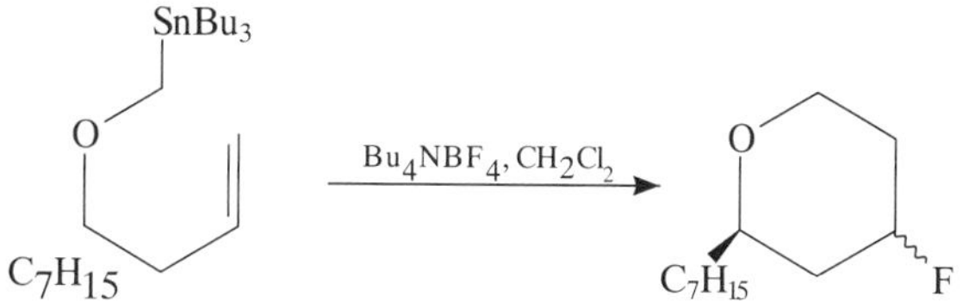

(14)

Reaction in methanol yields the substituted diether **13** whereas, in methylene chloride, intramolecular cyclization to an ether cation takes place and attack by the nucleophilic F^- from the electrolyte yields a fluorinated cyclic ether **14** (cf also Reference 197).

Further evidence for the above-mentioned mechanism of HOMO elevation by group 14 elements is provided by studies of thioethers. The decrease in oxidation potential of silyl ethers as compared to ethers is not realized in the case of α-silylthioethers whereas α-stannyl substituents in thioethers cause a considerable cathodic shift in oxidation potential. Moreover, the effect is geometry-dependent. Values for substituted cyclic dithianes **15** are summarized in Table 21. The difference between Si and Sn in this case is illustrative. The lone nonbonding pair in the 3p orbital of sulfur is much too low in energy compared to

TABLE 21.   Peak potentials of oxidation (vs Ag$^+$/AgNO$_3$
0.1 M/AN), of substituted 1,3-dithianes (**15**)^a

Compound	R	R′	E_p (V)
15a	SiMe$_3$	H	0.99
15b	SiMe$_3$	t-Bu	0.95
15c	SiMe$_3$	Ph	0.85
15d	SiMe$_3$	SiMe$_3$	0.70
15e	SnMe$_3$	H	0.75
15f	SnMe$_3$	t-Bu	0.54
15g	SnMe$_3$	Ph	0.81
15h	SnMe$_3$	SnMe$_3$	0.19
15i	SiMe$_3$	SnMe$_3$	0.44

^{a} From CV on Pt in 0.1 M LiClO$_4$, in AN[199].

the C—Si σ-orbital and overlap is not as effective as with the higher-energy 2p oxygen
orbital. The C—Sn is much closer in energy to the lone pair of sulfur, hence overlap is
possible[199]. The geometrical constraint on the extent of overlap is reflected in results with
4,6-*cis*-dimethyl-1,3-dithianes **16**. These are conformationally locked (not necessarily in
chair conformation).

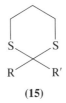

(15)

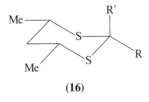

(16)

From the results in Table 22 it is apparent that the neighboring tin effect is high when
—SnMe$_3$ is in an axial conformation and lower for equatorial —SnMe$_3$. The results in
Table 21 may be similarly interpreted. In **15e** and **15g**, —SnMe$_3$ is mainly equatorial;

TABLE 22.   Peak potentials of oxidation (vs Ag$^+$/AgNO$_3$
0.1 M/AN), of substituted 4,6-*cis*-dimethyl-1,3-dithianes (**16**)^a

Compound	R	R′	E_p (V)
16a	SnMe$_3$	H	0.75
16b	H	SnMe$_3$	0.40
16c	SnMe$_3$	SnMe$_3$	0.35

^{a} From CV on Pt in 0.1 M LiClO$_4$, in AN[199].

however, **15f** and **15i** have a considerable proportion of $-SnMe_3$ in axial position and accordingly oxidation potentials are lower. The considerable cathodic shift in **15h** also fits the expectation that one $-SnMe_3$ group is axial in the predominant chair conformation. The resulting oxidation potential of 0.19 V is by 1 V lower than that of 1,3-dithiane. Peak potentials measured on Pt were reportedly board, and better defined measurements (for **16a**) were obtained on glassy carbon anodes. To ascertain that potential measurements were not affected by surface effects, compound **15h** was examined by photoelectron spectroscopy which, too, showed a lowering of 1 eV in the ionization potential for a nonbonding electron as compared to 1,3-dithiane.

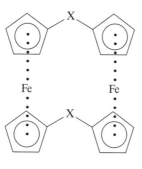

(**17a**) X = $(n\text{-}Bu)_2Sn$
(**17b**) X = CHMe

Studies of ET in 1,1,12,12-tetra-*n*-butyl[1,12]stannaferrocenophane, **17a**, add another perspective to this aspect, in a comparison between this compound and 1,12-dimethylferrocenophane, **17b**. CV of both compounds show two reversible single-electron oxidations to the mono- and dication, respectively. Each step is shown by polarography to be a single ET. The separation of the peak potentials ΔE_p of the oxidation waves should reflect the extent of interaction between the two Fe atoms[200]. ET interaction between the two ferrocene units in these cyclophanes is possible either through the bridge bonds or through space (field interaction). The value of ΔE_p should decrease with increasing Fe–Fe distance[201]. The absence of change in ΔE_p from **17a** to **17b** is taken as an indication of a through-bond interaction which plays a role in a possibly d–p overlap of Sn with cyclopentadiene rings though field interaction is not altogether excluded.

B. Applications of Group 14 Derivatives in Organic Electrosynthesis

Ge like Sn shows the β effect, namely promotion of electron energy levels in an oxygen atom once removed. The effect is apparent in ethers, i.e. alkoxygermanes as well as acylgermanes. Voltammetry by RDE shows considerable cathodic shifts in the oxidation

TABLE 23.   Half-wave oxidation potentials (vs SCE), separation of peak potentials and Fe–Fe distances in compounds **17a** and **17b**[200]

Compound	$E^1_{1/2}$	$E^2_{1/2}$	ΔE_p	Fe–Fe (Å)
17a	0.23	0.43	0.20	4.6
17b	0.50	0.70	0.20	5.50

TABLE 24.   Decomposition potentials E_d and half-wave potentials $E_{1/2}$ (V), of menthyloxymethyl-R (**18**) and methoxymethl-R, R′ (**19**) as determined by RDE on glassy carbon 0.1 M LiClO$_4$/AN, vs Ag/AgCl/sat. KCl[202]

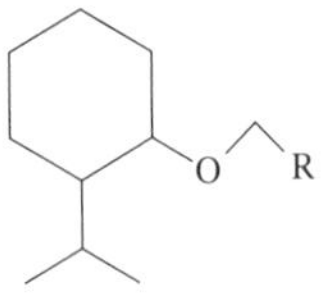

(**18**) 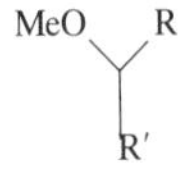   (**19**)

Compound	R=	R′ =	E_d	$E_{1/2}$	E_d	$E_{1/2}$
18a	GeMe$_3$	—	1.49	1.67		
18b	SiMe$_3$	—	1.65	1.83		
18c	CH$_3$	—	2.16			
19a	SiMe$_3$	H			1.62	1.85
19b	GeMe$_3$	SiMe$_3$			1.20	1.31
19c	SiMe$_3$	SiMe$_3$			1.35	1.54
20	C$_{12}$H$_{25}^a$	GeMe$_3$			1.93	—

^{a} Dodecanoyl-

potentials of (menthyloxymethyl)trimethylgermanes **18**, (trimethylsilyl)methoxymethyltrimethylgermanes **19** and dodecanoyltrimethylgermane **20** (Table 24). Follow-up by preparative electrolysis shows cleavage of the C—Ge bond as the main reaction, and its preferred cleavage in competition with C—Si when these are also present[202]. Table 24 gives the decomposition potentials E_d and half-wave potentials. Compound **20** has a decomposition value (E_d) only.

Utilization of the electroreduction of aryltrimethylsilanes and aryltrimethylgermanes for the preparation of silyl- and germyl-substituted cyclohexadienes was extensively studied[203], for example:

$$PhM(Me)_3 + e \longrightarrow C_6H_7M(Me)_3 + C_6H_8 + Me_3MH$$

$$M = Si, Ge;   C_6H_7 = cyclohexadienyl\text{-};   C_6H_8 = cyclohexadiene$$

Aryltrimethylstananes, on the other hand, undergo cleavage of the Sn—C bond under the described conditions:

$$PhSn(Me)_3 + e \longrightarrow PhH + Me_3SnH + C_6H_8$$

These studies, however, involve exhaustive electrolysis probably at excess energy, and not CPE. It would be interesting to follow up the potential differentiation of the possible hydrogenation reactions and the three M—C bonds, to check whether the differences in the outcome of the reduction stem from the LUMO level being highest in the tin compound. More informative is the study on electroreduction of substituted cyclooctatetraenes RCOT[204], where R = Me, Me$_3$C, Me$_3$Si, Me$_3$Ge, Me$_3$Sn. Polarography in THF and CV in HMPA (vs AgClO$_4$/Ag$^+$) showed two reduction waves, to radical anion and to anion.

XIV. POLYMERS

Recently increasing interest is evident in polymers bearing group 14 atoms. Polygermane and germane-silane polymers show properties such as semiconductivity, photoconductivity and photoluminescence[205,206]. Electropolymerization is often the preferred method of preparation, as the thin polymer layer forms directly on the electrode surface. Extent of polymerization, layer thickness and several other properties can be controlled by common electrodynamic analysis.

Generally, it would appear that R_3M- groups (M = Ge, Sn, Pb) should be good leaving groups for many synthones. In polymerizations, the presence of trimethylgermyl and trimethylplumbyl pendant groups on thiophene was instrumental in facilitating chain growth by easy M$-$C bond cleavage[207]. This way, substituted thiophene monomers such as 3-trimethylgermylthiophene and 3-trimethylplumbylthiophene and corresponding pyrroles were polymerized by controlled potential electrooxidation on indium-doped SnO_2 electrodes in 0.2 M Bu_4NPF_6–nitrobenzene solution. The films were electrochromic, with a conductivity of the order of 0.05–77 S/cm. The possibility that the group 14 elements play a role in bond fission and monomer activation is indicated by the negligible amounts of group 14 elements in the films produced, which is sometimes not even measurable. The electrooxidation of other aromatic heterocycles containing group 14 elements yields oxidized polymers[208]. In other cases, however, no evidence of M$-$C bond fission was brought forward. Trimethylgermyl pyrrole has been electropolymerized to give an electrochromic polymer[209]. Electropolymerization to produce N-trimethylgermylpyrrole polymers has also been studied together with polymerization of several silylpyrroles[158]. The polymers obtained in this case show a different electrochromism than polypyrrole. Unfortunately, they were not analyzed for elemental constitution, but from electrochromism and repeated CV tests it was concluded that the germyl and silyl pendant groups were not cleaved by oxidation.

Incidentally, oxidation data of the pyrrole monomers show an interesting increase in oxidation potentials when containing heavier substituents (Table 25). However, the ionization potential of N-methylpyrrole (7.95 V) is smaller than that of pyrrole (8.21 V). The accepted linear relationship between ionization potential and oxidation potential[210] would have it the other way round. Considering, however, that trimethylsilyl and trimethylgermyl groups are weak electron donors[211], it is plausible that a nonelectronic effect is responsible for the observed trend and the potential shifts are associated with steric effects.

The electrolysis in DMF of halogen derivatives of distannyls, such as $(R_2XSn)_2(CH_2)_m$, yields cyclocarbotin compounds. Halides prepared from bis(triphenylstannyl)methane $(Ph_3Sn)_2CH_2$ were electrolyzed to yield polymers, cyclic products and polymers containing mercury from the cathode[212]:

$$(Ph_2ClSn)_2CH_2 + 2e \quad (at - 1.35 \text{ V}) \longrightarrow (Ph_2Sn-CH_2-SnPh_2)_m$$

The reaction occurs with intermediate formation of $(Ph_2Sn-CH_2-SnPh_2-Hg-)_n$. Also produced is the compound 1,1,3,3,4,4,6,6-octaphenyl-1,3,4,6-tetrastannacyclohexane.

Sn$-$C bond formation is achieved by electrochemical alkylation of stannyl chlorides. The polarogram of dichlorodiphenyltin Ph_2SnCl_2, on mercury cathode, shows four waves having $E_{1/2}$ values which are dependent on the electrolyte (see Table 26). Of these four,

TABLE 25.   Oxidation peak potentials of substituted pyrroles[158]

R-, in pyrrole-R	E_p^{ox}
H	$+0.47 \pm 0.01$
Me	$+0.50 \pm 0.01$
SiMe$_3$	$+0.53 \pm 0.01$
GeMe$_3$	$+0.57 \pm 0.01$

TABLE 26. Polarography and coulometry of Ph_2SnCl_2: half-wave potentials and coulometric reading[213]

Electrolyte	Wave no.:	$E_{1/2}$ (V vs SCE)			
		1	2	3	4
Meoh, LiCl[0.1 M]		−0.52	−1.07	NA	−1.40
coulometric value n:			1.6		2.1
MeOH, $NH_4Cl[0.1$ M]		−0.50	−0.94	NA	−1.27
coulometric value n:			1.7		3.0

only the first (an absorption wave) and the last are persistent in dilute solutions. CPE was carried out at the plateau of the second and the fourth wave, and the number of electrons n involved in each step is given in Table 26[213]. CPE on the fourth wave yields polymers $[Ph_2Sn]_n$ and small amounts of Ph_2SnH_2. Up to 7% tin is deposited on the cathode. This implies a 2e reduction with the following consequent reactions:

1. $Ph_2SnCl_2 + 2e \longrightarrow 2Cl^- + Ph_2Sn_{absorbed} \longrightarrow (Ph_2Sn)_n$

2. $Ph_2Sn_{absorbed} + Ph_2SnCl_2 \longrightarrow Ph_2ClSnSnClPh_2$

Further insertion of the absorbed diphenyltin unit into the Sn−Cl bond of the dimer explains the mixed polymeric Sn−Hg product:

3. $Ph_2Sn_{(absorbed)} + Ph_2ClSnSnClPh_2 \longrightarrow Ph_2ClSn(HgSnPh_2)_n SnClPh_2$

CPE on the second wave yields $(Ph_3Sn)_2Hg$ and Ph_3SnH, which is taken as evidence for the formation of Ph_3SnCl as an intermediate. A 4e reduction at the fourth wave is also assumed possible, involving the dimer formed in reaction 2:

4. $Ph_2ClSnSnClPh_2 + 4e \longrightarrow 2PhH + Sn + (Ph_2Sn)_n$

5. $Ph_2ClSnSnClPh_2 + 4e \longrightarrow (Ph_2Sn)_n + Ph_2Sn^{2-} \longrightarrow (Ph_2Sn)H_2$

All this shows the complicated array of reduction routes, ET processes and surface reactions of absorbed intermediates which must be taken into account.

The method for preparation of polygermanes by reduction of organodichlorogermanes with alkali metals was replaced by electroreduction of organodichlorogermanes and silanes on magnesium cathodes[214]. The conditions were mild and safe and the product had a satisfactorily defined weight distribution. Ge−Ge and Ge−Si bonds are successfully created. The following test reactions have been performed to show that conditions are indeed favorable for metal−metal bond formation:

$$2(CH_3)_3GeCl + 2e \xrightarrow{\text{Mg}} (CH_3)_6Ge_2 + 2Cl^- \quad 84\%$$

$$(CH_3)_3GeCl + (CH_3)_2PhSiCl + 2e \xrightarrow{\text{Mg}} (CH_3)_3GeSi(CH_3)_2Ph + 2Cl^-$$

$$(CH_3)_3GeCl + (Ph)_3SiCl + 2e \xrightarrow{\text{Mg}} (CH_3)_3GeSi(Ph)_3 + 2Cl^-$$

Polygermanes and germane-silane polymers were prepared similarly. Polygermanes from the electroreduction of dichlorophenylbutylgermane showed typically a molecular weight of $M_n = 19900$ (10% yield). Germane-silane copolymers showed similar M_n values. The Ge/Si ratio in mixed polymers was roughly proportional to the ratio of monomers

TABLE 27. Oxidation of polygermanes and reduction of the oxidized form[215]

Sample		E_p^{ox} (V)	E_p^{red} (V)
$Me(GeMe_2)_n Me$	$n = 2$	1.28	−0.46
	$n = 3$	0.93	−0.56
	$n = 4$	0.72	−0.53
	$n = 5$	0.618	−0.48
	$n = 6$	0.53	−0.53
$(GeMe_2)_n$	$n = 6^a$	0.66	−0.34
By comparison:			
$Me(SiMe_2)_n Me$	$n = 2$	1.36	−0.59
	$n = 3$	1.10	−0.58

^{a} Cyclic compound.

employed. The method is relatively simple, using cells with two Mg electrodes. However, electrode passivation apparently occurs since sonication and frequent alternations between electrodes are required.

The CV of several permethylpolygermanes was measured[215]. This was carried out in AN solutions with TBAP on Pt with Ag wire reference and ferrocene in solution. Potentials are relative to Fc/Fc$^+$. The CV is irreversible, shows oxidation of the polygermane and reduction of the oxidized form (Table 27). It is concluded from these data that, with growing molecular size, the increase in HOMO level of the polymer is larger than the decrease in the LUMO level. The chemistry of polygermanes was studied further also by flash photolysis, which was used to generate germylenes and polygermyl radicals from the polymer[216].

Substitution of carbon by silicone or germanium in polymers like poly[(germylene)di-acetylenes] has been described. Polymers of the type $-[R_2MC{\equiv}C{-}C{\equiv}C{-}]_n$ and $-[RR'MC{\equiv}C{-}C{\equiv}C{-}]_n$ where M = Si or Ge and R or R' = CH$_3$−; CH$_3$CH$_2$− or Ph− were prepared by the reactions of LiC≡C−C≡CLi or BrMgC≡C−C≡CMgBr with SiCl$_2$ or GeBr$_2^{217}$. The resulting insulating polymers (10^{-12}–10^{-13} S cm^{-1}) became conducting when thin films were doped with FeCl$_3$. Values of 10^{-5}–10^{-3} S cm^{-1} were obtained and specific values are dependent on the type of R substituent rather than on the presence of Ge or Si. Sample values are given in Table 28. The doping is reversible and iron salts can be easily removed.

Another type of polymeric structure is based macrocyclics or on stacks of macro-cyclic units. An exhaustive electrodynamic investigation of the polymer derived from μ-oxo-(tetra-*tert*-butylphthalocyaninato)germanium and of the monomer was carried out using CPE, CV, voltammetry by RDE, rotating ring-disk electrode and differential pulse

TABLE 28. Conductivity values for poly(germylene)diacetylenes and poly(silylene)diacetylenes doped with FeCl$_3^{217}$

Polymer	Conductivity (S cm^{-1})
$-[Ph_2 GeC{\equiv}C{-}C{\equiv}C{-}]_n$	10^{-4}
$-[CH_3PhGeC{\equiv}C{-}C{\equiv}C{-}]_n$	3×10^{-5}
$-[(CH_3CH_2)PhGeC{\equiv}C{-}C{\equiv}C{-}]_n$	2×10^{-5}
$-[Ph_2SiC{\equiv}C{-}C{\equiv}C{-}]_n$	3×10^{-3}
$-[CH_3PhSiC{\equiv}C{-}C{\equiv}C{-}]_n$	10^{-4}

voltammetry[218]. The polymer, itself known to be electroconductive in the solid state, can be reduced in solution to any fractional degree between the 0 to -1 oxidation states. The voltammograms, though featureless, suggest a reversible Nernstian behavior. Reduction exceeding one F-equivalent, namely one electron per phthalocyanine unit, causes disintegration of the polymer to monomer. It is possible to follow the potential and the rate of this decomposition by rotating ring-disk electrode voltammetry. A rate constant of 0.2 s^{-1} was observed for decomposition by reduction.

Phthalocyanine-based molecular metals and conductive polymers have been prepared and investigated by ^{13}C solid state NMR. Among others, Ge serves as a central atom[219,220]. However, no discussion of a special role of the metal is presented in this work.

XV. TECHNICAL APPLICATIONS OF SPECIAL MATERIALS

A series of salts, $[TTF][SnEt_2Cl_3]$, $[TTF]_2[SnPh_2Cl_4]$, $[TTF]_3[SnEt_2Cl_4]$, $[TTF]_{3.3}[SnPh_2Cl_4]$, $[TSF]_2[SnPh_2Cl_4]$ and $[TSF]_{3.3}[SnPh_2Cl_4]$ has been prepared by electrocrystallization (electrolysis in acetonitrile for 11 days) of $SnEt_2Cl_2$ and $SnPh_2Cl_2$ with $[TTF]_3[BF_3]_2$ and $[TSF]_3[BF_3]_2$ (TTF = tetrathiafulvalen; TSF = tetraselenafulvalen). These behave as semiconductors showing electronic interactions of the TTF and TST cation radicals with the anions[221,222].

A. Sensor Electrodes

Recently, considerable activity has been aimed at the development of ion-selective membrane electrodes by incorporating organometallic compounds in the polymer matrix. Bis(p-chlorobenzyl)tin dichloride, dibenzyltin dichloride and bis(p-methylbenzyl)tin dichloride incorporated into a PVC liquid membrane gave electrodes with high selectivity for dibasic phosphates, HPO_4^{2-} as well as adenosine-5′-diphosphate and adenosine-5′-triphosphate. The first tin compound was most effective, though several other tin compounds were not effective at all. At present, the influence of these compounds and cationic sites in general on the permeation and selectivity of membranes is not well undersood[223].

A lead phthalocyanine complex together with LaF_3 are the elements of an oxygen sensor electrode[224]. Electrodes are based on thin film structures of conducting materials capable of electron conduction, hole conduction and ion conduction in thin layers of SnO_2, AgI and Pb phthalocyanine[225-227]. A tin phthalocyanine polymer has been described as a constituent in an anode coating of an oxygen sensor[228]. Organotin compounds, such as R_2SnX_2 and R_3SnX (R = alkyl- and aryl-), were used in PVC liquid membranes under the influence of an electric field as neutral carriers for anions by complex formation with oxoanions. Some selectivity to dihydrogenphosphate has been observed[229]. Heat-resistant electrodes were formed on substrates, by irradiation of gas plasma containing organic compounds of Fe and Sn with a high-frequency source[230].

Another aspect of tin as a constituent of electrode material is shown by tin(IV)TPP complexes incorporated into PVC membrane electrodes. These increase the selectivity to salicylate over anions such as Cl^-, Br^- I^-, IO_4^-, ClO_4^-, citrate, lactate and acetate. The specificity is attributed to the oxophilic character of the Sn ion in TPP at the axial coordination sites. Indeed, carboxyl groups incorporated into the membrane polymer compete for these binding sites. The complete complex structure is important. Substitution of TPP with octaethylporphirine results in loss of salicylate selectivity[231]. Preparation and analytical evaluation of a lead-selective membrane electrode, containing lead diethyldithiocarbamate chelate, has also been described[232].

XVI. REFERENCES

1. On-line *STN International*. The author wishes to thank Dr. Y. Wolman of The Hebrew University of Jerusalem for his contribution and expert advice.
2. P. G. Harrison (Ed.), *The Chemistry of Tin*, Blackie, Glasgow, 1989; P. G. Harrison, *Organometallic Chemistry of Germanium, Tin and Lead*, Chapman & Hall, London, 1985.
3. *The Gmelin Handbook of Inorganic and Organometallic Chemistry*, 8th ed., Springer, Berlin, Organometallic compounds of Pb, **1**, 1987; **2**, 1990; **3**, 1992 (Ed. F. Huber)
4. Exhaustive listings of monographs, reviews and major articles are found in: *The Gmelin Handbook of Inorganic and Organometallic Chemistry*, 8th ed., Springer, Berlin, Organometallic compounds of Ge, **1**, 1988; **3**, 1990; **5**, 1993 (Ed. F. Glockling); Organometallic compounds of Sn, **18**, 1989; **19**, 1991; **20** 1993 (Eds. H. Schumann and I. Schumann). Only few deal with electrochemistry.
5. N. N. Greenwood and A. Earnshaw, *Chemistry of the Elements*, Pergamon Press, Oxford, 1986: (a) p.460; (b) p. 432.
6. (a) Ullman's *Encyclopedia of Chemical Technology* (Eds. B. Elvers, S. Hawkins and G. Schulz) VCH, Series A; (b) **A16** (1990), 729; **A1**, 216; **A15**, 228 and 254; **A18**, 39 and 220.
7. L. Walder, in *Organic Electrochemistry* (Eds. H. Lund and M. M. Baizer), 3rd ed., Dekker, New York, 1991: (a) p. 809; (b) p. 861.
8. D. A. White, in *Organic Electrochemistry* (Eds. M. M. Baizer and H. Lund), 2nd ed., Dekker, New York, 1983, p. 591.
9. H. Lehmkuhl, *Synthesis*, 377 (1973).
10. M. Devaud, *Rev. Silicon, Germanium, Tin and Lead Compd.* (Ed. M. Gielen), **2**, 87 (1975–77).
11. G. Mengoli, *Rev. Silicon, Germanium, Tin and Lead Compd.* (Ed. M. Gielen), **4**, 59 (1979).
12. A. P. Tomilov, S. G. Mairanovskii, M. Ya. Fioshin and V. A. Smirnov, *Electrochemistry of Organic Compounds* (Engl. edition by Israel Program for Scientific Translations, Jerusalem), Halstead Press, New York, 1972, p. 479.
13. L. Eberson, *Electron Transfer Reactions in Organic Chemistry*, Springer-Verlag, Berlin, 1987.
14. A. Bard and L. R. Faukner, *Electrochemical Methods*, Wiley, New York, 1980; see, for a general discussion, also References 7 and 8.
15. This point is amply discussed in *Handbook Series in Organic Electrochemistry* (Eds. L. Meites, P. Zuman and E. Rupp), Vol. V, CRC Press, 1982 and other volumes in this series.
16. J. E. Huheey, *Inorganic Chemistry*, Harper & Row, New York, 1983, p. 842.
17. J. M. Savéant, *Acc. Chem. Res.*, **26**, 456 (1993).
18. L. L. Bott, *Hydrocarbon Processing*, **44**, 115 (1965).
19. D. Danley, in *Organic Electrochemistry* (Eds. M. M. Baizer and H. Lund), 2nd ed., Dekker, New York, 1983, p. 959.
20. F. Goodridge and C. J. H. King, in *Techniques of Electroorganic Synthesis*, Part 1 (Ed. N. L. Winberg), Wiley-Interscience, New York, 1974, p. 137.
21. W. J. Settineri and L. D. McKeever, in *Techniques of Electroorganic Synthesis*, (Ed. N. L. Winberg), Vol. 5, Part 2, Wiley-Interscience, New York, 1975, p. 397.
22. K. Yang, J. D. Reedy and W. H. Harwood, U.S. Pat. 3,622,476 to Continental Oil Co.; *Chem. Abstr.*, **76**, 46306b (1972).
23. W. P. Banks and W. H. Harwood, U.S. Pat. 3,649,481 to Continental Oil Co.; *Chem. Abstr.*, **76**, 161698t (1972).
24. W. H. Harwood, U.S. Pat. 3,655,536 to Continental Oil Co.; *Chem. Abstr.*, **77**, 13296r (1972).
25. C. L. Baimbridge, J. R. Minderhout, R. W. Bearman and D. E. Carpenter, U.S. Pat. 3,925,169 to Nalco Chem. Co.; *Chem. Abstr.*, **84**, 81745f (1976).
26. J. C. Shepard Jr., E. E. Johnson and R. W. Bearman, U.S. Pat. 3,853,735, 1974, to Nalco Chem. Co.; *Chem. Abstr.*, **82**, 147137b (1975).
27. C. L. Baimbridge, J. R. Minderhout, R. W. Bearman, and D. E. Carpenter, U.S. Pat. 4,002,548, 11 Jan. 1977, to Nalco Chemical Co.; *Chem. Abstr.*, **86**, 80915r (1977); G. E. Blackmar, U.S. Pat. 3,573,178, March 1971, to Nalco Chemical Co.; Chem. Abstr. 74, 150413 (1971).
28. J. B. Ganci and P. Manos, U.S. Pat. 3,630,858 to duPont; *Chem. Abstr.*, **76**, 80308p (1972).
29. E. A. Mayerle and J. R. Minderhout, U.S. Pat. 3,696,009 to Nalco Chem. Co.; *Chem. Abstr.* **77**, 171997z (1972).
30. E. H. McDonald and W. P. Banks, U.S. Pat. 3,707,456 to Continental Oil Co.; *Chem. Abstr.*, **78**, 79056q (1973); E. H. McDonald and W. P. Banks, U.S. Pat. 3,640,802; *Chem. Abstr.*, **76**, 107346p (1972).

31. C. L. Baimbridge, J. R. Minderhout, R. W. Bearman and D. E. Carpenter, U.S. Pat. 4,002,548 to Nalco Chem. Co.; *Chem. Abstr.*, **86**, 80915r (1977).
32. D. R. Lunam, *The Past, Present and Future of Lead in Gasoline*, Ethyl Corp., Baton Rouge, LA, 1988.
33. *Ullmann's Encyclopedia of Industrial Chemistry* (Ed. W. Gerhartz), 5th ed., 1985, p. 216.
34. *Kirk Othmer Encyclopedia of Chemical Technology*, 3rd ed. (Ed. M. Grayson), Vol. 8, Wiley, New York, 1979, pp. 702–714; J. H. Wagenknecht, *J. Chem. Educ.*, **60**, 271 (1983).
35. K. Ziegler, Belg. Pat. 617,628; *Chem. Abstr.*, **60**, 3008 (1964).
36. H. Lehmkuhl, *Chem.-Ing.-Tech.*, **36**, 612 (1964).
37. G. Mengoli and S. Daolio, *J. Organomet. Chem.*, **131**, 409 (1977).
38. G. E. Coates, M. L. H. Green and K. Wade, *Organometallic Compounds, The Main Group Elements*, 3rd ed., Vol. 1, Methuen, London, 1967, p. 487.
39. G. Mengoli, *J. Electrochem. Soc.*, **124**, 364 (1977).
40. G. Mengoli and S. Daolio, *Electrochim. Acta*, **21**, 889 (1976).
41. N. S. Hush and K. B. Oldham, *J. Electroanal. Chem.*, **6**, 34 (1963).
42. G. Mengoli and S. Daolio, *J. Chem. Soc., Chem. Commun.*, 96 (1976); G. Mengoli and F. Furlanetoo, *J. Organometal. Chem.*, **73**, 119 (1976).
43. G. Mengoli and S. Daolio, *J. Appl. Electrochem.*, **6**, 521 (1976).
44. K. Uneyama, H. Matsuda and S. Torii, *Tetrahedron Lett.*, **25**, 6017 (1984).
45. J. J. Habeeb and D. G. Tuck, *J. Organometal Chem.*, **134**, 363 (1977).
46. A. P. Tomilov, S. M. Makarochkina, Yu. I. Rozin, V. B. Busse-Machukas, K. M. Samarin, L. V. Zhitareva, V. M. Burmakov, SU 706420 30 Dec 1979 *Otkrytiya, Izobret., Prom. Obraztsy, Tovarnye Znaki*, 89. (1979). *Chem. Abstr.*, **92**, 111157 (1980).
47. Yu. I. Rozin, S. M. Makarochkina, K. M Samarin, L. Z. Zhitareva, F. A. Gorina and A. P. Tomilov, *Elektrokhimiya*, **20** (6), 849 (1984); *Chem. Abstr.*, **101**, 74739y (1984).
48. L. S. Feoktistov, in *Organic Electrochemistry*, (Eds. M. M. Baizer and H. Lund), 2nd ed., Marcel Dekker, New York, 1983, p. 259.
49. J. Casanova and V. Prakash Reddy, *The Chemistry of Halides*, Supplement D2 (Eds. S. Patai and Z. Rappoport), Wiley Interscience, Chichester, 1995, in press.
50. J. Y. Becker, in *The Chemistry of Halides, Pseudohalides and Azides*, Supplement D (Eds. S. Patai and Z. Rappoport), Wiley Interscience, Chichester, 1983, p. 203.
51. G. M. F. Calingaert, U.S. Pat. 1,539,297; *Chem. Abstr.*, **20**, 607 (1926); B. Mead, U.S. Pat. 1,567,159; *Chem. Abstr.*, **19**, 2210 (1925).
52. H. E. Ulery, *J. Electrochem. Soc.*, **116**, 1201 (1969).
53. R. Galli, *J. Electroanal. Chem.*, **22**, 75 (1969).
54. R. Galli and F. Olivani, *J. Electroanal. Chem.*, **25**, 331 (1970).
55. M. Fleischmann, D. Pletcher and C. J. Vance, *J. Electroanal. Chem.*, **29**, 325 (1971).
56. O. R. Brown, K. Taylor and H. R. Thirsk, *Electroanal. Chem., Interfac. Electrochem.*, **53**, 261 (1974).
57. M. Schuler, unpublished, cited in Reference 52.
58. A. P. Tomilov, S. M. Makarochkina, Yu. I. Rozin, V. F. Pavlichenko, K. M. Samarin and L. V. Zhitareva, U.S.S.R. SU 1081165, A1, 23 Mar. 1984, from: *Otkrytiya, Izobret., Prom. Obraztsy, Tovarnye Znaki*, (11), 81; 1984 *Chem. Abstr.*, **101** (3), 23727k (1984).
59. Yu. I. Rozin, S. M. Makarochkina, L. V. Zhitareva and A. P. Tomilov, *Elektrokhimiya*, **25** (11), 1540 (1989); *Chem. Abstr.*, **112** (4), 27210m (1990).
60. S. M. Makarochkina, Yu. I. Rozin, K. M. Samarin, V. F. Pavlichenko, L. V. Zhitareva and A. P. Tomilov, *Elektrokhimiya*, **21**, 1617 (1985); *Chem. Abstr.*, **104**, 77565m (1986).
61. A. P. Tomilov, M. Ya. Fioshin and V. A. Smirnov, *Electrochemical Synthesis of Organic Compounds*, Chimia, Moscow, (1976).
62. A. P. Tomilov, S. M. Makarochkina, Yu. I. Rozin, V. A. Klimov, K. M. Samarin and L. V. Zhitareva, U.S.S.R. SU 833976, 30 May 1981; *Chem. Abstr.*, **95**, 115753b (1981), From: *Obraztsy, Tovarnye Znaki* x, (20), 105 (1981).
63. 1T. A. Kharlamova, I. N. Chernykh, and A. P. Tomilov, *Khim. Promst. (Moscow)* (10), 589 1979; *Chem. Abstr.*, **92**, 111120h (1980).
64. K. M. Samarin, S. M. Makarochkina, A. P. Tomilov and L. V. Zhitareva, *Elektrokhimiya*, **16** (3), 326 (1980); *Chem. Abstr.*, **92**, 205967s (1980).
65. J. S. Blais and W. D. Marshall, *Appl. Radiat. Isot.*, **39**, 1259 (1988).
66. J. S. Blais and W. D. Marshall, *Appl. Organomet. Chem.*, **1**, 251 (1987).

67. E. R. Gonzalez, L. A. Avaca and M. D. Capelato, *An. Simp. Bras. Eletroquim. Eletroanal.*, 2nd (Ed. T. Rabokai), 179 (1980); *Chem. Abstr.*, **96**, 132014z (1982).
68. G. Mengoli, S. Daolio and F. Furlanetto, *Ann. Chim. (Rome)*, **68**, 455 (1978); *Chem. Abstr.*, **91**, 5308v (1979).
69. I. N. Chernykh and A. P. Tomilov, *Sov. Electrochem.*, **10**, 1363 (1974).
70. M. Fleischmann, G. Mengoli and D. Pletcher, *J. Electroanal. Chem.*, **43**, 308 (1973).
71. I. N. Brago, L. V. Kaabak and A. P. Tomilov, *Zh. Vses. Khim. Ova.*, **12** (4), 472 (1967); *Chem. Abstr.*, **67**, 104513u (1967).
72. S. M. A. Jorge and N. R. Stradiotto, *Anal. Lett.*, **22**, 1709 (1989).
73. S. Daolio, G. Mengoli and D. Pletcher, *Ann. Chim.*, **72**, 263 (1982).
74. D. A. Ashurov and Z. F. Mamedov, *Azerb. Khim. Zh.*, 115 (1981); *Chem. Abstr.*, **96**, 170970r. (1982).
75. J. Yoshida, H. Funahashi, H. Iwasaki and N. Kawabata, *Tetrahedron Lett.*, **27**, 4469 (1986).
76. B. Fleet and N. B. Fouzder, *J. Electroanal. Chem. Interfacial Electrochem.*, **99**, 215 (1979).
77. B. Fleet and N. B. Fouzder, *J. Electroanal. Chem. Interfacial Electrochem.*, **99**, 227 (1979).
78. R. E. Dessy, W. Kitching and T. Chivers, *J. Am. Chem. Soc.*, **88**, 453 (1966).
79. D. A. Kochkin, T. L. Shkorbatova, L. I. Ryzhova and G. E. El'Khanov, *Zh. Obshch. Khim.*, **41**, 74 (1971).
80. J. P. Colliard and M. Devaud, *Bull. Soc. Chim. Fr.*, 4068 (1972).
81. M. Devaud and A. Nezel, *J. Chem. Res.(S)*, 370 (1986); (*M*), 3201 (1986).
82. G. Chobert and M. Devaud, *J. Chem. Res.(S)*, (7), 228 (1980).
83. V. G. Kumar Das., S. W. Ng and L. H. Gan, *J. Organometal. Chem.*, **157**, 219 (1978); J. P. Colliard and M. Devaud, *Bull. Soc. Chim. Fr.*, 1541 (1973).
84. A. M. Bond, R. T. Gettar, N. M. McLachlan and G. B. Deacon, *Inorg. Chim. Acta*, **166**, 279 (1989).
85. M. Engel and M. Devaud, *J. Organomet. Chem.*, **307**, 15 (1986).
86. M. Devaud and J. L. Lecat, *Bull. Soc. Chim. Fr.*, 1187 (1985).
87. C. Feasson and M. Devaud, *J. Chem. Res.(S)*, (7), 248 (1986).
88. A. M. Bond and N. M. McLachlan *J. Electroanal. Chem. Interfacial Electrochem.*, **227**, 29 (1987).
89. M. Engel and M. Devaud, *J. Chem. Res.(S)*, (5), 152 (1984).
90. C. Feasson and M. Devaud, *Bull. Soc. Chim. Fr.* (1–2), 40 (1983).
91. C. Feasson and M. Devaud, *J. Chem. Res.(S)*, 152 (1982); (*M*), 1161 (1982).
92. G. A. Mazzocchin, R. Seeber and G. Bontempelli, *J. Organomet. Chem.*, **121**, 55 (1976).
93. M. Devaud, M. Engle, C. Feasson and J. L. Lecat, *J. Organomet. Chem.*, **281**, 181 (1985).
94. M. Devaud and P. Lepousez, *J. Chem. Res.(S)*, 100 (1982); (*M*), 1121 (1982).
95. F. A. Abeed, T. A. K. Al-Allaf and K. S. Ahmed, *Appl. Organomet. Chem.*, **4**, 133 (1990).
96. C. Feasson and M. Devaud, *J. Chem. Res.(S)*, 6 (1986); (*M*), 101 (1986).
97. R. J. P. Corriu, G. Dabosi and M. Martineau, *J. Organometal. Chem.*, **188**, 63 (1980).
98. R. J. P. Corriu, G. Dabosi and M. Martineau, *J. Organometal. Chem.*, **188**, 19 (1980).
99. R. J. P. Corriu, G. Dabosi and M. Martineau, *J. Chem. Soc., Chem., Commun.*, 457 (1979).
100. C. Combes, R. J. P Corriu, G. Dabosi, B. J. L. Henner and M. Martineau, *J. Organomet. Chem.*, **270**, 141 (1984).
101. C. Combes, R. J. P Corriu, G. Dabosi, B. J. L Henner and M. Martineau, *J. Organomet. Chem.*, **270**, 131 (1984).
102. K. Mochida, S. Okui, K. Ichikawa, O. Kanakubo, T. Tsuchiya and K. Yamamoto, *Chem. Lett.*, 805 (1986).
103. B. Fleet and N. B. Fouzder, *J. Electroanal. Chem. Interfacial Electrochem.*, **101**, 375 (1979).
104. R. J. Boczkowski and R. S. Bottei, *J. Organomet. Chem.*, **49**, 389 (1973).
105. M. Okano, T. Kugita and K. Mochida, *J. Electroanal. Chem.*, **356**, 303 (1993).
106. M. Okano and K. Mochida, *Bull. Chem. Soc. Jan.*, **64**, 1381 (1991).
107. (a) N. B. Fouzder and B. Fleet, in *Polarography of Molecules of Biological Significance* (Ed. W. F. Smyth), Academic Press, London, 1979; (b) p. 261.
108. P. J. Hayes and R. M. Smyth, *Anal. Proc. (London)*, **23**, 34 (1986).
109. G. N. Howell, M. J. O'Connor, A. M. Bond, H. A. Hudson, P. J. Hanna and S. Strother, *Aust. J. Chem.*, **39**, 1167 (1986).
110. M. P. Colombini, G. Corbini, R. Fuoco and P. Papoff, *Sci. Total Environ.*, **37**, 61 (1984).
111. M. P. Colombini, R. Fuoco and P. Papoff, *Ann. Chim. (Rome)*, **72**, 547 (1982).

112. A. M. Bond, J. R. Bradbury, G. N. Howell, H. A. Hudson, P. J. Hanna and S. Strother, *J. Electroanal. Chem. Interfacial Electrochem.*, **154**, 217 (1983).
113. For discussion of these methods see: R. Greef, R. Peat, L. M. Peter, D. Pletcher and J. Robinson, *Instrumental Methods in Electrochemistry*, Southampton Electrochemistry Group, Ellis Horwood Ltd., Chichester, 1985.
114. M. P. Colombini, G. Corbini, R. Fuoco and P. Papoff, *Ann. Chim. (Rome)*, **71**, 609 (1981).
115. S. E. Birnie and D. J. Hodges, *Environ. Technol. Lett.*, **2**, 433 (1981).
116. A. M. Bond and N. M. McLachlan, *Anal. Chem.*, **58**, 756 (1986).
117. D. Jagner, L. Renman and Y. Wang, *Anal. Chim. Acta*, **267**, 165 (1992).
118. M. Robecke and K. Cammann, *Fresenius. Z. Anal. Chem.*, **341**, 555 (1991).
119. N. Mikac and M. Branica, *Anal. Chim. Acta*, **212**, 349 (1988).
120. A. M. Bond and N. M. McLachlan, *J. Electroanal. Chem. Interfacial Electrochem.*, **194**, 37 (1985).
121. A. M. Bond and G. G. Wallace, *Anal. Chem.*, **56**, 2085 (1984).
122. A. M. Bond and N. M. McLachlan, *J. Electroanal. Chem. Interfacial Electrochem.*, **182**, 367 (1985).
123. K. Meng and D. Wu, *Fenxi Huaxue*, **9**, 708 (1981); *Chem. Abstr.*, **97**, 60059p (1982).
124. A. M. Bond and F. Scholz, *Langmuir*, **7**, 3197 (1991).
125. A. M. Bond and F. Scholz, *J. Phys. Chem.*, **95**, 7460 (1991).
126. C. M. G. Van den Berg, S. H. Khan and J. P. Riley, *Anal. Chim. Acta*, **222**, 43 (1989).
127. H. Kitamura, A. Sugimae and M. Nakamoto, *Bull. Chem. Soc. Jpn.*, **58**, 2641 (1985).
128. H. Kitamura, Y. Yamada and M. Nakamoto, *Chem. Lett., Chem. Soc. Jpn.*, 837 (1984).
129. P. Kenis and A. Zirino, *Anal. Chim. Acta*, **149**, 157 (1983).
130. B. Fleet and N. B. Fouzder, *J. Electroanal. Chem.*, **63**, 69 (1975).
131. B. Fleet and N. B. Fouzder, *J. Electroanal. Chem.*, **63**, 59 (1975).
132. C. B. Pascual and V. A. Vincente - Beckett, *Anal. Chim. Acta*, **224**, 97 (1989).
133. M. D. Booth and B. Fleet, *Anal. Chem.*, **42**, 825 (1970).
134. T. A. Annan, C. Peppe and D. G. Tuck, *Can. J. Chem.*, **68**, 1598 (1990).
135. T. A. Annan, C. Peppe and D. G. Tuck, *Can. J. Chem.*, **68**, 423 (1990).
136. H. E. Mabrouk and D. G. Tuck, *J. Chem. Soc., Dalton Trans.*, 2539 (1988).
137. A. T. Casey and A. M. Vecchio, *Inorg. Chim. Acta*, **131**, 191 (1987).
138. I. C. Hamilton and R. Woods, *Langmuir*, **2**, 770 (1986).
139. Y. P. Pena, A. J. Mora, O. J. Marquez, J. M. Ortega and J. Marquez, *Acta Cien. Venez.*, **38**, 37 (1987).
140. F. Capitan, P. Espinosa, F. Molina and L. F. Capitan-Vallvey, *An. Asoc. Quim. Argent.*, **74**, 517 (1986); *Chem. Abstr.*, **107**, 50722k (1987).
141. A. Ichimura, Y. Morimoto, H. Kitamura and T. Kitagawa, *Bunseki Kagaku*, **33** (12), E503 (1984); *Chem. Abstr.*, **102** (16), 139647j (1985).
142. S. Chandra, B. D. James and R. J. Magee, *Proc. Indian Acad. Sci., Chem. Sci.*, **99**, 317 (1987).
143. F. S. Holland, Eur. Pat. EP 84932, 1983; *Chem. Abstr.*, **99**, 112950k (1983).
144. R. E. Dessy, P. M. Weissman and R. L. Pohl, *J. Am. Chem. Soc.*, **88**, 5117 (1966).
145. R. E. Dessy and P. M. Weissman, *J. Am. Chem. Soc.*, **88**, 5124 (1966).
146. R. E. Dessy, R. L. Pohl and R. Bruce King, *J. Am. Chem. Soc.*, **88**, 5121 (1966).
147. R. E. Dessy and R. L. Pohl, *J. Am. Chem. Soc.*, **90**, 2005 (1968).
148. G. A. Russell, E. G. Jansen and T. Strom, *J. Am. Chem. Soc.*, **86**, 1807 (1964).
149. L. Doretti and G. Tagliavini, *J. Organometal. Chem.*, **13**, 195 (1968).
150. L. Doretti and G. Tagliavini, *J. Organometal. Chem.*, **12**, 203 (1968).
151. J. P. Bullock, M. C. Pallazzotto and K. R. Mann, *Inorg. Chem.*, **29**, 4413 (1990).
152. M. Okano and K. Mochida, *Chem. Lett.*, **64**, 1381 (1991).
153. M. Okano A. Toda and K. Mochida, *Bull. Chem. Soc. Jpn.*, **63**, 1716 (1990).
154. K. Mochida and T. Kugita, *Main Group Metal Chem.*, **XI**, 215 (1988).
155. R. Guilard and K. M. Kadish, *Chem. Rev.*, **88**, 1121 (1988).
156. K. M. Kadish, Q. Y. Xu and J. E. Anderson, *Electrochem. Surf. Sci. Mol. Phenom. Electrode Surf.*, ACS Symposium Series, **378**, 451 (1988).
157. K. M. Kadish, D. Dubois, J. -M. Barbe and R. Guilard, *Inorg. Chem.*, **30**, 4498 (1991).
158. K. M. Kadish, D. Dubois, S. Koeller, J. -M. Barbe and R. Guilard, *Inorg. Chem.*, **31**, 3292 (1992).
159. R. Guilard, J. -M. Barbe, M. Fahim, A. Atmani, G. Moninot and K. M. Kadish, *New J. Chem.*, **16**, 815 (1992).

160. R. Guilard, C. Ratti, J. -M. Barbe, D, Dubois and K. M. Kadish, *Inorg. Chem.*, **30**, 1537 (1991).
161. R. Guilard, J. M. Barbe, A. Boukhris, C. Lecomte, J. E. Anderson, Q. Y. Xu and K. M. Kadish, *J. Chem. Soc., Dalton Trans.* (5), 1109 (1988).
162. Q. Y. Xu, J. M. Barbe and K. M. Kadish, *Inorg. Chem.*, **27**, 2373 (1988).
163. K. M. Kadish, Q. Y. Xu and J. M. Barbe, *Inorg. Chem.*, **26**, 2565 (1987)
164. K. M. Kadish, Q. Y. Xu, J. M. Barbe, J. E. Anderson, E. Wang, and R. Guilard, *J. Am. Chem. Soc.*, **109**, 7705 (1987).
165. K. M. Kadish, Q. Y. Xu, J. M. Barbe, J. E. Anderson, E. Wang and R. Guilard, *Inorg. Chem.*, **27**, 691 (1988).
166. K. M. Kadish, C. Swistak, B. Boisselier-Cocolios, J. M. Barbe and R. Guilard, *Inorg. Chem.*, **25**, 4336 (1986).
167. J. H. Fuhrhop, K. M. Kadish and D. G. Davis. *J. Am. Chem. Soc.*, **95**, 5140 (1973).
168. K. M. Kadish, B. Boisselier-Cocolios, C. Swistak, J. M. Barbe and R. Guilard, *Inorg. Chem.*, **25**, 121 (1986).
169. A. Harriman, M. C. Richoux and P. Neta, *J. Phys. Chem.*, **87**, 4957 (1983).
170. J. A. Ferguson, T. J. Meyer and D. G. Whitten, *Inorg. Chem.*, **11**, 2767 (1972).
171. D. G. Whitten, J. C. Yau and F. A. Carroll, *J. Am. Chem. Soc.*, **93**, 2291 (1971).
172. M. El Meray, A. Louati, J. Simon, A. Giraudeau, M. Gross, T. Malinski and K. M. Kadish, *Inorg. Chem.*, **23**, 2606 (1984).
173. A. B. P. Lever, P. C. Minor and J. P. Wilshire, *Inorg. Chem.*, **20**, 2550 (1981).
174. C. C. Leznoff and A. B. P. Lever (Eds.), *Phthalocyanines, Properties and Applications* Vol. 3, VCH Publ., New York, 1993, p. 20.
175. L. F. LeBlevenec and J. G. Gaudiello, *J. Electroanal. Chem.*, **312**, 97 (1991).
176. G. T. Burns, E. Colomer, R. J. P. Corriu, M. Lheureux, J. Dubac, A. Laporterie and H. Iloughmane, *Organometallics*, **6**, 1398 (1987).
177. A. M. Bonny, .J. Crane N. A. P. Kane-Maguire, *J. Organomet. Chem.*, **289**, 157 (1985).
178. D. Miholová and A. A. Vlcek, *Inorg. Chim. Acta*, **73**, 249 (1983).
179. D. Miholová and A. A. Vlcek, *Inorg. Chim. Acta*, **43**, 43 (1980).
180. P. N. Lindsay, B. M. Peake, B. H. Robinson, J. Simpson, U. Honrath, H. Vahrenkamp and A. M. Bond, *Organometallics*, **3**, 413 (1984).
181. G. K. Campbell, P. B. Hitchcock, M. F. Lappert and M. C. Misra, *J. Organomet. Chem.*, **289**, C1 (1985).
182. J. C. Luong, R. A. Faltynek and M. S. Wrighton, *J. Am. Chem. Soc.*, **102**, 7892 (1980).
183. W. A. Herrmann, H. J. Kneuper and E. Herdtweck, *Chem. Ber.*, **122**, 445 (1989).
184. J. P. Zebrowski, R. K. Hayashi and L. F. Dahl, *J. Am. Chem. Soc.*, **115**, 1142 (1993).
185. E. Gileadi, *Electrode Kinetics*, VCH Publ., New York, 1993.
186. E. Steckhan, in *Topics in Current Chemistry*, **142**, Electrochemistry I (Ed. E. Steckhan), Springer-Verlag, Berlin, 1987, p. 1.
187. G. Dabosi, M.È. Martineau and J. Simonet, *J. Electronal. Chem. Interfacial Electrochem.*, **139**, 211 (1982).
188. C. Amatore, D. J. Kuchinka and J. K. Kochi, *J. Electroanal. Chem. Interfacial Electrochem.*, **241**, 181 (1988).
189. H. Tanaka, H. Suga, H. Ogawa, A. K. M. Abdul Hai, S. Torii, A. Jutand and C. Amatore, *Tetrahedron Lett.*, **33**, 6495 (1992).
190. B. A. Narayanan and J. K. Kochi, *J. Organomet. Chem.*, **272**, C49 (1984).
191. H. C. Gardner and J. K. Kochi, *J. Am. Chem. Soc.*, **97**, 855 (1975).
192. R. J. Klingler and J. K. Kochi, *J. Phys. Chem.*, **85**, 1731 (1981).
193. S. Fukuzumi, C. L. Wong and J. K. Kochi, *J. Am. Chem. Soc.*, **102**, 2928 (1980).
194. L. Eberson, *Adv. Phys. Org. Chem.*, **18**, 79 (1982).
195. K. Mochida, A. Itani, M. Yokoyama, T. Tsuchiya, S. D. Worley and J. K. Kochi, *Bull. Chem. Soc. Jpn.*, **58**, 2149 (1985).
196. J. Yoshida, Y. Ishichi and S. Isoe, *J. Am. Chem. Soc.*, **114**, 7594 (1992).
197. J. Yoshida, Y. Ishichi, K. Nishiwaki, S. Shiozawa and S. Isoe, *Tetrahedron Lett.*, **33**, 2599 (1992).
198. M. Pereyre, J. -P. Quintard and A. Rahm, *Tin in Organic Synthesis*, Butterworth, London, 1987.
199. R. S. Glass, A. M. Radspinner and W. P. Singh, *J. Am. Chem. Soc.*, **114**, 4921 (1992).
200. T Dong, M. Hwang, Y. Wen and W. Hwang, *J. Organomet. Chem.*, **391**, 377 (1990).
201. W. H. Morrison Jr., S. Krogsrud and D. N. Hendrickson, *Inorg. Chem.*, **12**, 1998 (1973).
202. J. I. Yoshida, Y. Morita, M. Itoh, Y. Ishichi and S. Isoe, *Synlett*, 843 (1992).

203. C. Eaborn, R. A. Jackson and R. Pearce, *J. Chem. Soc., Perkin Trans. 1*, 2055 (1974).
204. L. A. Paquette, C. D. Wright III, S. G. Traynor, D. L. Taggart and G. D. Ewing, *Tetrahedron*, **32**, 1885 (1976).
205. M. Stokla, M. A. Abkowitz, F. E. Knier, R. J. Weagley and K. M. McGrane, *Synth. Metals*, **37**, 295 (1990).
206. H. Isaka, M. Fujiki, M. Fujino and N. Matsumoto, *Macromolecules*, **24**, 2647 (1991).
207. S. K. Ritter and R. E. Noftle, *Chem, Mater.*, **4**, 872 (1992).
208. G. Casalbore-Miceli, G. Beggiato, N. Camaioni, L. Favaretto, D, Pietropaolo and G. Poggi, *Ann. Chim. (Rome)*, **82**, 161 (1992); *Chem. Abstr.*, **117**, 27337m (1992).
209. M. Okano and K. Mochida, Jpn. Kokai Tokyo Koho JP 01,201,323; *Chem. Abstr.*, **112**, 57069g (1990).
210. R. J. Klingler and J. K. Kochi, *J. Am. Chem. Soc.*, **102**, 4790 (1980).
211. C. Eaborn and S. B. Parker, *J. Chem. Soc.*, 939 (1954).
212. M. Engel and M. Devaud. *J. Chem. Res. (S)*, 152 (1984); (*M*), 1544 (1984).
213. C. Feasson and M. Devaud, *J. Chem. Res. (S)*, 228 (1987); (*M*), 1869 (1987).
214. T. Shono, S. Kashimura and H. Murase, *J. Chem. Soc., Chem. Commun.*, 896 (1992).
215. M. Okano, A. Toda and K. Mochida, *Chem. Lett.*, 701 (1990).
216. K. Mochida, K. Kimijima, H. Chiba, M. Wakasa and H. Hayashi, *Organometallics*, **13**, 404 (1994).
217. J. L. Bréfort, R. J. P. Coriu, Ph. Gerbier, C. Guérin, B. J. L. Henner, A. Jean, Th. Kuhlmann, F. Garnier and A. Yassar, *Organometallics*, **11**, 2500 (1992).
218. L. F. LeBlevenec and J. G. Gaudiello, *J. Electroanal. Chem. Interfacial Chem.*, **312**, 97 (1991).
219. P. J. Toscano and T. J. Marks, *J. Am. Chem. Soc.*, **108**, 437 (1986).
220. J. G. Gaudiello, G. E. Kellogg, S. M. Tetrick and T. J. Marks, *J. Am. Chem. Soc.*, **111**, 5259 (1989).
221. K. Ueyama, G. Matsubayashi, R. Shimizu and T. Tanaka, *Polyhedron*, **4**, 1783 (1985).
222. G. Matsubayashi, K. Miyake K. Ueyama and T. Tanaka, *Inorg. Chim. Acta*, **105**, 9 (1985).
223. S. A. Glazier and M. A. Arnold, *Anal. Chem.*, **63**, 754 (1991).
224. J. P. Lukaszewicz, N. Miura and N. Yamazoe, *Sens. Actuators*, B 1392, B9 55 (1992); *Chem. Abstr.*, **117**, 204064t (1992).
225. H. D. Wiemhoefer, D. Schmeisser and W. Goepel, *Solid State Ionics*, **40–41**, 421 (1990).
226. W. Goepel, K. D. Schierbaum, D. Schmeisser and H. D. Wiemhoefer, *Sens. Actuators*, **17**, 377 (1989), *Chem. Abstr.*, **112**, 47749x (1990).
227. W. R. Barger, H. Wohltjen, A. W. Snow, J. Lint and N. L. Jarvis, in *Fundamental Applications of Chemical Sensors*, ACS series **309**, 155 (1986); *Chem. Abstr.*, **105**, 53611t (1986).
228. Jpn. Kokai Tokkyo Koho JP 59,138,943 to Nippon Telegraph; *Chem. Abstr.*, **102**, 39107q (1985).
229. K. Fluri, J. Koudelka and W. Simon, *Helv. Chim. Acta*, **75**, 1012 (1992).
230. K. Ozaki, Jpn. Kokai Tokkyo, JP 02050965; *Chem. Abstr.*, **113**, 220203k (1990).
231. N. A. Chaniotakis, S. B. Park and M. E. Meyerhof, *Anal. Chem.*, **61**, 566 (1989).
232. W. Szczepaniak, J. Malicka and K. Ren, *Chem. Anal. (Warsaw)*, **20**, 1141 (1975).

CHAPTER **14**

The photochemistry of organometallic compounds of germanium, tin and lead

CHARLES M. GORDON and CONOR LONG

School of Chemical Sciences, Dublin City University, Dublin 9, Ireland
Fax: 353-1-704-5503; e-mail: LONGC@VAX1.DCU.IE

I. INTRODUCTION

In comparison with the photochemistry of organosilicon compounds, the photochemistry of organometallic compounds of the heavier Group 14 elements has received little attention. In recent years, however, a number of synthetically useful photoreactions of such species have been developed, and it would seem opportune to review this area of research. In particular, organogermanium and organotin reagents are becoming

The chemistry of organic germanium, tin and lead compounds
Edited by S. Patai   © 1995 John Wiley & Sons Ltd

increasingly important. The photochemical behaviour of such compounds is in many aspects similar to that of the equivalent silicon compounds, although there are some important differences which arise as a result of the weakening of the E—C bond (E = Group 14 atom) as one moves down the group. The photochemical behaviour of organolead compounds, however, remains little studied.

The initial photochemical step in almost all of the reactions described in this chapter is formation of either trivalent radicals of the type $R_3E\cdot$, or else the divalent analogues of carbenes, R_2E:. Such species are obviously very reactive, and are only observed as intermediates or in experiments in the presence of trapping agents. The relative stability of the intermediates depends greatly on the nature of the substituents R, and this can influence the type of reaction products ultimately formed. Where appropriate, comparisons with the behaviour of the analogous silicon species are made.

II. METAL ALKYL COMPOUNDS

Organostannyl radicals are important intermediates in many synthetically useful systems[1]. Such radicals are generated by hydrogen abstraction from the metal hydride[2], attack of alkoxy radicals on ditin compounds[3], or by photolysis of organotin compounds[4–7]. Consequently, laser flash photolysis has been used in the study of such radicals; this can provide direct kinetic information on their reactivity[8]. This work confirmed that photolysis of tetraalkyltin or hexaalkylditin compounds produces the trialkyl tin radicals (reactions 1 and 2; R = Ph or Bu).

$$R_4Sn \longrightarrow R_3Sn^{\bullet} + R^{\bullet} \tag{1}$$

$$R_3SnSnR_3 \longrightarrow 2R_3Sn^{\bullet} \tag{2}$$

The photochemistry of tetramethylstannane has also been investigated in low-temperature matrices[9]. Evidence was found for the production of CH_4 and a mononuclear tin compound, tentatively identified as the tin carbene $MeSn=CH_2$. Prolonged photolysis produced cyclic tin compounds, identified by Mössbauer spectroscopy. Stable trivalent germyl and stannyl alkyls can be synthesized photochemically from the appropriate MCl_2 salt and the alkyllithium reagent RLi^{10}. The alkyl groups in this work are bulky $CH(Me_3Si)_2$ ligands, and the stability of the trivalent product is the result of steric crowding around the central metal atom. Persistent trivalent tin radicals are also produced following photolysis of either R_3SnH or $R_3Sn—SnR_3$ (R = $PhMe_2CCH_2$), resulting from homolytic cleavage of either the Sn—H or Sn—Sn bonds as appropriate[11]. Irradiation of 1 in n-pentane results in both the cleavage of the carbon–tin bond and also that of the C—Br bond; the product of the latter process, 2, is obtained in low yield (4%)[12]. Longer irradiation with a medium pressure mercury lamp produced some of the stannyl migration product 3 along with some polymeric material.

| (1) | (2) | (3) |

In contrast to the observed photochemistry of adamantyltris(trimethylsilyl)silane which efficiently yields the appropriate silene compound, similar photolysis of the germanium analogue provided no evidence for the production of germene[13]. However, photolysis of the germanium compound in CCl_4 did result in a Norrish type 1 cleavage of the Ge—C bond to produce tris(trimethylsilyl)chlorogermane, adamantoylchloride and trimethylchlorosilane. The absolute rate constants for the reactions of $(t\text{-Bu})_3\text{Sn}^\bullet$ and $(t\text{-Bu})_3\text{Ge}^\bullet$ radicals with carbonyl compounds have been determined by laser flash photolysis techniques[14]. The reactivity of the $R_3M^\bullet$ radicals towards a carbonyl group follows the order[15,16]:

$$R_3Si^\bullet > R_3Ge^\bullet \sim R_3Sn^\bullet > R_3Pb^\bullet.$$

In this study, the metal centred radicals were formed indirectly by the photolysis of $(\text{Me}_3\text{CO})_2$ yielding $\text{Me}_3\text{CO}^\bullet$ radicals which then abstracted a hydrogen atom from the $(n\text{-Bu})_3\text{MH}$ (M = Ge or Sn). The resulting metal based radical reacts with the carbonyl compound by adding to the oxygen atom (reaction 3).

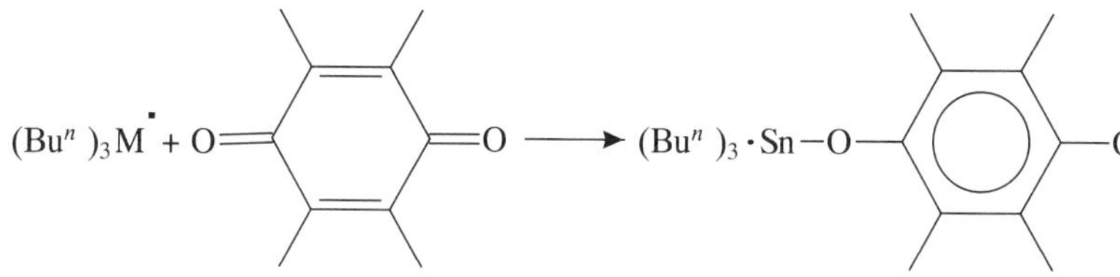

$$(\text{Bu}^n)_3M^\bullet + O = \text{...} = O \longrightarrow (\text{Bu}^n)_3 \cdot Sn - O - \text{...} - O^\bullet \quad (3)$$

The rate constants for these reactions are outlined in Table 1, and in all cases indicate that the tin and germanium radicals are less reactive than the silicon analogue. The rate constants for the addition of these radicals with other unsaturated materials are also presented.

It is generally accepted that photolysis of R_3SnX (R = alkyl group; X = halogen) results in the homolytic cleavage of the Sn—X bond. However, under certain conditions, for example in polar solvents such as EtOH which can act as a Lewis base, the photochemistry can switch to heterolytic cleavage of the Sn—X bond followed by formation of solvent adducts such as $R_3(X)\text{Sn(Sol)}$ (Sol = solvent)[17].

Ring closure reactions, with both 1,3 and 1,5 mechanisms, have been observed following the photolysis of trimethyl(3-methoxypropyl)tin in the presence of a variety of H abstractors such as Ph_2CO, $(t\text{-BuO})_2$, FeCl_3 or PhCOMe in benzene solution[18]. The mechanism involves the competitive H abstraction route (reactions 4 and 5), and the Sn—C cleavage process (reaction 6) which ultimately results in the formation of the various radical coupling products like $(\text{Me}_3\text{Sn})_2$. The similarity of the yields of acyclic and cyclic ethers (reactions 7 and 8) suggest that the intermolecular substitution reaction at tin and the hydrogen abstraction from the α-carbons on the ether occur

TABLE 1.   The rate constants for the reaction of $(n\text{-Bu})_3M^\bullet$ (M = Ge or Sn) with various substrates at 300 ± 3 K

Substrate	$k \times 10^{-8}$ $(\text{M}^{-1}\,\text{s}^{-1})$	Metal
Duroquinone	7.4	Ge
Duroquinone	14	Sn
Benzil	0.96	Ge
Benzil	1.3	Sn
Styrene	0.86	Ge
Styrene	0.99	Sn
1,4-Pentadiene	0.46	Ge
1,4-Pentadiene	0.68	Sn
t-Butyl bromide	0.86	Ge
t-Butyl bromide	1.7	Sn

with similar efficiency.

$$Me_3SnCH_2CH_2CH_2OMe + Ph_2CO^*(T) \longrightarrow Me_3SnCH_2CH_2CH_2O^\bullet CH_2 + Ph_2C^\bullet OH$$

(4)

$$Me_3SnCH_2CH_2CH_2OMe + Ph_2CO^*(T) \longrightarrow Me_3SnCH_2CH_2C^\bullet HOCH_3 + Ph_2C^\bullet OH$$

(5)

$$Me_3SnCH_2CH_2CH_2OCH_3 + Ph_2CO^*(T) \longrightarrow \overset{\bullet}{C}H_2CH_2CH_2OCH_3 + Ph_2\overset{\bullet}{C}OSnMe_3$$

(6)

$$Me_3Sn\overset{\bullet}{C}H_2CH_2CHOCH_3 \longrightarrow \text{[triangle]}-O\text{-}CH_3 + Me_3\overset{\bullet}{S}n$$

(7)

$$Me_3SnCH_2CH_2CH_2O\overset{\bullet}{C}H_2 \longrightarrow \text{[cyclic O]} + Me_3\overset{\bullet}{S}n$$

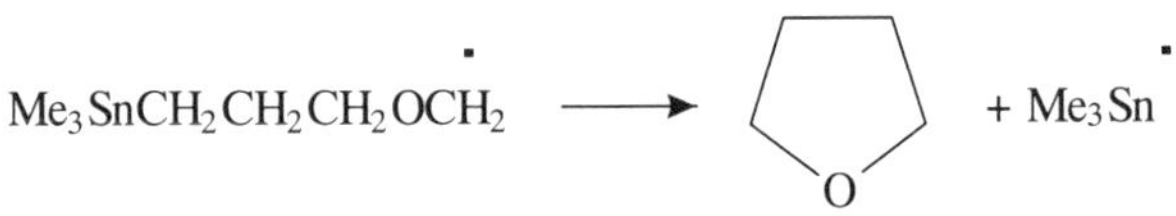

(8)

While C−Sn bond cleavage is a common feature of the photochemistry of organotin derivatives, the organic products are often influenced by the nature of organic ligand on the tin atom. For example, the photochemistry of organotin compounds containing a neighbouring carbonyl group can be characterized as follows:
(i) when the C−Sn bond is α to the carbonyl group, photolysis results in cleavage of the C−Sn bond;
(ii) the product types are dependent on the substitution pattern of β-stannyl ketones;
(iii) solvent molecules often are involved in the reaction sequences.
In the case of compound **4** photolysis with a medium pressure mercury lamp results in exclusive cleavage of the C−Sn bond β to the carbonyl group, while the terminal C−Sn bond remains intact. The cyclic ketone **5** and the α, β-unsaturated ketone **6** are produced in equal if low yields (18%)[19]. The reduced reactivity of remote C−Sn bonds in such compounds appears to be a general feature of their chemistry, and is also observed in acyclic systems. Thus the photochemistry of organotin compounds with carbonyl groups in the β-position can be classified as follows:
(i) hydrogen abstraction reactions following C−Sn cleavage;
(ii) β-elimination following C−Sn cleavage;
(iii) acyl migration;
(iv) incorporation of a solvent molecule.
Germanium porphyrins containing σ-alkyl or aryl ligands have potential antitumor activity. This is because of their tendency to accumulate in malignant tissue and their ability to act as alkylating agents under conditions of visible irradiation. A range of octaethyl or tetraphenyl porphyrins has been synthesized[20]. The photochemistry of these materials is consistent with an efficient cleavage of the Ge−C bond using visible light. On the basis of UV/visible, [1]H n.m.r. and e.s.r. data, the light-induced cleavage of the metal alkyl bond in (TPP)GeR(R′) metalloporphyrins (**7**) is achieved albeit with low quantum yields[21]. If the R group is ferrocenyl, the quantum yields for Ge−R rupture are significantly reduced, presumably because of efficient intramolecular triplet quenching processes.

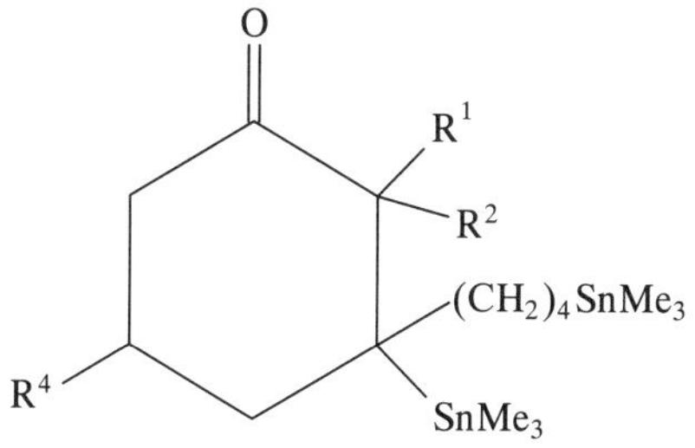

(4)

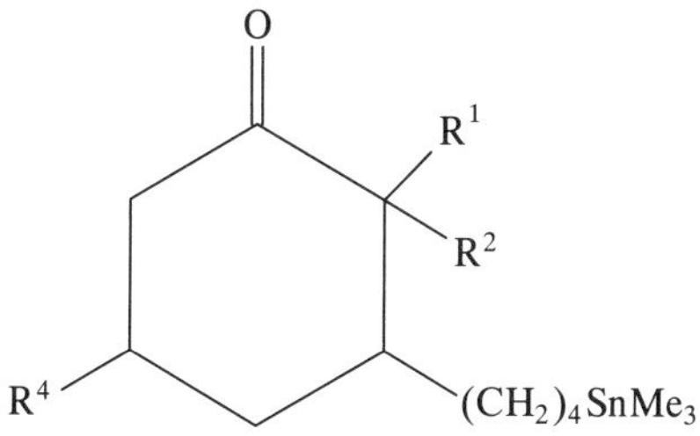

(5)

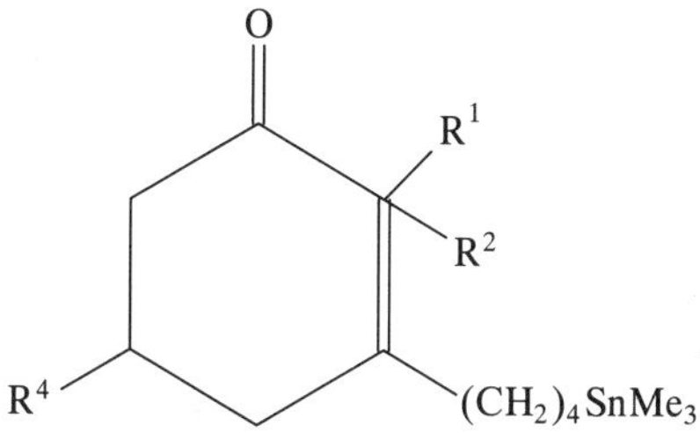

(6)

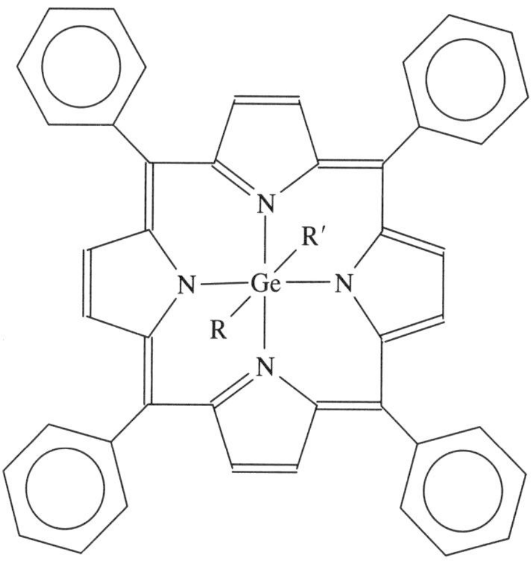

(7)

The bond dissociation energies (BDE) of several alkyl- and aryl-substituted germanium hydrides were measured by the laser-induced photoacoustic effect[22]. The alkyl-substituted compounds exhibited similar BDEs to that of GeH_4 (81.6–82.6 kcal mol^{-1}) while aryl substitution results in a slight weakening of the Ge—H bond (BDEs 79.2–80.2 kcal mol^{-1}).

The photochemical addition of diphenyl(trimethylstannyl)phosphine to either alkynes or allenes has been investigated[23]. The E-isomer is usually predominant (reaction 9) except when such a product is sterically impossible. Addition to the amide $BuC≡CCONMe_2$, however, gives a mixture of all four isomers, the Z-isomer being predominant. Addition of Me_3SnPPh_2 to allenes results in the production of the E-isomer predominantly, with the phosphine attack occurring at the central carbon atom (reaction 10). This topic is covered in greater detail in Section III.

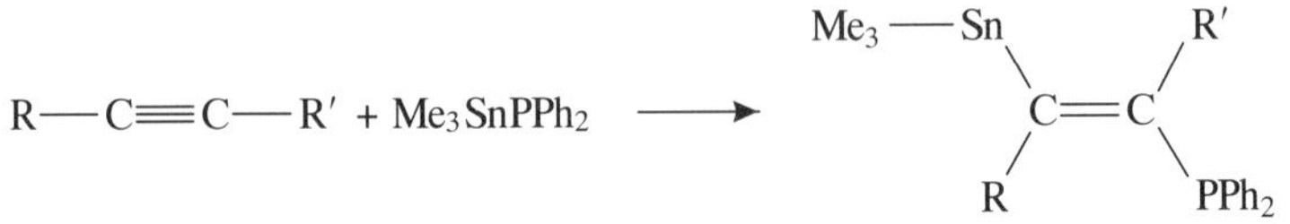

$$R—C≡C—R' + Me_3SnPPh_2 \longrightarrow \quad \begin{array}{c} Me_3—Sn \\ \diagdown \\ C=C \\ \diagup \quad \diagdown \\ R \qquad PPh_2 \end{array} \begin{array}{c} R' \end{array} \qquad (9)$$

$$Me_3SnPPh_2 + RCH{=}C{=}CH_2 \longrightarrow \quad \begin{array}{c} R \\ \diagup \\ C \\ \diagup \quad \diagdown \\ Me_3Sn \qquad PPh_2 \end{array} \qquad (10)$$

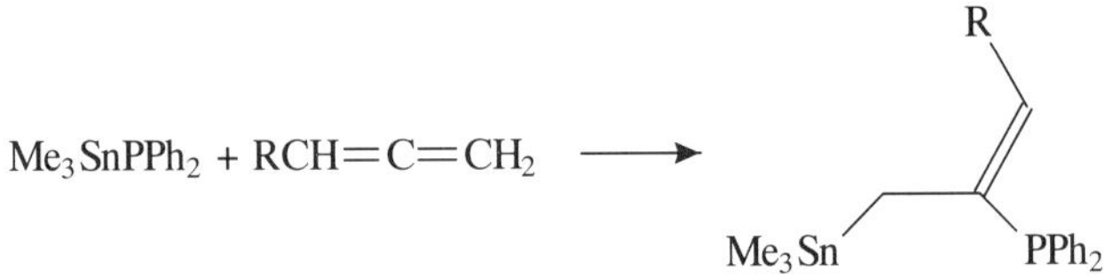

The photoreductions of a number of carbonyl compounds with either lowest $n\pi^*$ or $\pi\pi^*$ triplet states in the presence of tributyltinhydride are reported[24]. The carbonyl compounds include cyclohexanone and acetone which possess $n\pi^*$ lowest-energy triplets, and 2-acetonaphthone, 1-naphthaldehyde and 2-naphthaldehyde which possess lowest-energy $\pi\pi^*$ triplets. In the case of the two $n\pi^*$ triplets, a simple mechanism is proposed which involves the abstraction of a hydrogen atom from the tributyltinhydride by the triplet state of the ketone followed by a radical chain process (reactions 11–16).

$$R_2CO^T + Bu_3SnH \longrightarrow R_2C^•OH + Bu_3Sn^• \qquad (11)$$

$$R_2C^•OH + Bu_3SnH \longrightarrow R_2CHOH + Bu_3Sn^• \qquad (12)$$

$$Bu_3Sn^• + R_2CO \longrightarrow R_2C^•OSnBu_3 \qquad (13)$$

$$R_2C^•OSnBu_3 + Bu_3SnH \longrightarrow R_2CHOSnBu_3 + Bu_3Sn^• \qquad (14)$$

$$2Bu_3Sn^• \longrightarrow Bu_3SnSnBu_3 \qquad (15)$$

$$2Bu_3SnOC^•R_2 \longrightarrow Bu_3SnOCR_2CR_2OSnBu_3 \qquad (16)$$

Termination is principally via radical coupling forming hexabutylditin, or to a lesser degree via the coupling of ketyl radicals. In the case of the $\pi\pi^*$ ketones a different mechanism is proposed. The rate of abstraction of H from the tributyltinhydride by benzylic radicals is slower than the corresponding abstraction by alkyl radicals. Since the rate at which the tributyltin radical will add to aromatic carbonyls is similar to the addition rate to aliphatic carbonyls, the dominant radical species for the $\pi\pi^*$ systems is the ketyl radical. The primary termination process involves the coupling of the predominant radical species resulting in pinacol formation.

The photoinduced regio- and stereoselective $[2\pi + 2\pi]$ cycloadditions of electron-deficient alkenes to 1- and 2-naphthylmethylgermanes have been reported[25]. The photore-action of $CH_2{=}CH{-}CN$ with 1-naphthylmethyltrimethylgermanium yielded two products **8** and **9** in the ratio 15:1. Irradiation of 1-naphthylmethyltributylstannane in the presence of CH_2=CH—CN gave a complex mixture with 1,2-bis(1-naphthyl)ethane as the major product, presumably arising from an initial homolysis of the C—Sn bond followed by radical coupling.

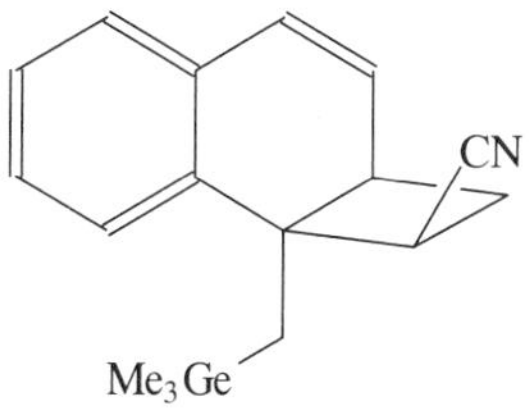

(8) 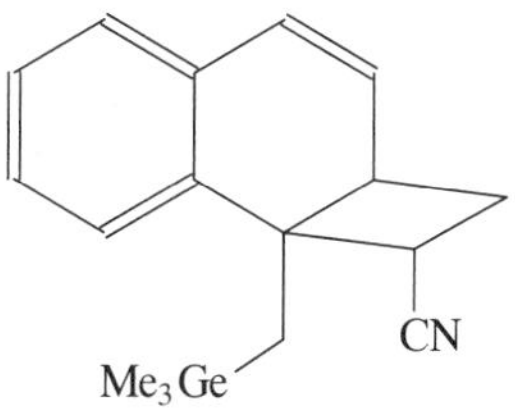   (9)

III ALLYL-SUBSTITUTED COMPOUNDS

The photochemistry of allyl-substituted Group 14 metal compounds can be divided into two sections, namely photoisomerization reactions, and those involving some kind of intermolecular reaction.

A. Photoismerization Reactions

A study of the photochemistry of 2-trimethylstannyl-1,3-butadiene (**10** in Scheme 1) shows that only isomerization is observed, with no Sn—C cleavage or polymerization[26]. Three different isomers are formed in approximately equal proportions, as shown in Scheme 1. In the case of the formation of 3-trimethylstannyl-1,2-butadiene (**11**), an intermediate is observed, but not characterized. The intermediate subsequently either reforms **10**, or forms **11**. The quantum yield of the reaction shows a strong dependence on the concentration of **10**. At $[10] = 10^{-2}$ mol l^{-1},

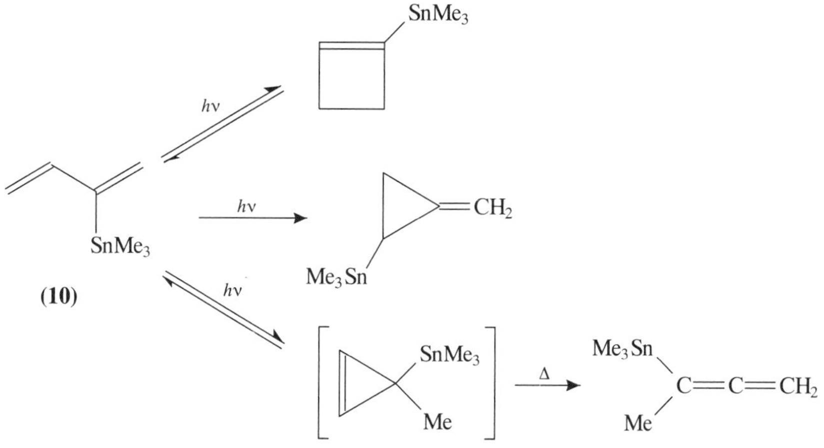

SCHEME 1

$\Phi_{(10,\text{disappearance})}$ is *ca* 8×10^{-3}, while at the much lower concentration of 1.5×10^{-4} mol l^{-1}, $\Phi_{(10,\text{disappearance})}$ is 0.12. It is suggested that these differences arise because of an unusually long-lived singlet excited state.

N.m.r. studies have shown that the amount of photochemical isomerization of Z-Me$_3$SnC(R)=CHSnMe$_3$ to the equivalent E-isomer depends to a large degree on the nature of the R group[27]. For example, when R = n-Bu, 15 hours irradiation results in 38% conversion Z to E, while with R = MeOCH$_2$, 22 hours photolysis gives 100% conversion. Extended photolysis of Me$_3$SnC(HOCH$_2$)=CHSnMe$_3$, however, results in polymerization. Photochemical Z to E isomerization was also observed on irradiation of Me$_3$Sn(Ph)C=CHPh at 313 nm[28]. This system reaches a photostationary state on extended photolysis, with a Z:E ratio of 27:73. No *trans*-stilbene is formed, showing that Sn−C cleavage does not occur. The higher yield of the E-form is attributed to the slightly higher extinction coefficient of the Z-form at 313 nm. The same publication reports similar results for Me$_3$Sn(Ph)C=CMePh, but gives no further details. The isomerization is also studied by transfer of energy from triplet sensitizers, such as benzophenone; some variation in the relative quantum yields is observed. A further example of isomerization of Me$_3$Sn-substituted alkene complexes is the smooth E to Z rearrangement of **12** on UV irradiation (reaction 17)[29]. Compound **13** is the first example of an alkene derivative containing a stannyl and boryl group *trans* to each other.

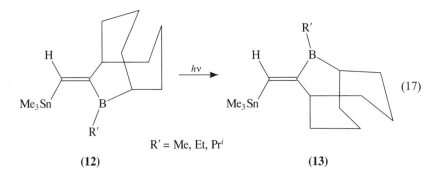

$$R' = \text{Me, Et, Pr}^i$$

(12)   **(13)**

B. Intermolecular Reactions

A series of symmetrically 1,3-disubstituted 2-propyl radicals was generated by photolysis of a mixture of a Group 14 substituted olefin and the corresponding Group 14 hydride, as illustrated in reaction 18. The structures of these radicals were then investigated using E.S.R. measurements.

$$\text{H}_2\text{C=C(X)CH}_2\text{ER}_3 + \text{R}_3\text{E·} \longrightarrow \text{R}_3\text{ECH}_2\text{C·(X)CH}_2\text{ER}_3 \qquad (18)$$

E = Si, Ge, Sn; R = Me.

The photochemical dissociation of Me$_2$Ge: from 7,7-dimethyl-1,4,5,6-tetraphenyl-2,3-benzo-7-germanorbornadiene (**14**) has been studied by flash photolysis, low-temperature matrix isolation and CIDNP ^{1}H NMR techniques[30]. The results suggest that a biradical (**15**) is formed as an intermediate species in the photoreaction. The biradical is initially formed in the singlet state, which undergoes conversion to the triplet state before irreversible decomposition to form Me$_2$Ge: and tetraphenylnaphthalene (TPN) (reaction 19).

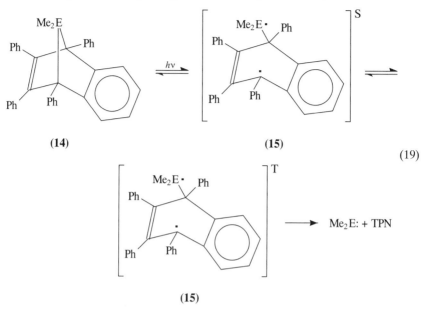

(14)   **(15)**

(19)

(15)

The photogenerated Me_2Ge: can subsequently insert into the C—Br bond of $PhCH_2Br$, and the Sn—Cl bond of Me_3SnCl. The reaction is thought to occur by way of radical pair intermediates; see reactions 20 and 21 (see also Scheme 2).

$$14 \longrightarrow Me_2Ge: \longrightarrow PhCH_2 \cdot \; \cdot GeMe_2Br \longrightarrow PhCH_2GeMe_2Br \qquad (20)$$

$$14 \longrightarrow Me_2Ge: \longrightarrow Me_3Sn \cdot \; \cdot GeMe_2Cl \longrightarrow Me_3SnGeMe_2Cl \qquad (21)$$

$$14 \xrightarrow{h\nu} Me_2Ge: $$

PhCH$_2$Br

$$\overline{PhCH_2{}^{\cdot} \; {}^{\cdot}GeMe_2Br} \longrightarrow PhCH_2GeMe_2Br$$

Me$_3$SnCl

$$\overline{Me_3Sn^{\cdot\cdot} {}^{\cdot}GeMe_2Cl} \longrightarrow Me_3SnGeMe_2Cl$$

SCHEME 2

UV irradiation of a saturated, O_2-flushed solution of trimethyl(triphenylethyl)tin, containing a catalytic quantity of I_2, results in cyclization, forming 9-phenyl-10-trimethylstannylphenanthrene (reaction 22)[31].

$$\xrightarrow[I_2/O_2]{h\nu}$$

(22)

Me$_3$SnPPh$_2$ will add to allenes under photochemical conditions, giving two regioisomeric products[32]. The predominant species is that in which the phosphine residue attaches to the central carbon atom (reaction 23). The overall yield, and relative proportions of **16** and **17** produced, depends on the nature of the substituent R. For R = H, yield = 78%, ratio **16** : **17** = 89:11; for R = Me, yield = 67%, ratio = 73:27; for R = Bu, yield = 58%, ratio = 88:12.

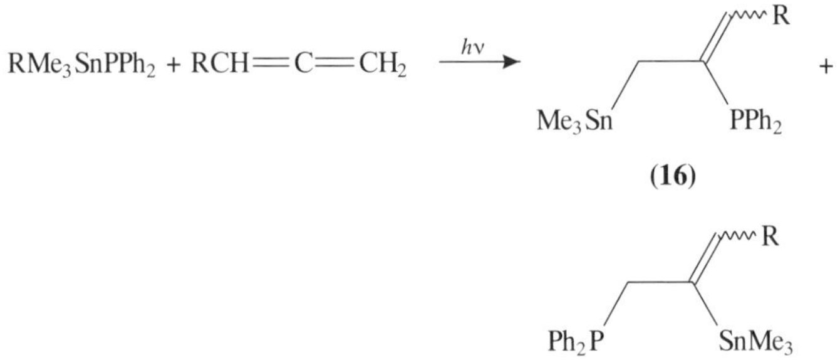

$$RMe_3SnPPh_2 + RCH{=}C{=}CH_2 \xrightarrow{h\nu}$$

(16)

(17)

(23)

The same stannyl phosphine will also add to both terminal and non-terminal alkynes[32], giving stannyl-substituted alkenes. In the case of terminal alkynes E- and Z-isomers are formed, with a mixture of the two possible regioisomers. Total yields are 60–80%, with 60–90% preference for the E-isomer, depending on the substituents on the alkyne, although exact experimental details are not given (reaction 24).

$$Me_3SnPPh_2 + RC{\equiv}CH \longrightarrow (E/Z)\text{-}Me_3SnCH{=}C(R)PPh_2 \qquad (24)$$

(R = Bu, Ph, Et$_2$NCH$_2$)

For reactions with non-terminal alkynes, no regioisomers are obtained, but a mixture of the E- and Z-isomers is present. Yields range from 50–90%, and the E:Z ratio once again varies, with the E-isomer predominating (reaction 25).

$$Me_3SnPPh_2 + RC{\equiv}CR' \longrightarrow (E/Z)\text{-}Me_3Sn(R)C{=}C(R')PPh_2 \qquad (25)$$

Russell and coworkers have made an extensive study of the photolytically initiated substitution reactions of a variety of reagents with 1-alkenyl derivatives of SnBu$_3$[33,34], the general reaction being as shown in reaction 26. The process is thought to involve addition–elimination in a free radical chain mechanism, illustrated in Scheme 3.

$$R(R')C{=}CHSnBu_3 + Q{-}Y \longrightarrow R(R')C{=}CHQ + YSnBu_3 \qquad (26)$$

R,R' = H, Me, Ph, MeO$_2$C; Q–Y = ArS–SAr, PhSO$_2$–Cl, PhSe–SO$_2$Ar.

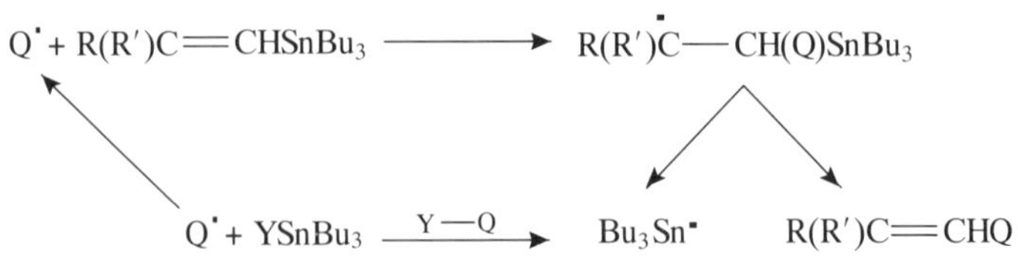

SCHEME 3

The photoreactions of $R(R')C{=}CHSnBu_3$ with mercurial compounds of the type Q_2Hg or $QHgCl$ also result in substitution[35]. Once again a radical chain mechanism is postulated, as outlined in Scheme 4.

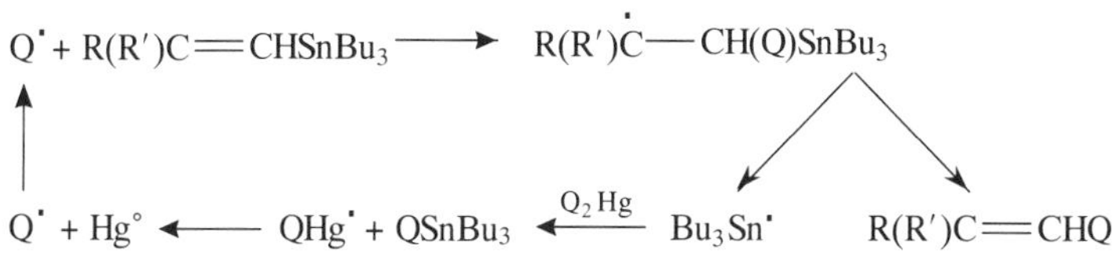

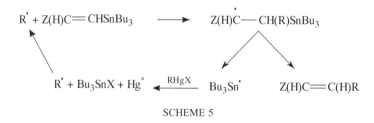

SCHEME 4

Irradiation of $PhC{\equiv}CSnBu_3$ in the presence of PhSSPh and $(PhS)_2Hg$ proved also to result in substitution of $SnBu_3$. The $PhC{\equiv}CSPh$ formed readily undergoes further photo-stimulated reactions. The photosubstitution reactions of 1-alkenylstannanes with RHgCl ($R = i\text{-Pr}$, $t\text{-Bu}$, $c\text{-C}_5H_9$, and $c\text{-C}_6H_{11}$) were also studied (Scheme 5)[34]. From this it was noted that the rates of reaction increase in the order primary < secondary < tertiary alkyl; this was attributed to the fact that the rate of reaction is controlled by the stability of the $R^{.}$ radical formed in the chain reaction. Related to this is the observation that ethenyl and ethynyl stannanes are much more reactive than $Bu_3SnCH_2CH{=}CH_2$, presumably as Bu_3Sn stabilizes the adduct radical. It is also observed that the substitution reaction is generally stereospecific, with good retention of configuration.

$$R^{.} + Z(H)C{=}CHSnBu_3 \longrightarrow Z(H)\overset{.}{C}{-}CH(R)SnBu_3$$

$$R^{.} + Bu_3SnX + Hg^{\circ} \xleftarrow{RHgX} Bu_3Sn^{.} \qquad Z(H)C{=}C(H)R$$

SCHEME 5

These reactions are compared with those of the equivalent 1-alkenyl HgCl compounds. Some differences are observed, notably that the mercury complexes will react with $PhSe{-}SePh$, while the tin complexes undergo no reaction at all. The reactions with Q_2Hg, QHgCl and RHgCl, however, are broadly similar to that shown in reaction 26.

The photochemistry of vinyl- and styryl-substituted digermanes has been extensively studied by Mochida and coworkers[35], and the results compared with the better known photochemistry of the analogous disilanes. Irradiation of vinylpentamethyldigermane (**18**) in cyclohexane gives a large excess of a high boiling point product, along with $Me_3GeCH_2CH_2Ge_2Me_5$ (8% yield) as the major identifiable product. Formation of traces of Me_3GeH, $CH_2{=}CHMe_2GeH$ and Me_6Ge_2 (along with other trace products) suggests that germyl radicals ($Me_3Ge^{.}$) are intermediate species. This was tested by carrying out the photolysis in cyclohexane containing CCl_4, as $Me_3Ge^{.}$ rapidly abstracts a chlorine atom. The expected chlorogermanes Me_3GeCl and $CH_2{=}CHMe_2GeCl$ were obtained, along with $ClCH_2CH_2Ge_2Me_5$ and C_2Cl_6.

The formation of Me_5Ge_2H in cyclohexane, and Me_5Ge_2Cl in the presence of CCl_4, implies that not only does $Ge{-}Ge$ bond cleavage occur on irradiation, but also $Ge{-}C$ cleavage to form digermyl radicals. The main photochemical paths are summarized in Scheme 6.

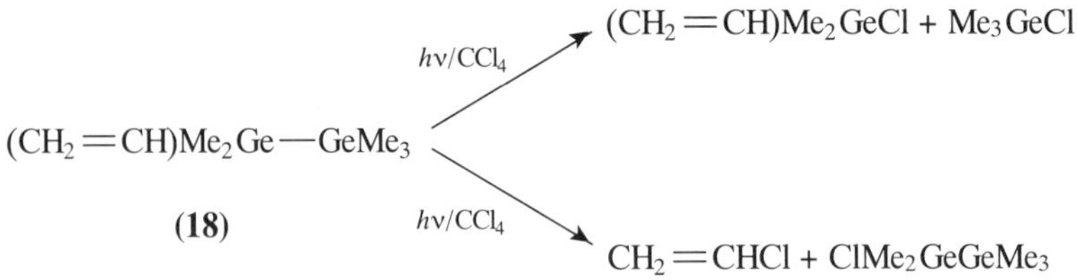

$$(CH_2\!=\!CH)Me_2Ge\!-\!GeMe_3 \xrightarrow[h\nu/CCl_4]{} (CH_2\!=\!CH)Me_2GeCl + Me_3GeCl$$

$$\xrightarrow[h\nu/CCl_4]{} CH_2\!=\!CHCl + ClMe_2GeGeMe_3$$

(18)

SCHEME 6

Another trace product observed is $CH_2\!=\!CHGeMe_3$, suggesting that Me_2Ge: is also formed. This was tested by photolysis in hexane containing a large excess of 2,3-dimethyl-1,3-butadiene as a trapping agent, which resulted in the formation of a small yield of the trapped germylene, shown in reaction 27 in Scheme 7.

$$(CH_2\!=\!CH)Me_2Ge\!-\!GeMe_3 \xrightarrow{h\nu} CH_2\!=\!CHGeMe_3 + [:GeMe_2] \qquad (27)$$

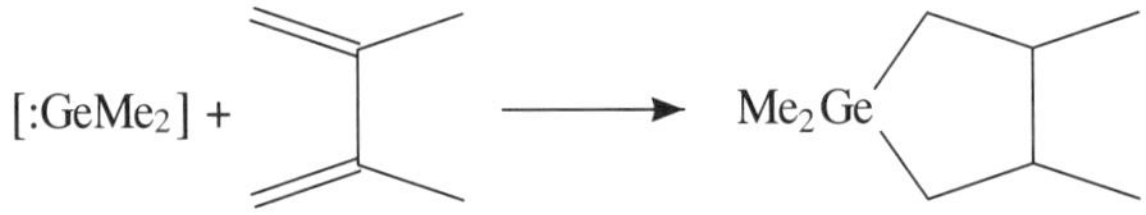

$$[:GeMe_2] + \quad \longrightarrow \quad Me_2Ge$$

SCHEME 7

Styrylpentamethyldigermane, $PhCH\!=\!CHGeMe_2GeMe_3$ (**19**), was similarly irradiated in cyclohexane, giving $PhCH\!=\!CHGeMe_3$ (**20**) as the primary photoproduct, along with traces of Me_3GeH (**21**) and Me_5Ge_2H (**22**). As in the photolysis of **18**, the observation of **20** indicated that germylene was formed; this was once more confirmed by photolysis in the presence of trapping agents. Similarly, the observation of traces of hydrogermanes **21** and **22** indicates that germyl radicals are involved as intermediates once again, and this was confirmed by the formation of the equivalent chlorogermanes on photolysis in the presence of CCl_4.

$$PhCH\!=\!CHGeMe_2GeMe_3 \qquad PhCH\!=\!CHGeMe_3 \qquad Me_3GeH \qquad Me_5Ge_2H$$

$$\textbf{(19)} \qquad\qquad\qquad \textbf{(20)} \qquad\qquad \textbf{(21)} \qquad\quad \textbf{(22)}$$

One photoproduct identified on photolysis of **19** whose analogue was not observed on photolysis of **18** is the trigermane $PhCH\!=\!CHGeMe_2GeMe_2GeMe_3$. It is suggested that this product arises from the insertion of germylene into a Ge—Ge bond, possibly via a germacyclopropane intermediate (by analogy with equivalent reactions of styryldisilanes[36]). An attempt made to trap any germacyclopropane by photolysis in the presence of methanol, which has been used in the analogous silane systems, failed to produce the expected 1,1-digermyl-2-phenylethanes[37].

The photochemistry of benzyl-substituted digermanes was examined in the same study[35]; these results are discussed in Section IV. Overall, however, the photolysis of all of these digermanes results in photolytic homolysis of Ge—C and Ge—Ge bonds as a major reaction path, with extrusion of dimethylgermylene as a minor path. This contrasts with the behaviour of the analogous silicon compounds, in which 1,3-silyl migration is observed on photolysis of vinyldisilanes[38], and 1,2-silyl migration for styryldisilanes[36].

In addition, no formation of silylenes has been observed to date[39]. The differences in photochemistry are attributed to the lower energy of Ge−Ge and Ge−C bonds relative to Si−Si and Si−C bonds.

The irradiation of two isomeric triethylstannane-substituted pyrazolenines (23 and 24) results in the evolution of N_2, and the formation of the same substituted cyclopropene[40], as illustrated in reaction 28.

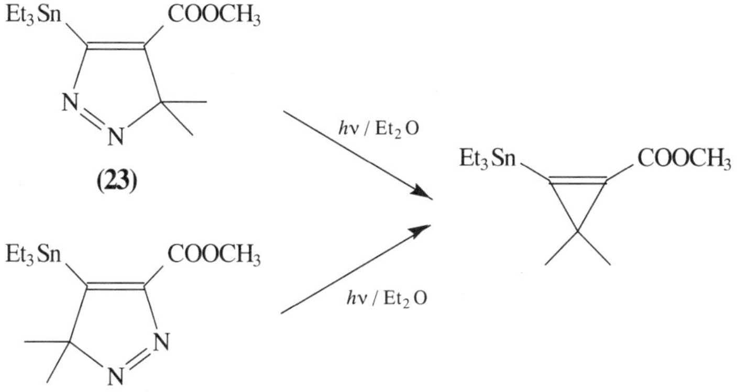

(28)

IV. PHENYL AND AROMATIC SYSTEMS

Simple Group 14 tetraaryl complexes (Ar_4E) appear to exhibit no particular photochemical behaviour. In the majority of aryl-substituted Group 14 complexes, the principal photochemical step is the loss of a ligand other than an aryl group. The initial photoproducts are either the radical $Ar_3E\cdot$, the radical anion $Ar_3E^-\cdot$ or Ar_2E:.

A typical example of this is shown in the laser flash photolysis of NaphPhMeGeH, an optically active hydrogermane[41]. On irradiation H· is lost, to form NaphPhMeGe·, and the decay of this species is monitored. In benzene and di-t-butyl peroxide the decay of this radical is slow, but in pure CCl_4 it reacts very rapidly by abstracting Cl. In pure CCl_4 the configuration at the Ge centre is predominately retained; as the CCl_4 is diluted with cyclohexane, however, optical purity is reduced, to an increasing degree with increased dilution (Scheme 8).

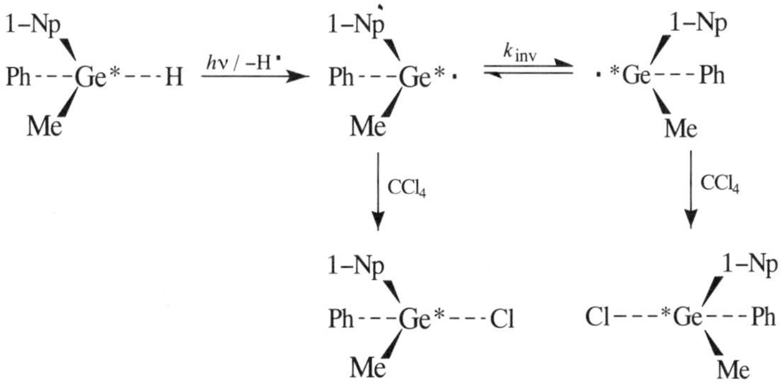

SCHEME 8

The Ph$_3$Sn· radical has also been generated by photolysis of Ph$_3$Sn$^-$ in THF and 2-methyl THF glasses at 93 K[42]. The radical was identified by ESR spectroscopy, and its signal was found to disappear on warming the glass to 150 K. A similar reaction has been studied by Mochida and coworkers, who looked at the photolysis of Ph$_n$Me$_{3-n}$E$^-$ ($n = 1-3$, E = Si, Ge; $n = 3$, E = Sn)[43]. Continuous photolysis of the anions in THF solution results in the formation of the respective dimers, presumably via the radical intermediate (reaction 29).

$$2 \ Ph_n Me_{3-n}E^- \longrightarrow 2 \ Ph_n Me_{3-n}E \cdot + 2 \ e^- \longrightarrow (Ph_n Me_{3-n}E)_2 \qquad (29)$$

An additional process was the formation of the equivalent hydrides, but this was shown to result from hydrolysis of unreacted anions by the observation of uptake of deuterium on hydrolysis of the starting reaction mixture with D$_2$O. The relative yields of the dimers increased in the order Si < Ge ~ Sn. The system was further examined[43] using laser flash photolysis, from which the radicals Ph$_n$Me$_{3-n}$Sn· were identified by their transient UV spectra, with λ_{max} in the region 315–330 nm, in good agreement with known values for such species[44]. In the same work the germyl radicals were identified by ESR measurements taken at 77 K, with g values once again in good agreement with those previously reported[45].

Haloarenes have been found to undergo nucleopilic substitution when irradiated with the triphenyl stannyl anion[46], reacting via a radical S$_{RN}$1 mechanism. In many cases the reaction will only occur under photochemical conditions. The reaction is found to proceed with chloro- and bromo-substituted arenes, but not iodo-compounds. The anion is produced either by treatment of triphenyltin chloride or hexaphenylditin with sodium metal in liquid ammonia, and will react with a wide variety of arenes (reaction 30).

$$p\text{-ClC}_6\text{H}_4\text{CH}_3 + Ph_3Sn^- \longrightarrow p\text{-CH}_3\text{C}_6\text{H}_4(\text{SnPh}_3) + \text{Cl}^- \qquad (30)$$

It is suggested[46] that the reaction mechanism involves initial electron transfer from Ph$_3$Sn$^-$ to the arene on irradiation to form the Ph$_3$Sn· radical and Ar$^-$. Disubstitution is also observed on irradiation of dihaloarenes, with a mechanism as outlined in Scheme 9.

Acyltriphenylgermanes react photochemically with styrene to form $_\sigma 2 + _\pi 2$ and $_\sigma 2 + _\pi 2 + _\pi 2$ adducts[47]. The yields of the two adducts **25** and **26** are *ca* 40% and *ca* 20%, respectively. The presence of GePh$_3$ in both products indicates that addition takes place before the germyl and acyl radical pair diffuse apart (reaction 31).

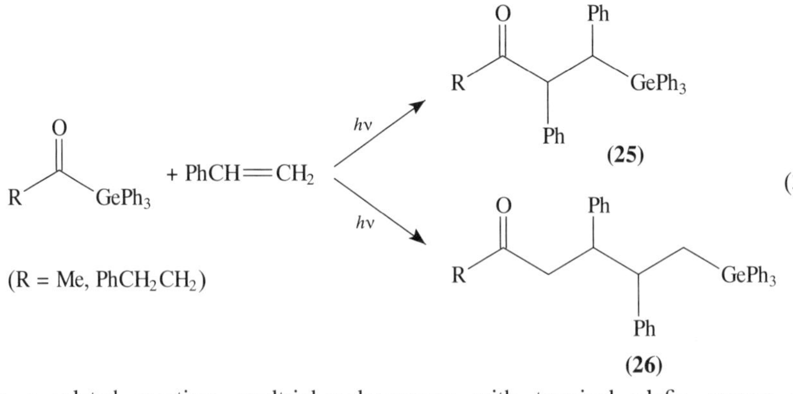

In a related reaction, acyltriphenylgermanes with terminal olefin groups can undergo photocyclization to form five- and six-membered cyclic ketones with an

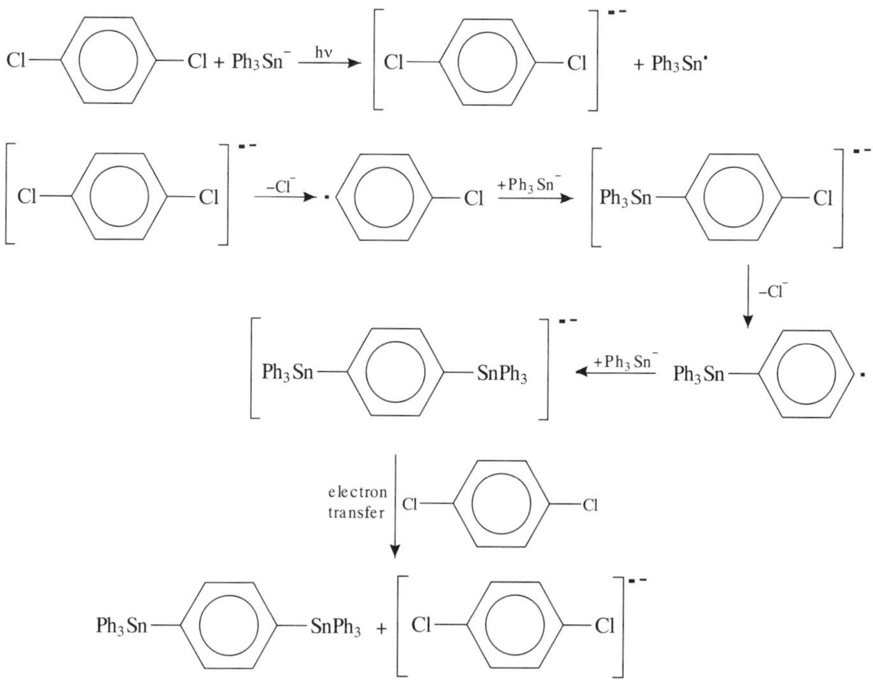

SCHEME 9

α-(triphenylgermyl)methyl group[48]. Twelve different systems are discussed, with cyclopentanones and cyclohexanones generally produced in good yield (> 75%). Starting materials with a non-terminal olefin did not cyclize, however, and it proved impossible to produce a seven-membered ring. Table 2 indicates a number of examples of the experiments attempted.

A rare example of photolytic arene ring loss is exhibited in the photolysis of Ph_3GeMe (27)[49]. Irradiation of a cyclohexane solution of 27 in the presence of CCl_4 results in the formation of $Ph_2MeGeCl$, suggesting that $Ph_2MeGe\cdot$ is an intermediate in the reaction. Laser flash photolysis of 27 at 266 nm, in micellar polyoxyethylene dodecyl ether, gives a radical pair (28) as the initial photoproduct. This can subsequently decay into a number of products as outlined in Scheme 10. It was observed that the rate of decay of 28 decreases with increasing magnetic field strength, suggesting that the magnetic field stabilizes the radical pair.

Photolysis of $(Me_3Si)_2GePh_2$ in cyclohexane gives the germylene Ph_2Ge: (29), which can then react with trapping agents such as substituted butadienes (reaction 32)[50]. This system was subsequently studied using laser flash photolysis[50]. On irradiation at 266 nm, an intermediate was observed with an absorption maximum at $\lambda = 445$ nm. This was assigned as the germylene intermediate 29 by comparison with matrix isolation data[51], and also with known values for the silylene analogue[52]. The rate of reaction of 29 with a variety of trapping agents (O_2, $EtMe_2SiH$, MeOH, 1,3-dienes) was measured, along with the simple dimerization reaction. In comparison with Ph_2Si:[52], the reactivity towards 1,3-dienes was somewhat lower (for the reaction with 2,3-dimethyl-1,3-butadiene, $k_2 = 2.75 \times 10^4$ $M^{-1}s^{-1}$ for 29 as against $k_2 = 8.26 \times 10^4$ $M^{-1}s^{-1}$ for Ph_2Si:), and the

TABLE 2. Examples of products formed following photolysis of some acyltriphenylger-manes containing terminal olefin groups

Acylgermane	Product	% Yield
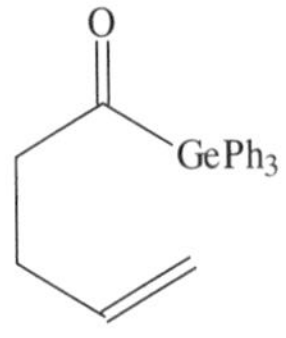	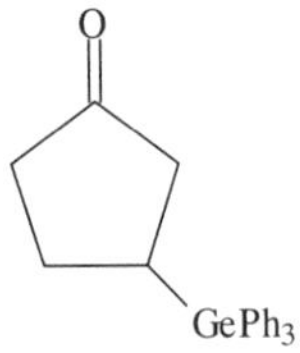	27
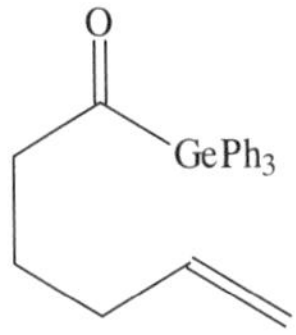	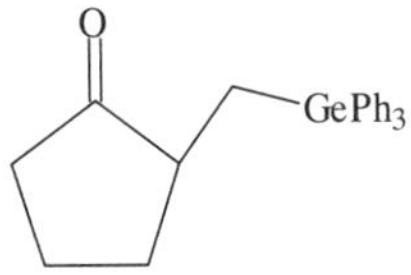	92
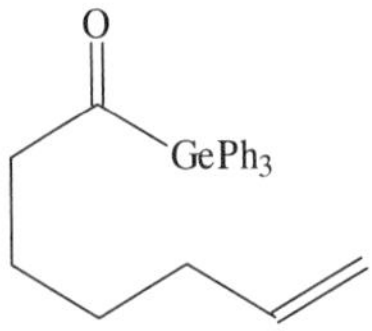	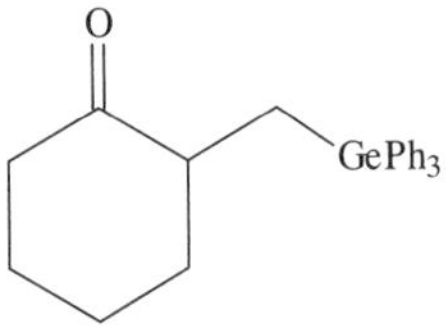	86
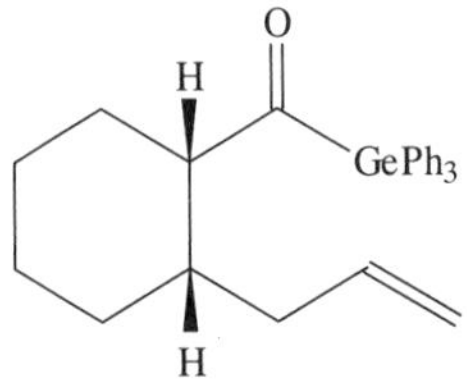	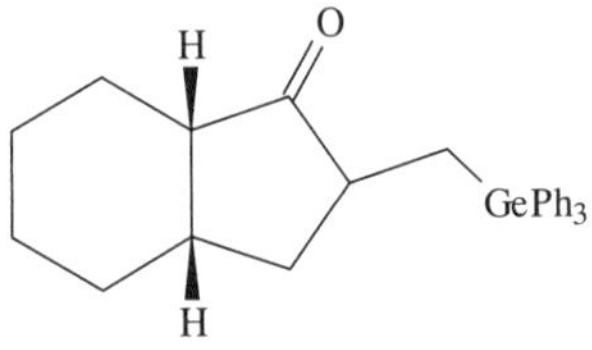	73
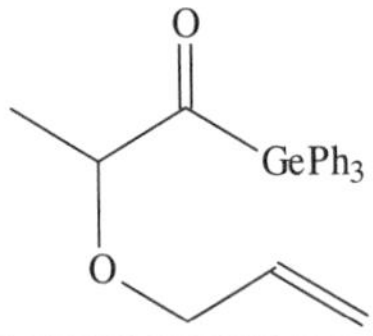	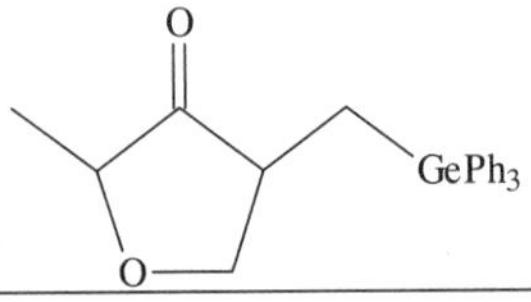	75

reactivity towards Si—H insertion somewhat greater (reaction with EtMe$_2$SiH, $k_2 = 1.01 \times 10^4$ M^{-1} s^{-1} for **29**, $k_2 = 1.54 \times 10^3$ M^{-1} s^{-1} for Ph$_2$Si:). In addition it was found that **29** does not react with MeOH, while Ph$_2$Si: reacts rapidly. In the absence of trapping agents the transient band assigned to **29** decays in a second-order process with a half-life of 270 ms; this is accompanied by the growth of a second transient centred at $\lambda = 320$ nm which is assigned to the digermene Ph$_2$Ge=GePh$_2$.

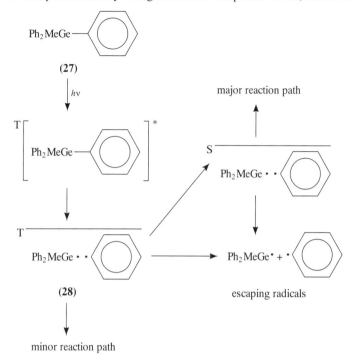

(27)

hv

(28)

major reaction path

escaping radicals

minor reaction path

SCHEME 10

$$(Me_3Si)_2GePh_2 \xrightarrow[-(Me_3Si)_2]{h\nu} [Ph_2Ge:] \xrightarrow{\quad} Ph_2Ge \qquad (32)$$

(29)

Another compound which loses R_2E: on irradiation is **30** (reaction 33), which forms perylene with the loss of Me_2Sn[53]. The authors report no more than the basic details of the reaction, however.

$$Me_2Sn \quad SnMe_2 \xrightarrow{h\nu} \qquad (33)$$

(30)

Mochida and coworkers have studied the photochemistry of the benzyl-substituted digermanes $PhCH_2Me_2GeGeMe_3$ (**31**) and $(PhCH_2Me_2Ge)_2$ (**32**)[35]. Irradiation of **31** in cyclohexane gave Me_3GeH, Me_5Ge_2H and $(Ph_2CH_2)_2$ as the major products, along with a number of trace products. These products suggested the involvement of germyl radicals as intermediates, and this was confirmed by the formation of the equivalent chloroger- manes Me_3GeCl and Me_5Ge_2Cl on photolysis in the presence of CCl_4. Analogous results were obtained on photolysis of **32**. All of this was interpreted as implying that homoly- sis of both the Ge—C and Ge—Ge bonds is the principal photoreaction, as indicated in Scheme 11.

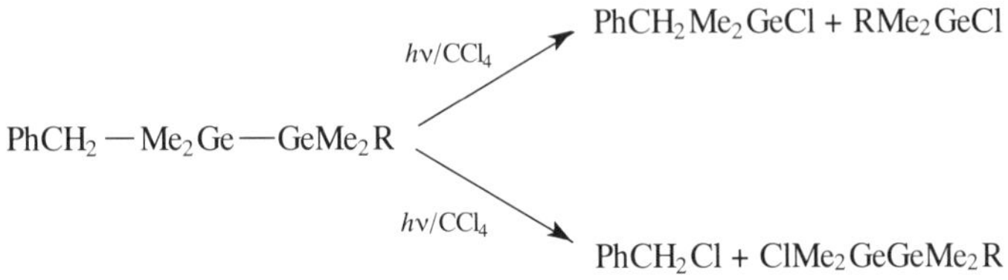

$$PhCH_2—Me_2Ge—GeMe_2R \quad \overset{h\nu/CCl_4}{\nearrow} \quad PhCH_2Me_2GeCl + RMe_2GeCl$$
$$\underset{h\nu/CCl_4}{\searrow} \quad PhCH_2Cl + ClMe_2GeGeMe_2R$$

R = Me (**31**), $PhCH_2$ (**32**)

SCHEME 11

The observation of $PhCH_2GeMe_3$ as a minor product in the photolysis of **31** and $(PhCH_2)_2GeMe_2$ in the photolysis of **32** suggests that loss of Me_2Ge: may be occurring as a competing reaction path (see Scheme 12). This was confirmed by the trapping of Me_2Ge: when photolysis was carried out in the presence of 2,3-dimethyl-1,3-butadiene.

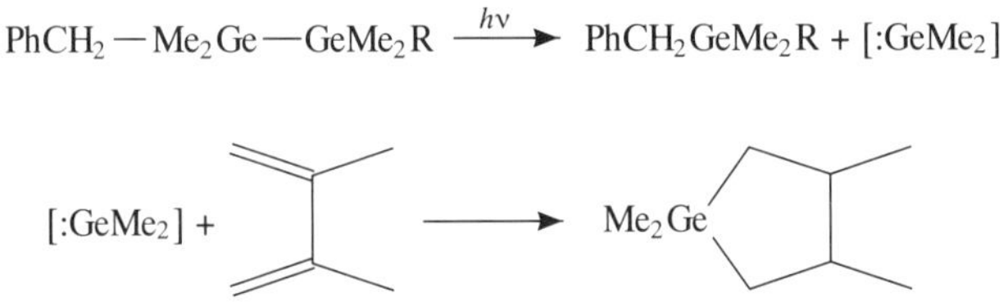

$$PhCH_2—Me_2Ge—GeMe_2R \quad \overset{h\nu}{\longrightarrow} \quad PhCH_2GeMe_2R + [:GeMe_2]$$

$$[:GeMe_2] + \quad \text{(diene)} \quad \longrightarrow \quad Me_2Ge \text{(ring)}$$

SCHEME 12

The photochemistry of benzyl-substituted digermanes is thus basically the same as that observed for vinyl- and styryl-substituted digermanes reported in Section III. In compari- son, the equivalent disilane compounds exhibit only Si—Si bond homolysis, with no Si—C cleavage or silylene formation[39].

V. MULTIHAPTIC SYSTEMS

Compounds of the type $CpSnX_3$ (X = alkyl, Cp, Cl) are very photosensitive com- pared with alkyltin compounds, and on irradiation show strong E.S.R. spectra of the Cp· radical[54]. The equivalent silicon and germanium compounds do not show this reactivity (reaction 34).

$$CpSnX_3 \longrightarrow Cp· + ·SnX_3 \qquad (34)$$

These reactions provide an efficient route to the formation of tin-centred radicals carrying a variety of ligands. Their reactions with a variety of ligands such as alkyl halides, alkenes and 1,2-diones have been investigated[55], and are summarized in Scheme 13.

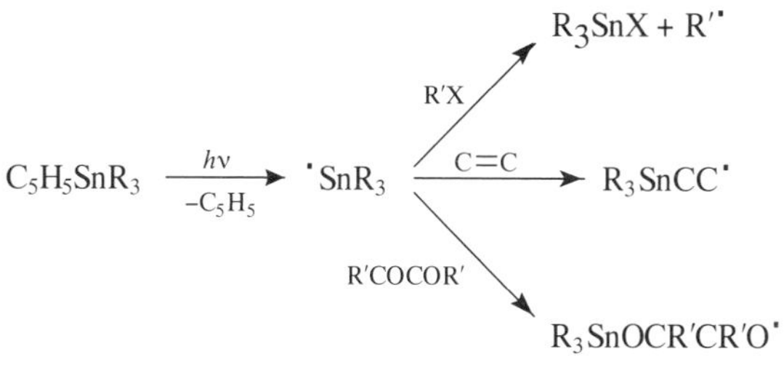

$$C_5H_5SnR_3 \xrightarrow[-C_5H_5]{h\nu} \cdot SnR_3$$

with products:

$$\xrightarrow{R'X} R_3SnX + R''\cdot$$

$$\xrightarrow{C=C} R_3SnCC\cdot$$

$$\xrightarrow{R'COCOR'} R_3SnOCR'CR'O\cdot$$

SCHEME 13

Photolysis of CpPbR$_3$ (R = Me, Ph) and Cp$_2$PbR$_2$ gives only signals for Cp·, when monitored by E.S.R. spectroscopy[56]. The yields of Cp· are particularly good for the latter two compounds. When R = Et on the other hand, above $-50\,^\circ$C a weak Cp· signal is seen by E.S.R., while below this temperature an Et· signal is also seen. When the temperature is below $-100\,^\circ$C only a weak Et· signal is seen. This behaviour is attributed to competitive cleavage of the Cp—Pb and Cp—Et bonds.

The photolysis of CpPbR$_3$ in the presence of various reagents was also investigated. Some differences were observed compared with the analogous tin compounds. The chemistry observed is summarized in Scheme 14, the main differences from CpSnR$_3$ being the lack of reaction with alkyl halides or alkenes.

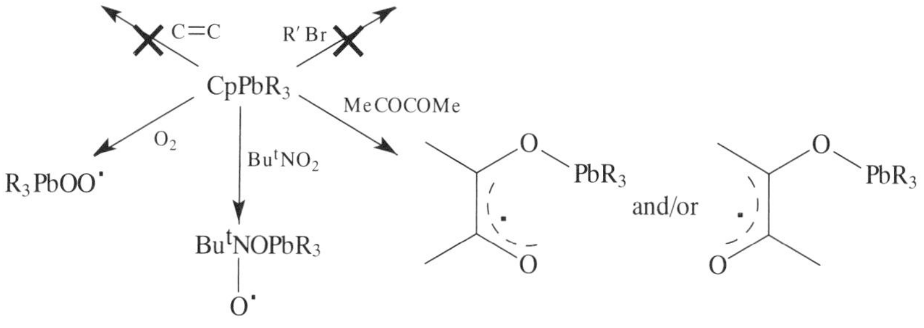

SCHEME 14

Photolysis of dicyclopentadienyltin results in formation of the Cp· radical (again detected by ESR), along with the precipitation of some unidentified yellow solid[54]. In contrast, photolysis of dicyclopentadienyllead produces no Cp·, unless di-t-butyl peroxide or biacetyl are added to the reaction mixture. The trimethylstannylcyclopentadienyl radical was produced by photolysis of bis(trimethylstannyl)cyclopentadiene (reaction 35), and was detected using ESR spectroscopy[57].

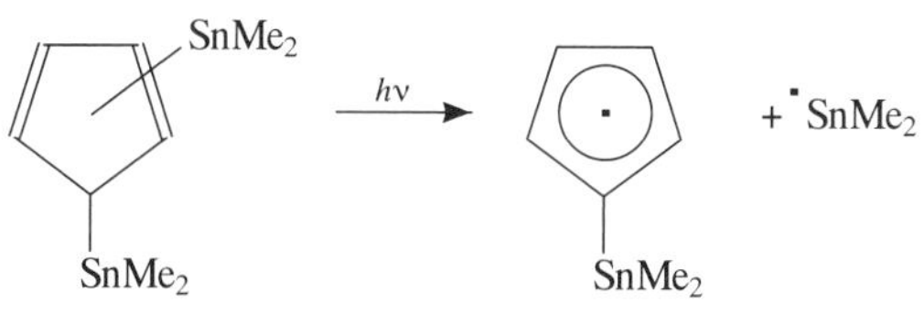

(35)

One method of producing penta(methoxycarbonyl)cyclopentadiene radical is by irradiation of $[SnBu_3(OH_2)_2]^+[C_5(CO_2Me)_5]^-$ (reaction 36)[58]. The radical thus produced is remarkably stable, with the ESR signal not decreasing in intensity one hour after cessation of photolysis.

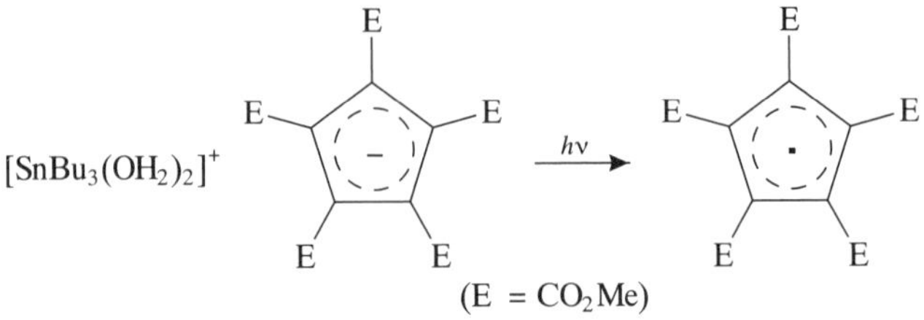

$$[SnBu_3(OH_2)_2]^+ \qquad \xrightarrow{h\nu} \qquad (36)$$

$$(E = CO_2Me)$$

VI. METALLOKETONES

Irradiation of metalloketones of silicon or germanium in t-BuOH results in the incorporation of a molecule of the solvent in the product[59]. For example, irradiation of **33** (M = Si) resulted in the formation of the cyclic acetyl compound while the germanium analogue gave the acyclic t-butyl-2-methyl-2-germa-6-hexanoate (**34**). The mechanism for the formation of the latter compound is thought to involve the initial cleavage of the Ge—C bond followed by an intramolecular hydrogen abstraction, producing a keto intermediate which then adds one solvent molecule (Scheme 15).

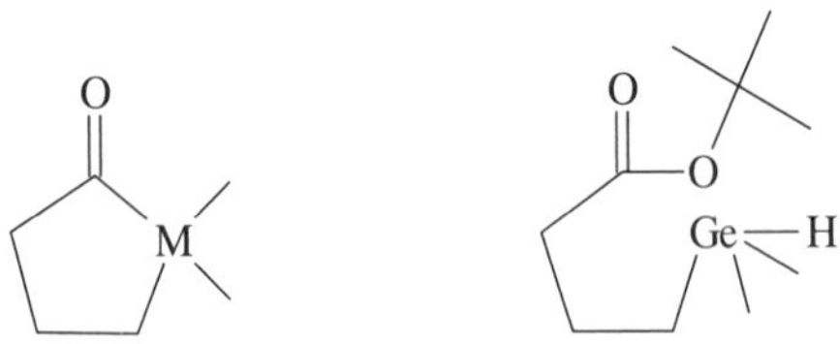

(33)   (34)

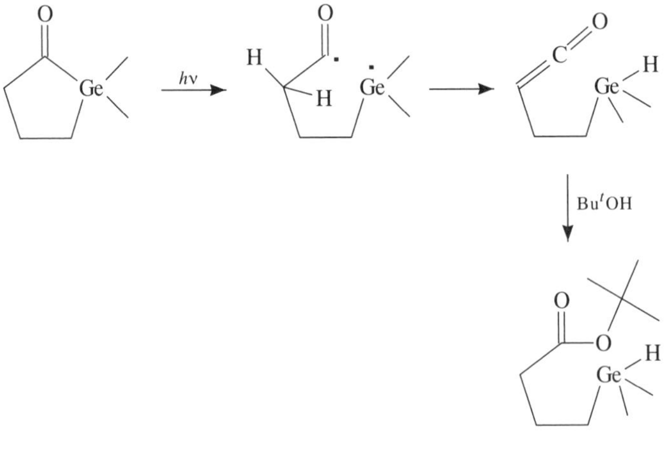

SCHEME 15

VII. AMIDES, AZIDES AND DIAZO COMPOUNDS

A convenient method of producing α-germylcarbene compounds has been published which utilizes the diazomethane precursor **35**. This yields the desired germene compound **36** under either thermal or photochemical conditions[60]. The chemistry of the germene parallels that of the silicon derivative in that it reacts almost quantitatively with CH_3OD to produce **37**. Triethylgermanium azide was irradiated ($\lambda = 253$ nm) in benzene solution in the presence of pinacol as a trapping agent (reaction 37)[61]. The mechanism of this reaction is thought to involve the intermediate germa-imine compound $Et_2Ge=NEt$ which polymerizes to produce $(Et_2Ge-NEt)_n$ in the absence of trapping agents. The triphenylgermanium azide, however, yields a more complex mixture of products upon photolysis. Two schemes are proposed to explain this photochemistry. The first involves the intermediate formation of the germaimine (Scheme 16), while the second (reaction 38) invokes a dimerization of the imine.

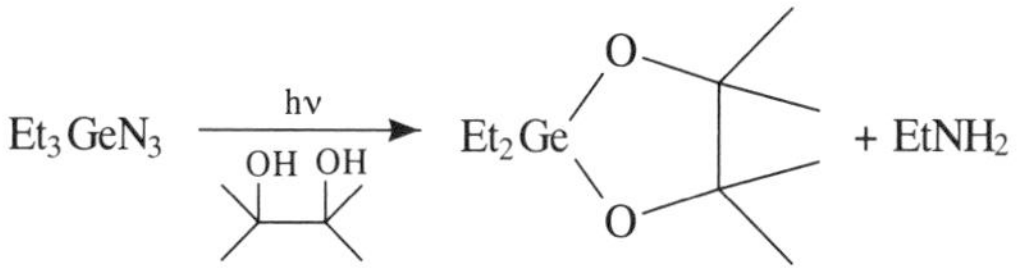

$$Et_3GeN_3 \xrightarrow[\text{OH OH}]{h\nu} Et_2Ge \overset{O}{\underset{O}{\diagdown\diagup}} + EtNH_2 \qquad (37)$$

$$\begin{array}{c} Ph_2Ge \!\!-\!\! NPh \\ | \qquad | \\ PhN \!\!-\!\! GePh_2 \end{array} \xrightarrow{h\nu} (Ph_2Ge)_n + PhNH_2 + Ph^-Ph \qquad (38)$$

$$\begin{array}{c} N_2 \\ \| \\ Me_3Ge\!\!-\!\!C\!\!-\!\!GeMe_3 \end{array} \qquad Me_2Ge\!=\!C\overset{Me}{\underset{GeMe_3}{\diagup}} \qquad \begin{array}{c} Me_2Ge\!\!-\!\!C\overset{Me}{\underset{GeMe_3}{\diagup}} \\ | \qquad | \\ Me\!\!-\!\!O \quad D \end{array}$$

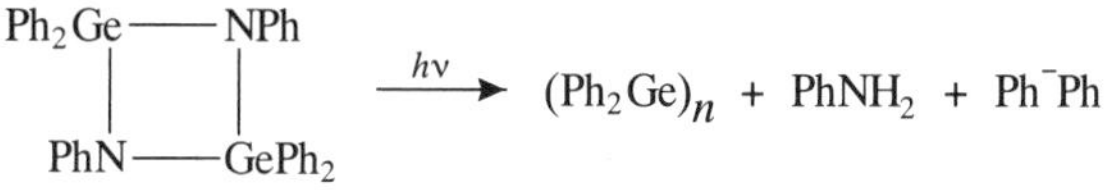

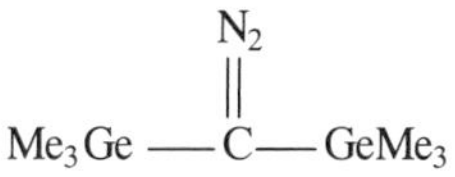

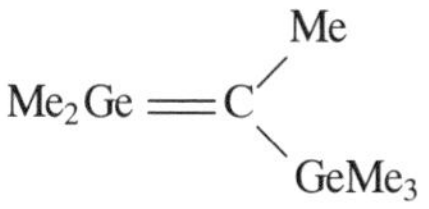

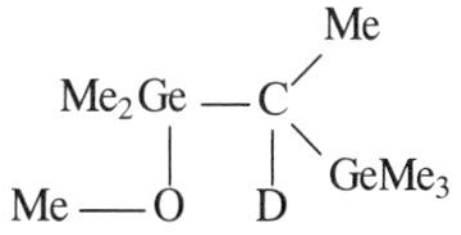

$$(35) \qquad\qquad (36) \qquad\qquad (37)$$

$$Ph_3GeN_3 \xrightarrow{h\nu} [\,Ph_2Ge\!=\!NPh\,] \longrightarrow \begin{cases} [\,Ph_2Ge\!:\,] + [PhN] \\ \quad\searrow\!\!\nearrow \\ \qquad h\nu \\ (Ph_2Ge^-NPh)_{2,n} \end{cases}$$

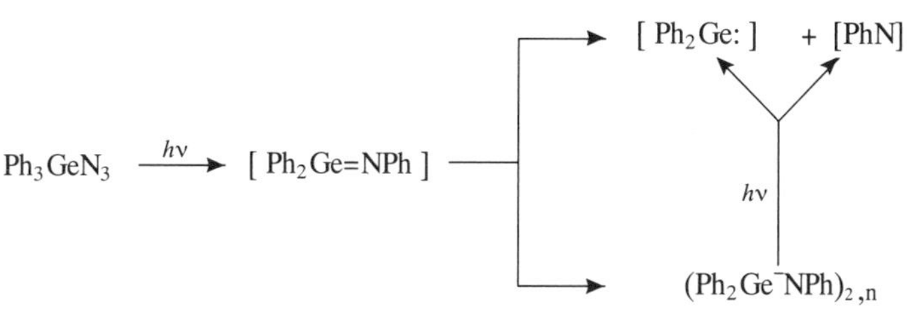

SCHEME 16

Treatment of the metal(II) halide ($GeCl_2$·dioxan, $SnCl_2$, or $PbCl_2$) with a lithium amide $Li(NR^1R^2)$ affords the appropriate metal diamide $M(NR^1R^2)_2$[62]. Some controversy exists as to the exact nature of these diamides, with some workers proposing that they are dimeric in nature[63]. However, with bulky R groups these compounds appear to exist as mononuclear species. For M = Ge or Sn, photolysis of the diamine produces a metal-centred radical species $M\cdot(NM'R^2)_3$ (M' = $SiMe_3$, $GeMe_3$, or $GeEt_3$). These radicals persist for several minutes at room temperature.

VIII. HETEROCYCLIC SYSTEMS

Photolysis of the four-membered heterocycle **38** yields the cyclotrisilane compound **39** by the extrusion of germylene[64]. The production of germylene was confirmed by trapping experiments. Germylene was again produced following photolysis of 2,3-dimethylgerma-1,4-dithiocyclohexane **40**, while photolysis of **41** gave **42** in almost quantitative yield[65]. Photolysis of the dithiagermolane (**43**) produces a range of heterocyclic products as outlined in Scheme 17[66], in which a homolytic cleavage of the S—S bond is proposed as the primary event. The diradical species thus formed then loses $CH_2{=}S$ producing the 1-dimethylgerma-2-thiocyclopropane intermediate.

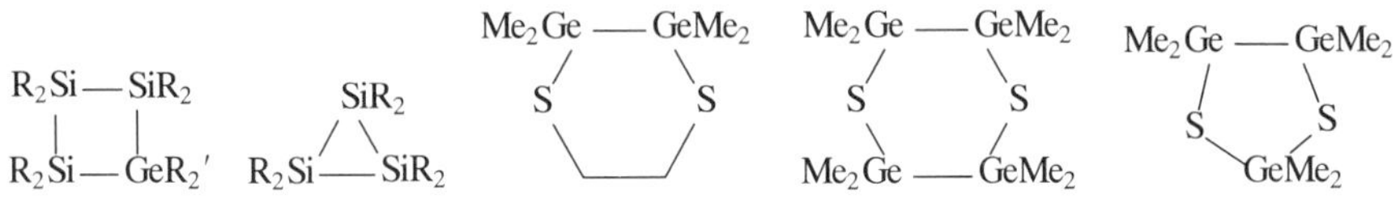

(38)   **(39)**   **(40)**   **(41)**   **(42)**

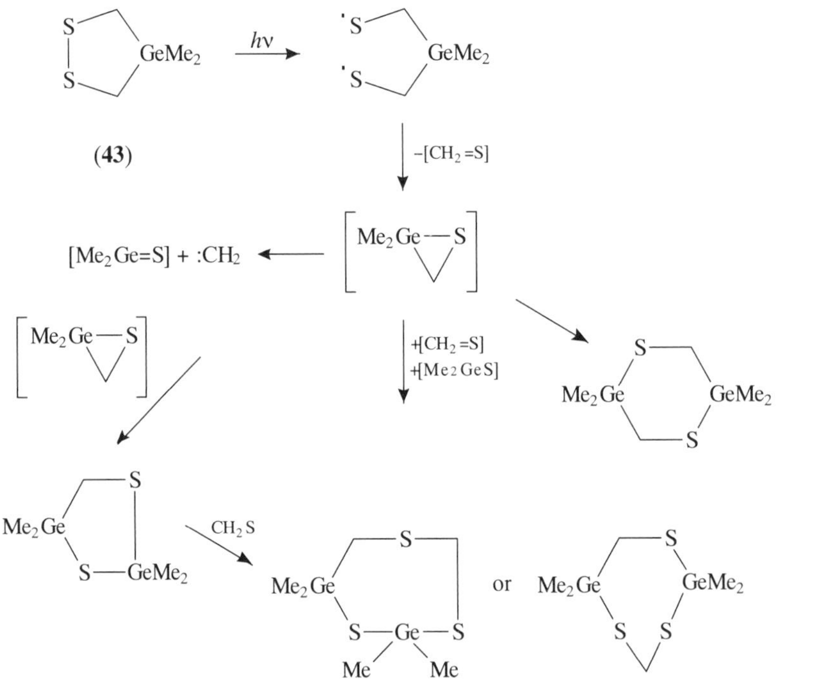

SCHEME 17

IX. METAL–METAL BONDED SPECIES

Laser flash photolysis of phenylated Group 14 catenates followed by trapping of the radical intermediates indicates that homolytic cleavage of the metal–metal bond is the

TABLE 3. The product distribution obtained following photolysis of some digermanes in the presence of various trapping agents

Digermane	Trapping agent	Main products (yield %)
$(PhMe_2Ge)_2$	None	$PhMe_2GeH$ (11), Ph_2GeMe_2 (17), $(Me_2Ge)_4$ (4)
	CCl_4	$PhMe_2GeCl$ (94)
	$DMBD^a$	$PhMe_2GeH$ (21), Ph_2GeMe_2 (5), $(Me_2Ge)_4$ (4)
$(Ph_2MeGe)_2$	None	Ph_2MeGeH (20), Ph_3GeMe (15)
	CCl_4	$Ph_2MeGeCl$ (95)
	$DMBD^a$	$PhMeGeCH_2C(Me)C(Me)CH_2$ (1)
$Me_3GeGePh_3$	None	Me_3GeH (12), Ph_3GeH (32), $PhGeMe_3$ (19)
	CCl_4	Me_3GeCl^b, Ph_3GeCl^b

^{a} Dimethylbutadiene.
^{b} Trace.

dominant process[67]. The generation of silicon–carbon double-bonded compounds results from the photolysis of aryldisilane compounds[68]. Trapping experiments as well as laser flash photolysis techniques have again been used to investigate the photochemistry of the analogous germanium systems. Laser flash photolysis of a number of $(Ph_nMe_{3-n}Ge)_2$ ($n = 1-3$) compounds has been undertaken in both THF and hydrocarbon solutions[69]. Two types of transient species were detected by their UV/visible spectra. The shorter-lived products were assigned to the germanium-centred radical species while the longer-lived products were tentatively assigned to either a germene species ($R_2Ge=CH_2$) or a coupling product of the caged radical pair (**44**). The products obtained following photolysis of a number of phenylated digermanes, along with the trapping agents used, are presented in Table 3[70]. The results of the flash photolysis experiments can be best rationalized in terms of the reactions outlined in Scheme 18. Similarly, the existence of germyl radical intermediates has been suggested from studies of phenylpentamethyldigermane compounds[71]. The photochemical homolysis of the Ge—Ge bond again appears to be the dominant process, and the products were identified by GC MS following steady-state photolysis. The product distribution is outlined in Scheme 19.

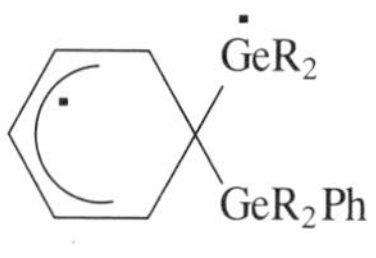

(**44**)

More recently, the photochemistry of polymeric germanes has been reported[72]. These polymers were prepared by the action of sodium metal on the appropriate dialkylgermanedichloride. Photolysis results in either homolytic cleavage of the Ge—Ge bond or germylene extrusion. The photogenerated germylenes can then insert into the C—Cl bonds of the halocarbon solvent producing a haloalkylgermane, while the germyl radicals abstract chlorine atoms from the solvent. Similar chemistry was observed for the analogous tin polymers[73], where the photoinduced reactions of $(Bu_2Sn)_n$ with alkyl halides (RX) gave the appropriate Bu_2RSnX compounds. The mechanism is again thought to involve the extrusion of stannylene Bu_2Sn: which inserts into the C—X bond of the alkyl halide.

The photolysis of phenyl-substituted trigermanes yields both the digermanes and germylenes, as determined by trapping, matrix isolation and flash photolysis experiments[74]. The data presented indicate two independent photochemical processes as outlined in

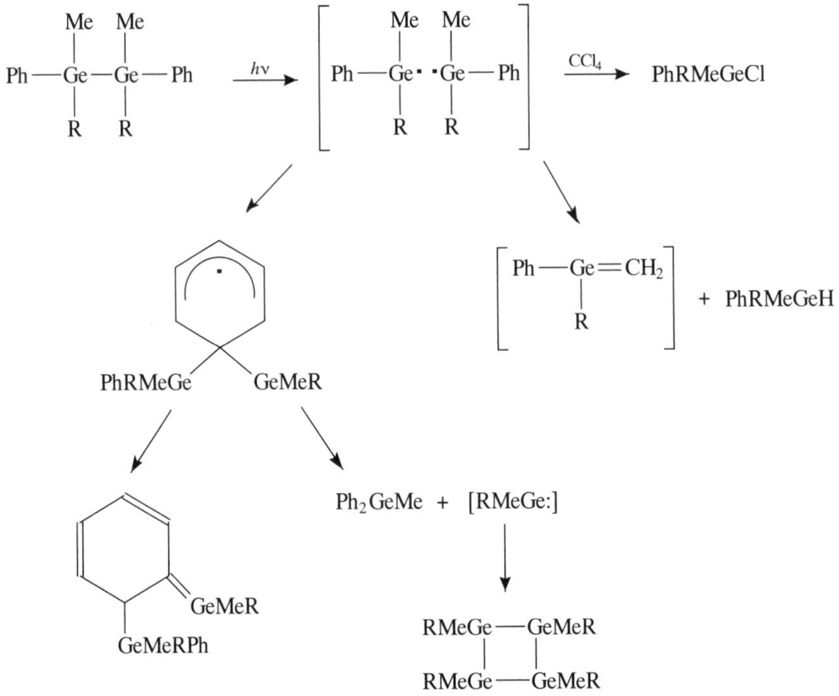

(58)

SCHEME 18

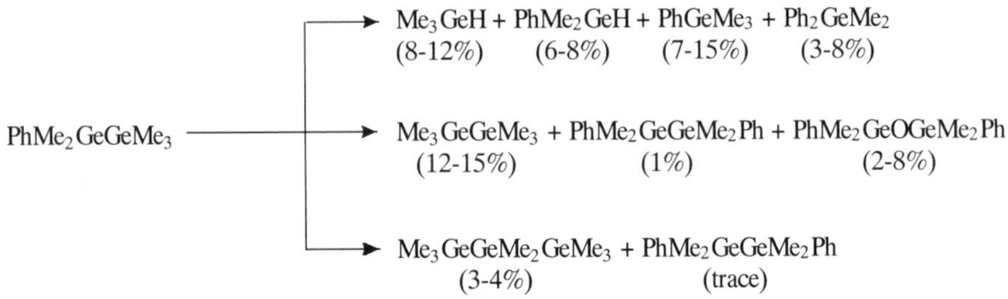

SCHEME 19

Scheme 20. The steric bulk of the alkyl or aryl substituents on hexa-substituted cyclotriger-manes influences the stability of the photoproducts formed, whether digermenes or ger-mylenes[75-77]. For instance, on photolysis of $(Ar_2Ge)_3$, when Ar = 2,6-dimethylphenyl, the resulting digermene undergoes photoinduced polymerization, while the equivalent compound with Ar = 2,6-diethylphenyl appears to be photostable. Treatment of the digermene **45** with CH_2N_2 produced the digermirane compound **46**, while photolysis in the presence of 2,3-dimethylbutadiene gave the appropriate germacyclopentene, which confirmed the intermediacy of germylene in the photochemistry[78]. Photolysis of **47** in the presence of adamanthanethione **48** resulted in the formation of the germathiirane compound **49** in reasonable yield (53%)[79]. Compound **49** is both air and moisture stable,

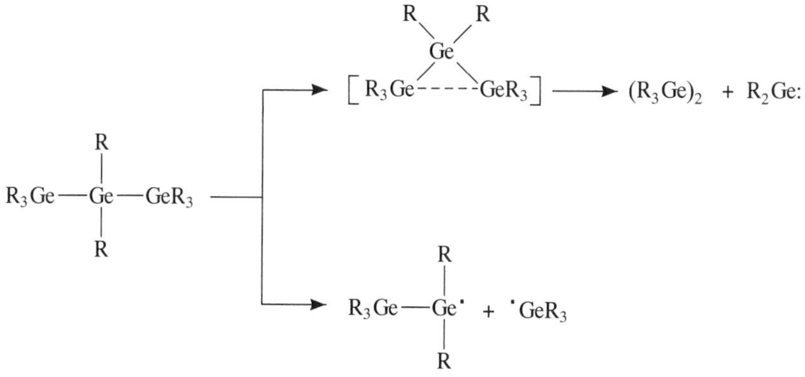

SCHEME 20

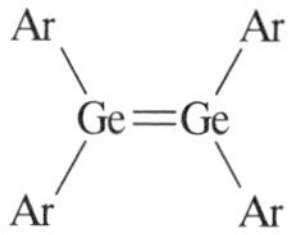

Ar = 2,6-diethylphenyl

(45)

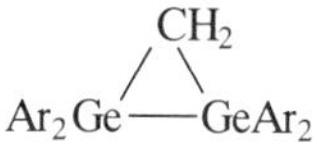

(46)

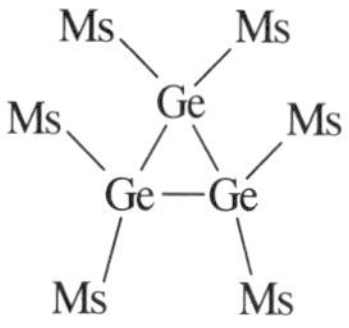

Ms = Me$_3$C$_6$H$_2$

(47)

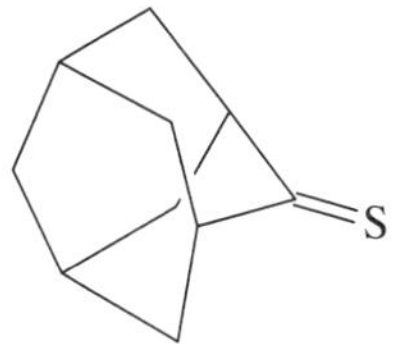

(48)

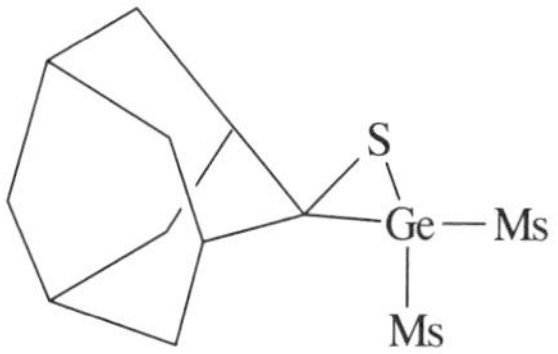

Ms = Me$_3$C$_6$H$_2$

(49)

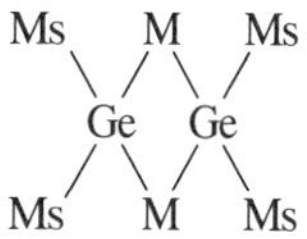

M = S or Se

(50)

and in general the stability of germathiiranes is greatly influenced by the bulk of the substituents. The related digermathiirane and digermaepiselenide were also synthesized from the hexamesitylcyclotrigermane precursor in the presence of elemental Group 16 elements (Scheme 21), and these in turn undergo a photochemical transformation to the doubly bridged species **50**[80]. The hexa(2,4,6-triisopropylphenyl)cyclotristannane compound has also been synthesized[81]. In this case the tin compound is in thermal equilibrium with the tetra(2,4,6-triisopropylphenyl)distannene. This equilibrium can be displaced in favour of the distannene upon irradiation with near-visible light.

Metallocyclopentenes are frequently formed in photochemical reactions of the Group 14 metal alkyls or catenates in the presence of dimethylbutadiene. This class of compound also has an extensive photochemistry[82]. For example, photolysis of **51** (R = H or Me) produced the allylic alcohols **52** and **53** and, for R = H, **54**. These alcohols could be dehydrated over Al$_2$O$_3$ to give the germole **55** along with other diene compounds.

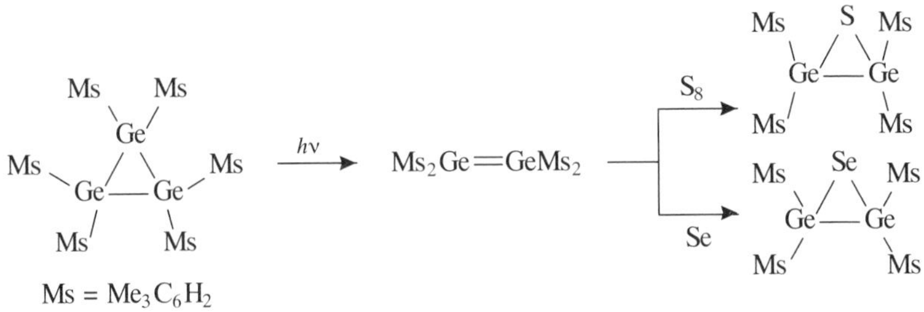

$Ms = Me_3C_6H_2$

SCHEME 21

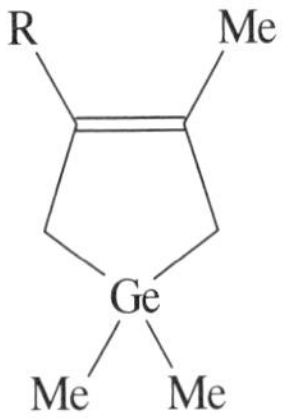

(51) 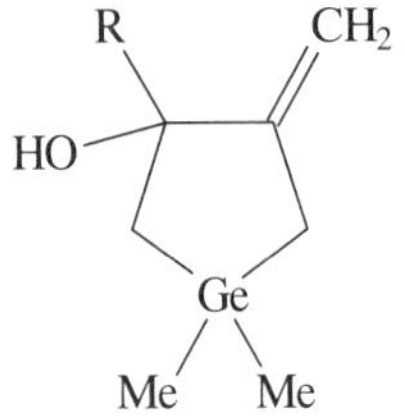   (52) 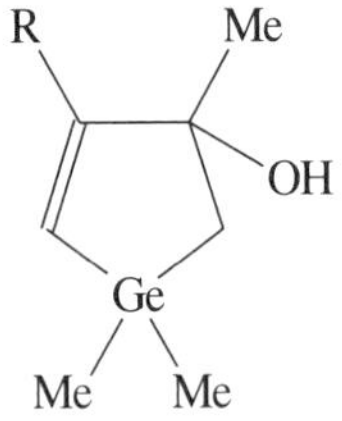   (53) 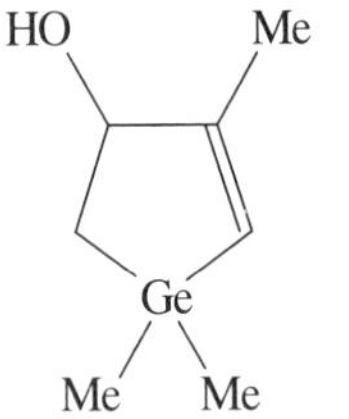   (54)

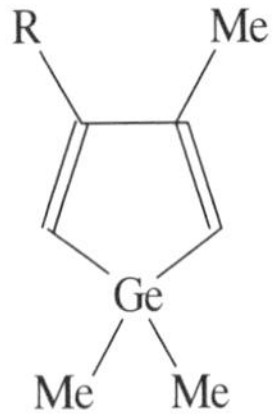

(55) 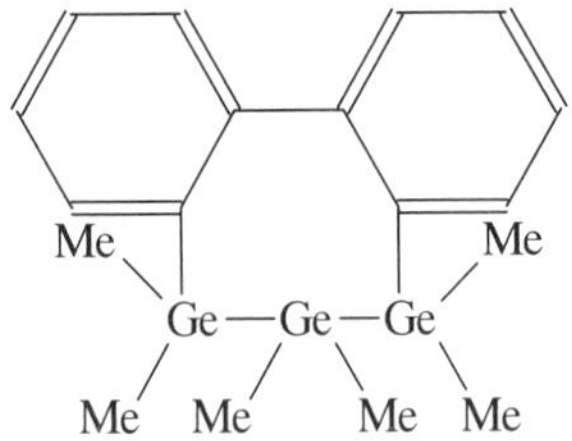   (56) 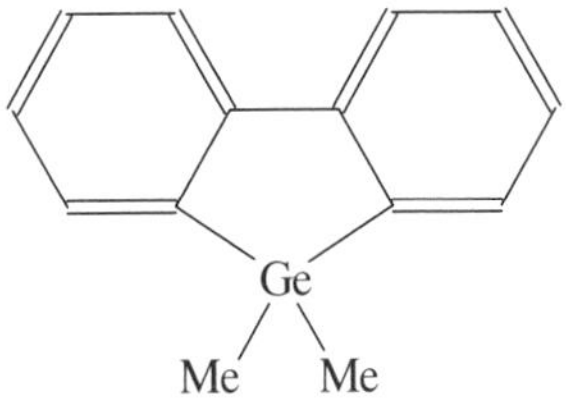   (57)

The photochemistry of unsaturated cycloheptatrigermanium compounds has also been investigated[83]. Photolysis of **56** in the presence of 2,3-dimethylbutadiene yielded the germafluorene **57** quantitatively, along with the expected germacyclopentene. At liquid nitrogen temperature the cyclic tetragermane **58** (R = Me; cf Scheme 18) was also formed. In general photolysis of polygermanes, cyclogermanes and polygermylmercury compounds leads to the formation of polymetallated chains, containing one or two germanium-centred radicals[84]. These radicals may then undergo a variety of processes to produce germylenes, germanium-centred radicals, α-digermyl radicals, or β- or γ-polygermyl diradical species. All of these intermediates have been inferred by trapping experiments with dimethylbutadiene, dimethyldisulphide or biacetyl. For instance, photolysis of acyclic tetragermanes such as $(PhX_2Ge)_3GePh$ (X = Cl or Me) yields the appropriate germylene compounds $Ph(PhX_2Ge)Ge$: along with the digermane $PhX_2Ge-GeX_2Ph$ formed via a radical coupling mechanism[85]. The methyl derivative was also prepared by UV photolysis of the polymeric mercury derivative (reaction 39); the germylene thus produced can be trapped by the addition of 2,3-dimethyl-1,3-butadiene[86].

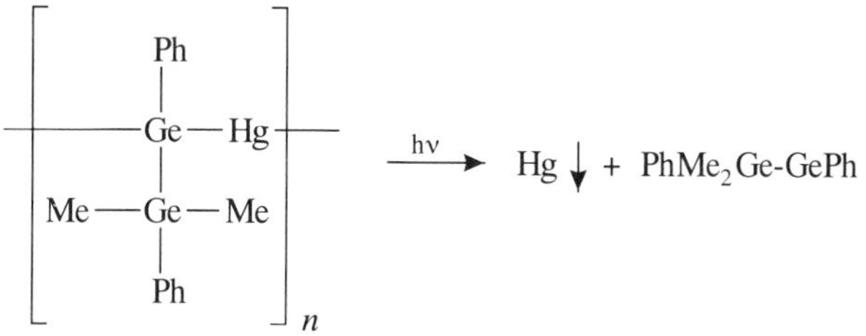

$$\left[\begin{array}{c} \overset{\displaystyle Ph}{\underset{\displaystyle Me-Ge-Me}{\overset{|}{\underset{|}{Ge}}}-Hg} \\ \overset{|}{Ph} \end{array}\right]_n \xrightarrow{\;h\nu\;} Hg\downarrow + PhMe_2Ge\text{-}GePh \qquad (39)$$

Heterocyclic derivatives such as **59** provide a convenient route to the synthesis of many cyclic germanium compounds[87]. For instance, photolysis of **59** in the presence of alkynes **60** or **61**, or allenes **62**, produces the unsaturated heterocyclic compounds **63** or **64**, respectively.

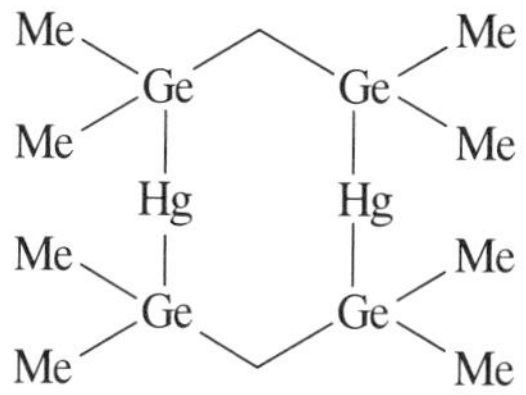

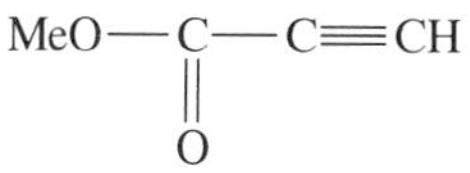

$$MeO-\underset{\underset{\displaystyle O}{\|}}{C}-C\equiv CH \qquad Me_2NCH_2C\equiv CH$$

(59)   **(60)**   **(61)**

$$Me_2C=C=CH_2$$

(62)   **(63)** 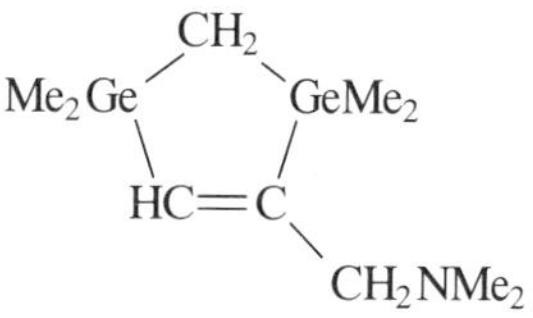   **(64)**

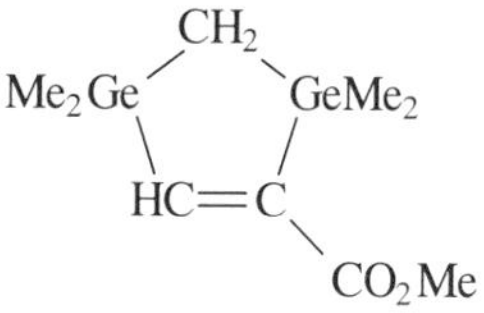

X. HETEROMETALLIC SYSTEMS

The majority of such systems involves the coordination of a Group 14 organometallic to a transition metal, in most cases a carbonyl complex. This section falls into two categories — the formation of such complexes by photochemical means, and their photochemistry once formed.

A. Formation of Transition Metal–Group 14 Organometallic Compounds

The photochemical formation of these complexes generally occurs from initial loss of CO or some similarly photolabile substituent from the transition metal centre. A common mode of attack of the Group 14 organometallic on the unsaturated species thus formed is by oxidative addition. There are many examples of such reactions, the most common involve E–H cleavage[88]; equations 40 and 41 show typical reactions.

$$Cp^*M(CO)_2 + R_3SnH \longrightarrow Cp^*M(CO)H(SnR_3) \qquad (40)$$

$$Cp^*M(C_2H_4)_2 + 2R_3SnH \longrightarrow Cp^*M(H)_2(SnR_3)_2 \qquad (41)$$

$$(M = Ir, \; Rh; \; R = n\text{-Bu}, \; Me)$$

A number of mechanistic studies of the reaction of stannanes with metal carbonyls have been carried out[89]. Continuous photolysis of a mixture of $Cp_2Fe_2(CO)_4$ and $HSnBu_3$ results in the formation of $CpFe(CO)_2H$, $CpFe(CO)_2SnBu_3$ and $CpFe(CO)(H)(SnBu_3)_2$. The relative yields of these products vary considerably, depending on the relative concentrations of starting material and type of protective gas atmosphere used for the reaction. For example, it was observed that increasing the concentration of CO results in inhibition of the overall reaction. A flash photolysis study of the reaction was also carried out, and in combination with the results obtained from continuous photolysis suggested that the reaction mechanism was as shown in Scheme 22.

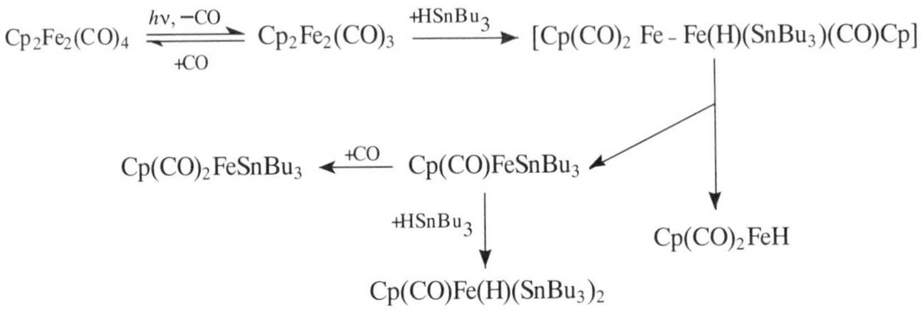

$$Cp_2Fe_2(CO)_4 \; \underset{+CO}{\overset{h\nu,\,-CO}{\rightleftharpoons}} \; Cp_2Fe_2(CO)_3 \; \xrightarrow{+HSnBu_3} \; [Cp(CO)_2\,Fe\text{-}Fe(H)(SnBu_3)(CO)Cp]$$

$$Cp(CO)_2FeSnBu_3 \; \xleftarrow{+CO} \; Cp(CO)FeSnBu_3$$

$$Cp(CO)_2FeH$$

$$Cp(CO)Fe(H)(SnBu_3)_2$$

SCHEME 22

Analogous conclusions were reached for the reaction of $Mn_2(CO)_{10}$ with $HSnBu_3$[89]. Oxidative addition of germanes has been carried out in a similar manner[90]. In some cases a state of 'arrested oxidation' can be achieved, with a $M-E-H$ 3-centre, 2-electron interaction (analogous to the $M-C-H$ agostic bond) being formed (reaction 42)[91].

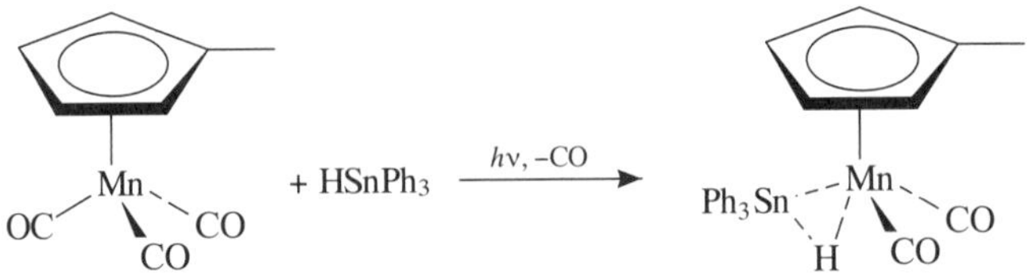

$$\text{(structure with Mn, OC, CO, CO)} \; + \; HSnPh_3 \; \xrightarrow{h\nu,\,-CO} \; \text{(structure with Ph}_3Sn, Mn, H, CO, CO)} \qquad (42)$$

A number of related stannane compounds have now been characterized, although to date no plumbanes. In addition, there is no definite evidence for the formation of $M-Ge-H$ interactions, although Carre and coworkers[90] have suggested that one may exist in $MeCpMn(CO)_2(H)GePh_3$. For a full coverage of this subject see the review by Schubert[92].

B. Reaction of Transition Metal–Group 14 Organometallic Compounds

Barrau and coworkers have synthesized a series of iron and ruthenium complexes by irradiation of $Me_2HGe(CH)_nGeMe_2H$ and $Me_2HGe(CH)_nSiMe_2H$ ($n = 1, 2$) in the presence of $Fe(CO)_5$ and $Ru_3(CO)_{12}$[93]. In each case irradiation causes CO loss, with the formation of the $M(CO)_4$ species (reaction 43). When $n = 2$ the products are photostable; with $n = 1$ (**65**) a mixture of products (**66–69**) are obtained due to secondary photolysis (reaction 44). The mechanism, outlined in Scheme 23, is presented to explain these observations.

$$\text{Me}_2\text{HGe(CH}_2)_n\text{GeHMe}_2 + \text{Fe(CO)}_5 \xrightarrow[\text{pentane}]{h\nu\,/\,-\text{CO}} \begin{array}{c} \text{Me}_2\text{Ge(CH}_2)_n\text{-GeMe}_2 \\ \diagdown\diagup \\ \text{Fe(CO)}_4 \end{array} \quad (43)$$

62%

$$\begin{array}{c} \text{Me}_2\text{Ge}\diagup\diagdown\text{GeMe}_2 \\ \diagdown\diagup \\ \text{Fe(CO)}_4 \end{array} \xrightarrow[\text{C}_6\text{H}_6]{h\nu\,/\,20\,\text{h}} \begin{array}{c} \text{Me}_2\text{Ge}\diagup\diagdown\text{GeMe}_2 \\ | \qquad | \\ \text{H} \qquad \text{H} \end{array} +$$

(65)   (66)

(CO)$_4$ (44)

$$\begin{array}{c} \text{Me}_2\text{Ge}\diagup\diagdown\text{GeMe}_2 \\ | \qquad\qquad | \\ \text{Me}_2\text{Ge}\diagdown\diagup\text{GeMe}_2 \end{array} + \begin{array}{c} \text{Me}_2\text{Ge}\diagup\diagdown\text{GeMe}_2 \\ \diagdown\diagup \\ \text{Fe(CO)}_4 \end{array} + \begin{array}{c} \text{Fe}\diagdown\text{GeMe}_2 \\ \text{Me}_2\text{Ge} \\ \text{Fe}\diagup\text{GeMe}_2 \\ \text{(CO)}_4 \end{array}$$

(67)   (68)   (69)

There is evidence that free carbene (:CH$_2$) is formed in these reactions; photolysis in the presence of excess Me$_3$GeH results in the formation of Me$_4$Ge. When 65 and 68 (in Scheme 23) are photolysed in the presence of PPh$_3$, Fe(CO)$_4$(PPh$_3$) and Fe(CO)$_3$(PPh$_3$)$_2$ are formed. This indicates that Fe–Ge cleavage is occurring as well as the expected CO loss.

Oxidative addition across Sn–C bonds has also been observed[94], as demonstrated in reaction 45. This reaction also occurs thermally, but is accelerated by photolysis. A related series of reactions has been observed by Pannell and Kapoor[95], looking at the photochemical decomposition of CpM(CO)$_n$PbR$_3$ ($n = 2$, M = Fe; $n = 3$, M = Cr, Mo, W; R = Me, Et, Ph). In the case of the phenyl analogues the products are metallic lead, PbPh$_4$, and CpM(CO)$_n$Ph (reaction 46).

$$\begin{array}{c} \text{Ph}_2\text{P}\diagup\diagdown\text{SnPh}_3 \\ | \\ \text{Fe(CO)}_4 \end{array} \xrightarrow{h\nu\,/\,-\text{CO}} \begin{array}{c} \text{OC} \quad \overset{\text{Ph}}{\underset{|}{\,}} \quad \text{CO} \\ \diagdown \; \diagup \\ \text{Fe} \\ \text{OC}\diagup \; | \;\diagdown\text{SnPh}_2 \\ \text{Ph}_2\text{P}\diagdown\diagup \end{array} \quad (45)$$

$$2\text{CpFe(CO)}_2\text{PbPh}_3 \longrightarrow 2\text{CpFe(CO)}_2\text{Ph} + \text{Pb} + \text{PbPh}_4 \quad (46)$$

When M = Fe, the methyl analogue undergoes an identical reaction[95]. When M = W, the reaction is similar, but no PbMe$_4$ was isolated. When the ethyl-substituted compounds are photolysed, a rather different reaction occurs (reaction 47):

$$2\text{CpMo(CO)}_3\text{PbEt}_3 \longrightarrow [\text{CpMo(CO)}_3]_2\text{PbEt}_2 + \text{PbEt}_4 \quad (47)$$

No mechanisms are suggested for these reactions.

Pannell and Sharma have studied the photochemistry of digermyl and isomeric silylgermyl and germylsilyl complexes of the [CpFe(CO)$_2$] system[96]. The photolysis of CpFe(CO)$_2$GeMe$_2$GePh$_3$ gave CpFe(CO)$_2$GeMe$_2$Ph (10%), CpFe(CO)$_2$GeMePh$_2$ (82%) and CpFe(CO)$_2$GePh$_3$ (8%). This product distribution is almost identical to that obtained with the equivalent Si analogue[97], suggesting an identical reaction mechanism.

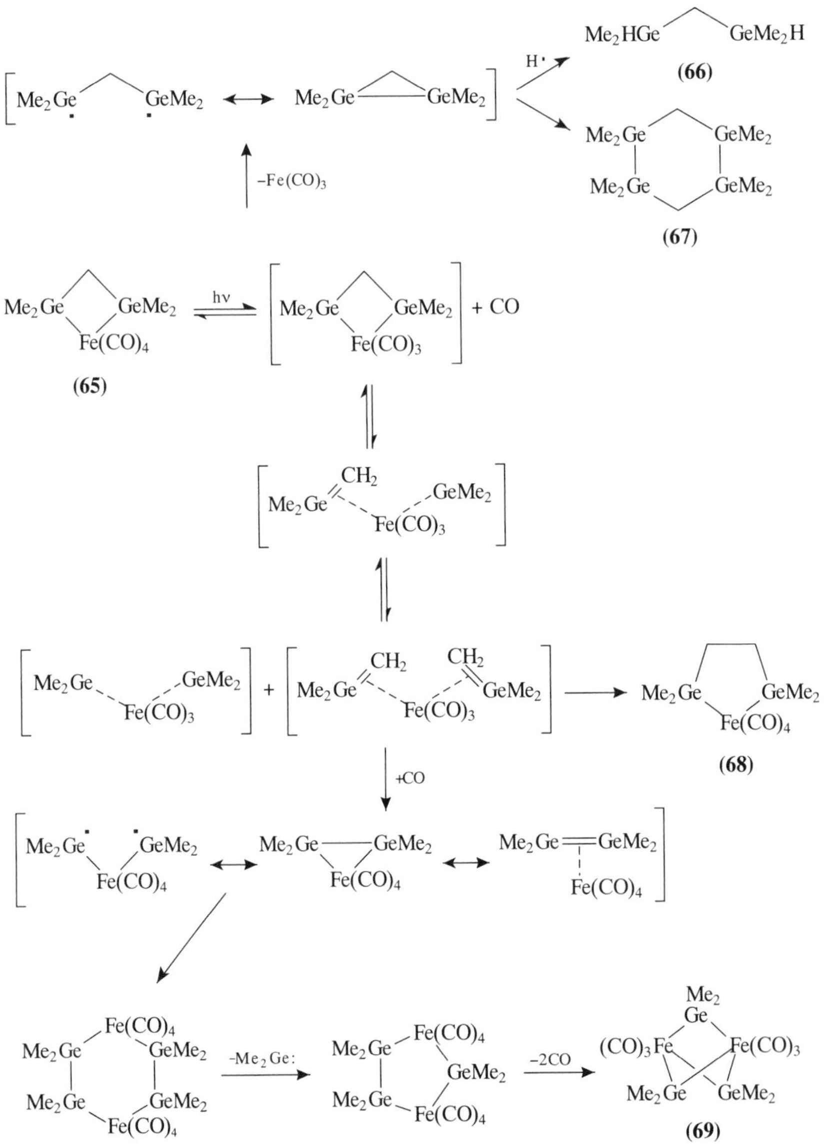

SCHEME 23

When the isomeric compounds $CpFe(CO)_2SiMe_2GeMe_3$ and $CpFe(CO)_2GeMe_2SiMe_3$ are photolysed, identical results are obtained, namely $CpFe(CO)_2SiMe_3$ as the principal (> 95%) product, along with a small amount of $CpFe(CO)_2GeMe_3$ (< 5%). Photolysis of $CpFe(CO)_2SiMe_2GePh_3$ gave only Si-containing $CpFe(CO)_2$ products (reaction 48).

$$CpFe(CO)_2SiMe_2GePh_3 \longrightarrow CpFe(CO)_2SiMe_2Ph + CpFe(CO)_2SiMePh_2$$

$$+ CpFe(CO)_2SiPh_3 + Ge\ polymer \tag{48}$$

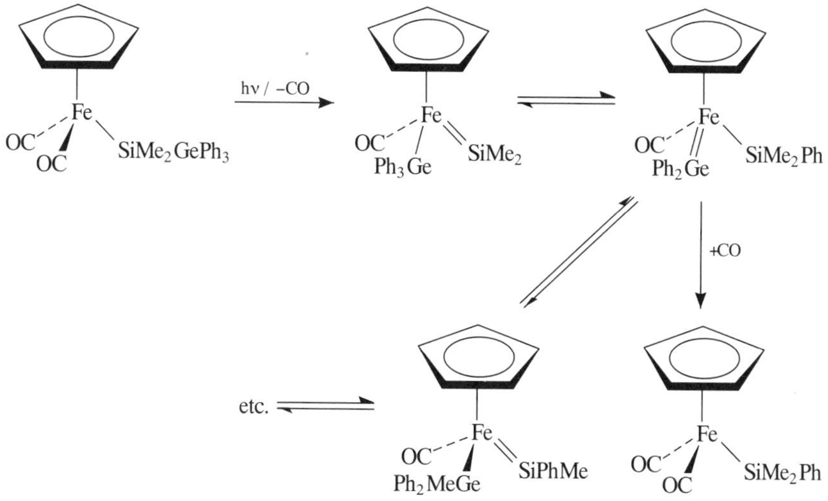

SCHEME 24

The mechanism of this reaction is suggested to involve silylene and germylene intermediates, as shown in Scheme 24.

The isomeric species $CpFe(CO)_2GeMe_2SiPh_3$ gives the same products, but also 10–15% of the germane complexes $CpFe(CO)_2GeR_3$ (R = Me_nPh_{3-n}). The authors suggest that this arises from a photochemical Fe—Ge cleavage reaction competing with the Fe—CO cleavage reaction, which is the primary photochemical step in the other molecules. Note that a similar effect was observed in the photolysis of **65** and **68** mentioned above[93]. In all of this work there was no observation of free germylene or silylene fragments.

Job and Curtis investigated the photochemistry of σ-bonded $(CH_3)_2GeCH=CH_2$ derivatives of metal carbonyls[98], in an attempt to produce species with π-allyl coordination. Three different metal complexes were examined, giving different types of product in each case, as shown in reactions 49–51. No π-allyl products were formed, however.

$$(CH_3)_2C_2H_3GeMo(CO)_3Cp \longrightarrow Cp(CO)_2Mo\equiv Mo(CO)_2Cp \quad (49)$$

$$(CH_3)_2C_2H_3GeMn(CO)_5 \longrightarrow Mn_2(CO)_{10} \quad (50)$$

$$(CH_3)_2C_2H_3GeFe(CO)_2Cp \longrightarrow Cp(CO)Fe(\eta\text{-}CO)(\eta\text{-}GeMe_2)Fe(CO)Cp \quad (51)$$

The ionic complex $Me_5C_5Ge^+[Cl_3-W(CO)_5]^-$ is formed on photolysis of $W(CO)_6$ with two different Ge complexes (reactions 52 and 53)[99]:

$$Me_5C_5Ge^+GeCl^- + W(CO)_6 \longrightarrow Me_5C_5Ge^+[Cl_3Ge-W(CO)_5]^- \quad (52)$$

$$Me_5C_5GeCl + W(CO)_6 \longrightarrow Me_5C_5Ge^+[Cl_3Ge-W(CO)_5]^- \quad (53)$$

Germylenes, stabilized by coordination to a transition metal centre, have also been produced by irradiation of $M(CO)_6$ in the presence of the appropriate free carbene homologue[100], as shown in reaction 54.

$$Ge[CH(SiMe_3)_2]_2 + Cr(CO)_6 \longrightarrow Cr(CO)_5Ge[CH(SiMe_3)_2]_2 \quad (54)$$

The photochemistry of compounds of the type $R_3E-M(CO)_3L$ (R = Me, Ph; M = Mn, Re; L = 1,10-phenanthroline, 2,2'-bipyridyl, 2,2'-biquinoline) has been studied by

Wrighton and coworkers[101]. The basic photochemistry is thought to involve the formation of an excited state charge transfer complex $[R_3E^+-Re(CO)_3L^-]^*$. This in turn cleaves to form $R_3E\cdot$ and $[Re(CO)_3L]\cdot$. In the presence of suitable quenching agents, the excited state can be quenched.

XI. MERCURIC SYSTEMS

A subgroup of heterometallic Group 14 systems worthy of special note is those containing an E−Hg linkage. Complexes basically fall into two different categories, the first being simple $(R_3E)_2Hg$ systems, and the second $(R_3E)HgX$, where X can be either a halogen or a more complicated substituent.

A. (R₃E)₂Hg Compounds

Photolysis of compounds of the type $(R_3E)_2Hg$ results in the formation of Hg metal and $R_3E\cdot$ radicals. The reaction has been reported for $[(CF_3)_3Ge]_2Hg^{102}$ and $(Me_3Ge)_2Hg^{103,104}$. When the reaction is carried out in the presence of aromatic species, such as benzene, toluene and anisole, two competing reaction paths are observed[103]. The $Me_3Ge\cdot$ radicals can dimerize to form $(Me_3Ge)_2$, or else aromatic substitution may occur, forming (in the case of the reaction with benzene) phenyltrimethylgermane (**70**) and 2,5-cyclohexadienyltrimethylgermane (**71**). A reaction scheme is suggested (Scheme 25). In this system, dimerization occurs in much greater yield ($> 95\%$) than the aromatic substitution, and the yield of substitution products decreases still further with increasing temperature. In contrast, with the equivalent silanes there is a much higher yield of substitution products. This is explained in terms of the greater strength of the Si−C bond as compared with the Ge−C bond[104]. With mono-substituted arene rings, the position of substitution is in an approximate 2:2:1 ratio for o-: m-: p-.

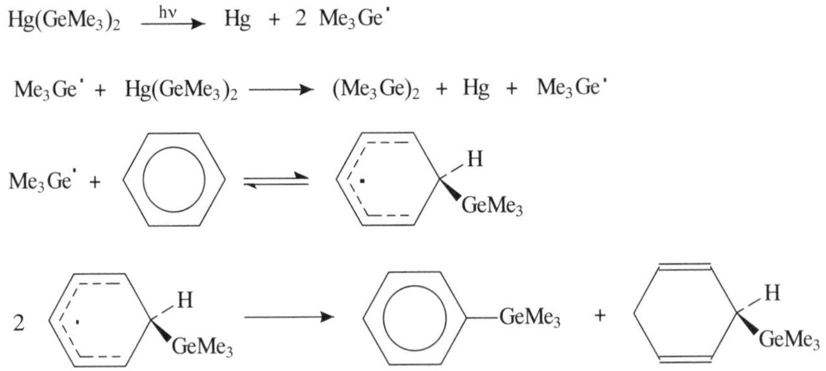

SCHEME 25

Exposure of $(Me_3Ge)_2Hg$ to visible light in the presence of alkyl halides results in the rapid formation of Me_3GeHal and Me_3GeHgR^{104}. This reaction is thought to occur via a radical chain mechanism (Scheme 26). When the reaction is instead carried out in the presence of aryl halides, the products are R_3GeHal and Ar−Hg−Ar; this reaction has a quantum yield > 1. The mechanism is thought to be similar to that for the alkyl systems, except that $R_3Ge-Hg-Ar$ formed initially goes on to react further (Scheme 27).

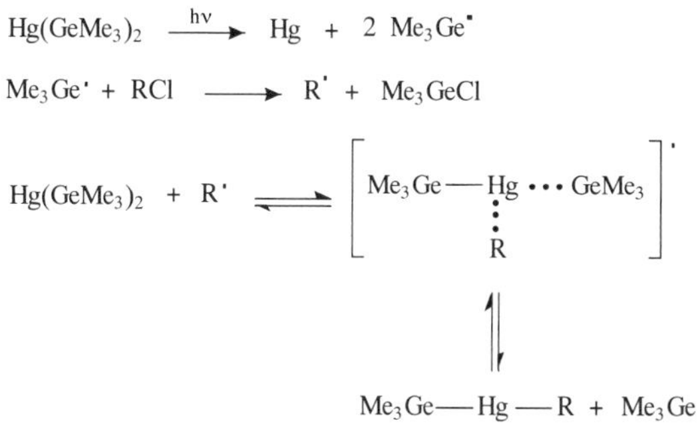

$$Hg(GeMe_3)_2 \xrightarrow{h\nu} Hg + 2\ Me_3Ge^{\bullet}$$

$$Me_3Ge^{\bullet} + RCl \longrightarrow R^{\bullet} + Me_3GeCl$$

$$Hg(GeMe_3)_2 + R^{\bullet} \rightleftharpoons \left[\begin{array}{c} Me_3Ge\text{---}Hg \cdots GeMe_3 \\ \vdots \\ R \end{array} \right]^{\bullet}$$

$$Me_3Ge\text{---}Hg\text{---}R + Me_3Ge$$

SCHEME 26

$$Hg(GeMe_3)_2 \xrightarrow{h\nu} Hg + 2\ Me_3Ge^{\bullet}$$

$$Me_3Ge^{\bullet} + ArBr \longrightarrow Ar^{\bullet} + Me_3GeBr$$

$$Ar^{\bullet} + (Me_3Ge)_2Hg \rightleftharpoons \left[\begin{array}{c} Me_3Ge\text{---}Hg \cdots GeMe_3 \\ \vdots \\ Ar \end{array} \right]^{\bullet} \xrightarrow{-Me_3Ge^{\bullet}} Me_3Ge\text{---}Hg\text{---}Ar$$

$$Me_3Ge\text{---}Hg\text{---}Ar + Ar^{\bullet} \rightleftharpoons \left[\begin{array}{c} Me_3Ge \cdots Hg \cdots Ar \\ \vdots \\ Ar \end{array} \right]^{\bullet} \xrightarrow{-Me_3Ge} Ar\text{---}Hg\text{---}Ar$$

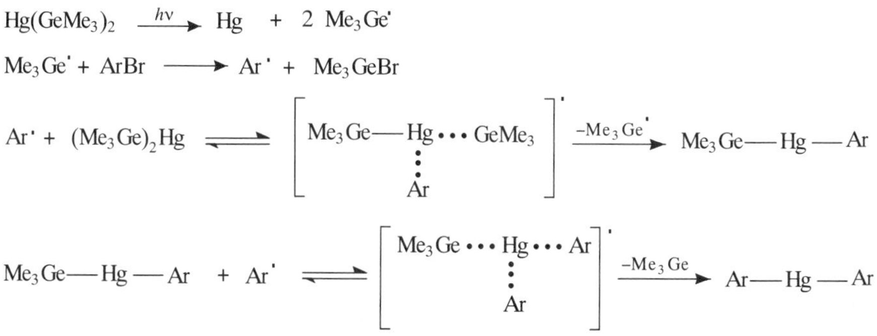

SCHEME 27

Irradiation of $[(C_6F_5)_3Ge]_2Hg$ in toluene gives Hg, $(C_6F_5)_6Ge_2$ and $(C_6F_5)_3GeH$ as the main products[105]. In addition, however, dibenzyl and $(C_6F_5)_3GeCH_2C_6H_5$ are also formed.

Bochkarev and coworkers studied the photochemistry of $(C_6F_5)_3GeHgPt(PPh_3)_2Ge$-$(C_6F_5)_3$[106]. As with many of these systems, Hg metal was evolved in almost 100% yield, along with a mixture of products. Of these, $(C_6F_5)_3GeH$ (50% yield), $(C_6F_5)_6Ge_2$ (3% yield) and a polymer of the form $[(C_6F_5)_3GePt(PPh_3)_2]_n$ were identified.

B. Asymmetric Systems

The photochemical decomposition of $Mes_3GeHgCl$ was studied by Castel and coworkers[107]. The main photoreaction is the ejection of Hg, and the formation of Mes_3GeCl, presumably via an intramolecular mechanism. A number of other reactions are observed, however, all occurring via formation of $Mes_3Ge\cdot$. The photochemistry is summarized in Scheme 28, with the reactions being followed by E.S.R. and N.M.R. spectroscopies.

On irradiation of $RHgSnR_3'$, the mixture of products formed depended on the nature of R and R'[108]. For example, photolysis of t-$BuHgSnMe_3$ gave Hg, t-Bu· and Me_6Sn_2, while with other substituents R_2Hg, $R_6'Sn_2$, Hg and $R_3'SnR$ were formed. The relative yields

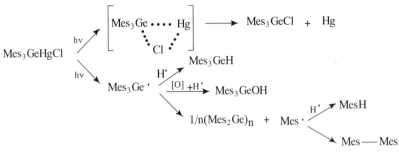

SCHEME 28

of products were not consistent from experiment to experiment, however. It is suggested that the primary photochemical step is Hg—C cleavage.

XII. REFERENCES

1. H. Sakurai, in *Free Radicals* (Ed. J. F. Kochi), Vol. II, Wiley, New York, 1973.
2. H. G. Kuivilla, *Acc. Chem. Res.*, **1**, 299 (1968).
3. J. Cooper, A. Hudson and R. A. Jackson, *J. Chem. Soc., Perkin Trans.*, 2, 1056 (1973).
4. A. G. Davies and N. -W. Tse, *J. Chem. Soc., Chem. Commun.*, 353 (1978).
5. P. J. Barker and A. G. Davies, *J. Chem. Soc., Chem. Commun.*, 815 (1979).
6. G. A. Razuvaev, N. S. Vyazankin and O. A. Shchepetkova, *Tetrahedron*, **18**, 667 (1962).
7. A. G. Davies, in *Comprehensive Organometallic Chemistry* (Eds. G. Wilkinson, F. G. A. Stone and E. D. Abel), Pergamon Press, Oxford, 1982.
8. K. Mochido, M. Wakasa, Y. Sakaguchi and H. Hayashi, *Chem. Lett.*, **10**, 1793 (1986).
9. C. Obayashi, H. Sato and T. Tominaga, *J. Radioanal. Nucl. Chem.*, **164**, 365 (1992).
10. J. D. Cotton, C. S. Cundy, D. H. Harris, A. Hudson, M. Lappert and P. Lednor, *J. Chem. Soc., Chem. Commun*, 651 (1974).
11. H. U. Buschhaus, M. Lehnig and W. P. Neumann, *J. Chem. Soc., Chem. Commun.*, 129 (1977).
12. H. J. R de Boer, O. S. Akkerman and F. Bickelhaupt, *Angew. Chem., Int. Ed. Engl.*, **27**, 687 (1988).
13. A. G. Brook, F. Abdesaken and H. Soellradl, *J. Organomet. Chem.*, **299**, 9 (1986).
14. K. U. Ingold, J. Lusztyk and J. C. Scaiano, *J. Am. Chem. Soc.*, **106**, 343 (1984).
15. H. Sakurai, *J. Organomet. Chem. Libr.*, **12**, 267 (1981).
16. J. Cooper, A. Hudson and R. A. Jackson, *J. Chem. Soc., Perkin Trans.*, 1933 (1973).
17. E. A. Mendoza and H. D. Gafney, *Inorg. Chem.*, **29**, 4853 (1990).
18. D. D. Davis and F. U. Ahmed, *J. Am. Chem. Soc.*, **103**, 7653 (1981).
19. T. Sato and K. Takezoe, *Tetrahedron Lett.*, **32**, 4003 (1991).
20. K. M. Kadish, Q. Y. Xu, J. -M. Barbe, J. E. Anderson, E. Wang and R. Guillard, *J. Am. Chem. Soc.*, **109**, 7705 (1987).
21. G. B. Maiya, J. -M. Barbe and K. M. Kadish, *Inorg. Chem.*, **28**, 2524 (1989).
22. K. B. Clark and D. Griller, *Organometallics*, **10**, 746 (1991).
23. T. N. Mitchell and H-J. Belt, *J. Organomet. Chem.*, **386**, 167 (1990).
24. M. H. Fisch, J. J. Dannenberg, M. Pereyre, W. G. Anderson, J. Rens and W. E. L. Grossman, *Tetrahedron*, **40**, 293 (1984).
25. K. Mizuno, K. Nakanishi, M. Yasueda, H. Miyata and Y. Otsuji, *Chem. Lett.*, **11**, 2001 (1991).
26. P. Vanderlinden and S. Boue, *J. Organomet. Chem.*, **87**, 183 (1975).
27. T. N. Mitchell, A. Amamria, H. Killing and D. Rutschow, *J. Organomet. Chem.*, **304**, 257 (1986).
28. J. M. Kelly and R. J. Trautman, *J. Chem. Soc., Dalton Trans.*, 909 (1984).
29. B. Wrackmeyer and S. T. Abu-Orabi, *Chem. Ber.*, **120**, 1603 (1987).
30. M. P. Egorov, A. S. Dvornikov, S. P. Kolesnikov, V. A. Kuzmin and O. M. Nefedov, *Izv. Akad. Nauk SSSR, Ser. Khim.*, 1200 (1987); S. P. Kolesnikov, M. P. Egorov, A. S. Dvornikov, V. A. Kuzmin and O. M. Nefedov, *Metalloorgan. Khim.*, **2**, 2735 (1989); S. P. Kolesnikov,

M. P. Egorov, A. M. Galminas, M. B. Ezhova, O. M. Nefedov, T. V. Leshina, M. B. Taraban, A. I. Kruppa and V. I. Maryasova, *J. Organomet. Chem.*, **391**, C1 (1990).

31. C. J. Cardin, D. J. Cardin, J. M. Kelly, D. J. H. L. Kirwan, R. J. Norton and A. Roy, *Proc. Royal Irish Acad.*, 365 (1977).
32. T. N. Mitchell and H. -J. Belt, *J. Organomet. Chem.*, **345**, C28 (1988).
33. G. A. Russell, P. Ngoviwatchai, H. I. Tashtoush and J. Hershberger, *Organometallics*, **6**, 1414 (1987).
34. G. A. Russell, P. Ngoviwatchai and H. I. Tashtoush, *Organometallics*, **7**, 696 (1988).
35. K. Mochida, H. Kikkawa and Y. Nakadaira, *Bull. Chem. Soc. Jpn.*, **64**, 2772 (1991).
36. R. L. Lambert Jr. and J. Seyferth, *J. Am. Chem. Soc.*, **94**, 9246 (1972); D. Seyferth, C. K. Haas and D. C. Annarelli, *J. Organomet. Chem.*, **56**, C7 (1973); D. Seyferth, D. P. Duncan and S. C. Vicks, *J. Organomet. Chem.*, **125**, C5 (1977).
37. W. Ando and T. Tsumuraya, *Organometallics*, **7**, 1882 (1988).
38. H. Sakurai, Y. Kamiyama and Y. Nakadaira, *J. Am. Chem. Soc.*, **98**, 7424 (1976); M. Ishikawa, T. Fuchigami and M. Kumada, *J. Organomet. Chem.*, **117**, C58 (1976).
39. M. Kira, H. Sakurai and H. Yoshida, *J. Am. Chem. Soc.*, **107**, 7767 (1985).
40. G. Guillerm, A. L'Honore, L. Veniard, G. Pourcelot and J. Benaim, *Bull. Soc. Chim. France*, **9–10**, 2739 (1973).
41. K. Mochida, T. Yamauchi and H. Sakurai, *Bull. Chem. Soc. Jpn.*, **62**, 1982 (1989).
42. B. A. King and F. B. Bramwell, *J. Inorg. Nucl. Chem.*, **43**, 1479 (1981).
43. K. Mochida, M. Wakasa, Y. Sakaguchi and H. Hayashi, *J. Am. Chem. Soc.*, **109**, 7942 (1987).
44. H. Hayashi and K. Mochida, *Chem. Phys. Lett.*, **101**, 307 (1983); C. Chatgilialoglu, K. U. Ingold, J. Lusztyk, A. S. Narzen and J. C. Scaiano, *Organometallics*, **2**, 1332 (1983).
45. H. Sakurai, K. Mochida and M. Kira, *J. Am. Chem. Soc.*, **97**, 929 (1975).
46. C. C. Yammal, J. C. Podesta and R. A. Rossi, *J. Org. Chem.*, **57**, 5720 (1992).
47. S. Kiyooka, M. Hamada, H. Matsue and R. Fujiyama, *Chem. Lett.*, 1385 (1989).
48. S. Kiyooka, Y. Kaneko, H. Matsue, M. Hamada and R. Fujiyama, *J. Org. Chem.*, **55**, 5562 (1990).
49. M. Wakasa, Y. Sakaguchi and H. Hayashi, *Chem. Phys. Lett.*, **176**, 541 (1991).
50. S. Konieczny, S. J. Jacobs, J. K. Braddock Wilking and P. P. Gaspar, *J. Organomet. Chem.*, **341**, C17 (1988).
51. W. Ando, T. Tsumuraya and A. Sekiguchi, *Chem. Lett.*, 317 (1987).
52. P. P. Gaspar, D. Holten, S. Konieczny and J. Y. Corey, *Acc. Chem. Res.*, **20**, 329 (1987).
53. J. Meinwald, S. Knapp, T. Tatsuoka, J. Finer and J. Clardy, *Tetrahedron Lett.*, **26**, 2247 (1977).
54. P. J. Barker, A. G. Davies and M. W. Tse, *J. Chem. Soc., Perkin Trans. 2*, 941 (1980).
55. P. J. Barker, A. G. Davies, J. A. -A. Hawari and M. W. Tse, *J. Chem. Soc., Perkin Trans. 2*, 1488 (1980).
56. A. G. Davies, J. A. -A. Hawari, C. Gaffney and P. G. Harrison, *J. Chem. Soc., Perkin Trans. 2*, 631 (1982).
57. P. J. Barker, A. G. Davies, R. Henriques and J. -Y. Nedelec, *J. Chem. Soc., Perkin Trans. 2*, 745 (1982).
58. A. G. Davies, J. P. Goddard, M. B. Hursthouse and N. P. C. Walker, *J. Chem. Soc., Dalton Trans.*, 1873 (1986).
59. A. Hassner and J. A. Soderquist, *Tetrahedron Lett.*, **21**, 429 (1980).
60. T. J. Barton and S. K. Hoekman, *J. Am. Chem. Soc.*, **102**, 1584 (1980).
61. A. Baceiredo, G. Bertrand and P. Mazerolles, *Tetrahedron Lett.*, **22**, 2553 (1981).
62. M. J. S. Gynane, D. H. Harris, M. F. Lappert, P. P. Power, P. Rivière and M. Rivière-Baudet, *J. Chem. Soc., Dalton Trans.*, 2005 (1977).
63. C. D. Schaeffer and J. J. Zuckerman, *J. Am. Chem. Soc.*, **96**, 7160 (1974).
64. H. Suzuki, K. Okabe, R. Kato, N. Sato, Y. Fukuda and H. Watanabe, *J. Chem. Soc., Chem. Commun.*, 1298 (1991).
65. J. Barrau, M. El Amine, G. Rima and J. Satgé, *J. Organomet. Chem.*, **277**, 323 (1984).
66. J. Barrau, G. Rima, M. El Amine and J. Satgé, *J. Organomet. Chem.*, **345**, 39 (1988).
67. K. Mochida, H. Kikkawa and Y. Nakadaira, *J. Organomet. Chem.*, **412**, 9 (1991).
68. P. Boudjouk, J. R. Roberts, C. M. Gollino and L. H. Sommer, *J. Am. Chem. Soc.*, **94**, 7926 (1972).
69. K. Mochida, M. Wakasa, Y. Nakadaira, Y. Sakaguchi and H. Hayashi, *Organometallics*, **7**, 1869 (1988).

70. K. Mochida, M. Wakasa, Y. Sakaguchi and H. Hayashi, *Bull. Chem. Soc. Jpn.*, **64**, 1889 (1991).
71. K. Mochida, H. Kikkawa and Y. Nakadaira, *Chem. Lett.*, **7**, 1089 (1988).
72. K. Mochido and H. Chiba, *J. Organomet. Chem.*, **473**, 45 (1994).
73. S. Kozima, K. Kobayashi and M. Kawanisi, *Bull. Chem. Soc. Jpn.*, **49**, 2837 (1976).
74. M. Wakasa, I. Yoneda and K. Mochida, *J. Organomet. Chem.*, **366**, C1 (1989).
75. J. T. Snow, S. Murakami, S. Masamune and D. J. Williams, *Tetrahedron Lett*, **25**, 4191 (1984).
76. S. Collins, S. Murakami, J. T. Snow and S. Masamune, *Tetrahedron Lett.*, **26**, 1281 (1985).
77. S. Masamune, Y. Hanzawa and D. J. Williams, *J. Am. Chem. Soc.*, **104**, 6136 (1982).
78. W. Ando and T. Tsumuraya, *Organometallics*, **7**, 1882 (1988).
79. K. Tomioka, K. Yasuda, H. Kawasaki, and K. Koga, *Tetrahedron Lett.*, **27**, 3247 (1986).
80. T. Tsumuraya, S. Sato, and W. Ando, *Organometallics*, **7**, 2015 (1988).
81. S. Masamune and L. R. Sita, *J. Am. Chem. Soc.*, **107**, 6390 (1985).
82. A. Laporterie, G. Manuel, J. Dubac and P. Mazerolles, *Nouv. J. Chim.*, **6**, 67 (1982).
83. H. Sakurai, K. Sakamoto and M. Kira, *Chem. Lett.*, **8**, 1379 (1984).
84. P. Rivière, A. Castel, J. Satgé and D. Guyot, *J. Organomet. Chem.*, **264**, 193 (1984).
85. P. Rivière, J. Satgé and D. Soula, *J. Organomet. Chem.*, **63**, 167 (1973).
86. P. Rivière, A. Castel and J. Satgé, *J. Organomet. Chem.*, **212**, 351 (1981).
87. J. Barrau, N. B. Hamida, A. Agrebi and J. Satgé, *Organometallics*, **6**, 659 (1987).
88. J. Ruiz, C. M. Spencer, B. E. Mann, B. F. Taylor and P. M. Maitlis, *J. Organomet. Chem.*, **325**, 253 (1987).
89. R. J. Sullivan and T. L. Brown, *J. Am. Chem. Soc.*, **113**, 9155 (1991); S. Zhang and T. L. Brown, *Organometallics*, **11**, 2122 (1992).
90. F. Carre, E. Colomer, R. J. P. Corriu and A. Vioux, *Organometallics*, **3**, 1272 (1984).
91. U. Schubert, E. Kunz, B. Harkers, J. Willnecker and J. Meyer, *J. Am. Chem. Soc.*, **111**, 2572 (1989).
92. U. Schubert, *Adv. Organomet. Chem.*, **30**, 151 (1990).
93. J. Barrau, N. B. Hamida, A. Agrebi and J. Satgé, *Organometallics*, **8**, 1585 (1989); J. Barrau, N. B. Hamida and J. Satgé, *J. Organomet. Chem.*, **387**, 65 (1990); J. Barrau, N. B. Hamida and J. Satgé, *J. Organomet. Chem.*, **395**, 27 (1990).
94. U. Schubert, S. Grubert, U. Schultz and S. Mock, *Organometallics*, **11**, 3163 (1992).
95. K. H. Pannell and R. N. Kapoor, *J. Organomet. Chem.*, **214**, 47 (1981); K. H. Pannell and R. N. Kapoor, *J. Organomet. Chem.*, **269**, 59 (1984).
96. K. H. Pannell, S. Sharma, *Organometallics*, **10**, 1655 (1991).
97. K. H. Pannell, J. Cervantes, C. Hernandez, C. Cassias and S. P. Vincenti, *Organometallics*, **5**, 1056 (1986).
98. R. C. Job and M. D. Curtis, *Inorg. Chem.*, **12**, 2510 (1973).
99. P. Jutzi and B. Hampel, *J. Organomet. Chem.*, **301**, 283 (1986).
100. M. F. Lappert, S. J. Miles, P. P. Power, A. J. Carty and N. J. Taylor, *J. Chem. Soc., Chem. Commun.*, 458 (1977).
101. J. C. Luong, R. A. Faltynek and M. S. Wrighton, *J. Am. Chem. Soc.*, **102**, 7892 (1980).
102. M. N. Bochkarev, N. L. Ermolaev, G. A. Razuvaev, Y. K. Grishin and Y. A. Ustynyuk, *J. Organomet. Chem.*, **229**, C1 (1982).
103. S. W. Bennett, C. Eaborn, R. A. Jackson and R. Pearce, *J. Organomet. Chem.*, **28**, 59 (1971).
104. F. Werner, W. P. Neumann and H. P. Becker, *J. Organomet. Chem.*, **97**, 389 (1975).
105. V. V. Bashilov, V. I. Sokolov and O. A. Reutov, *Dokl. Akad. Nauk SSSR*, **228**, 603 (1976); V. V. Bashilov and O. A. Reutov, *J. Organomet. Chem.*, **111**, C13 (1976).
106. M. N. Bochkarev, G. A. Razuvaev, L. P. Maiorova, N. P. Makarenko, V. I. Sokolov, V. V. Bashilov and O. A. Reutov, *J. Organomet. Chem.*, **131**, 399 (1977).
107. A. Castel, P. Rivière, J. Satgé, Y. H. Ko and D. Desor, *J. Organomet. Chem.*, **397**, 7 (1990).
108. T. N. Mitchell, *J. Organomet. Chem.*, **71**, 27 (1974).

CHAPTER **15**

Syntheses and uses of isotopically labelled organic derivatives of Ge, Sn and Pb

KENNETH C. WESTAWAY and HELEN JOLY

Department of Chemistry, Laurentian University, Sudbury, Ontario P3E 2C6, Canada
Fax: 705-675-4844; e-mail: KWESTAWA@NICKEL.LAURENTIAN.CA and
HJOLY@NICKEL.LAURENTIAN.CA

I. SYNTHESIS AND USES OF ISOTOPICALLY LABELLED ORGANOMETALLIC DERIVATIVES OF GERMANIUM (Ge)

A. Isotopes of Germanium

There are five naturally occurring isotopes of germanium ranging in mass from 70 to 76. The percent natural abundance for these isotopes is presented in Table 1. ^{73}Ge and

The chemistry of organic germanium, tin and lead compounds
Edited by S. Patai © 1995 John Wiley & Sons Ltd

TABLE 1. Percent natural abundance
of germanium isotopes

Isotope	% Natural abundance
^{70}Ge	20.52
^{72}Ge	27.43
^{73}Ge	7.76
^{74}Ge	36.54
^{76}Ge	7.76

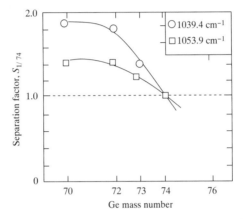

FIGURE 1. Separation factors for specified germanium isotopes with repect to ^{74}Ge (reproduced from
Reference 2)

^{76}Ge, which are found in equal quantities (*ca* 8%), are the least abundant of the naturally
occurring isotopes of germanium while the remaining three range in abundance from
21–37% with ^{74}Ge being the most abundant[1].

Molecular laser isotope separation (MLIS) was investigated as a means of separating
germanium isotopes[2]. The decomposition of germanium tetramethoxide, Ge(OCH$_3$)$_4$, a
relatively volatile compound, was induced by a TEA CO$_2$ laser. Determination of the
isotopic composition of the residual gas sample by mass spectrometry indicated that
this technique could selectively separate the isotopes of Ge. The separation factors for
the various naturally occurring Ge isotopes (with respect to ^{74}Ge) based on the specific isotopic composition before and after irradiation are shown graphically (Figure 1)
for two wavelengths, 1039.4 and 1053.9 cm^{-1}, respectively. The optimum wavelength
for separation of the isotopes was found to be 1039.4 cm^{-1}. In addition, the separation factors show that irradiating Ge(OCH$_3$)$_4$ resulted in a sample significantly enriched
in ^{70}Ge.

The ^{73}Ge nucleus has proven to be useful in studying organogermyl radicals by EPR
spectroscopy because it possesses a spin of 9/2. For instance, the Ge hyperfine coupling
constants derived from the EPR spectra of organogermyl radicals generated by treating
R$_3$GeH with *t*-butoxyl radical or by the photolysis of (R$_3$Ge)$_2$Hg, suggested that germyl
radicals are pyramidal. The s-character of the singly occupied orbital (SOMO) is related
to the isotropic hyperfine coupling constant[3].

B. The Synthesis of Labelled Organogermanium Compounds

1. Preparation of organogermanium compounds

The first reported preparation of an organogermanium compound dates back to 1887 when Winkler[4] synthesized tetraethylgermanium. There has, since that time, been numerous publications describing the preparation of organogermanium compounds. However, they remain primarily of academic interest since a practical application for germanium organyls has yet to be found.

Several methods, similar to those used to produce organosilicon compounds, are available for the preparation of germanium organyls. Organogermanium halides, which are the source of practically all other types of germanium organyls, can be formed by the direct reaction of aryl and alkyl halides with Ge/Cu alloys (equation 1).

$$\text{Ge/Cu alloy} + \text{RX} \longrightarrow \text{R}_x\text{GeX}_{4-x} \tag{1}$$

Alternatively, organogermanium monobromides can be prepared by the reaction of bromine or hydrogen bromide with tetra-alkyl or -arylgermanes in the presence of aluminium tribromide (equation 2).

$$
\text{R}_4\text{Ge} + \text{AlBr}_3 \quad
\begin{array}{c}
\xrightarrow{\text{HBr}} \quad \text{R}_3\text{GeBr} \\[2ex]
\xrightarrow[\text{Br}_2]{} \quad \text{R}_3\text{GeBr}
\end{array}
\tag{2}
$$

Other mono-halo derivatives are prepared indirectly from the corresponding bromide. For instance, $(\text{CH}_3\text{CH}_2)_3\text{GeX}$ (where $X = F$, Cl or I) is synthesized by treating the hydrolysis product of triethylbromogermane, bis(triethylgermanium) oxide, with HX (equation 3).

$$2\text{Et}_3\text{GeBr} \xrightarrow{\text{H}_2\text{O}} \text{Et}_3\text{GeOGeEt}_3 + 2\text{HBr} \xrightarrow{2\text{HX}} 2\text{Et}_3\text{GeX} + \text{H}_2\text{O} \tag{3}$$

Addition of an excess of organolithium or organomagnesium reagents to germanium halides results in the displacement of the halide ions and the formation of tetra-alkyl and -arylgermanium compounds. Representative examples of these reactions are shown in equations 4 and 5, respectively.

$$\text{GeCl}_4 + 4\text{LiR} \longrightarrow \text{R}_4\text{Ge} + 4\text{LiCl} \tag{4}$$

$$\text{Me}_2\text{GeCl}_2 + 2\text{RMgX} \longrightarrow \text{Me}_2\text{GeR}_2 + 2\text{MgXCl} \tag{5}$$

Organogermanes $\text{R}_n\text{GeH}_{4-x}$ are conveniently prepared by the substitution of the halogens of organohalogermanes by the hydride ion of reducing agents such as LAH and sodium borohydride (equation 6).

$$\text{R}_n\text{GeX}_{4-n} \xrightarrow[\text{or NaBH}_4]{\text{LiAlH}_4} \text{R}_n\text{GeH}_{4-n} \tag{6}$$

2. Preparation of isotopically labelled organogermanes

A number of isotopically labelled organogermanium compounds have been prepared and subjected to spectroscopic investigation (microwave, infrared and Raman) with a view to obtaining structural information about the compound. More specifically, these studies have sought information on bond strengths, bond lengths, dipole moments, quadrupolar coupling constants, etc.

Most of the labelled organogermanium compounds contain an isotope in the organic moiety. These labels are introduced in a variety of ways. For instance, Laurie[5] prepared a number of isotopically substituted methylgermanes to determine the structure and dipole moment of the parent compound by microwave spectroscopy. Methylgermane-d_3 was synthesized by the $LiAlD_4$ reduction of triiodomethylgermane. A mixture of CH_3GeH_2D and CH_3GeHD_2 was prepared by using a mixture of $LiAlH_4$ and $LiAlD_4$ as the reducing agent. The introduction of a CD_3, CH_2D, or $^{13}CH_3$ group to form CD_3GeH_3, CH_2DGeH_3 and $^{13}CH_3GeH_3$ was accomplished by first reacting the appropriate methyl halide (CD_3Br, CH_2DBr and $^{13}CH_3Br$, respectively) with GeI_4 to form isotopically substituted methyl-triiodogermanes. The methyltriiodogermanes were subsequently reduced to the desired products by $LiAlH_4$, (Scheme 1).

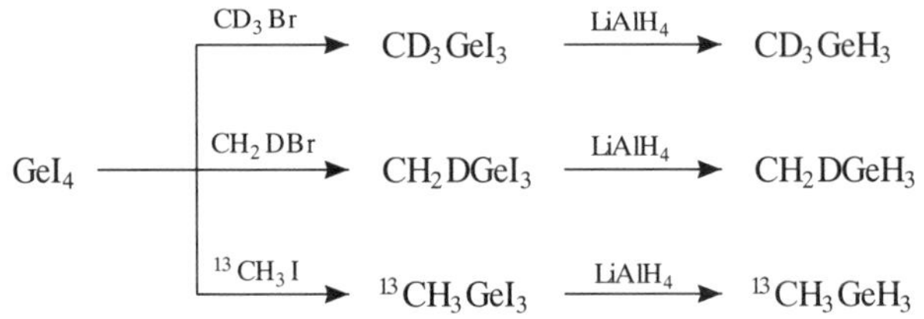

SCHEME 1

Eujen and coworkers[6-8] prepared trifluoromethylgermane-d_3 for an infrared and Raman study, by treating trifluoromethyltriiodogermane with sodium borodeuteride in a phosphoric acid-d_3–deuterium oxide mixture at room temperature (equation 7).

$$CF_3GeI_3 + D_3PO_4 \xrightarrow[D_2O]{NaBD_4} CF_3GeD_3 \qquad (7)$$

After repeated fractional condensation at $-196\,°C$, the purity of the deuterated germane was determined to be $> 90\%$. The isotopomers, CF_3GeHD_2 and CF_3GeH_2D, were also synthesized using an analogous reaction scheme employing $NaBD_4$ in a mixture of H_3PO_4 and H_2O and $NaBH_4$ in a mixture of D_3PO_4 and D_2O, respectively.

Vinylgermane-d_3 was formed[9] by adding vinyltrichlorogermane to a suspension of $LiAlD_4$ in diglyme at $115\,°C$ under an inert atmosphere. The desired material was collected in a trap cooled by liquid nitrogen. The purity of the compound was confirmed by IR spectroscopy.

A study of the vibrational spectra and normal coordinate analysis of trimethylgermane[10] necessitated the preparation of trimethylgermane-d, trimethyl-d_9-germane and trimethyl-d_9-germane-d. Tetramethylgermane and tetramethyl-d_{12}-germane were converted into the respective monoiodides in a reaction with iodine in a sealed glass tube at $50\,°C$. The resulting trimethyliodogermane and trimethyl-d_9-iodogermane were reduced with $LiAlH_4$ and $LiAlD_4$, respectively, to produce the desired isotopomers (Scheme 2).

$$(CH_3)_4Ge \xrightarrow[50\,°C]{I_2} (CH_3)_3GeI \xrightarrow{LiAlD_4} (CH_3)_3GeD$$

$$(CD_3)_4Ge \xrightarrow[50\,°C]{I_2} (CD_3)_3GeI \xrightarrow{LiAlD_4} (CD_3)_3GeD$$

SCHEME 2

The effect of trifluoromethyl groups on the structure of organogermanes was studied by vibrational spectroscopic techniques. Eujen and Burger[11] prepared a series of $(CF_3)_n Ge(CH_3)_{4-n}$ ($n = 1, 2, 3$) and their perdeuterated analogues. Trifluoromethyl-trimethyl-d_9 germane was prepared by a ligand exchange reaction between dimethyl-d_6 zinc and trifluoromethyltriiodogermane (equation 8).

$$2CF_3GeI_3 + 3(CD_3)_2Zn \longrightarrow 3ZnI_2 + 2CF_3Ge(CD_3)_3 \qquad (8)$$

Similarly, the two remaining perdeuterated analogues $(CF_3)_n Ge(CD_3)_{4-n}$ ($n = 2$ and 3) form on treating bis(trifluoromethyl)diiodo- and tris(trifluoromethyl)iodo-germane, respectively, with dimethyl-d_6 cadmium (equations 9 and 10).

$$(CF_3)_2GeI_2 + (CD_3)_2Cd \longrightarrow (CF_3)_2Ge(CD_3)_2 + CdI_2 \qquad (9)$$

$$2(CF_3)_3GeI + (CD_3)_2Cd \longrightarrow 2(CF_3)_3Ge(CD_3) + CdI_2 \qquad (10)$$

3. Preparation of isotopically labelled organohalogermanes

Trimethyl-d_9-chlorogermane was prepared for an infrared study by reacting tetramethyl-d_{12}-germane with acetyl chloride in the presence of a Lewis acid such as aluminium trichloride[12] (equation 11).

$$(CD_3)_4Ge + CH_3COCl \xrightarrow{AlCl_3} (CD_3)_3GeCl \qquad (11)$$

In the same study[12] trimethyl-d_9-bromogermane was synthesized by treating trimethyl-d_9-germane with mercuric bromide (equation 12).

$$(CD_3)_3GeH + HgBr_2 \longrightarrow (CD_3)_3GeBr \qquad (12)$$

Durig and coworkers[13] analysed the microwave spectra of eighteen isotopic species of methylgermyl bromide in order to determine the $Ge-H$ bond lengths for this molecule. The isotopic species include $^{12}CH_3GeH_2Br$, $^{12}CH_3GeD_2Br$ and $^{13}CH_3GeH_2Br$ with $Ge = {}^{70}Ge$, ^{72}Ge or ^{74}Ge, and $Br = {}^{79}Br$ or ^{81}Br.

Methylbromogermane-d_2 was prepared by first reducing trichloromethylgermane with $LiAlD_4$ at $0\,°C$ to give methylgermane-d_3, a volatile gas which can be trapped at $-160\,°C$. This was followed by the reaction of boron tribromide (BBr_3) and methylgermane-d_3 at $-78\,°C$ for 1 h, yielding methylbromogermane-d_2 after subsequent fractionation of the product (equation 13).

$$CH_3GeD_3 + BBr_3 \xrightarrow{-78\,°C} CH_3GeD_2Br \qquad (13)$$

The resulting product is a mixture of mainly six isotopomers of methylbromogermane-d_2, namely $^{12}CH_3{}^{70}GeD_2{}^{79}Br$, $^{12}CH_3{}^{72}GeD_2{}^{79}Br$, $^{12}CH_3{}^{74}GeD_2{}^{79}Br$, $^{12}CH_3{}^{70}GeD_2{}^{81}Br$, $^{12}CH_3{}^{72}GeD_2{}^{81}Br$ and $^{12}CH_3{}^{74}GeD_2{}^{81}Br$. The ^{13}C-labelled trimethylgermane is not available commercially. Hence another reaction scheme, equation 14,

$$GeH_4 + {}^{13}CH_3I \xrightarrow[DME]{KOH} {}^{13}CH_3GeH_3 \xrightarrow[-78\,°C]{BBr_3} {}^{13}CH_3GeH_2Br \qquad (14)$$

$$(\mathbf{1})$$

was used to prepare the isotopic species corresponding to $^{13}CH_3GeH_2Br$ (**1**). The synthesis first involved the generation of ^{13}C-labelled methylgermane by adding $^{13}CH_3I$ to germane in the presence of a base like potassium hydroxide in dimethoxyethane at $-78\,^{\circ}C$. The six isotopomers of methyl-^{13}C-bromogermane, $^{13}CH_3{}^{70}GeD_2{}^{79}Br$, $^{13}CH_3{}^{72}GeD_2{}^{79}Br$, $^{13}CH_3{}^{74}GeD_2{}^{79}Br$, $^{13}CH_3{}^{70}GeD_2{}^{81}Br$, $^{13}CH_3{}^{72}GeD_2{}^{81}Br$ and $^{13}CH_3{}^{74}GeD_2{}^{81}Br$ were obtained by reacting methyl-^{13}C-germane with boron tribromide as described above.

In an earlier study, Roberts and coworkers[14] synthesized isotopically labelled methyl-fluorogermane. These workers showed that ^{13}C-labelled methylfluorogermane could be prepared by reacting 1 with freshly prepared PbF_2 supported on layers of pyrex glass wool. Methyl-d_3-fluorogermane was synthesized in a similar fashion. More specifically, methyl-d_3 iodide was added to germyl anion ($GeH_3{}^-$) generated *in situ* by treating germane with a strong base. The resulting methyl-d_3-germane was subsequently allowed to react with bromine at $-196\,^{\circ}C$. Organobromogermane formed in this reaction was converted into the corresponding methyl-d_3-fluorogermane with PbF_2 as described above.

4. Preparation of isotopically labelled organo pseudohalogenogermanes

The structure[15,16] of methylcyanogermane was determined by using a combination of microwave, IR and Raman spectroscopic techniques. This study required that a number of isotopic species, namely $^{13}CH_3GeH_2CN$, $CH_3GeH_2C^{15}N$, $CH_3GeH_2{}^{13}CN$, CH_3GeD_2CN and CD_3GeH_2CN be prepared. The synthesis of the various isotopic samples was based on the exchange between bromide ion and cyanide ion that occurs when methylbromogermane is treated with silver cyanide (equation 15).

$$CH_3GeH_2Br + AgCN \longrightarrow CH_3GeH_2CN + AgBr \qquad (15)$$

The preparation of isotopically labelled methylbromogermane has already been discussed. The ^{13}C-labelled and ^{15}N-labelled germyl cyanides were synthesized from the ^{13}C-labelled and ^{15}N-labelled silver cyanides which are commercially available.

Durig and Attia[17] synthesized methylisocyanatogermane and its deuterated analogues CD_3GeH_2NCO and CH_3GeD_2NCO, for study by IR and Raman spectroscopy. As for the cyanides, the labelled methylisocyanatogermanes were produced by passing the appropriate isotopically labelled methylbromogermane (CD_3GeH_2Br or CH_3GeD_2Br) through a column packed with a mixture of glass wool and the silver cyanate salt (equation 16).

$$\begin{matrix} CH_3GeD_2Br \\ \text{or} \\ CD_3GeH_2Br \end{matrix} + AgNCO \longrightarrow \begin{matrix} CH_3GeD_2NCO \\ \text{or} \\ CD_3GeH_2NCO \end{matrix} \qquad (16)$$

5. Preparation of miscellaneous isotopically labelled organogermanium compounds

^{13}C-labelled triethylgermylacetylene and bis(triethylgermyl)diacetylene were prepared[18] for use in a spectroscopic study (Raman, IR and ^{13}C-NMR). The synthetic scheme was based on that developed for the silyl and stannyl analogues[19]. Lithiation of acetylene-1,2-$^{13}C_2$ and reaction with chlorotriethylgermane resulted in the formation of triethylgermyl-acetylene-1,2-$^{13}C_2$ (2) in 81% yield (equation 17).

$$H-{}^{13}C\equiv{}^{13}C-H \xrightarrow[\text{ii. Et}_3\text{GeCl}]{\text{i. Li}} Et_3Ge-{}^{13}C\equiv{}^{13}CH \qquad (17)$$
$$\textbf{(2)}$$

Bis(triethylgermyl)diacetylene-1,2,3,4-$^{13}C_4$ (3) forms in a 66% yield on treatment of 2 with with a mixture of copper(I) chloride and N,N,N',N'-tetramethylethylenediamine (equation 18).

$$2 + (CH_3)_2NCH_2CH_2N(CH_3)_2-CuCl \longrightarrow Et_3Ge-^{13}C\equiv^{13}C-^{13}C\equiv^{13}C-GeEt_3$$
$$\textbf{(3)}$$

(18)

In a study involving trapping reactions of dimethylgermylene, Shusterman and coworkers[20] showed that the reaction of 7,7-dimethyl-1,4,5,6-tetraphenyl-7-germabenzonorbornadiene with (*E*)-2-deuteriostyrene at 70 °C for 3 hours afforded the *cis* (**4a**) and *trans* (**4b**) isomers of 2,5-dideutero-1,1-dimethyl-3,4-diphenylgermacyclopentane in addition to the undeuterated analogues (equation 19). The (*E*)-2-deuteriostyrene was prepared by stirring phenylacetylene with zirconocene hydride chloride and treating the orange-red glass which formed upon solvent evaporation with D_2SO_4 (equation 20).

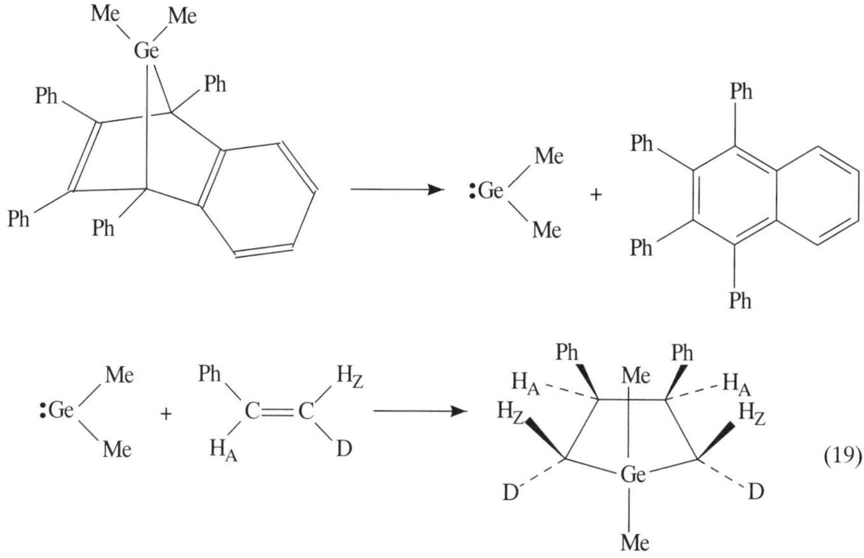

(19)

(**4a**)

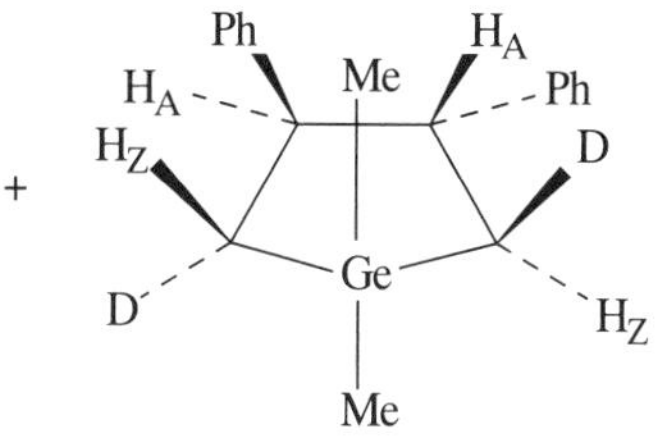

(**4b**)

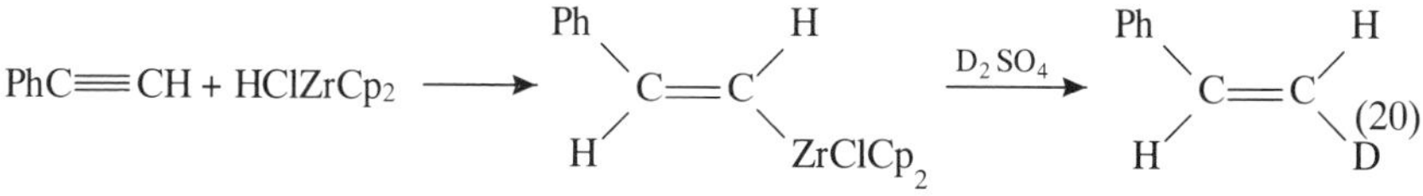

$$PhC\equiv CH + HClZrCp_2 \longrightarrow \underset{H}{\overset{Ph}{\diagdown}}C=C\underset{ZrClCp_2}{\overset{H}{\diagup}} \xrightarrow{D_2SO_4} \underset{H}{\overset{Ph}{\diagdown}}C=C\underset{D}{\overset{H}{\diagup}}$$

(20)

$$Cp = C_5H_5$$

TABLE 2. A summary of the organogermanium compounds and isotopic analogues used in spectroscopic studies

Parent compound	Isotopic species	Spectroscopy	Structural information	Reference
CH_3GeH_3	CH_3GeH_3, $^{13}CH_3GeH_3$, CH_3GeD_3, CD_3GeH_3, CH_2DGeH_3, CH_3GeH_2D, CH_3GeD_2H, Ge $= ^{70}$Ge, ^{72}Ge, ^{74}Ge, ^{76}Ge	Microwave	r_s structural parameters Barrier to internal rotation about the C–Ge bond Dipole moment Quadrupole coupling constant	5
CF_3GeH_3	CF_3GeH_3, CF_3GeH_2D, CF_3GeHD_2, CF_3GeD_3	IR Raman	Assignment of all fundamental bands Force constants for the Ge–H, C–F and Ge–C bonds using normal coordinate analysis	6
CF_3GeH_3	CF_3GeD_3, Ge $= ^{70}$Ge, ^{72}Ge, ^{73}Ge, ^{74}Ge, ^{76}Ge, CF_3GeDH_2, Ge $= ^{70}$Ge, ^{72}Ge, ^{74}Ge, CF_3GeD_2H, Ge $= ^{70}$Ge, ^{72}Ge, ^{74}Ge	Microwave	r_0 structural parameters r_s structural parameters for the GeH$_3$ group	8
CH_2CHGeH_3	CH_2CHGeH_3, CH_2CHGeD_3	IR Raman	Assignment of the vibrational modes	
CH_2CHGeH_3	CH_2CHGeH_3, CH_2CHGeD_3 Ge $= ^{70}$Ge, ^{72}Ge, ^{73}Ge, ^{74}Ge, ^{76}Ge	Microwave	Calculated r_s structural parameters Barrier to internal rotation of the germyl group Dipole moment Quadrupole coupling constant < 2.8 MHz	9 21
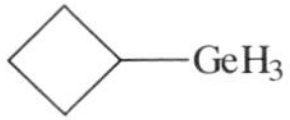 cyclobutyl–GeH$_3$	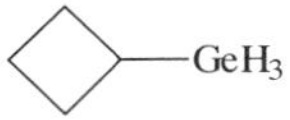 Ge $= ^{70}$Ge, ^{72}Ge, ^{74}Ge	Microwave	Equatorial and axial conformers have been detected Rotational constants for the 2 conformations of each isotopic species	22

Compound	Isotopologues	Method	Information obtained	Ref.
CH_2FGeH_3	CH_2FGeH_2, $Ge = {}^{74}Ge$, ${}^{72}Ge$, ${}^{70}Ge$	Microwave	Barrier to internal rotation of the GeH_3 group Dipole moment	23
Me_2GeH_2	$(CH_3)_2GeH_2$, $X = {}^{70}Ge$, ${}^{72}Ge$, ${}^{73}Ge$, ${}^{74}Ge$, ${}^{76}Ge$	Microwave	Structural parameters Dipole moment	24
CF_3GeMe_3 $(CF_3)_2GeMe_2$ $(CF_3)_3GeMe$	$CF_3Ge(CD_3)_3$ $(CF_3)_2Ge(CD_3)_2$ $(CF_3)_3GeCD_3$	IR Raman	Normal coordinate analysis Force constants	11
Me_3GeH	Me_3GeD $(CD_3)_3GeH$ $(CD_3)_3GeD$	IR Raman	Assignment of fundamental bands Normal coordinate analysis	10
Me_3GeX $X = Cl$, Br	Me_3GeX $(CD_3)_3GeX$	IR Raman	Assignment of fundamental bands with the exception of internal torsional bands Normal coordinate analysis	12
$MeGeH_2Br$	$MeGeH_2Br$ $MeGeD_2Br$ ${}^{13}CH_3GeH_2Br$ $Ge = {}^{70}Ge$, ${}^{72}Ge$, ${}^{74}Ge$ $Br = {}^{79}Br$, ${}^{81}Br$	Microwave	Calculation of r_s structural parameters r_0 structural parameters Quadrupolar coupling constants	13
$MeGeH_2F$	$MeGeH_2F$ ${}^{13}CH_3GeH_2F$ CD_3GeH_2F $Ge = {}^{70}Ge$, ${}^{72}Ge$, ${}^{74}Ge$, ${}^{76}Ge$	Microwave	Calculation of r_s structural parameters Dipole moment Barriers to internal rotation of the CH_3 group	14
$MeGeH_2CN$	$MeGeH_2CN$ ${}^{13}CH_3GeH_2CN$ $MeGeH_2C^{15}N$ $MeGeH_2{}^{13}CN$ $MeGeD_2CN$ CD_3GeH_2CN $Ge = {}^{70}Ge$, ${}^{72}Ge$, ${}^{74}Ge$, ${}^{76}Ge$	Microwave Raman IR	r_s structural parameters Dipole moment Threefold barrier to internal rotation Vibrational assignments of IR and Raman bands	16
$MeGeH_2NCO$	$MeGeH_2NCO$ CD_3GeH_2NCO CH_3GeD_2NCO	IR Raman	Vibrational assignment of bonds	17

C. The Use of Isotopically Labelled Organogermanium Compounds

1. The use of isotopically labelled organogermanium compounds in spectroscopic studies

Isotopically (natural and synthetic) labelled organogermanium compounds have been studied extensively by microwave spectroscopy. A summary of the organogermanium compounds subjected to microwave spectroscopy is found in Table 2. In general these studies determine the ground-state rotational transitions from the microwave spectra. The rotational transitions are subsequently used to calculate rotational constants for the various species. Molecular structure can be deduced using an iterative process which compares the experimental rotational constants to values calculated for an assumed structure. Refinement of bond lengths, bond angles, etc. are made, and the rotational constants are recalculated. The process continues until the best least-squares fit of the experimental rotational constants is obtained. A summary of the structural information obtained for several organogermanium compounds is presented in Table 3.

TABLE 3. A summary of the structural information obtained for a number of organogermanium compounds using microwave spectroscopy

(i) *Organogermanes*

Information	$MeGeH_3$	CF_3GeH_3	$H_2C{=}CHGeH_3$	Me_2GeH_2
r_{cx} (Å)	1.083 (X = H)	1.352 (X = F)	1.347 (X = C)	1.083 (X = H)
r_{GeH} (Å)	1.529	1.499	1.521	—
r_{CGe} (Å)	1.9453	1.997	1.926	1.950
∠HCH (deg)	108.4	—	—	108.5
∠HGeH (deg)	109.3	—	—	—
∠CGeH (deg)	—	107.1	110.7	—
∠XCGe (deg)	—	111.6	122.9	—
∠CGeC (deg)	—	—	—	110.0
Barrier to rotation (kcal mol^{-1})	1.24	1.3	1.24	1.18
Dipole moment (*D*)	0.635	—	0.50	0.616
Quadrupolar coupling constant (*Q* in MHz)	3	—	< 2.8	< 3.0
Reference	5	8	21	24

(ii) *Halogeno- and pseudohalogenogermanes*

Information	$MeGeH_2Br$	$MeGeH_2F$	$MeGeH_2CN$
r_{GeX} (Å)	2.308 (X = Br)	1.751 (X = F)	1.927 (X = CN)
r_{GeH} (Å)	1.520	1.525	1.521
r_{CGe} (Å)	1.933	1.925	1.933 (Me)
			1.927 (CN)
r_{CH} (Å)	1.100 (s)	1.094	1.092 (s)
	1.097 (a)		1.092 (a)
∠CGeX (deg)	107.0(X = Br)	106.3 (X = F)	107.4 (X = CN)
∠HGeH (deg)	111.5	110	111.4
∠HGeC (deg)	112.2	—	112.9 (Me)
∠GeCH$_s$ (deg)	110.3	—	109.5
∠GeCH$_a$ (deg)	111.9	—	109.3
∠HCH (deg)	—	108.9	—
Barrier to rotation (kcal mol^{-1})	—	0.941	1.15
Dipole moment (*D*)	—	2.59	4.22
Reference	13	14	16

There have been a number of gas-phase IR and Raman studies of organogermanium compounds. Eujen and Burger[6] investigated the gas-phase IR and liquid-phase Raman spectra of CF_3GeH_3, CF_3GeH_2D, CF_3GeHD_2 and CF_3GeD_3. More specifically, the deuterated species were used to assign the 18 normal vibrations of the CF_3GeH_3 IR spectrum. Generally, the trifluoromethyl and the germyl moiety are independent of one another, i.e. the CF_3 modes are not significantly perturbed by deuteration. The only example of substantial mixing is found for the pair of perpendicular vibrations v_7/v_{11} of CF_3GeH_3 which are at approximately 490 and 590 cm^{-1}.

The vibrational spectra and normal coordinate calculations[10] of Me_3GeH using deuterated analogues showed that the asymmetric methyl deformations are not greatly influenced by the adjacent germanium atom. For Me_3GeH and Me_3GeD, the asymmetric methyl deformations are observed at approximately 1410 cm^{-1}. The absorptions due to the symmetric methyl deformations are relatively strong and sharp and the germanium atom causes the band to shift to 1250 cm^{-1}. The symmetric and asymmetric methyl deformation bands for $(CD_3)_3GeH$ and $(CD_3)_3GeD$ are observed at 970 cm^{-1} and 1040 cm^{-1}, respectively. In addition, deuteration of the methyl groups causes the GeC_3 stretching mode observed in the Raman spectra at 550 cm^{-1} to shift to 520 cm^{-1}. The GeD and the GeH bending vibrations are observed at 480 cm^{-1} and 780 cm^{-1}, respectively.

Kamienska-Trela and Luettke[19] determined the carbon–carbon spin–spin coupling constants, J(CC), for ^{13}C-enriched mono(triethylgermyl)acetylene and bis(triethylgermyl)diacetylene by NMR spectroscopy. The acetylenic-1,2-$^{13}C_2$ fragment represents an AB-type spin system and the ^1J(CC) coupling constant was measured directly from the NMR spectrum. On the other hand, the proton-decoupled spectrum of the diacetylene moiety of bis(triethylgermyl)diacetylene is an example of an AA'BB' spin system. Computer-assisted analysis of the spectrum yielded the four possible spin–spin couplings between the carbon nuclei, namely ^1J(C_1–C_2), ^1J(C_2–C_3) and two long-range couplings ^2J(C_1–C_3) and ^3J(C_1–C_4). The ^1J(C_1–C_2) values for mono(triethylgermyl)acetylene (132.5 Hz) and bis(triethylgermyl)diacetylene (146.8 Hz) are significantly smaller than those reported for t-butylacetylene (168.7 Hz) and bis(t-butylacetylene) (188.3 Hz). The ^1J(C_2–C_3) value for coupling across the central C–C single bond of bis(triethylgermyl)diacetylene (137.7 Hz) is also smaller than the corresponding ^1J(C_2–C_3) value reported for diacetylene (154.8 Hz). Similar results have been obtained for triethylsilyl derivatives of acetylene and diacetylene. The large variations observed for the carbon–carbon spin–spin coupling constants have been attributed to the electronegativity of the triethylgermyl and triethylsilyl groups. Finally, the J(CC) across two and three bonds for the bis(triethylgermyl)diacetylene were found to be 12.4 Hz and 14.7 Hz, respectively.

2. The use of isotopically labelled organogermanium compounds in mechanistic studies

A solvent kinetic isotope effect, k_{H_2O}/k_{D_2O}[25,26], of 1.71 was found for the acid-catalysed cleavage of the C–Ge bond in protodegermylation of p-methoxyphenyltriethylgermane in aqueous dioxane (25% H_2O or D_2O: 75% dioxane). The magnitude of the k_{H_2O}/k_{D_2O} indicates there is transfer of a proton from the solvent to the substrate in the rate-determining step of the reaction. The mechanism which has been suggested for the protodegermylation reaction is shown in Scheme 3. The first step which involves the formation of a σ-complex via electrophilic attack of the *ipso* carbon by H$^+$ is rate-determining. In the transition state, the O–H bond of a hydronium ion breaks as a C–H bond forms. In

the second step, the complex decomposes rapidly by loosing the germyl moiety to form
p-methoxybenzene.

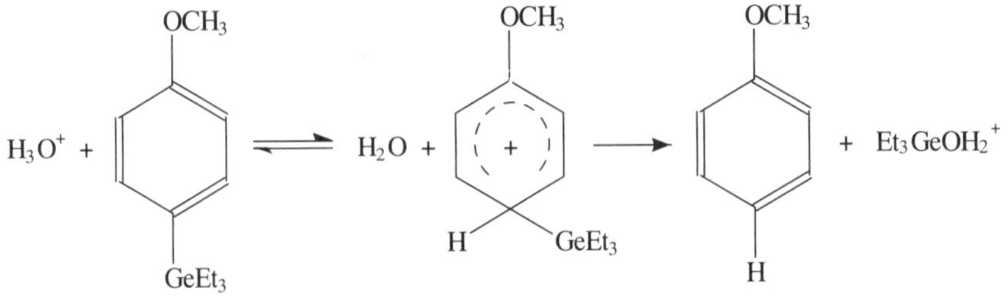

SCHEME 3

The base-catalysed detritiation of arylgermanes-t_1 in methanol is thought to proceed by
a two-step mechanism (Scheme 4). More specifically, methoxide ion abstracts the triton
to produce an arylgermyl anion in the first step, which is thought to be rate-determining.
The germyl anion removes a proton from methanol in a subsequent fast step. Eaborn and
Singh[27] studied the detritiation of two families of arylgermanes (equations 21 and 22) to
determine the degree of conjugative delocalization of the lone pair on the forming germyl
anion into the aromatic ring. The authors made several very interesting discoveries:

$$(XC_6H_4)_3GeT \xrightarrow[20-40\,°C]{MeONa/MeOH} (XC_6H_4)_3GeH \qquad (21)$$

$$X = m\text{-Cl},\ p\text{-Cl},\ m\text{-CH}_3,\ p\text{-CH}_3,\ o\text{-CH}_3,\ p\text{-CH}_3O,\ o\text{-CH}_3O$$

$$(XC_6H_4)Ph_2GeT \xrightarrow[20-40\,°C]{MeONa/MeOH} (XC_6H_4)Ph_2GeH \qquad (22)$$

$$X = m\text{-Cl},\ p\text{-NO}_2,\ p\text{-CN},\ p\text{-F}$$

$$(XC_6H_4)_3GeT + CH_3O^- \xrightarrow{slow} (XC_6H_4)_3Ge^- + CH_3OT$$

$$(XC_6H_4)_3Ge^- + CH_3OH \xrightarrow{fast} (XC_6H_4)_3GeH + CH_3O^-$$

SCHEME 4

First, the rate increases when a more electron-withdrawing group is on the benzene
ring. For example, the *meta*-chloro substituted substrate, $[(m\text{-ClC}_6H_4)_3GeT]$, is approximately 380 times as reactive as the parent compound, Ph_3GeT. The rate increase occurs
because the electron-withdrawing group stabilizes the germyl anion. However, a difference in the degree of activation of the *meta*-chloro substituent was observed between the
two families of substrates, i.e. $(m\text{-ClC}_6H_4)_3GeT$ and $(m\text{-ClC}_6H_4)Ph_2GeT$. More specifically, the degree of activation for $(m\text{-ClC}_6H_4)_3GeT$ is approximately 1000/380 or 2.6
times smaller than that predicted by the data collected for $(m\text{-ClC}_6H_4)Ph_2GeT$ where k_{rel}
was found to be 10. In addition, a good Hammett-ρ correlation was found between the
$\log(k_X/k_H)$ and the corresponding substituent constant σ_X, for $(XC_6H_4)_3GeT$. The large
ρ of 2.2 per phenyl group suggested a substantial dispersal of the negative charge from
the forming anionic centre into the aromatic rings. However, for the second family of
germanes, $(XC_6H_4)Ph_2GeT$, a better correlation was obtained when σ constants were

used instead of σ. The need to use σ constants for the p-CN and p-NO$_2$ substituents, which possess powerful electron-withdrawing resonance effects, is strong evidence for conjugative delocalization of charge into the aromatic rings from the germyl anion centre.

An inverse solvent kinetic isotope effect, k_{MeOD}/k_{MeOH}, of 1.7 was also determined[27] for hydrogen deuterium exchange reactions of both Ph$_3$GeH and $(m\text{-ClC}_6\text{H}_4)_3$GeH. Although this value is within the normal range for reactions catalysed by sodium methoxide in methanol, it is significantly lower than the value of 2.0–2.2 reported for analogous organosilane reactions where the methoxide ion is thought to be virtually free of solvating methanol molecules. The smaller k_{MeOD}/k_{MeOH} in the germyl reaction suggests that the methoxide ion is partially solvated, and as a result abstraction of the proton by methoxide is thought to be less advanced in the transition state than in the case of unsolvated methoxide ion. The solvent isotope effect is therefore an indicator of the degree of methoxide ion solvation. The near-zero values for $\Delta S^{\ddagger}$ calculated for a number of $(XC_6H_4)_3$GeH compounds is also consistent with the liberation of solvating methanol from the methoxide ion as the transition state is approached. The release of solvent compensates for the loss of freedom associated with the transition state of a bimolecular reaction.

Germylenes (**5**), which are isoelectronic to carbenes, are highly reactive species. Hence, germylenes are useful intermediates in the synthesis of organogermanium compounds. However, because the mechanism of their formation is not well understood, their usefulness in synthesis is somewhat limited. For instance, the thermal decomposition of 7-germanorbornadiene (**6**) results in the formation of germylene. Some controversy exists as to whether germylene formation occurs via a concerted fragmentation (cleavage of both germanium–carbon bonds in a single step) of the norbornadiene precursor or by a stepwise mechanism involving the formation of a diradical intermediate (**7**) which can subsequently undergo germanium carbon bond cleavage to yield germylene (equation 23).

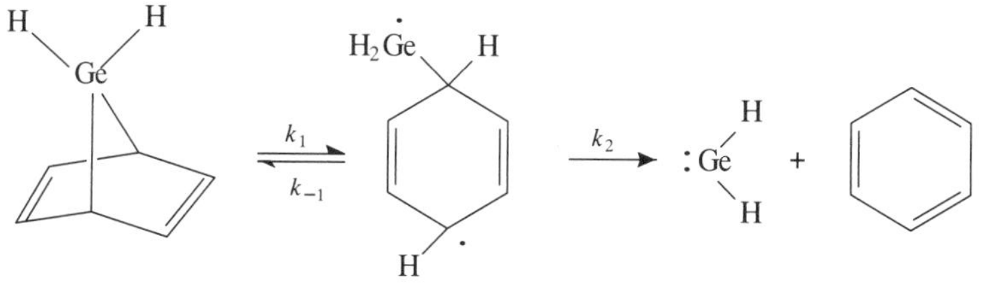

(23)

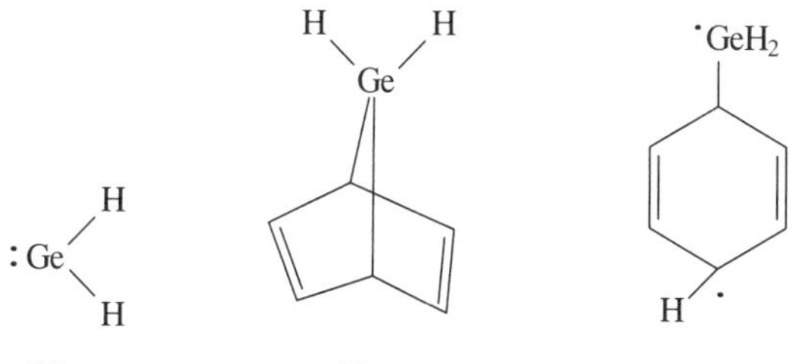

(**5**)   (**6**)   (**7**)

Shusterman and coworkers[20] studied the decomposition of 7,7-dimethyl-1,4,5,6-tetraphenyl-7-germabenzonorbornadiene (**8**) by monitoring the disappearance of the absorbance at 320 nm in the UV. They found that the rate of formation of dimethylgermylene was not affected by the addition of trapping agents such as styrene, 2,3-dimethylbutadiene or carbon tetrachloride. This suggested to the authors that either dimethylgermylene formation was occurring in a concerted fashion or via the rate-determining formation of the diradical (Scheme 5). Activation parameters of $\Delta H^{\ddagger} = 27.8$ kcal mol^{-1} and $\Delta S^{\ddagger} = 6.8$ e.u. determined for the decomposition reaction are similar to those found when norbornadienes undergo cleavage of the bridgehead bond. However, concerted fragmentation of **8** is also consistent with the observed activation parameters.

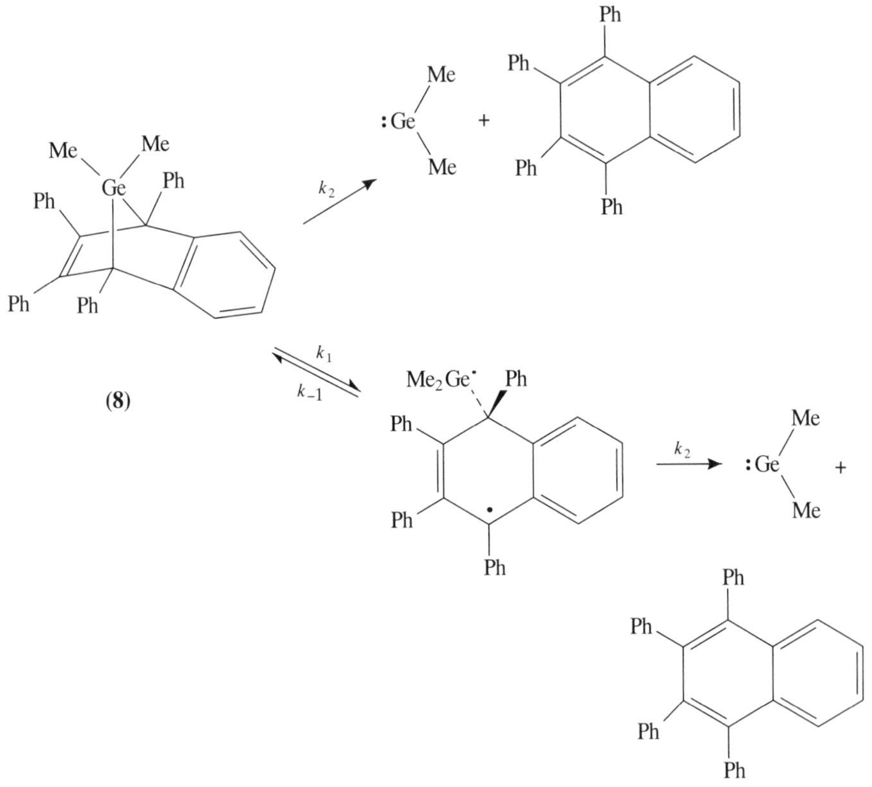

SCHEME 5

When dimethylgermylene is formed by the decomposition of **8** in the presence of excess styrene, *cis*- and *trans*-1,1-dimethyl-3,4-diphenylgermacyclopentane forms from the apparent cycloaddition of dimethylgermylene to two alkene molecules. This is thought to occur by a stepwise mechanism where dimethylgermylene adds in a concerted fashion to styrene to yield a phenyl-substituted germacyclopropane, in the first step. In a second step, concerted addition of the germacyclopropane to a second styrene molecule has been proposed (Scheme 6).

A NMR analysis of the reaction product resulting from the thermolysis of **8** in the presence of (*E*)-2-deuteriostyrene indicated an equal ratio of the *cis* and *trans* isomers of 2,5-dideuterio-1,1-dimethyl-3,4-diphenylgermacyclopentane. In addition, integration of the resonances for the ring protons indicated that the deuterium label was partially scrambled. It was concluded that the scrambling occurred during the cycloaddition, since

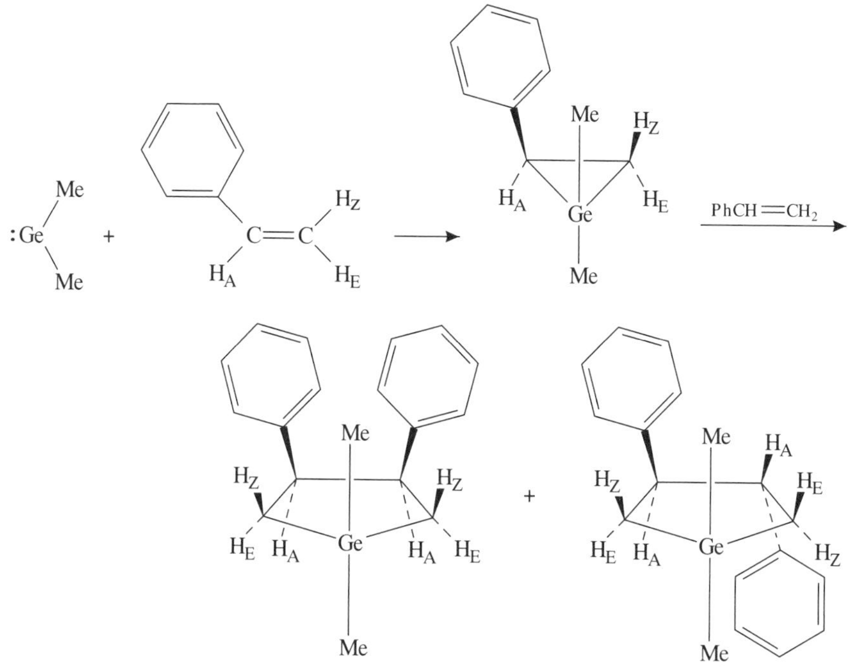

SCHEME 6

stability studies showed that the *cis* and *trans* isomers do not interconvert upon heating, exposure to room lighting or when subjected to repeated TLC. The cycloaddition mechanism proposed to explain the randomization of the deuterium label is given in Scheme 7. Cleavage of the germacyclopropane (**9**) gives the diradical **10**, which reacts with a second styrene molecule to form **11**. The direction of the ring cleavage and subsequent styrene addition are governed by the stabilizing effect of the phenyl groups on the radical centres. Scrambling of the deuterium label could occur in either **10** or **11** due to rotation about the carbon–carbon bond. Incomplete scrambling could result from ring closure of **11** at a rate that is competitive with bond rotation.

Nagasawa and Saito[28] prepared ^{14}C- and ^{2}H-labelled acetylacetone to study the ligand isotopic exchange of tris(acetylacetonato)germanium(IV) perchlorate in 1,1,2,2-tetrachloroethane, nitromethane and acetonitrile. The ^{14}C-labelled acetylacetone was prepared with a specific activity of 10 μCi g^{-1} by the Claisen condensation between ethyl acetate and ^{14}C-labelled acetone (equation 24). Incorporation of the ^{14}C-labelled acetylacetone into the tris(acetylacetonato)germanium(IV) perchlorate occurred only if the organic solvent contained water or acid, suggesting that proton transfer played an important role in the ligand exchange. This lead Nagasawa and Saito to present the mechanism in Scheme 8 for the ligand isotopic exchange of tris(acetylacetonato)germanium(IV) perchlorate.

$$\underset{\text{O}}{\overset{\text{O}}{\underset{\|}{CH_3\overset{\text{O}}{C}CH_3}}} + CH_3\overset{\overset{\text{O}}{\|}}{C}\underset{OCH_2CH_3}{} \xrightarrow{\text{base}} CH_3\overset{\text{O}}{\overset{\|}{C}}CH_2\overset{\text{O}}{\overset{\|}{C}}CH_3 \qquad (24)$$

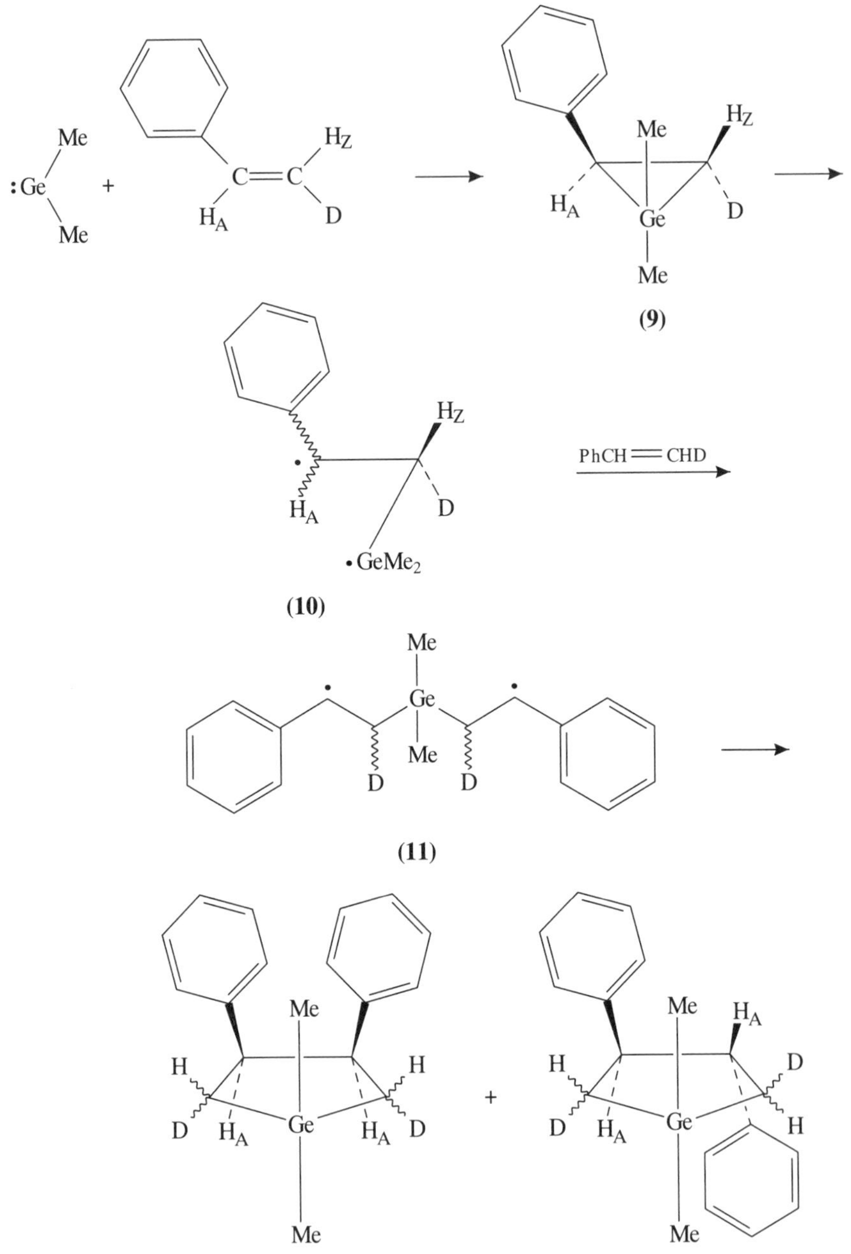
Me
:Ge
Me
+
C=C
HA
HZ
D
Me
HZ
HA
Ge
D
Me
(9)
HZ
HA
D
•GeMe2
(10)
PhCH=CHD
Me
Ge
Me
D
D
(11)
Me
H
H
D
HA
Ge
HA
D
Me
+
Me
HA
H
D
D
HA
Ge
H
Me

SCHEME 7

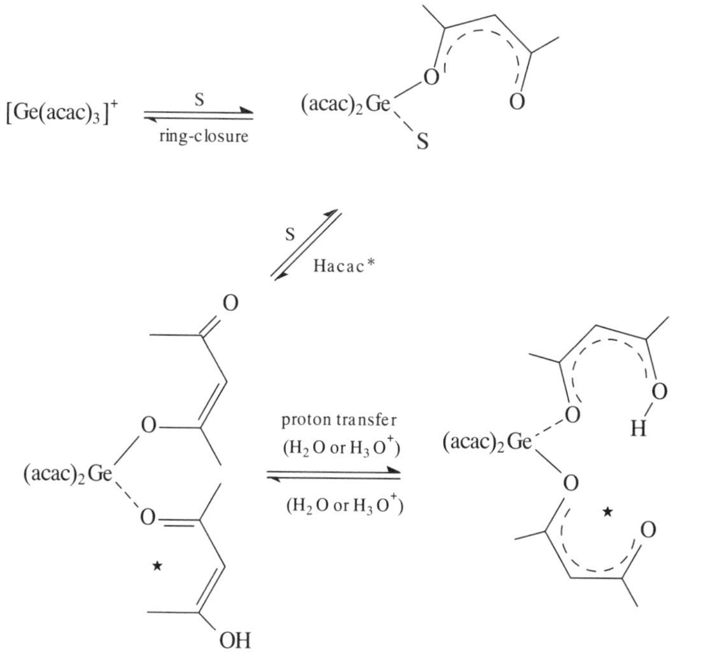

$[Ge(acac)_3]^+$ $\xrightleftharpoons[\text{ring-closure}]{S}$ $(acac)_2 Ge$ ...

$\Big\uparrow$ Hacac*

$(acac)_2 Ge$ ... $\xrightleftharpoons[(H_2O \text{ or } H_3O^+)]{\text{proton transfer} (H_2O \text{ or } H_3O^+)}$ $(acac)_2 Ge$ ...

(12)

SCHEME 8

In the first step, solvent coordinates to germanium and assists in breaking one of the GeO bonds present in the Δ-complex. In the second step, the solvent molecule is displaced by the ^{14}C-labelled acetylacetone (Hacac*) to give intermediate 12. A proton is next transferred from the ^{14}C-labelled ligand to the originally coordinated acetylacetone with the aid of water or acid added to the solvent. The deprotonated ^{14}C-labelled acetylacetone undergoes ring closure and the originally coordinated acetylacetone is released, yielding the ligand exchange product $[(acac)_2Ge(acac^*)]^+$. It is worth noting that all the processes are reversible. When deuterated acetylacetone, prepared by refluxing acetylacetone with deuterium oxide in dichloromethane, was added to tris(acetylacetonato)germanium(IV) perchlorate in acetonitrile, a decrease in the ligand exchange rate was observed. The primary deuterium isotope effect of 1.4 calculated for this process confirmed that the water or acid in the solvent helps transfer the proton between ligands in the rate-determining step of the reaction.

3. The use of organogermanium compounds in preparing labelled substrates

The production of radiopharmaceuticals by the radiohalogenation of biomolecules is an active area of research. The incorporation of radiohalogens into organic moieties is synthetically demanding for a number of reasons. It is essential that the radiohalogen be incorporated at a specific molecular site to ensure that the radiolabelled analogue remains pharmacologically active. In addition, the radiohalogenation must be rapid to allow for the incorporation of short-lived positron-emitting radionuclides. It is desirable to produce substrates with high radiochemical yields, which usually means that mild halogenation conditions are necessary. A route which seems to satisfy most of the criteria listed above is a halogenodemetallation reaction in which an electrophilic radiohalogen displaces a labile moiety from an aromatic ring. Moerlein and coworkers[29,30] have demonstrated that regiospecific incorporation (> 97%) of non-carrier added radiobromine and radioiodine into aromatic rings is possible via halogenodegermylation. More specifically, the radioiodination and radiobromination of a series of *para*-substituted aryltrimethylgermanium compounds (equation 25) have been investigated in several solvents, and at various oxidant concentrations and pHs, etc.

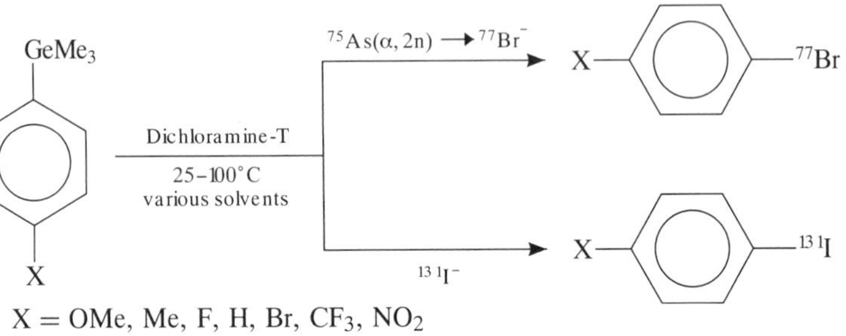

$$X = OMe, Me, F, H, Br, CF_3, NO_2$$

Table 4 shows the radiochemical yields for regiospecific halogenodegermylation of a series of *para*-substituted phenyltrimethylgermanium compounds. The authors found that the maximum product yield was obtained within 10 min in most substrates for both radioiodination and radiobromination. Radiobromination yields were notably lower. For example, the radiobromination of the *para*-H substrate in $MeCO_2H$ gives a 46% yield

TABLE 4. Radiochemical yields for regiospecific halogenodegermylation of *para*-substituted phenyltrimethylgermanium compounds.[a]

| Solvent | X | Radiochemical yield (%)[b] | |
		^{77}Br	^{131}I
$MeCO_2H$	OCH_3	68	87
	CH_3	54	85
	F	51	88
	H	46	79
	Br	20	59
	CF_3	*ca* 10	13
MeOH	OCH_3	53	95
	CH_3	39	87
	F	35	86
	H	25	86
	Br	*ca* 10	35
	CF_3	*ca* 10	16
CCl_4	OCH_3	60	77
	CH_3	28	63
	F	21	53
	H	12	42
	Br	*ca* 10	12

[a] Reaction conditions: 100 μCi dry $^{77}Br^-$ or 50 μCi dry $^{131}I^-$, 10 μl $XC_6H_4GeMe_3$, 5 mg dichloramine-T, 1 ml solvent, 25 °C, reaction time 30 min.
[b] Percentage of total reactivity in solution.

of product as compared to 79% in the radioiodination reaction. In general, radiochemical yields were higher for reactions carried out in hydrogen-bonding, high-dielectric-constant solvents such as methanol. They also noted that low concentrations of oxidant (0.1–1% of dichloramine-T) led to the highest percentage of halogen incorporation. This means that halogenodegermylation can even be used with oxidation-sensitive substrates. It was concluded that organogermanium compounds are superior to organosilicon and organotin compounds for radiohalogenation reactions due to (a) their ability to undergo both rapid bromo- and iododemetallation (unlike organosilicon compounds) and (b) their chemical stability and low toxicity (unlike the organotin compounds). Finally, the sensitivity of the halogenodegermylation reaction to the nature of the electrophile (I^- vs Br^-), the substituents on the benzene ring, and the solvent acidity and polarity means that this synthetic approach can be used to tailor the chemical synthesis to the specific radiopharmaceutical.

The radiochemical yields of the aryl halides formed in the halogenodegermylation reaction are intermediate between those obtained for analogous reactions involving arylorganosilanes and arylorganostannanes (*vide infra*). This finding is consistent with the suggestion that the halogenodegermylation involves the formation of a σ-complex similar to that proposed by Eaborn[25,26] for the protodemetallation of aryltrimethyl group IVb organometallics (*vide supra*, Scheme 3). From a steric point of view, the electrophilic attack at the *ipso* carbon to form the σ-complex should be dependent on the carbon–metal bond length (where C–Si, C–Ge and C–Sn are 1.31, 1.36 and 1.54 Å), i.e. a longer carbon–metal bond facilitates the approach of the electrophile and formation of the σ-complex. In addition, the decomposition of the σ-complex to form the aryl halide will require less energy as the carbon–metal bond weakens. This is predicted by the mechanism, because the radiochemical yield increases as one proceeds down group IVb (C–Si, C–Ge and C–Sn with bond strengths of 352, 308 and 257 kJ mol^{-1}, respectively). The σ-complex which is formed in the rate-limiting step of the halogenodegermylation reaction is stabilized by σ-pi conjugative electron release from the C–metal bond.

The study was subsequently extended to include the radiofluorodemetallation of organogermanium compounds[31]. Both fluorine-18 and ^{18}F-acetylhypofluorite were used to displace the trimethylgermyl moiety from a series of *para*-substituted aryltrimethylgermanes (equation 26). The ^{18}F was produced in a cyclotron by bombarding Ne atoms with deuterium ions (equation 27).

$$X\!-\!\!\left\langle\!\!\bigcirc\!\!\right\rangle\!\!-\!GeMe_3 \quad \xrightarrow[\text{[^{18}F]-CH}_3\text{CO}_2\text{F, 0}^\circ\text{C}]{\text{[^{18}F]-F}_2,\,-78^\circ\text{C}} \quad X\!-\!\!\left\langle\!\!\bigcirc\!\!\right\rangle\!\!-\!^{18}F \tag{26}$$

$$X = CH_3O,\ CH_3,\ H,\ F,\ Br,\ CF_3,\ O_2N$$

$$^{20}_{10}Ne + {}^{2}_{1}D \longrightarrow {}^{4}_{2}\alpha + {}^{18}_{9}F \tag{27}$$

A comparison of the yields of *para*-substituted [^{18}F]fluoroarenes indicate that: (i) acetyl hypofluorite is an inferior fluorination agent for the fluorodegermylation reaction and (ii) the aromatic substituents have considerable influence over the reactivity of the fluorination. A decrease in the fluorodegermylation yield was observed with electron-withdrawing aromatic substituents. The electrophilic aromatic degermylation reaction is thought to proceed via a σ-complex intermediate (Scheme 3). It has been hypothesized that the yield of aryl fluoride is influenced to some extent by the aromatic substituents' ability to stabilize the σ-complex intermediate.

II. THE SYNTHESIS AND USES OF LABELLED ORGANOMETALLIC DERIVATIVES OF TIN

A. Isotopes of Tin

There are ten, naturally occurring isotopes of tin and some artificial isotopes have been synthesized in cyclotrons and nuclear reactors. The naturally occurring tin isotopes range in mass from 112 to 124. The natural abundance (percentage of each isotope) also varies widely. For example, tin-115 accounts for only 0.35% while tin-120, the most abundant of the naturally occurring tin isotopes, accounts for almost 35% of the tin isotopes (Table 5)[32].

TABLE 5. The isotopic composition of naturally occurring tin

Isotope	% Natural abundance
^{112}Sn	0.96
^{114}Sn	0.66
^{115}Sn	0.35
^{116}Sn	14.30
^{117}Sn	7.61
^{118}Sn	24.03
^{119}Sn	8.58
^{120}Sn	34.72
^{122}Sn	4.72
^{124}Sn	5.94

Highly enriched samples of isotopic purities ranging from 84% to 99% have been prepared for most of the isotopes of tin. Typical percentage compositions for enriched samples of tin(IV) oxide are given in Table 6[33].

TABLE 6. The composition of isotopically enriched samples of tin(IV) oxide

Isotope	^{112}Sn	^{116}Sn	^{117}Sn	^{118}Sn	^{119}Sn	^{120}Sn	^{122}Sn	^{124}Sn
^{112}Sn	**98.93**	<0.02	0.08	<0.01	<0.02	<0.02	<0.03	0.04
^{114}Sn	1.04	<0.02	0.04	<0.01	<0.02	<0.02	<0.03	0.04
^{115}Sn	0.05	0.07	0.06	<0.02	<0.02	<0.02	<0.03	0.02
^{116}Sn		**96.68**	2.34	0.37	0.40	0.13	0.86	0.21
^{117}Sn		1.23	**89.2**	0.79	0.85	0.11	0.46	0.17
^{118}Sn		1.18	4.5	**95.75**	3.63	0.61	1.64	0.43
^{119}Sn		0.22	1.12	1.22	**84.48**	0.66	0.71	0.31
^{120}Sn		0.52	2.16	1.61	9.98	**98.05**	3.91	1.07
^{122}Sn		0.05	0.26	0.15	0.44	0.34	**91.24**	1.00
^{124}Sn		0.05	0.28	0.07	0.20	0.10	1.18	**96.71**

The availability of a large number of tin isotopes has made these isotopes useful in the analyses of organotin compounds in the environment and as tracers in mechanistic and environmental studies (*vide infra*).

B. The Synthesis of Labelled Tin Compounds

Two major types of isotopically labelled tin compounds have been synthesized. The first class of labelled organotin compounds has been prepared using specific isotopes of tin, while the second class have the isotope in one of the groups bonded to the tin atom. Both types of syntheses are discussed in the following paragraphs.

1. The synthesis of compounds with specific isotopes of tin

Sloop and coworkers[33] have prepared eight different isotopically substituted tin hydrides. All their syntheses started with a tin oxide enriched in one of the isotopes of tin. An example of their synthetic scheme is presented using tin-112.

$$^{112}SnO_2 + NaC{\equiv}N \xrightarrow{\text{red heat}} {^{112}Sn} \xrightarrow{\text{Mg}} {^{112}SnMg_2} \xrightarrow[\substack{160\,°C \\ 18\,h}]{\text{EtBr}} Et_3{^{112}}SnBr \xrightarrow{\text{LiAlH}_4} Et_3{^{112}}Sn{-}H$$

This synthetic method was used to prepare organotin hydrides highly enriched in ^{112}Sn, ^{116}Sn, ^{117}Sn, ^{118}Sn, ^{119}Sn, ^{120}Sn, ^{122}Sn and ^{124}Sn. These labelled tin hydrides were subsequently converted into N-succinimidyl-3-(triethylstannyl)propanoate, which was attached to oligonucleotides as a mass label. The N-succinimidyl-3-(triethylstannyl)propanoate was synthesized in two ways, via the 3-(triethylstannyl)propanoic acid (equation 28), and via the N-succinamidyl ester of 2-propenoic acid (equation 29).

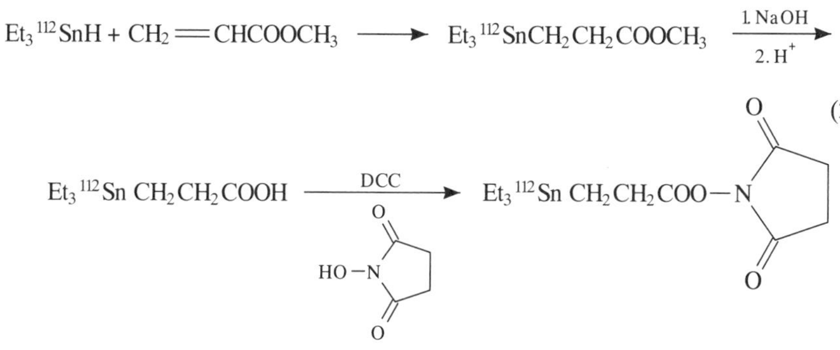

$$Et_3{^{112}}SnH + CH_2{=}CHCOOCH_3 \longrightarrow Et_3{^{112}}SnCH_2CH_2COOCH_3 \xrightarrow[\text{2. H}^+]{\text{1. NaOH}}$$

(28)

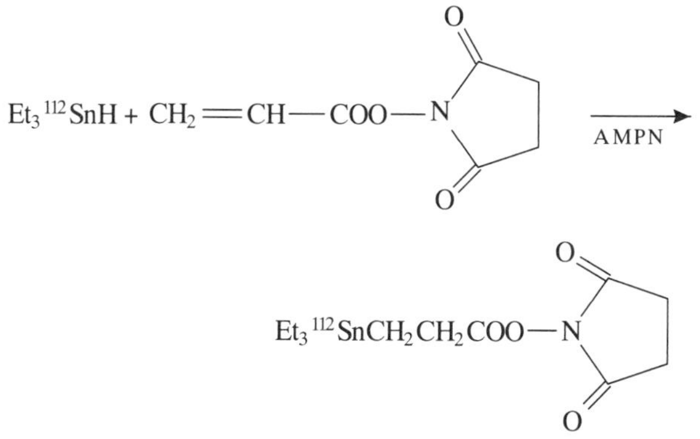

$$Et_3{^{112}}Sn\,CH_2CH_2COOH \xrightarrow{\text{DCC}} Et_3{^{112}}Sn\,CH_2CH_2COO{-}N$$

(29)

$$Et_3{^{112}}SnH + CH_2{=}CH{-}COO{-}N \xrightarrow{\text{AMPN}}$$

$$Et_3{^{112}}SnCH_2CH_2COO{-}N$$

AMPN = 2,2′–azobis–(2–methylpropionitrile)

Testa and Dooley[34] have also prepared several organotin compounds labelled with specific isotopes of tin. Tetrabutyltin-124 was prepared in 80% yield by reacting tin tetrachloride enriched to 94% purity in tin-124 with butylmagnesium bromide (equation 30).

$$^{124}SnCl_4 + 4BuMgBr \longrightarrow Bu_4{}^{124}Sn + 4MgClBr \qquad (30)$$

The synthesis was carried out by dissolving the tin-124 tetrachloride in hexane and adding the resulting solution to an excess of the Grignard reagent in diethyl ether at 0 °C. After hydrolysis with dilute HCl, the organic layer from the reaction was washed with potassium fluoride to precipitate the various butyltin chlorides as insoluble fluorides. The tetrabutyltin-124, which was soluble in the ether layer, was used as a tracer for organotin compounds in the environment. The advantage of using the specifically labelled organotin compound was that it simplified the mass spectrum of the tin compounds, thereby increasing the detection level for these compounds in the environment.

Tributyltin-124 bromide has been prepared in a yield of 80% by reacting tetrabutyltin-124 in anhydrous methanol with one equivalent of bromine (equation 31).

$$Bu_4{}^{124}Sn + Br_2 \longrightarrow Bu_3{}^{124}SnBr \qquad (31)$$

This compound has also been prepared by reacting three equivalents of butylmagnesium bromide with tin-124 tetrachloride[35] (equation 32).

$$^{124}SnCl_4 + 3BuMgBr \longrightarrow Bu_3{}^{124}SnBr \qquad (32)$$

This material was used to show that tributyltin halides decompose into dibutyltin dihalides and butyltin trihalides in the environment.

Imura and Suzuki[36] have prepared labelled organotin compounds from artificial tin isotopes produced in a cyclotron. The carrier-free tin-113 radioisotope was produced by irradiating indium-115 oxide with 40-MeV protons (equation 33).

$$^{115}In_2O + p \longrightarrow 3n + {}^{113}Sn \qquad (33)$$

An iodine extraction method was used to isolate the tin-113 from the irradiation mixture as $^{113}SnI_4$. The desired organotin compound, butyltin-113 trichloride, was prepared by first converting the $^{113}SnI_4$ into $^{113}SnCl_4$ by treatment with anhydrous tin tetrachloride. Then, heating the tin-113 tetrachloride with an excess of tetrabutyltin for three hours at 110 °C (equation 34) yielded butyltin-113 trichloride.

$$^{113}SnI_4 + SnCl_4 \longrightarrow {}^{113}SnCl_4 \xrightarrow[110\,°C,\,3\,h]{Bu_4Sn} Bu{}^{113}SnCl_3 \qquad (34)$$

When the reaction with tetrabutyltin was carried out at 220 °C for four hours, the product was dibutyltin-113 dichloride (equation 35).

$$^{113}SnCl_4 \xrightarrow[220\,°C]{Bu_4Sn} Bu_2{}^{113}SnCl_2 \qquad (35)$$

Other workers have also made tributyltin-113 labelled compounds for environmental and metabolic studies. For instance, Brown and coworkers[37] prepared bis(tributyltin-113) oxide by first refluxing tin-113, which was produced by neutron irradiation of metallic tin, in a bromine–chloroform solution for four hours. The resulting tin-113 tetrabromide was subsequently converted into tributyltin-113 bromide by reaction with three equivalents of unlabelled tetrabutyltin for four hours at 220 °C. The bis(tributyltin-113) oxide was finally obtained by hydrolysing the tributyltin-113 bromide with a KOH–95% ethanol solution

$$^{113}Sn + 2Br_2 \longrightarrow {}^{113}SnBr_4$$

$$^{113}SnBr_4 + 3Bu_4Sn \xrightarrow[220\,°C]{4\,h} Bu_3{}^{113}SnBr$$

$$2\,Bu_3{}^{113}SnBr + 2KOH \longrightarrow (Bu_3{}^{113}Sn)_2O$$

SCHEME 9

(Scheme 9). This labelled tin oxide was used to determine the clearance of bis(tributyltin) oxide from mice. The study was undertaken because bis(tributyltin) oxide is used to control schistomiasis. Many other labelled tin compounds have been used as tracers for tin compounds in fish, in animals and in the environment[34].

Tenny and Tenny[38] also prepared bis(tributyltin-113) oxide, although their yield was lower than that of Brown and coworkers. The Tenny and Tenny synthesis involved reacting tetrabutyltin with tin-113 tetrachloride. The tributyltin-113 chloride obtained in this reaction was subsequently hydrolysed to bis(tributyltin-113) oxide.

Otto and coworkers[39] have prepared tributyltin-113 benzoate from tin-113 in a three-step synthesis (Scheme 10).

$$^{113}Sn + 2Cl_2 \longrightarrow {}^{113}SnCl_4$$

$$^{113}SnCl_4 + 3Bu_4Sn \longrightarrow Bu_3{}^{113}SnCl$$

$$Bu_3{}^{113}SnCl + PhCOONa \longrightarrow Bu_3{}^{113}SnOOCPh$$

SCHEME 10

Tributyltin chloride labelled with a metastable tin-117 isotope has been prepared by Imura and Suzuki[36]. This artificial isotope was made by irradiating tin-118 labelled bis(tributyltin) oxide with 30-MeV γ-rays for six hours (equation 36).

$$(Bu_3{}^{118}Sn)_2O + \gamma\text{- rays} \longrightarrow (Bu_3{}^{117m}Sn)_2O + n \qquad (36)$$

The bis(tributyltin-117m) oxide was converted into the tributyltin-117m chloride in a reaction with sodium chloride (equation 37).

$$(Bu_3{}^{117m}Sn)_2O + NaCl \longrightarrow Bu_3{}^{117m}SnCl \qquad (37)$$

The tributyltin-117m chloride was purified by extraction into benzene. This isotope of tin, which has a half-life of 14 days, was used in the isotope dilution analysis of environmental organotin compounds.

A bis(tributyltin-125) oxide has been made by irradiating the corresponding tin-124 compound with a high neutron flux for one minute (equation 38).

$$(Bu_3{}^{124}Sn)_2O + n \longrightarrow (Bu_3{}^{125}Sn)_2O + \gamma \qquad (38)$$

The radioactive tin-125 isotope in the bis(tributyltin-125) oxide enabled Klotzer and Gorner[40] to use a neutron activation analysis to determine the percent tin in the organotin compounds formed in the photochemical reaction between bis(tributyltin-125) oxide and glucose (equation 39).

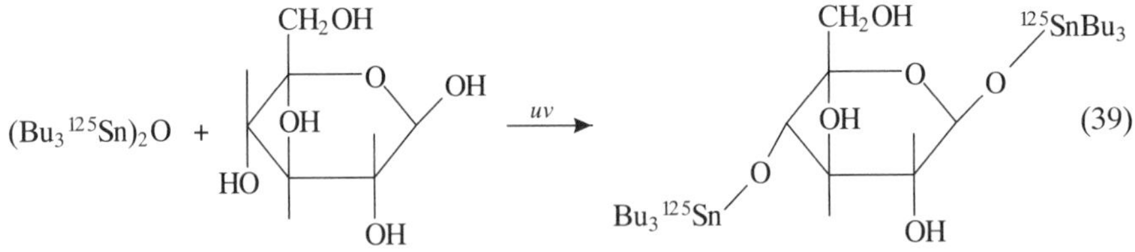

$$(Bu_3{}^{125}Sn)_2O \; + \qquad \xrightarrow{uv} \qquad \qquad (39)$$

Finally, Podoplelov and coworkers[41] have prepared trimethyltin hydride and benzyltrimethyltin labelled with tin-117 from a tin sample enriched to 92% in ^{117}Sn. The photochemical reaction between dibenzyl ketone and trimethyltin-117 hydride was used to investigate the chemically induced dynamic nuclear polarization of tin-containing radicals.

2. Preparation of compounds with the isotope in a group bonded to the tin atom

Most compounds in this category are tin hydrides or stannanes where a hydrogen isotope is bonded to the tin atom. The synthesis of these labelled stannanes will be presented in detail. However, the synthesis of other labelled tin compounds is also discussed.

a. The synthesis of tin compounds a hydrogen isotope is bonded to the tin atom. Although tributyltin deuteride is available from chemical suppliers, a thorough discussion of the methods that have been used for the preparation of isotopically labelled trialkyl- and triarylstannanes with a hydrogen isotope bonded to the tin atom is given in the following paragraphs.

The most commonly used method for synthesizing organotin hydrides with a hydrogen bonded to the tin atom is to reduce the appropriate chlorostannanes with labelled hydride reagents, such as lithium aluminium deuteride or sodium borodeuteride. For example, tributylchlorostannane can be reduced with lithium aluminium deuteride[42-45] or deuterated or tritiated sodium borohydride[46] to give tributyltin deuteride and tritiated tributyltin hydride, respectively (equations 40 and 41).

$$Bu_3SnCl \xrightarrow[\text{or NaBD}_4]{\text{LiAlD}_4} Bu_3SnD \qquad (40)$$

$$Bu_3SnCl \xrightarrow{\text{NaBT}_4} Bu_3SnT \qquad (41)$$

Another synthetic route which gives a good yield of the labelled tin hydride involves the hydrolysis of an organometallic intermediate such as a trialkylstannyllithium[46] with deuterated or tritiated water. The trialkylstannyllithium can be prepared by treating the trialkyltin chloride with lithium metal in THF[46]. This process is shown in equation 42.

$$Bu_3SnCl \xrightarrow[\text{THF}]{\text{Li}} Bu_3SnLi \begin{array}{c} \nearrow^{D_2O} \; Bu_3SnD \\ \searrow_{T_2O} \; Bu_3SnT \end{array} \qquad (42)$$

Alternatively, trialkyltinmagnesium halides[47], prepared by a Grignard reaction using either alkyl bromides or chlorides, can be hydrolysed to give deuterated or tritiated tin hydrides (equation 43).

$$Bu_3SnCl + RMgCl \longrightarrow Bu_3SnMgCl + RH \qquad (43)$$

$$Bu_3SnMgCl + T_2O \longrightarrow Bu_3SnT + MgClOT$$

A wide variety of alkyl groups such as i-propyl, s-butyl, t-butyl and cyclohexyl have been used in the synthesis of labelled tin hydrides.

A comparison of the trialkylstannyllithium–^3H_2O procedure (equation 42) and the tritiated sodium borohydride method (equation 41) indicated that the latter method was cleaner and gave higher yields than the former procedure[46].

One modification to the Grignard method has involved adding a radical initiator, galvinoxyl, to the reaction mixture containing tributyltin hydride and the sterically crowded Grignard reagent, cyclohexylmagnesium bromide[48] (equation 44). In the reaction where this has been used, the tributyltin deuteride was obtained in 89% yield.

$$\xrightarrow{\text{D}_2\text{O}} \text{Bu}_3\text{SnD} + \text{MgBrOD}$$

Finally, Szammer and Otvos have prepared labelled tributyltinhydrides by reducing bis-(tributyltin) oxide with deuterated or tritiated sodium borohydride (equation 45)[49,50].

$$(\text{Bu}_3\text{Sn})_2\text{O} + \text{NaBD}_4 \longrightarrow \text{Bu}_3\text{SnD} \tag{45}$$

However, the yields from these reactions are significantly lower than those from the other methods.

b. Synthesis of other labelled tin compounds. Several other organotin compounds with isotopic labels in the groups on the tin atom have been prepared. For example, McInnis and Dobbs[51] have prepared carbon-14 labelled tetramethyltin and trimethyltin chloride. The actual synthesis involved first reacting methyl-[^{14}C]-magnesium iodide with unlabelled trimethyltin chloride to form the carbon-14 labelled methyl-[^{14}C]-trimethyltin. This compound was then converted into the methyl-[^{14}C]-dimethyltin chloride with tin tetrachloride (equation 46).

$$^{14}\text{CH}_3\text{MgI} + \text{Me}_3\text{SnCl} \longrightarrow {}^{14}\text{CH}_3\text{SnMe}_3 \xrightarrow{\text{SnCl}_4} {}^{14}\text{CH}_3\text{SnMe}_2\text{Cl} \tag{46}$$

The methyl-[^{14}C]-dimethyltin chloride was used to compare the performance of packed and megabore capillary columns in a gas chromatographic analysis for separating mixtures of a carbon-14 labelled trimethyllead chloride, tetramethyltin, dimethyltin dichloride and methyltin trichloride. The megabore column was able to separate all four methyltin compounds quickly, i.e., before the tetramethyltin decomposed into trimethyltin chloride and dimethyltin dichloride (equation 47), a reaction which did occur on the packed columns. Thus, the megabore column enabled the determination of the precise distribution of the various methyltin compounds in an environmental sample. The packed columns, on the other hand, could not separate dimethyltin dichloride and the methyltin trichloride and allowed significant decomposition of the tetramethyltin during the 15 minutes the analysis required.

$$\text{Me}_4\text{Sn} + \text{CH}_3\text{SnCl}_3 \longrightarrow \text{Me}_3\text{SnCl} + \text{Me}_2\text{SnCl}_2 \tag{47}$$

The formation of methyltin derivatives in the environment has been of concern because methyltin compounds are quite toxic. As a result, several workers have attempted to determine how the methyltin compounds are generated in the environment. In one study, a carbon-13 label was used to demonstrate that alkyltin derivatives can be formed in an aquatic environment. Shugui and coworkers[52] mixed methyl-[^{13}C] iodide with tin(II) and/or tin(IV) chlorides in artificial sea water at 20 °C for 30 days in sealed tubes. When the incubation period was complete, the organotin compounds were extracted with a benzene–tropolone mixture and reacted with an excess of butylmagnesium bromide to form the various tetraalkyltin compounds. Examples of these reactions using tin(II) compounds are shown in equations 48 and 49.

$$^{13}CH_3SnCl_3 + 3BuMgBr \longrightarrow Bu_3{}^{13}CH_3Sn \tag{48}$$

$$(^{13}CH_3)_2Sn + 2BuMgBr \longrightarrow Bu_2(^{13}CH_3)_2Sn \tag{49}$$

The various tetraalkyltin compounds were then identified and quantitated by GC MS and gas chromatography–atomic absorption analyses. The results indicated that tin(II) chloride in the simulated sea water was converted into methyltin trichloride and dimethyltin dichloride. The tin(IV) chloride, on the other hand, only formed methyltin trichloride. No trace of trimethyltin chloride was found from either tin(II) or tin(IV). The maximum amount of methyltin trichloride was formed near pH = 6 and at a salinity of 28%. The rate expression for the reaction is

$$d[^{13}CH_3SnCl_3]/dt = (k + k'\ [^{13}CH_3I])[Sn(II)] \tag{50}$$

Since the reaction is first order in both methyl iodide and tin(II), the dimethyltin dichloride is thought to be formed in a disproportionation reaction (Scheme 11).

$$^{13}CH_3I + SnCl_2 \longrightarrow {}^{13}CH_3SnCl_2I$$

$$2\ ^{13}CH_3SnCl_2I \longrightarrow (^{13}CH_3)_2SnCl_2 + SnCl_2I_2$$

SCHEME 11

Wells and coworkers[53] have prepared a series of deuterated tetramethyltin compounds, which they used to study the long-range deuterium isotope effects on the proton chemical shifts of tetramethyltin. The various deuterated tetramethyltin compounds, with one to four trideuteromethyl groups on the tin atom, were prepared by a series of methyl group exchanges beginning with tri-trideuteromethyltin chloride and undeuterated tetramethyltin (Scheme 12).

$$(CD_3)_3SnCl + (CH_3)_4Sn \rightleftharpoons (CD_3)_3SnCH_3 + (CH_3)_3SnCl$$

$$(CD_3)_3SnCl + (CD_3)_3SnCH_3 \rightleftharpoons (CD_3)_4Sn + (CD_3)_2SnCH_3Cl$$

$$(CD_3)_2SnCH_3Cl + (CH_3)_4Sn \rightleftharpoons (CD_3)_2Sn(CH_3)_2 + (CH_3)_3SnCl$$

$$(CH_3)_3SnCl + (CD_3)_3SnCH_3 \rightleftharpoons CD_3Sn(CH_3)_3 + (CD_3)_2SnCH_3Cl$$

SCHEME 12

The NMR spectra of the various deuterated tetramethyltin compounds demonstrated that the long-range isotope effects on the chemical shift are additive and operate in the direction of increased shielding. However, there is no significant isotope effect on the H−Sn coupling constant.

Zhang, Prager and Weiss[54] also synthesized four different deuterated tetramethyltin compounds. These workers prepared the $(CD_3)_4Sn$, $(CD_3)_3CH_3Sn$, $(CD_3)_2(CH_3)_2Sn$ and

$CD_3(CH_3)_3Sn$ by reacting an excess of trideuteromethylmagnesium iodide with tin tetra-chloride, methyltin tribromide, dimethyltin dibromide and trimethyltin bromide, respec-tively. In each reaction, CD_3 groups replace all the halogen atoms on the tin forming the various deuterated tetramethyltin compounds. Equation 51 illustrates these reactions using methyltin tribromide.

$$3CD_3MgI + CH_3SnBr_3 \longrightarrow (CD_3)_3SnCH_3 \qquad (51)$$

The inelastic neutron-scattering technique was used to study the rotational tunnelling of the methyl groups in these deuterated tetramethyltin compounds. The decrease in the tunnel splittings in the inelastic neutron-scattering spectra for the deuterated compounds is evidence of a difference in the inter- and intramolecular association of the deuterated and undeuterated methyl groups. These differences are attributed to the smaller crystal lattice (the C—D bonds are shorter than C—H bonds[55]) and the greater dipole moment of the CD_3 groups.

Stanislawski and coworkers[56] have prepared di-(trideuteromethyl)tin dichloride and (trideuteromethyl)methyltin dichloride for an investigation of the isotope effect on tun-nelling.

Imai and coworkers synthesized tri-trideuteromethyltin bromide and chloride for an investigation of the infra-red spectra of the trimethylgermyl and trimethyltin halides[12]. The tetra-trideuteromethyltin chloride was prepared by reacting tetra-trideuteromethyltin with tin tetrachloride (equation 52), while the tri-trideuteromethyltin bromide was obtained by reacting the tetra-trideuteromethyltin with molecular bromine[57] (equation 53). The tetra-trideuteromethyltin was presumably obtained by reacting an excess of trideuteromethyl-magnesium iodide with tin tetrachloride.

$$(CD_3)_4Sn + SnCl_4 \longrightarrow (CD_3)_3SnCl + CD_3SnCl_3 \qquad (52)$$

$$(CD_3)_4Sn + Br_2 \longrightarrow (CD_3)_3SnBr \qquad (53)$$

Veith and Huch[58] have made a tin dimer containing a deuterium atom on one of the groups bonded to the tin atom (Figure 2). This unusual tin compound was obtained when the acid–base adduct formed between 1,3-di-*tert*-butyl-2,2-dimethyl-1,3,2,4λ^2-diazasilastannetidine with triphenylmethylene-1,1-d$_2$ phosphorane was heated at $120\,^\circ$C in toluene (equation 54).

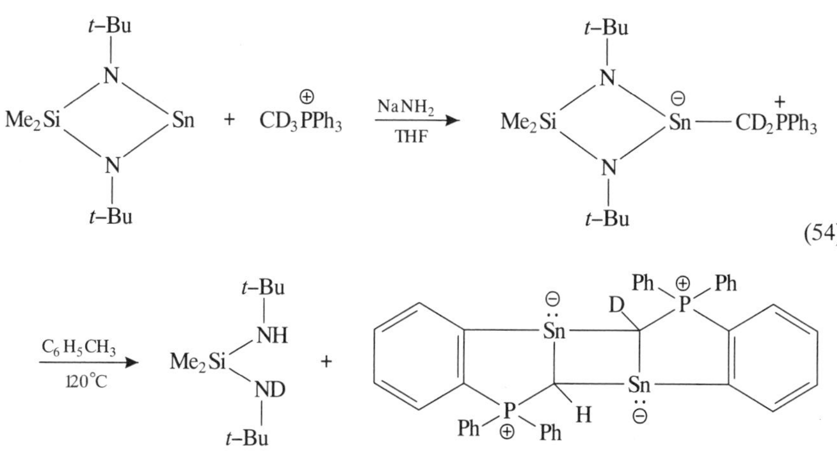

(54)

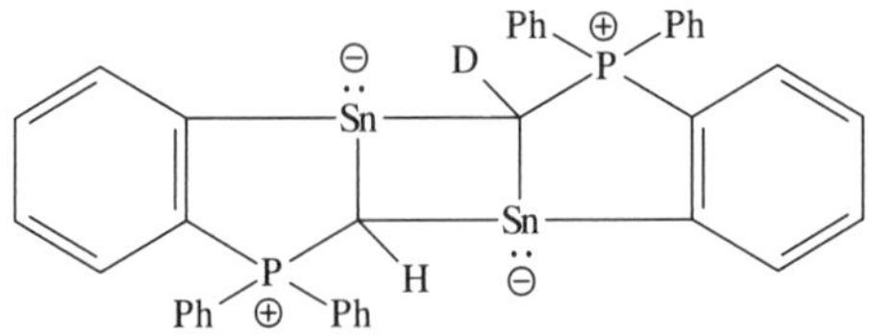

FIGURE 2. The deuterated tin compound formed when the acid–base adduct formed between 1,3-di-*tert*-butyl-2,2-dimethyl-1,3,2,4λ^2-diazasilastannetidine and triphenylmethylene-1,1-d$_2$ phosphorane was heated at 120 °C in toluene

A primary hydrogen–deuterium kinetic isotope effect of 5.34, found for the thermal decomposition of the acid–base adduct, indicated that the alpha-hydrogen (deuterium) of the phosphorane is transferred to the amino nitrogen in the slow step of this first-order reaction. Analysis of the products formed when the deuterated acid–base adduct was heated in toluene showed that only one deuterium is transferred to the amino nitrogens. This occurs because only one hydrogen (deuterium) [H(1) in Scheme 13] is near an amino nitrogen in the crystal structure of the acid–base adduct. The hydrogen that is transferred to the second amino group [H(22) in Scheme 13] is thought to come from a phenyl group on the phosphorus that is over the nitrogen in the crystal structure of the acid–base adduct.

Finally, tin compounds (mainly stannous chloride) have been widely used in the preparation of technetium-99 labelled compounds which are used as tracers for medical purposes. In these syntheses, the tin is used to reduce the metastable technetium-99 from the $+7$ oxidation state to the $+4$ oxidation state (equation 55)[59,60].

$$^{99m}T_cO_4{}^- + SnCl_2 \longrightarrow {}^{99m}TcO^{+2} + Sn^{+4} \tag{55}$$

The technitium $+4$ is then used to form the labelled organometallic technetium complexes that are absorbed in specific organs in the body. Many examples of this use of tin in the preparation of labelled technetium derivatives are given in a review by Noronha[60].

C. Syntheses Using Labelled Tin Compounds

Two important classes of reactions use labelled tin compounds to prepare labelled compounds for mechanistic and analytical purposes. In the first type of reaction, labelled trialkyl- or triaryl tin hydrides (stannanes) are used to reduce (replace) several different groups such as halogen, $-NO_2$, $-N{\equiv}C$, $-N{=}C{=}Se$, $-COOR$, $-SR$ or an acetal group with a deuterium or a tritium atom.

In the second class of reactions, stannanes are added to carbon–carbon π bonds. This enables either (i) the direct addition of a labelled tin atom to an organic substrate when the tin atom is isotopically labelled, or (ii) the introduction of an unlabelled tin group that is subsequently replaced by another isotopically labelled atom such as ^{36}Cl, to give a labelled compound. These two types of reactions are discussed separately in the following paragraphs.

1. Using labelled tin compounds to produce labelled substrates via reduction reactions

The first type of reaction uses a deuterated or tritiated tin hydride (R_3Sn-L) to reduce (replace) a group (LG) with deuterium or tritium, respectively (equation 56),

$$R_3Sn-L + R-LG \longrightarrow R-L + R_3Sn-LG \tag{56}$$

where $L = D$ or T, and LG is the group that is displaced in the reaction.

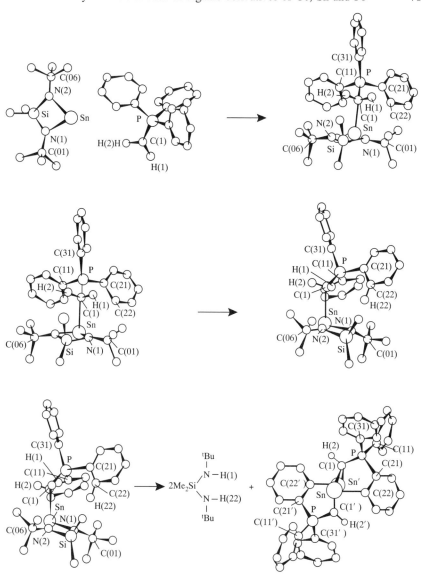

SCHEME 13

Although by far the largest number of these reduction reactions have involved reducing carbon–halogen bonds, a wide variety of groups can be reduced with labelled tin hydrides. Examples of all these reactions are presented in the following paragraphs. This type of reaction has also been recently reviewed by Kotora, Svata, and Leseticky[61].

a. Reduction of alkyl and aryl halides with labelled tin hydrides. Although all the alkyl halides can be reduced with labelled trialkyl- and triaryltin hydrides, bromides are the usual substrates in these reductions. This is because the ease of removing halogen atoms from alkyl halides decreases in the order I > Br > Cl > F. The carbon–fluorine bond is difficult to reduce and is often unreactive when tin hydrides are used[62]. Iodides, on the other hand, are so reactive that they are reduced at room temperature even without an initiator and solvent[63]. In some cases, the reactions of alkyl bromides are also very fast[63–65] although most of the reductions of alkyl and aryl bromides require higher temperatures and initiators[64–73]. Although a few exceptions involving especially reactive chlorides have been reported[74,75], the reductions of alkyl chlorides almost always require initiators and elevated temperatures.

The initiators which are commonly used in these reactions range from azo-bis-isobutyronitrile (AIBN) to organic peroxides such as dibenzoyl peroxide, to ultra-violet light. In some cases, the reaction has been done with two different initiators[42]. A few illustrative examples of this reaction are presented in equations 57–59.

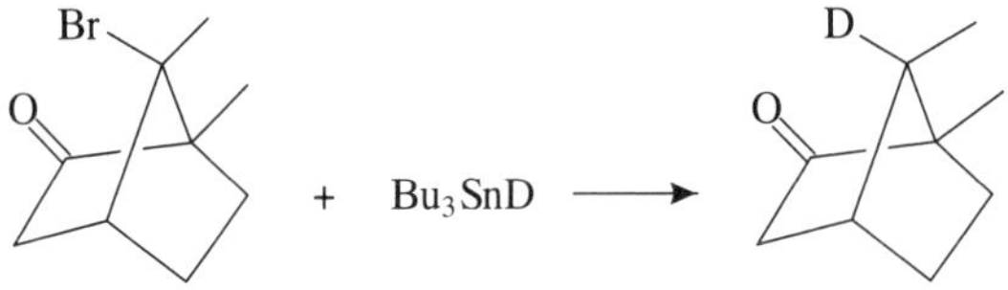

$$+ \quad Bu_3SnD \quad \longrightarrow \qquad\qquad\qquad (57)$$

$$+ \quad Bu_3SnD \quad \xrightarrow{UV} \qquad\qquad\qquad (58)$$

$$+ \quad Bu_3SnD \quad \xrightarrow{AIBN} \qquad\qquad\qquad (59)$$

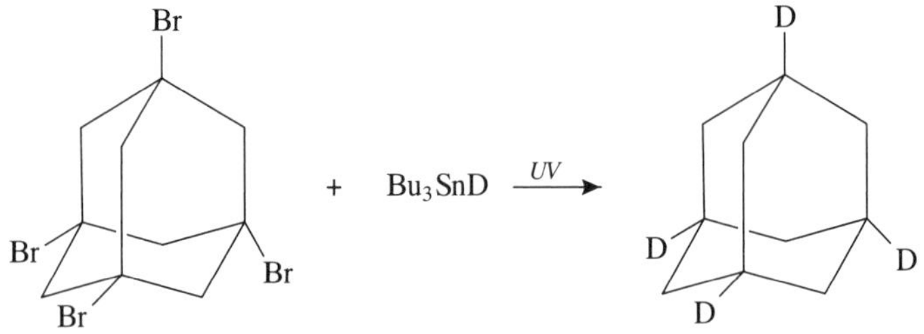

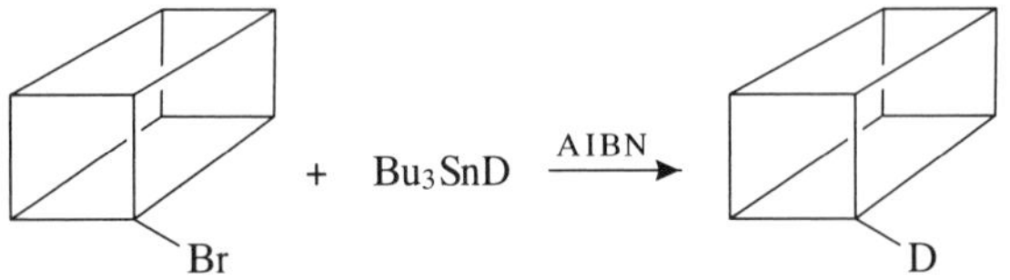

Aryl halides can also be reduced by tin hydrides[76,77], although these reactions always require initiators because the stronger C–X bonds in aryl halides are less reactive than the C–X bonds in alkyl halides. In fact, a series of *meta*- and *para*-substituted bromobenzenes, where X is either *meta*- or *para*-CH$_3$O-, C≡N, Cl, F, CF$_3$, CH$_3$, Bu-*t* or 2,6-dichloro, have been reduced by tributyltin deuteride (equation 60). It is worth noting that the more reactive bromide is reduced selectively in the presence of the less reactive chloride and fluoride groups (equation 61).

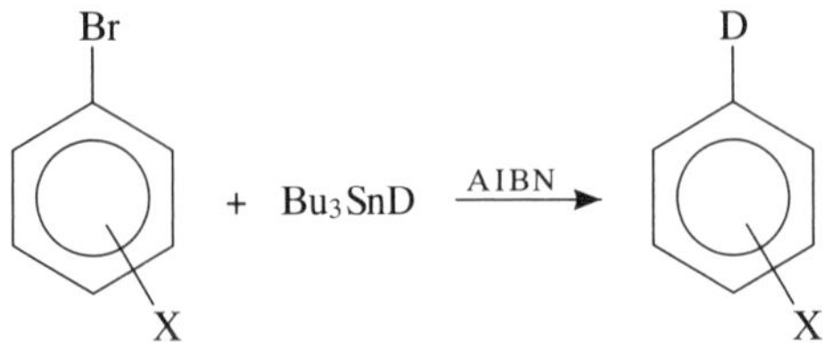

$$+ \quad Bu_3SnD \quad \xrightarrow{AIBN} \qquad\qquad\qquad (60)$$

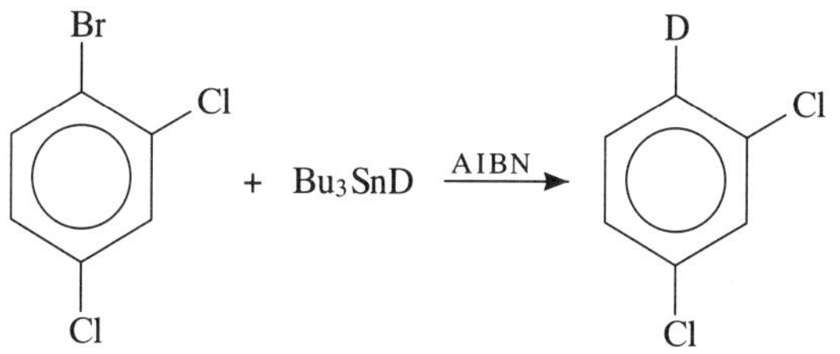

$$(61)$$

Although the tin hydride reductions of alkyl halides seem simple, one must be careful because these reactions occur by a free radical mechanism. This is important, because the carbon radical produced in the reaction can isomerize[68,78] and one often obtains two different stereoisomers from the synthesis. Another problem is that chiral centres can be lost in tin hydride reductions when an optically active halide is reduced. One example of this is the reduction of benzyl-6-isocyanopenicillanate with tributyltin deuteride[78] (Scheme 14). The amount of isomerization depends on the temperature, the concentration of the tin hydride and the presence of β- and γ-substituents[78-82]. However, some authors have reported tin hydride reductions where no racemization was observed[78].

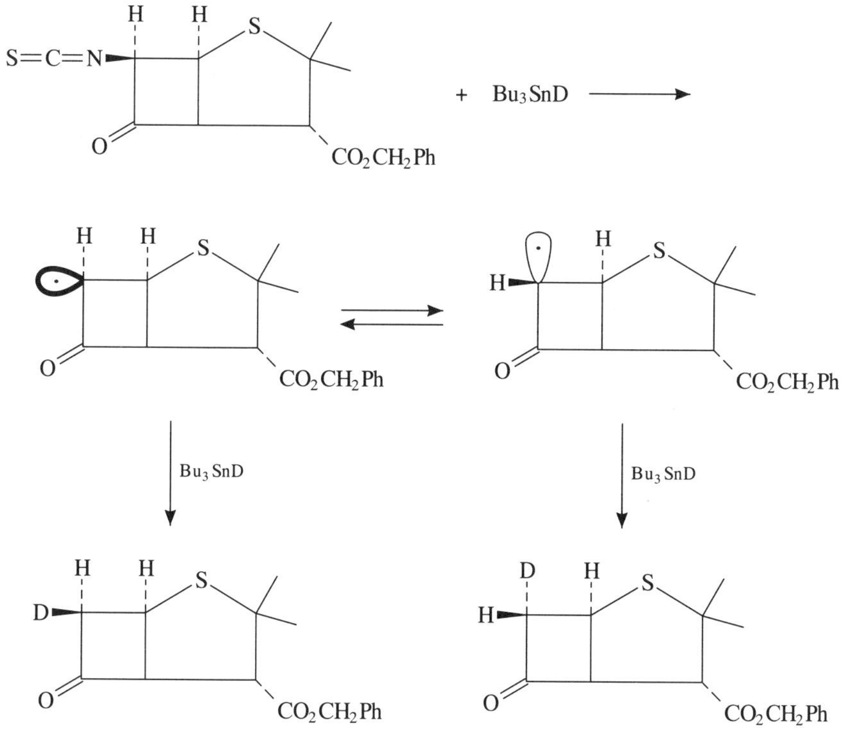

SCHEME 14

A similar problem can occur when vicinal dihalides are reduced. Although both aliphatic and benzylic dihalides can be fully reduced to the labelled hydrocarbon[83,84] (equations 62 and 63), a problem arises when the dihalide is only partially reduced because two products are obtained[74,84–86] (equation 64).

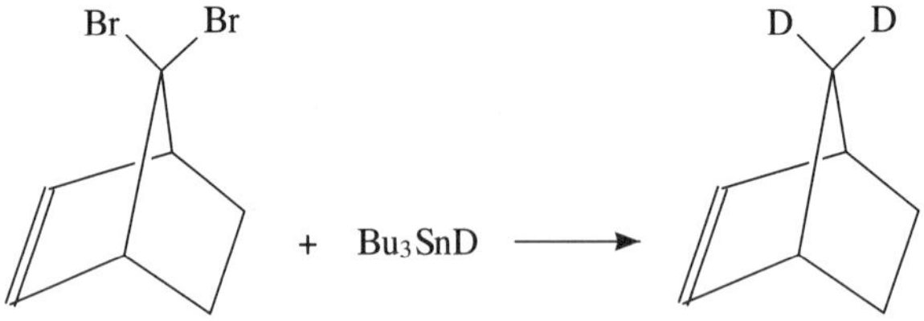

(62)

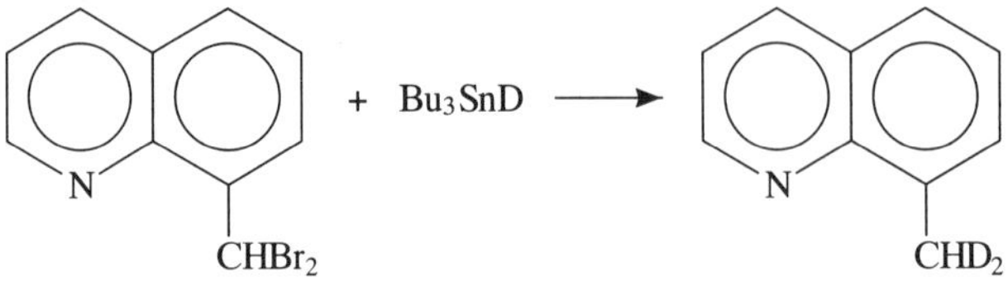

(63)

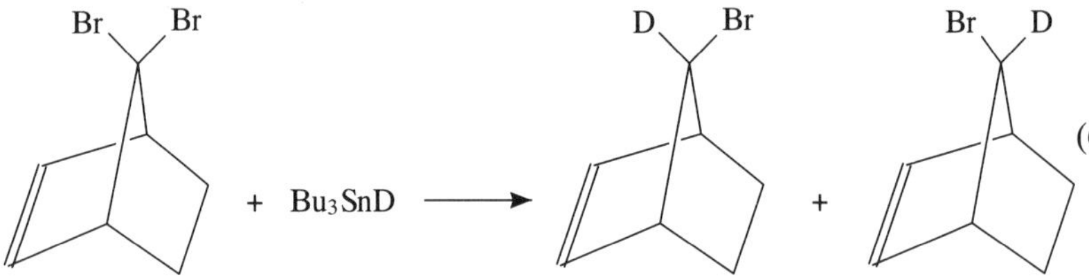

(64)

Trialkylstannanes can be used selectively to reduce halogen atoms in compounds containing different functional groups, e.g. the halogen is reduced preferentially to the carbonyl group of a haloketone[44,87,88] (equation 65) or an ester group of a haloester[89–92] (equation 66). It is interesting that the reduction reaction in equation 66 goes with retention of configuration at the chiral centre.

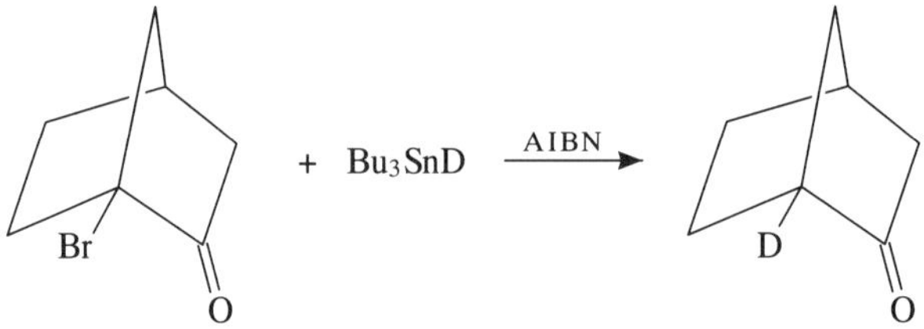

(65)

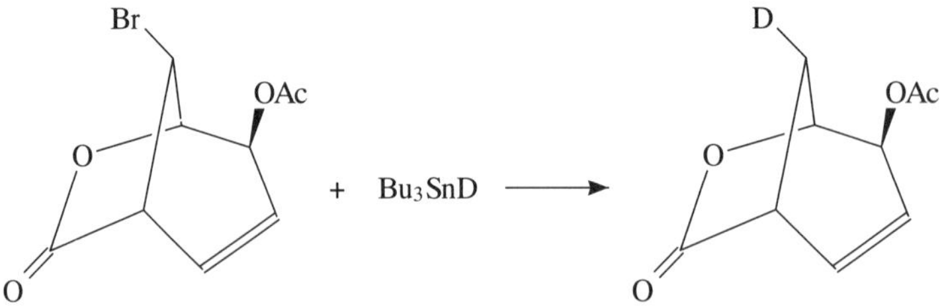

(66)

Halides in heterocyclic compounds can also be reduced selectively by tin hydrides. For example, the more reactive chlorine in a fluorinated aziridine can be reduced readily using very mild conditions and without a catalyst[75]. Again, the more reactive chlorine is reduced preferentially (equation 67).

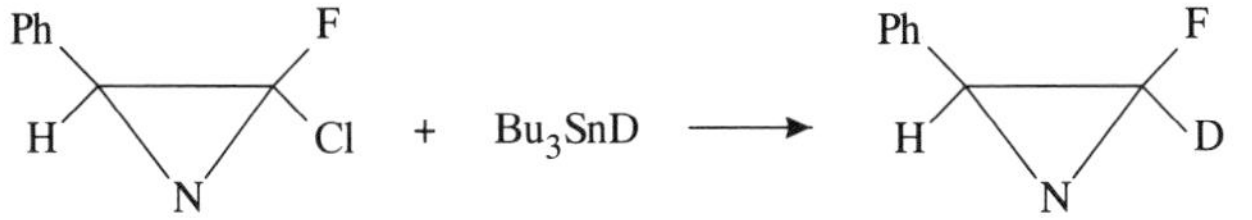

$$(67)$$

Deuterated and tritiated tin hydrides have been used to prepare deuterated saccharides[93] and tritiated steroids[46] from alkyl bromides, (equations 68 and 69). It is important to note that isomerization has occurred at the chiral reaction centre in the saccharide reaction (equation 68). For the steroid, the tin hydride reaction is regiospecific, i.e. it only reacts at the more reactive bromide rather than the less reactive chloride site and does not react with the keto group, the hydroxyl group or the acetal group.

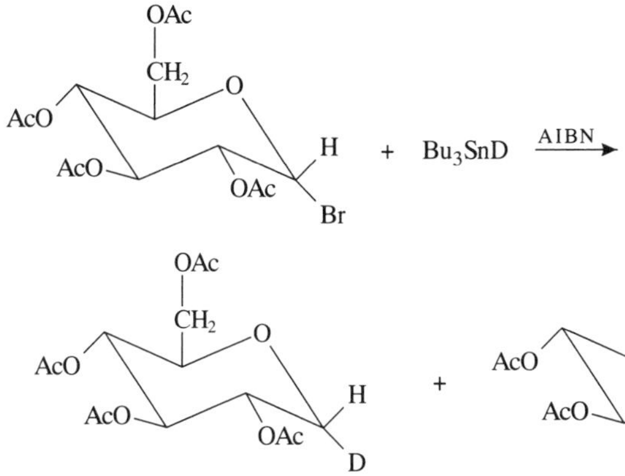

$$(68)$$

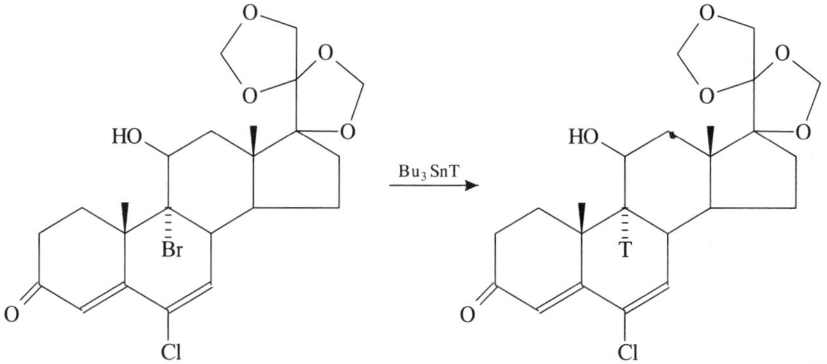

$$(69)$$

Several workers have investigated the reduction of halogens β to a carbon–carbon π bond. For example, Fantazier and Poutsma[94] have found that a halogen β to a carbon–carbon triple bond can be reduced easily with a trialkyltin hydride (equation 70).

$$HC\equiv CCR_2Cl + Bu_3SnH \xrightarrow{AIBN} HC\equiv CCR_2H \qquad (70)$$

Although the alkyne is the major product of these reactions, some rearranged allene is also formed. The allene is produced when the hydrogen on the tin atom is transferred to the terminal carbon of the triple bond of the conjugated radical intermediate $HC\equiv C\dot{C}R_2$ (equation 71).

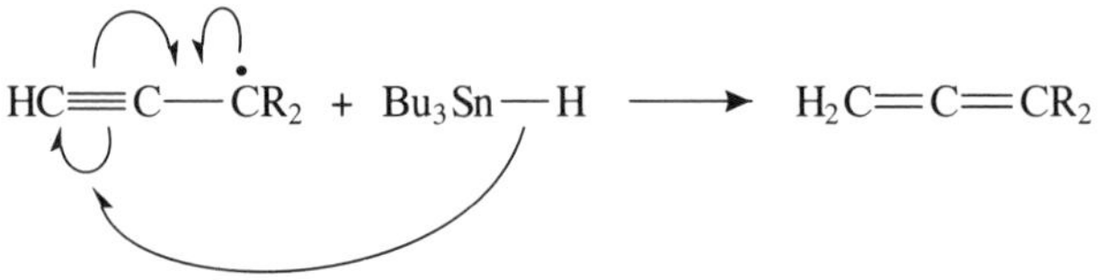

$$HC\equiv C-CR_2 + Bu_3Sn-H \longrightarrow H_2C=C=CR_2 \tag{71}$$

In another study, Hoyte and Denney[95] have demonstrated that alkyl halides β to a carbon–carbon double bond can be reduced cleanly (equation 72).

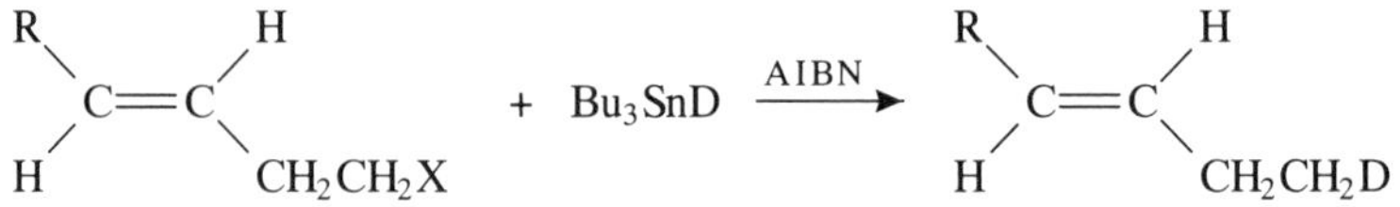

$$\begin{array}{c} R \\ \diagdown \\ C=C \\ H CH_2CH_2X \end{array} + Bu_3SnD \xrightarrow{AIBN} \begin{array}{c} R H \\ \diagdown \diagup \\ C=C \\ \diagup \diagdown \\ H CH_2CH_2D \end{array} \tag{72}$$

In some cases, however, the reactions of β-haloolefins are not clean. For example, cis–trans isomerization has been observed[95] when the halogen atom β to the double bond is reduced (equation 73).

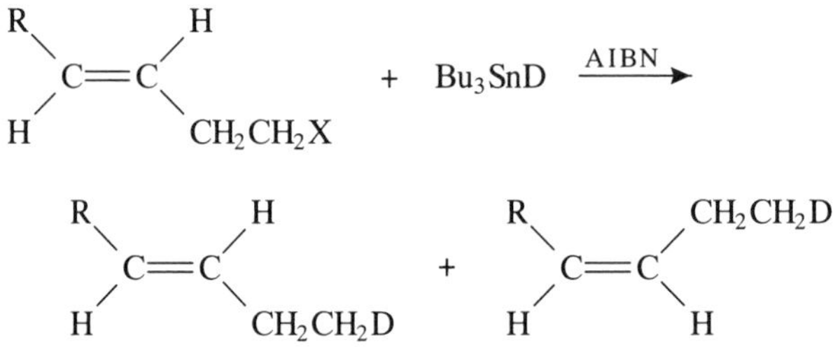

$$\begin{array}{c} R H \\ \diagdown \diagup \\ C=C \\ \diagup \diagdown \\ H CH_2CH_2X \end{array} + Bu_3SnD \xrightarrow{AIBN}$$

$$\begin{array}{c} R H \\ \diagdown \diagup \\ C=C \\ \diagup \diagdown \\ H CH_2CH_2D \end{array} + \begin{array}{c} R CH_2CH_2D \\ \diagdown \diagup \\ C=C \\ \diagup \diagdown \\ H H \end{array} \tag{73}$$

Another complication that can arise in these reactions is that a cycloaddition reaction of the radical intermediate formed in the tin hydride reduction, to an adjacent carbon–carbon double bond, can compete with the simple reduction reaction. This occurs when a five- or six-membered ring can be formed in an intramolecular cycloaddition reaction. For example, Beckwith and Lawrence[96] found both five- and six-membered rings in the product when 1-bromo-2,2,5-trimethylhex-1-ene was treated with tributyltin hydride (Scheme 15).

John and coworkers[78] also observed cyclization to a five-membered ring heterocycle in their hydride reduction of methyl 6β-isothiocyanatopenicillanate with both tributyltin hydride and triphenyltin hydride (Scheme 16). John and coworkers also found this type of ring closure when an isocyanide was reduced with tributyltin deuteride. The mechanism of this arrangement (Scheme 17) has been confirmed by deuterium labelling.

b. Reduction of other groups with tin hydrides. Several other groups can be reduced with tin hydrides to give labelled compounds. For example, acyl halides have been reduced with tin deuterides under mild conditions in the presence of Pd complexes to give deuterated aldehydes in yields of almost 100% (equation 74)[61,97]. Tosylates[98] (equation 75) and S-methyldithiocarbonate groups[99–102] (equation 76) can be removed by treatment with

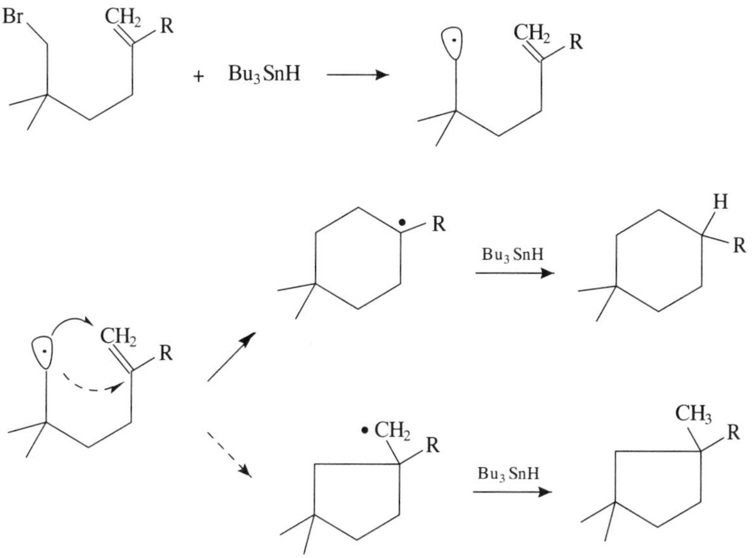

SCHEME 15

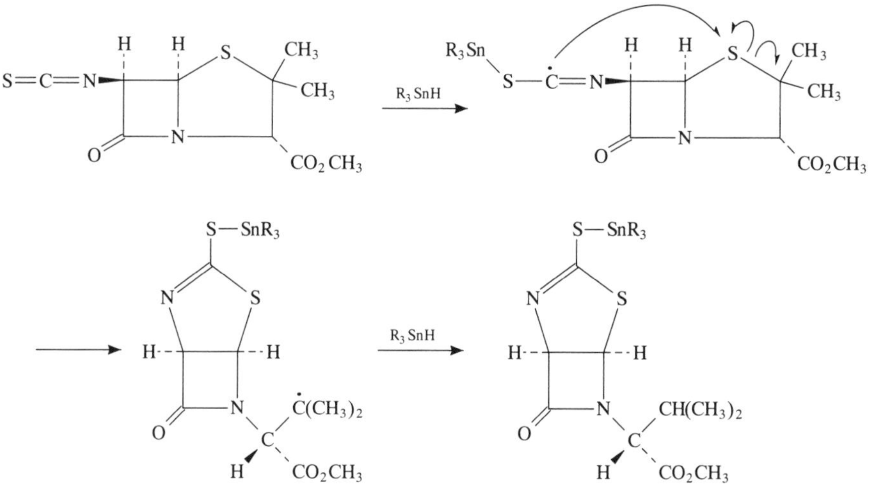

SCHEME 16

organotin hydrides. Unfortunately, some isomerization occurs during the reduction shown in equation 76.

$$CH_3(CH_2)_5\overset{O}{\underset{Cl}{C}} + Bu_3SnD \longrightarrow CH_3(CH_2)_5\overset{O}{\underset{D}{C}} \tag{74}$$

$$CH_3(CH_2)_6CH_2OTs + Bu_3SnD \xrightarrow{\text{AIBN}} CH_3(CH_2)_6CH_2D \tag{75}$$

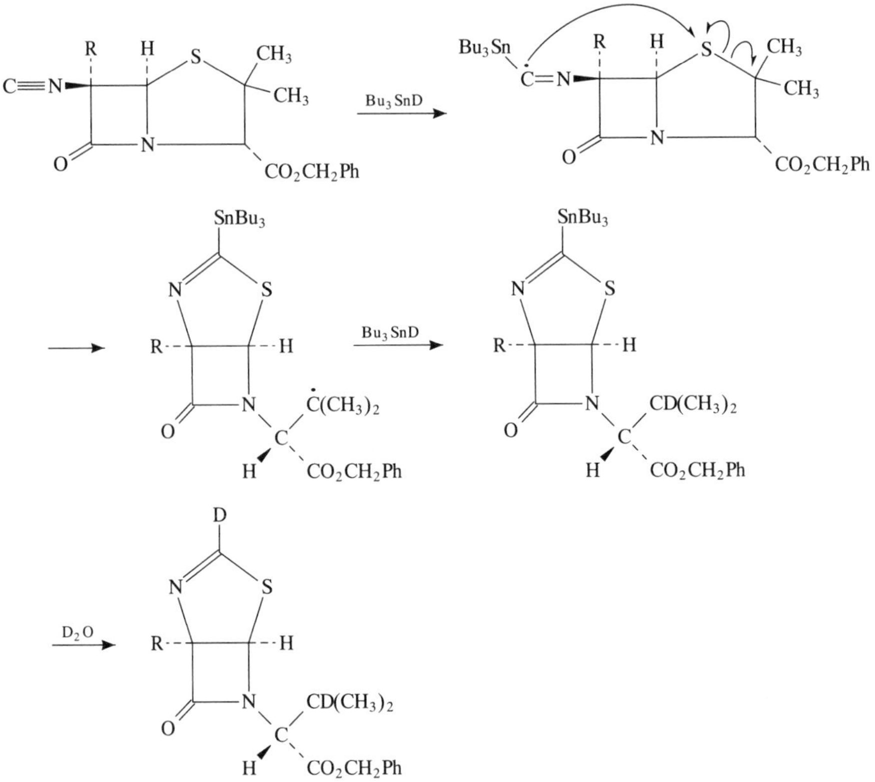

SCHEME 17

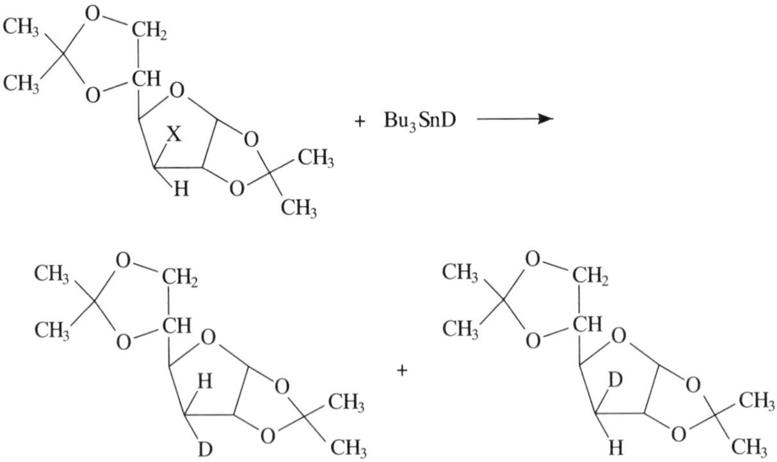

where X is $-O-\overset{\overset{\displaystyle S}{\|}}{C}-SCH_3$

(76)

Organotin hydrides can also reduce a nitro group on either a secondary (equation 77) or a tertiary (equation 78) carbon[103-106]. Although the organotin hydride is not expected to react with the keto, the ester or the cyano groups in the substrates in equations 77 and 78, the reaction in equation 77 demonstrates that the nitro group is more readily reduced than the $-SPh$ (sulphide) group.

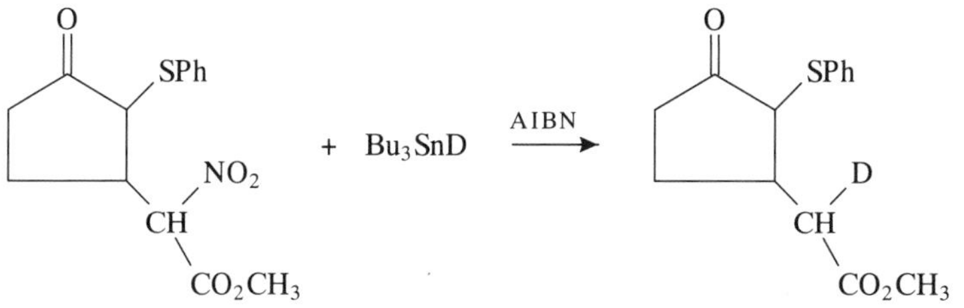

$$(77)$$

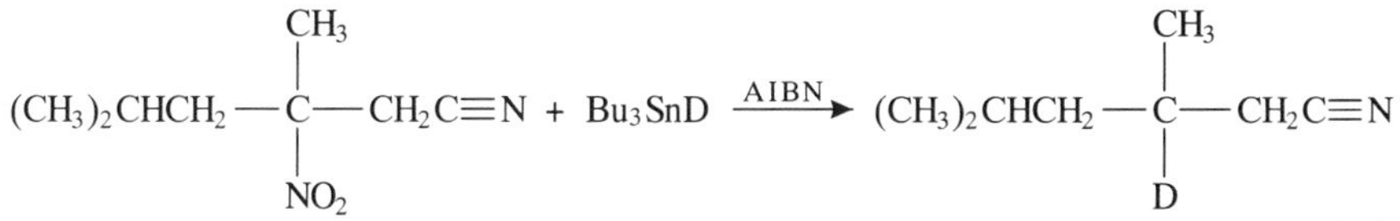

$$(78)$$

Many other groups can be selectively reduced by organotin hydrides. Although these reactions have not been used to date to form labelled compounds, the reactions are included here using tributyltin deuteride as the reducing agent to illustrate the synthetic possibilities for this reagent.

Clive and coworkers have investigated the reduction of both selenide and telluride groups with organotin hydrides[107]. Both the selenide and selenoacetal groups are easily reduced by organotin hydrides without initiators at 120 °C (equations 79 and 80). These groups, which are easily reduced, are useful because they can be reduced selectively in the presence of the less reactive carbonyl and thioacetal groups that are not reduced under these mild conditions.

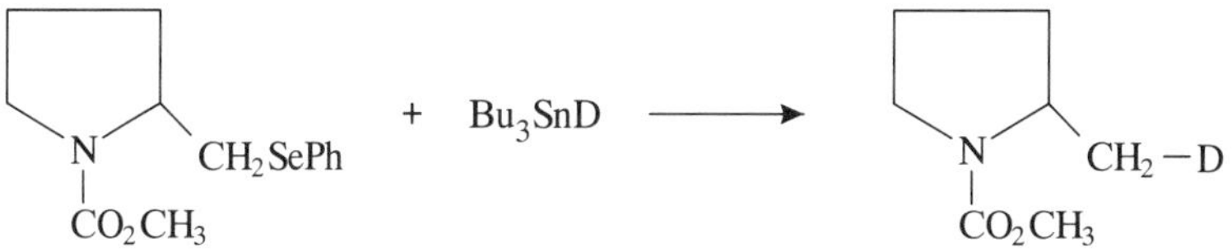

$$(79)$$

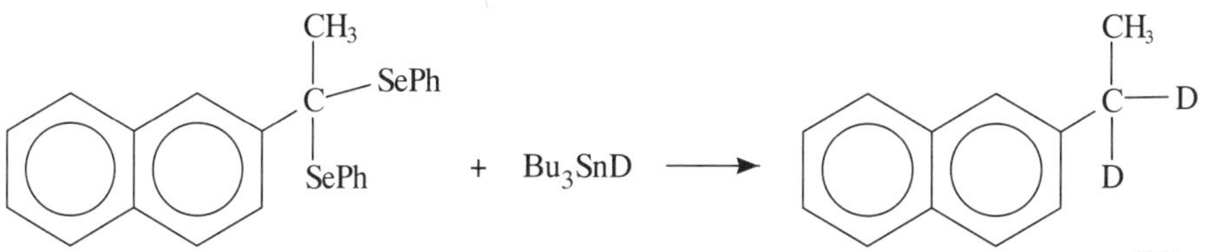

$$(80)$$

The same reaction can be performed under even milder conditions with tellurides, i.e. they react at 80 °C without a catalyst[107] (equation 81).

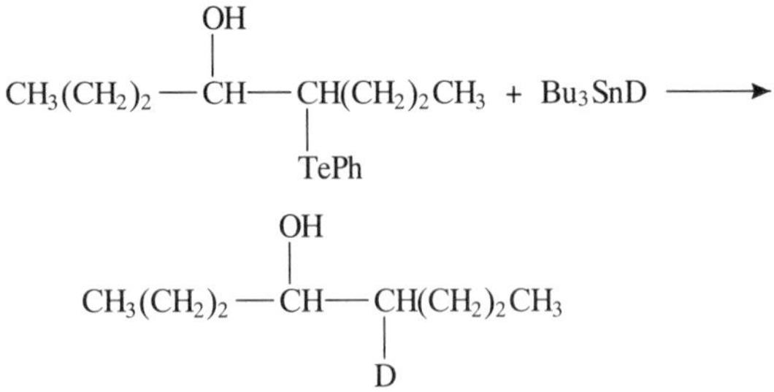

$$CH_3(CH_2)_2 \overset{\overset{\displaystyle OH}{|}}{CH} \overset{}{CH(CH_2)_2CH_3} + Bu_3SnD \longrightarrow$$

$$\underset{\overset{|}{TePh}}{}$$

$$CH_3(CH_2)_2 \overset{\overset{\displaystyle OH}{|}}{CH} \overset{}{CH(CH_2)_2CH_3}$$

$$\underset{\overset{|}{D}}{}$$

(81)

Aldehydes are selectively reduced under mild conditions to alcohols even in the presence of a keto group (equation 82)[108].

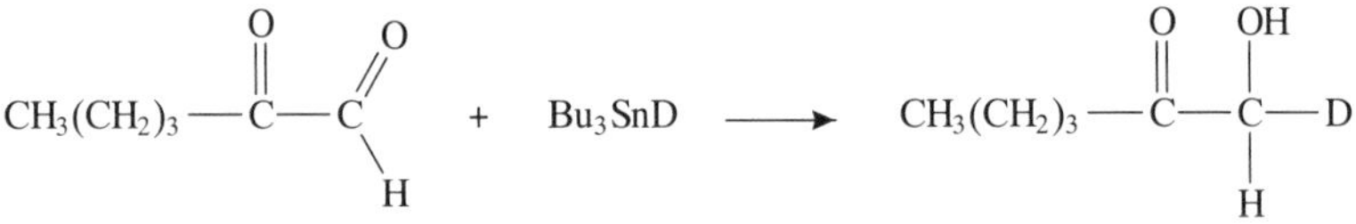

$$CH_3(CH_2)_3 \overset{\overset{\displaystyle O}{\|}}{C} \overset{\overset{\displaystyle O}{\|}}{C} + Bu_3SnD \longrightarrow CH_3(CH_2)_3 \overset{\overset{\displaystyle O}{\|}}{C} \overset{\overset{\displaystyle OH}{|}}{C} D$$

(82)

John and coworkers[78] and Barton and coworkers[109] have successfully reduced the isocyano, isothiocyano and isoselenato groups to the corresponding hydrocarbon when the reaction is initiated with AIBN. The general reaction is shown in equation 83.

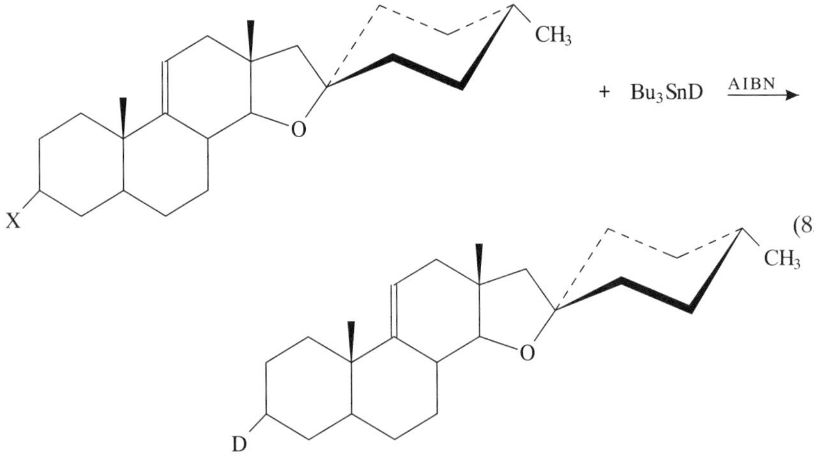

+ Bu₃SnD $\xrightarrow{\text{AIBN}}$

(83)

where X = $-N\equiv C$, $-N=C=S$ or $-N=C=Se$

Although these groups can be reduced with organotin hydrides, John and coworkers[78] found that cyclic products accompanied the reduced (hydrocarbon) products in some reactions; see Schemes 16 and 17 (*vide supra*).

Finally, some important functional groups, notably the keto group and the carboxylic acid group, cannot be reduced with organotin hydrides. However, these groups can be reduced provided they are converted into a more reactive group.

For example, it is possible to reduce the normally unreactive keto group to a hydrocarbon if the ketone is first converted into the more reactive thioacetal[110] (equation 84).

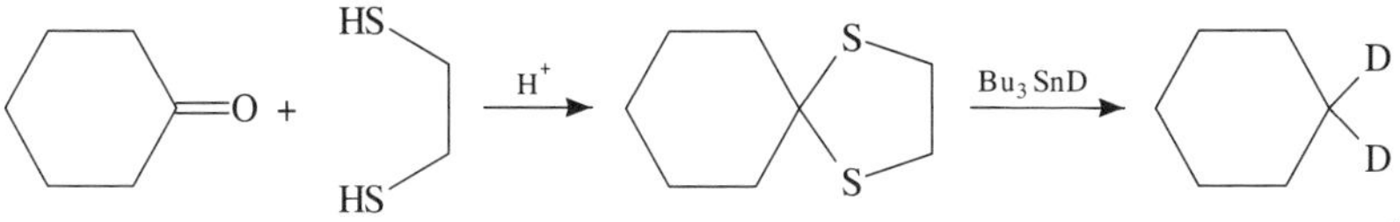

$$(84)$$

A carboxylic acid group that does not react with an organotin hydride can be reduced to a hydrocarbon if one converts the acid group into the more reactive ester group. The ester, however, is not very reactive so the reduction has to be done with an initiator and a very good leaving group. This reaction is illustrated with *trans*-9-hydroxy-10-thiophenyl-9,10-dihydrophenanthrene (Scheme 18). These unusual esters only react because the phenanthrene leaving group, that is released in the step that generates the carboxylate radical, is so stable. It is noteworthy that the reduction does not affect either of the less reactive acetoxy, sulphide or keto groups[111].

2. Using labelled tin compounds to produce labelled substrates via addition to π bonds

In this type of reaction, organotin hydrides or stannanes are added across carbon–carbon π bonds. The addition of an organotin hydride to an alkene or alkyne is important because it opens up a wide range of synthetic possibilities for the formation of labelled tin compounds. When an organotin hydride is added to a carbon–carbon π bond, an isotopically labelled compound can be formed if the organotin hydride is labelled at either (i) the tin atom, (ii) an organic group on the tin atom or (iii) with an isotope of hydrogen on the tin atom (equation 85).

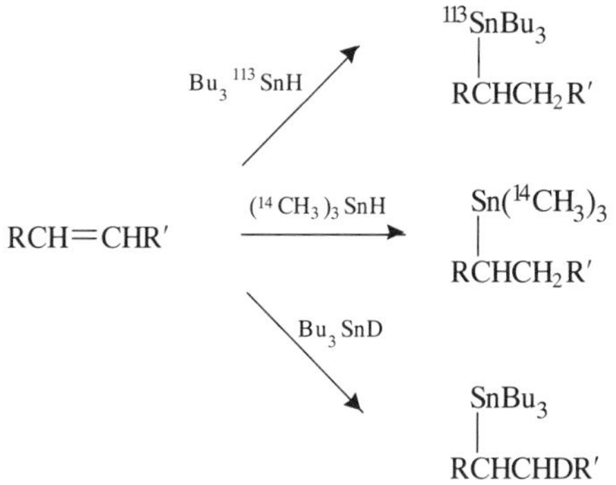

$$(85)$$

Another synthetic route to a labelled compound involves adding an unlabelled tin hydride to a carbon–carbon π bond and then replacing the tin moiety with an isotopically labelled atom, such as ^{125}I[112] (equation 86).

$$(86)$$

All of these syntheses begin with the addition of an organotin hydride to a carbon–carbon π bond. Therefore, it is important to understand the scope of these addition reactions. Organotin hydrides add without initiators to olefinic bonds, which are activated with an electron-withdrawing group. The hydride of the organotin hydride always adds to the more positive carbon in the addition reaction (equation 87).

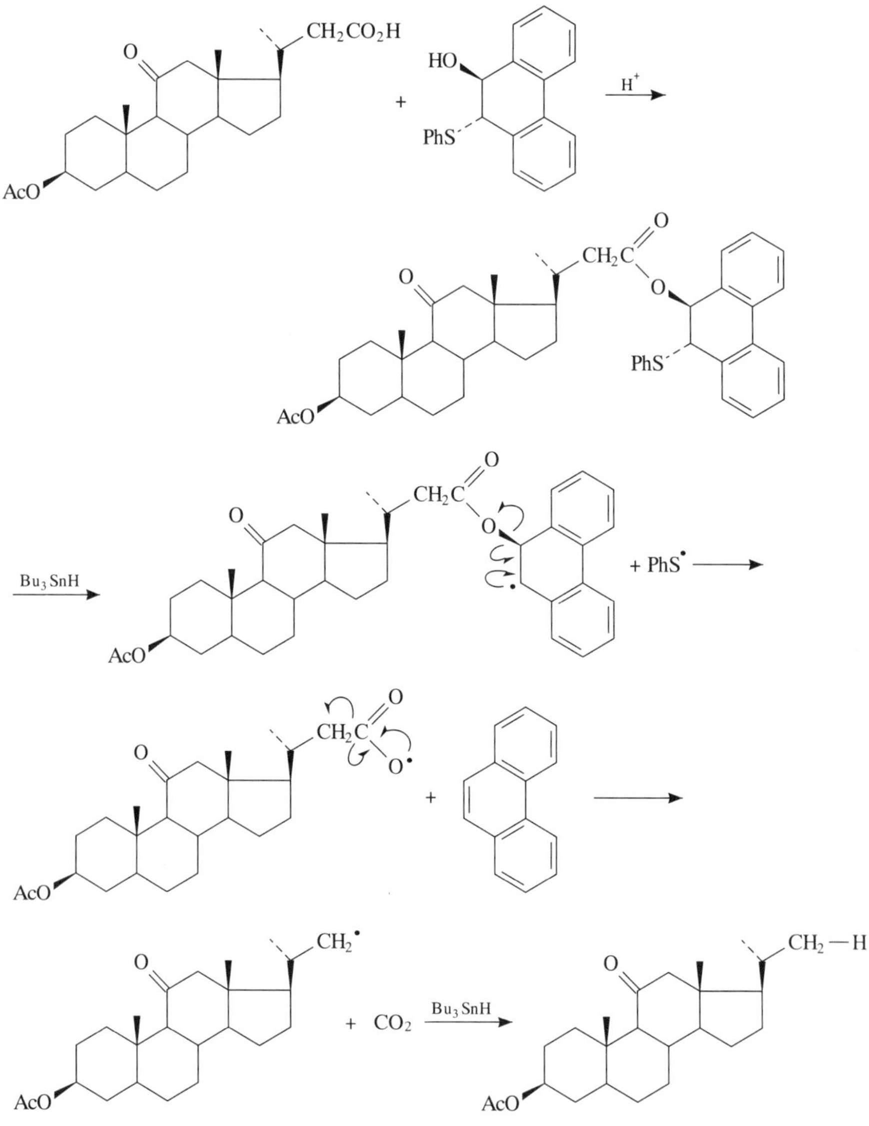

SCHEME 18

$$RCH{=}CHX + Bu_3SnD \longrightarrow \overset{\overset{\displaystyle SnBu_3}{\displaystyle |}}{RCHCHDX}$$

(87)

where $X = -CO_2H, -CO_2H, -C{\equiv}N$ or $-\overset{\displaystyle O}{\overset{\displaystyle \|}{C}}\text{-}R$

Organotin hydrides can also be added to unactivated olefinic (without an electron-withdrawing group on one carbon) bonds provided the reaction is initiated by UV radiation, AIBN or γ-rays[113-116]. In one instance, an organotin hydride has even been added to an unactivated olefin without an initiator. This unexpected addition reaction occurred at high pressure[117].

Organotin hydrides add across carbon–carbon triple bonds even more easily than they add across carbon–carbon double bonds. For example, the addition to the unactivated C≡C bond of phenylacetylene proceeds spontaneously at 20 °C without an initiator (equation 88), whereas the addition to an unactivated olefin requires the presence of an initiator[118].

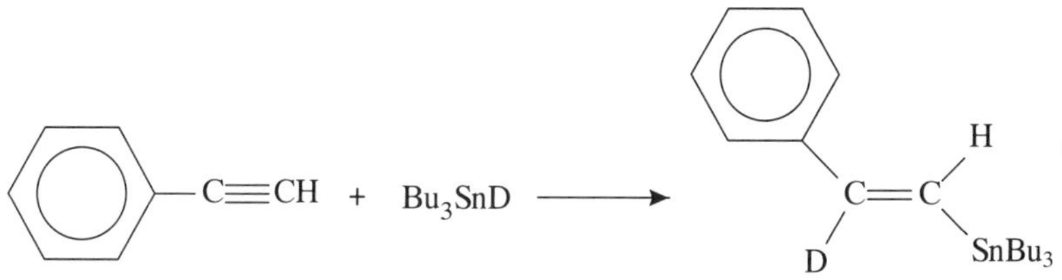

$$\text{Ph}-C\equiv CH \;+\; Bu_3SnD \longrightarrow \quad (88)$$

Although acetylenic bonds are more reactive than C=C bonds, the reactions are often initiated by AIBN or UV radiation. Baldwin and Barden[119] have used the latter method to treat a doubly labelled phenylacetylene with triphenyltin deuteride (Scheme 19). The addition of the triphenyltin deuteride was both regiospecific and gave a stereochemically pure product. A five-step synthesis (Scheme 20) converted this product into an optically pure trideuterophenylcyclopropane, which was used to study the thermal stereomutations that these compounds undergo.

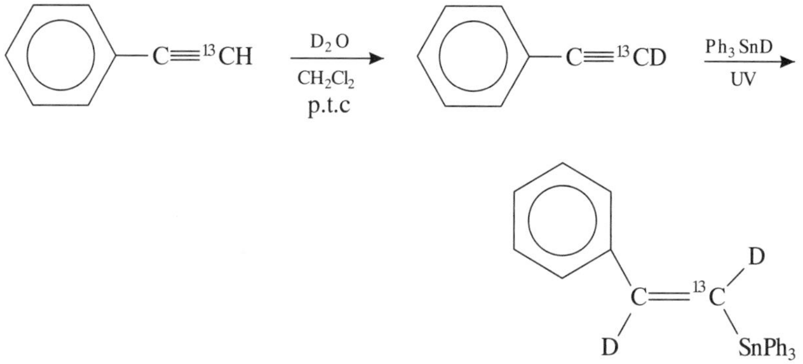

SCHEME 19

While Baldwin and Barden[119] found that triphenyltin deuteride added to the carbon–carbon triple bond of phenylacetylene in a stereospecific reaction, several workers[120-124] have found that the addition of a trialkyltin hydride to a carbon–carbon triple bond gives a mixture of the *cis* and *trans* isomers (equation 89). The more stable *trans* isomer is produced in the highest yield.

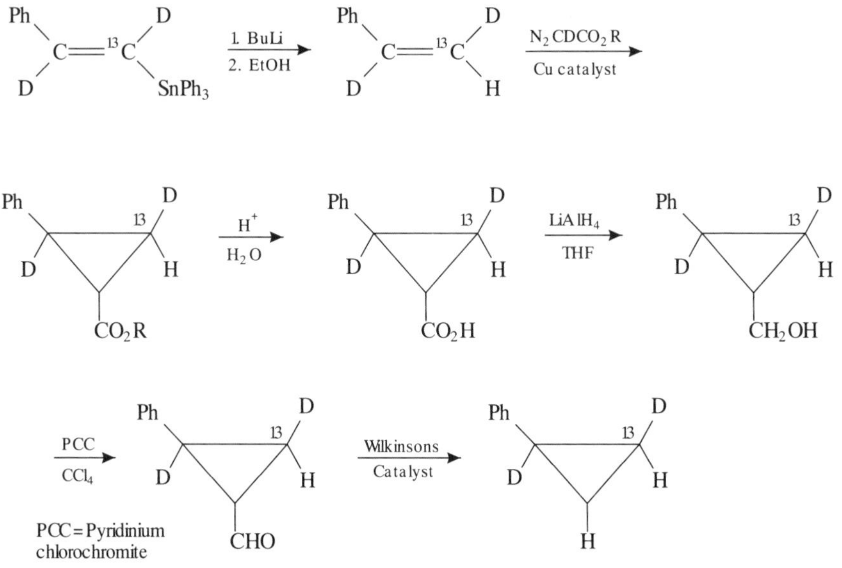

SCHEME 20

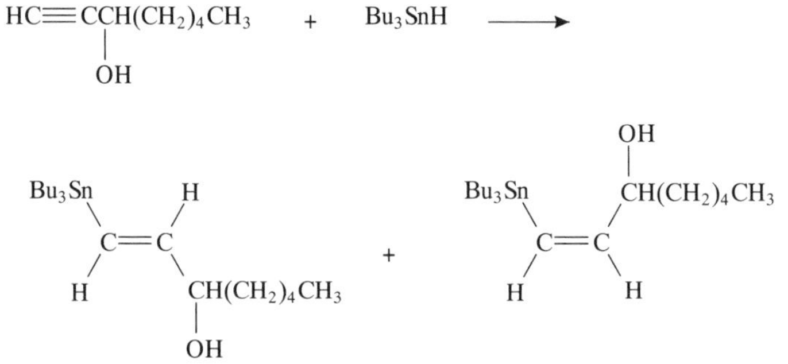

$$HC\equiv CCH(CH_2)_4CH_3 \quad + \quad Bu_3SnH \longrightarrow$$

with OH below the propargyl carbon

(89)

3   :   1

Obviously, the stereochemistry of these addition reactions is controlled by other factors. For instance, the mixture of *cis* and *trans* products obtained depends on the amount of the trialkyltin hydride and the temperature. Generally, a greater excess of the trialkyltin hydride and a higher temperature increase the yield of the more stable *trans* isomer[123].

In one case, the addition reaction was neither regiospecific nor stereospecific[124] (equation 90). Obviously, one will have to choose the reaction conditions for the addition reactions to alkynes carefully when stereochemistry is important.

D. Using Tin Compounds to Prepare Labelled Compounds

A wide variety of labelled compounds can be synthesized by a two-step process involving the formation of a trialkyl- or triarylstannyl derivative. In the first step, a trialkyl- or triaryltin group is added to (i) a π bond of an alkene or an alkyne, or (ii) an aryl group in the substrate. Then, in the second step, the trialkyl- or triarylstannyl group is replaced by either (i) an isotope of hydrogen or (ii) an isotopically labelled group[78,125,126].

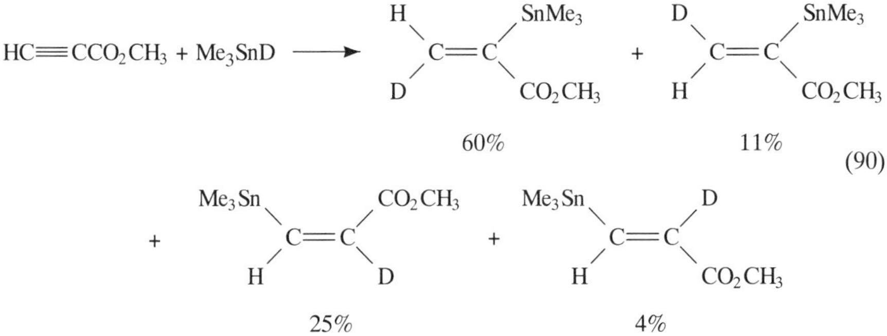

$$HC{\equiv}CCO_2CH_3 + Me_3SnD \longrightarrow \underset{D}{\overset{H}{>}}C{=}C\underset{CO_2CH_3}{\overset{SnMe_3}{<}} + \underset{H}{\overset{D}{>}}C{=}C\underset{CO_2CH_3}{\overset{SnMe_3}{<}}$$

60%   11%

$$+ \underset{H}{\overset{Me_3Sn}{>}}C{=}C\underset{D}{\overset{CO_2CH_3}{<}} + \underset{H}{\overset{Me_3Sn}{>}}C{=}C\underset{CO_2CH_3}{\overset{D}{<}}$$

25%   4%

(90)

An example of the first type of reaction, where a trialkylstannyl group is replaced by an isotope of hydrogen, is shown in equation 91. In this example, methanolysis using methanol–O–D replaces the tributylstannyl group with a deuterium atom[127,128].

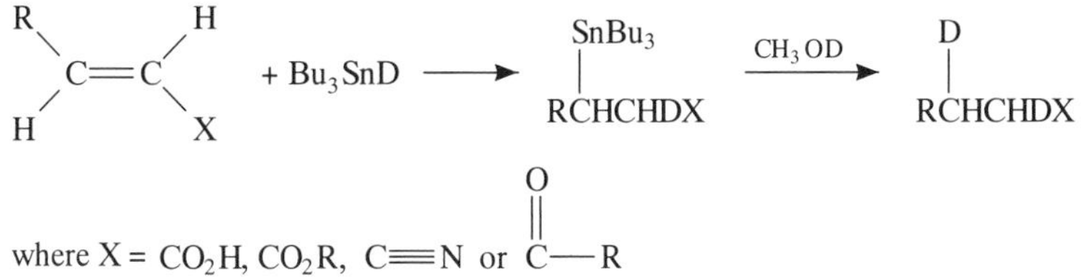

$$\underset{H}{\overset{R}{>}}C{=}C\underset{X}{\overset{H}{<}} + Bu_3SnD \longrightarrow \underset{RCHCHDX}{\overset{SnBu_3}{|}} \xrightarrow{CH_3OD} \underset{RCHCHDX}{\overset{D}{|}}$$

(91)

$$\text{where } X = CO_2H, \ CO_2R, \ C{\equiv}N \ \text{ or } \ \overset{\overset{O}{\parallel}}{C}{-}R$$

The second type of reaction, which involves displacing a trialkylstannyl group with a radioisotope, is illustrated by the radiohalodestannylation shown in equation 92. In fact, this latter synthetic strategy has been widely used for preparing radiohalogen compounds, which in turn have been used as therepeutic agents and/or radiotracers in medical and biochemical research. The general procedure involves preparing a vinyl- or an aryltrialkyltin compound and then displacing the tin moiety with a radioactive halogen (equations 93 and 94).

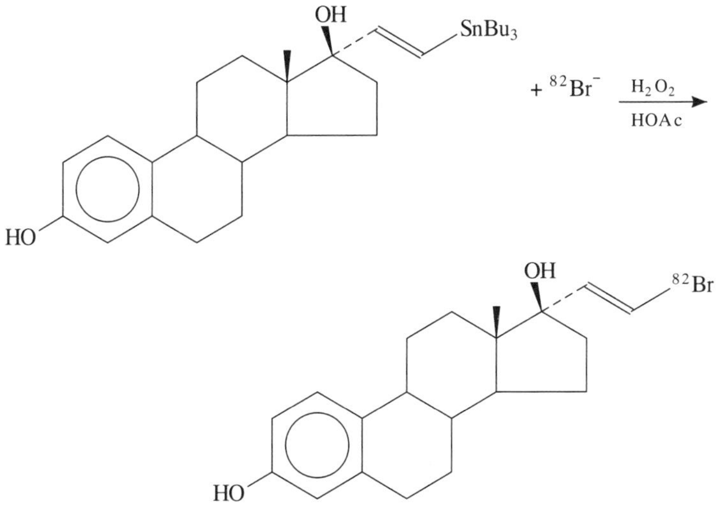

(92)

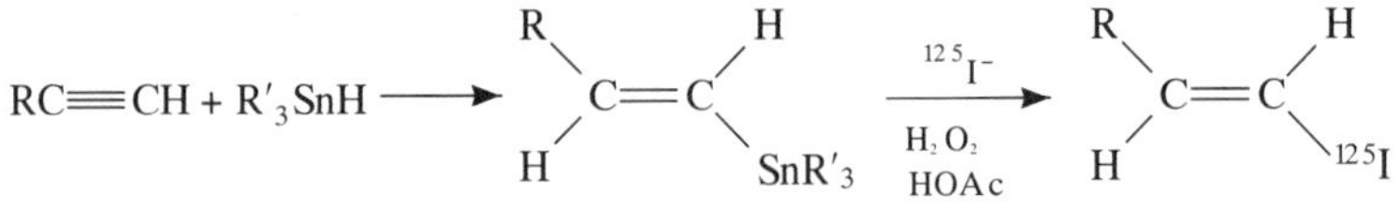

$$RC\equiv CH + R'_3SnH \longrightarrow \underset{H}{\overset{R}{\diagdown}}C=C\underset{SnR'_3}{\overset{H}{\diagup}} \xrightarrow[\substack{H_2O_2 \\ HOAc}]{^{125}I^-} \underset{H}{\overset{R}{\diagdown}}C=C\underset{^{125}I}{\overset{H}{\diagup}} \quad (93)$$

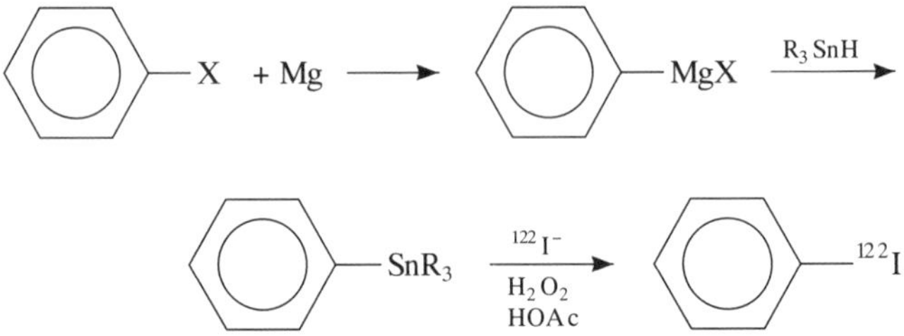

(94)

The trialkylstannyl intermediates required in this synthetic sceme to prepare labelled compounds can be obtained in several ways. One method is the addition of the organotin hydride to the carbon–carbon triple bond of an alkyne (equation 93). These reactions have already been discussed in detail above. A second approach is to add a trialkylstan-nylvinyllithium to a ketone (equation 95), and a third method involves adding trialkylstan-nyllithium to a β-halo, α, β-unsaturated ester (equation 96). Although this last reaction gives a suitable trialkylstannane, these stannanes have proven to be inert in the destanny-lation reaction and, therefore, have not been used extensively to prepare radiolabelled compounds.

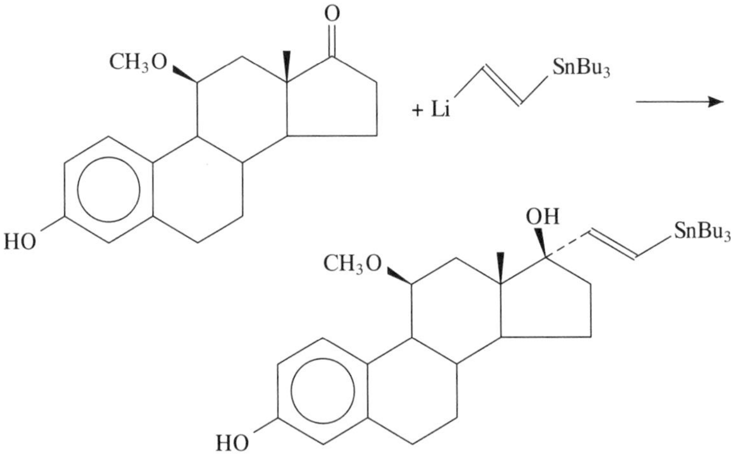

(95)

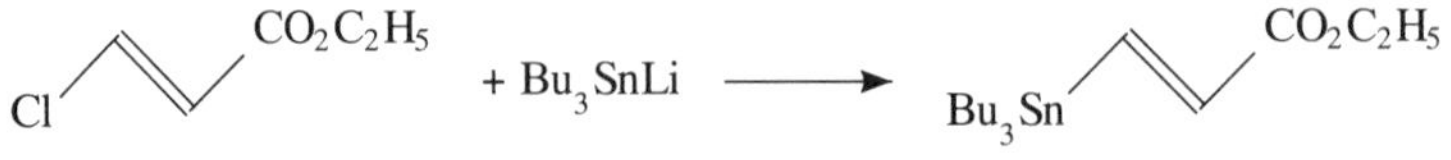

(96)

A trialkyl- or triarylstannyl group can also be added to an aromatic ring. One way this can be accomplished is by treating an aromatic or heteroaromatic compound possessing an active hydrogen(s) with an alkyl lithium and then reacting the lithium salt with a trialkyltin halide (equation 97). Another general method that has been used to attach a trialkyl- or triarylstannyl group to an aromatic ring is to form an organolithium or a Grignard reagent from an aryl halide, and then treat the carbanionic intermediate with a trialkyltin halide (equations 98 and 99).

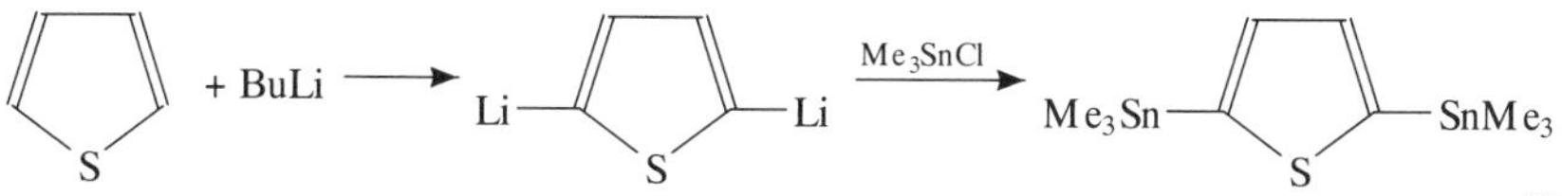

$$(97)$$

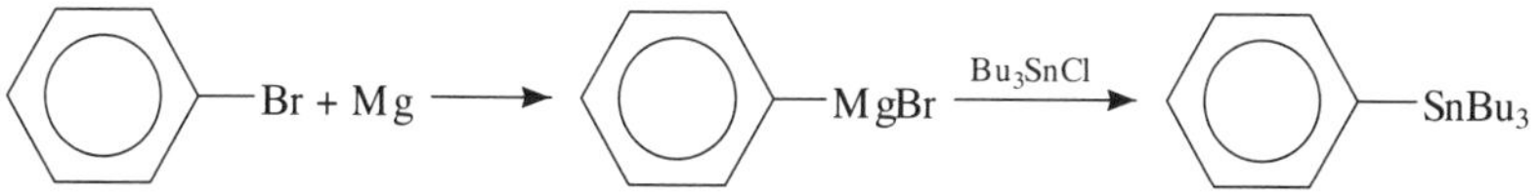

$$(98)$$

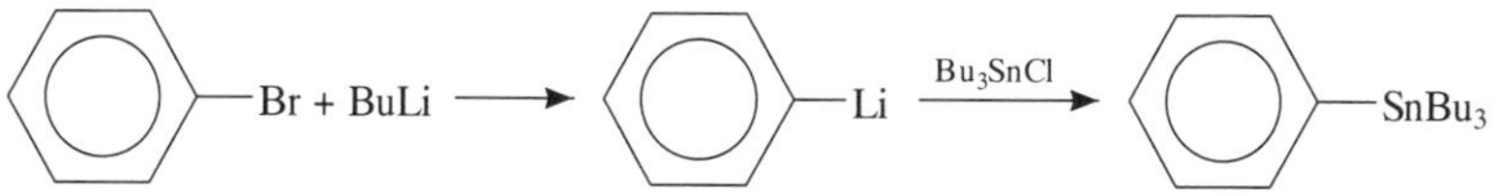

$$(99)$$

The trialkyl- and triarylstannyl derivatives obtained using the above reactions have been used extensively in the preparation of radiolabelled halides (equations 93 and 94). This methodology has been particularly useful for labelling aromatic and vinyl groups, because the trialkyl- or triarylstannyl group on a benzene ring or a vinyl group undergoes *ipso* substitution almost exclusively in the radiohalodestannylation reaction. Although a variety of oxidizing agents have been used in the radiohalodestannylation, this reaction is usually accomplished by reacting a radioactive halide ion with the trialkylstannyl derivative in a hydrogen peroxide–acetic acid medium[129]. These reactions are usually completed in 2–5 minutes and the isolated yields are high, i.e. they are generally between 80 and 90% at the no-carrier-added level[123,130–135]. This general technique has been used to prepare labelled iodides, bromides, fluorides and selenides[136–138,128,129], although the reaction using fluorine-18 had to be done under carrier-added conditions and had relatively low yields[138]. Finally, it is worth noting that the reactions are clean and that the unreacted substrates and tin moiety are invariably easily separated from the product by column chromatography.

This synthetic approach is attractive because the trialkyltin group can be removed rapidly and cleanly under mild reaction conditions and because the expensive label, which invariably has a short half-life, is not added until the last step in the synthesis. Because of these advantages, this synthetic approach has been widely used by several research groups to prepare radiohalogenated labelled hormones, neurohormones, neurotransmitter receptors, etc. for radiotracer and therepeutic uses[123].

Several examples illustrating the scope of this synthetic procedure are presented in Table 7. The use of these radiolabelled halogen compounds has become important because many of the halogen radioisotopes have short half-lifes (Table 8). The short half-life of these radiohalogens is important because one can prepare a specifically labelled radiohalogen compound of high specific activity, inject it into a living species where it can be absorbed into a specific organ and be easily detected when it decays. Then, due to the short half-life, the radioactive material is rapidly lost from the body when the radiohalide decays into a stable species. Having several radiohalogens available is also important, because it allows one to select the radiohalogen with the appropriate half-life for the experiment.

TABLE 7. Some of the labelled compounds that have been formed using the radiohalodestannylation reaction with vinyl- or aryltrialkyl- or triarylstannanes

Substrate	Product	Reference
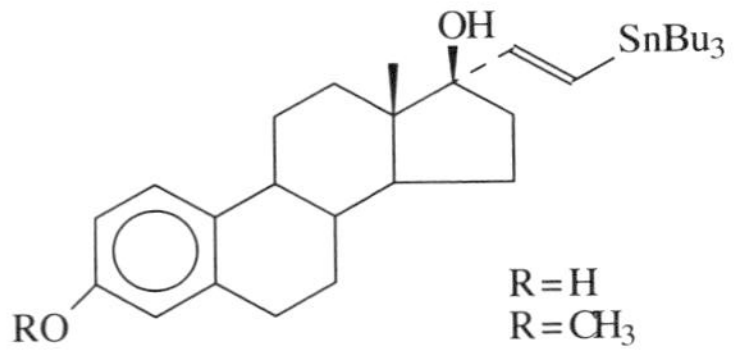 R=H, R=CH₃	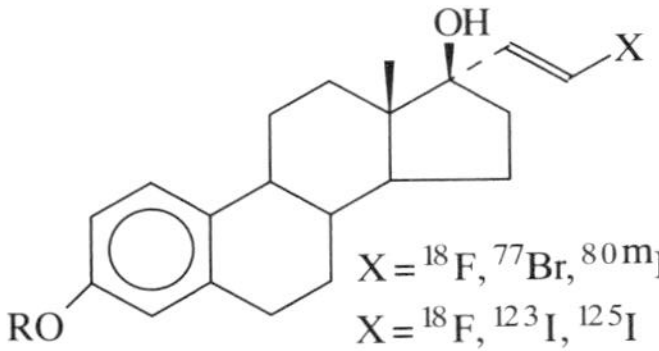 X = ^{18}F, ^{77}Br, ^{80m}Br, ^{82}Br, ^{122}I, ^{123}I, ^{125}I, ^{131}I, ^{211}At, D X = ^{18}F, ^{123}I, ^{125}I	123 136 139
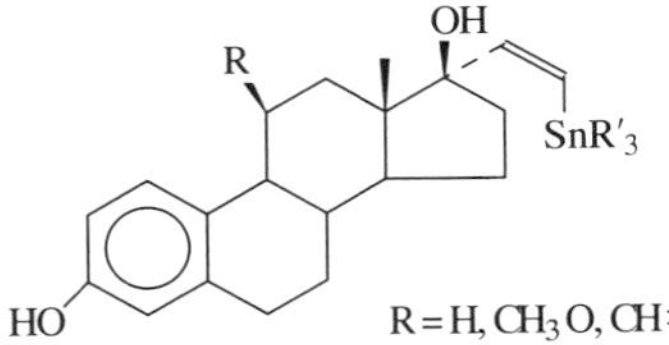 R=H, CH₃O, CH=CH₂	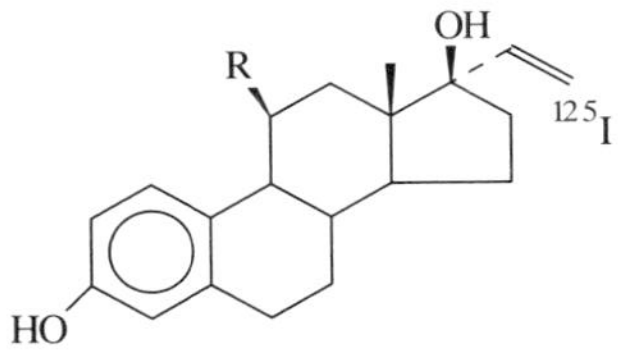	123 137
R¹=OCH₃, R²=H R¹=OCH₃, R²=CH₃ R¹=C₂H₅ R²=H R¹=CH=CH₂ R²=H	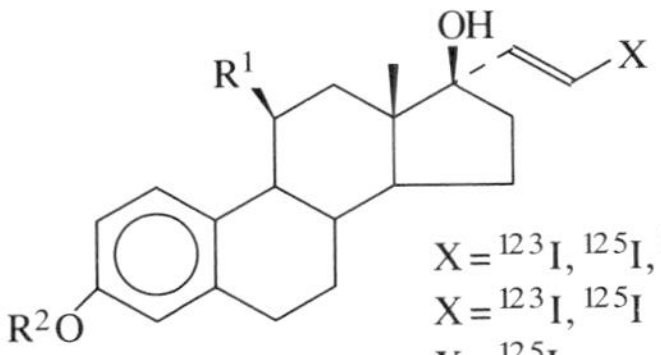 X = ^{123}I, ^{125}I, ^{80m}Br, ^{211}At X = ^{123}I, ^{125}I X = ^{125}I X = ^{125}I	123

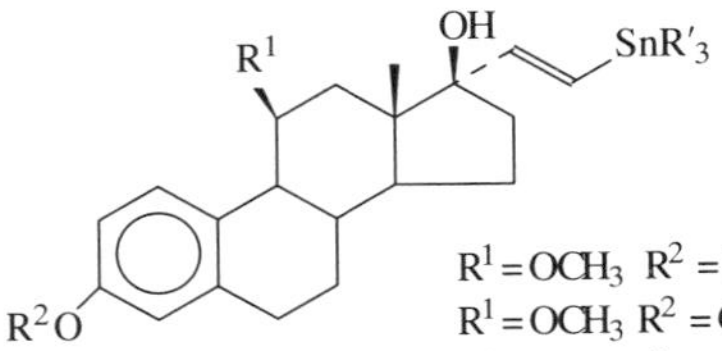

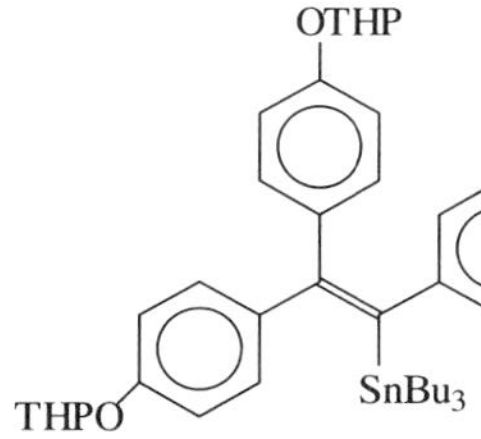

123

132
140
141

142

X = ^{125}I, ^{82}Br, ^{211}At

OH

HO

80m Br

O(CH$_2$)$_2$NMe$_2$

X

C$_2$H$_5$

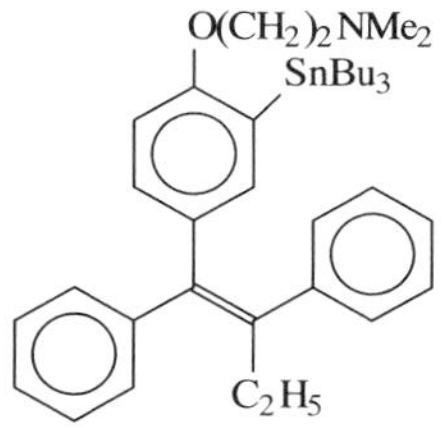

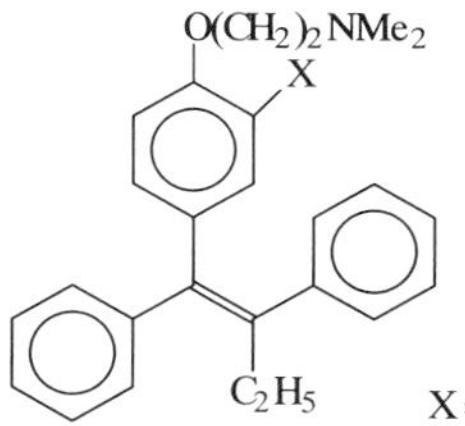

O(CH$_2$)$_2$—N◁

^{125}I

OTHP

THPO

SnBu$_3$

THP = Tetrahydropyranyl

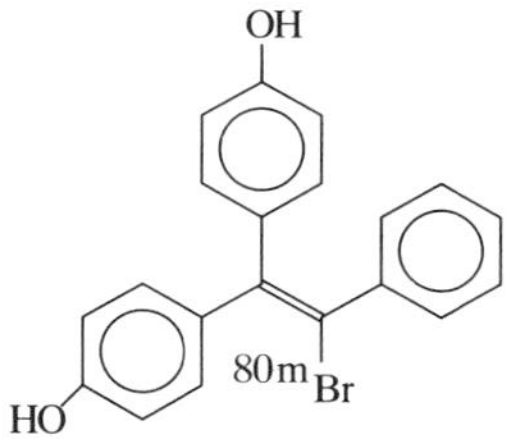

O(CH$_2$)$_2$NMe$_2$

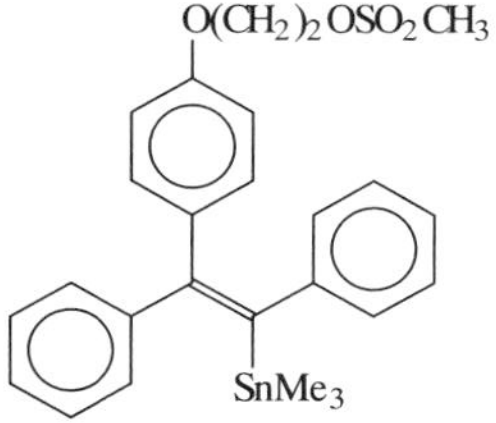

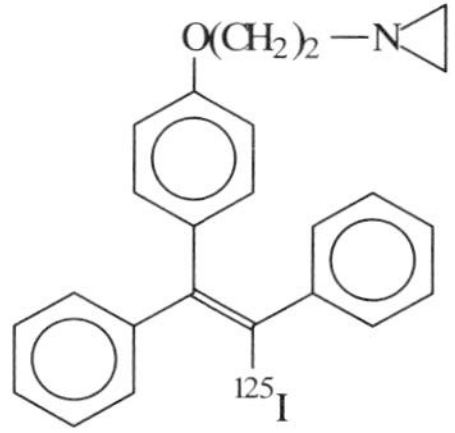

SnBu$_3$

C$_2$H$_5$

O(CH$_2$)$_2$OSO$_2$CH$_3$

SnMe$_3$

(continued overleaf)

TABLE 7. (continued)

Substrate	Product	Reference

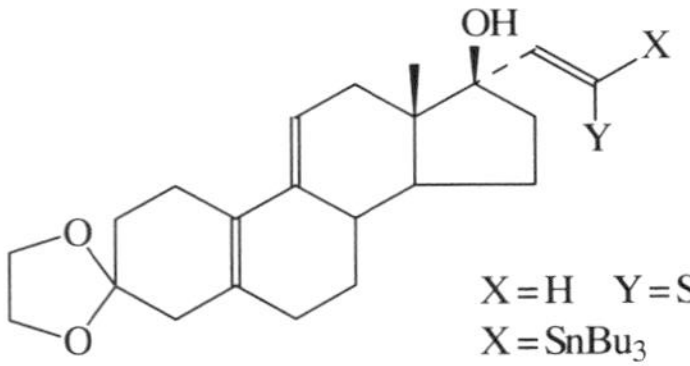

X = H   Y = SnBu₃
X = SnBu₃   Y = H

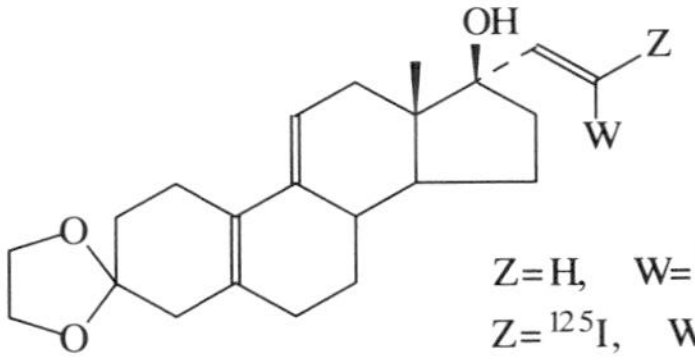

Z = H,   W = ¹²⁵I
Z = ¹²⁵I,   W = H

123

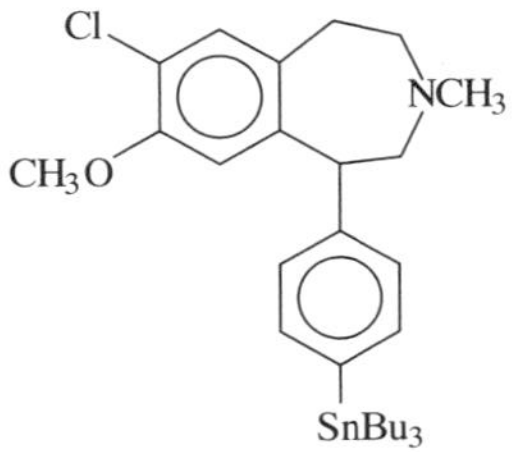

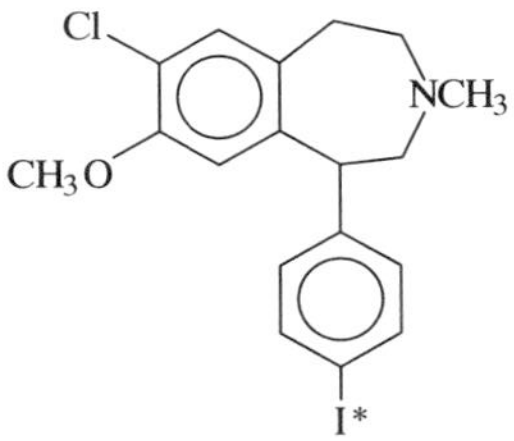

143
144

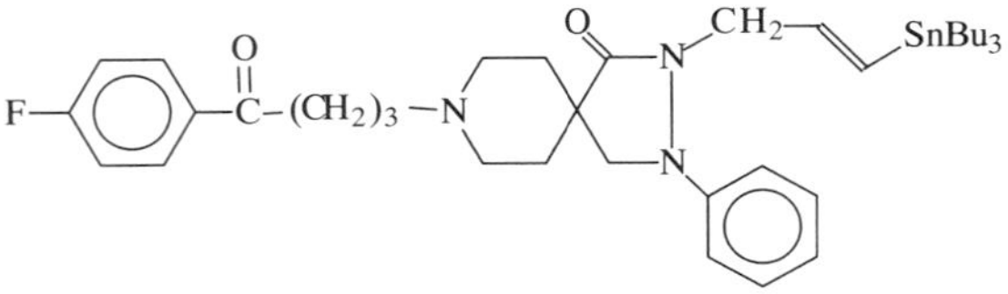

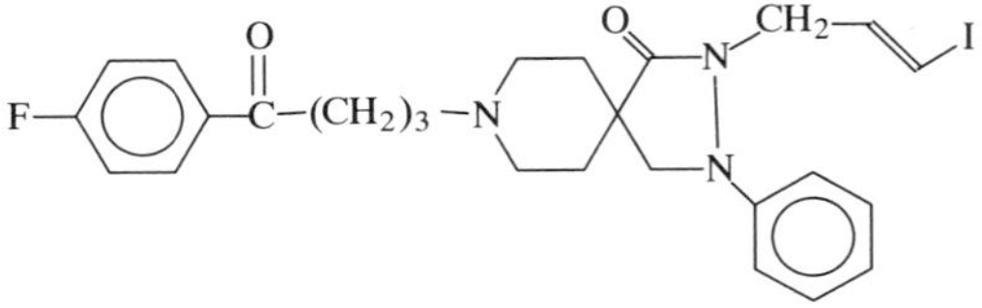

147
148
149

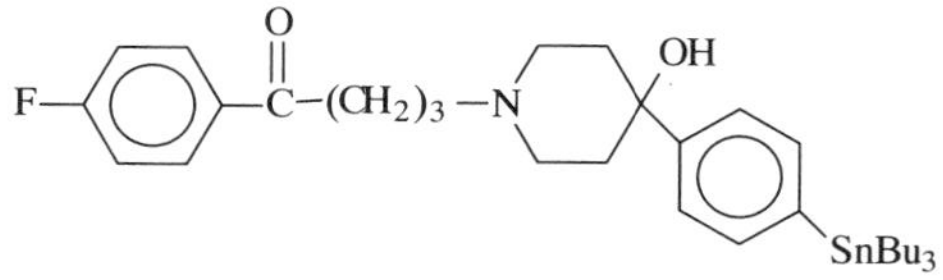

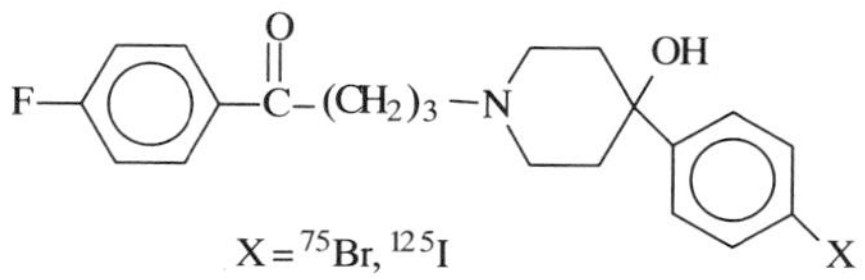

145
146

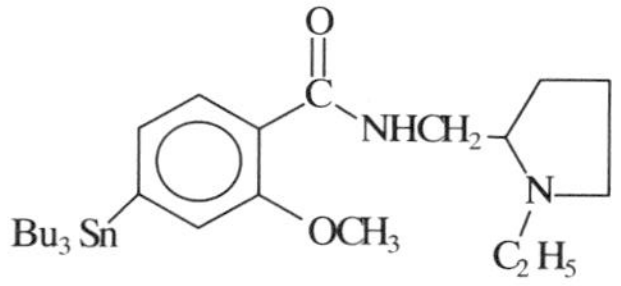

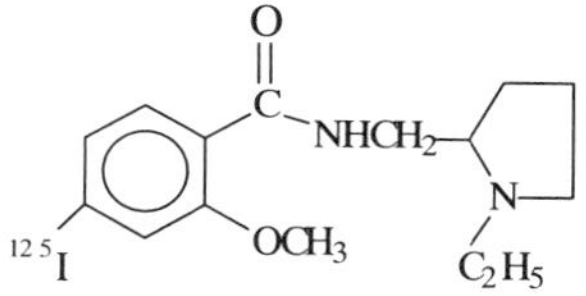

150

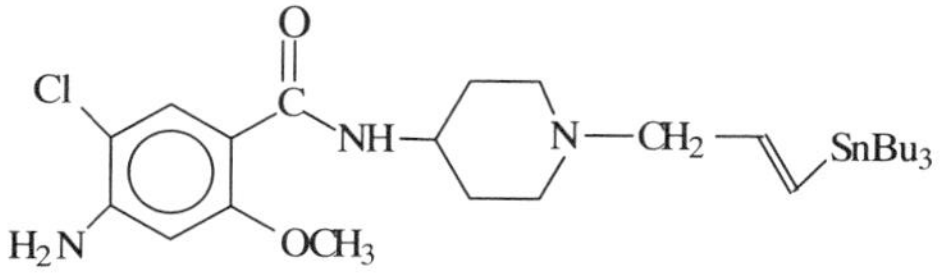

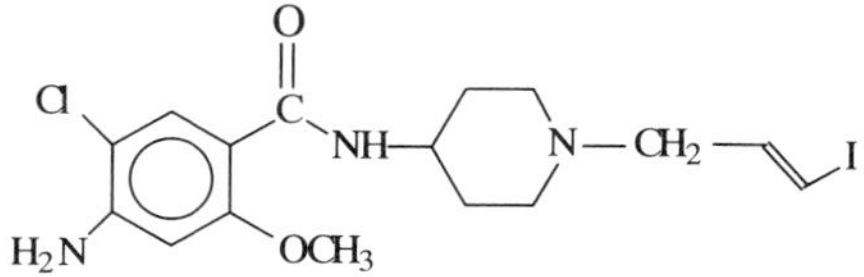

151

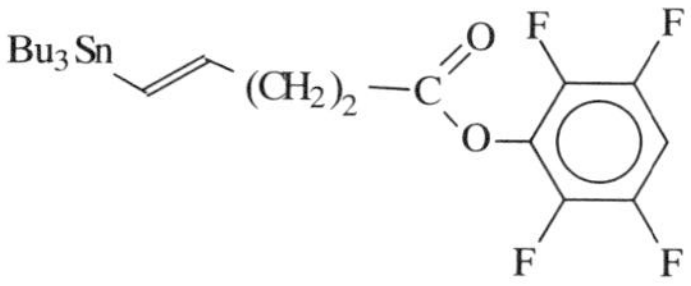

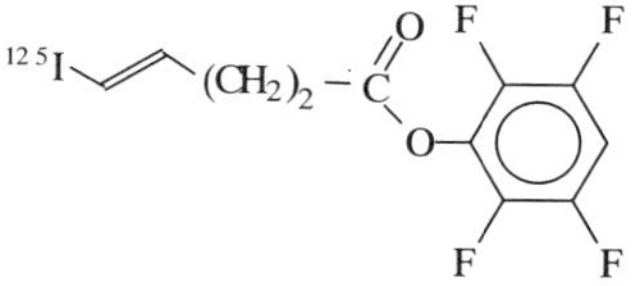

152

X = ^{75}Br, ^{125}I

OH

F—C(=O)—(CH$_2$)$_3$—N ... 4-(4-X-phenyl)-4-hydroxypiperidine

F ... C(=O)—(CH$_2$)$_3$—N ... OH ... SnBu$_3$

NHCH$_2$ ... C$_2$H$_5$ ... OCH$_3$ ... ^{125}I

NHCH$_2$ ... C$_2$H$_5$ ... OCH$_3$ ... Bu$_3$Sn

N—CH$_2$ ... I ... NH ... OCH$_3$ ... Cl ... H$_2$N

N—CH$_2$ ... SnBu$_3$ ... NH ... OCH$_3$ ... Cl ... H$_2$N

F ... F ... F ... F ... O—C(=O)—(CH$_2$)$_2$ ... ^{125}I

F ... F ... F ... F ... O—C(=O)—(CH$_2$)$_2$ ... Bu$_3$Sn

(continued overleaf)

TABLE 7. (*continued*)

Substrate	Product	Reference
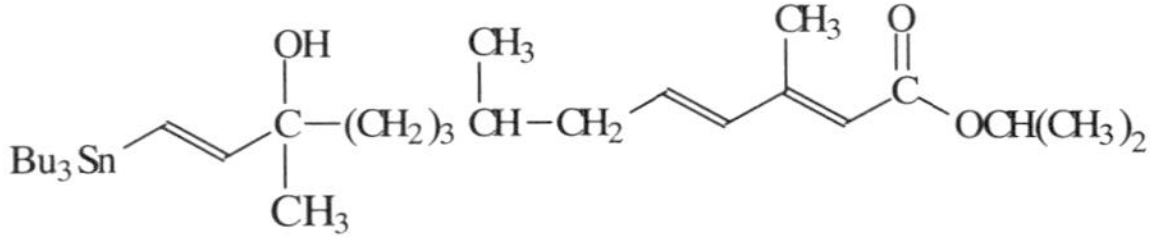	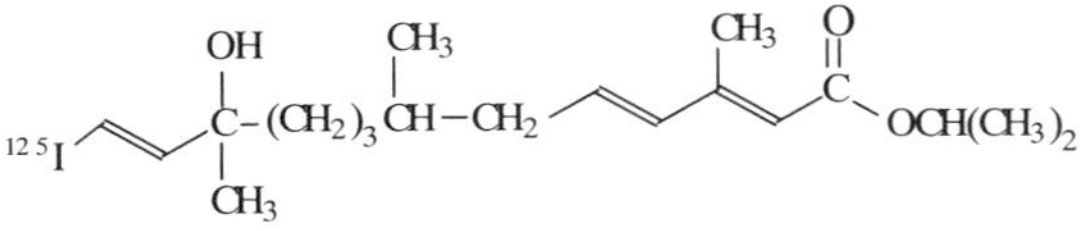	153 154
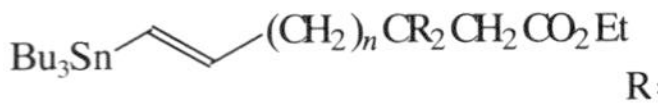	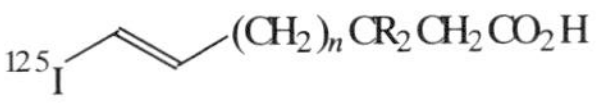	155
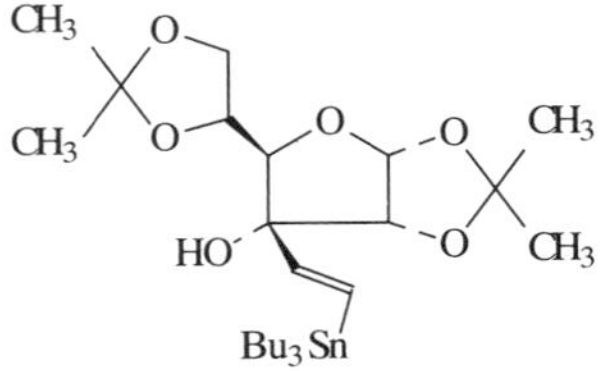	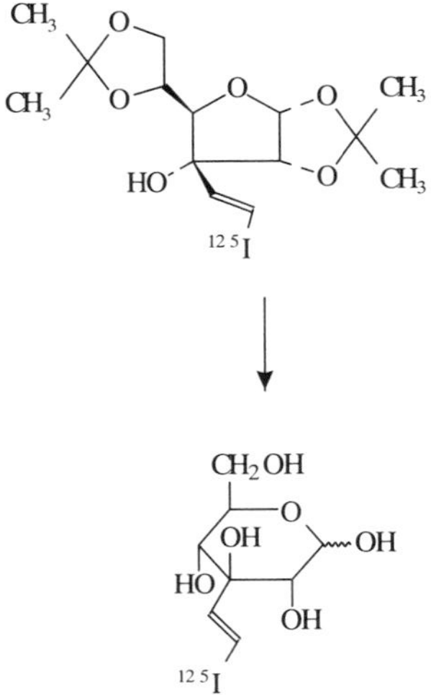	156 157

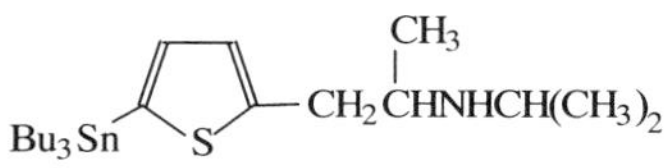

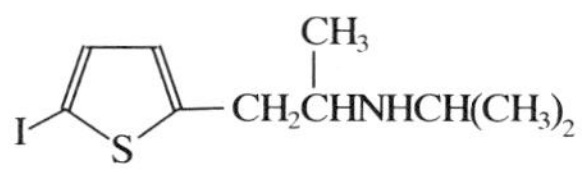

123

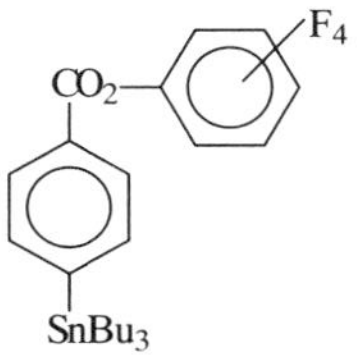

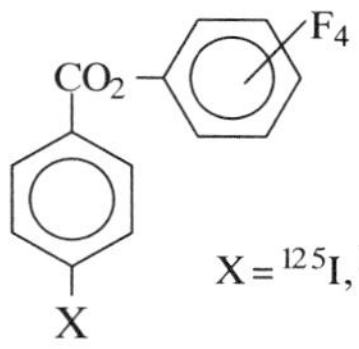

123

137

CH$_3$
|
CH$_2$CHNHCH(CH$_3$)$_2$

Bu$_3$Sn — S

CH$_3$
|
CH$_2$CHNHCH(CH$_3$)$_2$

I — S

F$_4$

CO$_2$ — SnBu$_3$

F$_4$

CO$_2$ — X

X = ^{125}I, ^{211}At

SnMe$_3$

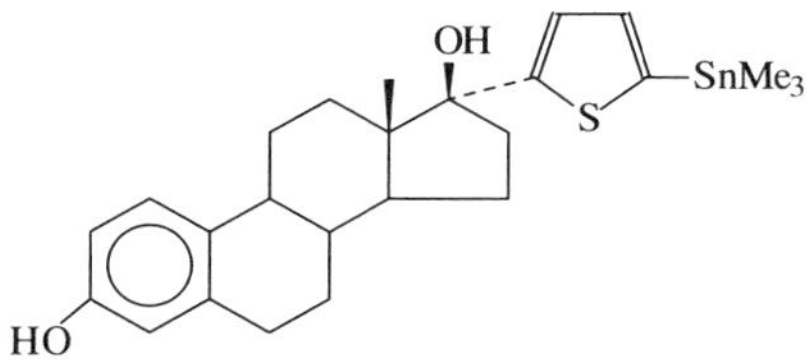

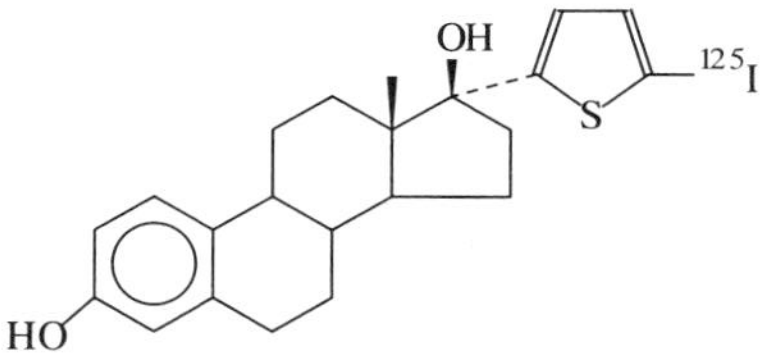

S

OH

HO

^{125}I

S

OH

HO

TABLE 8. The half-lives, the types of particles emitted and the energies of the particles emitted from several radiohalogens used in the syntheses of labelled compounds

Isotope	Half-life	Particle emitted	Energy emitted (keV)
^{18}F	110 min	positron	511
^{75}Br	101 min	positron	511; E.C.[a], 287
^{76}Br	15.9 h	positron	511; E.C.[a], 559
^{77}Br	56 h	E.C.[a]	E.C.[a], 239, 521
^{80m}Br	4.4 h	I.T.[b]	I.T.[b] 39
^{122}I	3.6 min	positron	511
^{123}I	13.3 h	gamma	159
^{125}I	60.2 days	gamma	27
^{131}I	8.1 days	gamma	364
^{211}At	7.2 h	alpha	5.86 meV

[a] E.C. is orbital electron capture.
[b] I.T. is isomeric transition from an upper to a lower isomeric state.

E. Investigation of Reaction Mechanisms Using Labelled Tin Compounds

Vlcek and Gray[157] have investigated the hydrogen atom abstraction reaction between trialkyl- and triphenyltin and germanium hydrides with triplet $d\sigma^*$–$p\sigma$ excited states of d^8–d^8 binuclear platinum complexes, $^3Pt_2(P_2O_5H_2)_4{}^{4-}$ (equation 100). Finding $Pt_2(P_2O_5H_2)_4{}^{4-}H_2$ as a product and observing a primary hydrogen–deuterium kinetic isotope effect of 1.7 for the photochemical reaction between $^3Pt_2{}^*(P_2O_5H_2)_4{}^{4-}$ and tributyltin hydride in acetonitrile at 25 °C led the authors to suggest that the reactions with both the tin and germanium hydrides occur via a slow hydrogen abstraction from the tin hydride to one of the two open axial coordination sites of the triplet $d\sigma^*$–$p\sigma$ excited states of d^8–d^8 binuclear platinum. Because the primary hydrogen–deuterium isotope effect of 1.7 measured for the tributyltin hydride reaction is in the range of the isotope effects found for other tin hydride reactions, i.e. 2.3 to 1.2 with various organic radicals, the authors concluded that the Pt- - -H(D)- - -M transition state is linear. It is interesting that rates of these reactions are in the order Sn > Ge, which is consistent with the strengths of the M—H bonds.

$$R_3M-H + {}^3Pt_2{}^*(P_2O_5H_2)_4{}^{4-} \longrightarrow \bullet Pt_2(P_2O_5H_2)_4{}^{4-}H + R_3M\bullet \qquad (100)$$

where M = Sn or Ge, and R = alkyl or aryl.

Hannon and Traylor[158] used a specifically labelled organotin hydride, *threo*-3-deutero-2-trimethylstannylbutane, to determine the mechanism and stereochemistry of the hydride abstraction from an organostannane by a carbocation (equation 101).

$$Ph_3C^+ + H-\overset{|}{\underset{|}{C}}-\overset{|}{\underset{|}{C}}-SnMe_3 \longrightarrow Ph_3CH + \diagdown C=C\diagup + \overset{+}{S}nMe_3 \quad (101)$$

Three mechanisms have been proposed for this reaction (Scheme 21). The reaction is first order in each of the reactants. In another study, Reutov and coworkers[159] found a large primary hydrogen–deuterium kinetic isotope effect of 3.8 for the reaction of tri-(*para*-methylphenyl)methyl carbocation with tetrabutyltin. This isotope effect clearly demonstrates that the hydride ion is transferred in the slow step of the reaction. This means that the first step must be rate-determining if the reaction proceeds by either of the stepwise mechanisms in Scheme 21. The primary hydrogen–deuterium kinetic isotope effect is, of course, consistent with the concerted mechanism shown in Scheme 21.

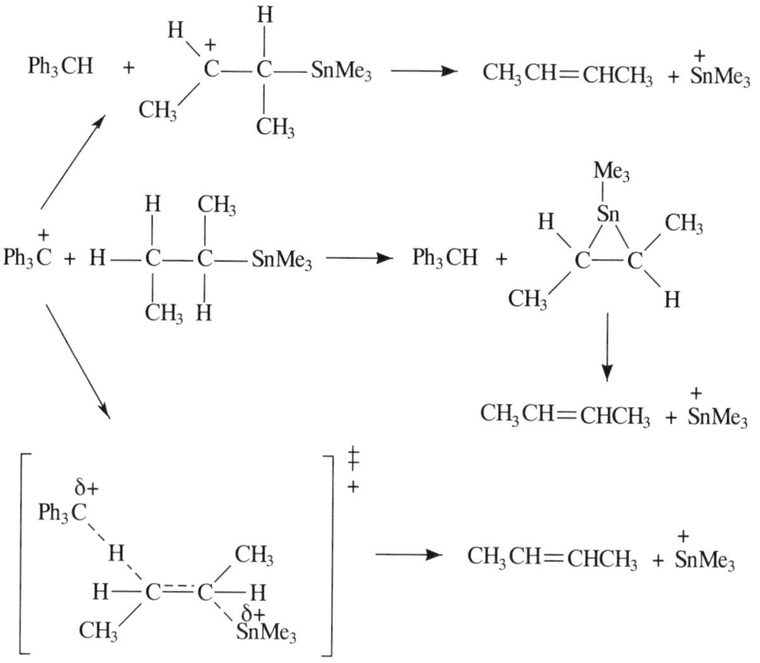

$$Ph_3CH + \quad \underset{CH_3}{\overset{H}{\underset{|}{\overset{|}{C}}}}\!-\!\underset{CH_3}{\overset{H}{\underset{|}{\overset{|}{C}}}}\!-\!SnMe_3 \longrightarrow CH_3CH\!=\!CHCH_3 + \overset{+}{SnMe_3}$$

$$\overset{+}{Ph_3C} + H\!-\!\underset{CH_3}{\overset{H}{\underset{|}{\overset{|}{C}}}}\!-\!\underset{H}{\overset{CH_3}{\underset{|}{\overset{|}{C}}}}\!-\!SnMe_3 \longrightarrow Ph_3CH + \underset{CH_3}{\overset{H}{\overset{}{C}}}\!-\!\underset{H}{\overset{Me_3\ Sn}{\overset{}{C}}}CH_3$$

$$\longrightarrow CH_3CH\!=\!CHCH_3 + \overset{+}{SnMe_3}$$

$$\left[\ \begin{array}{c} \overset{\delta+}{Ph_3C} \\ \quad H \qquad\qquad CH_3 \\ H\!-\!C\!\cdots\!C\!-\!H \\ CH_3 \qquad\qquad \overset{\delta+}{SnMe_3} \end{array}\ \right]^{\ddagger}_{+} \longrightarrow CH_3CH\!=\!CHCH_3 + \overset{+}{SnMe_3}$$

SCHEME 21

The *threo*-3-deutero-2-trimethylstannylbutane that Hannon and Traylor[158] used to determine the stereochemistry of the hydride transfer reaction and to shed light on the mechanism of this reaction was synthesized using the reactions in Scheme 22. Each of the reactions in Scheme 22 is stereospecific and the analysis showed that the product was at least 97% *threo*-3-deutero-2-trimethylstannylbutane. If the elimination reaction from *threo*-3-deutero-2-trimethylstannylbutane occurs with an *anti*-periplanar stereochemistry, the products shown in Scheme 23 will be obtained. Thus, if the elimination occurs by an *anti*-periplanar stereochemistry, all the *trans*-2-butene will be monodeuterated while the *cis*-2-butene will not be deuterated. A *syn*-periplanar elimination from *threo*-3-deutero-2-trimethylstannylbutane, on the other hand, would give the products shown in Scheme 24. If this occurs, the *cis*-2-butene will contain one deuterium atom and the *trans*-2-butene will contain none.

The results of a mass spectrometric investigation of the products, after correcting for the ^{13}C content and the M-1 fractionation of the molecular ion, showed that the *trans*-2-butene was > 97% deuterated and that less than 1% of the *cis*-2-butene was deuterated. This means that at least 97% of the elimination reaction to form the *trans*-2-butene and > 99% of the elimination to form the *cis*-2-butene occurred by an *anti*-periplanar mechanism (Scheme 23).

The results from these experiments also allowed Hannon and Traylor to determine the primary and secondary hydrogen deuterium kinetic isotope effects for the hydride abstraction reaction. If one assumes that there is no kinetic isotope effect associated with the formation of 3-deutero-1-butene, i.e. that $CH_2\!=\!CHCHDCH_3$ is formed at the same rate (k') from both the deuterated and undeuterated substrate (Scheme 25), then one can obtain both the primary (where a deuteride ion is abstracted) and the secondary deuterium

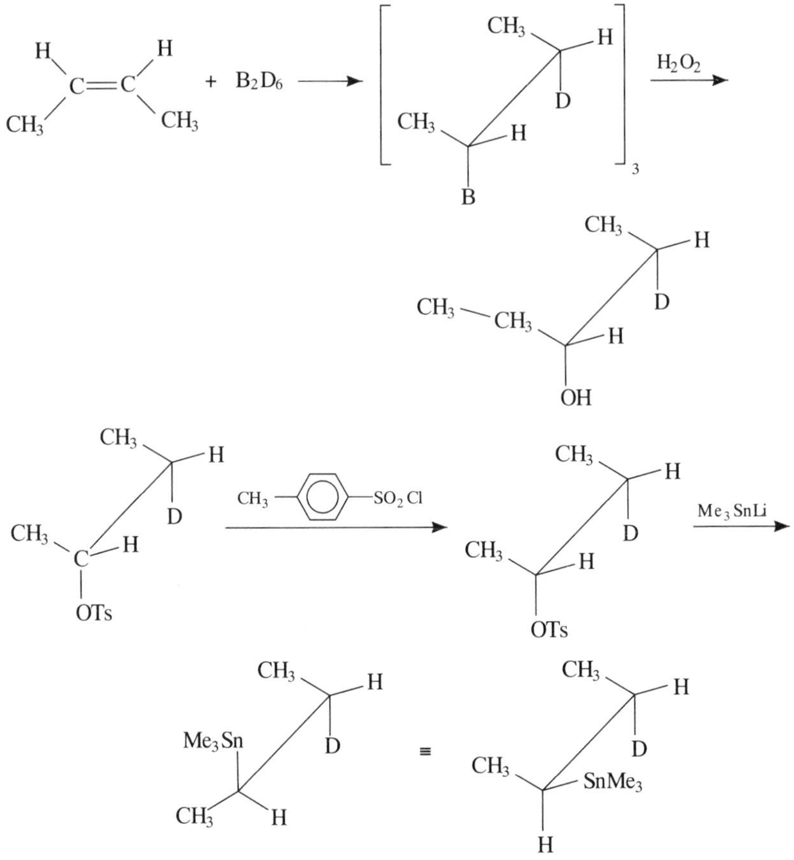

SCHEME 22

(where a hydride ion is abstracted) kinetic isotope effects for the reaction from the product ratios from the reactions of the undeuterated and deuterated substrates (equations 102 and 103).

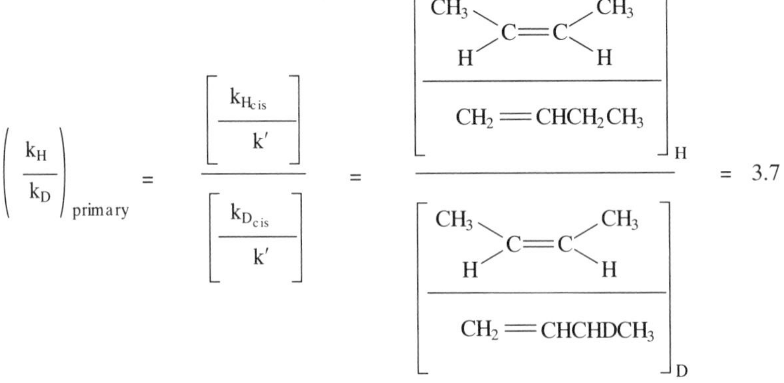

$$\left(\frac{k_H}{k_D}\right)_{primary} = \frac{\left[\dfrac{k_{H_{cis}}}{k'}\right]}{\left[\dfrac{k_{D_{cis}}}{k'}\right]} = \frac{\left[\dfrac{CH_3\,{>}C{=}C{<}\,CH_3}{\substack{H\qquad H}}\atop{CH_2{=}CHCH_2CH_3}\right]_H}{\left[\dfrac{CH_3\,{>}C{=}C{<}\,CH_3}{\substack{H\qquad H}}\atop{CH_2{=}CHCHDCH_3}\right]_D} = 3.7 \quad (102)$$

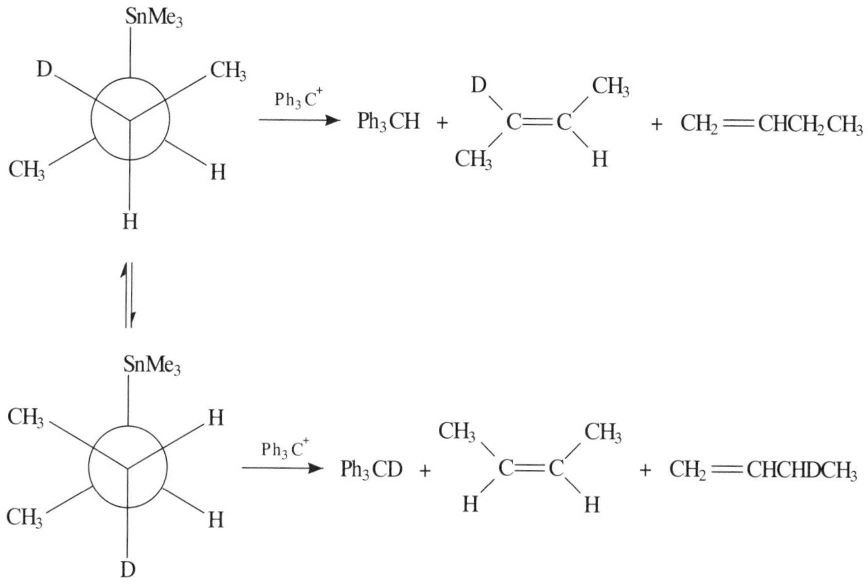

SCHEME 23

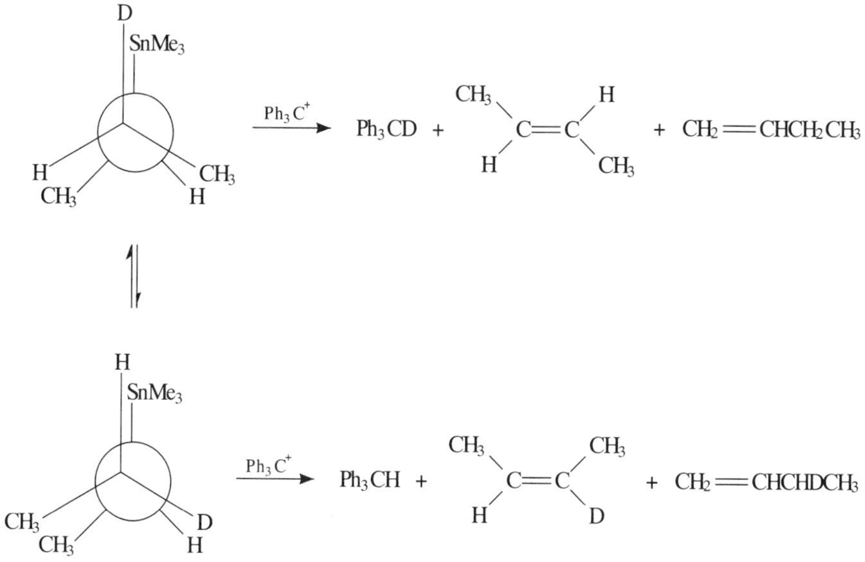

SCHEME 24

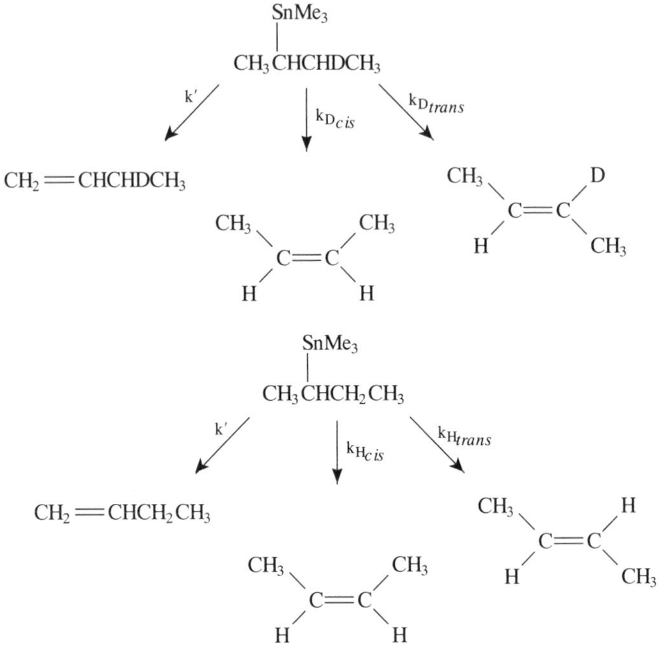

SCHEME 25

$$\left(\frac{k_H}{k_D}\right)_{secondary} = \frac{\left[\dfrac{k_{H_{trans}}}{k'}\right]}{\left[\dfrac{k_{D_{trans}}}{k'}\right]} = \frac{\left[\dfrac{\substack{CH_3 \diagdown \hspace{1em} \diagup H \\ C=C \\ H \diagup \hspace{1em} \diagdown CH_3}}{CH_2=CHCH_2CH_3}\right]_H}{\left[\dfrac{\substack{CH_3 \diagdown \hspace{1em} \diagup D \\ C=C \\ H \diagup \hspace{1em} \diagdown CH_3}}{CH_2=CHCHDCH_3}\right]_D} = 1.1 \quad (103)$$

The primary hydrogen–deuterium kinetic isotope effect is obtained from the percent *cis*-2-butene obtained from the deuterated and undeuterated stannanes. This is possible because a hydride and a deuteride are transferred to the carbocation when the undeuterated and deuterated stannane, respectively, forms *cis*-2-butene. The secondary deuterium kinetic isotope effect for the hydride transfer reaction is obtained from the relative amounts of *trans*-2-butene in each reaction. This is because a hydride is transferred from a deuterated and undeuterated stannane when *trans*-2-butene is formed.

The primary hydrogen–deuterium kinetic isotope effect for the reaction was 3.7 and the secondary alpha-deuterium kinetic isotope effect was found to be 1.1. It is worth noting that the primary hydrogen–deuterium kinetic isotope effect of 3.7 is in excellent agreement with that found by Reutov and coworkers in a similar reaction[159]. The secondary alpha-deuterium kinetic isotope effect of 1.1 is large and is characteristic of a carbocation reaction. Hence, Hannon and Traylor concluded on the basis of this isotope effect, and other evidence[160], that the reaction occurred by way of a carbocation intermediate (equation 104).

$$Ph_3C^+ + \underset{\underset{CH_3CH_2CHCH_3}{|}}{SnMe_3} \longrightarrow Ph_3CH + \underset{\underset{CH_3\overset{+}{C}HCHCH_3}{|}}{SnMe_3} \longrightarrow \tag{104}$$

$$CH_3CH{=}CHCH_3 + \overset{+}{S}nMe_3$$

The unusual stereospecific elimination via a carbocation was rationalized by suggesting that extensive C-metal $\sigma-\pi$ conjugation is present in the carbocation intermediate (Figure 3). The *anti*-periplanar stereochemistry is found because the $\sigma-\pi$ stabilization to the SnMe$_3$ group is strongest when the metal is involved in a vertical stabilization of the carbocation.

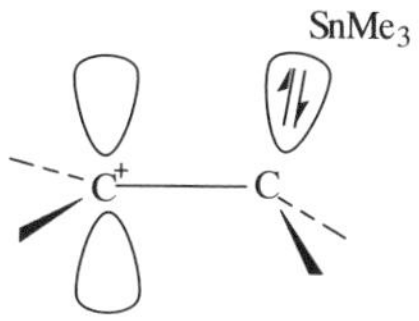

FIGURE 3. The C-metal $\sigma-\pi$ conjugation in the carbocation intermediate formed in the hydride transfer reaction

Song and Beak[161] have used intramolecular and intermolecular hydrogen–deuterium kinetic isotope effects to investigate the mechanism of the tin tetrachloride catalysed ene-carbonyl enophile addition reaction between diethyloxomalonate and methylenecyclohexane (equation 105). These ene reactions with carbonyl enophiles can occur by a concerted (equation 106) or a stepwise mechanism (equation 107), where the formation of the intermediate is either fast and reversible and the second step is slow ($k_{-1} > k_2$), or where the formation of the intermediate (the k_1 step) is rate-determining.

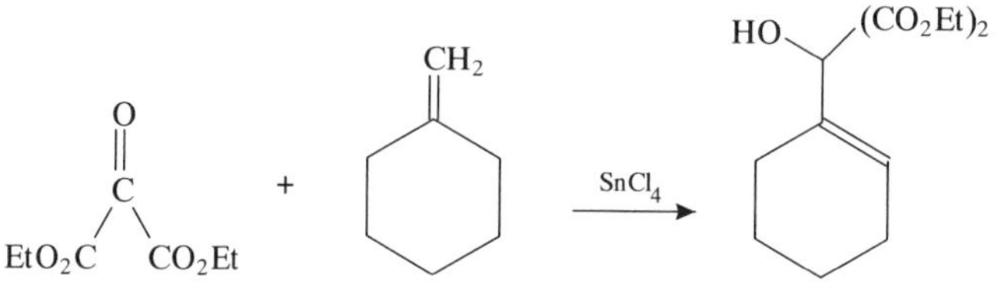

$$\tag{105}$$

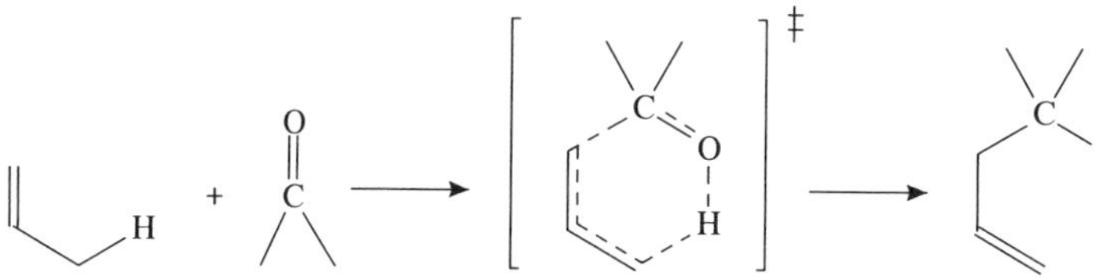

$$\tag{106}$$

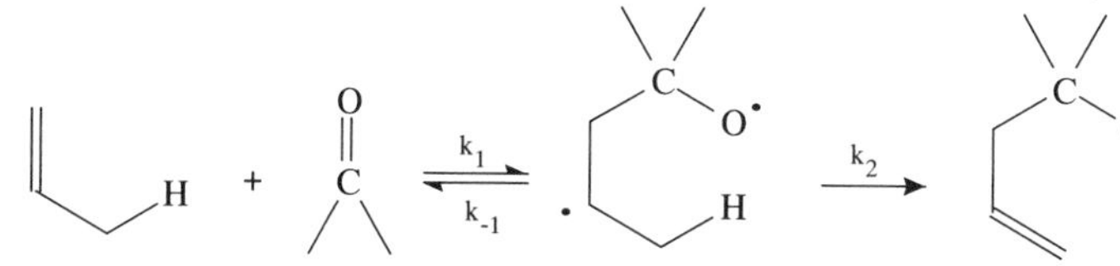

$$\tag{107}$$

The intermolecular hydrogen–deuterium kinetic isotope effect was determined from the rates of the undeuterated and tetradeuterated methylenecyclohexanes (equation 108), and the intramolecular isotope effect was determined from the reaction of the dideuterated methylenecyclohexane where the isotopic competition is between reaction at the CH_2 and the CD_2 allylic hydrogens (equation 109). A concerted reaction will have significant intermolecular and intramolecular kinetic isotope effects because the bond to the allylic hydrogen (deuterium) of the ene is broken in the slow step of the reaction. While large intermolecular and intramolecular isotope effects would also be found for the two-step reaction if the hydrogen (deuterium) is transferred in the slow k_2 step, small intermolecular and intramolecular isotope effects are expected when the hydrogen transfer does not occur in the slow step of the reaction, i.e. when the k_1 step is rate-determining. Thus, although observing large intramolecular and intermolecular isotope effects does not establish the mechanism, observing small isotope effects would indicate that the reaction occurs by a stepwise mechanism where the intermediate is formed in the slow step.

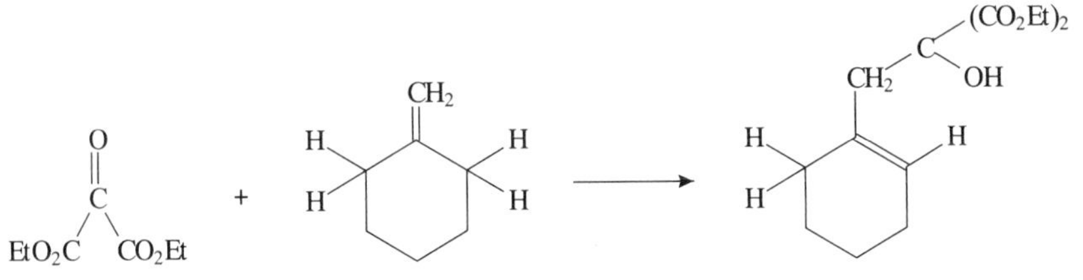

(108)

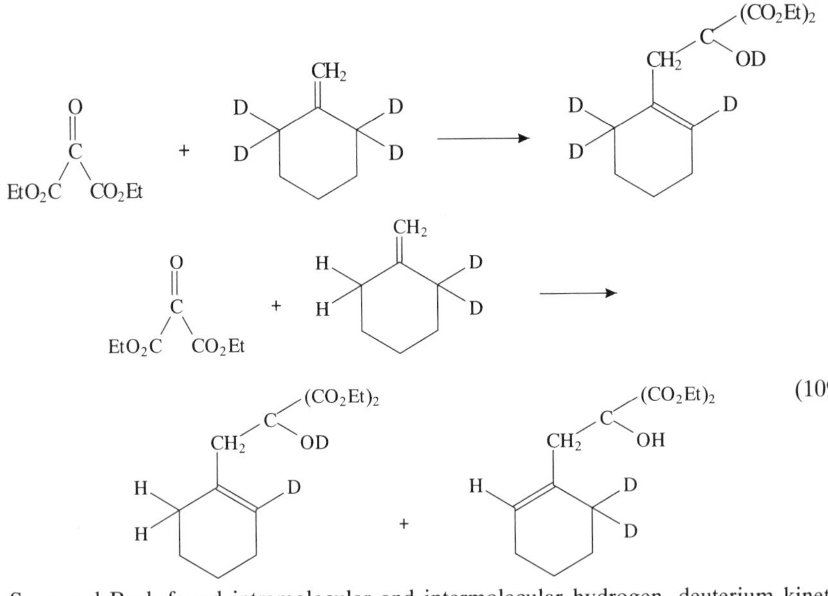

(109)

Song and Beak found intramolecular and intermolecular hydrogen–deuterium kinetic isotope effects of 1.1 ± 0.2 and 1.2 ± 0.1, respectively, for the tin tetrachloride catalysed ene reaction. Since significant intramolecular and intermolecular primary deuterium kinetic isotope effects of between two and three have been found for other concerted ene addition reactions[161], the tin-catalysed reaction must proceed by the stepwise pathway with the k_1 rate determining step (equation 107).

The isotope effects have been interpreted in terms of a mechanism involving two equilibrating zwitterionic intermediates (equation 110). In this instance, the k_1 step is partially reversible and both the intermolecular and intramolecular isotope effects are a composite of the isotope effects in several steps (Schemes 26 and 27).

Intermolecular

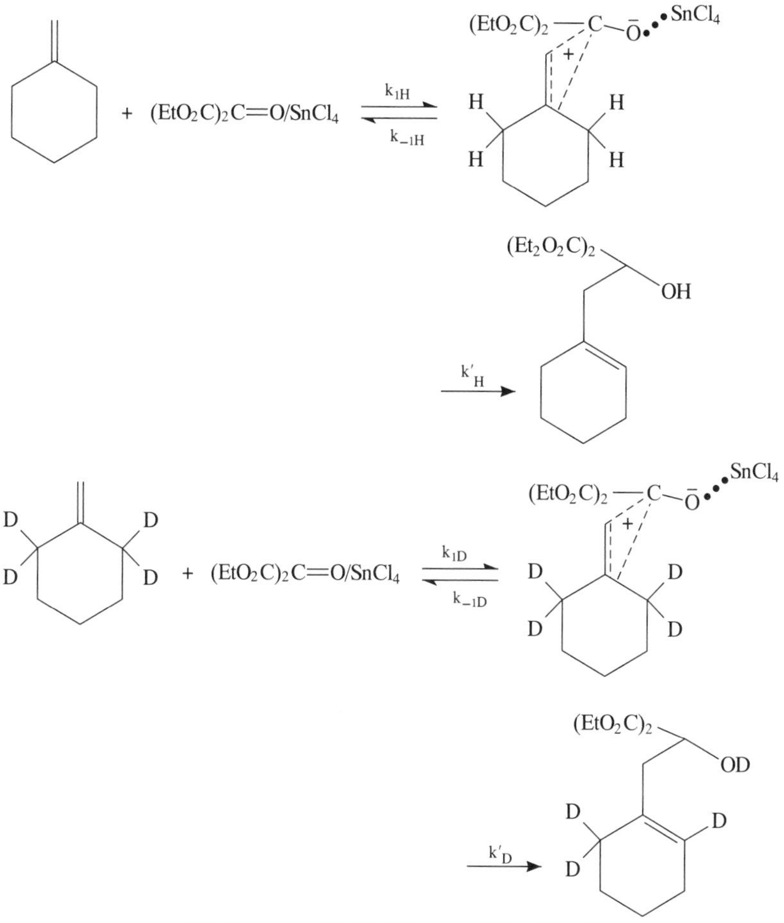

$$\left(\frac{k_H}{k_D}\right)_{inter} = \left(\frac{k_H}{k'_D}\right)\left(\frac{k_{-1D} + k'_D}{k_{-1H} + k'_H}\right)\left(\frac{k_{1H}}{k_{1D}}\right)$$

SCHEME 26

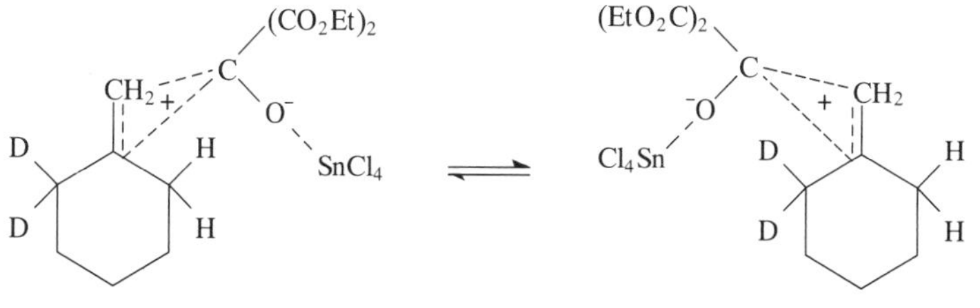

(110)

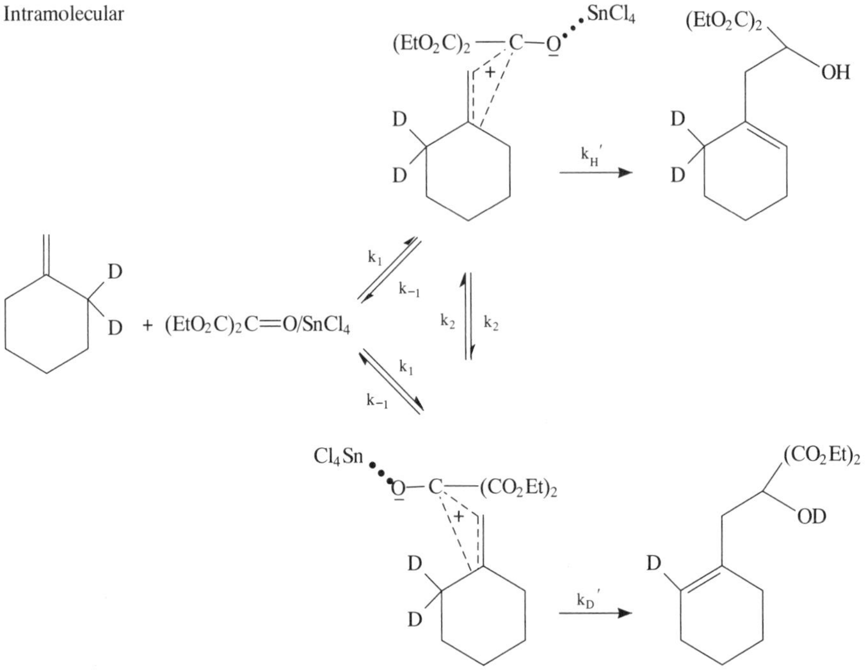

Intramolecular

$$\left(\frac{k_H}{k_D}\right)_{intra} = \left(\frac{k'_H}{k'_D}\right)\left(\frac{k_{-1} + 2k_2 + k'_D}{k_{-1} + 2k_2 + k'_H}\right)$$

SCHEME 27

Abeywickrema and Beckwith[162] have measured the primary hydrogen–deuterium kinetic isotope effect for the reaction between an aryl radical and tributyltin hydride. The actual isotope effect was determined by reacting tributyltin hydride and deuteride with the *ortho*-alkenylphenyl radical generated from 2-(3-butenyl)bromobenzene (equation 111).

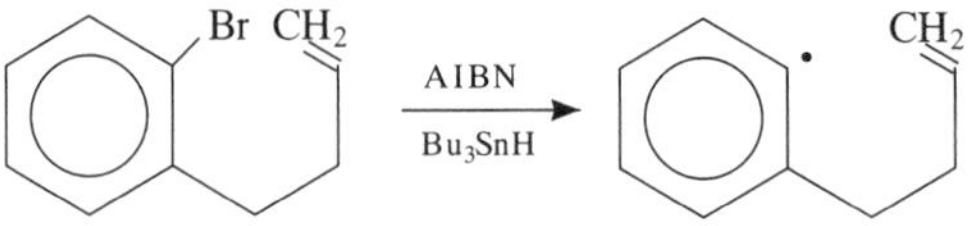

(111)

The 3-butenylphenyl radical can either react with tributyltin hydride to form an alkane or undergo a ring closure to form cyclic products (Scheme 28).

The *exo* and the *endo* ring closures (the k_C reactions) are in competition with the aryl radical-tributyltin hydride transfer (the k_H or k_D reaction). These workers[162] used this competition to determine the primary hydrogen–deuterium kinetic isotope effect in the hydride transfer reaction between the aryl radical and tributyltin hydride and deuteride.

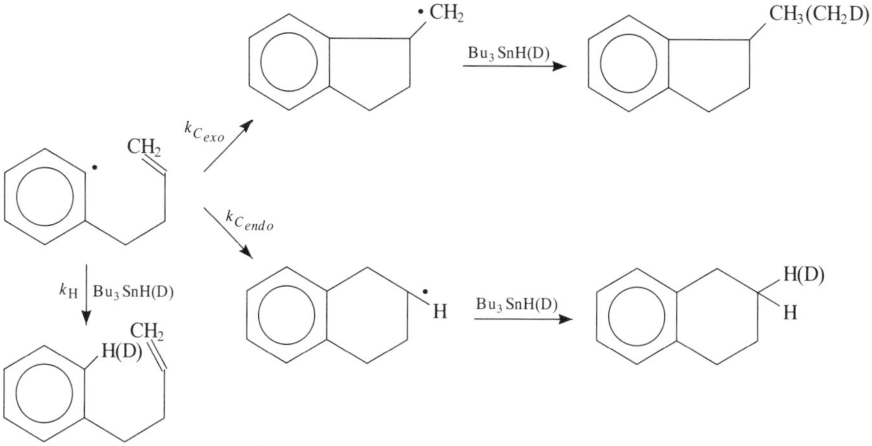

SCHEME 28

There is no hydrogen transfer in either of the ring cyclization (k_C) reactions, so their rates are not affected by the hydrogen isotope on the tin atom. The k_H reaction, on the other hand, will display a primary hydrogen–deuterium kinetic isotope effect because the hydrogen is transferred from the tin atom to the aryl radical in this reaction. Since the $\Sigma k_C/k_H$ ratio, where $\Sigma k_C = (k_{C_{endo}} + k_{C_{exo}})$, is directly proportional to the product ratio (ring closure/aryl radical-tin hydride product), dividing the product ratio from the reaction with the tributyltin deuteride ($\Sigma k_C/k_D$) by that from the tributyltin hydride reaction ($\Sigma k_C/k_H$) gives the primary hydrogen–deuterium kinetic isotope effect (k_H/k_D) for the reaction between the aryl radical and tributyltin hydride (equation 112).

$$k_H/k_D = \frac{\dfrac{\Sigma k_C}{k_D}}{\dfrac{\Sigma k_C}{k_H}} =$$

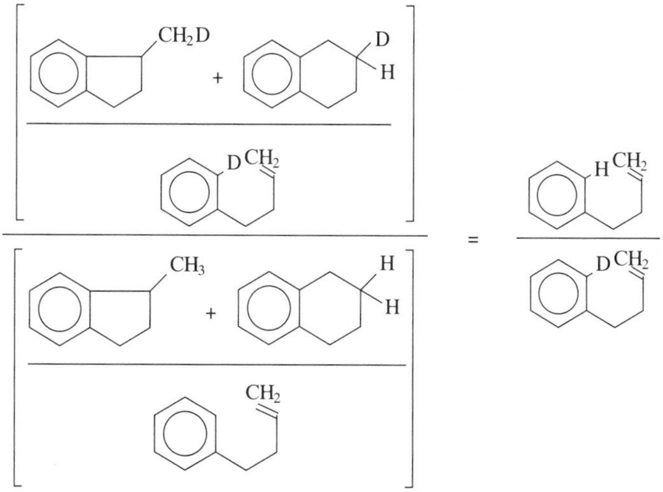

This method gave a primary hydrogen–deuterium kinetic isotope effect of 1.3 for the reaction between the aryl radical and tributyltin hydride. This isotope effect is smaller than the isotope effect of 1.9 which San Filippo and coworkers reported for the reaction between the less reactive alkyl radicals and tributyltin hydride[163] (*vide infra*). The smaller isotope effect of 1.3 in the aryl radical reaction is reasonable, because an earlier transition state with less hydrogen transfer, and therefore a smaller isotope effect[164], should be observed for the reaction with the more reactive aryl radicals.

Tributyltin deuteride has also been used to help determine the mechanism of the electro-chemical oxidation of carbonylmanganese phosphites and carbonylmanganese phosphines (equation 113).

$$Mn(CO)_3P_2^- \xrightarrow{Bu_3SnD} DMn(CO)_3P_2 + Bu_3Sn-Mn(CO)_3P_2 \qquad (113)$$

where P = tri-isopropylphosphite or triphenylphosphine.

Kochi and coworkers[165] found that all of the product was labelled with a deuterium atom when the anion was oxidized in the presence of tributyltin deuteride. This demonstrated that the hydrogen attached to the manganese atom in the product was supplied entirely by the tin atom. They also found that the reaction was first order in both the manganese radical and tributyltin hydride and concluded that the k_2 step of the three-step mechanism shown in Scheme 29 was rate-determining. Then, these workers measured the primary hydrogen–deuterium kinetic isotope effect for the reaction in an effort to confirm that the hydrogen was transferred in the slow (k_2) step of the reaction. Their results indicated that the reaction proceeded at the same rate within experimental error when tributyltin hydride and deuteride were used, i.e. that there was no isotope effect in this oxidation. Since the hydrogen atom must be transferred in the slow (k_2) step of the reaction and no isotope effect is observed, the authors concluded that the hydrogen transfer was virtually complete, i.e. that the reaction had a very product-like transition state[164].

$$Mn(CO)_3P_2^- \xrightleftharpoons{k_1} Mn(CO)_3P_2^\bullet + e^-$$

$$Mn(CO)_3P_2^\bullet + Bu_3SnD \xrightleftharpoons{k_2} DMn(CO)_3P_2 + Bu_3Sn^\bullet$$

$$Bu_3Sn^\bullet + Mn(CO)_3P_2^\bullet \xrightarrow{k_3} Bu_3Sn-Mn(CO)_3P_2$$

SCHEME 29

Several workers have measured the primary hydrogen–deuterium kinetic isotope effects for the reaction between organic radicals and tributyltin hydrides (equation 114).

$$R^\bullet + Bu_3SnH \longrightarrow RH + Bu_3Sn^\bullet \qquad (114)$$

In one study, Ingold and coworkers[166] measured the rate constants for the reactions of several alkyl radicals with tributyltin hydride using a laser flash photolytic technique and direct observation of the tributyltin radical. They also used this technique with tributyltin deuteride to determine the primary hydrogen–deuterium kinetic isotope effects for three of these reactions. The isotope effects were 1.9 for reaction of the ethyl radical, and 2.3 for reaction of the methyl and *n*-butyl radicals with tributyltin hydride at 300 K.

Other primary hydrogen–deuterium kinetic isotope effects have been measured for radical reactions with tributyltin hydride. For example, Carlsson and Ingold[167] found primary hydrogen–deuterium kinetic isotope effects of 2.7 and 2.8, respectively, for the reaction of cyclohexyl and *tert*-butyl radicals with tributyltin hydrides. These isotope effects were determined from the deuterium content of the alkane formed when the radicals were reacted in the presence of equimolar amounts of tributyltin hydride and tributyltin deuteride (equation 115).

$$R^\bullet + Bu_3SnH + Bu_3SnD \longrightarrow RH + RD + Bu_3Sn^\bullet \qquad (115)$$

where $k_H/k_D = ([RH]/[RD])$.

San Filippo and coworkers[163] have determined the temperature dependence of the primary hydrogen–deuterium kinetic isotope effects for the hydrogen transfer reactions between several organic radicals and tributyltin hydride (deuteride); see equation 116.

$$R^\bullet + Bu_3Sn-H(D) \longrightarrow R-H(D) + Bu_3Sn^\bullet \qquad (116)$$

Some of their isotope effects are presented in Table 9.

TABLE 9. The primary hydrogen–deuterium kinetic isotope effects for the hydrogen transfer reactions between alkyl radicals and tributyltin hydride (deuteride)

Substrate	Temperature ($^\circ$C)	k_H/k_D
$CH_3(CH_2)_6CH_2-Cl$	80.4	1.84
$CH_3(CH_2)_6CH_2-Br$	80.5	1.81
$CH_3(CH_2)_6CH_2-I$	80.5	1.78
$CH_3(CH_2)_5CHCH_3-Br$	85.2	1.92
$(C_2H_5)_2CH_3CBr$	80.0	1.83
C_6H_5-Br	78.1	1.41
$C_6H_5CH_2-Br$	70.0	2.42

The first observation is that the isotope effect is effectively independent of the leaving group. This is expected because the leaving group has departed before the radical reacts with the tributyltin hydride. Secondly, the primary hydrogen–deuterium kinetic isotope effects decrease slightly as the radical is changed from benzyl, to secondary, to tertiary, to primary, to phenyl. It is important to note that the temperature effects on the isotope effects in Table 9 are very small and do not affect the observed trends. However, regardless of the structure of the radical, all of the isotope effects are similar, i.e. they range from 1.4 to 2.4 at between 70 and 80 $^\circ$C. Moreover, the Arrhenius A factor ratio for the undeuterated and deuterated reactions, A_H/A_D, were all normal, i.e. approximately unity, indicating there was no tunnelling in these hydrogen transfer reactions[168]. The larger hydrogen–deuterium kinetic isotope effects that were found with the benzyl radicals suggest that the benzyl radical–tributyltin hydride transition states are more symmetrical than those for the less stable alkyl radicals, i.e. that the hydrogen transfer is more complete in the reactions with the larger isotope effect[164].

One interesting aspect of their results[163] was that the magnitude of the isotope effect for a series of primary and secondary radicals was directly related to the bond dissociation energy of the carbon–hydrogen bond formed in the reaction. This suggests that the primary hydrogen–deuterium kinetic isotope effects in radical reactions might be used to determine the strengths of C—H bonds. However, the magnitude of the isotope effect was not related to the C—H bond strength when a tertiary radical was used. Presumably this is for steric reasons. Finally, the hydrogen–deuterium isotope effects for the reactions of a series of *para*-substituted benzyl radicals with tributyltin hydride decreased when a more electron-withdrawing group was in the *para*-position on the benzene ring (Table 10). This decrease in the isotope effect with a more electron-withdrawing substituent (the Hammett

TABLE 10. The primary hydrogen–deuterium kinetic isotope effects for the reactions of a series
of *para*-substituted benzyl radicals with tributyltin
hydride[a]

Para-substituent	Reaction temperature (°C)	Primary k_H/k_D
CH_3	60.8	2.71
H	55.0	2.61
F	54.2	2.55
CF_3	60.5	2.44

[a] The differences between the temperatures where the isotope effects were measured do not affect the magnitude of
the isotope effect significantly.

$\rho = -0.189$ with a correlation coefficient of 0.996) clearly demonstrates that polar effects do affect radical reactions. Finally, it is interesting that the reaction with the most stable radical has the largest isotope effect and most symmetrical transition state. It is noteworthy that this trend in isotope effect with radical stability is the same as that observed for the benzyl and the primary alkyl radicals (*vide supra*).

In the most comprehensive study of the primary hydrogen kinetic isotope effects for the reactions between tin hydrides and organic radicals, Kozuka and Lewis[169] measured the primary hydrogen–tritium kinetic isotope effect for the reactions between several alkyl radicals and tributyltin hydride at 298 K. The method used by these workers was identical to that used by Carlsson and Ingold[167], i.e. the radicals were reacted with an excess of both tributyltin hydride and tributyltin hydride-t and the product composition, which is equal to the isotope effect, was determined by mass spectrometric analysis of the product. The tributyltin hydride-t used in this study was obtained by reacting tributyltin chloride with sodium borohydride-t_4. The primary tritium isotope effects found by Kozuka and Lewis are presented in Table 11. The primary hydrogen–deuterium kinetic isotope effects in Table 11 were calculated from the tritium isotope effects using the Swain–Schaad equation[170].

The results indicate that the isotope effect is dependent on the halide that is used, the temperature and the type of radical. The isotope effects are slightly larger with the chlorides than with bromides, and the iodide reactions have still smaller isotope effects. The form of the radical also affects the magnitude of the isotope effect. Except for the cyclohexyl and the tertiary, 2-methyl-2-pentyl radicals which have smaller isotope effects, possibly for steric reasons, all the reactions with primary and secondary alkyl radicals have isotope effects around 2.6. The larger isotope effects found in the benzyl radical reactions suggest a more symmetrical (more advanced) transition state for the benzyl radical reaction[164]. Changing the substitutent on the benzyl radical also affects the isotope effect. A smaller isotope effect is observed when a more electron-withdrawing substituent is on the phenyl ring. The smaller isotope effects are thought to represent an earlier transition state with less Sn- - - -H(T) bond rupture. Finally, this study also showed that there was a significant increase in the isotope effect with decreasing temperature, as one would expect. It is important to note that all the trends in these hydrogen–tritium isotope effects have been confirmed by San Filippo and coworkers' more recent study using hydrogen–deuterium kinetic isotope effects[163].

In addition, Kozuka and Lewis measured the tritium isotope effect for the reaction between the *n*-hexyl, the 2-hexyl and the 2-methyl-2-pentyl radicals with triphenyltin hydride and triphenyltin hydride-t; see the last three entries in Table 11. The isotope effect of 2.55 found for the triphenyltin hydride–*n*-hexyl radical reaction was slightly smaller

TABLE 11. The primary hydrogen–tritium kinetic isotope effects found in the reactions between various alkyl radicals and tributyltin hydride and tributyltin hydride-t

R	X	Temp ($^\circ$C)	k_H/k_T	Estimated[a] k_H/k_D
n-Hexyl	Br	80	2.65	2.0
2-Hexyl	Br	80	2.72	2.0
2-Methyl-2-pentyl	Br	80	2.53	1.9
n-Hexyl	Cl	80	2.96	2.1
n-Hexyl	Cl	25	3.07	2.2
Cyclopentyl	Br	80	2.71	2.0
Cyclohexyl	Br	80	2.38	1.8
Cyclohexyl	Br	25	2.60	1.9
Benzyl	Br	80	4.01	2.6
Benzyl	Cl	80	4.12	2.7
Benzyl	Cl	4	6.32	3.6
Benzyl	I	80	3.86	2.6
p-Methylbenzyl	Cl	80	3.92	2.6
p-Chlorobenzyl	Cl	80	3.76	2.5
m-Chlorobenzyl	Cl	80	3.68	2.5
n-Hexyl	Br	80	2.55^b	1.9^b
2-Hexyl	Br	80	2.30^b	1.8^b
2-Methyl-2-pentyl	Br	80	2.14^b	1.7^b

^{a} Calculated using the expression $k_H/k_T = (k_H/k_D)^{1.442}$ (Reference 170).
^{b} Measured using triphenyltin hydride rather than tributyltin hydride.

than the isotope effect of 2.65 found for the tributyltin hydride–n-hexyl radical reaction. Although smaller isotope effects were also found when triphenyltin hydride reacted with the 2-hexyl and the 2-methyl-2-pentyl radicals, the corresponding differences in the isotope effects were significantly larger, i.e. 2.30 versus 2.72, and 2.14 versus 2.53, for the 2-hexyl and 2-methyl-2-pentyl radical reactions, respectively. Again, the smaller isotope effects are thought to represent an earlier transition state with less Sn- - - -H(T) bond rupture[164]. This is expected because the triphenyltin reaction is faster than the tributyltin reactions.

Finally, this study of the tritium kinetic isotope effects demonstrated that the primary hydrogen–deuterium kinetic isotope effect of 2.7 reported by Carlsson and Ingold[167] for the tributyltin hydride–cyclohexyl radical reaction was incorrect. In fact, the hydrogen–deuterium isotope effect calculated from Lewis's work suggests that the isotope effect reported by Carlsson and Ingold should only be approximately 1.9. The isotope effect of 2.8 reported by Carlsson and Ingold for the t-butyl radical reaction[167] also seems to be too large because Kozuka and Lewis found a k_H/k_T of 2.5 ($k_H/k_D = 1.9$) for the reaction of the tertiary, 2-methyl-2-pentyl, radical. This conclusion is supported by the excellent agreement between Lewis's estimated hydrogen-deuterium isotope effects and the more recent isotope effects reported by Ingold and coworkers for the reactions of other alkyl radicals with tributyltin hydrides[166] and those found for the reactions between the benzyl radicals and tributyltin hydrides by San Filippo and coworkers[163]. The error in the Carlsson–Ingold isotope effect for the cyclohexyl radical reaction was confirmed when Kozuka and Lewis found a primary hydrogen–deuterium isotope effect of 1.6 in this reaction.

Finally, Franz and coworkers[171] measured the rate constants and primary hydrogen–deuterium kinetic isotope effects for the radical reactions between tributyltin hydride and the neophyl and the 2-allylbenzyl radical in diphenyl ether. The isotope effect in the first reaction was 1.64 at 192.5 $^\circ$C and that in the second reaction was 1.91 at 236 $^\circ$C. These values compare well with those predicted from Kozuka and Lewis's primary hydrogen–tritium isotope effects considering the temperature difference. In another study, Franz and coworkers measured a primary hydrogen–deuterium kinetic isotope effect of 2.26 for the reaction between benzyl radicals and tributyltin hydride and deuteride at 25 °C[172]. This isotope effect is much smaller than that found by other workers. The isotope effect calculated from Kozuka and Lewis's tritium isotope effects using the Swain–Schaad equation[170] was 3.2 and that reported by San Filippo and coworkers was 2.9[163]. The problem may be that Franz and coworkers used rate constants determined in different solvents to determine their isotope effect. Franz and coworkers used these isotope effects to help determine the rates of competing ring closure and rearrangement reactions that could be used as 'radical clocks'.

III. SYNTHESIS AND USES OF LABELLED ORGANOMETALLIC DERIVATIVES OF LEAD

A. Isotopes of Lead

The four, naturally occurring isotopes of lead are listed in Table 12 along with their percent natural abundance[173].

^{207}Pb with a spin of 1/2 has been used to study organolead compounds by NMR spectroscopy[174]. The lead-207 chemical shifts are of the order of 1300 ppm. Coupling constants, which range from 62–155 Hz for J(^{207}Pb-^{1}H), provide an insight into bonding. Lead-207 NMR chemical shifts and coupling constants, J (^{207}Pb-^{13}C), were determined for phenyl-substituted lead anions as a function of solvent and counter ion (Table 13). The lead-207 resonances were sensitive to a change in the anion–alkali metal interaction. When the lead-207 chemical shifts were compared for lithium and potassium salts, it was found that in methyltetrahydrofuran, where anion–cation interaction is significant, the lead-207 chemical shift for the lithium species was shifted slightly upfield relative to that for the potassium species. In the more polar solvents tetrahydrofuran, dimethoxyethane and 1,3-dimethyl-3,4,5,6-tetrahydro-2(H)-pyrimidinone, where solvent-separated ion pairs are formed more easily because the cation–anion interactions are weak, the lead-207 resonances for the lithium and potassium salts are shifted downfield. However, this effect

TABLE 12. The percent natural abundance of lead isotopes

Isotope	% Natural abundance
^{204}Pb	1.48
^{206}Pb	23.6
^{207}Pb	22.6
^{208}Pb	52.3

TABLE 13. ^{207}Pb NMR chemical shifts and J couplings of phenyl-substituted Pb anions[a]

	MTHF		THF		DME		DMPU	
	Li	K	Li	K	Li	K	Li	K
Ph$_3$Pb X[b]	1036.6	1045.7	1062.6	1045	1060.1	1055.1	1047.3	1048.1
^1J (^{207}Pb-^{13}C)	1020	1021	1030	1030	1038		1030	
^2J (^{207}Pb-^{13}C)	60	57	59	56	59		59	
^3J (^{207}Pb-^{13}C)	29	29	31	25	31			

[a] Abbreviations for solvents; MTHF, 2-methyltetrahydrofuran; THF, tetrahydrofuran; DME, 1,2-dimethoxyethane; DMPU, 1,3-dimethyl-3,4,5,6-tetrahydro-2(H)-pyrimidinone.
[b] Referenced relative to external hexaphenyldilead/CS$_2$/hexane.

is most pronounced for the lithium salts. This observation supports the theory that lithium salts form solvent separated ion pairs more easily than the potassium analogues. The degree of anion–cation interaction appears to be reflected in the magnitude of the shift in the lead-207 resonances. Large $J(^{207}Pb-^{13}C)$ of 1020–1038 Hz were noted for the phenyl-substituted lead anions. The relative magnitude of the one-, two- and three-bond $^{207}Pb-^{13}C$ coupling constants were of the order $^1J(^{207}Pb-^{13}C) > {}^2J(^{207}Pb-^{13}C) > {}^3J(^{207}Pb-^{13}C)$, which is different from that reported for Ph_4Pb where $^1J(^{207}Pb-^{13}C) > {}^3J(^{207}Pb-^{13}C) > {}^2J(^{207}Pb-^{13}C)$.

^{206}Pb, which is radiogenic, is used to determine the lead content of biological materials by a technique known as isotope dilution. Samples are spiked with a known quantity of ^{206}Pb. The ratio of ^{208}Pb and ^{206}Pb can be measured directly by flow injection ICP-MS and used to determine the ^{208}Pb concentration. Accurate determination of the lead content of biological materials can be made at the 0.1 to 10.0 μg/g level using the isotope dilution technique[175].

B. The Synthesis of Labelled Organolead Compounds

Alkyllead salts have been found to be 10 to 100 times more toxic than the corresponding inorganic lead salts. The toxicity arises because trialkyllead chlorides affect the central nervous system by inhibiting oxidative phosphorylation. The LD_{50} value of tetraalkyllead is comparable to that of the trialkyl derivatives because tetraalkyllead is easily dealkylated in the liver to form the trialkyl species. In spite of their toxicity, alkyllead compounds have a wide range of applications, and have been used as catalysts, stabilizers in vinyl resins and transformer oils, antioxidants, biocidal agents and, until recently, as antiknocking agents for gasoline.

In order to trace the fate of these compounds in the environment and in biological samples, the synthesis and detection of radiolabelled organolead compounds has been investigated. Blais and Marshall[176] showed that C-14 labelled Me_3PbCl could be prepared by the elaborate seven-step synthesis summarized in Scheme 30. This synthesis was accomplished as follows. Firstly, a Grignard reaction was initiated by adding a long-chain alkyl halide, such as n-decyl bromide, to an excess of magnesium metal. Then, C-14 labelled CH_3I was added at this point yielding a mixture of $(C_{10}H_{21})MgBr$ and $^{14}CH_3MgI$. Once the excess magnesium was removed, the mixture of organomagnesium reagents was added dropwise to chlorotrimethylplumbane to give n-decyltrimethylplumbane and C-14 labelled tetramethylplumbane. The excess chlorotrimethylplumbane was extracted into an aqueous phase containing the sodium salt of ethylenediaminetetraacetic acid (Na_4EDTA). Adding gaseous HCl converted the tetraalkyllead compounds into chlorodecyldimethylplumbane and C-14 labelled chlorotrimethylplumbane, respectively. After 1 hour the solution was neutralized. Extraction of the reaction mixture with a 1% aqueous solution of Na_4EDTA resulted in the selective complexation of the $(^{14}CH_3)(CH_3)_2Pb^+$ cation.

The C-14 labelled trimethyllead cation was recovered from the aqueous phase by adding dimethyldithiocarbamate (DMDTC) in hexane to form a trimethyllead–DMDTC complex. After the latter had been purified on preparative polyamide-6 TLC plates, the $(^{14}CH_3)(CH_3)_2PbCl$ was obtained by treating the pure C-14 labelled $(CH_3)_3Pb$–DMDTC complex with HCl.

Blais and Marshall found that C-14 labelled Me_4Pb could be prepared by an electrochemical route and could be used to prepare C-14 labelled Me_3PbCl. The electrochemical reactor shown in Figure 4 was designed by the authors to prepare the desired product. It consists of two compartments filled with 5% $NaClO_4$ in anhydrous DMF. The anode, a silver wire, is introduced through a silicon plug in the lower compartment. The cathode, a 4 cm × 4 cm × 0.13 mm piece of polished lead foil, is suspended by a stainless steel wire in the upper chamber. Stainless steel needles, inserted through a rubber septum sealing the

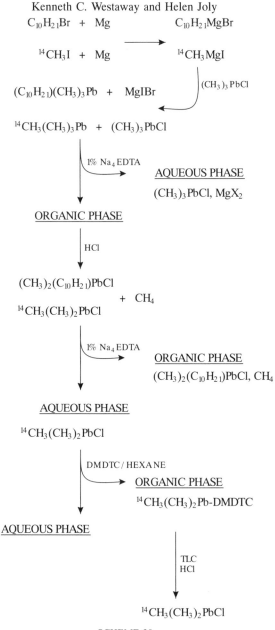

$$C_{10}H_{21}Br + Mg \longrightarrow C_{10}H_{21}MgBr$$

$$^{14}CH_3I + Mg \longrightarrow ^{14}CH_3MgI$$

$$(C_{10}H_{21})(CH_3)_3Pb + MgIBr$$

$$^{14}CH_3(CH_3)_3Pb + (CH_3)_3PbCl$$

SCHEME 30

upper compartment, are used to deliver and remove N_2 so the cell can be flushed during electrolysis. When a mixture of $^{14}CH_3I$ and CH_3I in DMF was transferred to the cathodic compartment of the electrochemical reactor and subjected to electrolysis, the alkyl halides were reduced at the sacrificial lead cathode giving tetraalkyllead (equation 117).

$$7CH_3I/^{14}CH_3I + Pb^0 \longrightarrow (CH_3)_4Pb/^{14}CH_3(CH_3)_3Pb \qquad (117)$$

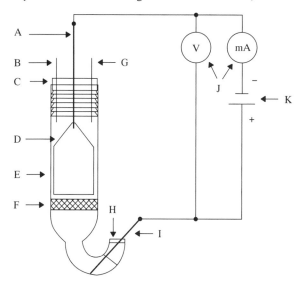

FIGURE 4. Electrochemical reactor composed of a 2 cm^3 capacity filter funnel and (A) stainless steel wire, (B, G) stainless steel needles for N$_2$ entry and exit ports, (C) rubber septum, (D) lead foil cathode, (F) fine porosity glass filter, (H) plug of silicone glue and (I) silver wire anode. (Taken from Reference 176.)

Maximum yield was obtained when the applied voltage was 13.1 V, the cathode surface area was 4 cm^2 and the total CH$_3$I concentration was 35.2 μmol. The C-14 labelled tetramethyllead was isolated by the extraction procedures developed for the Grignard route (*vide supra*). It was next converted to ^{14}CH$_3$(CH$_3$)$_2$PbCl by controlled oxidation with HCl.

Blais and Marshall[177] adapted the electrochemical technique to the synthesis of tetramethyllead-210, Me$_4$^{210}Pb. In the first step of the synthesis, ^{210}Pb was electrodeposited on a 10-cm copper coil made of 27 gauge wire when a potential of 3 V was applied between the copper cathode and a silver wire anode immersed in a solution containing 0.3 M HNO$_3$, 0.1% polyoxyethylene lauryl ether and 3 μmol of Pb(NO$_3$)$_2$ [407 MBq ^{210}Pb/mmol]. The lead-coated copper coil was next inserted into the electrochemical reactor (Figure 4) containing 5 M CH$_3$I and 2% (w/v) NaClO$_4$ in DMF and the cell was operated at a potential of 22 V. The tetramethyllead-210 formed by this process was extracted into diethyl ether.

The lead-210 labelled chlorotrimethylplumbane was prepared by adding HCl to the tetramethyllead-210 at 0 °C. The crude product was treated with the complexometric agent dimethyldithiocarbamate and the trimethyllead-210–DMDTC complex was purified by preparative TLC. Addition of HCl to the pure lead-210 labelled Me$_3$Pb–DMDTC complex yielded the desired lead-210 labelled Me$_3$PbCl.

A more efficient synthesis of Me$_4$^{210}PbCl was subsequently proposed by Marshall and coworkers[178]. The key step in this improved synthesis involved the reaction of methylmagnesium bromide with ^{210}Pb(DMDTC)$_2$ in the presence of excess CH$_3$I. This is thought to occur by the chemical reactions shown in Scheme 31. Initially ^{210}Pb(DMDTC)$_2$, prepared by shaking an aqueous solution of ^{210}Pb(NO$_3$)$_2$ with NaDMDTC, reacted with the methymagnesium bromide to give (CH$_3$)$_2$^{210}Pb(NO$_3$)$_2$. The disproportionation of the dimethyllead-210 resulted in the formation of tetramethyllead-210 and metallic lead-210. Excess CH_3I was used to recycle the $^{210}Pb^0$ and to form dimethyllead-210 iodide, $(CH_3)_2{}^{210}PbI_2$, which was subsequently reacted with methylmagnesium bromide to give tetramethyllead. The latter was mono-demethylated by gaseous HCl to give $Me_3{}^{210}PbCl$ which was recovered complexometrically with the aid of DMDTC, as $Me_3{}^{210}Pb-DMDTC$. Controlled oxidation of $Me_3{}^{210}Pb-DMDTC$ with anhydrous hydrochloric acid resulted in the formation of lead-210 labelled chlorotrimethylplumbane. The crude product was purified by reversed-phase high pressure liquid chromatography giving $Me_3{}^{210}PbCl$ in 71% radiochemical yield.

$$4CH_3MgBr + 2Pb(DMDTC)_2 \longrightarrow 2[(CH_3)_2Pb(II)] + 4Mg(Br)(DMDTC)$$

$$2[(CH_3)_2Pb(II)] \longrightarrow (CH_3)_4Pb + Pb^0$$

$$2CH_3I + Pb^0 \longrightarrow (CH_3)_2PbI_2$$

$$(CH_3)_2PbI_2 + 2CH_3MgBr \longrightarrow (CH_3)_4Pb + 2Mg(Br)(I)$$

SCHEME 31

A mixture of $Me_3{}^{210}PbCl$ and $^{210}Pb(NO_3)_2$ was used to study the rate of ionic trimethyllead uptake by exposed plant surfaces. More specifically, the mean cumulative activity of the lead toxicants transferred across tomato cuticle was measured daily over a six-day period. Reversed-phase HPLC was used to separate and identify the lead species crossing the plant cuticle. It was found that appreciably more trimethyllead(I) (75% of the theoretical) than inorganic lead(II) (39%) was transferred. The apparent rate constants derived from the first-order plot of time in days versus the difference in observed activity were 0.0788 and 0.0346 day^{-1} for transfer of the trimethyllead(I) and inorganic lead(II), respectively.

McInnis and Dobbs[179] showed that both a DB-5 megabore capillary-GC column and a packed (20% SP2100 on 60–80 mesh Supelcoport) GC column could be used for the analysis of C-14 labelled chlorotrimethylplumbane. Analysis times using the DB-5 megabore column were approximately one-half those required on the packed column.

C. The Use of Labelled Organolead Compounds in Spectroscopic Studies

Wei and collaborators[180] studied the long-range deuterium isotope effects on the proton chemical shifts of tetramethyllead and chlorotrimethylplumbane. The perdeuterated analogues of these organolead compounds were generated *in situ* by reaction of equimolar amounts of $(CD_3)_3SnCl$ and $(CH_3)_4Pb$ in methanol-d, in equation 118.

$$(CH_3)_4Pb + (CD_3)_3SnCl \rightleftharpoons (CH_3)_3PbCl + CH_3Sn(CD_3)_3 \qquad (118)$$

The exchange of chloride for methyl occurs relatively rapidly and the equilibrium favours formation of tetramethylstannane and chlorotrimethylplumbane. Subsequently, exchange occurs between tetramethylstannane and chlorotrimethylplumbane (equation 119).

$$(CH_3)_3PbCl + CH_3Sn(CD_3)_3 \rightleftharpoons (CH_3)_3PbCD_3 + CH_3Sn(CD_3)_2Cl \qquad (119)$$

With time, the CD_3 group is distributed throughout the system resulting in the formation of $(CH_3)_n(CD_3)_{4-n}Pb$ ($n = 3, 2, 1$) and $(CH_3)_n(CD_3)_{3-n}PbCl$ ($n = 2, 1$), as shown in Scheme 32. These species were used to determine the effect that a deuteron positioned four bonds away, **13**, would have on the proton chemical shifts of a methyl group. These long-range four-bond deuterium isotope effects, $^4\Delta H(D_3)$, are summarized in Table 14.

TABLE 14.   Summary of isotope shifts and ^2J coupling constants for organolead compounds

Species	$\delta^{a,b}$	$^4\Delta H(D_3)^a$	$^2J(^1H,^{207}Pb)^c$
$(CH_3)_4Pb$	0.723	0	—
$(CH_3)_3Pb(CD_3)$		-3.8×10^{-3}	62.2
$(CH_3)_2Pb(CD_3)_2$		-7.7×10^{-3}	—
$(CH_3)Pb(CD_3)_3$		-10.0×10^{-3}	—
$(CH_3)_3PbCl$	1.473	0	79.16
$(CH_3)_2(CD_3)PbCl$		-3.9×10^{-3}	79.04
$(CH_3)(CD_3)_2PbCl$		-7.8×10^{-3}	78.90

^{a} In ppm.
^{b} Relative to TMS.
^{c} In Hz.

Deuteration of the methyl groups of organolead compounds caused the chemical shifts of the organolead compounds to move downfield. The effect of isotope substitution was found to be additive, with $^4\Delta H(D_3)$ changing from -3.8×10^{-3} to -7.7×10^{-3} to -10.0×10^{-3} as the CH_3 groups are sequentially substituted by CD_3 groups. The magnitude of $^4\Delta H(D_3)$ for $M(CH_3)_{4-n}(CD_3)_n$ where M = Sn and Hg was similar to that for Pb, indicating the small influence exerted by the metal centre. A comparison of the two-bond H–Pb coupling constants $^2J(^1H,\ ^{207}Pb)$ (cf Table 14, column 4) for the isotopomers of chlorotrimethylplumbane indicated that isotopic substitution has little effect.

$$(CH_3)_4Pb + (CD_3)_3SnCl \rightleftharpoons (CH_3)_3PbCl + CH_3Sn(CD_3)_3$$
$$\rightleftharpoons (CH_3)_3PbCD_3 + CH_3Sn(CD_3)_2Cl$$
$$(CH_3)_3PbCD_3 + CH_3Sn(CD_3)_2Cl \rightleftharpoons (CH_3)_2PbCD_3Cl + (CH_3)_2Sn(CD_3)_2$$
$$\rightleftharpoons (CH_3)_2Pb(CD_3)_2 + (CH_3)_2Sn(CD_3)Cl$$
$$(CH_3)_2Pb(CD_3)_2 + (CH_3)_2Sn(CD_3)Cl \rightleftharpoons (CH_3)Pb(CD_3)_2Cl + (CH_3)_3Sn(CD_3)$$
$$\rightleftharpoons (CH_3)Pb(CD_3)_3 + (CH_3)_3SnCl$$

SCHEME 32

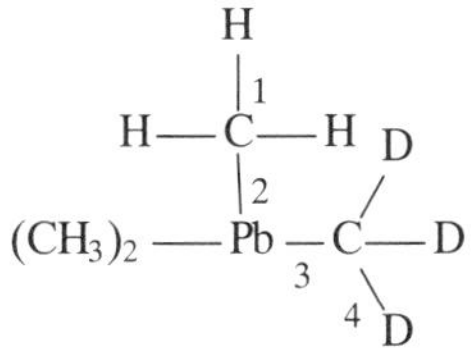

(13)

D. The Use of Labelled Organolead Compounds in Mechanistic Studies

Banerjee and coworkers[181–184] have been interested in elucidating the reaction mechanism of the oxidation of mandelic acid and its derivatives by lead tetraacetate $[Pb(OAc)_4]$.

They determined the activation parameters for this reaction in benzene and in benzene containing pyridine (equation 120).

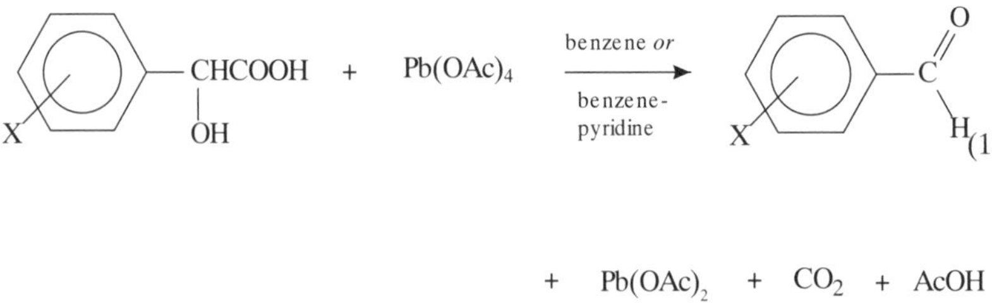

$$X = p\text{-H},\ m\text{-NO}_2,\ p\text{-NO}_2,\ m\text{-Cl},\ p\text{-Cl},\ p\text{-Br},\ p\text{-CH}_3,\ p\text{-CH}_3\text{CH}_2$$

The oxidation product in both media was primarily (85±5%) benzaldehyde. When compared to the rate data obtained for related substrates such as $PhCR_2COOH$ and $PhCR_2OH$ with $Pb(OAc)_4$, the oxidation of mandelic acid was found to be much faster, indicating that there was anchimeric assistance by both the hydroxy and the carboxy group in the rate-determining step of the reaction. The oxidation proved to be first order in Pb(IV) and mandelic acid in both benzene and benzene–pyridine. However, the activation parameters proved to be significantly different (Table 15), indicating that the mechanism of the oxidation is different in these two solvents. In addition, the oxidations were carried out with deuterated analogues of mandelic acid, namely $PhCD(OH)COOH$, $PhCH(OD)COOD$ and $PhCD(OD)COOD$. The resulting ratios of the second-order rate constants for the oxidation of deuterated and undeuterated (k_H/k_D) substrates undergoing oxidation are presented in Table 16.

The absence of an isotope effect when the reaction was carried out in benzene suggested to the authors that no significant change in bonding occurred to the α-H, the hydroxyl H or the carboxyl H in the rate-determining step of the reaction. Hence they proposed that $Pb(OAc)_4$ undergoes rapid anion exchange with mandelic acid in the non-polar medium

TABLE 15. The activation parameters for the oxidation of mandelic acid[a] by $Pb(OAc)_4$ in benzene and in benzene–pyridine

[pyridine] M	$\Delta H^{\ddagger}$ kcal mol^{-1}	$\Delta S^{\ddagger}$ e.u.
0	8.4	−34
5×10^{-4}	20.9	+24

[a] The concentration of mandelic acid was 2.5×10^{-3} M.

TABLE 16. The hydrogen–deuterium kinetic isotope effects measured for the oxidation of mandelic acid[a] by $Pb(OAc)_4$ in benzene and in benzene–pyridine

Substrate	$(k_H/k_D)_{benz}$	$(k_H/k_D)_{benz-pyr}$
PhCD(OH)COOH	1.04	1.01
PhCH(OD)COOD	1.01	1.19
PhCD(OD)COOD	1.02	1.21

[a] The concentration of mandelic acid was 1.0×10^{-3} M at 25 °C.

to form either **14** or **15** (Scheme 33) in the rate-limiting step of the oxidation. The subsequent step involves homolytic rupture of the Pb—O bond followed by rapid loss of the COO$^{\bullet}$/$^{\bullet}$COOH (Scheme 33). The homolytic cleavage of the Pb—O bond in **14** and **15** yields radicals or radical anions which would rapidly collapse to product.

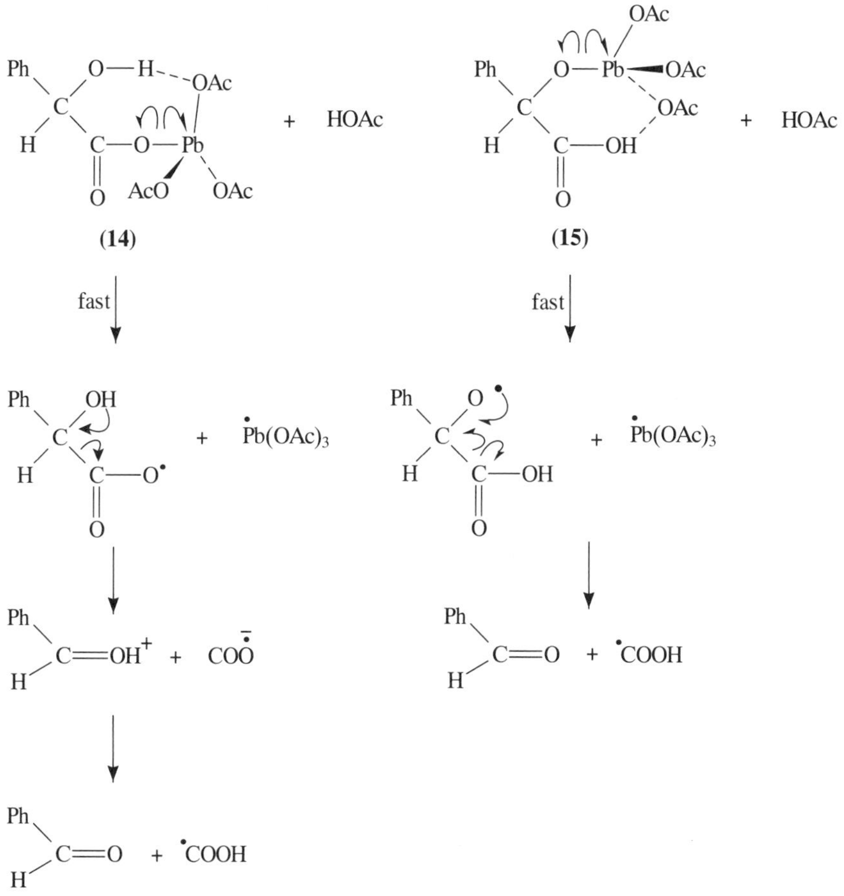

SCHEME 33

Although the reaction could proceed via intermediate **14** or **15**, the authors favour a mechanism where the formation of **14** is rate-determining because the displacement of the acetate at Pb by carboxylate anions is known to be rapid. The large negative $\Delta S^{\ddagger}$ (-34 e.u./mol) observed for the oxidation reaction is consistent with formation of the 'pseudo-cyclic' intermediate **14**. Also, the small Hammett ρ value of 0.4 determined for a series of *meta-* and *para*-substituted mandelic acids indicates that there is very little charge development on the benzyl carbon in the transition state of the rate-determining step. This is also consistent with the proposed mechanism.

However, the formation of intermediate **14** requires at least two steps, (i) a proton trans-fer and (ii) the formation of the cyclic intermediate. If formation of the intermediate, **14**, is rate-determining, the carboxy hydrogen must be lost in a pre-equilibrium step because no deuterium kinetic isotope effect is observed for this reaction (Scheme 34). Alternatively, the mandelic acid could displace an acetate ligand in a slow step and the proton could be transferred to the acetate ion in a fast, subsequent step (Scheme 35). Unfortunately, the results do not indicate which step in the formation of the cyclic intermediate, **14**, is rate-determining.

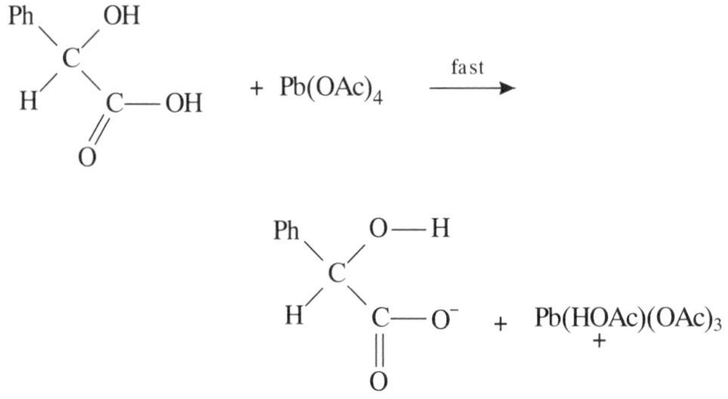

SCHEME 34

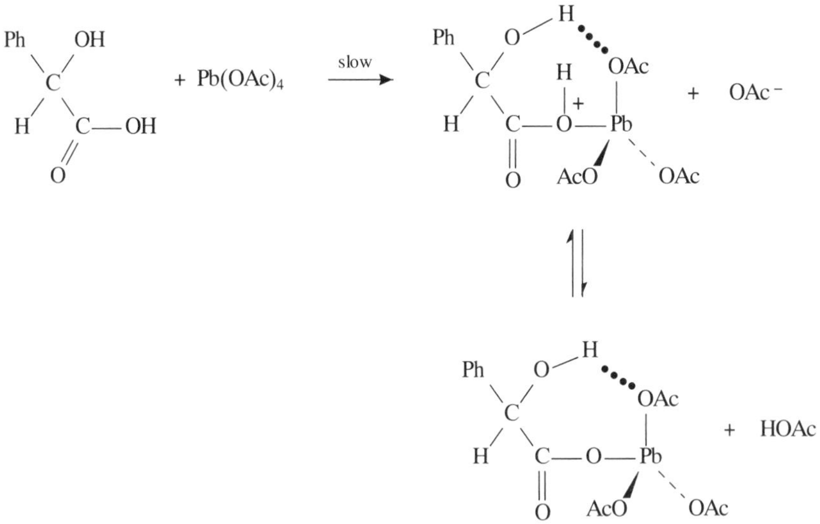

SCHEME 35

For the reaction carried out in the presence of pyridine, α-deuteration did not affect the second-order rate constant significantly. However, small isotope effects of 1.19 and 1.21 were observed for the lead tetraacetate oxidation of PhCH(OD)COOD and

PhCD(OD)COOD, respectively. This indicated to the authors that changes in the bonding to the hydroxy or carboxy hydrogen were occurring in the slow step of the reaction. Therefore, after the lead tetraacetate undergoes ligand exchange with the mandelic acid, abstraction of the hydroxy or carboxy proton by pyridine becomes rate-determining (Scheme 36).

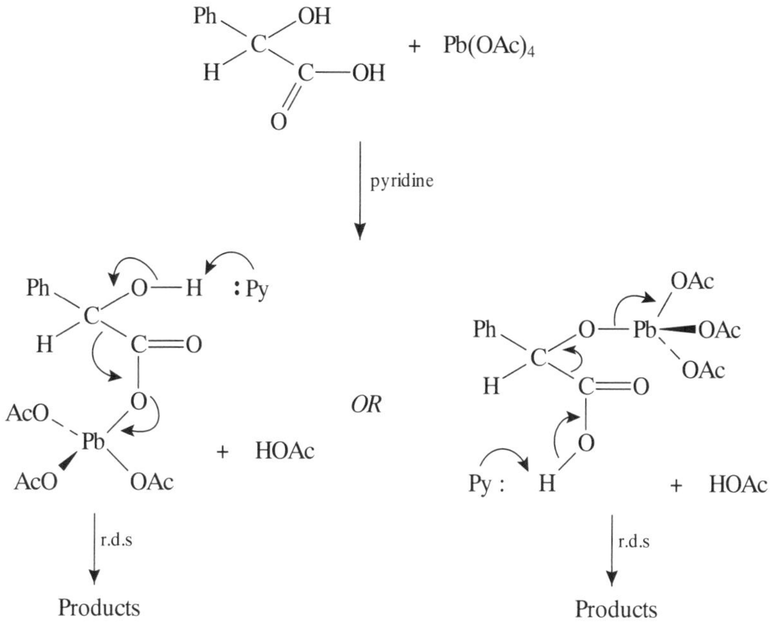

SCHEME 36

The small Hammett ρ value of $+0.16$ observed for a series of related *meta*- and *para*-substituted mandelic acids indicates that there is a very small negative charge development on the benzyl carbon in the transition state of the rate-determining step of the pyridine catalysed oxidation of mandelic acid. The large positive $\Delta S^{\ddagger}$ value ($+24$ e.u./mol) found for the catalysed reaction led Banerjee and coworkers to conclude that the transition state (Figure 5) is 'product-like'. This conclusion is consistent with the small k_H/k_D that is observed in this reaction[164]. The Pb$-$O bond is shown to rupture in a heterolytic fashion because Partch and Monthony[185] have demonstrated that pyridine diverts the reaction from a homolytic to a heterolytic mechanism.

This work was extended to include the lead tetraacetate oxidation of methyl esters of *meta*- and *para*-substituted mandelic acids[183,184] shown in equation 121. A kinetic study by Banerjee and collaborators showed the kinetic dependence on the ester concentration changed from second order in 1% (v/v) acetic acid in benzene to first order when the solvent contained more than 10% (v/v) acetic acid. These workers observed a significant decrease in $\Delta H^{\ddagger}$ (from 82.9 to 53.6 kcal mol^{-1}) and in $\Delta S^{\ddagger}$ (from -5.84 to -35.6 e.u.) when the solvent composition was changed from 1% acetic acid to greater than or equal to 10% acetic acid in benzene.

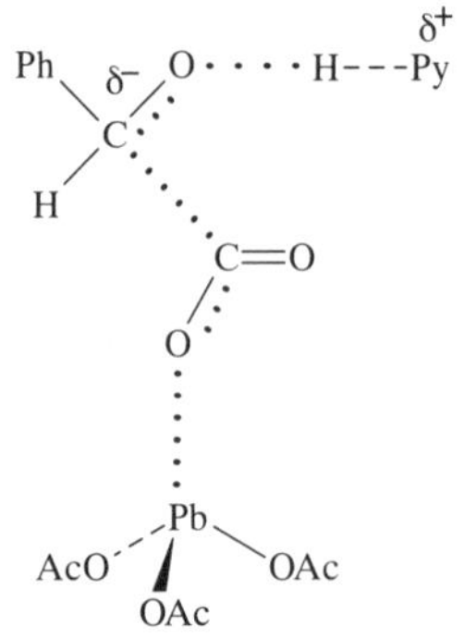

FIGURE 5. The transition state for the pyridine-catalysed oxidation of mandelic acid with lead tetraacetate

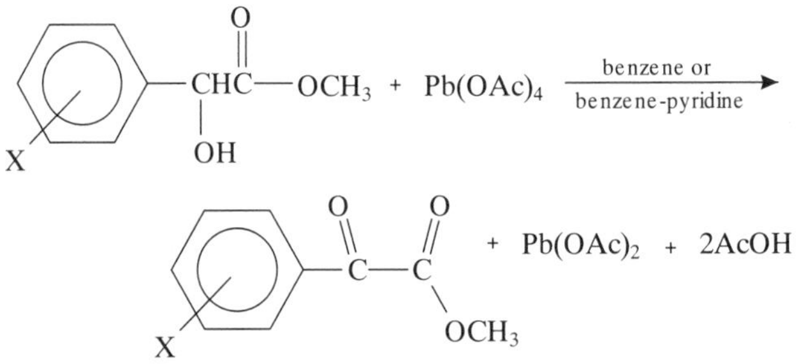

(121)

These observations led the authors to propose the mechanism shown in Scheme 37. The acetic acid is mainly in the form of dimers when the solvent is 1% (v/v) acetic acid–benzene. As a result, the equilibrium concentration of $Pb(OAc)_3(OR)$ is small and the formation of $Pb(OAc)_2(OR)_2$ is favoured. Hence the reaction proceeds through the intermediate, $Pb(OAc)_2(OR)_2$, and the k_2 step is rate-determining. At higher percent acetic acid, the increased concentration of the monomeric form of acetic acid prevents the formation of $Pb(OAc)_2(OR)_2$. As a result, the product is formed via the rate-limiting disproportionation of $Pb(OAc)_3OR$, i.e. the k_1 step is rate-determining.

$$2AcOH \xrightleftharpoons{K_{DM}} (AcOH)_2$$

$$Pb(OAc)_4 + ROH \rightleftharpoons Pb(OAc)_3(OR) + AcOH \xrightarrow{k_1} products$$

$$Pb(OAc)_3(OR) + ROH \rightleftharpoons Pb(OAc)_2(OR)_2 + AcOH \xrightarrow{k_2} products$$

SCHEME 37

The negative $\Delta S^{\ddagger}$ was explained in terms of the solvation of the transition state by the reaction medium. Dimerisation reduces the polar nature of acetic acid. Hence, in 1% acetic acid the 'polar' transition state is thought to 'freeze' the benzene molecules, thus producing a negative $\Delta S^{\ddagger}$ (−5.8 e.u./mol). When the acetic acid concentration is increased, the monomers of acetic acid solvate the transition state preferentially and the benzene molecules remain virtually unaffected. The authors believe this solvation accounts for the very negative $\Delta S^{\ddagger}$ (-35.6 e.u./mol). One should also add at this point that the oxidation changes from a rate-determining disproportionation of $Pb(OAc)_2(OR)_2$ to a rate-determining disproportionation of $Pb(OAc)_3(OR)$ when the acetic acid concentration increases. No reason for the variation in $\Delta H^{\ddagger}$ with the concentration of acetic acid was offered.

Banerjee's group also observed a large primary kinetic isotope effect of 4.2 for the oxidation of methyl mandelate, $PhCD(OH)COOCH_3$. This suggested that $C_\alpha-H$ bond cleavage occurs in the rate-determining step of the oxidation. In addition, a systematic study of the effect of substituents on the oxidation showed the rate constant was correlated with the Hammett σ. The positive Hammett ρ value of $+0.75$ suggested the development of only a small negative charge on the benzyl carbon in the transition state. The authors believe this information is consistent with a rate-limiting disproportionation of the dialkoxy Pb(IV) derivative. The Hammett ρ value and primary deuterium isotope effect suggest there is significant proton transfer in the transition state, **16**.

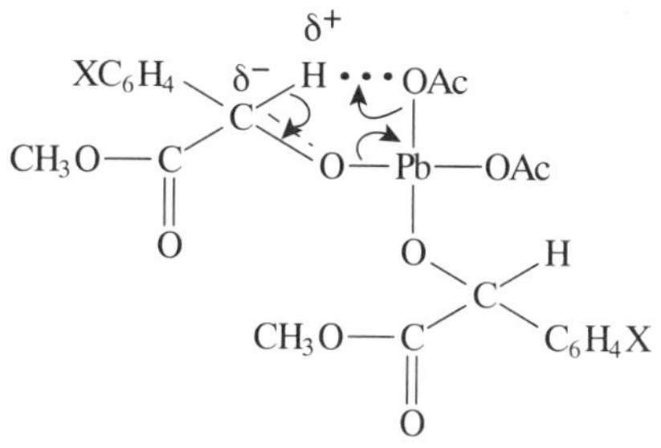

(16)

The oxidation of methyl mendalates is also catalysed by pyridine. Again, the activation parameters, $\Delta S^{\ddagger}$ and $\Delta H^{\ddagger}$, for the pyridine-catalysed reactions are significantly different from those for the uncatalysed oxidations. The $\Delta G^{\ddagger}$, however, is the same in the catalysed and uncatalysed reactions. The pyridine-catalysed reaction shows a first-order dependence on [pyridine], [ester] and [Pb(IV)]. The first-order dependence on ester concentration suggests that one of the unused coordination sites of Pb(IV) is blocked by the pyridine and that the oxidation involves rate-limiting disproportionation of a monoalkoxy–Pb(IV) derivative (Scheme 38). The pyridine is thought to catalyse the reaction by abstracting the α-proton from the methyl mandelate in the rate-determining step of the reaction. This is supported by the primary kinetic isotope effect of 1.8 observed for the oxidation of $PhCD(OH)COOCH_3$ in the presence of pyridine. The magnitude of the kinetic isotope effect was thought to reflect the product-like nature of the transition state[164].

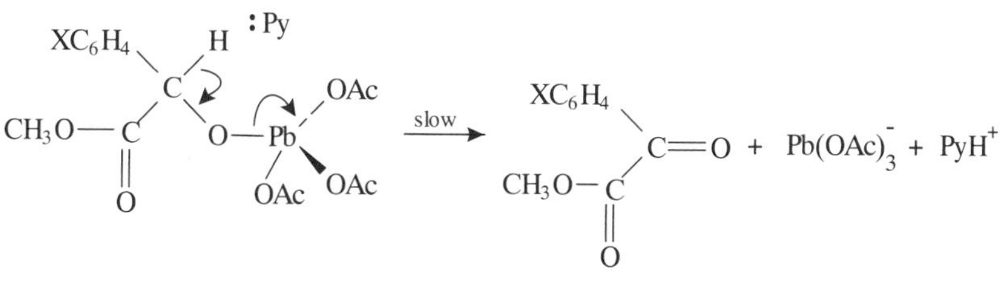

SCHEME 38

Lead tetraacetate also catalyses the oxidation of benzyl alcohols to benzaldehyde. Bhatia and Banerji[186] found that the reaction was first order in lead tetraacetate and second order in benzyl alcohol. They also found a primary hydrogen–deuterium kinetic isotope effect of 2.36 and a Hammett ρ value of -1.00 for this oxidation. This led Banerjee and Shanker[187] to propose that in the first step, benzyl alcohol exhanges with two acetate ligands on the lead tetraacetate to form $(PhCH_2O)_2Pb(OAc)_2$. The second step is a rate-determining hydride ion transfer of one of the benzylic hydrogens to the second $PhCH_2O^-$ entity coordinated to Pb. The intermediate then decomposes in a fast step to give benzaldehyde, $Pb(OAc)_2$ and benzyl alcohol. The primary hydrogen–deuterium isotope effect indicates that the $C_\alpha-H$ bond is broken in the rate-determining step and the large Hammett ρ value shows that there is considerable positive charge on the alpha carbon in the transition state. Although Banerjee and Shanker have suggested the formation of an intermediate, it would seem more likely that hydride ion is transferred and benzaldehyde is formed in a single step (Scheme 39). The negative ρ value indicates the development of a δ^+ charge on the reaction centre in the transiton state of the rate-determining step. This suggests there is more $C_\alpha-H$ bond rupture than C=O bond formation in the transition state.

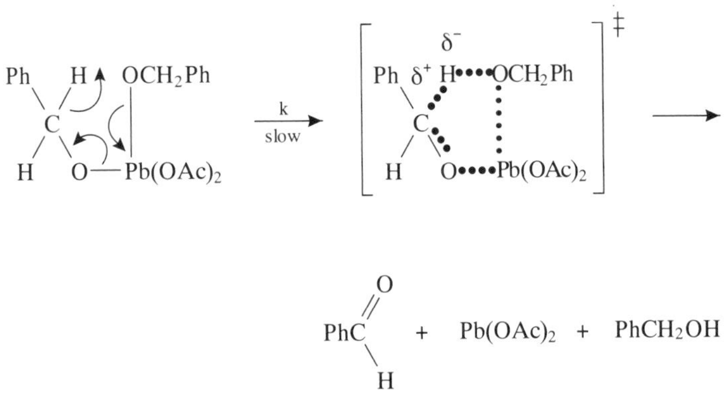

SCHEME 39

The pyridine-catalysed lead tetraacetate oxidation of benzyl alcohols shows a first-order dependence in $Pb(OAc)_4$, pyridine and benzyl alcohol concentration. An even larger primary hydrogen kinetic isotope effect of 5.26 and a Hammett ρ value of -1.7 led Banerjee and Shanker[187] to propose that benzaldehyde is formed by the two concurrent pathways shown in Schemes 40 and 41. Scheme 40 describes the hydride transfer mechanism consistent with the negative ρ value. In the slow step of the reaction, labilization of the Pb–O bond resulting from the coordination of pyridine occurs as the $C_\alpha-H$ bond is broken. The loss of $Pb(OAc)_2$ completes the reaction with transfer of ^+OAc to an anion.

In view of the large hydrogen–deuterium isotope effect of 5.26, Banerjee and coworkers proposed that the proton transfer mechanism (Scheme 41) is also operating. In this mechanism, pyridine behaves as a base and abstracts a proton in the rate-determining step.

It is interesting that the authors did not observe any curvature in their Hammett-ρ plots, which would have been expected if the two proposed mechanisms where competing because they would have ρ values of opposite sign, i.e. ρ should be negative for the proton transfer mechanism and positive for the hydride transfer mechanism. The authors proposed the two mechanisms because of the seemingly conflicting information given by (i) the ρ value which favours the hydride transfer mechanism and (ii) the large hydrogen–deuterium kinetic isotope effect which is greater than that for the uncatalysed

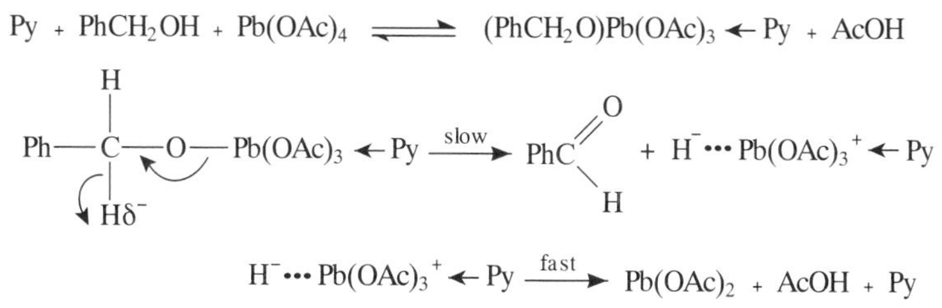

$$Py + PhCH_2OH + Pb(OAc)_4 \rightleftharpoons (PhCH_2O)Pb(OAc)_3 \leftarrow Py + AcOH$$

$$H^- \cdots Pb(OAc)_3{}^+ \leftarrow Py \xrightarrow{fast} Pb(OAc)_2 + AcOH + Py$$

SCHEME 40

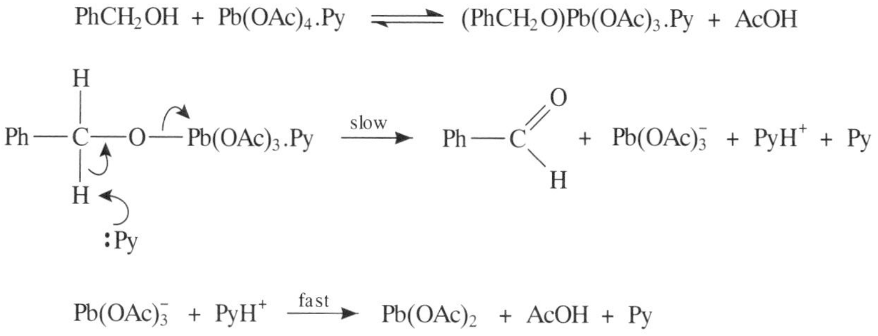

$$PhCH_2OH + Pb(OAc)_4 \cdot Py \rightleftharpoons (PhCH_2O)Pb(OAc)_3 \cdot Py + AcOH$$

$$Pb(OAc)_3^- + PyH^+ \xrightarrow{fast} Pb(OAc)_2 + AcOH + Py$$

SCHEME 41

reaction. This larger isotope effect led the authors to suggest the proton transfer mechanism. However, the data seem to be more consistent with one mechanism operating in both the catalysed and uncatalysed reaction, namely the hydride transfer mechanism. If this is the case, the larger deuterium kinetic isotope effect found for the pyridine-catalysed reaction simply indicates that the transfer of hydride ion is more complete in the transition state.

IV. REFERENCES

1. R. C. Weast (Ed.), *Handbook of Chemistry and Physics*, 66th Edition, CRC Press, Boca Raton, 1985, pp. B270–B271.
2. Y. Okada, S. Katu, S. Satooka and K. Takeuchi, *Spectrochim. Acta*, **46A**(4), 643 (1990).
3. C. Elschenbrioch and A. Salzer, *Organometallics: A Concise Introduction*, VCH, New York, 1992.
4. C. Winkler, *J. Prakt. Chem.*, **36**, 177 (1887).
5. V. W. Laurie, *J. Chem. Phys.*, **30**(5), 1210 (1959).
6. R. Eujen and H. Burger, *Spectrochim. Acta*, **37A**(12), 1029 (1981).
7. R. J. Lagow, R. Eujen, L. L. Gerchman and J. A. Morrison, *J. Am. Chem. Soc.*, **100**, 1722 (1978).
8. J. F. Sullivan, C. M. Whang, J. R. Durig, H. Burger, R. Eujen and S. Cradock, *J. Mol. Struct.*, **223**, 457 (1990).
9. J. R. Durig and J. B. Turner, *Spectrochim. Acta*, **27A**, 1623 (1971).
10. Y. Imai and K. Aida, *Bull. Chem. Soc. Jpn.*, **54**, 3323 (1981).
11. R. Eujen and H. Burger, *Spectrochim. Acta*, **35A**, 1135, (1979).
12. Y. Imai, K. Aida, K. -I. Sohma and F. Watari, *Polyhedron*, **1**(4), 397 (1982).
13. J. R. Durig, J. F. Sullivan, A. B. Mohamad, S. Cradock and Y. S. Li, *J. Chem. Phys.*, **84**(10), 5796 (1986).
14. R. F. Roberts, R. Varma and J. F. Nelson, *J. Chem. Phys.*, **64**, 5035 (1976).
15. J. E. Drake and R. T. Hemmings, *Can. J. Chem.*, **51**, 302 (1973).

16. J. R. Durig, A. B. Mohamad, G. M. Attia, Y. S. Li and S. Cradock, *J. Chem. Phys.*, **83**, 9 (1985).
17. J. R. Durig and G. M. Attia, *Spectrochim. Acta*, **44A**, 517 (1988).
18. K. Kamienska-Trela, H. Ilcewicz, H. Baranska and A. Labudzinska, *Bull. Pol. Acad. Sci.*, **32**(3–6), 144 (1984).
19. K. Kamienska-Trela and W. Luettke, *Pol. J. Chem.*, **54**(3), 611 (1980).
20. A. J. Shusterman, B. E. Landrum and R. L. Miller, *Organometallics*, **8**, 1851 (1989).
21. J. R. Durig, K. L. Kizer and Y. S. Li, *J. Am. Chem. Soc.*, **96**, 7400 (1974).
22. J. R. Durig, T. J. Geyer, P. Groner and M. Dakkouri, *Chem. Phys.*, **125**, 299 (1988).
23. L. C. Krisher, W. A. Watson and J. A. Morrison, *J. Phys. Chem.*, **60**(9), 3417 (1974).
24. E. C. Thomas and V. W. Laurie, *J. Chem. Phys.*, **50**(8), 3512 (1969).
25. R. W. Bott, C. Eaborn and P. M. Greasley, *J. Chem. Soc.*, 4804 (1964).
26. C. Eaborn, *J. Organometal. Chem.*, **100**, 43 (1975).
27. C. Eaborn and B. Singh, *J. Organomet. Chem.*, **177**, 333, (1979).
28. A. Nagasawa and K. Saito, *Bull. Chem. Soc. Jpn.*, **51**(7), 2015 (1978).
29. S. M. Moerlein, *J. Chem. Soc., Perkin Trans. 1*, 1687 (1985).
30. S. M. Moerlein and H. H. Coenen, *J. Chem. Soc., Perkin Trans. 1*, 1941 (1985).
31. H. H. Coenen and S. M. Moerlein, *J. Fluorine Chem.*, **36**(1), 63 (1987).
32. R. C. Weast (Ed.), *Handbook of Chemistry and Physics*, 66th Edition, CRC Press, Boca Raton, 1985, pp. B-300, B-301.
33. F. V. Sloop, G. M. Brown, R. S. Foote, K. B. Jacobson and R. A. Sachleben, *Biconjugate Chem.*, **4**, 406 (1993).
34. J. F. Testa, Jr. and C. A. Dooley, *J. Labelled Compd. Radiopharm.*, **27**, 753 (1989).
35. K. J. Meyers-Schulte and C. A. Dooley, *Marine Chem.*, **29**, 339 (1990).
36. H. Imura and N. Suzuki, *Anal. Chem.*, **55**, 1107 (1983).
37. R. A. Brown, C. M. Nazario, R. S. de Tirado, J. Castillon and E. T. Agard, *Environ. Res.*, **13**, 56 (1977).
38. K. S. Tenny and A. M. Tenny, *J. Labelled Compd.*, **4**, 54 (1968).
39. P. Ph. H. L. Otto, H. M. J. C. Creemers and J. G. Luijten, *J. Labelled Compd.*, **39**(2), 339 (1966).
40. D. Klotzer and W. Gorner, *Isotopenpraxis*, **20**, 58 (1984).
41. A. A. V. Podoplelov, S. C. Su, R. Z. Sagdeev, M. S. Shtein, V. M. Moralev, V. I. Goldanskii and Y. N. Molin, *Izv. Akad. Nauk SSSR, Ser. Khim.*, **10**, 2207 (1985).
42. E. W. Della and H. K. Patney, *Aust. J. Chem.*, **29**, 2469 (1976).
43. W. Kitching, A. R. Atkins, G. Wickham and V. Alberts. *J. Org. Chem.*, **46**, 563 (1981).
44. L. A. Paquette, C. W. Doeche, F. R. Kearney, A. F. Drake and S. F. Mason, *J. Am. Chem. Soc.*, **102**, 7228 (1980).
45. D. J. Kuchynka, C. Amatore and J. K. Kochi, *J. Organomet. Chem.*, **328**, 133 (1987).
46. H. Parnes and J. Pease, *J. Org. Chem.*, **44**, 151 (1979).
47. J. -C. Lahournere and J. Valade, *J. Organomet. Chem.*, **22**, C-3 (1970).
48. H. -J. Albert and W. P. Neumann, *Synthesis*, 942 (1980).
49. J. Szammer and L. Otvos, *Chem. Abstr.*, **108**, 150719g (1988).
50. J. Szammer and L. Otvos, *Hung. Teljes HU*, **41**, 412 (1987).
51. B. L. McInnis and T. K. Dobbs, *LC-GC*, **4**, 450 (1986); *Chem. Abstr.*, **105**, 53849b (1986).
52. D. Shugui, H. Guolan and C. Yong, *Appl. Organomet. Chem.*, **3**, 115 (1989).
53. Y. -C. Wei, P. R. Wells and L. K. Lambert, *Magn. Res. Chem.*, **24**, 659 (1986).
54. D. Zhang, M. Prager and A. Weiss, *J. Chem. Phys.*, **94**, 1765 (1991).
55. K. C. Westaway, in *Isotopes in Organic Reactions*, Vol. 7 (Eds. E. Buncel and C. C. Lee), Elsevier, New York, 1987, pp. 288–290.
56. J. Stanislawski, M. Prager and W. Hausler, *Physica B*, **156–157**, 356 (1989).
57. C. A. Kraus and W. V. Sessions, *J. Am. Chem. Soc.*, **47**, 2361 (1925).
58. M. Veith and V. Huch, *J. Organomet. Chem.*, **308**, 263 (1986).
59. K. Horiuchi, A. Yokoyama, Y. Fujibayashi, H. Tanaka, T. Odori, H. Saji, R. Morita and K. Torizuka, *Int. J. Appl. Radiat. Isot.*, **32**, 47 (1981).
60. O. P. D. Noronha, *Nuklearmedizin (Stuttgart)*, **17**, 110 (1978).
61. M. Kotora, V. Svata and L. Leseticky, *Radioisotopy*, **30**, 319 (1989).
62. H. G. Kuivila, *Synthesis*, 499 (1970).
63. H. G. Kuivila and C. W. Menapace, *J. Org. Chem.*, **28**, 2165 (1963).
64. J. E. Leibner and J. Jacobus, *J. Org. Chem.*, **44**, 449 (1979).
65. R. C. Fort, Jr. and J. J. Hiti, *J. Org. Chem.*, **42**, 3968 (1977).

66. W. Kitching, A. R. Atkins, G. Wickham and V. Alberts, *J. Org. Chem.*, **46**, 563 (1981).
67. T. -Y. Luh and L. M. Stock, *J. Org. Chem.*, **42**, 2790 (1977).
68. S. J. Cristol and A. L. Noreen, *J. Am. Chem. Soc.*, **91**, 3969 (1969).
69. G. A. Russell and G. W. Holland, *J. Am. Chem. Soc.*, **91**, 3968 (1969).
70. E. V. Blackburn and D. D. Tanner, *J. Am. Chem. Soc.*, **102**, 692 (1980).
71. H. R. Rogers, C. L. Hill, Y. Fujiwara, R. J. Rogers, H. L. Mitchell and G. M. Whitesides, *J. Am. Chem. Soc.*, **102**, 217 (1980).
72. J. J. Barber and G. M. Whitesides, *J. Am. Chem. Soc.*, **102**, 239 (1980).
73. W. M. Dadson and T. Money, *J. Chem. Soc., Chem. Commun.*, 112 (1982).
74. L. A. Hull and P. D. Bartlett, *J. Org. Chem.*, **40**, 824 (1975).
75. H. Yamanaka, I. Kikui and K. Teramura, *J. Org. Chem.*, **41**, 3794 (1976).
76. W. P. Neumann and H. Hillgartner, *Synthesis*, 537 (1971).
77. H. R. Rogers, R. J. Rogers, H. L. Mitchell and G. M. Whitesides, *J. Am. Chem. Soc.*, **102**, 231 (1980).
78. I. D. John, N. D. Tyrrel and E. J. Thomas, *Tetrahedron*, **39**, 2477 (1983).
79. J. San Filippo, Jr. and G. M. Anderson, *J. Org. Chem.*, **39**, 473 (1974).
80. T. Ando, H. Yamanaka, F. Nagimata and W. Fumasaka, *J. Org. Chem.*, **35**, 33 (1970).
81. T. Ando, T. Ishikava, E. Ohtani and M. Sawada, *J. Org. Chem.*, **46**, 4446 (1981).
82. T. Ando, F. Nagimata, H. Yamanaka and W. Fumasaka, *J. Am. Chem. Soc.*, **89**, 5719 (1967).
83. L. K. Sydnes, *Acta Chem. Scand., Ser. B***32**, 47 (1978).
84. J. B. Lambert, B. T. Ziemnicka-Merchant, M. A. Hayden and A. T. Hjelmfelt, *J. Chem. Soc., Perkin Trans. 2*, 1553 (1991).
85. N. I. Yakushina, G. A. Zakharova, C. S. Surmina and I. G. Bolesov, *Zh. Org. Khim.*, **16**, 1834 (1980).
86. D. B. Ledlie, T. Swan, L. Bowers and J. Pile, *J. Org. Chem.*, **41**, 419 (1976).
87. D. P. G. Hammon and K. R. Richards, *Aust. J. Chem.*, **36**, 2243 (1983).
88. I. A. McDonald, A. S. Dreiding, H. M. Hartmacher and H. Musso, *Helv. Chim. Acta*, **56**, 1385 (1973).
89. K. E. Coblens, V. B. Muralidharan and B. Ganem, *J. Org. Chem.*, **47**, 5041 (1982).
90. W. Heller and C. Tamm, *Helv. Chim. Acta*, **57**, 1766 (1974).
91. W. Bolland and L. Jaenicke, *Chem. Ber.*, **110**, 1823 (1977).
92. D. E. Applequist, M. R. Johnston and F. Fisher, *J. Am. Chem. Soc.*, **92**, 4614 (1970).
93. J. P. Praly, *Tetrahedron Lett.*, **24**, 3075 (1983).
94. R. M. Fantazier and M. L. Poutsma, *J. Am. Chem. Soc.*, **90**, 5940 (1968).
95. R. M. Hoyte and D. B. Denney, *J. Org. Chem.*, **39**, 2607 (1974).
96. A. L. J. Beckwith and T. Lawrence, *J. Chem. Soc., Perkin Trans. 2*, 1535 (1979).
97. M. Kotora, Thesis, Faculty of Natural Sciences, Charles Univ., Prague (1986).
98. Y. Ueno, C. Tanaka and M. Okawara, *Chem. Lett.*, 795 (1983).
99. J. J. Patroni and R. V. Stick, *J. Chem. Soc., Chem. Commun.*, 449 (1978).
100. J. J. Patroni and R. V. Stick, *Aust. J. Chem.*, **32**, 411 (1979).
101. T. S. Fuller and R. V. Stick, *Aust. J. Chem.*, **33**, 2509 (1980).
102. D. H. R. Barton, W. B. Motherwell and A. Stange, *Synthesis*, 743 (1981).
103. D. D. Tanner, E. V. Blackburn and G. E. Diaz, *J. Am. Chem. Soc.*, **103**, 1557 (1981).
104. A. G. M. Barrett, D. Dauzonne, I. A. O'Neil and A. Renaud, *J. Org. Chem.*, **49**, 4409 (1984).
105. N. Ono, H. Miyake, R. Tamura and A. Kaji, *Tetrahedron Lett.*, **22**, 1705 (1981).
106. N. Ono, H. Miyake, H. Fujii and A. Kaji, *Tetrahedron Lett.*, **24**, 3477 (1983).
107. D. L. J. Clive, G. J. Chittattu, V. Farina, W. A. Kiel, S. M. Menchen, C. G. Russell, A. Singh, C. K. Wong and N. J. Curtis, *J. Am. Chem. Soc.*, **102**, 4438 (1980).
108. N. Y. M. Fung, P. De Mayo, J. H. Schauble and A. C. Weedon, *J. Org. Chem.*, **43**, 3977 (1978).
109. D. H. R. Barton, G. Bringmann, G. Lamotte, W. B. Motherwell, R. S. M. Motherwell and A. E. A. Porter, *J. Chem. Soc., Perkin Trans. 1*, 2657 (1980).
110. C. G. Gutierrez, R. A. Stringham, T. Nitasaka and K. G. Glasscock, *J. Org. Chem.*, **45**, 3393 (1980).
111. D. H. R. Barton, M. A. Dowlatshahi, W. B. Motherwell and D. Villemin, *J. Chem. Soc., Chem. Commun.*, 732 (1980).
112. R. N. Hanson and L. A. Franke, *J. Nucl. Med.*, **25**, 998 (1984).
113. J. P. Quintard, M. Degueil-Castaing, B. Barbe and M. Petraud, *J. Organomet. Chem.*, **234**, 41 (1982).

114. M. Gielen and Y. Tonduer, *J. Organomet. Chem.*, **169**, 265 (1979).
115. V. S. Lopatina, N. I. Sheverdina and K. A. Kocheshkov, *Zh. Obshch. Khim.*, **47**, 359 (1977).
116. V. S. Lopatina, N. I. Sheverdina, N. V. Fomina, K. A. Kocheshkov and E. M. Panov, *Izv. Akad. Nauk SSSR, Ser. Khim.*, 378 (1980).
117. A. Rahm, M. Degueil-Castaing and M. Pereyre, *J. Organomet. Chem.*, **232**, C-29 (1982).
118. J. E. Baldwin and C. G. Carter, *J. Am. Chem. Soc.*, **104**, 1362 (1982).
119. J. E. Baldwin and T. C. Barden, *J. Am. Chem. Soc.*, **106**, 5312 (1984).
120. J. P. Quintard and M. Pereyre, *J. Labelled Compd. Radiopharm.*, **14**, 633 (1978).
121. R. Fosty, M. Gielen, M. Pereyre and J. P. Quintard, *Bull. Soc. Chim. Belg.*, **85**, 523 (1976).
122. J. P. Quintard, M. Degueil-Castaing, G. Dumartin, B. Barbe and M. Petraud, *J. Organomet. Chem.*, **234**, 27 (1982).
123. R. N. Hanson, *Isot. Phys. Biomed. Sci.*, **1**, 285 (1991).
124. J. P. Quintard, M. Degueil-Castaing, G. Dumartin, A. Rahm and M. Pereyre, *J. Chem. Soc., Chem. Commun.*, 1004 (1980).
125. H. E. Ensley, R. R. Buescher and K. Lee, *J. Org. Chem.*, **47**, 404 (1982).
126. P. W. Collins, C. J. Jung, A. Gasiecki and R. Pappo, *Tetrahedron Lett.*, 3187 (1978).
127. M. Pereyre and J. Valade, *Tetrahedron Lett.*, 489 (1969).
128. H. R. Wolf and M. P. Zink, *Helv. Chim. Acta*, **56**, 1062 (1973).
129. R. N. Hanson, *Isot. Phys. Biomed. Sci.*, **1**, 285 (1991).
130. W. F. Goure, M. E. Wright, P. D. Davis, S. S. Labadie and J. K. Stille. *J. Am. Chem. Soc.*, **106**, 6417 (1984).
131. J. G. M. Vanderkerk and J. G. Noltres, *J. Appl. Chem.*, **9**, 179 (1959).
132. G. L. Tonnesen, R. N. Hanson and D. E. Seitz, *Int. J. Appl. Radiat. Isot.*, **32**, 171 (1980).
133. R. N. Hanson, D. E. Seitz and G. L. Tonnesen, *J. Labelled Compd. Radiopharm.*, **18**, 99 (1981).
134. R. N. Hanson, G. L. Tonnesen, W. H. McLaughlin, W. D. Bloomer and D. E. Seitz, *J. Labelled Compd. Radiopharm.*, **18**, 128 (1981).
135. M. J. Adam, B. D. Pate, T. J. Ruth, J. M. Berry and L. D. Hall, *J. Chem. Soc., Chem. Commun.*, 733 (1981).
136. R. N. Hanson and H. El-Wakil, *J. Org. Chem.*, **52**, 3687 (1987).
137. R. N. Hanson, in *Proceedings of the Third International Symposium on the Synthesis and Applications of Isotopically Labelled Compounds* (Eds. T. A. Ballie and J. R. Jones), Elsevier, Amsterdam, 1989, pp. 275–281.
138. J. A. Balatano, M. J. Adams and D. Hall, *J. Nucl. Med.*, **27**, 972 (1986).
139. D. E. Seitz, G. L. Tonnesen, S. Hellman, R. N. Hanson and S. J. Adelstein, *J. Organomet. Chem.*, **186**, C33, (1980).
140. R. A. Milius, W. H. McLaughlin, R. M. Lambrect and A. P. Wolf, *Int. J. Appl. Radiat. Isot.*, **37**, 799 (1986).
141. F. G. Salituro, K. E. Carlson, J. F. Ellison, B. S. Katzenellenbogen and J. A. Katzenellenbogen, *Steroids*, **48**, 287 (1986).
142. S. Chumpradit, J. Billings, M. P. Kung, S. Pan and H. F. Kung, *J. Nucl. Med.*, **30**, 803 (1989).
143. S. Chumpradit, H. F. Kung, J. Billings, M. P. Kung and S. Pan., *J. Med. Chem.*, **32**, 1431 (1989).
144. L. L. Iversen, M. A. Rogawski and R. J. Miller, *Mol. Pharmacol.*, **12**, 251 (1976).
145. L. L. Iversen, *Science*, **188**, 1084 (1975).
146. J. R. Lever, J. L. Musachio, U. A. Scheffel, M. Stathis and N. H. Wagner, Jr., *J. Nucl. Med.*, **30**, 803 (1989).
147. J. L. Musachio and J. R. Lever, *Tetrahedron Lett.*, **30**, 3613 (1989).
148. H. F. Kung, J. J. Billings, Y. Z. Guo and R. H. Mach., *Nucl. Med. Biol.*, **15**, 203 (1988).
149. T. dePaulis, A. Janowsky, R. M. Kessler, J. A. Clanton and H. E. Smith, *J. Med. Chem.*, **31**, 2017 (1988).
150. R. N. Hanson, *J. Labelled Compd. Radiopharm.*, **26**, 3 (1989).
151. D. S. Wilbur, S. W. Hadley, M. D. Hylarides, P. A. Beaumier and A. R. Fritzberg, *J. Nucl. Med.*, **29**, 777 (1988).
152. G. D. Prestwich, *Science*, **237**, 999 (1987).
153. G. D. Prestwich, W. Eng, S. Robles, R. G. Voigt. J. R. Wisniewski and C. Wawzenczyk, *J. Biol. Chem.*, **263**, 1398 (1988).
154. M. M. Goodman, K. H. Neff, K. R. Ambrose and F. F. Knapp, Jr., *J. Nucl. Med.*, **28**, 775 (1987).
155. M. M. Goodman and F. F. Knapp. Jr., *Chem. Abstr.*, **111**, 78541g (1989).
156. M. M. Goodman, G. W. Kagalka, R. Marks, F. F. Knapp, Jr. and S. Truelove, *J. Nucl. Med.*, **29**, 777 (1988).

157. A. Vlcek, Jr. and H. B. Gray, *J. Am. Chem. Soc.*, **109**, 286 (1987).
158. S. J. Hannon and T. G. Traylor, *J. Org. Chem.*, **46**, 3645 (1981).
159. E. V. Uglova, I. G. Brodskaya and O. A. Reutov, *J. Org. Chem. USSR (Engl. Transl.)*, **12**, 1357 (1976).
160. T. G. Traylor and G. S. Koerner, *J. Org. Chem.*, **46**, 3651 (1981).
161. Z. Song and P. Beak, *J. Am. Chem. Soc.*, **112**, 8126 (1990).
162. A. N. Abeywickrema and A. L. J. Beckwith, *J. Chem. Soc., Chem. Commun.*, 464 (1986).
163. H. L. Strong, M. L. Brownawell and J. San Filippo, Jr., *J. Am. Chem. Soc.*, **105**, 6526 (1983).
164. F. Westheimer, *Chem. Rev.*, **61**, 265 (1961).
165. D. J. Kuchynka, C. Amatore and J. K. Kochi, *J. Organomet. Chem.*, **328**, 133 (1987).
166. C. Chatgiliaoglu, K. U. Ingold and J. C. Scaiano, *J. Am. Chem. Soc.*, **103**, 7739 (1981).
167. D. J. Carlsson and K. U. Ingold, *J. Am. Chem. Soc.*, **90**, 7047 (1968).
168. L. Melander and W. H. Saunders, Jr., *Reaction Rates of Isotopic Molecules*, Wiley, Interscience, New York, 1980.
169. S. Kozuka and E. S. Lewis, *J. Am. Chem. Soc.*, **98**, 2254 (1976).
170. C. G. Swain, E. C. Stivers, J. F. Reuwer, Jr. and L. J. Schaad, *J. Am. Chem. Soc.*, **80**, 5885 (1958).
171. J. A. Franz, R. D. Barrows and D. M. Camaioni, *J. Am. Chem. Soc.*, **106**, 3964 (1984).
172. J. A. Franz, N. K. Seleman and M. S. Alnajjar, *J. Org. Chem.*, **51**, 19 (1986).
173. R. C. Weast (Ed.), *Handbook to Chemistry and Physics*, 66th Edition, CRC Press, Boca Raton, 1985, pp. B318–B319.
174. U. Edlund, T. Lejon, P. Pyykko, T. K. Venkatachalam and E. Buncel, *J. Am. Chem. Soc.*, **109**, 5982 (1987).
175. J. R. Dean, L. Ebdon, H. M. Crews and R. C. Massey, *J. Anal. Atom. Spec.*, **3**, 349 (1988).
176. J. S. Blais and W. D. Marshall, *Appl. Organomet. Chem.*, **1**, 251 (1987).
177. J. S. Blais and W. D. Marshall, *Appl. Radiat. Isot.*, **39**(12), 1259 (1988).
178. J. S. Blais, G. M. Momplaisir and W. D. Marshall, *Appl. Organomet. Chem.*, **3**, 89 (1989).
179. B. L. McInnis and T. K. Dobbs, *LC-GC*, **4**, 450 (1986).
180. Y. -C. Wei, P. R. Wells and L. K. Lambert, *Magn. Res. Chem.*, **24**, 659 (1986).
181. R. Shanker, S. K. Banerjee and O. P. Sachdeo, *Z. Naturforsch.*, **28**(6), 375 (1973).
182. S. K. Banerjee, S. Singh, R. Shanker and O. P. Sachdeva, *Bull. Soc. Kinet. Ind.*, **11**(1), 1 (1989).
183. S. K. Banerjee, R. Shanker, and O. P. Sachdeva, *Z. Phys. Chem. (Leipzig)*, **267**(5), 931 (1986).
184. O. P. Sachdeva, S. K. Banerjee and R. Shanker, *Z. Naturforsch.*, **41b**, 467 (1986).
185. R. E. Partch and J. Monthony, *Tetrahedron Lett.*, 4427 (1967).
186. I. Bhatia and K. K. Banerji, *J. Chem. Res. (S)*, 97 (1982).
187. S. K. Banerjee and R. Shanker, *Z. Phys. Chem.* (Leipzig), **268** (2), 409 (1987).

CHAPTER **16**

The environmental methylation of germanium, tin and lead

P. J. CRAIG and J. T. VAN ELTEREN

Department of Chemistry, De Montfort University, The Gateway, Leicester LE1 9BH, UK
Fax: 0116-2577135

I. THE METHYLATION OF GERMANIUM COMPOUNDS UNDER ENVIRONMENTAL OR MODEL ENVIRONMENTAL CONDITIONS

A. Methods of Analysis

Reduction of inorganic and methylgermanium species using sodium borohydride ($NaBH_4$) is successful in view of the stability to disproportionation and water of

---

The chemistry of organic germanium, tin and lead compounds
Edited by S. Patai © 1995 John Wiley & Sons Ltd

the germanium carbon and hydrogen bonds[1]. The true divalent state for germanium is unknown in the environment and germanium chemistry resembles that of silicon[2,3]. Reduction without disproportionation allows separation of derivatized mixed organogermanium hydrides by purge and trap or gas chromatographic methodologies in the normal way[4,5]. Graphite furnace AA has been used for detection[4-6]. Absolute detection limits for environmental samples are around 150 pg for each species and potential interference effects from carbonate and sulphide have been eliminated by preliminary acidification and helium purging. Use of sodium tetraethylborate (NaBEt$_4$) for the production of volatile organogermanium species for analysis has been investigated but dismutation of Me$_3$GeCl to a mixed series of Me$_4$Ge ($\sim$ 20%), Me$_3$GeCl (40%) and the expected Me$_3$GeEt (40%) occurred[7].

It is not inconceivable that re-investigation under different conditions might lead to viable results with this method. The reaction of Grignard reagents with organogermanium halides is known to produce volatile R$_3$GeR[1] species[8] but this has not been attempted for samples from the natural environment[2]. Hence there is a reliable analytical method that could detect organogermanium compounds if they exist in the natural environment.

B. Organogermanium Compounds Found in the Environment

Only methylgermanium species of the theoretically possible organic series have been detected. Unlike the cases for tin and lead there are no large-scale higher alkyl industrial products for germanium that would produce higher alkylgermaniums in the environment (compare, for example, the butyltins and the alkylleads).

However, small concentrations of methylgermaniums (Me$_2$Ge^{2+} and MeGe^{3+}) have been detected in most of the oceans (though only on few and small-scale surveys). Levels of MeGe^{3+} were around 330 pmol dm^{-3} and Me$_2$Ge^{2+} at 100 pmol dm^{-3}, respectively[9,10]. This is an organic proportion of over 70% compared to total germanium. Me$_3$Ge$^+$ has not been detected, possibly owing to low concentration, insolubility in ocean water or liposolubility in marine species. The observed concentration order MeGe^{3+} > Me$_2$Ge^{2+} > Me$_3$Ge$^+$ argues for sequential natural methylation of an initial inorganic germanium precursor.

Me$_2$Ge^{2+} and MeGe^{3+} behave conservatively in the oceans, having little or no concentration variation with depth[11]. There is no evidence for production or removal of methylgermanium species in the euphotic zone or in the upper thermocline, suggesting that germanium species do not methylate or demethylate in the oceans[12].

Methylgermanium species also behave conservatively in estuaries. As with the oceans, concentrations vary directly with salinity, concentrations being very low in freshwaters, whether or not the river is polluted. Again, it appears that estuarine processes do not methylate or demethylate germanium species[9].

Non-polluted rivers surveyed so far have very low levels of methylgermanium species in the 1–10 pmol dm^{-3} range[12] (i.e. low ppb). In view of the conservative oceanic and estuarine behaviour, this suggests a terrestrial source for germanium methylation, such as methanogenic swamps etc.[13]. However, such a methylating source on land has not yet been established. Sea salt aerosols as a source of methylgermaniums in rivers appear to be unlikely[13]. The ratio of MeGe^{3+} to Me$_2$Ge^{2+} in rivers is < 0.5 to 1, compared to a 3:1 ratio in seawater. The estimated residence time for methylgermaniums in the oceans is about 10^6 years, at a removal rate of < 10^{-4}% pa[13].

Contaminated rivers contain much higher levels of inorganic germanium, e.g. from steel or coal industries locally. In such rivers, methylgermanium concentrations up to 100 times higher than in unpolluted locations may be found, e.g. 10–20 pmol dm^{-3} region for MeGe^{3+} and Me$_2$Ge^{2+}[11,12,14]. The source of this inorganic germanium is likely to arise from coal fly ash, with the methylation taking place in local sewage treatment works (i.e. a bacterial methylation process).

In view of the lack of success in discovering a plausible natural methylating source for the methylgermanium species[14], some efforts have been made to investigate model methylating systems in laboratories.

C. Microorganism Culture Experiments to Demonstrate Methylation of Germanium

To date there appear to have been no positive results for the methylation of germanium by marine algal species. Sediment cultures from the River Zenne (Belgium) have been shown to produce methylgermanium species from added GeO$_2$. The sediments were anaerobic and probably were methanogens. 100 nmol dm^{-3} concentrations of GeO$_2$ were incubated at 30 °C at pH 7.7 in the dark for 10 months. All three methylgermanium species were found to have been produced[12]. No methylgermanium species were found in a sterile control experiment, indicating that the methylation observation in the microbe-rich systems was a real phenomenum. Percentage conversion was at a maximum down to a depth of 20 cm of sediment and decreased below that depth. Maximum total percentage conversion of germanium to methylated products was 29% at a depth of 13 cm, decreasing to 4.5% at 28 cm depth. In each case production was Me$_2$Ge^{2+} > Me$_3$Ge$^+$ > MeGe^{3+}. However, a second series of experiments the next year with this system did not produce methylated species[13]. Further experiments[13] were carried out with enriched cultures from the Zenne system, with the cultures variously selected for methanogens, sulphate reducers or iron reducers. However, there was little clear evidence of methylation in any of these selected systems.

Conversion of natural, unknown inorganic germanium species to all three methylgermanium species has been measured in an anaerobic sewage digester system. Production of organic germanium was matched with loss of inorganic germanium. Although small quantities of methylgermanium were present at the start of the experiment, the ratio of final methylgermanium concentration to control was up to 6.6:1, suggesting a genuine biological methylation had occurred. Concentration levels of all the germanium species were at the pmol dm^{-3} level. No germanium was added in those experiments[12].

D. Model Experiments to Demonstrate Methylation of Germanium

Chemical experiments with methyl cobalamium (Me$_3$CoB$_{12}$) or iodomethane (Me$_3$I) have been attempted. Reaction of Me$_3$CoB$_{12}$ with GeI$_2$ at pH 1 gave a 1.3% yield of a methylated product (Me$_3$Ge^{3+} only); no methylation occurred at pH 7. Concentrations of GeI$_2$ were 300 μmol dm^{-3}, and of Me$_3$CoB$_{12}$ 600 μmol dm^{-3}. With Me$_3$I, 6% of Me$_3$Ge^{3+} was produced at pH 7.6 in artificial seawater. Me$_3$CoB$_{12}$ did not react with Ge^{IV} (GeO$_2$). Although well known as a synthetic reaction on a larger scale[15], the oxidative addition reaction of Me$_3$I with Ge^{II} is not inherently impossible in the environment as Me$_3$I is a natural product. The reaction of Me$_3$CoB$_{12}$ with GeI$_2$ probably occurs by a free radical route. From the natural environmental results noted above, the methylating source molecule needs to be available terrestrially. Me$_3$I and DMSP (dimethyl sulphopropiothetin) are mainly, but not exclusively, marine products; no experiments appear to date to have been carried out with SAM (S-adenosyl methionine).

II. THE METHYLATION OF TIN COMPOUNDS UNDER ENVIRONMENTAL OR MODEL ENVIRONMENTAL CONDITIONS

It should be noted for tin species that derivatization with $NaBH_4$, $NaBEt_4$ or Grignard reagents have all been attempted with apparent success, and that volatile derivatized organotins can be separated and identified using interfaced GC FID, GC AA and GC MS etc., or similar HPLC systems.

A. Organotin Compounds Found in the Environment

There were several reports of the existence of methyltin compounds in the natural environment in the late 1970s and early 1980s[16−20]. It is usually assumed that they were formed there, since dispersion of methyltin products to those locations seemed unlikely. However, the case of methyltins as stabilizers in PVC water pipes and the possibility of methyltin impurities in butyltin-containing paints should be borne in mind. The examples above were found in harbours, estuaries, rivers and ocean waters and were at the $ng\,dm^{-3}$ level (viz $\mu mol\,dm^{-3}$). At about this time, methyltin hydrides were also found as such (i.e. without reaction with $NaBH_4$) in an anoxic harbour mud[18]. In aerobic environments, or at higher concentrations, such species would not be expected to be stable. Methyl(butyl)tins have also been reported in harbour sediments[21] and waters[22].

More recent work has demonstrated that methyltin compounds are found widely in sediment[23], shellfish[24], algae[25] and marsh grass[26]. *S. alterniflora* (marsh grass) from the Great Bay Estuary, NH, USA has been shown to contain inorganic tin [*ca* 30 $ng\,g^{-1}$ level (viz approximately $pmol\,dm^{-3}$)], $MeSn^{3+}$ (2 $ng\,g^{-1}$) and Me_2Sn^{2+} (10 $ng\,g^{-1}$)[27]. The tin species were analysed by hydride generation ($NaBH_4$) and GC AA. The concentration of inorganic tin in the estuary water containing the grass was in the region of 1.5–2.0 $ng\,cm^{-3}$, $MeSn^{3+}$ (0.16 $ng\,cm^{-3}$) and Me_2Sn^{2+} (1.4 $ng\,cm^{-3}$) was also present.

The role of the decaying marsh grass *S. alterniflora*, when exposed to inorganic tin, has recently been investigated[27] and the authors concluded that formation of methyltin may have occurred. Although methyltin species were present naturally in the estuary waters used (Great Bay, NH, USA), incubation of the estuarine water without plant leaves in the presence of added inorganic tin ($SnCl_4/5H_2O$) (added at 75 $ng\,cm^{-3}$) produced enhanced net methylation (from a mean of 152 $ng\,cm^{-3}$ in the flask to 174 $ng\,cm^{-3}$ over three replicate experiments). The identification of the methylator in the estuarine water was not attempted. The same group has also detected methyltin compounds in other sediments[28].

Recent work has demonstrated the existence of methyltin species in a sediment core in an inorganic environment (Sepetiba Bay, Brazil)[29]. The ultimate source of tin is thought to be anthropogenic. The environment was rich in organic material, rich in microbes and anaerobic. In general, total concentrations increased from bottom to top in the core (i.e. anthropogenic input). Between 30 and 50 cm depth no tin compounds could be detected. At the surface inorganic tin was 164 $ng\,g^{-1}$, $MeSn^+$ 98 $ng\,g^{-1}$, Me_2Sn^{2+} 83 $ng\,g^{-1}$ and Me_3Sn^+ 26$ng\,g^{-1}$. Again this order parallels a hypothetical carbonium ion methylation starting from inorganic tin. (In general though, in this core, $Me_2Sn^{2+} > MeSn^{3+}$ — but Me_3Sn^+ was always the methyltin species in least concentration.) Natural levels of methyltins in Baltimore Harbour sediments were found to be *ca* 8 $ng\,g^{-1}$ ($MeSn^{3+}$), *ca* 1 $ng\,g^{-1}$ (Me_2Sn^{2+}) and 0.3 $ng\,g^{-1}$ (Me_3Sn^+) (wet weight)[30,31]. This has been compared to levels of $MeSn^{3+}$ measured at a polluted site in Chesapeake Bay (0.6 $ng\,g^{-1}$ dry weight)[32]. Estimates of methyltin levels in the Patuxent River, USA are 1 $ng\,g^{-1}$ $MeSn^{3+}$ and 0.1 ng Me_2Sn^{2+}. Me_3Sn^+ was not detectable[30].

Methyltins have also been isolated from seaweeds, and bacteria isolated from the seaweed-produced methyltins[33].

Recently methyltin compounds have been detected in unfiltered waters from Antwerp Harbour in Belgium[34]. The methyltins were derivatized by a Grignard method (n-$C_5H_{11}MgBr$) and detection of the volatile derivative was by GC QFAAS or GC AED. Me_2Sn^{2+} and $MeSn^{3+}$ were found to be around 3–5 $ng\,dm^{-3}$ and 2–10 $ng\,dm^{-3}$, respectively, in the water. Me_3Sn^+ was not detected. The butyltins found in the same locations were ascribed as being derived from anti-fouling paints — the methyltins could have arisen as an anti-fouling impurity or by methylation of decomposed butyltins (i.e. inorganic tin) in the environment. Methyl(butyl)tins were not found in this study.

B. Microorganism Culture Experiments to Demonstrate Methylation of Tin

There were two early reports of the incubation of an inorganic tin compound with a biological medium to produce a methyltin product[18,35]. In the first case[18], $SnCl_4/5H_2O$ was converted to Me_3Sn^+ and Me_2Sn^{2+} species by incubation with aliquots of natural sediments from Chesapeake Bay, USA. The source of methyl groups was identified as sediment microflora but no individual identifications were made. Only small traces of monomethyltin species were found. In the other report[35] inorganic tin(II) and tin(IV) compounds were incubated with an aerobic strain of Pseudomas 244 (Ps244) from Chesapeake Bay. Only small amounts of methyltins were produced from tin(II), but tin(IV) produced various stannanes after hydride generation, viz $(Me)_4SnH_{4-n}(n = 2-4)$.

Methylation of several tin compounds by strains of yeast (*Saccharomyces cerevisiae*) has been reported[36]. Tin(0), $SnCl_4$ and SnS produced little evidence of methylation. Various tin(II) derivatives of amino acids [e.g. tin(II) pencillamine] produced mainly $MeSn^{3+}$, but some Me_2Sn^{2+} species was obtained. The yields based on the tin species were small (up to 0.05%) over periods up to 50 hours. The methyltins were characterized by reduction of the liquid phase by $NaBH_4$ and identification by GC FID and GC MS. The absolute amounts of Me_3SnH_3 produced after reaction with $NaBH_4$ were up to 48.3 μg, well within the analytical range of the equipment used. Methylation of tin(II) in preference to tin(IV) suggests oxidative methylation by S adenosylmethionine (i.e. attack by Me_3^+).

Decomposition of the alga Enteromorpha in the presence of inorganic tin produces all the methyltin species[25]. Hydroponically grown *S. alterniflora* plants have been shown to accumulate Me_3Sn^+ from tin(IV)-treated media[37].

Work on estuarine sediment systems has demonstrated that these mixed cultures, and other single microorganisms, can methylate inorganic tin ($SnCl_4$) in oxygen-free sediment slurries[30,31]. Although microbial methylation of tin had been shown to occur in estuarine sediments, there was no demonstration of the microbial species involved or of mechanistic factors or pathways[38]. In a more recent study[30] more quantitative results were obtained, showing that tin methylation in estuarine sediments is a microbially mediated process favoured under anoxic conditions. In this case[15], the percentage methylation on tin was never greater than 0.02% over 60 days. This could be caused by the formation of insoluble tin sulphide species in the media; the yield on *available* tin is probably much greater. (This is a common observation in this type of experiment.) The sediments in this case came from Baltimore Harbour and the Patuxent River (USA). $MeSn^{3+}$ was always the main product with smaller amounts of Me_2Sn^{2+} being produced (at about 10% of the $MeSn^{3+}$ production); Me_3Sn^+ was barely detectable. There was no evidence in this series of experiments for tin methylation without microbial activity. Most of the methylating isolates were found to be sulphate reducers[30]. Autoclaving prevented methylation.

Several workers have reported results on the methylation of *methyltin* compounds. Clearly such methylations in principle could amount to no more than *redistribution* of methyl groups about tin, and the yields in such experiments would allow for this. Recent experiments with mixed cultures derived from Boston Harbour, USA sediments did give evidence of methylating $MeSnCl_3$ to form Me_3Sn^+ in the culture[39]. In this case methylation took place with a co-culture of two organisms isolated from the sediments; both were *Bacillus* sp. When these were grown in the presence of $MeSnCl_3$, Me_3SnH was detected after $NaBH_4$ reduction after 21 days. After 35 days, 55% of the $MeSnCl_3$ had disappeared and 35% of the original amount was detected as Me_3SnH. (If this calculation is based on *tin*, then the extra methyl groups could not have arisen by dismutation.) Me_2Sn^{2+} was not detected. Neither organism *alone* produced methylation.

This study parallels an earlier one made on oxygen-deficient sediments from San Francisco Bay, USA[38,40]. These, on incubation with Me_3Sn^+ under anaerobic conditions, produced Me_4Sn in both cultures and sterilized controls (redistribution?), but about three times as much Me_4Sn was produced in the live cultures. However, the sterilization process could have reduced an ability of the system to promote or catalyse dismutation of methyl groups. Similarly, Canadian Lake sediments methylated both inorganic tin(II) and tin(IV), and also methyltins — the methyltin compounds were methylated more readily than inorganic tin compounds[41].

There have been several reports of a pseudomonas sp in aerobic incubation with inorganic tin(II) and tin(IV) yielding Me_2Sn^{2+} and probably $Me_4Sn^{42,43}$.

Very recent work has shown that two microorganisms can convert $BuSnCl_3$ to methyltin products[44]. *Pseudomonas fluorescens* degraded 10% of the $BuSnCl_3$ substrate quantitatively to Me_2Sn^{2+} in solution. *Schizossacharomyces pombe* metabolized 13% of $BuSnCl_3$ to Me_3Sn^{2+} (10%) and Me_3Sn^+ (3%). Yields are based on tin. The probable intermediate, inorganic tin(IV) was not detected. In this study, all the microorganisms used gave methylation of inorganic tin(IV) when incubated[44]. The only product detected was Me_3Sn^+ and percentage conversion ranged from 1 to 16 (up to 155 ng of Me_3Sn^+ being produced by *S. cerevisiae*). This conversion occurred under aerobic conditions with fungi, yeasts and bacteria, including *Chaetomium globosum, Penicillium citrinum, Aspergillas tomarri, Schizosaccharomyces pombe, Saccharomyces cerevisiae* and *Pseudomonas fluorescens*. Analysis of products was by hydride generation GC and purge and trapping, and also by QF AA. Tin concentrations in the medium were 10 $\mu g\,dm^{-3}$ ($BuSnCl_3$) and 20 $\mu g\,dm^{-3}$ ($SnCl_4$). The basic mechanism for $BuSnCl_3$ proposed was enzymatic debutylation of the $BuSnCl_3$ followed by methylation of the tin(IV) species formed. It is not easy to specify the methylating agent for tin(IV) unless it is Me^- (as a carbanion donor). These experiments are better rationalized on the basis of tin(II) oxidative methylation by carbonium ion donors (viz attack essentially of Me^+). Seidal and coworkers reported that bacteria from seaweed produced Me_4Sn which was characterized by its GC retention time[33]. They also found methyltins in the seaweeds.

In general, the incubation experiments are judged against abiotic control experiments in which no, or reduced, tin concentrations are methylated. In some experiments potential biological methylating systems have been augmented by likely chemical methylating agents — which themselves may have been produced biologically. These include MeI, $MeAsO(OH)_2$, $(Me)_3Pb^+$. $SnCl_2$ has been methylated by MeI in porewater and Me_3Sn^{3+} only has been found[45]. Fulvic acid was able to reduce the yields of $MeSn^{3+}$ significantly and yields were significantly higher under anaerobic compared to aerobic conditions. This is rationalized on the basis that tin(II) is being methylated by oxidative addition of a carbonium ion type methylating agent — this does not occur with tin(IV). In the environment it is proposed that sulphate reducers convert tin(IV) to tin(II) to allow this to take place. The authors suggested that methylation in sediments occurs essentially in the anoxic zone rather than in the oxic surface layer. (This gives problems in findings where methyltins are highest in top sediment layers.) Reduction of yields by fulvic acid was ascribed to sequestration of the MeI methylating agent. Maximum yields for the methylation were 12.0% based on tin.

A similar augmented system has also been studied. Added Me_3Pb^+ and Me_2Pb^{2+} were able to transfer methyl groups to tin(II) and tin(IV) to form varying yields of all the methyltins including Me_4Sn^{46}. The presence of sediment increased yields. Methylarsenic species did not transfer methyl groups. The ratio of methyl donor to tin was typically 100:1. Incubations with sediments were abiotic, at 20 °C and carried out in the presence of light.

Earlier work on general biotransformation properties of tin and tin compounds has been reviewed by Cooney[47].

C. Model Experiments to Demonstrate Methylation of Tin

Transmethylation reactions have been frequently reported in organometallic chemistry. Only three relevant to environmental methylation will be discussed here. Essential features include feasible methylating agents and aqueous solutions. $MeCoB_{12}$, which has been frequently shown to methylate mercury(II) under these conditions, has also been shown to methylate tin(II). A transient role for tin(III) was suggested[48,49] in a free radical process and aerobic conditions were required. Similarly, $SnCl_2$ and tin amino acid complexes [e.g. Sn(II) penicillamine] have been methylated by $MeCoB_{12}$ to produce $MeSn^{3+}$ and small amounts of Me_2Sn^{2+}. No attempt was made to exclude oxygen from the solutions, and the assumption was made that the process is a carbonium ion transfer. Alternatively, a carbanion substitution followed by disproportionation of unstable Me_2Sn is feasible[50].

Disproportionation reactions are well known in tin chemistry and can account for the occurrence of several methyltin compounds, assuming a methyltin species was present in the first place. Similarly, transmethylation can occur between inorganic tin species and methylated forms of other elements. Examples of this nature, some of which may take place in water, are given in Section V, Reference 2. Their environmental significance is unknown. Disproportionation reactions are also covered in Section V, Reference 1. A general theory of methylation by $MeCoB_{12}$ was proposed, based on redox potentials[51].

Abiotic formation of $MeSnI_3$ by reaction of SnS and MeI in water has also been observed[52].

Tin(II) compounds have been converted to methyltins in water in model reactions by a series of reagents[53]. Methylating agents used were MeI, $(Me)_3SI$, $(Me)_3^+NCH_2COO^-$. MeI was successful with both tin(0) and tin(II) compounds with yields varying from around 0.1 to 2.0%. Me_2Sn^{2+}, Me_3Sn^+ and Me_4Sn were also detected. Hydride generation and GC FID were used to detect tin-containing products. Very small yields of methyltin products were found from use of $(Me)_3SI$ and $(Me)_3NCH_2COO^-$ as methylating agents.

Relative yields were such as to suggest sequential methylation from inorganic tin by an initial oxidation–addition process involving free radicals[53].

MeI has also been studied as a model methylator for tin(II) in aqueous solution, either alone or in combination with MnO_2 and/or a cationic model $MeCoB_{12}$ substance, based on a MeCo tetrazacyclotetradecatetraene system $\{(Me)_2Co(N_4)^+\}^{36}$. The full range of methyltin products could be produced by various experimental combinations of the methylators. Yields ranged as follows: $MeSn^+$ (0–14.4%), Me_2Sn^{2+} (0.8–8.8%), Me_3Sn^+ (0–2.65%), Me_4Sn (0–3.32%), indicative of an initial $MeSn^{2+}$ product followed by disproportionation or further methylation. Analysis was by GC FID with use of $NaBH_4$ to volatilize the ionic methyltins.

III. THE METHYLATION OF LEAD COMPOUNDS UNDER ENVIRONMENTAL OR MODEL ENVIRONMENTAL CONDITIONS

A. Methods of Analysis

Hydride generation techniques for the volatilization and separation of alkyl leads have been infrequently used as the organolead hydrides are not very stable. The commonest derivatization reagents are $NaB(C_2H_5)_4$ or Grignard reagents. The normal separation systems and methods of detection are used.

B. Methyllead Compounds Found in the Natural Environment and Arising from Biomethylation

The vast majority of measurements of organolead compounds in the environment do not constitute evidence for biomethylation of lead. Most environmental organic lead comes from incomplete combustion or spillage of methyl- or ethyl-lead gasoline additives (viz tetraalkylleads or TALs). A literature search will produce several hundred TAL or ionic alkyllead results, but few of them are evidence for methylation *in* or *by* the environment.

Discovery of methylleads in the absence of accompanying ethylleads is also not really evidence for biomethylation. Methylleads are sometimes used alone in petrol and ethylleads, being less stable, may decay faster than co-existing methylleads. Monomethyllead species are not observed as these are essentially unstable in aquatic media.

In a similar manner to tin, the production of Me_4Pb arising from an incubation or similar experiment involving Me_3Pb^+ cannot be taken as conclusive evidence for a lead methylation.

It is also significant that recent studies tend to show methyllead contents for common species (e.g. wine) which relate to local and temporal conditions for alkyllead use. A recent study has shown that some French wines have higher methyl- and ethyl-lead contents than Californian wines[55]. In the latter case there is almost no use of TALs in gasoline in California, while TALs are still heavily used in France. The vintage of a wine is, of course, a tangible record of local conditions at the time of bottling (and sometimes of quality). Both Californian and French wines show much higher levels of alkylleads in the 1970s and 1980s compared with now, corresponding to periods of time when lead was more heavily used in gasoline. All this points to contamination being the cause of the organic leads observed, with atmospheric transport being the agency of distribution. The conclusions of this local study can be applied generally.

Recent studies also parallel this, with decreasing concentrations of alkylleads being found in rainwater over the last 10 years[56]. However, in a recent case it has been pointed out that the decrease does not fully reflect the amount of reduction of TAL to gasoline over that period and that the *proportion* of TAL to inorganic lead in rainwater has actually increased[56]. The reasons for this are unclear, although in the location studied, whereas the amount of TAL in gasoline may have been reduced, there has been more vehicular traffic over the last 10 years. However, gross emission rates of lead declined. In general these results can be explained by atmospheric transport.

Variation in the proportions of ethylleads to methylleads found in samples can also be used to argue a gasoline-based origin for methylleads. Where mixed ethyl- and methylleads are used in gasoline in the region, mixed ethyl- and methyl-leads are found in local biota. Where Et_4Pb only is used in petrol, the biota tends to contain less methylleads (which occur from other sources via atmospheric transport). Results also reflect changes in the methyl/ethyl lead usage in gasoline. A recent study on Canadian Great Lake fish has shown these effects[57]. The evidence is that the methyllead in the fish came, via atmospheric or water transport, from commercial alkyllead production and not from biomethylation, although the authors did not exclude biomethylation.

Et$_4$Pb has been the exclusive TAL used in gasoline in Canada. Biota from urban locations (Montreal) was shown to contain high proportions (mainly Et$_3$Pb$^+$). of ethyl-lead compared to methyllead[58]. However, in mallard ducks from a sanctuary in eastern Ontario, the major toxicant observed was Me$_3$Pb$^+$. This observation was ascribed to an environmental methylation process. However, long-range transport of methylleads in the atmosphere might account for the observations. Concentrations of both ethyl- and methyl-leads in the birds was in the ng g^{-1} range. Similar conclusions were postulated after work on alkylleads in Canadian herring gulls[59]. Me$_3$Pb$^+$ was ubiquitous, despite the lack of an automotive source for the methyllead. Methyl- and ethyl-lead in the birds did not correlate and different sources for methyl and ethyl groups were proposed.

Harrison and coworkers noted unusually high ratios of alkyl to total lead in the atmosphere at several UK rural sites[60,61]. Normally the ratio is about 0.5–8.0%, but in these cases a maximum of 33% was found. Analysis of air-mass trajectories revealed that these elevated ratios were associated with air that had passed over the open sea and estuarine and coastal areas.

The work of the Harrison group on alkylleads in air is perhaps the most persuasive of the reports as to the advocation of an environmental methylating process. Some of the sites whose atmospheric lead was sampled were very remote, such as Harris Island, Outer Hebrides, UK. Air reaching this site should not have any anthropogenic content but concentrations of alkylleads were found to be in the 3–7 ng m^{-3} range[61]. There was a higher than normal ratio of alkyl to total lead present here also (from 10–30%). A maritime source of volatile alkyllead was proposed[61,62].

Measurement in pristine prehistoric Antarctic ice has given lead concentrations which, in order to be accounted for, require a natural input of lead in prehistoric times of the order of 10^5 tonnes per year to the atmosphere. Biomethylation could be responsible for this input[63].

C. Microorganism Culture Experiments to Demonstrate Methylation of Lead

The first report that sediments will incubate inorganic lead(II) to produce Me$_4$Pb was made in 1975[64]. Although lead (including TAL) was already present naturally in the sediments, certain sediments produced *more* Me$_3$Pb$^+$ when lead(II) was added. This could have simply been a displacement of weakly coordinated Me$_4$Pb on the sediment to head space by the stronger Lewis acid lead(II). Similarly, it was reported that Me$_4$Pb could be formed from lead(II) acetate after incubation in a culture[64]. There is the possibility that the methyl groups arose from the acetate moiety. Conversion of lead(II) {Pb(OAc)$_2$} to Me$_4$Pb in seeded water and sediment has been reported[65]. Me$_3$PbOAc has been incubated in marine sediments to produce Me$_4$Pb quantitatively, but inorganic lead conversion was low[66]. Me$_4$Pb from Pb(NO$_3$)$_2$ was produced in 0.026% yield after incubation in the sediments for 600 hours.

Hewitt and Harrison[61] incubated inorganic^{210} Pb(NO$_3$)$_2$ with sediments and measured ^{210}Pb alkyllead after a 14-day period. The conversion rate was very low, viz the ratio of ^{210}Pb alkyllead to ^{210}Pb added to sediment after 14 days was between 0.9 and 2.6 × 10^{-7}.

Incubation of lead(II) (as nitrate or acetate) with marine algae and a S-adenosylmethionine rich yeast produced methyl leads in the culture solution[62]. Marine macrophyte cultures produced mainly Me$_3$Pb$^+$. Production with the yeast was much less efficient. Concentration levels of methyllead produced in the cultures for the algae were of the order of 10–20 ng dm^{-3}.

Me$_4$Pb has also been produced by incubation of inorganic lead salts [Pb(NO$_3$)$_2$, PbCl$_2$, Pb(OAc)$_2$] with biologically active sediments and waters from the Tamar Estuary, UK[67].

For two lead salts, about 0.03% of added lead was converted to Me_4Pb. The nitrate and chloride salts produced 6 times as much Me_4Pb as did the acetate. No Me_4Pb was detected in sterile control experiments. The concentration of inorganic lead in the media was 10 mg $Pb\,dm^{-3}$ and Me_4Pb was evolved to a head space. There was a strong seasonal variation noted in the production of Me_4Pb in these experiments, winter sediments producing no detectable Me_4Pb.

A number of negative reports have also arisen from incubation experiments. It has been pointed out that methylation of Me_3Pb^+ species to Me_4Pb may arise through a sulphide-mediated disproportionation[68-70]. Me_3PbOAc has been incubated with both sterilized and unsterilized lake sediments and in all cases similar amounts of Me_4Pb were evolved, i.e. disproportionation without biomethylation can account for the results[71]. Use of labelled carbon and lead in a series of attempted biomethylation in cultures produced no evidence of biomethylation but did confirm that sulphide-promoted disproportionation is possible[72].

D. Model Experiments to Demonstrate Methylation of Lead

As for tin, two potential routes to methyllead compounds might appear to exist: (1) Me^- carbanion attack on lead(II) (e.g. by Me^- from $MeCoB_{12}$) followed by dismutation and (2) oxidative addition by a Me^+ carbonium ion from e.g. S adenosylmethionine or MeI etc.

Direct methylation of lead(II) salts by $MeCoB_{12}$ has not been detected[68,73-76]. Methylation of $(Me)_2Pb^{2+}$ by $MeCoB_{12}$ has been reported[74]. Methylation of lead(II) salts with more active model methyl donor systems modelled on $MeCoB_{12}$ have been reported, but their environmental significance is not conclusive. This aspect has been extensively reviewed[77]. Methylation of $Pb(NO_3)_2$ by a dimethylcobalt(III) macrocyclic complex in water to give Me_4Pb occurs[78]. Me_4Pb was the only alkyllead detectable by the GC FID or GC ECD systems used — the non-volatile ionic alkylleads were not sought.

Lead(II) compounds have been converted to Me_4Pb, Me_3Pb^+, Me_2Pb^{2+} by incubation with MeI in aqueous systems[53]. For lead(0) 1.18% of methylated product was produced, for $Pb(NO_3)_2$ and $Pb(OAc)_2$ smaller yields were found. Me_3Pb^+ and Me_2Pb^{2+} were produced in much higher yields than Me_4Pb, suggesting sequential methylation with diminishing yields at each step. A radical one-equivalent oxidative addition of alkyl halide to the lead(0) or lead(II) species in a non-chain process was suggested. Fully methylated species then occur by dismutation. A small amount of Me_2Pb^{2+} was produced from the reaction of lead(0) with Me_3SI (5.67×10^{-3}% yield).

$MeCoB_{12}$ reacted in aqueous solution with insoluble PbO_2 and the rate of decay of $MeCoB_{12}$ was followed, but no methylleads were detected[79].

Using the active dimethylcobalt systems modelled on $MeCoB_{12}$, viz $Me_2Co(N_4)^4$, Weber and coworkers demonstrated the effect on methyllead production of methylation alone and in combination with sediments[80]. Total methyllead yields were 0.037 to 0.11%. Me_4Pb, Me_3Pb^{2+} and Me_2Pb^+ were detected. $MeCoB_{12}$ alone could not methylate the lead. The methylation was carried out in a matrix of Great Bay, NH, USA sediments and μg quantities of organoleads were produced. Presumably the dimethylcobalt cationic complex transferred methyl carbonium ions to lead(II) in an oxidative addition step. Low yields of Me_4Pb have also been found after reactions of $Me_2Co(N_4)^+$ with lead(II) in aqueous KNO_3. Use of deuterated methyl groups showed that methylation of Me_3Pb^+ and Me_2Pb^{2+} occurred by methyl transfer from $Me_2Co(N_4)^+$[81].

Several other dimethylcobalt complexes based on $MeCoB_{12}$ have been reported to methylate lead(II), but no methylleads were characterized[82,83]. Reaction between lead(II) and $(Me)_2Co(CDo)(DOH)pn$ in MeCN yielded Me_4Pb^{84}. Weber and coworkers have investigated a number of systems of this nature and have certainly demonstrated that methyl groups bound to cobalt may transfer to lead.

IV. GENERAL CONCLUSIONS

It is the view of this writer that for certain Main Group metallic elements, methylation in or by the environment has been demonstrated conclusively. Tin joins mercury and arsenic in this category. An insufficient number of observations has been made in the case of germanium (and also of antimony). To date, the lead case seems hopelessly bound up with the use of lead in gasoline and the final decision is still awaited in that case. It is more difficult if this comes down to proving the negative.

V. REVIEWS AND EARLIER WORK

1. J. S. Thayer and F. E. Brinckman, *Adv. Organomet. Chem.*, **20**, 313 (1982).
2. E. W. Abel, F. G. A. Stone and G. Williamson (Eds.), *Comprehensive Organometallic Chemistry*, Pergamon, Oxford, (1982).
3. F. E. Brinckman and J. M. Bellama (Eds.), *Organometals and Organometalloids: Occurrence and Fate in the Environment*, ACS Symp. Ser. No. 82, ACS, Washington, DC, (1978).
4. P. J. Craig (Ed.), *Organometallic Compounds in the Environment*, Longman, London, (1986).
5. J. S. Thayer, *Organometallic Compounds and Living Organisms*, Academic Press, New York, (1984).
6. P. J. Craig, 'Biological and environmental methylation of metals', in *The Chemistry of the Metal Carbon Bond, Volume 5* (F. R. Hartley Ed.), Chap. 10, Wiley, Chichester, (1989), p. 437.

VI. REFERENCES

1. P. J. Craig (Ed.), *Organometallic Compounds in the Environment*, Longman, London, 1986, p. 24 (Occurrence and Pathways — General Considerations).
2. R. C. Poller, in *Comprehensive Organic Chemistry: The Synthesis and Reactions of Organic Compounds* (Ed. D. N. Jones), Pergamon, New York, 1979, p. 1061.
3. P. Riviere, M. Riviere-Baudet and J. Satge, in *Comprehensive Organometallic Chemistry* (Eds. E. W. Abel, F. G. A. Stone and G. Wilkinson), Vol. 2, Pergamon, Oxford, 1982, p. 399.
4. M. O. Andreae and P. N. Froelich, *Anal. Chem.*, **53**, 1766 (1981).
5. G. A. Hambrick, P. N. Froelich, M. O. Andreae and B. L. Lewis, *Anal. Chem.*, **56**, 421 (1984).
6. H. P. Mayer and S. Rapsomanikis, *Appl. Organometal. Chem.*, **6**, 173 (1992).
7. S. Clark 'A new method for the analysis of metal and metalloidal compounds', Thesis, Leicester Polytechnic, 1988, pp. 204–206.
8. Reference 3, p. 404.
9. B. L. Lewis, P. N. Froelich and M. O. Andreae, *Nature*, **313**, 303 (1985).
10. B. L. Lewis, M. O. Andreae, P. N. Froelich and R. A. Mortlock, *Sci. Total. Environ.*, **73**, 107 (1988).
11. P. N. Froelich, G. A. Hambrick, M. O. Andreae, R. A. Mortlock and J. M. Edmond, *J. Geophys. Res.*, **90**, 1122 (1985).
12. B. L. Lewis, M. O. Andreae and P. N. Froelich, *Marine Chem.*, **27**, 179 (1989).
13. B. L. Lewis and H. P. Mayer, in *Metal Ions in Biological Systems* (Eds. H. Sigel and A. Sigel), Dekker, New York, 1993, p. 79.
14. M. O. Andreae and P. N. Froelich, *Tellus*, **36B**, 101 (1984).
15. Reference 3, p. 403.
16. R. S. Braman and M. A. Tompkins, *Anal. Chem.*, **51**, 12 (1979).
17. V. F. Hodge, S. L. Seidel and E. D. Goldberg, Anal. Chem., **51**, 1256 (1979).

18. J. -A. A. Jackson, W. R. Blair, F. E. Brinckman and W. P. Iverson, *Environ. Sci. Technol.*, **16**, 110 (1982).
19. J. T. Byrd and M. O. Andreae, *Science*, **218**, 565 (1982).
20. Y. -K. Chau, P. T. S. Wong and G. A. Bengert, *Anal. Chem.*, **54**, 246 (1982).
21. R. J. Maguire, *Environ. Sci. Technol.*, **18**, 291 (1984).
22. R. J. Maguire, R. J. Tcacz, Y. K. Chau, G. A. Bengert and P. T. S. Wong, *Chemosphere*, **15**, 253 (1986).
23. J. A. J. Thompson, M. G. Shelfer, R. C. Pierce, Y. K. Chau, J. J. Cooney, W. R. Cullen and R. J. Maguire, in *Organotin Compounds in the Aquatic Environment; Scientific Criteria for Assessing Effects on Environmental Quality*, NRCC Publication No. 22494, NRC, Canada, 1985.
24. R. J. Maguire, *Water Pollut. Res. J. Can.*, **26**, 243 (1991).
25. O. F. X. Donard, F. T. Short and J. H. Weber, *Can. J. Fish Aquat. Sci.*, **44**, 140 (1987).
26. A. M. Falke and J. H. Weber, *Environ. Technol.*, **14**, 851 (1993).
27. A. M. Falke and J. H. Weber, *Appl. Organomet. Chem.*, **8**, 351 (1994).
28. L. Randall, J. S. Han and J. H. Weber, *Environ. Technol. Lett.*, **7**, 571 (1986).
29. P. Quevauviller, O. F. X. Donard, J. C. Wasserman, F. M. Martin and J. Schneider, *Appl. Organomet. Chem.*, **6**, 221 (1992).
30. C. C. Gilmour, J. H. Tuttle and J. C. Means, *Microbiol. Ecol.*, **14**, 233 (1987).
31. C. C. Gilmour, J. H. Tuttle and J. C. Means, *Anal. Chem.*, **58**, 1848 (1986).
32. S. Tugral, T. I. Balkas and E. Goldberg, *Marine Poll. Bull.*, **14**, 297 (1983).
33. S. L. Seidal, V. F. Hodge and E. D. Goldberg, *Thalassia Jugosl.*, **16**, 209 (1980).
34. W. Dirkxx, R. Lobinski, M. Ceulemans and F. Adams, *Sci. Total Environ.*, **136**, 279 (1993).
35. L. E. Hallas, J. C. Means and J. J. Cooney, *Science*, **215**, 1505 (1982).
36. J. Ashby and P. J. Craig, *Appl. Organomet. Chem.*, **1**, 275 (1987).
37. J. H. Weber and J. J. Alberts, *Environ. Technol.*, **11**, 3 (1990).
38. L. E. Hallas, J. C. Means and J. J. Cooney, *Science*, **215**, 1505 (1982).
39. N. S. Makkor and J. J. Cooney, *Geomicrobiol. J.*, **8**, 101 (1990).
40. H. E. Guard, A. B. Cobet and W. M. Coleman, III, *Science*, **213**, 770 (1981).
41. Y. K. Chau, P. T. S. Wong, O. Kramer and G. A. Bergert, Proc 3rd Int. Conf. Heavy Metals Env., Amsterdam, 1981, CEP, Edinburgh, p. 6421.
42. F. E. Brinckman, J. A. Jackson, W. R. Blair, G. J. Olson and W. P. Iverson, in *Trace Metals in Sea Water* (NATO Conf Ser 4:9; Eds. C. S. Wong *et al.*), Plenum Press, New York, 1983, p. 39.
43. C. Huey, F. E. Brinckman, S. Grim and W. P. Iverson, Proc. Int. Conf. Transport Persist Chemicals Aquat. Ecosystems, Ottawa, 1974, N.R.C. Canada, p. II-73.
44. O. Errecaulde, M. Astruc, G. Maury and R. Pinel, *Appl. Organomet. Chem.*, in press (1994).
45. D. S. Lee and J. H. Weber, *Appl. Organomet. Chem.*, **2**, 435 (1988).
46. Y. K. Chau, P. T. S. Wong, C. A. Mojesky and A. J. Carty, *Appl. Organomet. Chem.*, **1**, 235 (1987).
47. J. J. Cooney, *J. Ind. Microbiol.*, **3**, 195 (1988).
48. Y. -T. Fanchiang and J. M. Wood, *J. Am. Chem. Soc.*, **103**, 5100 (1981).
49. L. J. Dizikes, W. P. Ridley and J. M. Wood, *J. Am. Chem. Soc.* **100**, 1010 (1978).
50. J. R. Ashby and P. J. Craig, *Sci. Total Environ.*, **100**, 337 (1991).
51. W. P. Ridley, L. J. Dizikes and J. M. Wood, *Science*, **197**, 329 (1977).
52. W. F. Manders, G. J. Olson, F. E. Brinckman and J. M. Billama, *J. Chem. Soc., Chem. Commun.*, 538 (1984).
53. P. J. Craig and S. Rapsomanikis, *Environ. Sci. Technol.*, **19**, 726 (1985).
54. S. Rapsomanikis and J. H. Weber, *Environ. Sci. Technol.*, **19**, 352 (1985).
55. P. -L. Teissedre, R. Lobinski, M. -T. Cabanis, J. Szpunar-Lobinska, J. -C. Cabanis and F. C. Adams, *Sci. Total Environ.*, **153**, 247 (1994).
56. A. B. Turnbull, Y. Wang and R. M. Harrison, *Appl. Organomet. Chem.*, **7**, 567 (1993).
57. D. S. Forsyth, R. W. Dabeka and C. Cleroux, *Appl. Organomet. Chem.*, **4**, 591 (1990).
58. D. S. Forsyth, W. D. Marshall and M. C. Collette, *Appl. Organomet. Chem.*, **2**, 233 (1988).
59. D. S. Forsyth and W. D. Marshall, *Environ. Sci. Technol.*, **20**, 1033 (1986).
60. R. M. Harrison and D. P. Laxen, *Nature*, **275**, 738 (1978).
61. C. N. Hewitt and R. M. Harrison, *Environ. Sci. Technol.*, **21**, 260 (1987).
62. R. M. Harrison and A. G. Allen, *Appl. Organomet. Chem.*, **2**, 49 (1989).
63. C. F. Boutron and C. C. Patterson, *Geochim. Cosmochim. Acta*, **47**, 1355 (1983).
64. P. T. S. Wong, Y. K. Chau and P. L. Luxon, *Nature*, **253**, 263 (1975).

65. U. Schmidt and F. Huber, *Nature*, **259**, 157 (1976).
66. J. A. J. Thompson and J. A. Crerar, *Marine Pollut. Bull.*, **11**, 251 (1980).
67. A. P. Walton, L. Ebdon and G. E. Millnard, *Appl. Organomet. Chem.*, **2**, 87 (1988).
68. A. W. P. Jarvie, R. N. Markall and H. R. Potter, *Nature*, **255**, 217 (1975).
69. A. W. P. Jarvie, A. P. Whitmore, R. N. Markall and H. R. Potter, *Environ. Poll. (B)*, **6**, 69 (1983).
70. A. W. P. Jarvie, A. P. Whitmore, R. N. Markall and H. R. Potter, *Environ. Poll. (B)*, **6**, 81 (1983).
71. P. J. Craig, *Environ. Technol. Lett.*, **1**, 17 (1980).
72. K. Reisinger, M. Stoeppler and H. W. Nurnberg, *Nature*, **291**, 228 (1981).
73. R. T. Taylor and M. L. Hanna, *J. Environ. Sci. Health*, **A11**, 201 (1976).
74. W. P. Ridley, L. J. Dizikes and J. M. Wood, *Science*, **197**, 329 (1977).
75. G. Agnes, S. Bendle, H. A. O. Hill, F. R. Williams, R. J. P. Williams, *Chem, Commun.*, 850 (1971).
76. J. Lewis, R. H. Prince and D. A. Stotter, *J. Inorg. Nucl. Chem.*, **35**, 341 (1973).
77. S. Rapsomanikis and J. H. Weber, in *Organometallic Compounds in the Environment* (Ed. P. J. Craig), Longman, London, 1986.
78. S. F. Rhode and J. H. Weber, *Environ. Technol. Lett.*, **5**, 63 (1984).
79. J. S. Thayer, *Appl. Organomet. Chem.*, **1**, 545 (1987).
80. S. Rapsomanikis, O. F. X. Donard and J. H. Weber, *Appl. Organomet. Chem.*, **1**, 115 (1987).
81. S. Rapsomanikis, J. J. Ciejka and J. H. Weber, *Inorg. Chim. Acta*, **89**, 179 (1984).
82. M. W. Witman and J. H. Weber, *Inorg. Chem.*, **15**, 2375 (1976).
83. M. W. Witman and J. H. Weber, *Inorg. Chem.*, **16**, 2512 (1977).
84. J. H. Dimmit and J. H. Weber, *Inorg. Chem.*, **21**, 1554 (1982).

CHAPTER 17

Toxicity of organogermanium compounds

E. LUKEVICS and L. M. IGNATOVICH

Latvian Institute of Organic Synthesis, Riga, LV 1006 Latvia
Fax: 371-7-821038

I. INTRODUCTION

Numerous organogermanium compounds possessing antitumour, immunomodulating interferon-inducing, radio-protective, hypotensive and neurotropic properties have been synthesized[1,2]. Two of them, spirogermanium[1,3−6] and 2-carboxyethylgermsesquioxane (Ge-132, proxigermanium, rexagermanium)[1,7−11], have been tested in clinics as antitumour remedies.

The toxicological studies performed demonstrated that most organogermanium compounds tried were less toxic than the corresponding organosilicon and organotin analogues[1,12]. On the other hand, in some cases nephrotoxicity caused by long-term administration of germanium-containing organic preparations in large doses was documented[13−16].

The acute toxicity of organogermanium compounds depends strongly on the structure of substituents at the germanium atom, thus ranging them from non-toxic compounds (tetraalkylgermanes, germanols, germoxanes, adamantyl derivatives of germanium) with LD_{50} more than 3000–5000 mg kg^{-1} to highly toxic (thienylgermatranes) having LD_{50} about 15–20 mg kg^{-1}.

II. TETRAORGANOGERMANIUM COMPOUNDS

Tetraorganylgermanes R_4Ge ($R = Et$, Pr, Bu, Am, Hs, PhCH$_2$, Ph), administered *p.o.*, *i.p.* or *s.c.* to mice, exhibit low toxicity. The mean lethal dose (LD_{50}) for *n*-alkyl

The chemistry of organic germanium, tin and lead compounds
Edited by S. Patai   © 1995 John Wiley & Sons Ltd

derivatives varies from 10,000 to 2000 mg kg^{-1}[17-19]. Tetraisopropylgermane is more toxic (620 mg kg^{-1}) than tetra(n-propyl)germane (5690 mg kg^{-1}) while the unsaturated triethylallylgermane (114 mg kg^{-1}) appears to be the most toxic in this series of compounds[19]. Rats have been found to be more sensitive to these germanes[18,19].

Adamantyl derivatives of germanium are considerably less toxic than the corresponding derivatives of silicon. LD$_{50}$ of AdCH$_2$CH$_2$GeMe$_3$ equals 1480 mg kg^{-1}. The introduction of one more methylene group between the adamantane ring and the germanium atom dramatically decreases the toxicity of the compound[20].

The toxicity of the majority of carbofunctional tetraorganogermanes administered *i.p.* to mice lies within the 3000–1500 mg kg^{-1} range[1,2,20]. The germanium derivatives of titanocene dichloride belong to the moderate toxic compounds[21,22]. The introduction of nitrogen-containing substituents may increase their toxicity[2].

LD$_{50}$ of spirogermanium-2-(3-dimethylaminopropyl)-8,8-diethyl-2-azaspiro[4,5]decane administered to white mice equals 224 mg kg^{-1} (*p.o.*), 150 mg kg^{-1} (*i.p.*), 134 mg kg^{-1} (*i.m.*) and 75 mg kg^{-1} (*i.v.*)[1,23,24]. This compound appeared more toxic for rats and dogs[23,24]. The neurotoxicity of spirogermanium at doses of 32–60 mg m^{-2} has been observed in clinical trials as well[1].

The iodomethylammonium salts of 5-trimethylgermyl-2-furfurylamine have been shown to possess the highest toxicity (82 mg kg^{-1}). 5-Trimethylgermylfurfurylidenethiocarbazide, 1-(5-trimethylgermyl-2-furfurylidene)hydantoin, 2-(2-pyridyl)ethyltrimethylgermane hydrochloride and 2-trimethylgermylisobutyrohydroxamic acid have mean LD$_{50}$ values — within the 205–355 mg kg^{-1} range[2,20,25].

Substitution of an isobutyrohydroxamic group for the propiohydroxamic one decreases the toxicity of the germanium derivative more than twofold, whereas the introduction of the triethylgermyl group instead of the trimethylgermyl one in position 2 in the isobutyrohydroxamic acid molecule increases the acute toxicity twofold[25]. The heteroatom essentially affects the toxicity values of β-trimethylsilyl-, β-trimethylgermyl- and β-trimethylstannylpropiohydroxamic acids. The acute toxicity of the tin-substituted acid (20.5 mg kg^{-1}) is 100 times higher than that of the acid-containing germanium (2000 mg kg^{-1}), and 40 times higher than that of silicon derivative[25].

III. COMPOUNDS WITH Ge—N BONDS

Germylation of imidazoline lowers its toxicity. Dimethylbis(imidazolino)germane has LD$_{50}$ = 500 mg kg^{-1} (imidazoline hydrochloride, 200 mg kg^{-1})[26]. 2-Naphthylmethylimidazoline and its N-diisoamylgermyl and N-dihexylgermyl derivatives possess comparable toxicity (50–75 mg kg^{-1})[26].

IV. COMPOUNDS WITH Ge—O BONDS

Tricyclohexylgermanol and hexaphenyldigermoxane administered *s.c.* have been shown to be non-toxic for mice[17]. Tricyclohexylgermanol exhibits low toxicity also at *i.p.* administration, its LD$_{50}$ value exceeding 5000 mg kg^{-1} [20].

The toxicity of hexaorganodigermoxanes (R$_3$Ge)$_2$O for rats depends strongly on the nature of substituents at the germanium. Hexaethyldigermoxane (LD$_{50}$ = 240 mg kg^{-1} *p.o.* and 14.7 mg kg^{-1} *i.p.*) is more toxic than hexabutyldigermoxane (6000 and 3000 mg kg^{-1})[27]. Digermoxanes are less toxic for mice, LD$_{50}$ values being within 130–240 mg kg^{-1} for R = Me, 30–650 mg kg^{-1} for R = Et and 2270–7380 mg kg^{-1} for R = Pr. The higher homologues (R = Bu, Am, Hs, Ph) are even less toxic[28].

The water solution of octamethylcyclotetragermoxane (Me$_2$GeO)$_4$ has been found to be non-toxic for rabbits[29], its toxicity being low also for mice at oral (LD$_{50}$ = 4470 mg kg^{-1}) and intraperitoneal (2720 mg kg^{-1}) administration. Hexabutylcyclotrigermoxane is more toxic for rats (1310 mg kg^{-1}, *p.o.*) than for mice (4640 mg kg^{-1}, *p.o.*)[30].

Octamethylcyclotetragermoxane reveals higher embryotoxicity for chicken embryos than acetone[31,32]. Oral toxicity of hexabutylcyclotrigermoxane and diethylpolygermoxane is low for chickens (2000–4000 mg kg^{-1})[27].

Most of the organylgermsesquioxanes (RGeO$_{1.5}$)$_n$ tested are low-toxic compounds[1,10,11,17,20,33–45].

2-Carboxyethylgermsesquioxane exhibits acute toxicity for mice and rats with a mean value of LD$_{50}$ = 6000–10,000 mg kg^{-1} (*p.o.*, *i.p.*) and 4500–5700 mg kg^{-1} (*i.v.*)[34]. It also shows low tocixity in subacute[36,37] and chronic[38,40] experiments.

A single oral administration of proxigermanium at 2000 and 4000 mg kg^{-1} is not lethal for dogs, but it induces diarrhea and vomiting. About 50% of rats which received proxigermanium orally in 4000 mg kg^{-1}/day for 3 months died. Oral administration of proxigermanium for a year at 750 mg kg^{-1}/day induced diarrhea in rats. However, it has been found that a dose of 83 mg kg^{-1} is not toxic. Proxigermanium administered in a dose of 15–240 mg/body has not affected physiological function in healthy volunteers[11].

The compound is not embryotoxic, teratogenic[35,39,40,43], mutagenic or antigenic[44]. The oral administration of proxigermanium does not affect fertility at 350, 700 and 1400 mg kg^{-1}/day during 60 days before and at mating in male rats and during 14 days before, and at and during 7 days after mating in female rats[11].

Hydroxamic acid (O$_{1.5}$GeCH$_2$CH$_2$CONHOH)$_n$, its sodium salt and 1-(2-pyrrolidonyl)ethylgermsesquioxane appear to be low-toxic compounds as well, their LD$_{50}$ values exceeding 5000 mg kg^{-1}. 3,5-Dimethylpyrazolylmethylgermsesquioxane exhibits acute toxicity with a mean value of LD$_{50}$ = 708 mg kg^{-1} [20].

All derivatives of heterylgermatranes RGe(OCH$_2$CH$_2$)$_3$N (see Table 1) are several times less toxic than their silicon analogues[12,46]. Thus, 2-thienylgermatrane (the most toxic among all studied germatranes) is 55 times less toxic than 2-thienylsilatrane.

Furylgermatranes are low-toxic compounds, their LD$_{50}$ exceeding 1000 mg kg^{-1} [47]. The derivatives of thienylgermatranes on the other hand, are highly toxic compounds with LD$_{50}$ values within the 16–89 mg kg^{-1} range. 5-Ethyl-2-thienylgermatrane (> 1000 mg kg^{-1}) appears to be an exception in this series of compounds. Comparison of 5-methyl-, 5-ethyl- and 5-bromo-2-thienylgermatranes demonstrates that the substitution of the methyl group for the ethyl one reduces noticeably the acute toxicity of the compound. The introduction of a bromine atom instead of a methyl group does not change the toxicity value. 2-Isomers belonging to the thiophene series appear to be the most toxic, while in the furan series the 2-derivatives are less toxic than the 3-isomers[47–49]. The insertion of a CH$_2$ group between the furan ring and the germanium atom decreases the toxicity[49].

Vinylgermatrane (5600 mg kg^{-1}), 1-hydroxygermatrane (8400 mg kg^{-1}), germatranyl derivatives of adamantane (> 5000 mg kg^{-1}), pyrrolidone (6500–10,000 mg kg^{-1}) and hexabarbital (10,000 mg kg^{-1}) are low-toxic compounds[2,20,33,50].

LD$_{50}$ values of chloromethyl- and methoxycarbonylpropylgermatranes exceed 3000 mg kg^{-1}. Bromomethylgermatrane appears considerably more toxic (355 mg kg^{-1})[2].

Siloxygermatranes[51] and nitrogen-containing germatranes[2] are low-toxic compounds, their mean lethal doses exceeding 1000 mg kg^{-1}. Diethylaminomethylgermatrane is a sole exception having LD$_{50}$ = 355 mg kg^{-1}. Hydrogermatrane reveals a similar toxicity (320 mg kg^{-1})[2].

TABLE 1.   Acute toxicity of germatranes $RGe(OCH_2CH_2)_3N$ (*i.p.* administration to white mice)

R	LD_{50} (mg kg^{-1})	R	LD_{50} (mg kg^{-1})	R	LD_{50} (mg kg^{-1})
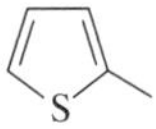 (thiophene)	16.5	 (3-methylfuran)	1630	Et_2N–C$_6$H$_4$– 	3250
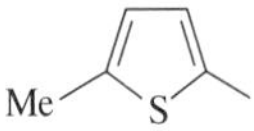 Me-thiophene	20.5	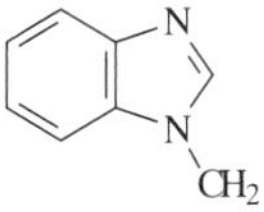 benzimidazole-CH$_2$	1780	Me_3SiO	3500
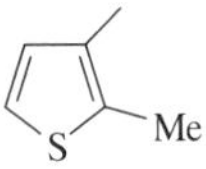 thiophene-Me	20.5	Ph_3GeO	~2000	Me_2N–C$_6$H$_4$– 	3680
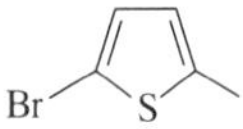 Br-thiophene	20.5	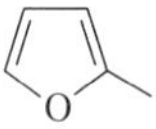 furan-methyl	2050	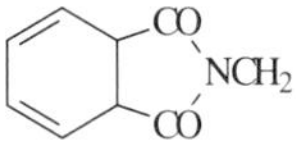 phthalimide-NCH$_2$	4100
 thiophene	89	p-ClC$_6$H$_4$CONHCH$_2$	2050	CH_2CH_2CN	4300
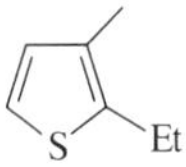 thiophene-Et	89	CH_2CH_2COOEt	2400	p-FC$_6$H$_4$CONHCH$_2$	>5000
H	320	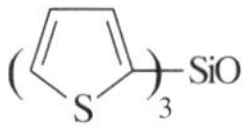 (thiophene)$_3$SiO	~2500	1-Ad	>5000
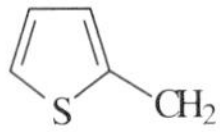 thiophene-CH$_2$	325	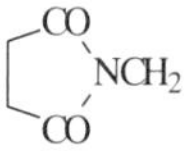 succinimide-NCH$_2$	2500	CH_2=CH	5600
BrCH$_2$	355	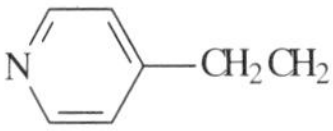 pyridine-CH$_2$CH$_2$	2580	 NCHMe	6500
Et$_2$NCH$_2$	355	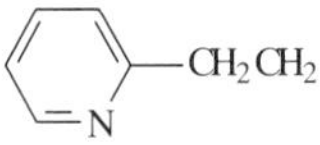 pyridine-CH$_2$CH$_2$	2820	$CH_2CH(CH_3)COOMe$	6820
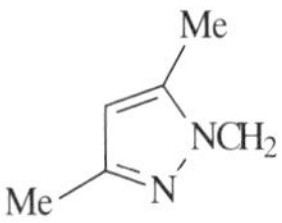 Me,Me-pyrazole-NCH$_2$	708	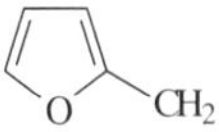 furan-CH$_2$	2960	OH	8400
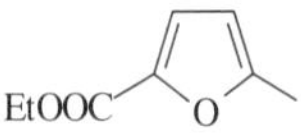 EtOOC-furan	1090	ClCH$_2$	2960	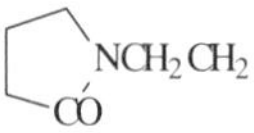 NCH$_2$CH$_2$	10,000

In the series of trialkylacetoxygermanes R_3GeOAc the ethyl derivative is the most toxic for rats $(125-250$ mg kg^{-1}, *p.o.*$)^{27,52}$.

V. COMPOUNDS WITH Ge—S BONDS

The germanium derivative of cysteine $Ge[SCH_2CH(NH_2)COOH]_4$ shows low toxicity in acute $(3402$ mg kg$^{-1})$, subacute and chronic experiments[53]. No teratogenic effects were noticed in rats and mice after subcutaneous administration of the compound[54].

Cyclic[55] and acyclic[56] organogermanium derivatives of cysteamine and methylcysteamine exhibit toxicity within the $300-800$ mg kg^{-1} range. Introduction of the germatranyl group into aminothiol molecules considerably decreases the toxicity of the compound; radioprotective properties remain at the same level. The same can be observed in the case of the ring closure of trithiagermatrane $(550$ mg kg$^{-1})$ and dithiagermocane $(200$ mg kg$^{-1})^{57}$.

The tolerable *s.c.* dose of $[(PhCH_2)_3Ge]_2S$ in mice equals 5000 mg kg^{-1}, that of sesquithianes $(RGeS_{1.5})_n$ 1250 mg kg^{-1} $(R = Ph)$ and 2500 mg kg^{-1} $(R = p-Me_2NC_6H_4)$; LD_{50} of the compound with $R = CHPhCH_2CONH_2$ is 500 mg kg^{-1} [1].

VI. ORGANOHALOGENOGERMANES

Tricyclohexyl as well as tribenzylchloro-, -bromo and iodogermanes administered subcutaneously to white mice have low toxicity $(1250-5000$ mg kg$^{-1})^1$. The toxicity of butylchlorogermanes $Bu_{4-n}GeCl_n$ administered *i.p.* grows with the increase in the number of the chlorine atoms in the molecule $(1280, 96, 50$ mg kg^{-1} for mice and 1970, 100, 48 mg kg^{-1} for rats). Methyl-, ethyl- and propyltriiodogermanes have similar toxicity (about 200 mg kg$^{-1})^{19}$.

Triphenylchlorogermane does not inhibit the growth of *Tribolium castaneum* larvae[58]. On the other hand, trialkylchlorogermanes are stronger inhibitors of the growth of *Escherichia coli* bacteria and *Selenastrum capricornutum* algae than the analogous chlorosilanes; but their activity is inferior to that of the tin derivatives[59].

VII. REFERENCES

1. E. Lukevics, T. K. Gar, L. M. Ignatovich and V. F. Mironov, *Biological Activity of Germanium Compounds*, Zinatne, Riga, 1990 (in Russian).
2. E. Lukevics, S. K. Germane and L. M. Ignatovich, *Appl. Organomet. Chem.*, **6**, 543 (1992).
3. J. J. Kavanagh, P. B. Saul, L. J. Copeland, D. M. Gershenson and I. H. Krakoff, *Cancer Treat. Rep.*, **69**, 139 (1985).
4. J. H. Saiers, M. Slavik, R. L. Stephens and E. D. Crawford, *Cancer Treat. Rep.*, **71**, 207 (1987).
5. F. H. Dexeus, C. Logothetis, M. L. Samuels and B. Hassan, *Cancer Treat. Rep.*, **70**, 1129 (1986).
6. N. Vogelzang, D. Gesme and B. Kennedy, *Am. J. Clin. Oncol.*, **8**, 341 (1985).
7. K. Asai, *Organic Germanium: A Medical Godsend*, L. Kagakusha, Tokyo, 1977.
8. K. Asai, *Miracle Cure: Organic Germanium*, Japan Publ. Inc., Tokyo, 1980.
9. K. Miyao and N. Tanaka, *Drugs of the Future*, **13**, 441 (1988).
10. Asai Germanium Research Institute, *Drugs of the Future*, **18**, 472 (1993).
11. A. Hoshi, *Drugs of the Future*, **18**, 905 (1993).
12. E. Lukevics and L. M. Ignatovich, *Appl. Organomet. Chem.*, **6**, 113 (1992).
13. M. Nagata, T. Yoneyama, K. Yanagida, K. Ushio, S. Yanagihara, O. Matsubara and Y. Eishi, *J. Toxicol. Sci.*, **10**, 333 (1985).
14. S. Okada, S. Kijama, Y. Oh, K. Shimatsu, N. Oochi, K. Kobayashi, F. Nanishi, S. Fujimi, K. Onoyama and M. Fujishima, *Curr. Ther. Res.*, **141**, 265 (1987).
15. O. Wada and M. Nagahashi, *Nippon Ishikai Zasshi*, **99**, 1929 (1988); *Chem. Abstr.*, **109**, 66198 (1988).

16. A. G. Schauss, *Biol. Trace Element. Res.*, **29**, 267 (1991).
17. M. Rothermundt and K. Burschkies, *Z. Immunitätsforsch. Exp. Ther.*, **87**, 445 (1936).
18. F. Caujolle, D. Caujolle and H. Bouissou, *C.R. Seances Acad. Sci.*, **257**, 551 (1963).
19. F. Caujolle, D. Caujolle, Dao-Huy-Giao, J. L. Foulquier and E. Maurel, *C.R. Seances Acad. Sci.*, **262**, 1302 (1966).
20. E. Lukevics, S. K. Germane, M. A. Trushule, V. F. Mironov, T. K. Gar, N. A. Viktorov and D. N. Chernysheva, *Khim.-Farm.Zh.*, **21**, 1070 (1987).
21. P. Köpf-Maier, W. Kahl, N. Klouras, G. Hermann and H. Köpf, *Eur. J. Med. Chem. Chim. Ther.*, **16**, 275 (1981).
22. P. Köpf-Maier and H. Köpf, *J. Organomet. Chem.*, **342**, 167 (1988).
23. M. C. Henry, E. Rosen, C. D. Port and B. S. Levine, *Cancer Treat. Rep.*, **64**, 1207 (1980).
24. L. M. Rice, J. W. Wheeler and C. F. Geschicter, *J. Heterocycl. Chem.*, **11**, 1041 (1974).
25. E. Lukevics, S. K. Germane, M. S. Trushule, A. E. Feoktistov and V. F. Mironov, *Latv.PSR Zināt. Akad. Vēstis*, **5**, 79 (1988).
26. G. Rima, J. Satgé, H. Sentenac-Roumanou, M. Fatome, C. Lion and J. D. Laval, *Eur. J. Med. Chem.*, **28**, 761 (1993).
27. F. Rijkens and G. J. van der Kerk; *Investigations in the Field of Organogermanium Chemistry*, Germanium Research Committee, 1964, p.95.
28. H. Bouissou, F. Caujolle, D. Caujolle and M. C. Voisin, *C.R. Seances Acad. Sci.*, **259**, 3408 (1964).
29. E. S. Rochow and B. M. Sindler, *J. Am. Chem. Soc.*, **72**, 1218 (1950).
30. D. Caujolle, Dao-Huy-Giao, J. L. Foulquier and M. -C. Voisin, *Ann. Biol. Clin. (Paris)*, **24**, 479 (1966).
31. F. Caujolle, D. Caujolle, S. Cros, Dao-Huy-Giao, F. Moulas, Y. Tollon and J. Caylus, *Bull. Trav. Soc. Pharm. Lyon*, **9**, 221 (1965).
32. F. Caujolle, R. Huron, F. Moulas and S. Cros, *Ann. Pharm.Fr.*, **24**, 23 (1966).
33. E. Lukevics, S. K. Germane, A. A. Zidermane, A. Zh. Dauvarte, I. M. Kravchenko, M. A. Trushule, V. F. Mironov, T. K. Gar, N. Yu. Khromova, N. A. Viktorov and V. I. Shiryaev, *Khim.-Pharm.Zh.*, **18**, 154 (1984).
34. S. Nakayama, T. Tsuji and K. Usami, *Showa Igakkai Zasshi*, **46**, 227 (1986); *Chem. Abstr.*, **106**, 334 (1987).
35. Y. Sugiya, K. Eda, K. Yoshida, S. Sakamaki and H. Satoh, *Oyo Yakuri*, **32**, 113 (1986); *Chem. Abstr.*, **105**, 164632 (1986).
36. Y. Sugiya, S. Sakamaki and H. Satoh, *Oyo Yakuri*, **31**, 1191 (1986); *Chem. Abstr.*, **105**, 126951 (1986).
37. Y. Sugiya, S. Sakamaki, T. Sugita, Y. Abo and H. Satoh, *Oyo Yakuri*, **31**, 1181 (1986); *Chem. Abstr.*, **105**, 126950 (1986).
38. Y. Sugiya, T. Sugita, S. Sakamaki, Y. Abo and H. Satoh, *Oyo Yakuri*, **32**, 93 (1986); *Chem. Abstr.*, **105**, 164631 (1986).
39. Y. Sugiya, K. Yoshida, K. Eda, S. Sakamaki and H. Satoh, *Oyo Yakuri*, **32**, 139 (1986); *Chem. Abstr.*, **105**, 164634 (1986).
40. Y. Sugiya, K. Yoshida, S. Sakamaki, K. Eda and H. Satoh, *Oyo Yakuri*, **32**, 123 (1986); *Chem. Abstr.*, **105**, 164633 (1986).
41. S. Tomizawa, R. Sato, H. Sato and A. Ishikawa, *Rep. Asai Germanium Research Inst.*, **1**, 5 (1972).
42. S. Tomizawa, R. Sato, H. Sato and A. Ishikawa, *Rep. Asai Germanium Research Inst.*, **1**, 7 (1972).
43. H. Nagai, K. Hasegawa and K. Shimpo, *Oyo Yakuri*, **20**, 271 (1980); *Chem. Abstr.*, **96**, 62662 (1980).
44. M. Kagoshima and M. Suzuki, *J. Med. Pharm. Sci.*, **15**, 1497 (1986).
45. S. Tomizawa, N. Suguro and M. Kagoshima, *Oyo Yakuri*, **16**, 671 (1978); *Chem. Abstr.*, **90**, 162125 (1979).
46. E. Lukevics and L. M. Ignatovich, *Chem. Heterocycl. Compd.*, **28**, 603 (1992).
47. E. Lukevics, L. M. Ignatovich, N. Porsyurova and S. K. Germane, *Appl. Organomet. Chem.*, **2**, 115 (1988).
48. E. Lukevics and L. M. Ignatovich, *Metalloorg. Khim.*, **2**, 184 (1989).
49. E. Lukevics and L. M. Ignatovich, *Main Group Metal. Chem.*, **7**, 133 (1994).
50. E. Lukevics, S. K. Germane, M. A. Trushule, V. F. Mironov, T. K. Gar, O. A. Dambrova and N. A. Victorov, *Khim.-Pharm.Zh.*, **22**, 163 (1988).
51. E. Lukevics, L. M. Ignatovich, N. Shilina and S. K. Germane, *Appl. Organomet. Chem.*, **6**, 261 (1992).

52. J. E. Cremer and W. N. Aldridge, *Brit. J. Ind. Med.*, **21**, 214 (1964).
53. R. Ho, T. Tanihata and T. Hidano, *Toho Igakkai Zasshi*, **20**, 633 (1973); *Chem. Abstr.*, **83**, 53372 (1975).
54. S. Hosokawa and K. Makabe, *Yamaguchi Igaku*, **22**, 107 (1973); *Chem. Abstr.*, **82**, 118883 (1975).
55. J. Satgé, A. Gazes, M. Bouchaut, M. Fatome, H. Sentenac-Roumanou and C. Lion, *Eur. J. Med. Chem. Chim. Ther.*, **17**, 433 (1982).
56. M. Fatome, H. Sentenac-Roumanou, C. Lion, F. Satgé and G. Rima, *Eur. J. Med. Chem.*, **23**, 257 (1988).
57. J. Satgé, G. Rima, M. Fatome, H. Sentenac-Roumanou and C. Lion, *Eur. J. Med., Chem.*, **24**, 48 (1989).
58. I. Ishaaya, R. L. Holmstead and J. E. Casida, *Pestic. Biochem. Physiol.*, **7**, 573 (1977).
59. G. Eng, E. J. Tierney, G. J. Olson, F. E. Brinckman and J. M. Bellama, *Appl. Organomet. Chem.*, **5**, 33 (1991).

CHAPTER **18**

Organotin toxicology

LARRY R. SHERMAN

Department of Chemistry, University of Scranton, Scranton, PA 18510–4626, USA
Fax: 011-717-941-7510;   e-mail: WPGATE::IN%"shermanl1@saguar.vefs.edu"

ABBREVIATIONS

ANP	atrial natriuretic peptide	TaET	tetraethyltin
DBT	dibutyltin	TaMT	tetramethyltin
DBTCl2	dibutyltin dichloride	TaMPb	tetramethyllead
DBTL	dibutyltinlaurate	TBT	tributyltin
DMT	dimethyltin	TBTO	bistri-*n*-butyltin oxide
DOT	dioctyltin	TcHT	tricyclohexyltin
DOTCl2	dioctyltin dichloride	TEPb	tetraethyl lead
DPhT	diphenyltin	TET	triethyltin
cGMP	cyclo guanidiene monophosphate	THT	trihexyltin
MET	monoethyltin	TMT	trimethyltin
NCI	National Cancer Institute	TPT	tripropyltin
NOEL	no observable effect level	TPhT	triphenyltin
OTC	organotin compounds		

I. GENERAL

Few clinical or epidemiological studies exist concerning inorganic tin intoxication primarily because inorganic tin(IV) compounds are poorly absorbed by mammals. The absorption from the gut for most compounds is less than 2%. However, when inorganic tin compounds are injected iv, they have a higher absorption and are usually stored in the bones with a half-life of approximately 400 days[1].

Organotin compounds (OTC), on the other hand, are considerably more toxic than inorganic compounds. Their toxicity is species-dependent, dependent upon the length

The chemistry of organic germanium, tin and lead compounds
Edited by S. Patai   © 1995 John Wiley & Sons Ltd

of the carbon chain and dependent upon the number of Sn—C bonds in the molecule. Adsorption is generally much higher, being about 8% for TET and 2% for TcHT in rats and 10% for TPhT compounds in cows; however, the absorption was even higher for TPhT compounds in guinea pigs. Fortunately, the half-life for OTC is quite short, being 30 days or less with most of the absorption occurring in the soft tissue[2]. The di- and tri-substituted methyl and ethyl compounds are the most toxic to humans. As the carbon chain length of the organic constituents increases, the compounds become more toxic to aquatic life but less toxic to humans and other mammals. Except for cyclic compounds, organotin derivatives with five or more carbons in the organic group have little mammalian toxicity except for premature atrophy of the thymus gland[3]. In fact, data on the LD_{50} for dioctyltin compounds and higher homologs is poor because it is difficult to dope animals with sufficient OTC to cause death (the LD_{50} is greater than 2000 mg/kg for some of the compounds)[1]. The biochemistry of OTC is basically dependent upon the lipid solubility of the compound[4] and its partition coefficient[5]. Longer-chain compounds as well as short-chain organotins can produce acute skin burns that may appear 1–8 hours after exposure. Exposure to organotin vapors or direct eye contact causes lachrymation and severe suffusion of the conjunctiva which can persist for days[6]. The compounds readily interact through insertion into the plasma membrane resulting in disturbances to membrane function, including signal transduction and ion movements. In many cases the OTC affect ion transport processes by acting as ionophores, inhibiting specific membrane-bound transport enzymes and interfering with receptor-mediated transport.

Even though general trends in OTC toxicity have been well established, there are inconsistencies in the literature concerning the various observed toxic effects of many OTC. A recent paper[7] reported a severe synergistic effect between polysorbate 80 and TBTCl. The work leads to speculation that much of the inconsistencies in the literature may be due to the unknown synergism and not to poor data collection. Because of their select toxicity, OTC exhibit excellent anti-tumor potentials when using the NCI P388 leukemia mouse test. More than 590 organotin compounds yielded in vivo anti-tumor properties with this test. Even so, no OTC has yet reached clinical testing except a tin porphyrin which is a photoactivating agent. This has been a disappointment for many OTC scientists, but the work still continues and bis(adeninato-N9) diphenyltin(IV) has recently been selected by NCI for further testing against a number of tumor cells. However, the initial anti-tumor action corresponds to the standard response to DPhT moieties[8]. This recent work seems to illustrate again that OTC toxicity toward either normal or malignant cells is primarily dependent upon the number and type of functional groups attached to the tin atom, which was carefully documented in the early 1980s[9].

In summary, OTC toxicity either toward animal organs or as drugs appears in four broad target areas, namely neurotoxicity, hepatoxicity, immunotoxicity and cutaneous toxicity[10]. Although other effects occur in select species or at high dose levels, their importance is minor compared to the four areas listed above.

II. HUMAN MORBIDITY

In 1954, a proprietary formulation, 'Stalinon', was marketed in France for oral adminis-tration of boils. The formulation was primarily linoleic acid but was contaminated with triethyltin (TET)iodide (LD_{50} in rats, ca 0.7 mg/kg[1]) and may have contained the mono-, di- and tetraethyltin compounds. The formulation led to the death of 102 people and the intoxication of more than 200 others. The TET caused altopic cerebral edema of the white matter of the brain. Since the 'Stalinon Affair', a great deal of control has been exerted in the manufacture and marketing of organotins and few deaths have occurred[11]. The acci-dental exposure of six industrial workers to TMT led to one death and to two seriously disabled workers[12]. Attempted suicides with OTC as well as accidental poisoning rarely lead to death, primarily because of the slow destruction of vital organs (most acute studies with OTC are performed over three days rather than one day). Yet, when TPhT was used as a suicidal agent, the victim exhibited abdominal pain, diarrhoea and vomiting with severe ataxia, dysmetria, nystagmus and blurring vision which persisted for many days. It was two months before the patient totally recovered[13].

III. IMMUNOTOXICITY

Although little epidemiological data exist for humans, most organotins, especially the dibutyl- and dioctyltins, have severe effects on the immune system of animals including premature atrophy of the thymus gland. The first observation was made by Seinen and Willems[3] and a great deal of work has since been performed in this area.

DBTCl2 and DOTCl2 exhibit severe depletion of small lymphocytes in the thymic cortex. Following a single ip dose of 1 mg of OTC/kg, DOTCl2 exhibited maximum cortex depletion after 96 hours. After 120 hours some repopulation occurred; yet a maximum dose suppression of the T-cell-dependent immune response was observed two days after dosing due to a selective reduction in the number of rapidly proliferating lymphoblasts in the thymus gland. As a consequence, the large pool of small lymphocytes declined in the following two days. On the fourth day, the atrophy was most pronounced with the frequency of the lymphoblasts increasing faster than in the controls[14,15]. At 1 to 10 mg/kg, TPhTCl significantly reduced the weight of the spleen and thymus gland without significantly affecting the animal's body weight when the OTC was administered at a constant dose for 14 days. At higher doses, the immunoglobulin E antibody exhibited severe suppression but little other effect was observed upon the secondary immune response[16]. Action of TPhT is related to the same immunotoxicity observed with other organotins[17].

TBT at the μM level hyperpolarized thymocytes and depolarized mouse thymocytes at higher concentrations. At μM levels, TBT caused a rapid increase in cytosolic free Ca^{+2} influx through membrane channels. The elevation of Ca^{+2} was associated with extensive DNA fragmentation. Yet other OTC, primarily TMT, TPhT and DBT, had minimal effects on Ca^{+2}, DNA fragmentation and cell viability. This is consistent with a greater susceptibility of thymocytes to trialkyltins containing four carbons[18]. OTC-stimulated human lymphocytes resulted in statistically significant increase in frequencies of hyperdiploid cells. A recent study indicates that OTC are able to induce aneuploidy, probably affecting spindle function of the cells[19]. The more lipophilic compounds of TPT, TBT, THT and TPhT are the most cytotoxic, with reduced thymidine incorporation at concentrations as low as 0.05–1.0 μM that can lead to membrane damage and Cr release, especially at higher levels[20].

A comparison of the immunotoxic effects between adult and pre-weanling rats subacutely dosed with TBTO exhibited alteration in body and lymphoid organ weights, mitogen and mixed lymphocyte reaction, lymphoproliferative responses, natural killer cell activity, cytotoxic T lymphocyte responses and primary antibody plaque-forming cell responses[21]. At low TBTO concentration, it showed only marginal loss of viability in isolated thymocytes. However, these changes included nuclear chromatin condensation associated with increased DNA fragmentation, cytoplasmic contraction and formation of membrane-bound apoptotic bodies. Comparable morphological changes and cleavage of DNA into oligonucleosomal fragments were evident in thymocytes. The work illustrates that, even at levels where TBTO is not overtly cytotoxic, it is capable of inducing programmed cell death in rat thymocytes[22]. Basically, the di- and tri-OTC with three to six carbons have a rather pronounced effect on the primary and secondary immune systems.

IV. NEUROTOXINS

The second most documented toxicological effect occurs between OTC and the central nervous system. TMT causes cell loss in the central nervous system and TET causes brain and spinal cord edema. Brain retention of the organotin moiety increased as follows:

$$TcHT < TPhT < TBT < TPT < TMT < TET$$

Furthermore, TMT, TET, TPT and TBT decreased susceptibility of mice to electroshock seizures[23] with retention in the brain about 4 days longer for low molecular weight OTC than in other parts of the body[23]. Ca^{2+} ATPase activity in the brain was significantly reduced in a dose-dependent manner in the presence of TMT and TET, but the effect only occurs at the highest levels for TBT. The order of inhibition was TET > TMT > TBT, a nice correlation with the LD_{50} for the chemicals[18].

TMTCl and TETBr exhibited acute effects on cochlea function following ip injection and generated the release of neurotransmitters from the inner hair cells and the subsequent depolarization of spiral ganglion cells. TMT impaired compound action thresholds at all frequencies within 30 minutes. At low doses, impairment became more noticeable after 60 minutes. Both OTC initially disrupt the functional integrity of either inner hair cells or spiral ganglion cells within the cochlea[24].

TET-treated rats showed muscular weakness. Tremors and especially the dragging of hind limbs with hyper-excitability were observed in rats treated with TMT. At high doses of TBT (> 2.5 mg/kg/d) all rats exhibited tremors[25] whereas administration of DBTL by gavage(GAVAGE-forced feeding of animals by tube to administer a known quantity of chemical to the animal's stomach) produced significant increase in polyamine levels in select rat brain areas and at higher doses spermidine levels were raised in pons-medulla, hypothalamus and frontal cortex. The observed induction in regional brain polyamines with DBTL-treated rats may lead to disturbances in synaptic function and further enhance its neurotoxic potential[26]. The fungicides, TBT and TPhT acetate at 297 mg/kg and 402 mg/kg produced severe central nervous and respiratory depression in mammals. The findings showed pulmonary, peptic and renal congestion, brain hemorrhages and destruction of the intestinal mucosa[27].

DMT, TMT, DBT, TBT and DPhT chlorides exhibited in vitro spindle disturbance in V79 Chinese hamster cells of brain tubulin. The V79 cells lose stainable spindles at higher concentration. The cell mitosis activity effect at low concentration increased with the lipophilicity of the OTC, but all compounds showed a concentration dependence on microtubules. The OTC seem to act through two different cooperative mechanisms: inhibition of microtubule assembly and interaction with hydrophobic sites. The latter mechanism might involve Cl/OH ion exchange[28].

TBTCl produced a dose-dependent inhibition of ANP on vascular smooth muscle responses with an effect on norepinephrine, nitroprusside and atrial natriuretic peptide in isolated aortic rings of rats. The inhibition of vasorelaxation was accompanied by a parallel inhibition of ANP-induced cGMP generation[29].

Calmodulin, a calcium binding protein, is involved in Ca^{2+}-dependent regulation of several synaptic functions of the brain: synthesis, uptake and release of neurotransmitters, protein phosphorylation and Ca^{+2} transport. It reacts with TET, TMT and TBT which then inactivates enzymes like Ca^{+2}-ATPase and phosphodiesterase. *In vitro* studies indicated TBT was greater at inhibiting calmodulin activity than TET and TMT, whereas *in vivo* the order was TET > TMT > TBT. This may be due to the greater detoxification of TBT (66%) in the liver before moving to other organs[30,31].

Other organs. Yallapragada and collaborators plotted the change in body weight versus time for TMT, TET and TBT administration over a six-day period and showed significant weight loss at high doses[25,30]. It has been postulated that OTC exhibit a strong antifeeding effect even at very low levels[1] and can lead to significant malnutrition[31]. Antifeeding may cause animals to succumb to malnutrition rather than direct toxicity of the OTC. A strong gender effect was evident in OTC studies with male rats succumbing faster than females to the adverse effects of the compounds[31-33].

Long exposure of workers to DBT moieties caused bile-duct damage and exposure to TPhT compounds caused liver damage[34]. This is probably due to high biliary excretion rate of butylin compounds.

Rat hemoglobin has a high affinity for TMT and TET but, because it is highly selective to only two amino acids in the correct sequence, the affinity is quite low in other animals.

TBT < TET < TMT inhibit the cardiovascular system in a concentration-dependent manner which affects the Ca^{+2} pump as well as protein phosphorylation. The compounds also inhibited the Ca^{+2}-Atpase function similar to that observed with nerve cells[35].

Seven days after injecting DBT (o-hydroxyacetophenone S-methyldithiocarbamate) into adult rat testes, the following was observed: seminiferous tubules, arrest of spermato-genesis, disorganization of interstital epitheliums and infiltration of polymorphonuclear leukocytes indicating advance necrosis and possible sterilization through action on the spermatocytes and spermatids[36].

Moderate but prolonged exposure of rats to TBT and TPhT acetate at subchronic levels (< 20 mg/kg OTC) brought about histopathologic lesions in lungs, liver, intestines and kidneys besides reduction in lymphocyte count at higher concentrations[37].

V. BIODEGRADATION

Lower alkyltins are very rapidly biotransformed in mammals with some conversion occurring within 15 minutes[38]. Administration of TaET, TEPb, TaMT and TaMPb yield toxicities about the same as their trialkyl compounds, because of the rapid dealkylation to the more toxic trialkyl compounds. However, gross intoxication is usually less with the tetra compounds because they are poorly assimilated into the animal's system. Further dealkylation of the trialkyltin to dialkyltins occurs at slower rates than with the tetra com-pounds; the mechanism is probably through hydroxylation of a beta carbon[23] leading to the less toxic mono species, which seems to be the end product in most animals and is rapidly excreted by the kidneys. The dealkylation of the OTC is further enhanced when animals are pretreated with phenobarbital, a typical synergism is found with all organometallic com-pounds. The mechanism in the latter case involves a hepatic microsomal P-450 mediated mono-oxygenated system with the dealkylation occurring in the liver microsomes[39].

VI. ENVIRONMENTAL TOXICITY

The most significant environmental effects have been observed with the TBT compounds[11] but other OTC have a select effect with the relative toxicity being a function of the hydrophobicity of the OTC rather than electronic or steric effects of the compounds[29].

Beginning with yolk sac fry, trout were continuously exposed for 110 days to TBT, TPhT, TcHTCl, DBTCl and DPhTCl at concentrations of 0.12–0.15 nM. The diorganotin compounds were about 3 orders of magnitude less toxic than the tri-OTC compounds: the di-OTC had a NOEL near 160 nM. The TcHT was the most toxic chemical studied (3 nM) causing 100% morbidity in one week. Histopathological examination revealed depletion of glycogen in the liver cells without atrophy of the thymus[40].

Guppies exposed to TPhTCl died as soon as a body burden 20 ± 10 nM was reached. Accumulation of TPhTCl can be predicted using kinetic parameters[41]. The bio-concentration factors of both TBTCl and TPhTCl via gill intake of goldfishes reached a plateau after 21 days of exposure[42].

VII. REFERENCES

1. L. R. Sherman, *Rev. Silicon, Germanium, Tin, Lead Compd.*, **9**, 323 (1986).
2. WHO Working Group, 'TBT', Environment Health Criteria, **116**, 1 (1990).
3. W. Seinen and W. Willems, *Toxicol. Appl. Pharmacol.*, **35**, 63 (1976).
4. R. M. Zucker, K. H. Elstein, R. E. Easterling and E. J. Massaro, *Gov. Rep. Announce. Index*, **13** (1990).
5. K. A. Winship, *Adverse Drug Reaction Acute Poisoning Review*, **7**, 19 (1988).
6. Int. Prog. on Chemical Safety, 'TBT', World Health Organization, Geneva 27, Switzerland (1990).
7. L. R. Sherman and G. L. Kellner, *Appl. Organometal. Chem.*, **4**, 379 (1990).
8. R. Barbieri, G. Ruisi and G. Atassi, *J. Inorg. Biochem.*, **41**, 25 (1991).
9. G. Atassi, *Rev. Silicon, Germanium, Tin, Lead Compd.*, **8**, 219 (1985).
10. N. J. Snoeij, A. H. Penninks and W. Seinen, *Environ. Res.*, **44**, 335 (1987).
11. J. J. Cooney, J. H. Weber, and L. R. Sherman, in *Biological Diversity Problems and Challenges* (Eds. S. K. Majumdar *et al.*), Pa. Acad. Sci., Easton, PA 18042, 1994.
12. R. Besser, G. Kramer, R. Thumler, J. Bohl, L. Gutmann and H. C. Hopf, *Neurology*, **37**, 945 (1987).
13. R. M. Wu, Y. C. Chang and H. C. Chiu, *J. Neurol. Neurosurg. Psychiatry*, **53**, 356 (1990).
14. N. J. Snoeij, A. H. Penninks and W. Seinen, *Int. J. Immunopharmacol.*, **10**, 891 (1989).
15. A. Penninks, F. Kuper, B. J. Spit and W. Seinen, *Immunopharmacology*, **10**, 1 (1985).
16. H. Nishida, H. Matsui, H. Sugiura, K. Kitagaki, M. Fuchigami, N. Inagaki, H. Nagai and A. Koda, *J. Pharmacol. Biodyn.*, **13**, 543 (1990).
17. Y. Oyama, L. Chikahisa, F. Tomiyoshi and H. Hayashi, *Jpn. J. Pharmacol.*, **57**, 419 (1991).
18. T. Y. Aw, P. Nicotera, L. Manzo and S. Orrenius, *Arch. Biochem. Biophys.*, **283**, 46 (1990).
19. K. G. Jensen, O. Andersen and M. Ronne, *Mutat. Res.*, **246**, 109 (1991).
20. N. J. Snoeij, A. A. van Iersel, A. H. Penninks and W. Seinen, *Toxicology*, **39**, 71 (1986).
21. R. J. Smialowicz, M. M. Riddle, R. R. Rogers, R. W. Luebke and C. B. Copeland, *Toxicology* **57**, 97 (1989).
22. M. Raffray and G. M. Cohen, *Arch. Toxicol.*, **65**, 135 (1991).
23. S. S. Brown and J. Savory, *Chemical Toxicology and Clinical Chemical of Metals*, Academic Press, New York, 1983.
24. W. J. Clerici, B. Ross and L. D. Fechter, *Toxicol. Appl. Pharmacol.*, **109**, 547 (1991).
25. P. R. Yallapragada, P. J. Vig, P. R. Kodavanti and D. Desaiah, *J. Toxicol. Environ. Heal.*, **34**, 229 (1991).
26. M. A. Khaliq, R. Husain, P. K. Seth and S. P. Srivastava, *Toxicol. Lett.*, **55**, 179 (1991).
27. U. S. Attahiru, T. T. Iyaniwura, A. O. Adaudi and J. J. Bonire, *Vet. Hum. Toxicol.*, **33**, 554 (1991).
28. K. G. Jensen, A. Onfelt, M. Wallin, V. Lidums and O. Andersen, *Mutagenesis*, **6**, 409 (1991).
29. R. Solomon and V. Krishnamurty, *Toxicology*, **76**, 39 (1992).
30. P. R. Yallapragada, P. J. S. Vig and D. Desaiah, *J. Toxicol. Environ. Health*, **29**, 317 (1990).
31. G. L. Kellner and L. R. Sherman, *Microchem. J.*, **47**, 67 (1993).
32. A. L. Boyd and J. M. Jones, *Toxicol. Lett.*, **30**, 253 (1986).
33. E. I. Krajnc, J. G. Vos, P. W. Wester, J. G. Loeber and C. A. Van Der Heijden in *Toxicology and Analysis of the Tributytins — the Present Status* (Ed. J. A. Jonker), ORTEP, 4380 AB Vlissingen-Oost, The Netherlands.
34. S. C. Srivastava, *Toxicol. Lett.*, **52**, 287 (1990).
35. P. R. Kodavanti, J. A. Cameron, P. R. Yallapragada, P. J. Vig and D. Desaiah, *Arch. Toxicol.*, **65**, 311 (1991).
36. A. Saxena, J. K. Koacher and J. P. Tandon, *J. Toxicol. Environ. Health*, **15**, 503 (1985).
37. U. S. Attahiru, T. T. Iyaniwura, A. O. Adaudi and J. J. Bonire, *Vet. Hum. Toxicol.*, **33**, 499 (1991).
38. O. H. Wada and Y. Arakawa, *J. Anal. Toxicol.*, **5**, 300 (1981).
39. L. D. Hamilton, W. H. Medeiros, P. D. Meskowitz and K. Tybicka, *Gov. Rep. Announce. Index*, **8** (1989).
40. H. DeVries, A. Penninks, N. J. Snoeij and W. Seinen, *Sci. Total Environ.*, **103**, 229 (1991).
41. J. W. Tas, W. Seinen and A. Opperhuizen, *Comp. Biochem. Physiol. C*, **100**, 59 (1991).
42. T. Tsuda, S. Aoki, M. Kojima and H. Harada, *Comp. Biochem. Physiol. C*, **99**, 69 (1991).

CHAPTER **19**

Safety and environmental effects

SHIGERU MAEDA

Department of Applied Chemistry and Chemical Engineering, Faculty of Engineering, Kagoshima University, 1-21-40 Korimoto, Kagoshima 890, Japan
Fax: 81-992-85-8339;   e-mail: maeda@apc.eng.Kagoshima-u.ac.jp

The chemistry of organic germanium, tin and lead compounds
Edited by S. Patai   © 1995 John Wiley & Sons Ltd

871

I. INTRODUCTION

A considerable number of organometallic species of germanium, tin and lead have been detected in the natural environment. A number of these are nonmethyl compounds which have entered the environment after manufacture and use (e.g. butyltin and phenyltin compounds by diffusion from antifouling paints on boats, and ethylleads from leaded gasoline). Only a few methyl compounds are now manufactured and used (e.g. some methyltin compounds are used as oxide film precursors on glass)[1].

It is now well established that organometallic compounds are formed in the environment from mercury, arsenic, selenium, tellurium and tin and hence were also deduced on the basis of analytical evidence for lead, germanium, antimony and thallium. Biological methylation of tin has been demonstrated by the use of experimental organisms. Methylgermanium and methyllead were widely found in the environment but it is debatable whether germanium and lead are directly methylated by biological activity in natural environment.

The main interest in methylation (or alkylation) is the change in properties resulting from the attachment of methyl groups (or other alkyl groups) to the inorganic elements or compounds. Lipid solubility, volatility and persistence of metals in biological systems may be increased in the methyl (or alkyl) derivatives. Most organometallic compounds are more toxic than the inorganic ones, but sometimes the reverse is the case (particularly for arsenic). When impact of toxic organometallics exceeds the capacity of the resistance mechanism of an organism and/or of an environment, the organism and/or the environment may suffer toxic effects.

This review describes factors concerning the safety and environmental effects of organic germanium, tin and lead compounds. The factors involve the production and use of the elements, alkylation, degradation, toxicity, health effect assessment and so on.

II. GERMANIUM

A. Introduction

Germanium is a grayish-white lustrous crystalline metal that belongs to group IVA of the periodic table of elements, so that its physical and chemical properties resemble those of the non-metal silicon and, to a lesser extent, tin[2]. Germanium is ubiquitous in the earth's crust in an abundance of $1.4 \times 10^{-4}\%$ (Clarke's number, 6.7×10^{-4}). In nature, germanium is widely, albeit sparsely, distributed. It is associated with sulfide ores of other elements, particularly with those of copper, zinc, lead, tin and antimony[3]. Minerals in which germanium is concentrated are germanite, a sulfoarsenite of copper, germanium and iron with an average content of 5% germanium; argyrodite, a double sulfide of germanium and silver containing 5% to 7% germanium; renirite, a complex sulfide of arsenic, copper, germanium, iron, tin and zinc with 6% to 8% germanium; and several other minerals such as canfieldite, itoite, stottite and ultrabasite[4].

Germanium enters aquatic environment indirectly from germanium-rich residues, mainly zinc base metal smelting operations. Sea water contains 0.05 μg Ge l^{-1}.

There is no known biological requirement for germanium, germanates or any organogermanium compounds. Germanium deficiency has not been demonstrated in any animal[5].

The properties of 299 organic germanium compounds have been summarized by Harrison[6].

B. Production and Use of Germanium and Germanium Compounds

1. Production

The data in Table 1 represent the annual production *capacity* for refineries on December 31, 1990, i.e. the maximum quantity of product which these refineries are able to produce.

TABLE 1. World capacity of annual germanium refinery production, December, 1990 (ton yr^{-1})[4]

Area	Capacity (ton)
North America	
Canada	10
USA	60
Total	70
Europe	
Belgium	50
Centrally planned economy countries	40
Other	65
Total	155
Asia	
China	10
Japan	35
Total	45
World Total	270

In 1990, the *actual* world refinery production of germanium was estimated at 76 ton, a decrease of about 7% compared with the 1989 level[4]. This decline is attributed to an oversupply and to a lower level of demand for the metal. The main producers of germanium products are located in the United States, Belgium, France, Germany and Japan.

Germanium metal production in Japan was 3368 kg in 1990. Dioxide production decreased from 13,302 kg in 1989 to 12,350 kg in 1990[4].

2. Use

The major use of germanium is as optical materials. The US Bureau of Mines estimated that the consumption pattern for the use of germanium in 1990 was as follows: infrared systems, 60%; fiber optics, 8%; gamma-ray, X-ray and infrared detectors, 9%; semiconductors (including transistors, diodes and rectifiers), 10%; and other applications (catalysts, phosphors, metallurgy and chemotherapy), 13%[4].

Germanium lenses and filters have been used in instruments which operate in the infrared region of the spectrum. Windows and lenses of germanium are vital components of some laser and infrared guidance or detection systems. Glasses prepared with germanium dioxide have a higher refractivity and dispersion than do comparable silicate glasses, and may be used in wide-angle camera lenses and microscopes. The GeO_2–TiO_2–P_2O_5-type glasses have excellent infrared transmission characteristics that make them ideal for use as windows for the protection of ultrasensitive infrared detectors used in the space programs[3]. In the United States, infrared optics are mainly used for military guidance and weapon-sighting systems[4].

The second major use of germanium is as catalyst in the production of polyesters [e.g. poly(ethylene terephthalate)] and synthetic textile fibers (especially those produced in Europe and Japan).

The remaining interest for germanium in electronics is based on the advantageous mobility characteristics of charge carriers in this material. Typical applications are high-power divices with low energy loss and photodetectors[7]. During World War II germanium was investigated for its use in the rectification of microwaves for radar applications, and several types of diodes were developed. The sale of germanium diodes and transistors peaked in 1966, and then this demand declined because germanium was replaced by

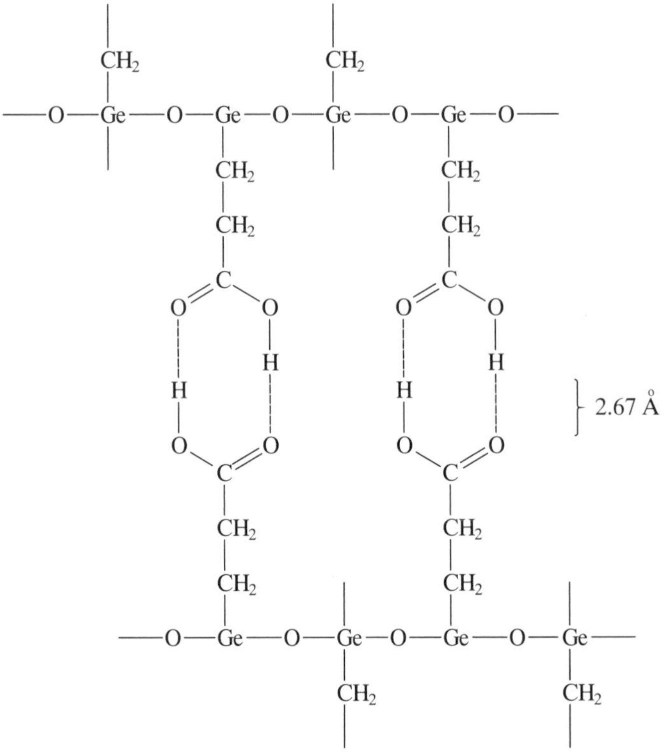

FIGURE 1. Structure proposed for carboxyethyl-germanium sesquioxide (modified from Reference 12)

electronic-grade silicon[3]. Little technical or scientific effort has been expended on germanium integrated circuits. However, when ultrahigh-speed switching circuitry is required, germanium is inherently better than silicon since its mobility values for electrons and holes are twice as great as those for silicon[3]. The reduced demand for germanium in the electronics field was offset by a dramatic increase in demand for germanium in both infrared night vision systems and fiber-optic communication networks in the USA[4].

A small amount of organogermanium compounds is used for medical applications. Thus, carboxyethyl-germanium sesquioxide [$O_3(GeCH_2CH_2COOH)_2$], trade-name Ge-132 (Figure 1), originally synthesized at Asai Germanium Research Institute, Tokyo, Japan, is an immuno-potentiating agent with interferon (IFN)-inducing and antitumour activities[8].

C. Concentration and Speciation of Organogermanium Compounds in the Natural Environment

The concentration of Ge in sea water and in the earth's crust have been estimated at 0.05 $\mu g\,l^{-1}$ and 2 $mg\,kg^{-1}$, respectively[9]. Germanium is generally found in the +4 oxidation state as the oxide or in solution as germanic acid under various environmental conditions.

Braman and Tompkins[10] developed the atomic emission spectrometric determination method for the determination of inorganic and organic germanium compounds in the environment. They analyzed a number of natural offshore waters, in and around Tampa Bay, Florida, USA. The average germanium concentrations found in fresh waters (13 sites), saline waters (10 sites) and estuarine waters (5 sites) were 0.016, 0.079 and 0.029 μg l^{-1}, respectively. In rain and tap waters the average germanium content found was 0.045 and 0.0088 μg Ge l^{-1}, respectively. Several waters in deep wells in Oregon were found to contain unusually high levels of germanium(IV) but no methylgermanium compounds. The average value reported was 0.47 μg Ge l^{-1}. Methylgermanium compounds were not detected in any environmental samples analyzed at a detection limit of 10 ng l^{-1}, i.e. their concentrations must be less than this value.

Andreae and Froelich[11] determined arsenic, antimony and germanium species concentrations from five hydrographic stations along the central axis of the Baltic Sea from the Bornholm Basin to the Gulf of Finland. The Baltic Sea is a brackish, landlocked sea surrounded by highly industrialized countries and is considered to be one of the most seriously polluted marine areas in the world, receiving pollutants from domestic and industrial sources, as well as from river inputs, atmospheric deposition and via inflow through the Danish Straits (Belt Sea). Seawater samples were collected at every 10 m depth from five stations during June 10–15, 1981. The hydrographic conditions in the central Baltic Sea during the study period were discussed. Depth profiles of germanium compounds in the Baltic Sea are shown in Figure 2[11]. Three dissolved germanium species were observed in the Baltic Sea (Figure 2): inorganic germanium, which is thought to exist in seawater as germanic acid [Ge(OH)$_4$], and the organogermanium species, monomethylgermanium (CH$_3$Ge^{3+}) and dimethylgermanium [(CH$_3$)$_2$Ge^{2+}]. The methylated species are likely to be present in the form of the uncharged hydroxide complexes rather than the free ions. Trimethylgermanium was not found at a detection limit of 10 pM. Germanium acid concentrations are about ten times higher than in the ocean and much higher than can be accounted for by fluvial input. The methylated species, on the other hand, show only a small degree of vertical structure. This rules out the possibility of production of organogermanium species in anoxic basins of the Baltic. The vertical distributions of germanium within the Baltic Sea are controlled by biogeochemical cycling, involving biogenic uptake, particulate acavenging and partial regeneration. A mass balance including river and atmospheric inputs, exchange with the Atlantic through the Belt Sea and

TABLE 2. Germanium concentration in selected plants for herb medicines[12]

Plants	Germanium (μg Ge kg^{-1})
Polypore	800–2000
Ginseng (Korea, 20 years or more)	2000–4000
Ginseng	250–320
Litchi	800–2000
Garlic	745–756
Acanthopanax senticosus	310–400
Chebulae fructus	260
Bandai mushroom	255
Sophorae subprostratae	250
Water caltrop	230–257
Chinese maltrimony vine	120–124
Comfrey	76–152
Lithospermi radix	58
Adlay	50

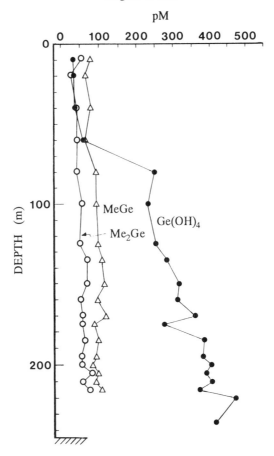

FIGURE 2. Depth profile of germanium compounds in Baltic Sea (modified from Reference 11): MeGe, monomethylgermanium; Me₂Ge, dimethylgermanium; Asi, total dissolved inorganic arsenic

removal by sediment deposition suggests that anthropogenic inputs make a significant contribution to the budgets of germanium, with atmospheric fluxes dominating the input to the Baltic Sea[11].

Edible plant foods usually contain less than 1 μg g^{-1}, although some plants, such as Shiitake mushroom, pearly barley and garlic, contain appreciably higher amounts[5].

Omae[12] collected the data on germanium concentration in selected plants used for herb medicines and also some foods and summarized them in Table 2, which shows that these contain germanium at an extremely high level. It is found that the germanium in the plants is present in organic form including a Ge−O bond. However, the complete chemical structure of the germanium compounds is not known[12].

The mean concentrations of germanium in normal human tissues were: lymph node, 0.9 μg kg^{-1} wet wt.; skeletal muscle, 3 μg kg^{-1} wet wt.; liver, 0.04 mg kg^{-1} wet wt.; lung, 0.09 mg kg^{-1} wet wt.; brain, 0.1 mg kg^{-1} wet wt.; blood, 0.2 mg kg^{-1} wet wt.; testes, 0.5 mg kg^{-1} wet wt; and kidney, 9.0 mg kg^{-1} wet wt.[5].

D. Methylation of Germanium Compounds

While both monomethylgermanium and dimethylgermanium have been detected in natural waters, there is no evidence for biological methylation. According to the biological methylation test using diatoms, aerobic bacteria and fungi, methylated germanium was not produced.

Mayer and Rapsomanikis[13] studied chemical methylation of germanium(II) in aqueous solution. In these experiments, inorganic germanium was reacted with methyl iodide as carbanion donor and methylcobalamin (CH_3CoB_{12}) as carbonium donor. The results showed that CH_3CoB_{12} and CH_3I are able to methylate Ge(II) forming monomethylgermanium at pH 1 and pH 7.6, respectively. No dimethylgermanium or trimethylgermanium was produced. For the reaction of CH_3CoB_{12} with germanium(II) a free-radical mechanism is assumed, whereas methylation by CH_3I is likely an oxidative addition mechanism.

Methylation experiments with CH_3I in artificial seawater indicate that methylation of germanium(II) to monomethylgermanium could occur in the ocean. However, germanium is considered to exist in the +IV state in the ocean. A methylation of Ge(IV) is not possible by oxidative addition of CH_3I. Although the experiments described above have shown that a chemical methylation of germanium(II) to monomethylgermanium by CH_3I is possible, it is not clear whether this reaction contributes to the methylgermanium compounds found in natural waters[13].

E. Toxicity and Environmental Effects of Organogermaniums

1. Inorganic germanium compounds

a. GeH₄, germanium hydride. GeH_4 is a colorless toxic gas of low stability with a characteristic unpleasant odor. GeH_4 (70 mg m^{-3}, minimally effective concentration) for 2–15 days caused nonspecific and nonpersistent changes in the nervous system, kidney and blood composition[7]. The maximum time-weighted average 8-h safe exposure limit is only 0.2 ppm[14]. Like other metal hydrides such as AsH_3, it shows hemolytic action in animals. The lethal concentration in air is 150 ppm[7].

b. GeO₂, germanium dioxide. The LD_{50} of GeO_2 for mice ranges from 2025 mg kg^{-1} (female, intraperitoneally) to 6300 mg kg^{-1} (male, per os), with corresponding values for rats of 1620–3700 mg kg^{-1}, respectively[5]. Although GeO_2 has been believed to have low toxicity in mammals owing to its diffusibility and rapid elimination, more recent evidence suggests that it may be involved in the pathogenesis of Ge-induced nephrotoxicities in humans following long-term Ge administration[5].

Slawson and coworkers reviewed early studies on the toxicity of GeO_2 to microorganisms, which focused on diatoms because of the unique use of SiO_2 in shell material[15]. According to the review, concentrations of only 1 mg l^{-1} of GeO_2 could significantly inhibit diatom growth. The diatom *Phaeodactylum tricornutum* was the least sensitive to GeO_2 inhibition and also contained the least amount of silicon in its shell. The growth inhibitory effect of GeO_2 could be reversed by adding SiO_2. These results, along with the chemical similarity between germanium and silicon, suggest that toxic effects of germanium may be due to inhibition of silica shell formation in diatoms[15].

The toxic effect and bioaccumulation of GeO_2 on 21 bacterial and 13 yeast strains were investigated by Van Dyke and coworkers[16,17]. Bacteria were more tolerant than yeasts to the growth inhibitory effects of germanium. Some examples of germanium accumulation in bacteria and in a yeast strain are provided in Figure 3. Bacillus strains accumulated the highest levels, ranging from 1.3 to 1.5 mg Ge g^{-1} dry wt.[15]. Lee and coworkers found that germanium accumulation was temperature and pH dependent, with high levels being

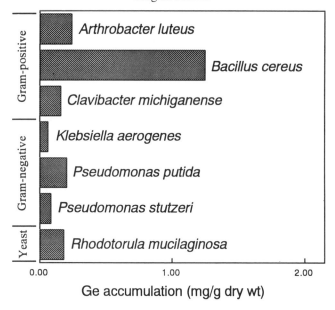

FIGURE 3. Germanium accumulation by selected microorganisms (modified from Reference 15). Cultures were incubated in a medium containing 10 g glucose l^{-1} and 0.5 g GeO$_2$ l^{-1}, final pH 7, at 28°C for 20 h

accumulated at pH 11 and incubation at 50 °C, conditions under which *P. stutzeri* cells were no longer viable. At pH 11, about 17 mg Ge g^{-1} dry cell wt. was accumulated[2].

Chmielowsky and Klapcinska studied the bioaccumulation of germanium by *Pseudomonas putida* to provide evidence that germanium is transported into bacterial cells as a complex with an aromatic substrate (catechol)[18]. They reached the conclusion that the uptake of germanium from a medium containing germanium complexed with readily dissimilated substrates might be considered an example of a nonspecific intracellular accumulation of a metal by microorganisms[18].

2. Organogermanium compounds

The main organogermanium compounds produced are shown in Figure 4. Therapeutic effects of organic germanium as shown below were reviewed by Goodman[19] and other investigators[20]. Compounds **2** and **3** in Figure 4 have therapeutic effects, such as analgesic potentiating activity for morphine, as well as antitumor activity.

a. Alkylgermanium oxides. Dimethylgermanium oxide is only slightly toxic to rats, whereas triethylgermanium acetate shows considerable toxicity[7].

*b. Carboxyethyl germanium sesquioxide (Ge-132, **1**).* Ge-132 (**1**) ([O$_3$(GeCH$_2$CH$_2$-COOH)$_2$]) was originally synthesized by Asai in Japan in 1967. Its solubility in water is 1.19 mg 100 ml^{-1} at 31 °C.

In pharmocokinetic C14-labelled studies of its absorption, excretion, distribution and metabolism, Ge-132 administered orally was absorbed about 30%, distributed evenly, with

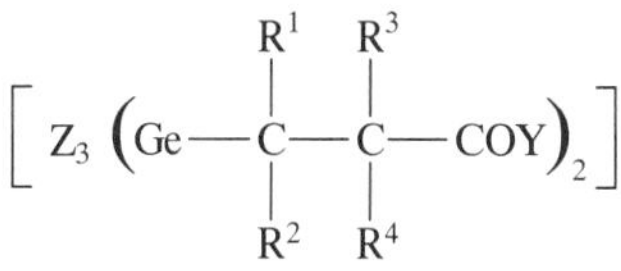

$$\left[Z_3 \left(Ge - \underset{R^2}{\overset{R^1}{C}} - \underset{R^4}{\overset{R^3}{C}} - COY \right)_2 \right]$$

R¹~R⁴ = H, alkyl, aryl, etc.
Z = O, S
Y = OH, OR, NH₂, NR₂, etc.

(2)

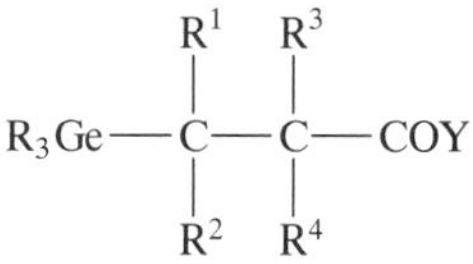

$$R_3 Ge - \underset{R^2}{\overset{R^1}{C}} - \underset{R^4}{\overset{R^3}{C}} - COY$$

R¹~R⁴ = H, alkyl, aryl, NHCOCH₃, etc.
R = H, alkyl, aryl, halogen
R₃ = N(CH₂CH₂O-)₃, N(CH₂CH₂S-)₃
Y = OH, OR¹, OM, NH₂, NR¹₂
R¹ = alkyl

(3)

Et₂Ge ⋯⋯ N(CH₂)₃NMe₂ · HCl

(4)

Spirogermanium

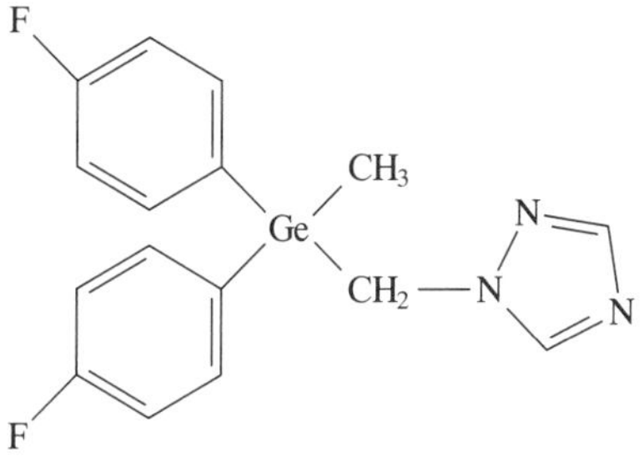

(5)

Bis(4-fluorophenyl)methyl(1H-1,2,4-triazo-1-yl-methyl)-german

FIGURE 4. Main organogermanium compounds (modified from Reference 20)

almost no residual concentration after 12 hours. It was excreted, unchanged metabolically, in the urine in 24 hours[19]. Toxicities of the compound were determined in Wistar rats (acute and chronic) or beagle dogs (chronic) by intravenous ($125-500$ mg kg^{-1}) or oral ($30-3000$ mg kg^{-1}) administration. At all doses examined, no significant toxicity was detected[21,22].

Aso and coworkers described the induction of interferon by oral administration of Ge-132[23]. The metabolic fate of Ge-132 was investigated by Kagoshima and Onishi[24].

c. Sanumgerman (lactate-citrate-germanium). For sanumgerman the LD$_{50}$ in mice (CWF/Bog strain) and Wistar rats was determined to be 275 mg kg^{-1} 24 h^{-1} and 250 mg kg^{-1} 48 h^{-1}, respectively[19]. Sanumgerman exerted a positive inhibiting effect on the central nervous system of mice.

d. Spirogermanium (4). Spirogermanium is an azaspirane compound containing nitrogen linked to a dimethyl amino propyl substituent (2-aza-8-germanspiro decane-2-propamine-8,8-diethyl-N,N-dimethyl hydrochloride). This drug inhibited DNA and RNA synthesis, as measured by radioactive amino acid uptake and analysis[5].

The LD_{10} value in mice (strain CDF1) was determined to be $105-147$ mg kg^{-1}. The highest nontoxic dose in beagle dogs was 12.5 mg kg^{-1}, the lowest toxic dose was 25 mg kg^{-1} and the lethal dose was 800 mg kg^{-1}. Toxic effects included focal necrosis of lymph nodes, inflammation and necrosis of the gastrointestinal mucosa and abnormal liver function[19].

e. Germatranes (tricyclic organogermanium derivatives of triethanolamine-1-germa-2,8,9-trioxa-5-azatricyclo[3.3.3.0^{1,5}]undecane) and their derivatives. Neurotrophic activity of 62 germatrane derivatives (e.g. germanols, germsesquioxanes, germyladamantanes, germylamides, germylimides and germyl-substituted amines, imines and hydroxamic acids and so on) were reviewed by Lukevics and coworkers[25]. The data obtained were processed statistically and the mean effective (ED_{50}) and mean lethal (LD_{50}) doses for mice were determined. Minimum LD_{50} througout 62 compounds tested was 16.5 mg kg^{-1}, six out of 62 compounds were below 100 mg kg^{-1} and 45 compounds out of 62 compounds were above 1000 mg kg^{-1}.

f. Bis(4-fluorophenyl)methyl(1H-1,2,4-triazo-1-yl-methyl)-german (5). The compound (**5**, shown in Figure 4) is a germanium analog of the leading agricultural fungicide flusilazole, which is a highly potent Si-based ergosterol biosynthesis-inhibiting fungicide. The biological properties of flusilazole and of 5 were compared and found to show similar fungicidal properties, as shown in Table 3[26].

[b]**5** is bis(4-fluorophenyl)methyl(1H-1,2,4-triazo-1-yl-methyl)-german.

The NADH- and oxygen-dependent microsomal metabolism of the di-, tri- and tetra-ethyl substituted derivatives of germanium, tin and lead was shown to give rise to ethylene as a major product and ethane as a minor product[27]. These reactions were shown to be catalyzed by the liver microsomal fractions.

F. Health Effect Assessment and Safety of Organogermaniums

Germanium is widely distributed in edible plant foods as described above in Section II.C. The estimated average dietary[5] intake of germanium in humans is 1.5 mg day^{-1} (range, $0.40-3.40$ mg day^{-1}), of which 96% or more is absorbed.

Germanium and most germanium compounds are comparatively low in toxicity because of pharmacological inertness, slow diffusibility and rapid excretion[7]. However, some exceptions exist, the most important being germanium hydride (germane). Surprisingly,

TABLE 3. Fungicidal activity (MIC values[a]) of the Si/Ge analogues Flusilazole and **5** against plant and human pathogenic fungi (modified from Reference 26)

Fungus	MIC value (μg ml^{-1})	
	Flusilazole	**5**[b]
Saccharomycopsis lipolytica	1.7	1.6
Pyricularia oryzae	0.9	1.0
Pseudocercosporella herpotrichoides	0.3	0.5
Fusarium culmorum	9.2	1.9
Botrytis cinerea	3.1	8.9
Pyrenophora teres	1.2	0.7

[a] Minimal inhibition concentrations.

soluble germanium compounds are more toxic by oral than by parenteral uptake. Industrial exposure is due mainly to germanium fumes and dusts generated during production[7].

Gastrointestinal absorption of germanium oxides and cationic salts is poor. No reports of germanium accumulation in human or animal tissue exist.

High exposure levels of germanium salts may disturb the water balance leading to dehydration, hemoconcentration, decrease in blood pressure and hypothermia, without showing gross tissue damage[7].

Poisoning by germanium compounds has occurred frequently during medical therapy. Some Japanese people have ingested orally Ge elixirs containing germanium dioxide (GeO_2) or Ge-132[28]. Furthermore, Ge-132 and lactate-citrate-germanate (Ge lactate citrate) have been sold as nutritional supplements in some countries for their purported immunomodulatory effects or as health-producing elixirs, causing intakes of germanium significantly exceeding the estimated average dietary intake[5]. Since 1982, there have been 18 reported cases of acute renal dysfunction or failure, including 2 deaths, linked to oral intake of elixirs containing germanium oxide or Ge-132. In 17 of the 18 cases, accumulated elemental germanium intakes reportedly ranged between 1666 to 328 g over a 4–36 month period, or between 100 to 2000 times the average estimated dietary intake for humans[5].

The U.S. Food and Drug Administration (FDA) has recently begun to take action against Ge-containing supplements as nonconforming food additives, while official warnings have been released in Germany and other countries[5].

A TLV(Threshold Limit Value) for germane (germanium hydride) of 0.2 ppm, equivalent to 0.64 mg m^{-3}, was fixed by the TLV Committee of the American Conference of Governmental Industrial Hygienists in 1973 as an average limit of permissible exposure during an 8-h working day[7]. TLV-STEL (Short Term Exposure Limit) in Japan is 0.6 ppm, equivalent to 1.8 mg m^{-3}.

G. Summary

Since plants used as herb medicines contain high amounts of germanium compounds, the latter have been regarded as having immunopotentiation and antitumour activities. At present, organogermanium compounds are mainly used for medical applications. For example, Ge-132 is well known to have not only an antitumour effect, but also to induce interferon (IFN) production in vivo[8].

However, poisoning by germanium and germanium compounds has occurred more frequently during medical therapy than by exposure at the work place. It was reported that a significantly high intake level of Ge elixirs containing GeO_2 or Ge-132 caused renal failure[5].

Obviously any medicine has some side effects, if it is taken in significantly high doses and/or for longer periods. Therefore, we need always to pay attention to dose levels and duration of intake.

III. TIN

A. Introduction

Tin is an essential trace element for animals. It is soft, pliable and colorless and belongs to group IV of the periodic table, and is corrosion-resistant to many media. Tin occurs in nature mostly as the oxide mineral cassiterite and is ubiquitous in the earth' crust in an abundance of $2.5 \times 10^{-4}\%$ (Clarke's number, 4×10^{-3}). It is one of the earliest metals known to mankind, and evidence of its use dates back over 4000 years. The ancients found that tin has unique properties, and realized that tin alloys readily with copper to produce bronze[29]. Tin metal is commonly used as a protective coating or as an alloy with other metals. It finds applications in products and processes as diverse as tin cans, solder for electronics, tin chemicals, bronze fittings and flat-glass production[29,30]. The element reacts with both strong acids and strong bases, but it is relatively resistant to solutions that are nearly neutral[29].

Most usual coordination numbers for its tetravalent compounds are 4,5 and 6, although examples of 7 and 8 coordination are known. While bivalent organotin compounds such as $(C_5H_5)_2Sn$ and $[(Me_3Si)_2CH]_2Sn$ are 2-coordinate, coordination numbers of 6 and 7 have been observed for arene–tin(II) compounds[30].

The properties of 300 organotin compounds have been summarized by Harrison[30].

B. Production and Use of Tin and Tin Compounds

1. Production

At least 35 countries mine or smelt tin. Virtually every continent has an important tin-mining country. In 1990, the leading countries in tin mining and smelting were Brazil, China, Malaysia, Thailand and Bolivia. World mine and smelter production of tin by country are shown in Table 4[31].

2. Consumption and use

In the United States, primary tin consumption in 1990 remained about the same as in the prior year. Only the category of solder and tinning increased significantly, and only the category of brass and bronze declined significantly. The USA consumption of finished products of tin is shown in Table 5[31].

Today's major use of tin is for tin cans used for preserving foods and beverages. Other important uses are solder alloys, bearing metals, bronzes, pewter and miscellaneous industrial alloys.

a. Principal inorganic compounds. Stannous oxide, SnO, is a blue-black crystalline product which is soluble in common acids and strong alkalis. It is used in making stannous salts for plating and glass manufacture.

TABLE 4. World tin mine and smelter production in 1990 (modified from Reference 31)

Country	Mine production (metric tons)	Smelter production (metric tons)
Bolivia	18,000	13,400
China	40,000	40,000
Indonesia	30,200	30,389
Japan	—	816
Malaysia	28,468	50,000
Thailand	14,635	15,512
USSR	15,000	19,700
United Kingdom	4,200	12,000
United States	W[a]	W[a]
Other	29,681	27,737
Total	219,333	249,804

[a] Withheld to avoid disclosing company proprietary data; not included in total.

TABLE 5.   USA consumption of tin, by finished product (modified from Reference 31)

Product	Consumption in 1990 (ton yr^{-1})		
	Primary ('virgin')	Secondary (recovered)	Total
Alloys (miscellaneous)	W^a	W^a	W^a
Babbitt	552	211	763
Bar tin	603	—	603
Bronze and brass	1,160	2,003	3,163
Chemicals	6,275	W^a	6,275
Solder	11,567	4,011	15,578
Tinning	1,707	W^a	1,707
Tinplate	11,750	W^a	11,750
Tin powder	563	W^a	563
White metal	1,045	W^a	1,045
Other	1,394	1,522	2,916
Total	36,616	7,747	44,363

^{a} Withheld to avoid disclosing company proprietary data; included in 'Other'.

Stannic oxide, SnO$_2$, is a white powder, insoluble in acids and alkalis. It is an excellent glaze opacifier, a component of pink, yellow and maroon ceramic stains and of dielectric and refractory bodies.

Stannous chloride, SnCl$_2$, is the major ingredient in acid electrotinning electrolyte and is an intermediate for tin chemicals.

Stannic chloride, SnCl$_4$, a fuming liquid, is used in the preparation of organic compounds and chemicals to weight silk and to stabilize perfumes and colors in soap.

Sodium or potassium stannate is used in alkaline electrotinning baths. Heavy-metal stannates are important in the manufacture of capacitor bodies.

b. Principal organotin compounds. Organotin compounds contain at least one tin–carbon bond and the tin is usually present in the +IV oxidation state with the general formulae R$_4$Sn, R$_3$SnX, R$_2$SnX$_2$, and RSnX$_3$: R is an organic group, while X is an inorganic substituent, commonly chloride.

Based on a report from the London International Tin Institute, Publication No. 665 (1986), the annual consumption of tin metal as organotin chemicals in the U.S.A., Europe and Japan is summarized in Table 6[32]. Chemical formulae of commonly used organotin chemicals are summarized in Table 7[32].

Most of the production of organotin compounds is for the stabilization of polyvinyl chloride (PVC) plastics[29]. Tin stabilizers are effective in preventing the degradation of the plastic during processing or during prolonged exposure to light or heat. The U.S.A. production and consumption of organotin compounds foreseen in 1990 indicate that the organotin compounds produced a little more than 60% will be used as PVC stabilizers[33].

TABLE 6.   Annual consumption of tin metal as organotin chemicals in the USA, Europe and Japan (modified from Reference 32)

Uses	Annual consumption	(ton in Sn base)
Polyvinyl chloride stabilizers	5400	(59%)
Biocidal chemicals	2800	(30%)
Catalysts	400	(4%)
Others	600	(7%)
Total	9200	(100%)

TABLE 7. Industrial applications of organotin compounds (modified from Reference 32)

Application	Organotin compound
	R_3SnX
Agriculture	
fungicides	Ph_3SnX (X = OH, OAc)
antifeedants	Ph_3SnX (X = OH, OAc)
acaricides	$(Cycl\text{-}C_6H_{11})_3SnX$ (X = OH, $-N-C=NC=N$)
	$[(Ph(CH_3)_2CCH_2)_3Sn]_2O$
Antifouling paint biocides	Ph_3SnX (X = OH, OAc, F, Cl, SCS, $N(CH_3)_2$, $OCOCH_2Cl$)
	$Ph_3SnOCOCH_2CBr_2COOSnPh_3$
	$(Bu_3Sn)_2O$
Wood preservative fungicide	$(Bu_3Sn)_2O$, $Bu_3Sn(naphthalenate)$, $(Bu_3Sn)_3PO_4$
Stone preservation	$Bu_3SnOCOPh$, $(Bu_3Sn)_2O$
Molluscicides	Bu_3SnF, $(Bu_3Sn)_2O$
	R_2SnX_2
Heat and light stabilizers	$R_2Sn(SCH_2COO\text{-}Oct)_2$ (R = Me, Bu, Oct, etc.)
for rigid PVC	$(R_2SnOCOCH=CHCOO)_n$ (R = Bu, Oct), etc.
Homogeneous catalysts for	$Bu_2Sn(OCOCH_3)_2$, $Bu_2Sn(OCO\text{-}Oct)_2$,
RTV silicon, polyurethanefoams	$Bu_2Sn(OCOC_{12}H_{25})_2$, $Bu_2Sn(OCOC_{11}H_{23})_2$
and transesterification reaction	
Anthelmintics	$Bu_2Sn(OCOC_{11}H_{23})_2$
	$RSnX_3$
Heat stabilizers for PVC	$RSn(SCH_2COO\text{-}Oct)_3$ (R = Me, Bu, Oct, etc.)
Catalyzers	$(BuSn(O)OH)_n$, $BuSn(OH)_2Cl$

Certain dioctyltin compounds are used for clear PVC materials. Dialkyltin carboxylate and dialkyltin mercaptide heat stabilizers are used in the PVC industry in many application[34].

Trialkyltin and triaryltin compounds possess powerful biocidal properties. These are manifested to a high degree only when the tin atom is combined directly with three carbon atoms, as in trialkyl compounds (R_3SnX); biocidal effects are at a maximum when the total number of carbon atoms attached to Sn is 12. These compounds are used as fungicides, insecticides and as pest control in agricultural applications[29].

Tributyltin acetate, $(C_4H_9)_3SnOOCCH_3$, and bis(tributyltin) oxide, $(C_4H_9)_3Sn\text{-}O\text{-}Sn(C_4H_9)_3$, have been commercialized as antimicrobial agents in the paper, wood preservation, plastics and textile industries[29].

Marine antifouling paints containing tributyltin oxide or tributyltin fluoride are widely used all over the world[29]. However, tributyltin from marine antifouling paints has been shown to have a major impact on the oyster industry in Australia and in France. Concern for these effects has now led to the banning in many countries of the use of tributyltin-based antifouling paints on pleasure boats[35].

Recently, salicylaldoxime with dibutyltin oxide[36], diorganotin bisxanthates [R_2Sn $(S_2COR')_2$] (R = Me, Et, Bu, Ph; R' = Et, i-Pr, Hex)[37], diaminoalkyl complexes of tin halides[38], dibutyltin and diethyltin monofluorobenzoates[39], diorganotin(IV) dipeptide complexes[40] and dicyclohexyltin derivatives of dipeptides[41] were investigated for antitumor activities.

Since the discovery of the anticancer and antitumor activity of platinum complexes, there has been an effort to identify organic complexes of other metals that possess similar activities. Gielen and coworkers[39] state that the antitumor activities of the di-n-butyltin(IV) and diethyltin(IV) fluorobenzoates ($FC_6H_4COO)_2SnR_2$ and $(FC_6H_4\text{-}COOSnR_2)_4O_2$, R = Et, Bu are satisfactory.

C. Concentration and Speciation of Organotin Compounds in the Natural Environment

1. Total tin concentration

The abundance of tin in sea water is below $3 \mu g l^{-1}$. Typical abundance of tin in the human (adult) body is 17–130 mg: tin distribution is 25% in skin and lipo-tissues; 3.2% in red blood cell; 0.8% in blood plasma; the remainder in soft tissues[9].

Tin concentrations in algae collected near the Scripps Institution of Oceanography, Calif. and in coastal marine sediments from Narragansett Bay, USA, were determined by Hodge and coworkers[42]. Tin concentrations of the blades of *Pelagophycus porra, Macrocystis pyrifera* and *Eisenia arborea* were 0.71 ± 0.01, 0.83 ± 0.01 and $1.06 \pm 0.02 \mu g Sn g^{-1}$ dry wt., respectively. Tin concentrations in the core of sediments are shown in Table 8.

It is evident from Table 8 that tin concentration in sedimentary material deposited in the last 50 years in the Narragansett Bay core are significantly higher than those in pre-1900 sediments. This is probably a consequence of the increased use and subsequent dispersion of tin by human activity.

2. Organotin concentration

Braman and Tompkins first reported methylated tin compounds in environmental materials[43]. Saline water, estuary water, fresh water, rain water and tap water were analyzed for methyltin compounds: tin levels were at $ng l^{-1}$. Average total tin concentration of human urine (11 samples) was $1 \mu g Sn l^{-1}$, and those of methyltin, dimethyltin and trimethyltin were 90, 73 and $42 ng Sn l^{-1}$, respectively. Methyltin compounds were also observed in shell samples at the $0.1 ng g^{-1}$ level. About 17–60% of the total tin was present in monomethyltin form[43].

Ashby and coworkers[44] reviewed experimental data on organotin compounds from environmental matrices found in the world during 1978–1988. According to this excellent review, reported tin levels (in $ng kg^{-1}$ or $ng l^{-1}$ units) in environmental matrices are as follows. In sea and lake waters and sediments: < 18 MeSn, < 63 Me_2Sn, 22–1220 BuSn and 10–1600 Bu_2Sn[42]; fresh and sea waters and human urine: below ppt level[43]; sea water: 200 Me_2SnH_2, 400 Me_3SnH, 480 Me_4Sn and 50–100 $BuSnH_3$[45]; water: 300–1200 MeSn, 100–400 Me_2Sn and not detected Me_3Sn[46]; lake, river and harbor

TABLE 8. Tin in sediments from Narragansett Bay, USA (modified from Reference 42)

Depth in core (cm)	Depositional period[a]	$\mu g Sn g^{-1}$, dry wt.[b]
1–2	1972–1973	20
4–5	1969–1970	16
7–8	1966–1967	15
11–12	1962–1963	14
14–15	1959–1960	13
25–27	1935–1947	13
39–41	1900	6
50–53	pre-1900	2
79–84	pre-1900	1

[a] Determined by unsupported [210]Pb.
[b] Precision of 5%.

waters: 200–8480 BuSn, 100–7300 Bu_2Sn and 100–2910 Bu_3Sn[47]; sediment: 4.7×10^5 Bu_3Sn and 1.13×10^6 Bu_3Sn[48]; water and sediment: 1–10 and 2,000–16,000 Bu_3Sn, respectively[49]; sediment: 100–500 Bu_3Sn[50]; seawater: 10–60 BuSn, 50–150 Bu_2Sn and 20–160 Bu_3Sn[51]; harbor water and sediment: 10–200 and 80–1280 Bu_3Sn, respectively[40]; seawater: 100–800 organotin[52]; estuary water: not detected Bu_3Sn and < 212 BuSn[53]; sediment and fish: $< 10,000$ and 10–240 Bu_3Sn, respectively[54]; water: 68 Bu_3Sn and 108 Bu_2Sn[55]; estuary water: 66 Bu_3Sn[56]; sediment: 12–28 BuSn, 1.2–15 Bu_2Sn and 3.0–30 Bu_3Sn[57]; seawater: < 930 (surface) and < 550 (bottom)[58]; water, sediment and sewage sludge: < 15, 140 and 300–6000 (dry), respectively[59].

Organotin levels of the sediments, especially in tributyltin, are found to be very high compared with those of waters. It has been shown that Bu_3Sn^+, like other organometallics, is adsorbed rapidly by marine and other sediments. The sediment therefore represents a sink for Bu_3Sn^+ leached from the antifouling coating of boats.

Concentrations of inorganic tin and organotin compounds in some environmental waters in the USA are shown in Table 9[42].

Butyltin and dibutyltin compounds are found in Lake Michigan water at $0–1,600$ ng l^{-1}, also generally higher than the methyl species or inorganic tin. The higher concentrations of inorganic tin and the butyltin compounds in the upper 20 m may be indicative of atmospheric input. Butyltin compounds are not detected in surface waters in San Diego Bay. No organotin compounds were detected in both San Francisco Bay and California Coast sea off waters[42].

Ashby and Craig analyzed some U.K. sediments for tributyltin (TBT) and dibutyltin (DBT) by their newly developed analytical method[60]. TBT and DBT concentrations (μg g^{-1}) observed in river sediments were as follows: 0.46–5.7 and 0.78–3.95 in River Hamble, 0.21–1.12 and ND–1.20 in River Beaulieu, 0.28–4.30 and ND–2.62 in River Lymington, 0.74–1.06 and 0.99–1.23 in Plymouth Sutton Harbor, 0.63–1.83 and 0.16–5.25 top sediment in River Dart, 1.35–4.84 and 2.55–3.40 bottom sediment in River Dart, 0.81–0.48 and 0.96 top sediment in River Teign and 0.67–0.83 and 0.48–0.66 bottom sediment in River Teign[60].

TABLE 9. Tin(IV) and organotin compounds[a] in some environmental waters (modified from Reference 42)

Location	Collection date	ng Sn l^{-1}				
		Sn(IV)	MeSnCl$_3$	Me$_2$SnCl$_2$	BuSnCl$_3$	Bu$_2$SnCl$_2$
Lake Michigan off	9/14/78					
Grand Haven, Mich.						
10 m		490 ± 25	13 ± 2	0	1220 ± 60	1600 ± 80
18 m		290 ± 15	12 ± 2	10 ± 2	840 ± 40	1500 ± 70
32 m		112 ± 6	18 ± 2	63 ± 2	22 ± 3	10 ± 3
62 m		84 ± 4	6 ± 1	7 ± 1	56 ± 4	64 ± 4
San Diego Bay, Calif.	10/5/78					
(surface water)[b]		16 ± 1	4 ± 1	31 ± 1	0	0
San Francisco Bay, Calif.,						
15 m[c]	10/14/78	2.7 ± 0.3	0	0	0	0
California Coast	10/11–12/78					
off, San Francisco[d]		0.5 ± 0.3	0	0	0	0

[a] Organotin compounds are assumed to be chlorides for the calculation; 0 = not detected.
[b] Average in five locations.
[c] Average in two locations.
[d] Average in three locations.

In 1987, in collaboration with the New South Wales (NSW) State Pollution Control Commission, measurements were made on water samples from Sydney Harbor and the nearby Georges River estuary. Oyster culture (*Saccostrea commercialis*) in eastern Australia is a multimillion dollar industry. They showed concentrations in water ranging from 8 to 220 ng Sn l^{-1} in areas of high boating activity, including a naval base, while in the majority of samples concentrations were below 45 ng Sn l^{-1} and nearer 10 ng Sn l^{-1} in uncontaminated sites[35].

Batley and Scammell[35] undertook to survey tributyltin in oysters from tin-contaminated and slightly contaminated sites. While the the latter had tissue concentrations below 2 ng Sn g^{-1}, values between 80 and 130 ng Sn g^{-1} were obtained in areas of high boat density. A set of oysters from Sand Brook Inlet contained 350 ng Sn g^{-1} (Table 10)[35]. These data added strength to the case for government action to restrict TBT usage. A ban on the sale and usage of TBT-based antifouling paints on boats under 25 m in length was instituted in NSW in 1989, and similar bans are in place in most States in Australia.

Batley and Scammell[35] measured tributyltin in other Australian waters. In most instances, concentrations below 20 ng Sn l^{-1} have been obtained, except when the samples had been collected in close proximity to large surface areas of TBT-antifouled boats. Since the banning of tributyltin-based antifouling paints in New South Wales in 1989, sites where concentrations near 45 ng Sn l^{-1} were previously obtained now show much lower values. Sediments have also been investigated from similar sites. Whilst high concentrations of TBT (2–40 mg Sn g^{-1}) have been obtained from samplings close to marinas, it was typical to find values nearer 1 ng Sn g^{-1} in sandy sediments and 50 ng Sn g^{-1} in silty material. It is likely that the very high values might include paint flakes from the hydroblasting of paint on marina slipways where waste waters were not being contained[35].

Maguire[61] reviewed studies concerning the aquatic chemistry, fate and toxicity of tributyltin. He summarized investigations of the occurrence and persistence of tributyltin and its degradation products in water and sediments in Canada. Tributyltin was mainly found in areas of heavy boating or shipping traffic, consistent with its use as an antifouling agent. In about 8% of the 269 locations across Canada at which samples were collected, tributyltin was found in water at concentrations which would cause chronic toxicity in a sensitive species, such as rainbow trout. Estimations of the half-life of biological degradation of tributyltin in fresh water and in sediments in Canada are from a few weeks to 4–5 months[61].

Schebe and coworkers[62] also reported experimental results on concentration of methyltin and butyltin compounds in environmental waters and sediment samples.

TABLE 10. Tributyltin in Australian aquatic biota (1989) (modified from Reference 35)

Species	Site	TBT (ng Sn g^{-1}, fresh wt.)
Saccostrea	Upper Georges River, NSW	40–128
commercialis	Lower Georges River, NSW	14–44
	Coba Bay, Hawkesbury River, NSW	7
	Sand Brook Inlet, Hawkesbury River, NSW	350
	Wallis Lake, NSW	2
	Botany Bay, NSW	15
Crassostrea gigas	Upper Georges River, NSW	175
Ostrea angasi	Port Phillips Bay, VIC	<1
Mytilus edulis	Cockburn Sound, WA	18
	Near slipway, Cockburn, WA	166
Pecten alba	Port Phillips Bay, VIC	5

Forsyth and coworkers[63] found organotin compounds in fruit juices. Thus juices from apple and passion fruit contained low or undetectable levels of butyl-, phenyl- and octyltin compounds. Octyltins were present in juices sold in containers constructed of poly(vinyl chloride) but not in those made from poly(ethyleneterephthalate). Therefore, the likely source of the octyltin was the PVC container material[63].

High tributyltin (TBT) concentrations were observed in sediments and selected shell-fish from Suva Harbor, Fiji. Sediments in the immediate vicinity of foreshore slipways and boat yards were exceedingly contaminated, with a maximum observed level of 38 μg TBT-Sn g^{-1}. Concentrations were much lower in surfacial sediments from commercial docks and yacht mooring areas, namely 16–83 ng TBT-Sn g^{-1}. Concentrations as high as 3180 ng TBT-Sn g^{-1} were found in mangrove oysters[64].

Morita[65] determined the concentration of tributyltin and triphenyltin compounds in shellfish tissue (shortnecked clam) collected in Tokyo Bay, Japan in 1980–1987. The results are illustrated in Figure 5. The contamination by organotin compounds reflected in the shellfish tissue has been gradually increasing since 1980. In Japan, the production and use of triphenyltin have been stopped in 1990 and that of tributyltin has been restricted since 1991. The question may be raised as to whether the regulations effectively reduce the environmental level of these compounds or whether their effect appears delayed. It seems to be necessary to make observations for longer periods in order to identify the decline of these compounds in the environment[65].

In order to observe the decline of triphenyltin level in fish, The Environment Agency of the Japanese government collected fishes from 35 sites all over the Japanese coasts and carried out a survey of the triphenyltin concentration[66]. Experimental results are illustrated in Figure 6.

The average limit of a tolerable daily intake was fixed by the World Health Organization (WHO) as 0.5 μg kg^{-1} day^{-1} for triphenyltin and this figure corresponds to an allowed concentration in fish of 0.25 μg g^{-1}. About half the triphenyltin concentrations shown in Figure 6 are higher than the limit of 0.25 μg g^{-1}. The decline of these triphenyltin concentrations is expected since the banning.

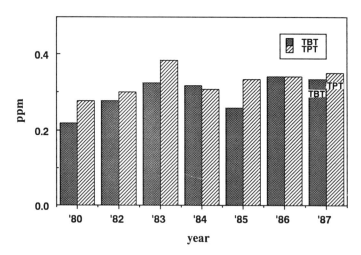

FIGURE 5. Tributyltin (TBT) and triphenyltin (TPT) levels in shortnecked clam collected in Tokyo Bay (1980–1987) (modified from Reference 65)

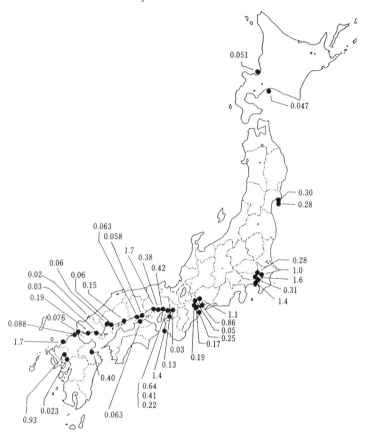

FIGURE 6. Triphenyltin chloride (TPTCl) level in fish (gray mullet, sea bass, young bass, floun-
der, greenling, crusian carp, flatfish and dace) collected from all over Japanese coasts (modified from
Reference 66) (μg TPTCl g^{-1}, wet wt.)

D. Methylation of Tin Compounds

There have been numerous reports that organic tin compounds were found in natural
waters, sediments, aquatic plants and shellfish. It is likely that the organic tin compounds
in the environment originate from methylation (biological or abiological) of inorganic
tin compounds, and from anthropogenic antifouling paints, PVC stabilizers and biocide
inputs[35,67,68]. Tin is similar to most organometallics in that the methylated derivatives
are more toxic than the inorganic metal substrates; arsenic is an exception in this respect.
Fortunately, organic tin compounds may be converted to other tin species. For instance,
tributyltins in aqueous solution are decomposed under UV irradiation, and furthermore, the
persistence of tributyltin compounds in freshwater ecosystems is controlled by microbial
degradation[40]. This suggests that organotin compounds are degraded, and perhaps turned
into less toxic inorganic tin species in the environment. On the other hand, methyltin
compounds, which are rarely used in industry, have been found in marine sediments

and natural waters as described before, so there might be a possibility of formation of methyltin species in the environment[69].

There is compelling evidence for the existence of methyltins in the environment, but it is not certain that methylation of naturally occurring inorganic tin compounds account for their presence. However, various biological or chemical processes have been described under laboratory conditions in which methyltin products have been formed from inorganic tin starting materials[70].

Incubation experiments carried out with marine sediment converted $SnCl_4$ to methyl tin species[71]. Me_2Sn^{2+} and Me_3Sn^+ products were identified from all cultures whereas $MeSn^{3+}$ was occasionally seen. Sterile controls and sediments did not produce methyltins[71], suggesting that inorganic tin(II) was biologically methylated.

Ashby and Craig[72] reported that $MeSn^{3+}$ and small amounts of Me_2Sn^{2+} are also produced when a baker's yeast (*Saccharomyces cerevisiae*) is incubated with tin(II) compounds including tin(II) oxalate, tin(II) sulfide and various tin amino acid complexes. Tin(II) chloride and tin(II) amino acid complexes were methylated by methyl-cobalamin, under conditions of chloride ion concentrations and pH relevant to the natural environment[73]. The main identified product of all reactions was monomethyltin.

Rapsomanikis and coworkers observed methylation of tin(II) and lead(II) in sediments by carbanion donors[74]. A factorial experimental design determined separate and combined effects of $MeCoB_{12}$ (methylcobalamin) and $Me_2Co(N_4)^+$ (a methylcobalamin model) (6) on methylation of Sn(II) in sediment matrices. Experimental results for methylation of tin are shown in Table 11.

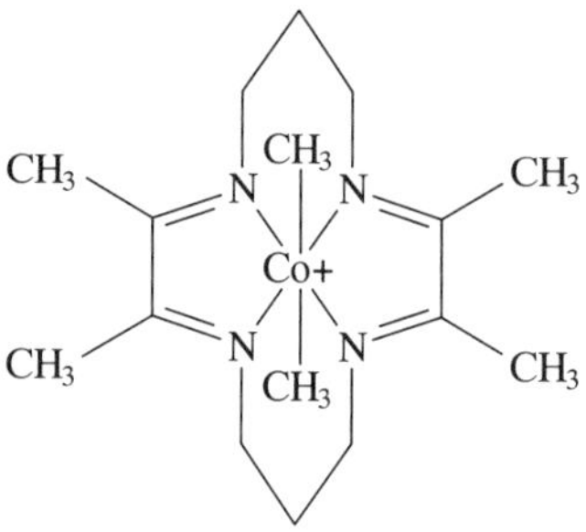

(6)

$[Me_2Co(N_4)]^+$

TABLE 11. Experimental design and yields of methyltin compounds[a] (modified from Reference 74)

Carbanion donor (μmol)		Percent yield[b] (μg)[c]				
$Me_2Co(N_4)^+$	$MeCoB_{12}$	$MeSn^{3+}$	Me_2Sn^{2+}	Me_3Sn^+	Me_4Sn	Total
10	10	0.03(0.15)	0.08(0.50)	2.86(17.0)	0.19(1.13)	3.16(18.8)
10	0	nd	nd	1.23(7.30)	0.19(1.13)	1.42(8.53)
0	10	nd	nd	nd	nd	nd
0	0	nd	nd	nd	nd	nd

[a] Each experiment contained 5 μmol (595 μg) Sn(II) as $SnCl_2$.
[b] Based on Sn(II) added.
[c] Limit of detection is 20 pg; nd means not detected.

As shown in Table 11, total methyltin yields ranged from 1.4% to 3.2%; no methyltin products occurred in the absence of $Me_2Co(N_4)^+$. Rapsomanikis and Weber also observed methylation of tin by methyl iodide[75].

Quevauviller and coworkers[76] studied the occurrence of methylated tin and dimethylmercury compounds in a sediment core from Sepetibaa Bay, Brazil, in order to investigate possible methylation pathways in a mangrove environment. The results have revealed that the physicochemical conditions existing in this type of environment (high organic inputs, anaerobic conditions, microbial activity, etc.) account for high methyltin concentrations (mono-, di- and trimethyltin) in the sediments, which are dependent upon the total load of metal released (e.g. anthropogenic sources).

Weber and Alberts[67] studied methylation of tin(IV) chloride by hydroponically incubated *Spartina alterniflora* plants which had been collected from salt marshes, 10–25 cm in height. Incubation of *S. alterniflora* plants for five days in tin-amended Hoaglund's solution increased their inorganic tin and methyltin concentrations. Compared to control plants, roots, but not leaves, accumulated considerable inorganic tin from solution. Mono- and dimethyltin concentrations in leaves of treated plants were not significantly enhanced from concentrations in the control plants. However, trimethyltin, which was not detected in control leaves, had the highest concentration of any methyltin compound in treated plants. Weber and Alberts proposed that methyl iodide is very likely involved in the methylation process. The reaction (equation 1), which is well known for methyl iodide, readily occurs in model systems[75]. However, further methylation to di- or trimethyltin is more difficult to explain because methylation by carbocation donors would require preceding reduction of methyltin(IV) compounds, however methyltin(II) compounds were not found. The most likely mechanism by which this can occur is a rearrangement process (equations 2 and 3).

$$Sn(II)^{2+} + MeI \longrightarrow MeSn(IV)^{3+} + I^- \tag{1}$$

$$2\ MeSn(IV)^{3+} \longrightarrow Me_2Sn(IV)^{2+} + Sn(IV)^{4+} \tag{2}$$

$$2\ Me_2Sn(IV)^{2+} \longrightarrow Me_3Sn(IV)^+ + MeSn(IV)^{3+} \tag{3}$$

E. Bioaccumulation, Toxicity and Environmental Effects of Organotins

Tin is comparable in its toxicological behavior to lead and mercury. Bivalent tin compounds generally are more toxic than the tetravalent compounds. Furthermore, organic tin compounds are more toxic than inorganic ones and the trialkyl analogs (triethyltin, trimethyltin, tributyltin) are the most toxic. As the number of carbon atoms attached to tin increases, the toxicity of the organic tin compounds rapidly declines[9,77].

Since the tragic human exposure to diethyltin salts for the therapy of an infectious skin disease by Staphylococcus in France in the 1950s, the toxic and biochemical effects of many of these derivatives have been explored. Di- and tri-ethyltin salts have been demonstrated to have pronounced effects on intermediary metabolism in brain and liver. These effects have been suggested to be due to inhibition of the mitochondrial functions[9,27].

Rey and coworkers[78] reported methyltin intoxication in six chemical workers exposed to Me_2SnCl_2 and Me_3SnCl. After a latent period of 1–3 days, the first symptoms occurred, including headache, tinnitus, deafness, impair of memory, disorientation, aggressiveness, psychotic behavior, syncope, loss of consciousness and, in the most severe cases, respiratory depression requiring ventilatory assistance. Increased tin excretion was detected in the urine of all patients, particularly those most ill. The patient with the highest tin levels died 12 days after the initial exposure.

Oral LD_{50} values against experimental animals for organotin compounds were summarized by Zuckerman[79].

1. Effect on microorganims

Uptake of inorganic tin ($SnCl_4$) by an alga, *Ankistrodesmus falcatus*, occurred very rapidly[80]. The alga accumulated 2.3 and 4.1 mg Sn g^{-1} dry weight cells after 2 and 60 min of incubation, respectively, in the presence of 46 μg Sn l^{-1} of medium.

Huang and coworkers[81] investigated toxicity and bioaccumulation of organotin compounds. Toxicity (96-h-EC_{50}) and relative toxicity of eight organotin compounds, as well as organolead and organoarsenic compounds for growth of freshwater alga, *Scenedesmus obliquus*, are shown in Table 12.

Accumulation of organotin compounds in freshwater alga, Sc (*Scenedesmus obliquus*), and marine alga, Du (*Dunaliella salina*), is shown in Table 13. It was found that freshwater and marine algae accumulated organotin at a very high level[81].

Röderer compared biological effects of inorganic and organic compounds of mercury, lead, tin and arsenic upon the freshwater alga, *Poteriochromonas malhamensis*[82]. The order of toxicity (LC_{72h}) was determined as follows: MePbCl (0.2 μM) > $HgCl_2$(25 μM) > Me_3PbAc(50 μM) > Me_3SnCl(80 μM) > $PbCl_2$(1,500 μM) > $SnCl_4$(3,500 μM) > K_2HAsO_4(50,000 μM) > $NaMe_2AsO_2$(400,000 μM). The organometallics proved to be considerably more toxic to the alga than their inorganic forms, whereas the reverse was found for the arsenic.

The relative toxicity of organotin compounds to alga *Scenedesmus obliquus* is shown in Table 12. Pettibone and Cooney investigated the toxicity of methyltins (MMT: $MeSnCl_3$; DMT: Me_2SnCl_2; and TMT: Me_3SnCl) to microbial populations in estuary sediments in

TABLE 12. 96-h-EC_{50} and relative toxicity of organotin compounds for growth of *S. obliquus* (modified from Reference 81)[a]

Organotin	TBT	TPT	DBT	Cy₂MTA	TML	Cy₂MTB	DPT	TMT	DMA	DMT
EC50 (μg l^{-1})	3.4	5.6	16.7	24.3	24.3	33.2	256	389	822	1118
Relative toxicity	329	200	67	46	46	34	4.4	2.9	1.4	1

[a] TBT (tributyltin), TPT (triphenyltin), DBT (dibutyltin), Cy_2MTA (dicyclohexylmethyltin acetate), TML (trimethyllead acetate), Cy_2MTB (dicyclohexylmethyltin isobutyrate), DPT (diphenyltin), TMT (trimethyltin), DMA (dimethylarsine), DMT (dimethyltin).
96-h-EC_{50} = median effective concentration in 96 h culture.

TABLE 13. Accumulation of organotin compounds in algae over seven days (modified from Reference 81)[a]

Organotin	Algae	Initial conc. (μg l^{-1})	Tin in algae (μg g^{-1})	Residue in supernatant (μg l^{-1})	BCF ($\times 10^5$)
TBT	Sc	1.0	16.6	not detected	>3.32
TPT	Sc	3.0	114	1.0	1.14
TBT	Du	1.0	17.4	not detected	>3.48

[a] Detection limit for organotins in supernatant is 0.05 $\mu g l^{-1}$.
BCF = (conc. of TBT in algae [$\mu g g^{-1}$])/(conc. of TBT in supernatant medium [$\mu g g^{-1}$]). Sc denotes *Scenedesmus obliquus*; Du denotes *Dunaliella salina*.

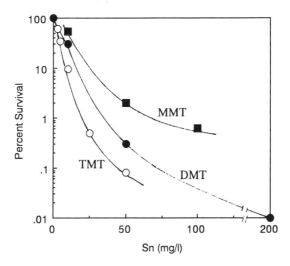

FIGURE 7. Effect of methyltins on total viable counts of natural populations from sediments in Boston Harbor (modified from Reference 68). Tin added as: ■, CH_3SnCl_3(MMT); •, $(CH_3)_2SnCl_2$(DMT); ○, $(CH_3)_3SnCl$(TMT)

Boston Harbor, MA, U.S.A.[68]. All three were toxic to organisms in these sediments, and the order of the toxicity was TMT > DMT > MMT as shown in Figure 7.

Shinoda[83] collected water samples from 10 different stations including fresh, brackish and saline water environments in Okayama Prefecture, Japan. The samples were diluted, inoculated on agar plates containing different concentrations of tributyltin (TBT) and incubated. The plaque-forming-unit (PFU) was counted. Addition of TBT less than 1 μM did not reduce PFU (bacterial growth). Decrease of PFU was observed by addition of more than 10 μM TBT, whereas some bacterial colonies could survive even at 100 μM TBT. Dominant genera of TBT-resistant bacteria were *Enterobacteriaceae, Pseudomonas* and *Alcaligenes*.

Kumari and coworkers[84] synthesized a trimethyltin(IV) derivative of the type Me_3Sn (SCZ) (where SCZ is the anion of a semicarbazone ligand) and evaluated its antimicrobial effects on different species of pathogenic fungi and bacteria. The results showed that the methyltin compound is highly active against these pathogens even below 200 $\mu g\,cm^{-3}$.

Structure–toxicity relationships of organotin compounds on algae were summarized by Wong and coworkers[85].

2. Effect on aquatic organisms

Recently, Horiguchi and coworkers studied the effects of organotin compounds [bis(tributyltin)oxide, tributyltin chloride, tributyltin fluoride, tributyltin acetate, triphenyltin chloride and triphenyltin acetate] on marine (algae, crustacea, bivalves, gastropods and sea urchins) and freshwater (algae, molluscs, benthos and amphibia) organisms[86]. Acute toxicity of tributyltin compounds to the above organisms, and subacute and chronic toxicities to marine crustacea were reported. They also studied the effect of tributyltin and triphenyltin on imposex (impotence for propagation), in gastropods. In Japan, imposex has so far been confirmed in 16 species of gastropods including *T. clavigera* and *T. bronni*. Judging from flow-through exposure experiments carried out for 3 months, imposex was induced in adult females in *T. clavigera* at environmental concentrations of approximately 1 ng l^{-1} of TBT[86].

3. Effect on nucleic acids and genotoxicity

Organic tin compounds have the effects of neurotoxicity, immunotoxicity and genotoxicity; however, studies on the genotoxicity of organic tin compounds are comparatively scarce. It is important to study the genotoxicity of organic tin compounds, because those that show genotoxicity have the potency to induce mutations or cancer.

Hamasaki and coworkers[87] investigated the genotoxicity of 14 organic tin compounds (methyl-, butyl- and phenyltins) and inorganic tin ($SnCl_4$) on *Escherichia coli* and on *Bacillus subtilis*. Dibutyltin dichloride, tributyltin chloride, tributyltin chloride, bis(tributyltin)oxide, dimethyltin dichloride and trimethyltin chloride were all found to be genotoxic.

Piro and coworkers observed that R_2SnCl_2 and R_3SnCl (R=Me, Et, Bu, Oct, Ph) reacted with aqueous native DNA. They suggested that the genotoxic effects of these chemicals might be ascribed to coordination by the phosphodiester group of the nucleic acid[88]. The effect of trimethyltin on RNA was described by Veronesi and coworkers[89], and by Usta and Griffiths[90].

Effects of dibutyltin compounds on ATPase were investigated by Griffiths and coworkers[91,92].

4. Neurotoxicity

The neurotoxicity of trimethyltin has been investigated extensively in recent years. According to these investigations, trimethyltin has toxic effects on the central nervous system (CNS) including the pyriform cortex, amygdaloid nucleus, neocortex, olfactory bulbs and hippocampus. In the hippocampus there occurs prominent destruction of the hippocampus subfield CA1 region pyramidal cells, focal damage to the dentate gyrus, mild necrosis of the CA3 region and sparing of the CA2 region. Trimethyltin has been shown to produce neuronal necrosis in the CNS, while triethyltin has been shown to produce intramyelinic vacuolation and oedema in the CNS[77,93−102]. Trimethyltin-induced gross behavioral changes include hyperactivity, aggression, tremor, self-mutilation and spontaneous motor seizure activity[93].

The biochemical mechanism through which organic tin compounds induce cell damages remains unclear. However, several reports suggest that trimethyltin acts as a Cl-/OH-exchanger in mitochondrial membranes and that it inhibits mitochondrial ATP synthesis, increases Ca^{2+} concentration and reduces neurotransmitter uptake. Since oxygen reactive species, such as superoxide anion, hydrogen peroxide and hydroxyl radicals, are believed to be initiators of peroxidative cell damage, LeBel and coworkers[103] studied the organometal-induced increase in oxygen reactive species in mice and found that the levels of oxygen reactive species were elevated by trimethyltin injection. This suggests that oxidative damage may be a mechanism underlying the toxicity of organic tin compounds.

Effects of triethyltin on brain octamines and their metabolism in the rat has been investigated[104].

The following acute toxicological data are available for organic tin compounds: triethyltin, Et_3Sn, LD_{50} 10–12.5 mg kg^{-1} (BALB/c mice, oral)[105], trimethyltin, Me_3Sn, LD_{50} 2.25 mg kg^{-1} (CFW mice, oral)[105], dimethyltin dichloride, Me_2SnCl_2, LD_{50} 74 mg kg^{-1} (rat, oral)[34], dibutyltin dichloride, Bu_2SnCl_2, LD_{50} 126 mg kg^{-1} (rat, oral)[34], methyltin trichloride, LD_{50} 1370 mg kg^{-1} (rat, oral)[34], butyltin trichloride, LD_{50} 2300 mg kg^{-1} (rat, oral)[34], octyltin trichloride, LD_{50} 3800 mg kg^{-1} (rat, oral)[34], dioctyltin dichloride, LD_{50} 7000 mg kg^{-1} (rat, oral)[34].

F. Fate of Organotin Compounds in the Environment

Bis(tributyltin)oxide in antifouling paint was found to change to tributyltin chloride and an unknown organotin species by Allen and coworkers[106]. The organotin species appear to be held strongly within the paint film. In the case of triphenyltin chloride and triphenyltin acetate, evidence of dephenylation to form diphenyltin and monophenyltin compounds has been obtained.

As mentioned before, the diethyl-, triethyl- and tetraethyltin compounds were dealkylated to give rise to ethylene as a major product and ethane as a minor product with liver microsomal fractions, NADH and oxygen[27]. Tetraethyltin was found to be converted into triethyltin salts in significant concentrations.

Dowson and coworkers studied partitioning and sorptive behavior of tributyltin (TBT) and its degradation products, dibutyltin (DBT) and monobutyltin (MBT) in the aquatic environment[107]. The determination of the sorptive behavior of TBT is necessary in order to understand its fate in freshwater and estuary environments. The results indicate that MBT and TBT in freshwater will be partitioned to a lesser extent towards the particulate phase, whereas DBT exhibits a 50:50 partitioning between the particulate and solution phases. In estuary waters, MBT will almost exclusively be adsorbed on the particulates, while TBT will be predominantly in the solid-phase fractions but 10–30% may remain in solution. DBT, in contrast, is solubilized in estuary waters. The order of adsorption to particulate matter for butyltins is MBT > TBT > DBT[107].

Contamination of water and sediment in Toronto Harbor, Canada by the highly toxic tributyltin species (Bu_3Sn^+) and its less toxic degradation products, the dibutyltin species (Bu_2Sn^{2+}), butyltin species ($BuSn^{3+}$) and inorganic tin was demonstrated[40]. The main factors limiting the persistence of Bu_3Sn^+ in aquatic ecosystems were photolysis in water and biological degradation in water and sediment. The half-life of Bu_3Sn^+ was likely to be at least a few to several months.

Soderquist and Crosby[41] reported that aqueous triphenytin hydroxide (Ph_3SnOH) (solubility in water: $1.2 \ mg \ l^{-1}$ at pH 7–9 and $6.6 \ mg \ l^{-1}$ at pH 4.2) was readily degraded by homolytic cleavage of the tin–carbon bond to diphenyltin oxide when exposed to sunlight or to UV light in a laboratory photoreactor. While neither tetraphenyltin, monophenyltin species nor inorganic tin were detected as products, the formation of a water-soluble, nonextractable organotin polymer was indicated[41].

In Japan, the amounts of production and discharge to the environment of tributyltin compounds were considered much bigger than those of triphenyltin compounds. Nevertheless, the observed concentration in marine animals is higher in triphenyltin than tributyltin as mentioned before (Figure 5)[65]. One possible reason for this may be the long life-time of triphenyltin compounds in the environment. Morita[65] examined the photochemical degradation of both compounds using a UV light source. The tributyltin compound was easily decomposed giving rise to debutylated products, while the triphenyltin compound showed less decomposition.

G. Health Effect Assessment and Safety of Organotins

Recently, pollution by organotin compounds was extensive in the environment and its effects on human health are feared. Organotin compounds such as tributyltin and triphenyltin, which are widely used in antifouling paints for ships and fishing nets, are very toxic to aquatic organisms[38]. These compounds may be accumulated in various organisms through the food chain.

Penninks[108] assessed the evaluation of a safety factor to determine a Tolerable Daily Intake (TDI) value for the environmental contaminant bis(tributyltin)oxide (TBTO). The most sensitive parameter of TBTO toxicity was shown to be on lymphoid organs and lymphoid function. Subsequently, safety factors were derived from the published data on interspecies and interindividual differences in both kinetics and dynamics of TBTO. Lack of information on human data concerning the nature of toxicity as well as kinetics and dynamics of TBTO finally resulted in an estimated safety factor of 100. A TDI of 5 or 0.25 mg kg^{-1} body weight per day was assessed based on reductions of lymphoid organ weights or lymphoid function, respectively.

Generally, tin compounds of 3–4 mg day^{-1} are ingested in the human body, and are absorbed poorly from the digestive tracts. 40% of the tin are excreted via urine and feces. Tin compounds tend to accumulate in liver, kidney and bone. By the oral route, high levels of tin compounds cause gastrointestinal symptoms such as nausea, vomiting and abdominal pain and headache.

The average limit of a tolerable daily intake (TDI) fixed by the World Health Organization (WHO) is 0.5 μg kg^{-1} day^{-1} for triphenyltin and this figure corresponds to a permissive concentration in fish of 0.25 μg g^{-1}. TDI of 1.6 μg kg^{-1} day^{-1} for tributyltin oxide is assessed by the Ministry of Health and Welfare, Japan[66].

H. Summary

Tin is shown to be methylated by organisms and other environmental factors. Organic tin compounds are much more toxic than the inorganic ones. This increase in toxicity due, e.g., to methylation is attributable to the increasing membrane permeability of the methylated metal compounds and to the remaining sensibility to react with sulfhydryl groups of proteins, especially of enzyme proteins.

Tin is comparable in its toxicological behavior to lead. Trialkyl species are the most toxic in both cases. While the toxicity of alkyltin species declines with alkyl chain length, that of alkyllead species increases with alkyl chain length.

Organic tin compounds have been used in various fields, such as antifouling paints, stabilizers of PVC, agricultural chemicals and so on, while lately anticancer and antitumor activities of organic tin compounds are also studied[38,39].

Pollution by organic tin compounds in the environment has become extensive, and the effects on various ecosystems, including humans, are feared. Fortunately, the use and production of tributyltin and triphenyltin compounds as antifouling paints have been restricted in several countries. These organotin compounds are degraded sooner or later by both biotic and abiotic processes. In many countries, environmental monitoring of tributyltins and phenyltins has been carried out. We need to observe carefully the trend of the decline of organotin level in the environment. On the other hand, the development of an alternative to organotin compounds is also required.

IV. LEAD

A. Introduction

Lead is universally present in plants, but is a nonessential element for plants and animals. Lead is a soft, heavy metal of bluish color, which tarnishes to dull gray and is the most corrosion resistant of the common metals[109]. Lead has relatively high resistance to attack by sulfuric and hydrochloric acids but dissolves slowly in nitric acid. Lead forms many salts, oxides and organometallic compounds[110]. Lead is one of the oldest known metals, dating from about 3000 B.C. The Romans used lead for water pipes, while in ancient Egypt lead compounds were employed for glazing pottery and ornamental objects, a purpose for which they are still used today.

TABLE 14. Some typical organolead compounds[111]

Compounds	MW	mp (°C)	Uses
PhPb(OAc)₃	461.43	102	Polyurethane foam catalyst
Bu₂Pb(OAc)₂	439.50	103	Anthelmintic for tapeworm
Ph₃Pb(OAc)	497.55	205	Toxicant for ship bottom paints
Et₄Pb	323.45	−137	Anti-knock agent in gasoline

Organolead compounds comprise the broad class of structures that are characterized by at least one carbon atom bonded directly to a lead atom.

In general, organolead compounds are well-characterized substances that are quite stable at room temperature. However, they are thermally the least stable of the organometallic compounds of group IV. Most organolead compounds are liable to severe decomposition on heating to 100–200 °C, but they are not explosive[111].

Stable organolead compounds are mostly derived from tetravalent lead and most fall within the four basic categories shown in Table 14, namely $RPbX_3$, R_2PbX_2, R_3PbX and R_4Pb (where R is an alkyl or aryl group, and X is a halogen, OH or an acid radical).

Tetraethyllead was the first known organolead compound, synthesized in 1853. Its antiknock properties were discovered in 1922, and since then its consumption has rapidly increased.

The properties of 216 organolead compounds have been summarized by Harrison[112].

B. Production and Use of Lead and Lead Compounds

According to the International Lead and Zinc Study Group (ILZSG) statistics, world refinery production was 5.9 million tons in 1990[109]. Annual productions of the main countries in 1987 and 1990 are shown in Table 15.

The leading countries in lead mining in 1990 were Australia (563,000 ton), the United States (495,000 ton), Russia (450,000 ton), China (315,000 ton) and Canada (236,000 ton)[109].

The largest single use of lead compounds is for the manufacture of lead-acid storage batteries. Other industrially important uses of lead compounds range from antiknock additives in gasoline to lead crystal glasses and stabilizers for plastics during thermal processing[111]. Demand for lead is surpassed only by iron, copper, aluminum and zinc.

TABLE 15. World refinery production of lead (thousand metric tons; modified from Reference 109)

Country	1987	1990
Australia	216.7	227.0
Canada	230.7	224.0
France	245.9	260.0
Germany (West and East)	389.5	396.5
Italy	173.7	172.0
Japan	338.3	329.0
Mexico	212.0	214.0
Spain	122.7	120.0
U.S.S.R.	750.0	700.0
United Kingdom	338.2	360.0
United States	1083.8	1326.6
Other	1616.2	1612.6
Total	5717.7	5941.7

Table 16 shows the consumption of major countries in the world (United States, Japan, Germany, France, United Kingdom and Italy) in 1977 and 1987. According to Table 16, the 1987 level shows a decrease of 6.71% compared with the 1977 level. This decline is attributable to environmental regulations[113].

Today's major use of lead is in storage batteries that consist of a negative plate of porous lead, a positive plate of lead peroxide and an electrolyte of sulfuric acid solution. Other important applications of inorganolead compounds are for the manufacture of pigments, cable covering, solder and ammunition. $Pb(CN)_2$, Pb_3O_4 and $2PbCO_3 \cdot Pb(OH)_2$ are extensively used as pigments, and lead azide, $Pb(N_3)_2$, is the standard detonator for explosives[110,113].

The only large-scale industrial use of organolead compounds utilizes the antiknock properties of the ethyl- and methylleads. When 2–3 ml of tetraethyllead is added to 1 gal of gasoline, the octane number of the fuel is raised by about 10 octane numbers. The higher this number, the greater is the gasoline's resistance to knocking in combustion[110,114].

Data from the International Lead and Zinc Study Group (ILZSG) in London indicate a decline of *ca* 66% in the world consumption of refined lead for gasoline additives, from a peak consumption of 365,200 ton in 1970 (excluding Eastern-bloc countries) to 124,000 ton in 1986. The limits in 1988 on the lead content of gasoline in various countries are given in Table 17. Consumption of gasoline and organolead compounds in 1989 are also given in Table 17. These limits were imposed as a means of meeting automotive emission standards. For catalyst-equipped cars, unleaded gasoline must be used to avoid poisoning of the catalytic converter[111].

The only non-antiknock use of organolead compounds has been in the preparation of RHgX for application as seed disinfectants and for the control of fungi[114]. Thiomethyl

TABLE 16.   Consumption of lead by end-use (modified from Reference 113)

Use	1977		1987	
	million ton	(%)	million ton	(%)
Batteries	1.472	50.1	1.717	62.7
Cables	0.175	6.0	0.105	3.8
Semifinished	0.324	11.0	0.292	10.7
Pigments and chemicals	0.648	22.1	0.407	14.9
Alloys	0.171	5.8	0.110	4.0
Others	0.146	5.0	0.108	3.9
Total	2.936	100.0	2.739	100.0

TABLE 17.   Limit of lead content of gasoline (in 1988)[111], and consumption of gasoline and organolead compounds (in 1989) (modified from References 111 and 113)

Country	Maximum lead in gasoline (g l^{-1})	Consumption of gasoline (million tons)	Consumption of organoleads (thousand tons)
USA	0.026	334.1	11.7
Canada	0.291	26.9	10.5
Italy, France, Spain, Portugal	0.399	—	—
Others in European Community	0.151	152.4	56.4
Australia	0.304–0.840	14.4	11.1
Korea	0.301	—	—
Japan	0.00	49.5	0.0

triphenyllead is used as an antifungal agent, cotton preservative and lubricant additive; thiopropyl triphenyllead as a rodent repellent; tributyllead imidazole as a lubricant additive and cotton preservative; and tributyllead cyanamide and tributyllead cyanoguanidine as lubricant additives. Triphenyllead acetate has been commercialized in Germany for antifouling paint. A number of other uses has been examined but none has been commercialized[111,114].

C. Concentration and Speciation of Organolead Compounds in the Natural Environment

It is now well established that environmental samples may contain organolead compounds, originating from their use as antiknock additives in gasoline and possibly from natural alkylation of inorganolead compounds[115]. Tetraalkyllead antiknock additives in gasoline are volatile and water-insoluble but environmentally labile. They gradually undergo degradation to inorganolead via various highly toxic ionic alkyllead compounds. From the gas phase of the atmosphere these compounds are washed into the hydrosphere, and are ultimately adsorbed into sediments; these compounds also become incorporated into the food chain[116,117].

Neves and coworkers[115] determined the concentration of alkyllead compounds in fish and water. Water was sampled at 11 stations distant from industrial activities producing or using alkyllead in Essex, U.K.; sediments or sands were collected at the same location; fishes caught within 5 miles of the coast were obtained direct from fishermen in Eastern England. The experimental results are shown in Tables 18 and 19. The River Thames, Thames Lagoon, Colchester, Ardleigh and Dedham are fresh water sites, Rowhedge, St. Osyth and Wivenhoe estuary water and Walton, Frinton and Clacton saline water.

According to Table 18, the freshwater river sites (River Thames, Colchester and Dedham) exhibit the lowest concentrations of alkyllead compounds. The highest total alkyllead concentrations were observed at St. Osyth, in an estuary region experiencing significant marine pleasure-craft traffic, attributable largely to fully alkylated forms and suggesting spillage of leaded gasoline[115]. Samples from the lagoon contained elevated levels of both fully alkylated and ionic species, as well as the highest mean Pb^{2+} concentration measured. This site is near a roundabout carrying a heavy traffic load, where car exhaust and road runoff clearly influence levels[115].

TABLE 18. Mean concentration of organolead and inorganic lead compounds ($ng \, Pb \, l^{-1}$) in water in UK (modified from Reference 115)

Site	Organolead compounds									Total alkyl-lead	Inorg. Pb
	Me_4Pb	Me_3EtPb	Me_2Et_2Pb	$MeEt_3Pb$	Et_4Pb	Me_3Pb^+	Me_2Pb^{++}	Et_3Pb^+	Et_2Pb^{++}		
River Thames	0.09	0.18	0.18	*	*	*	*	*	*	0.45	3.4
Thames Lagoon	0.70	0.60	0.30	0.24	0.02	0.30	0.19	0.3		2.65	6.6
Colchestel	0.05	0.03	0.02	0.02	0.20	0.02	*	*	*	0.34	3.0
Rowhedge	0.02	0.12	0.06	*	*	0.06	0.11	*	*	0.37	4.7
Ardleigh	0.30	0.19	0.16	0.02	0.05	0.05	0.09	0.09	0.1	1.05	5.2
Dedham	0.02	*	*	*	*	*	*	*	*	0.02	2.5
Walton	0.3	0.17	0.11	0.15	0.04	0.04	0.03	*	*	0.84	3.6
Frinton	0.3	0.2	0.10	0.14	0.03	0.02	0.03	*	*	0.82	3.3
Clacton	0.4	0.3	0.15	0.06	0.40	0.20	*	*	*	1.51	3.5
St. Osyth	0.7	0.6	1.10	*	*	0.10	0.02	0.20	0.05	2.77	5.4
Wivenhoe	0.09	0.12	0.04	0.07	0.10	0.09	0.03	0.10	0.04	0.68	4.8

*denotes below detection limit.

TABLE 19. Mean concentration in each species of fish/wet weight in UK (ng Pb g^{-1}) (modified from Reference 115)

Species	n^a	Me$_4$Pb	Me$_3$EtPb	Me$_2$Et$_2$Pb	MeEt$_3$Pb	Et$_4$Pb	Me$_3$Pb$^+$	Me$_2$Pb^{++}	Et$_3$Pb$^+$	Et$_2$Pb^{++}	Inorg.
Herring	46	0.2	0.01	0.04	0.008	0.008	*	*	*	*	2.0
Dabs	5	*	*	0.23	*	*	*	*	*	*	3.9
Squid	4	0.06	*	*	*	*	*	*	*	*	1.8
Coley	7	*	*	*	*	*	*	*	*	*	2.3
Sprats	8	*	0.13	0.13	0.08	*	*	*	*	*	4.1
Skate	4	0.2	*	1.1	*	*	*	*	*	*	0.9
Mackerel	3	*	*	*	*	*	*	*	*	*	2.6
Haddock	4	*	*	*	*	*	*	*	*	*	0.6
Plaice	1	*	*	*	*	*	*	*	*	*	*
Whiting	4	*	*	*	*	*	*	*	*	*	1.2
Cod	29	0.06	0.06	0.05	*	*	*	*	*	*	3.1
Sardine	9	*	*	*	*	*	*	*	*	*	0.3

$^a n$ = number of samples,
*denotes below detection limit.

As shown in Table 19, amounts of organolead were frequently below the detection limits and levels of both organic and inorganic lead were found to be subject to wide fluctuations which did not correspond closely to the water body.

The concentration factor may also be expressed as ng Pb kg^{-1} (fish)/ng Pb l^{-1} (water), for the individual tetraalkyllead compounds and for the various species of fish in which they were detected, and data for inorganic lead were derived in a similar manner. The concentration factors are summarized in Table 20.

Massive enrichment of organolead relative to inorganic lead is observed in fish tissues, while incorporation of inorganic forms appears to be very low (mean ratio 0.76). The highest alkyllead ratios are observed for Me$_2$Et$_2$Pb (mean of all fish species, 2580)[115].

Wong and coworkers[118] analyzed fish and other environmental samples (clam, macrophytes, sediments and waters) from areas upstream and downstream from alkyllead manufacturing sites beside the St. Lawrence and St. Clair Rivers, Ontario, and found a clear indication of elevated alkyllead levels in samples near the industries. Most species of fish contained alkyllead compounds with tetraethyllead and triethyllead as the predominant forms. Most fish from the contaminated areas contained 70% or more of the total lead as alkyllead. Average alkyllead levels varied from year to year but declined steadily after 1981. For example, the mean value of alkyllead in carp from the St. Lawrence River decreased from 4207 µg kg^{-1} in 1981 to 2000 µg kg^{-1} in 1982 and to 49 µg kg^{-1} in

TABLE 20. Concentration factor of organotin compounds (modified from Reference 115)

Species	Me$_4$Pb	Me$_3$EtPb	Me$_2$Et$_2$Pb	MeEt$_3$Pb	Et$_4$Pb	Inorg. Pb
Herring	610	40	330	70	50	0.59
Dabs	—	—	1920	—	—	1.12
Squid	180	—	—	—	—	0.52
Sprats		590	1080	670	—	1.18
Skate	610	—	9170	—	—	0.26
Cod	180	270	420	—	—	0.89
All species	400	300	2580	370	50	0.76

1987, reflecting the reduction of alkylleads in the effluents and the closure of one of the factories in 1985[118].

Chakraborti and coworkers[117] surveyed ionic alkyllead compounds in environmental water and sediments collected at an artificial lake within the University of Antwerp campus, and in rivers and the sea in Belgium. In the lake water, abundance of the species is in the order $PbMe_3^+ > PbEt_3^+ > PbEt_2^+$, $PbMe_2^+$, while in the river, the order is reversed.

Turnbull and coworkers[119] measured concentrations of tetra-, tri- and dialkyllead compounds in rain collected at rural and urban sites during November 1992 and April 1993 in England. The measurements are compared with similar data collected in the early 1980s, prior to a 72% reduction in the emission of lead from combustion of leaded petroleum. While concentrations of inorganic lead have fallen broadly in line with emissions of automotive lead, alkyllead concentrations in rain have fallen by only 50% or less, and thus the ratio of alkyllead to inorganic lead in rain has increased appreciably. The data suggest that lead in rainwater would fall to approximately $2 \ \mu g \ l^{-1}$ if automotive lead emission fell to zero[119].

D. Methylation of Lead Compounds

The major organolead compounds found in the environment are the tetraalkyllead compounds and their di- and trialkyl decomposition products. Elevated levels of tetraalkylleads have two possible sources: either (i) anthropogenic leaded petroleum inputs or (ii) environmental methylation of natural lead compounds. While the former is well established, the latter is the subject of some controversy in the literature. Interest in the environmental methylation process derives from the increased toxicity of methyllead compounds compared to their inorganic analogs.

Wong and coworkers[120] investigated in 1975 whether microorganisms in lake sediments can transform certain inorganic and organic lead compounds into the volatile tetramethyllead. The sediments were collected from Hamilton Harbor (Lake Ontario), Mitchell Bay (Lake St Clair) and Erieau Harbor (Lake Erie). The authors concluded from the results of 50 experiments that incubation of some lead-containing sediments generates Me_4Pb; that Me_3Pb^+ salts are readily converted to Me_4Pb by microorganisms in lake waters or in nutrient media, with or without the sediment, and both in the presence or absence of light; that conversion of inorganic lead to Me_4Pb occurred in the presence of certain sediments; and that the conversion is purely a biological process[120].

Chau and Wong[121] observed biological methylation of inorganic lead and of trimethyllead acetate in the presence of eleven different sediments as shown in Table 21. The conversion of Me_3PbOAc was observed in all experiments but that of lead nitrate was only sporadic.

Thompson and Crerar[122] reported the methylation of lead in marine sediments.

Huber and coworkers[123] also reported biomethylation of Pb^{2+} and of Me_3PbX. They followed the redistribution of Me_3PbX in anaerobic cultures (bacteria from the surface of a natural lake gron under N_2, or from the anaerobic sediment of a small pond), and observed a rate increase, but less Pb^{2+} and more Me_4Pb were obtained than were expected from equation 4:

$$3 \ Me_3PbX \longrightarrow 2 \ Me_4Pb + PbX_2 + MeX \qquad (4)$$

A blank and also a sterile solution containing Pb^{2+} of methyllead compounds showed no Pb content in the methanolic solution after the same treatment. The author concluded that Me_4Pb was produced in the biomethylation of Pb^{2+} by bacteria[123]. Thompson and Crerar[122] also observed that about 0.03% of lead as $Pb(NO_3)_2$ underwent methylation and trimethyllead acetate, $(CH_3)_3PbOAc$, was methylated nearly quantitatively in incubation experiments with marine sediments from the British Columbia coastline.

TABLE 21. Methylation of lead in lake or harbor (or bay) sediments in Canada (modified from Reference 121)[a]

Lake or harbor	Total Pb conc. (mg kg^{-1} dry wt.)	μg Me$_4$Pb generated from sediment supplemented with		
		no addition	Pb(NO$_3$)$_2$	Me$_3$PbOAc
Mitchel Bay	110	1.20	2.20	4.7
Erieau Harbor	60	0	0	2.9
Port Stanley	69	0	0	5.2
Long Lake	116	0	0.09	8.6
Kelly Lake	285	1.80	1.10	4.7
Kunch Lake	47	0.16	0.14	4.0
Robinson Lake	48	0	2.10	21.0
Dill Lake	47	0.71	0.55	7.6
Norway Lake	48	0	0	2.4
Babine Lake	43	0	0	1.4
Hamilton Harbor	273	0.01	0.13	6.4

[a] 50 mg sediment (wet wt.), 150 cm^3 lake water, 0.5% nutrient broth and 0.1% glucose with and without addition of 1 mg Pb as Pb(NO$_3$)$_2$ or Me$_3$PbOAc. Final Pb concentration, 5 μg cm^{-3}.

Harrison and Allen[124] presented evidence of a natural alkylation process of lead as a source of atmospheric alkyllead.

Berdicevsky and coworkers[125] reported the conversion of inorganic lead into organic derivatives by marine microorganisms.

Some claims have, however, been made that the methylation of lead was not biologically mediated. In 1980, Craig[126] studied the methylation of trimethyllead acetate in a lake sediment (Lake Minetonka, Minn., U.S.A.), and concluded that it is not necessary to invoke a biological route to the methylation and that the results in this case can be explained entirely by a disproportionation process[126]. Craig and Wood also proposed an abiotic methylation route of lead[127].

Jarvie and coworkers conducted an unsuccessful search for lead biomethylation using sediments. They studied modified and abnormal sediments and culture systems, but could not find any definite evidence for lead biomethylation in any of the systems investigated[128].

Walton and coworkers[129] studied the methylation of inorganic lead by the use of biologically active sediments, collected from the low-salinity region of the Tamar Estuary, SW England, which showed the presence of the microorganisms Genus bacillus and Pseudomonas. Sediments from this estuary are biologically active and have been shown to convert an inorganic arsenic to dimethylarsenic species. In these biologically active sediment systems, tetramethyllead was produced from Pb(NO$_3$)$_2$, PbCl$_2$ and Pb(OAc)$_2$, in a two-stage process involving an initial lag phase of 80–150 hours followed by the exponential appearance of tetramethyllead. The first stage is the slow formation of the (CH$_3$)Pb^{3+} intermediate followed by more rapid methylation, through dismutation. It is thought that (CH$_3$)Pb^{3+} may be produced by oxidative addition of carbocation.

Forsyth and coworkers[116] postulated that the ubiquity of trimethyllead in ducks can be accounted for if the methylation of Pb^{2+} was environmentally mediated.

Rapsomanikis and coworkers[74,75] observed methylation of tin(II) and lead(II) in sediments by carbanion donors as described before. A factorial experimental design determined separate and combined effects of (i) [CH$_3$)$_2$Co(N$_4$)]ClO$_4$ (methylcobalamin model and carbanion donor), CH$_3$I and MnO$_2$[130]; and (ii) MeCoB$_{12}$ (methylcobalamin) and

TABLE 22. Experimental design and yields of methyllead compounds (modified from Reference 74)

Carbanion donor (μmol)		Percent yield^b $\times 10^2 (\mu g)^c$			
Me$_2$Co(N$_4$)$^+$	MeCoB$_{12}$	Me$_2$Pb^{2+}	Me$_3$Pb$^+$	Me$_4$Pb	Total
10	10	1.4 (0.14)	1.9 (0.20)	0.4 (0.04)	3.7 (0.38)
10	0	6.2 (0.65)	4.4 (0.46)	0.5 (0.05)	11.4 (1.18)
0	10	nd	nd	nd	nd
0	0	nd	nd	nd	nd

^{a} Each experiment contained 5 μmol (1040 μg) Pb(II) as Pb(NO$_3$)$_2$.
^{a} Based on Pb(II) added.
^{c} Limit of detection is 20 pg; nd = not detected.

Me$_2$Co(N$_4$)$^+$ (methylcobalamin model)(**6**)[74] on the methylation of Pb(II) in aqueous systems and sediment matrices, respectively. Results obtained for methylation of tin in the latter experiment are shon in Table 11.

Methylation experiments of lead(II) under the same conditions as used for tin resulted in very low yields of di-, tri- and tetramethyllead (Table 22). Yields in the presence of Me$_2$Co(N$_4$)$^+$ and MeCoB$_{12}$ are lower than with Me$_2$Co(N$_4$)$^+$ alone, but the decrease is insignificant at the 95% confidence level. No methyllead compound as detected with MeCoB$_{12}$ alone. Methyllead products probably form via disproportionation of MePb(II) (equation 7) and Me$_2$Pb(II) (equation 8) followed by successive methylation of di- and trimethyllead (equation 9, steps c and d)[74].

$$M(II) + Me \longrightarrow MeM(II) \tag{5}$$

$$MeM(II) + Me \longrightarrow Me_2M(II) \tag{6}$$

$$2\ MeM(II) \longrightarrow Me_2M(IV) + M^0 \tag{7}$$

$$2\ Me_2M(II) \longrightarrow Me_4M + M^0 \tag{8}$$

$$M(IV) \xrightarrow[a]{Me^-} MeM(IV) \xrightarrow[b]{Me^-} Me_2M(IV) \xrightarrow[c]{Me^-} Me_3(IV) \xrightarrow[d]{Me^-} Me_4M(IV) \tag{9}$$

However, the low yields of methyllead products are insufficient to clarify the controversy over their possible formation in sediments[74].

E. Bioaccumulation, Toxicity and Environmental Effects of Organoleads

1. Bioaccumulation

Satake and coworkers[131] studied lead accumulation and location in the shoots of the aquatic liverwort, *Scapania undulata*, in stream water (Pb content, 20 μg l^{-1}) at a mine in England. The lead concentration in the shoots ranged from 7 to 24 mg Pb g^{-1} on a dry weight basis, giving an enrichment ratio of $3.5 \times 10^5 - 1.2 \times 10^6$. Lead was localized in the cell wall[131].

Wong and coworkers[132] and Chau and coworkers[133] investigated the bioaccumulation of alkylleads. A freshwater green alga, *Ankistrodesmus falcatus*, exposed to solutions of trialkyllead, dialkyllead and inorganic lead(II) compounds (1 mg l^{-1}) for 24 hours accumulated these compounds with concentration factors of about 100, 2,000 and 20,000, respectively, as shown in Table 23[132]. They found in experiments of long-term (28 days) incubation of this alga with trimethyllead that the latter followed a dealkylation sequence with the formation of dimethyllead and lead(II) compounds[133].

TABLE 23.   Accumulation of alkyllead and lead(II) compounds by *A. Falcatus* after incubation in solutions of 1 mg l^{-1} of the lead compounds for 24 hours (modified from Reference 132)

Alkyllead	Concentration of lead compounds		
	Supernatant (mg l^{-1})	Cells (μg g^{-1})	Conc. factor[a]
Pb^{2+}	0.31	6,242	20,135
Me$_2$Pb^{2+}	0.71	1,978	2,786
Me$_3$Pb$^+$	1.02	100	98
Et$_2$Pb^{2+}	0.94	1,821	1,937
Et$_3$Pb$^+$	1.00	170	170

[a] Concentration factor = conc. in algae/conc. in supernatant.

Chau and coworkers[133] investigated the bioaccumulation of alkyllead compounds from water and from contaminated sediments by freshwater mussels, *Elliptio complanata*. Higher levels of trimethyllead than triethyllead species were accumulated after the same exposure period. In vivo transformation of the trialkyllead species by a series of dealkylation reactions giving dialkyllead and inorganic lead(II) species appears to take place. Rates of accumulation are higher for the more contaminated sediments[133].

Vighi[134] set up a laboratory continuous-flow system to breed a simple trophic chain (*Selenastrum capricornutum* ⟶ *Daphnia magna* ⟶ *Poecilia reticulata*) in order to evaluate both the accumulation of lead at various trophic levels and the transfer between trophic levels. The various organisms were bred in separate, but communicating, vessels and results are summarized in Table 24[134]. The experiments demonstrated that lead accumulates in the trophic chain with a decreasing concentration from the lowest to the highest levels as shown in Table 24. *P. reticulata* took up lead at a higher lead level via food than via water[134].

2. Comparison of toxicity in alkylleads

As shown before in Section III.E., the toxic effects upon the freshwater alga, *Poterioochromonas malhamensis*, of the trimethyllead compound (Me$_3$PbAc) was thirtyfold larger than that of the inorganic lead compound (PbCl$_2$)[82].

Babich and Borenfreund[135] compared in vitro cytotoxicities of inorganic lead and tri- and dialkyllead compounds in fish cells (bluegill sunfish, *Lepomis macrochiru*).

TABLE 24.   Bioaccumulation (μg Pb g^{-1} dry weight), concentration factor and biological half-time ($t_{1/2}$, day) for lead uptake in three different trophic levels of organisms (modified from Reference 134)

Organisms	Initial Pb conc: 5 μg l^{-1}			Initial Pb conc: 50 μg l^{-1}		
	uptake (μg g^{-1})	conc. factor	$t_{1/2}$ (day)	uptake (μg g^{-1})	conc. factor	$t_{1/2}$ (day)
Selenastrum	460	100,000	5.3	1350	29,000	5.3
Daphnia	25	5,000	7.7	74	1,900	7.7
Poecilia						
total	30	3,600	25.7	70	1,000	25.7
via H$_2$O	4.5	800	7.7	13	250	7.7
via food	25.5	5,000	33	57	900	33

The sequence of cytotoxicity was $(C_2H_5)_3Pb^+$ > $(CH_3)_3Pb^+$ > $(CH_3)_2Pb^{2+}$ > $(C_2H_5)_2Pb^{2+}$ > Pb^{2+}. The organolead compounds were more toxic than inorganic Pb^{2+} and the trialkyllead compounds were more toxic than the dialkyllead compounds. Whereas triethyllead was more toxic than trimethyllead, the dimethyllead was more toxic than diethylead[135].

Jarvie and Marshall[136] investigated the effects of alkyllead compounds on freshwater and marine algae. A number of researchers have found R_4Pb to be comparable in toxicity to the R_3Pb^+ compounds and they suggested that the R_4Pb compounds themselves are completely nontoxic and their toxicity is due to R_3Pb^+ breakdown products. Two algal species were also cultured in the presence of Me_4Pb, Me_3PbCl, Et_3PbCl, Bu_3PbCl and Et_2PbCl_2 to assess relationships between alkyl chain length and degree of substitution around the lead on algal activity. The results showed that the trialkylleads were the most toxic of the several series studied, and within the trialkyl series the toxicity increased with alkyl chain length[136]. Similar results were obtained on the sedimental organism *Scrobiculavia plana*, by Marshall and Jarvie[137].

Somewhat different results were obtained by Zuckerman[79] on the comparative bactericidal effects of group IVa di- and triorganoderivatives.

3. Mechanism of toxicity

The long-term consequences of neonatal exposure to triethyllead were examined with respect to the development of the central nervous system of rats[138]. The studies of the developmental exposure to triethyllead lead to the conclusion that this compound causes permanent hippocampus damage (neurotoxicity) in rats.

4. Metabolism

As mentioned before, trimethyllead was found by Wong[132] to be dealkylated to form dimethyllead and inorganic lead(II) by freshwater algae.

Krishnan and Marshall[139] investigated the metabolism of ethyllead salts by the Japanese quail *Coturnix coturnix japonica*, which were provided with drinking water containing 0.0 or 250 mg l^{-1} of $Pb(NO_3)_2$, 25 mg l^{-1} of Et_2PbCl_2 or 2.5 or 0.25 mg l^{-1} of Et_3PbCl for 9 weeks. Eggs as well as soft tissues (liver, kidney, brain and breast muscle) were recovered at the termination of the trials and analyzed for alkyllead salts. Using the inorganic lead salt, no evidence for hostmediated methylation was observed in any of the samples. With each of the alkyllead compounds the toxicant was rapidly transferred to the eggs. Et_2Pb^{2+} was metabolized to Et_3Pb^+, Me_2Pb^{2+} and Me_3Pb^+ in low yields which accumulated in the eggs. The major toxicant in the soft tissues was Et_3Pb^+. Metabolic dealkylation of Et_3Pb^+ was a minor process, and Et_2Pb^{2+} accumulated mainly in the eggs. Only traces of mixed alkyllead cations (Et_2MePb^+ and $EtMePb^{2+}$) were detected in the liver or the kidney if either Et_3PbCl or Et_2PbCl_2 served as test toxicant[139].

Arai and Yamamura[140] investigated the excretion of tetramethyllead, trimethyllead, dimethyllead and inorganic lead after injection of either 9.9 mg kg^{-1} or 39.7 mg kg^{-1} of tetramethyllead into rabbits, and urinary and fecal excretions of lead compounds were analyzed. In the former administration, urinary total lead excretion was 73% Me_2Pb^{2+}, 19% Me_3Pb^+, 6% Pb^{2+} and 2% Me_4Pb. In the latter administration, urinary total excretion was 67% Me_2Pb^{2+}, 14% Me_3Pb^+, 17% Pb^{2+} and 2% Me_4Pb. During 7 days, 1–3% of the administered dose was excreted in urine, 7–9% in feces. Some tetramethyllead was found to be metabolized to trimethyl-, dimethyl- and inorganic lead compounds and excreted by rabbit[140].

Tetraalkylleads are highly lipid, soluble, rapidly metabolized and readily cross the blood–brain barrier. These physicochemical properties make the central nervous system the main site of the toxic action of the alkyllead species[138]. The earliest symptoms include insomnia, which may be followed by lack of appetite, nausea, vomiting, diarrhea hallucinations, delusions, delirium and ultimately death[141]. Continuing exposure may lead to complaints associated with central nervous system disturbances[141].

Booze and Mactutus[138] examined the long-term consequences of neonatal exposure to triethyllead. The results demonstrated that triethyllead preferentially damages a select area of the developing rat central nervous system namely, the hippocampus. Furthermore, pharmacological studies of the neurotransmitter system suggested functional and dose-dependent relationships between the behavioral hyperactivity and hippocampus damage via cholinergic, but not dopaminergic pathways. Other studies indicate that trialkyl-lead compounds inhibit mitochondrial respiration, oxidative phosphorylation and ATP synthesis[142,143].

F. Health Effect Assessment and Safety of Organoleads

The estimated average dietary intake of lead in humans is 300 $\mu g\, day^{-1}$. Fortunately, only 8% of ingested lead is retained in the body. About 90% of the accumulated lead is localized in bone[9]. Organolead compounds, however, scarcely accumulate in bone[138].

Lead is toxic in all forms, but to different degrees depending upon the chemical nature and solubility of the lead compound. In general, organolead compounds are more toxic than inorganic lead salts if ingested, inhaled or absorbed by humans. Children are more sensitive than adults to the effects of lead exposure. Such exposure may be acute (large single dose) or chronic (repeated low doses)[111]. The main site of the toxic action of organolead compounds is the central nervous system[141,114].

Sources of lead in the environment include soils near smelters, lead pigments, automobile exhaust emissions, food and water and industrial exposure. Each of these sources is being reduced or eliminated through environmental and occupational safety regulations. For example, organolead levels in reinwater collected in the early 1990s declined compared with similar data in the early 1980s in samples taken at a number of sites in England[119].

In the United States, the ambient airborne lead limit is 1.5 $\mu g\, m^{-3}$, which applies to the exterior of plant environments and emissions. Moreover, lead-based paints are banned for residential use; leaded gasoline is being phased out; also lower limits of lead in drinking water have been proposed[111]: in the USA below 20 $\mu g\, Pb\, l^{-1}$, while in other countries as follows[9]: Japan 0.01 $mg\, l^{-1}$ (since 1992), France and Yugoslavia 0.05 $mg\, l^{-1}$, Netherlands and Mexico 0.1 $mg\, l^{-1}$, Germany 0.5 $mg\, l^{-1}$. The limits of lead content in gasoline in 1988 are shown in Table 17[111].

Lead-exposed employees in the United States[111], for example, must be removed from work if the average of their last three blood lead determinations is at or above 0.50 $\mu g\, Pb\, g^{-1}$ whole blood and if the airborne lead level is at or above 30 $\mu g\, m^{-3}$.

G. Summary

According to the results of numerous investigations using biologically active sediments and micro-organisms, it seems likely that lead compounds may undergo alkylation under natural environmental conditions by either biological or chemical processes. Methylation of divalent inorganic lead (Pb^{2+}) is theoretically unlikely due to the difficulty, indicated by thermodynamic considerations, of the initial oxidative conversion of lead(II) to lead(IV)[130]. Alkylation of organic lead(IV) compounds proceeds rapidly, however, although it is difficult to ascribe the process to purely biological or chemical pathways[126−128]. Of particular concern is the possibility that lead may be biotransformed into more toxic organolead compounds.

The most toxic organoleads are the trialkylleads, whose toxicity increases with alkyl chain length. Tetraalkylleads themselves are completely nontoxic and their apparent toxicity is due to their trialkyllead breakdown products[136].

Triethyllead was suggested to be one of the factors causing progressive damage of European forests[144]. The toxic effect of triethylleads on cells was attributable to inhibition of microtubule assembly. Triethylleads react with SH groups present in tubulin, whereupon the latter loses its capability for microtubule assembly[144].

Lipid solubility and volatility of lead in biological systems may be increased in the alkyl derivatives, making the central nervous system a critical target for organolead toxicity. Furthermore, it is well known that the toxic effect of organolead may be further exacerbated for specific subpopulations[138]. The usual therapeutic treatment of inorganolead poisoning involves the use of chelating agents such as ethylenediaminetetraacetic acid (EDTA). However, since EDTA does not bind Et_3Pb^+ or Et_2Pb^{2+}, it may not be very useful[138].

Organolead compounds have been handled quite safely in industry for many years. As in the case of inorganic lead, control measures are well defined, and if they are properly applied, neither inorganic nor organic lead intoxication should develop.

Sources of organoleads in the environment are being reduced or eliminated through environmental and occupational safety regulations. Even no, we need to further observe and control the decline of organolead levels in our environment.

V. ACKNOWLEDGMENT

The author is sincerely grateful to Professor A. Inoue (Director of the Research Center for the South Pacific, Kagoshima University) and Professor K. Arai (Department of Environmental Science and Technology, Kagoshima University) for reading the manuscript and for valuable discussions and useful suggestions, and to Associate Professors A. Ohki, K. Naka, M. Nakazaki and T. Kuroiwa (Department of Applied Chemistry and Chemical Engineering, Kagoshima University) for their help in the preparation of this chapter.

VI. REFERENCES

1. P. J. Craig in *The Chemistry of the Metal–Carbon Bond*, Vol. 5 (Ed. F. R. Hartley), Chap. 10, Wiley, Chichester 1989, pp. 437–463.
2. H. Lee, J. T. Trevors and M. I. Van Dyke, *Biotech. Adv.*, **8**, 539 (1990).
3. P. S. Gleim, in *McGraw-Hill Encyclopedia of Science & Technology*, Vol. 8, 6th ed., McGraw-Hill, New York, 1987, pp. 76–83.
4. T. O. Llewellyn, in *Minerals Yearbook—1990*, Vol. 1 (Metals and Minerals), US Department of the Interior Bureau of Mines, US Government Printing Office, Washington, 1993, pp. 491–494.
5. A. G. Schauss, *Biological Trace Element Research*, **29**, 267 (1991).
6. P. G. Harrison, in *Dictionary of Organometallic Compounds*, Chapman and Hall, London, 1984, pp. 966–972.
7. J. Scoyer, H. Guislain and H. U. Wolf, in *Ullman's Encyclopedia of Industrial Chemistry*, Vol. A12, 5th ed., VCH, Weinheim, 1989, pp. 351–363.
8. F. Suzuki, R. R. Brutkiewicz and R. B. Pollard, *Br. J. Cancer*, **52**, 757 (1985).
9. T. Oka., *Mizu to kenkou (Water and Health)*, **17**, 80 (1991).
10. R. S. Braman and M. A. Tompkins, *Anal. Chem.*, **50**, 1088 (1978).
11. M. O. Andreae and P. N. J. Froelich, *Tellus*, **36B**, 101 (1984).
12. I. Omae, *Kagaku-Kogyou (Chemicals Industry)*, **1991**, 237 (1991a).
13. H. P. Mayer and S. Rapsomanikis, *Appl. Organomet. Chem.*, **6**, 173 (1992).

14. J. H. Adams, in *Encyclopedia of Chemical Technology*, Vol. 11, 3rd ed., Wiley, New York, 1978, pp. 791–802.
15. R. M. Slawson, M. I. Van Dyke, H. Lee and J. T. Trevors, *Plasmid*, **27**, 72 (1992).
16. M. I. Van Dyke, H. Lee and J. T. Trevors, *Arch. Microbiol.*, **152**, 533 (1989).
17. M. I. Van Dyke, H. Lee and J. T. Trevors, *J. Ind. Microbiol.*, **4**, 299 (1989).
18. J. Chmielowski and B. Klapcinska, *Appl. Environ. Microbiol.*, **51**, 1099 (1986).
19. S. Goodman, *Medical Hypotheses*, **26**, 207 (1988).
20. I. Omae, *Kagaku-Kogyou (Chemicals Industry)*, **1991**, 392 (1991).
21. K. Miyao, T. Ohnishi, K. Asai, S. Tomizawa and F. Suzuki, *Current Chemother. Inf. Discuss. (Proceedings of the 11th ICC & 19th ICAAC American Society of Microbiology)*, **2**, 1527 (1980).
22. T. Nagata, Y. Aramaki, M. Enomoto, H. Isaka and J. Otuka, *Pharmacometrics*, **16**, 613 (1978).
23. H. Aso, F. Suzuki, T. Yamaguchi, Y. Hayashi, T. Ebina and N. Ishida, *Microbiol. Immunol.*, **29**, 65 (1985).
24. M. Kagoshima and T. Onishi, *Ouyou Yakuri (Applied Pharmacology)*, **32**, 89 (1986); *Chem. Abst.*, **105**, 164451 (1993).
25. E. Lukevics, S. Germane and L. Ignatovich, *Appl. Organomet. Chem.*, **6**, 543 (1992).
26. R. Tacke, B. Becker, D. Berg, W. Brandes, S. Dutzmann and K. Schaller, *J. Organomet. Chem.*, **438**, 45 (1992).
27. R. A. Prough, M. A. Stalmach, P. Wiebkin and J. W. Bridges, *Biochem. J.*, **196**, 763 (1981).
28. T. Sanai, N. Oochi, S. Okuda, S. Osato, S. Kiyama, T. Komota, K. Onoyama and M. Fujishima, *Toxicology and Applied Pharmacology*, **103**, 345 (1990).
29. J. B. Long, in *McGraw-Hill Encyclopedia of Science & Technology*, Vol. 18, 6th ed., McGraw-Hill, New York, 1987, pp. 368–371.
30. P. G. Harrison, in *Dictionary of Organometallic Compounds*, Vol. 2, Chapman and Hall, London, 1984, pp. 2152–2156..
31. J. F. J. Carlin, in *Minerals Yearbook — 1990*, Vol. 1 (Metals and Minerals), US Department of the Interior Bureau of Mines, US Government Printing Office, Washington, 1993, pp. 1157–1175.
32. I. Omae, *Kagaku-kougyou (Chemicals Industry)*, **1990**, 944 (1990).
33. P. Quevauviller, A. Bruchet and O. F. X. Donard, *Appl. Organomet. Chem.*, **5**, 125 (1991).
34. K. A. Mesch and T. G. Kugele, *J. Vinyl Technol.*, **14**, 131 (1992).
35. G. E. Batley and M. S. Scammell, *Appl. Organomet. Chem.*, **5**, 99 (1991).
36. M. Boualam, M. Biesemans, J. Meunir-Piret, R. Willem and M. Gielen, *Appl. Organomet. Chem.*, **6**, 197 (1992).
37. N. Donoghue, E. R. T. Tiekink and L. Webster, *Appl. Organomet. Chem.*, **7**, 109 (1993).
38. G. Eng and T. W. Engle, *Bull. Soc. Chim. Belg.*, **96**, 69 (1987).
39. M. Gielen, A. E. Khloufi, B. Monique and R. Willem, *Appl. Organomet. Chem.*, **7**, 119 (1993).
40. R. J. Maguire and J. Tkacz, *J. Agric. Food Chem.*, **33**, 947 (1985).
41. C. J. Soderquist and D. G. Crosby, *J. Agric. Food Chem.*, **28**, 111 (1980).
42. V. F. Hodge, S. L. Seidel and E. D. Golberg, *Anal. Chem.*, **51**, 1256 (1979).
43. R. S. Braman and M. A. Tompkins, *Anal. Chem.*, **51**, 12 (1979).
44. J. Ashby, S. Clark and P. J. Craig, *J. Anal. At. Spectrom.*, **3**, 735 (1988).
45. J. A. Jackson, W. R. Blair, F. E. Brickman and W. P. Iverson, *Environ. Sci. Technol.*, **16**, 110 (1982).
46. Y. K. Chau, P. T. S. Wong and G. A. Bengert., *Anal. Chem.*, **54**, 256 (1982).
47. R. J. Maguire, Y. K. Chan, G. A. Bengert, E. J. Hale, P. T. S. Wong and O. Kramer, *Environ. Sci. Technol.*, **16**, 698 (1982).
48. Y. Hattori, A. Kobayashi, S. Takemoto, K. Takami, Y. Kuge, A. Sugimae and M. Nakamoto, *J. Chromatogr.*, **315**, 341 (1984).
49. M. D. Mueller, *Fresenius Z. Anal. Chem.*, **317**, 32 (1984).
50. R. J. Maguire, *Environ. Sci. Technol.*, **18**, 291 (1984).
51. A. O. Valkirs, P. F. Seligman, C. Vafa, P. M. Stang, V. Homer and S. H. Lieberman., *NOSC Technical Report*, **1037**, 1 (1985).
52. J. J. Cleary and A. R. D. Stebbing, *Mar. Pollut. Bull.*, **16**, 350 (1985).
53. O. F. X. Donard, S. Rapsomanikis and J. H. Weber, *Anal. Chem.*, **58**, 772 (1986).
54. R. J. Maguire, R. J. Tkacz and Y. K. Chau., G. A. Bengert and P. T. S. Wong, *Chemosphere*, **15**, 253 (1986).
55. C. L. Matthias, J. M. Bellama, G. J. Olson and F. E. Brinckman, *Environ. Sci. Technol.*, **20**, 609 (1986).

56. M. A. Unger, W. G. Mackintyre, J. Greaves and P. J. Hugett., *Chemosphere*, **15**, 461 (1986).
57. L. Randall, J. S. Han and J. M. Weber, *Environ. Technol. Lett.*, **7**, 571 (1986).
58. A. D. Valkirs, P. F. Seligman, P. M. Stanf, V. Homer, S. H. Lieberman, G. Vafa and C. A. Dooley, *Mar. Pollut. Bull.*, **17**, 319 (1986).
59. M. D. Mueller, *Anal. Chem.*, **59**, 617 (1987).
60. J. R. Ashby and P. J. Craig, *The Science of the Total Environment*, **78**, 219 (1989).
61. R. J. Maguire, *Water Sci. Tecnol.*, **25**, 125 (1992).
62. L. Schebe, M. O. Andreae and H. J. Tobschall., *Int. J. Environ. Anal. Chem.*, **45**, 257 (1991).
63. D. S. Forsyth, D. Weber and L. Barlow, *Appl. Organomet. Chem.*, **6**, 579 (1992).
64. C. Stewart and S. J. de Mora, *Appl. Organomet. Chem.*, **6**, 507 (1992).
65. M. Morita, in *Circulation and Control of Man-Made Substances in Environment (1990–1992)*, **G083-N10b** (Ed. N. Soga), The Ministry of Education, Culture and Science, Japan, 1993, p. 357.
66. M. Morita, *Bunseki (Analysis)*, **1991**, 785 (1991).
67. J. H. Weber and J. J. Alberts, *Environ. Technol.*, **11**, 3 (1990).
68. G. W. Pettibone and J. J. Cooney, *J. Ind. Microbiol.*, **2**, 373 (1988).
69. T. Hamasaki, H. Nagase, T. Sato, H. Kito and Y. Ose, *Appl. Organomet. Chem.*, **5**, 83 (1991).
70. J. Ashby, S. Clark and P. J. Craig, *Spec. Publ. R. Soc. Chem.*, **66** (Biological Alkylation of Heavy Elements), 263 (1988).
71. L. E. Hallas, J. C. Means and J. J. Cooney, *Science*, **215**, 1505 (1982).
72. J. R. Ashby and P. J. Craig, *Appl. Organomet. Chem.*, **1**, 275 (1987).
73. J. R. Ashby and P. J. Craig, *The Science of the Total Environment*, **100**, 337 (1991).
74. S. Rapsomanikis, O. F. X. Donard and J. H. Weber, *Appl. Organomet. Chem.*, **1**, 115 (1987).
75. S. Rapsomanikis and J. H. Weber, *Environ. Sci. Technol.*, **19**, 352 (1985).
76. P. Quevauviller, O. F. X. Donard, J. C. Wasserman, F. M. Martin and J. Schneider, *Appl. Organomet. Chem.*, **6**, 221 (1992).
77. B. Earley, A. Biegon and B. E. Leonard, *Neurochem. Int.*, **15**, 475 (1989).
78. C. Rey, H. J. Reinecke and R. Besser, *Vet. Hum. Toxicol.*, **26**, 121 (1984).
79. J. J. Zuckerman, *ACS Symp. Ser.*, **82**, 388 (1978).
80. P. T. S. Wong, R. J. Maguire, Y. K. Chau and O. Kramar, *Can. J. Fish. Aquat. Sci.*, **41**, 1570 (1984).
81. G. Huang, Z. Bai, S. Dai and Q. Xie, *Appl. Organomet. Chem.*, **7**, 373 (1993).
82. G. Röderer, *Trace Subst. Environ. Health*, **16**, 137 (1982).
83. S. Shinoda, In *Circulation and Control of Man-Made Substances in Environment (1990–1992)*, **G083-N10b** (Ed. N. Soga), The Ministry of Education, Culture and Science, Japan, 341 (1993).
84. A. Kumari, J. P. Tandon and R. V. Singh, *Appl. Organomet. Chem.*, **7**, 655 (1993).
85. P. T. S. Wong, Y. K. Chau, O. Kramar and G. A. Bengert, *Can. J. Fish. Aquat. Sci.*, **39**, 483 (1982).
86. T. Horiguchi, H. Shiraishi, M. Shimizu, S. Yamazaki and M. Morita, *Main Group Metal Chemistry*, **17**, 81 (1994).
87. T. Hamasaki, T. Sato, H. Nagase and H. Kito, *Mutat. Res.*, **280**, 195 (1992).
88. V. Piro, F. D. Simone, G. Madonia, A. Silvestri, A. M. Giuliani, G. Ruisi and R. Barbieri, *Appl. Organomet. Chem.*, **6**, 537 (1992).
89. B. Veronesi, K. Jones, S. Gupta, J. Pringle and C. Mezei, *Neurotoxicology*, **12**, 265 (1991).
90. J. Usta and D. E. Griffiths, *Appl. Organomet. Chem.*, **7**, 193 (1993).
91. D. E. Griffiths, J. Usta and Y. M. Tian, *Appl. Oganomet. Chem.*, **7**, 401 (1993).
92. D. E. Griffiths, *Appl. Organomet. Chem.*, **8**, 149 (1994).
93. D. L. Whittington, M. L. Woodruff and R. H. Baisden, *Neurotoxicology and Teratology*, **11**, 21 (1989).
94. D. G. Robertson, R. H. Gray and F. A. De La Iglesia, *Toxicologic Pathology*, **15**, 7 (1987).
95. M. G. Simpson, S. L. Allen and R. Sheldon, *Marine Environmental Research*, **28**, 437 (1989).
96. L. U. Naalsund and F. Fonnum, *Neurotoxicology*, **7**, 53 (1986).
97. W. E. Wilson, P. M. Hudson, T. Kanamatsu, T. J. Walsh, H. A. Tilson, J. S. Hong, R. R. Marenpot and M. Thomson, *Neurotoxicology*, **7**, 63 (1986).
98. M. L. Woodruff, R. H. Baisden and A. J. Nonneman, *Neurotoxicology*, **12**, 427 (1991).
99. M. E. Brodie, J. Opacka-Juffry, D. W. Peterson and A. W. Brown, *Neurotoxicology*, **11**, 35 (1990).
100. M. E. Stanton, K. F. Jensen and C. V. Pickens, *Neurotoxicology and Teratology*, **13**, 525 (1991).
101. P. J. Bushnell and K. E. Angell, *Neurotoxicology*, **13**, 429 (1992).

102. A. D. Toews, R. B. Ray, N. D. Goines and T. W. Bouldin, *Brain Res.*, **398**, 298 (1986).
103. C. P. LeBel, S. F. Ali, M. McKee and S. C. Bondy, *Toxicol. Appl. Pharmacol.*, **104**, 17 (1990).
104. J. F. Coulon, P. Lacoix, P. Linee and J. C. David, *Eur. J. Pharmacol.*, **135**, 53 (1987).
105. G. R. Wenger, D. E. McMillan and L. W. Chang, *Neurobehavioral Toxicology and Teratology*, **8**, 659 (1986).
106. D. W. Allen, J. S. Brooks and S. J. Campbell, *Appl. Organomet. Chem.*, **7**, 531 (1993).
107. P. H. Dowson, J. M. Bubb and J. N. Lester, *Appl. Organomet. Chem.*, **7**, 623 (1993).
108. A. H. Penninks, *Food Addit. Contam.*, **10**, 351 (1993).
109. W. D. Woodbury, *Minerals Yearbook—1990*, Vol. 1 (Metals and Minerals), US Department of the Interior Bureau of Mines, US Government Printing Office, Washington, 1993, pp. 657–682.
110. H. Shapiro and J. D. Johnston, *McGraw-Hill Encyclopedia of Science & Technology*, Vol. 9, 6th ed., McGraw-Hill, New York, 1987, pp. 623–625.
111. D. S. Carr, in *Ullman's Encyclopedia of Industrial Chemistry*, 5th ed., Vol. A15, VCH, Weinheim, 1989, pp. 249–257.
112. P. G. Harrison, in *Dictionary of Organometallic Compounds*, Chapman and Hall, London, 1984, p. 1456.
113. I. Omae, *Kagaku-Kogyo (Chemicals Industry)*, **1991**, 579 (1991).
114. W. B. McCormack, R. Moore and C. A. Sandy, in *Encyclopedia of Chemical Technology*, Vol. 14, 3rd ed., Wiley, New York, 1978, pp. 180–195.
115. A. G. Neves, A. G. Allen and R. M. Harrison, *Environmental Technology*, **11**, 877 (1990).
116. D. S. Forsyth, W. D. Marshall and M. C. Collette, *Appl. Organomet. Chem.*, **2**, 233 (1988).
117. D. Chakraborti, R. J. A. V. Cleuvenbergen and F. C. Adams, *Hydrobiologia*, **176/177**, 151 (1989).
118. P. T. S. Wong, Y. K. Chau, J. Yaromich, P. Hodson and M. Whittle, *Appl. Organomet. Chem.*, **3**, 59 (1989).
119. A. B. Turnbull, Y. Wang and R. M. Harrison, *Appl. Organomet. Chem.*, **7**, 567 (1993).
120. P. T. S. Wong, Y. K. Chau and P. L. Luxon, *Nature*, **253**, 263 (1975).
121. Y. K. Chau and P. T. S. Wong, *ACS Symp. Ser.*, **82**, 39 (1978).
122. J. A. J. Thompson and J. A. Crerar, *Mar. Pollut. Bull.*, **11**, 251 (1980).
123. F. Huber, U. Schmidt and H. Kirshmann, *ACS Symp. Ser.*, **82**, 65 (1978).
124. R. M. Harrison and A. G. Allen, *Appl. Organomet. Chem.*, **2**, 49 (1989).
125. I. Berdicevsky, M. Shacar and S. Yannai, *Arch. Toxicol. Suppl.*, **6**, 285 (1983).
126. P. J. Craig, *Environmental Technology Letters*, **1**, 17 (1980).
127. P. J. Craig and J. M. Wood, *Environ. Pollut. Ser. B*, **6** (*Environmental Lead*), 333 (1981).
128. A. W. P. Jarvie, A. P. Whitmore, R. N. Markall and H. R. Potter, *Environmental Pollution (Ser. B)*, **6**, 81 (1983).
129. A. P. Walton, L. Ebdon and G. E. Millward, *Appl. Organomet. Chem.*, **2**, 87 (1988).
130. S. Rapsomanikis, J. J. Ciejka and J. H. Weber, *Inorg. Chim. Acta*, **89**, 179 (1984).
131. K. Satake, T. Takamatsu, M. Soma, K. Shibata, M. Nishikawa, P. J. Say and B. A. Whitton, *Aquat. Bot.*, **33**, 111 (1989).
132. P. T. S. Wong, Y. K. Chau, J. L. Yaromich and O. Klamar, *Can. J. Fish. Aquat. Sci.*, **44**, 1257 (1987).
133. Y. K. Chau, P. T. S. Wong, G. A. Bengert and J. Wasslen, *Appl. Organomet. Chem.*, **2**, 427 (1988).
134. M. Vighi, *Ecotoxicology and Environmental Safety*, **5**, 177 (1981).
135. H. Babich and E. Borenfreund, *Bull. Environ. Contam. Toxicol.*, **44**, 456 (1990).
136. A. W. P. Jarvie and S. J. Marshall, *Appl. Organomet. Chem.*, **1**, 29 (1987).
137. S. J. Marshall and A. W. P. Jarvie, *Appl. Organomet. Chem.*, **2**, 143 (1988).
138. R. M. Booze and C. F. Mactutus, *Experientia*, **46**, 292 (1990).
139. K. Krishnan and W. D. Marshall, *Environ. Sci. Technol.*, **22**, 1038 (1988).
140. F. Arai and Y. Yamamura, *Ind. Health*, **28**, 63 (1990).
141. J. F. Cole, in *McGraw-Hill Encyclopedia of Science & Technology*, Vol. 9, 6th ed., McGraw-Hill, New York, 1987, pp. 636–637.
142. D. J. Minnema, G. P. Cooper and M. M. Schamer, *Neurotoxicology and Teratology*, **13**, 257 (1991).
143. H. Komulainen and S. C. Bondy, *Toxicol. Appl. Pharmacol.*, **88**, 77 (1987).
144. A. Hager, I. Moser and W. Berthold, *Z. Naturforsch.*, **42c**, 1116 (1987).

Author index

This author index is designed to enable the reader to locate an author's name and work with the aid of the reference numbers appearing in the text. The page numbers are printed in normal type in ascending numerical order, followed by the reference numbers in parentheses. The numbers in *italics* refer to the pages on which the references are actually listed.

Index compiled by K. Raven

Subject index

975

Index compiled by P. Raven